D0321655

The Finite Element Method: Its Basis and Fundamentals

Sixth edition

Professor O.C. Zienkiewicz, CBE, FRS, FREng is Professor Emeritus at the Civil and Computational Engineering Centre, University of Wales Swansea and previously Director of the Institute for Numerical Methods in Engineering at the University of Wales Swansea, UK. He holds the UNESCO Chair of Numerical Methods in Engineering at the Technical University of Catalunya, Barcelona, Spain. He was the head of the Civil Engineering Department at the University of Wales Swansea between 1961 and 1989. He established that department as one of the primary centres of finite element research. In 1968 he became the Founder Editor of the *International Journal for Numerical Methods in Engineering* which still remains today the major journal in this field. The recipient of 27 honorary degrees and many medals, Professor Zienkiewicz is also a member of five academies – an honour he has received for his many contributions to the fundamental developments of the finite element method. In 1978, he became a Fellow of the Royal Society and the Royal Academy of Engineering. This was followed by his election as a foreign member to the U.S. Academy of Engineering (1981), the Polish Academy of Science (1985), the Chinese Academy of Sciences (1998), and the National Academy of Science, Italy (Academia dei Lincei) (1999). He published the first edition of this book in 1967 and it remained the only book on the subject until 1971.

Professor R.L. Taylor has more than 40 years' experience in the modelling and simulation of structures and solid continua including two years in industry. He is Professor in the Graduate School and the Emeritus T.Y. and Margaret Lin Professor of Engineering at the University of California at Berkeley. In 1991 he was elected to membership in the US National Academy of Engineering in recognition of his educational and research contributions to the field of computational mechanics. Professor Taylor is a Fellow of the US Association of Computational Mechanics – USACM (1996) and a Fellow of the International Association of Computational Mechanics – IACM (1998). He has received numerous awards including the Berkeley Citation, the highest honour awarded by the University of California at Berkeley, the USACM John von Neumann Medal, the IACM Gauss–Newton Congress Medal and a Dr.-Ingenieur ehrenhalber awarded by the Technical University of Hannover, Germany. Professor Taylor has written several computer programs for finite element analysis of structural and non-structural systems, one of which, *FEAP*, is used world-wide in education and research environments. A personal version, *FEAPpv*, available from the publisher's website, is incorporated into the book.

Dr J.Z. Zhu has more than 20 years' experience in the development of finite element methods. During the last 12 years he has worked in industry where he has been developing commercial finite element software to solve multi-physics problems. Dr Zhu read for his Bachelor of Science degree at Harbin Engineering University and his Master of Science at Tianjin University, both in China. He was awarded his doctoral degree in 1987 from the University of Wales Swansea, working under the supervision of Professor Zienkiewicz. Dr Zhu is the author of more than 40 technical papers on finite element methods including several on error estimation and adaptive automatic mesh generation. These have resulted in his being named in 2000 as one of the highly cited researchers for engineering in the world and in 2001 as one of the top 20 most highly cited researchers for engineering in the United Kingdom.

Plate 1 Contours of the deviatoric stress invariants for the nave of Gothic cathedral

Courtesy of Prof. Miguel Cervera, CIMNE, Barcelona, Source: P. Roca, M. Cervera, L. Pellegrini and J. Torrent, 'Studies on the Structure of Gothic Cathedrals', Int. Conf. 40th Anniversary of the Int. Ass. Shells and Spatial Structures, Madrid, Spain 1999.

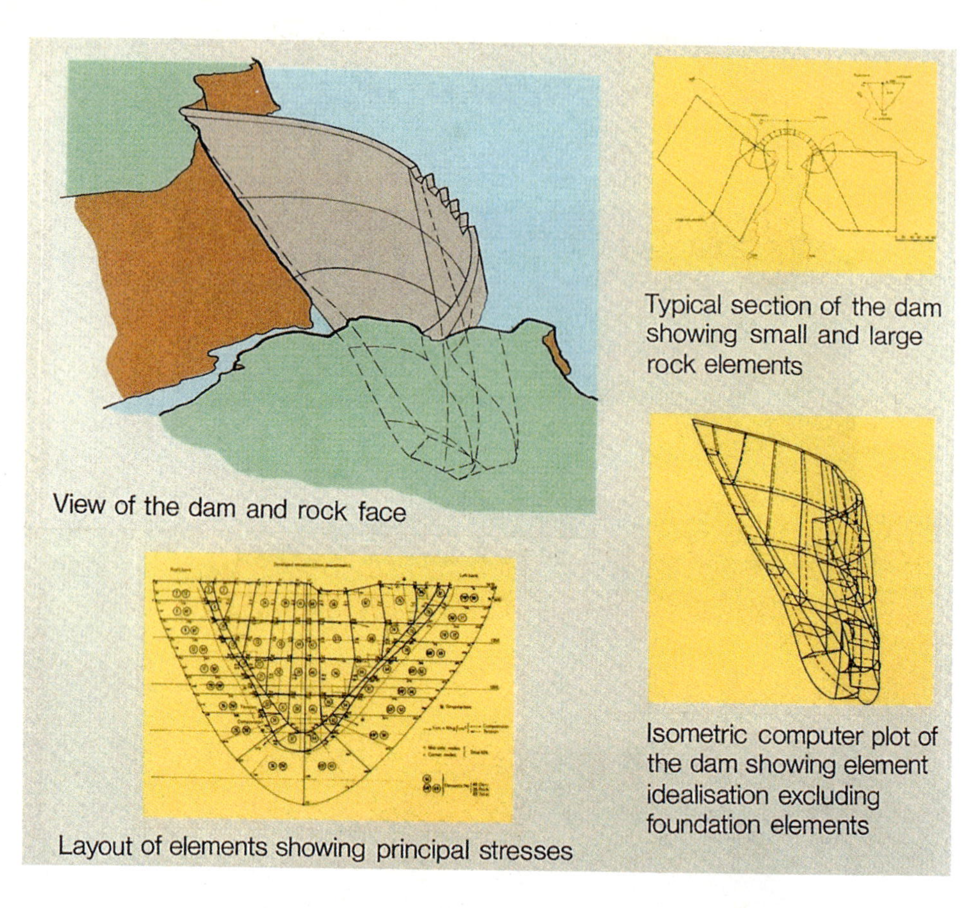

Plate 2 Analysis of an arch dam in China

An early three dimensional analysis (1970). Analysis by OCZ and Cedric Taylor, Department of Civil Engineering, University of Wales Swansea.

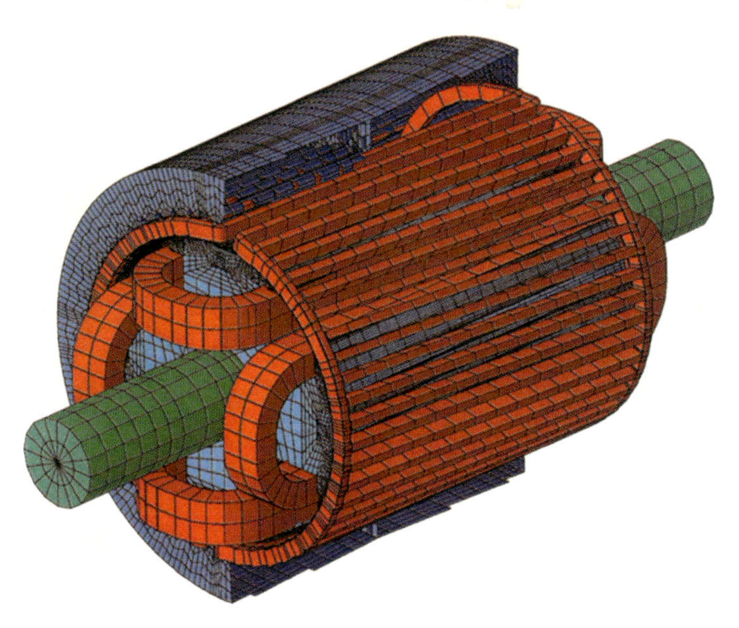

(a) Stator slots have been 'skewed' by one slot pitch in order to reduce cogging torque, i.e. torque associated with the alignment and misalignment of rotor poles and stator teeth.

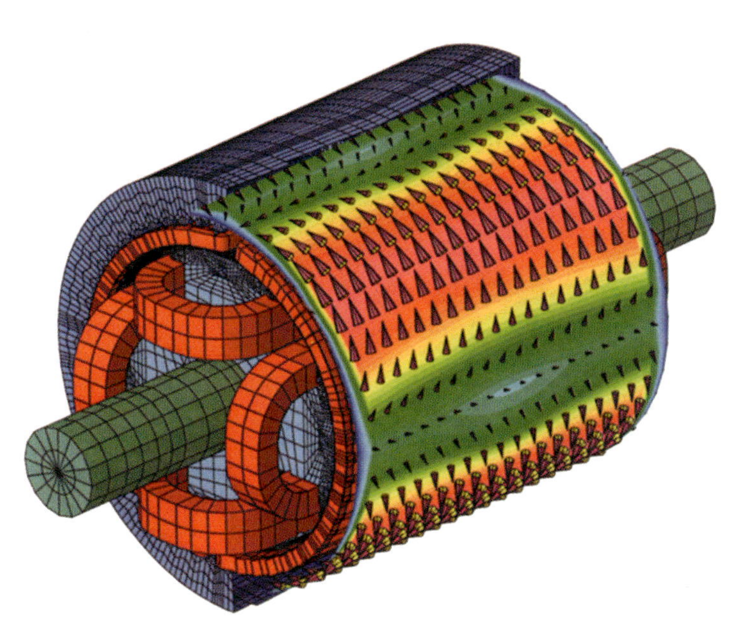

(b) Contours and vectors that indicate the strength and direction of the magnetic fields in the stator core back.

Plate 3 4 pole generator

Courtesy of Mr. William Trowbridge, Vector Fields, Kidlington, Oxfordshire. Source: J. Simkin and C.W. Trowbridge 'Three dimensional nonlinear electromagnetic field computations, using scalar potentials', *Proc. IEE*, **127**, Pt. B, No. 6, Nov. 1980.

Plate 4 Scattering of a plane electromagnetic wave by a perfectly conducting aircraft
Courtesy of Prof. Ken Morgan, Department of Civil Engineering, University of Wales Swansea. Source: K. Morgan, P.J. Brookes, O. Hassan and N. P. Wetherill, 'Parallel processing for the simulation of problems involving scattering of electromagnetic waves, *Comp. Meth. Appl. Eng.,* **152, 157-174,** 1998

The Finite Element Method: Its Basis and Fundamentals

Sixth edition

O.C. Zienkiewicz, CBE, FRS
UNESCO Professor of Numerical Methods in Engineering
International Centre for Numerical Methods in Engineering, Barcelona
Previously Director of the Institute for Numerical Methods in Engineering
University of Wales, Swansea

R.L. Taylor
Professor in the Graduate School
Department of Civil and Environmental Engineering
University of California at Berkeley
Berkeley, California

J.Z. Zhu
Senior Scientist
ESI US R & D Inc.
5850 Waterloo Road, Suite 140
Columbia, Maryland

ELSEVIER
BUTTERWORTH
HEINEMANN

AMSTERDAM • BOSTON • HEIDELBERG • LONDON • NEW YORK • OXFORD
PARIS • SAN DIEGO • SAN FRANCISCO • SINGAPORE • SYDNEY • TOKYO

Butterworth-Heinemann is an imprint of Elsevier
Linacre House, Jordan Hill, Oxford OX2 8DP, UK
30 Corporate Drive, Suite 400, Burlington, MA 01803, USA

First edition published by McGraw-Hill 1967
Fifth edition published by Butterworth Heinemann 2000
Reprinted 2002
Sixth edition 2005
Reprinted 2005, 2006 (three times)

British Library Cataloguing in Publication Data
A catalogue record for this book is available from the British Library

Library of Congress Cataloging-in-Publication Data
A catalog record for this book is available from the Library of Congress

ISBN–13: 978-07506-6320-5
ISBN–10: 0-7506-6320-0

Published with the cooperation of CIMNE, the international Centre for Numerical Methods in Engineering, Barcelona, Spain (www.cimne.upc.es)

Printed and bound in *Great Britain*

06 07 08 09 10 10 9 8 7 6 5 4

Dedication

This book is dedicated to our wives Helen, Mary Lou and Song and our families for their support and patience during the preparation of this book, and also to all of our students and colleagues who over the years have contributed to our knowledge of the finite element method. In particular we would like to mention Professor Eugenio Oñate and his group at CIMNE for their help, encouragement and support during the preparation process.

Contents

Preface

It is thirty-eight years since the *The Finite Element Method in Structural and Continuum Mechanics* was first published. This book, which was the first dealing with the finite element method, provided the basis from which many further developments occurred. The expanding research and field of application of finite elements led to the second edition in 1971, the third in 1977, the fourth as two volumes in 1989 and 1991 and the fifth as three volumes in 2000. The size of each of these editions expanded geometrically (from 272 pages in 1967 to the fifth edition of 1482 pages). This was necessary to do justice to a rapidly expanding field of professional application and research. Even so, much filtering of the contents was necessary to keep these editions within reasonable bounds.

In the present edition we have decided not to pursue the course of having three contiguous volumes but rather we treat the whole work as an assembly of three separate works, each one capable of being used without the others and each one appealing perhaps to a different audience. Though naturally we recommend the use of the whole ensemble to people wishing to devote much of their time and study to the finite element method.

In particular the first volume which was entitled *The Finite Element Method: The Basis* is now renamed *The Finite Element Method: Its Basis and Fundamentals*. This volume has been considerably reorganized from the previous one and is now, we believe, better suited for teaching fundamentals of the finite element method. The sequence of chapters has been somewhat altered and several examples of worked problems have been added to the text. A set of problems to be worked out by students has also been provided.

In addition to its previous content this book has been considerably enlarged by including more emphasis on use of higher order shape functions in formulation of problems and a new chapter devoted to the subject of automatic mesh generation. A beginner in the finite element field will find very rapidly that much of the work of solving problems consists of preparing a suitable mesh to deal with the whole problem and as the size of computers has seemed to increase without limits the size of problems capable of being dealt with is also increasing. Thus, meshes containing sometimes more than several million nodes have to be prepared with details of the material interfaces, boundaries and loads being well specified. There are many books devoted exclusively to the subject of mesh generation but we feel that the essence of dealing with this difficult problem should be included here for those who wish to have a complete 'encyclopedic' knowledge of the subject.

The chapter on computational methods is much reduced by transferring the computer source program and user instructions to a web site.† This has the very substantial advantage of not only eliminating errors in program and manual but also in ensuring that the readers have the benefit of the most recent version of the program available at all times.

The two further volumes form again separate books and here we feel that a completely different audience will use them. The first of these is entitled *The Finite Element Method in Solid and Structural Mechanics* and the second is a text entitled *The Finite Element Method in Fluid Dynamics*. Each of these two volumes is a standalone text which provides the full knowledge of the subject for those who have acquired an introduction to the finite element method through other texts. Of course the viewpoint of the authors introduced in this volume will be continued but it is possible to start at a different point.

We emphasize here the fact that all three books stress the importance of considering the finite element method as a unique and whole basis of approach and that it contains many of the other numerical analysis methods as special cases. Thus, imagination and knowledge should be combined by the readers in their endeavours.

The authors are particularly indebted to the International Center of Numerical Methods in Engineering (CIMNE) in Barcelona who have allowed their pre- and post-processing code (GiD) to be accessed from the web site. This allows such difficult tasks as mesh generation and graphic output to be dealt with efficiently. The authors are also grateful to Professors Eric Kasper and Jose Luis Perez-Aparicio for their careful scrutiny of the entire text and Drs Joaquim Peiró and C.K. Lee for their review of the new chapter on mesh generation.

Resources to accompany this book

Worked solutions to selected problems in this book are available online for teachers and lecturers who either adopt or recommend the text. Please visit http://books.elsevier.com/manuals and follow the registration and log in instructions on screen.

OCZ, RLT and JZZ

† Complete source code and user manual for program *FEAPpv* may be obtained at no cost from the publisher's web page: http://books.elsevier.com/companions or from the authors' web page: http://www.ce.berkeley.edu/~rlt

1

The standard discrete system and origins of the finite element method

1.1 Introduction

The limitations of the human mind are such that it cannot grasp the behaviour of its complex surroundings and creations in one operation. Thus the process of subdividing all systems into their individual components or 'elements', whose behaviour is readily understood, and then rebuilding the original system from such components to study its behaviour is a natural way in which the engineer, the scientist, or even the economist proceeds.

In many situations an adequate model is obtained using a finite number of well-defined components. We shall term such problems *discrete*. In others the subdivision is continued indefinitely and the problem can only be defined using the mathematical fiction of an infinitesimal. This leads to differential equations or equivalent statements which imply an infinite number of elements. We shall term such systems *continuous*.

With the advent of digital computers, *discrete* problems can generally be solved readily even if the number of elements is very large. As the capacity of all computers is finite, *continuous* problems can only be solved exactly by mathematical manipulation. The available mathematical techniques for exact solutions usually limit the possibilities to oversimplified situations.

To overcome the intractability of realistic types of continuous problems (continuum), various methods of *discretization* have from time to time been proposed by engineers, scientists and mathematicians. All involve an *approximation* which, hopefully, approaches in the limit the true continuum solution as the number of discrete variables increases.

The discretization of continuous problems has been approached differently by mathematicians and engineers. Mathematicians have developed general techniques applicable directly to differential equations governing the problem, such as finite difference approximations,[1–3] various weighted residual procedures,[4, 5] or approximate techniques for determining the stationarity of properly defined 'functionals'.[6] The engineer, on the other hand, often approaches the problem more intuitively by creating an analogy between real discrete elements and finite portions of a continuum domain. For instance, in the field of solid mechanics McHenry,[7] Hrenikoff,[8] Newmark,[9] and Southwell[2] in the 1940s, showed that reasonably good solutions to an elastic continuum problem can be obtained by replacing small portions of the continuum by an arrangement of simple elastic bars. Later, in the same context, Turner *et al.*[10] showed that a more direct, but no less intuitive, substitution of properties can be made much more effectively by considering that small portions or 'elements' in a continuum behave in a simplified manner.

It is from the engineering 'direct analogy' view that the term 'finite element' was born. Clough[11] appears to be the first to use this term, which implies in it a direct use of a *standard methodology applicable to discrete systems* (see also reference 12 for a history on early developments). Both conceptually and from the computational viewpoint this is of the utmost importance. The first allows an improved understanding to be obtained; the second offers a unified approach to the variety of problems and the development of standard computational procedures.

Since the early 1960s much progress has been made, and today the purely mathematical and 'direct analogy' approaches are fully reconciled. It is the object of this volume to present a view of the finite element method as *a general discretization procedure of continuum problems posed by mathematically defined statements*.

In the analysis of problems of a discrete nature, a standard methodology has been developed over the years. The civil engineer, dealing with structures, first calculates force–displacement relationships for each element of the structure and then proceeds to assemble the whole by following a well-defined procedure of establishing local equilibrium at each 'node' or connecting point of the structure. The resulting equations can be solved for the unknown displacements. Similarly, the electrical or hydraulic engineer, dealing with a network of electrical components (resistors, capacitances, etc.) or hydraulic conduits, first establishes a relationship between currents (fluxes) and potentials for individual elements and then proceeds to assemble the system by ensuring continuity of flows.

All such analyses follow a standard pattern which is universally adaptable to discrete systems. It is thus possible to define a *standard discrete system*, and this chapter will be primarily concerned with establishing the processes applicable to such systems. Much of what is presented here will be known to engineers, but some reiteration at this stage is advisable. As the treatment of elastic solid structures has been the most developed area of activity this will be introduced first, followed by examples from other fields, before attempting a complete generalization.

The existence of a unified treatment of 'standard discrete problems' leads us to the first definition of the finite element process as a method of approximation to continuum problems such that

(a) the continuum is divided into a finite number of parts (elements), the behaviour of which is specified by a finite number of parameters, and
(b) the solution of the complete system as an assembly of its elements follows precisely the same rules as those applicable to *standard discrete problems*.

The development of the standard discrete system can be followed most closely through the work done in structural engineering during the nineteenth and twentieth centuries. It appears that the 'direct stiffness process' was first introduced by Navier in the early part of the nineteenth century and brought to its modern form by Clebsch[13] and others. In the twentieth century much use of this has been made and Southwell,[14] Cross[15] and others have revolutionized many aspects of structural engineering by introducing a relaxation iterative process. Just before the Second World War matrices began to play a larger part in casting the equations and it was convenient to restate the procedures in matrix form. The work of Duncan and Collar,[16–18] Argyris,[19] Kron[20] and Turner[10] should be noted. A thorough study of direct stiffness and related methods was recently conducted by Samuelsson.[21]

It will be found that most classical mathematical approximation procedures as well as the various direct approximations used in engineering fall into this category. It is thus difficult to determine the origins of the finite element method and the precise moment of its invention.

Table 1.1 shows the process of evolution which led to the present-day concepts of finite element analysis. A historical development of the subject of finite element methods has been presented by the first author in references 34–36. Chapter 3 will give, in more detail, the mathematical basis which emerged from these classical ideas.[1, 22–27, 29, 30, 32]

1.2 The structural element and the structural system

To introduce the reader to the general concept of discrete systems we shall first consider a structural engineering example with linear elastic behaviour.

Figure 1.1 represents a two-dimensional structure assembled from individual components and interconnected at the nodes numbered 1 to 6. The joints at the nodes, in this case, are pinned so that moments cannot be transmitted.

As a starting point it will be assumed that by separate calculation, or for that matter from the results of an experiment, the characteristics of each element are precisely known. Thus, if a typical element labelled (1) and associated with nodes 1, 2, 3 is examined, the forces acting at the nodes are uniquely defined by the displacements of these nodes, the distributed loading acting on the element (p), and its initial strain. The last may be due to temperature, shrinkage, or simply an initial 'lack of fit'. The forces and the corresponding

Table 1.1 History of approximate methods

ENGINEERING | MATHEMATICS

Trial functions
Rayleigh 1870[22]
Ritz 1908[23]

Finite differences
Richardson 1910[1]
Liebman 1918[24]
Southwell 1946[2]

Variational methods
Rayleigh 1870[22]
Ritz 1908[23]

Weighted residuals
Gauss 1795[25]
Galerkin 1915[26]
Biezeno–Koch 1923[27]

Structural analogue substitution
Hrenikoff 1941[8]
McHenry 1943[28]
Newmark 1949[9]

Piecewise continuous trial functions
Courant 1943[29]
Prager–Synge 1947[30]
Argyris 1955[19]
Zienkiewicz 1964[31]

Direct continuum elements
Turner *et al.* 1956[10]

Variational finite differences
Varga 1962[32]
Wilkins 1964[33]

PRESENT-DAY FINITE ELEMENT METHOD

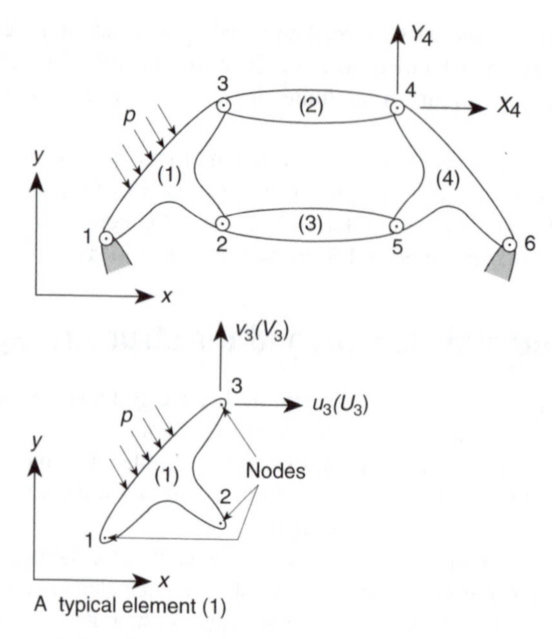

Fig. 1.1 A typical structure built up from interconnected elements.

displacements are defined by appropriate components (U, V and u, v) in a common co-ordinate system (x, y).

Listing the forces acting on all the nodes (three in the case illustrated) of the element (1) as a matrix† we have

$$\mathbf{q}^1 = \begin{Bmatrix} \mathbf{q}_1^1 \\ \mathbf{q}_2^1 \\ \mathbf{q}_3^1 \end{Bmatrix} \qquad \mathbf{q}_1^1 = \begin{Bmatrix} U_1 \\ V_1 \end{Bmatrix}, \qquad \text{etc.} \tag{1.1}$$

and for the corresponding nodal displacements

$$\mathbf{u}^1 = \begin{Bmatrix} \mathbf{u}_1^1 \\ \mathbf{u}_2^1 \\ \mathbf{u}_3^1 \end{Bmatrix} \qquad \mathbf{u}_1^1 = \begin{Bmatrix} u_1 \\ v_1 \end{Bmatrix}, \qquad \text{etc.} \tag{1.2}$$

Assuming linear elastic behaviour of the element, the characteristic relationship will always be of the form

$$\mathbf{q}^1 = \mathbf{K}^1\mathbf{u}^1 + \mathbf{f}^1 \tag{1.3}$$

in which $\mathbf{f}^1$ represents the nodal forces required to balance any concentrated or distributed loads acting on the element. The first of the terms represents the forces induced by displacement of the nodes. The matrix $\mathbf{K}^e$ is known as the *stiffness matrix* for the element (e).

Equation (1.3) is illustrated by an example of an element with three nodes with the interconnection points capable of transmitting only two components of force. Clearly, the

†A limited knowledge of matrix algebra will be assumed throughout this book. This is necessary for reasonable conciseness and forms a convenient book-keeping form. For readers not familiar with the subject a brief appendix (Appendix A) is included in which sufficient principles of matrix algebra are given to follow the development intelligently. Matrices and vectors will be distinguished by bold print throughout.

same arguments and definitions will apply generally. An element (2) of the hypothetical structure will possess only two points of interconnection; others may have quite a large number of such points. Quite generally, therefore,

$$\mathbf{q}^e = \begin{Bmatrix} \mathbf{q}_1^e \\ \mathbf{q}_2^e \\ \vdots \\ \mathbf{q}_m^e \end{Bmatrix} \quad \text{and} \quad \mathbf{u}^e = \begin{Bmatrix} \mathbf{u}_1 \\ \mathbf{u}_2 \\ \vdots \\ \mathbf{u}_m \end{Bmatrix} \tag{1.4}$$

with each $\mathbf{q}_a^e$ and $\mathbf{u}_a$ possessing the same number of components or *degrees of freedom*.

The stiffness matrices of the element will clearly always be square and of the form

$$\mathbf{K}^e = \begin{bmatrix} \mathbf{K}_{11}^e & \mathbf{K}_{12}^e & \cdots & \mathbf{K}_{1m}^e \\ \mathbf{K}_{21}^e & \ddots & & \vdots \\ \vdots & \vdots & & \vdots \\ \mathbf{K}_{m1}^e & \cdots & \cdots & \mathbf{K}_{mm}^e \end{bmatrix} \tag{1.5}$$

in which $\mathbf{K}_{11}^e$, $\mathbf{K}_{12}^e$, etc., are submatrices which are again square and of the size $l \times l$, where l is the number of force and displacement components to be considered at each node.

The element properties were assumed to follow a simple linear relationship. In principle, similar relationships could be established for non-linear materials, but discussion of such problems will be postponed at this stage. In most cases considered in this volume the element matrices $\mathbf{K}^e$ will be symmetric.

1.3 Assembly and analysis of a structure

Consider again the hypothetical structure of Fig. 1.1. To obtain a complete solution the two conditions of

(a) displacement compatibility and
(b) equilibrium

have to be satisfied throughout.

Any system of nodal displacements $\mathbf{u}$:

$$\mathbf{u} = \begin{Bmatrix} \mathbf{u}_1 \\ \vdots \\ \mathbf{u}_n \end{Bmatrix} \tag{1.6}$$

listed now for the whole structure in which all the elements participate, automatically satisfies the first condition.

As the conditions of overall equilibrium have already been satisfied *within* an element, all that is necessary is to establish equilibrium conditions at the nodes (or assembly points) of the structure. The resulting equations will contain the displacements as unknowns, and once these have been solved the structural problem is determined. The internal forces in elements, or the stresses, can easily be found by using the characteristics established *a priori* for each element.

If now the equilibrium conditions of a typical node, a, are to be established, the sum of the component forces contributed by the elements meeting at the node are simply accumulated. Thus, considering *all* the force components we have

$$\sum_{e=1}^{m} \mathbf{q}_a^e = \mathbf{q}_a^1 + \mathbf{q}_a^2 + \cdots = \mathbf{0} \tag{1.7}$$

in which $\mathbf{q}_a^1$ is the force contributed to node a by element 1, $\mathbf{q}_a^2$ by element 2, etc. Clearly, only the elements which include point a will contribute non-zero forces, but for conciseness in notation all the elements are included in the summation.

Substituting the forces contributing to node a from the definition (1.3) and noting that nodal variables $\mathbf{u}_a$ are common (thus omitting the superscript e), we have

$$\left(\sum_{e=1}^{m} \mathbf{K}_{a1}^e\right)\mathbf{u}_1 + \left(\sum_{e=1}^{m} \mathbf{K}_{a2}^e\right)\mathbf{u}_2 + \cdots + \sum_{e=1}^{m} \mathbf{f}_i^e = \mathbf{0} \tag{1.8}$$

The summation again only concerns the elements which contribute to node a. If all such equations are assembled we have simply

$$\mathbf{K}\mathbf{u} + \mathbf{f} = \mathbf{0} \tag{1.9}$$

in which the submatrices are

$$\mathbf{K}_{ab} = \sum_{e=1}^{m} \mathbf{K}_{ab}^e \quad \text{and} \quad \mathbf{f}_a = \sum_{e=1}^{m} \mathbf{f}_a^e \tag{1.10}$$

with summations including all elements. This simple rule for assembly is very convenient because as soon as a coefficient for a particular element is found it can be put immediately into the appropriate 'location' specified in the computer. *This general assembly process can be found to be the common and fundamental feature of all finite element calculations and should be well understood by the reader.*

If different types of structural elements are used and are to be coupled it must be remembered that at any given node the rules of matrix summation permit this to be done only if these are of identical size. The individual submatrices to be added have therefore to be built up of the same number of individual components of force or displacement.

1.4 The boundary conditions

The system of equations resulting from Eq. (1.9) can be solved once the prescribed support displacements have been substituted. In the example of Fig. 1.1, where both components of displacement of nodes 1 and 6 are zero, this will mean the substitution of

$$\mathbf{u}_1 = \mathbf{u}_6 = \begin{Bmatrix} 0 \\ 0 \end{Bmatrix}$$

which is equivalent to reducing the number of equilibrium equations (in this instance 12) by deleting the first and last pairs and thus reducing the total number of unknown displacement components to eight. It is, nevertheless, often convenient to assemble the equation according to relation (1.9) so as to include all the nodes.

Clearly, without substitution of a minimum number of prescribed displacements to prevent rigid body movements of the structure, it is impossible to solve this system, because the displacements cannot be uniquely determined by the forces in such a situation. This physically obvious fact will be interpreted mathematically as the matrix $\mathbf{K}$ being singular, i.e., not possessing an inverse. The prescription of appropriate displacements after the assembly stage will permit a unique solution to be obtained by deleting appropriate rows and columns of the various matrices.

If all the equations of a system are assembled, their form is

$$
\begin{aligned}
&\mathbf{K}_{11}\mathbf{u}_1 + \mathbf{K}_{12}\mathbf{u}_2 + \cdots + \mathbf{f}_1 = \mathbf{0} \\
&\mathbf{K}_{21}\mathbf{u}_1 + \mathbf{K}_{22}\mathbf{u}_2 + \cdots + \mathbf{f}_2 = \mathbf{0} \qquad (1.11) \\
&\text{etc.}
\end{aligned}
$$

and it will be noted that if any displacement, such as $\mathbf{u}_1 = \bar{\mathbf{u}}_1$, is prescribed then the total 'force' $\mathbf{f}_1$ cannot be simultaneously specified and remains unknown. The first equation could then be *deleted* and substitution of known values $\bar{\mathbf{u}}_1$ made in the remaining equations.

When all the boundary conditions are inserted the equations of the system can be solved for the unknown nodal displacements and the internal forces in each element obtained.

1.5 Electrical and fluid networks

Identical principles of deriving element characteristics and of assembly will be found in many non-structural fields. Consider, for instance, the assembly of electrical resistances shown in Fig. 1.2.

If a typical resistance element, ab, is isolated from the system we can write, by Ohm's law, the relation between the currents (J) *entering* the element at the ends and the end voltages (V) as

$$J_a^e = \frac{1}{r^e}(V_a - V_b) \qquad \text{and} \qquad J_b^e = \frac{1}{r^e}(V_b - V_a) \qquad (1.12)$$

or in matrix form

$$\begin{Bmatrix} J_a^e \\ J_b^e \end{Bmatrix} = \frac{1}{r^e}\begin{bmatrix} 1 & -1 \\ -1 & 1 \end{bmatrix}\begin{Bmatrix} V_a \\ V_b \end{Bmatrix}$$

which in our standard form is simply

$$\mathbf{J}^e = \mathbf{K}^e\mathbf{V}^e \qquad (1.13)$$

This form clearly corresponds to the stiffness relationship (1.3); indeed if an external current were supplied along the length of the element the element 'force' terms could also be found.

To assemble the whole network the continuity of the voltage (V) at the nodes is assumed and a current balance imposed there. With no external input of current at node a we must

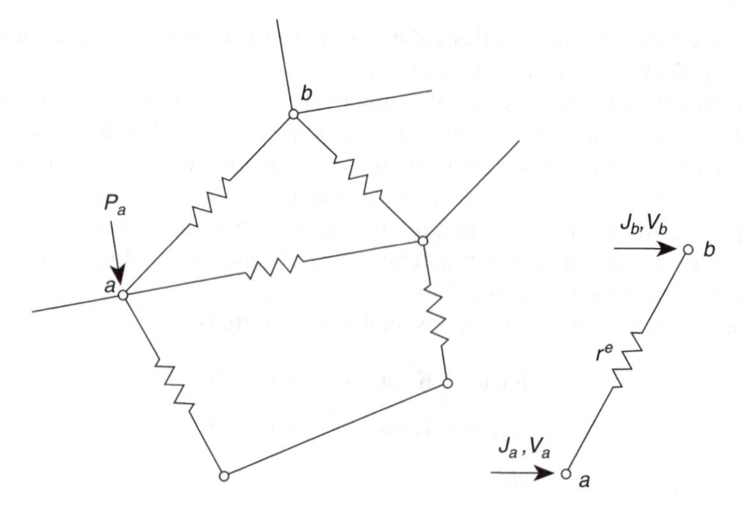

Fig. 1.2 A network of electrical resistances.

have, with complete analogy to Eq. (1.8),

$$\sum_{b=1}^{n} \sum_{e=1}^{m} K_{ab}^{e} V_b = 0 \tag{1.14}$$

where the second summation is over all 'elements', and once again for all the nodes

$$\mathbf{KV} = \mathbf{0} \tag{1.15}$$

in which

$$K_{ab} = \sum_{e=1}^{m} K_{ab}^{e}$$

Matrix notation in the latter has been dropped since the quantities such as voltage and current, and hence also the coefficients of the 'stiffness' matrix, are scalars.

If the resistances were replaced by fluid-carrying pipes in which a laminar regime pertained, an identical formulation would once again result, with V standing for the hydraulic head and J for the flow.

For pipe networks that are usually encountered, however, the linear laws are in general not valid and non-linear equations must be solved.

Finally it is perhaps of interest to mention the more general form of an electrical network subject to an alternating current. It is customary to write the relationships between the current and voltage in *complex arithmetic form* with the resistance being replaced by complex impedance. Once again the standard forms of (1.13)–(1.15) will be obtained but with each quantity divided into real and imaginary parts.

Identical solution procedures can be used if the equality of the real and imaginary quantities is considered at each stage. Indeed with modern digital computers it is possible to use standard programming practice, making use of facilities available for dealing with complex numbers. Reference to some problems of this class will be made in the sections dealing with vibration problems in Chapter 15.

1.6 The general pattern

An example will be considered to consolidate the concepts discussed in this chapter. This is shown in Fig. 1.3(a) where five discrete elements are interconnected. These may be of structural, electrical, or any other linear type. In the solution:

The first step is the determination of element properties from the geometric material and loading data. For each element the 'stiffness matrix' as well as the corresponding 'nodal

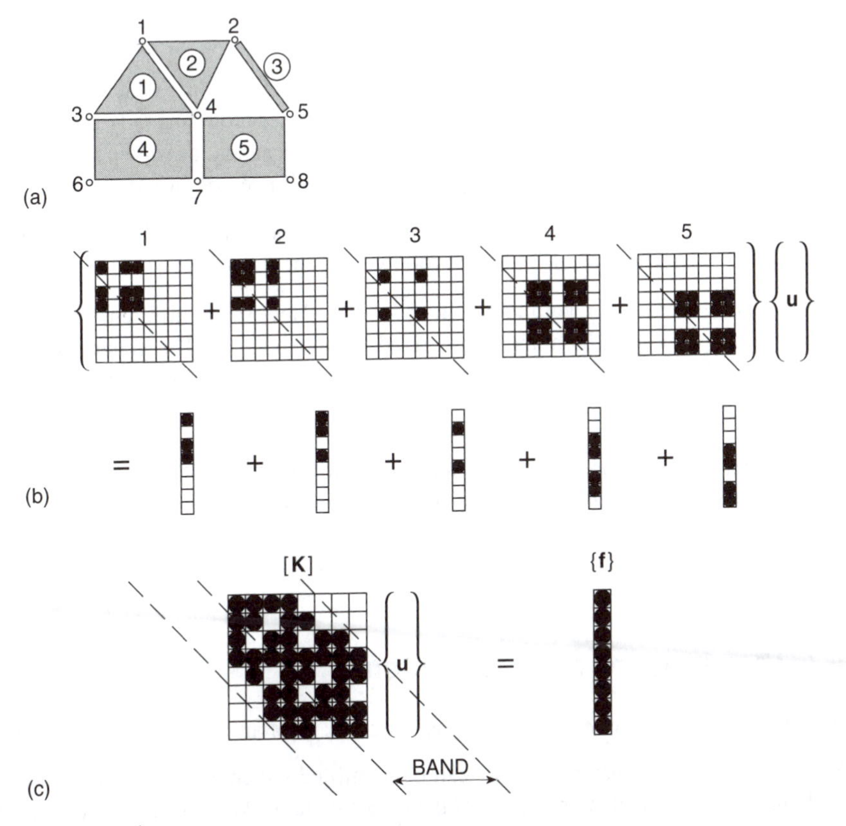

Fig. 1.3 The general pattern.

loads' are found in the form of Eq. (1.3). Each element shown in Fig. 1.3(a) has its own identifying number and specified nodal connection. For example:

element	1	connection	1	3	4	
	2		1	4	2	
	3		2	5		
	4		3	6	7	4
	5		4	7	8	5

Assuming that properties are found in global coordinates we can enter each 'stiffness' or 'force' component in its position of the global matrix as shown in Fig. 1.3(b). Each shaded square represents a single coefficient or a submatrix of type $\mathbf{K}_{ab}$ if more than one quantity is being considered at the nodes. Here the separate contribution of each element is shown and the reader can verify the position of the coefficients. Note that the various types of 'elements' considered here present no difficulty in specification. (All 'forces', including nodal ones, are here associated with elements for simplicity.)

The second step is the assembly of the final equations of the type given by Eq. (1.9). This is accomplished according to the rule of Eq. (1.10) by *simple addition* of all numbers in the appropriate space of the global matrix. The result is shown in Fig. 1.3(c) where the non-zero coefficients are indicated by shading.

If the matrices are symmetric only the half above the diagonal shown needs, in fact, to be found.

All the non-zero coefficients are confined within a *band* or *profile* which can be calculated *a priori* for the nodal connections. Thus in computer programs only the storage of the elements within the profile (or sparse structure) is necessary, as shown in Fig. 1.3(c). Indeed, if $\mathbf{K}$ is symmetric only the upper (or lower) half need be stored.

The third step is the insertion of prescribed boundary conditions into the final assembled matrix, as discussed in Sec. 1.3. This is followed by the final step.

The final step solves the resulting equation system. Here many different methods can be employed, some of which are summarized in Appendix C. The general subject of equation solving, though extremely important, is in general beyond the scope of this book.

The final step discussed above can be followed by substitution to obtain stresses, currents, or other desired *output* quantities. All operations involved in structural or other network analysis are thus of an extremely simple and repetitive kind. We can now define *the standard discrete system* as one in which such conditions prevail.

1.7 The standard discrete system

In the *standard discrete system*, whether it is structural or of any other kind, we find that:

1. A set of discrete parameters, say $\mathbf{u}_a$, can be identified which describes simultaneously the behaviour of each element, e, and of the whole system. We shall call these the *system parameters*.

2. For each element a set of quantities $\mathbf{q}_a^e$ can be computed in terms of the system parameters $\mathbf{u}_a$. The general function relationship can be non-linear, for example

$$\mathbf{q}_a^e = \mathbf{q}_a^e(\mathbf{u}) \tag{1.16}$$

but in many cases a linear form exists giving

$$\mathbf{q}_a^e = \mathbf{K}_{a1}^e \mathbf{u}_1 + \mathbf{K}_{a2}^e \mathbf{u}_2 + \cdots + \mathbf{f}_a^e \tag{1.17}$$

3. The final *system equations* are obtained by a simple addition

$$\mathbf{r}_a = \sum_{e=1}^{m} \mathbf{q}_a^e = \mathbf{0} \tag{1.18}$$

where $\mathbf{r}_a$ are system quantities (often prescribed as zero). In the linear case this results in a system of equations

$$\mathbf{Ku} + \mathbf{f} = \mathbf{0} = 0 \tag{1.19}$$

such that

$$\mathbf{K}_{ab} = \sum_{e=1}^{m} \mathbf{K}_{ab}^e \quad \text{and} \quad \mathbf{f}_a = \sum_{e=1}^{m} \mathbf{f}_a^e \tag{1.20}$$

from which the solution for the system variables $\mathbf{u}$ can be found after imposing necessary boundary conditions.

The reader will observe that this definition includes the structural, hydraulic, and electrical examples already discussed. However, it is broader. In general neither linearity nor symmetry of matrices need exist – although in many problems this will arise naturally. Further, the narrowness of interconnections existing in usual elements is not essential.

While much further detail could be discussed (we refer the reader to specific books for more exhaustive studies in the structural context[37, 38]), we feel that the general exposé given here should suffice for further study of this book.

Only one further matter relating to the change of discrete parameters need be mentioned here. The process of so-called transformation of coordinates is vital in many contexts and must be fully understood.

1.8 Transformation of coordinates

It is often convenient to establish the characteristics of an individual element in a coordinate system which is different from that in which the external forces and displacements of the assembled structure or system will be measured. A different coordinate system may, in fact, be used for every element, to ease the computation. It is a simple matter to transform the coordinates of the displacement and force components of Eq. (1.3) to any other coordinate system. Clearly, it is necessary to do so before an assembly of the structure can be attempted.

Let the local coordinate system in which the element properties have been evaluated be denoted by a prime suffix and the common coordinate system necessary for assembly have no embellishment. The displacement components can be transformed by a suitable matrix of direction cosines $\mathbf{L}$ as

$$\mathbf{u}' = \mathbf{Lu} \tag{1.21}$$

As the corresponding force components must perform the same amount of work in either system†

$$\mathbf{q}^{\mathrm{T}}\mathbf{u} = \mathbf{q}'^{\mathrm{T}}\mathbf{u}' \tag{1.22}$$

On inserting (1.21) we have

$$\mathbf{q}^{\mathrm{T}}\mathbf{u} = \mathbf{q}'^{\mathrm{T}}\mathbf{L}\mathbf{u}$$

or

$$\mathbf{q} = \mathbf{L}^{\mathrm{T}}\mathbf{q}' \tag{1.23}$$

The set of transformations given by (1.21) and (1.23) is called *contravariant.*

To transform 'stiffnesses' which may be available in local coordinates to global ones note that if we write

$$\mathbf{q}' = \mathbf{K}'\mathbf{u}' \tag{1.24}$$

then by (1.23), (1.24), and (1.21)

$$\mathbf{q} = \mathbf{L}^{\mathrm{T}}\mathbf{K}'\mathbf{L}\mathbf{u}$$

or in global coordinates

$$\mathbf{K} = \mathbf{L}^{\mathrm{T}}\mathbf{K}'\mathbf{L} \tag{1.25}$$

In many complex problems an external constraint of some kind may be imagined, enforcing the requirement (1.21) with the number of degrees of freedom of $\mathbf{u}$ and $\mathbf{u}'$ being quite different. Even in such instances the relations (1.22) and (1.23) continue to be valid.

An alternative and more general argument can be applied to many other situations of discrete analysis. We wish to replace a set of parameters $\mathbf{u}$ in which the system equations have been written by another one related to it by a transformation matrix $\mathbf{T}$ as

$$\mathbf{u} = \mathbf{T}\mathbf{v} \tag{1.26}$$

In the linear case the system equations are of the form

$$\mathbf{K}\mathbf{u} = -\mathbf{f} \tag{1.27}$$

and on the substitution we have

$$\mathbf{K}\mathbf{T}\mathbf{v} = -\mathbf{f} \tag{1.28}$$

The new system can be premultiplied simply by $\mathbf{T}^{\mathrm{T}}$, yielding

$$(\mathbf{T}^{\mathrm{T}}\mathbf{K}\mathbf{T})\mathbf{v} = \mathbf{T}^{\mathrm{T}} - \mathbf{T}^{\mathrm{T}}\mathbf{f} \tag{1.29}$$

which will preserve the symmetry of equations if the matrix $\mathbf{K}$ is symmetric. However, occasionally the matrix $\mathbf{T}$ is not square and expression (1.26) represents in fact *an approximation* in which a larger number of parameters $\mathbf{u}$ is *constrained*. Clearly the system of equations (1.28) gives more equations than are necessary for a solution of the reduced set of parameters $\mathbf{v}$, and the final expression (1.29) presents a reduced system which in some sense approximates the original one.

We have thus introduced the basic idea of approximation, which will be the subject of subsequent chapters where infinite sets of quantities are reduced to finite sets.

† With $(\ )^{\mathrm{T}}$ standing for the transpose of the matrix.

1.9 Problems

1.1 A simple fluid network to transport water is shown in Fig. 1.4. Each 'element' of the network is modelled in terms of the flow, $\mathbf{J}$, and head, $\mathbf{V}$, which are approximated by the linear relation

$$\mathbf{J}^e = -\mathbf{K}^e \mathbf{V}^e$$

where $\mathbf{K}^e$ is the coefficient array for element (e). The individual terms in the flow vector denote the total amount of flow entering (+) or leaving (−) each end point. The properties of the elements are given by

$$\mathbf{K}^e = c^e \begin{bmatrix} 3 & -2 & -1 \\ -2 & 4 & -2 \\ -1 & -2 & 3 \end{bmatrix}$$

for elements 1 and 4, and for elements 2 and 3 by

$$\mathbf{K}^e = c^e \begin{bmatrix} 1 & -1 \\ -1 & 1 \end{bmatrix}$$

where c^e is an element related parameter. The system is operating with a known head of 100 m at node 1 and 20 m at node 6. At node 2, 30 cubic metres of water per hour are being used and at node 4, 10 cubic metres per hour.

(a) For all $c^e = 1$, assemble the total matrix from the individual elements to give

$$\mathbf{J} = \mathbf{K}\,\mathbf{V}$$

N.B. $\mathbf{J}$ contains entries for the specified usage and connection points.

(b) Impose boundary conditions by modifying $\mathbf{J}$ and $\mathbf{K}$ such that the known heads at nodes 1 and 6 are recovered.

(c) Solve the equations for the heads at nodes 2 to 5. (Result at node 4 should be $V_4 = 30.8133$ m.)

(d) Determine the flow entering and leaving each element.

1.2 A plane truss may be described as a standard discrete problem by expressing the characteristics for each member in terms of end displacements and forces. The behaviour of the elastic member shown in Fig. 1.5 with modulus E, cross-section A and length L is given by

$$\mathbf{q}' = \mathbf{K}'_e\,\mathbf{u}'$$

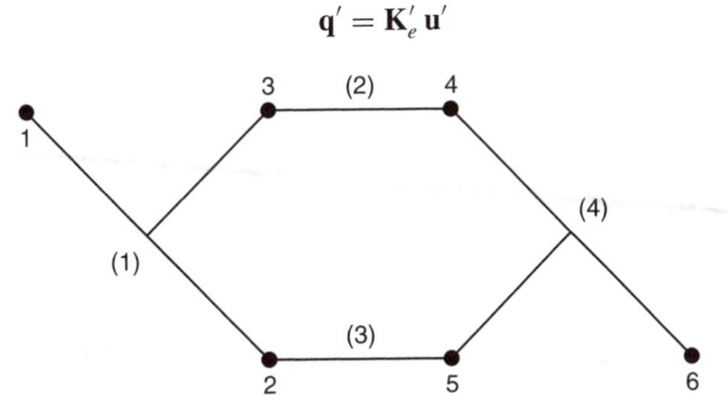

Fig. 1.4 Fluid network for Problem 1.1.

where

$$\mathbf{q}' = \begin{Bmatrix} U_1' \\ U_2' \end{Bmatrix}; \quad \mathbf{u}' = \begin{Bmatrix} u_1' \\ u_2' \end{Bmatrix} \quad \text{and} \quad \mathbf{K}_e' = \frac{EA}{L} \begin{bmatrix} 1 & -1 \\ -1 & 1 \end{bmatrix}$$

To obtain the final assembled matrices for a standard discrete problem it is necessary to transform the behaviour to a global frame using Eqs 1.23 and 1.25 where

$$\mathbf{L} = \begin{bmatrix} \cos\theta & \sin\theta & 0 & 0 \\ 0 & 0 & \cos\theta & \sin\theta \end{bmatrix}; \quad \mathbf{q} = \begin{Bmatrix} U_1 \\ V_1 \\ U_2 \\ V_2 \end{Bmatrix} \quad \text{and} \quad \mathbf{u} = \begin{Bmatrix} u_1 \\ v_1 \\ u_2 \\ v_2 \end{Bmatrix}$$

(a) Compute relations for $\mathbf{q}$ and $\mathbf{K}$ in terms of $\mathbf{L}$, $\mathbf{q}'$ and $\mathbf{K}_e'$.

(b) If the numbering for the end nodes is reversed what is the final form for $\mathbf{K}$ compared to that given in (a)? Verify your answer when $\theta = 30^\circ$.

1.3 A plane truss has nodes numbered as shown in Fig. 1.6(a).

(a) Use the procedure shown in Fig. 1.3 to define the non-zero structure of the coefficient matrix $\mathbf{K}$. Compute the maximum bandwidth.

(b) Determine the non-zero structure of $\mathbf{K}$ for the numbering of nodes shown in 1.6(b). Compute the maximum bandwidth.

Which order produces the smallest band?

1.4 Write a small computer program (e.g., using MATLAB[39]) to solve the truss problem shown in Fig. 1.6(b). Let the total span of the truss be 2.5 m and the height 0.8 m and use steel as the property for each member with $E = 200$ GPa and $A = 0.001\ \text{m}^2$. Restrain node 1 in both the u and v directions and the right bottom node in the v direction only. Apply a vertical load of 100 N at the position of node 6 shown in Fig. 1.6(b). Determine the maximum vertical displacement at any node. Plot the undeformed and deformed position of the truss (increase the magnitude of displacements to make the shape visible on the plot).

You can verify your result using the program *FEAPpv* available at the publisher's web site (see Chapter 18).

1.5 An axially loaded elastic bar has a variable cross-section and lengths as shown in Fig. 1.7(a). The problem is converted into a standard discrete system by considering each prismatic section as a separate member. The array for each member segment is given as

$$\mathbf{q}^e = \mathbf{K}^e \mathbf{u}^e$$

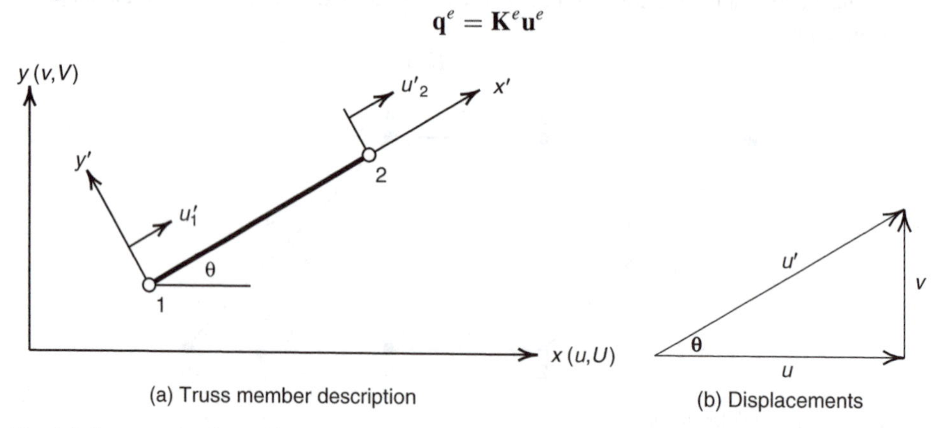

Fig. 1.5 Truss member for Problem 1.2.

where

$$\mathbf{K}^e = \frac{EA_e}{h}\begin{bmatrix} 1 & -1 \\ -1 & 1 \end{bmatrix} \mathbf{q}^e = \begin{Bmatrix} q_e^e \\ q_{e+1}^e \end{Bmatrix} \quad \text{and} \quad \mathbf{u}^e = \begin{Bmatrix} u_e \\ u_{e+1} \end{Bmatrix}$$

Equilibrium for the standard discrete problem at joint e is obtained by combining results from segment $e-1$ and e as

$$q_e^{e-1} + q_e^e + U_e = 0$$

where U_e is any external force applied to a joint. Boundary conditions are applied for any joint at which the value of u_e is known *a priori.*

Solve the problem shown in Fig. 1.7(b) for the joint displacements using the data $E_1 = E_2 = E_3 = 200$ GPa, $A_1 = 25$ cm^2, $A_2 = 20$ cm^2, $A_3 = 12$ cm^2, $L_1 = 37.5$ cm, $L_2 = 25.0$ cm, $L_3 = 12.5$ cm, $P_2 = 10$ kN, $P_3 = -3.5$ kN and $P_4 = 6$ kN.

1.6 Solve Problem 1.5 for the boundary conditions and loading shown in Fig. 1.7(c). Let $E_1 = E_2 = E_3 = 200$ GPa, $A_1 = 30$ cm^2, $A_2 = 20$ cm^2, $A_3 = 10$ cm^2, $L_1 = 37.5$ cm, $L_2 = 30.0$ cm, $L_3 = 25.0$ cm, $P_2 = -10$ kN and $P_3 = 3.5$ kN.

1.7 A tapered bar is loaded by an end load P and a uniform loading b as shown in Fig. 1.8(a). The area varies as $A(x) = A\,x/L$ when the origin of coordinates is located as shown in the figure.

The problem is converted into a standard discrete system by dividing it into equal length segments of constant area as shown in Fig. 1.8(b). The array for each segment is determined from

$$\mathbf{q}^e = \mathbf{K}^e\mathbf{u}^e + \mathbf{f}^e$$

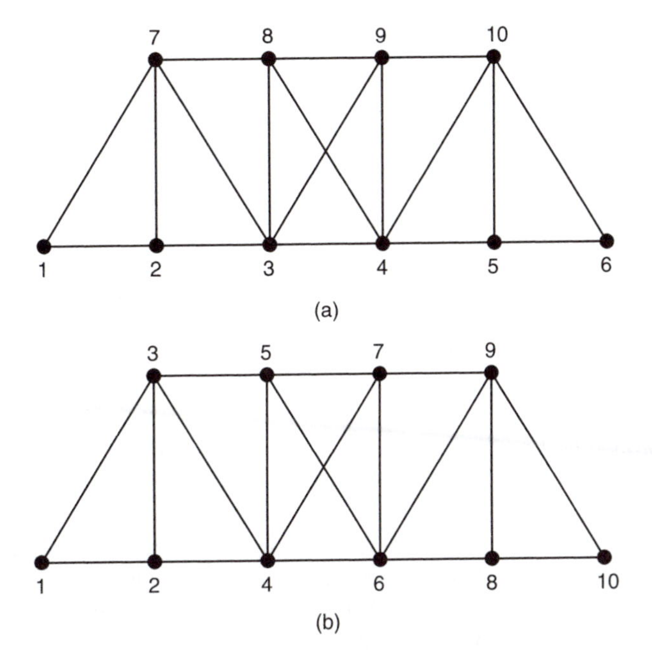

Fig. 1.6 Truss for Problems 1.3 and 1.4.

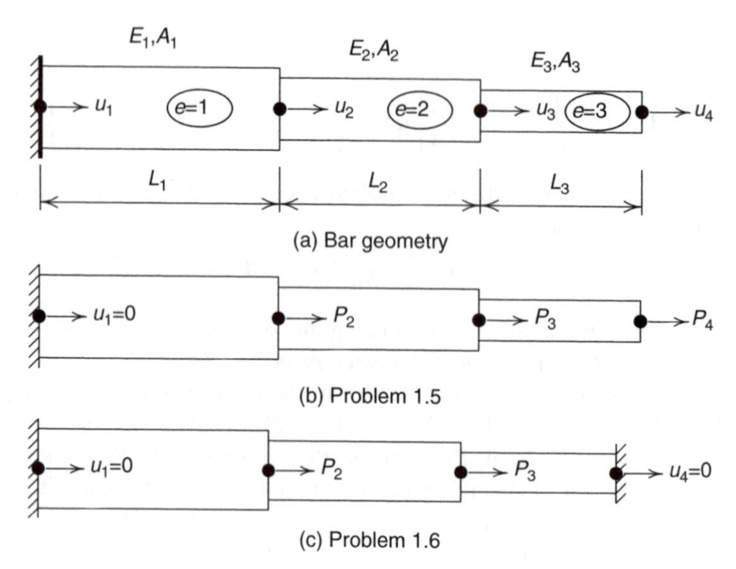

Fig. 1.7 Elastic bars. Problems 1.5 and 1.6.

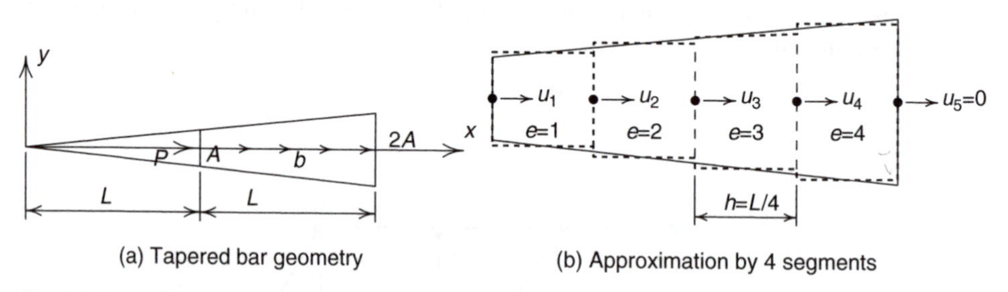

Fig. 1.8 Tapered bar. Problem 1.7.

where $\mathbf{K}^e$ and $\mathbf{u}^e$ are defined in Problem 1.5 and

$$\mathbf{f}^e = \tfrac{1}{2}\, b\, h \begin{Bmatrix} 1 \\ 1 \end{Bmatrix}$$

For the properties $L = 100$ cm, $A = 2$ cm^2, $E = 10^4$ kN/cm^2, $P = 2$ kN, $b = -0.25$ kN/cm and $u(2L) = 0$, the displacement from the solution of the differential equation is $u(L) = -0.03142513$ cm.

Write a small computer program (e.g., using MATLAB[39]) that solves the problem for the case where $e = 1, 2, 4, 8, \cdots$ segments. Continue the solution until the absolute error in the tip displacement is less than 10^{-5} cm (let error be $E = |u(L) - u_1|$ where u_1 is the numerical solution at the end).

References

1. L.F. Richardson. The approximate arithmetical solution by finite differences of physical problems. *Trans. Roy. Soc. (London)*, A210:307–357, 1910.
2. R.V. Southwell. *Relaxation Methods in Theoretical Physics*. Clarendon Press, Oxford, 1st edition, 1946.
3. D.N. de G. Allen. *Relaxation Methods*. McGraw-Hill, London, 1955.
4. S. Crandall. *Engineering Analysis*. McGraw-Hill, New York, 1956.
5. B.A. Finlayson. *The Method of Weighted Residuals and Variational Principles*. Academic Press, New York, 1972.
6. K. Washizu. *Variational Methods in Elasticity and Plasticity*. Pergamon Press, New York, 3rd edition, 1982.
7. D. McHenry. A lattice analogy for the solution of plane stress problems. *J. Inst. Civ. Eng.*, 21:59–82, 1943.
8. A. Hrenikoff. Solution of problems in elasticity by the framework method. *J. Appl. Mech., ASME*, A8:169–175, 1941.
9. N.M. Newmark. Numerical methods of analysis in bars, plates and elastic bodies. In L.E. Grinter, editor, *Numerical Methods in Analysis in Engineering*. Macmillan, New York, 1949.
10. M.J. Turner, R.W. Clough, H.C. Martin, and L.J. Topp. Stiffness and deflection analysis of complex structures. *J. Aero. Sci.*, 23:805–823, 1956.
11. R.W. Clough. The finite element method in plane stress analysis. In *Proc. 2nd ASCE Conf. on Electronic Computation*, Pittsburgh, Pa., Sept. 1960.
12. R.W. Clough. Early history of the finite element method from the view point of a pioneer. *Int. J. Numer. Meth. Eng.*, 60:283–287, 2004.
13. R.F. Clebsch. *Théorie de l'elasticité des corps solides*. Dunod, Paris, 1883.
14. R.V. Southwell. Stress calculation in frame works by the method of systematic relaxation of constraints, Part I & II. *Proc. Roy. Soc. London (A)*, 151:56–95, 1935.
15. Hardy Cross. *Continuous Frames of Reinforced Concrete*. John Wiley & Sons, New York, 1932.
16. W.J. Duncan and A.R. Collar. A method for the solution of oscillation problems by matrices. *Phil. Mag.*, 17:865, 1934. Series 7.
17. W.J. Duncan and A.R. Collar. Matrices applied to the motions of damped systems. *Phil. Mag.*, 19:197, 1935. Series 7.
18. R.R. Frazer, W.J. Duncan, and A.R. Collar. *Elementary Matrices*. Cambridge University Press, London, 1960.
19. J.H. Argyris and S. Kelsey. *Energy Theorems and Structural Analysis*. Butterworths, London, 1960. Reprinted from a series of articles in *Aircraft Eng.*, 1954–55.
20. G. Kron. *Equivalent Circuits of Electrical Machinery*. John Wiley & Sons, New York, 1951.
21. A. Samuelsson. Personal communication, 2003.
22. Lord Rayleigh (J.W. Strutt). On the theory of resonance. *Trans. Roy. Soc. (London)*, A161:77–118, 1870.
23. W. Ritz. Über eine neue Methode zur Lösung gewisser variationsproblem der mathematischen physik. *Journal für die reine und angewandte Mathematik*, 135:1–61, 1908.
24. H. Liebman. Die angenäherte Ermittlung: harmonishen, functionen und konformer Abbildung. *Sitzber. Math. Physik Kl. Bayer Akad. Wiss. München*, 3:65–75, 1918.
25. C.F. Gauss. *Werke*. Dietrich, Göttingen, 1863–1929. See: Theoretishe Astronomie, Bd. VII.
26. B.G. Galerkin. Series solution of some problems in elastic equilibrium of rods and plates. *Vestn. Inzh. Tech.*, 19:897–908, 1915.
27. C.B. Biezeno and J.J. Koch. Over een Nieuwe Methode ter Berekening van Vlokke Platen. *Ing. Grav.*, 38:25–36, 1923.
28. D. McHenry. A new aspect of creep in concrete and its application to design. *Proc. ASTM*, 43:1064, 1943.

29. R. Courant. Variational methods for the solution of problems of equilibrium and vibration. *Bull. Am. Math Soc.*, 49:1–61, 1943.
30. W. Prager and J.L. Synge. Approximation in elasticity based on the concept of function space. *Quart. J. Appl. Math*, 5:241–269, 1947.
31. O.C. Zienkiewicz and Y.K. Cheung. The finite element method for analysis of elastic isotropic and orthotropic slabs. *Proc. Inst. Civ. Eng.*, 28:471–488, 1964.
32. R.S. Varga. *Matrix Iterative Analysis*. Prentice-Hall, Englewood Cliffs, N.J., 1962.
33. M.L. Wilkins. Calculation of elastic-plastic flow. In B. Alder, editor, *Methods in Computational Physics*, volume 3, pages 211–263. Academic Press, New York, 1964.
34. O.C. Zienkiewicz. Origins, milestones and directions of the finite element method. *Arch. Comp. Meth. Eng.*, 2:1–48, 1995.
35. O.C. Zienkiewicz. Origins, milestones and directions of the finite element method. A personal view. In P.G. Ciarlet and J.L Lyons, editors, *Handbook of Numerical Analysis*, volume IV, pages 3–65. North Holland, 1996.
36. O.C. Zienkiewicz. The birth of the finite element method and of computational mechanics. *Int. J. Numer. Meth. Eng.*, 60:3–10, 2004.
37. J.S. Przemieniecki. *Theory of Matrix Structural Analysis*. McGraw-Hill, New York, 1968.
38. R.K. Livesley. *Matrix Methods in Structural Analysis*. Pergamon Press, New York, 2nd edition, 1975.
39. MATLAB. www.mathworks.com, 2003.

2

A direct physical approach to problems in elasticity: plane stress

2.1 Introduction

The process of approximating the behaviour of a continuum by 'finite elements' which behave in a manner similar to the real, 'discrete', elements described in the previous chapter can be introduced through the medium of particular physical applications or as a general mathematical concept. We have chosen here to follow the first path, narrowing our view to a set of problems associated with structural mechanics which historically were the first to which the finite element method was applied. In Chapter 3 we shall generalize the concepts and show that the basic ideas are widely applicable.

In many phases of engineering the solution of stress and strain distributions in elastic continua is required. Special cases of such problems may range from two-dimensional plane stress or strain distributions, axisymmetric solids, plate bending, and shells, to fully three-dimensional solids. In all cases the number of interconnections between any 'finite element' isolated by some imaginary boundaries and the neighbouring elements is continuous and therefore infinite. It is difficult to see at first glance how such problems may be discretized in the same manner as was described in the preceding chapter for simpler systems. The difficulty can be overcome (and the approximation made) in the following manner:

1. The continuum is separated by imaginary lines or surfaces into a number of 'finite elements'.
2. The elements are assumed to be interconnected at a discrete number of nodal points situated on their boundaries and occasionally in their interior. The displacements of these nodal points will be the basic unknown parameters of the problem, just as in simple, discrete, structural analysis.
3. A set of functions is chosen to define uniquely the state of displacement within each 'finite element' and on its boundaries in terms of its nodal displacements.
4. The displacement functions now define uniquely the state of strain within an element in terms of the nodal displacements. These strains, together with any initial strains and the constitutive properties of the material, define the state of stress throughout the element and, hence, also on its boundaries.
5. A system of 'equivalent forces' concentrated at the nodes and equilibrating the boundary stresses and any distributed loads is determined, resulting in a stiffness relationship

of the form of Eq. (1.3). The determination of these equivalent forces is done most conveniently and generally using the *principle of virtual work* which is a particular mathematical relation known as a *weak form of the problem.*

Once this stage has been reached the solution procedure can follow the standard discrete system pattern described in Chapter 1.

Clearly a series of approximations has been introduced. First, it is not always easy to ensure that the chosen displacement functions will satisfy the requirement of displacement continuity between adjacent elements. Thus, the compatibility condition on such lines may be violated (though within each element it is obviously satisfied due to the uniqueness of displacements implied in their continuous representation). Second, by concentrating the equivalent forces at the nodes, equilibrium conditions are satisfied in the overall sense only. Local violation of equilibrium conditions within each element and on its boundaries will usually arise.

The choice of element shape and of the form of the displacement function for specific cases leaves many opportunities for the ingenuity and skill of the analyst to be employed, and obviously the degree of approximation which can be achieved will strongly depend on these factors.

The approach outlined here is known as the *displacement formulation.*[1,2]

The use of the principle of virtual work (weak form) is extremely convenient and powerful. Here it has only been justified intuitively though in the next chapter we shall see its mathematical origins. However, we will also show the determination of these equivalent forces can be done by minimizing the *total potential energy*. This is applicable to situations where elasticity predominates and the behaviour is reversible. While the virtual work form is always valid, the principle of minimum potential energy is not and care has to be taken. The recognition of the equivalence of the finite element method to a minimization process was late.[2,3] However, Courant[4] in 1943† and Prager and Synge[5] in 1947 proposed minimizing methods that are in essence identical.

This broader basis of the finite element method allows it to be extended to other continuum problems where a variational formulation is possible. Indeed, general procedures are now available for a finite element discretization of any problem defined by a properly constituted set of differential equations. Such generalizations will be discussed in Chapter 3, and throughout the book application to structural and some non-structural problems will be made. It will be found that the process described in this chapter is essentially an application of trial-function and Galerkin-type approximations to the particular case of solid mechanics.

2.2 Direct formulation of finite element characteristics

The 'prescriptions' for deriving the characteristics of a 'finite element' of a continuum, which were outlined in general terms, will now be presented in more detailed mathematical form.

† It appears that Courant had anticipated the essence of the finite element method in general, and of a triangular element in particular, as early as 1923 in a paper entitled 'On a convergence principle in the calculus of variations.' Kön. Gesellschaft der Wissenschaften zu Göttingen, Nachrichten, Berlin, 1923. He states: 'We imagine a mesh of triangles covering the domain ... the convergence principles remain valid for each triangular domain.'

It is desirable to obtain results in a general form applicable to any situation, but to avoid introducing conceptual difficulties the general relations will be illustrated with a very simple example of plane stress analysis of a thin slice. In this a division of the region into triangular-shaped elements may be used as shown in Fig. 2.1. Alternatively, regions may be divided into rectangles or, indeed using a combination of triangles and rectangles. In later chapters we will show how many other shapes also may be used to define elements.

2.2.1 Displacement function

A typical finite element, e, with a triangular shape is defined by local nodes 1, 2 and 3, and straight line boundaries between the nodes as shown in Fig. 2.2(a). Similarly, a rectangular element could be defined by local nodes 1, 2, 3 and 4 as shown in Fig. 2.2(b). The choice of displacement functions for each element is of paramount importance and in Chapters 4 and 5 we will show how they may be developed for a wide range of types; however, in the rest of this chapter we will consider only the 3-node triangular and 4-node rectangular element shapes.

Let the displacements $\mathbf{u}$ at any point within the element be approximated as a column vector, $\hat{\mathbf{u}}$:

$$\mathbf{u} \approx \hat{\mathbf{u}} = \sum_a \mathbf{N}_a \tilde{\mathbf{u}}_a^e = \begin{bmatrix} \mathbf{N}_1, & \mathbf{N}_2, & \ldots \end{bmatrix} \begin{Bmatrix} \tilde{\mathbf{u}}_1 \\ \tilde{\mathbf{u}}_2 \\ \vdots \end{Bmatrix}^e = \mathbf{N}\tilde{\mathbf{u}}^e \tag{2.1}$$

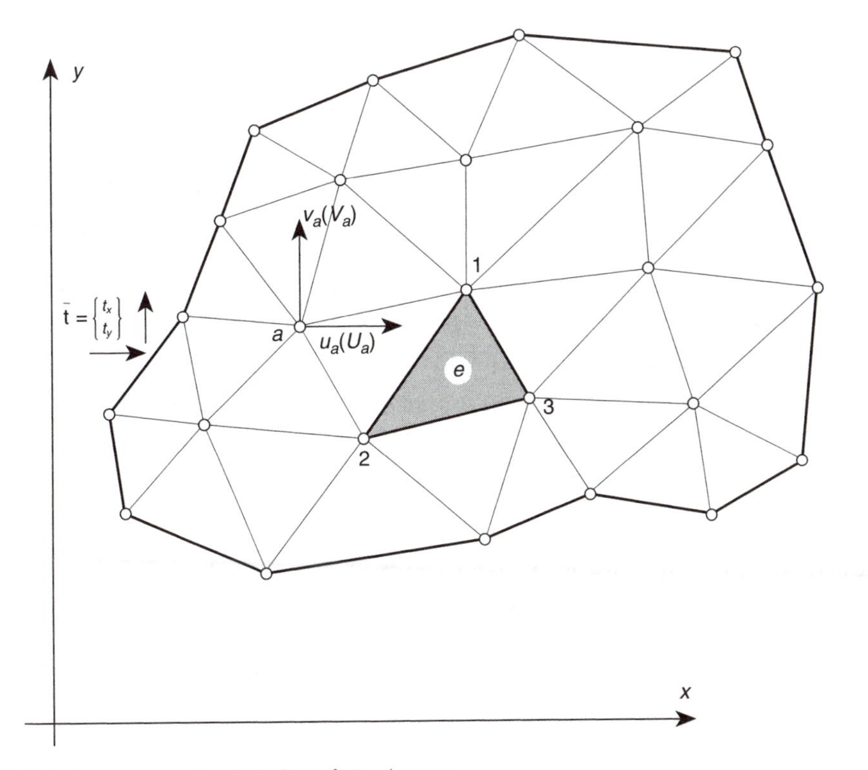

Fig. 2.1 A plane stress region divided into finite elements.

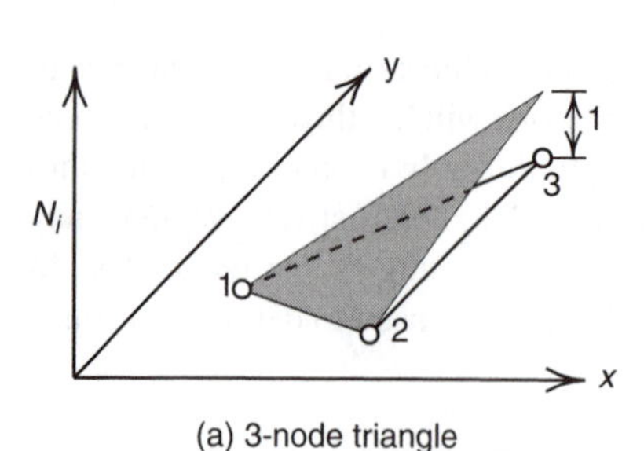

(a) 3-node triangle

(b) 4-node rectangle

Fig. 2.2 Shape function N_3 for one element.

In the case of plane stress, for instance,

$$\mathbf{u} = \begin{Bmatrix} u(x, y) \\ v(x, y) \end{Bmatrix}$$

represents horizontal and vertical movements (see Fig. 2.1) of a typical point within the element and

$$\tilde{\mathbf{u}}_a = \begin{Bmatrix} \tilde{u}_a \\ \tilde{v}_a \end{Bmatrix}$$

the corresponding displacements of a node a.

The functions $\mathbf{N}_a$, $a = 1, 2, \ldots$ are called *shape functions* (or basis functions, and, occasionally interpolation functions) and must be chosen to give appropriate nodal displacements when coordinates of the corresponding nodes are inserted in Eq. (2.1). Clearly in general we have

$$\mathbf{N}_a(x_a, y_a) = \mathbf{I} \quad \text{(identity matrix)}$$

while

$$\mathbf{N}_a(x_b, y_b) = \mathbf{0}, \quad a \neq b$$

If both components of displacement are specified in an identical manner then we can write

$$\mathbf{N}_a = N_a \mathbf{I} \tag{2.2}$$

and obtain N_a from Eq. (2.1) by noting that $N_a(x_a, y_a) = 1$ but is zero at other vertices. The shape functions $\mathbf{N}$ will be seen later to play a paramount role in finite element analysis.

Triangle with 3 nodes

The most obvious linear function in the case of a triangle will yield the shape of N_a of the form shown in Fig. 2.2(a). Writing, the two displacements as

$$\begin{aligned} u &= \alpha_1 + \alpha_2 x + \alpha_3 y \\ v &= \alpha_4 + \alpha_5 x + \alpha_6 y \end{aligned} \tag{2.3}$$

we may evaluate the six constants by solving two sets of three simultaneous equations which arise if the nodal coordinates are inserted and the displacements equated to the appropriate nodal values. For example, the u displacement gives

$$\begin{aligned} \tilde{u}_1 &= \alpha_1 + \alpha_2 x_1 + \alpha_3 y_1 \\ \tilde{u}_2 &= \alpha_1 + \alpha_2 x_2 + \alpha_3 y_2 \\ \tilde{u}_3 &= \alpha_1 + \alpha_2 x_3 + \alpha_3 y_3 \end{aligned} \tag{2.4}$$

We can easily solve for α_1, α_2 and α_3 in terms of the nodal displacements $\tilde{u}_1$, $\tilde{u}_2$ and $\tilde{u}_3$ and obtain finally

$$u = \frac{1}{2\Delta}\left[(a_1 + b_1 x + c_1 y)\,\tilde{u}_1 + (a_2 + b_2 x + c_2 y)\,\tilde{u}_2 + (a_3 + b_3 x + c_3 y)\,\tilde{u}_3\right] \qquad (2.5)$$

in which

$$\begin{aligned} a_1 &= x_2 y_3 - x_3 y_2 \\ b_1 &= y_2 - y_3 \\ c_1 &= x_3 - x_2 \end{aligned} \qquad (2.6)$$

with other coefficients obtained by cyclic permutation of the subscripts in the order 1, 2, 3, and where

$$2\Delta = \det \begin{vmatrix} 1 & x_1 & y_1 \\ 1 & x_2 & y_2 \\ 1 & x_3 & y_3 \end{vmatrix} = 2 \cdot (\text{area of triangle } 123) \qquad (2.7)$$

From (2.5) we see that the shape functions are given by

$$N_a = (a_a + b_a\, x + c_a\, y)/(2\Delta); \quad a = 1, 2, 3 \qquad (2.8)$$

Since displacements with these shape functions vary linearly along any side of a triangle the interpolation (2.5) guarantees continuity between adjacent elements and, with identical nodal displacements imposed, the same displacement will clearly exist along an interface between elements. We note, however, that in general the derivatives will not be continuous between elements.

Rectangle with 4 nodes

An alternative subdivision can use rectangles of the form shown in Fig. 2.3. The rectangular element has side lengths of a and b in the x and y directions, respectively. For the derivation

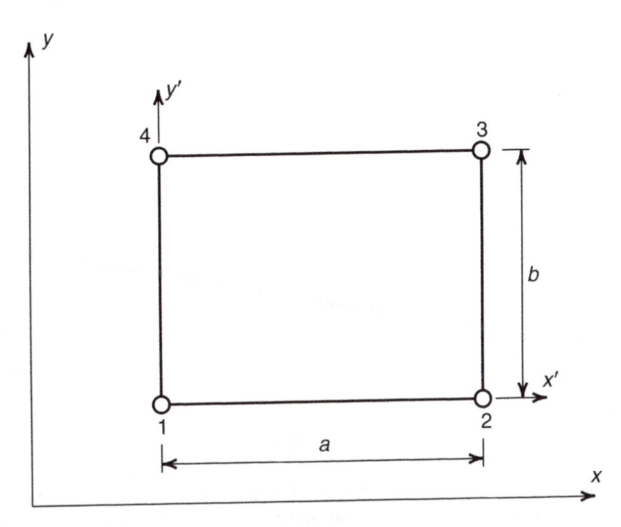

Fig. 2.3 Rectangular element geometry and local node numbers.

of the shape functions it is convenient to use a local cartesian system x', y' defined by

$$x' = x - x_1$$
$$y' = y - y_1$$

We now need four functions for each displacement component in order to uniquely define the shape functions. In addition these functions must have linear behaviour along each edge of the element to ensure interelement continuity. A suitable choice is given by

$$\begin{aligned} u &= \alpha_1 + x'\,\alpha_2 + y'\,\alpha_3 + x'y'\,\alpha_4 \\ v &= \alpha_5 + x'\,\alpha_6 + y'\,\alpha_7 + x'y'\,\alpha_8 \end{aligned} \tag{2.9}$$

The coefficients α_a may be obtained by expressing (2.9) at each node, giving for u

$$\begin{aligned} \tilde{u}_1 &= \alpha_1 \\ \tilde{u}_2 &= \alpha_1 + a\,\alpha_2 \\ \tilde{u}_3 &= \alpha_1 + a\,\alpha_2 + b\,\alpha_3 + ab\,\alpha_4 \\ \tilde{u}_4 &= \alpha_1 + b\,\alpha_3 \end{aligned} \tag{2.10}$$

We can again easily solve for α_a in terms of the nodal displacements to obtain finally

$$u = \frac{1}{ab}[(a - x')(b - y')\,\tilde{u}_1 + x'\,(b - y')\,\tilde{u}_2 + x'\,y'\,\tilde{u}_3 + (a - x')\,y'\,\tilde{u}_4] \tag{2.11}$$

An identical expression is obtained for v by replacing $\tilde{u}_a$ by $\tilde{v}_a$.

From (2.11) we obtain the shape functions

$$\begin{aligned} N_1 &= (a - x')(b - y')/(ab) \\ N_2 &= \;x'\,(b - y')/(ab) \\ N_3 &= \;x'\,y'\,/(ab) \\ N_4 &= (a - x')\,y'\,/(ab) \end{aligned} \tag{2.12}$$

2.2.2 Strains

With displacements known at all points within the element the 'strains' at any point can be determined. These will always result in a relationship that can be written in matrix notation as†

$$\boldsymbol{\varepsilon} = \boldsymbol{\mathcal{S}}\mathbf{u} \tag{2.13}$$

where $\boldsymbol{\mathcal{S}}$ is a suitable linear differential operator. Using Eq. (2.1), the above equation can be approximated by

$$\boldsymbol{\varepsilon} \approx \hat{\boldsymbol{\varepsilon}} = \mathbf{B}\tilde{\mathbf{u}}^e \tag{2.14}$$

with

$$\mathbf{B} = \boldsymbol{\mathcal{S}}\mathbf{N} \tag{2.15}$$

† It is known that strain is a second rank tensor by its transformation properties; however, in this book we will normally represent quantities using matrix (Voigt) notation. The interested reader is encouraged to consult Appendix B for the relations between tensor forms and the matrix quantities.

For the plane stress case the relevant strains of interest are those occurring in the plane and are defined in terms of the displacements by well-known relations[6] which define the operator $\boldsymbol{\mathcal{S}}$

$$\boldsymbol{\varepsilon} = \begin{Bmatrix} \varepsilon_x \\ \varepsilon_y \\ \gamma_{xy} \end{Bmatrix} = \begin{Bmatrix} \dfrac{\partial u}{\partial x} \\ \dfrac{\partial v}{\partial y} \\ \dfrac{\partial u}{\partial y} + \dfrac{\partial v}{\partial x} \end{Bmatrix} = \begin{bmatrix} \dfrac{\partial}{\partial x}, & 0 \\ 0, & \dfrac{\partial}{\partial y} \\ \dfrac{\partial}{\partial y}, & \dfrac{\partial}{\partial x} \end{bmatrix} \begin{Bmatrix} u \\ v \end{Bmatrix}$$

With the shape functions $\mathbf{N}_1$, $\mathbf{N}_2$ and $\mathbf{N}_3$ already determined for a triangular element, the matrix $\mathbf{B}$ can easily be obtained using (2.15). If the linear form of the shape functions is adopted then, in fact, the strains are constant throughout the element (i.e., the $\mathbf{B}$ matrix is constant).

A similar result may be obtained for the rectangular element by adding the results for $\mathbf{N}_4$; however, in this case the strains are not constant but have linear terms in x and y.

2.2.3 Stresses

In general, the material within the element boundaries may be subjected to initial strains such as those due to temperature changes, shrinkage, crystal growth, and so on. If such strains are denoted by $\boldsymbol{\varepsilon}_0$ then the stresses will be caused by the difference between the actual and initial strains.

In addition it is convenient to assume that at the outset of the analysis the body is stressed by some known system of initial residual stresses $\boldsymbol{\sigma}_0$ which, for instance, could be measured, but the prediction of which is impossible without the full knowledge of the material's history. These stresses can simply be added on to the general definition. Thus, assuming linear elastic behaviour, the relationship between stresses and strains will be linear and of the form

$$\boldsymbol{\sigma} = \mathbf{D}(\boldsymbol{\varepsilon} - \boldsymbol{\varepsilon}_0) + \boldsymbol{\sigma}_0 \tag{2.16}$$

where $\mathbf{D}$ is an elasticity matrix containing the appropriate material properties.

Again for the particular case of plane stress three components of stress corresponding to the strains already defined have to be considered. These are, in familiar notation,

$$\boldsymbol{\sigma} = \begin{Bmatrix} \sigma_x \\ \sigma_y \\ \tau_{xy} \end{Bmatrix}$$

and for an isotropic material the $\mathbf{D}$ matrix may be simply obtained from the usual stress–strain relationship[6]

$$\begin{aligned} \varepsilon_x - \varepsilon_{x0} &= \frac{1}{E}(\sigma_x - \sigma_{x0}) - \frac{\nu}{E}(\sigma_y - \sigma_{y0}) \\ \varepsilon_y - \varepsilon_{y0} &= -\frac{\nu}{E}(\sigma_x - \sigma_{x0}) + \frac{1}{E}(\sigma_y - \sigma_{y0}) \\ \gamma_{xy} - \gamma_{xy0} &= \frac{2(1+\nu)}{E}(\tau_{xy} - \tau_{xy0}) \end{aligned}$$

i.e., on solving,

$$\mathbf{D} = \frac{E}{1-\nu^2}\begin{bmatrix} 1 & \nu & 0 \\ \nu & 1 & 0 \\ 0 & 0 & (1-\nu)/2 \end{bmatrix}$$

2.2.4 Equivalent nodal forces

Let

$$\mathbf{q}^e = \begin{Bmatrix} \mathbf{q}_1^e \\ \mathbf{q}_2^e \\ \vdots \end{Bmatrix}$$

define the nodal forces which are statically equivalent to the boundary stresses and distributed body forces acting on the element. Each of the forces $\mathbf{q}_a^e$ must contain the same number of components as the corresponding nodal displacement $\tilde{\mathbf{u}}_a$ and be ordered in the appropriate, corresponding directions.

The distributed body forces $\mathbf{b}$ are defined as those acting on a unit volume of material within the element with directions corresponding to those of the displacements $\mathbf{u}$ at that point.

In the particular case of plane stress the nodal forces are, for instance,

$$\mathbf{q}_a^e = \begin{Bmatrix} U_a^e \\ V_a^e \end{Bmatrix}$$

with components U and V corresponding to the directions of u and v, respectively (viz. Fig. 2.1), and the distributed body forces are

$$\mathbf{b} = \begin{Bmatrix} b_x \\ b_y \end{Bmatrix}$$

in which b_x and b_y are the 'body force' components per unit of volume.

In the absence of body forces equivalent nodal forces for the 3-node triangular element can be computed directly from equilibrium considerations. In Fig. 2.4(a) we show a triangular element together with the geometric properties which are obtained by the linear interpolation of the displacements using (2.1) to (2.8). In particular we note from the figure [and (2.6)] that

$$b_1 + b_2 + b_3 = 0 \quad \text{and} \quad c_1 + c_2 + c_3 = 0$$

The stresses in the element are given by (2.16) in which we assume that ε_0 and $\boldsymbol{\sigma}_0$ are constant in each element and strains are computed from (2.14) and, for the 3-node triangular element, are also constant in each element. To determine the nodal forces resulting from the stresses, the boundary tractions are first computed from

$$\mathbf{t} = \begin{Bmatrix} t_x \\ t_y \end{Bmatrix} = t\begin{bmatrix} n_x & 0 & n_y \\ 0 & n_y & n_x \end{bmatrix}\begin{Bmatrix} \sigma_x \\ \sigma_y \\ \tau_{xy} \end{Bmatrix} \tag{2.17}$$

where t is a constant thickness of the plane strain slice and n_x, n_y are the direction cosines of the outward normal to the element boundary. For the triangular element the tractions

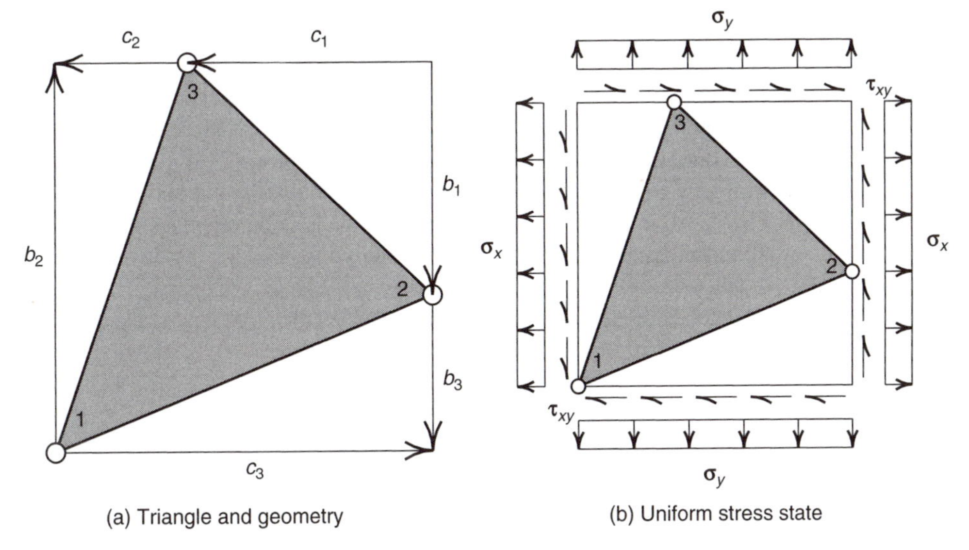

(a) Triangle and geometry (b) Uniform stress state

Fig. 2.4 3-node triangle, geometry and constant stress state.

are constant. The resultant for each side of the triangle is the product of the triangle side length (l_a) times the traction. Here l_a is the length of the side opposite the triangle node a and we note from Fig. 2.4(a) that

$$l_a\, n_x = -\, b_a \quad \text{and} \quad l_a\, n_y = -\, c_a \tag{2.18}$$

Therefore,

$$l_a \mathbf{t} = \begin{Bmatrix} l_a t_x \\ l_a t_y \end{Bmatrix} = t \begin{bmatrix} -b_a & 0 & -c_a \\ 0 & -c_a & -b_a \end{bmatrix} \begin{Bmatrix} \sigma_x \\ \sigma_y \\ \tau_{xy} \end{Bmatrix}$$

The resultant acts at the middle of each side of the triangle and, thus, by sum of forces and moments is equivalent to placing half at each end node. Thus, by static equivalence the nodal forces at node 1 are given by

$$\begin{aligned} \mathbf{q}_1 &= \frac{t}{2}\left(\begin{bmatrix} -b_2 & 0 & -c_2 \\ 0 & -c_2 & -b_2 \end{bmatrix} + \begin{bmatrix} -b_3 & 0 & -c_3 \\ 0 & -c_3 & -b_3 \end{bmatrix} \right) \boldsymbol{\sigma} \\ &= \frac{t}{2} \begin{bmatrix} b_1 & 0 & c_1 \\ 0 & c_1 & b_1 \end{bmatrix} \boldsymbol{\sigma} = \mathbf{B}_1^{\mathrm{T}} \boldsymbol{\sigma}\, \Delta\, t \end{aligned} \tag{2.19a}$$

Similarly, the forces at nodes 2 and 3 are given by

$$\begin{aligned} \mathbf{q}_2 &= \mathbf{B}_2\, \boldsymbol{\sigma}\, \Delta\, t \\ \mathbf{q}_3 &= \mathbf{B}_3\, \boldsymbol{\sigma}\, \Delta\, t \end{aligned} \tag{2.19b}$$

Combining with the expression for stress and strain for each element we obtain

$$\begin{aligned} \mathbf{q} &= \mathbf{B}^{\mathrm{T}} \left[\mathbf{D} \left(\mathbf{B}\tilde{\mathbf{u}}^e - \boldsymbol{\varepsilon}_0 \right) + \boldsymbol{\sigma}_0 \right] \Delta\, t \\ &= \mathbf{K}^e \tilde{\mathbf{u}}^e + \mathbf{f}^e \end{aligned} \tag{2.20a}$$

where

$$\mathbf{K}^e = \mathbf{B}^{\mathrm{T}}\mathbf{D}\,\mathbf{B}\,\Delta t \quad \text{and} \quad \mathbf{f}^e = \mathbf{B}^{\mathrm{T}}(\boldsymbol{\sigma}_0 - \mathbf{D}\,\boldsymbol{\varepsilon}_0)\,\Delta t \tag{2.20b}$$

The above gives a result which is now in the form of the standard discrete problem defined in Sec. 1.2. However, when body forces are present or we consider other element forms the above procedure fails and we need a more general approach. To make the nodal forces statically equivalent to the actual boundary stresses and distributed body forces, the simplest general procedure is to impose an arbitrary (virtual) nodal displacement and to equate the external and internal work done by the various forces and stresses during that displacement.

Let such a virtual displacement be $\delta\tilde{\mathbf{u}}^e$ at the nodes. This results, by Eqs (2.1) and (2.14), in virtual displacements and strains within the element equal to

$$\delta\mathbf{u} = \mathbf{N}\,\delta\tilde{\mathbf{u}}^e \quad \text{and} \quad \delta\boldsymbol{\varepsilon} = \mathbf{B}\,\delta\tilde{\mathbf{u}}^e \tag{2.21}$$

respectively.

The external work done by the nodal forces is equal to the sum of the products of the individual force components and corresponding displacements, i.e., in matrix form

$$\delta\tilde{\mathbf{u}}_1^{e\mathrm{T}}\mathbf{q}_1^e + \delta\tilde{\mathbf{u}}_2^{e\mathrm{T}}\mathbf{q}_2^e \ldots = \delta\tilde{\mathbf{u}}^{e\mathrm{T}}\mathbf{q}^e \tag{2.22}$$

Similarly, the internal work per unit volume done by the stresses and distributed body forces subjected to a set of virtual strains and displacements is

$$\delta\boldsymbol{\varepsilon}^{\mathrm{T}}\boldsymbol{\sigma} - \delta\mathbf{u}^{\mathrm{T}}\mathbf{b} \tag{2.23}$$

or, after using (2.21),†

$$\delta\tilde{\mathbf{u}}^{e\mathrm{T}}\left(\mathbf{B}^{\mathrm{T}}\boldsymbol{\sigma} - \mathbf{N}^{\mathrm{T}}\mathbf{b}\right) \tag{2.24}$$

Equating the external work with the total internal work obtained by integrating (2.24) over the volume of the element, Ω_e, we have

$$\delta\tilde{\mathbf{u}}^{e\mathrm{T}}\mathbf{q}^e = \delta\tilde{\mathbf{u}}^{e\mathrm{T}}\left(\int_{\Omega_e}\mathbf{B}^{\mathrm{T}}\boldsymbol{\sigma}\,\mathrm{d}\Omega - \int_{\Omega_e}\mathbf{N}^{\mathrm{T}}\mathbf{b}\,\mathrm{d}\Omega\right) \tag{2.25}$$

As this relation is valid for any value of the virtual displacement, the multipliers must be equal. Thus

$$\mathbf{q}^e = \int_{\Omega_e}\mathbf{B}^{\mathrm{T}}\boldsymbol{\sigma}\,\mathrm{d}\Omega - \int_{\Omega_e}\mathbf{N}^{\mathrm{T}}\mathbf{b}\,\mathrm{d}\Omega \tag{2.26}$$

This statement is valid quite generally for any stress–strain relation. With (2.14) and the linear law of Eq. (2.16) we can write Eq. (2.26) as

$$\mathbf{q}^e = \mathbf{K}^e\tilde{\mathbf{u}}^e + \mathbf{f}^e \tag{2.27}$$

where

$$\mathbf{K}^e = \int_{\Omega_e}\mathbf{B}^{\mathrm{T}}\mathbf{D}\,\mathbf{B}\,\mathrm{d}\Omega \tag{2.28a}$$

and

$$\mathbf{f}^e = -\int_{\Omega_e}\mathbf{N}^{\mathrm{T}}\mathbf{b}\,\mathrm{d}\Omega - \int_{\Omega_e}\mathbf{B}^{\mathrm{T}}\mathbf{D}\,\boldsymbol{\varepsilon}_0\,\mathrm{d}\Omega + \int_{\Omega_e}\mathbf{B}^{\mathrm{T}}\boldsymbol{\sigma}_0\,\mathrm{d}\Omega \tag{2.28b}$$

† Note that by the rules of matrix algebra for the transpose of products $(\mathbf{A}\,\mathbf{B})^{\mathrm{T}} = \mathbf{B}^{\mathrm{T}}\mathbf{A}^{\mathrm{T}}$.

For the plane stress problem

$$\int_{\Omega_e} (\cdot)\, \mathrm{d}\Omega = \int_{A_e} (\cdot)\, t\, \mathrm{d}A$$

where A_e is the area of the element. Here t now can be allowed to vary over the element. In the last equation the three terms represent forces due to body forces, initial strain, and initial stress respectively. The relations have the characteristics of the discrete structural elements described in Chapter 1.

If the initial stress system is self-equilibrating, as must be the case with normal residual stresses, then the forces given by the initial stress term of Eq. (2.28b) are identically zero after assembly. Thus frequent evaluation of this force component is omitted. However, if for instance a machine part is manufactured out of a block in which residual stresses are present or if an excavation is made in rock where known tectonic stresses exist a removal of material will cause a force imbalance which results from the above term.

For the particular example of the plane stress triangular element these characteristics will be obtained by appropriate substitution. It has already been noted that the **B** matrix in that example was not dependent on the coordinates; hence the integration will become particularly simple and, in the absence of body forces, $\mathbf{K}^e$ and $\mathbf{f}^e$ are identical to those given in (2.20b).

The interconnection and solution of the whole assembly of elements follows the simple structural procedures outlined in Chapter 1. This gives

$$\mathbf{r} = \sum_e \mathbf{q}^e = \mathbf{0} \tag{2.29}$$

A note should be added here concerning elements near the boundary. If, at the boundary, displacements are specified, no special problem arises as these can be satisfied by specifying some of the nodal parameters $\tilde{\mathbf{u}}$. Consider, however, the boundary as subject to a distributed external loading, say $\bar{\mathbf{t}}$ per unit area (traction). A loading term on the nodes of the element which has a boundary face Γ_e will now have to be added. By the virtual work consideration, this will simply result in

$$\mathbf{f}^e \rightarrow \mathbf{f}^e - \int_{\Gamma_e} \mathbf{N}^{\mathrm{T}} \bar{\boldsymbol{t}}\, \mathrm{d}\Gamma \tag{2.30}$$

with integration taken over the boundary area of the element. It will be noted that $\bar{\mathbf{t}}$ must have the same number of components as $\mathbf{u}$ for the above expression to be valid. Such a boundary element is shown again for the special case of plane stress in Fig. 2.1.

Once the nodal displacements have been determined by solution of the overall 'structural'-type equations, the stresses at any point of the element can be found from the relations in Eqs (2.14) and (2.16), giving

$$\sigma = \mathbf{D}\left(\mathbf{B}\tilde{\mathbf{u}}^e - \varepsilon_0\right) + \sigma_0 \tag{2.31}$$

Example 2.1: Stiffness matrix for 3-node triangle. The stiffness matrix for an individual element is computed by evaluating Eq. (2.28a). For a 3-node triangle in which the moduli and thickness are constant over the element the solution for the stiffness becomes

$$\mathbf{K}^e = \mathbf{B}^{\mathrm{T}}\, \mathbf{D}\, \mathbf{B}\, \Delta\, t \tag{2.32}$$

where Δ is the area of the triangle computed from (2.7). Evaluating (2.15) using the shape functions in (2.8) gives

$$\mathbf{B}_a = \frac{1}{2\Delta} \begin{bmatrix} b_a & 0 \\ 0 & c_a \\ c_a & b_a \end{bmatrix} \tag{2.33}$$

Thus, the expression for the stiffness of the triangular element is given by

$$\mathbf{K}_{ab} = \frac{t}{4\Delta} \begin{bmatrix} b_a & 0 & c_a \\ 0 & c_a & b_a \end{bmatrix} \begin{bmatrix} D_{11} & D_{12} & D_{13} \\ D_{21} & D_{22} & D_{23} \\ D_{31} & D_{32} & D_{33} \end{bmatrix} \begin{bmatrix} b_b & 0 \\ 0 & c_b \\ c_b & b_b \end{bmatrix} \tag{2.34}$$

where $D_{ij} = D_{ji}$ are the elastic moduli.

Example 2.2: Nodal forces for boundary traction. Let us consider a problem in which a traction boundary condition is to be imposed along a vertical surface located at $x = x_b$. A triangular element has one of its edges located along the boundary as shown in Fig. 2.5 and is loaded by a specified traction given by

$$\bar{\mathbf{t}} = \begin{Bmatrix} t_x \\ t_y \end{Bmatrix} = t \begin{Bmatrix} \sigma_x \\ \tau_{xy} \end{Bmatrix}$$

The normal stress σ_x is given by a linearly varying stress in the y direction and the shearing stress τ_{xy} is assumed zero, thus, to compute nodal forces we use the expressions

$$\sigma_x = k_x y \quad \text{and} \quad \tau_{xy} = 0$$

in which k_x is a specified constant.

Along the boundary the shape functions for either a triangular element or a rectangular element are linear functions in y and are given by

$$N_1 = (y_2 - y)/(y_2 - y_1) \quad \text{and} \quad N_2 = (y - y_1)/(y_2 - y_1)$$

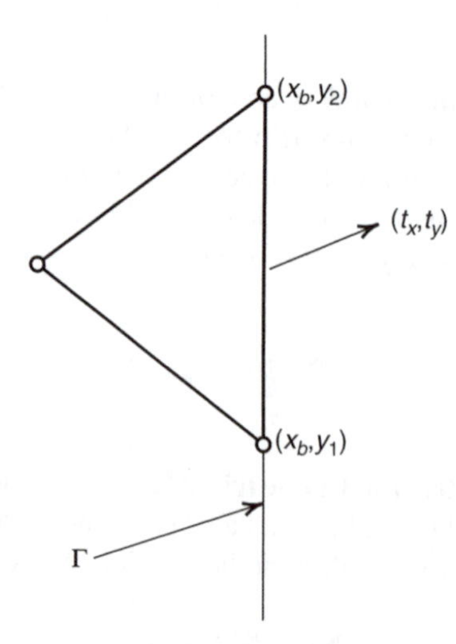

Fig. 2.5 Traction on vertical face.

thus, the nodal forces for the element shown are computed from Eq. (2.30) and given by

$$\mathbf{f}_1 = -\int_{y_1}^{y_2} t\,N_1 \begin{Bmatrix} \sigma_x \\ \tau_{xy} \end{Bmatrix} \mathrm{d}y = -\begin{Bmatrix} k_x\,t(2y_1 + y_2)(y_2 - y_1)/6 \\ 0 \end{Bmatrix}$$

and

$$\mathbf{f}_2 = -\int_{y_1}^{y_2} tN_2 \begin{Bmatrix} \sigma_x \\ \tau_{xy} \end{Bmatrix} \mathrm{d}y = -\begin{Bmatrix} k_x\,t(y_1 + 2y_2)(y_2 - y_1)/6 \\ 0 \end{Bmatrix}$$

2.3 Generalization to the whole region – internal nodal force concept abandoned

In the preceding section the virtual work principle was applied to a single element and the concept of equivalent nodal force was retained. The assembly principle thus followed the conventional, direct equilibrium, approach.

The idea of nodal forces contributed by elements replacing the continuous interaction of stresses between elements presents a conceptual difficulty. However, it has a considerable appeal to ‘practical’ engineers and does at times allow an interpretation which otherwise would not be obvious to the more rigorous mathematician. There is, however, no need to consider each element individually and the reasoning of the previous section may be applied directly to the whole continuum.

Equation (2.1) can be interpreted as applying to the whole structure, that is,

$$\mathbf{u} = \bar{\mathbf{N}}\,\tilde{\mathbf{u}} \quad \text{and} \quad \delta\mathbf{u} = \bar{\mathbf{N}}\,\delta\tilde{\mathbf{u}} \tag{2.35}$$

in which $\tilde{\mathbf{u}}$ and $\delta\tilde{\mathbf{u}}$ list all the nodal points and

$$\bar{\mathbf{N}}_a = \sum_e \mathbf{N}_a^e \tag{2.36}$$

when the point concerned is within a particular element e and a is a node point associated with the element. If a point does not occur within the element (see Fig. 2.6)

$$\bar{\mathbf{N}}_a = \mathbf{0} \tag{2.37}$$

A matrix $\bar{\mathbf{B}}$ can be similarly defined and we shall drop the bar, considering simply that the shape functions, etc., are always defined over the whole domain, Ω.

For any virtual displacement $\delta\tilde{\mathbf{u}}$ we can now write the sum of internal and external work for the whole region as

$$\delta\tilde{\mathbf{u}}^{\mathrm{T}}\mathbf{r} = \int_{\Omega} \delta\boldsymbol{\varepsilon}^{\mathrm{T}}\boldsymbol{\sigma}\,\mathrm{d}\Omega - \int_{\Omega} \delta\mathbf{u}^{\mathrm{T}}\,\mathbf{b}\,\mathrm{d}\Omega - \int_{\Gamma} \delta\mathbf{u}^{\mathrm{T}}\,\bar{\mathbf{t}}\,\mathrm{d}\Gamma = 0 \tag{2.38}$$

In the above equation, $\delta\tilde{\mathbf{u}}$, $\delta\mathbf{u}$ and $\delta\boldsymbol{\varepsilon}$ can be completely arbitrary, providing they stem from a continuous displacement assumption. If for convenience we assume they are simply variations linked by relations (2.35) and (2.14) we obtain, on substitution of the constitutive relation (2.16), a system of algebraic equations

$$\mathbf{K}\,\tilde{\mathbf{u}} + \mathbf{f} = \mathbf{0} \tag{2.39}$$

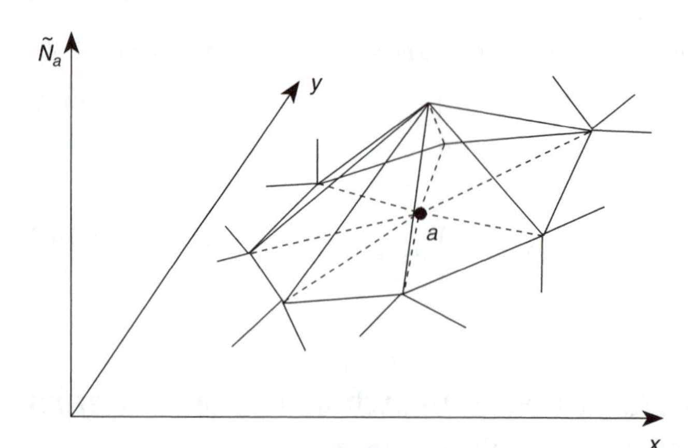

Fig. 2.6 Shape function $\tilde{N}_a$ for whole domain.

where

$$\mathbf{K} = \int_{\Omega} \mathbf{B}^{\mathrm{T}} \mathbf{D} \mathbf{B} \, \mathrm{d}\Omega \tag{2.40a}$$

and

$$\mathbf{f} = -\int_{\Omega} \mathbf{N}^{\mathrm{T}} \mathbf{b} \, \mathrm{d}\Omega - \int_{\Gamma} \mathbf{N}^{\mathrm{T}} \bar{\mathbf{t}} \, \mathrm{d}\Gamma + \int_{\Omega} \mathbf{B}^{\mathrm{T}} (\boldsymbol{\sigma}_0 - \mathbf{D} \boldsymbol{\varepsilon}_0) \, \mathrm{d}\Omega \tag{2.40b}$$

The integrals are taken over the whole domain Ω and over the whole surface area Γ on which tractions are given.

It is immediately obvious from the above that

$$\mathbf{K}_{ab} = \sum_{e} \mathbf{K}_{ab}^{e} \quad \text{and} \quad \mathbf{f}_a = \sum_{e} \mathbf{f}_a^{e} \tag{2.41}$$

by virtue of the property of definite integrals requiring that the total be the sum of the parts:

$$\int_{\Omega} (\cdot) \mathrm{d}\Omega = \sum_{e} \int_{\Omega_e} (\cdot) \mathrm{d}\Omega \quad \text{and} \quad \int_{\Gamma} (\cdot) \mathrm{d}\Gamma = \sum_{e} \int_{\Gamma_e} (\cdot) \mathrm{d}\Gamma \tag{2.42}$$

The same is obviously true for the surface integrals in Eq. (2.40b). We thus see that the 'secret' of the approximation possessing the required behaviour of a 'standard discrete system' of Chapter 1 lies simply in the requirement of writing the relationships in integral form.

The assembly rule as well as the whole derivation has been achieved without involving the concept of 'interelement forces' (i.e., $\mathbf{q}^e$). In the remainder of this book the element superscript will be dropped unless specifically needed. Also no differentiation between element and system shape functions will be made.

However, an important point arises immediately. In considering the virtual work for the whole system [Eq. (2.38)] and equating this to the sum of the element contributions it is implicitly assumed that *no discontinuity in displacement between adjacent elements develops*. If such a discontinuity developed, a contribution equal to the work done by the stresses in the separations would have to be added.

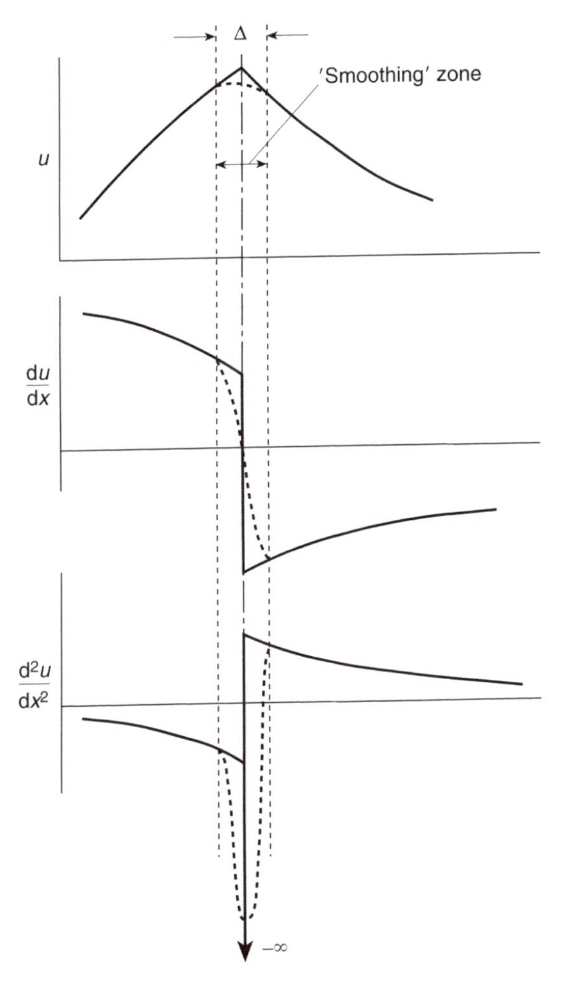

Fig. 2.7 Differentiation of function with sloped discontinuity (C_0 continuous).

Put in other words, we require that the terms integrated in Eq. (2.42) be finite. These terms arise from the shape functions N_a used in defining the displacement **u** [by Eq. (2.35)] and its derivatives associated with the definition of strain [viz. Eq. (2.14)]. If, for instance, the 'strains' are defined by first derivatives of the functions **N**, the displacements must be continuous. In Fig. 2.7 we see how first derivatives of continuous functions may involve a 'jump' but are still finite, while second derivatives may become infinite. Such functions we call C_0 continuous.

In some problems the 'strain' in a generalized sense may be defined by second derivatives. In such cases we shall obviously require that both the function **N** and its slope (first derivative) be continuous. Such functions are more difficult to derive but are used in the analysis of thin plate and shell problems (e.g., see volume on solid mechanics[7]). The continuity involved now is called C_1.

2.4 Displacement approach as a minimization of total potential energy

The principle of virtual displacements used in the previous sections ensured satisfaction of equilibrium conditions within the limits prescribed by the assumed displacement pattern. Only if the virtual work equality for all, arbitrary, variations of displacement was ensured would the equilibrium be complete.

As the number of parameters of $\tilde{\mathbf{u}}$ which prescribes the displacement increases without limit then ever closer approximation of all equilibrium conditions can be ensured.

The virtual work principle as written in Eq. (2.38) can be restated in a different form if the virtual quantities $\delta\tilde{\mathbf{u}}$, $\delta\mathbf{u}$, and $\delta\boldsymbol{\varepsilon}$ are considered as *variations* of the real quantities.[8, 9]

Thus, for instance, we can write the first term of Eq. (2.38), for elastic materials, as

$$\delta U = \int_{\Omega} \delta\boldsymbol{\varepsilon}^{\mathrm{T}}\boldsymbol{\sigma}\, \mathrm{d}\Omega \tag{2.43}$$

where U is the *strain energy* of the system. For the linear elastic material described by Eq. (2.16) the strain energy is given by

$$U = \frac{1}{2}\int_{\Omega} \boldsymbol{\varepsilon}^{\mathrm{T}}\mathbf{D}\,\boldsymbol{\varepsilon}\, \mathrm{d}\Omega + \int_{\Omega} \boldsymbol{\varepsilon}^{\mathrm{T}}(\boldsymbol{\sigma}_0 - \mathbf{D}\,\boldsymbol{\varepsilon}_0)\, \mathrm{d}\Omega \tag{2.44}$$

and will, after variation, yield the correct expression providing $\mathbf{D}$ is a symmetric matrix (this is a necessary condition for a single-valued U to exist).[8, 9]

The last two terms of Eq. (2.38) can be written as

$$\delta W = -\delta\left(\int_{\Omega} \mathbf{u}^{\mathrm{T}}\mathbf{b}\, \mathrm{d}\Omega + \int_{\Gamma} \mathbf{u}^{\mathrm{T}}\bar{\mathbf{t}}\, \mathrm{d}\Gamma\right) \tag{2.45}$$

where W is the *potential energy of the external loads*. The above is certainly true if $\mathbf{b}$ and $\bar{\mathbf{t}}$ are conservative (or independent of displacement) where we obtain simply

$$W = -\int_{\Omega} \mathbf{u}^{\mathrm{T}}\mathbf{b}\, \mathrm{d}\Omega - \int_{\Gamma} \mathbf{u}^{\mathrm{T}}\bar{\mathbf{t}}\, \mathrm{d}\Gamma \tag{2.46}$$

Thus, instead of Eq. (2.38), we can write the *total potential energy*, Π, as

$$\Pi = U + W \tag{2.47}$$

in which U is given by (2.44) and W by (2.46) and require

$$\delta\Pi = \delta(U + W) = 0 \tag{2.48}$$

In this form Π is known as a *functional* and (2.48) is a requirement which renders the functional *stationary*.

The above statement means that for equilibrium to be ensured the *total potential energy must be stationary* for variations of the admissible displacements. The finite element equations for the total potential energy are obtained by substituting the approximation for displacements [viz. Eq. (2.35)] into Eqs (2.44) and (2.46) giving

$$\Pi = \frac{1}{2}\tilde{\mathbf{u}}^{\mathrm{T}}\mathbf{K}\,\tilde{\mathbf{u}} + \tilde{\mathbf{u}}^{\mathrm{T}}\mathbf{f} \tag{2.49}$$

in which $\mathbf{K}$ (where $\mathbf{K} = \mathbf{K}^T$) and $\mathbf{f}$ are given by Eqs (2.40a) to (2.41). The variation with respect to displacements with the finite number of parameters $\tilde{\mathbf{u}}$ is now written as

$$\frac{\partial \Pi}{\partial \tilde{\mathbf{u}}} = \begin{Bmatrix} \dfrac{\partial \Pi}{\partial \tilde{\mathbf{u}}_1} \\ \dfrac{\partial \Pi}{\partial \tilde{\mathbf{u}}_2} \\ \vdots \end{Bmatrix} = \mathbf{K}\,\tilde{\mathbf{u}} + \mathbf{f} = \mathbf{0} \tag{2.50}$$

It can also be shown that in stable elastic situations the total potential energy is not only stationary but is a minimum.[8] *Thus the finite element process seeks such a minimum within the constraint of an assumed displacement pattern.*

The greater the number of degrees of freedom, the more closely the solution will approximate the true one, ensuring complete equilibrium, providing the true displacement can, in the limit, be represented. The necessary convergence conditions for the finite element process could thus be derived. Discussion of these will, however, be deferred to subsequent sections.

It is of interest to note that if true equilibrium requires an absolute minimum of the total potential energy, Π, a finite element solution by the displacement approach will always provide an approximate Π greater than the correct one. *Thus a bound on the value of the total potential energy is always achieved.*

If the functional Π could be specified, *a priori*, then the finite element equations could be derived directly by the differentiation specified by Eq. (2.50).

The well-known Rayleigh[10]–Ritz[11] process of approximation frequently used in elastic analysis is based precisely on this approach. The total potential energy expression is formulated and the displacement pattern is assumed to vary with a finite set of undetermined parameters. A set of simultaneous equations minimizing the total potential energy with respect to these parameters is set up. Thus the finite element process as described so far can be considered to be the Rayleigh–Ritz procedure. The difference is only in the manner in which the assumed displacements are prescribed. In the traditionally used Ritz process the functions are usually given by expressions valid throughout the whole region, thus leading to simultaneous equations in which the coefficient matrix is full. In the finite element process this specification is usually piecewise, each nodal parameter influencing only adjacent elements, and thus a sparse and usually banded matrix of coefficients is found.

By its nature the conventional Ritz process is limited to relatively simple geometrical shapes of the total region while this limitation only occurs in finite element analysis in the element itself. Thus complex, realistic, configurations can be assembled from relatively simple element shapes.

A further difference is in the usual association of the undetermined parameter $\tilde{\mathbf{u}}_a$ with a particular nodal displacement. This allows a simple physical interpretation invaluable to an engineer. Doubtless much of the early popularity of the finite element process is due to this fact.

2.4.1 Bound on strain energy in a displacement formulation

While the approximation obtained by the finite element displacement approach always overestimates the true value of the total potential energy Π (the absolute minimum corresponding to the exact solution), this is not directly useful in practice. It is, however, possible to obtain a more useful limit in special cases.

Consider the problem in which no initial strains ε_0 or initial stresses $\boldsymbol{\sigma}_0$ exist. Now by the principle of energy conservation the strain energy will be equal to the work done by the external loads which increase uniformly from zero.[12] This work done is equal to $-W/2$ where W is the potential energy of the loads. Thus,

$$U + \frac{1}{2}W = 0 \tag{2.51}$$

or

$$\Pi = U + W = -U \tag{2.52}$$

whether an exact or approximate displacement field is assumed.

If only one external concentrated load R_a is present, the strain energy bound immediately informs us that the finite element deflection under this load has been underestimated (as $U = -W/2 = -R_a u_a/2$, where u_a is the deflection at the load point). In more complex loading cases the usefulness of this bound is limited as neither local displacements nor local stresses, i.e., the quantities of real engineering interest, can be bounded. It is also important to remember that this bound on strain energy is only valid in the absence of any initial stresses or strains.

The expression for U in this case can be obtained from Eq. (2.44) as

$$U = \frac{1}{2}\int_\Omega \boldsymbol{\varepsilon}^{\mathrm{T}} \mathbf{D}\, \boldsymbol{\varepsilon}\, \mathrm{d}\Omega \tag{2.53}$$

which becomes by Eq. (2.14) simply

$$U = \frac{1}{2}\tilde{\mathbf{u}}^{\mathrm{T}} \int_\Omega \mathbf{B}^{\mathrm{T}}\mathbf{D}\,\mathbf{B}\,\mathrm{d}\Omega\, \tilde{\mathbf{u}} = \frac{1}{2}\tilde{\mathbf{u}}^{\mathrm{T}}\mathbf{K}\,\tilde{\mathbf{u}} \tag{2.54}$$

a *quadratic matrix form* in which **K** is the stiffness matrix previously discussed.

When sufficient supports are provided to prevent rigid body motion and only linear elastic materials are considered, the above energy expression is always *positive* from physical considerations. It follows therefore that the matrix **K** occurring in all the finite element assemblies is not only symmetric but is *positive definite* (a property defined in fact by the requirement that the quadratic form should always be greater than zero).

This feature is of importance when the numerical solution of the simultaneous equations is considered, as simplifications arise in the case of symmetric positive definite equations.[13]

2.4.2 Direct minimization

The fact that the finite element approximation reduces to the problem of minimizing the total potential energy Π defined in terms of a finite number of nodal parameters led us to the formulation of the simultaneous set of equations given symbolically by Eq. (2.50). This is the most usual and convenient approach, especially in linear solutions, but other search procedures, now well developed in the field of optimization, could be used to estimate the lowest value of Π. In this text we shall continue with the simultaneous equation process but the interested reader could well bear the alternative possibilities in mind.[14, 15]

2.5 Convergence criteria

The assumed shape functions limit the infinite degrees of freedom of the real system, and the true minimum of the energy may never be reached, irrespective of the fineness of subdivision. To ensure convergence to the correct result certain simple requirements must be satisfied. Obviously, for instance, the displacement function should be able to represent the true displacement distribution as closely as desired. It will be found that this is not so if the chosen functions are such that straining is possible when the element is subjected to rigid body displacements at the nodes. Thus, the first criterion that the displacement function must obey is:

> *Criterion* 1. The displacement shape functions chosen should be such that they do not permit straining of an element to occur when the nodal displacements are caused by a rigid body motion.

This self-evident condition can be violated easily if certain types of function are used; care must therefore be taken in the choice of displacement functions.

A second criterion stems from similar requirements. Clearly, as elements get smaller nearly constant strain conditions will prevail in them. If, in fact, constant strain conditions exist, it is most desirable for good accuracy that a finite size element is able to reproduce these exactly. It is possible to formulate functions that satisfy the first criterion but at the same time require a strain variation throughout the element when the nodal displacements are compatible with a constant strain solution. Such functions will, in general, not show good convergence to an accurate solution and cannot, even in the limit, represent the true strain distribution. The second criterion can therefore be formulated as follows:

> *Criterion* 2. The displacement shape functions have to be of such a form that if nodal displacements are compatible with a constant strain condition such constant strain will in fact be obtained.

It will be observed that Criterion 2 in fact incorporates the requirement of Criterion 1, as rigid body displacements are a particular case of constant strain – with a value of zero. This criterion was first stated by Bazeley *et al.*[16] in 1966. *Strictly, both criteria need only be satisfied in the limit as the size of the element tends to zero.* However, the imposition of these criteria on elements of finite size leads to improved accuracy, although in certain situations (such as in axisymmetric analysis) the imposition of the second one is not possible or essential.

Lastly, as already mentioned in Sec. 2.3, it is implicitly assumed in this derivation that no contribution to the virtual work arises at element interfaces. It therefore appears necessary that the following criterion be included:

> *Criterion* 3. The displacement shape functions should be chosen such that the strains at the interface between elements are finite (even though they may be discontinuous).

This criterion implies a certain continuity of displacements between elements. In the case of strains being defined by first derivatives, as in the plane stress example quoted here, the displacements only have to be continuous (C_0 continuity). If, however, the 'strains' are defined by second derivatives, first derivatives of these have also to be continuous (C_1 continuity).[2]

The above criteria are included mathematically in a statement of 'functional completeness' and the reader is referred elsewhere for full mathematical discussion.[17–22] The 'heuristic' proof of the convergence requirements given here is sufficient for practical purposes and we shall generalize all of the above criteria in Sec. 3.6 and more fully in Chapter 9. Indeed in the latter we shall show a universal test which justifies convergence even if some of the above criteria are violated.

2.6 Discretization error and convergence rate

In the foregoing sections we have assumed that the approximation to the displacement as represented by Eq. (2.1) will yield the exact solution in the limit as the size h of elements decreases. The arguments for this are simple: if the expansion is capable, in the limit, of exactly reproducing any displacement form conceivable in the continuum, then as the solution of each approximation is unique it must approach, in the limit of $h \rightarrow 0$, the unique exact solution. In some cases the exact solution is indeed obtained with a finite number of subdivisions (or even with one element only) if the *polynomial expansion used in that element fits the exact solution*. Thus, for instance, if the exact solution is of the form of a quadratic polynomial *and* the shape functions include all the polynomials of that order, the approximation will yield the exact answer.

The last argument helps in determining the order of convergence of the finite element procedure as the exact solution can always be expanded in a Taylor series in the vicinity of any point (or node) a as a polynomial:

$$\mathbf{u} = \mathbf{u}_a + \left(\frac{\partial \mathbf{u}}{\partial x}\right)_a (x - x_a) + \left(\frac{\partial \mathbf{u}}{\partial y}\right)_a (y - y_a) + \cdots \tag{2.55}$$

If within an element of 'size' h a polynomial expansion complete to degree p is employed, this can fit locally the Taylor expansion up to that degree and, as $x - x_a$ and $y - y_a$ are of the order of magnitude h, the error in $\mathbf{u}$ will be of the order $O(h^{p+1})$. Thus, for instance, in the case of the plane elasticity problem discussed, we used a complete linear expansion and $p = 1$. We should therefore expect a *convergence* rate of order $O(h^2)$, i.e., the error in displacement being reduced to $1/4$ for a halving of the mesh spacing.

By a similar argument the strains (or stresses) which are given by the mth derivatives of displacement should converge with an error of $O(h^{p+1-m})$, i.e., as $O(h)$ in the plane stress example, where $m = 1$. The strain energy, being given by the square of the stresses, will show an error of $O(h^{2(p+1-m)})$ or $O(h^2)$ in the plane stress example.

The arguments given here are perhaps 'heuristic' from a mathematical viewpoint – they are, however, true[21, 22] and correctly give the orders of convergence, which can be expected to be achieved asymptotically as the element size tends to zero and *if the exact solution does not contain singularities*. Such singularities may result in infinite values of the coefficients in terms omitted in the Taylor expansion of Eq. (2.55) and invalidate the arguments. However, in many well-behaved problems the mere determination of the order of convergence often suffices to extrapolate the solution to the correct result. Thus, for instance, if the displacement converges at $O(h^2)$ and we have two approximate solutions u^1 and u^2 obtained with meshes of size h and $h/2$, we can write, with u being the exact solution,

$$\frac{u^1 - u}{u^2 - u} = \frac{O(h^2)}{O(h/2)^2} \approx 4 \tag{2.56}$$

From the above an (almost) exact solution u can be predicted. This type of extrapolation was first introduced by Richardson[23] and is of use if convergence is monotonic and nearly asymptotic.

We shall return to the important question of estimating errors due to the discretization process in Chapter 13 and will show that much more precise methods than those arising from convergence rate considerations are possible today. Indeed automatic mesh refinement processes can be introduced so that the specified accuracy can be achieved (viz. Chapters 8 and 14).

Discretization error is not the only error possible in a finite element computation. In addition to obvious mistakes which can occur when introducing data into computers, errors due to *round-off* are always possible. With the computer operating on numbers rounded off to a finite number of digits, a reduction of accuracy occurs every time differences between 'like' numbers are being formed. In the process of equation solving many subtractions are necessary and accuracy decreases. Problems of matrix conditioning, etc., enter here and the user of the finite element method must at all times be aware of accuracy limitations which simply do not allow the exact solution ever to be obtained. Fortunately in many computations, by using modern machines which carry a large number of significant digits, these errors are often small.

Another error that is often encountered occurs in approximation of curved boundaries by polynomials on faces of elements. For example, use of linear triangles to approximate a circular boundary causes an error of $O(h^2)$ to be introduced.

2.7 Displacement functions with discontinuity between elements – non-conforming elements and the patch test

In some cases considerable difficulty is experienced in finding displacement functions for an element which will automatically be continuous along the whole interface between adjacent elements.

As already pointed out, the discontinuity of displacement will cause infinite strains at the interfaces, a factor ignored in this formulation because the energy contribution is limited to the elements themselves.

However, if, in the limit, as the size of the subdivision decreases continuity is restored, then the formulation already obtained will still tend to the correct answer. This condition is always reached if

(a) constant strain condition automatically ensures displacement continuity, and
(b) the constant strain criteria of the previous section are satisfied.

To test that such continuity is achieved for any mesh configuration when using such *non-conforming* elements it is necessary to impose, on an arbitrary patch of elements, nodal displacements corresponding to any state of constant strain. *If nodal equilibrium is simultaneously achieved without the imposition of external, nodal, forces and if a state of constant stress is obtained, then clearly no external work has been lost through interelement discontinuity.*

Elements which pass such a *patch test* will converge, and indeed at times non-conforming elements will show a superior performance to conforming elements.

The patch test was first introduced by Irons[16] and has since been demonstrated to give a sufficient condition for convergence.[22, 24–28] The concept of the patch test can be generalized to give information on the rate of convergence which can be expected from a given element.

We shall return to this problem in detail in Chapter 9 where the test will be fully discussed.

2.8 Finite element solution process

The finite element solution of a problem follows a standard methodology. The steps in any solution process are always performed by the following steps:

1. Define the problem to be solved in terms of differential equations. Construct the integral form for the problem as a virtual work, variational or weak formulation.
2. Select the type and order of finite elements to be used in the analysis.
3. Define the mesh for the problem. This involves the description of the node and element layout, as well as the specification of boundary conditions and parameters for the formulation used. The process for mesh generation will be described in more detail in Chapter 8.
4. Compute and assemble the element arrays. The particular virtual work, variational or weak form provide the basis for computing specific relationships of each element.
5. Solve the resulting set of linear algebraic equations for the unknown parameters. See Appendix C for a brief discussion on solution of linear algebraic equations.
6. Output the results for the nodal and element variables. Graphical outputs also are useful for this step. An accurate procedure to project element values to nodes is described in Chapter 6.

Much of the discussion in the following chapters is concerned with the development of the theory needed to compute element arrays. For a steady-state problem the two arrays are a coefficient array $\mathbf{K}$, which we refer to as a 'stiffness' matrix, and a force array $\mathbf{f}$.

In the next section, however, we first illustrate the solution steps for two problems for which the exact solution is available.

2.9 Numerical examples

Let us now consider the solution to a set of problems for which an exact solution is known. This will enable us to see how the finite element results compare to the known solution and also to demonstrate the convergence properties for different element types. Of course, the power of the finite element method is primarily for use on problems for which no alternative solution is possible using results from classical books on elasticity and in later chapters we will include results for several such example problems.

2.9.1 Problems for accuracy assessment

Example 2.3: Beam subjected to end shear. We consider a rectangular beam in a state of plane stress. The geometric properties are shown in Fig. 2.8(a). The solution to the problem is given in Timoshenko and Goodier based on use of a stress function solution.[6] The solution for stresses is given by

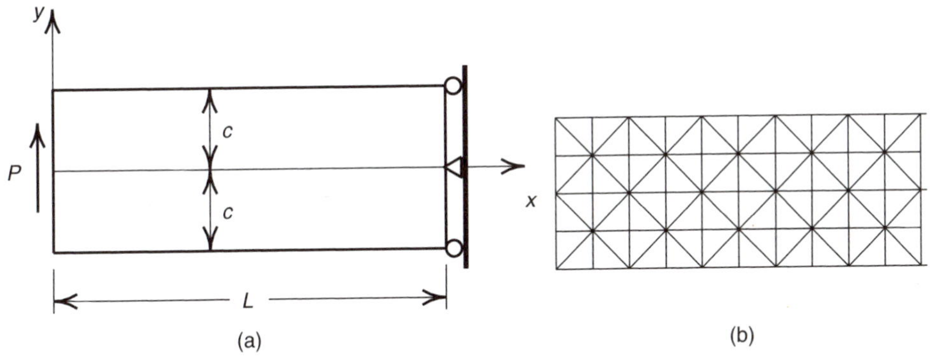

Fig. 2.8 End loaded beam: (a) Problem geometry and (b) coarse mesh.

$$\sigma_x = -\frac{3}{2}\frac{Pxy}{c^3}$$
$$\sigma_y = 0$$
$$\tau_{xy} = -\frac{3P}{4c}\left[1 - \left(\frac{y}{c}\right)^2\right]$$

where P is the applied load and c the half-depth of the beam. For the displacement boundary conditions

$$u(L, 0) = v(L, 0) = 0 \quad \text{and} \quad u(L, c) = u(L, -c) = 0$$

As shown in Fig. 2.8(a), the solution for displacements is given by

$$u = -\frac{P(x^2 - L^2)y}{2EI} - \frac{\nu Py(y^2 - c^2)}{6EI} + \frac{Py(y^2 - c^2)}{6GI}$$
$$v = \frac{\nu Pxy^2}{2EI} + \frac{P(x^3 - L^3)}{6EI} - \left(\frac{PL^2}{2EI} + \frac{\nu Pc^2}{6EI} + \frac{Pc^2}{3GI}\right)(x - L)$$

In the above E and ν are the elastic modulus and Poisson ratio, G is the shear modulus given by $E/[2(1+\nu)]$ and I is the area moment of inertia which is equal to $2tc^3/3$ where t is a constant beam thickness.

For this solution the tractions on the boundaries become

$$\begin{Bmatrix} t_x \\ t_y \end{Bmatrix} = t \begin{Bmatrix} 0 \\ -\tau_{xy} \end{Bmatrix} \quad \text{for } x = 0; \quad -c \le y \le c$$
$$\begin{Bmatrix} t_x \\ t_y \end{Bmatrix} = t \begin{Bmatrix} \sigma_x \\ \tau_{xy} \end{Bmatrix} \quad \text{for } x = L; \quad -c \le y \le c$$

For the numerical solution we choose the properties

$$c = 10; \ L = 100; \ t = 1; \ P = 80; \ E = 1000 \text{ and } \nu = 0.25$$

In order to perform a finite element solution to the problem we need to compute the nodal forces for the tractions using Eq. 2.30. When many elements are used in an analysis this

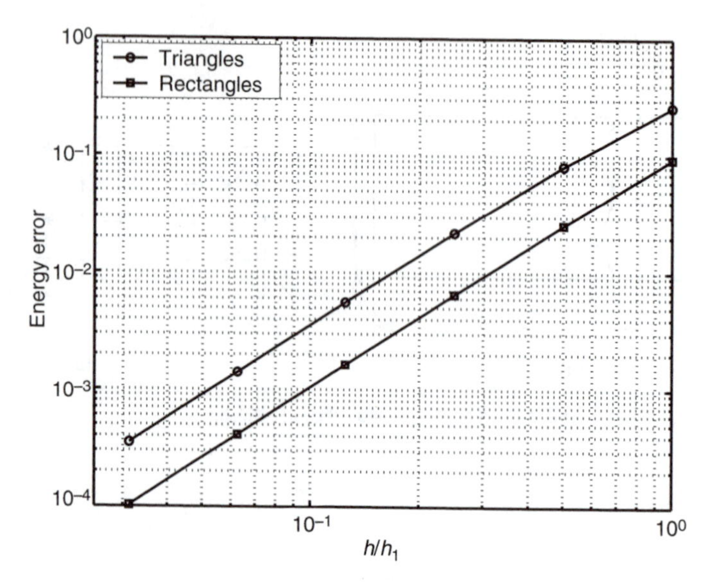

Fig. 2.9 Convergence in energy error for 3-node triangles and 4-node rectangular elements.

step can be quite tedious and it is best to write a small computer program to carry out the integrations (e.g., using MATLAB[29] or any other programming language).†

The solution to the problem is carried out using a uniform mesh of (a) 3-node triangular elements and (b) 4-node rectangular elements and the results for the error in energy given by

$$\eta_E = \frac{|E_{ex} - E_{fe}|}{E_{ex}} \approx C\, h^2$$

is plotted versus the log in element size h in Fig. 2.9. Here E_{ex} is the energy of the exact solution and E_{fe} that of the finite element solution. Results for the energy are given in Table 2.1 and the exact value for the geometry and properties selected is 3296 (energy here is work done which is twice the stored elastic strain energy). The element size is normalized to that of the coarsest mesh [h_1 shown in Fig. 2.8(b)] and the energy error is computed using the exact value. The expected slope of 2 is achieved for both element types with the 4-node element giving a smaller constant C due to the presence of the xy term in each shape function.

The stresses in each triangle are constant. The values for σ_x, σ_y and τ_{xy} obtained in the elements at the right end of the beam (where σ_x is largest) are shown in Fig. 2.10(a). The distribution of σ_x for $x = 90$ is shown in Fig. 2.10(b) where we also include values computed by a nodal averaging method. In Chapter 6 we will show how more accurate stresses may be obtained at nodes.

Example 2.4: Circular beam subjected to end shear. We consider a circular beam in a state of plane stress. The geometric properties are shown in Fig. 2.11. The solution to the problem is given in Timoshenko and Goodier based on use of a stress function.[6] The geometry and loading for the problem are shown in Fig. 2.11. The solution for stresses is

† For the triangular elements discussed in this chapter, the program *FEAPpv* available as a companion to this book includes automatic computation of nodal forces for this type of loading.[30]

Table 2.1 Mesh size and energy for end loaded beam

	(a) 3-node triangles		(b) 4-node rectangles	
Nodes	Elmts	Energy	Elmts	Energy
55	80	2438.020814633	40	2984.863896144
189	320	3027.225730752	160	3212.088124234
697	1280	3223.959303515	640	3274.561666674
2673	5120	3277.628191064	2560	3290.607667261
10465	20480	3291.381304522	10240	3294.649630776
41409	81920	3294.843527071	40960	3295.662249951
Exact	–	3296.000000000	–	3296.000000000

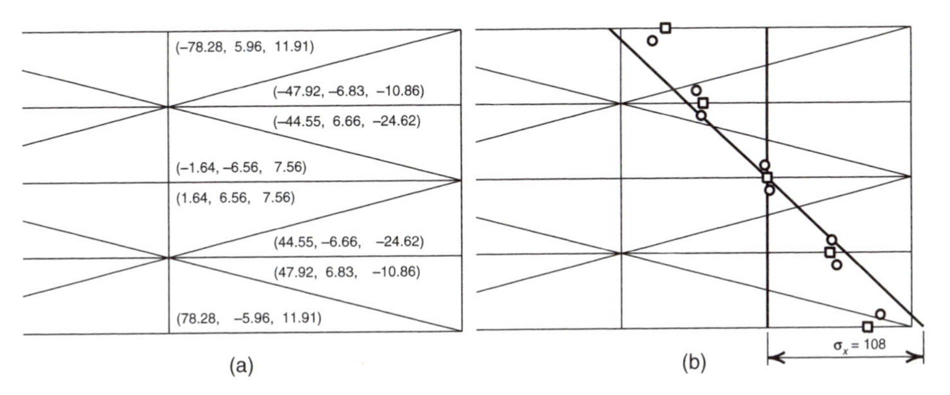

Fig. 2.10 End loaded beam: (a) Element stresses (σ_x, σ_y, τ_{xy}) and (b) σ_x stress distribution for $x = 90$, ○ element values and □ nodal average values.

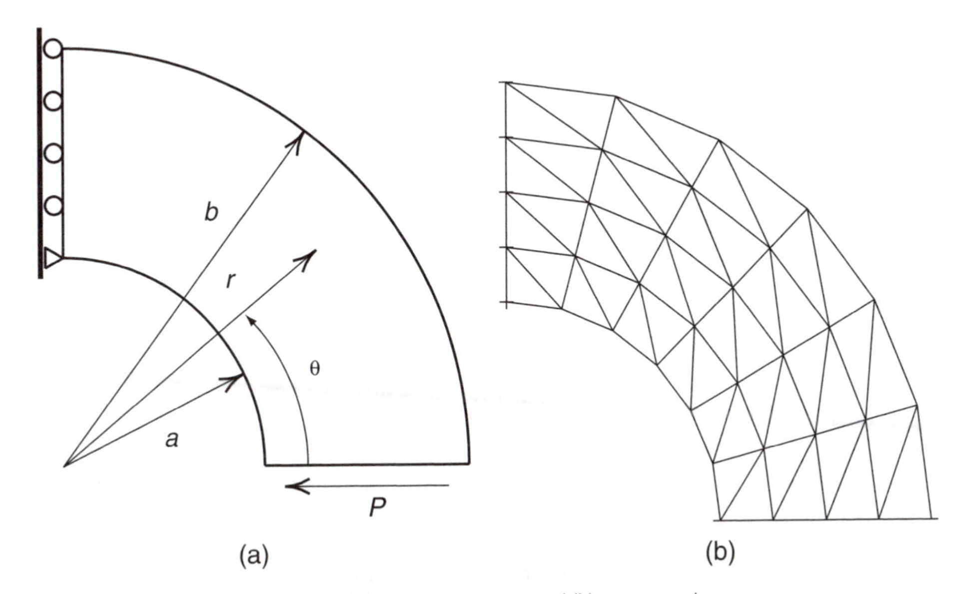

Fig. 2.11 End loaded circular beam: (a) Problem geometry and (b) coarse mesh.

given by

$$\sigma_{rr} = \frac{P}{N}\left[r + \frac{a^2b^2}{r^3} - \frac{a^2+b^2}{r}\right] \sin\theta$$

$$\sigma_{\theta\theta} = \frac{P}{N}\left[3r - \frac{a^2b^2}{r^3} - \frac{a^2+b^2}{r}\right] \sin\theta$$

$$\tau_{r\theta} = -\frac{P}{N}\left[r + \frac{a^2b^2}{r^3} - \frac{a^2+b^2}{r}\right] \cos\theta$$

where $N = a^2 - b^2 + (a^2+b^2)\log b/a$. For the restraints shown in Fig. 2.11(a) the solution for displacements is given by

$$u_r = \frac{P}{NE}\left\{ \left[\frac{1}{2}(1-3\nu)r^2 - \frac{a^2b^2(1+\nu)}{2r^2} - (a^2+b^2)(1-\nu)\log r\right] \sin\theta \right.$$
$$\left. +(a^2+b^2)(\theta-\pi)\cos\theta \right\} - K\sin\theta$$

$$u_\theta = \frac{P}{NE}\left\{ \left[\frac{1}{2}(5+\nu)r^2 - \frac{a^2b^2(1+\nu)}{2r^2} + (a^2+b^2)[(1-\nu)\log r - (1+\nu)]\right] \cos\theta \right.$$
$$\left. -(a^2+b^2)(\theta-\pi)\sin\theta \right\} - K\cos\theta$$

where for $u_r(a, \pi/2) = 0$ we obtain

$$K = \frac{P}{NE}\left[\frac{1}{2}(1-3\nu)a^2 - \frac{b^2(1+\nu)}{2} - (a^2+b^2)(1-\nu)\log a\right]$$

In the above E and ν are the elastic modulus and Poisson ratio; a and b are the inner and outer radii, respectively (see Fig. 2.11).

For this solution the displacement u_r for $\theta = 0$ is constant and given by

$$u_r(r, 0) = -\frac{\pi P}{EN}(a^2+b^2) = u_0$$

Thus, instead of computing the nodal forces for the traction on this boundary we merely set all the nodal displacements in the x direction to a constant value.

For the numerical solution we choose the properties

$$a = 5;\ b = 10;\ t = 1;\ u_0 = -0.01;\ E = 10\,000 \text{ and } \nu = 0.25$$

In addition the displacements on the boundaries are prescribed as

$$u(x, 0) = u_0 \text{ and } u(0, y) = v(0, a) = 0$$

The finite element solution to the problem is carried out using a uniform mesh of 3-node triangular elements oriented as shown in Fig. 2.11(b). Results for the energy are given in Table 2.2 and the 'exact' value is computed from

$$E_{ex} = \frac{1}{\pi}\left[\log 2 - 0.6\right] = 0.02964966844238$$

for the geometry and properties selected. The element size is normalized to that of the coarsest mesh [shown in Fig. 2.11(b)] and the energy error given in Table 2.2 again has the expected slope of 2.

Finally, in Fig. 2.13 we compare the u_r and u_θ displacements from the finite element solution of the coarsest mesh to the exact values. We observe that even with this coarse distribution of elements the solution is quite good. Unfortunately, the stress distribution is not as accurate and quite fine meshes are needed to obtain good values. In Chapter 6 we will show how use of higher order elements can significantly improve both the displacements and stresses obtained.

2.9.2 A practical application

Obviously, the practical applications of the finite element method are limitless, and it has superseded experimental technique for plane problems because of its high accuracy, low cost, and versatility. The ease of treatment of general boundary shapes and conditions, material anisotropy, thermal stresses, or body force problems add to its practical advantages.

Stress flow around a reinforced opening

An example of an actual early application of the finite element method to complex problems of engineering practice is a steel pressure vessel or aircraft structure in which openings are introduced in the stressed skin. The penetrating duct itself provides some reinforcement round the edge and, in addition, the skin itself is increased in thickness to reduce the stresses due to concentration effects.

Analysis of such problems treated as cases of plane stress present no difficulties. The elements are chosen so as to follow the thickness variation, and appropriate values of this are assigned.

The narrow band of thick material near the edge can be represented either by special bar-type elements, or by very thin triangular elements of the usual type, to which appropriate thickness is assigned. The latter procedure was used in the problem shown in Fig. 2.14 which gives some of the resulting stresses near the opening itself. The fairly large extent of the region introduced in the analysis and the grading of the mesh should be noted.

Table 2.2 Mesh size and energy for curved beam

	3-node triangles		
Nodes	Elmts	Energy	Error (%)
35	48	0.04056964168222	36.830
117	192	0.03245261212845	9.454
425	768	0.03035760000738	2.388
1617	3072	0.02982725603614	0.598
6305	12 288	0.02969411302439	0.150
24 897	49 152	0.02966078320581	0.037
Exact	–	0.02964966844238	–

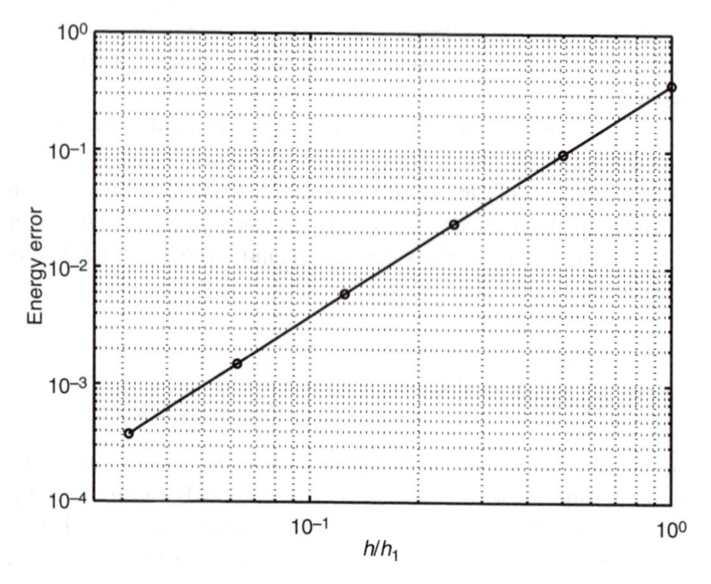

Fig. 2.12 Curved beam: Convergence in energy error for 3-node triangles.

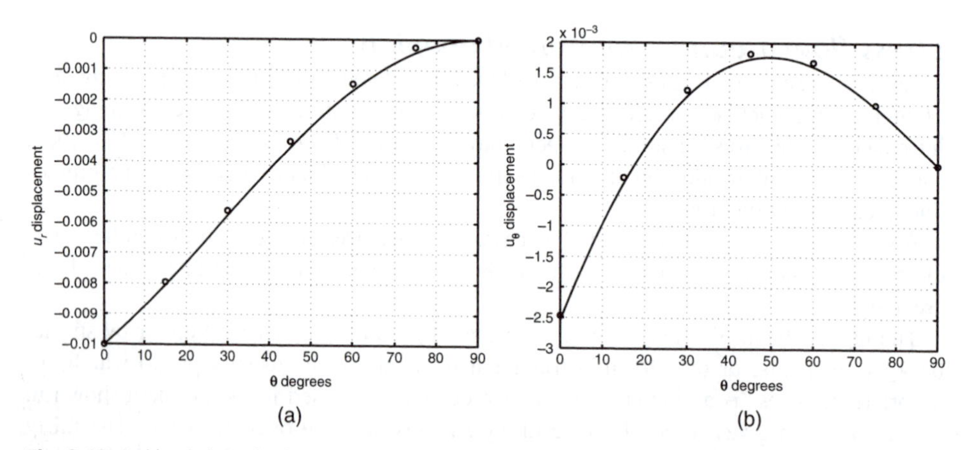

Fig. 2.13 End loaded circular beam: (a) u_r displacement and (b) u_θ displacement for $r = a$.

2.10 Concluding remarks

The 'displacement' approach to the analysis of elastic solids is still undoubtedly the most popular and easily understood procedure. In many of the following chapters we shall use the general formulae developed here in the context of linear elastic analysis (e.g., in Chapter 6). These are also applicable in the context of non-linear analysis, the main variants being the definitions of the stresses, generalized strains, and other associated quantities.[7]

In Chapter 3 we shall show that the procedures developed here are but a particular case of finite element discretization applied to the governing differential equations written in terms of displacements.[31] Clearly, alternative starting points are possible. Some of these will be mentioned in Chapters 10 and 11.

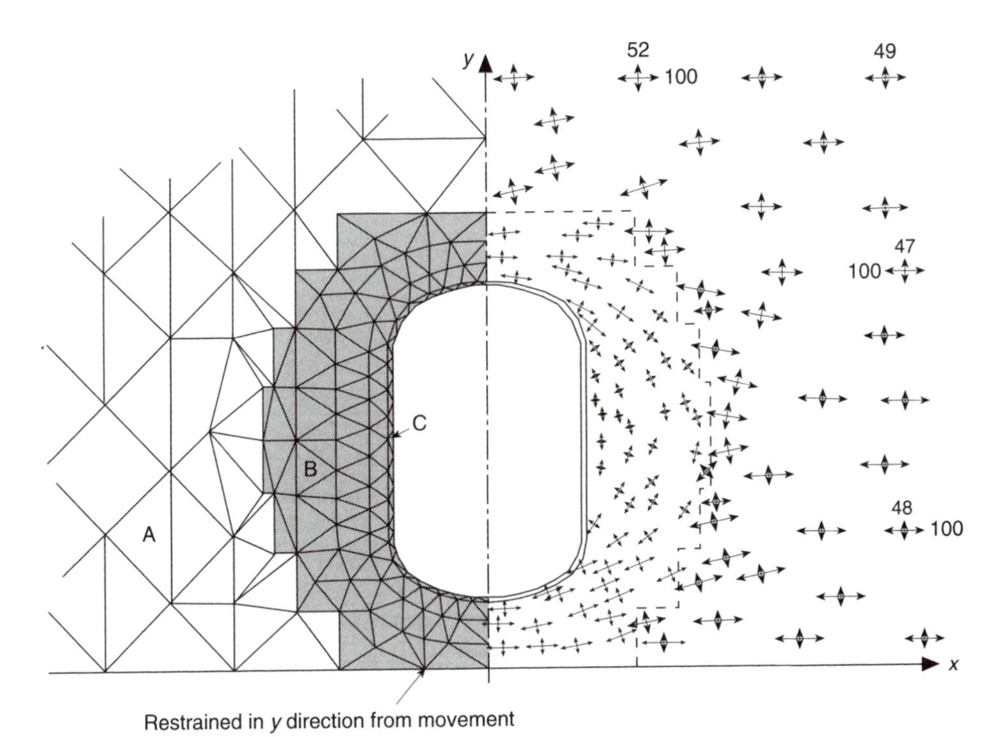

Fig. 2.14 A reinforced opening in a plate. Uniform stress field at a distance from opening $\sigma_x = 100$, $\sigma_y = 50$. Thickness of plate regions A, B, and C is in the ratio of 1 : 3 : 23.

2.11 Problems

2.1 For the triangular element shown in Fig. 2.15(a), the dimensions are: $a = 3$ cm and $b = 4$ cm. Compute the shape functions $\mathbf{N}$ for the three nodes of the element.

2.2 For the rectangular element shown in Fig. 2.15(b), the dimensions are: $a = 6$ cm and $b = 4$ cm. Compute the shape functions $\mathbf{N}$ for the four nodes of the element.

2.3 Use the results from Problem 2.1 to compute the strain-displacement matrix $\mathbf{B}$ for the triangular element shown in Fig. 2.15(a).

2.4 Use the results from Problem 2.2 to compute the strain-displacement matrix $\mathbf{B}$ for the rectangular element shown in Fig. 2.15(b). The body force vector in a plane stress problem is given by $b_x = 5$ and $b_y = 0$. Using the shape functions determined in Problem 2.1 compute the body force vector for the triangular element shown in Fig. 2.15(a).

2.5 Repeat Problem 2.4 using $b_x = 0$ and $b_y = -30$.

2.6 The body force vector in a plane stress problem is given by $b_x = 5$ and $b_y = 0$. Using the shape functions determined in Problem 2.2 compute the body force vector for the rectangular element shown in Fig. 2.15(b).

2.7 Repeat Problem 2.6 using $b_x = 0$ and $b_y = -30$.

2.8 The edge of the triangular element defined by nodes 2–3 shown in Fig. 2.15(a) is to be assigned boundary conditions $u_n = 0$ and $t_s = 0$ where n is a direction normal to the edge and s tangential to the edge. Determine the transformation matrix $\mathbf{L}$

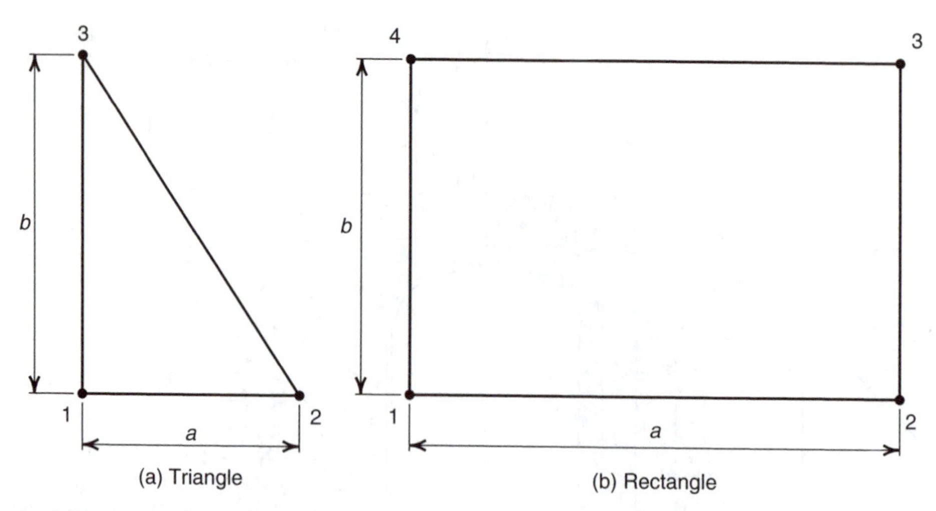

Fig. 2.15 Elements for Problems 2.1 to 2.4.

[viz. Eq. (1.21)] required to transform the nodal degrees of freedom at node 2 and 3 to be able to impose the boundary conditions.

2.9 A concentrated load, F, is applied to the edge of a two-dimensional plane strain problem as shown in Fig. 2.16(a).
   (a) Use equilibrium conditions to compute the statically equivalent forces acting at nodes 1 and 2.
   (b) Use virtual work to compute the equivalent forces acting on nodes 1 and 2.

2.10 A triangular traction load is applied to the edge of a two-dimensional plane strain problem as shown in Fig. 2.16(b).
   (a) Use equilibrium conditions to compute the statically equivalent forces acting at nodes 1 and 2.
   (b) Use virtual work to compute the equivalent forces acting on nodes 1 and 2.

2.11 For the rectangular and triangular element shown in Fig. 2.17, compute and assemble the stiffness matrices associated with nodes 2 and 5 (i.e., $\mathbf{K}_{22}$, $\mathbf{K}_{25}$ and $\mathbf{K}_{55}$). Let $E = 1000$, $\nu = 0.25$ for the rectangle and $E = 1200$, $\nu = 0$ for the triangle. The thickness for the assembly is constant with $t = 0.2$ cm.

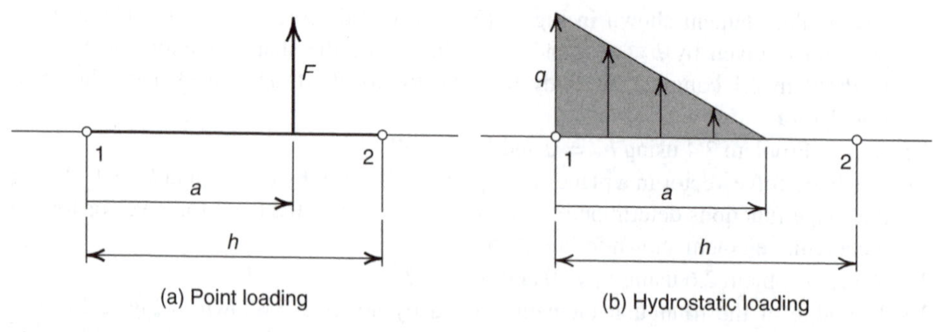

Fig. 2.16 Traction loading on boundary for Problems 2.9 and 2.10.

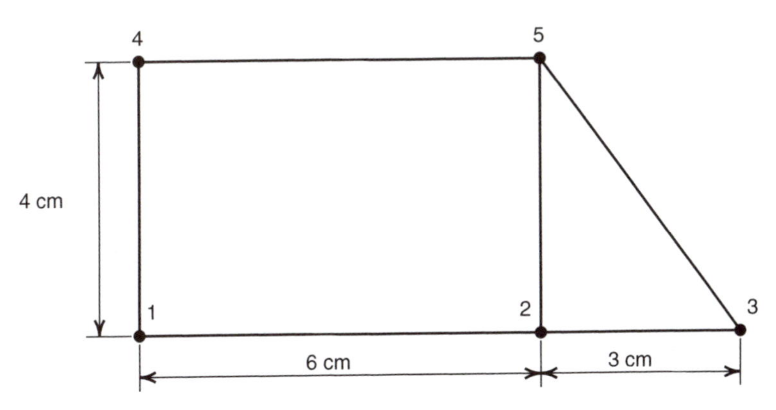

Fig. 2.17 Element assembly for Problem 2.11.

2.12 The formulation given in Sec. 2.3 may be specialized to one dimension by using displacement functions

$$u^e = \alpha_1 + \alpha_2 x = N_1(x)\,\tilde{u}_1 + N_2(x)\,\tilde{u}_2$$
$$\delta u^e = N_1(x)\,\delta\tilde{u}_1 + N_2(x)\,\delta\tilde{u}_2$$

and simplifying the equation to:

$$\text{Strain:} \quad \varepsilon = \frac{\mathrm{d}u}{\mathrm{d}x} = \mathbf{B}\tilde{\mathbf{u}}$$
$$\text{Stress:} \quad \sigma = E\,\varepsilon = E\mathbf{B}\tilde{\mathbf{u}}$$

where E is the modulus of elasticity and we assume initial stress σ_0 and initial strain ε_0 are zero.

(a) For a two-node element with coordinates located at x_1^e and x_2^e compute the shape functions N_a which satisfy the linear approximation given above.
(b) Compute the strain matrix $\mathbf{B}$ for the shape functions determined in (a).
(c) Using Eqs (2.25) to (2.28b) compute the element stiffness and force vector. Assume the body force b in the x direction is constant in each element.
(d) For the two element problem shown in Fig. 2.18 let each element have length $a = 5$, the end traction $t = 4$, the body force $b = 2$ and the modulus of elasticity $E = 200$.
   i. Generalize the element formulation given in (b) to form the whole problem.
   ii. Impose the boundary condition $u(0) = \tilde{u}_1 = 0$.
   iii. Determine the solution for $\tilde{u}_2$ and $\tilde{u}_3$.
   iv. Plot the computed finite element displacement u and stress σ vs x.

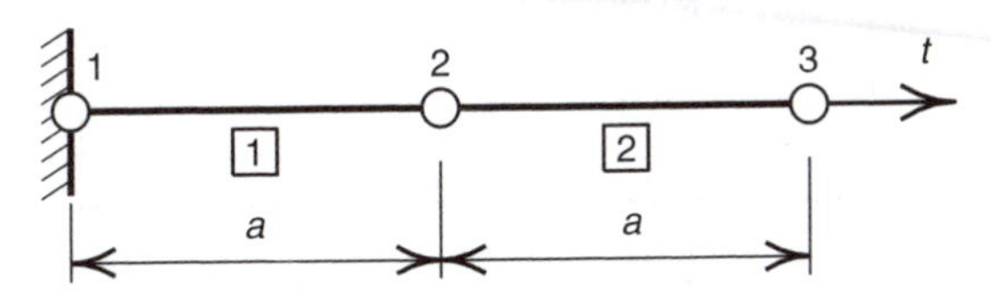

Fig. 2.18 One-dimensional elasticity. Problem 2.12.

(e) The exact solution to the problem satisfies the equilibrium equation

$$\frac{d\sigma}{dx} + b = 0$$

and boundary conditions $u(0) = 0$ and $t(10) = \sigma(10) = 4$. Compute and plot the exact solution for $u(x)$ and $\sigma(x)$.

(f) What is the maximum error in the finite element solutions for u and σ?

(g) Subdivide the mesh into four elements and repeat the above solution steps.

2.13 Download the program *FEAPpv* and user manual from a web site given in Chapter 19. Note that both source code for the program and an executable version for Windows-based systems are available at the site.

If source code is used it is necessary to compile the program to obtain an executable version.

2.14 Use *FEAPpv* (or any available program) to solve the rectangular beam problem given in Example 2.3 – verify results shown in Table 2.1.

2.15 Use *FEAPpv* (or any available program) to solve the curved beam problem given in Example 2.4 – verify results shown in Table 2.2.

2.16 The uniformly loaded cantilever beam shown in Fig. 2.19 has properties

$$L = 2 \text{ m}; \quad h = 0.4 \text{ m}; \quad t = 0.05 \text{ m} \text{ and } q_0 = 100 \text{ N/m}$$

Use *FEAPpv* (or any available program) to perform a plane stress analysis of the problem assuming linear isotropic elastic behaviour with $E = 200$ GPa and $\nu = 0.3$.

In your analysis:

(a) Use 3-node triangular elements with an initial mesh of two elements in the depth and ten elements in the length directions.

(b) Compute consistent nodal forces for the uniform loading.

(c) Compute nodal forces for a parabolically distributed shear traction at the restrained end which balances the uniform loading q_0.

(d) Report results for the centreline displacement in the vertical direction and the stored energy in the beam.

(e) Repeat the analysis three additional times using meshes of 4×20, 8×40 and 16×80 elements. Tabulate the tip vertical displacement and stored energy for each solution.

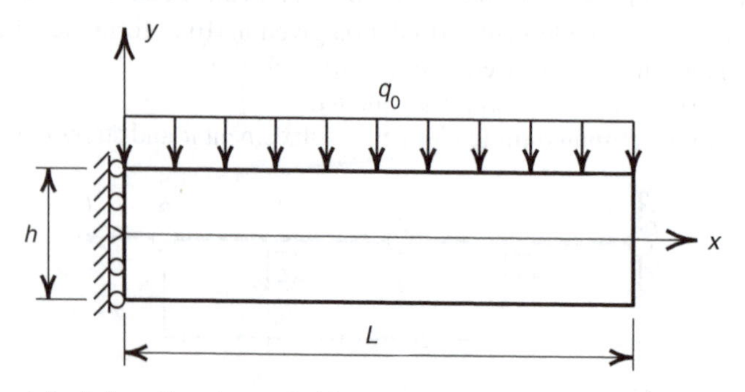

Fig. 2.19 Uniformly loaded cantilever beam. Problem 2.16.

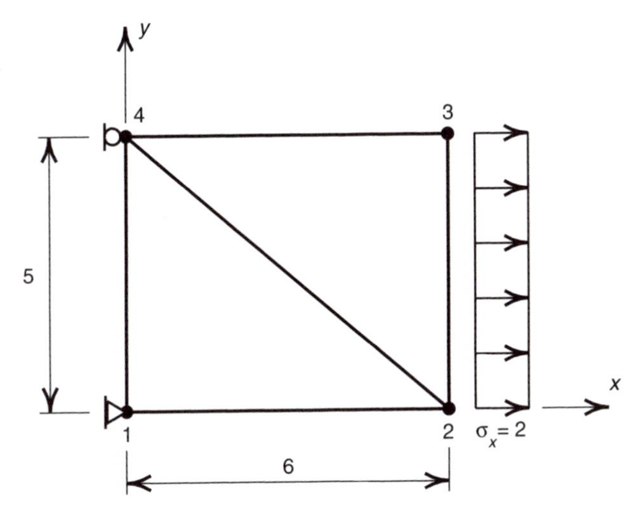

Fig. 2.20 Patch test for triangles. Problem 2.17.

(f) If the energy error is given by

$$\Delta E = E_n - E_{n-1} = Ch^p$$

estimate C and p for your solution.

(g) Repeat the above analysis using rectangular 4-node elements.

2.17 Program development project:† Write a MATLAB[29] program‡ to solve plane stress problems.

Your program system should have the following features:

(a) Input module which describes:
   i. Nodal coordinate values, $\mathbf{x}_a$;
   ii. Nodes connected to each element and material properties of each element;
   iii. Node and degree-of-freedom (dof) for each applied nodal forces;
   iv Node and dof for fixed (essential) boundary condition – also value if non-zero.

(b) Module to compute the stiffness matrix for a 3-node triangular element [use Eq. (2.34)].

(c) Module to assemble element arrays into global arrays and specified nodal forces and displacements.

(d) Module to solve $\mathbf{K}\tilde{\mathbf{u}} + \mathbf{f} = \mathbf{0}$.

(e) Module to output nodal displacements and element stress and strains.

Use your program to solve the *patch test* problem shown in Fig. 2.20. Use the properties: $E = 2 \cdot 10^5$, $\nu = 0.3$ and $t = 1$ (t is thickness of slab). You can verify the correctness of your answer by computing an exact solution to the problem. The correctness of computed arrays may be obtained using results from *FEAPpv* (or any available plane stress program).

† If programming is included as a part of your study, it is recommended that this problem be solved. Several extensions will be suggested later to create a solution system capable of performing all steps of finite element analysis.

‡ Another programming language may be used; however, MATLAB offers many advantages to write simple programs and is also useful to easily complete later exercises.

2.18 Program development project: Add a graphics capability to the program developed in Problem 2.17 to plot contours of the computed finite element displacements. (Hint: MATLAB has *contour* and *surf* options to easily perform this operation.)

Solve the curved beam problem for the mesh shown in Fig. 2.11. Plot contours for u and v displacements. (Hint: Write a separate MATLAB program to generate the nodal coordinates and element connections for the simple geometry of the curved beam.) Refine the mesh by increasing the number of segments in each direction by a factor of 2 and repeat the solution of the curved beam problem.

References

1. R.W. Clough. The finite element method in plane stress analysis. In *Proc. 2nd ASCE Conf. on Electronic Computation*, Pittsburgh, Pa., Sept. 1960.
2. R.W. Clough. The finite element method in structural mechanics. In O.C. Zienkiewicz and G.S. Holister, editors, *Stress Analysis*, Chapter 7. John Wiley & Sons, Chichester, 1965.
3. J. Szmelter. The energy method of networks of arbitrary shape in problems of the theory of elasticity. In W. Olszak, editor, *Proc. IUTAM Symposium on Non-Homogeneity in Elasticity and Plasticity*. Pergamon Press, 1959.
4. R. Courant. Variational methods for the solution of problems of equilibrium and vibration. *Bull. Am. Math Soc.*, 49:1–61, 1943.
5. W. Prager and J.L. Synge. Approximation in elasticity based on the concept of function space. *Quart. J. Appl. Math.*, 5:241–269, 1947.
6. S.P. Timoshenko and J.N. Goodier. *Theory of Elasticity*. McGraw-Hill, New York, 3rd edition, 1969.
7. O.C. Zienkiewicz and R.L. Taylor. *The Finite Element Method for Solid and Structural Mechanics*. Butterworth-Heinemann, Oxford, 6th edition, 2005.
8. K. Washizu. *Variational Methods in Elasticity and Plasticity*. Pergamon Press, New York, 3rd edition, 1982.
9. F.B. Hildebrand. *Methods of Applied Mathematics*. Prentice-Hall (reprinted by Dover Publishers, 1992), 2nd edition, 1965.
10. Lord Rayleigh (J.W. Strutt). On the theory of resonance. *Trans. Roy. Soc. (London)*, A161:77–118, 1870.
11. W. Ritz. Über eine neue Methode zur Lösung gewisser variationsproblem der mathematischen physik. *Journal für die reine und angewandte Mathematik*, 135:1–61, 1908.
12. B. Fraeijs de Veubeke. Displacement and equilibrium models in finite element method. In O.C. Zienkiewicz and G.S. Holister, editors, *Stress Analysis*, Chapter 9, pages 145–197. John Wiley & Sons, Chichester, 1965.
13. J. Demmel. *Applied Numerical Linear Algebra*. Society for Industrial and Applied Mathematics, Philadelphia, PA, 1997.
14. R.L. Fox and E.L. Stanton. Developments in structural analysis by direct energy minimization. *J. AIAA*, 6:1036–1044, 1968.
15. F.K. Bogner, R.H. Mallett, M.D. Minich, and L.A. Schmit. Development and evaluation of energy search methods in non-linear structural analysis. In *Proc. 1st Conf. Matrix Methods in Structural Mechanics*, volume AFFDL-TR-66-80, Wright Patterson Air Force Base, Ohio, Oct. 1966.
16. G.P. Bazeley, Y.K. Cheung, B.M. Irons, and O.C. Zienkiewicz. Triangular elements in bending – conforming and non-conforming solutions. In *Proc. 1st Conf. Matrix Methods in Structural Mechanics*, volume AFFDL-TR-66-80, pages 547–576, Wright Patterson Air Force Base, Ohio, Oct. 1966.

17. S.C. Mikhlin. *The Problem of the Minimum of a Quadratic Functional.* Holden-Day, San Francisco, 1966.
18. M.W. Johnson and R.W. McLay. Convergence of the finite element method in the theory of elasticity. *J. Appl. Mech., ASME*, 274–278, 1968.
19. P.G. Ciarlet. *The Finite Element Method for Elliptic Problems.* North-Holland, Amsterdam, 1978.
20. T.H.H. Pian and P. Tong. The convergence of finite element method in solving linear elastic problems. *Int. J. Solids Struct.*, 4:865–880, 1967.
21. E.R. de Arantes e Oliveira. Theoretical foundations of the finite element method. *Int. J. Solids Struct.*, 4:929–952, 1968.
22. G. Strang and G.J. Fix. *An Analysis of the Finite Element Method.* Prentice-Hall, Englewood Cliffs, N.J., 1973.
23. L.F. Richardson. The approximate arithmetical solution by finite differences of physical problems. *Trans. Roy. Soc. (London)*, A210:307–357, 1910.
24. B.M. Irons and A. Razzaque. Experience with the patch test for convergence of finite elements. In A.K. Aziz, editor, *The Mathematics of Finite Elements with Application to Partial Differential Equations*, pages 557–587. Academic Press, New York, 1972.
25. B. Fraeijs de Veubeke. Variational principles and the patch test. *Int. J. Numer. Meth. Eng.*, 8:783–801, 1974.
26. R.L. Taylor, O.C. Zienkiewicz, J.C. Simo, and A.H.C. Chan. The patch test – a condition for assessing FEM convergence. *Int. J. Numer. Meth. Eng.*, 22:39–62, 1986.
27. O.C. Zienkiewicz, S. Qu, R.L. Taylor, and S. Nakazawa. The patch test for mixed formulations. *Int. J. Numer. Meth. Eng.*, 23:1873–1883, 1986.
28. O.C. Zienkiewicz and R.L. Taylor. The finite element patch test revisited: a computer test for convergence, validation and error estimates. *Comp. Meth. Appl. Mech. Eng.*, 149:523–544, 1997.
29. MATLAB. www.mathworks.com, 2003.
30. R.L. Taylor. *FEAP – A Finite Element Analysis Program: Personal version, User Manual.* University of California, Berkeley. http://www.ce.berkeley.edu/~rlt/feappv.
31. O.C. Zienkiewicz and K. Morgan. *Finite Elements and Approximation.* John Wiley & Sons, London, 1983.

3

Generalization of the finite element concepts. Galerkin-weighted residual and variational approaches

3.1 Introduction

We have so far dealt with one possible approach to the approximate solution of the particular problem of linear elasticity. Many other continuum problems arise in engineering and physics and usually these problems are posed by appropriate differential equations and boundary conditions to be imposed on the unknown function or functions. It is the object of this chapter to show that all such problems can be dealt with by the finite element method.

Posing the problem to be solved in its most general terms we find that we seek an unknown function $\mathbf{u}$ such that it satisfies a certain differential equation set

$$\mathcal{A}(\mathbf{u}) = \begin{Bmatrix} A_1(\mathbf{u}) \\ A_2(\mathbf{u}) \\ \vdots \end{Bmatrix} = \mathbf{0} \tag{3.1}$$

in a 'domain' (volume, area, etc.), Ω, together with certain boundary conditions

$$\mathcal{B}(\mathbf{u}) = \begin{Bmatrix} B_1(\mathbf{u}) \\ B_2(\mathbf{u}) \\ \vdots \end{Bmatrix} = \mathbf{0} \tag{3.2}$$

on the boundaries, Γ, of the domain as shown in Fig. 3.1.

The function sought may be a scalar quantity or may represent a vector of several variables. Similarly, the differential equation may be a single one or a set of simultaneous equations and does not need to be linear. It is for this reason that we have resorted to matrix notation in the above.

The finite element process, being one of approximation, will seek the solution in the approximate form†

$$\mathbf{u} \approx \hat{\mathbf{u}} = \sum_{a=1}^{n} \mathbf{N}_a \tilde{\mathbf{u}}_a = \mathbf{N}\tilde{\mathbf{u}} \tag{3.3}$$

where $\mathbf{N}_a$ are shape functions prescribed in terms of independent variables (such as the coordinates x, y, etc.) and all or most of the parameters $\tilde{\mathbf{u}}_a$ are unknown.

† In the sequel we will also use summation convention for any repeated index. Thus $\mathbf{N}_a\tilde{\mathbf{u}}_a \equiv \sum_a \mathbf{N}_a\tilde{\mathbf{u}}_a$, etc.

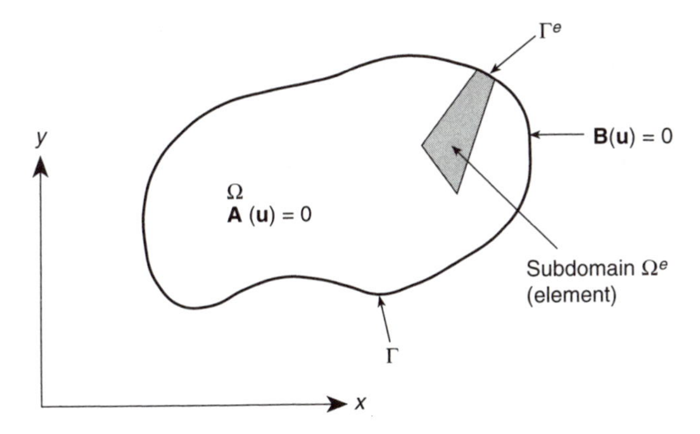

Fig. 3.1 Problem domain Ω and boundary Γ.

In the previous chapter we have seen that precisely the same form of approximation was used in the displacement approach to elasticity problems. We also noted there that (a) the shape functions were usually defined locally for elements or subdomains and (b) the properties of discrete systems were recovered if the approximating equations were cast in *an integral form* [viz. Eqs (2.38)–(2.42)]. With this object in mind we shall seek to cast the equation from which the unknown parameters $\tilde{\mathbf{u}}_a$ are to be obtained in the integral form

$$\int_\Omega \mathbf{G}_b(\hat{\mathbf{u}})\mathrm{d}\Omega + \int_\Gamma \mathbf{g}_b(\hat{\mathbf{u}})\mathrm{d}\Gamma = \mathbf{0} \qquad b = 1 \text{ to } n \tag{3.4}$$

in which $\mathbf{G}_b$ and $\mathbf{g}_b$ prescribe known functions or operators.

These integral forms will permit the approximation to be obtained element by element and an assembly to be achieved by the use of the procedures developed for *standard discrete systems* in Chapter 1, since, providing the functions $\mathbf{G}_b$ and $\mathbf{g}_b$ are integrable, we have

$$\int_\Omega \mathbf{G}_b\mathrm{d}\Omega + \int_\Gamma \mathbf{g}_b\mathrm{d}\Gamma = \sum_{e=1}^{m}\left(\int_{\Omega_e} \mathbf{G}_b\mathrm{d}\Omega + \int_{\Gamma_e} \mathbf{g}_b\mathrm{d}\Gamma\right) = \mathbf{0} \tag{3.5}$$

where Ω_e is the domain of each element and Γ_e its part of the boundary.

Two distinct procedures are available for obtaining the approximation in such integral forms. The first is the *method of weighted residuals* (known alternatively as the Galerkin procedure); the second is the determination of *variational functionals* for which stationarity is sought. We shall deal with both approaches in turn.

If the differential equations are linear, i.e., if we can write (3.1) and (3.2) as

$$\begin{aligned} \mathcal{A}(\mathbf{u}) &\equiv \mathcal{L}\mathbf{u} + \mathbf{b} = \mathbf{0} \qquad \text{in } \Omega \\ \mathcal{B}(\mathbf{u}) &\equiv \mathcal{M}\mathbf{u} + \mathbf{t} = \mathbf{0} \qquad \text{on } \Gamma \end{aligned} \tag{3.6}$$

then the approximating equation system (3.4) will yield a set of linear equations of the form

$$\mathbf{K}\,\tilde{\mathbf{u}} + \mathbf{f} = \mathbf{0} \tag{3.7}$$

with

$$\mathbf{K}_{ab} = \sum_{e=1}^{m} \mathbf{K}^e_{ab} \qquad \mathbf{f}_a = \sum_{e=1}^{m} \mathbf{f}^e_a$$

The reader not used to abstraction may well now be confused about the meaning of the various terms. We shall introduce here some typical sets of differential equations for which we will seek solutions (and which will make the problems a little more definite).

Example 3.1: Steady-state heat conduction equation in a two-dimensional domain. Here the equations are written as

$$A(\phi) = -\frac{\partial}{\partial x}\left(k\frac{\partial \phi}{\partial x}\right) - \frac{\partial}{\partial y}\left(k\frac{\partial \phi}{\partial y}\right) + Q = 0 \quad \text{in } \Omega$$

$$B(\phi) = \begin{cases} \phi - \bar{\phi} = 0 & \text{on } \Gamma_\phi \\ k\dfrac{\partial \phi}{\partial n} + \bar{q} = 0 & \text{on } \Gamma_q \end{cases} \tag{3.8}$$

where $\mathbf{u} \equiv \phi$ indicates temperature, k is the conductivity, Q is a heat source, $\bar{\phi}$ and $\bar{q}$ are the prescribed values of temperature and heat flow on the boundaries and n is the direction normal to Γ. In the context of this equation the boundary condition for Γ_ϕ is called a Dirichlet condition and the one on Γ_q a Neumann one.

In the above problem k and Q can be functions of position and, if the problem is non-linear, of ϕ or its derivatives.

Example 3.2: Steady-state heat conduction–convection equation in two dimensions. When convection effects are added the differential equation becomes

$$A(\phi) = -\frac{\partial}{\partial x}\left(k\frac{\partial \phi}{\partial x}\right) - \frac{\partial}{\partial y}\left(k\frac{\partial \phi}{\partial y}\right) + u_x\frac{\partial \phi}{\partial x} + u_y\frac{\partial \phi}{\partial y} + Q = 0 \quad \text{in } \Omega \tag{3.9}$$

with boundary conditions as in the first example. Here u_x and u_y are known functions of position and represent velocities of an incompressible fluid in which heat transfer occurs.

Example 3.3: A system of three first-order equations equivalent to Example 3.1. The steady-state heat equation in two dimensions may also be split into the three equations

$$\boldsymbol{\mathcal{A}}(\mathbf{u}) = \begin{Bmatrix} \dfrac{\partial q_x}{\partial x} + \dfrac{\partial q_y}{\partial y} + Q \\ q_x + k\dfrac{\partial \phi}{\partial x} \\ q_y + k\dfrac{\partial \phi}{\partial y} \end{Bmatrix} = 0 \quad \text{in } \Omega \tag{3.10}$$

and

$$\boldsymbol{\mathcal{B}}(\mathbf{u}) = \begin{cases} \phi - \bar{\phi} = 0 & \text{on } \Gamma_\phi \\ q_n - \bar{q} = 0 & \text{on } \Gamma_q \end{cases}$$

where q_n is the flux normal to the boundary.

Here the unknown function vector $\mathbf{u}$ corresponds to the set

$$\mathbf{u} = \begin{Bmatrix} \phi \\ q_x \\ q_y \end{Bmatrix}$$

This last example is typical of a so-called *mixed formulation*. In such problems the number of dependent unknowns can always be reduced in the governing equations by suitable algebraic operation, still leaving a solvable problem [e.g., obtaining Eq. (3.8) from (3.10) by eliminating q_x and q_y].

If this cannot be done [viz. Eq.(3.8)] we have an *irreducible formulation*.

Problems of mixed form present certain complexities in their solution which we shall discuss in Chapter 11.

In Chapter 7 we shall return to detailed examples of the first problem and other examples will be introduced throughout the book. The above three sets of problems will, however, be useful in their full form or reduced to one dimension (by suppressing the y variable) to illustrate the various approaches used in this chapter.

Weighted residual methods

3.2 Integral or 'weak' statements equivalent to the differential equations

As the set of differential equations (3.1) has to be zero at each point of the domain Ω, it follows that

$$\int_{\Omega} \mathbf{v}^{\mathrm{T}} \boldsymbol{\mathcal{A}}(\mathbf{u}) \mathrm{d}\Omega \equiv \int_{\Omega} [v_1 A_1(\mathbf{u}) + v_2 A_2(\mathbf{u}) + \cdots] \mathrm{d}\Omega \equiv 0 \tag{3.11}$$

where

$$\mathbf{v} = \begin{Bmatrix} v_1 \\ v_2 \\ \vdots \end{Bmatrix}$$

is a set of arbitrary functions equal in number to the number of equations (or components of $\mathbf{u}$) involved.

The statement is, however, more powerful. *We can assert that if (3.11) is satisfied for all* $\mathbf{v}$ *then the differential equations (3.1) must be satisfied at all points of the domain.* The proof of the validity of this statement is obvious if we consider the possibility that $\boldsymbol{\mathcal{A}}(\mathbf{u}) \neq \mathbf{0}$ at any point or part of the domain. Immediately, a function $\mathbf{v}$ can be found which makes the integral of (3.11) non-zero, and hence the point is proved.

If the boundary conditions (3.10) are to be simultaneously satisfied, then we require that

$$\int_{\Gamma} \bar{\mathbf{v}}^{\mathrm{T}} \boldsymbol{\mathcal{B}}(\mathbf{u}) \mathrm{d}\Gamma \equiv \int_{\Gamma} [\bar{v}_1 B_1(\mathbf{u}) + \bar{v}_2 B_2(\mathbf{u}) + \cdots] \mathrm{d}\Gamma = 0 \tag{3.12}$$

for any set of arbitrary functions $\bar{\mathbf{v}}$.

Indeed, the integral statement that

$$\int_{\Omega} \mathbf{v}^{\mathrm{T}} \boldsymbol{\mathcal{A}}(\mathbf{u}) \mathrm{d}\Omega + \int_{\Gamma} \bar{\mathbf{v}}^{\mathrm{T}} \boldsymbol{\mathcal{B}}(\mathbf{u}) \mathrm{d}\Gamma = 0 \tag{3.13}$$

is satisfied for all $\mathbf{v}$ and $\bar{\mathbf{v}}$ is equivalent to the satisfaction of the differential equations (3.1) and their boundary conditions (3.2).

In the above discussion it was implicitly assumed that integrals such as those in Eq. (3.13) are capable of being evaluated. This places certain restrictions on the possible families to

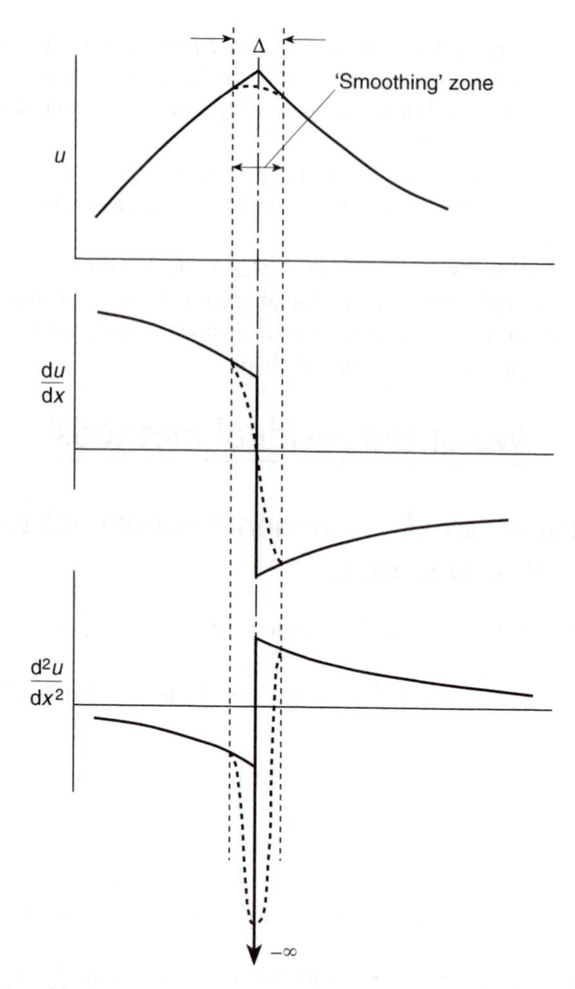

Fig. 3.2 Differentiation of function with slope discontinuity (C_0 continuity).

which the functions **v** or **u** must belong. *In general we shall seek to avoid functions which result in any term in the integrals becoming infinite.* Thus, in Eq. (3.13) we generally limit the choice of **v** and $\bar{\mathbf{v}}$ to bounded functions without restricting the validity of previous statements.

What restrictions need to be placed on the functions? The answer obviously depends on the order of differentiation implied in the equations $\mathcal{A}(\mathbf{u})$ [or $\mathcal{B}(\mathbf{u})$]. Consider, for instance, a function **u** which is continuous but has a discontinuous slope in the x direction, as shown in Fig. 3.2 which is identical to Fig. 2.7 but is reproduced here for clarity. We imagine this discontinuity to be replaced by a continuous variation in a very small distance Δ (a process known as 'molification') and study the behaviour of the derivatives. It is easy to see that although the first derivative is not defined here, it has finite value and can be integrated easily but the second derivative tends to infinity. This therefore presents difficulties if integrals are to be evaluated numerically by simple means, even though the integral is finite. If such derivatives are multiplied by each other the integral does not exist and the function is known as *non-square integrable. Such a function is said to be* C_0 *continuous.*

In a similar way it is easy to see that if nth-order derivatives occur in any term of $\mathcal{A}$ or $\mathcal{B}$ then the function has to be such that its $n-1$ derivatives are continuous (C_{n-1} continuity).

On many occasions it is possible to perform an integration by parts on Eq. (3.13) and replace it by an alternative statement of the form

$$\int_\Omega \mathcal{C}(\mathbf{v})^{\mathrm{T}}\mathcal{D}(\mathbf{u})\mathrm{d}\Omega + \int_\Gamma \mathcal{E}(\bar{\mathbf{v}})^{\mathrm{T}}\mathcal{F}(\mathbf{u})\mathrm{d}\Gamma = 0 \tag{3.14}$$

In this the operators $\mathcal{C}$ to $\mathcal{F}$ usually contain lower order derivatives than those occurring in operators $\mathcal{A}$ or $\mathcal{B}$. Now a lower order of continuity is required in the choice of the $\mathbf{u}$ function at a price of higher continuity for $\mathbf{v}$ and $\bar{\mathbf{v}}$.

The statement (3.14) is now more 'permissive' than the original problem posed by Eq. (3.1), (3.2), or (3.13) and is called a *weak form* of these equations. It is a somewhat surprising fact that often this weak form is more realistic physically than the original differential equation which implied an excessive 'smoothness' of the true solution.

Integral statements of the form of (3.13) and (3.14) will form the basis of finite element approximations, and we shall discuss them later in fuller detail. Before doing so we shall apply the new formulation to an example.

Example 3.4: Weak form of the heat conduction equation – forced and natural boundary conditions. Consider now the integral form of Eq. (3.8). We can write the statement (3.13) as

$$\int_\Omega v\left[-\frac{\partial}{\partial x}\left(k\frac{\partial\phi}{\partial x}\right) - \frac{\partial}{\partial y}\left(k\frac{\partial\phi}{\partial y}\right) + Q\right]\mathrm{d}x\,\mathrm{d}y + \int_{\Gamma_q}\bar{v}\left[k\frac{\partial\phi}{\partial n} + \bar{q}\right]\mathrm{d}\Gamma = 0 \tag{3.15}$$

noting that v and $\bar{v}$ are scalar functions† and presuming that one of the boundary conditions, i.e.,

$$\phi - \bar{\phi} = 0$$

is automatically satisfied by the choice of the functions ϕ on Γ_ϕ. This type of boundary condition is often called a 'forced' or an 'essential' one.

Equation (3.15) can now be integrated by parts to obtain a weak form similar to Eq. (3.14). We shall make use here of general formulae for such integration (Green's formulae) which we derive in Appendix G and which on many occasions will be useful, i.e.,

$$\begin{aligned}\int_\Omega v\frac{\partial}{\partial x}\left(k\frac{\partial\phi}{\partial x}\right)\mathrm{d}x\,\mathrm{d}y &\equiv -\int_\Omega \frac{\partial v}{\partial x}\left(k\frac{\partial\phi}{\partial x}\right)\mathrm{d}x\,\mathrm{d}y + \oint_\Gamma v\left(k\frac{\partial\phi}{\partial x}\right)n_x\mathrm{d}\Gamma\\ \int_\Omega v\frac{\partial}{\partial y}\left(k\frac{\partial\phi}{\partial y}\right)\mathrm{d}x\,\mathrm{d}y &\equiv -\int_\Omega \frac{\partial v}{\partial y}\left(k\frac{\partial\phi}{\partial y}\right)\mathrm{d}x\,\mathrm{d}y + \oint_\Gamma v\left(k\frac{\partial\phi}{\partial y}\right)n_y\mathrm{d}\Gamma\end{aligned} \tag{3.16}$$

We have thus in place of Eq. (3.15)

$$\begin{aligned}\int_\Omega \left(\frac{\partial v}{\partial x}k\frac{\partial\phi}{\partial x} + \frac{\partial v}{\partial y}k\frac{\partial\phi}{\partial y} + vQ\right)\mathrm{d}x\,\mathrm{d}y - \oint_\Gamma vk\left(\frac{\partial\phi}{\partial x}n_x + \frac{\partial\phi}{\partial y}n_y\right)\mathrm{d}\Gamma&\\ + \int_{\Gamma_q}\bar{v}\left[k\frac{\partial\phi}{\partial n} + \bar{q}\right]\mathrm{d}\Gamma = 0&\end{aligned} \tag{3.17}$$

† Two functions are introduced such that simplifications are possible at later stages of the development.

Noting that the derivative along the normal is given as

$$\frac{\partial \phi}{\partial n} \equiv \frac{\partial \phi}{\partial x} n_x + \frac{\partial \phi}{\partial y} n_y \tag{3.18}$$

and, further, making

$$\bar{v} = v \qquad \text{on } \Gamma \tag{3.19}$$

without loss of generality (as both functions are arbitrary), we can write Eq. (3.17) as

$$\int_{\Omega} (\nabla v)^{\mathrm{T}} (k \nabla \phi) \mathrm{d}\Omega + \int_{\Omega} v\, Q \, \mathrm{d}\Omega + \int_{\Gamma_q} v\, \bar{q} \, \mathrm{d}\Gamma - \int_{\Gamma_\phi} v\, k \frac{\partial \phi}{\partial n} \mathrm{d}\Gamma = 0 \tag{3.20}$$

where the operator ∇ is simply

$$\nabla = \begin{Bmatrix} \dfrac{\partial}{\partial x} \\ \dfrac{\partial}{\partial y} \end{Bmatrix}$$

We note that

(a) the variable ϕ has disappeared from the integrals taken along the boundary Γ_q and that the boundary condition

$$B(\phi) = k \frac{\partial \phi}{\partial n} + \bar{q} = 0$$

on that boundary is automatically satisfied – such a condition is known as a *natural boundary condition* – and

(b) if the choice of ϕ is restricted so as to satisfy the *forced boundary conditions* $\phi - \bar{\phi} = 0$, we can omit the last term of Eq. (3.20) by restricting the choice of v to functions which give $v = 0$ on Γ_ϕ.

The form of Eq. (3.20) is the *weak form* of the heat conduction statement equivalent to Eq. (3.8). It admits discontinuous conductivity coefficients k and temperature ϕ which show discontinuous first derivatives, a real possibility not easily admitted in the differential form.

3.3 Approximation to integral formulations: the weighted residual-Galerkin method

If the unknown function $\mathbf{u}$ is approximated by the expansion (3.3), i.e.,

$$\mathbf{u} \approx \hat{\mathbf{u}} = \sum_{a=1}^{n} \mathbf{N}_a \tilde{\mathbf{u}}_a = \mathbf{N}\tilde{\mathbf{u}}$$

then it is clearly impossible to satisfy both the differential equation and the boundary conditions in the general case. The integral statements (3.13) or (3.14) allow an approximation to be made if, in place of *any function* $\mathbf{v}$, we put a finite set of approximate functions

$$\mathbf{v} \approx \sum_{b=1}^{n} \mathbf{w}_b \, \delta\tilde{\mathbf{u}}_b \quad \text{and} \quad \hat{\mathbf{v}} = \sum_{b=1}^{n} \bar{\mathbf{w}}_b \, \delta\tilde{\mathbf{u}}_b \tag{3.21}$$

in which $\delta\tilde{\mathbf{u}}_b$ are arbitrary parameters. Inserting the approximations into Eq. (3.13) we have the relation

$$\delta\tilde{\mathbf{u}}_b^{\mathrm{T}}\left[\int_\Omega \mathbf{w}_b^{\mathrm{T}}\mathcal{A}(\mathbf{N}\,\tilde{\mathbf{u}})\,\mathrm{d}\Omega + \int_\Gamma \bar{\mathbf{w}}_b^{\mathrm{T}}\mathcal{B}(\mathbf{N}\,\tilde{\mathbf{u}})\,\mathrm{d}\Gamma\right] = 0$$

and since $\delta\tilde{\mathbf{u}}_b$ is arbitrary we have a set of equations which is sufficient to determine the parameters $\tilde{\mathbf{u}}$ as

$$\int_\Omega \mathbf{w}_b^{\mathrm{T}}\mathcal{A}(\mathbf{N}\,\tilde{\mathbf{u}})\,\mathrm{d}\Omega + \int_\Gamma \bar{\mathbf{w}}_b^{\mathrm{T}}\mathcal{B}(\mathbf{N}\,\tilde{\mathbf{u}})\,\mathrm{d}\Gamma = \mathbf{0}; \quad b = 1, 2, \ldots, n \tag{3.22}$$

Performing similar steps using Eq. (3.14) gives the set

$$\int_\Omega \mathcal{C}^{\mathrm{T}}(\mathbf{w}_b)\mathcal{D}(\mathbf{N}\,\tilde{\mathbf{u}})\,\mathrm{d}\Omega + \int_\Gamma \mathcal{E}^{\mathrm{T}}(\bar{\mathbf{w}}_b)\mathcal{F}(\mathbf{N}\,\tilde{\mathbf{u}})\,\mathrm{d}\Gamma = \mathbf{0}; \quad b = 1, 2, \ldots, n \tag{3.23}$$

If we note that $\mathcal{A}(\mathbf{N}\tilde{\mathbf{u}})$ represents the *residual or error* obtained by substitution of the approximation into the differential equation [and $\mathcal{B}(\mathbf{N}\tilde{\mathbf{u}})$, the residual of the boundary conditions], then Eq. (3.22) is a *weighted integral of such residuals*. The approximation may thus be called the *method of weighted residuals*.

In its classical sense it was first described by Crandall,[1] who points out the various forms used since the end of the nineteenth century. Later a very full exposé of the method was given by Finlayson.[2] Clearly, almost any set of independent functions $\mathbf{w}_b$ could be used for the purpose of weighting and, according to the choice of function, a different name can be attached to each process. Thus the various common choices are:

1. *Point collocation.*[3] $\mathbf{w}_b = \boldsymbol{\delta}_b$ in (3.22), where $\boldsymbol{\delta}_b$ is such that for $x \neq x_b$; $y \neq y_b$, $\mathbf{w}_b = 0$ but $\int_\Omega \mathbf{w}_b \mathrm{d}\Omega = \mathbf{I}$ (unit matrix). This procedure is equivalent to simply making the residual zero at n points within the domain and integration is 'nominal' (incidentally although $\mathbf{w}_b$ defined here does not satisfy all the criteria of Sec. 3.2, it is nevertheless admissible in view of its properties). Finite difference methods are particular cases of this weighting.
2. *Subdomain collocation.*[4] $\mathbf{w}_b = \mathbf{I}$ in subdomain Ω_b and zero elsewhere. This essentially makes the integral of the error zero over the specified subdomains. When used with (3.23) this is one of the many finite volume methods.[5]
3. *The Galerkin method* (Bubnov–Galerkin).[4, 6] $\mathbf{w}_b = \mathbf{N}_b$. Here simply the original shape (or basis) functions are used as weighting. This method, as we shall see, frequently (but by no means always) leads to symmetric matrices and for this and other reasons will be adopted in this book almost exclusively.

The name of 'weighted residuals' is clearly much older than that of the 'finite element method'. The latter uses mainly locally based (element) functions in the expansion of Eq. (3.3) but the general procedures are identical. As the process always leads to equations which, being of integral form, can be obtained by summation of contributions from various subdomains, we choose to embrace all weighted residual approximations under the name of *generalized finite element method*. On occasion, simultaneous use of both local and 'global' trial functions will be found to be useful.

In the literature the names of Petrov and Galerkin[6] are often associated with the use of weighting functions such that $\mathbf{w}_b \neq \mathbf{N}_b$. It is important to remark that the well-known *finite difference method* of approximation is a particular case of collocation with locally defined

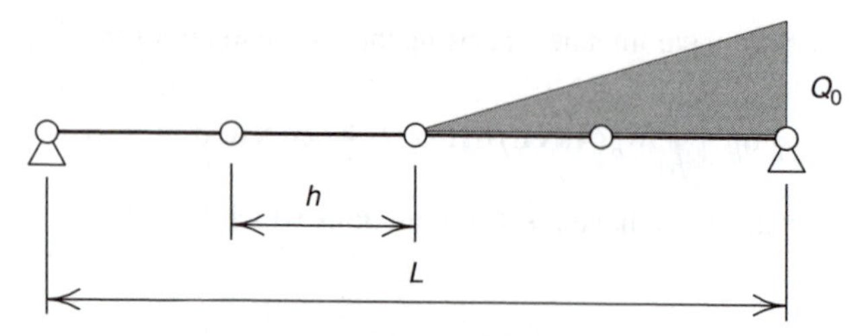

Fig. 3.3 Problem description and loading for 1-d heat conduction example.

basis functions and is thus a case of a Petrov–Galerkin scheme. We shall return to such unorthodox definitions in more detail in Chapter 15.

To illustrate the procedure of weighted residual approximation and its relation to the finite element process let us consider some specific examples.

Example 3.5: One-dimensional equation of heat conduction. The problem here will be a one-dimensional representation of the heat conduction equation [Eq. (3.8)] with unit conductivity. (This problem could equally well represent many other physical situations, e.g., deflection of a loaded string with unit tension.) Here we have (see Fig. 3.3)

$$A(\phi) = -\frac{\mathrm{d}^2\phi}{\mathrm{d}x^2} + Q(x) = 0 \quad (0 < x < L) \tag{3.24a}$$

with $Q(x)$ given by

$$Q(x) = \begin{cases} 0 & 0 < x \leq L/2 \\ -2\,Q_0\,(x/L - 1/2) & L/2 < x < L \end{cases} \tag{3.24b}$$

The boundary conditions assumed will be simply $\phi = 0$ at $x = 0$ and $x = L$.

The problem is solved by a Galerkin-weighted residual method in which the field $\phi(x)$ is approximated by piecewise defined (locally based) functions. Here we use the equivalent of Eq. (3.14) which results from an integration by parts of

$$\int_0^L w_b \left[-\frac{\mathrm{d}^2}{\mathrm{d}x^2} \left(\sum_a N_a\, \tilde{\phi}_a \right) + Q \right] \mathrm{d}x = 0$$

to obtain

$$\int_0^L \left[\frac{\mathrm{d}w_b}{\mathrm{d}x} \left(\sum_a \frac{\mathrm{d}N_a}{\mathrm{d}x}\, \tilde{\phi}_a \right) + w_b Q \right] \mathrm{d}x = 0 \tag{3.25}$$

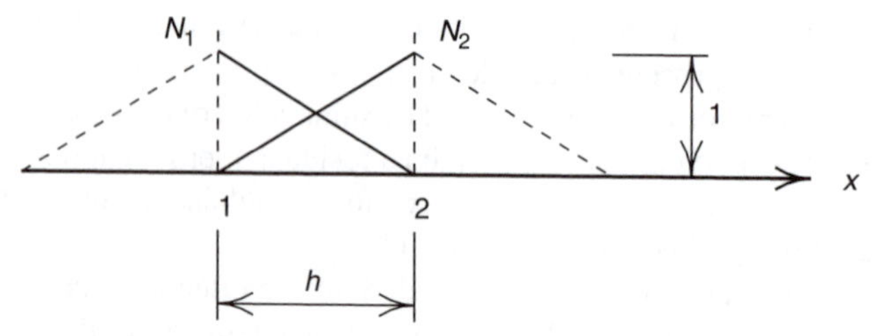

Fig. 3.4 Linear locally based one-dimensional shape functions.

in which the boundary terms disappear if we require $w_b = 0$ at the two ends.

For the Galerkin solution we use $w_b = N_b$; hence, the above equation can be written as

$$\mathbf{K}\,\tilde{\phi} + \mathbf{f} = \mathbf{0} \tag{3.26}$$

where for each element of length h,

$$\begin{aligned} K^e_{ba} &= \int_0^h \frac{\mathrm{d}N_b}{\mathrm{d}x}\frac{\mathrm{d}N_a}{\mathrm{d}x}\,\mathrm{d}x \\ f^e_b &= \int_0^h N_b\, Q(x)\,\mathrm{d}x \end{aligned} \tag{3.27}$$

with the usual rules of adding pertaining, i.e.,

$$K_{ba} = \sum_e K^e_{ba} \quad \text{and} \quad f_b = \sum_e f^e_b$$

The reader will observe that the matrix $\mathbf{K}$ is symmetric, i.e., $K_{ba} = K_{ab}$.

As the shape functions need only be of C_0 continuity, a piecewise linear approximation is conveniently used, as shown in Fig. 3.4. Considering a typical element 1–2 shown, we can write (translating the cartesian origin of x to point 1)

$$N_1 = 1 - x/h \quad \text{and} \quad N_2 = x/h \tag{3.28}$$

giving for a typical element

$$\begin{aligned} \mathbf{K}^e &= \frac{1}{h}\begin{bmatrix} 1 & -1 \\ -1 & 1 \end{bmatrix} \\ \mathbf{f}^e &= -\frac{h}{6}\begin{Bmatrix} 2\,Q_1 + Q_2 \\ Q_1 + 2\,Q_2 \end{Bmatrix} \end{aligned} \tag{3.29}$$

where Q_1 and Q_2 are load intensities at the x_1 and x_2 coordinates, respectively.

Assembly of four equal size elements results, after inserting the boundary conditions $\tilde{\phi}_1 = \tilde{\phi}_5 = 0$, in the equation set

$$\frac{4}{L}\begin{bmatrix} 2 & -1 & 0 \\ -1 & 2 & -1 \\ 0 & -1 & 2 \end{bmatrix}\begin{Bmatrix} \tilde{\phi}_2 \\ \tilde{\phi}_3 \\ \tilde{\phi}_4 \end{Bmatrix} = \frac{Q_0 L}{48}\begin{Bmatrix} 0 \\ 1 \\ 6 \end{Bmatrix} \tag{3.30}$$

The solution is shown in Fig. 3.5 along with the exact solution to the problem. For comparison purposes we also show a finite difference solution in which simple collocation is used in a weighted residual equation together with the approximation for the second derivative given by a Taylor expansion

$$\left.\frac{\mathrm{d}^2\phi}{\mathrm{d}x^2}\right|_{x_a} \approx \frac{1}{h^2}\left(\tilde{\phi}_{a-1} - 2\tilde{\phi}_a + \tilde{\phi}_{a+1}\right) \tag{3.31}$$

which yields the approximation for each node point

$$\frac{1}{h^2}\left(-\tilde{\phi}_{a-1} + 2\tilde{\phi}_a - \tilde{\phi}_{a+1}\right) + Q_a = 0 \tag{3.32}$$

Again after including the boundary conditions a set of three equations for the points 2, 3 and 4 is expressed as

$$\frac{16}{L^2}\begin{bmatrix} 2 & -1 & 0 \\ -1 & 2 & -1 \\ 0 & -1 & 2 \end{bmatrix}\begin{Bmatrix} \tilde{\phi}_2 \\ \tilde{\phi}_3 \\ \tilde{\phi}_4 \end{Bmatrix} = Q_0 \begin{Bmatrix} 0 \\ 0 \\ 1/2 \end{Bmatrix} \tag{3.33}$$

The reader will note that the coefficient matrix for the finite element and finite difference methods differ by only a constant multiplier (for the boundary conditions assumed in this one-dimensional problem); however, the right sides differ significantly. We also plot the solution to (3.33) (and one for half the mesh spacing) in Fig. 3.5. Here we note that the nodal results for the finite element method are *exact* whereas those for the finite difference solution are all in error (although convergence can be observed for the finer subdivision). The nodal exactness is a property of the particular equation being solved and unfortunately does not carry over to general problems.[7] (See also Appendix H.) However, based on the above result and other experiences we can say that the finite element method always achieves (the same or) better results than classical finite difference methods. In addition, the finite element method permits an approximation of the solution *at all points in the domain* as indicated by the dashed lines in Fig. 3.5 for the one-dimensional problem.

The problem is repeated using 4-quadratic order finite elements and results are shown in Fig. 3.6. It is evident that the use of quadratic order greatly increases the accuracy of the results obtained. Indeed, if cubic order elements were used results would be exact, since for linear varying Q the solution over the loaded portion will only contain polynomials up to cubic order.

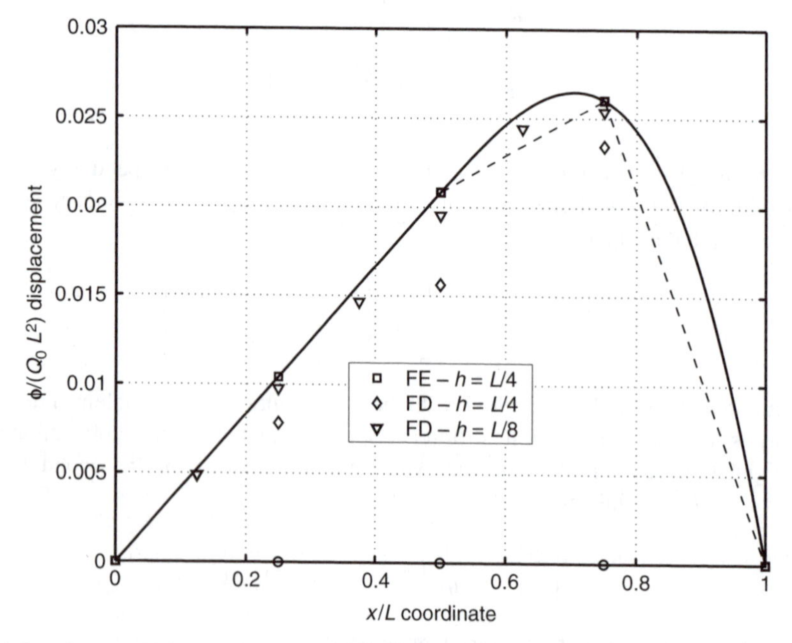

Fig. 3.5 One-dimensional heat conduction. Solution by finite element method with linear elements and $h = L/4$; finite difference method with $h = L/4$ and $h = L/8$.

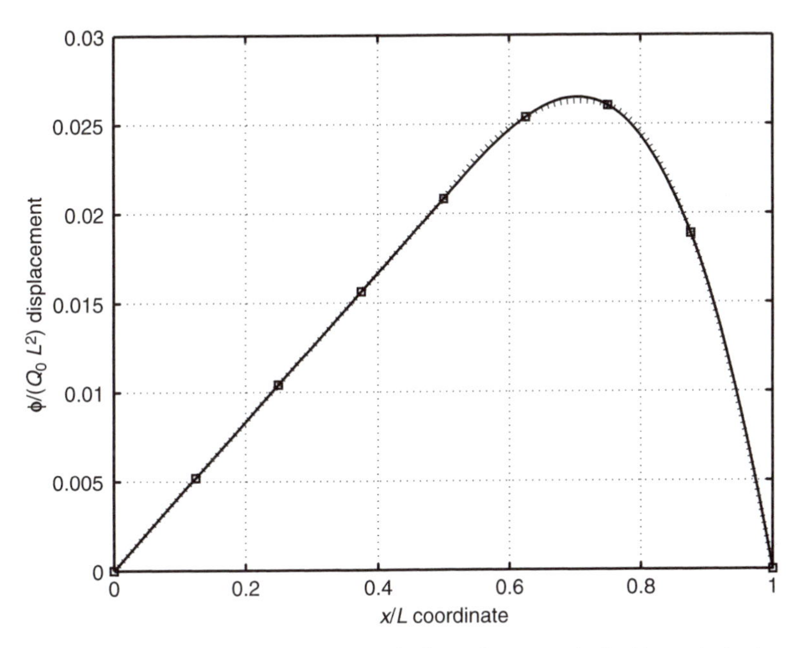

Fig. 3.6 One-dimensional heat conduction. Solution by finite element method with quadratic elements and $h = L/4$.

Example 3.6: Steady-state heat conduction in two dimensions: Galerkin formulation with triangular elements. We have already introduced the problem in Sec. 3.1 and defined it by Eq. (3.8) with appropriate boundary conditions. The weak form has been obtained in Eq. (3.20). Approximating the weight by $v = \sum N_b \delta\tilde{\phi}_b$ and solution by $\phi = \sum N_a \tilde{\phi}_a$ we have immediately that

$$\mathbf{K}\,\tilde{\boldsymbol{\phi}} + \mathbf{f} = \mathbf{0}$$

where

$$\begin{aligned} K^e_{ba} &= \int_\Omega \left(\frac{\partial N_b}{\partial x}\, k\, \frac{\partial N_a}{\partial x} + \frac{\partial N_b}{\partial y}\, k\, \frac{\partial N_a}{\partial y} \right) \mathrm{d}\Omega \\ f^e_b &= \int_\Omega N_b\, Q\, \mathrm{d}\Omega + \int_\Gamma N_b\, \bar{q}\, \mathrm{d}\Gamma \end{aligned} \tag{3.34}$$

Once again the components of K_{ba} and f_b can be evaluated for a typical element or subdomain and the system of equations built by standard methods.

For instance, considering the set of nodes and elements shown shaded in Fig. 3.7(b), to compute the equation for node 1 it is only necessary to compute the K^e_{ba} for two element shapes as indicated in Fig. 3.7. For the Type 1 element (left element in Fig. 3.7(c)) the shape functions evaluated from (2.8) using (2.6) and (2.7) gives

$$N_1 = 1 - \frac{y}{h}; \quad N_2 = \frac{x}{h}; \quad N_3 = \frac{y - x}{h}$$

thus, the derivatives are given by:

$$\frac{\partial \mathbf{N}}{\partial x} = \begin{Bmatrix} \dfrac{\partial N_1}{\partial x} \\ \dfrac{\partial N_2}{\partial x} \\ \dfrac{\partial N_3}{\partial x} \end{Bmatrix} = \begin{Bmatrix} 0 \\ \dfrac{1}{h} \\ -\dfrac{1}{h} \end{Bmatrix} \quad \text{and} \quad \frac{\partial \mathbf{N}}{\partial y} = \begin{Bmatrix} \dfrac{\partial N_1}{\partial y} \\ \dfrac{\partial N_2}{\partial y} \\ \dfrac{\partial N_3}{\partial y} \end{Bmatrix} = \begin{Bmatrix} -\dfrac{1}{h} \\ 0 \\ \dfrac{1}{h} \end{Bmatrix}$$

Similarly, for the Type 2 element the shape functions are expressed by

$$N_1 = 1 - \frac{x}{h}; \quad N_2 = \frac{x-y}{h}; \quad N_3 = \frac{y}{h}$$

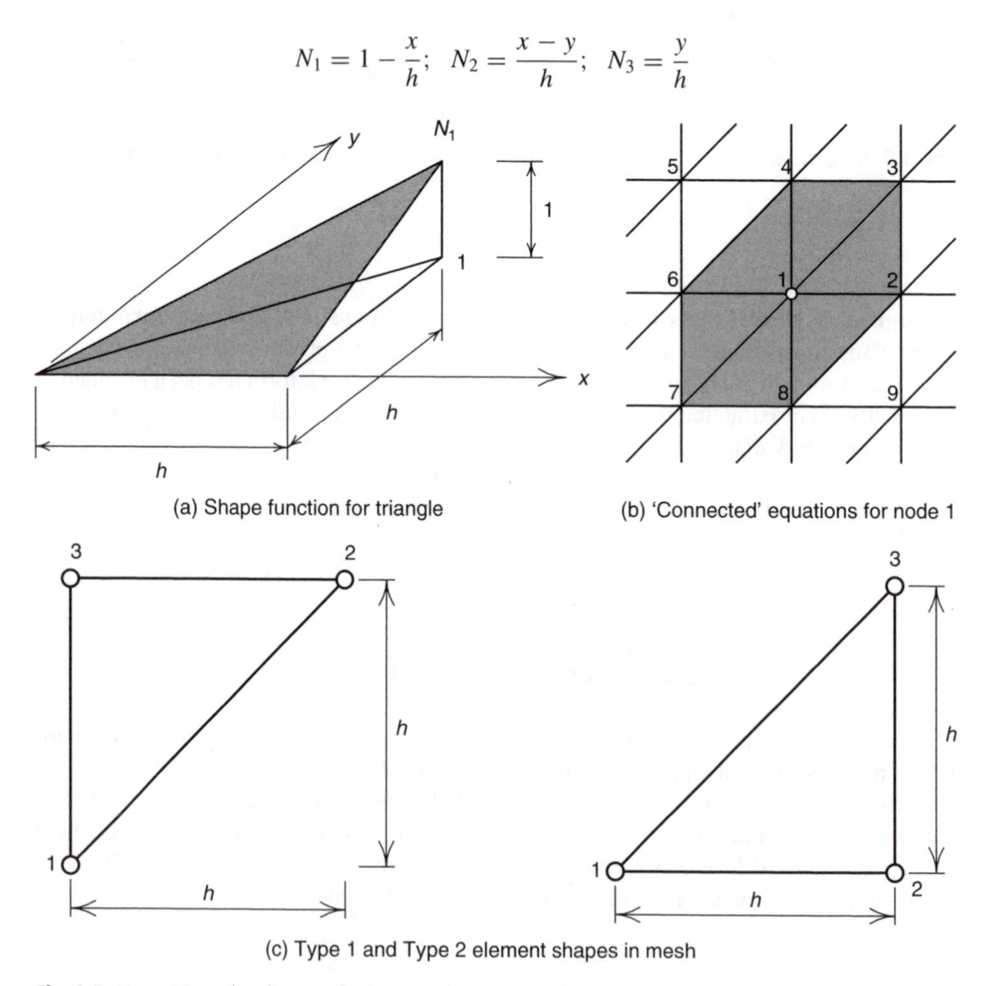

Fig. 3.7 Linear triangular elements for heat conduction example.

and their derivatives by

$$\frac{\partial \mathbf{N}}{\partial x} = \begin{Bmatrix} \dfrac{\partial N_1}{\partial x} \\ \dfrac{\partial N_2}{\partial x} \\ \dfrac{\partial N_3}{\partial x} \end{Bmatrix} = \begin{Bmatrix} -\dfrac{1}{h} \\ \dfrac{1}{h} \\ 0 \end{Bmatrix} \quad \text{and} \quad \frac{\partial \mathbf{N}}{\partial y} = \begin{Bmatrix} \dfrac{\partial N_1}{\partial y} \\ \dfrac{\partial N_2}{\partial y} \\ \dfrac{\partial N_3}{\partial y} \end{Bmatrix} = \begin{Bmatrix} 0 \\ -\dfrac{1}{h} \\ \dfrac{1}{h} \end{Bmatrix}$$

Evaluation of the matrix K_{ba}^e and f_b^e for Type 1 and Type 2 elements gives

$$\mathbf{K}^e \tilde{\phi}^e = \frac{1}{2} k \begin{bmatrix} 1 & 0 & -1 \\ 0 & 1 & -1 \\ -1 & -1 & 2 \end{bmatrix} \begin{Bmatrix} \tilde{\phi}_1^e \\ \tilde{\phi}_2^e \\ \tilde{\phi}_3^e \end{Bmatrix} \quad \text{and} \quad \mathbf{K}^e \tilde{\phi}^e = \frac{1}{2} k \begin{bmatrix} 1 & -1 & 0 \\ -1 & 2 & -1 \\ 0 & -1 & 1 \end{bmatrix} \begin{Bmatrix} \tilde{\phi}_1^e \\ \tilde{\phi}_2^e \\ \tilde{\phi}_3^e \end{Bmatrix},$$

respectively. Note that the stiffness matrix does not depend on the size h of the element. This is a property of all two-dimensional elements in which the **B** matrix depends *only* on first derivatives of C_0 shape functions. The force vector for a constant Q over each element is given by

$$\mathbf{f}^e = \frac{1}{6} Q h^2 \begin{Bmatrix} 1 \\ 1 \\ 1 \end{Bmatrix}$$

for both types of elements. Assembling the patch of elements shown in Fig. 3.7(b) gives the equation with non-zero coefficients for node 1 as

$$k[4 \quad -1 \quad -1 \quad -1 \quad -1] \begin{Bmatrix} \tilde{\phi}_1 \\ \tilde{\phi}_2 \\ \tilde{\phi}_4 \\ \tilde{\phi}_6 \\ \tilde{\phi}_8 \end{Bmatrix} = Q h^2$$

The reader should note that the final stiffness for node 1 does not depend on nodes 3 and 7 of the patch, whereas there are non-zero stiffness coefficients in the individual elements. Thus, the final result is only true when the arrangement of the nodes is regular. If the location of any of the nodes lies on an irregular pattern then final stiffness coefficients will remain for these nodes also.

Repeating the construction of the stiffness terms using a finite difference approximation [as given in Eq. (3.31)] directly in the differential equation (3.8) gives the approximation

$$\frac{k}{h^2}[4 \quad -1 \quad -1 \quad -1 \quad -1] \begin{Bmatrix} \tilde{\phi}_1 \\ \tilde{\phi}_2 \\ \tilde{\phi}_4 \\ \tilde{\phi}_6 \\ \tilde{\phi}_8 \end{Bmatrix} = Q$$

and once again the assembled node is identical to the finite difference approximation to within a constant multiplier. If all the boundary conditions are forced (i.e., $\phi = \bar{\phi}$) no differences arise between a finite element and a finite difference solution for the regular mesh assumed. However, if any boundary conditions are of natural type or the mesh is irregular differences will arise. Indeed, no restrictions on shape of elements or assembly type are imposed by the finite element approach.

Example 3.7: Steady-state heat conduction–convection in two dimensions: Galerkin formulation. We have already introduced the problem in Sec. 3.1 and defined it by Eq. (3.9) with appropriate boundary conditions. The equation differs only in the convective terms from that of simple heat conduction for which the weak form has already been obtained in Eq. (3.20). We can write the weighted residual equation immediately from this, substituting $v = w_b \delta\tilde{\phi}_b$ and adding the convective terms. Thus we have

$$\int_\Omega (\nabla w_b)^{\mathrm{T}} k \, \nabla\hat{\phi} \mathrm{d}\Omega + \int_\Omega w_b \left(u_x \frac{\partial \hat{\phi}}{\partial x} + u_y \frac{\partial \hat{\phi}}{\partial y} \right) \mathrm{d}\Omega + \int_\Omega w_b Q \, \mathrm{d}\Omega + \int_{\Gamma_q} w_b \bar{q} \, d\Gamma = 0 \tag{3.35}$$

with $\hat{\phi} = \sum_a N_a \tilde{\phi}_a$ being such that the prescribed values of $\bar{\phi}$ are given on the boundary Γ_ϕ and that $\delta\tilde{\phi}_b = 0$ on that boundary [ignoring that term in (3.35)].

Specializing to the Galerkin approximation, i.e., putting $w_b = N_b$, we have immediately a set of equations of the form

$$\mathbf{K}\tilde{\boldsymbol{\phi}} + \mathbf{f} = \mathbf{0} \tag{3.36}$$

with

$$\begin{aligned} K_{ba} &= \int_\Omega (\nabla N_b)^{\mathrm{T}} k \nabla N_a \mathrm{d}\Omega + \int_\Omega \left(N_b u_x \frac{\partial N_a}{\partial x} + N_b u_y \frac{\partial N_a}{\partial y} \right) \mathrm{d}\Omega \\ &= \int_\Omega \left(\frac{\partial N_b}{\partial x} k \frac{\partial N_a}{\partial x} + \frac{\partial N_b}{\partial y} k \frac{\partial N_a}{\partial y} \right) \mathrm{d}\Omega + \int_\Omega \left(N_b u_x \frac{\partial N_a}{\partial x} + N_b u_y \frac{\partial N_a}{\partial y} \right) \mathrm{d}\Omega \\ f_b &= \int_\Omega N_b Q \mathrm{d}\Omega + \int_{\Gamma_q} N_b \bar{q} \mathrm{d}\Gamma \end{aligned} \tag{3.37}$$

Once again the components K_{ba} and f_b can be evaluated for a typical element or subdomain and systems of equations built up by standard methods.

At this point it is important to mention that to satisfy the boundary conditions some of the parameters $\tilde{\phi}_a$ have to be prescribed and the number of approximation equations must be equal to the number of unknown parameters. It is nevertheless often convenient to form all equations for all parameters and prescribe the fixed values at the end using precisely the same techniques as we have described in Chapter 1 for the insertion of prescribed boundary conditions in standard discrete problems.

A further point concerning the coefficients of the matrix **K** should be noted here. The first part, corresponding to the pure heat conduction equation, is symmetric ($K_{ab} = K_{ba}$) but the second is not and thus a system of non-symmetric equations needs to be solved. There is a basic reason for such non-symmetries which will be discussed in Sec. 3.9.

To make the problem concrete consider the domain Ω to be divided into regular square elements of side h (Fig. 3.8(b)). To preserve C_0 continuity with nodes placed at corners, shape functions given as the product of the linear expansions can be written. For instance, for node 1, as shown in Fig. 3.8(a),

$$N_1 = \frac{x}{h} \frac{y}{h}$$

and for node 2,

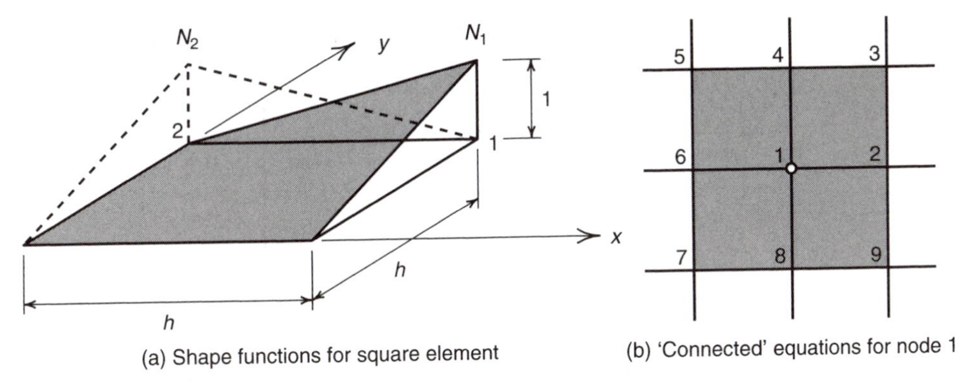

(a) Shape functions for square element (b) 'Connected' equations for node 1

Fig. 3.8 Linear square elements for heat conduction–convection example.

$$N_2 = \frac{(h-x)}{h}\frac{y}{h}, \qquad \text{etc.}$$

With these shape functions the reader is invited to evaluate typical element contributions and to assemble the equations for point 1 of the mesh numbered as shown in Fig. 3.8(b). If Q is assumed to be constant, the result will be

$$\begin{aligned}\frac{8k}{3}\tilde{\phi}_1 &- \left(\frac{k}{3} - \frac{u_x h}{3}\right)\tilde{\phi}_2 - \left(\frac{k}{3} - \frac{u_x h}{12} - \frac{u_y h}{12}\right)\tilde{\phi}_3 - \left(\frac{k}{3} - \frac{u_y h}{3}\right)\tilde{\phi}_4 \\ &- \left(\frac{k}{3} + \frac{u_x h}{12} - \frac{u_y h}{12}\right)\tilde{\phi}_5 - \left(\frac{k}{3} + \frac{u_x h}{3}\right)\tilde{\phi}_6 - \left(\frac{k}{3} + \frac{u_x h}{12} + \frac{u_y h}{12}\right)\tilde{\phi}_7 \\ &- \left(\frac{k}{3} + \frac{u_y h}{3}\right)\tilde{\phi}_8 - \left(\frac{k}{3} - \frac{u_x h}{12} + \frac{u_y h}{12}\right)\tilde{\phi}_9 = -Qh^2\end{aligned} \tag{3.38}$$

This equation is similar to those that would be obtained by using finite difference approximations to the same equations in a fairly standard manner.[8,9] In the example discussed some difficulties arise when the convective terms are large. In such cases the Galerkin weighting is not acceptable and other forms have to be used. For problems dealing with fluid dynamics this is discussed in detail in reference 10.

3.4 Virtual work as the 'weak form' of equilibrium equations for analysis of solids or fluids

In Chapter 2 we introduced a finite element by way of an application to the solid mechanics problem of linear elasticity. The integral statement necessary for formulation in terms of the finite element approximation was supplied via the principle of *virtual work*, which was assumed to be so basic as not to merit proof. Indeed, to many this is so, and the virtual work principle is considered by some as a statement of mechanics more fundamental than the traditional equilibrium conditions of Newton's laws of motion. Others will argue with this view and will point out that all work statements are derived from the classical laws pertaining to the equilibrium of the particle. We shall therefore show in this section that the virtual work statement is simply a 'weak form' of equilibrium equations.

In a general three-dimensional continuum the equilibrium equations of an elementary volume can be written in terms of the components of the symmetric cartesian stress tensor as[11, 12]

$$\left\{\begin{array}{c} A_1 \\ A_2 \\ A_3 \end{array}\right\} = -\left\{\begin{array}{c} \dfrac{\partial \sigma_x}{\partial x} + \dfrac{\partial \tau_{xy}}{\partial y} + \dfrac{\partial \tau_{xz}}{\partial z} + b_x \\ \dfrac{\partial \tau_{xy}}{\partial x} + \dfrac{\partial \sigma_y}{\partial y} + \dfrac{\partial \tau_{yz}}{\partial z} + b_y \\ \dfrac{\partial \tau_{xz}}{\partial x} + \dfrac{\partial \tau_{yz}}{\partial y} + \dfrac{\partial \sigma_z}{\partial z} + b_z \end{array}\right\} = \mathbf{0} \tag{3.39}$$

where $\mathbf{b} = \begin{bmatrix} b_x & b_y & b_z \end{bmatrix}^{\mathrm{T}}$ stands for the body forces acting per unit volume (which may well include acceleration effects by the d'Alembert principle).

In solid mechanics the six stress components will be some general functions of the six components of strain (ε) which are computed from the displacement

$$\mathbf{u} = \begin{bmatrix} u & v & w \end{bmatrix}^{\mathrm{T}} \tag{3.40}$$

and in fluid mechanics of the velocity vector $\mathbf{u}$, which has identically named components. Thus Eq. (3.39) can be considered as a general equation of the form Eq. (3.1), i.e., $\mathcal{A}(\mathbf{u}) = \mathbf{0}$. To obtain a weak form we shall proceed as before, introducing an arbitrary weighting function vector defined as

$$\mathbf{v} \equiv \delta\,\mathbf{u} = \begin{bmatrix} \delta\,u, & \delta\,v, & \delta\,w \end{bmatrix}^{\mathrm{T}} \tag{3.41}$$

We can now write the integral statement of Eq. (3.11) as

$$\begin{aligned} \int_\Omega \delta\mathbf{u}^{\mathrm{T}} \mathcal{A}(\mathbf{u})\,\mathrm{d}\Omega &= -\int_\Omega \left[\delta u \left(\frac{\partial \sigma_x}{\partial x} + \frac{\partial \tau_{xy}}{\partial y} + \frac{\partial \tau_{xz}}{\partial y} + b_x \right) + \delta v(A_2) + \delta w(A_3) \right] \mathrm{d}\Omega \\ &= 0 \end{aligned} \tag{3.42}$$

where the volume, Ω, is the problem domain.

Integrating each term by parts and rearranging we can write this as

$$\begin{aligned} &\int_\Omega \left[\frac{\partial \delta u}{\partial x} \sigma_x + \left(\frac{\partial \delta u}{\partial y} + \frac{\partial \delta v}{\partial x} \right) \tau_{xy} + \cdots - \delta u\, b_x - \delta v\, b_y - \delta w\, b_z \right] \mathrm{d}\Omega \\ &- \int_\Gamma \left[\delta u\, t_x + \delta v\, t_y + \delta w\, t_z \right] \mathrm{d}\Gamma = 0 \end{aligned} \tag{3.43}$$

where

$$\mathbf{t} = \left\{\begin{array}{c} t_x \\ t_y \\ t_z \end{array}\right\} = \left\{\begin{array}{c} n_x \sigma_x + n_y \tau_{xy} + n_z \tau_{xz} \\ n_x \tau_{xy} + n_y \sigma_y + n_z \tau_{yz} \\ n_x \tau_{xz} + n_y \tau_{yz} + n_z \sigma_z \end{array}\right\} \tag{3.44}$$

are *tractions* acting per unit area of external boundary surface Γ of the solid [in (3.43) Green's formulae of Appendix G are again used].

In the first set of bracketed terms in (3.43) we can recognize immediately the small strain operators acting on $\delta\mathbf{u}$, which can be termed a virtual displacement (or virtual velocity).

We can therefore introduce a virtual strain (or strain rate) defined as

$$\delta\boldsymbol{\varepsilon} = \left\{ \begin{array}{c} \dfrac{\partial \delta u}{\partial x} \\ \dfrac{\partial \delta v}{\partial y} \\ \dfrac{\partial \delta w}{\partial z} \\ \dfrac{\partial \delta u}{\partial y} + \dfrac{\partial \delta v}{\partial x} \\ \vdots \end{array} \right\} = \boldsymbol{\mathcal{S}}\delta\mathbf{u} \tag{3.45}$$

where the strain operator is defined as in Chapter 2 [Eqs (2.13)–(2.15)].

Arranging the six stress components in a vector $\boldsymbol{\sigma}$ in an order corresponding to that used for $\delta\boldsymbol{\varepsilon}$, we can write Eq. (3.43) simply as

$$\int_\Omega \delta\boldsymbol{\varepsilon}^{\mathrm{T}}\boldsymbol{\sigma}\,\mathrm{d}\Omega - \int_\Omega \delta\mathbf{u}^{\mathrm{T}}\mathbf{b}\,\mathrm{d}\Omega - \int_\Gamma \delta\mathbf{u}^{\mathrm{T}}\mathbf{t}\,\mathrm{d}\Gamma = 0 \tag{3.46}$$

which is the three-dimensional equivalent virtual work statement used in Eqs (2.25) and (2.38) of Chapter 2.

We see from the above that the virtual work statement is precisely the weak form of equilibrium equations and is valid for non-linear as well as linear stress–strain (or stress–strain rate) relations.

The finite element approximation which we have derived in Chapter 2 *is in fact a Galerkin formulation of the weighted residual process applied to the equilibrium equation.* Thus if we take $\delta\mathbf{u}$ as the shape function times arbitrary parameters

$$\delta\mathbf{u} = \sum_b \mathbf{N}_b\delta\tilde{\mathbf{u}}_b \tag{3.47}$$

where the displacement field is discretized, i.e.,

$$\mathbf{u} = \sum_a \mathbf{N}_a\tilde{\mathbf{u}}_a \tag{3.48}$$

together with the strain-displacement relations

$$\boldsymbol{\varepsilon} = \sum_a \boldsymbol{\mathcal{S}}\mathbf{N}_a\tilde{\mathbf{u}}_a = \sum_a \mathbf{B}_a\tilde{\mathbf{u}}_a \tag{3.49}$$

and constitutive relation of Eq. (2.16), we shall determine once again all the basic expressions of Chapter 2 which are so essential to the solution of elasticity problems. We shall consider this class of problems further in Chapter 6.

Similar expressions are vital to the formulation of equivalent fluid mechanics problems as discussed in reference 10.

3.5 Partial discretization

In the approximation to the problem of solving the differential equation (3.1) by an expression of the standard form of Eq. (3.3), we have assumed that the shape functions $\mathbf{N}$ include *all* independent coordinates of the problem and that $\tilde{\mathbf{u}}$ was simply a set of constants. The final approximation equations were thus always of an algebraic form, from which a unique set of parameters could be determined.

In some problems it is convenient to proceed differently. Thus, for instance, if the independent variables are x, y and z we could allow the parameters $\tilde{\mathbf{u}}$ to be functions of z and do the approximate expansion only in the domain of x, y, say $\bar{\Omega}$. Thus, in place of Eq. (3.3) we would have

$$\mathbf{u} = \mathbf{N}(x, y)\,\tilde{\mathbf{u}}(z) \tag{3.50}$$

Clearly the derivatives of $\tilde{\mathbf{u}}$ with respect to z will remain in the final discretization and the result will be a set of *ordinary differential equations* with z as the independent variable. In linear problems such a set will have the appearance

$$\mathbf{K}\tilde{\mathbf{u}} + \mathbf{C}\dot{\tilde{\mathbf{u}}} + \cdots + \mathbf{f} = \mathbf{0} \tag{3.51}$$

where $\dot{\tilde{\mathbf{u}}} \equiv \mathrm{d}\tilde{\mathbf{u}}/\mathrm{d}z$, etc.

Such a partial discretization can obviously be used in different ways, but is particularly useful when the domain $\bar{\Omega}$ is not dependent on z, i.e., when the *problem is prismatic*. In such a case the coefficient matrices of the ordinary differential equations, (3.51), are independent of z and the solution of the system can frequently be carried out efficiently by standard analytical methods.

This type of partial discretization has been applied extensively by Kantorovich[13] and is frequently known by his name. Semi-analytical treatments are presented in reference 14 for prismatic solids where the final solution is obtained in terms of Fourier (or other) series. However, the most frequently encountered 'prismatic' problem is one involving the time variable, where the space domain $\bar{\Omega}$ is not subject to change. We shall address such problems in Chapter 16 of this volume. It is convenient by way of illustration to consider here heat conduction in a two-dimensional equation in its transient state. This is obtained from Eq. (3.8) by addition of the heat storage term $c(\partial\phi/\partial t)$, where c is the specific heat per unit volume. We now have a problem posed in a domain $\Omega(x, y, t)$ in which the following equation holds:

$$A(\phi) \equiv -\frac{\partial}{\partial x}\left(k\frac{\partial \phi}{\partial x}\right) - \frac{\partial}{\partial y}\left(k\frac{\partial \phi}{\partial y}\right) + Q + c\frac{\partial \phi}{\partial t} = 0 \tag{3.52}$$

with boundary conditions identical to those of Eq. (3.8) and the temperature taken as zero at time zero. Taking

$$\phi \approx \hat{\phi} = \sum_a N_a(x, y)\,\tilde{\phi}_a(t) \tag{3.53}$$

and using the Galerkin weighting procedure we follow precisely the steps outlined in Eqs (3.35)–(3.37) and arrive at a system of ordinary differential equations

$$\mathbf{K}\tilde{\boldsymbol{\phi}} + \mathbf{C}\frac{\mathrm{d}\tilde{\boldsymbol{\phi}}}{\mathrm{d}t} + \mathbf{f} = \mathbf{0} \tag{3.54}$$

Here the expression for K_{ba} and f_b are identical with that of Eq. (3.34) and the reader can verify that the matrix $\mathbf{C}$ is defined by

$$C_{ab} = \int_{\Omega} N_a c N_b \,\mathrm{d}x\,\mathrm{d}y \tag{3.55}$$

Once again the matrix **C** can be assembled from its element contributions. Various analytical and numerical procedures can be applied simply to the solution of such transient, ordinary, differential equations which, again, we shall discuss in detail in Chapters 16 and 17. However, to illustrate the detail and the possible advantage of the process of partial discretization, we shall consider a very simple problem.

Example 3.8: Heat equation with heat generation. Consider a long bar with a square cross-section of size $L \times L$ in which the transient heat conduction equation (3.52) applies and assume that the rate of heat generation varies with time as

$$Q = Q_0 \mathrm{e}^{-\alpha t} \tag{3.56}$$

(this might approximate a problem of heat development due to hydration of concrete). We assume that at $t = 0$, $\phi = 0$ throughout. Further, we shall take $\phi = 0$ on all boundaries for all times.

An approximation for the solution is taken:

$$\begin{aligned} \phi &= \sum_{m=1}^{M} \sum_{n=1}^{N} N_{mn}(x, y)\, \tilde{\phi}_{mn}(t) \\ N_{mn} &= \cos \frac{m\pi x}{L} \cos \frac{n\pi y}{L}; \quad m, n = 1, 3, 5, \cdots \end{aligned} \tag{3.57}$$

with x and y measured from the centre (Fig. 3.9). The even components of the Fourier series are omitted due to the required symmetry of solution. Evaluating the coefficients (only diagonal terms exist in **K**), we have

$$\begin{aligned} K_{mn} &= \int_{-L/2}^{L/2} \int_{-L/2}^{L/2} \left[k \left(\frac{\partial N_{mn}}{\partial x} \right)^2 + k \left(\frac{\partial N_{mn}}{\partial y} \right)^2 \right] \mathrm{d}x\, \mathrm{d}y = \frac{\pi^2 k}{4} (m^2 + n^2) \\ C_{mn} &= \int_{-L/2}^{L/2} \int_{-L/2}^{L/2} c N_1^{mn} \mathrm{d}x\, \mathrm{d}y = \frac{L^2 c}{4} \\ f_{mn} &= \int_{-L/2}^{L/2} \int_{-L/2}^{L/2} N_{mn} Q_0 \mathrm{e}^{-\alpha t} \mathrm{d}x\, \mathrm{d}y = \frac{4 Q_0 L^2}{mn\pi^2} (-1)^{(m+3)/2} (-1)^{(n+3)/2} \mathrm{e}^{-\alpha t} \end{aligned} \tag{3.58}$$

This leads to an ordinary differential equation with parameters $\tilde{\phi}_{mn}$:

$$K_{mn} \tilde{\phi}_{mn} + C_{mn} \frac{\mathrm{d}\tilde{\phi}_{mn}}{\mathrm{d}t} + f_{mn} = 0 \tag{3.59}$$

with $\tilde{\phi}_{mn} = 0$ when $t = 0$. The exact solution of this is easy to obtain, as is shown in Fig. 3.9 for specific values of the parameters M, N, α and $k/L^2 c$.

The remarkable accuracy of the approximation with $M = N = 3$ in this example should be noted. In this example we have used trigonometric functions in place of the more standard polynomials used in the finite element method. In Chapter 7 we recalculate the solution using a standard finite element method in which the solution to the time problem is computed using a finite difference method.

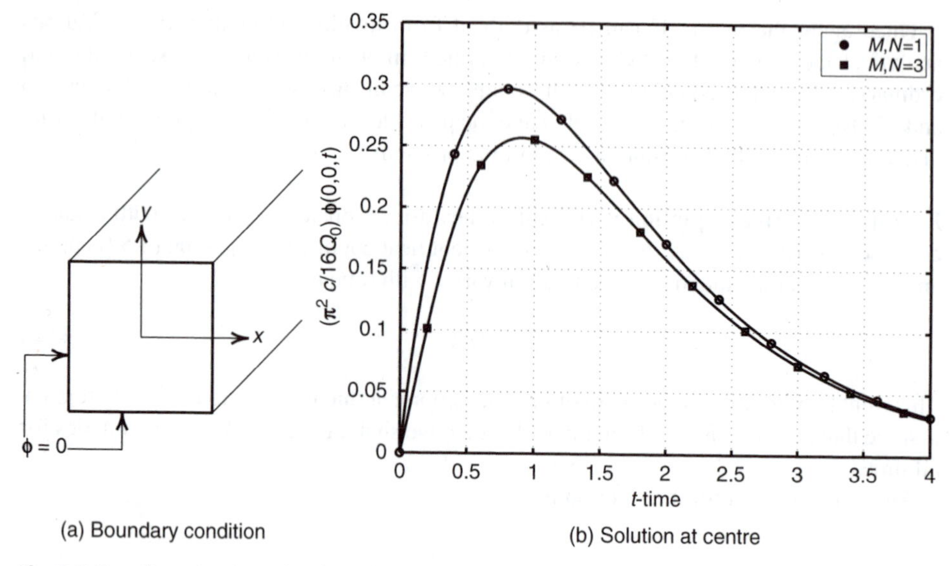

Fig. 3.9 Two-dimensional transient heat development in a square prism – plot of temperature at centre.

3.6 Convergence

In the previous sections we have discussed how approximate solutions can be obtained by use of an expansion of the unknown function in terms of trial or shape functions. Further, we have stated the necessary conditions that such functions have to fulfil in order that the various integrals can be evaluated over the domain. Thus if various integrals contain only the values of $\mathbf{N}$ and its first derivatives then $\mathbf{N}$ has to be C_0 continuous. If second derivatives are involved, C_1 continuity is needed, etc. The problem which we have not yet addressed ourselves consists of the questions of *just how good the approximation is* and *how it can be systematically improved to approach the exact answer*. The first question is more difficult to answer and presumes knowledge of the exact solution (see Chapter 13). The second is more rational and can be answered if we consider some systematic way in which the number of parameters $\tilde{\mathbf{u}}$ in the standard expansion of Eq. (3.3),

$$\hat{\mathbf{u}} = \sum_{a=1}^{n} \mathbf{N}_a \tilde{\mathbf{u}}_a$$

is presumed to increase.

In some examples we have assumed, in effect, a trigonometric Fourier-type series limited to a finite number of terms with a single form of trial function assumed over the whole domain. Here addition of new terms would be simply an extension of the number of terms in the series included in the analysis, and as the Fourier series is known to be able to represent any function within any accuracy desired as the number of terms increases, we can talk about *convergence* of the approximation to the true solution as the number of terms increases.

In other examples of this chapter we have used locally based polynomial functions which are fundamental in the finite element analysis. Here we have tacitly assumed that *convergence occurs as the size of elements decreases and, hence, the number of $\tilde{\mathbf{u}}$ parameters specified at nodes increases*. It is with such convergence that we need to be concerned and we have already discussed this in the context of the analysis of elastic solids in Chapter 2 (Sec. 2.6).

We have now to determine

(a) that, as the number of elements increases, the unknown functions can be approximated as closely as required, and
(b) how the error decreases with the size, h, of the element subdivisions (h is here some typical dimension of an element).

The first problem is that of *completeness* of the expansion and we shall here assume that all trial functions are polynomials (or at least include certain terms of a polynomial expansion).

Clearly, as the approximation discussed here is to the weak, integral form typified by Eq. (3.11) or (3.14) it is necessary that every term occurring under the integral be in the limit capable of being approximated as nearly as possible and, in particular, giving a constant value over any arbitrary infinitesimal part of the domain Ω.

If a derivative of order m exists in any such term, then it is obviously necessary for the local polynomial to be at least of the order m so that, in the limit, such a constant value can be obtained.

We will thus state that a necessary condition for the expansion to be convergent is the *criterion of completeness*: that, if mth derivatives occur in the integral form, a constant value of all derivatives up to order m be attainable in the element domain when the size of any element tends to zero.

This criterion is automatically ensured if the polynomials used in the shape function $\mathbf{N}$ are complete to mth order. This criterion is also equivalent to the one of constant strain postulated in Chapter 2 (Sec. 2.5). This, however, has to be satisfied only in the limit $h \to 0$.

If the actual order of a complete polynomial used in the finite element expansion is $p \geq m$, then *the order of convergence* can be ascertained by seeing how closely such a polynomial can follow the local Taylor expansion of the unknown $\mathbf{u}$. Clearly the order of error will be simply $O(h^{p+1})$ since only terms of order p can be rendered correctly.

Knowledge of the order of convergence helps in ascertaining how good the approximation is if studies on several decreasing mesh sizes are conducted. Though, in Chapter 14, we shall see the asymptotic convergence rate is seldom reached if singularities occur in the problem. Once again we have re-established some of the conditions discussed in Chapter 2.

We shall not discuss, at this stage, approximations which do not satisfy the postulated continuity requirements except to remark that once again, in many cases, convergence and indeed improved results can be obtained (see Chapter 9).

In the above we have referred to the convergence of a given element type as its size is reduced. This is sometimes referred to as *h convergence*.

On the other hand, it is possible to consider a subdivision into elements of a given size and to obtain convergence to the exact solution by increasing the polynomial order p of each element. This is referred to as *p convergence*, which is obviously assured. In general p convergence is more rapid per degree of freedom introduced. We shall discuss both types further in Chapter 14; although we have already noted in some examples how improved accuracy occurs with higher term polynomials being added at each element level.

Variational principles

3.7 What are 'variational principles'?

What are variational principles and how can they be useful in the approximation to continuum problems? It is to these questions that the following sections are addressed.

First a definition: a 'variational principle' specifies a scalar quantity (functional) Π, which is defined by an integral form

$$\Pi = \int_\Omega F\left(\mathbf{u}, \frac{\partial \mathbf{u}}{\partial x}, \cdots\right) \mathrm{d}\Omega + \int_\Gamma E\left(\mathbf{u}, \frac{\partial \mathbf{u}}{\partial x}, \cdots\right) \mathrm{d}\Gamma \tag{3.60}$$

in which $\mathbf{u}$ is the unknown function and F and E are specified differential operators. The solution to the continuum problem is a function $\mathbf{u}$ which makes Π *stationary* with respect to arbitrary changes $\delta\mathbf{u}$. Thus, for a solution to the continuum problem, the 'variation' is

$$\delta\Pi = 0 \tag{3.61}$$

for any $\delta\mathbf{u}$, which defines the condition of stationarity.[15]

If a 'variational principle' can be found, then means are immediately established for obtaining approximate solutions in the standard, integral form suitable for finite element analysis.

Assuming a trial function expansion in the usual form [Eq. (3.3)]

$$\mathbf{u} \approx \hat{\mathbf{u}} = \sum_{a=1}^{n} \mathbf{N}_a \tilde{\mathbf{u}}_a$$

we can insert this into Eq. (3.60) and write

$$\delta\Pi = \frac{\partial \Pi}{\partial \tilde{\mathbf{u}}_1}\delta\tilde{\mathbf{u}}_1 + \frac{\partial \Pi}{\partial \tilde{\mathbf{u}}_2}\delta\tilde{\mathbf{u}}_2 + \cdots + \frac{\partial \Pi}{\partial \tilde{\mathbf{u}}_n}\delta\tilde{\mathbf{u}}_n = 0 \tag{3.62}$$

This being true for any variations $\delta\tilde{\mathbf{u}}$ yields a set of equations

$$\frac{\partial \Pi}{\partial \tilde{\mathbf{u}}} = \begin{Bmatrix} \dfrac{\partial \Pi}{\partial \tilde{\mathbf{u}}_1} \\ \vdots \\ \dfrac{\partial \Pi}{\partial \tilde{\mathbf{u}}_n} \end{Bmatrix} = \mathbf{0} \tag{3.63}$$

from which parameters $\tilde{\mathbf{u}}_a$ are found. The equations are of an integral form necessary for the finite element approximation as the original specification of Π was given in terms of domain and boundary integrals.

The process of finding stationarity with respect to trial function parameters $\tilde{\mathbf{u}}$ is an old one and is associated with the names of Rayleigh[16] and Ritz.[17] It has become extremely important in finite element analysis which, to many investigators, is typified as a 'variational process'.

If the functional Π is 'quadratic', i.e., if the function $\mathbf{u}$ and its derivatives occur in powers not exceeding 2, then Eq. (3.63) reduces to a standard linear form similar to Eq. (3.7), i.e.,

$$\frac{\partial \Pi}{\partial \tilde{\mathbf{u}}} \equiv \mathbf{K}\tilde{\mathbf{u}} + \mathbf{f} = \mathbf{0} \tag{3.64}$$

It is easy to show that the matrix $\mathbf{K}$ will now always be symmetric. To do this let us consider a linearization of the vector $\partial \Pi / \partial \tilde{\mathbf{u}}$. This we can write as

$$\Delta \left(\frac{\partial \Pi}{\partial \tilde{\mathbf{u}}} \right) = \left\{ \begin{matrix} \dfrac{\partial}{\partial \tilde{\mathbf{u}}_1} \left(\dfrac{\partial \Pi}{\partial \tilde{\mathbf{u}}_1} \right) \Delta \tilde{\mathbf{u}}_1 + \dfrac{\partial}{\partial \tilde{\mathbf{u}}_2} \left(\dfrac{\partial \Pi}{\partial \tilde{\mathbf{u}}_1} \right) \Delta \tilde{\mathbf{u}}_2 + \cdots \\ \vdots \end{matrix} \right\} \equiv \mathbf{K}_\mathrm{T}\, \Delta \tilde{\mathbf{u}} \tag{3.65}$$

in which $\mathbf{K}_\mathrm{T}$ is generally known as the tangent matrix, of significance in non-linear analysis, and $\Delta \tilde{\mathbf{u}}$ are small incremental changes to $\tilde{\mathbf{u}}$. Now it is easy to see that

$$\mathbf{K}_{\mathrm{T}ab} = \frac{\partial^2 \Pi}{\partial \tilde{\mathbf{u}}_a \partial \tilde{\mathbf{u}}_b} = \mathbf{K}_{\mathrm{T}ba}^\mathrm{T} \tag{3.66}$$

Hence $\mathbf{K}_\mathrm{T}$ is symmetric.

For a quadratic functional we have, from Eq. (3.64),

$$\Delta \left(\frac{\partial \Pi}{\partial \tilde{\mathbf{u}}} \right) = \mathbf{K} \Delta \tilde{\mathbf{u}} \qquad \text{with} \qquad \mathbf{K} = \mathbf{K}^\mathrm{T} \tag{3.67}$$

and hence symmetry must exist.

The fact that *symmetric matrices will arise whenever a variational principle exists is one of the most important merits of variational approaches for discretization.* However, symmetric forms will frequently arise directly from the Galerkin process. In such cases we simply conclude that the variational principle exists but we shall not need to use it directly. Further, the discovery of symmetry from a weighted residual process leads directly to known (or previously unknown) variational principles.[18]

How then do 'variational principles' arise and is it always possible to construct these for continuous problems?

To answer the first part of the question we note that frequently the physical aspects of the problem can be stated directly in a variational principle form. Theorems such as minimization of total potential energy to achieve equilibrium in mechanical systems, least energy dissipation principles in viscous flow, etc., may be known to the reader and are considered by many as the basis of the formulation. We have already referred to the first of these in Sec. 2.4 of Chapter 2.

Variational principles of this kind are 'natural' ones but unfortunately they do not exist for all continuum problems for which well-defined differential equations may be formulated.

However, there is another category of variational principles which we may call 'contrived'. Such contrived principles can always be constructed for any differentially specified problem, either by extending the number of unknown functions $\mathbf{u}$ by additional variables known as Lagrange multipliers, or by procedures imposing a higher degree of continuity requirements such as in least squares problems. In subsequent sections we shall discuss, respectively, such 'natural' and 'contrived' variational principles.

Before proceeding further it is worth noting that, in addition to symmetry occurring in equations derived by variational means, sometimes further motivation arises. When 'natural' variational principles exist the quantity Π may be of specific interest itself. If this arises a variational approach possesses the merit of easy evaluation of this functional.

The reader will observe that if the functional is 'quadratic' and yields Eq. (3.64), then we can write the approximate 'functional' Π simply as

$$\Pi = \tfrac{1}{2}\tilde{\mathbf{u}}^{\mathrm{T}}\mathbf{K}\tilde{\mathbf{u}} + \tilde{\mathbf{u}}^{\mathrm{T}}\mathbf{f} \tag{3.68}$$

By simple differentiation

$$\delta\Pi = \tfrac{1}{2}\delta(\tilde{\mathbf{u}}^{\mathrm{T}})\mathbf{K}\tilde{\mathbf{u}} + \tfrac{1}{2}\tilde{\mathbf{u}}^{\mathrm{T}}\mathbf{K}\,\delta\tilde{\mathbf{u}} + \delta\tilde{\mathbf{u}}^{\mathrm{T}}\mathbf{f} = \mathbf{0}$$

As $\mathbf{K}$ is symmetric,

$$\delta\tilde{\mathbf{u}}^{\mathrm{T}}\mathbf{K}\tilde{\mathbf{u}} \equiv \tilde{\mathbf{u}}^{\mathrm{T}}\mathbf{K}\delta\tilde{\mathbf{u}}$$

Hence

$$\delta\Pi = \delta\tilde{\mathbf{u}}^{\mathrm{T}}(\mathbf{K}\tilde{\mathbf{u}} + \mathbf{f}) = 0$$

which is true for all $\delta\tilde{\mathbf{u}}$ and hence

$$\mathbf{K}\tilde{\mathbf{u}} + \mathbf{f} = \mathbf{0}$$

when inserted into (3.68) we obtain

$$\Pi = \frac{1}{2}\tilde{\mathbf{u}}^{\mathrm{T}}\mathbf{f} = -\frac{1}{2}\tilde{\mathbf{u}}^{\mathrm{T}}\mathbf{K}\,\tilde{\mathbf{u}}$$

3.8 'Natural' variational principles and their relation to governing differential equations

3.8.1 Euler equations

If we consider the definitions of Eqs (3.60) and (3.61) we observe that for stationarity we can write, after performing some differentiations and integrations by parts,

$$\delta\Pi = \int_{\Omega} \delta\mathbf{u}^{\mathrm{T}}\mathcal{A}(\mathbf{u})\mathrm{d}\Omega + \int_{\Gamma} \delta\mathbf{u}^{\mathrm{T}}\mathcal{B}(\mathbf{u})\mathrm{d}\Gamma = 0 \tag{3.69}$$

As the above has to be true for any variations $\delta\mathbf{u}$, we must have

$$\mathcal{A}(\mathbf{u}) = \mathbf{0} \quad \text{in } \Omega \quad \text{and} \quad \mathcal{B}(\mathbf{u}) = \mathbf{0} \quad \text{on } \Gamma \tag{3.70}$$

If $\mathcal{A}$ corresponds precisely to the differential equations governing the problem of interest and $\mathcal{B}$ to its boundary conditions, then the variational principle is a *natural* one. Equations (3.70) are known as the Euler differential equations corresponding to the variational principle requiring the stationarity of Π. It is easy to show that for any variational principle a corresponding set of Euler equations can be established. The reverse is unfortunately not true, i.e., only certain forms of differential equations are Euler equations of a variational functional. In the next section we shall consider the conditions necessary for the existence of variational principles and give a prescription for the establishing Π from a set of suitable linear differential equations. In this section we shall continue to assume that the form of the variational principle is known.

To illustrate the process let us now consider a specific example. Suppose we specify a problem by requiring the stationarity of a functional

$$\Pi = \int_\Omega \left[\tfrac{1}{2} k \left(\frac{\partial \phi}{\partial x} \right)^2 + \tfrac{1}{2} k \left(\frac{\partial \phi}{\partial y} \right)^2 + Q\phi \right] \mathrm{d}\Omega + \int_{\Gamma_q} \bar{q}\phi \, \mathrm{d}\Gamma \qquad (3.71)$$

in which k and Q depend only on position and we assume $\phi = \bar{\phi}$ is satisfied on Γ_ϕ.

We now perform the variation.[15] This can be written following the rules of differentiation as

$$\delta\Pi = \int_\Omega \left[k \frac{\partial \phi}{\partial x} \delta \left(\frac{\partial \phi}{\partial x} \right) + k \frac{\partial \phi}{\partial y} \delta \left(\frac{\partial \phi}{\partial y} \right) + Q\delta\phi \right] \mathrm{d}\Omega + \int_{\Gamma_q} (\bar{q}\,\delta\phi) \, \mathrm{d}\Gamma = 0 \qquad (3.72)$$

As

$$\delta \left(\frac{\partial \phi}{\partial x} \right) = \frac{\partial}{\partial x} (\delta\phi) \qquad (3.73)$$

we can integrate by parts (as in Sec. 3.3) and, since $\delta\phi = 0$ on Γ_ϕ, obtain

$$\begin{aligned} \delta\Pi = & \int_\Omega \delta\phi \left[-\frac{\partial}{\partial x} \left(k \frac{\partial \phi}{\partial x} \right) - \frac{\partial}{\partial y} \left(k \frac{\partial \phi}{\partial y} \right) + Q \right] \mathrm{d}\Omega \\ & + \int_{\Gamma_q} \delta\phi \left(k \frac{\partial \phi}{\partial n} + \bar{q} \right) \mathrm{d}\Gamma = 0 \end{aligned} \qquad (3.74a)$$

This is of the form of Eq. (3.69) and we immediately observe that the Euler equations are

$$\begin{aligned} A(\phi) &= -\frac{\partial}{\partial x} \left(k \frac{\partial \phi}{\partial y} \right) - \frac{\partial}{\partial y} \left(k \frac{\partial \phi}{\partial y} \right) + Q = 0 \qquad \text{in } \Omega \\ B(\phi) &= k \frac{\partial \phi}{\partial n} + \bar{q} = 0 \qquad \text{on } \Gamma_q \end{aligned} \qquad (3.74b)$$

If ϕ is prescribed so that $\phi = \bar{\phi}$ on Γ_ϕ and $\delta\phi = 0$ on that boundary, then the problem is precisely the one we have already discussed in Sec. 3.2 and the functional (3.71) specifies the *two-dimensional heat conduction* problem in an alternative way.

In this case we have 'guessed' the functional but the reader will observe that the variation operation could have been carried out for any functional specified and corresponding *Euler* equations could have been established.

Let us continue the process to obtain an approximate solution of the linear heat conduction problem. Taking, as usual,

$$\phi \approx \hat{\phi} = \sum_a N_a \tilde{\phi}_a = \mathbf{N}\tilde{\boldsymbol{\phi}} \qquad (3.75)$$

we substitute this approximation into the expression for the functional Π [Eq. (3.71)] and obtain

$$\begin{aligned} \Pi = & \int_\Omega \frac{1}{2} k \left(\sum_a \frac{\partial N_a}{\partial x} \tilde{\phi}_a \right)^2 \mathrm{d}\Omega + \int_\Omega \frac{1}{2} k \left(\sum_a \frac{\partial N_a}{\partial y} \tilde{\phi}_a \right)^2 \mathrm{d}\Omega \\ & + \int_\Omega Q \sum_a N_a \tilde{\phi}_a \, \mathrm{d}\Omega + \int_{\Gamma_q} \bar{q} \sum_a N_a \tilde{\phi}_a \, \mathrm{d}\Gamma \end{aligned} \qquad (3.76)$$

On differentiation with respect to a typical parameter $\tilde{\phi}_b$ we have

$$\frac{\partial \Pi}{\partial \tilde{\phi}_b} = \int_\Omega k\left(\sum_a \frac{\partial N_a}{\partial x}\tilde{\phi}_a\right)\frac{\partial N_b}{\partial x}\,\mathrm{d}\Omega + \int_\Omega k\left(\sum_a \frac{\partial N_a}{\partial y}\tilde{\phi}_a\right)\frac{\partial N_b}{\partial y}\,\mathrm{d}\Omega + \int_\Omega Q N_b\,\mathrm{d}\Omega + \int_{\Gamma_q} \bar{q} N_b\,\mathrm{d}\Gamma = 0 \tag{3.77}$$

and a system of equations for the solution of the problem is

$$\mathbf{K}\tilde{\boldsymbol{\phi}} + \mathbf{f} = \mathbf{0} \tag{3.78}$$

with

$$\begin{aligned} K_{ab} = K_{ba} &= \int_\Omega k\frac{\partial N_a}{\partial x}\frac{\partial N_b}{\partial x}\,\mathrm{d}\Omega + \int_\Omega k\frac{\partial N_a}{\partial y}\frac{\partial N_b}{\partial y}\,\mathrm{d}\Omega \\ f_b &= \int_\Omega N_b Q\,\mathrm{d}\Omega + \int_{\Gamma_q} N_b \bar{q}\,\mathrm{d}\Gamma \end{aligned} \tag{3.79}$$

The reader will observe that the approximation equations are here identical with those obtained in Sec. 3.5 for the same problem using the Galerkin process. No special advantage accrues to the variational formulation here, and indeed we can predict now that *Galerkin and variational procedures must give the same answer for cases where natural variational principles exist.*

3.8.2 Relation of the Galerkin method to approximation via variational principles

In the preceding example we have observed that the approximation obtained by the use of a natural variational principle and by the use of the Galerkin weighting process proved identical. That this is the case follows directly from Eq. (3.69), in which the variation was derived in terms of the original differential equations and the associated boundary conditions.

If we consider the usual trial function expansion [Eq. (3.3)]

$$\mathbf{u} \approx \hat{\mathbf{u}} = \mathbf{N}\tilde{\mathbf{u}}$$

we can write the variation of this approximation as

$$\delta\hat{\mathbf{u}} = \mathbf{N}\,\delta\tilde{\mathbf{u}} \tag{3.80}$$

and inserting the above into (3.69) yields

$$\delta\Pi = \delta\tilde{\mathbf{u}}^{\mathrm{T}} \int_\Omega \mathbf{N}^{\mathrm{T}} \boldsymbol{\mathcal{A}}(\mathbf{N}\tilde{\mathbf{u}})\,\mathrm{d}\Omega + \delta\tilde{\mathbf{u}}^{\mathrm{T}} \int_\Gamma \mathbf{N}^{\mathrm{T}} \boldsymbol{\mathcal{B}}(\mathbf{N}\tilde{\mathbf{u}})\,\mathrm{d}\Gamma = 0 \tag{3.81}$$

The above form, being true for all $\delta\tilde{\mathbf{u}}$, requires that the expression under the integrals should be zero. The reader will immediately recognize this as simply the Galerkin form of the weighted residual statement discussed earlier [Eq. (3.22)], and identity is hereby proved.

We need to underline, however, that this is only true if the Euler equations of the variational principle coincide with the governing equations of the original problem. The Galerkin process thus retains its greater range of applicability.

3.9 Establishment of natural variational principles for linear, self-adjoint, differential equations

General rules for deriving natural variational principles from non-linear differential equations are complicated and even the tests necessary to establish the existence of such variational principles are not simple. Much mathematical work has been done in this context by Vainberg,[19] Tonti,[18] Oden,[20, 21] and others.

For linear differential equations the situation is much simpler and a thorough study is available in the works of Mikhlin,[22, 23] and in this section a brief presentation of such rules is given.

We shall consider here only the establishment of variational principles for a linear system of equations with *forced* boundary conditions, implying only variation of functions which yield $\delta\mathbf{u} = \mathbf{0}$ on their boundaries. The extension to include natural boundary conditions is simple and will be omitted.

Writing a linear system of differential equations as

$$\boldsymbol{\mathcal{A}}(\mathbf{u}) \equiv \boldsymbol{\mathcal{L}}\mathbf{u} + \mathbf{b} = \mathbf{0} \tag{3.82}$$

in which $\boldsymbol{\mathcal{L}}$ is a linear differential operator it can be shown that natural variational principles require that the operator $\boldsymbol{\mathcal{L}}$ be such that

$$\int_\Omega \boldsymbol{\psi}^{\mathrm{T}}(\boldsymbol{\mathcal{L}}\boldsymbol{\gamma})\,\mathrm{d}\Omega = \int_\Omega \boldsymbol{\gamma}^{\mathrm{T}}(\boldsymbol{\mathcal{L}}\boldsymbol{\psi})\,\mathrm{d}\Omega + \text{b.t.} \tag{3.83}$$

for any two function sets $\boldsymbol{\psi}$ and $\boldsymbol{\gamma}$. In the above, 'b.t.' stands for boundary terms which we disregard in the present context. The property required in the above operator is called that of *self-adjointness* or *symmetry*.

If the operator $\boldsymbol{\mathcal{L}}$ is self-adjoint, the variational principle can be written immediately as

$$\Pi = \int_\Omega \left[\tfrac{1}{2}\mathbf{u}^{\mathrm{T}}\,(\boldsymbol{\mathcal{L}}\mathbf{u}) + \mathbf{u}^{\mathrm{T}}\mathbf{b}\right]\mathrm{d}\Omega + \text{b.t.} \tag{3.84}$$

To prove the veracity of the last statement a variation needs to be considered. We thus write (omitting boundary terms)

$$\delta\Pi = \int_\Omega [\tfrac{1}{2}\delta\mathbf{u}^{\mathrm{T}}\boldsymbol{\mathcal{L}}\mathbf{u} + \tfrac{1}{2}\mathbf{u}^{\mathrm{T}}\delta(\boldsymbol{\mathcal{L}}\mathbf{u}) + \delta\mathbf{u}^{\mathrm{T}}\mathbf{b}]\,\mathrm{d}\Omega = 0 \tag{3.85}$$

Noting that for any linear operator

$$\delta(\boldsymbol{\mathcal{L}}\mathbf{u}) \equiv \boldsymbol{\mathcal{L}}\,\delta\mathbf{u} \tag{3.86}$$

and that $\mathbf{u}$ and $\delta\mathbf{u}$ can be treated as any two independent functions, by identity (3.83) we can write Eq. (3.85) as

$$\delta\Pi = \int_\Omega \delta\mathbf{u}^{\mathrm{T}}[\boldsymbol{\mathcal{L}}\mathbf{u} + \mathbf{b}]\,\mathrm{d}\Omega = 0 \tag{3.87}$$

We observe immediately that the term in the brackets, i.e., the Euler equation of the functional, is identical with the original equation postulated, and therefore the variational principle is verified.

The above gives a very simple test and a prescription for the establishment of natural variational principles for differential equations of the problem.

Example 3.9: Helmholz problem in two dimensions. A Helmholz problem is governed by a differential equation similar to the heat conduction equation, e.g.,

$$\nabla^2\phi + c\phi + Q = 0 \tag{3.88}$$

with c and Q being dependent on position only.

The above can be written in the general form of Eq. (3.82), with

$$\mathcal{L} = \left[\frac{\partial^2}{\partial x^2} + \frac{\partial^2}{\partial y^2} + c\right]; \qquad \mathbf{b} = Q \qquad \text{and} \qquad \mathbf{u} = \phi \tag{3.89}$$

Verifying that self-adjointness applies (which we leave to the reader as an exercise), we immediately have a variational principle

$$\Pi = \int_\Omega \left[\frac{1}{2}\phi\left(\frac{\partial^2\phi}{\partial x^2} + \frac{\partial^2\phi}{\partial y^2} + c\phi\right) + \phi Q\right] \mathrm{d}x\,\mathrm{d}y \tag{3.90}$$

with ϕ satisfying the forced boundary condition, i.e., $\phi = \bar{\phi}$ on Γ_ϕ. Integrating by parts of the first two terms results in

$$\Pi = -\int_\Omega \left[\frac{1}{2}\left(\frac{\partial\phi}{\partial x}\right)^2 + \frac{1}{2}\left(\frac{\partial\phi}{\partial y}\right)^2 - \frac{1}{2}c\phi^2 - \phi Q\right] \mathrm{d}x\,\mathrm{d}y \tag{3.91}$$

on noting that boundary terms with prescribed ϕ do not alter the principle.

Example 3.10: First-order form of heat equation. This problem concerns the one-dimensional heat conduction equation (Example 3.5, Sec. 3.3) written in first order form as

$$\mathcal{A}(\mathbf{u}) = \begin{Bmatrix} -q - \dfrac{\mathrm{d}\phi}{\mathrm{d}x} \\ \dfrac{\mathrm{d}q}{\mathrm{d}x} + Q \end{Bmatrix} = \mathbf{0}$$

or, using Eq. (3.82), as

$$\mathcal{L} \equiv \begin{bmatrix} -1, & -\dfrac{\mathrm{d}}{\mathrm{d}x} \\ \dfrac{\mathrm{d}}{\mathrm{d}x}, & 0 \end{bmatrix}; \quad \mathbf{b} = \begin{Bmatrix} 0 \\ Q \end{Bmatrix} \quad \text{and} \quad \mathbf{u} = \begin{Bmatrix} q \\ \phi \end{Bmatrix}$$

Again self-adjointness of the operator can be tested and found to be satisfied. We now write the functional as

$$\begin{aligned} \Pi &= \int_\Omega \left[\frac{1}{2}\begin{Bmatrix} q \\ \phi \end{Bmatrix}^{\mathrm{T}} \left(\begin{bmatrix} -1, & -\dfrac{\mathrm{d}}{\mathrm{d}x} \\ \dfrac{\mathrm{d}}{\mathrm{d}x}, & 0 \end{bmatrix} \begin{Bmatrix} q \\ \phi \end{Bmatrix}\right) + \begin{Bmatrix} q \\ \phi \end{Bmatrix}^{\mathrm{T}} \begin{Bmatrix} 0 \\ Q \end{Bmatrix}\right] \mathrm{d}x \\ &= \int_\Omega \left[\frac{1}{2}\left(-q^2 - q\frac{\mathrm{d}\phi}{\mathrm{d}x} + \phi\frac{\mathrm{d}q}{\mathrm{d}x}\right) + \phi Q\right] \mathrm{d}x \end{aligned} \tag{3.92}$$

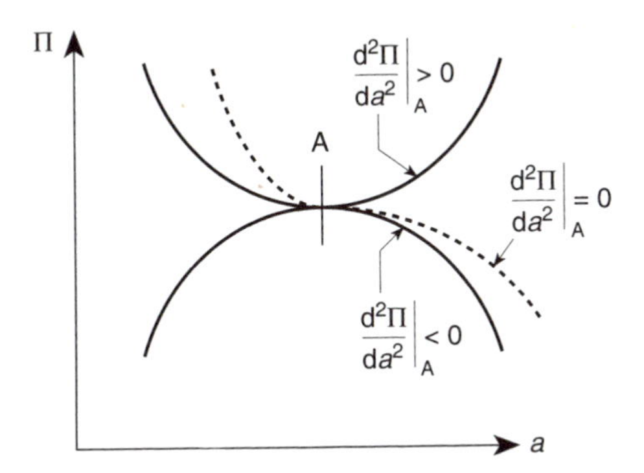

Fig. 3.10 Maximum, minimum and a 'saddle' point for a functional Π of one variable.

The verification of the correctness of the above, by executing a variation, is left to the reader.

These two examples illustrate the simplicity of application of the general expressions. The reader will observe that self-adjointness of the operator will generally exist if even orders of differentiation are present. For odd orders self-adjointness is only possible if the operator is a 'skew'-symmetric matrix such as occurs in the second example.

3.10 Maximum, minimum, or a saddle point?

In discussing variational principles so far we have assumed simply that at the solution point $\delta\Pi = 0$, that is the functional is stationary. It is often desirable to know whether Π is at a maximum, minimum, or simply at a 'saddle point'. If a maximum or a minimum is involved, then the approximation to Π will always be 'bounded', i.e., will provide approximate values of Π which are either smaller or larger than the correct ones.† The bound in itself may be of practical significance in some problems.

When, in elementary calculus, we consider a stationary point of a function Π of one variable u, we investigate the rate of change of $\mathrm{d}\Pi$ with $\mathrm{d}u$ and write

$$\mathrm{d}(\mathrm{d}\Pi) = \mathrm{d}\left(\frac{\partial \Pi}{\partial u}\mathrm{d}u\right) = \frac{\partial^2 \Pi}{\partial u^2}(\mathrm{d}u)^2 \tag{3.93}$$

The sign of the second derivative determines whether Π is a minimum, maximum, or simply stationary (saddle point), as shown in Fig. 3.10. By analogy in the calculus of variations we shall consider changes of $\delta\Pi$. Noting the general form of this quantity given by Eq. (3.62) and the notion of the second derivative of Eq. (3.65) we can write, in terms of discrete parameters,

$$\delta(\delta\Pi) \equiv \delta\left(\frac{\partial \Pi}{\partial \tilde{\mathbf{u}}}\right)^{\mathrm{T}} \delta\tilde{\mathbf{u}} = \delta\tilde{\mathbf{u}}^{\mathrm{T}}\delta\left(\frac{\partial \Pi}{\partial \tilde{\mathbf{u}}}\right) = \delta\tilde{\mathbf{u}}^{\mathrm{T}}\left(\frac{\partial^2 \Pi}{\partial \tilde{\mathbf{u}}\,\partial \tilde{\mathbf{u}}}\delta\tilde{\mathbf{u}}\right) = \delta\tilde{\mathbf{u}}^{\mathrm{T}}\mathbf{K}_{\mathrm{T}}\,\delta\tilde{\mathbf{u}} \tag{3.94}$$

If, in the above, $\delta(\delta\Pi)$ is always negative then Π is obviously reaching a maximum, if it is always positive then Π is a minimum, but if the sign is indeterminate this shows only the existence of a saddle point.

† Provided all integrals are exactly evaluated.

As $\delta\tilde{\mathbf{u}}$ is an arbitrary vector this statement is equivalent to requiring the matrix $\mathbf{K}_T$ to be negative definite for a maximum *or* positive definite for a minimum. The form of the matrix $\mathbf{K}_T$ (or in linear problems of $\mathbf{K}$ which is identical to it) is thus of great importance in the solution of variational problems.

3.11 Constrained variational principles. Lagrange multipliers

3.11.1 Lagrange multipliers

Consider the problem of making a functional Π stationary, subject to the unknown $\mathbf{u}$ obeying some set of additional differential relationships

$$\mathbf{C}(\mathbf{u}) = \mathbf{0} \quad \text{in } \Omega \tag{3.95}$$

We can introduce this constraint by forming another functional

$$\bar{\Pi}(\mathbf{u}, \boldsymbol{\lambda}) = \Pi(\mathbf{u}) + \int_\Omega \boldsymbol{\lambda}^T \mathbf{C}(\mathbf{u})\, d\Omega \tag{3.96}$$

in which $\boldsymbol{\lambda}$ is some set of functions of the independent coordinates in the domain Ω known as *Lagrange multipliers*. The variation of the new functional is now

$$\delta\bar{\Pi} = \delta\Pi + \int_\Omega \boldsymbol{\lambda}^T \delta\mathbf{C}(\mathbf{u})\, d\Omega + \int_\Omega \delta\boldsymbol{\lambda}^T \mathbf{C}(\mathbf{u})\, d\Omega = 0 \tag{3.97}$$

which immediately gives $\mathbf{C}(\mathbf{u}) = \mathbf{0}$ and, simultaneously, an added contribution to the original $\delta\Pi$ involving $\boldsymbol{\lambda}$.

In a similar way, constraints can be introduced at some points or over boundaries of the domain. For instance, if we require that $\mathbf{u}$ obey

$$\mathbf{E}(\mathbf{u}) = \mathbf{0} \quad \text{on } \Gamma \tag{3.98}$$

we would add to the original functional the term

$$\int_\Gamma \boldsymbol{\lambda}^T \mathbf{E}(\mathbf{u})\, d\Gamma \tag{3.99}$$

with $\boldsymbol{\lambda}$ now being an unknown function defined only on Γ. Alternatively, if the constraint $\mathbf{C}$ is applicable only at one or more points of the system, then the simple addition of $\boldsymbol{\lambda}^T\mathbf{C}(\mathbf{u})$ at these points to the general functional Π will introduce a discrete number of constraints.

It appears, therefore, possible to always introduce additional functions $\boldsymbol{\lambda}$ and modify a functional to include any prescribed constraints. In the 'discretization' process we shall now have to use trial functions to describe both $\mathbf{u}$ and $\boldsymbol{\lambda}$.

Writing, for instance,

$$\hat{\mathbf{u}} = \sum_a \mathbf{N}_a \tilde{\mathbf{u}}_a = \mathbf{N}\tilde{\mathbf{u}} \qquad \hat{\boldsymbol{\lambda}} = \sum_b \bar{\mathbf{N}}_b \tilde{\boldsymbol{\lambda}}_b = \bar{\mathbf{N}}\tilde{\boldsymbol{\lambda}} \tag{3.100}$$

we shall obtain a set of equations

$$\frac{\partial \Pi}{\partial \mathbf{w}} = \begin{Bmatrix} \dfrac{\partial \Pi}{\partial \tilde{\mathbf{u}}} \\ \dfrac{\partial \Pi}{\partial \tilde{\boldsymbol{\lambda}}} \end{Bmatrix} = \mathbf{0} \qquad \text{where} \qquad \mathbf{w} = \begin{Bmatrix} \tilde{\mathbf{u}} \\ \tilde{\boldsymbol{\lambda}} \end{Bmatrix} \tag{3.101}$$

from which both the sets of parameters $\tilde{\mathbf{u}}$ and $\tilde{\boldsymbol{\lambda}}$ can be obtained. It is somewhat paradoxical that the 'constrained' problem has resulted in a larger number of unknown parameters than the original one and, indeed, has complicated the solution. We shall, nevertheless, find practical use for Lagrange multipliers in formulating some physical variational principles, and will make use of these in a more general context in Chapters 10 and 11.

Before proceeding further it is of interest to investigate the form of equations resulting from the modified functional Π of Eq. (3.96). If the original functional Π gave as its Euler equations a system

$$\boldsymbol{\mathcal{A}}(\mathbf{u}) = \mathbf{0} \tag{3.102}$$

then we have (omitting the boundary terms)

$$\delta \bar{\Pi} = \int_\Omega \delta \mathbf{u}^{\mathrm{T}} \boldsymbol{\mathcal{A}}(\mathbf{u})\, \mathrm{d}\Omega + \int_\Omega \delta \mathbf{C}^{\mathrm{T}} \boldsymbol{\lambda}\, \mathrm{d}\Omega + \int_\Omega \delta \boldsymbol{\lambda}^{\mathrm{T}} \mathbf{C}(\mathbf{u})\, \mathrm{d}\Omega = 0 \tag{3.103}$$

Substituting the trial functions (3.100) we can write for a linear set of constraints

$$\mathbf{C}(\mathbf{u}) = \boldsymbol{\mathcal{L}}_1 \mathbf{u} + \mathbf{C}_1$$

that

$$\begin{aligned} \delta \bar{\Pi} = \delta \tilde{\mathbf{u}}^{\mathrm{T}} &\left[\int_\Omega \mathbf{N}^{\mathrm{T}} \boldsymbol{\mathcal{A}}(\hat{\mathbf{u}})\, \mathrm{d}\Omega + \int_\Omega (\boldsymbol{\mathcal{L}}_1 \mathbf{N})^{\mathrm{T}} \hat{\boldsymbol{\lambda}}\, \mathrm{d}\Omega \right] \\ &+ \delta \tilde{\boldsymbol{\lambda}}^{\mathrm{T}} \int_\Omega \bar{\mathbf{N}}^{\mathrm{T}} (\boldsymbol{\mathcal{L}}_1 \hat{\mathbf{u}} + \mathbf{C}_1)\, \mathrm{d}\Omega = 0 \end{aligned} \tag{3.104}$$

As this has to be true for all variations $\delta \tilde{\mathbf{u}}$ and $\delta \tilde{\boldsymbol{\lambda}}$, we have a system of equations

$$\begin{aligned} \int_\Omega \mathbf{N}^{\mathrm{T}} \mathbf{A}(\hat{\mathbf{u}})\, \mathrm{d}\Omega + \int_\Omega (\boldsymbol{\mathcal{L}}_1 \mathbf{N})^{\mathrm{T}} \hat{\boldsymbol{\lambda}}\, \mathrm{d}\Omega &= \mathbf{0} \\ \int_\Omega \bar{\mathbf{N}}^{\mathrm{T}} (\boldsymbol{\mathcal{L}}_1 \hat{\mathbf{u}} + \mathbf{C}_1)\, \mathrm{d}\Omega &= \mathbf{0} \end{aligned} \tag{3.105}$$

For linear equations $\boldsymbol{\mathcal{A}}$, the first term of the first equation is precisely the ordinary, unconstrained, variational approximation

$$\mathbf{K}_{uu} \tilde{\mathbf{u}} + \mathbf{f}_u \tag{3.106}$$

and inserting again the trial functions (3.100) we can write the approximated Eq. (3.105) as a linear system:

$$\mathbf{K}_w \mathbf{w} = \begin{bmatrix} \mathbf{K}_{uu}, & \mathbf{K}_{u\lambda} \\ \mathbf{K}_{u\lambda}^{\mathrm{T}}, & \mathbf{0} \end{bmatrix} \begin{Bmatrix} \tilde{\mathbf{u}} \\ \tilde{\boldsymbol{\lambda}} \end{Bmatrix} + \begin{Bmatrix} \mathbf{f}_u \\ \mathbf{f}_\lambda \end{Bmatrix} = \mathbf{0} \tag{3.107}$$

with

$$\mathbf{K}_{u\lambda}^{\mathrm{T}} = \int_\Omega \bar{\mathbf{N}}^{\mathrm{T}} (\boldsymbol{\mathcal{L}}_1 \mathbf{N})\, \mathrm{d}\Omega; \qquad \mathbf{f}_\lambda = \int_\Omega \bar{\mathbf{N}}^{\mathrm{T}} \mathbf{C}_1\, \mathrm{d}\Omega \tag{3.108}$$

Clearly the system of equations is symmetric but now possesses zeros on the diagonal, and therefore the variational principle Π is merely stationary. Further, computational difficulties may be encountered unless the solution process allows for zero diagonal terms.

Example 3.11: Constraint enforcement using Lagrange multiplier. The point about increasing the number of parameters to introduce a constraint may perhaps be best illustrated in a simple algebraic situation in which we require a stationary value of a quadratic function of two variables u_1 and u_2:

$$\Pi = 2u_1^2 - 2u_1u_2 + u_2^2 + 18u_1 + 6u_2 \tag{3.109}$$

subject to a constraint

$$u_1 - u_2 = 0 \tag{3.110}$$

The obvious way to proceed would be to insert directly the equality 'constraint' and obtain

$$\Pi = u_1^2 + 24u_1 \tag{3.111}$$

and write, for stationarity,

$$\frac{\partial \Pi}{\partial u_1} = 0 = 2u_1 + 24 \qquad u_1 = u_2 = -12 \tag{3.112}$$

Introducing a Lagrange multiplier λ we can alternatively find the stationarity of

$$\bar{\Pi} = 2u_1^2 - 2u_1u_2 + u_2^2 + 18u_1 + 6u_2 + \lambda(u_1 - u_2) \tag{3.113}$$

and write *three* simultaneous equations

$$\begin{aligned}
\frac{\partial \bar{\Pi}}{\partial u_1} &= 4u_1 - 2u_2 + \lambda + 18 = 0 \\
\frac{\partial \bar{\Pi}}{\partial u_2} &= -2u_1 + 2u_2 - \lambda + 6 = 0 \\
\frac{\partial \bar{\Pi}}{\partial \lambda} &= u_1 - u_2 = 0
\end{aligned} \tag{3.114}$$

The solution of the above system again yields the correct answer

$$u_1 = u_2 = -12 \qquad \lambda = 6$$

but at considerably more effort. Unfortunately, in most continuum problems direct elimination of constraints cannot be so simply accomplished.†

† In the finite element context, Szabo and Kassos[24] use such direct elimination; however, this involves considerable algebraic manipulation.

3.11.2 Identification of Lagrange multipliers. Forced boundary conditions and modified variational principles

Although the Lagrange multipliers were introduced as a mathematical concept necessary for the enforcement of certain external constraints required to satisfy the original variational principle, we shall find that in many situations they can be identified with certain physical quantities of importance to the original mathematical model. Such an identification will follow immediately from the definition of the variational principle established in Eq. (3.96) and through the first of the Euler equations in (3.105) corresponding to it. The variation $\delta\bar{\Pi}$, written in Eq. (3.97), supplies through its third term the constraint equation. The first two terms can always be rewritten as

$$\int_{\Omega} \delta\mathbf{C}(\mathbf{u})^{\mathrm{T}}\boldsymbol{\lambda}\,\mathrm{d}\Omega + \int_{\Omega} \delta\mathbf{u}^{\mathrm{T}}\boldsymbol{\mathcal{A}}(\mathbf{u})\,\mathrm{d}\Omega = \mathbf{0} \tag{3.115a}$$

or

$$\int_{\Gamma} \delta\mathbf{E}(\mathbf{u})^{\mathrm{T}}\boldsymbol{\lambda}\,\mathrm{d}\Gamma + \int_{\Gamma} \delta\mathbf{u}^{\mathrm{T}}\boldsymbol{\mathcal{B}}(\mathbf{u})\,\mathrm{d}\Gamma = \mathbf{0} \tag{3.115b}$$

This supplies the identification of $\boldsymbol{\lambda}$.

In the literature of variational calculation such identification arises frequently and the reader is referred to the excellent text by Washizu[25] for numerous examples.

Example 3.12: Identification of Lagrange multiplier for boundary condition. Here we shall introduce this identification by means of the example considered in Sec. 3.8.1. As we have noted, the variational principle of Eq. (3.71) established the governing equation and the natural boundary conditions of the heat conduction problem providing the forced boundary condition

$$\mathbf{E}(\phi) = \phi - \bar{\phi} = 0 \tag{3.116}$$

was satisfied on Γ_ϕ in the choice of the trial function for ϕ.

The above forced boundary condition can, however, be considered as a constraint on the original problem. We can write the constrained variational principle as

$$\bar{\Pi} = \Pi + \int_{\Gamma_\phi} \lambda(\phi - \bar{\phi})\,\mathrm{d}\Gamma \tag{3.117}$$

where Π is given by Eq. (3.71).

Performing the variation we have

$$\delta\bar{\Pi} = \delta\Pi + \int_{\Gamma_\phi} \delta\phi\lambda\,\mathrm{d}\Gamma + \int_{\Gamma_\phi} \delta\lambda(\phi - \bar{\phi})\,\mathrm{d}\Gamma = 0 \tag{3.118}$$

$\delta\Pi$ is now given by the expression (3.74a) augmented by an integral

$$\int_{\Gamma_\phi} \delta\phi\, k\,\frac{\partial\phi}{\partial n}\,\mathrm{d}\Gamma \tag{3.119}$$

which was previously disregarded (as we had assumed that $\delta\phi = 0$ on Γ_ϕ). In addition to the conditions of Eq. (3.74b), we now require that

$$\int_{\Gamma_\phi} \delta\lambda(\phi - \bar{\phi})\,\mathrm{d}\Gamma + \int_{\Gamma_\phi} \delta\phi\left(\lambda + k\frac{\partial\phi}{\partial n}\right)\mathrm{d}\Gamma = 0 \tag{3.120}$$

which must be true for all variations $\delta\lambda$ and $\delta\phi$. The first simply reiterates the constraint

$$\phi - \bar{\phi} = 0 \qquad \text{on } \Gamma_\phi \tag{3.121}$$

The second *defines* λ as

$$\lambda = -k\frac{\partial \phi}{\partial n} \tag{3.122}$$

Noting that $k(\partial\phi/\partial n)$ is the negative to the flux q_n on the boundary Γ_ϕ, the physical identification of the multiplier has been achieved – that is, $\lambda \equiv q_n$.

The identification of the Lagrange variable leads to the possible establishment of a modified variational principle in which λ is replaced by the identification.

We could thus write a new principle for the above example:

$$\bar{\Pi} = \Pi - \int_{\Gamma_\phi} k\frac{\partial \phi}{\partial n}(\phi - \bar{\phi})\, \mathrm{d}\Gamma \tag{3.123}$$

in which once again Π is given by the expression (3.71) but ϕ is not constrained to satisfy any boundary conditions. Use of such modified variational principles can be made to restore interelement continuity and appears to have been first introduced for that purpose by Kikuchi and Ando.[26] In general these present interesting new procedures for establishing useful variational principles.

A further extension of such principles has been made use of by Chen and Mei[27] and Zienkiewicz *et al.*[28] Washizu[25] discusses many such applications in the context of structural mechanics. The reader can verify that the variational principle expressed in Eq. (3.123) leads to automatic satisfaction of all the necessary boundary conditions in the example considered.

The use of modified variational principles restores the problem to the original number of unknown functions or parameters and is often computationally advantageous.

3.12 Constrained variational principles. Penalty function and perturbed lagrangian methods

In the previous section we have seen how the process of introducing Lagrange multipliers allows constrained variational principles to be obtained at the expense of increasing the total number of unknowns. Further, we have shown that even in linear problems the algebraic equations which have to be solved are now complicated by having zero diagonal terms. In this section we shall consider alternative procedures of introducing constraints which do not possess these drawbacks.

3.12.1 Penalty functions

Considering once again the problem of obtaining stationarity of Π with a set of constraint equations $\mathbf{C}(\mathbf{u}) = \mathbf{0}$ in domain Ω, we note that the product

$$\mathbf{C}^{\mathrm{T}}\mathbf{C} = C_1^2 + C_2^2 + \cdots \tag{3.124}$$

where $\mathbf{C}^{\mathrm{T}} = [C_1, C_2, \ldots]$ must always be a quantity which is positive or zero. Clearly, the latter value is found when the constraints are satisfied and also clearly the variation

$$\delta(\mathbf{C}^{\mathrm{T}}\mathbf{C}) = 0 \tag{3.125}$$

as the product reaches that minimum.

We can now write a new functional

$$\bar{\bar{\Pi}} = \Pi + \frac{1}{2}\alpha \int_\Omega \mathbf{C}^{\mathrm{T}}(\mathbf{u})\mathbf{C}(\mathbf{u})\,\mathrm{d}\Omega \tag{3.126}$$

in which α is a 'penalty number' and then require the stationarity for the constrained solution. If Π is itself a minimum of the solution then α should be a positive number. The solution obtained by the stationarity of the functional $\bar{\bar{\Pi}}$ will satisfy the constraints only approximately. The larger the value of α the better will be the constraints achieved. Further, it seems obvious that the process is best suited to cases where Π is a minimum (or maximum) principle, but success can be obtained even with purely saddle point problems. The process is equally applicable to constraints applied on boundaries or simple discrete constraints. In this latter case integration is dropped.

3.12.2 Perturbed lagrangian

We consider once again the problem of obtaining stationarity of Π with a set of constraint equations $\mathbf{C}(\mathbf{u}) = \mathbf{0}$ in domain Ω. The Lagrange multiplier form to embed the constraint is given in Eq. (3.96). Here we modify the expression by appending a quadratic term of the form $\boldsymbol{\lambda}^{\mathrm{T}}\boldsymbol{\lambda}$ scaled by a parameter α. The form of the final equation is given by

$$\breve{\Pi}(\mathbf{u}, \boldsymbol{\lambda}) = \Pi(\mathbf{u}) + \int_\Omega \boldsymbol{\lambda}^{\mathrm{T}}\mathbf{C}(\mathbf{u})\,\mathrm{d}\Omega - \frac{1}{2\alpha}\int_\Omega \boldsymbol{\lambda}^{\mathrm{T}}\boldsymbol{\lambda}\,\mathrm{d}\Omega \tag{3.127}$$

We note that as the parameter α tends toward infinity the form approaches a Lagrange multiplier form. Accordingly, this form is called a *perturbed lagrangian* functional. Taking the variation we obtain the result

$$\delta\breve{\Pi} = \delta\Pi + \int_\Omega \boldsymbol{\lambda}^{\mathrm{T}}\delta\mathbf{C}(\mathbf{u})\,\mathrm{d}\Omega + \int_\Omega \delta\boldsymbol{\lambda}^{\mathrm{T}}\mathbf{C}(\mathbf{u})\,\mathrm{d}\Omega - \frac{1}{\alpha}\int_\Omega \delta\boldsymbol{\lambda}^{\mathrm{T}}\boldsymbol{\lambda}\,\mathrm{d}\Omega = 0 \tag{3.128}$$

If the constraints are a linear form given by

$$\mathbf{C}(\mathbf{u}) = \mathbf{C}_0\mathbf{u}$$

we can introduce the approximations (3.100) into (3.128) to obtain the set of equations

$$\begin{bmatrix} \mathbf{K}_{uu} & \mathbf{K}_{u\lambda} \\ \mathbf{K}_{\lambda u} & -\frac{1}{\alpha}\mathbf{K}_{\lambda\lambda} \end{bmatrix} \begin{Bmatrix} \tilde{\mathbf{u}} \\ \tilde{\boldsymbol{\lambda}} \end{Bmatrix} = \begin{Bmatrix} \mathbf{f} \\ \mathbf{0} \end{Bmatrix} \tag{3.129}$$

where $\mathbf{K}_{uu}$ is the coefficient array from $\delta\Pi$ and

$$\mathbf{K}_{u\lambda} = \int_\Omega \bar{\mathbf{N}}^{\mathrm{T}}\mathbf{C}_0\,\mathrm{d}\Omega \quad \text{and} \quad \mathbf{K}_{\lambda\lambda} = \int_\Omega \bar{\mathbf{N}}^{\mathrm{T}}\bar{\mathbf{N}}\,\mathrm{d}\Omega$$

The second equation of (3.129) may be solved for $\tilde{\boldsymbol{\lambda}}$ in terms of $\tilde{\mathbf{u}}$ and substituted into the first equation to obtain

$$\bar{\mathbf{K}}_{uu}\tilde{\mathbf{u}} = \left[\mathbf{K}_{uu} + \alpha\,\mathbf{K}_{u\lambda}\mathbf{K}_{\lambda\lambda}^{-1}\mathbf{K}_{\lambda u}\right]\tilde{\mathbf{u}} = \mathbf{f}$$

It is now apparent that the perturbed lagrangian and penalty forms are closely related. The perturbed lagrangian uses

$$\mathbf{K}_{u\lambda}\mathbf{K}_{\lambda\lambda}^{-1}\mathbf{K}_{\lambda u}$$

to impose the constraint whereas the penalty approach uses

$$\int_\Omega \mathbf{C}_0^{\mathrm{T}}\mathbf{C}_0\,\mathrm{d}\Omega$$

When the constraint is a simple scalar relation the two methods are identical; however, when any other form is considered the methods will yield different approximations unless the shape functions for $\boldsymbol{\lambda}$ include all the terms contained in $\delta\mathbf{C}(\mathbf{u})$.

Example 3.13: Constraint enforcement by penalty method. To clarify ideas let us once again consider the algebraic problem of Sec. 3.11.1, in which the stationarity of a functional given by Eq. (3.109) was sought subject to a constraint. With the penalty function approach we now seek the minimum of a functional

$$\bar{\bar{\Pi}} = 2u_1^2 - 2u_1u_2 + u_2^2 + 18u_1 + 6u_2 + \tfrac{1}{2}\,\alpha\,(u_1 - u_2)^2 \tag{3.130}$$

with respect to the variation of both parameters u_1 and u_2. Writing the two simultaneous equations

$$\frac{\partial\bar{\bar{\Pi}}}{\partial u_1} = 0, \quad \frac{\partial\bar{\bar{\Pi}}}{\partial u_2} = 0$$

we find

$$\begin{bmatrix} (4+\alpha) & -(2+\alpha) \\ -(2+\alpha) & (2+\alpha) \end{bmatrix} \begin{Bmatrix} u_1 \\ u_2 \end{Bmatrix} + \begin{Bmatrix} 18 \\ 6 \end{Bmatrix} = \begin{Bmatrix} 0 \\ 0 \end{Bmatrix} \tag{3.131}$$

and note as α is increased we approach the correct solution. In Table 3.1 the results are set out demonstrating the convergence.

The reader will observe that in a problem formulated in the above manner the constraint introduces no additional unknown parameters – but neither does it decrease their original number. The process will always result in strongly positive definite matrices if the original variational principle is one of a minimum and, similarly, negative definite matrices are obtained for a maximum principle if α is negative.

In practical applications the method of penalty functions has proved to be quite effective,[29] and indeed is often introduced intuitively.

Table 3.1 Convergence of two-term solution

α	1/2	1	3	5	50	500
u_1	−12.000	−12.000	−12.000	−12.000	−12.000	−12.000
u_2	−13.500	−13.000	−12.429	−12.273	−12.030	−12.003

In the example presented next the forced boundary conditions are not introduced *a priori* and the problem gives, on assembly, a singular system of equations

$$\mathbf{K}\tilde{\mathbf{u}} + \mathbf{f} = \mathbf{0} \tag{3.132}$$

which can be obtained from the functional (providing $\mathbf{K}$ is symmetric)

$$\Pi = \tfrac{1}{2}\tilde{\mathbf{u}}^{\mathrm{T}}\mathbf{K}\tilde{\mathbf{u}} + \tilde{\mathbf{u}}^{\mathrm{T}}\mathbf{f} \tag{3.133}$$

Introducing a prescribed value of u_1, i.e., writing

$$u_1 - \bar{u}_1 = 0 \tag{3.134}$$

the functional can be modified to

$$\bar{\bar{\Pi}} = \Pi + \tfrac{1}{2}\alpha(u_1 - \bar{u}_1)^2 \tag{3.135}$$

yielding

$$\bar{\bar{K}}_{11} = K_{11} + \alpha \qquad \bar{\bar{f}}_1 = f_1 - \alpha\bar{u}_1 \tag{3.136}$$

and giving no change in any of the other matrix coefficients. Many applications of such a 'discrete' kind are discussed by Campbell.[30]

It is easy to show in another context[29, 31] that the use of a high Poisson's ratio ($\nu \to 0.5$) for the study of incompressible solids or fluids is in fact equivalent to the introduction of a penalty term to suppress any compressibility allowed by an arbitrary displacement variation.

The use of the penalty function in the finite element context presents certain difficulties. *First*, the constrained functional of Eq. (3.126) leads to equations of the form

$$(\mathbf{K}_1 + \alpha\mathbf{K}_2)\tilde{\mathbf{u}} + \bar{\bar{\mathbf{f}}} = \mathbf{0} \tag{3.137}$$

where $\mathbf{K}_1$ derives from the original functional and $\mathbf{K}_2$ from the constraints. As α increases the above equation degenerates to:

$$\mathbf{K}_2\tilde{\mathbf{u}} = -\mathbf{f}/\alpha \to \mathbf{0}$$

and $\tilde{\mathbf{u}} = \mathbf{0}$ unless the matrix $\mathbf{K}_2$ is singular. The phenomenon where $\tilde{\mathbf{u}} \Rightarrow \mathbf{0}$ is known as *locking* and has often been encountered by researchers who failed to recognize its source. This singularity in the equations does not always arise and we shall discuss means of its introduction in Chapters 10 and 11.

Second, with large but finite values of α numerical difficulties will be encountered. Noting that discretization errors can be of comparable magnitude to those due to not *satisfying* the constraint, we can make

$$\alpha = \text{constant}(1/h)^n$$

ensuring a limiting convergence to the correct answer. Fried[32, 33] discusses this problem in detail.

A more general discussion of the whole topic is given in reference 34 and in Chapter 11 where the relationship between Lagrange constraints and penalty forms is made clear.

3.13 Least squares approximations

A general variational principle also may be constructed if the constraints described in the previous section are simply the governing equations of the problem

$$\mathbf{C}(\mathbf{u}) = \mathbf{A}(\mathbf{u}) \tag{3.138}$$

Obviously the same procedure can be used in the context of the penalty function approach by setting $\Pi = 0$ in Eq. (3.126). We can thus write a 'variational principle'

$$\bar{\bar{\Pi}} = \tfrac{1}{2}\int_\Omega (A_1^2 + A_2^2 + \cdots)\,\mathrm{d}\Omega = \tfrac{1}{2}\int_\Omega \boldsymbol{\mathcal{A}}^\mathrm{T}(\mathbf{u})\boldsymbol{\mathcal{A}}(\mathbf{u})\,\mathrm{d}\Omega \tag{3.139}$$

for any set of differential equations. In the above equation the boundary conditions are assumed to be satisfied by $\mathbf{u}$ (forced boundary condition) and the parameter α is dropped as it becomes a multiplier.

Clearly, the above statement is a requirement that the sum of the squares of the residuals of the differential equations should be a minimum at the correct solution. This minimum is obviously zero at that point, and the process is simply the well-known *least squares method* of approximation.

It is equally obvious that we could obtain the correct solution by minimizing any functional of the form

$$\bar{\bar{\Pi}} = \tfrac{1}{2}\int_\Omega (p_1 A_1^2 + p_2 A_2^2 + \cdots)\,\mathrm{d}\Omega = \tfrac{1}{2}\int_\Omega \boldsymbol{\mathcal{A}}^\mathrm{T}(\mathbf{u})\mathbf{p}\boldsymbol{\mathcal{A}}(\mathbf{u})\,\mathrm{d}\Omega \tag{3.140}$$

in which $p_1, p_2, \ldots$, etc., are positive valued weighting functions or constants and $\mathbf{p}$ is a diagonal matrix:

$$\mathbf{p} = \begin{bmatrix} p_1 & & & 0 \\ & p_2 & & \\ & & p_3 & \\ 0 & & & \ddots \end{bmatrix} \tag{3.141}$$

The above alternative form is sometimes convenient as it puts different importance on the satisfaction of individual components of the equation set and allows additional freedom in the choice of the approximate solution. Once again this weighting function could be chosen so as to ensure a constant ratio of terms contributed by various equations.

A least squares method of the kind shown above is a very powerful alternative procedure for obtaining integral forms from which an approximate solution can be started, and has been used with considerable success.[35, 36] As a least squares variational principle can be written for *any* set of differential equations without introducing additional variables, we may well enquire what is the difference between these and the *natural variational principles* discussed previously. On performing a variation in a specific case the reader will find that the Euler equations which are obtained no longer give the original differential equations but give higher order derivatives of these. This introduces the possibility of spurious solutions if incorrect boundary conditions are used. Further, higher order continuity of trial functions is now generally needed. This may be a serious drawback but frequently can be by-passed by stating the original problem as a set of lower order equations.

We shall now consider the general form of discretized equations resulting from the least squares approximation for linear equation sets (again neglecting boundary conditions which are assumed forced). Thus, if we take

$$\mathcal{A}(\mathbf{u}) = \mathcal{L}\mathbf{u} + \mathbf{b} \tag{3.142}$$

and take the usual trial function approximation

$$\hat{\mathbf{u}} = \mathbf{N}\tilde{\mathbf{u}} \tag{3.143}$$

we can write, substituting into (3.140),

$$\bar{\bar{\Pi}} = \tfrac{1}{2}\int_\Omega [(\mathcal{L}\mathbf{N})\tilde{\mathbf{u}} + \mathbf{b}]^{\mathrm{T}}\mathbf{p}[(\mathcal{L}\mathbf{N})\tilde{\mathbf{u}} + \mathbf{b}]\,\mathrm{d}\Omega \tag{3.144}$$

and obtain

$$\delta\bar{\bar{\Pi}} = \tfrac{1}{2}\int_\Omega \delta\tilde{\mathbf{u}}^{\mathrm{T}}(\mathcal{L}\mathbf{N})^{\mathrm{T}}\mathbf{p}[(\mathcal{L}\mathbf{N})\tilde{\mathbf{u}}+\mathbf{b}]\,\mathrm{d}\Omega + \tfrac{1}{2}\int_\Omega [(\mathcal{L}\mathbf{N})\tilde{\mathbf{u}}+\mathbf{b}]^{\mathrm{T}}\mathbf{p}(\mathcal{L}\mathbf{N})\,\delta\tilde{\mathbf{u}}\,\mathrm{d}\Omega = 0 \tag{3.145}$$

or, as $\mathbf{p}$ is symmetric,

$$\delta\bar{\bar{\Pi}} = \delta\tilde{\mathbf{u}}^{\mathrm{T}}\left\{\left[\int_\Omega (\mathcal{L}\mathbf{N})^{\mathrm{T}}\mathbf{p}(\mathcal{L}\mathbf{N})\,\mathrm{d}\Omega\right]\tilde{\mathbf{u}} + \int_\Omega (\mathcal{L}\mathbf{N})^{\mathrm{T}}\mathbf{p}\mathbf{b}\,\mathrm{d}\Omega\right\} = 0 \tag{3.146}$$

This immediately yields the approximation equation in the usual form:

$$\mathbf{K}\tilde{\mathbf{u}} + \mathbf{f} = \mathbf{0} \tag{3.147}$$

and the reader can observe that the matrix $\mathbf{K}$ is symmetric and positive definite.

Example 3.14: Least squares solution for Helmholz equation. To illustrate an actual example, consider the Helmholz problem governed by Eq. (3.88) for which we have already obtained a *natural* variational principle [Eq. (3.91)] in which only first derivatives were involved requiring C_0 continuity for $\mathbf{u}$. Now, if we use the operator $\mathcal{L}$ and term $\mathbf{b}$ defined by Eq. (3.89), we have a set of approximating equations with

$$\begin{aligned} K_{ab} &= \int_\Omega (\nabla^2 N_a + cN_a)(\nabla^2 N_b + cN_b)\,\mathrm{d}x\,\mathrm{d}y \\ f_a &= \int_\Omega (\nabla^2 N_a + cN_a)Q\,\mathrm{d}x\,\mathrm{d}y \end{aligned} \tag{3.148}$$

The reader will observe that due to the presence of second derivatives C_1 continuity is now needed for the trial functions $\mathbf{N}$.

Example 3.15: Least squares solution for Helmholz equation in first-order form. An alternative, avoiding the requirement of C_1 functions, is to write Eq. (3.88) as a first-order system. This can be written as

$$\mathbf{A}(\mathbf{u}) = \left\{\begin{matrix} \dfrac{\partial q_x}{\partial x} + \dfrac{\partial q_y}{\partial y} + c\phi + Q \\ \dfrac{\partial \phi}{\partial x} - q_x \\ \dfrac{\partial \phi}{\partial y} - q_y \end{matrix}\right\} = 0 \tag{3.149}$$

or, introducing the vector **u**,

$$\mathbf{u} = [\phi, q_x, q_y]^{\mathrm{T}} = \mathbf{N}\tilde{\mathbf{u}} \tag{3.150}$$

as the unknown we can write an approximation as

$$\mathbf{u} \approx \hat{\mathbf{u}} = \begin{bmatrix} \mathbf{N}_\phi & \mathbf{0} & \mathbf{0} \\ \mathbf{0} & \mathbf{N}_q & \mathbf{0} \\ \mathbf{0} & \mathbf{0} & \mathbf{N}_q \end{bmatrix} \begin{Bmatrix} \tilde{\phi} \\ \tilde{\mathbf{q}}_x \\ \tilde{\mathbf{q}}_y \end{Bmatrix} = \mathbf{N}\tilde{\mathbf{u}} \tag{3.151}$$

where $\mathbf{N}_\phi$ and $\mathbf{N}_q$ are C_0 shape functions for the ϕ and q_x, q_y variables, respectively. The least squares approximation is now given by

$$\delta\bar{\bar{\Pi}} = \delta\tilde{\mathbf{u}}^{\mathrm{T}} \int_\Omega (\mathcal{L}\mathbf{N})^{\mathrm{T}} \left[(\mathcal{L}\mathbf{N})\tilde{\mathbf{u}} + \mathbf{b}\right] \mathrm{d}\Omega = 0 \tag{3.152a}$$

where

$$\mathcal{L}\mathbf{N} = \begin{bmatrix} c\,\mathbf{N}_\phi, & \dfrac{\partial \mathbf{N_q}}{\partial x}, & \dfrac{\partial \mathbf{N_q}}{\partial y} \\ \dfrac{\partial \mathbf{N}_\phi}{\partial x}, & -\mathbf{N_q}, & \mathbf{0} \\ \dfrac{\partial \mathbf{N}_\phi}{\partial y}, & \mathbf{0}, & -\mathbf{N_q} \end{bmatrix} \qquad \mathbf{b} = \begin{Bmatrix} Q \\ 0 \\ 0 \end{Bmatrix} \tag{3.152b}$$

The reader can now perform the final steps to obtain the **K** and **f** matrices. The approximation equations in a form requiring only C_0 continuity are obtained, however, at the expense of additional variables. Use of such forms has been made extensively in the finite element context.[35–41]

3.13.1 Galerkin least squares, stabilization

It is interesting to note that the concept of penalty formulation introduced in the previous section was anticipated as early as 1943 by Courant[42] in a somewhat different manner. He used the original variational principle augmented by the differential equations of the problem employed as least squares constraints. In this manner he claimed, though never proved, that the convergence rate could be accelerated.

The suggestion put forward by Courant has been used effectively by others though in a somewhat different manner. Noting that the Galerkin process is, for self-adjoint equations, equivalent to that of minimizing a functional, the least squares formulation using the original equation is simply added to the Galerkin form. Here it allows non-self-adjoint operators to be used, for instance, and this feature has been exploited with success. Consider, for instance, an equation of the form

$$\frac{\mathrm{d}^2\phi}{\mathrm{d}x^2} + \alpha\frac{\mathrm{d}\phi}{\mathrm{d}x} + Q = 0$$

The first order term multiplying α is a *convective* term and, due to its presence, no natural variational equation is available as the differential equation is non-self-adjoint. However, Galerkin methods have been successfully used in its solution providing the convection term $(\alpha\mathrm{d}\phi/\mathrm{d}x)$ remains relatively small compared to the second derivative term (the diffusion term). However, it is found that as the convection term increases the solution becomes highly oscillatory. Here we only consider the problem in a preliminary manner and refer the reader to references on fluid dynamics for further study (e.g., see reference 10). Suppose in a Galerkin form given by

$$\int_\Omega \left\{ \frac{dv}{dx}\frac{d\phi}{dx} - v\left(\alpha \frac{d\phi}{dx} + Q\right)\right\} dx = 0 \tag{3.153}$$

we add a multiple of the minimization of the least squares of the total equation. The result is

$$\begin{aligned}\int_\Omega &\left\{ \frac{dv}{dx}\frac{d\phi}{dx} - v\left(\alpha \frac{d\phi}{dx} + Q\right)\right\} dx \\ &+ \int_\Omega \left(\frac{d^2 v}{dx^2} + \alpha \frac{dv}{dx}\right) \tau \left(\frac{d^2\phi}{dx^2} + \alpha \frac{d\phi}{dx} + Q\right) dx = 0\end{aligned} \tag{3.154}$$

and we see immediately that an additional diffusive term has been added which depends on the parameter τ, though at the expense of having higher derivatives appearing in the integrals. If only linear elements are used and the discontinuities ignored at element interfaces, the process of adding the diffusive terms can *stabilize* the oscillations which would otherwise occur. The idea appears to have first been used by Hughes[43–45] and later studied by Codina.[46] This process in the view of the authors is somewhat unorthodox as discontinuity of derivatives is ignored, and alternatives to this are discussed at length in reference 10.

It is interesting to note also that another application of the same Galerkin least squares process can be made to the mixed formulation with two variables **u** and p for incompressible problems. We shall discuss such problems in Chapter 11 of this volume and show how this process can be made applicable there.

Finally, it is of interest to note that the simple procedure introduced by Courant can also be effective in the prevention of locking of other problems. The treatment for beams has been studied by Freund and Salonen[47] and it appears that quite an effective process can be reached.

3.14 Concluding remarks – finite difference and boundary methods

This very extensive chapter presents the general possibilities of using the finite element process in almost any mathematical or mathematically modelled physical problem. The essential approximation processes have been given in as simple a form as possible, at the same time presenting a fully comprehensive picture which should allow the reader to understand much of the literature and indeed to experiment with new permutations. In the chapters that follow we shall apply to various physical problems a limited selection of the methods to which allusion has been made. In some we shall show, however, that certain extensions of the process are possible (Chapters 11 and 15) and in another (Chapter 9) how a violation of some of the rules here expounded can be accepted.

The numerous approximation procedures discussed fall into several categories. To remind the reader of these, we present in Table 3.2 a comprehensive catalogue of the methods used here and in Chapter 2. The only aspect of the finite element process mentioned in

Table 3.2 Finite element approximation

Integral forms of continuum problems trial functions $\mathbf{u} = \sum_a \mathbf{N}_a \tilde{\mathbf{u}}_a$

Direct physical model

Variational principles

Weighted integrals of partial differential equation governing (weak formulations)

Global physical statements (e.g. virtual work)

Meaningful physical principles

Constrained lagrangian forms

Penalty function forms

Least square forms

Miscellaneous weight functions

Collocation (point or subdomain)

Galerkin ($\mathbf{W}_b = \mathbf{N}_b$)

this table that has not been discussed here is that of a *direct physical method.* In such models an 'atomic' rather than continuum concept is the starting point. While much interest exists in the possibilities offered by such models, their discussion is outside the scope of this book.

In all the continuum processes discussed the first step is always the choice of suitable shape or trial functions. A few simple forms of such functions have been introduced as the need demanded and many new forms will be introduced in the next two chapters. Indeed, the reader who has mastered the essence of the present chapter will have little difficulty in applying the finite element method to any suitably defined physical problem. For further reading references 48–52 could be consulted.

The methods listed do not include specifically two well-known techniques, i.e., *finite difference* methods and *boundary solution* methods (sometimes known as boundary elements). In the general sense these belong under the category of the *generalized finite element* method discussed here.[48]

1. Boundary solution methods choose the trial functions such that the governing equation is automatically satisfied in the domain Ω. Thus starting from the general approximation equation (3.22), we note that only boundary terms remain to be satisfied. We shall return to such approximations in Chapter 12.
2. Finite difference procedures can be interpreted as an approximation based on local, discontinuous, shape functions with collocation weighting applied (although usually the derivation of the approximation algorithm is based on a Taylor expansion).

   As Galerkin or variational approaches give, in the energy sense, the best approximation, this method has only the merit of computational simplicity and occasionally a loss of accuracy.

To illustrate this process we recall the approximation carried out for the one-dimensional equation (3.24a) (viz. p. 62). We now represent a localized approximation through equally spaced nodal points by

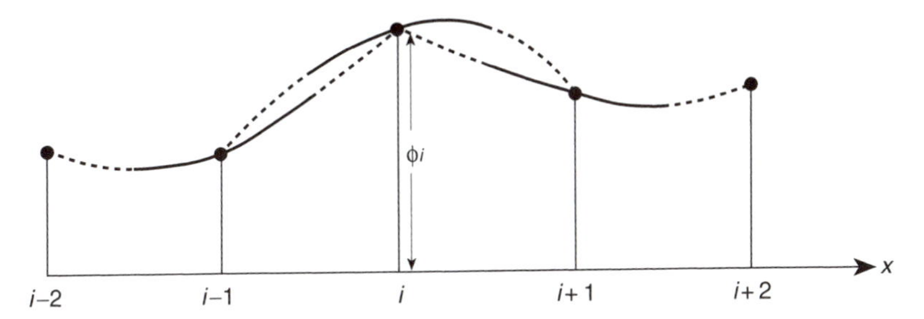

Fig. 3.11 A local, discontinuous shape function by parabolic segments used to obtain a finite difference approximation.

$$\phi(x) = \left[\frac{1}{2}\left(\frac{(x-x_a)^2}{h^2} - \frac{x-x_a}{h}\right), \left(1 - \frac{(x-x_a)^2}{h^2}\right), \frac{1}{2}\left(\frac{(x-x_a)^2}{h^2} + \frac{x-x_a}{h}\right)\right] \times \begin{Bmatrix} \tilde{\phi}_{a-1} \\ \tilde{\phi}_a \\ \tilde{\phi}_{a+1} \end{Bmatrix} \tag{3.155}$$

where $h = x_{a+1} - x_a$ (shown in Fig. 3.11). It is now clear that adjacent parabolic approximations in this case are discontinuous between the nodes. Values of the function and its first two derivatives at a typical node i are given by

$$\begin{aligned} \phi(x_a) &= \tilde{\phi}_a \\ \left.\frac{\partial \phi}{\partial x}\right|_{x=x_a} &= \frac{1}{2h}(\tilde{\phi}_{a+1} - \tilde{\phi}_{a-1}) \\ \left.\frac{\partial^2 \phi}{\partial x^2}\right|_{x=x_a} &= \frac{1}{h^2}(\tilde{\phi}_{a+1} - 2\tilde{\phi}_a + \tilde{\phi}_{a-1}) \end{aligned} \tag{3.156}$$

If we insert these into the governing equation at node i, we note immediately that the approximating equation at the node becomes

$$-\frac{1}{h^2}(\tilde{\phi}_{a-1} - 2\tilde{\phi}_a + \tilde{\phi}_{a+1}) + Q_a = 0 \tag{3.157}$$

This is identical to the result based on Taylor expansion given by Eq. (3.31). This is indeed one of the cases in which the finite difference approximation is identical to the finite element one rather than different. In Chapter 15 we shall be discussing such finite difference and point approximations in more detail. However, the reader will note the present exercise is simply given to underline the similarity of finite element and finite difference processes.

Many textbooks deal exclusively with these types of approximations. References 8, 53–55 discuss finite difference approximation and references 56–59 relate to boundary methods.

3.15 Problems

3.1 Write weak forms for the following differential equations and boundary conditions. For each form state appropriate continuity conditions for approximations to the dependent variable u and the weighting function v. The domain for each one-dimensional differential equation is $0 < x < 1$.

(a) $a\dfrac{\mathrm{d}u}{\mathrm{d}x} + cu + q = 0; \quad u(0) = \bar{g}$

(b) $\dfrac{\mathrm{d}}{\mathrm{d}x}\left(a\dfrac{\mathrm{d}u}{\mathrm{d}x}\right) + q = 0; \quad u(0) = \bar{g}_0 \;\&\; a\dfrac{\mathrm{d}u}{\mathrm{d}x} + ku = \bar{g}_1 \text{ at } x = 1$

(c) $-\dfrac{\mathrm{d}}{\mathrm{d}x}\left(a\dfrac{\mathrm{d}u}{\mathrm{d}x}\right) + b\dfrac{\mathrm{d}u}{\mathrm{d}x} + q = 0; \quad u(0) = \bar{g}_0; \; u(1) = \bar{g}_2$

(d) $\dfrac{\mathrm{d}}{\mathrm{d}x}\left(a\dfrac{\mathrm{d}^2u}{\mathrm{d}x^2}\right) + f = 0; \quad u(0) = \bar{g}_0; \; \left.\dfrac{\mathrm{d}u}{\mathrm{d}x}\right|_{x=0} = \bar{h}_0 \;\&\; u(1) = \bar{g}_1$

(e) $-\boldsymbol{\nabla}^{\mathrm{T}}(\mathbf{k}\boldsymbol{\nabla}u) + c\,\mathbf{b}^{\mathrm{T}}(\boldsymbol{\nabla}u) + q = 0 \text{ in } \Omega; \quad u = \bar{g} \text{ on } \Gamma$

The differential equations for bending of a beam are given by

$$(1)\;\frac{\mathrm{d}V}{\mathrm{d}x} + q = 0 \qquad (2)\;\frac{\mathrm{d}M}{\mathrm{d}x} + V = 0$$

$$(3)\;\frac{\mathrm{d}\theta}{\mathrm{d}x} - \frac{M}{EI} = 0 \quad (4)\;\frac{\mathrm{d}w}{\mathrm{d}x} - \theta - \frac{V}{GA} = 0$$

in which V is shear force, M is moment, θ is section rotation, w is displacement, EI is bending stiffness, GA is shear stiffness and q is load as shown in Fig. 3.12. Boundary conditions are given by

$$(1)\; V = \bar{V} \quad \text{or} \quad w = \bar{w}$$

$$(2)\; M = \bar{M} \quad \text{or} \quad \theta = \bar{\theta}$$

3.2 Construct a weak form for the beam equations by multiplying (1) by $\pm\delta w$, (2) by $\pm\delta\theta$, (3) by δM and (4) by δV.

Choose the correct sign for δw and $\delta\theta$ to give symmetry.

3.3 Add all boundary conditions to the weak form obtained in Problem 3.2.

3.4 Construct a variational theorem which gives the weak form obtained in Problems 3.2 and 3.3 as the first variation.

3.5 For $GA = \infty$ (no shear deformation) deduce the irreducible differential equation in terms of w. Express all boundary conditions in terms of w.

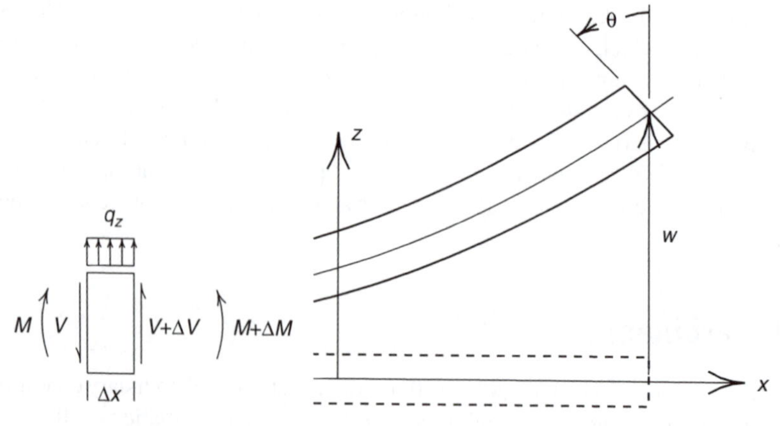

Fig. 3.12 Beam bending description.

3.6 Construct a weak form for Problem 3.5. What is the required continuity of the dependent variable needed for approximation by a finite element method? What are the natural and essential boundary conditions for the weak form?

3.7 Construct a variational theorem which has Problem 3.6 as its first variation.

3.8 For $GA = \infty$ (no shear deformation) deduce the differential equations in terms of w and M. Express all boundary conditions in terms of these variables.

3.9 Deduce a weak form for Problem 3.8 that permits approximation using C_0 functions to approximate w and M. Let

$$w = \sum_{a=1}^{2} N_a \tilde{w}_a \quad \text{and} \quad M = \sum_{a=1}^{2} N_a \tilde{M}_a$$

where N_a are given by (3.28). Ensure your weak form gives a symmetric coefficient matrix for these approximations.

Compute typical element matrices **K** and **f** for an element of length h with constant EI and q in the element.

3.10 For a simply supported beam of length 10 and constant cross-section $EI = 3$ compute the solution for a uniform load of $q = 1$. The boundary conditions at each end of the beam for a simple support are $w = M = 0$. Obtain a solution using 2, 4, and 8 elements. It is recommended that a small computer program be written using a high level language, e.g. MATLAB,[60] to perform the numerical calculations. Compare your results to an exact solution.

3.11 Solve the one-dimensional heat equation given in Example 3.5 by enforcing the boundary conditions by the penalty formulation described in Sec. 3.12.1. How large must each penalty parameter be taken to make the boundary error less than $10^{-6}|\phi_{max}|$?

3.12 Deduce the Euler differential equation and boundary conditions for the variational principle expressed as

$$\Pi(u) = \int_a^b \left[EI\,(\frac{\mathrm{d}u}{\mathrm{d}x})^2 - P\,u^2 \right] \mathrm{d}x - ug\Big|_{x=b} \; ; \quad u(a) = 0$$

Classify Π as a minimum, maximum or saddle point form.

3.13 Deduce the Euler differential equation and boundary conditions for the variational principle expressed as

$$\Pi(u) = \int_a^b \left[EA\,(\frac{\mathrm{d}u}{\mathrm{d}x})^2 + ku^2 - 2\,qu \right] \mathrm{d}x + \alpha \left[(u(a))^2 + (u(b))^2 \right]$$

where EA and k are constant parameters and α is a penalty parameter.

3.14 Deduce the Euler equations and boundary conditions for the variational principle expressed as

$$\Pi(u, \lambda_a, \lambda_b) = \int_a^b \left[EA\,(\frac{\mathrm{d}u}{\mathrm{d}x})^2 + ku^2 - 2\,qu \right] \mathrm{d}x + \lambda_a u(a) + \lambda_b u(b)$$

where EA, k and q are constant parameters and λ_a, λ_b are Lagrange multipliers.

3.15 The transient heat equation in one dimension is given by

$$-\frac{\partial}{\partial x}\left(k\,\frac{\partial \phi}{\partial x} \right) + Q + c\,\frac{\partial \phi}{\partial t} = 0$$

where ϕ is temperature, k thermal conductivity, Q heat generation per unit length and c specific heat.

Boundary conditions may be given as

$$\phi = \bar{\phi} \text{ on } \Gamma_1 \quad \text{or} \quad q = -k\,\frac{\partial \phi}{\partial x} = \bar{q} \text{ on } \Gamma_2$$

where q is the heat flux and $\bar{\phi}$, $\bar{q}$ are specified values. Initial conditions are given as $\phi(x, 0) = \bar{\phi}_0(x)$.

(a) Construct a weak form for the problem.

(b) Using the shape functions given in Eq. (3.28) and the approximation

$$u^e = N_1(x)\tilde{u}_1(t) + N_2(x)\tilde{u}_2(t)$$
$$\delta u^e = N_1(x)\delta\tilde{u}_1 + N_2(x)\delta\tilde{u}_2$$

construct the semi-discrete form for a typical element of length h.

(c) Consider a region of length 10, with properties $k = 5$, $c = 1$, $Q = 0$. Divide the region into four equal length elements and establish the set of global semi-discrete equations.

(d) Consider a set of discrete times t_n. Approximate time derivatives of nodal values by $\mathrm{d}\phi/\mathrm{d}t(t_n) \approx (\phi_n - \phi_{n-1})/\Delta t$ where ϕ_n is the approximation to $\phi(t_n)$ and $\Delta t = t_n - t_{n-1}$ and write the fully discrete equations.

Write a computer program (e.g., using MATLAB) to solve the problem. Assume the initial temperature of the region is zero and boundary conditions $\phi(0) = 0$ and $\phi(10) = 1$ are applied at time zero and held constant. Solve the problem using 10 steps with $\Delta t = 0.01$, followed by 9 steps with $\Delta t = 0.1$ and finally 9 steps with $\Delta t = 1$. Plot the finite element solution for ϕ vs x at times 0.01, 0.1, 1.0 and 10.0.

Replace the element matrix associated with c by a diagonal (lumped) form with $ch/2$ on each diagonal ($h = x_2^e - x_1^e$). Repeat the above solution and compare results with the consistent form for the matrix.

References

1. S. Crandall. *Engineering Analysis*. McGraw-Hill, New York, 1956.
2. B.A. Finlayson. *The Method of Weighted Residuals and Variational Principles*. Academic Press, New York, 1972.
3. R.A. Frazer, W.P. Jones, and S.W. Sken. Approximations to functions and to the solution of differential equations. Technical Report 1799, Aero. Research Committee Report, 1937.
4. C.B. Biezeno and R. Grammel. *Technishe Dynamik*. Springer-Verlag, Berlin, 1933, p. 142.
5. O.C. Zienkiewicz and E. Oñate. Finite elements versus finite volumes. Is there a choice? In P. Wriggers and W. Wagner, editors, *Nonlinear Computational Mechanics. State of the Art*. Springer, Berlin, 1991.
6. B.G. Galerkin. Series solution of some problems in elastic equilibrium of rods and plates. *Vestn. Inzh. Tech.*, 19:897–908, 1915.
7. P. Tong. Exact solution of certain problems by the finite element method. *J. AIAA*, 7:179–180, 1969.
8. R.V. Southwell. *Relaxation Methods in Theoretical Physics*. Clarendon Press, Oxford, 1st edition, 1946.

9. R.S. Varga. *Matrix Iterative Analysis*. Prentice-Hall, Englewood Cliffs, N.J., 1962.
10. O.C. Zienkiewicz, R.L. Taylor, and P. Nithiarasu. *The Finite Element Method for Fluid Dynamics*. Butterworth-Heinemann, Oxford, 6th edition, 2005.
11. S.P. Timoshenko and J.N. Goodier. *Theory of Elasticity*. McGraw-Hill, New York, 3rd edition, 1969.
12. I.S. Sokolnikoff. *The Mathematical Theory of Elasticity*. McGraw-Hill, New York, 2nd edition, 1956.
13. L.V. Kantorovich and V.I. Krylov. *Approximate Methods of Higher Analysis*. John Wiley & Sons (International), New York, 1964. English translation by Curtis D. Benster.
14. O.C. Zienkiewicz and R.L. Taylor. *The Finite Element Method for Solid and Structural Mechanics*. Butterworth-Heinemann, Oxford, 6th edition, 2005.
15. F.B. Hildebrand. *Methods of Applied Mathematics*. Prentice-Hall (reprinted by Dover Publishers, 1992), 2nd edition, 1965.
16. Lord Rayleigh (J.W. Strutt). On the theory of resonance. *Trans. Roy. Soc. (London)*, A161:77–118, 1870.
17. W. Ritz. Über eine neue Methode zur Lösung gewisser variationsproblem der mathematischen physik. *J. Reine angew. Math.*, 135:1–61, 1908.
18. E. Tonti. Variational formulation of non-linear differential equations. *Bull. Acad. Roy. Belg. Classe Sci.*, 55:137–165 & 263–278, 1969.
19. M.M. Vainberg. *Variational Methods for the Study of Nonlinear Operators*. Holden-Day Inc., San Francisco, CA, 1964.
20. J.T. Oden. A general theory of finite elements. Part I. Topological considerations. *Int. J. Numer. Meth. Eng.*, 1:205–246, 1969.
21. J.T. Oden. A general theory of finite elements. Part II. Applications. *Int. J. Numer. Meth. Eng.*, 1:247–254, 1969.
22. S.C. Mikhlin. *Variational Methods in Mathematical Physics*. Macmillan, New York, 1964.
23. S.C. Mikhlin. *The Problem of the Minimum of a Quadratic Functional*. Holden-Day, San Francisco, 1966.
24. B.A. Szabo and T. Kassos. Linear equation constraints in finite element approximations. *Int. J. Numer. Meth. Eng.*, 9:563–580, 1975.
25. K. Washizu. *Variational Methods in Elasticity and Plasticity*. Pergamon Press, New York, 3rd edition, 1982.
26. F. Kikuchi and Y. Ando. A new variational functional for the finite element method and its application to plate and shell problems. *Nucl. Eng. Des.*, 21(1):95–113, 1972.
27. H.S. Chen and C.C. Mei. Oscillations and water forces in an offshore harbour. Technical Report 190, Ralph M. Parsons Laboratory for Water Resources and Hydrodynamics, Massachusetts Institute of Technology, Cambridge, MA, 1974.
28. O.C. Zienkiewicz, D.W. Kelley, and P. Bettess. The coupling of the finite element and boundary solution procedures. *Int. J. Numer. Meth. Eng.*, 11:355–375, 1977.
29. O.C. Zienkiewicz. Constrained variational principles and penalty function methods in finite element analysis. In *Lecture Notes in Mathematics*, No. 363, pages 207–214, Springer-Verlag, Berlin, 1974.
30. J. Campbell. *A finite element system for analysis and design*. Ph.D. thesis, Department of Civil Engineering, University of Wales, Swansea, 1974.
31. D.J. Naylor. Stresses in nearly incompressible materials for finite elements with application to the calculation of excess pore pressures. *Int. J. Numer. Meth. Eng.*, 8:443–460, 1974.
32. I. Fried. Shear in c^0 and c^1 bending finite elements. *Int. J. Solids Struct.*, 9:449–460, 1973.
33. I. Fried. Finite element analysis of incompressible materials by residual energy balancing. *Int. J. Solids Struct.*, 10:993–1002, 1974.
34. O.C. Zienkiewicz and E. Hinton. Reduced integration, function smoothing and non-conformity in finite element analysis. *J. Franklin Inst.*, 302:443–461, 1976.

35. P.P. Lynn and S.K. Arya. Finite elements formulation by the weighted discrete least squares method. *Int. J. Numer. Meth. Eng.*, 8:71–90, 1974.
36. O.C. Zienkiewicz, D.R.J. Owen, and K.N. Lee. Least square finite element for elasto-static problems – use of reduced integration. *Int. J. Numer. Meth. Eng.*, 8:341–358, 1974.
37. B.-N. Jiang. Optimal least-squares finite element method for elliptic problems. *Comp. Meth. Appl. Mech. Eng.*, 102:199–212, 1993.
38. B.-N. Jiang. On the least-squares method. *Comp. Meth. Appl. Mech. and Eng.*, 152:239–257, 1998.
39. B.-N. Jiang. *Least Squares Finite Element Method: Theory and Applications in Computational Fluid Dynamics and Electromagnetics*. Springer, New York, 1998.
40. B.-N. Jiang. The least-squares finite element method in elasticity. I. Plane stress or strain with drilling degrees of freedom. *Int. J. Numer. Meth. Eng.*, 53:621–636, 2002.
41. B.-N. Jiang. The least-squares finite element method in elasticity. II. Bending of thin plates. *Int. J. Numer. Meth. Eng.*, 54:1459–1475, 2002.
42. R. Courant. Variational methods for the solution of problems of equilibrium and vibration. *Bull. Am. Math Soc.*, 49:1–61, 1943.
43. T.J.R. Hughes, L.P. Franca, and M. Balestra. A new finite element formulation for computational fluid dynamics: V. Circumventing the Babuška-Brezzi condition: a stable Petrov-Galerkin formulation of the Stokes problem accommodating equal-order interpolations. *Comp. Meth. Appl. Mech. Eng.*, 59:85–99, 1986.
44. T.J.R. Hughes and L.P. Franca. A new finite element formulation for computational fluid dynamics: VII. The Stokes problem with various well-posed boundary conditions: symmetric formulation that converge for all velocity/pressure spaces. *Comp. Meth. Appl. Mech. Eng.*, 65:85–96, 1987.
45. T.J.R. Hughes, L.P. Franca, and G.M. Hulbert. A new finite element formulation for computational fluid dynamics: VIII. The Galerkin/least-squares method for advective-diffusive equations. *Comp. Meth. Appl. Mech. Eng.*, 73:173–189, 1989.
46. R. Codina, M. Vázquez, and O.C. Zienkiewicz. General algorithm for compressible and incompressible flows, Part III – a semi-implicit form. *Int. J. Numer. Meth. Fluids*, 27:13–32, 1998.
47. Jouni Freund and Eero-Matti Salonen. Sensitizing according to Courant the Timoshenko beam finite element solution. *Int. J. Numer. Meth. Eng.*, x:129–160, 1999.
48. O.C. Zienkiewicz and K. Morgan. *Finite Elements and Approximation*. John Wiley & Sons, London, 1983.
49. E.B. Becker, G.F. Carey, and J.T. Oden. *Finite Elements: An Introduction*, volume 1. Prentice-Hall, Englewood Cliffs, N.J., 1981.
50. B. Szabo and I. Babuška. *Finite Element Analysis*. John Wiley & Sons, New York, 1991.
51. T.J.R. Hughes. *The Finite Element Method: Linear Static and Dynamic Analysis*. Dover Publications, New York, 2000.
52. C.A.T. Fletcher. *Computational Galerkin Methods*. Springer-Verlag, Berlin, 1984.
53. D.N. de G. Allen. *Relaxation Methods*. McGraw-Hill, London, 1955.
54. F.B. Hildebrand. *Introduction to Numerical Analysis*. Dover Publishers, 2nd edition, 1987.
55. A.R. Mitchell and D. Griffiths. *The Finite Difference Method in Partial Differential Equations*. John Wiley & Sons, London, 1980.
56. P.K. Banerjee. *The Boundary Element Methods in Engineering*. McGraw-Hill, London, 1994.
57. Prem K. Kythe. *An Introduction to Boundary Element Methods*. CRC Press, 1995.
58. G. Beer and J.O. Watson. *Programming the Boundary Element Method: An Introduction for Engineers*. John Wiley & Sons, Chichester, 2001.
59. L. Gaul. *Boundary Element Methods for Engineers and Scientists*. Springer, Berlin, 2003.
60. MATLAB. www.mathworks.com, 2003.

4

'Standard' and 'hierarchical' element shape functions: some general families of C_0 continuity

4.1 Introduction

In Chapters 2 and 3 the reader was shown in some detail how linear elasticity and other problems could be formulated and solved using very simple element forms. Although the detailed algebra was only concerned with shape functions which arose from triangular or rectangular shapes, it should by now be obvious that other element forms could equally well be used. Indeed, once the element and the corresponding shape functions are determined, subsequent operations follow a standard, well-defined path. It will be seen later that it is possible to program a computer to deal with wide classes of problems by specifying the shape functions only. The choice of these is, however, a matter to which intelligence has to be applied and in which the human factor remains paramount. In this chapter some rules for the generation of several families of one-, two-, and three-dimensional elements will be presented.

In the problems of elasticity illustrated in Chapters 2 and 3 the displacement variable was a vector with two or three components and the shape functions were written in matrix form. They were, however, derived for each component separately and the matrix expressions in these were derived by multiplying a scalar function by an identity matrix [e.g., Eq. (2.2)]. In this chapter we shall concentrate on the scalar shape function forms, calling these simply N_a.

The shape functions used in the displacement formulation of elasticity problems were such that they satisfy the convergence criteria of Chapter 2:

1. The continuity of the *unknown only* had to occur between elements (i.e., slope continuity is not required), or, in mathematical notation, C_0 continuity was needed;
2. The function has to allow any arbitrary linear form to be taken so that the constant strain (constant first derivative) criterion could be observed in each element.

The shape functions described in this chapter will require the satisfaction of these two criteria. They will thus be applicable to all the problems requiring C_0 continuity (i.e., all problems governed by first or second order differential equations). Indeed they are applicable to any situation where the functional Π or $\delta\Pi$ (see Chapter 3) is defined by derivatives of first order only.

The element families discussed will progressively have an increasing number of degrees of freedom. The question may well be asked as to whether any economic or other advantage is gained by increasing the complexity of an element. The answer here is not an easy one although it can be stated as a general rule that as the order of an element increases so the total number of unknowns in a problem can be reduced for a given accuracy of representation. Economic advantage requires, however, a reduction of total computation and data preparation effort, and this does not follow automatically for a reduced number of total variables.

However, an overwhelming economic advantage in the case of three-dimensional analyses occurs. The same kind of advantage arises on occasion in other problems but in general the optimum element may have to be determined from case to case.

In Sec. 2.6 of Chapter 2 we have shown that the order of error in the approximation to the unknown function is $O(h^{p+1})$, where h is the element 'size' and p is the degree of the complete polynomial present in the expansion. Clearly, as the element shape functions increase in degree so will the order of error increase, and convergence to the exact solution becomes more rapid. While this says nothing about the magnitude of error at a particular subdivision, it is clear that we should seek element shape functions with the highest complete polynomial for a given number of degrees of freedom.

4.2 Standard and hierarchical concepts

The essence of the finite element method already stated in Chapters 2 and 3 is in approximating the unknown (displacement) by an expansion given in Eqs (2.1) and (3.3). This, for a scalar variable u, can be written as

$$u \approx \hat{u} = \sum_{a=1}^{n} N_a \tilde{u}_a = \mathbf{N}\tilde{\mathbf{u}}^e \tag{4.1}$$

where n is the total number of functions used and $\tilde{u}_a$ are the unknown parameters to be determined.

We have explicitly chosen to identify such variables with the values of the unknown function at element nodes, thus making

$$\tilde{u}_a = \hat{u}(x_a) \tag{4.2}$$

The shape functions so defined will be referred to as 'standard' ones and are the basis of most finite element programs. If polynomial expansions are used and the element satisfies Criterion 1 of Chapter 2 (which specifies that rigid body displacements cause no strain), it is clear that a constant value of $\tilde{u}_a$ specified at all nodes must result in a constant value of $\hat{u}$:

$$\hat{u} = \left(\sum_{a=1}^{n} N_a \right) u_0 = u_0 \tag{4.3}$$

when $\tilde{u}_a = u_0$. It follows that

$$\sum_{a=1}^{n} N_a = 1 \tag{4.4}$$

at all points of the domain. This important property is known as a *partition of unity*[1] which we will make extensive use of here and in Chapter 15. The first part of this chapter will deal with such *standard shape functions*.

A serious drawback exists, however, with 'standard' functions, since when element refinement is made totally new shape functions have to be generated and hence all calculations repeated. It would be of advantage to avoid this difficulty by considering the expression (4.1) as a *series* in which the shape function N_a does not depend on the number of nodes in the mesh n. This indeed is achieved with *hierarchic shape functions* to which the second part of this chapter is devoted.

The hierarchic concept is well illustrated by the one-dimensional (elastic bar) problem of Fig. 4.1. Here for simplicity elastic properties are taken as constant ($D = E$) and the body force b is assumed to vary in such a manner as to produce the exact solution shown on the figure (with zero displacements at both ends).

Two meshes are shown and a linear interpolation between nodal points assumed. For both standard and hierarchic forms the coarse mesh gives

$$K_{11}^c \tilde{u}_1^c = f_1 \tag{4.5}$$

For a fine mesh two additional nodes are added and with the standard shape function the equations requiring solution are

$$\begin{bmatrix} K_{11}^F & K_{12}^F & K_{13}^F \\ K_{21}^F & K_{22}^F & 0 \\ K_{31}^F & 0 & K_{33}^F \end{bmatrix} \begin{Bmatrix} \tilde{u}_1 \\ \tilde{u}_2 \\ \tilde{u}_3 \end{Bmatrix} = \begin{Bmatrix} f_1 \\ f_2 \\ f_3 \end{Bmatrix} \tag{4.6}$$

In this form the zero matrices have been automatically inserted due to element interconnection which is here obvious, and we note that as no coefficients are the same, the new equations have to be resolved [Eq. (2.28a) shows how these coefficients are calculated and the reader is encouraged to work these out in detail].

With the 'hierarchic' form using the shape functions shown, a similar form of equation arises and an identical approximation is achieved (being simply given by a series of straight segments). The *final* solution is identical but the meaning of the parameters $\tilde{u}_a^\star$ is now different, as shown in Fig. 4.1.

Quite generally,

$$K_{11}^F = K_{11}^c \tag{4.7}$$

as an identical shape function is used for the first variable. Further, in this particular case the off-diagonal coefficients are zero and the final equations become, for the fine mesh,

$$\begin{bmatrix} K_{11}^c & 0 & 0 \\ 0 & K_{22}^F & 0 \\ 0 & 0 & K_{33}^F \end{bmatrix} \begin{Bmatrix} \tilde{u}_1^\star \\ \tilde{u}_2^\star \\ \tilde{u}_3^\star \end{Bmatrix} = \begin{Bmatrix} f_1 \\ f_2 \\ f_3 \end{Bmatrix} \tag{4.8}$$

The 'diagonality' feature is only true in the one-dimensional problem, but in general it will be found that the matrices obtained using hierarchic shape functions are more nearly diagonal and hence usually imply better conditioning than those with standard shape functions.

Although the variables are now not subject to the obvious interpretation (as local displacement values), they can be easily transformed to those if desired. Though it is not usual

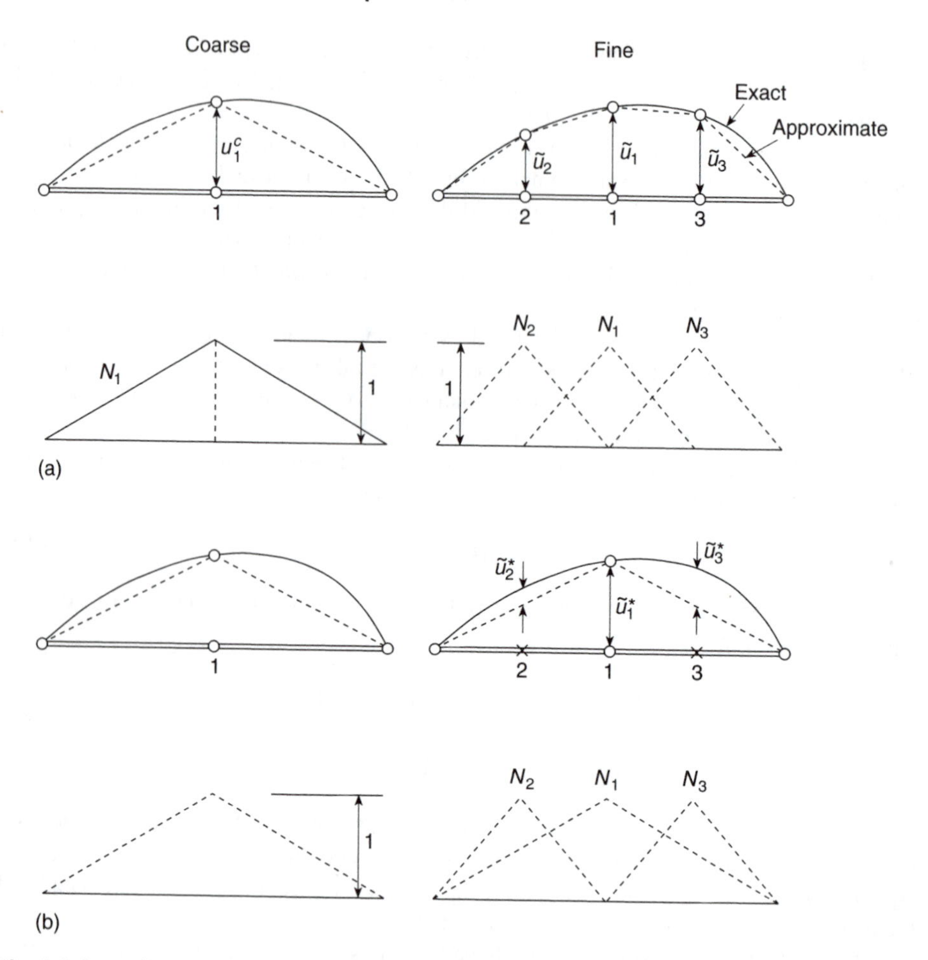

Fig. 4.1 A one-dimensional problem of stretching of a uniform elastic bar by prescribed body forces.

to use hierarchic forms in linearly interpolated elements their derivation in polynomial form is simple and very advantageous.

The reader should note that with hierarchic forms it is convenient to consider the finer mesh as still using the same, coarse, elements but now adding additional refining functions.

Hierarchic forms provide a link with other approximate (orthogonal) series solutions. Many problems solved in classical literature by trigonometric, Fourier series, expansion are indeed particular examples of this approach.

In the next sections of this chapter we shall consider the development of shape functions for high order elements with many boundary and internal degree of freedoms. Such development will generally be made on simple geometric forms and the reader may well question the wisdom of using increased accuracy for such simple shaped domains – having already observed the advantage of generalized finite element methods in fitting arbitrary domain shapes. This concern is well founded, but in the next chapter we shall show a general method to map high order elements into quite complex shapes.

Part 1. 'Standard' shape functions

Two-dimensional elements

4.3 Rectangular elements – some preliminary considerations

Conceptually (especially if the reader is conditioned by education to thinking in the cartesian coordinate system) the simplest element form of a two-dimensional kind is that of a rectangle with sides parallel to the x and y axes. Consider, for instance, the rectangle shown in Fig. 4.2 with nodal points numbered 1 to 8, located as shown, and at which the values of an unknown function u (here representing, for instance, one of the components of displacement) form the element parameters. How can suitable C_0 continuous shape functions for this element be determined?

Let us first assume that u is expressed in polynomial form in x and y. To ensure interelement continuity of u along the top and bottom sides the variation must be linear. Two points at which the function is common between elements lying above or below exist, and as two values uniquely determine a linear function, its identity all along these sides is ensured with that given by adjacent elements. Use of this fact was already made in specifying linear expansions on edges for a triangle and a rectangle.

Similarly, if a cubic variation along the vertical sides is assumed, continuity will be preserved there as four values determine a unique cubic polynomial. Conditions for satisfying the first criterion are now obtained.

To ensure the existence of constant values of the first derivative it is necessary that all the linear polynomial terms of the expansion be retained.

Finally, as eight points are to determine uniquely the variation of the function, only eight coefficients of the expansion can be retained and thus we could write

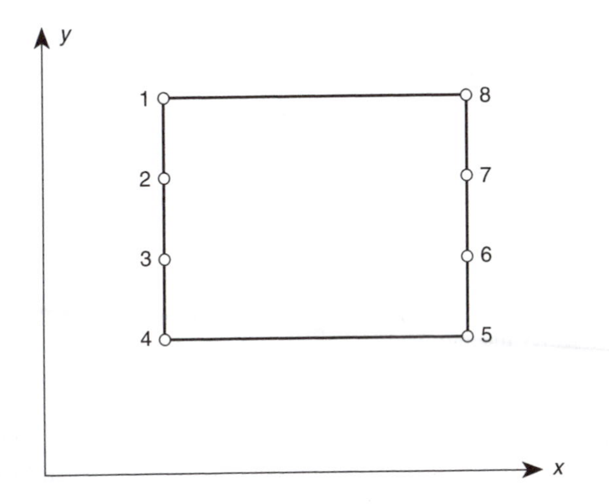

Fig. 4.2 A rectangular element.

$$u = \alpha_1 + \alpha_2 x + \alpha_3 y + \alpha_4 xy + \alpha_5 y^2 + \alpha_6 xy^2 + \alpha_7 y^3 + \alpha_8 xy^3 \tag{4.9}$$

The choice can in general be made unique by retaining the lowest possible expansion terms, though in this case apparently no such choice arises.† The reader will easily verify that all the requirements have now been satisfied.

Substituting coordinates of the various nodes a set of simultaneous equations will be obtained. This can be written in exactly the same manner as was done for a triangle in Eq. (2.4) as

$$\begin{Bmatrix} \tilde{u}_1 \\ \vdots \\ \tilde{u}_8 \end{Bmatrix} = \begin{bmatrix} 1, & x_1, & y_1, & x_1 y_1, & y_1^2, & x_1 y_1^2, & y_1^3, & x_1 y_1^3 \\ \vdots & \vdots & & & & & & \vdots \\ 1, & x_8, & y_8, & \dots & & & \dots & x_8 y_8^3 \end{bmatrix} \begin{Bmatrix} \alpha_1 \\ \vdots \\ \alpha_8 \end{Bmatrix} \tag{4.10}$$

or simply as

$$\tilde{\mathbf{u}}^e = \mathbf{C}\boldsymbol{\alpha}. \tag{4.11}$$

Formally,

$$\boldsymbol{\alpha} = \mathbf{C}^{-1}\tilde{\mathbf{u}}^e \tag{4.12}$$

and we could write Eq. (4.9) as

$$u = \mathbf{P}(x, y)\boldsymbol{\alpha} = \mathbf{P}(x, y)\mathbf{C}^{-1}\tilde{\mathbf{u}}^e \tag{4.13}$$

in which

$$\mathbf{P}(x, y) = [1, x, y, xy, y^2, xy^2, y^3, xy^3] \tag{4.14}$$

Thus the shape functions for the element defined by

$$u = \mathbf{N}\tilde{\mathbf{u}}^e = [N_1, N_2, \dots, N_8]\,\tilde{\mathbf{u}}^e \tag{4.15}$$

can be found as

$$\mathbf{N}(x, y) = \mathbf{P}(x, y)\mathbf{C}^{-1} \tag{4.16}$$

This process has, however, some considerable disadvantages. Occasionally an inverse of $\mathbf{C}$ may not exist[2, 3] and *always* considerable algebraic difficulty is experienced in obtaining an expression for the inverse in general terms suitable for all element geometries. It is therefore worthwhile to consider whether shape functions $N_a(x, y)$ can be written down directly. Before doing this some general properties of these functions have to be mentioned.

Inspection of the defining relation, Eq. (4.15), reveals immediately some important characteristics. First, as this expression is valid for all components of $\tilde{\mathbf{u}}^e$,

$$N_a(x_b, y_b) = \delta_{ab} = \begin{cases} 1; & a = b \\ 0; & a \neq b \end{cases}$$

where δ_{ab} is known as the Kronecker delta. Further, the basic type of variation along boundaries defined for continuity purposes (e.g., linear in x and cubic in y in the above example) must be retained. The typical form of the shape functions for the elements considered is illustrated isometrically for two typical nodes in Fig. 4.3. It is clear that these could have been written down directly as a product of a suitable linear function in x with a

† Retention of a higher order term of expansion, ignoring one of lower order, will usually lead to a poorer approximation though still retaining convergence,[2] providing the linear terms are always included.

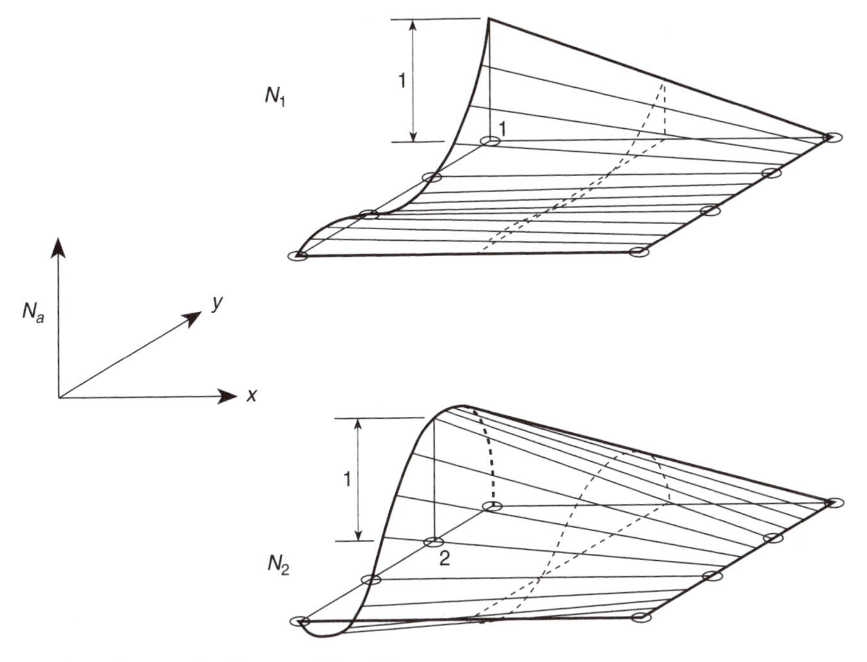

Fig. 4.3 Shape functions for elements of Fig. 4.2.

cubic function in y. The easy solution of this example is not always as obvious but given sufficient ingenuity, a direct derivation of shape functions is always preferable.

It will be convenient to use normalized coordinates in our further investigation. Such normalized coordinates are shown in Fig. 4.4 and are chosen so that their values are ± 1 on the faces of the rectangle:†

$$\begin{aligned} \xi &= \frac{x - x_c}{a} & \qquad \mathrm{d}\xi &= \frac{\mathrm{d}x}{a} \\ \eta &= \frac{y - y_c}{b} & \qquad \mathrm{d}\eta &= \frac{\mathrm{d}y}{b} \end{aligned} \tag{4.17}$$

Once the shape functions are known in the normalized coordinates, translation into actual coordinates or transformation of the various expressions occurring, for instance, in the stiffness derivation is trivial for rectangular shapes. Consideration of other more convenient 'mapping' methods will be addressed in Chapter 5.

4.4 Completeness of polynomials

The shape function derived in the previous section was of a rather special form [viz. Eq. (4.9)]. Only a linear variation with the coordinate x was permitted, while in y a full cubic was available. The complete polynomial contained in it was thus of order 1. In general use, a convergence order corresponding to a linear variation would occur despite an increase of the total number of variables. Only in situations where the linear variation

† In Chapter 5 we will show that this is convenient for purposes of numerical integration.

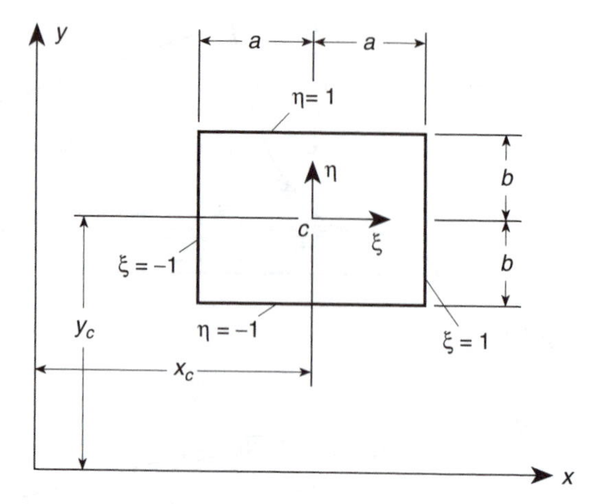

Fig. 4.4 Normal coordinates for a rectangle.

in x corresponded closely to the exact solution would a higher order of convergence occur, and for this reason elements with such 'preferential' directions should be restricted to special use, e.g., in narrow beams or strips. Usually, we will seek element expansions which possess the highest order of a complete polynomial for a minimum of degrees of freedom. In this context it is useful to recall the Pascal triangle (Fig. 4.5) from which the number of terms occurring in a polynomial in two variables x, y can be readily ascertained. For instance, first-order polynomials require three terms, second order require six terms, third order require ten terms, etc.

4.5 Rectangular elements – Lagrange family

Consider the element shown in Fig. 4.6 in which a series of nodes, external and internal, is placed on a regular grid. It is required to determine a shape function for the point indicated

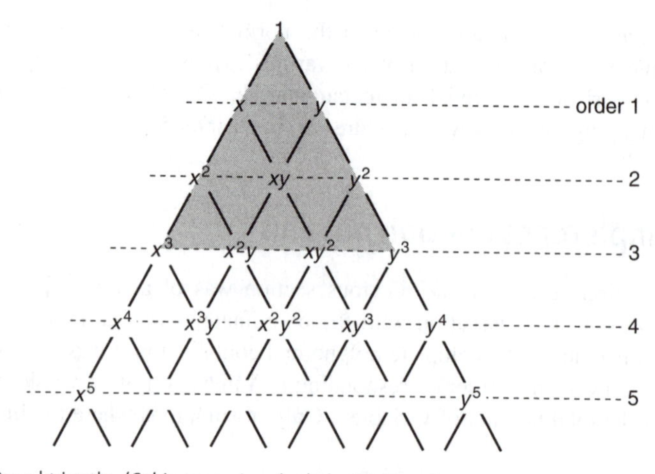

Fig. 4.5 The Pascal triangle. (Cubic expansion shaded – 10 terms.)

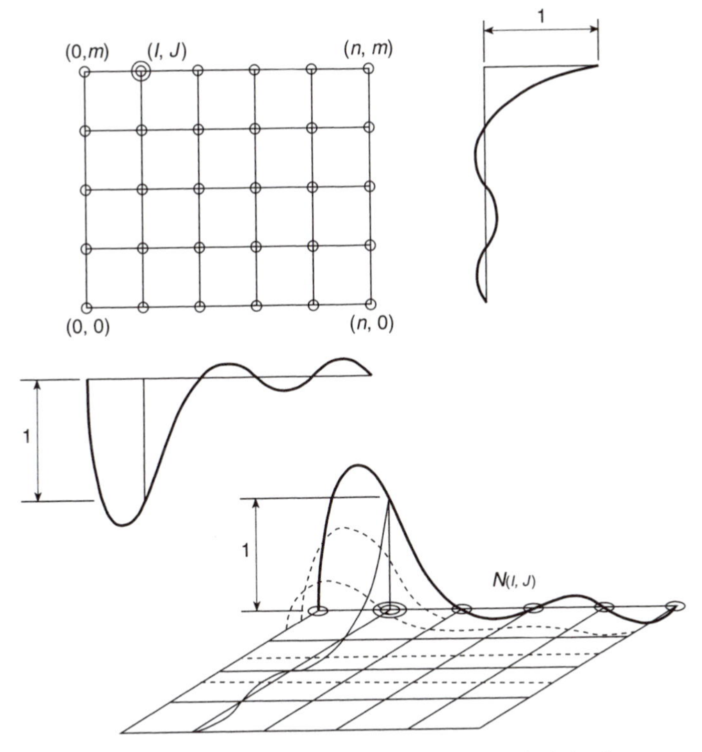

Fig. 4.6 A typical shape function for a lagrangian element ($n = 5$, $m = 4$, $I = 1$, $J = 4$).

by the heavy circle. Clearly the product of a fifth-order polynomial in ξ which has a value of unity at points of the second column of nodes and zero at the other nodal columns and that of a fourth-order polynomial in η having unity on the coordinate corresponding to the top row of nodes and zero at other nodal rows satisfies all the interelement continuity conditions and gives unity at the nodal point concerned. Polynomials in one coordinate having this property are known as *Lagrange polynomials* and can be written down directly as

$$l_k^n(\xi) = \frac{(\xi - \xi_0)(\xi - \xi_1)\cdots(\xi - \xi_{k-1})(\xi - \xi_{k+1})\cdots(\xi - \xi_n)}{(\xi_k - \xi_0)(\xi_k - \xi_1)\cdots(\xi_k - \xi_{k-1})(\xi_k - \xi_{k+1})\cdots(\xi_k - \xi_n)} = \prod_{\substack{i=0 \\ i\neq k}}^{n} \frac{\xi - \xi_i}{\xi_k - \xi_i} \tag{4.18}$$

giving unity at ξ_k and passing through zero at the remaining n points.

An easy and systematic method of generating shape functions of any order now can be achieved by simple products of Lagrange polynomials in the two coordinates.[4–6] Thus, in two dimensions, if we label the node by its column and row number, I, J, we have

$$N_a \equiv N_{IJ} = l_I^n(\xi) l_J^m(\eta) \tag{4.19}$$

where n and m stand for the number of subdivisions in each direction. Figure 4.7 shows a few members of this unlimited family where $m = n$. For $m = n = 1$ we obtain the simple result

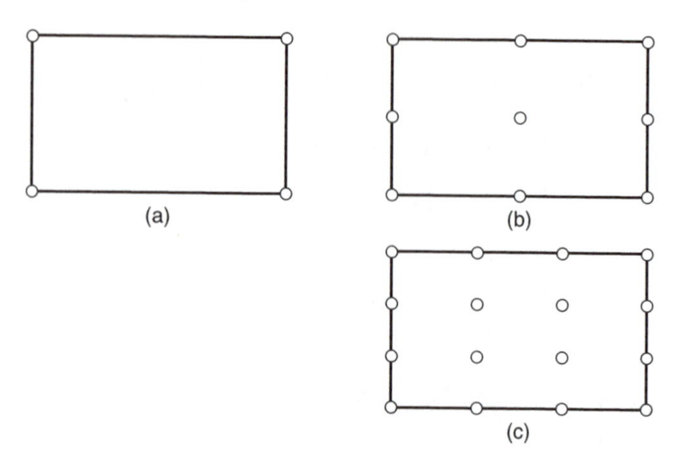

Fig. 4.7 Three elements of the Lagrange family: (a) linear, (b) quadratic, (c) cubic.

$$N_a = \tfrac{1}{4}(1 + \xi_a \xi)(1 + \eta_a \eta) \tag{4.20}$$

in which ξ_a, η_a are the normalized coordinates at node a.

Indeed, if we examine the polynomial terms present in a situation where $n = m$ we observe in Fig. 4.8, based on the Pascal triangle, that a large number of polynomial terms is present above those needed for a complete expansion.[7] However, when mapping of shape functions is considered (viz. Chapter 5) some advantages occur for this family.

4.6 Rectangular elements – 'serendipity' family

It is often more efficient to make the functions dependent on nodal values placed on the element boundary. Consider, for instance, the first three elements of Fig. 4.9. In each a progressively increasing and equal number of nodes are placed on the element boundary.

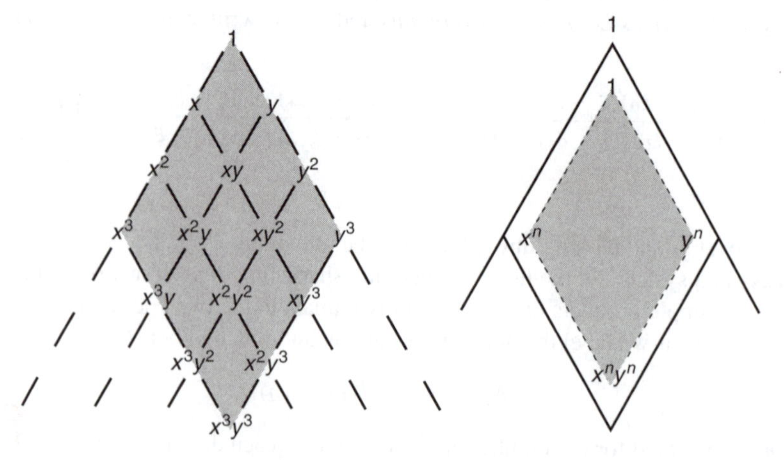

Fig. 4.8 Terms generated by a lagrangian expansion of order 3 × 3 (or $m \times n$). Complete polynomials of order 3 (or n).

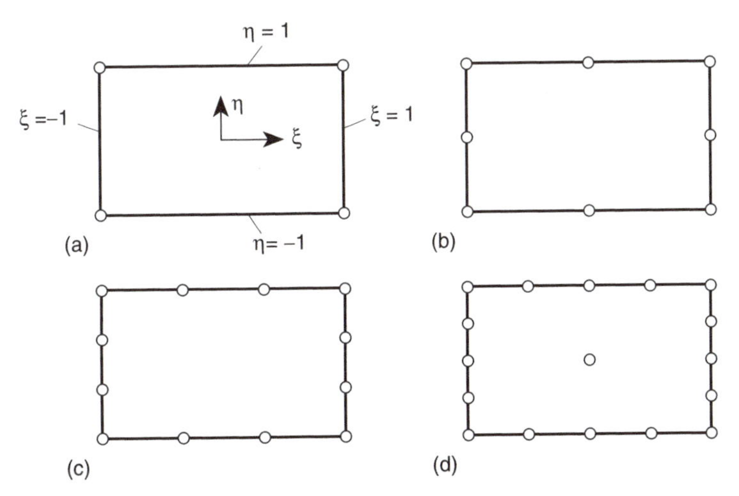

Fig. 4.9 Rectangles of boundary node (serendipity) family: (a) linear, (b) quadratic, (c) cubic, (d) quartic.

The variation of the function on the edges to ensure continuity is linear, parabolic, and cubic in increasing element order.

To achieve the shape function for the first element it is obvious that a product of linear lagrangian polynomials of the form

$$\tfrac{1}{4}(1+\xi)(1+\eta) \tag{4.21}$$

gives unity at the top right corner where $\xi = \eta = 1$ and zero at all the other corners. Further, a linear variation of the shape function of all sides exists and hence continuity is satisfied. Indeed this element is identical to the lagrangian one with $n = 1$ and again all the shape functions may be written as one expression:

$$N_a = \tfrac{1}{4}(1+\xi_a\xi)(1+\eta_a\eta)$$

As a linear combination of these shape functions yields any arbitrary linear variation of u, the second convergence criterion is satisfied.

The reader can verify that the following functions satisfy all the necessary criteria for quadratic and cubic members of the family.

'Quadratic' element

Corner nodes:

$$N_a = \tfrac{1}{4}(1+\xi_a\xi)(1+\eta_a\eta)(\xi_a\xi+\eta_a\eta-1) \tag{4.22a}$$

Mid-side nodes:

$$\begin{aligned} \xi_a &= 0 \qquad N_a = \tfrac{1}{2}(1-\xi^2)(1+\eta_a\eta) \\ \eta_a &= 0 \qquad N_a = \tfrac{1}{2}(1+\xi_a\xi)(1-\eta^2) \end{aligned} \tag{4.22b}$$

'Cubic' element

Corner nodes:

$$N_a = \tfrac{1}{32}(1+\xi_a\xi)(1+\eta_a\eta)[9(\xi^2+\eta^2)-10] \tag{4.23a}$$

Mid-side nodes:

$$\begin{aligned} \xi_a &= \pm 1 \quad \text{and} \quad \eta_a = \pm \tfrac{1}{3} \\ N_a &= \tfrac{9}{32}(1+\xi_a\xi)(1-\eta^2)(1+9\eta_a\eta) \end{aligned} \tag{4.23b}$$

and

$$\begin{aligned} \xi_a &= \pm \tfrac{1}{3} \quad \text{and} \quad \eta_a = \pm 1 \\ N_a &= \tfrac{9}{32}(1-\xi^2)(1+9\xi_a\xi)(1+\eta_a\eta) \end{aligned} \tag{4.23c}$$

which all satisfy the requirement

$$N_a(\xi_b, \eta_b) = \delta_{ab} = \begin{cases} 1; a = b \\ 0; a \neq b \end{cases} \tag{4.23d}$$

The above functions were originally derived by inspection, and progression to yet higher members is difficult and requires some ingenuity.[4, 5] It was therefore appropriate to name this family 'serendipity' after the famous princes of Serendip noted for their chance discoveries (Horace Walpole, 1754).

However, a quite systematic way of generating the 'serendipity' shape functions can be devised, which becomes apparent from Fig. 4.10 where the generation of a quadratic shape function is presented.[7, 8]

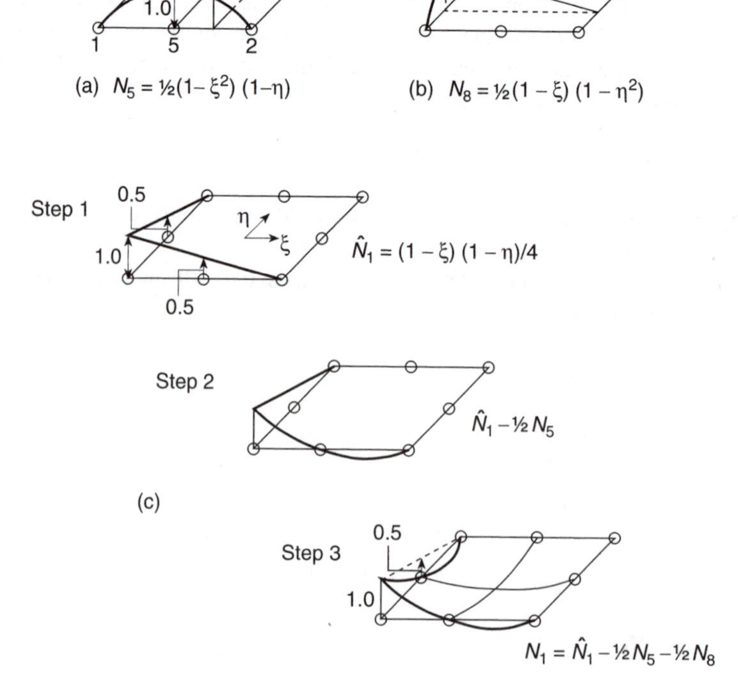

Fig. 4.10 Systematic generation of 'serendipity' shape functions.

As a starting point we observe that for *mid-side* nodes a lagrangian interpolation of a quadratic × linear type suffices to determine N_a at nodes 5 to 8. N_5 and N_8 are shown in Fig. 4.10(*a*) and (*b*). For a *corner* node, such as Fig. 4.10(*c*), we start with a bilinear lagrangian family $\hat{N}_1$ and note immediately that while $\hat{N}_1 = 1$ at node 1, it is not zero at nodes 5 or 8 (step 1). Successive subtraction of $1/2\, N_5$ (step 2) and $1/2\, N_8$ (step 3) ensures that a zero value is obtained at these nodes. The reader can verify that the final expressions obtained coincide with those of Eqs (4.22a) and (4.22b).

Indeed, it should now be obvious that for all higher order elements the *mid-side* and *corner shape* functions can be generated by an identical process. For the former a simple multiplication of mth-order and first-order lagrangian interpolations suffices. For the latter a combination of bilinear corner functions, together with appropriate fractions of mid-side shape functions to ensure zero at appropriate nodes, is necessary.

It also is quite easy to generate shape functions for elements with different numbers of nodes along each side by a similar systematic algorithm. This may be very desirable if a transition between elements of different order is to be achieved, enabling a different order of accuracy in separate sections of a large problem to be studied. Figure 4.11 illustrates the necessary shape functions for a cubic/linear transition. Use of such special elements was first introduced in reference 8, but the simpler formulation used here is that of reference 7.

With the mode of generating shape functions for this class of elements available it is immediately obvious that fewer degrees of freedom are now necessary for a given complete

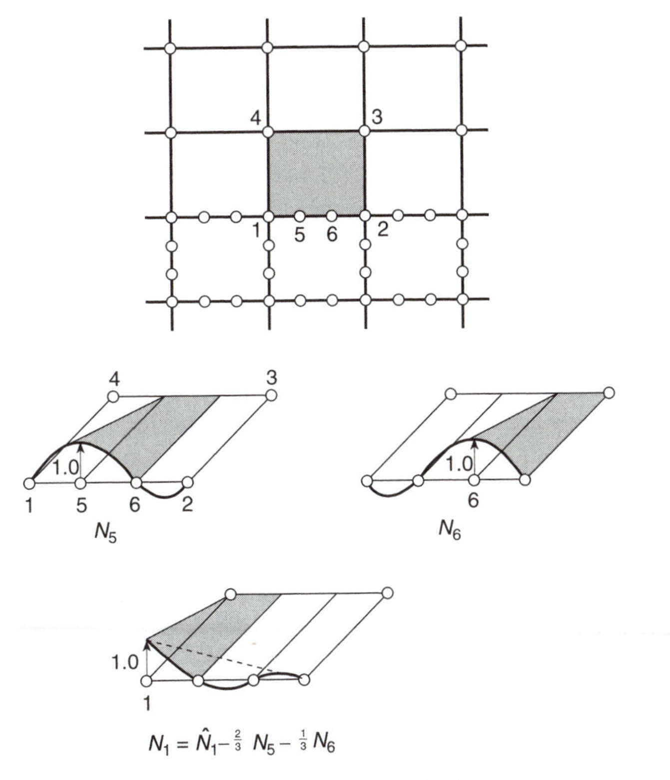

Fig. 4.11 Shape functions for a transition 'serendipity' element, cubic/linear.

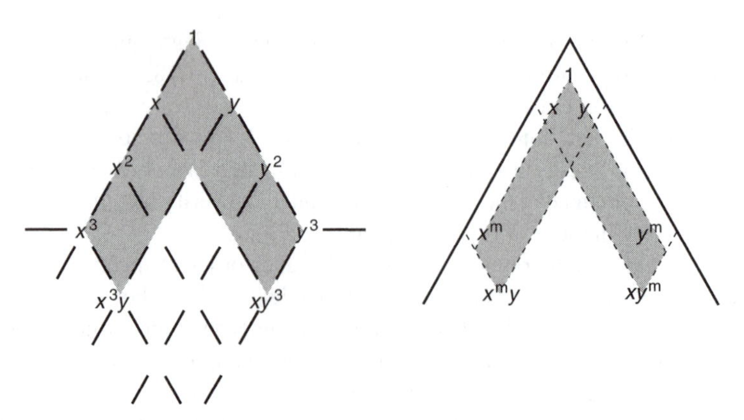

Fig. 4.12 Terms generated by edge shape functions in serendipity-type elements (3×3 and $m \times m$).

polynomial expansion. Figure 4.12 shows this for a cubic element where only two surplus terms arise (as compared with six surplus terms in a lagrangian of the same degree). However, when mapping to general quadrilateral shape is introduced (Chapter 5) some of these advantages are lost rendering the lagrangian form of interpolation advantageous.

It is immediately evident, however, that the functions generated by nodes placed only along the edges will not generate complete polynomials beyond cubic order. For higher order ones it is necessary to supplement the expansion by internal nodes or by the use of 'nodeless' variables which contain appropriate polynomial terms. For example, in the next, quartic, member[9] of this family a central node is added [viz. Fig. 4.9(d)] so that all terms of a complete fourth-order expansion will be available. This central node adds a shape function $(1 - \xi^2)(1 - \eta^2)$ which is zero on all outer boundaries and coincides with the internal function used in the quadratic lagrangian element. Once interior nodes are added it is necessary to modify the corner and mid-side shape functions to preserve the Kronnecker delta property (4.23d).

4.7 Triangular element family

The advantage of an arbitrary triangular shape in approximating to any boundary configuration has been amply demonstrated in earlier chapters. Its apparent superiority here over rectangular shapes needs no further discussion. However, the question of generating more elaborate higher order elements needs to be further developed.

Consider a series of triangles generated on a pattern indicated in Fig. 4.13. The number of nodes in each member of the family is now such that a complete polynomial expansion, of the order needed for interelement compatibility, is ensured. This follows by comparison with the Pascal triangle of Fig. 4.5 in which we see the number of nodes coincides exactly with the number of polynomial terms required. This particular feature puts the triangle family in a special, privileged position, in which the inverse of the **C** matrices of Eq. (4.11) will always exist.[3] However, once again a direct generation of shape functions will be preferred – and indeed will be shown to be particularly easy.

Before proceeding further it is useful to define a special set of normalized coordinates for a triangle.

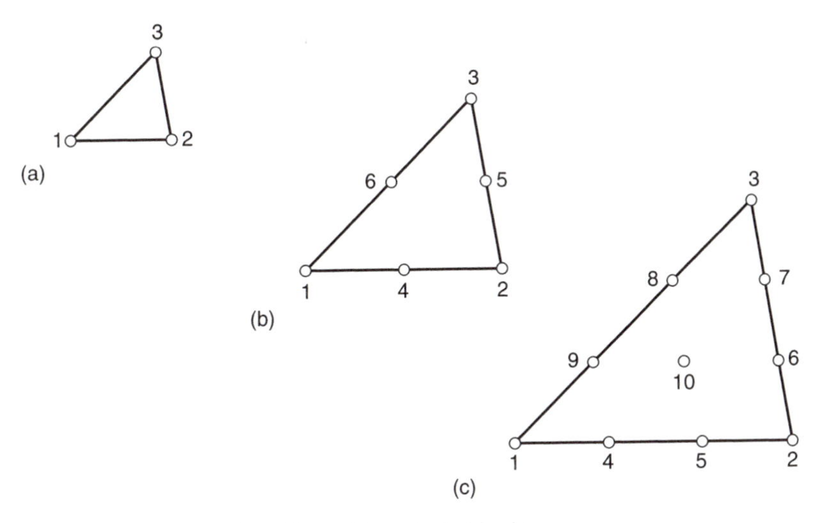

Fig. 4.13 Triangular element family: (a) linear, (b) quadratic, (c) cubic.

4.7.1 Area coordinates

While cartesian directions parallel to the sides of a rectangle were a natural choice for that shape, in the triangle these are not convenient.

A new set of coordinates, L_1, L_2, and L_3 for a triangle 1, 2, 3 (Fig. 4.14), is defined by the following linear relation between these and the cartesian system:

$$
\begin{aligned}
x &= L_1 x_1 + L_2 x_2 + L_3 x_3 \\
y &= L_1 y_1 + L_2 y_2 + L_3 y_3 \\
1 &= L_1 + L_2 + L_3
\end{aligned}
\tag{4.24}
$$

To every set, L_1, L_2, L_3 (which are not independent, but are related by the third equation), there corresponds a unique set of cartesian coordinates. At point 1, $L_1 = 1$ and $L_2 = L_3 = 0$, etc. A linear relation between the new and cartesian coordinates implies that contours of L_1 are equally placed straight lines parallel to side 2–3 on which $L_1 = 0$, etc.

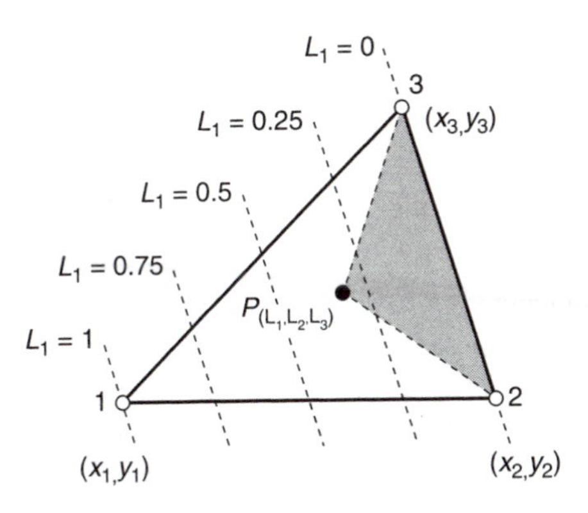

Fig. 4.14 Area coordinates.

Indeed it is easy to see that an alternative definition of the coordinate L_1 of a point P is by a ratio of the area of the shaded triangle to that of the total triangle:

$$L_1 = \frac{\text{area } P23}{\text{area } 123} \tag{4.25}$$

Hence the name *area coordinates.*

Solving Eq. (4.24) gives

$$L_a = \frac{a_a + b_a x + c_a y}{2\Delta}; \quad a = 1, 2, 3 \tag{4.26}$$

in which

$$\Delta = \tfrac{1}{2} \det \begin{bmatrix} 1 & x_1 & y_1 \\ 1 & x_2 & y_2 \\ 1 & x_3 & y_3 \end{bmatrix} = \text{area } 123 \tag{4.27}$$

and

$$a_1 = x_2 y_3 - x_3 y_2 \qquad b_1 = y_2 - y_3 \qquad c_1 = x_3 - x_2 \tag{4.28}$$

etc., with cyclic rotation of indices 1, 2, and 3.

The identity of expressions with those derived in Chapter 2 [Eqs (2.6) and (2.7)] is worth noting.

4.7.2 Shape functions

For the first element of the series [Fig. 4.13(a)], the shape functions are simply the area coordinates. Thus

$$N_1 = L_1 \qquad N_2 = L_2 \qquad N_3 = L_3 \tag{4.29}$$

This is obvious as each individually gives unity at one node, zero at others, and varies linearly everywhere.

To derive shape functions for other elements a simple recurrence relation can be derived.[3] However, it is very simple to write functions for an arbitrary triangle of order M in a manner similar to that used for the lagrangian element of Sec. 4.5.

Denoting a typical node a by three numbers I, J, and K corresponding to the position of coordinates L_{1a}, L_{2a}, and L_{3a} we can write the shape function in terms of three lagrangian interpolations as [see Eq. (4.18)]

$$N_a = l_I^I(L_1) l_J^J(L_2) l_K^K(L_3) \tag{4.30}$$

In the above l_I^I, etc., are given by expression (4.18), with L_1 taking the place of ξ, etc.

It is easy to verify that the above expression gives

$$N_a = 1 \quad \text{at} \quad L_1 = L_{1I}, \quad L_2 = L_{2J}, \quad L_3 = L_{3K}$$

and zero at all other nodes.

The highest term occurring in the expansion is $L_1^I L_2^J L_3^K$ and as $I + J + K \equiv M$ for all points the polynomial is also of order M.

Expression (4.30) is valid for quite arbitrary distributions of nodes of the pattern given in Fig. 4.15 and simplifies if the spacing of the nodal lines is equal (i.e., $1/m$). The formula was first obtained by Argyris *et al.*[10] and formalized in a different manner by others.[7, 11]

The reader can verify the shape functions for the second- and third-order elements as given below and indeed derive ones of any higher order easily.

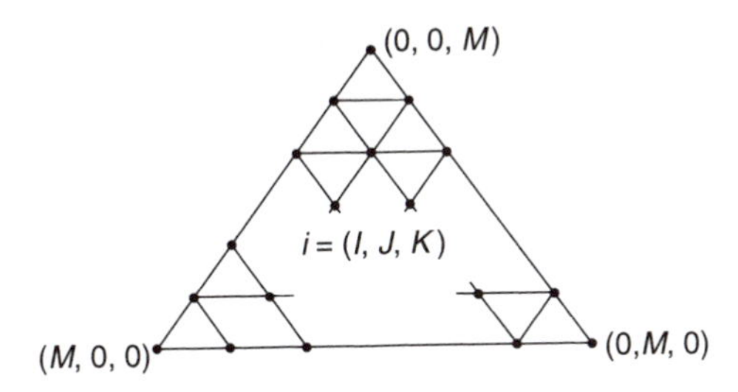

Fig. 4.15 A general triangular element.

Quadratic triangle [Fig. 4.13(b)]

Corner nodes:

$$N_a = (2L_a - 1)L_a, \qquad a = 1, 2, 3$$

Mid-side nodes:

$$N_4 = 4L_1L_2, \qquad N_5 = 4L_2L_3, \qquad N_6 = 4L_3L_1$$

Cubic triangle [Fig. 4.13(c)]

Corner nodes:

$$N_a = \tfrac{1}{2}(3L_a - 1)(3L_a - 2)L_a, \qquad a = 1, 2, 3$$

Mid-side nodes:

$$N_4 = \tfrac{9}{2}L_1L_2(3L_1 - 1), \qquad N_5 = \tfrac{9}{2}L_1L_2(3L_2 - 1), \qquad \text{etc.}$$

and for the internal node:

$$N_{10} = 27L_1L_2L_3$$

The last shape again is a 'bubble' function giving zero contribution along boundaries – and this will be found to be useful in other contexts (see the mixed forms in Chapter 11).

The quadratic triangle was first derived by Veubeke[12] and used later in the context of plane stress analysis by Argyris.[13]

When element matrices have to be evaluated it will follow that we are faced with integration of quantities defined in terms of area coordinates over the triangular region. It is useful to note in this context the following exact integration expression:

$$\iint_\Delta L_1^a L_2^b L_3^c \, \mathrm{d}x \, \mathrm{d}y = \frac{a!\,b!\,c!}{(a+b+c+2)!} 2\Delta \qquad (4.31)$$

One-dimensional elements

4.8 Line elements

So far in this book the continuum was considered generally in two or three dimensions. 'One-dimensional' members, being of a kind for which exact solutions are generally available, were treated only as trivial examples in Chapter 3 and in Sec. 4.2. In many practical two- or three-dimensional problems such elements do in fact appear in conjunction with the more usual continuum elements – and a unified treatment is desirable. In the context of elastic analysis these elements may represent lines of reinforcement (plane and three-dimensional problems) or sheets of thin lining material in axisymmetric bodies. In the context of heat conduction and other field problems similar effects occur.

Once the shape of such a function as displacement is chosen for an element of this kind, its properties can be determined, noting, however, that derived quantities such as strain, etc., have to be considered only in one dimension.

Figure 4.16 shows such an element sandwiched between two adjacent quadratic-type elements. Clearly for continuity of the function a quadratic variation of the unknown with the one variable ξ is all that is required. Thus the shape functions are given directly by the Lagrange polynomial as defined in Eq. (4.18).

Three-dimensional elements

4.9 Rectangular prisms – Lagrange family

In a precisely analogous way to that given in previous sections equivalent lagrangian family elements of three-dimensional type can be described.

Shape functions for such elements will be generated by a direct product of three Lagrange polynomials. Extending the notation of Eq. (4.19) we now have

$$N_a \equiv N_{IJK} = l_I^n(\xi) l_J^m(\eta) l_K^p(\zeta) \tag{4.32}$$

for n, m, and p subdivisions along each side and

$$\xi = \frac{x - x_c}{a}; \quad \eta = \frac{y - y_c}{b} \quad \text{and} \quad \zeta = \frac{z - z_c}{c}$$

This element again is suggested by Zienkiewicz *et al.*[5] and elaborated upon by Argyris *et al.*[6] All the remarks about internal nodes and the properties of the formulation with mappings (to be described in the next chapter) are applicable here. The first three members of the three-dimensional Lagrange family are shown in Fig. 4.17(a).

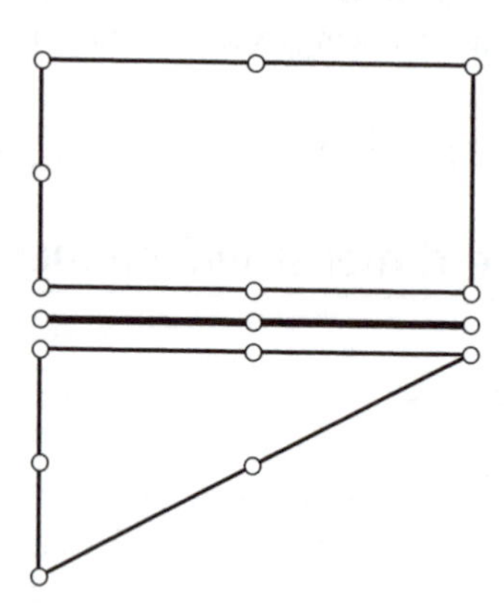

Fig. 4.16 A line element sandwiched between two-dimensional elements.

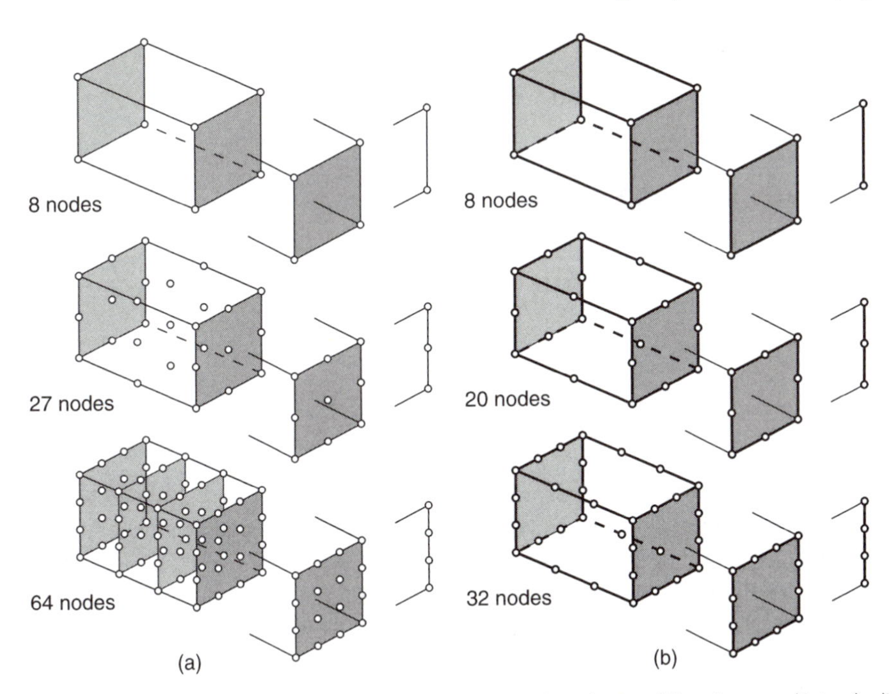

Fig. 4.17 Linear, quadratic and cubic right prisms with corresponding sheet and line elements. (Extra shading on 64-node element to show node location more clearly.)

For interelement continuity the simple rules given previously have to be modified. What is necessary to achieve such continuity is that *along a whole face of an element the nodal values define a unique variation of the unknown function*. It is obvious on a face that one of the l_I^i will be unity and the remaining product defines the two-dimensional form given by (4.19), thus ensuring continuity.

4.10 Rectangular prisms – 'serendipity' family

The serendipity family of elements shown in Fig. 4.17(b) is precisely equivalent to that of Fig. 4.9 for the two-dimensional case.[4, 8, 14] Using now three coordinates and otherwise following the terminology of Sec. 4.6 we have the following shape functions:

'Linear' element (8 nodes)

$$N_a = \tfrac{1}{8}(1 + \xi_a\xi)(1 + \eta_a\eta)(1 + \zeta_a\zeta)$$

which is identical with the linear lagrangian element.

'Quadratic' element (20 nodes)

Corner nodes:

$$N_a = \tfrac{1}{8}(1 + \xi_a\xi)(1 + \eta_a\eta)(1 + \zeta_a\zeta)(\xi_a\xi + \eta_a\eta + \zeta_a\zeta - 2)$$

Typical mid-side node:

$$\xi_a = 0 \qquad \eta_a = \pm 1 \qquad \zeta_a = \pm 1$$
$$N_a = \tfrac{1}{4}(1 - \xi^2)(1 + \eta_a \eta)(1 + \zeta_a \zeta)$$

'Cubic' elements (32 nodes)

Corner node:

$$N_a = \tfrac{1}{64}(1 + \xi_a \xi)(1 + \eta_a \eta)(1 + \zeta_a \zeta)[9(\xi^2 + \eta^2 + \zeta^2) - 19]$$

Typical mid-side node:

$$\xi_a = \pm \tfrac{1}{3} \qquad \eta_a = \pm 1 \qquad \zeta_a = \pm 1$$
$$N_a = \tfrac{9}{64}(1 - \xi^2)(1 + 9\xi_a \xi)(1 + \eta_a \eta)(1 + \zeta_a \zeta)$$

When $\zeta_a \zeta = \zeta^2 = 1$ the above expressions reduce to those of Eqs (4.20)–(4.23c). Indeed such elements of three-dimensional type can be joined in a compatible manner to sheet or line elements of the appropriate type as shown in Fig. 4.17.

Once again the procedure for generating the shape functions follows that described in Figs 4.10 and 4.11 and once again elements with varying degrees of freedom along the edges can be derived following the same steps.

The equivalent of a Pascal triangle is now a tetrahedron and again we can observe the small number of surplus degrees of freedom – a situation of even greater magnitude than in two-dimensional analysis.

4.11 Tetrahedral elements

The tetrahedral family shown in Fig. 4.18 not surprisingly exhibits properties similar to those of the triangle family.

First, once again complete polynomials in three coordinates are achieved at each stage. Second, as faces are divided in a manner identical with that of the previous triangles, the same order of polynomial in two coordinates in the plane of the face is achieved and element compatibility ensured. No surplus terms in the polynomial occur.

4.11.1 Volume coordinates

Once again special coordinates are introduced defined by (Fig. 4.19):

$$\begin{aligned} x &= L_1 x_1 + L_2 x_2 + L_3 x_3 + L_4 x_4 \\ y &= L_1 y_1 + L_2 y_2 + L_3 y_3 + L_4 y_4 \\ z &= L_1 z_1 + L_2 z_2 + L_3 z_3 + L_4 z_4 \\ 1 &= L_1 + L_2 + L_3 + L_4 \end{aligned} \tag{4.33}$$

Solving Eq. (4.33) gives

$$L_k = \frac{a_k + b_k x + c_k y + d_k z}{6V}; \quad k = 1, 2, 3, 4$$

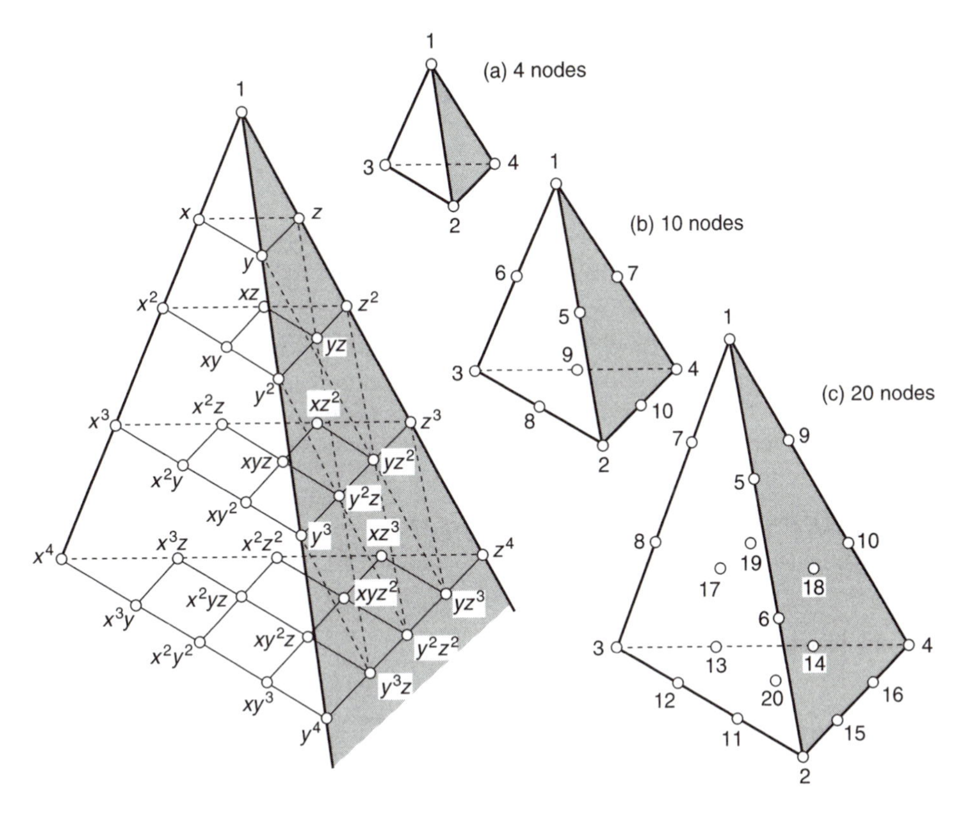

Fig. 4.18 The tetrahedral family: (a) linear, (b) quadratic, (c) cubic.

with

$$6V = \det \begin{bmatrix} 1 & x_1 & y_1 & z_1 \\ 1 & x_2 & y_2 & z_2 \\ 1 & x_3 & y_3 & z_3 \\ 1 & x_4 & y_4 & z_4 \end{bmatrix} \tag{4.34a}$$

in which, incidentally, the value V represents the volume of the tetrahedron. By expanding the other relevant determinants into their cofactors we have

$$\begin{aligned} a_1 &= \det \begin{bmatrix} x_2 & y_2 & z_2 \\ x_3 & y_3 & z_3 \\ x_4 & y_4 & z_4 \end{bmatrix} & \qquad b_1 &= -\det \begin{bmatrix} 1 & y_2 & z_2 \\ 1 & y_3 & z_3 \\ 1 & y_4 & z_4 \end{bmatrix} \\ c_1 &= -\det \begin{bmatrix} x_2 & 1 & z_2 \\ x_3 & 1 & z_3 \\ x_4 & 1 & z_4 \end{bmatrix} & \qquad d_1 &= -\det \begin{bmatrix} x_2 & y_2 & 1 \\ x_3 & y_3 & 1 \\ x_4 & y_4 & 1 \end{bmatrix} \end{aligned} \tag{4.34b}$$

with the other constants defined by cyclic interchange of the subscripts in the order 1, 2, 3, 4.

Again the physical nature of the coordinates can be identified as the ratio of volumes of tetrahedra based on an internal point P in the total volume, e.g., as shown in Fig. 4.19:

$$L_1 = \frac{\text{volume } P234}{\text{volume } 1234}, \qquad \text{etc.} \tag{4.35}$$

4.11.2 Shape functions

As the volume coordinates vary linearly with the cartesian ones from unity at one node to zero at the opposite face then shape functions for the linear element [Fig. 4.18(a)] are simply

$$N_a = L_a \qquad a = 1, 2, 3, 4 \tag{4.36}$$

Formulae for shape functions of higher order tetrahedra are derived in precisely the same manner as for the triangles by establishing appropriate Lagrange-type formulae similar to Eq. (4.30). The reader may verify the following shape functions for the quadratic and cubic order cases.

'Quadratic' tetrahedron [Fig. 4.18(b)]

For corner nodes:

$$N_a = (2L_a - 1)L_a \qquad a = 1, 2, 3, 4$$

For mid-edge nodes:

$$N_5 = 4L_1L_2, \qquad \text{etc.}$$

'Cubic' tetrahedron

Corner nodes:

$$N_1 = \tfrac{1}{2}(3L_a - 1)(3L_a - 2)L_a \qquad a = 1, 2, 3, 4$$

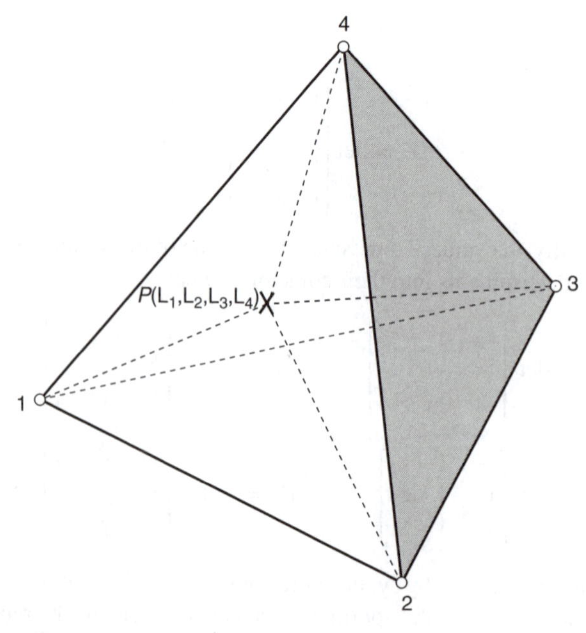

Fig. 4.19 Volume coordinates.

Mid-edge nodes:

$$N_5 = \tfrac{9}{2}L_1L_2(3L_1 - 1), \qquad \text{etc.}$$

Mid-face nodes:

$$N_{17} = 27L_1L_2L_3, \qquad \text{etc.}$$

A useful integration formula may again be quoted here:

$$\iiint_{\text{vol}} L_1^a L_2^b L_3^c L_4^d \; \mathrm{d}x \; \mathrm{d}y \; \mathrm{d}z = \frac{a!\,b!\,c!\,d!}{(a+b+c+d+3)!} 6V \tag{4.37}$$

4.12 Other simple three-dimensional elements

The possibilities of simple shapes in three dimensions are greater, for obvious reasons, than in two dimensions. A quite useful series of elements can, for instance, be based on triangular prisms (wedges) (Fig. 4.20). Here again variants of the product, Lagrange, approach or of the 'serendipity' type can be distinguished. The first element of both families, shown in Fig. 4.20(a), is identical and the shape functions are

$$N_a = \frac{1}{2} L_a (1 + \zeta_a \zeta) \qquad a = 1, 2, \ldots, 6$$

For the 'quadratic' element illustrated in Fig. 4.20(b) the shape functions are

Corner nodes

$$N_a = \tfrac{1}{2}L_a(2L_a - 1)(1 + \zeta_a\zeta) - \tfrac{1}{2}L_a(1 - \zeta^2) \qquad a = 1, 2, \ldots, 6$$

Mid-edge of rectangle:

$$N_7 = L_1(1 - \zeta^2), \qquad \text{etc.}$$

Mid-edge of triangles:

$$N_{10} = 2L_1L_2(1 + \zeta), \qquad \text{etc.}$$

Such elements are not purely esoteric but have a practical application as 'fillers' in conjunction with 20-noded serendipity elements.

Part 2. Hierarchical shape functions

4.13 Hierarchic polynomials in one dimension

The general ideas of hierarchic approximation were introduced in Sec.4.2 in the context of simple, linear, elements. The idea of generating higher order hierarchic forms is again simple. We shall start from a one-dimensional expansion as this has been shown to provide a basis for the generation of two- and three-dimensional forms in previous sections.

To generate a polynomial of order p along an element side we do not need to introduce nodes but can instead use parameters without any obvious physical meaning. We could use here a linear expansion specified by 'standard' functions N_1 and N_2 and add to this a

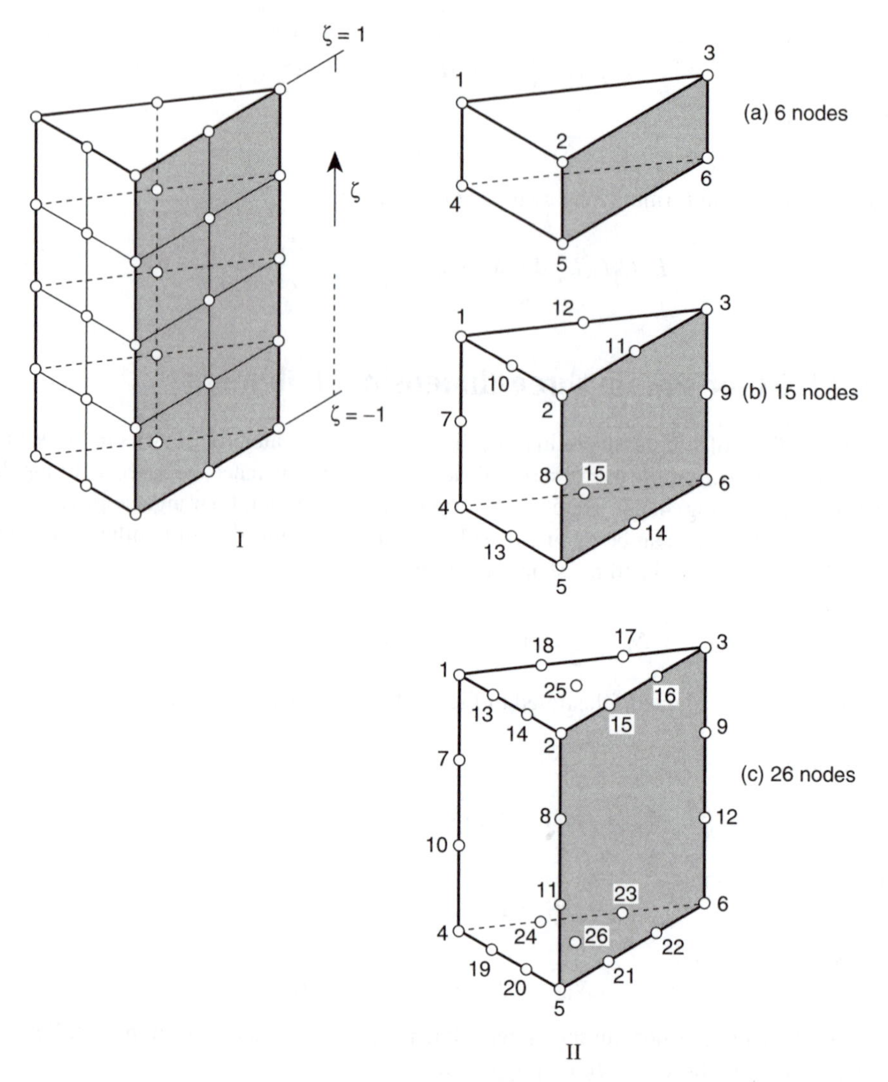

Fig. 4.20 Triangular prism elements (serendipity) family: (a) linear, (b) quadratic, (c) cubic.

series of polynomials always designed so as to have zero values at the ends of the range (i.e., points 1 and 2).

Thus for a quadratic approximation, we would write over the typical one-dimensional element, for instance,

$$\hat{u} = N_1\tilde{u}_1 + N_2\tilde{u}_2 + N_3\tilde{u}_3 \tag{4.38}$$

where

$$N_1 = \frac{1}{2}(1-\xi) \qquad N_2 = \frac{1}{2}(1+\xi) \qquad N_3 = (1-\xi^2) \tag{4.39}$$

using in the above the normalized x coordinate [viz. Eq. (4.17)].

We note that the parameter $\tilde{u}_3$ does in fact have a meaning in this case as it is the magnitude of the departure from linearity of the approximation $\hat{u}$ at the element centre, since N_3 has been chosen here to have the value of unity at that point.

In a similar manner, for a cubic element we simply have to add $N_4\tilde{u}_4$ to the quadratic expansion of Eq. (4.39), where N_4 is any cubic of the form

$$N_4 = \alpha_0 + \xi\alpha_1 + \xi^2\alpha_2 + \xi^3\alpha_3 \tag{4.40}$$

and which has zero values at $\xi = \pm 1$ (i.e., at nodes 1 and 2). Again an infinity of choices exists, and we could select a cubic of a simple form which has a zero value at the centre of the element and for which $\mathrm{d}N_4/\mathrm{d}\xi = 1$ at the same point. Immediately we can write

$$N_4 = \xi(1 - \xi^2) \tag{4.41}$$

as the cubic function with the desired properties. Now the parameter $\tilde{u}_4$ denotes the departure of the slope at the centre of the element from that of the linear approximation.

We note that we could proceed in a similar manner and define the fourth-order hierarchical element shape function as

$$N_5 = \xi^2(1 - \xi^2) \tag{4.42}$$

but a physical identification of the parameter associated with this now becomes more difficult (even though it is not strictly necessary).

As we have already noted, the above set is not unique and many other possibilities exist. An alternative convenient form for the hierarchical functions is defined by

$$N_{p+1}(\xi) = \begin{cases} \dfrac{1}{p!}(\xi^p - 1) & p \text{ even} \\ \dfrac{1}{p!}(\xi^p - \xi) & p \text{ odd} \end{cases} \tag{4.43}$$

where p (≥ 2) is the degree of the introduced polynomial.[16] This yields the set of shape functions:

$$\begin{aligned} N_3 &= \tfrac{1}{2}(\xi^2 - 1) \qquad N_4 = \tfrac{1}{6}(\xi^3 - \xi) \\ N_5 &= \tfrac{1}{24}(\xi^4 - 1) \qquad N_6 = \tfrac{1}{120}(\xi^5 - \xi), \qquad \text{etc.} \end{aligned} \tag{4.44}$$

We observe that all derivatives of N_{p+1} of second or higher order have the value zero at $\xi = 0$, apart from $\mathrm{d}^p N_{p+1}/\mathrm{d}\xi^p$, which equals unity at that point, and hence, when shape functions of the form given by Eq. (4.44) are used, we can identify the parameters in the approximation as

$$\tilde{u}_{p+1} = \left.\frac{\mathrm{d}^p \hat{u}}{\mathrm{d}\xi^p}\right|_{\xi=0} \qquad p \geq 2 \tag{4.45}$$

This identification gives a general physical significance but is by no means necessary.

In two- and three-dimensional elements a simple identification of the hierarchic parameters on interfaces will automatically ensure C_0 continuity of the approximation.

4.14 Two- and three-dimensional, hierarchical elements of the 'rectangle' or 'brick' type

In deriving 'standard' finite element approximations we have shown that all shape functions for the Lagrange family could be obtained by a simple multiplication of one-dimensional ones and those for serendipity elements by a combination of such multiplications. The situation is even simpler for hierarchic elements. Here *all* the shape functions can be obtained by a simple multiplication process.

Thus, for instance, in Fig. 4.21 we show the shape functions for a lagrangian nine-noded element and the corresponding hierarchical functions. The latter not only have simpler shapes but are more easily calculated, being simple products of linear and quadratic terms of Eq. (4.43) or (4.44). Using products of lagrangian polynomials the three functions illustrated are simply

$$\begin{aligned} N_1 &= (1-\xi)(1+\eta)/4 \\ N_2 &= (1-\xi)(1-\eta^2)/2 \\ N_3 &= (1-\xi^2)(1-\eta^2) \end{aligned} \tag{4.46}$$

The distinction between lagrangian and serendipity forms now disappears as for the latter in the present case the last shape function (N_3) is simply omitted.

Indeed, it is now easy to introduce interpolation for elements of the type illustrated in Fig. 4.11 in which a different expansion is used along different sides. This essential characteristic of hierarchical elements is exploited in adaptive refinement (viz. Chapter 14) where new degrees of freedom (or polynomial order increase) is made only when required by the magnitude of the error.

A similar process clearly applies to the three-dimensional family of hierarchical brick-type elements.

4.15 Triangle and tetrahedron family

Once again the concepts of multiplication can be introduced in terms of area or volume coordinates to define the triangle and tetrahedron family of elements.[15, 16] Starting from the linear shape functions for the corner nodes

$$N_a = L_a$$

hierarchical functions for mid-side and interior nodes can be added.

For the triangle shown in Fig. 4.14 we note that along the side 1–2, L_3 is identically zero, and therefore we have

$$(L_1 + L_2)_{1-2} = 1 \tag{4.47}$$

If ξ, measured along side 1–2, is the usual non-dimensional local element coordinate of the type we have used in deriving hierarchical functions for one-dimensional elements, we can write

$$L_1|_{1-2} = \tfrac{1}{2}(1-\xi) \qquad L_2|_{1-2} = \tfrac{1}{2}(1+\xi) \tag{4.48}$$

from which it follows that we have

$$\xi = (L_2 - L_1)_{1-2} \tag{4.49}$$

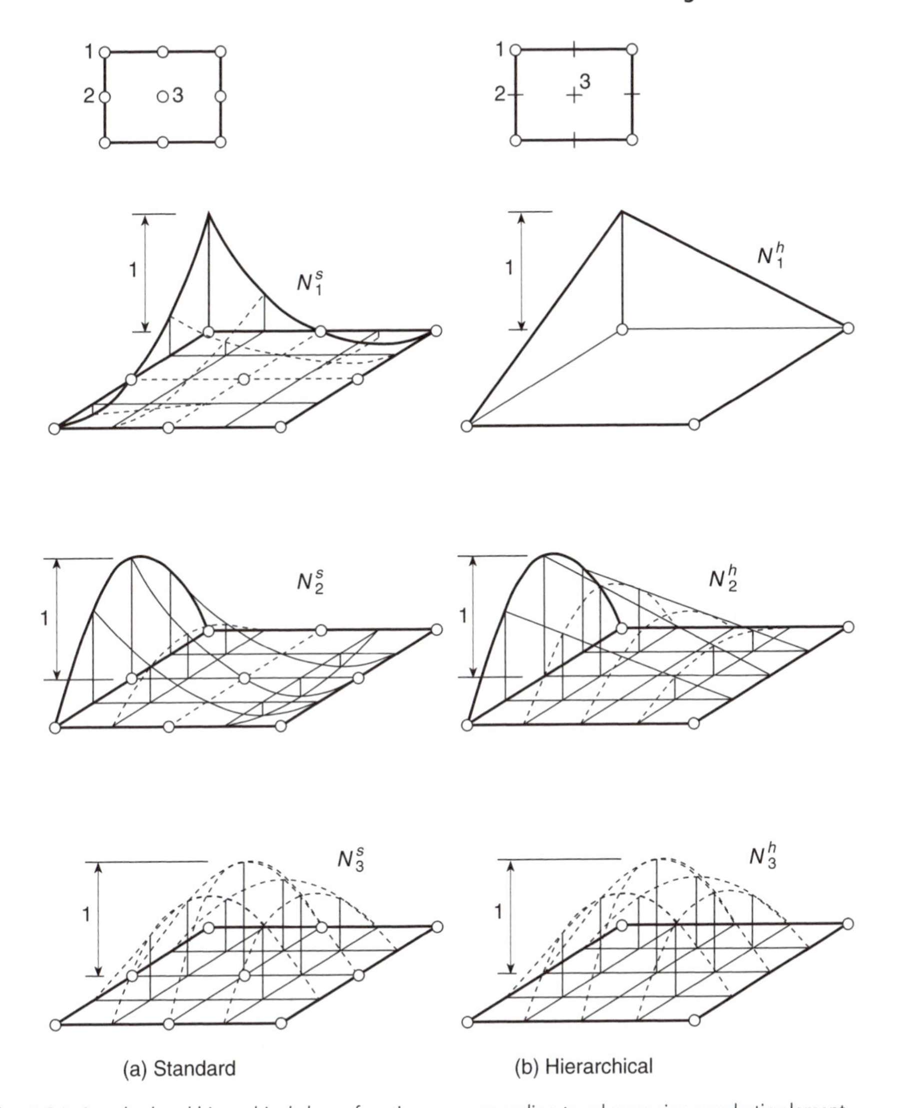

Fig. 4.21 Standard and hierarchical shape functions corresponding to a lagrangian quadratic element.

This suggests that we could generate hierarchical shape functions over the triangle by generalizing the one-dimensional shape function forms produced earlier. For example, using the expressions of Eq. (4.43), we associate with the side 1–2 the polynomial of degree p (≥ 2) defined by

$$N_{p(1-2)} = \begin{cases} \dfrac{1}{p!}[(L_2 - L_1)^p - (L_1 + L_2)^p] & p \text{ even} \\ \dfrac{1}{p!}[(L_2 - L_1)^p - (L_2 - L_1)(L_1 + L_2)^{p-1}] & p \text{ odd} \end{cases} \tag{4.50}$$

It follows from Eq. (4.48) that these shape functions are zero at nodes 1 and 2. In addition, it can easily be shown that $N_{p(1-2)}$ will be zero all along the sides 3–1 and 3–2 of the triangle, and so C_0 continuity of the approximation $\hat{u}$ is assured.

It should be noted that in this case for $p \geq 3$ the number of hierarchical functions arising from the element sides in this manner is insufficient to define a complete polynomial of degree p, and internal hierarchical functions, which are identically zero on the boundaries, need to be introduced; for example, for $p = 3$ the function $L_1L_2L_3$ could be used, while for $p = 4$ the three additional functions $L_1^2L_2L_3$, $L_1L_2^2L_3$, $L_1L_2L_3^2$ could be adopted.

In Fig. 4.22 typical, hierarchical, linear, quadratic, and cubic trial functions for a triangular element are shown. Identical procedures are obvious in the context of tetrahedra.

Hierarchical functions of other forms can be found in reference 23.

4.16 Improvement of conditioning with hierarchical forms

We have already mentioned that hierarchic element forms give a much improved equation conditioning for steady-state (static) problems due to their form which is more nearly diagonal. In Fig. 4.23 we show the 'condition number' (which is a measure of such diagonality and is defined in standard texts on linear algebra; see Appendix A) for a single cubic element and for an assembly of four cubic elements, using standard and hierarchic forms in their

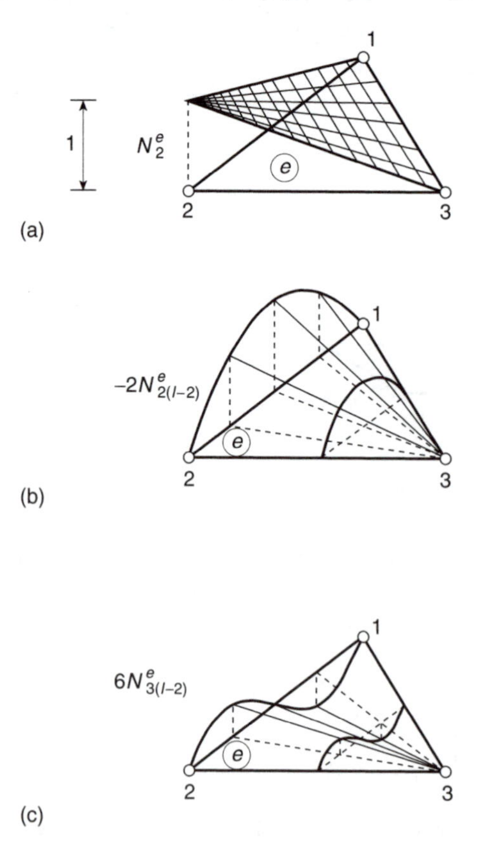

Fig. 4.22 Triangular elements and associated hierarchical shape functions of (a) linear, (b) quadratic, (c) cubic form.

Single element (Reduction of condition number = 10.7)

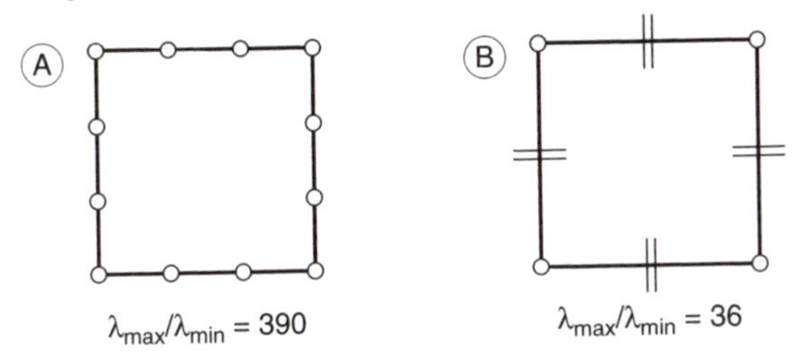

Four element assembly (Reduction of condition number = 13.2)

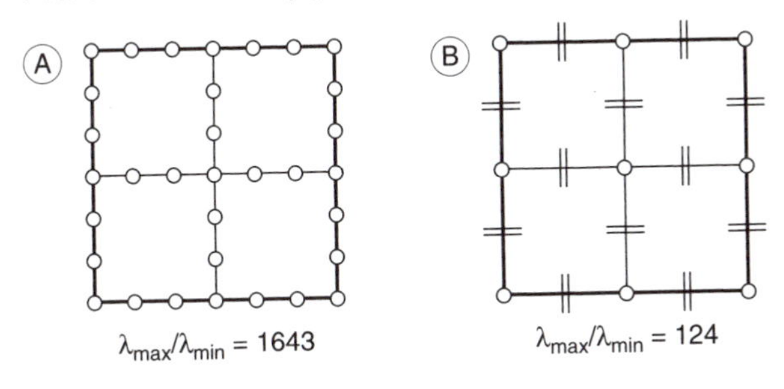

Cubic order elements

(A) Standard shape function

(B) Hierarchic shape function

Fig. 4.23 Improvement of condition number (ratio of maximum to minimum eigenvalue of the stiffness matrix) by use of hierarchical form (isotropic elasticity, $\nu = 0.15$).

formulation. The improvement of the conditioning is a distinct advantage of such forms and allows the use of iterative solution techniques to be more easily adopted.[17] Unfortunately much of this advantage disappears for transient analysis as the approximation must contain specific modes (see Chapter 16).

4.17 Global and local finite element approximation

The very concept of hierarchic approximations (in which the shape functions are not affected by the refinement) means that it is possible to include in the expansion

$$u = \sum_{a=1}^{n} N_a \tilde{u}_a \tag{4.51}$$

where functions N are not local in nature. Such functions may, for instance, be the exact solutions of an analytical problem which in some way resembles the problem dealt with, but do not satisfy some boundary or inhomogeneity conditions. The 'finite element', local, expansions would here be a device for correcting this solution to satisfy the real conditions.

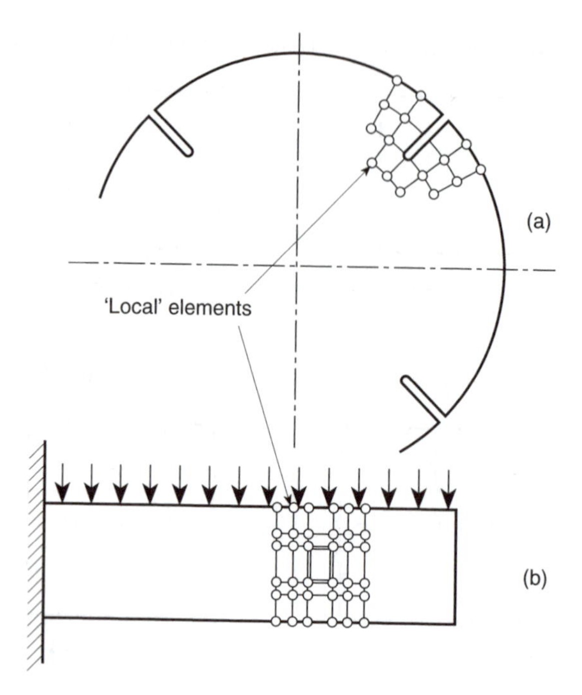

Fig. 4.24 Some possible uses of global–local approximation. (a) Rotating slotted disc. (b) Perforated beam.

This use of the global–local approximation was first suggested by Mote[18] in a problem where the coefficients of this function were fixed. The example involved here is that of a rotating disc with cutouts (Fig. 4.24). The global known solution is the analytical one corresponding to a disc without cutout, and finite elements are added locally to modify the solution. Other examples of such 'fixed' solutions may well be those associated with point loads, where the use of the global approximation serves to eliminate the singularity modelled badly by the discretization.

In some problems the singularity itself is unknown and the appropriate function can be added with an unknown coefficient. Some aspects of this are mentioned in Chapter 15 and, for waves, in the context of fluid dynamics, in reference 19.

4.18 Elimination of internal parameters before assembly – substructures

Internal nodes or nodeless internal parameters yield in the usual way the element properties

$$\frac{\partial \Pi^e}{\partial \tilde{\mathbf{u}}^e} = \mathbf{K}^e \tilde{\mathbf{u}}^e + \mathbf{f}^e \tag{4.52}$$

As $\tilde{\mathbf{u}}^e$ can be subdivided into parts which are common with other elements, $\tilde{\mathbf{u}}_1^e$, and others which occur in the particular element only, $\tilde{\mathbf{u}}_2^e$, we can immediately write

$$\frac{\partial \Pi}{\partial \tilde{\mathbf{u}}_2^e} = \frac{\partial \Pi^e}{\partial \tilde{\mathbf{u}}_2^e} = \mathbf{0}$$

and eliminate $\tilde{\mathbf{u}}_2^e$ from further consideration. Writing Eq. (4.52) in a partitioned form we have

$$\frac{\partial \Pi^e}{\partial \tilde{\mathbf{u}}^e} = \begin{Bmatrix} \dfrac{\partial \Pi^e}{\partial \tilde{\mathbf{u}}_1^e} \\ \dfrac{\partial \Pi^e}{\partial \tilde{\mathbf{u}}_2^e} \end{Bmatrix} = \begin{bmatrix} \mathbf{K}_{11}^e & \mathbf{K}_{12}^e \\ \mathbf{K}_{21}^e & \mathbf{K}_{22}^e \end{bmatrix} \begin{Bmatrix} \tilde{\mathbf{u}}_1^e \\ \tilde{\mathbf{u}}_2^e \end{Bmatrix} + \begin{Bmatrix} \mathbf{f}_1^e \\ \mathbf{f}_2^e \end{Bmatrix} = \begin{Bmatrix} \dfrac{\partial \Pi^e}{\partial \tilde{\mathbf{u}}_1^e} \\ \mathbf{0} \end{Bmatrix} \tag{4.53}$$

From the second set of equations given above we can write

$$\tilde{\mathbf{u}}_2^e = -(\mathbf{K}_{22}^e)^{-1}(\mathbf{K}_{21}^e \tilde{\mathbf{u}}_1^e + \mathbf{f}_2^e) \tag{4.54}$$

which on substitution yields

$$\frac{\partial \Pi^e}{\partial \tilde{\mathbf{u}}_1^e} = \bar{\mathbf{K}}_{11}^e \tilde{\mathbf{u}}_1^e + \bar{\mathbf{f}}_1^e \tag{4.55}$$

in which

$$\begin{aligned} \bar{\mathbf{K}}_{11}^e &= \mathbf{K}_{11}^e - \mathbf{K}_{12}^e (\mathbf{K}_{22}^e)^{-1} \mathbf{K}_{21}^e \\ \bar{\mathbf{f}}_1^e &= \mathbf{f}_1^e - \mathbf{K}_{12}^e (\mathbf{K}_{22}^e)^{-1} \mathbf{f}_2^e \end{aligned} \tag{4.56}$$

This process of partial solution is also known in the literature as 'static condensation'.[20]

Assembly of the total region then follows, by considering only the element boundary variables, thus giving a saving in the equation-solving effort at the expense of a few additional manipulations carried out at the element stage.[20]

Perhaps a structural interpretation of this elimination is desirable. What in fact is involved is the separation of a part of the structure from its surroundings and determination of its solution separately for any prescribed displacements at the interconnecting boundaries. $\bar{\mathbf{K}}_{11}^e$ is now simply the overall stiffness of the separated structure and $\bar{\mathbf{f}}_1^e$ the equivalent set of nodal forces.

If the triangulation of Fig. 4.25 is interpreted as an assembly of pin-jointed bars the reader will recognize immediately the well-known device of 'substructures' used frequently in structural engineering. Such a substructure is in fact simply a complex element from which the internal degrees of freedom have been eliminated. Immediately a new possibility for devising more elaborate, and presumably more accurate, elements is presented.

Figure 4.25(a) can be interpreted as a continuum field subdivided into linear triangular elements. The substructure results in fact in one complex element shown in Fig. 4.25(b) with a number of boundary nodes.

The only difference from elements derived in previous sections is the fact that the unknown **u** now is not approximated internally by one set of smooth shape functions but by a series of piecewise approximations. This presumably results in a slightly poorer approximation but an economic advantage may arise if the total computation time for such an assembly is saved.

Substructuring is an important device in complex problems, particularly where a repetition of complicated components arises.

In simple, small-scale finite element analysis, much improved use of simple triangular elements was found by the use of simple subassemblies of the triangles (or indeed tetrahedra). For instance, a quadrilateral based on four triangles from which the central node is eliminated was found to give an economic advantage over direct use of simple triangles (Fig. 4.26). This and other subassemblies based on triangles are discussed by Doherty *et al.*[21] and used by Nagtegaal *et al.*[22] and others.

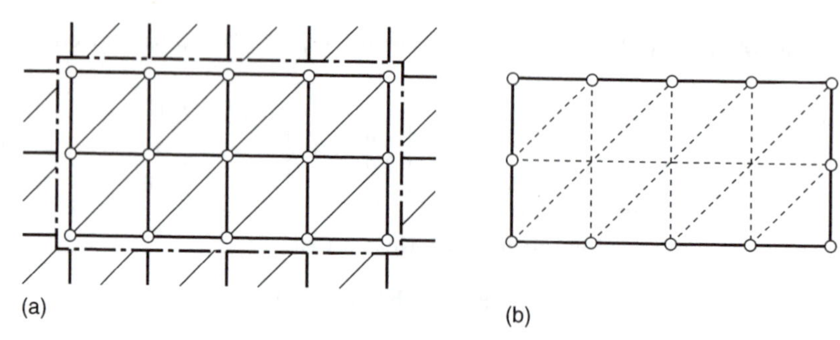

Fig. 4.25 Substructure of a complex element.

4.19 Concluding remarks

An unlimited selection of element types has been presented here to the reader – and indeed equally unlimited alternative possibilities exist.[4, 8] What is the use of such complex elements in practice? As presented so far the triangular and tetrahedral elements are limited to situations where the real region is of a suitable shape which can be represented as an assembly of flat facets and all other elements are limited to situations represented by an assembly of right prisms. Such a limitation would be so severe that little practical purpose would have been served by the derivation of such shape functions unless some way could be found of distorting these elements to fit realistic curved boundaries. In fact, methods for doing this are available and will be described in the next chapter.

4.20 Problems

4.1 Develop an explicit form of the standard shape functions at nodes 1, 3 and 6 for the element shown in Fig. 4.27(a).

Using a Pascal triangle in ξ and η show the polynomials included in the element.

4.2 Develop an explicit form of the standard shape functions at nodes 2, 3 and 9 for the element shown in Fig. 4.27(b).

Using a Pascal triangle in ξ and η show the polynomials included in the element.

4.3 Develop an explicit form of the standard shape functions at nodes 1, 2 and 5 for the element shown in Fig. 4.27(c).

Using a Pascal triangle in ξ and η show the polynomials included in the element.

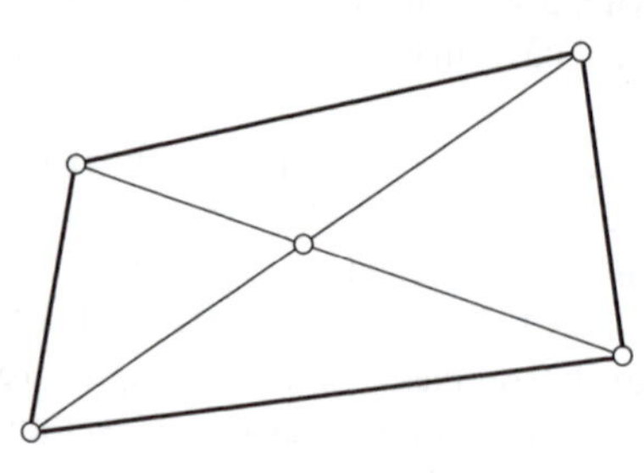

Fig. 4.26 A composite quadrilateral made from four simple triangles.

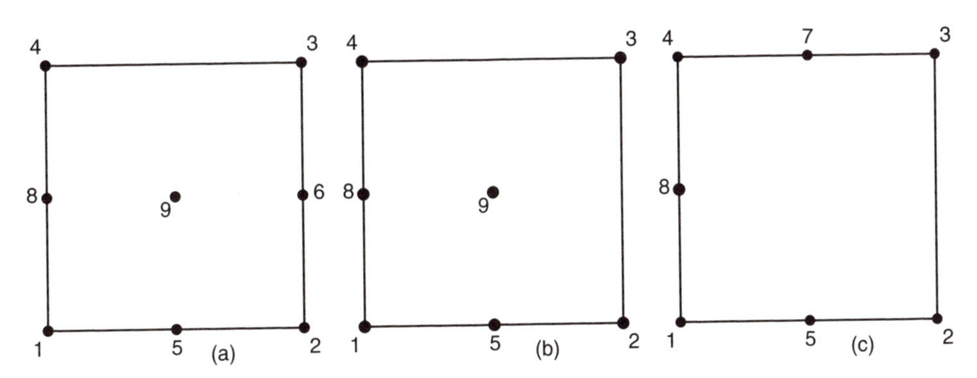

Fig. 4.27 Quadrilateral element for Problems 4.1 to 4.4.

4.4 Develop an explicit expression in hierarchical form for all nodes of the element shown in Fig. 4.27(c).

4.5 Develop an explicit form of the standard shape functions at nodes 1, 2 and 5 for the element shown in Fig. 4.28(a).

Using a Pascal triangle in ξ and η show the polynomials included in the element.

4.6 Develop an explicit form of the standard shape functions at nodes 1, 5 and 7 for the element shown in Fig. 4.28(b).

Using a Pascal triangle in ξ and η show the polynomials included in the element.

4.7 The mesh for a problem contains an 8-node quadratic serendipity rectangle adjacent to a 6-node quadratic triangle as shown in Fig. 4.29. Show that the coordinates computed from each element satisfy C^0 continuity along the edge 3–7–11.

4.8 Determine an explicit expression for the shape function of node 1 of the linear triangular prism shown in Fig. 4.20(a).

4.9 Determine an explicit expression for the hierarchical shape function of nodes 1, 7 and 10 of the quadratic triangular prism shown in Fig. 4.20(b).

4.10 Determine an explicit expression for the shape function of nodes 1, 7, 13 and 25 of the cubic triangular prism shown in Fig. 4.20(c).

4.11 On a sketch show the location of the nodes for the quartic member of the tetrahedron family. Construct an explicit expression for the shape function of the vertex node located at $(L_1, L_2, L_3, L_4) = (1, 0, 0, 0)$ and the mid-edge node located at $(0.25, 0.75, 0, 0)$.

4.12 On a sketch show the location of the nodes for the quartic member of the serendipity family. Construct an explicit expression for the shape function of the vertex node located at $(\xi, \eta, \zeta) = (1,\ 1,\ 1)$ and the mid-edge node located at $(0.5,\ 1,\ 1)$.

4.13 On a sketch show the location of the nodes for the quartic member of the triangular prism family shown in Fig. 4.20. Construct an explicit expression for the hierarchical shape function of a vertex node, an edge node of a triangular face and an edge node of a rectangular face.

4.14 On a sketch show the location of the nodes for the quadratic member of the triangular prism family in which lagrangian interpolation is used on rectangular faces (see Fig. 4.20). Construct an explicit expression for the shape function of a vertex node, an edge node of a triangular face and an edge node of a rectangular face.

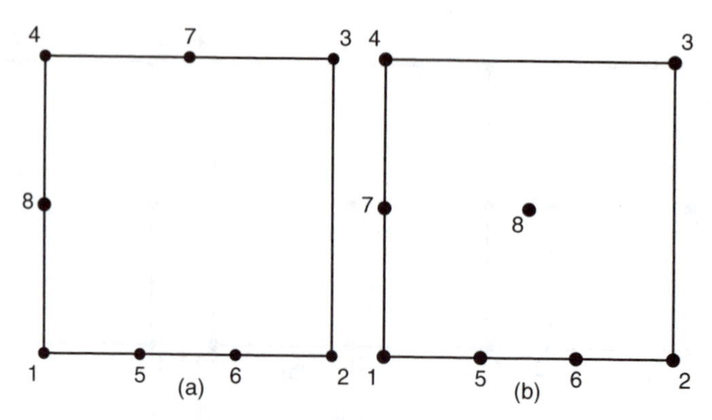

Fig. 4.28 Quadrilateral element for Problems 4.5 and 4.6.

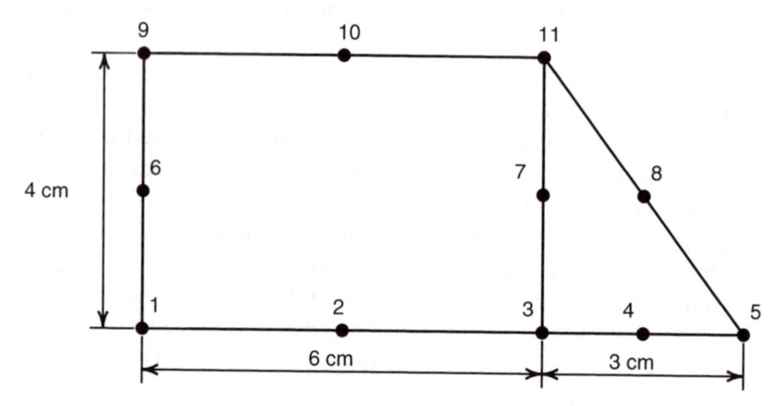

Fig. 4.29 Quadratic rectangle and triangle for Problem 4.7.

4.15 On a sketch show the location of the nodes for the cubic member of the triangular prism family in which lagrangian interpolation is used on rectangular faces (see Fig. 4.20). Construct an explicit expression for the shape function of a vertex node, an edge node of a triangular face, an edge node of a rectangular face, a mid-face node of a triangular face, a mid-face node of a rectangular face, and for any internal nodes.

References

1. W. Rudin. *Principles of Mathematical Analysis*. McGraw-Hill, 3rd edition, 1976.
2. P.C. Dunne. Complete polynomial displacement fields for finite element methods. *Trans. Roy. Aero. Soc.*, 72:245, 1968.
3. B.M. Irons, J.G. Ergatoudis, and O.C. Zienkiewicz. Comments on 'complete polynomial displacement fields for finite element method' (by P.C. Dunne). *Trans. Roy. Aeronaut. Soc.*, 72:709, 1968.
4. J.G. Ergatoudis, B.M. Irons, and O.C. Zienkiewicz. Curved, isoparametric, 'quadrilateral' elements for finite element analysis. *Int. J. Solids Struct*, 4:31–42, 1968.

5. O.C. Zienkiewicz, B.M. Irons, J.G. Ergatoudis, S. Ahmad, and F.C. Scott. Isoparametric and associated elements families for two and three dimensional analysis. In *Finite Element Methods in Stress Analysis*, Chapter 13. Tapir Press, Trondheim, 1969.
6. J.H. Argyris, K.E. Buck, H.M. Hilber, G. Mareczek, and D.W. Scharpf. Some new elements for matrix displacement methods. In *Proc. 2nd Conf. Matrix Methods in Structural Mechanics*, volume AFFDL-TR-68-150, Wright Patterson Air Force Base, Ohio, Oct. 1968.
7. R.L. Taylor. On completeness of shape functions for finite element analysis. *Int. J. Numer. Meth. Eng.*, 4:17–22, 1972.
8. O.C. Zienkiewicz, B.M. Irons, J. Campbell, and F.C. Scott. Three dimensional stress analysis. In *IUTAM Symposium on High Speed Computing in Elasticity*, Li'ege, 1970.
9. F.C. Scott. A quartic, two dimensional isoparametric element. Undergraduate Project, University of Wales, 1968.
10. J.H. Argyris, I. Fried, and D.W. Scharpf. The TET 20 and TEA 8 elements for the matrix displacement method. *Aero. J.*, 72:618–625, 1968.
11. P. Silvester. Higher order polynomial triangular finite elements for potential problems. *Int. J. Eng. Sci.*, 7:849–861, 1969.
12. B. Fraeijs de Veubeke. Displacement and equilibrium models in finite element method. In O.C. Zienkiewicz and G.S. Holister, editors, *Stress Analysis*, Chapter 9, pages 145–197. John Wiley & Sons, Chichester, 1965.
13. J.H. Argyris. Triangular elements with linearly varying strain for the matrix displacement method. *J. Roy. Aero. Soc. Tech. Note*, 69:711–713, 1965.
14. J.G. Ergatoudis, B.M. Irons, and O.C. Zienkiewicz. Three dimensional analysis of arch dams and their foundations. In *Proc. Symp. Arch Dams*, Inst. Civ. Eng., London, 1968.
15. A.G. Peano. Hierarchics of conforming finite elements for elasticity and plate bending. *Comp. Math. and Applications*, 2:3–4, 1976.
16. J.P. de S.R. Gago. *A posteri error analysis and adaptivity for the finite element method.* Ph.D. thesis, Department of Civil Engineering, University of Wales, Swansea, 1982.
17. O.C. Zienkiewicz, J.P. De S.R. Gago, and D.W. Kelly. The hierarchical concept in finite element analysis. *Comp. Struct.*, 16:53–65, 1983.
18. C.D. Mote. Global–local finite element. *Int. J. Numer. Meth. Eng.*, 3:565–574, 1971.
19. O.C. Zienkiewicz, R.L. Taylor, and P. Nithiarasu. *The Finite Element Method for Fluid Dynamics*. Butterworth-Heinemann, Oxford, 6th edition, 2005.
20. E.L. Wilson. The static condensation algorithm. *Int. J. Numer. Meth. Eng.*, 8:199–203, 1974.
21. W.P. Doherty, E.L. Wilson, and R.L. Taylor. Stress analysis of axisymmetric solids utilizing higher-order quadrilateral finite elements. Technical Report 69–3, Structural Engineering Laboratory, Univ. of California, Berkeley, Jan. 1969.
22. J.C. Nagtegaal, D.M. Parks, and J.R. Rice. On numerical accurate finite element solutions in the fully plastic range. *Comp. Meth. Appl. Mech. Eng.*, 4:153–177, 1974.
23. S.J. Sherwin and G.E. Karniadakis. A new triangular and tetrahedral basis for high-order (hp) finite element methods. *Int. J. Numer. Meth. Eng.* 38:3775–3802, 1995.

5

Mapped elements and numerical integration – 'infinite' and 'singularity elements'

5.1 Introduction

In the previous chapter we have shown how some general families of finite elements can be obtained for C_0 interpolations. A progressively increasing number of nodes and hence improved accuracy characterizes each new member of the family and presumably the number of such elements required to obtain an adequate solution decreases rapidly. To ensure that a small number of elements can represent a relatively complex form of the type that is liable to occur in real, rather than academic, problems, simple rectangles and triangles no longer suffice. This chapter is therefore concerned with the subject of distorting such simple forms into others of more arbitrary shape.

Elements of the basic one-, two-, or three-dimensional types will be 'mapped' into distorted forms in the manner indicated in Figs 5.1 and 5.2.

In these figures it is shown that the 'parent' ξ, η, ζ, or L_1, L_2, L_3, L_4 coordinates can be distorted to a new, curvilinear set when plotted in cartesian x, y, z space.

Not only can two-dimensional elements be distorted into others in two dimensions but the mapping of these can be taken into three dimensions as indicated by the flat sheet elements of Fig. 5.2 distorting into a three-dimensional space. This principle applies generally, providing a one-to-one correspondence between cartesian and curvilinear coordinates can be established, i.e., once the mapping relations of the type

$$\begin{Bmatrix} x \\ y \\ z \end{Bmatrix} = \begin{Bmatrix} f_x(\xi, \eta, \zeta) \\ f_y(\xi, \eta, \zeta) \\ f_z(\xi, \eta, \zeta) \end{Bmatrix} \quad \text{or} \quad \begin{Bmatrix} f_x(L_1, L_2, L_3, L_4) \\ f_y(L_1, L_2, L_3, L_4) \\ f_z(L_1, L_2, L_3, L_4) \end{Bmatrix} \tag{5.1}$$

can be established.

Once such coordinate relationships are known, shape functions can be specified in local (parent) coordinates and by suitable transformations the element properties established in the global coordinate system.

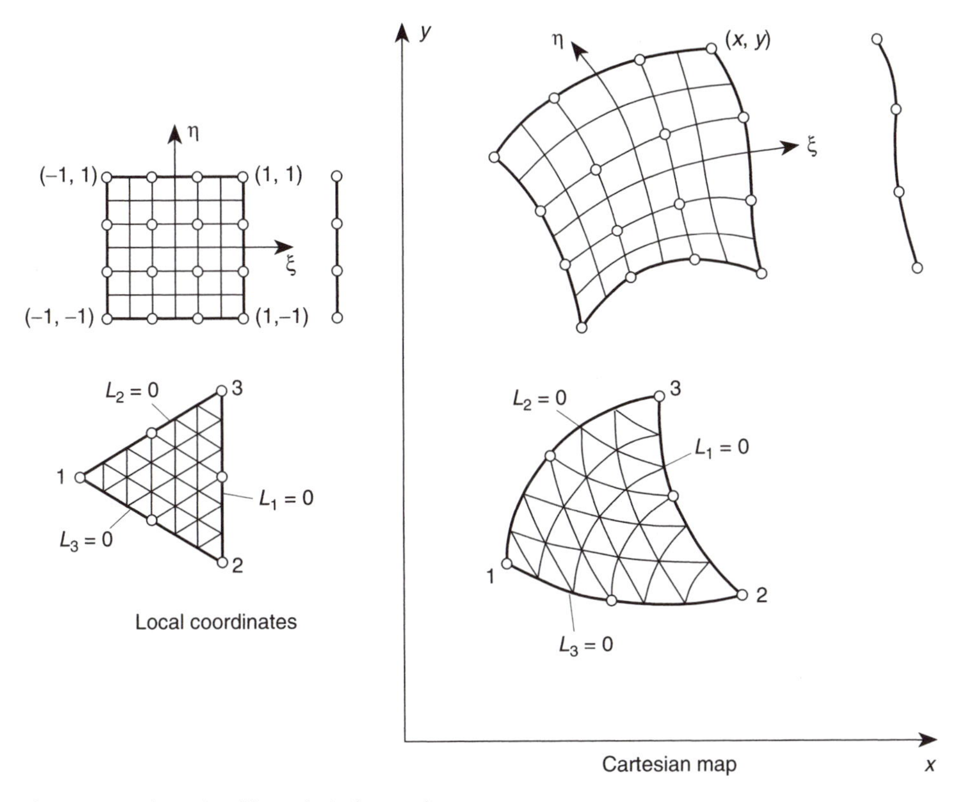

Fig. 5.1 Two-dimensional 'mapping' of some elements.

In what follows we shall first discuss the so-called isoparametric form of relationship (5.1) which has found a great deal of practical application. Full details of this formulation will be given, including the establishment of element matrices by numerical integration.

In later sections we shall show that many other coordinate transformations also can be used effectively.

Parametric curvilinear coordinates

5.2 Use of 'shape functions' in the establishment of coordinate transformations

A most convenient method of establishing the coordinate transformations is to use the 'standard' type of C_0 shape functions we have already derived to represent the variation of the unknown function.

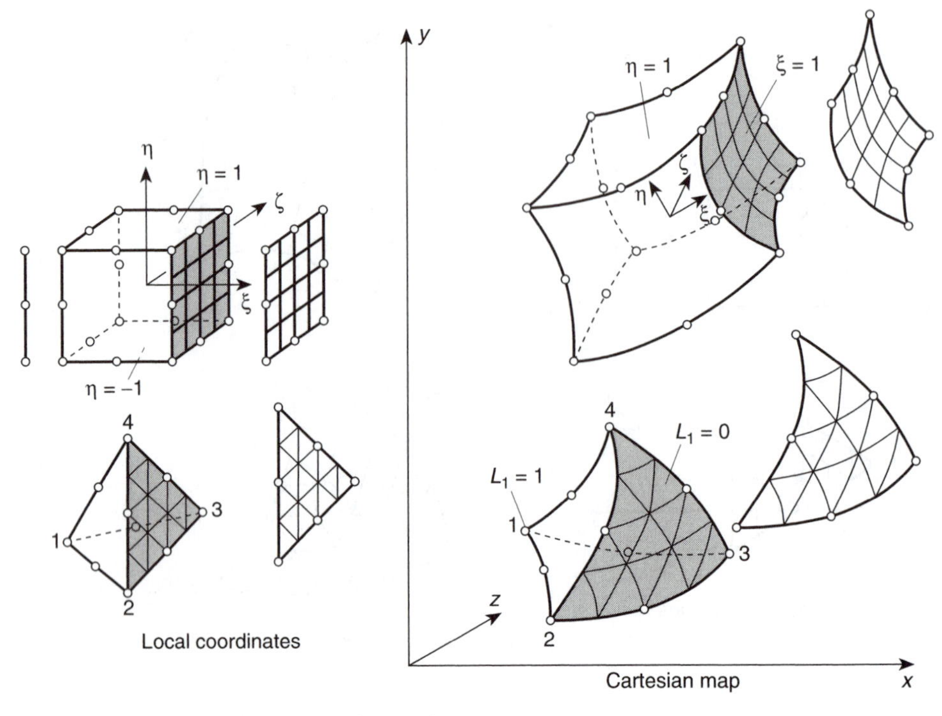

Fig. 5.2 Three-dimensional 'mapping' of some elements.

If we write, for instance, for each element

$$x = N'_1 x_1 + N'_2 x_2 + \cdots = \mathbf{N}' \begin{Bmatrix} x_1 \\ x_2 \\ \vdots \end{Bmatrix} = \mathbf{N}'\mathbf{x}$$

$$y = N'_1 y_1 + N'_2 y_2 + \cdots = \mathbf{N}' \begin{Bmatrix} y_1 \\ y_2 \\ \vdots \end{Bmatrix} = \mathbf{N}'\mathbf{y} \tag{5.2}$$

$$z = N'_1 z_1 + N'_2 z_2 + \cdots = \mathbf{N}' \begin{Bmatrix} z_1 \\ z_2 \\ \vdots \end{Bmatrix} = \mathbf{N}'\mathbf{z}$$

in which $\mathbf{N}'$ are standard shape functions given in terms of the local (parent) coordinates, then a relationship of the required form is immediately available. Further, the points with coordinates x_1, y_1, z_1, etc., will lie at appropriate points of the element boundary or interior (as from the general definitions of the standard shape functions we know that these have a value of unity at the point in question and zero elsewhere). These points can establish nodes *a priori*.

To each set of local coordinates there will correspond a set of global cartesian coordinates and in general only one such set. We shall see, however, that a non-uniqueness may arise if the nodal coordinates are placed such that a violent distortion occurs.

The concept of using such element shape functions for establishing curvilinear coordinates in the context of finite element analysis appears to have been introduced first by Taig.[1] In his first application basic linear quadrilateral relations were used. Irons generalized the idea for other elements.[2, 3]

Quite independently the exercises of devising various practical methods of generating curved surfaces for purposes of engineering design led to the establishment of similar definitions by Coons[4] and Forrest,[5] and indeed today the subjects of surface definitions and analysis are drawing closer together due to this activity.

In Fig. 5.3 an actual distortion of elements based on the quadratic and cubic members of the two-dimensional 'serendipity' family is shown. It is seen here that a one-to-one relationship exists between the local (ξ, η) and global (x, y) coordinates. If the fixed (nodal) points are such that a violent distortion occurs then a non-uniqueness can occur in the manner indicated for two situations in Fig. 5.4. Here at internal points of the distorted element two or more local coordinates correspond to the same cartesian coordinate and in addition to some internal points being mapped outside the element. Care must be taken in practice to avoid such gross distortion. Figure 5.5 shows two examples of a two-dimensional (ξ, η) element mapped into a three-dimensional (x, y, z) space.

We shall often refer to the basic element in undistorted, local, coordinates as a 'parent' element.

In Sec. 5.5 we shall define a quantity known as the jacobian determinant. The well-known condition for a *one-to-one* mapping (such as exists in Fig. 5.3 and does not in Fig. 5.4) is that the sign of this quantity should remain unchanged at all the points of the mapped element.

It can be shown that with a parametric transformation based on bilinear shape functions, the necessary condition is that no internal angle [such as α in Fig. 5.6(a)] be equal or greater than 180°.[6] In transformations based on quadratic 'serendipity' or 'lagrangian' functions,

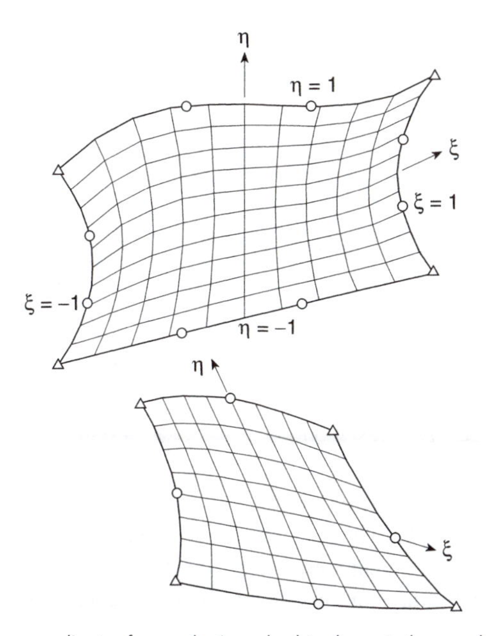

Fig. 5.3 Plots of curvilinear coordinates for quadratic and cubic elements (reasonable distortion).

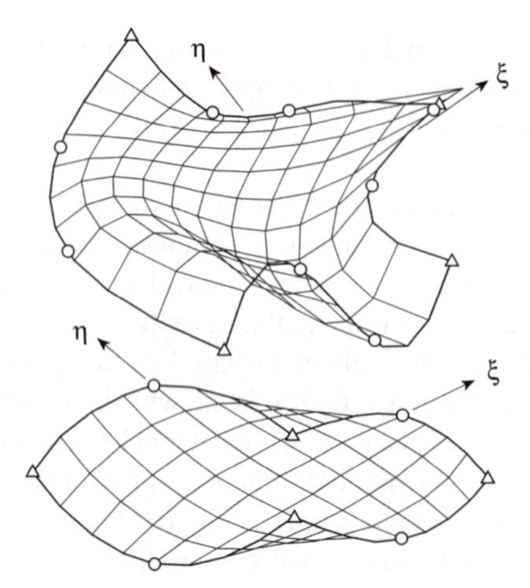

Fig. 5.4 Unreasonable element distortion leading to non-unique mapping and 'overspill'. Quadratic and cubic elements.

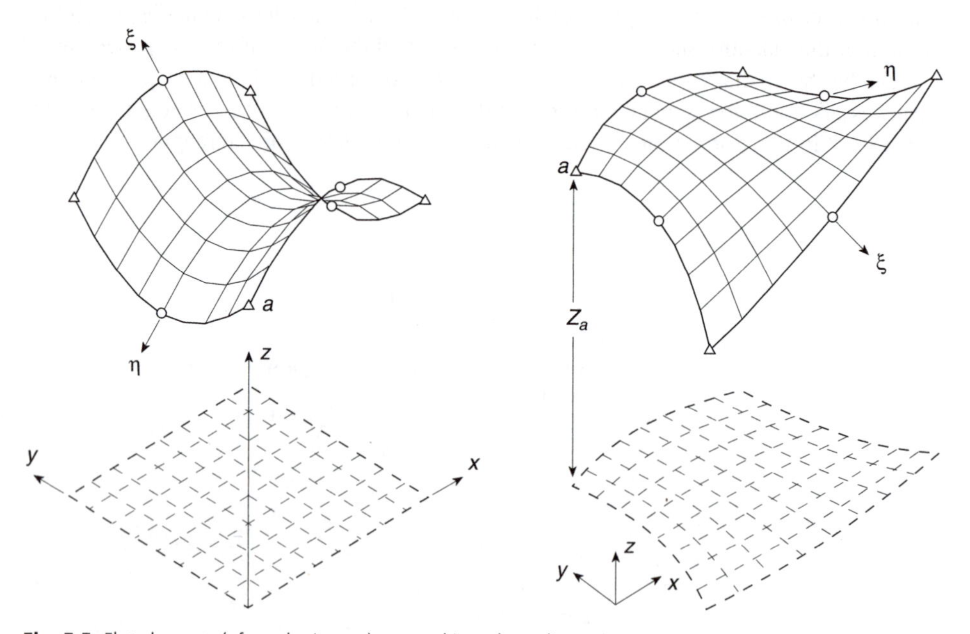

Fig. 5.5 Flat elements (of quadratic type) mapped into three dimensions.

it is necessary in addition to this requirement to ensure that the mid-side nodes are in the 'middle half' of the distance between adjacent corners but a 'middle third' shown in Fig. 5.6 is safer. For cubic functions such general rules are impractical and numerical checks on the sign of the jacobian determinant are necessary. In practice a quadratic distortion is usually sufficient.

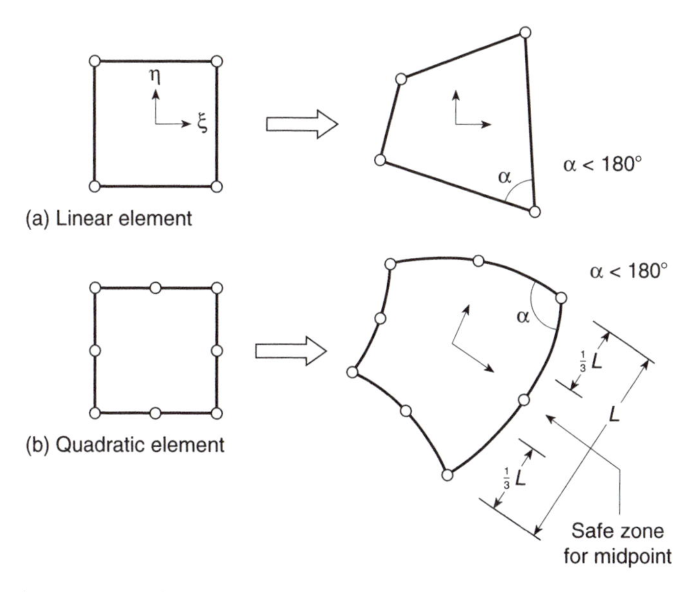

Fig. 5.6 Rules for uniqueness of mapping (a) and (b).

5.3 Geometrical conformity of elements

While it was shown that by the use of the shape function transformation each parent element maps uniquely a part of the real object, it is important that the subdivision of this into the new, curved, elements should leave no gaps. The possibility of such gaps is indicated by dotted lines in Fig. 5.7.

Theorem 1. *If two adjacent elements are generated from 'parents' in which the shape functions satisfy C_0 continuity requirements then the distorted elements will be continuous (compatible).*

This statement is obvious, as in such cases uniqueness of any function u required by continuity is simply replaced by that of uniqueness of the x, y, or z coordinate. As adjacent elements are given the same sets of coordinates at nodes, continuity is implied.

5.4 Variation of the unknown function within distorted, curvilinear elements. Continuity requirements

With the shape of the element now defined by the shape functions $\mathbf{N}'$ the variation of the unknown, u, has to be specified before we can establish element properties. This is most conveniently given in terms of local, curvilinear coordinates by the usual expression

$$u = \mathbf{N}\tilde{\mathbf{u}}^e \tag{5.3}$$

where $\tilde{\mathbf{u}}^e$ lists the nodal values.

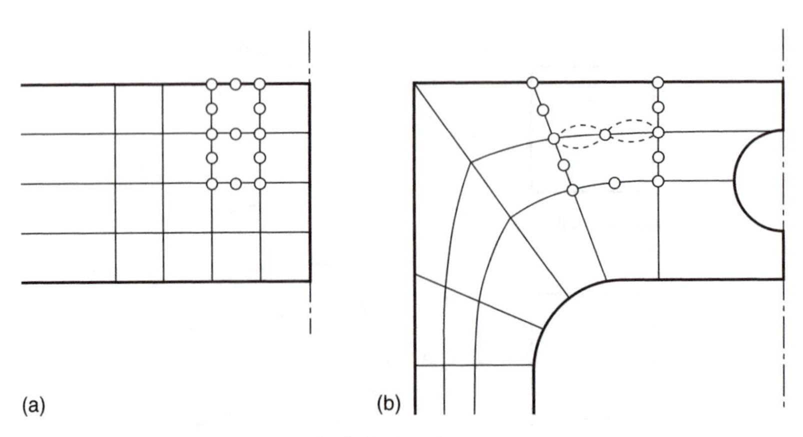

Fig. 5.7 Compatibility requirements in a real subdivision of space.

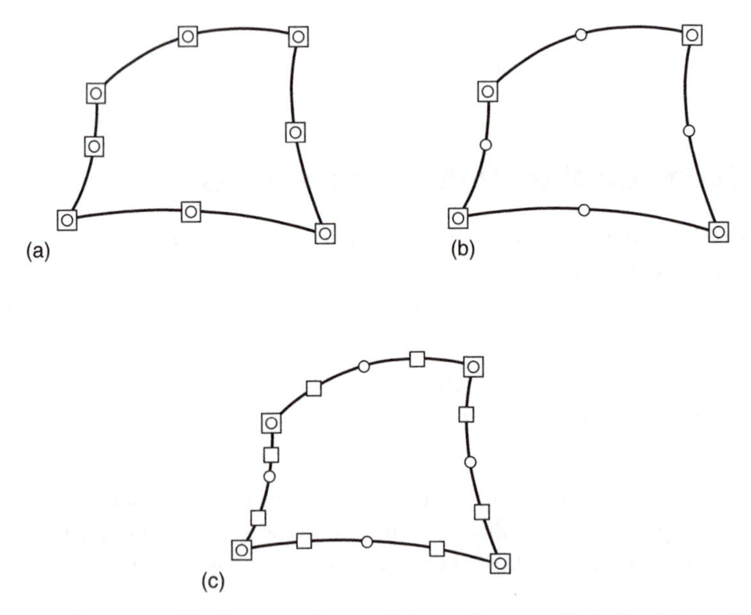

Fig. 5.8 Various element specifications: ○ point at which coordinate specified; □ point at which function parameter specified. (a) Isoparametric. (b) Superparametric. (c) Subparametric.

Theorem 2. *If the shape functions* **N** *used in (5.3) are such that* C_0 *continuity of* u *is preserved in the parent coordinates then* C_0 *continuity requirements will be satisfied in distorted elements.*

The proof of this statement follows the same lines as that in the previous section.

The nodal values may or may not be associated with the same nodes as used to specify the element geometry. For example, in Fig. 5.8 the points marked with a *circle* are used to define the element geometry. We could use the values of the function defined at nodes marked with a *square* to define the variation of the unknown.

In Fig. 5.8(a) the same points define the geometry and the finite element analysis points. If then

$$\mathbf{N} = \mathbf{N}' \tag{5.4}$$

i.e., the shape functions defining the geometry and the function are the same, the elements will be called *isoparametric*.

We could, however, use only the four corner points to define the variation of u [Fig. 5.8(b)]. We shall refer to such an element as *superparametric*, noting that the variation of geometry is more general than that of the actual unknown.

Similarly, if for instance we introduce more nodes to define u than are used to define the geometry, *subparametric* elements will result [Fig. 5.8(c)].

While for mapping it is convenient to use 'standard' forms of shape functions the interpolation of the unknown can, of course, use hierarchic forms defined in the previous chapter. Once again the definitions of sub- and superparametric variations are applicable.

Transformations

5.5 Evaluation of element matrices. Transformation in ξ, η, ζ coordinates

To perform finite element analysis the matrices defining element properties, e.g., stiffness, etc., have to be found. These will be of the form

$$\int_\Omega \mathbf{G}\, \mathrm{d}\Omega \tag{5.5}$$

in which the matrix $\mathbf{G}$ depends on $\mathbf{N}$ or its derivatives with respect to *global coordinates*. As an example of this we have the stiffness matrix

$$\mathbf{K} = \int_\Omega \mathbf{B}^{\mathrm{T}}\mathbf{D}\mathbf{B}\, \mathrm{d}\Omega \tag{5.6a}$$

and associated body force vectors

$$\mathbf{f} = \int_\Omega \mathbf{N}^{\mathrm{T}}\mathbf{b}\, \mathrm{d}\Omega \tag{5.6b}$$

For elastic problems the matrix for $\mathbf{B}$ is given explicitly by components [see the general form of Eq. (2.15)]. For plane stress problems we have

$$\mathbf{B}_a = \begin{bmatrix} \dfrac{\partial N_a}{\partial x}, & 0 \\ 0, & \dfrac{\partial N_a}{\partial y} \\ \dfrac{\partial N_a}{\partial y}, & \dfrac{\partial N_a}{\partial x} \end{bmatrix} \tag{5.7}$$

In elasticity problems the matrix **G** is thus a function of the first derivatives of **N** and this situation will arise in many other classes of problems. In all, C_0 continuity is needed and, as we have already noted, this is readily satisfied by the functions of Chapter 4, written now in terms of curvilinear coordinates.

To evaluate such matrices we note that two transformations are necessary. In the first place, as N_a is defined in terms of local (curvilinear) coordinates, it is necessary to devise some means of expressing the global derivatives of the type occurring in Eq. (5.7) in terms of local derivatives.

In the second place the element of volume (or surface) over which the integration has to be carried out needs to be expressed in terms of the local coordinates with an appropriate change in limits of integration.

5.5.1 Computation of global derivatives

Consider, for instance, the set of local coordinates ξ, η, ζ and a corresponding set of global coordinates x, y, z. By the usual rules of partial differentiation we can write, for instance, the ξ derivative as

$$\frac{\partial N_a}{\partial \xi} = \frac{\partial N_a}{\partial x}\frac{\partial x}{\partial \xi} + \frac{\partial N_a}{\partial y}\frac{\partial y}{\partial \xi} + \frac{\partial N_a}{\partial z}\frac{\partial z}{\partial \xi} \tag{5.8}$$

Performing the same differentiation with respect to the other two coordinates and writing in matrix form we have

$$\begin{Bmatrix} \dfrac{\partial N_a}{\partial \xi} \\ \dfrac{\partial N_a}{\partial \eta} \\ \dfrac{\partial N_a}{\partial \zeta} \end{Bmatrix} = \begin{bmatrix} \dfrac{\partial x}{\partial \xi}, & \dfrac{\partial y}{\partial \xi}, & \dfrac{\partial z}{\partial \xi} \\ \dfrac{\partial x}{\partial \eta}, & \dfrac{\partial y}{\partial \eta}, & \dfrac{\partial z}{\partial \eta} \\ \dfrac{\partial x}{\partial \zeta}, & \dfrac{\partial y}{\partial \zeta}, & \dfrac{\partial z}{\partial \zeta} \end{bmatrix} \begin{Bmatrix} \dfrac{\partial N_a}{\partial x} \\ \dfrac{\partial N_a}{\partial y} \\ \dfrac{\partial N_a}{\partial z} \end{Bmatrix} = \mathbf{J} \begin{Bmatrix} \dfrac{\partial N_a}{\partial x} \\ \dfrac{\partial N_a}{\partial y} \\ \dfrac{\partial N_a}{\partial z} \end{Bmatrix} \tag{5.9}$$

In the above, the left-hand side can be evaluated as the functions N_a are specified in local coordinates. Further, as x, y, z are explicitly given by the relation defining the curvilinear coordinates [Eq. (5.2)], the matrix **J** can be found explicitly in terms of the local coordinates. The array **J** is known as the *jacobian matrix* for the transformation.

To find now the global derivatives we invert **J** and write

$$\begin{Bmatrix} \dfrac{\partial N_a}{\partial x} \\ \dfrac{\partial N_a}{\partial y} \\ \dfrac{\partial N_a}{\partial z} \end{Bmatrix} = \mathbf{J}^{-1} \begin{Bmatrix} \dfrac{\partial N_a}{\partial \xi} \\ \dfrac{\partial N_a}{\partial \eta} \\ \dfrac{\partial N_a}{\partial \zeta} \end{Bmatrix} \tag{5.10}$$

In terms of the shape function defining the coordinate transformation $\mathbf{N}'$ (which as we have seen are only identical with the shape functions **N** when the isoparametric formulation is used) we have

$$\mathbf{J} = \begin{bmatrix} \sum_a \dfrac{\partial N'_a}{\partial \xi} x_a, & \sum_a \dfrac{\partial N'_a}{\partial \xi} y_a, & \sum_a \dfrac{\partial N'_a}{\partial \xi} z_a \\ \sum_a \dfrac{\partial N'_a}{\partial \eta} x_a, & \sum_a \dfrac{\partial N'_a}{\partial \eta} y_a, & \sum_a \dfrac{\partial N'_a}{\partial \eta} z_a \\ \sum_a \dfrac{\partial N'_a}{\partial \zeta} x_a, & \sum_a \dfrac{\partial N'_a}{\partial \zeta} y_a, & \sum_a \dfrac{\partial N'_a}{\partial \zeta} z_a \end{bmatrix} = \begin{bmatrix} \dfrac{\partial N'_1}{\partial \xi}, & \dfrac{\partial N'_2}{\partial \xi} & \cdots \\ \dfrac{\partial N'_1}{\partial \eta}, & \dfrac{\partial N'_2}{\partial \eta} & \cdots \\ \dfrac{\partial N'_1}{\partial \zeta}, & \dfrac{\partial N'_2}{\partial \zeta} & \cdots \end{bmatrix} \begin{bmatrix} x_1, & y_1, & z_1 \\ x_2, & y_2, & z_2 \\ \vdots & \vdots & \vdots \end{bmatrix} \tag{5.11}$$

For two-dimensional problems we drop all the terms containing z and/or ζ in Eqs (5.8) to (5.11).

5.5.2 Volume integrals

To transform the variables and the domain with respect to which the integration is made, a standard process will be used which involves the determinant of $\mathbf{J}$. Thus, for instance, a volume element becomes

$$\mathrm{d}x\,\mathrm{d}y\,\mathrm{d}z = J(\xi, \eta, \zeta)\,\mathrm{d}\xi\,\mathrm{d}\eta\,\mathrm{d}\zeta \tag{5.12}$$

where $J(\xi, \eta, \zeta) = \det \mathbf{J}$.

This type of transformation is valid irrespective of the number of coordinates used. For its justification the reader is referred to standard mathematical texts.† (See also Appendix F.)

Assuming that the inverse of $\mathbf{J}$ can be found we now have reduced the evaluation of the element properties to that of finding integrals of the form of Eq. (5.5).

More explicitly we can write this as

$$\int_{\Omega} \mathbf{G}(x, y, z)\,\mathrm{d}\Omega = \int_{-1}^{1}\int_{-1}^{1}\int_{-1}^{1} \bar{\mathbf{G}}(\xi, \eta, \zeta)\, J(\xi, \eta, \zeta)\,\mathrm{d}\xi\,\mathrm{d}\eta\,\mathrm{d}\zeta \tag{5.13}$$

where

$$\mathbf{G}(x, y, z) = \mathbf{G}(x(\xi, \eta, \zeta), y(\xi, \eta, \zeta), z(\xi, \eta, \zeta)) = \bar{\mathbf{G}}(\xi, \eta, \zeta)$$

and the curvilinear coordinates are of the normalized type based on the right prism. Indeed the integration *is carried out within such a prism* and not in the complicated distorted shape, thus accounting for the simple integration limits. One- and two-dimensional problems will similarly result in integrals with respect to one or two coordinates within simple limits.

While the limits of integration are simple in the above case, unfortunately the explicit form of $\bar{\mathbf{G}}$ is not. Apart from the simplest elements, algebraic integration usually defies our mathematical skill, and numerical integration has to be used. This, as will be seen from later sections, is not a severe penalty and has the advantage that algebraic errors are more easily avoided and that general programs, not tied to a particular element, can be written for various classes of problems.

† The determinant of the jacobian matrix is known in the literature simply as 'the jacobian' and is often written as

$$J(\xi, \eta, \zeta) \equiv \frac{\partial(x, y, z)}{\partial(\xi, \eta, \zeta)}$$

5.5.3 Surface integrals

In elasticity and other applications, surface integrals frequently occur. Typical here are the expressions for evaluating the contributions of surface tractions [see Chapter 2, Eq. (2.40b)]:

$$\mathbf{f} = -\int_{\Gamma} \mathbf{N}^{\mathrm{T}} \bar{\mathbf{t}} \, \mathrm{d}\Gamma$$

The element $\mathrm{d}\Gamma$ will generally lie on a surface where one of the coordinates (say ζ) is constant.

The most convenient process of dealing with the above is to consider $\mathrm{d}A$ as a vector oriented in the direction normal to the surface (see Appendix F). For three-dimensional problems we form the vector product

$$\mathbf{n}\,\mathrm{d}A = \mathrm{d}\mathbf{A} = \begin{Bmatrix} \dfrac{\partial x}{\partial \xi} \\ \dfrac{\partial y}{\partial \xi} \\ \dfrac{\partial z}{\partial \xi} \end{Bmatrix} \times \begin{Bmatrix} \dfrac{\partial x}{\partial \eta} \\ \dfrac{\partial y}{\partial \eta} \\ \dfrac{\partial z}{\partial \eta} \end{Bmatrix} \mathrm{d}\xi \, \mathrm{d}\eta$$

and on substitution integrate within the domain $-1 \leq \xi, \eta \leq 1$.

For two dimensions a line length $\mathrm{d}S$ arises and here the magnitude is simply

$$\mathbf{n}\,\mathrm{d}\Gamma = \mathrm{d}\boldsymbol{\Gamma} = \begin{Bmatrix} \dfrac{\partial x}{\partial \xi} \\ \dfrac{\partial y}{\partial \xi} \\ 0 \end{Bmatrix} \times \begin{Bmatrix} 0 \\ 0 \\ 1 \end{Bmatrix} \mathrm{d}\xi = \begin{Bmatrix} \dfrac{\partial y}{\partial \xi} \\ -\dfrac{\partial x}{\partial \xi} \\ 0 \end{Bmatrix} \mathrm{d}\xi$$

on constant η surfaces. This may now be reduced to two components for the two-dimensional problem.

5.6 Evaluation of element matrices. Transformation in area and volume coordinates

The general relationship (5.2) for coordinate mapping and indeed all the subsequent statements are equally valid for any set of local coordinates and could relate the local $L_1, L_2, \ldots$ coordinates used for triangles and tetrahedra in the previous chapter, to the global cartesian ones.

Indeed most of the discussion of the previous sections is valid if we simply rename the local coordinates suitably. However, two important differences arise.

The first concerns the fact that the local coordinates are not independent and in fact number one more than the cartesian system. The matrix $\mathbf{J}$ would apparently therefore become rectangular and would not possess an inverse. The second is simply the difference of integration limits which have to correspond with a triangular or tetrahedral 'parent' element.

A simple, though perhaps not the most elegant, way out of the first difficulty is to consider one variable as a dependent one. Thus, for example, we can introduce formally, in the case of the tetrahedra,

$$\begin{aligned} \xi &= L_1 \\ \eta &= L_2 \\ \zeta &= L_3 \\ 1-\xi-\eta-\zeta &= L_4 \end{aligned} \tag{5.14}$$

(by definition in the previous chapter) and thus preserve without change Eq. (5.8) and all the equations up to Eq. (5.12).

As the functions N_a are given in terms of L_1, L_2, etc., we must observe that

$$\frac{\partial N_a}{\partial \xi} = \frac{\partial N_a}{\partial L_1}\frac{\partial L_1}{\partial \xi} + \frac{\partial N_a}{\partial L_2}\frac{\partial L_2}{\partial \xi} + \frac{\partial N_a}{\partial L_3}\frac{\partial L_3}{\partial \xi} + \frac{\partial N_a}{\partial L_4}\frac{\partial L_4}{\partial \xi} \tag{5.15}$$

On using Eq. (5.14) this becomes simply

$$\frac{\partial N_a}{\partial \xi} = \frac{\partial N_a}{\partial L_1} - \frac{\partial N_a}{\partial L_4}$$

with the other derivatives obtainable by similar expressions.

The integration limits of Eq. (5.13) now change, however, to correspond with the tetrahedron limits, typically

$$\int_\Omega \mathbf{G}\,\mathrm{d}\Omega = \int_0^1 \int_0^{1-\zeta} \int_0^{1-\eta-\zeta} \bar{\mathbf{G}}(\xi,\eta,\zeta)\,\mathrm{d}\xi\,\mathrm{d}\eta\,\mathrm{d}\zeta \tag{5.16}$$

The same procedure will clearly apply in the case of triangular coordinates.

It must be noted that once again the expression $\bar{\mathbf{G}}$ will necessitate numerical integration which, however, is carried out over the simple, undistorted, parent region whether this be triangular or tetrahedral.

An alternative to the above is to express the coordinates and constraint as

$$\begin{aligned} r_x &= x - x_1 N_1' - x_2 N_2' - x_3 N_3' - \cdots = 0 \\ r_y &= y - y_1 N_1' - y_2 N_2' - y_3 N_3' - \cdots = 0 \\ r_z &= z - z_1 N_1' - z_2 N_2' - z_3 N_3' - \cdots = 0 \\ r_1 &= 1 - L_1 - L_2 - L_3 - L_4 = 0 \end{aligned} \tag{5.17}$$

where $N_a' = N_a'(L_1, L_2, L_3, L_4)$, etc. Now derivatives of the above with respect to x, y

and z may be written directly as

$$\begin{bmatrix} \frac{\partial r_x}{\partial x} & \frac{\partial r_x}{\partial y} & \frac{\partial r_x}{\partial z} \\ \frac{\partial r_y}{\partial x} & \frac{\partial r_y}{\partial y} & \frac{\partial r_y}{\partial z} \\ \frac{\partial r_z}{\partial x} & \frac{\partial r_z}{\partial y} & \frac{\partial r_z}{\partial z} \\ \frac{\partial r_1}{\partial x} & \frac{\partial r_1}{\partial y} & \frac{\partial r_1}{\partial z} \end{bmatrix} = \begin{bmatrix} 1 & 0 & 0 \\ 0 & 1 & 0 \\ 0 & 0 & 1 \\ 0 & 0 & 0 \end{bmatrix} - \begin{bmatrix} \sum_a x_a \frac{\partial N_a}{\partial L_1} & \sum_a x_a \frac{\partial N_a}{\partial L_2} & \sum_a x_a \frac{\partial N_a}{\partial L_3} & \sum_a x_a \frac{\partial N_a}{\partial L_4} \\ \sum_a y_a \frac{\partial N_a}{\partial L_1} & \sum_a y_a \frac{\partial N_a}{\partial L_2} & \sum_a y_a \frac{\partial N_a}{\partial L_3} & \sum_a y_a \frac{\partial N_a}{\partial L_4} \\ \sum_a z_a \frac{\partial N_a}{\partial L_1} & \sum_a z_a \frac{\partial N_a}{\partial L_2} & \sum_a z_a \frac{\partial N_a}{\partial L_3} & \sum_a z_a \frac{\partial N_a}{\partial L_4} \\ 1 & 1 & 1 & 1 \end{bmatrix}$$

$$\times \begin{bmatrix} \frac{\partial L_1}{\partial x} & \frac{\partial L_1}{\partial y} & \frac{\partial L_1}{\partial z} \\ \frac{\partial L_2}{\partial x} & \frac{\partial L_2}{\partial y} & \frac{\partial L_2}{\partial z} \\ \frac{\partial L_3}{\partial x} & \frac{\partial L_3}{\partial y} & \frac{\partial L_3}{\partial z} \\ \frac{\partial L_4}{\partial x} & \frac{\partial L_4}{\partial y} & \frac{\partial L_4}{\partial z} \end{bmatrix} = \mathbf{0} \tag{5.18}$$

The above may be solved for the partial derivatives of L_a with respect to the x, y, z coordinates and used directly with the chain rule written as

$$\frac{\partial N_a}{\partial x} = \frac{\partial N_a}{\partial L_1}\frac{\partial L_1}{\partial x} + \frac{\partial N_a}{\partial L_2}\frac{\partial L_2}{\partial x} + \frac{\partial N_a}{\partial L_3}\frac{\partial L_3}{\partial x} + \frac{\partial N_a}{\partial L_4}\frac{\partial L_4}{\partial x} \tag{5.19}$$

The above has advantages when the coordinates are written using mapping functions as the computation can still be more easily carried out. Also, the calculation of integrals will normally be performed numerically (as described in Sec. 5.11) where the points for integration are defined directly in terms of the volume coordinates.

Finally it should be remarked that any of the elements given in the previous chapter are capable of being mapped. In some, such as the triangular prism, both area and rectangular coordinates are used (Fig. 5.9). The remarks regarding the dependence of coordinates apply once again with regard to the former but the processes of the present section should make procedures clear.

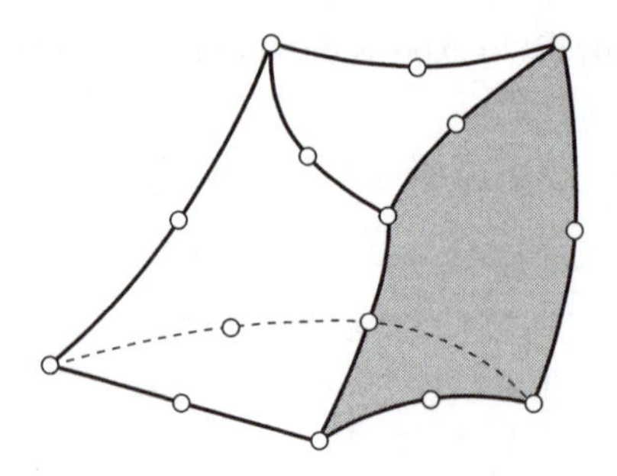

Fig. 5.9 A distorted triangular prism.

5.7 Order of convergence for mapped elements

If the shape functions are chosen in curvilinear coordinate space so as to observe the usual rules of convergence (continuity and presence of complete first-order polynomials in these coordinates), then convergence will occur. In the case of isoparametric (or subparametric) elements a complete linear field is always reproduced (i.e., 1, x, y) by the curvilinear coordinate expansion, and thus the lowest order patch test will be passed in the standard manner on such elements.

The proof of this is simple. Consider a standard isoparametric expansion

$$u = \sum_{a=1}^{n} N_a \tilde{u}_a \equiv \mathbf{N}\tilde{\mathbf{u}} \qquad \mathbf{N} = \mathbf{N}(\xi, \eta, \zeta) \tag{5.20}$$

with coordinates of nodes defining the transformation as

$$x = \sum N_a x_a \qquad y = \sum N_a y_a \qquad z = \sum N_a z_a \tag{5.21}$$

The question is under what circumstances is it possible for expression (5.20) to define a linear expansion in cartesian coordinates:

$$\begin{aligned} u &= \alpha_1 + \alpha_2 x + \alpha_3 y + \alpha_4 z \\ &\equiv \alpha_1 + \alpha_2 \sum N_a x_a + \alpha_3 \sum N_a y_a + \alpha_4 \sum N_a z_a \end{aligned} \tag{5.22}$$

If we take

$$\tilde{u}_a = \alpha_1 + \alpha_2 x_a + \alpha_3 y_a + \alpha_4 z_a$$

and compare expression (5.20) with (5.22) we note that identity is obtained between these providing

$$\sum N_a = 1$$

As this is the usual requirement of standard element shape functions [see Eq. (4.4)] we can conclude that the following theorem is valid.

Theorem 3. *The constant derivative condition will be satisfied for all isoparametric elements.*

As subparametric elements can always be expressed as specific cases of an isoparametric transformation this theorem is obviously valid here also.

It is of interest to pursue the argument and to see under what circumstances higher polynomial expansions in cartesian coordinates can be achieved under various transformations. The simple linear case in which we 'guessed' the solution has now to be replaced by considering in detail the polynomial terms occurring in expressions such as (5.20) and (5.22) and establishing conditions for equating appropriate coefficients.

Consider a specific problem: the circumstances under which the bilinear mapped quadrilateral of Fig. 5.10 can fully represent any quadratic cartesian expansion. We now have

$$x = \sum_{1}^{4} N'_a x_a \qquad y = \sum_{1}^{4} N'_a y_a \tag{5.23}$$

and we wish to be able to reproduce

$$u = \alpha_1 + \alpha_2 x + \alpha_3 y + \alpha_4 x^2 + \alpha_5 xy + \alpha_6 y^2 \tag{5.24}$$

Noting that the bilinear form of N'_a contains terms such as $1, \xi, \eta$ and $\xi\eta$, the above can be written as

$$u = \eta_1 + \eta_2\xi + \eta_3\eta + \eta_4\xi^2 + \eta_5\xi\eta + \eta_6\eta^2 + \eta_7\xi\eta^2 + \eta_8\xi^2\eta + \eta_9\xi^2\eta^2 \tag{5.25}$$

where η_1 to η_9 depend on the values of α_1 to α_6.

We shall now try to match the terms arising from the quadratic expansions of the serendipity kind shown in Fig. 5.10(b) where the interplation is

$$u = \sum_{a=1}^{8} N_a \tilde{u}_a \tag{5.26}$$

where the appropriate shape functions are of the kind defined in the previous chapter. We also can write (5.26) directly using polynomial coefficients b_a, $a = 1, \ldots, 8$, in place of the nodal variables $\tilde{u}_a$ (noting the terms occurring in the Pascal triangle) as

$$u = b_1 + b_2\xi + b_3\eta + b_4\xi^2 + b_5\xi\eta + b_6\eta^2 + b_7\xi\eta^2 + b_8\xi^2\eta \tag{5.27}$$

It is immediately evident that for arbitrary values of η_1 to η_9 it is impossible to match the coefficients b_1 to b_8 due to the absence of the term $\xi^2\eta^2$ in Eq. (5.27). [However, if higher order (quartic, etc.) expansions of the serendipity kind were used such matching would evidently be possible and we could conclude that for linearly distorted elements the serendipity family of order four or greater will always represent quadratic polynomials in x, y.]

For the 9-node, lagrangian, element [Fig. 5.10(c)] the expansion similar to (5.28) gives

$$u = \sum_{a=1}^{9} N_a \tilde{u}_a \tag{5.28}$$

which when expressed directly in polynomial coefficients b_a, $1 = 1, \ldots, 9$ yields

$$u = b_1 + b_2\xi + b_3\eta + b_4\xi^2 + \cdots + b_8\xi^2\eta + b_9\xi^2\eta^2 \tag{5.29}$$

and the matching of the coefficients of Eqs (5.29) and (5.25) can be made directly.

We can conclude therefore that 9-node elements better represent cartesian polynomials (when distorted linearly) and therefore are generally preferable in modelling smooth solutions. This matter was first presented by Wachspress[7] but the simple proof presented above is due to Crochet.[8] An example of this is given in Fig. 5.11 where we consider the results of a finite element calculation with 8- and 9-node elements respectively used to reproduce a simple beam solution in which we know that the exact answers are quadratic. With no distortion both elements give exact results but when distorted only the 9-node element does so, with the 8-node element giving quite wild stress fluctuation.

Similar arguments will lead to the conclusion that in three dimensions again only the lagrangian 27-node element is capable of reproducing fully a quadratic function in cartesian coordinates when trilinearly distorted (i.e., using the mapping for N'_a for the 8-node hexahedron).

Lee and Bathe[9] investigate the problem for cubic and quartic serendipity and lagrangian quadrilateral elements and show that under bilinear distortions the full order cartesian polynomial terms remain in lagrangian elements but not in serendipity ones. They also consider edge distortion and show that this polynomial order is always lost. Additional discussion of such problems is also given by Wachspress.[7]

5.8 Shape functions by degeneration

In the previous sections we have discussed the construction of shape functions for mapped elements of lagrangian and serendipity type, as well as those for triangular and tetrahedral type. We have also shown how mixtures of interpolation forms may be used to construct elements of prism type. One may ask what happens if we distort elements such that nodes for the lagrangian or serendipity type are coalesced – that is, they are assigned the same node number in the mesh. We call the approach where two or more nodes are common a *degenerate form*. In a degenerate form the shape function for a coalesced set of two or more nodes is obtained by adding together the shape functions of each individual node (in a hierarchic form, any mid-side and/or face functions are omitted).

Example 5.1: Quadrilateral degenerated into a triangular element. As a simple example we consider the degeneration of a 4-node quadrilateral in which nodes 3 and 4 are coalesced to form the third node of a triangular element as shown in Fig. 5.12. For an isoparametric form given in ξ, η coordinates, the shape functions for the degenerate triangular element are given by

$$\begin{aligned} N_1 &= \tfrac{1}{4}(1-\xi)(1-\eta) \\ N_2 &= \tfrac{1}{4}(1+\xi)(1-\eta) \\ N_3 &= \tfrac{1}{2}(1+\eta) \end{aligned} \tag{5.30}$$

where the last function results from adding together the standard shape functions for nodes 3 and 4 of the quadrilateral element. Computing now the global derivatives for the above functions we obtain [using (5.10)]

$$\frac{\partial N_a}{\partial x} = \frac{b_a(1-\eta)}{2\,\Delta(1-\eta)}; \quad \frac{\partial N_a}{\partial y} = \frac{c_a(1-\eta)}{2\,\Delta(1-\eta)} \tag{5.31}$$

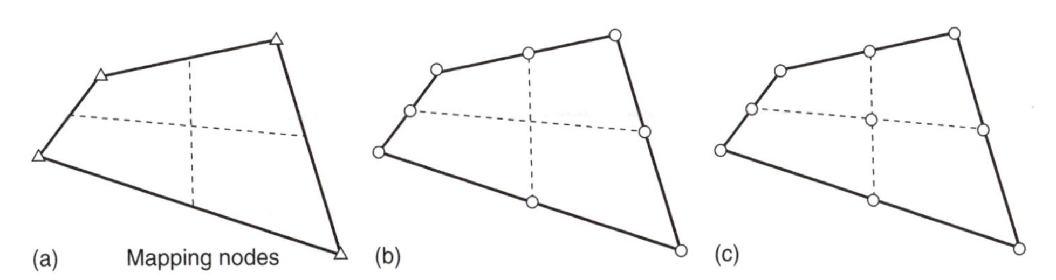

Fig. 5.10 Bilinear mapping of subparametric quadratic 8- and 9-node element.

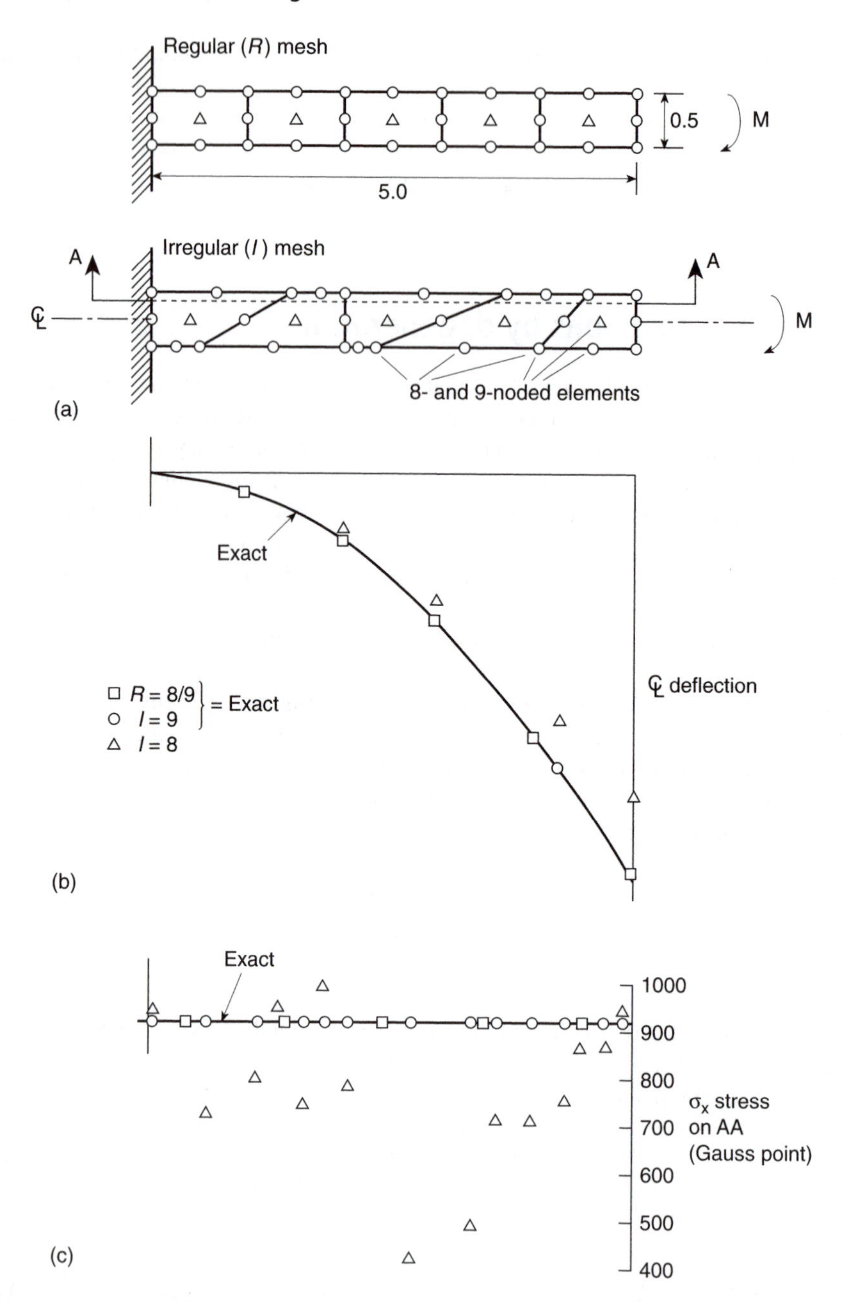

Fig. 5.11 Quadratic serendipity and lagrangian 8- and 9-node elements in regular and distorted form. Elastic deflection of a beam under constant moment. Note poor results of 8-node element.

where b_a and c_a coincide with results for the standard 3-node triangular element shape functions given in (2.6) and Δ is the area of the triangle as given in (2.7). Except for the point $\eta = 1$ (the point where the nodes are coalesced) the shape function derivatives are constant and identical to those obtained using area coordinates L_1, L_2, L_3. Thus, for the degeneration we have the identities

$$
\begin{aligned}
N_1 &= \tfrac{1}{4}(1-\xi)(1-\eta) &&= L_1 \\
N_2 &= \tfrac{1}{4}(1+\xi)(1-\eta) &&= L_2 \\
N_3 &= \tfrac{1}{2}(1+\eta) &&= L_3
\end{aligned}
\tag{5.32}
$$

and, provided we do not consider the point $\eta = 1$, we may compute the derivatives and integrals for 3-node triangular elements using the degeneration process.

A similar form to the above example holds when an 8-node brick element is degenerated into a 4-node tetrahedron. In addition, however, we can compute shape functions for other degenerate forms as indicated in Fig. 5.13. In all cases, the computation of derivatives gives a $0/0$ form at any point where nodes are coalesced. In addition, however, any faces which degenerate into an edge will also contain a $0/0$ in the derivative along that edge. The behaviour on any remaining face of a degenerate element is either the original quadrilateral one or a triangular one in which the shape functions are identical to the results given in (5.32).

5.8.1 Higher order degenerate elements

When nodes for higher order quadrilateral and hexahedral elements are coalesced to give a degenerate form it is necessary to modify the shape functions for some of the non-coalesced nodes in order to produce results which are consistent with those computed using area or volume coordinates, respectively. This aspect was first studied by Newton[10] and Irons[11] for serendipity-type elements. Here we extend the work reported in these references to include the lagrangian-type elements. Using lagrangian elements has a distinct advantage since all the degenerate elements preserve the properties of higher order approximation

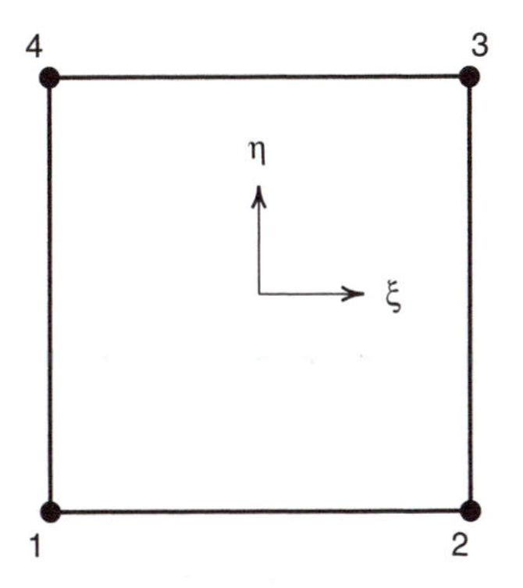

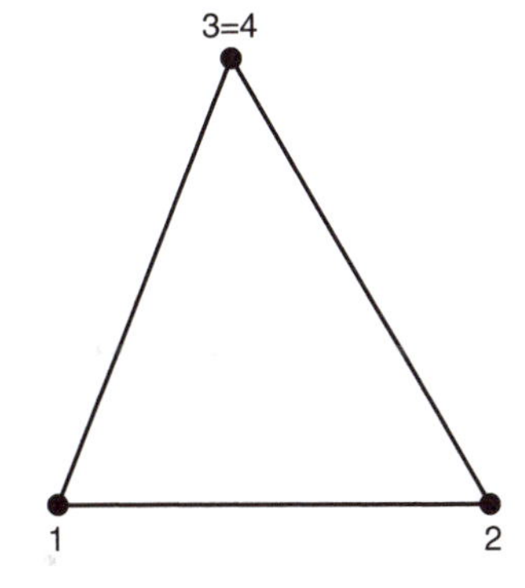

Fig. 5.12 Degeneration of a quadrilateral into a triangle.

in global coordinates when the element is mapped according to the trilinear form (i.e., a subparametric form using the 8-node hexahedron).

Example 5.2: Quadratic quadrilateral degenerated to a triangular element. As an example we consider the degeneration of a quadratic order quadrilateral to form a quadratic order triangular element. Expressing the shape functions in hierarchical form we have for

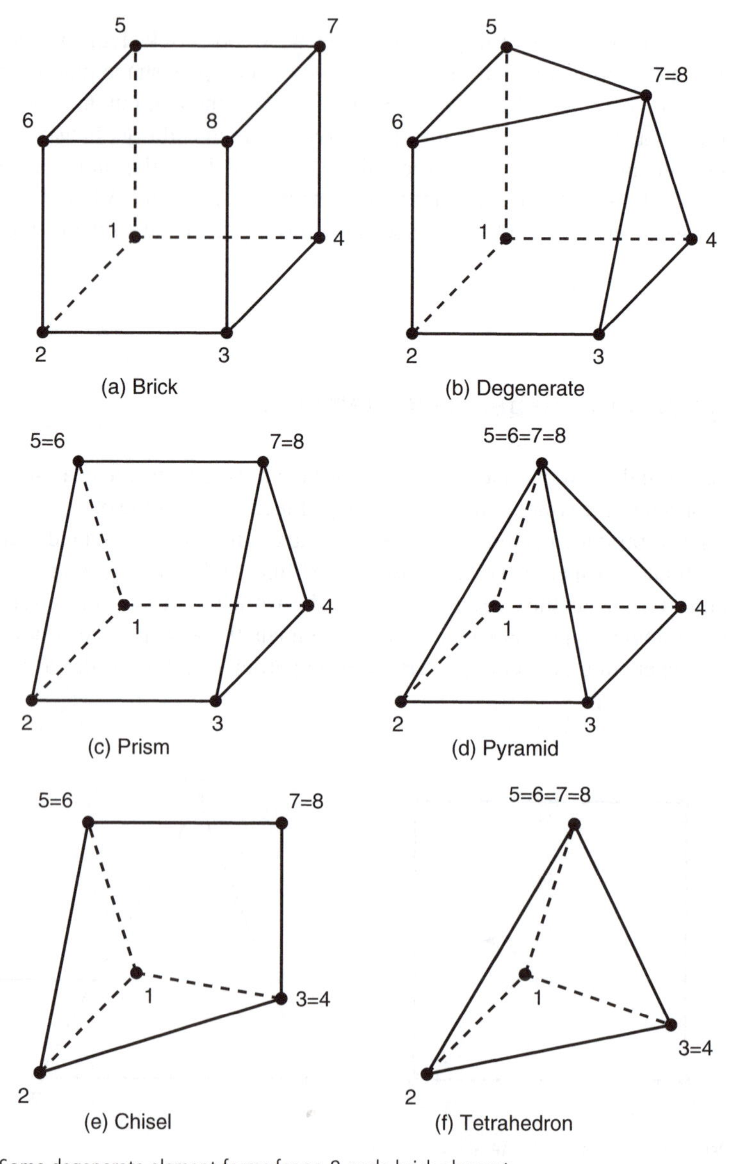

Fig. 5.13 Some degenerate element forms for an 8-node brick element.

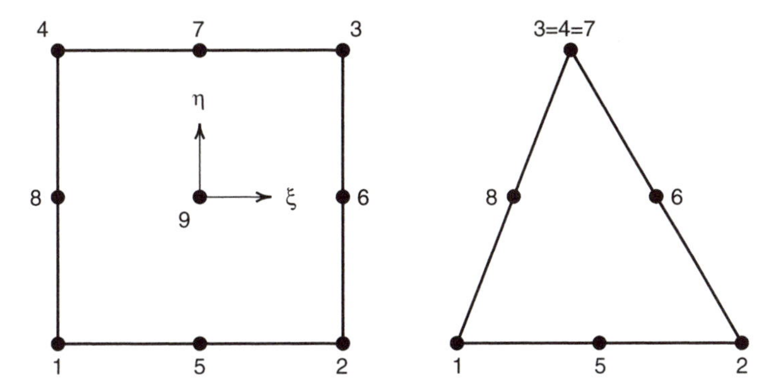

Fig. 5.14 Degeneration of a 8- or 9-node quadrilateral into a 6-node triangle.

the 8- or 9-node quadrilateral†

$$\begin{aligned} N_a^Q &= \tfrac{1}{4}(1+\xi_a\xi)(1+\eta_a\eta); \quad && a=1,2,3,4 \\ N_a^Q &= \tfrac{1}{2}(1+\xi_a\xi)(1-\eta^2); && a=6,8 \\ N_a^Q &= \tfrac{1}{2}(1+\eta_a\eta)(1-\xi^2); && a=5,7 \\ N_a^Q &= (1-\xi^2)(1-\eta^2); && a=9 \end{aligned} \tag{5.33}$$

for which a hierarchical lagrangian interpolation of any function is given by

$$f = \sum_{a=1}^{4} N_a^Q(\xi,\eta)\, f_a + \sum_{a=5}^{8} N_a^Q(\xi,\eta)\, \Delta f_a + N_9^Q(\xi,\eta)\, \Delta\Delta f_9 \tag{5.34}$$

where f_a are nodal values, Δf_a are departures from linear interpolation for mid-side nodes, and $\Delta\Delta f_9$ is the departure from the 8-node serendipity interpolation at the centre node. Thus, omitting the ninth function gives the serendipity form.

If we now coalesce the nodes 3, 4, 7 and use the above hierarchic form, the shape functions for the vertex nodes again are given by

$$\begin{aligned} N_1^Q &= \tfrac{1}{4}(1-\xi)(1-\eta) = L_1 = N_1^T \\ N_2^Q &= \tfrac{1}{4}(1+\xi)(1-\eta) = L_2 = N_2^T \\ N_3^Q &= \tfrac{1}{2}(1+\eta) \qquad\qquad = L_3 = N_3^T \end{aligned} \tag{5.35}$$

[note that $\Delta f_7 = 0$ in any interpolation and, thus, $N_7^T = 0$]. Also, for the 6-node form we omit the interior node 9 and thus, for the degenerate element, $N_9^T = 0$. If the resulting degenerate element is to be identical with the 6-node triangular element we require

$$\begin{aligned} N_5^T &= 4L_1L_2 \\ N_6^T &= 4L_2L_3 \\ N_8^T &= 4L_3L_1 \end{aligned} \tag{5.36}$$

† We use a superscript 'Q' for shape functions associated with the quadrilateral form and, later, 'T' to denote those for a triangular form.

Substituting the definitions for area coordinates given by (5.35) into (5.36) we find

$$\begin{aligned} N_5^T &= \tfrac{1}{4}(1-\xi^2)(1-\eta)^2 \\ N_6^T &= \tfrac{1}{2}(1+\xi)(1-\eta^2) \\ N_8^T &= \tfrac{1}{2}(1-\xi)(1-\eta^2) \end{aligned} \tag{5.37}$$

and, thus, comparing the forms given by (5.33) and (5.37) we obtain the result

$$N_5^T \neq N_5^Q; \quad N_6^T = N_6^Q; \quad N_8^T = N_8^Q \tag{5.38}$$

Thus, it only remains to correct the shape function for node 5. This is accomplished by noting

$$\begin{aligned} N_5^T &= \tfrac{1}{4}(1-\xi^2)(1-2\eta+\eta^2) \\ &= \tfrac{1}{4}(1-\xi^2)(2-2\eta-1+\eta^2) \\ &= \tfrac{1}{2}(1-\xi^2)(1-\eta) - \tfrac{1}{4}(1-\xi^2)(1-\eta^2) \end{aligned}$$

giving the 'corrected' degenerate function for node 5 as

$$N_5^T = N_5^Q - \tfrac{1}{4}N_9^Q \tag{5.39}$$

The hierarchical forms now can be converted to standard isoparametric form using the process given in Sec. 4.6.

Example 5.3: Degenerate forms for a quadratic 27-node hexahedron. The construction for quadratic degenerate three-dimensional forms follows a similar process and, when using the hierarchical form, the mid-side node opposite each coalesced node on a 'face' must be modified using a form similar to (5.39). Again, all the shapes shown in Fig. 5.13 are possible and permit the construction of meshes which use a mix of bricks, tetrahedra and degenerate transition forms. In addition to the 8 vertex nodes it is necessary to add 12 mid-edge nodes, 6 mid-face nodes and one internal node to form a lagrangian quadratic order hexahedron. For node numbers as given in Fig. 5.15 and using hierarchical interpolation, the shape functions are given by the following:

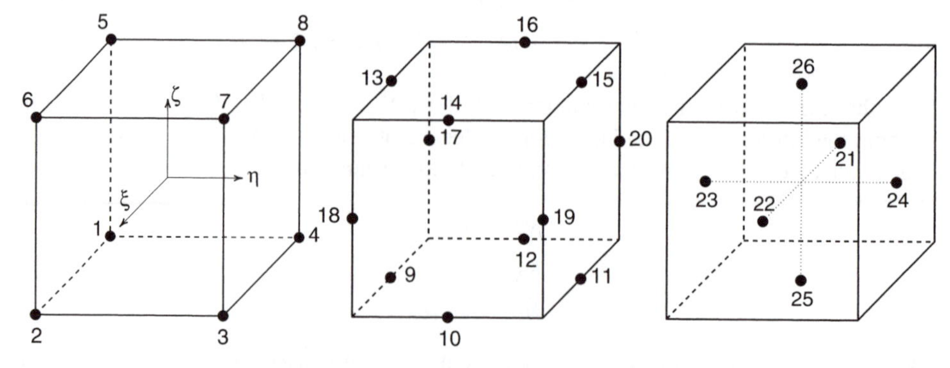

Fig. 5.15 Numbering for 27-node quadratic lagrangian hexagon. (Node 27 at origin of ξ, η, ζ coordinates).

Table 5.1 Degeneration modifications for 27-node hexahedron

Coalesced nodes	Modified nodes	
1 and 2 omit 9	11 by 25 omit 25	13 by 23 omit 23
2 and 3 omit 10	12 by 25 omit 25	14 by 22 omit 22
3 and 4 omit 11	9 by 25 omit 25	15 by 24 omit 24
4 and 1 omit 12	10 by 25 omit 25	16 by 21 omit 21
5 and 6 omit 13	15 by 26 omit 26	9 by 23 omit 23
6 and 7 omit 14	16 by 26 omit 26	10 by 22 omit 22
7 and 8 omit 15	13 by 26 omit 26	11 by 24 omit 24
8 and 5 omit 16	14 by 26 omit 26	12 by 21 omit 21
1 and 5 omit 17	18 by 23 omit 23	20 by 21 omit 21
2 and 6 omit 18	19 by 22 omit 22	17 by 23 omit 23
3 and 7 omit 19	20 by 24 omit 24	18 by 22 omit 22
4 and 8 omit 20	17 by 21 omit 21	19 by 24 omit 24

1. For vertex nodes

$$N_a = \tfrac{1}{8}(1+\xi_a\xi)(1+\eta_a\eta)(1+\zeta_a\zeta); \quad a = 1, 2, \ldots, 8 \tag{5.40a}$$

2. For mid-edge nodes

$$N_a = \tfrac{1}{4}\begin{cases} (1-\xi^2)(1+\eta_a\eta)(1+\zeta_a\zeta); & a = 9, 11, 13, 15 \\ (1+\xi_a\xi)(1-\eta^2)(1+\zeta_a\zeta); & a = 10, 12, 14, 16 \\ (1+\xi_a\xi)(1+\eta_a\eta)(1-\zeta^2); & a = 17, 18, 19, 20 \end{cases} \tag{5.40b}$$

3. For mid-face nodes

$$N_a = \tfrac{1}{2}\begin{cases} (1+\xi_a\xi)(1-\eta^2)(1-\zeta^2); & a = 21, 22 \\ (1-\xi^2)(1+\eta_a\eta)(1-\zeta^2); & a = 23, 24 \\ (1-\xi^2)(1-\eta^2)(1+\zeta_a\zeta); & a = 25, 26 \end{cases} \tag{5.40c}$$

4. For interior node

$$N_a = (1-\xi^2)(1-\eta^2)(1-\zeta^2); \quad a = 27 \tag{5.40d}$$

Table 5.1 indicates which shape functions are modified when vertex nodes are coalesced. The hierarchical shape functions to be omitted are also indicated. Note that shape functions should only be omitted (set zero) after all coalesced node pairs are considered. Also if a tetrahedral element is formed then all mid-face nodes are deleted and the interior node may also be omitted, giving the final tetrahedron as a 10-node element. Again, if any of the element forms is mapped using the degenerate subparametric form of the 8-node hexahedron for N'_a full quadratic behaviour in global coordinates is attained – showing the advantage of starting from lagrangian form elements.

Consideration of cubic and higher order forms are also possible and are left as an exercise for the interested reader.

5.9 Numerical integration – one dimensional

Some principles of numerical integration will be summarized here together with tables of convenient numerical coefficients.

To find numerically the integral of a function of one variable we can proceed in one of several ways as discussed next.

5.9.1 Newton–Cotes quadrature†

In the most obvious procedure, points at which the function is to be found are determined *a priori* – usually at equal intervals – and a polynomial passed through the values of the function at these points and exactly integrated [Fig. 5.16(a)].

As n values of the function define a polynomial of degree $n-1$, the errors will be of the order $O(h^n)$ where h is the element size. The well-known Newton–Cotes 'quadrature' formulae can be written as

$$I = \int_{-1}^{1} f(\xi)\,\mathrm{d}\xi = \sum_{i=1}^{n} f(\xi_i)\,w_i \tag{5.41}$$

for the range of integration between -1 and $+1$ [Fig. 5.16(a)]. For example, if $n = 2$, we have the well-known trapezoidal rule:

$$I = f(-1) + f(1) \tag{5.42a}$$

for $n = 3$, the Simpson 'one-third' rule:

$$I = \tfrac{1}{3}[f(-1) + 4f(0) + f(1)] \tag{5.42b}$$

and for $n = 4$:

$$I = \tfrac{1}{4}[f(-1) + 3f(-\tfrac{1}{3}) + 3f(\tfrac{1}{3}) + f(1)] \tag{5.42c}$$

Formulae for higher values of n are given in reference 12.

5.9.2 Gauss quadrature

If in place of specifying the position of sampling points *a priori* we allow these to be located at points to be determined so as to aim for best accuracy, then for a given number of sampling points increased accuracy can be obtained. Indeed, if we again consider

$$I = \int_{-1}^{1} f(\xi)\,\mathrm{d}\xi = \sum_{i=1}^{n} f(\xi_i)\,w_i \tag{5.43}$$

† 'Quadrature' is an alternative term to 'numerical integration'.

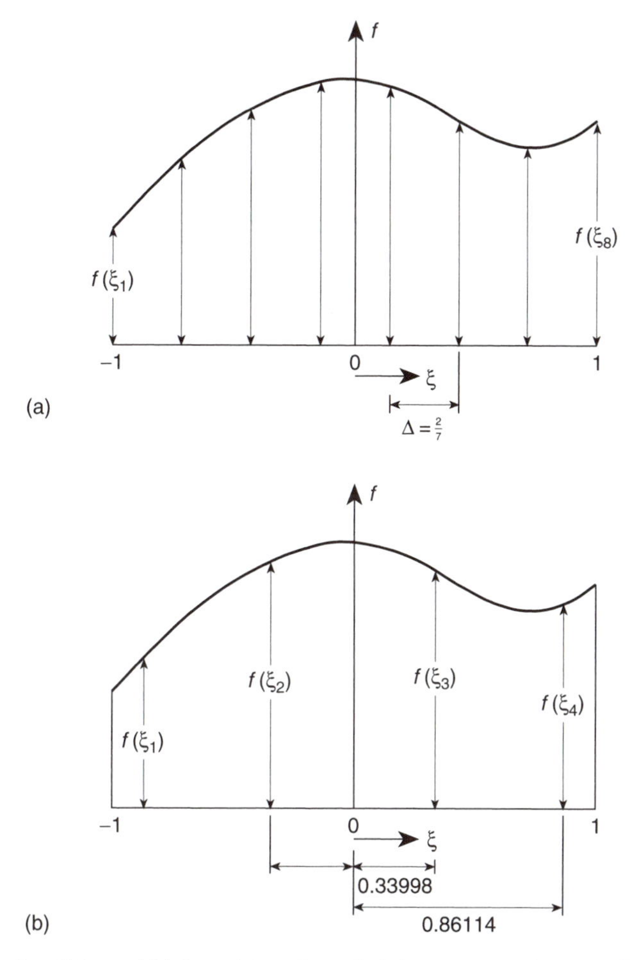

Fig. 5.16 (a) Newton–Cotes and (b) Gauss integrations. Each integrates exactly a seventh-order polynomial [i.e., error $O(h^8)$].

and again assume a polynomial expression, it is easy to see that for n sampling points we have $2n$ unknowns (w_i and ξ_i) and hence a polynomial of degree $2n-1$ could be constructed and exactly integrated [Fig. 5.16(b)]. The error is thus of order $O(h^{2n})$.

The simultaneous equations involved are difficult to solve, but some mathematical manipulation will show that the solution can be obtained explicitly in terms of Legendre polynomials. Thus this particular process is frequently known as Gauss–Legendre quadrature.[12]

Table 5.2 shows the positions and weighting coefficients for gaussian integration.

For purposes of finite element analysis complex calculations are involved in determining the values of f, the function to be integrated. Thus the Gauss-type processes, requiring the least number of such evaluations, are ideally suited and from now on will be used exclusively.

5.10 Numerical integration – rectangular (2D) or brick regions (3D)

The most obvious way of obtaining the integral

$$I = \int_{-1}^{1} \int_{-1}^{1} f(\xi, \eta)\, \mathrm{d}\xi\, \mathrm{d}\eta \tag{5.44}$$

is first to evaluate the inner integral keeping η constant, i.e.,

$$\int_{-1}^{1} f(\xi, \eta)\, \mathrm{d}\xi = \sum_{j=1}^{n} f(\xi_j, \eta)\, w_j = \psi(\eta) \tag{5.45}$$

Table 5.2 Gaussian quadrature abscissae and weights for $\int_{-1}^{1} f(x)\mathrm{d}x = \sum_{j=1}^{n} f(\xi_j)\, w_j$.

$\pm\xi_j$		w_j
	$n = 1$	
0		2.000 000 000 000 000
	$n = 2$	
$1/\sqrt{3}$		1.000 000 000 000 000
	$n = 3$	
$\sqrt{0.6}$		5/9
0.000 000 000 000 000		8/9
	$n = 4$	
0.861 136 311 594 053		0.347 854 845 137 454
0.339 981 043 584 856		0.652 145 154 862 546
	$n = 5$	
0.906 179 845 938 664		0.236 926 885 056 189
0.538 469 310 105 683		0.478 628 670 499 366
0.000 000 000 000 000		0.568 888 888 888 889
	$n = 6$	
0.932 469 514 203 152		0.171 324 492 379 170
0.661 209 386 466 265		0.360 761 573 048 139
0.238 619 186 083 197		0.467 913 934 572 691
	$n = 7$	
0.949 107 912 342 759		0.129 484 966 168 870
0.741 531 185 599 394		0.279 705 391 489 277
0.405 845 151 377 397		0.381 830 050 505 119
0.000 000 000 000 000		0.417 959 183 673 469
	$n = 8$	
0.960 289 856 497 536		0.101 228 536 290 376
0.796 666 477 413 627		0.222 381 034 453 374
0.525 532 409 916 329		0.313 706 645 877 887
0.183 434 642 495 650		0.362 683 783 378 362
	$n = 9$	
0.968 160 239 507 626		0.081 274 388 361 574
0.836 031 107 326 636		0.180 648 160 694 857
0.613 371 432 700 590		0.260 610 696 402 935
0.324 253 423 403 809		0.312 347 077 040 003
0.000 000 000 000 000		0.330 239 355 001 260
	$n = 10$	
0.973 906 528 517 172		0.066 671 344 308 688
0.865 063 366 688 985		0.149 451 349 150 581
0.679 409 568 299 024		0.219 086 362 515 982
0.433 395 394 129 247		0.269 266 719 309 996
0.148 874 338 981 631		0.295 524 224 714 753

Evaluating the outer integral in a similar manner, we have

$$
\begin{aligned}
I = \int_{-1}^{1} \psi(\eta)\, \mathrm{d}\eta &= \sum_{i=1}^{n} \psi(\eta_i)\, w_i \\
&= \sum_{i=1}^{n} w_i \sum_{j=1}^{n} f(\xi_j, \eta_i)\, w_j \\
&= \sum_{i=1}^{n} \sum_{j=1}^{n} f(\xi_j, \eta_i)\, w_i\, w_j
\end{aligned}
\tag{5.46}
$$

For a brick we have similarly

$$
\begin{aligned}
I &= \int_{-1}^{1} \int_{-1}^{1} \int_{-1}^{1} f(\xi, \eta, \zeta)\, \mathrm{d}\xi\, \mathrm{d}\eta\, \mathrm{d}\zeta \\
&= \sum_{k=1}^{n} \sum_{j=1}^{n} \sum_{i=1}^{n} f(\xi_i, \eta_j, \xi_k)\, w_i\, w_j\, w_k
\end{aligned}
\tag{5.47}
$$

In the above, the number of integrating points in each direction was assumed to be the same. Clearly this is not necessary and on occasion it may be an advantage to use different numbers in each direction of integration.

It is of interest to note that in fact the double summation can be readily interpreted as a single one over $(n \times n)$ points for a rectangle (or n^3 points for a cube). Thus in Fig. 5.17 we show the nine sampling points that result in exact integrals of order 5 in each direction. In the sequel when numerical integration is used we will denote the summation as a single sum over unique points, thus we will write

$$
I = \int_{-1}^{1} \int_{-1}^{1} f(\xi, \eta)\, \mathrm{d}\xi\, \mathrm{d}\eta = \sum_{l=1}^{m} f(\xi_l, \eta_l)\, W_l \tag{5.48}
$$

for two dimensions and

$$
I = \int_{-1}^{1} \int_{-1}^{1} \int_{-1}^{1} f(\xi, \eta, \zeta)\, \mathrm{d}\xi\, \mathrm{d}\eta\, \mathrm{d}\zeta = \sum_{l=1}^{m} f(\xi_l, \eta_l, \zeta_l)\, W_l \tag{5.49}
$$

for three dimensions. Here the weight W_l denotes the product of the appropriate one-dimensional weights.

We can also approach the problem directly and require an exact integration of a fifth-order polynomial in two dimensions. At any sampling point two coordinates and a value of f have to be determined in a weighting formula of type

$$
I = \int_{-1}^{1} \int_{-1}^{1} f(\xi, \eta)\, \mathrm{d}\xi\, \mathrm{d}\eta = \sum_{l=1}^{m} f(\xi_i, \eta_i)\, W_i \tag{5.50}
$$

It would appear that only seven points would suffice to obtain the same order of accuracy. Some formulae for three-dimensional bricks have been derived by Irons[13] and used successfully.[14]

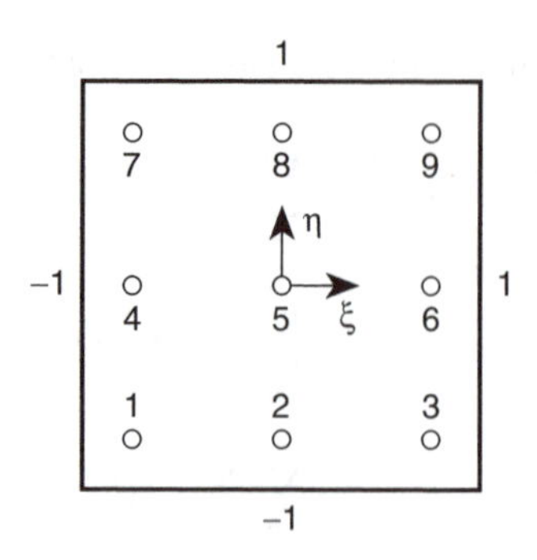

Fig. 5.17 Integrating points for $n = 3$ in a square region. (Exact for polynomial of fifth order in each direction).

5.11 Numerical integration – triangular or tetrahedral regions

For a triangle, in terms of the area coordinates the integrals are of the form

$$I = \int_0^1 \int_0^{1-L_1} f(L_1, L_2, L_3)\, \mathrm{d}L_2\, \mathrm{d}L_1 \qquad L_3 = 1 - L_1 - L_2 \tag{5.51}$$

Once again we could use n Gauss points and arrive at a summation expression of the type used in the previous section. However, the limits of integration now involve the variable itself and it is convenient to use alternative sampling points for the second integration by use of a special Gauss expression for integrals in which the integrand is multiplied by a linear function. These have been devised by Radau[15] and used successfully in the finite element context.[16] It is, however, much more desirable (and aesthetically pleasing) to use special formulae in which no bias is given to any of the natural coordinates L_a. Such formulae were first derived by Hammer *et al.*[17] and Felippa[18] and a series of necessary sampling points and weights is given in Table 5.3.[19] (A more comprehensive list of higher formulae derived by Cowper is given in reference 19.)

A similar extension for tetrahedra can obviously be made. Table 5.4 presents some formulae based on reference 17.

5.12 Required order of numerical integration

With numerical integration used in place of exact integration, an additional error is introduced into the calculation and the first impression is that this should be reduced as much as possible. Clearly the cost of numerical integration can be quite significant, and indeed in some early programs numerical formulation of element characteristics used a comparable amount of computer time as in the subsequent solution of the equations. It is of interest, therefore, to determine (a) the minimum integration requirement permitting convergence and (b) the integration requirements necessary to preserve the rate of convergence which would result if exact integration were used.

It will be found later (Chapters 9 and 11) that it is in fact often a disadvantage to use higher orders of integration than those actually needed under (b) as, for very good

Table 5.3 Numerical integration formulae for triangles

Order	Figure	Error	Points	Triangular coordinates	Weights
Linear		$R = O(h^2)$	a	$\frac{1}{3}, \frac{1}{3}, \frac{1}{3}$	1
Quadratic		$R = O(h^3)$	a	$\frac{1}{2}, \frac{1}{2}, 0$	$\frac{1}{3}$
			b	$\frac{1}{2}, 0, \frac{1}{2}$	$\frac{1}{3}$
			c	$0, \frac{1}{2}, \frac{1}{2}$	$\frac{1}{3}$
Cubic		$R = O(h^4)$	a	$\frac{1}{3}, \frac{1}{3}, \frac{1}{3}$	$-\frac{27}{48}$
			b	0.6, 0.2, 0.2	$\frac{25}{48}$
			c	0.2, 0.6, 0.2	$\frac{25}{48}$
			d	0.2, 0.2, 0.6	$\frac{25}{48}$
Quintic		$R = O(h^6)$	a	$\frac{1}{3}, \frac{1}{3}, \frac{1}{3}$	0.2250000000
			b	$\alpha_1, \beta_1, \beta_1$	0.1323941527
			c	$\beta_1, \alpha_1, \beta_1$	0.1323941527
			d	$\beta_1, \beta_1, \alpha_1$	0.1323941527
			e	$\alpha_2, \beta_2, \beta_2$	0.1259391805
			f	$\beta_2, \alpha_2, \beta_2$	0.1259391805
			g	$\beta_2, \beta_2, \alpha_2$	0.1259391805

with
$\alpha_1 = 0.059\,715\,871\,7$
$\beta_1 = 0.470\,142\,064\,1$
$\alpha_2 = 0.797\,426\,985\,3$
$\beta_2 = 0.101\,286\,507\,3$

reasons, a 'cancellation of errors' due to discretization and due to inexact integration can occur.

5.12.1 Minimum order of integration for convergence

In problems where the energy functional (or equivalent Galerkin integral statements) defines the approximation we have already stated that convergence can occur providing any arbitrary constant value of mth derivatives can be reproduced. In the present case $m = 1$ and we thus require that in integrals of the form (5.5) a constant value of $\mathbf{G}$ be correctly integrated. *Thus the volume of the element* $\int_\Omega d\Omega$ *needs to be evaluated correctly*

Table 5.4 Numerical integration formulae for tetrahedra

Order	Figure	Error	Points	Tetrahedral coordinates	Weights
Linear		$R = O(h^2)$	a	$\frac{1}{4}, \frac{1}{4}, \frac{1}{4}, \frac{1}{4}$	1
Quadratic		$R = O(h^3)$	a	$\alpha, \beta, \beta, \beta$	$\frac{1}{4}$
			b	$\beta, \alpha, \beta, \beta$	$\frac{1}{4}$
			c	$\beta, \beta, \alpha, \beta$	$\frac{1}{4}$
			d	$\beta, \beta, \beta, \alpha$	$\frac{1}{4}$
				$\alpha = 0.585\,410\,20$ $\beta = 0.138\,196\,60$	
Cubic		$R = O(h^4)$	a	$\frac{1}{4}, \frac{1}{4}, \frac{1}{4}, \frac{1}{4}$	$-\frac{4}{5}$
			b	$\frac{1}{2}, \frac{1}{6}, \frac{1}{6}, \frac{1}{6}$	$\frac{9}{20}$
			c	$\frac{1}{6}, \frac{1}{2}, \frac{1}{6}, \frac{1}{6}$	$\frac{9}{20}$
			d	$\frac{1}{6}, \frac{1}{6}, \frac{1}{2}, \frac{1}{6}$	$\frac{9}{20}$
			e	$\frac{1}{6}, \frac{1}{6}, \frac{1}{6}, \frac{1}{2}$	$\frac{9}{20}$

for convergence to occur. In curvilinear coordinates we can thus argue that $\int J \mathrm{d}\zeta \; \mathrm{d}\eta \; \mathrm{d}\xi$ has to be evaluated exactly.[3, 6]

5.12.2 Order of integration for no loss of convergence rate

In a general problem we have already found that the finite element approximate evaluation of energy (and indeed all the other integrals in a Galerkin-type approximation, see Chapter 3) was exact to the order $2(p-m)$, where p was the degree of the complete polynomial present and m the order of differential occurring in the appropriate expressions.

Providing the integration is exact to the order $2(p-m)$, or shows an error of $O(h^{2(p-m)+1})$, or less, then no loss of convergence order will occur.† If in curvilinear coordinates we take a curvilinear dimension h of an element, the same rule applies. For C_0 problems (i.e., $m = 1$) the integration formulae should be as follows:

$p = 1,$	linear elements	$O(h)$
$p = 2,$	quadratic elements	$O(h^3)$
$p = 3,$	cubic elements	$O(h^5)$

† For an energy principle use of quadrature may result in loss of a bound for $\Pi(\tilde{\mathbf{u}})$.

We shall make use of these results in practice, as will be seen later, but it should be noted that for a linear quadrilateral or triangle a single-point integration is adequate. For parabolic quadrilaterals (or bricks) 2×2 (or $2 \times 2 \times 2$), Gauss point integration is adequate and for parabolic triangles (or tetrahedra) three-point (and four-point) formulae of Tables 5.3 and 5.4 are needed.

The basic theorems of this section have been introduced and proved numerically in published work.[20–22]

5.12.3 Matrix singularity due to numerical integration

The final outcome of a finite element approximation in linear problems is an equation system

$$\mathbf{K}\tilde{\mathbf{u}} + \mathbf{f} = \mathbf{0} \tag{5.52}$$

in which the boundary conditions have been inserted and which should, on solution for the parameter $\tilde{\mathbf{u}}$, give an approximate solution for the physical situation. If a solution is unique, as is the case with well-posed physical problems, the equation matrix $\mathbf{K}$ should be non-singular. We have *a priori* assumed that this was the case with exact integration and in general have not been disappointed. With numerical integration, singularities may arise for low integration orders, and this may make such orders impractical. It is easy to show how, in some circumstances, a singularity of $\mathbf{K}$ must arise, but it is more difficult to prove that it will not. We shall, therefore, concentrate on the former case.

With numerical integration we replace the integrals by a weighted sum of independent linear relations between the nodal parameters $\tilde{\mathbf{u}}$. These linear relations supply the only information from which the matrix $\mathbf{K}$ is constructed. *If the number of unknowns* $\tilde{\mathbf{u}}$ *exceeds the number of independent relations supplied at all the integrating points, then the matrix* $\mathbf{K}$ *must be singular*.

To illustrate this point we shall consider two-dimensional elasticity problems using linear and parabolic serendipity quadrilateral elements with one- and four-point quadrature respectively.

Here at each integrating point *three* independent 'strain relations' are used and the total number of independent relations possible equals $3 \times$ (number of integration points). The number of unknowns $\tilde{\mathbf{u}}$ is simply $2 \times$ (number of nodes) less restrained degrees of freedom.

In Fig. 5.18(a) and (b) we show a single element and an assembly of two elements supported by a minimum number of specified displacements eliminating rigid body motion. The simple calculation shows that only in the assembly of the quadratic elements is elimination of singularities possible, all the other cases remaining strictly singular.

In Fig. 5.18(c) a well-supported block of both kinds of elements is considered and here for both element types non-singular matrices may arise although local, near singularity may still lead to unsatisfactory results (see Chapter 9).

The reader may well consider the same assembly but supported again by the minimum restraint of three degrees of freedom. The assembly of linear elements with a single integrating point *will* be singular while the quadratic ones will, in fact, usually be well behaved.

For the reason just indicated, linear single-point integrated elements are used infrequently in static solutions, although they do find wide use in 'explicit' dynamics codes – but needing

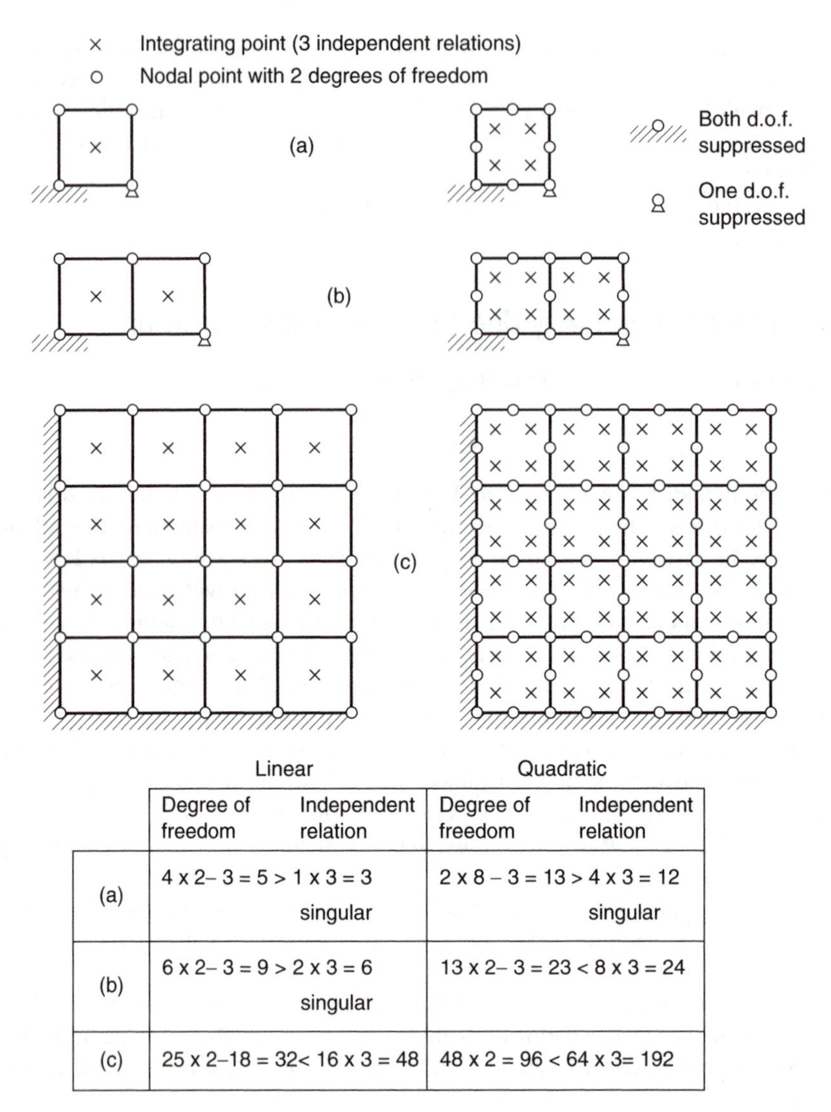

	Linear: Degree of freedom	Linear: Independent relation	Quadratic: Degree of freedom	Quadratic: Independent relation
(a)	4 x 2– 3 = 5 >	1 x 3 = 3 singular	2 x 8 – 3 = 13 >	4 x 3 = 12 singular
(b)	6 x 2– 3 = 9 >	2 x 3 = 6 singular	13 x 2– 3 = 23 <	8 x 3 = 24
(c)	25 x 2–18 = 32<	16 x 3 = 48	48 x 2 = 96 <	64 x 3= 192

Fig. 5.18 Check on matrix singularity in two-dimensional elasticity problems (a), (b), and (c).

certain remedial additions (e.g., hourglass control[23, 24]) – while four-point quadrature is often used for quadratic serendipity elements.†

In Chapter 9 we shall return to the problem of convergence and will indicate dangers arising from local element singularities.

However, it is of interest to mention that in Chapter 11 we shall in fact *seek* matrix singularities for special purposes (e.g., incompressibility) using similar arguments.

† Repeating the test for quadratic lagrangian elements indicates a singularity for 2 × 2 quadrature.

5.13 Generation of finite element meshes by mapping. Blending functions

It will be observed that it is an easy matter to obtain a coarse subdivision of the analysis domain with a small number of isoparametric elements. If second- or third-degree elements are used, the fit of these to quite complex boundaries is reasonable, as shown in Fig. 5.19(a) where four parabolic elements specify a sectorial region. This number of elements would be too small for analysis purposes *but a simple subdivision into finer elements* can be done automatically by, say, assigning new positions for the nodes at the mid-points of the curvilinear coordinates and thus deriving a larger number of similar elements, as shown in Fig. 5.19(b). Indeed, automatic subdivision could be carried out further to generate a field of triangular elements. The process thus allows us, with a small amount of original *input data*, to derive a finite element mesh of any refinement desirable. In reference 25 this type of mesh generation is developed for two- and three-dimensional solids and surfaces and is reasonably efficient. However, elements of predetermined size and/or gradation cannot be easily generated.

The main drawback of the mapping and generation suggested is the fact that the originally circular boundaries in Fig. 5.19(*a*) are approximated by simple parabolas and a geometric error can be developed there. To overcome this difficulty another form of mapping, originally developed for the representation of complex motor-car body shapes, can be adopted for this purpose.[26] In this mapping blending functions interpolate the unknown u in such a way as to satisfy *exactly* its variations along the edges of a square ξ, η domain. If the coordinates x and y are used in a parametric expression of the type given in Eq. (5.1), then any complex shape can be mapped by a single element. In reference 26 the region of Fig. 5.19 is in fact so mapped and a mesh subdivision obtained directly without any geometric error on the boundary.

The blending processes are of considerable importance and have been used to construct some interesting element families[27] (which in fact include the standard serendipity elements as a subclass). To explain the process we shall show how a function with prescribed variations along the boundaries can be interpolated.

Consider a region $-1 \leq \xi, \eta \leq 1$, shown in Fig. 5.20, on the edges of which an arbitrary function ϕ is specified [i.e., $\phi(-1, \eta)$, $\phi(1, \eta)$, $\phi(\xi, -1)$, $\phi(\xi, 1)$ are given]. The problem presented is that of interpolating a function $\phi(\xi, \eta)$ so that a smooth surface reproducing precisely the boundary values is obtained. Writing

$$\begin{aligned} N_1(\xi) &= \tfrac{1}{2}(1-\xi) \qquad & N_2(\xi) &= \tfrac{1}{2}(1+\xi) \\ N_1(\eta) &= \tfrac{1}{2}(1-\eta) \qquad & N_2(\eta) &= \tfrac{1}{2}(1+\eta) \end{aligned} \tag{5.53}$$

for our usual one-dimensional linear interpolating functions, we note that

$$P_\eta \phi \equiv N_1(\eta)\phi(\xi, -1) + N_2(\eta)\phi(\xi, 1) \tag{5.54}$$

interpolates linearly between the specified functions in the η direction, as shown in Fig. 5.20(b). Similarly,

$$P_\xi \phi \equiv N_1(\xi)\phi(\eta, -1) + N_2(\xi)\phi(\eta, 1) \tag{5.55}$$

interpolates linearly in the ξ direction [Fig. 5.20(c)]. Constructing a third function which is a standard bilinear interpolation of the kind we have already encountered [Fig. 5.20(d)],

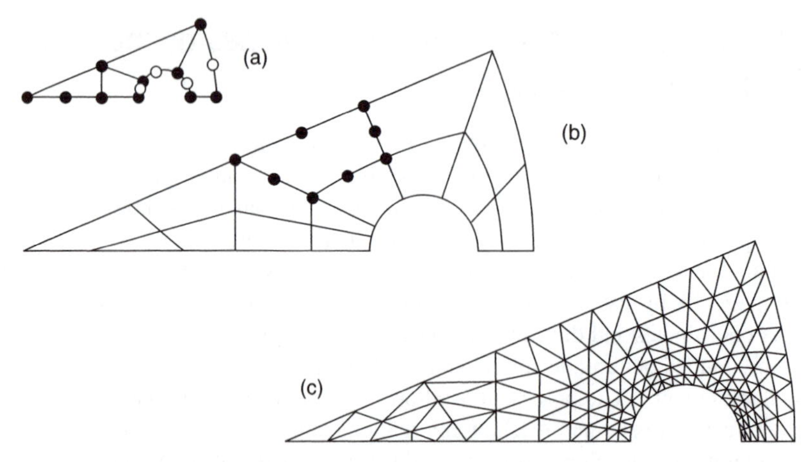

Fig. 5.19 Automatic mesh generation by quadratic isoparametric elements. (a) Specified mesh points. (b) Automatic subdivision into a small number of isoparametric elements. (c) Automatic subdivision into linear triangles.

i.e.,

$$P_\xi P_\eta \phi = N_1(\xi)N_1(\eta)\phi(-1,-1) + N_1(\xi)N_2(\eta)\phi(-1,1) + N_2(\xi)N_1(\eta)\phi(1,-1) + N_2(\xi)N_2(\eta)\phi(1,1) \tag{5.56}$$

we note by inspection that

$$\phi(\xi,\eta) = P_\eta\phi + P_\xi\phi - P_\xi P_\eta \phi \tag{5.57}$$

is a smooth surface interpolating exactly the boundary functions.

Extension to functions with higher order blending is almost evident, and immediately the method of mapping the quadrilateral region $-1 \le \xi, \eta \le 1$ to any arbitrary shape is obvious.

Though the above mesh generation method derives from mapping and indeed has been widely applied in two and three dimensions, we shall see in the chapter devoted to adaptivity (Chapter 14) that the optimal solution or specification of *mesh density* or *size* should guide the mesh generation. In Chapter 8 we will discuss in much more detail how meshes with prescribed density can be generated.

5.14 Infinite domains and infinite elements

5.14.1 Introduction

In many problems of engineering and physics infinite or semi-infinite domains exist. A typical example from structural mechanics may, for instance, be that of three-dimensional (or axisymmetric) excavation, illustrated in Fig. 5.21. Here the problem is one of determining the deformations in a semi-infinite half-space due to the removal of loads with the specification of zero displacements at infinity. Similar problems abound in electromagnetics and fluid mechanics but the situation illustrated is typical. The question arises as to how such

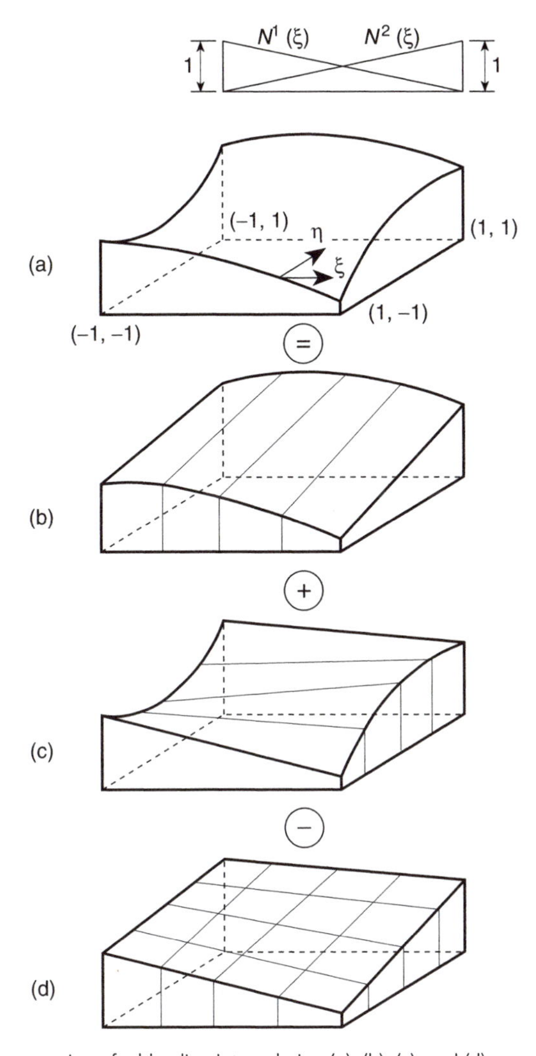

Fig. 5.20 Stages of construction of a blending interpolation (a), (b), (c), and (d).

problems can be dealt with by a method of approximation in which elements of decreasing size are used in the modelling process. The first intuitive answer is the one illustrated in Fig. 5.21(a) where the infinite boundary condition is specified at a finite boundary placed at a *large distance* from the object. This, however, begs the question of what is a 'large distance' and obviously substantial errors may arise if this boundary is not placed far enough away. On the other hand, pushing this out excessively far necessitates the introduction of a large number of elements to model regions of relatively little interest to the analyst.

To overcome such 'infinite' difficulties many methods have been proposed. In some a sequence of nesting grids is used and a recurrence relation derived.[28, 29] In others a boundary-type exact solution is used and coupled to the finite element domain.[30, 31] However, without doubt, the most effective and efficient treatment is the use of 'infinite elements'[32–35]

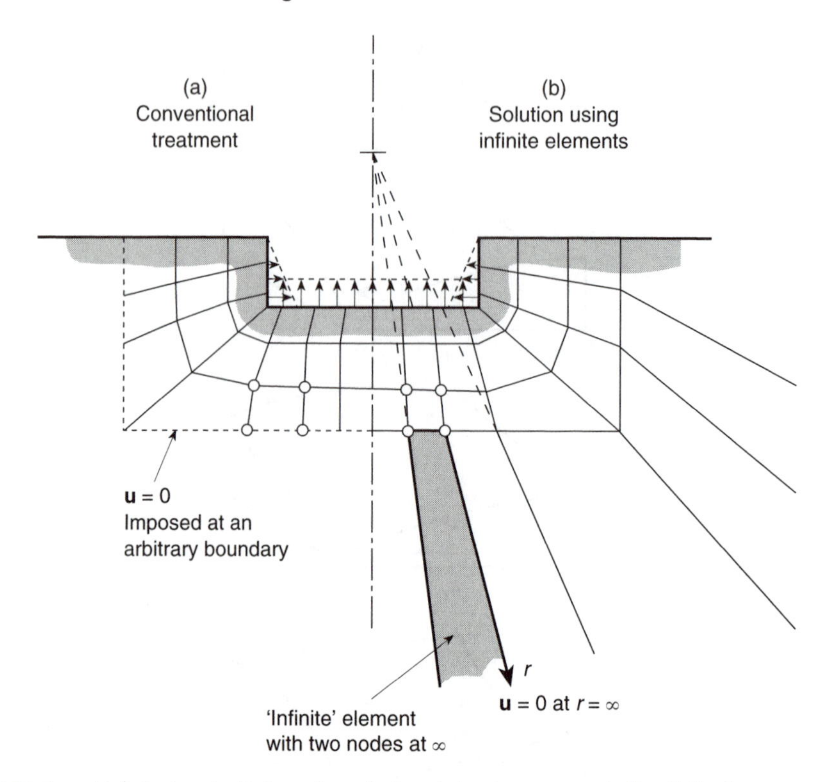

Fig. 5.21 A semi-infinite domain. Deformations of a foundation due to removal of load following an excavation. (a) Conventional treatment and (b) use of infinite elements.

pioneered originally by Bettess.[36] In this process the conventional, finite elements are coupled to elements of the type shown in Fig. 5.21(b) which model in a reasonable manner the material stretching to infinity.

The shape of such two-dimensional elements and their treatment is best accomplished by mapping[34–36] these onto a finite square (or a finite line in one dimension or cube in three dimensions). However, it is essential that the sequence of trial functions introduced in the mapped domain be such that it is complete and capable of modelling the true behaviour as the radial distance r increases. Here it would be advantageous if the mapped shape functions could approximate a sequence of the decaying form

$$\frac{C_1}{r} + \frac{C_2}{r^2} + \frac{C_3}{r^3} + \cdots \tag{5.58}$$

where C_a are arbitrary constants and r is the radial distance from the 'focus' of the problem.

In the next subsection we introduce a mapping function capable of doing just this.

5.14.2 The mapping function

Figure 5.22 illustrates the principles for generation of the derived mapping function.

We shall start with a one-dimensional mapping along a line CPQ coinciding with the x

direction. Consider the following function:

$$x = -\frac{\xi}{1-\xi}x_C + \left(1 + \frac{\xi}{1-\xi}\right)x_Q = \bar{N}_C x_C + \bar{N}_Q x_Q \tag{5.59a}$$

and we immediately observe that

$$\begin{aligned} \xi &= -1 \quad \text{corresponds to } x = \frac{x_Q + x_C}{2} \equiv x_P \\ \xi &= 0 \quad \text{corresponds to } x = x_Q \\ \xi &= 1 \quad \text{corresponds to } x = \infty \end{aligned}$$

where x_P is a point midway between Q and C.

Alternatively the above mapping could be written directly in terms of the Q and P coordinates by simple elimination of x_C. This gives, using our previous notation:

$$\begin{aligned} x &= N_Q x_Q + N_P x_P \\ &= \left(1 + \frac{2\xi}{1-\xi}\right)x_Q - \frac{2\xi}{1-\xi}x_P \end{aligned} \tag{5.59b}$$

Both forms give a mapping that is independent of the origin of the x coordinate as

$$N_Q + N_P = 1 = \bar{N}_C + \bar{N}_Q \tag{5.60}$$

The significance of the point C is, however, of great importance. It represents the centre from which the 'disturbance' originates and, as we shall now show, allows the expansion of the form of Eq. (5.58) to be achieved on the assumption that r is measured from C. Thus

$$r = x - x_C \tag{5.61}$$

If, for instance, the unknown function u is approximated by a polynomial function using, say, hierarchical shape functions and giving

$$u = \alpha_0 + \alpha_1\xi + \alpha_2\xi^2 + \alpha_3\xi^3 + \cdots \tag{5.62}$$

we can easily solve Eqs (5.59a) for ξ, obtaining

$$\xi = 1 - \frac{x_Q - x_C}{x - x_C} = 1 - \frac{x_Q - x_C}{r} \tag{5.63}$$

Substitution into Eq. (5.62) shows that a series of the form given by Eq. (5.58) is obtained with the linear shape function in ξ corresponding to $1/r$ terms, quadratic to $1/r^2$, etc.

In one dimension the objectives specified have thus been achieved and the element will yield convergence as the degree of the polynomial expansion, p, increases. Now a generalization to two or three dimensions is necessary. It is easy to see that this can be achieved by simple products of the one-dimensional infinite mapping with a 'standard' type of shape function in η (and ζ) directions in the manner indicated in Fig. 5.22.

First we generalize the interpolation of Eqs (5.59a) and (5.59b) for any straight line in x, y, z space and write (for such a line as $C_1P_1Q_1$ in Fig. 5.22)

$$\begin{aligned} x &= -\frac{\xi}{1-\xi}x_{C_1} + \left(1 + \frac{\xi}{1-\xi}\right)x_{Q_1} \\ y &= -\frac{\xi}{1-\xi}y_{C_1} + \left(1 + \frac{\xi}{1-\xi}\right)y_{Q_1} \\ z &= -\frac{\xi}{1-\xi}z_{C_1} + \left(1 + \frac{\xi}{1-\xi}\right)z_{Q_1} \quad \text{(in three dimensions)} \end{aligned} \tag{5.64}$$

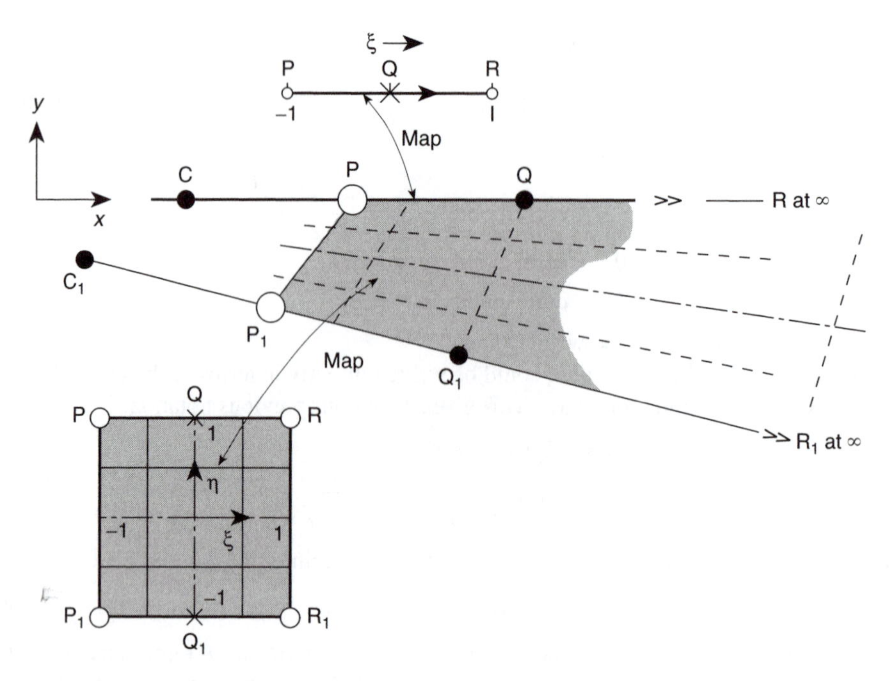

Fig. 5.22 Infinite line and element map. Linear η interpolation.

Second we complete the interpolation and map the whole $\xi\eta(\zeta)$ domain by adding a 'standard' interpolation in the $\eta(\zeta)$ directions. Thus for the linear interpolation shown we can write for elements $PP_1QQ_1RR_1$ of Fig. 5.22, as

$$\begin{aligned} x = N_1(\eta)\left[-\frac{\xi}{1-\xi}x_C + \left(1+\frac{\xi}{1-\xi}\right)x_Q\right] \\ + N_0(\eta)\left(-\frac{\xi}{1-\xi}x_{C_1} + \left(1+\frac{\xi}{1-\xi}\right)x_{Q_1}\right), \qquad \text{etc.} \end{aligned} \tag{5.65}$$

with

$$N_1(\eta) = \frac{1}{2}(1+\eta) \qquad N_0(\eta) = \frac{1}{2}(1-\eta)$$

and map the points as shown.

In a similar manner we could use quadratic interpolations and map an element as shown in Fig. 5.23 by using quadratic functions in η.

Thus it is an easy matter to create infinite elements and join these to a standard element mesh as shown in Fig. 5.21(b). In the generation of such element properties only the transformation jacobian matrix differs from standard forms, hence only this has to be altered in conventional programs. Moreover, integration is again over the usual 'parent' element.

The 'origin' or 'pole' of the coordinates C can be fixed arbitrarily for each radial line, as shown in Fig. 5.22. This will be done by taking account of the knowledge of the physical solution expected.

In Fig. 5.24 we show a solution of the Boussinesq problem (a point load on an elastic half-space). Here results of using a fixed displacement or infinite elements are compared

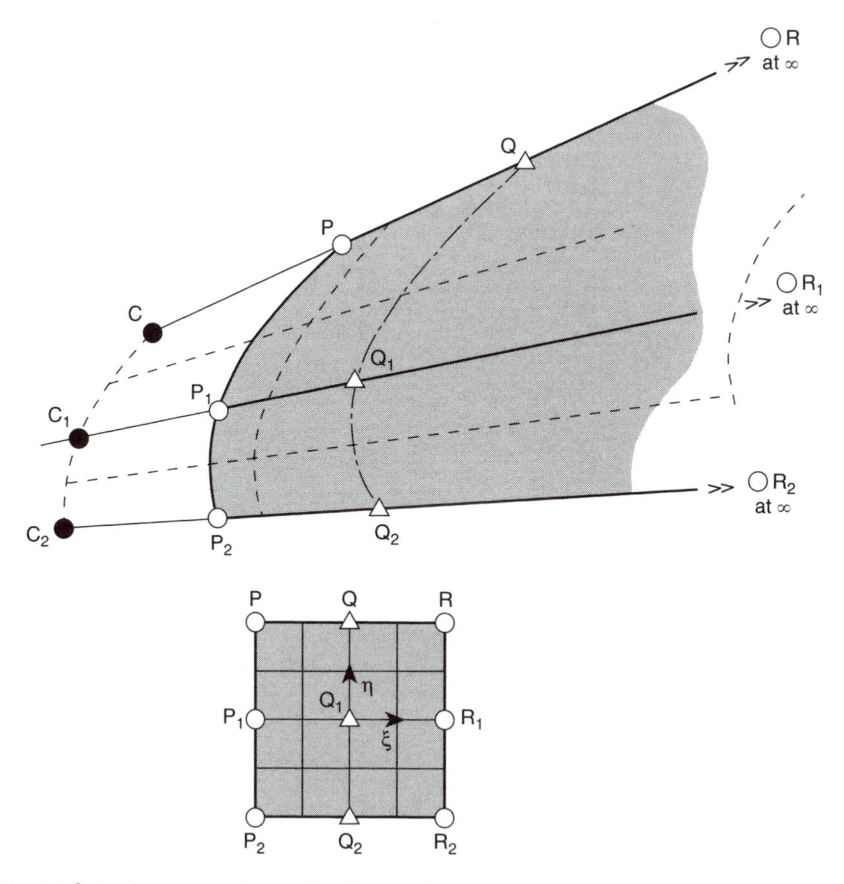

Fig. 5.23 Infinite element map. Quadratic η interpolation.

and the big changes in the solution noted. In this example the pole of each element was taken at the load point for obvious reasons.[35]

Figure 5.25 shows how similar infinite elements (of the linear kind) can give excellent results, even when combined with very few standard elements. In this example where a solution of the Laplace equation is used (see Chapter 3) for an irrotational fluid flow, the poles of the infinite elements are chosen at arbitrary points of the aerofoil centre-line.

In concluding this section it should be remarked that the use of infinite elements (as indeed of any other finite elements) must be tempered by background analytical knowledge and 'miracles' should not be expected. Thus the user should not expect, for instance, such excellent results as those shown in Fig. 5.24 for the displacement of a plane elasticity problem. It is 'well known' that in this case the displacements under any load which is not self-equilibrated will be infinite everywhere and the numbers obtained from the computation will not be, whereas for the three-dimensional or axisymmetric case it is infinite only at a point load.

Further use of infinite elements is made in the context of the solution of wave problems in fluids in reference 37.

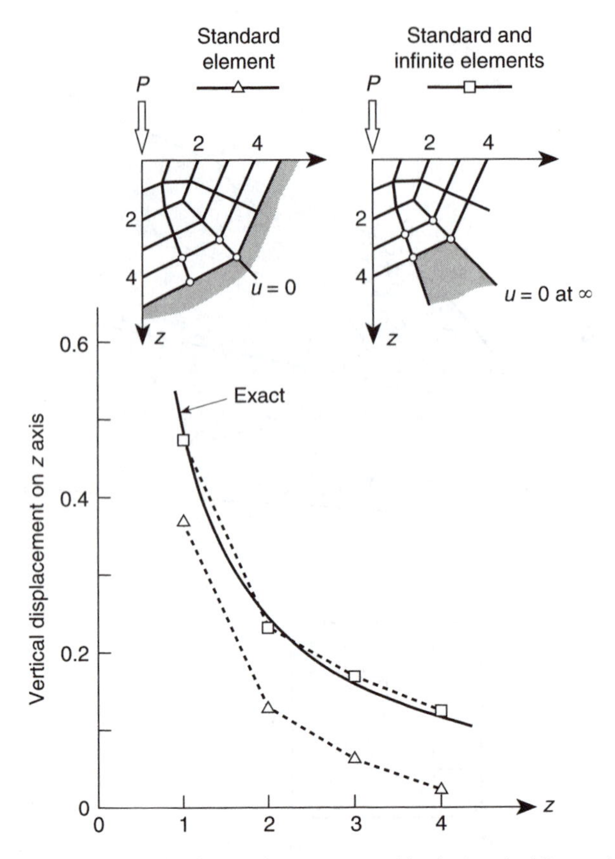

Fig. 5.24 A point load on an elastic half-space (Boussinesq problem). Standard linear elements and infinite line elements ($E = 1, \nu = 0.1, p = 1$).

5.15 Singular elements by mapping – use in fracture mechanics, etc.

In the study of fracture mechanics interest is often focused on the singularity point where quantities such as stress become (mathematically, but not physically) infinite. Near such singularities normal, polynomial-based, finite element approximations perform badly and attempts have frequently been made here to include special functions within an element which can model the analytically known singularity. References 38–53 give an extensive literature survey of the problem and finite element solution techniques. An alternative to the introduction of special functions within an element – which frequently poses problems of enforcing continuity requirements with adjacent, standard, elements – lies in the use of special mapping techniques.

An element of this kind, shown in Fig. 5.26(a), was introduced almost simultaneously by Henshell and Shaw[49] and Barsoum[50, 51] for quadrilaterals by a simple shift of the mid-side node to the quarter point in quadratic, isoparametric elements.

It can now be shown (and we leave this exercise to the curious reader) that along the element edges the derivatives $\partial u/\partial x$ (or strains) vary as $1/\sqrt{r}$ where r is the distance from

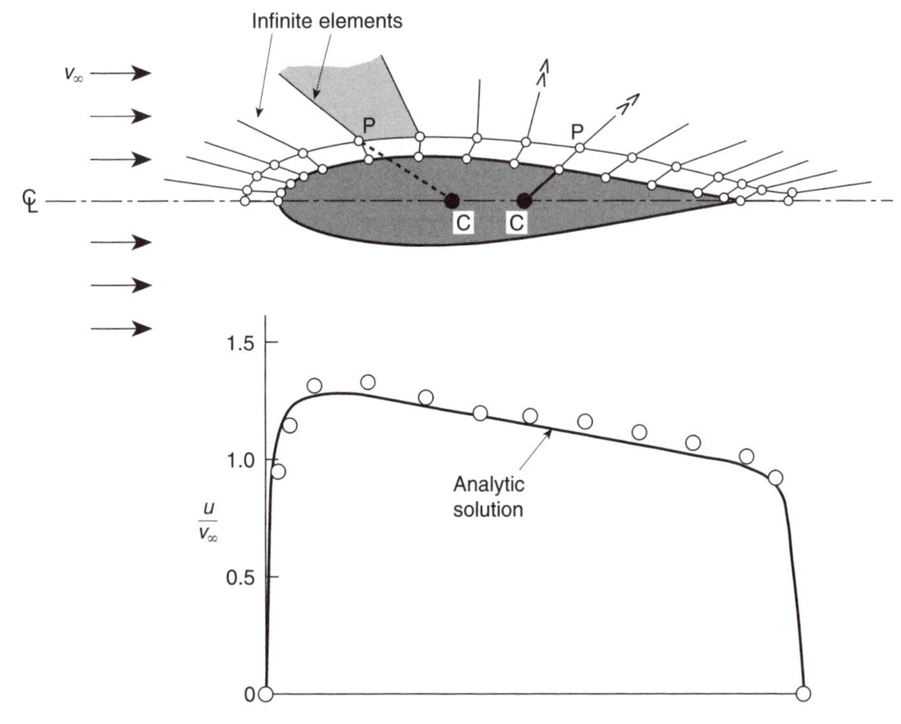

Fig. 5.25 Irrotational flow around NACA 0018 wing section.[31] (a) Mesh of bilinear isoparametric and infinite elements. (b) Computed ○ and analytical — results for velocity parallel to surface.

the corner node at which the singularity develops. Although good results are achievable with such elements the singularity is, in fact, not well modelled on lines other than element edges. A development suggested by Hibbitt[52] achieves a better result by using triangular second order elements for this purpose [Fig. 5.26(b)].

Indeed, the use of distorted or degenerate isoparametric elements is not confined to elastic singularities. Rice[43] shows that in the case of plasticity a shear strain singularity of $1/r$ type develops and Levy *et al.*[54] use an isoparametric, linear quadrilateral to generate such a singularity by the simple device of coalescing two nodes but treating these displacements independently. A variant of this is developed by Rice and Tracey.[39]

The elements just described are simple to implement without any changes in a standard finite element program. However, in Chapter 15 we introduce a method whereby any singularity (or other function) can be modelled directly. We believe the methods to be described there supersede the above described techniques.

5.16 Computational advantage of numerically integrated finite elements

One considerable gain that is possible in numerically integrated finite elements is the versatility that can be achieved in a single computer program.[55] It will be observed that for a

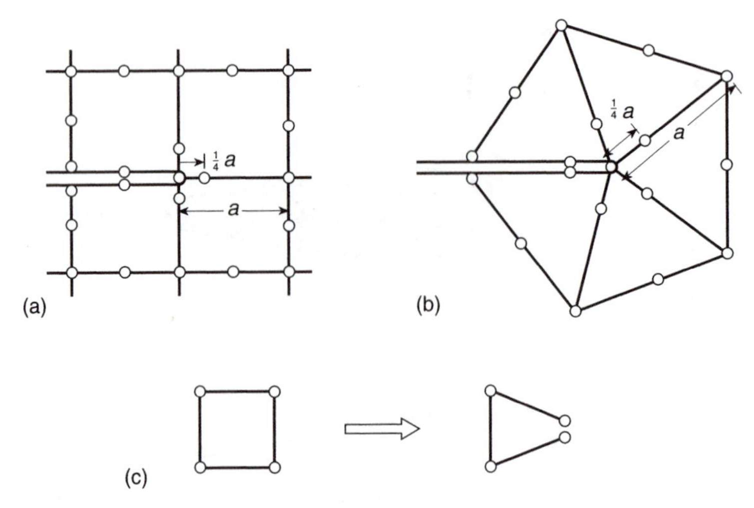

Fig. 5.26 Singular elements from degenerate isoparametric elements (a), (b), and (c).

given class of problems the general matrices are always of the same form [see the example of Eq. (5.7)] in terms of the shape function and its derivatives.

To proceed to evaluation of the element properties it is necessary first to *specify the shape function* and its derivatives and, second, to *specify the order of integration*. The computation of element properties is thus composed of three distinct parts as shown in Fig. 5.27. For a *given class of problems* it is only necessary to change the prescription of the shape functions to achieve a variety of possible elements. Conversely, the *same shape function* routines can be used in many different classes of problem.

Use of different elements, testing the efficiency of a new element in a given context, or extension of programs to deal with new situations can thus be readily achieved, and considerable algebra avoided (with its inherent possibilities of mistakes). The computer is thus placed in the position it deserves, i.e., of being the obedient slave capable of saving routine work.

The greatest practical advantage of the use of universal shape function routines is that they can be checked decisively for errors by a simple program with the patch test playing the crucial role (viz. Chapter 9).

The incorporation of simple, exactly integrable, elements in such a system is, incidentally, not penalized as the time of exact and numerical integration in many cases is almost identical.

5.17 Problems

5.1 A quadratic one-dimensional element is shown in Fig. 5.28 in parent form and in the mapped configuration. Let $a + b = h$ the total length of the mapped element.
   (a) Determine the shape functions $N_a(\xi)$ for the three nodes.
   (b) Plot ξ vs x for values of a ranging from $0.2h$ to $0.8h$ in increments of $0.1h$.
   (c) Plot N_a vs x for the range of a given in part (b).
   (d) Plot $\mathrm{d}N_a/\mathrm{d}x$ vs x for the range of a given in part (b).

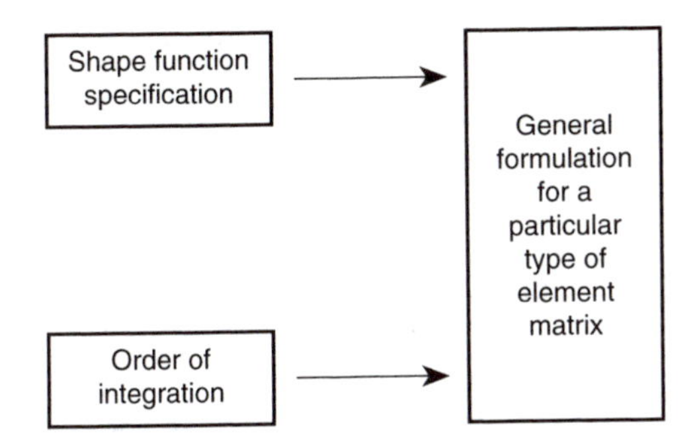

Fig. 5.27 Computational scheme for numerically integrated elements.

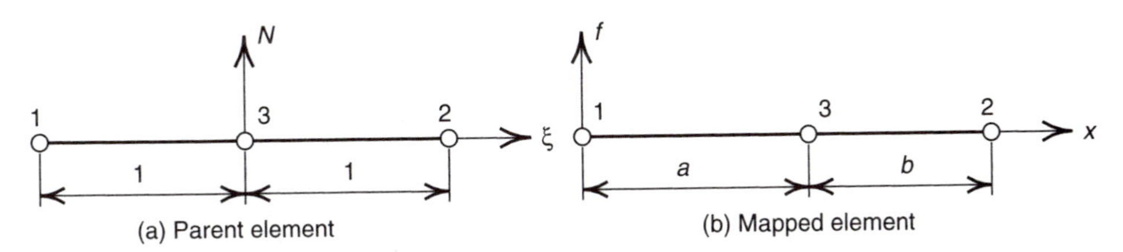

Fig. 5.28 Quadratic element for Problem 5.1.

5.2 Consider the one-dimensional problem for $0 \leq x \leq 1$ which is defined by the weak form

$$\int_0^1 \left[\frac{\mathrm{d}\delta u}{\mathrm{d}x} \frac{\mathrm{d}u}{\mathrm{d}x} - \delta u \, q \right] \mathrm{d}x - \delta u \, \sigma \Big|_{x=1} = 0 \;\; \text{with} \; u(0) = 0$$

with $q = \sigma = 1$.

(a) Deduce the Euler differential equation and boundary conditions for the problem.

(b) Construct an exact solution to the differential equation.

(c) Solve the weak form using a single quadratic order element with nodes placed at $x = 0, 5/16$ and 1 and shape functions N_a defined by:

i. Lagrange interpolation in x directly.

ii. Isoparametric interpolation for $N_a(\xi)$ with $x = N_a(\xi)\tilde{x}_a$.

Evaluate all integrals using two-point gaussian quadrature.

(d) Plot u and $\mathrm{d}u/\mathrm{d}x$ for the two solutions. Comment on differences in quality of the two solutions.

5.3 It is proposed to create transition elements for use with 4-node quadrilateral element meshes as shown in Fig. 5.29.

(a) Devise the shape functions for the transition element labelled A. The shape functions must maintain compatibility along all boundaries. (Hint: The element can be a composite form combining more than one 4-node element.)

(b) Devise the shape functions for the transition element labelled B.

(c) On a sketch show the location of quadrature points necessary to integrate each element form.

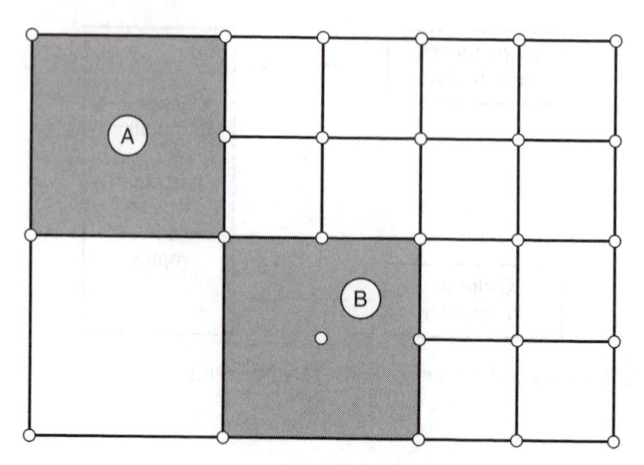

Fig. 5.29 Transition elements for use with 4-node quadrilaterals.

(d) As an alternative to transition elements, 4-node elements may be used for all elements and constraints imposed to maintain compatibility. For the mesh shown in the figure, number all nodes and write the constraint equations necessary to maintain compatibility. The interior node of element *B* is not needed and can be ignored.

5.4 Determine the hierarchical interpolation functions in ξ, η coordinates for the 16-node cubic order quadrilateral shown in Fig. 5.30(a). Express your hierarchic shape functions in a form such that interpolation is given by

$$f(\xi, \eta) = \sum_{a=1}^{4} N_a(\xi, \eta) f_a + \sum_{a=5}^{12} N_a(\xi, \eta) \Delta f_a + \sum_{a=13}^{16} N_a(\xi, \eta) \Delta\Delta f_a$$

5.5 Determine the hierarchical interpolation functions in L_1, L_2, L_3 area coordinates for the 10-node cubic order triangle shown in Fig. 5.30(b). Express your hierarchic shape functions in a form such that interpolation is given by

$$f(L_1, L_2, L_3) = \sum_{a=1}^{3} N_a(L_1, L_2, L_3) f_a + \sum_{a=5}^{8} N_a(L_1, L_2, L_3) \Delta f_a + \sum_{a=11}^{12} N_a(L_1, L_2, L_3) \Delta f_a + N_{13}(L_1, L_2, L_3) \Delta\Delta f_{13}$$

5.6 Using the shape functions developed in Problem 5.4, determine the modified shape functions to degenerate the cubic 16-node quadrilateral into the cubic 10-node triangular element using numbering as shown in Fig. 5.30. The final element must be completely consistent with the shape functions developed in Problem 5.5.

5.7 Degenerate an 8-node hexahedral element to form a pyramid form with a rectangular base. Write the resulting shape functions for the remaining 5 nodes.

5.8 For the triangular element shown in Fig. 5.31 show that the global coordinates may be expressed in local coordinates as

$$\mathbf{x} = \sum_{a=1}^{6} N_a(L_b)\, \tilde{\mathbf{x}}_a = 12L_2 + 18L_3$$

5.9 For the triangular element shown in Fig. 5.31 compute the integrals $\int_\Delta N_2 N_3 \, d\Delta$ and $\int_\Delta N_2 N_4 \, d\Delta$ using:
(a) Eq. (4.31) and
(b) an appropriate numerical integration using Table 5.3.

5.10 For the triangular element shown in Fig. 5.31 compute the integrals $\int_\Delta N_a \, d\Delta$; $a = 1, 2, \cdots, 6$ using:
(a) Eq. (4.31) and
(b) an appropriate numerical integration using Table 5.3.

5.11 The 4-node quadrilateral element shown in Fig. 5.32 is used in the solution of a problem in which the dependent variable is a scalar, u.
(a) Write the expression for an isoparametric mapping of coordinates in the element.
(b) Determine the location of the natural coordinates ξ and η which define the centroid of the element.
(c) Compute the expression for the jacobian transformation $\mathbf{J}$ of the element. Evaluate the jacobian at the centroid.
(d) Compute the derivatives of the shape function N_3 at the centroid.

5.12 A triangular element is formed by degenerating a 4-node quadrilateral element as shown in Fig. 5.33. If node 1 is located at $(x, y) = (10, 8)$ and the sides are $a = 20$ and $b = 30$:
(a) Write the expressions for x and y in terms of ξ and η.
(b) Compute the jacobian matrix $\mathbf{J}(\xi, \eta)$ for the element.
(c) Compute the jacobian $J(\xi, \eta)$.
(d) For a one-point quadrature formula given by

$$I = \int_{-1}^{1} \int_{-1}^{1} f(\xi, \eta) \, d\xi \, d\eta = f(\xi_i, \eta_i) W_i$$

determine the values of W_i, ξ_i and η_i which exactly integrate the jacobian J (and thus also any integral which is a constant times the jacobian).
(e) Is this the same point in the element as that using triangular coordinates L_a and the one-point formula from Table 5.3? If not, why?

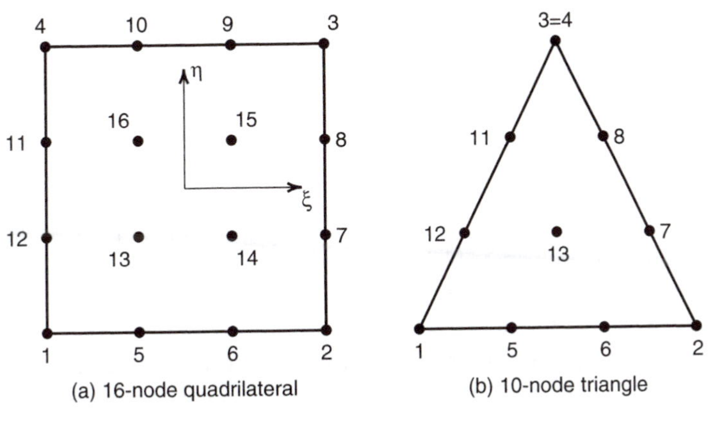

Fig. 5.30 Degeneration of cubic triangle for Problem 5.5.

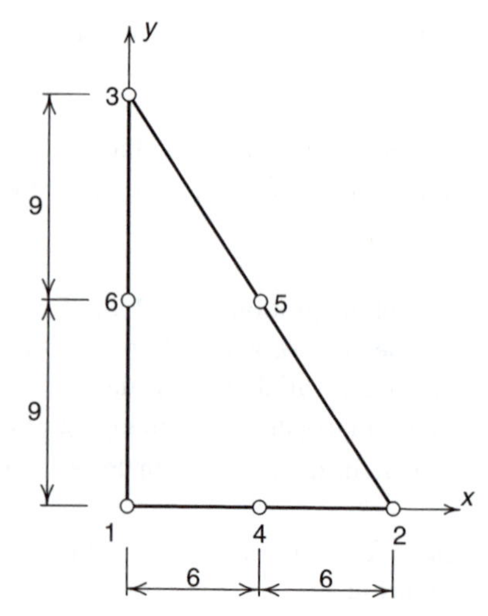

Fig. 5.31 Quadratic triangle. Problems 5.8 to 5.10.

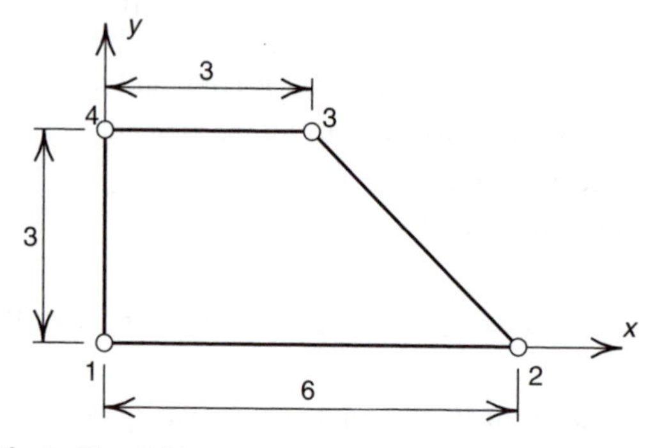

Fig. 5.32 Quadrilateral for Problem 5.11.

5.13 In some instances it is desirable to perform numerical integration in which quadrature points are located at the end points as well as at interior points. One such formula is the Newton–Cotes type shown in Fig. 5.16(a); however, a more accurate formula (known as Gauss–Lobatto quadrature) may be developed as

$$\int_{-1}^{1} f(\xi)\, \mathrm{d}\xi = [f(-1) + f(1)]W_0 + \sum_{i=1}^{n} f(\xi_i)W_i$$

Determine the location of the points ξ_i and the value of the weights W_i which exactly integrate the highest polynomial of f possible. Consider:

(a) The three-point formula ($n = 1$).

(b) The four-point formula ($n = 2$).

5.14 Write the blending function mapping for a two-dimensional quadrilateral region which has one circular edge and three straight linear edges. Make a clear sketch of the region defined by the function and a 3×3 division into 4-node quadrilateral elements.

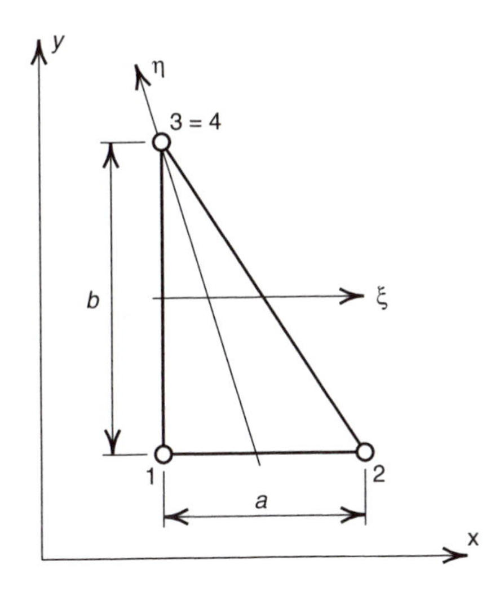

Fig. 5.33 Degenerate triangle for Problem 5.12.

5.15 Consider a 6-node triangular element with straight edges in which two of the mid-side nodes are placed at the quarter point. Show that the interpolation along the edge produces a derivative which varies as $1/\sqrt{r}$ where r is the distance measured from the vertex.

5.16 Compute the x and y derivatives for the shape function of nodes 1, 7 and 10 of the quadratic triangular prism shown in Fig. 4.20(b).

5.17 Program development project: Extend the program system started in Problem 2.17 to permit mesh generation using as input a 4-node isoparametric block and mapping as described in Sec. 5.13. The input data should be the coordinates of the block vertices and the number of subdivisions in each direction.

Include as an option generation of coordinates in r, θ coordinates that are then transformed to x, y cartesian form.

(Hint: Once coordinates for all node points are specified, MATLAB can generate a node connection list for 3-node triangles using DELAUNAY.† A plot of the mesh may be produced using TRIMESH.)

Use your program to generate a mesh for the rectangular beam described in Example 2.3 and the curved beam described in Example 2.4. Note the random orientation of diagonals which is associated with degeneracy in the Delaunay algorithm (viz. Chap. 8).

5.18 Program development project: Extend the mesh generation scheme developed in Problem 5.17 to permit specification of the block as a blending function. Only allow two cases: (i) Lagrange interpolation which is linear or quadratic; (ii) circular arcs with specified radius and end points.

Test your program for the beam problems described in Examples 2.3 and 2.4.

† In Chapter 8 we discuss mesh generation and some of the difficulties encountered with the Delaunay method.

References

1. I.C. Taig. Structural analysis by the matrix displacement method. Technical Report No. S.0.17, English Electric Aviation Ltd, April 1962. Based on work performed 1957–58.
2. B.M. Irons. Numerical integration applied to finite element methods. In *Proc. Conf. on Use of Digital Computers in Structural Engineering*, University of Newcastle, 1966.
3. B.M. Irons. Engineering applications of numerical integration in stiffness methods. *J. AIAA*, 4:2035–2037, 1966.
4. S.A. Coons. Surfaces for computer aided design of space forms. Technical Report MAC-TR-41, MIT Project MAC, 1967.
5. A.R. Forrest. Curves and surfaces for computer aided design. Technical Report, Computer Aided Design Group, Cambridge, England, 1968.
6. G. Strang and G.J. Fix. *An Analysis of the Finite Element Method.* Prentice-Hall, Englewood Cliffs, N.J., 1973.
7. E.L. Wachspress. High order curved finite elements. *Int. J. Numer. Meth. Eng.*, 17:735–745, 1981.
8. M. Crochet. Personal communication, 1988.
9. Nam-Sua Lee and K.-J. Bathe. Effects of element distortion on the performance of isoparametric elements. *Int. J. Numer. Meth. Eng.*, 36:3553–3576, 1993.
10. R.E. Newton. Degeneration of brick-type isoparametric elements. *Int. J. Numer. Meth. Eng.*, 7:579–581, 1974.
11. B.M. Irons. A technique for degenerating brick type isoparametric elements using hierarchical midside nodes. *Int. J. Numer. Meth. Eng.*, 8:209–211, 1974.
12. M. Abramowitz and I.A. Stegun, editors. *Handbook of Mathematical Functions.* Dover Publications, New York, 1965.
13. B.M. Irons. Quadrature rules for brick based finite elements. *Int. J. Numer. Meth. Eng.*, 3:293–294, 1971.
14. T.K. Hellen. Effective quadrature rules for quadratic solid isoparametric finite elements. *Int. J. Numer. Meth. Eng.*, 4:597–600, 1972.
15. R. Radau. Études sur les formules d'approximation qui servent à calculer la valeur d'une intégrale définie. *Journ. de math.,* 6:283–336, 1880.
16. R.G. Anderson, B.M. Irons, and O.C. Zienkiewicz. Vibration and stability of plates using finite elements. *Int. J. Solids Struct.*, 4:1033–1055, 1968.
17. P.C. Hammer, O.P. Marlowe, and A.H. Stroud. Numerical integration over symplexes and cones. *Math. Tables Aids Comp.*, 10:130–137, 1956.
18. C.A. Felippa. *Refined finite element analysis of linear and non-linear two-dimensional structures.* Ph.D. dissertation, Department of Civil Engineering, SEMM, University of California, Berkeley, 1966. Also: SEL Report 66–22, Structures Materials Research Laboratory.
19. G.R. Cowper. Gaussian quadrature formulas for triangles. *Int. J. Numer. Meth. Eng.*, 7:405–408, 1973.
20. I. Fried. Accuracy and condition of curved (isoparametric) finite elements. *J. Sound Vibration*, 31:345–355, 1973.
21. I. Fried. Numerical integration in the finite element method. *Comp. Struct.*, 4:921–932, 1974.
22. M. Zlamal. Curved elements in the finite element method. *SIAM J. Num. Anal.*, 11:347–362, 1974.
23. D. Kosloff and G.A. Frasier. Treatment of hour glass patterns in low order finite element codes. *Int. J. Numer. Anal. Meth. Geomech.*, 2:57–72, 1978.
24. T. Belytschko and W.E. Bachrach. The efficient implementation of quadrilaterals with high coarse mesh accuracy. *Comp. Meth. Applied Mech. Eng.*, 54:276–301, 1986.
25. O.C. Zienkiewicz and D.V. Phillips. An automatic mesh generation scheme for plane and curved surfaces by isoparametric coordinates. *Int. J. Numer. Meth. Eng.*, 3:519–528, 1971.

26. W.J. Gordon and C.A. Hall. Construction of curvilinear co-ordinate systems and application to mesh generation. *Int. J. Numer. Meth. Eng.*, 3:461–477, 1973.
27. W.J. Gordon and C.A. Hall. Transfinite element methods – blending-function interpolation over arbitrary curved element domains. *Numer. Math.*, 21:109–129, 1973.
28. R.W. Thatcher. On the finite element method for unbounded regions. *SIAMM J. Num. Anal.*, 15(3):466–476, 1978.
29. P. Silvester, D.A. Lowther, C.J. Carpenter, and E.A. Wyatt. Exterior finite elements for 2-dimensional field problems with open boundaries. *Proc. IEEE*, 123(12), Dec. 1977.
30. S.F. Shen. An aerodynamicist looks at the finite element method. In R.H. Gallagher *et al.*, editor, *Finite Elements in Fluids*, volume 2, pages 179–204. John Wiley & Sons, New York, 1975.
31. O.C. Zienkiewicz, D.W. Kelley, and P. Bettess. The coupling of the finite element and boundary solution procedures. *Int. J. Numer. Meth. Eng.*, 11:355–375, 1977.
32. P. Bettess. *Infinite Elements*. Penshaw Press, Cleadon, U.K., 1992.
33. P. Bettess and O.C. Zienkiewicz. Diffraction and refraction of surface waves using finite and infinite elements. *Int. J. Numer. Meth. Eng.*, 11:1271–1290, 1977.
34. G. Beer and J.L. Meek. Infinite domain elements. *Int. J. Numer. Meth. Eng.*, 17:43–52, 1981.
35. O.C. Zienkiewicz, C. Emson, and P. Bettess. A novel boundary infinite element. *Int. J. Numer. Meth. Eng.*, 19:393–404, 1983.
36. P. Bettess. Infinite elements. *Int. J. Numer. Meth. Eng.*, 11:53–64, 1977.
37. O.C. Zienkiewicz, R.L. Taylor, and P. Nithiarasu. *The Finite Element Method for Fluid Dynamics*. Butterworth-Heinemann, Oxford, 6th edition, 2005.
38. J.J. Oglesby and O. Lomacky. An evaluation of finite element methods for the computation of elastic stress intensity factors. *J. Eng. Ind.*, 95:177–183, 1973.
39. J.R. Rice and D.M. Tracey. Computational fracture mechanics. In S.J. Fenves *et al.*, editor, *Numerical and Computer Methods in Structural Mechanics*, pages 555–624. Academic Press, New York, 1973.
40. A.A. Griffiths. The phenomena of flow and rupture in solids. *Phil. Trans. Roy. Soc. (London)*, A221:163–198, 1920.
41. D.M. Parks. A stiffness derivative finite element technique for determination of elastic crack tip stress intensity factors. *Int. J. Fract.*, 10:487–502, 1974.
42. T.K. Hellen. On the method of virtual crack extensions. *Int. J. Numer. Meth. Eng.*, 9:187–208, 1975.
43. J.R. Rice. A path-independent integral and the approximate analysis of strain concentration by notches and cracks. *J. Applied Mech., ASME*, 35:379–386, 1968.
44. P. Tong and T.H.H. Pian. On the convergence of the finite element method for problems with singularity. *Int. J. Solids Struct.*, 9:313–321, 1972.
45. T.A. Cruse and W. Vanburen. Three dimensional elastic stress analysis of a fracture specimen with edge crack. *Int. J. Fract. Mech.*, 7:1–15, 1971.
46. P.F. Walsh. The computation of stress intensity factors by a special finite element technique. *Int. J. Solids Struct.*, 7:1333–1342, 1971.
47. P.F. Walsh. Numerical analysis in orthotropic linear fracture mechanics. *Inst. Eng. Australia, Civ. Eng. Trans.*, 15:115–119, 1973.
48. D.M. Tracey. Finite elements for determination of crack tip elastic stress intensity factors. *Eng. Fract. Mech.*, 3:255–265, 1971.
49. R.D. Henshell and K.G. Shaw. Crack tip elements are unnecessary. *Int. J. Numer. Meth. Eng.*, 9:495–509, 1975.
50. R.S. Barsoum. On the use of isoparametric finite elements in linear fracture mechanics. *Int. J. Numer. Meth. Eng.*, 10:25–38, 1976.
51. R.S. Barsoum. Triangular quarter point elements as elastic and perfectly elastic crack tip elements. *Int. J. Numer. Meth. Eng.*, 11:85–98, 1977.

52. H.D. Hibbitt. Some properties of singular isoparametric elements. *Int. J. Numer. Meth. Eng.*, 11:180–184, 1977.
53. S.E. Benzley. Representation of singularities with isoparametric finite elements. *Int. J. Numer. Meth. Eng.*, 8:537–545, 1974.
54. N. Levy, P.V. Marçal, W.J. Ostergren, and J.R. Rice. Small scale yielding near a crack in plane strain: a finite element analysis. *Int. J. Fract. Mech.*, 7:143–157, 1967.
55. B.M. Irons. Economical computer techniques for numerically integrated finite elements. *Int. J. Numer. Meth. Eng.*, 1:201–203, 1969.

6

Problems in linear elasticity

6.1 Introduction

In this and the next chapter we deal with the set of problems in elasticity and fields which are common in various engineering applications and will serve well to introduce practical examples of application of the various element forms discussed in the previous chapters.

Specifically, in this chapter we again consider the problem of stress analysis for linear elastic solids which was introduced in Chapter 2 and briefly discussed in Sec. 3.4.

The simplest two-dimensional continuum element is a triangle. In three dimensions its equivalent is a tetrahedron, an element with four nodal corners.†

Two-dimensional elastic problems were the first successful examples of the application of the finite element method.[1,2] Indeed, we have already used this situation to illustrate the basis of the finite element formulation in Chapter 2 where the general relationships were derived.

The first suggestions for the use of the simple tetrahedral element appear to be those of Gallagher *et al.*[3] and Melosh.[4] Argyris[5,6] elaborated further on the theme and Rashid and Rockenhauser[7] were the first to apply three-dimensional analysis to realistic problems.

It is immediately obvious, however, that the number of simple tetrahedral elements which has to be used to achieve a given degree of accuracy has to be very large and this will result in very large numbers of simultaneous equations in practical problems. This leads to large compute times when direct solution schemes based on Gauss elimination are used. Thus, in recent times there is increased interest in use of iterative solution methods.

To realize the order of magnitude of the problems presented let us assume that the accuracy of a triangle in two-dimensional analysis is comparable to that of a tetrahedron in three dimensions. If an adequate stress analysis of a square, two-dimensional region requires a mesh of some $20 \times 20 = 400$ nodes, the total number of simultaneous equations is around 800 given two displacement variables at a node (this is a fairly realistic figure). The bandwidth of the matrix involves 20 nodes, i.e., some 40 variables.

An equivalent three-dimensional region is that of a cube with $20 \times 20 \times 20 = 8000$ nodes. The total number of simultaneous equations is now some 24 000 as three displacement variables have to be specified. Further, the bandwidth now involves an interconnection of some $20 \times 20 = 400$ nodes or 1200 variables.

Given that with direct solution techniques the computation effort is roughly proportional to the number of equations and to the square of the bandwidth, the magnitude of the problems can be appreciated. It is not surprising therefore that efforts to improve accuracy by use

† The simplest polygonal shape which permits the approximation of the domain is known as the *simplex*. Thus a triangular and tetrahedral element constitutes the simplex forms in two and three dimensions, respectively.

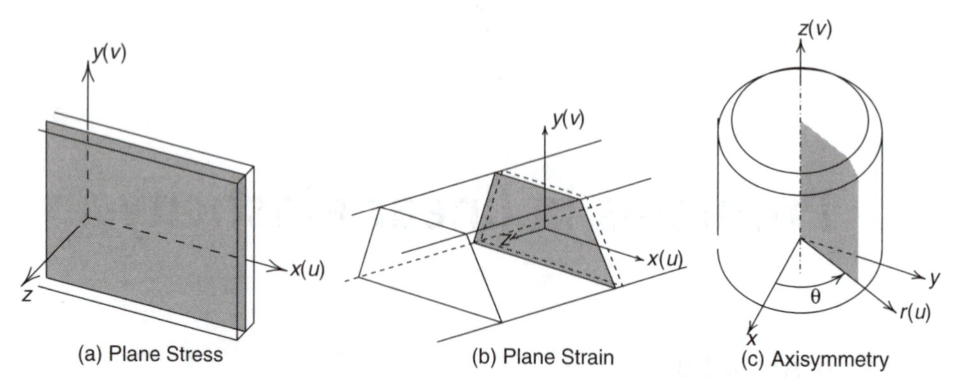

Fig. 6.1 Two-dimensional analysis types for plane stress, plane strain and axisymmetry.

of higher order elements was strongest in the area of three-dimensional analysis.[8–12] The development and practical application of such elements will be described in this chapter.

We shall deal with elasticity problems for general three-dimensional applications, as well as its simplification to some special two-dimensional situations. What we mean by two dimensions is that the total field should be capable of being defined by two components of displacement in the manner similar to that illustrated in Chapter 2.

The two-dimensional problems we consider are of three types:

(a) The *plane stress* case dealt with in Chapter 2 and shown in Fig. 6.1(a). In this problem the only non-zero stresses are those in the plane of the problem and normal to the lamina we have no stresses.
(b) The second case where again two displacement components exist is that of *plane strain* in which all straining normal to the plane considered is prevented. Such a situation may arise in a long prism which is being loaded in the manner shown in Fig. 6.1(b).
(c) The third and final case of two-dimensional analysis is that in which the situation is *axisymmetric*. Here the plane considered is one at constant θ in a cylindrical coordinate system r, z, θ [Fig. 6.1(c)] and again two displacements define the state of strain.

We assume the reader is familiar with the theory of linear elasticity; however, for completeness we will summarize the basic equations for the different problem classes to be considered. For a more general discussion the reader is referred to standard references on the subject (e.g., see references 13–18).

6.2 Governing equations

Although in some situations the use of indicial notation is advantageous as discussed in Appendix B, for simplicity we choose to continue here the matrix form of definitions.

6.2.1 Displacement function

For the three-dimensional problem the displacement field is given by

$$\mathbf{u} = \begin{Bmatrix} u(x, y, z) \\ v(x, y, z) \\ w(x, y, z) \end{Bmatrix} \tag{6.1}$$

where positions are denoted by the cartesian coordinates x, y, z.

For the two-dimensional cases considered the displacement field is given by

$$\mathbf{u} = \begin{Bmatrix} u(x, y) \\ v(x, y) \end{Bmatrix} \tag{6.2}$$

for plane stress and plane strain problems; and by

$$\mathbf{u} = \begin{Bmatrix} u(r, z) \\ v(r, z) \end{Bmatrix} \tag{6.3}$$

for problems with axisymmetric deformation. The only difference in the latter two is the coordinates used: x, y for cartesian coordinates and r, z for cylindrical coordinates. One may notice that in plane stress problems, changes of thickness occur; however, no explicit displacement assumption is given and the result will be included directly within the strain approximation.

6.2.2 Strain matrix

The strains for a problem undergoing small deformations are computed from the displacements. The form of the strain was given in Eq. (2.13) as

$$\boldsymbol{\varepsilon} = \boldsymbol{\mathcal{S}}\mathbf{u} \tag{6.4}$$

where $\boldsymbol{\mathcal{S}}$ is a differential operator and $\mathbf{u}$ the displacement field. We write the *six* independent components of strain in $\boldsymbol{\varepsilon}$ where

$$\boldsymbol{\varepsilon} = \begin{Bmatrix} \varepsilon_x \\ \varepsilon_y \\ \varepsilon_z \\ \gamma_{xy} \\ \gamma_{yz} \\ \gamma_{zx} \end{Bmatrix} = \begin{bmatrix} \frac{\partial}{\partial x} & 0 & 0 \\ 0 & \frac{\partial}{\partial y} & 0 \\ 0 & 0 & \frac{\partial}{\partial z} \\ \frac{\partial}{\partial y} & \frac{\partial}{\partial x} & 0 \\ 0 & \frac{\partial}{\partial z} & \frac{\partial}{\partial y} \\ \frac{\partial}{\partial z} & 0 & \frac{\partial}{\partial x} \end{bmatrix} \begin{Bmatrix} u \\ v \\ w \end{Bmatrix} \tag{6.5}$$

Note that in matrix form shear strain components are twice that given in tensor form in Appendix B (e.g., $\gamma_{xy} = 2\,\varepsilon_{xy}$).

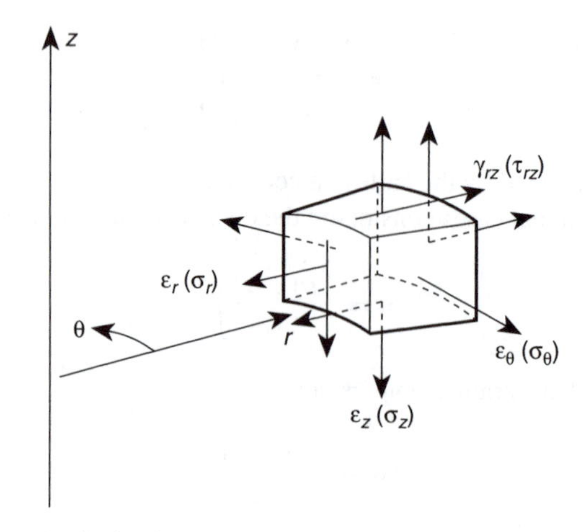

Fig. 6.2 Strains and stress involved in the analysis of axisymmetric solids.

For convenience in considering all three classes of two-dimensional problems in a unified manner, we include *four* components of strain in $\boldsymbol{\varepsilon}$ and write them as

$$\boldsymbol{\varepsilon} = \begin{Bmatrix} \varepsilon_x \\ \varepsilon_y \\ \varepsilon_z \\ \gamma_{xy} \end{Bmatrix} = \begin{bmatrix} \dfrac{\partial}{\partial x} & 0 \\ 0 & \dfrac{\partial}{\partial y} \\ 0 & 0 \\ \dfrac{\partial}{\partial y} & \dfrac{\partial}{\partial x} \end{bmatrix} \begin{Bmatrix} u \\ v \end{Bmatrix} + \begin{Bmatrix} 0 \\ 0 \\ \varepsilon_z \\ 0 \end{Bmatrix} = \boldsymbol{\mathcal{S}}\mathbf{u} + \boldsymbol{\varepsilon}_z \tag{6.6}$$

for plane problems (where ε_z is zero for plane strain but not for plane stress) and

$$\boldsymbol{\varepsilon} = \begin{Bmatrix} \varepsilon_r \\ \varepsilon_z \\ \varepsilon_\theta \\ \gamma_{rz} \end{Bmatrix} = \begin{bmatrix} \dfrac{\partial}{\partial r} & 0 \\ 0 & \dfrac{\partial}{\partial z} \\ \dfrac{1}{r} & 0 \\ \dfrac{\partial}{\partial z} & \dfrac{\partial}{\partial r} \end{bmatrix} \begin{Bmatrix} u \\ v \end{Bmatrix} = \boldsymbol{\mathcal{S}}\mathbf{u} \tag{6.7}$$

for the axisymmetric case (see Fig. 6.2).

The three problem types differ only by the presence of ε_z in the plane stress problem and the ε_θ component in the $\boldsymbol{\mathcal{S}}$ operator of the axisymmetric case.

6.2.3 Equilibrium equations

The equilibrium equations for the three-dimensional behaviour of a solid were presented in Sec. 3.4. They may be written in a matrix form as

$$\boldsymbol{\mathcal{S}}^{\mathrm{T}}\boldsymbol{\sigma} + \mathbf{b} = \mathbf{0} \tag{6.8}$$

where $\mathcal{S}$ is the same differential operator as that given for strains in (6.5); $\boldsymbol{\sigma}$ is the array of stresses which are ordered as

$$\boldsymbol{\sigma} = \begin{bmatrix}\sigma_x , & \sigma_y , & \sigma_z , & \tau_{xy} , & \tau_{yz} , & \tau_{zx}\end{bmatrix}^{\mathrm{T}} \tag{6.9}$$

and $\mathbf{b}$ is the vector of body forces given as

$$\mathbf{b} = \begin{bmatrix}b_x , & b_y , & b_z\end{bmatrix}^{\mathrm{T}} \tag{6.10}$$

In two-dimensional plane problems we omit τ_{yz}, τ_{zx} and b_z. For axisymmetric problems the stress is replaced by

$$\boldsymbol{\sigma} = \begin{bmatrix}\sigma_r , & \sigma_z , & \sigma_\theta , & \tau_{rz}\end{bmatrix}^{\mathrm{T}} \tag{6.11}$$

and body force by

$$\mathbf{b} = \begin{bmatrix}b_r , & b_z\end{bmatrix}^{\mathrm{T}} \tag{6.12}$$

6.2.4 Boundary conditions

Boundary conditions must be specified for each point on the surface of the solid. Here we consider two types of boundary conditions: (a) displacement boundary conditions and (b) traction boundary conditions. Thus we assume that the boundary may be divided into two parts, Γ_u for the displacement conditions and Γ_t for the traction conditions.

Displacement boundary conditions are specified at each point of the boundary Γ_u as

$$\mathbf{u} = \bar{\mathbf{u}} \tag{6.13}$$

where $\bar{\mathbf{u}}$ are known values.

Traction boundary conditions are specified for each point of the boundary Γ_t and are given in terms of stresses by

$$\mathbf{t} = \mathbf{G}^{\mathrm{T}}\boldsymbol{\sigma} = \bar{\mathbf{t}} \tag{6.14}$$

in which $\mathbf{G}^{\mathrm{T}}$ is the matrix

$$\mathbf{G}^{\mathrm{T}} = \begin{bmatrix} n_x & 0 & 0 & n_y & 0 & n_z \\ 0 & n_y & 0 & n_x & n_z & 0 \\ 0 & 0 & n_z & 0 & n_y & n_x \end{bmatrix} \tag{6.15}$$

where n_x, n_y, n_z are the direction cosines for the outward pointing normal to the boundary Γ_t. It should be noticed that the matrices $\mathbf{G}$ and $\mathcal{S}$ have identical non-zero structure. In two dimensions $\mathbf{G}$ reduces to

$$\mathbf{G}^{\mathrm{T}} = \begin{bmatrix} n_x & 0 & 0 & n_y \\ 0 & n_y & 0 & n_x \end{bmatrix} \tag{6.16}$$

with n_x, n_y the components of an outward pointing unit normal vector of the boundary. In axisymmetry $n_x = n_r$ and $n_y = n_z$.

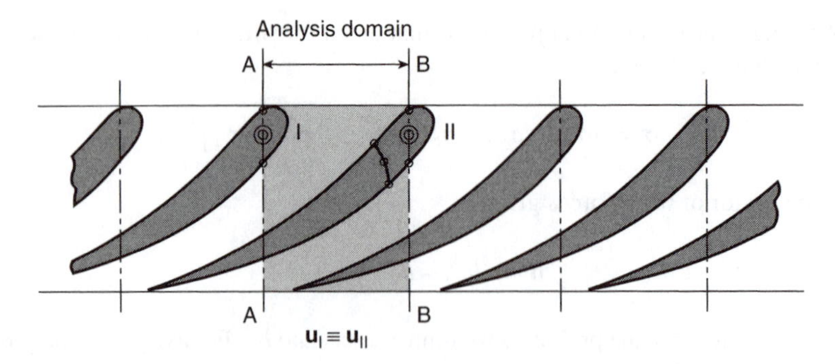

Fig. 6.3 Repeatability segments and analysis domain (shaded).

Boundary conditions on inclined coordinates

Each of the two boundary condition types are vectors and, at any point on the boundary, it is possible to specify some components by displacement conditions and others by traction ones. It is also possible that the conditions are given with respect to coordinates x', y', z' which are oriented with respect to the global axes x, y, z by

$$\mathbf{x}' = \mathbf{T}\mathbf{x} \tag{6.17}$$

where $\mathbf{T}$ is an orthogonal matrix of direction cosines given by

$$\mathbf{T} = \begin{bmatrix} \cos(x', x) & \cos(x', y) & \cos(x', z) \\ \cos(y', x) & \cos(y', y) & \cos(y', z) \\ \cos(z', x) & \cos(z', y) & \cos(z', z) \end{bmatrix} = \begin{bmatrix} t_{11} & t_{12} & t_{13} \\ t_{21} & t_{22} & t_{23} \\ t_{31} & t_{32} & t_{33} \end{bmatrix} \tag{6.18}$$

in which $\cos(x', x)$ is the cosine of the angle between the x' direction and the x direction. For two dimensions

$$\begin{aligned} \cos(x', z) = \cos(z', x) &= 0 \\ \cos(y', z) = \cos(z', y) &= 0 \\ \cos(z', z) &= 1 \end{aligned}$$

The displacement and traction vectors transform to the prime system in exactly the same way as the coordinates; hence, we have

$$\begin{aligned} \mathbf{u}' &= \mathbf{T}\mathbf{u} = \bar{\mathbf{u}}' \\ \mathbf{t}' &= \mathbf{T}\mathbf{t} = \mathbf{T}\mathbf{G}^{\mathrm{T}}\boldsymbol{\sigma} = \bar{\mathbf{t}}' \end{aligned} \tag{6.19}$$

Symmetry and repeatability

In many problems, the advantage of symmetry in loading and geometry can be considered when imposing the boundary conditions, thus reducing the whole problem to more manageable proportions. The use of symmetry conditions is so well known to the engineer and physicist that no statement needs to be made about it explicitly. Less known, however, appears to be the use of *repeatability*[19] when an identical structure (and) loading is continuously repeated, as shown in Fig. 6.3 for an infinite blade cascade. Here it is evident

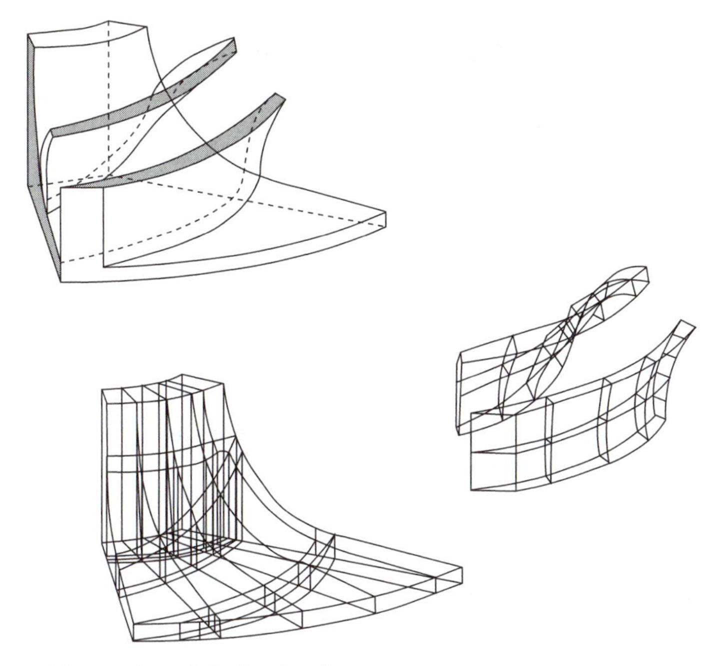

Fig. 6.4 Repeatable sector in analysis of an impeller.

that a typical segment shown shaded behaves identically to the next one, and thus functions such as velocities and displacements at corresponding points of AA and BB are simply identified, i.e.,

$$\mathbf{u}_{\mathrm{I}} = \mathbf{u}_{\mathrm{II}}$$

Similar repeatability, in radial coordinates, occurs very frequently in problems involving turbine or pump impellers. Figure 6.4 shows a typical three-dimensional analysis of such a repeatable segment.

Proper use of symmetry and repeatability can reduce the required compute effort significantly. Similar conditions can obviously be imposed to enforce conditions of 'asymmetry' also.

Normal pressure loading

When a pressure loading is applied normal to a surface the traction may be specified as

$$\mathbf{t} = p\,\mathbf{n} \tag{6.20}$$

where p is the magnitude of the traction and $\mathbf{n}$ again is the unit outward normal to the boundary. This is a condition which is often encountered in practical situations in which p arises from a fluid or gas loading in which tangential components are zero.

6.2.5 Transformation of stress and strain

The transformation of coordinates to a prime system may also be used to define transformations for stresses and strains. Expressing the stress in the cartesian tensor form

$$\boldsymbol{\sigma} = \begin{bmatrix} \sigma_{xx} & \sigma_{xy} & \sigma_{xz} \\ \sigma_{yx} & \sigma_{yy} & \sigma_{yz} \\ \sigma_{zx} & \sigma_{zy} & \sigma_{zz} \end{bmatrix} \tag{6.21}$$

these transformations are given by

$$\boldsymbol{\sigma}' = \mathbf{T}\boldsymbol{\sigma}\mathbf{T}^{\mathrm{T}} \tag{6.22}$$

In a matrix form, however, we must transform the quantities using the forms identified in Appendix B. Using this we obtain the relations

$$\begin{Bmatrix} \sigma_{x'} \\ \sigma_{y'} \\ \sigma_{z'} \\ \tau_{x'y'} \\ \tau_{y'z'} \\ \tau_{z'x'} \end{Bmatrix} = \begin{bmatrix} t_{11}t_{11} & t_{12}t_{12} & t_{13}t_{13} & 2\,t_{11}t_{12} & 2\,t_{12}t_{13} & 2\,t_{13}t_{11} \\ t_{21}t_{21} & t_{22}t_{22} & t_{23}t_{23} & 2\,t_{21}t_{22} & 2\,t_{22}t_{23} & 2\,t_{23}t_{21} \\ t_{31}t_{31} & t_{32}t_{32} & t_{33}t_{33} & 2\,t_{31}t_{32} & 2\,t_{32}t_{33} & 2\,t_{33}t_{31} \\ t_{11}t_{21} & t_{12}t_{22} & t_{13}t_{23} & (t_{11}t_{22}+t_{12}t_{21}) & (t_{12}t_{23}+t_{13}t_{22}) & (t_{13}t_{21}+t_{11}t_{23}) \\ t_{21}t_{31} & t_{22}t_{32} & t_{23}t_{33} & (t_{21}t_{32}+t_{22}t_{31}) & (t_{22}t_{33}+t_{23}t_{32}) & (t_{23}t_{31}+t_{21}t_{33}) \\ t_{31}t_{11} & t_{32}t_{12} & t_{33}t_{13} & (t_{31}t_{12}+t_{32}t_{11}) & (t_{32}t_{13}+t_{33}t_{12}) & (t_{33}t_{11}+t_{31}t_{13}) \end{bmatrix} \begin{Bmatrix} \sigma_{x} \\ \sigma_{y} \\ \sigma_{z} \\ \tau_{xy} \\ \tau_{yz} \\ \tau_{zx} \end{Bmatrix}$$

$$\boldsymbol{\sigma}' = \mathbf{T}_{\sigma}\boldsymbol{\sigma} \tag{6.23}$$

for stresses and

$$\begin{Bmatrix} \varepsilon_{x'} \\ \varepsilon_{y'} \\ \varepsilon_{z'} \\ \gamma_{x'y'} \\ \gamma_{y'z'} \\ \gamma_{z'x'} \end{Bmatrix} = \begin{bmatrix} t_{11}t_{11} & t_{12}t_{12} & t_{13}t_{13} & t_{11}t_{12} & t_{12}t_{13} & t_{13}t_{11} \\ t_{21}t_{21} & t_{22}t_{22} & t_{23}t_{23} & t_{21}t_{22} & t_{22}t_{23} & t_{23}t_{21} \\ t_{31}t_{31} & t_{32}t_{32} & t_{33}t_{33} & t_{31}t_{32} & t_{32}t_{33} & t_{33}t_{31} \\ 2\,t_{11}t_{21} & 2\,t_{12}t_{22} & 2\,t_{13}t_{23} & (t_{11}t_{22}+t_{12}t_{21}) & (t_{12}t_{23}+t_{13}t_{22}) & (t_{13}t_{21}+t_{11}t_{23}) \\ 2\,t_{21}t_{31} & 2\,t_{22}t_{32} & 2\,t_{23}t_{33} & (t_{21}t_{32}+t_{22}t_{31}) & (t_{22}t_{33}+t_{23}t_{32}) & (t_{23}t_{31}+t_{21}t_{33}) \\ 2\,t_{31}t_{11} & 2\,t_{32}t_{12} & 2\,t_{33}t_{13} & (t_{31}t_{12}+t_{32}t_{11}) & (t_{32}t_{13}+t_{33}t_{12}) & (t_{33}t_{11}+t_{31}t_{13}) \end{bmatrix} \begin{Bmatrix} \varepsilon_{x} \\ \varepsilon_{y} \\ \varepsilon_{z} \\ \gamma_{xy} \\ \gamma_{yz} \\ \gamma_{zx} \end{Bmatrix}$$

$$\boldsymbol{\varepsilon}' = \mathbf{T}_{\varepsilon}\boldsymbol{\varepsilon} \tag{6.24}$$

for strains. The differences between $\mathbf{T}_{\sigma}$ and $\mathbf{T}_{\varepsilon}$ occur from the use of the engineering definition of shearing strain where we have introduced

$$\gamma_{xy} = 2\,\varepsilon_{xy}, \quad \text{etc.}$$

If the principal material axes are oriented at angle of β with respect to the coordinate axes of the problem (Fig. 6.5), the two-dimensional representation of $\mathbf{T}_{\varepsilon}$ is given by

$$\mathbf{T}_{\varepsilon} = \begin{bmatrix} \cos^2\beta & \sin^2\beta & 0 & \sin\beta\cos\beta \\ \sin^2\beta & \cos^2\beta & 0 & -\sin\beta\cos\beta \\ 0 & 0 & 1 & 0 \\ -2\sin\beta\cos\beta & 2\sin\beta\cos\beta & 0 & \cos^2\beta-\sin^2\beta \end{bmatrix} \tag{6.25}$$

Table 6.1 Relations between isotropic elastic parameters

Parameters	E, ν	K, G	λ, μ
$E =$	–	$9KG/(3K+G)$	$\mu(3\lambda+2\mu)/(\lambda+\mu)$
$\nu =$	–	$(3K-2G)/(6K+2G)$	$\lambda/(\lambda+\mu)/2$
$K =$	$E/(1-2\nu)/3$	–	$\lambda+2\mu/3$
$G=\mu=$	$E/(1+\nu)/2$	G	μ
$\lambda =$	$\nu E/(1+\nu)/(1-2\nu)$	$K-2G/3$	–

6.2.6 Elasticity matrix

The stress–strain equations, also known as *constitutive relations*, for a linearly elastic material may be expressed by Eq. (2.16)

$$\boldsymbol{\sigma} = \mathbf{D}\,(\boldsymbol{\varepsilon} - \boldsymbol{\varepsilon}_0) + \boldsymbol{\sigma}_0 \tag{6.26a}$$

or by

$$\boldsymbol{\varepsilon} = \mathbf{D}^{-1}\,(\boldsymbol{\sigma} - \boldsymbol{\sigma}_0) + \boldsymbol{\varepsilon}_0 \tag{6.26b}$$

The $\mathbf{D}$ matrix is known as the *elasticity matrix of moduli* and the $\mathbf{D}^{-1}$ matrix as the *elasticity matrix of compliances*.[17] Without loss in generality, in the sequel we ignore the $\boldsymbol{\varepsilon}_0$ and $\boldsymbol{\sigma}_0$ terms. After obtaining final results they may be again added by replacing $\boldsymbol{\varepsilon}$ and $\boldsymbol{\sigma}$ by $\boldsymbol{\varepsilon}-\boldsymbol{\varepsilon}_0$ and $\boldsymbol{\sigma}-\boldsymbol{\sigma}_0$, respectively.

Isotropic materials

We write a general expression for isotropic materials in terms of the six stress and strain terms. We may use any two independent elastic constants for an isotropic material.[15, 17] Here we use Young's modulus of elasticity, E, and Poisson's ratio, ν. In Table 6.1 we indicate relationships between E, ν and other parameters frequently encountered in the literature. Using cartesian coordinates, for example, the expression is given by

$$\begin{Bmatrix} \varepsilon_x \\ \varepsilon_y \\ \varepsilon_z \\ \gamma_{xy} \\ \gamma_{yz} \\ \gamma_{zx} \end{Bmatrix} = \frac{1}{E}\begin{bmatrix} 1 & -\nu & -\nu & 0 & 0 & 0 \\ -\nu & 1 & -\nu & 0 & 0 & 0 \\ -\nu & -\nu & 1 & 0 & 0 & 0 \\ 0 & 0 & 0 & 2(1+\nu) & 0 & 0 \\ 0 & 0 & 0 & 0 & 2(1+\nu) & 0 \\ 0 & 0 & 0 & 0 & 0 & 2(1+\nu) \end{bmatrix}\begin{Bmatrix} \sigma_x \\ \sigma_y \\ \sigma_z \\ \tau_{xy} \\ \tau_{yz} \\ \tau_{zx} \end{Bmatrix} \tag{6.27}$$

Inverting to obtain the appropriate $\mathbf{D}$ matrix yields the result

$$\begin{Bmatrix} \sigma_x \\ \sigma_y \\ \sigma_z \\ \tau_{xy} \\ \tau_{yz} \\ \tau_{zx} \end{Bmatrix} = \frac{E}{d}\begin{bmatrix} (1-\nu) & \nu & \nu & 0 & 0 & 0 \\ \nu & (1-\nu) & \nu & 0 & 0 & 0 \\ \nu & \nu & (1-\nu) & 0 & 0 & 0 \\ 0 & 0 & 0 & (1-2\nu)/2 & 0 & 0 \\ 0 & 0 & 0 & 0 & (1-2\nu)/2 & 0 \\ 0 & 0 & 0 & 0 & 0 & (1-2\nu)/2 \end{bmatrix}\begin{Bmatrix} \varepsilon_x \\ \varepsilon_y \\ \varepsilon_z \\ \gamma_{xy} \\ \gamma_{yz} \\ \gamma_{zx} \end{Bmatrix} \tag{6.28}$$

where $d = (1 + \nu)(1 - 2\nu)$. The form of d places restrictions on the admissible values of ν to keep the material parameters positive; thus, we have

$$-1 < \nu < \tfrac{1}{2}$$

The limiting values of the parameters (namely -1 and $1/2$) are permitted by rewriting the material model with Lagrange multipliers replacing the terms which are indefinite. The case where $\nu = 1/2$ is associated with materials which are *incompressible* and we shall devote special attention to this problem in Chapter 11 since it has relevance for many applications in solid and fluid mechanics. Generally, of course, no material can be incompressible and we are only interested in the case where $\nu \to 1/2$. However, even for this case we will find that care must be used when developing a finite element form.

For isotropic materials the expression for two-dimensional problems is written in terms of the four stress and strain terms by omitting the last two rows and columns in (6.27). Using cartesian coordinates the expression then becomes

$$\begin{Bmatrix} \varepsilon_x \\ \varepsilon_y \\ \varepsilon_z \\ \gamma_{xy} \end{Bmatrix} = \frac{1}{E} \begin{bmatrix} 1 & -\nu & -\nu & 0 \\ -\nu & 1 & -\nu & 0 \\ -\nu & -\nu & 1 & 0 \\ 0 & 0 & 0 & 2(1+\nu) \end{bmatrix} \begin{Bmatrix} \sigma_x \\ \sigma_y \\ \sigma_z \\ \tau_{xy} \end{Bmatrix} \tag{6.29}$$

For the plane stress case we must set σ_z zero to compute the appropriate **D** matrix. This yields the result

$$\varepsilon_z = -\frac{\nu}{E}\left(\sigma_x + \sigma_y\right) \tag{6.30}$$

and including this in the inverse of (6.29) we obtain

$$\begin{Bmatrix} \sigma_x \\ \sigma_y \\ \sigma_z \\ \tau_{xy} \end{Bmatrix} = \frac{E}{(1-\nu^2)} \begin{bmatrix} 1 & \nu & 0 & 0 \\ \nu & 1 & 0 & 0 \\ 0 & 0 & 0 & 0 \\ 0 & 0 & 0 & (1-\nu)/2 \end{bmatrix} \begin{Bmatrix} \varepsilon_x \\ \varepsilon_y \\ \varepsilon_z \\ \gamma_{xy} \end{Bmatrix} \tag{6.31}$$

We have filled in the third column and row of the **D** array so that correct stresses are obtained. Indeed, if we deleted these we would once again get the result given in Chapter 2 and one may ask why we include the extra terms. The main reason is to permit a single form to be used for plane stress, plane strain and axisymmetric problems and therefore minimize the amount of programming needed to implement these in a computer program.

For the plane strain and axisymmetric problems the inverse may be performed directly from (6.29), as all components of stress can exist. Accordingly, for these two cases we obtain (again writing for the cartesian coordinate form)

$$\begin{Bmatrix} \sigma_x \\ \sigma_y \\ \sigma_z \\ \tau_{xy} \end{Bmatrix} = \frac{E}{(1+\nu)(1-2\nu)} \begin{bmatrix} (1-\nu) & \nu & \nu & 0 \\ \nu & (1-\nu) & \nu & 0 \\ \nu & \nu & (1-\nu) & 0 \\ 0 & 0 & 0 & (1-2\nu)/2 \end{bmatrix} \begin{Bmatrix} \varepsilon_x \\ \varepsilon_y \\ \varepsilon_z \\ \gamma_{xy} \end{Bmatrix} \tag{6.32}$$

which is identical to the three-dimensional problem if the last two rows and columns of each array in (6.28) are omitted. Here we observe that the case where ε_z is zero is treated merely by inserting that value when computing stresses; however, σ_z will always exist unless the Poisson ratio, ν, is zero.

Anisotropic materials

We may write a general relationship for anisotropic linearly elastic materials as

$$\begin{Bmatrix} \sigma_x \\ \sigma_y \\ \sigma_z \\ \tau_{xy} \\ \tau_{yz} \\ \tau_{zx} \end{Bmatrix} = \begin{bmatrix} D_{11} & D_{12} & D_{13} & D_{14} & D_{15} & D_{16} \\ D_{21} & D_{22} & D_{23} & D_{24} & D_{25} & D_{26} \\ D_{31} & D_{32} & D_{33} & D_{34} & D_{35} & D_{36} \\ D_{41} & D_{42} & D_{43} & D_{44} & D_{45} & D_{46} \\ D_{51} & D_{52} & D_{53} & D_{54} & D_{55} & D_{56} \\ D_{61} & D_{62} & D_{63} & D_{64} & D_{65} & D_{66} \end{bmatrix} \begin{Bmatrix} \varepsilon_x \\ \varepsilon_y \\ \varepsilon_z \\ \gamma_{xy} \\ \gamma_{yz} \\ \gamma_{zx} \end{Bmatrix} \tag{6.33}$$

For an elastic material the **D** matrix must be symmetric and, hence, $D_{ij} = D_{ji}$. This results in a possibility of 21 elastic constants for the general problem.[16, 20, 21]

An important class of anisotropic materials is one for which three planes of symmetry exist and is called an *orthotropic material.* Here the principal axes are also in rectangular cartesian coordinates. For orthotropic materials it is common to define the elastic material parameters in terms of their Young's moduli, Poisson ratios and shear moduli.

If we let x', y' and z' be the three axes of material symmetry, the elastic strain–stress relations may be expressed as

$$\begin{Bmatrix} \varepsilon_{x'} \\ \varepsilon_{y'} \\ \varepsilon_{z'} \\ \gamma_{x'y'} \\ \gamma_{y'z'} \\ \gamma_{z'x'} \end{Bmatrix} = \begin{bmatrix} \dfrac{1}{E_{x'}} & -\dfrac{\nu_{x'y'}}{E_{y'}} & -\dfrac{\nu_{x'z'}}{E_{z'}} & 0 & 0 & 0 \\ -\dfrac{\nu_{y'x'}}{E_{x'}} & \dfrac{1}{E_{y'}} & -\dfrac{\nu_{y'z'}}{E_{x'}} & 0 & 0 & 0 \\ -\dfrac{\nu_{z'x'}}{E_{x'}} & -\dfrac{\nu_{z'y'}}{E_{y'}} & \dfrac{1}{E_{z'}} & 0 & 0 & 0 \\ 0 & 0 & 0 & \dfrac{1}{G_{x'y'}} & 0 & 0 \\ 0 & 0 & 0 & 0 & \dfrac{1}{G_{y'z'}} & 0 \\ 0 & 0 & 0 & 0 & 0 & \dfrac{1}{G_{z'x'}} \end{bmatrix} \begin{Bmatrix} \sigma_{x'} \\ \sigma_{y'} \\ \sigma_{z'} \\ \tau_{x'y'} \\ \tau_{y'z'} \\ \tau_{z'x'} \end{Bmatrix} \tag{6.34}$$

where $E_{x'}$, $E_{y'}$,$E_{z'}$ are elastic moduli; $\nu_{x'y'}$, $\nu_{x'z'}$, etc. are Poisson ratios; and $G_{x'y'}$, $G_{y'z'}$, $G_{z'x'}$ are elastic shear modulus. Again symmetry of the **D** matrix results in

$$\frac{\nu_{i'j'}}{E_{j'}} = \frac{\nu_{j'i'}}{E_{i'}} \tag{6.35}$$

thus reducing the number of independent components for the three-dimensional cases considered to nine parameters (three direct moduli, three Poisson ratios, and three shear moduli).

The elastic moduli for the $\mathbf{D}'$ matrix are computed by inverting the square matrix appearing in Eq. (6.34).

The inverse may be written as

$$\boldsymbol{\sigma}' = \mathbf{D}'\,\boldsymbol{\varepsilon}' \tag{6.36}$$

If the principal material axes are expressed by the directions given in (6.17), it is necessary to transform Eq. (6.36) to the form (6.26a) before proceeding with an analysis. This is most easily performed by noting an equality of work given by

$$\begin{aligned} \boldsymbol{\sigma}'^{\mathrm{T}}\boldsymbol{\varepsilon}' &= \boldsymbol{\sigma}^{\mathrm{T}}\boldsymbol{\varepsilon} \\ \boldsymbol{\varepsilon}'^{\mathrm{T}}\mathbf{D}'\,\boldsymbol{\varepsilon}' &= \boldsymbol{\varepsilon}^{\mathrm{T}}\mathbf{D}\,\boldsymbol{\varepsilon} \end{aligned} \tag{6.37}$$

and using the expressions for transformation given by Eq. (6.24) in Eq. (6.37) gives

$$\mathbf{D} = \mathbf{T}_\varepsilon^{\mathrm{T}} \mathbf{D}' \mathbf{T}_\varepsilon \tag{6.38}$$

Such transformation also has been used in a slightly different context in Chapter 1 [viz. Eq. (1.25)] to transform a stiffness matrix.

For treatment of anisotropic materials in the two-dimensional problems, it is necessary for the direction normal to the plane of deformation (i.e., the z direction for plane problems or θ direction for axisymmetric problems) to be a direction of *material symmetry*. For this case we may write a general relationship for linearly elastic deformation as (using the cartesian form)

$$\begin{Bmatrix} \sigma_x \\ \sigma_y \\ \sigma_z \\ \tau_{xy} \end{Bmatrix} = \begin{bmatrix} D_{11} & D_{12} & D_{13} & D_{14} \\ D_{21} & D_{22} & D_{23} & D_{24} \\ D_{31} & D_{32} & D_{33} & D_{34} \\ D_{41} & D_{42} & D_{43} & D_{44} \end{bmatrix} \begin{Bmatrix} \varepsilon_x \\ \varepsilon_y \\ \varepsilon_z \\ \gamma_{xy} \end{Bmatrix} \tag{6.39}$$

For an elastic material the **D** matrix must be symmetric and, hence, $D_{ij} = D_{ji}$. This results in a possibility of ten elastic constants for the plane or axisymmetric problem.

In order to consider the plane stress case it is again necessary to impose the constraint $\sigma_z = 0$. If we solve for ε_z using the third row of (6.39) we have

$$\varepsilon_z = -[D_{31}\,\varepsilon_x + D_{32}\,\varepsilon_y + D_{34}\,\gamma_{xy}]/D_{33} \tag{6.40}$$

which may be substituted into the remaining three equations in (6.39) to give

$$\begin{Bmatrix} \sigma_x \\ \sigma_y \\ \sigma_z \\ \tau_{xy} \end{Bmatrix} = \begin{bmatrix} \hat{D}_{11} & \hat{D}_{12} & 0 & \hat{D}_{14} \\ \hat{D}_{21} & \hat{D}_{22} & 0 & \hat{D}_{24} \\ 0 & 0 & 0 & 0 \\ \hat{D}_{41} & \hat{D}_{42} & 0 & \hat{D}_{44} \end{bmatrix} \begin{Bmatrix} \varepsilon_x \\ \varepsilon_y \\ \varepsilon_z \\ \gamma_{xy} \end{Bmatrix} \tag{6.41}$$

in which

$$\hat{D}_{ij} = D_{ij} - D_{i3} D_{33}^{-1} D_{3j} \tag{6.42}$$

are reduced elastic moduli.

For the plane stress and strain problems of an *orthotropic material* the principal axes are also in rectangular cartesian coordinates; however, for the axisymmetric problem the principal axes also must be axisymmetric (i.e., cylindrical orthotropy which is similar to rings of a tree). For orthotropic materials it is common to define their properties in terms of their Young's moduli, Poisson ratios and shear moduli.

If we let $x',\ y'$ (or $r',\ z'$) be the two axes of material symmetry in the plane of deformation, the elastic strain–stress relations may be expressed by (6.34), after omitting the last two rows and columns of each array. Accordingly,

$$\begin{Bmatrix} \varepsilon_{x'} \\ \varepsilon_{y'} \\ \varepsilon_{z'} \\ \gamma_{x'y'} \end{Bmatrix} = \begin{bmatrix} \dfrac{1}{E_{x'}} & -\dfrac{\nu_{x'y'}}{E_{y'}} & -\dfrac{\nu_{x'z'}}{E_{z'}} & 0 \\ -\dfrac{\nu_{y'x'}}{E_{x'}} & \dfrac{1}{E_{y'}} & -\dfrac{\nu_{y'z'}}{E_{x'}} & 0 \\ -\dfrac{\nu_{z'x'}}{E_{x'}} & -\dfrac{\nu_{z'y'}}{E_{y'}} & \dfrac{1}{E_{z'}} & 0 \\ 0 & 0 & 0 & \dfrac{1}{G_{x'y'}} \end{bmatrix} \begin{Bmatrix} \sigma_{x'} \\ \sigma_{y'} \\ \sigma_{z'} \\ \tau_{x'y'} \end{Bmatrix} \tag{6.43}$$

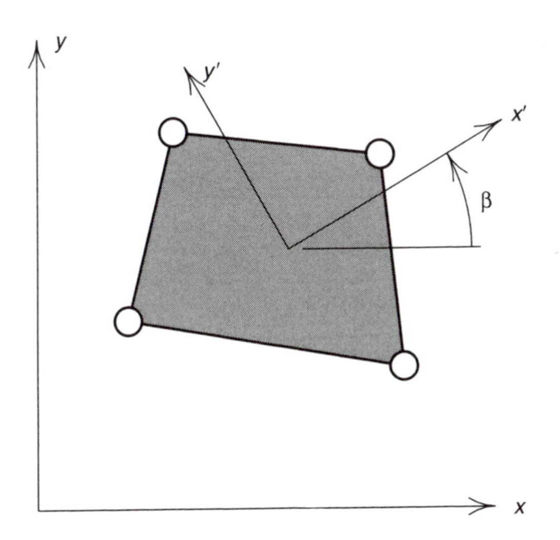

Fig. 6.5 Coordinate definition for transformation of material axes.

After considering symmetry the number of independent components for the two-dimensional cases considered reduces to seven parameters (three direct moduli, three Poisson ratios, and one shear modulus).

The elastic moduli for the $\mathbf{D}'$ matrix are computed by inverting the square matrix appearing in Eq. (6.43). If plane stress is considered, it is necessary to ensure $\sigma_{z'}$ is zero.

Inserting expression (6.24) into Eq. (6.37) again gives

$$\mathbf{D} = \mathbf{T}_\varepsilon^{\mathrm{T}} \mathbf{D}' \mathbf{T}_\varepsilon \tag{6.44}$$

Example 6.1: Anisotropic, stratified, material. With the z axis representing the normal to the planes of stratification, as shown in Fig. 6.6, we can rewrite (6.43) (again ignoring the initial strains and stresses for convenience) as

$$\begin{aligned}
\varepsilon_r &= \frac{\sigma_r}{E_1} - \frac{\nu_2 \sigma_z}{E_2} - \frac{\nu_1 \sigma_\theta}{E_1} \\
\varepsilon_z &= -\frac{\nu_2 \sigma_r}{E_1} + \frac{\sigma_z}{E_2} - \frac{\nu_2 \sigma_\theta}{E_1} \\
\varepsilon_\theta &= -\frac{\nu_1 \sigma_r}{E_1} - \frac{\nu_2 \sigma_z}{E_2} + \frac{\sigma_\theta}{E_1} \\
\gamma_{rz} &= \frac{\tau_{rz}}{G}
\end{aligned} \tag{6.45}$$

Writing the parameters as

$$\frac{E_1}{E_2} = n; \qquad \frac{G}{E_2} = m \qquad \text{and} \qquad d = (1 + \nu_1)(1 - \nu_1 - 2n\nu_2^2)$$

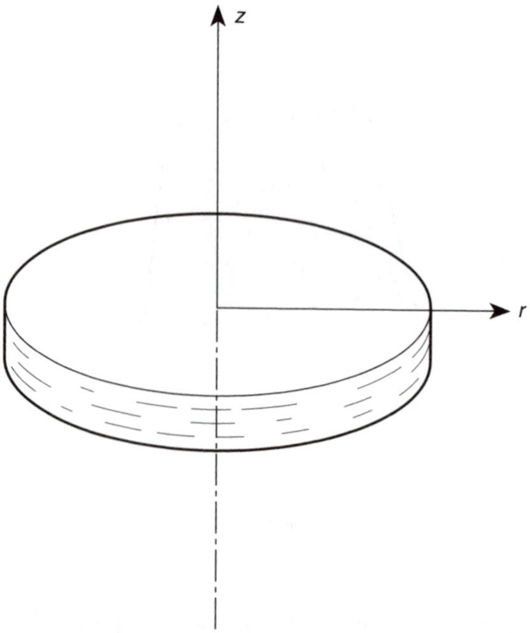

Fig. 6.6 Axisymmetrically stratified material.

we have on solving for the stresses that

$$\mathbf{D} = \frac{E_2}{d} \begin{bmatrix} n(1-n\nu_2^2), & n\nu_2(1+\nu_1), & n(\nu_1+n\nu_2^2), & 0 \\ n\nu_2(1+\nu_1), & 1-\nu_1^2, & n\nu_2(1+\nu_1), & 0 \\ n(\nu_1+n\nu_2^2), & n\nu_2(1+\nu_1), & n(1-n\nu_2^2), & 0 \\ 0, & 0, & 0, & md \end{bmatrix} \tag{6.46}$$

Initial strain – thermal effects

Initial strains may be due to many causes. Shrinkage, crystal growth, or temperature change will, in general, result in an initial strain vector

$$\boldsymbol{\varepsilon}_0 = \begin{bmatrix} \varepsilon_{x0} & \varepsilon_{y0} & \varepsilon_{z0} & \gamma_{xy0} & \gamma_{yz0} & \gamma_{zx0} \end{bmatrix}^{\mathrm{T}} \tag{6.47}$$

The initial strain will usually depend on position which may be included by interpolation using shape functions.

As an example, consider the effects of change in temperature in an isotropic material. The initial strain for a temperature change $\Delta T = T - T_0$ (with T_0 a temperature where no straining is caused) with a *linear* coefficient of thermal expansion α is given by

$$\boldsymbol{\varepsilon}_0 = \alpha\, \Delta T\, \mathbf{m} \tag{6.48}$$

where

$$\mathbf{m} = \begin{bmatrix} 1 & 1 & 1 & 0 & 0 & 0 \end{bmatrix}^{\mathrm{T}} \tag{6.49}$$

For an isotropic material normal strains ε_x, ε_y, ε_z are all equal and no shear strains are caused by a temperature change.

Anisotropic materials present no special problems with the coefficients of thermal expansion varying with direction in the material. For example, in an orthotropic material no shearing strains are caused for the principal material directions and we may replace (6.48) by

$$\varepsilon_0' = \Delta T \begin{bmatrix} \alpha_{x'} & \alpha_{y'} & \alpha_{z'} & 0 & 0 & 0 \end{bmatrix}^{\mathrm{T}} \tag{6.50}$$

It is now again necessary to use the transformation between principal material directions and those used for coordinates of the analysis using

$$\varepsilon_0 = \mathbf{T}_\varepsilon^{-1} \, \varepsilon_0' \tag{6.51}$$

where $\mathbf{T}_\varepsilon$ is given by (6.24). Now, in general, the $\gamma_{xy0}, \gamma_{yz0}, \gamma_{zx0}$ components are no longer equal to zero.

6.3 Finite element approximation

The above describes the governing equations for three-dimensional behaviour of solids. Except for the elastic constitutive equations, all the remaining equations are valid for general materials undergoing small deformations. A finite element solution process for the equations may be established by:

1. Using the virtual work (or weak form), equations for equilibrium given in Sec. 3.4.
2. Introducing an approximation for the displacement field $\mathbf{u}$ in terms of shape functions.
3. Computing strains from (6.5).
4. Computing stresses from (6.26a) where for linear elastic behaviour we use one of the forms given in Sec. 6.2.6 for the $\mathbf{D}$ matrix.
5. Performing the integrations over each element (usually by quadrature).
6. Assembling the element contributions to form the global stiffness and load arrays.
7. Imposing the known traction and displacement boundary conditions.
8. Solving the resulting stiffness and load matrices.
9. Reporting desired parts of the solution.

The above steps describe a general solution framework which we shall follow in all subsequent developments for solutions of problems in solid mechanics. Differences will occur later in the types of weak forms used and in the expressions for strains and constitutive equations. Otherwise the steps are standard. To illustrate the process we first consider the general three-dimensional problem in which the virtual work expression is given by [see (3.46) in Sec. 3.4]

$$\int_\Omega \delta\varepsilon^{\mathrm{T}} \boldsymbol{\sigma} \, \mathrm{d}\Omega - \int_\Omega \delta\mathbf{u}^{\mathrm{T}} \mathbf{b} \, \mathrm{d}\Omega - \int_\Gamma \delta\mathbf{u}^{\mathrm{T}} \mathbf{t} \, \mathrm{d}\Gamma = 0$$

To simplify the solution process we split the boundary into Γ_u and Γ_t and introduce the known traction boundary condition. We shall also impose the displacement boundary conditions in the approximations for $\mathbf{u}$ and assume the virtual displacement $\delta\mathbf{u}$ vanishes on Γ_u.

For linear elastic behaviour we also may introduce the constitutive equation given by (6.26a) and thus our virtual work equation simplifies to

$$\int_\Omega \delta\varepsilon^{\mathrm{T}} \left[\boldsymbol{\sigma}_0 + \mathbf{D}\left(\varepsilon - \varepsilon_0\right)\right] \mathrm{d}\Omega - \int_\Omega \delta\mathbf{u}^{\mathrm{T}} \mathbf{b} \, \mathrm{d}\Omega - \int_{\Gamma_t} \delta\mathbf{u}^{\mathrm{T}} \bar{\mathbf{t}} \, \mathrm{d}\Gamma = 0 \tag{6.52}$$

with the constraint $\mathbf{u} = \bar{\mathbf{u}}$ on Γ_u.

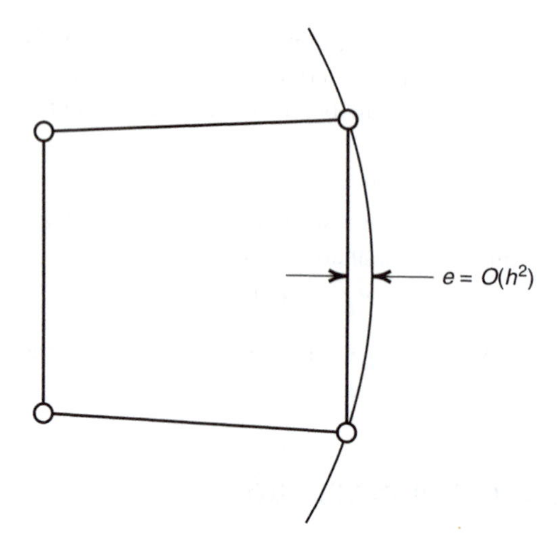

Fig. 6.7 Approximation of curved surface by linear element.

In the finite element solution we divide all integrals to be sums over individual elements and *approximate* the weak form by

$$\sum_e \int_{\Omega^e} \delta\boldsymbol{\varepsilon}^{\mathrm{T}} \left[\boldsymbol{\sigma}_0 + \mathbf{D}\left(\boldsymbol{\varepsilon} - \boldsymbol{\varepsilon}_0\right)\right] \mathrm{d}\Omega - \sum_e \int_{\Omega^e} \delta\mathbf{u}^{\mathrm{T}}\mathbf{b}\,\mathrm{d}\Omega - \sum_e \int_{\Gamma_t^e} \delta\mathbf{u}^{\mathrm{T}}\bar{\mathbf{t}}\,\mathrm{d}\Gamma = 0 \tag{6.53}$$

where Ω^e and Γ_t^e denote element domains and parts of boundaries of any element where tractions are specified, respectively. The 'approximation' in this step is associated with the fact that for curved boundary surfaces the sum of element domains Ω^e is not always exactly equal to Ω, nor is the sum of Γ_t^e equal to Γ_t. This is easily observed for approximations using linear elements as shown in Fig. 6.7. We observe that the error is $O(h^2)$ which is exactly the same as the error in displacement from shape functions using linear polynomials. Thus, the order of error in our solution is not increased by the boundary approximation.

Displacement and strain approximation

At this point we can introduce the finite element shape function expressions to define displacements. Accordingly, we have

$$\mathbf{u} \approx \hat{\mathbf{u}} = \begin{Bmatrix} \hat{u} \\ \hat{v} \\ \hat{w} \end{Bmatrix} = \sum_a N_a \begin{Bmatrix} \tilde{u}_a \\ \tilde{v}_a \\ \tilde{w}_a \end{Bmatrix} = \sum_a N_a \tilde{\mathbf{u}}_a \tag{6.54}$$

where $\tilde{u}_a$, $\tilde{v}_a$, $\tilde{w}_a$ are nodal values of the displacement. Any of the three-dimensional interpolations given in Chapter 4 may be used to define the shape functions N_a. Inserting

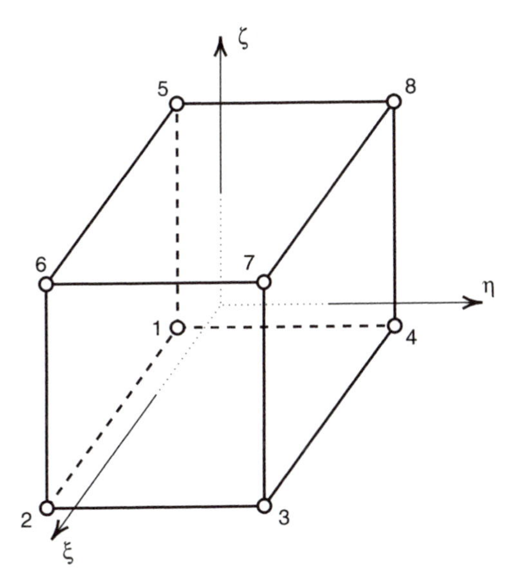

Fig. 6.8 8-node brick element. Local node numbering.

the interpolation into Eq. (6.5) gives

$$\boldsymbol{\varepsilon} = \begin{Bmatrix} \varepsilon_x \\ \varepsilon_y \\ \varepsilon_z \\ \gamma_{xy} \\ \gamma_{yz} \\ \gamma_{zx} \end{Bmatrix} \approx \hat{\boldsymbol{\varepsilon}} = \sum_a \begin{bmatrix} \dfrac{\partial N_a}{\partial x} & 0 & 0 \\ 0 & \dfrac{\partial N_a}{\partial y} & 0 \\ 0 & 0 & \dfrac{\partial N_a}{\partial z} \\ \dfrac{\partial N_a}{\partial y} & \dfrac{\partial N_a}{\partial x} & 0 \\ 0 & \dfrac{\partial N_a}{\partial z} & \dfrac{\partial N_a}{\partial y} \\ \dfrac{\partial N_a}{\partial z} & 0 & \dfrac{\partial N_a}{\partial x} \end{bmatrix} \begin{Bmatrix} \tilde{u}_a \\ \tilde{v}_a \\ \tilde{w}_a \end{Bmatrix} = \sum_a \mathbf{B}_a \tilde{\mathbf{u}}_a \tag{6.55}$$

A similar expression may be written for virtual strains.

Example 6.2: Strains for 8-node brick. As an example we consider the 8-node brick element shown in Fig. 6.8. The shape functions are given by

$$N_a = \tfrac{1}{8}\,(1 + \xi_a \xi)(1 + \eta_a \eta)(1 + \zeta_a \zeta)$$

for which the derivatives with respect to ξ, η, ζ are given by

$$\frac{\partial N_a}{\partial \xi} = \frac{1}{8}\,\xi_a\,(1+\eta_a\eta)(1+\zeta_a\zeta)$$
$$\frac{\partial N_a}{\partial \eta} = \frac{1}{8}\,\eta_a\,(1+\zeta_a\zeta)(1+\xi_a\xi)$$
$$\frac{\partial N_a}{\partial \zeta} = \frac{1}{8}\,\zeta_a\,(1+\xi_a\xi)(1+\eta_a\eta)$$

For an 8-node brick element, the jacobian matrix in (5.11) may be expressed as

$$\mathbf{J} = \frac{1}{8}\sum_{b=1}^{8}\begin{Bmatrix} \xi_b(1+\eta_b\eta)(1+\zeta_b\zeta) \\ \eta_b(1+\zeta_b\zeta)(1+\xi_b\xi) \\ \zeta_b(1+\xi_b\xi)(1+\eta_b\eta) \end{Bmatrix}\begin{bmatrix} x_b & y_b & z_b \end{bmatrix}$$

The shape function derivatives are now obtained from (5.10) as

$$\begin{Bmatrix} \dfrac{\partial N_a}{\partial x} \\ \dfrac{\partial N_a}{\partial y} \\ \dfrac{\partial N_a}{\partial z} \end{Bmatrix} = \frac{1}{8}\mathbf{J}^{-1}\begin{Bmatrix} \xi_a(1+\eta_a\eta)(1+\zeta_a\zeta) \\ \eta_a(1+\zeta_a\zeta)(1+\xi_a\xi) \\ \zeta_a(1+\xi_a\xi)(1+\eta_a\eta) \end{Bmatrix}$$

These may be used directly to define the $\mathbf{B}_a$ strain matrix given in Eq. (6.55).

For two-dimensional problems the finite element shape function expressions to define the displacements are given by

$$\mathbf{u} \approx \hat{\mathbf{u}} = \begin{Bmatrix} \hat{u} \\ \hat{v} \end{Bmatrix} = \sum_a N_a \begin{Bmatrix} \tilde{u}_a \\ \tilde{v}_a \end{Bmatrix} = \sum_a N_a \tilde{\mathbf{u}}_a \tag{6.56}$$

Inserting the interpolation into Eqs (6.6) and (6.7) gives

$$\boldsymbol{\varepsilon} = \begin{Bmatrix} \varepsilon_x \\ \varepsilon_y \\ \varepsilon_z \\ \gamma_{xy} \end{Bmatrix} \approx \hat{\boldsymbol{\varepsilon}} = \sum_a \begin{bmatrix} \dfrac{\partial N_a}{\partial x} & 0 \\ 0 & \dfrac{\partial N_a}{\partial y} \\ 0 & 0 \\ \dfrac{\partial N_a}{\partial y} & \dfrac{\partial N_a}{\partial x} \end{bmatrix} \begin{Bmatrix} \tilde{u}_a \\ \tilde{v}_a \end{Bmatrix} + \begin{Bmatrix} 0 \\ 0 \\ \varepsilon_z \\ 0 \end{Bmatrix} = \sum_a \mathbf{B}_a \tilde{\mathbf{u}}_a + \varepsilon_z \tag{6.57}$$

for the plane problems and

$$\boldsymbol{\varepsilon} = \begin{Bmatrix} \varepsilon_r \\ \varepsilon_z \\ \varepsilon_\theta \\ \gamma_{rz} \end{Bmatrix} = \sum_a \begin{bmatrix} \dfrac{\partial N_a}{\partial r} & 0 \\ 0 & \dfrac{\partial N_a}{\partial z} \\ \dfrac{N_a}{r} & 0 \\ \dfrac{\partial N_a}{\partial z} & \dfrac{\partial N_a}{\partial r} \end{bmatrix} \begin{Bmatrix} \tilde{u}_a \\ \tilde{v}_a \end{Bmatrix} = \sum_a \mathbf{B}_a \tilde{\mathbf{u}}_a \tag{6.58}$$

for the axisymmetric problem, respectively. In the form for axisymmetry the radius is computed from the parametric form given by Eq. (5.2). Accordingly, for this case we have

$$r = \sum_b N'_b r_b \tag{6.59}$$

where r_b are locations of the node points defining the N'_b functions.

Stiffness and load matrices

Introducing the above approximations into the weak form (6.53) results in

$$\sum_e \delta\tilde{\mathbf{u}}_a^{\mathrm{T}} \left[\int_{\Omega^e} \mathbf{B}_a^{\mathrm{T}} \left[\boldsymbol{\sigma}_0 + \mathbf{D} \left(\mathbf{B}_b \tilde{\mathbf{u}}_b - \boldsymbol{\varepsilon}_0 \right) \right] \mathrm{d}\Omega - \int_{\Omega^e} N_a \mathbf{b} \, \mathrm{d}\Omega - \int_{\Gamma_t^e} N_a \bar{\mathbf{t}} \, \mathrm{d}\Gamma \right] = 0 \tag{6.60}$$

which after summing the element integrals and noting that $\delta\tilde{\mathbf{u}}_a$ is arbitrary gives the system of linear equations

$$\mathbf{K}_{ab} \, \tilde{\mathbf{u}}_b + \mathbf{f}_a = \mathbf{0} \tag{6.61}$$

where

$$\begin{aligned} \mathbf{K}_{ab} &= \sum_e \int_{\Omega^e} \mathbf{B}_a^{\mathrm{T}} \mathbf{D} \mathbf{B}_b \, \mathrm{d}\Omega \\ \mathbf{f}_a &= \sum_e \int_{\Omega^e} \left[\mathbf{B}_a^{\mathrm{T}} \left(\boldsymbol{\sigma}_0 - \mathbf{D} \boldsymbol{\varepsilon}_0 \right) - N_a \mathbf{b} \right] \mathrm{d}\Omega - \sum_e \int_{\Gamma_t^e} N_a \bar{\mathbf{t}} \, \mathrm{d}\Gamma \end{aligned} \tag{6.62}$$

The integration over each element domain may be computed by quadrature in which

$$\begin{aligned} \int_{\Omega^e} (\cdot) \, \mathrm{d}\Omega &= \int_{\square} (\cdot) \; J \, \mathrm{d}\square \approx \sum_{l=1}^{L_3} (\cdot)_l \; J_l \, W_l \\ \int_{\Gamma_t^e} (\cdot) \, \mathrm{d}\Gamma &= \int_{\square} (\cdot) \; j \, \mathrm{d}\square \approx \sum_{l=1}^{L_2} (\cdot)_l \; j_l \, W_l \end{aligned} \tag{6.63}$$

where $J = \det \mathbf{J}$ is the determinant of the jacobian transformation between the global and local *volume* coordinate frames, $j = \det \mathbf{j}$ is the determinant of the jacobian transformation between the global and local *surface* coordinate frames, and subscript l is associated with each quadrature point with weight W_l. The points and weights are taken from the tables given in Chapter 5.

Example 6.3: Quadrature for 8-node brick element. For an 8-node brick element it is sufficient to perform volume integrals using a $2 \times 2 \times 2$ formula. Thus L_3 in (6.63) is equal to 8 and the points and weights may be given as†

l	1	2	3	4	5	6	7	8
ξ_l	$-c$	c	$-c$	c	$-c$	c	$-c$	c
η_l	$-c$	$-c$	c	c	$-c$	$-c$	c	c
ζ_l	$-c$	$-c$	$-c$	$-c$	c	c	c	c
W_l	1	1	1	1	1	1	1	1

where $c = 1/\sqrt{3}$.

† Note that ordering is unimportant and any other permutation for l is permissible.

Similarly for the surface integrals a 2×2 formula may be used and give L_2 equal 4 with points and weights ordered as

l	1	2	3	4
ξ_l	$-c$	c	$-c$	c
η_l	$-c$	$-c$	c	c
W_l	1	1	1	1

in which again $c = 1/\sqrt{3}$.

These formulae always have error equal to or less than that in the approximation of domain or in the shape functions. Hence, it is never necessary to use higher order quadrature than the above.[22]

A similar form holds for all the two-dimensional cases; however, the volume and surface elements are different for the plane and axisymmetric problems. For plane stress

$$\mathrm{d}\Omega = t\,\mathrm{d}x\,\mathrm{d}y \quad \text{and} \quad \mathrm{d}\Gamma = t\,\mathrm{d}s \tag{6.64}$$

where t is the thickness of the slab and may vary over the two-dimensional domain; for plane strain

$$\mathrm{d}\Omega = \mathrm{d}x\,\mathrm{d}y \quad \text{and} \quad \mathrm{d}\Gamma = \mathrm{d}s \tag{6.65}$$

where a *unit thickness* is considered; and for axisymmetry†

$$\mathrm{d}\Omega = 2\pi r\,\mathrm{d}r\,\mathrm{d}z \quad \text{and} \quad \mathrm{d}\Gamma = 2\pi r\,\mathrm{d}s \tag{6.66}$$

In the above

$$\mathrm{d}s = \left(\mathrm{d}x^2 + \mathrm{d}y^2\right)^{1/2} \tag{6.67}$$

for the plane problem with a similar expression for axisymmetry.

The finite element arrays may now be computed from (6.60) for quadrilateral-shaped elements of lagrangian or serendipity type; the stiffness and load matrices defined in (6.62) are computed using gaussian quadrature as

$$\begin{aligned}\mathbf{K}^e_{ab} &= \int_{-1}^{1}\int_{-1}^{1} \mathbf{B}_a^{\mathrm{T}}(\xi,\eta)\,\mathbf{D}\mathbf{B}_b(\xi,\eta)\,J(\xi,\eta)\,\mathrm{d}\xi\,\mathrm{d}\eta \\ &\approx \sum_l \mathbf{B}_a^{\mathrm{T}}(\xi_l,\eta_l)\,\mathbf{D}\mathbf{B}_b(\xi_l,\eta_l)\,J(\xi_l,\eta_l)\,W_l\end{aligned} \tag{6.68}$$

and

$$\begin{aligned}\mathbf{f}^e_a &= \int_{-1}^{1}\int_{-1}^{1} \left[\mathbf{B}_a^{\mathrm{T}}(\xi,\eta)\,(\boldsymbol{\sigma}_0 - \mathbf{D}\boldsymbol{\varepsilon}_0) - N_a(\xi,\eta)\mathbf{b}\right] J(\xi,\eta)\,\mathrm{d}\xi\,\mathrm{d}\eta \\ &\quad - \int_{-1}^{1} N_a(\xi)\,\bar{\mathbf{t}}\,j(\xi)\,\mathrm{d}\xi \\ &\approx \sum_l \left[\mathbf{B}_a^{\mathrm{T}}(\xi_l,\eta_l)\,(\boldsymbol{\sigma}_0 - \mathbf{D}\boldsymbol{\varepsilon}_0) - N_a(\xi_l,\eta_l)\mathbf{b}\right] J(\xi_l,\eta_l)\,W_l \\ &\quad - \sum_l N_a(\xi_l)\,\bar{\mathbf{t}}(\xi_l)\,j(\xi_l)\,w_l\end{aligned} \tag{6.69}$$

† Some programs omit the 2π in the definition of $\mathrm{d}\Omega$ and $\mathrm{d}\Gamma$ and compute matrices for one radian of arc.

in which

$$
\begin{aligned}
J &= t\ \det \mathbf{J}; \quad j = t\left[\left(\frac{\partial x}{\partial \xi}\right)^2 + \left(\frac{\partial y}{\partial \xi}\right)^2\right]^{1/2}; \text{ Plane stress} \\
J &= \det \mathbf{J}; \quad j = \left[\left(\frac{\partial x}{\partial \xi}\right)^2 + \left(\frac{\partial y}{\partial \xi}\right)^2\right]^{1/2}; \text{ Plane strain} \\
J &= 2\pi r\ \det \mathbf{J}; \quad j = 2\pi r\left[\left(\frac{\partial r}{\partial \xi}\right)^2 + \left(\frac{\partial z}{\partial \xi}\right)^2\right]^{1/2}; \text{ Axisymmetric}
\end{aligned}
\tag{6.70}
$$

for the domain and boundary. The quadrature points are denoted as ξ_l and η_l and weights as W_l. In (6.68) and (6.70) t, $\mathbf{D}$, $\mathbf{b}$, $\bar{\mathbf{t}}$ and initial stress and strain may vary in space in an arbitrary manner and $\det \mathbf{J}$ is computed as indicated in Sec. 5.5.

The simplicity of computation using shape functions and numerical integration should be especially noted. This permits easy consideration of different types of interpolations for the shape functions and different quadrature orders for numerical integration to be assessed.

6.4 Reporting of results: displacements, strains and stresses

The reader will now have observed that, while the finite element representation of displacements is in a sense optimal as it is the primary variable, both the strains and the stresses are not realistic. In particular, in ordinary engineering problems both strains and stresses tend to be continuous within a single material. The answers which are obtained by the finite element calculation result in discontinuities of both strains and stresses between adjacent elements. Thus if the direct calculation of these quantities were presented the answers would be deemed unrealistic. For this reason, from the beginning of the finite element method it was sought to establish these rather important quantities in a more realistic, and possibly more accurate, way. In the very early days of finite element calculation with simple C_0 continuity elements an averaging of element strains and stresses, which are constant in triangular elements, was made at each node. This of course gave improved results at most points – except at those which were on the boundary.

Since the simple days of averaging further attention was given to this subject and other methods were developed. The first of these methodologies was developed by Brauchli and Oden[23] in 1971 and consisted of assuming that a continuous representation of either strain or stress using the same C_0 functions as for displacements could be found by solving a least squares sense representation of the corresponding discontinuous (finite element) one. This method proved quite expensive but often gave results which were superior to the simple averaging – at least for some sets of problems. However, higher accuracy was not achieved despite the additional cost of solving a full set of algebraic equations. An alternative local procedure to improve results was proposed by Hinton and Campbell[24] and was once quite widely used.

The methods of recovery of strain and stress have progressed much further in recent years and in Chapter 13 we discuss these fully. We find that currently an optimal procedure, which generally gives higher order accuracy and has similar cost to simple averaging, is the patch recovery method. In this the process of determining values of recovered strains or stresses assumes that:

1. At some points of the domain or each element, the strains and stresses calculated by the direct differentiation of the shape functions are more accurate than elsewhere. Indeed on many occasions at such points 'superconvergence' is demonstrated which can make the accuracy at least one order higher than that of the finite element values computed from derivatives of shape functions.
2. A continuous representation of such strains and stresses can be given by finding nodal values which in the least squares sense approximate those computed by the optimal points. Now the increased accuracy will exist over the entire domain.

The discussion of the existence of such points at which higher order may exist is deferred to Chapter 13 but here we show how this can be easily incorporated into standard programs dealing with elasticity.

Basically in the procedure we will assume a strain exists for an element and can be expressed by

$$\varepsilon^{\star} = \sum_b N_b \tilde{\varepsilon}_b \tag{6.71}$$

where now ε is any component of strain. A similar expression may be written for a stress component. The goal is to find appropriate values for $\tilde{\varepsilon}_b$ which give improved results. To do this we use a least squares method in which the strain in a patch surrounding a *vertex* node a on elements may be expressed in *global coordinates* by a polynomial expression of higher order, suitable for the number of unknown parameters in the strain expression. This polynomial expression is given by

$$\varepsilon^{\star\star} = \begin{bmatrix} 1, & (x - x_a), & (y - y_a), & \cdots \end{bmatrix} \begin{Bmatrix} \tilde{\varepsilon}_a \\ \alpha_1 \\ \alpha_2 \\ \vdots \end{Bmatrix} = \mathbf{P}_a(\mathbf{x})\,\boldsymbol{\alpha}_a \tag{6.72}$$

For 3-node triangles or 4-node quadrilaterals in two dimensions and 4-node tetrahedrons or 8-node brick elements in three dimensions a linear interpolation is used. For the quadratic order elements the polynomial in $\mathbf{P}_a$ is also raised to quadratic order. Thus, for 6-node triangular and 8- or 9-node quadrilateral elements in two dimensions we use

$$\mathbf{P}_a = \begin{bmatrix} 1, & (x - x_a), & (y - y_a), & (x - x_a)^2, & (x - x_a)(y - y_a), & (y - y_a)^2 \end{bmatrix}$$

The parameters in $\varepsilon^{\star\star}$ are determined using the least squares problem given by

$$\Pi = \frac{1}{2} \sum_{e=1}^{n_a} \sum_{l} \left[\mathbf{P}_a(\mathbf{x}_l^e)\boldsymbol{\alpha}_a - \hat{\varepsilon}(\mathbf{x}_l^e) \right]^2 = \min \tag{6.73}$$

where n_a is the number of elements attached to node a and x_l^e are locations where strains are computed. The minimization condition results in

$$\mathbf{M}_a \boldsymbol{\alpha}_a = \mathbf{f}_a \tag{6.74}$$

where

$$\mathbf{M}_a = \sum_{e=1}^{n_a} \sum_{l} \mathbf{P}_a^{\mathrm{T}}(\mathbf{x}_l^e)\mathbf{P}_a(\mathbf{x}_l^e) \quad \text{and} \quad \mathbf{f}_a = \sum_{e=1}^{n_a} \sum_{l} \mathbf{P}_a^{\mathrm{T}}(\mathbf{x}_l^e)\hat{\varepsilon}(\mathbf{x}_l^e) \tag{6.75}$$

The values for the remaining nodes (e.g., at mid-side and boundary locations) may be computed by averaging the extrapolated values computed from (6.72). For example, from a patch, the result at node b ($b \neq a$) is given by

$$\varepsilon^{**}(\mathbf{x}_b) = \mathbf{P}_a(\mathbf{x}_b)\boldsymbol{\alpha}_a$$

and averaging the result from all patches which contain node b gives the final result for $\tilde{\varepsilon}_b$.

An identical process may be used to compute stress values. We recommend that a patch recovery method be used to report all strain or stress values. In addition, the method serves as the basis for error assessment and methods to efficiently construct adaptive solutions to a specified accuracy as we shall present in Chapter 14.

6.5 Numerical examples

To illustrate the application of the theory presented above we consider some example problems. Some of the problems we include can be solved by other analytic methods and thus serve to illustrate the accuracy of results obtained. Others, however, are from more practical situations where either no alternative solution method exists or the method is otherwise cumbersome to obtain thus rendering the finite element approach most useful.

Here we first solve again the two-dimensional plane stress problems considered in Chapter 2 to illustrate the advantages of using 4-, 9-, and 16-node quadrilateral isoparametric elements of lagrangian type.

Example 6.4: Beam subjected to end shear. The rectangular beam considered in Sec. 2.9.1 is solved again using lagrangian rectangular elements with 4 nodes (bilinear), 9 nodes (biquadratic) and 16 nodes (bicubic). The mesh for the bilinear model initially has six elements in the depth direction and 12 along the length for a total of 72 elements and 91 nodes. This is subsequently subdivided to form meshes with 12×24, 24×48, 48×96 and 96×192 elements. All other data are as defined in Sec. 2.9.1.

The analysis is repeated using 9-node biquadratic elements with an initial mesh of 3×6 elements, which gives the same number of nodes. Finally, the problem is solved with a mesh of 2×3 16-node bicubic elements which again gives a mesh with 91 nodes. Since the exact solution for displacements given in Sec. 2.9.1 contains all polynomial terms of degree 3 or less the solution with this coarse mesh is exact and no refinement is needed.

In Table 6.2 we present the results for the energy obtained from each mesh and in Fig. 6.9 we show the convergence behaviour for the 4-node and 9-node element forms. Again, the expected rates of convergence are attained as indicated by the slopes of 2 and 4 in the figure.

Example 6.5: Circular beam subjected to end shear. We consider next the circular beam problem described in Sec. 2.9.1. The solution to the problem is performed using isoparametric 4-node bilinear quadrilaterals, 9-node lagrangian quadrilaterals and 16-node lagrangian quadrilaterals. The geometric and material data for the problem is as given in Example 2.4.

The initial mesh for all element types uses a regular subdivision of the domain that produces initial element patterns with 6×12 4-node elements, 3×6 9-node elements and 2×4 16-node elements. The mesh for each element form is shown in Fig. 6.10.

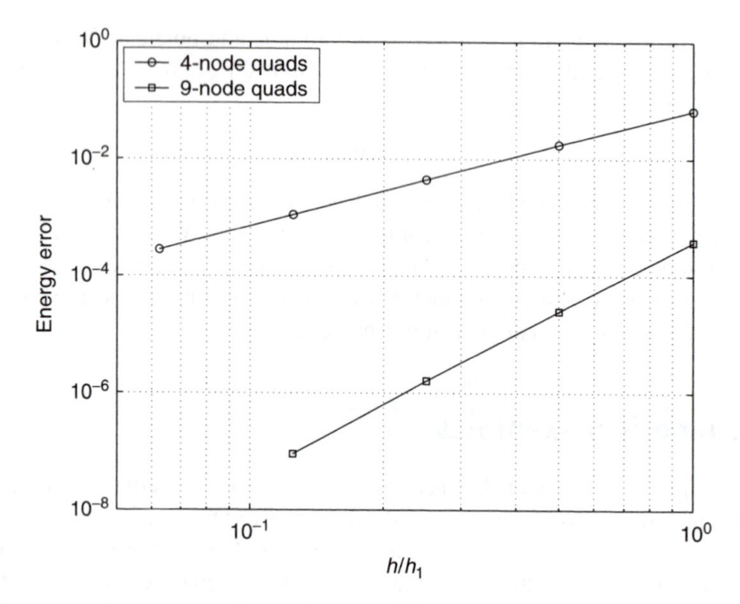

Fig. 6.9 Convergence in energy error for 4-node and 9-node rectangular elements.

Table 6.2 Mesh size and energy for end loaded beam

	4-node rectangles		9-node rectangles		16-node rectangles	
Nodes	Elmts	Energy	Elmts	Energy	Elmts	Energy
91	72	3077.4986	18	3294.7512	8	3296.0000
325	288	3238.2915	72	3295.9174		
1225	1152	3281.3465	288	3295.9947		
4753	4608	3292.3206	1152	3295.9997		
18721	18432	3295.0790	4608	3296.0000		
Exact	–	3296.0000	–	3296.0000	–	3296.0000

Results for the energy are given in Table 6.3 and compared to the exact value computed from

$$E_{ex} = 0.02964966844238$$

using the geometry and properties selected. The element size is normalized to that of the coarsest mesh [shown in Fig. 6.10(a)] and the energy error computed from Table 6.3 has the expected slope for 4-node elements, for 9-node elements and for cubic elements (viz. Fig. 6.11).

We now consider some practical examples for problems which have been solved using the finite element method.

Some simple typical examples are given which use both tetrahedral and isoparametric brick-type elements. The isoparametric examples are all performed using Gauss quadrature to approximate the necessary integrals.

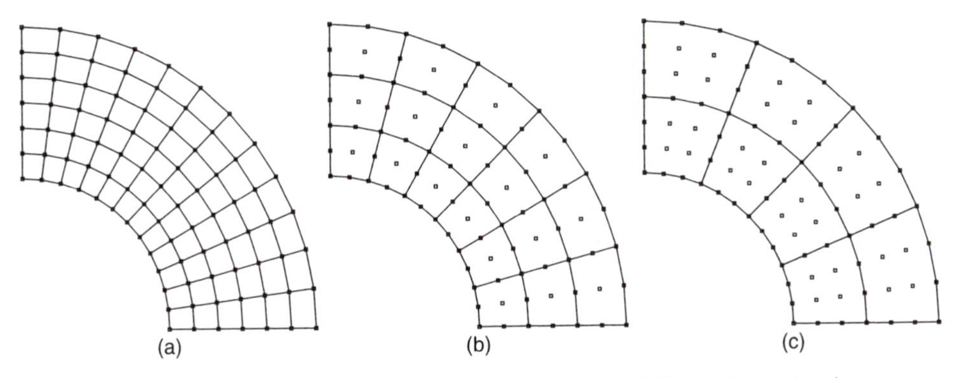

Fig. 6.10 End loaded circular beam: Coarse mesh for 4-node, 9-node and 16-node lagrangian elements.

Table 6.3 Mesh size and energy for curved beam

	4-node quadrilateral		9-node quadrilateral		16-node quadrilateral	
Nodes	Elmts	Energy	Elmts	Energy	Elmts	Energy
91	72	0.03042038175071	18	0.02970101373401	8	0.02965327376971
325	288	0.02984351371323	72	0.02965318188484	32	0.02964975296446
1225	1152	0.02969820784232	288	0.02964989418870	128	0.02964966996157
4753	4608	0.02966180825828	1152	0.02964968266120	512	0.02964966846707
18 721	18 432	0.02965270370808	4608	0.02964966933301	2048	0.02964966844276
Exact	–	0.02964966844238	–	0.02964966844238	–	0.02964966844238

6.5.1 A dam subject to external and internal water pressures

A buttress dam on a somewhat complex rock foundation is shown in Fig. 6.12 and analysed.[25, 26] This dam (completed in 1964) is of particular interest as it is the first to which the finite element method was applied during the design stage. The heterogeneous foundation region is subject to plane strain conditions while the dam itself is considered in a state of plane stress of variable thickness.

With external and gravity loading no special problems of analysis arise.

When pore pressures are considered, the situation, however, requires perhaps some explanation.

It is well known that in a porous material the water pressure is transmitted to the structure as a *body force* of magnitude

$$b_x = -\frac{\partial p}{\partial x} \qquad b_y = -\frac{\partial p}{\partial y} \tag{6.76}$$

and that now the external pressure need not be considered.

The pore pressure p is, in fact, now a body force potential which may be determined by solving a 'field problem' as described in the next chapter. Figure 6.12 shows the element subdivision of the region and the outline of the dam. Figure 6.13(a) and (b) shows the stresses resulting from gravity (applied to the dam only) and due to water pressure assumed to be acting as an external load or, alternatively, as an internal pore pressure. Both solutions indicate large tensile regions, but the increase of stresses due to the second assumption is important.

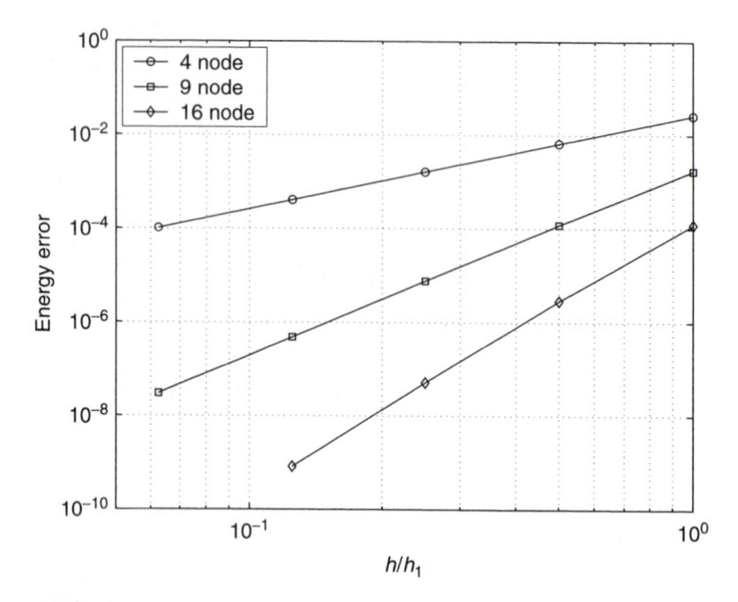

Fig. 6.11 Curved beam: Convergence in energy error for quadrilateral elements.

The stresses calculated here are the so-called 'effective' stresses. These represent the forces transmitted between the solid particles and are defined in terms of the *total* stresses $\boldsymbol{\sigma}$ and the pore pressures p by

$$\boldsymbol{\sigma}' = \boldsymbol{\sigma} + \mathbf{m}p \qquad \mathbf{m}^{\mathrm{T}} = [1, 1, 0] \tag{6.77}$$

i.e., simply by removing the hydrostatic pressure component from the *total* stress.[27, 28]

The effective stress is of particular importance in the mechanics of porous media such as those that occur in the study of soils, rocks, or concrete. The basic assumption in deriving the body forces of Eq. (6.76) is that only the effective stress is of any importance in deforming the solid phase. This leads immediately to another possibility of formulation.[29] If we examine the equilibrium conditions of Eq. (6.8) we note that this is written in terms of total stresses. Writing the constitutive relation, Eq. (6.26a), in terms of effective stresses, i.e.,

$$\boldsymbol{\sigma}' = \mathbf{D}'(\boldsymbol{\varepsilon} - \boldsymbol{\varepsilon}_0) + \boldsymbol{\sigma}'_0 \tag{6.78}$$

and substituting into the weak form we find that the stiffness matrix is given in terms of the matrix $\mathbf{D}'$ and the force terms are augmented by an additional force

$$-\int_{\Omega^e} \mathbf{B}^{\mathrm{T}}\mathbf{m}\, p\, \mathrm{d}\Omega \tag{6.79}$$

or, if p is interpolated by shape functions N'_b, the force becomes

$$-\int_{\Omega^e} \mathbf{B}^{\mathrm{T}}\mathbf{m}\mathbf{N}'\, \mathrm{d}\Omega\, \tilde{\mathbf{p}}^e \tag{6.80}$$

This alternative form of introducing pore pressure effects allows a discontinuous interpolation of p to be used [as in Eq. (6.79) no derivatives occur] and this is now frequently used in practice.

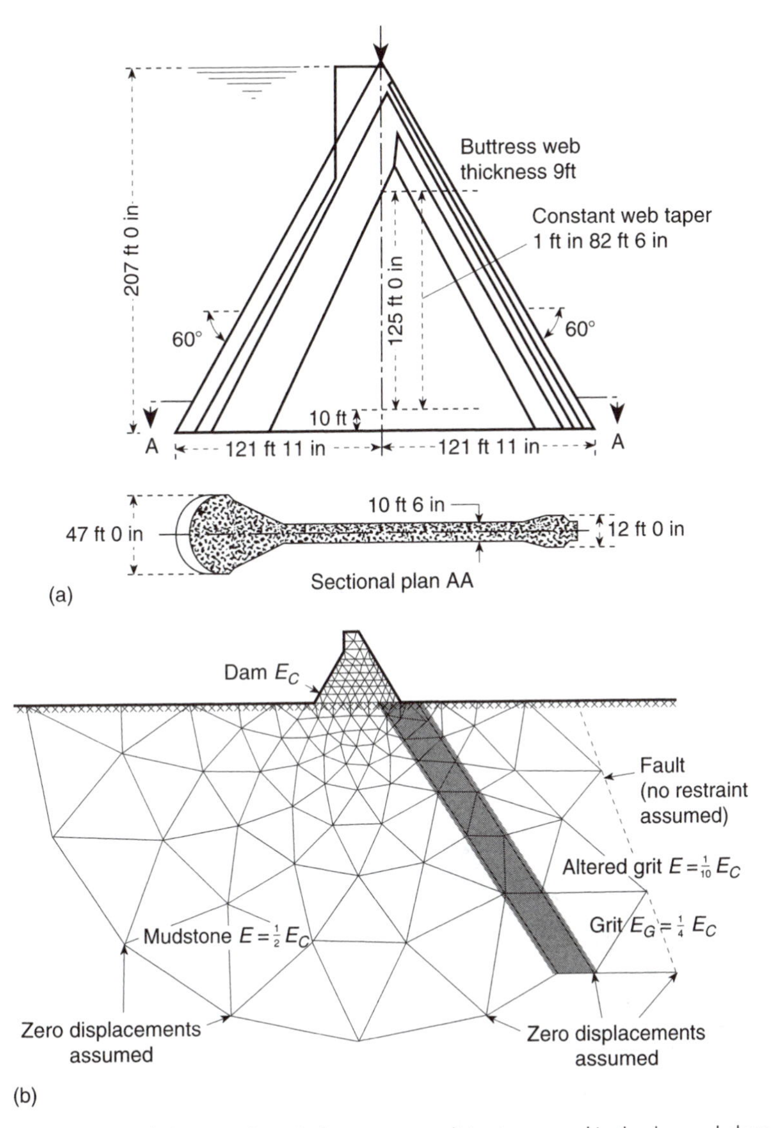

Fig. 6.12 Stress analysis of a buttress dam. A plane stress condition is assumed in the dam and plane strain in the foundation. (a) The buttress section analysed. (b) Extent of foundation considered and division into finite elements.

6.5.2 Rotating disc

Here only 18 cubic serendipity elements are needed to obtain an adequate solution, arranged as shown in Fig. 6.14. It is of interest to observe that all mid-side nodes of the cubic elements may be generated within the computer program and need not be specified. Also, the problem requires the specification of body forces caused by the centrifugal effects of the rotating disk. Here,

$$b_r = -\rho\, r\, \omega^2$$

where ρ is the mass density of the material and ω is the angular velocity.

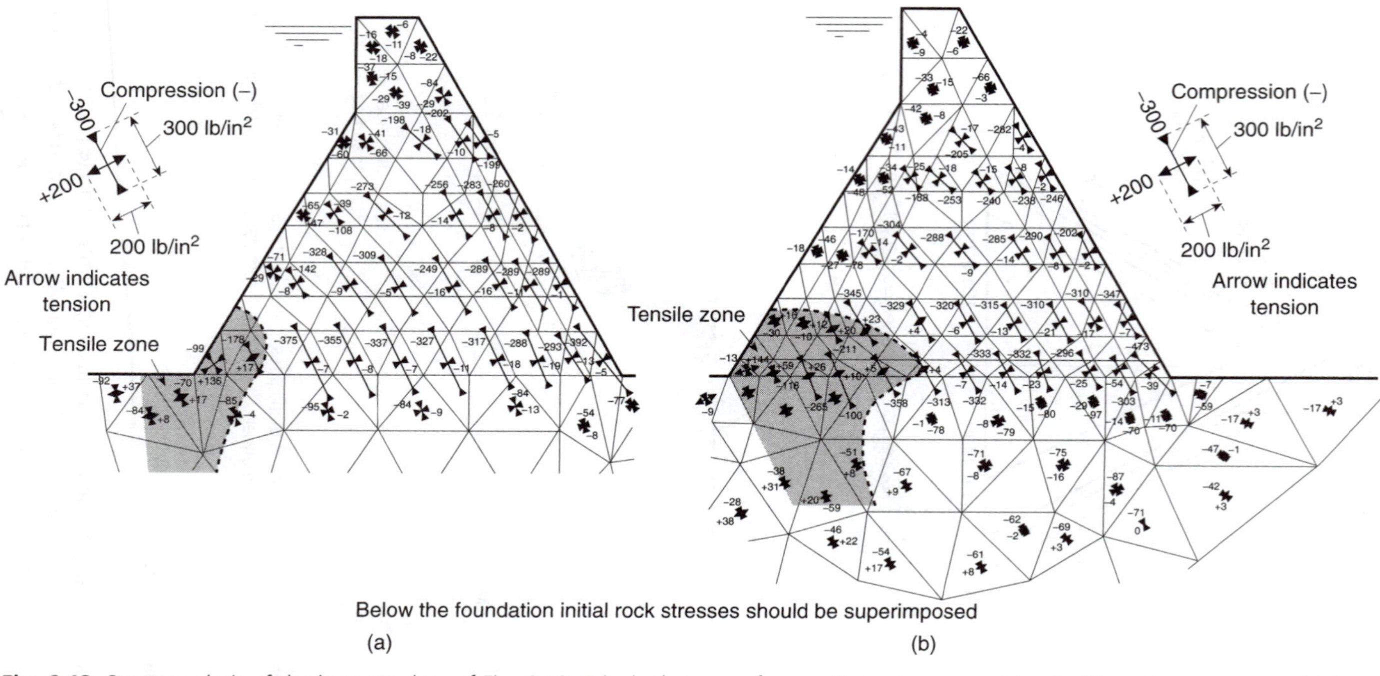

Fig. 6.13 Stress analysis of the buttress dam of Fig. 6.12. Principal stresses for gravity loads are combined with water pressures, which are assumed to act (a) as external loads, (b) as body forces due to pore pressure.

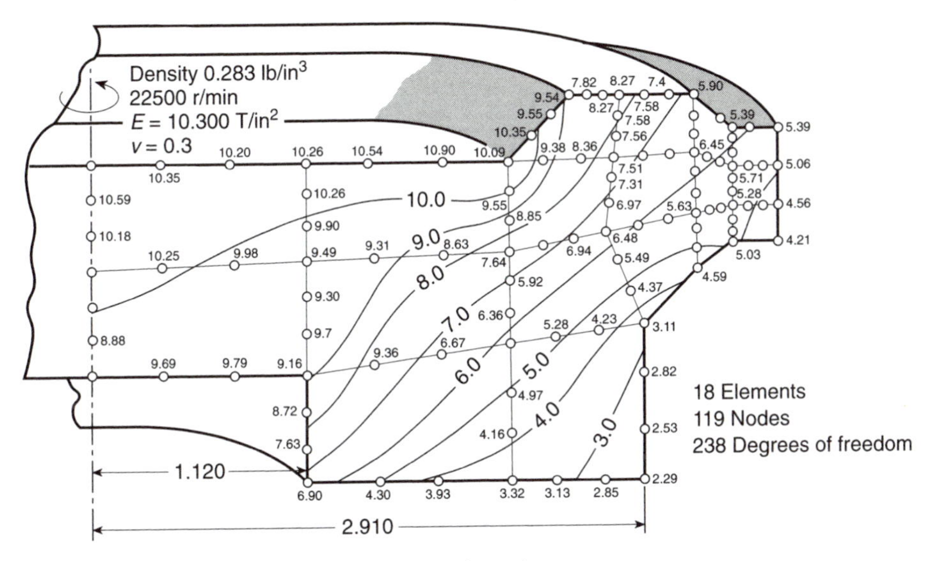

Fig. 6.14 A rotating disc – analysed with cubic serendipity elements.

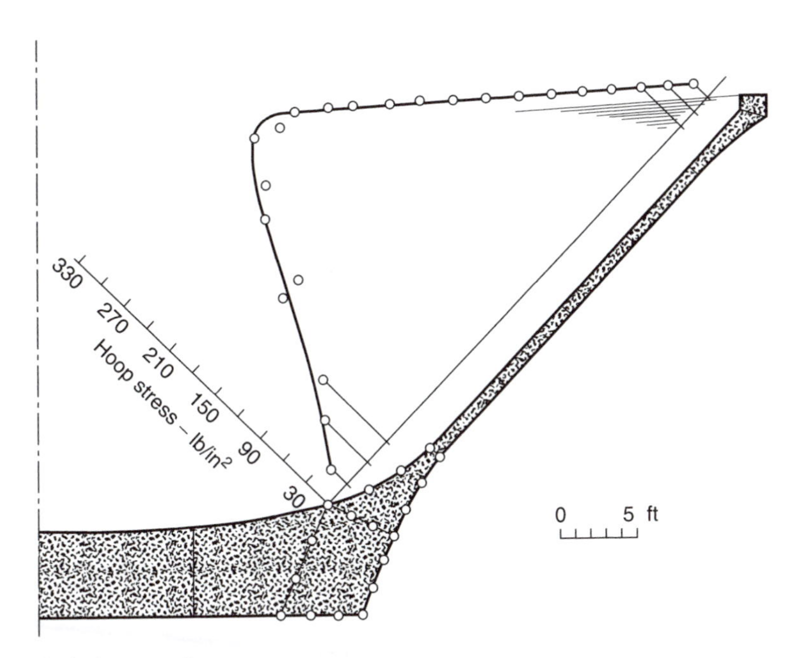

Fig. 6.15 Conical water tank.

6.5.3 Conical water tank

In this problem cubic serendipity elements are again used as shown in Fig. 6.15. It is worth noting that single-element thickness throughout is adequate to represent the bending effects in both the thick and thin parts of the container. With simple 3-node triangular elements, several layers of elements would have been needed to give an adequate solution.

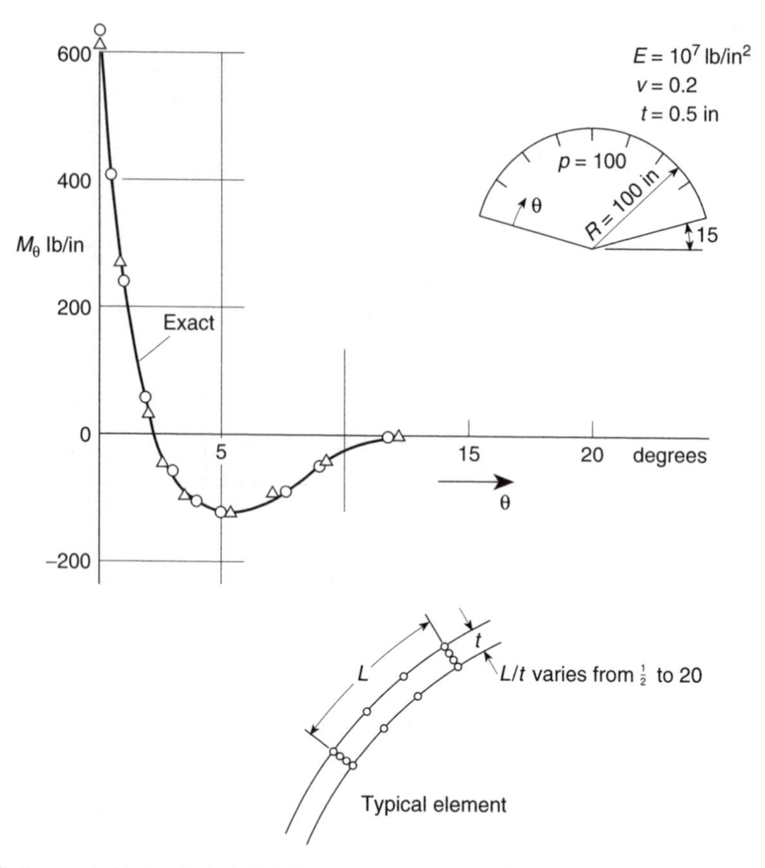

Fig. 6.16 *Encastré*, thin hemispherical shell. Solution with 15 and 24 cubic serendipity elements.

6.5.4 A hemispherical dome

The possibilities of dealing with shells approached in the previous example are here further exploited to show how a limited number of elements can adequately solve a thin shell problem as illustrated in Fig. 6.16. This type of solution can be further improved upon from the economy viewpoint by making use of the well-known shell assumptions involving a linear variation of displacements across the thickness. Thus the number of degrees of freedom can be reduced (e.g., see reference 30).

6.5.5 Arch dam in a rigid valley

This problem, perhaps a little unrealistic from the engineer's viewpoint, was the subject of a study carried out by a committee of the Institution of Civil Engineers and provided an excellent test for a convergence evaluation of three-dimensional analysis.[10] In Fig. 6.17 two subdivisions into quadratic and two into cubic elements are shown. In Fig. 6.18 the convergence of displacements in the centre-line section is shown, indicating that quite remarkable accuracy can be achieved with even one element.

The comparison of stresses in Fig. 6.19 is again quite remarkable, though showing a greater 'oscillation' with coarse subdivision. The finest subdivision results can be taken as 'exact' from checks by models and alternative methods of analysis.

6.5.6 Pressure vessel problem

A more ambitious problem treated with simple tetrahedra is given in reference 7. Figure 6.20 illustrates an analysis of a complex pressure vessel. Some 10 000 degrees of freedom are involved in this analysis. A similar problem using higher order isoparametric elements permits a sufficiently accurate analysis for a very similar problem to be performed with only 2121 degrees of freedom (Fig. 6.21).

6.6 Problems

6.1 Use the transformation array given by

$$\mathbf{T} = \begin{bmatrix} \cos\theta & \sin\theta & 0 \\ -\sin\theta & \cos\theta & 0 \\ 0 & 0 & 1 \end{bmatrix}$$

with $\theta = 45^\circ$ to transform stress and strain components from their x, y, z components to their x', y', z' components. Let the material be linearly elastic with material parameters given by E and ν. Show that $G = E/[2(1+\nu)]$.

6.2 For an isotropic material expressed in E and ν compute the mean stress $p = (\sigma_x + \sigma_y + \sigma_z)$. If the bulk modulus is given by

$$p = K\,\varepsilon_v$$

where $\varepsilon_v = \varepsilon_x + \varepsilon_y + \varepsilon_z$ is the volume strain, show that $K = E/[3(1-2\nu)]$.

6.3 The strain displacement equations for a one-dimensional problem in plane polar coordinates are given by

$$\boldsymbol{\varepsilon} = \begin{Bmatrix} \varepsilon_{rr} \\ \varepsilon_{\theta\theta} \\ \gamma_{r\theta} \end{Bmatrix} = \begin{Bmatrix} \dfrac{\partial u_r}{\partial r} \\ \dfrac{1}{r}\dfrac{\partial u_\theta}{\partial \theta} + \dfrac{u_r}{r} \\ \dfrac{1}{r}\left(\dfrac{\partial u_r}{\partial \theta} - u_\theta\right) + \dfrac{\partial u_\theta}{\partial r} \end{Bmatrix}$$

The displacements are expanded in a Fourier series as

$$u_r = \sum u^n(r)\cos n\theta \quad \text{and} \quad u_\theta = \sum v^n(r)\sin n\theta$$

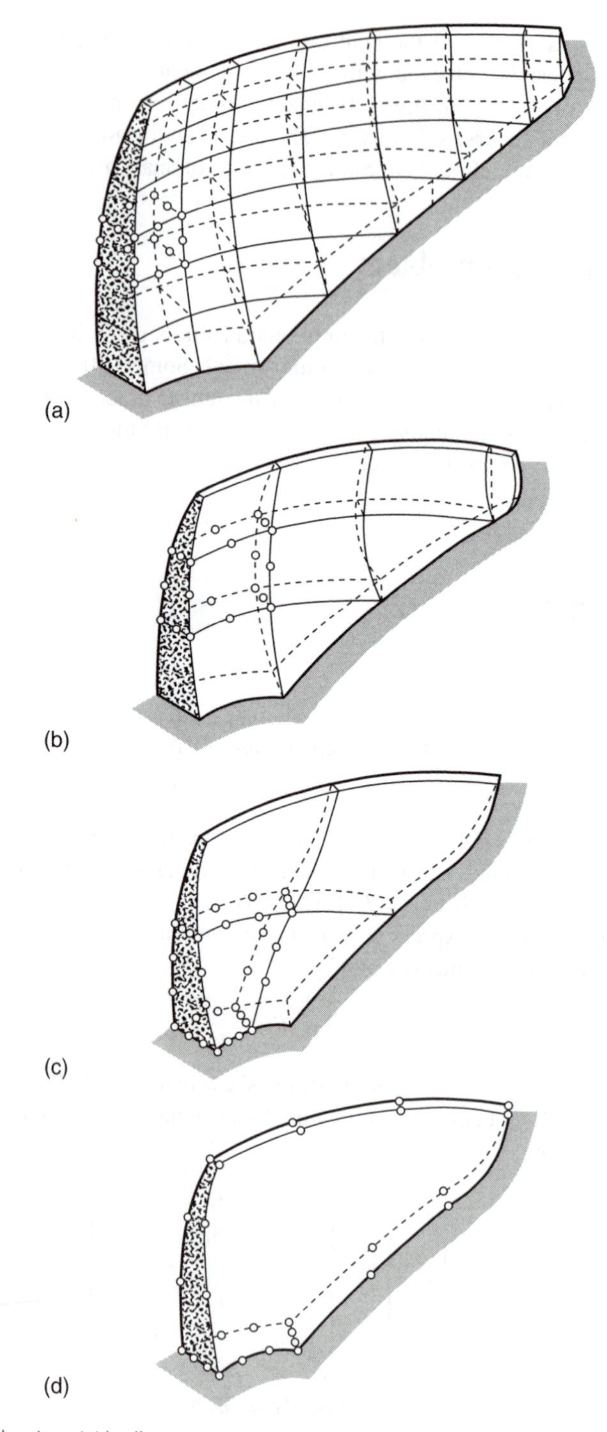

Fig. 6.17 Arch dam in a rigid valley – various element subdivisions.

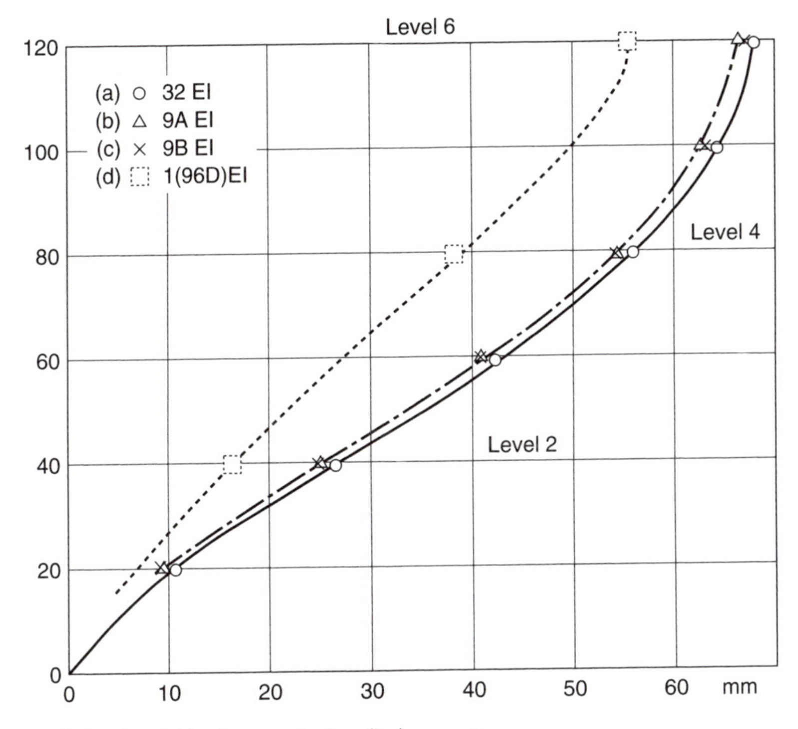

Fig. 6.18 Arch dam in a rigid valley – centre-line displacements.

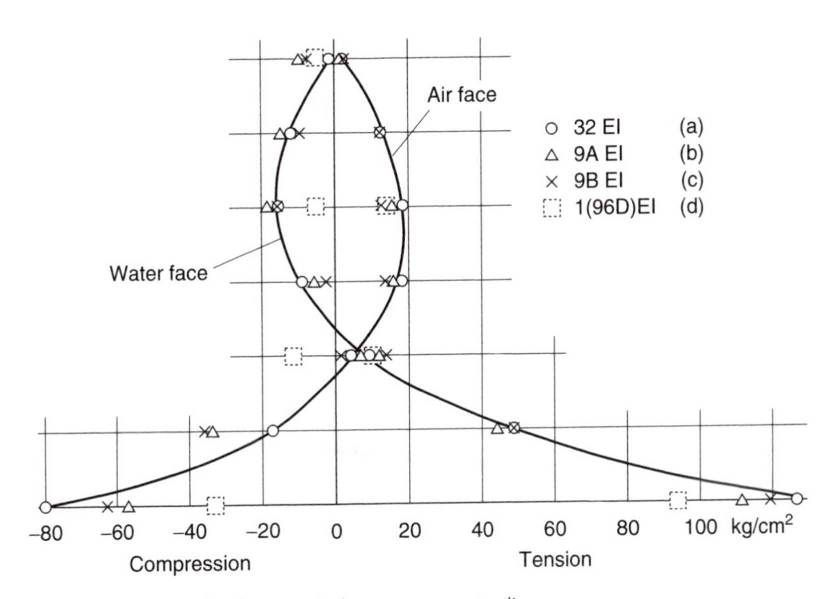

Fig. 6.19 Arch dam in a rigid valley – vertical stresses on centre-line.

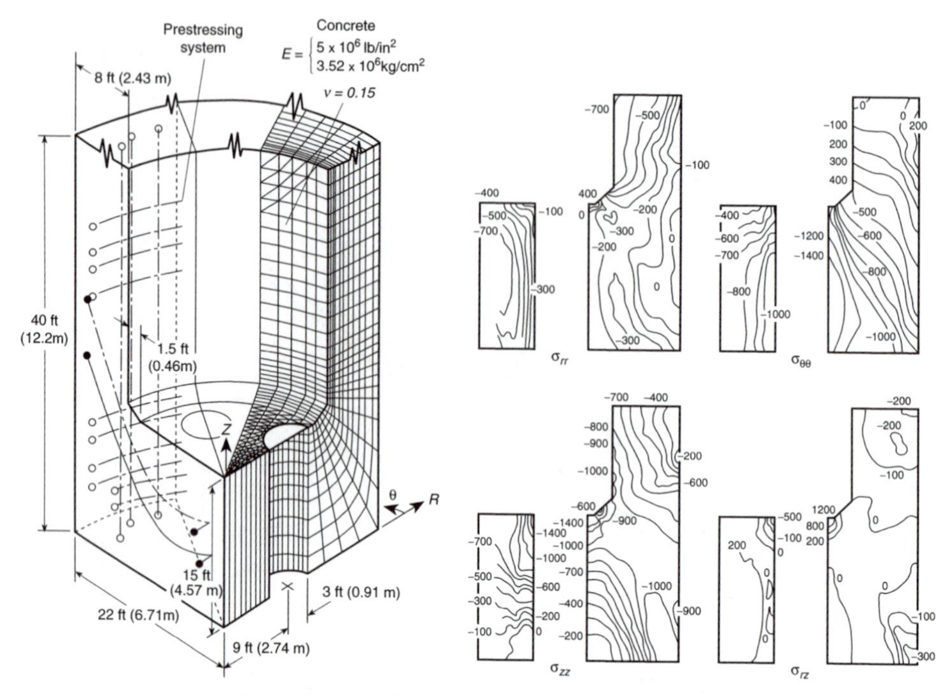

Fig. 6.20 A nuclear pressure vessel analysis using simple tetrahedral elements.[7] Geometry, subdivision, and some stress results. N.B. Not all edges are shown.

(a) Express $u^n(r)$ and $v^n(r)$ in terms of shape functions and parameter $\tilde{u}_a^n$ and $\tilde{v}_a^n$, respectively, and determine the strain displacement matrix for each harmonic n.

(b) For a linear elastic material show that the stiffness matrix for each harmonic is independent of other terms in the Fourier series. (Hint: Perform integrals in θ analytically.)

6.4 Cartesian coordinates may be expressed in terms of spherical components r, θ, and ϕ as

$$x = r\,\cos\theta\,\sin\phi; \quad y = r\,\sin\theta\,\sin\phi \text{ and } z = r\,\cos\phi$$

This form permits the solution of spherically symmetric problems for which displacements depend only on r and the strain-displacement equations are expressed as

$$\varepsilon_{rr} = \frac{\partial u_r}{\partial r}; \quad \varepsilon_{\theta\theta} = \varepsilon_{\phi\phi} = \frac{u_r}{r}$$
$$\gamma_{r\phi} = \frac{\partial u_\phi}{\partial r} - \frac{u_\phi}{r}; \quad \gamma_{r\theta} = \frac{\partial u_\theta}{\partial r} - \frac{u_\theta}{r}; \quad \gamma_{\theta\phi} = 0$$

(a) How many rigid body modes exist for this problem?

(b) Express the displacement components u_r, u_θ and u_ϕ in finite element form using one-dimensional shape functions in r.

(c) Determine the form of the strain-displacement matrix $\mathbf{B}_a$ for each shape function N_a.

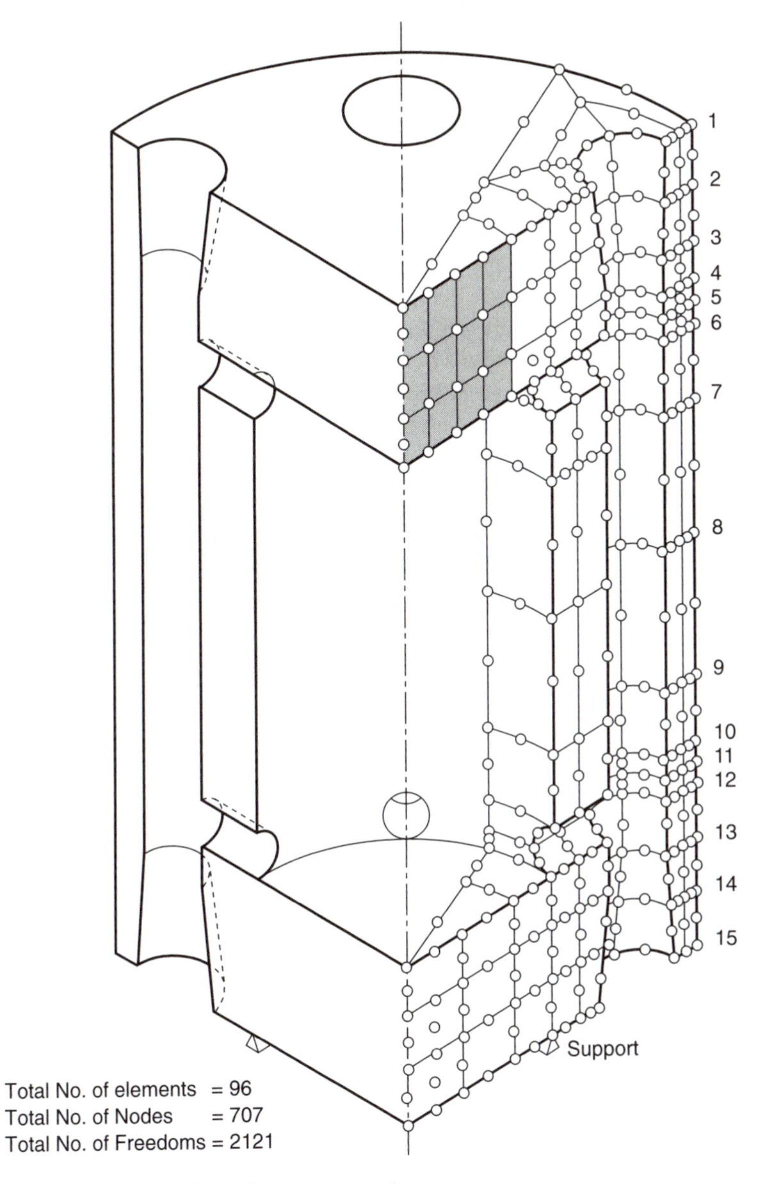

Fig. 6.21 Three-dimensional analysis of a pressure vessel.

(d) For a linear elastic isotropic material write the form of the stiffness matrix for the nodal pair a and b. Show that the problem decomposes into three separate problems in terms of each displacement component.

(e) For linear shape functions obtain an expression for the stiffness components corresponding to the u_r displacements using a one-point quadrature formula. Check if the resulting stiffness matrix has correct rank.

6.5 For a linear elastic isotropic material the stiffness matrix may be computed by numerical integration using Eq. (6.68). Alternatively, the stiffness matrix may be computed in indicial form as indicated in Appendix B.

Consider a plane strain problem which is modelled by 4-node quadrilateral elements. Assume the stiffness matrix is computed using a 2×2 gaussian quadrature formula.

(a) Compute separately the number of additions/subtractions and multiplications necessary to evaluate the stiffness using Eq. (6.68). Count only operations involving non-zero values in **B** or **D**.

(b) Repeat the above calculation using the method of Appendix B given by Eqs (B.52) and (B.54).

6.6 In the classical plane strain problem the strain normal to the plane of deformation (i.e., ε_z) was assumed to be zero. The problem may be 'generalized' by assuming ε_z is constant over the entire analysis domain. The constant strain may then be related to a resultant force F_z applied normal to the deformation plane.

(a) Following the steps given in Sec. 6.3, develop the virtual work expression (weak form) for the generalized plane strain problem.

(b) Write finite element approximations for all the terms in the weak form.

(c) Write the expression for an element stiffness in terms of nodal parameters and the strain ε_z.

(d) Show how the resultant force F_z is related to the constant strain ε_z.

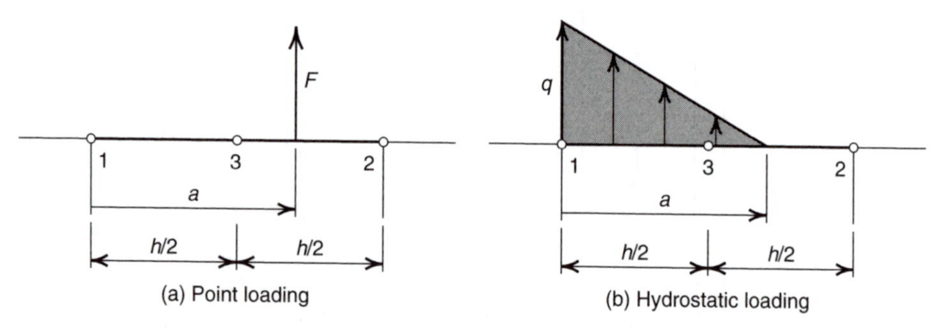

Fig. 6.22 Traction loading on boundary for Problems 6.7 and 6.8.

6.7 A concentrated load, F, is applied to the edge of a two-dimensional plane strain problem which is modelled using quadratic order finite elements as shown in Fig. 6.22(a). Compute the equivalent forces acting on nodes 1, 2 and 3.

6.8 A triangular traction load is applied to the edge of a two-dimensional plane strain problem as shown in Fig. 6.22(b).

(a) Compute the equivalent forces acting on nodes 1, 2 and 3 by performing the integrals exactly.

(b) Use numerical integration to compute the integrals which define the equivalent forces. Use the minimum number of points that integrate the integral exactly. What is the result if one-order lower is used?

6.9 An arc of 2θ for a circular boundary of radius R is approximated by the quadratic isoparametric interpolation as shown in Fig. 6.23. For this case $h = R \sin\theta$ and $c = R(1 - \cos\theta)$.

A concentrated load, F, is applied normal to the boundary at the point labelled $a(\xi)$. Let $F = 100$ N, $R = 10$ cm and $\theta = 15°$. For $\xi = 0, 0.25, 0.50, 0.75, 1.0$ determine:

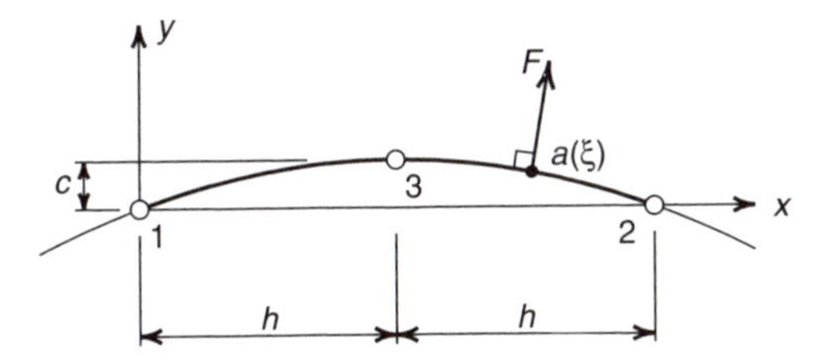

Fig. 6.23 Concentrated normal load on a curved boundary. Problem 6.9.

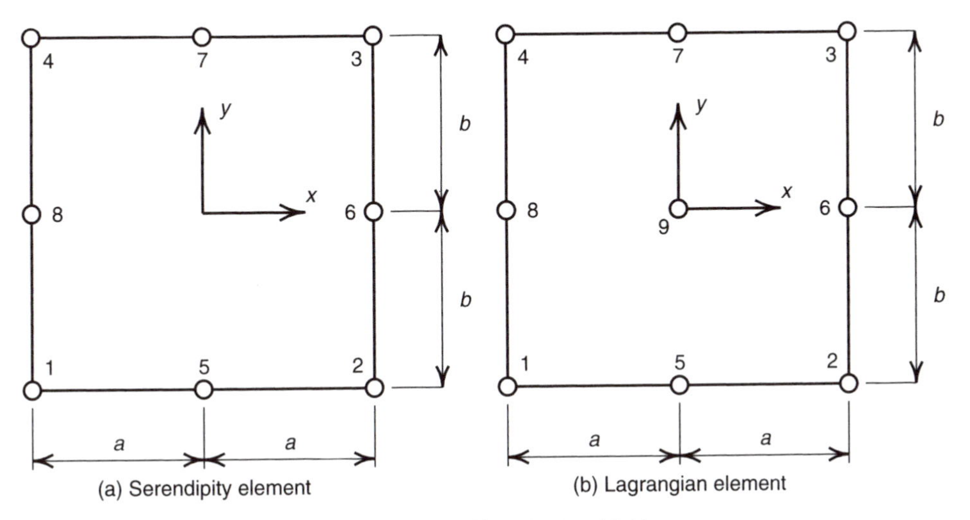

Fig. 6.24 Quadrilateral 8- and 9-node elements. Problems 6.10 and 6.11.

(a) The equivalent forces acting on nodes 1, 2 and 3 for the case when the normal is computed from the quadratic interpolation.
(b) The equivalent nodal forces using the normal to the circular boundary.
(c) The error between the two forms. Show on a sketch.

6.10 A mesh for a plane strain problem contains the quadratic order rectangular elements shown in Figs 6.24(a) and (b). The elements are subjected to a constant body force $\mathbf{b} = (0, -\rho\, g)^T$ where ρ is mass density and g is acceleration of gravity. For each element type:
(a) Use standard shape functions for N_a and develop a closed form expression for the nodal forces in terms of a, b and $\rho\, g$.
(b) Use hierarchical shape functions for N_a and develop a closed form expression for the nodal forces in terms of a, b and $\rho\, g$.

6.11 A mesh for a plane strain problem contains the quadratic order rectangular elements shown in Figs 6.24(a) and (b). The elements are subjected to a constant temperature change ΔT. Each element is made from an isotropic elastic material with constant properties E, ν and α.

For each element type:

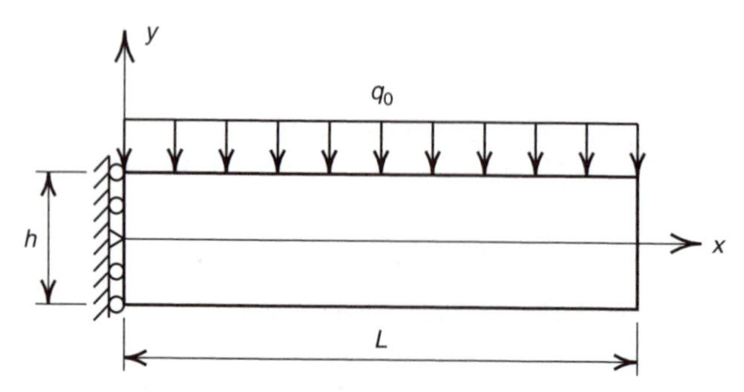

Fig. 6.25 Uniformly loaded cantilever beam. Problem 6.12.

(a) Use standard shape functions for N_a and develop a closed form expression for the nodal forces in terms of a, b and the elastic properties.
(b) Use hierarchical shape functions for N_a and develop a closed form expression for the nodal forces in terms of a, b and the elastic properties.

6.12 Use the program *FEAPpv* (or any other available program) to solve the rectangular beam problem given in Example 6.4 and verify the results shown in Table 6.2.

6.13 Use the program *FEAPpv* (or any other available program) to solve the curved beam problem given in Example 6.5 and verify the results shown in Table 6.3.

6.14 The uniformly loaded cantilever beam shown in Fig. 6.25 has properties

$$L = 2 \text{ m}; \quad h = 0.4 \text{ m}; \quad t = 0.05 \text{ m} \text{ and } q_0 = 100 \text{ N/m}$$

Use *FEAPpv* or any other available program to perform a plane stress analysis of the problem assuming linear isotropic elastic behaviour with $E = 200$ GPa and $\nu = 0.3$.
In your analysis:

(a) Use quadratic lagrangian elements with an initial mesh of 1 element in the depth and 5 elements in the length directions.
(b) Compute consistent nodal forces for the uniform loading.
(c) Compute nodal forces for a parabolically distributed shear traction at the restrained end which balances the uniform loading q_0.
(d) Report results for the centre-line displacement in the vertical direction and the stored energy in the beam.
(e) Repeat the analysis three additional times using meshes of 2×10, 4×20 and 8×40 elements. Tabulate the tip vertical displacement and stored energy for each solution.
(f) If the energy error is given by

$$\Delta E = E_n - E_{n-1} = Ch^q$$

estimate C and q for your solution. Is the convergence rate as expected? Explain your answer.

6.15 A circular composite disk is restrained at its inner radius and free at the outer radius. The disk is spinning at a constant angular velocity ω as shown in Fig. 6.26. The disk is manufactured by bonding a steel layer on top of an aluminium layer as shown in Fig. 6.26(b).

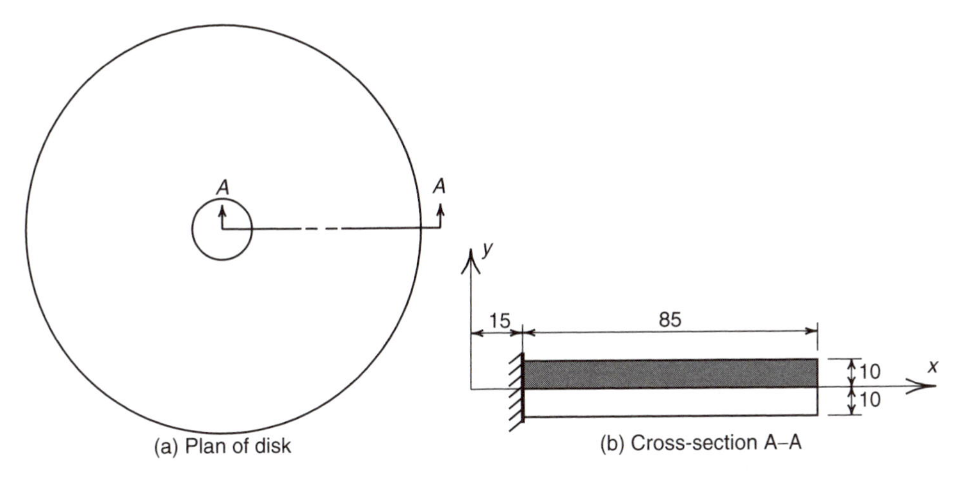

Fig. 6.26 Spinning composite disk. Problem 6.15.

When spinning at an angular velocity of 50 rpm it is desired that the top surface be flat. This will be accomplished by milling the initial shape of the top to a specified level. Your task is to determine the profile for milling. To accomplish this

(a) Perform an analysis for an initially flat top surface using the dimensions given in the figure (lengths given in mm). The elastic properties for steel are $E = 200$ GPa, $\nu = 0.3$ and $\rho = 7.8\ \mu\text{g/mm}^3$; those for aluminium are $E = 70$ GPa, $\nu = 0.35$ and $\rho = 2.6\ \mu\text{g/mm}^3$ (where $\mu = 10^{-6}$). Be sure to use consistent units (say, mm, sec, and μg).

The inner radius of the disk is to be restrained in the radial direction (i.e., $u(15, z) = 0$). Axial restraint is only applied at the centre of the disk (i.e., $v(15, 0) = 0$).

(b) Using the results for the vertical displacements computed in (a) reposition the top nodes to new values for which a reanalysis should give improved results.

(c) Reanalyse the problem for the new coordinates. How accurate does this analysis predict the desired result? What would you do to improve your answer?

6.16 A rectangular region with a circular hole is shown in Fig. 6.27. The traction on the circular hole is zero. The region is to be used for the solution of an infinitely extending

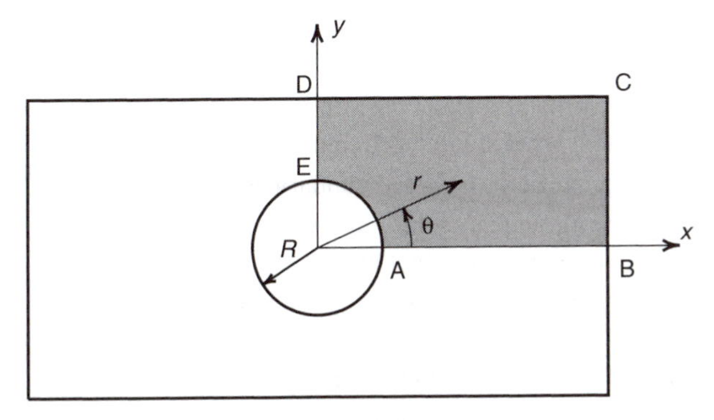

Fig. 6.27 Rectangular region with circular hole. Problem 6.16.

plane stress problem in which the stress at infinity is given by a uniformly distributed normal stress σ_0 acting in the x direction.

The stress distribution in polar coordinates for the problem is given by

$$\begin{aligned}
\sigma_r &= \tfrac{1}{2}\,\sigma_0 \left\{ \left[1 - (\frac{a}{r})^2\right] + \left[1 + 3(\frac{a}{r})^4 - 4\,(\frac{a}{r})^2\right] \cos 2\theta \right\} \\
\sigma_\theta &= \tfrac{1}{2}\,\sigma_0 \left\{ \left[1 + (\frac{a}{r})^2\right] - \left[1 + 3(\frac{a}{r})^4\right] \cos 2\theta \right\} \\
\tau_{r\theta} &= -\tfrac{1}{2}\,\sigma_0 \left\{ 1 - 3(\frac{a}{r})^4 + 2(\frac{a}{r})^2 \right\} \sin 2\theta
\end{aligned}$$

and the displacements by

$$\begin{aligned}
u_r = \frac{\sigma_0 r}{2E} & \left\{ \left[1 + (\frac{a}{r})^2\right] + \left[1 - (\frac{a}{r})^4 + 4(\frac{a}{r})^2\right] \cos 2\theta \right. \\
& \left. + \nu \left[1 - (\frac{a}{4})^2\right] - \nu \left[1 - (\frac{a}{r})^4\right] \cos 2\theta \right\} \\
u_\theta = \frac{\sigma_0 r}{2E} & \left\{ \left[1 + (\frac{a}{r})^4 + 2(\frac{a}{r})^2\right] + \nu \left[1 + (\frac{a}{4})^4 - 2(\frac{a}{r})^2\right] \right\} \sin 2\theta
\end{aligned}$$

In order for the region to satisfy the above solution it is necessary to:

(a) Enforce symmetry conditions along the boundaries AB and DE and

(b) Apply the tractions of the exact solution on the boundary BCD.

Program development project: Write a program that uses numerical integration to compute the consistent nodal forces on the boundary BCD. (Hint: This may be done by adding an element to *FEAPpv* which computes only the nodal forces for line elements defined on the boundary BCD or by writing a MATLAB program which, given the location of nodal coordinates on BCD, computes the nodal forces.) Your program should also compute

$$E = t \int_{\mathrm{BCD}} \left[u\, t_x + v\, t_y\right] \, \mathrm{d}\Gamma$$

which is twice the stored energy in a slice of thickness t. When accurately computed (e.g., to 9 or 10 digit accuracy) this may be used as the 'exact' solution for the region. Use your program and *FEAPpv* (or any other available program) to solve a plane stress problem.

Let the hole radius be $R = 10$ cm, the thickness of the slice be $t = 0.1R$ and take $E = 200$ GPa and $\nu = 0.3$ for the elastic properties. The boundary BC should be placed at about $3R$ and the boundary CD at 2 to $3R$. Assume a unit value for the stress σ_0.

(a) Use 4-node quadrilateral elements to solve the problem on a sequence of meshes in which element sizes are reduced in half for each succeeding mesh.

(b) Plot the displacement at the hole boundary and compare to the exact solution.

(c) Compute the work done by your finite element program. (Note: In *FEAPpv* this will be the 'energy' reported by the solver.)

(d) Compute the rate of convergence for your solution and plot on a figure similar to that given in Fig. 6.9.

(e) Repeat the solution using 8-node serendipity elements.

(f) Repeat the solution using 9-node lagrangian elements.

Write a short report discussing your findings.

6.17 Program development project: Extend the program developed in Problem 2.17 to consider plane strain and axisymmetric geometry.

6.18 Program development project: Extend the program developed in Problem 2.17 to compute nodal forces for specified boundary tractions which are normal or tangential to the element edge. Assume tractions can vary up to quadratic order (i.e., constant, linear and parabolic distributions) and use numerical integration to compute values.

Test your program for an edge with constant normal stress. Then test for linear normal and finally quadratic tangential values. Compare results with those computed by *FEAPpv* (or any available program).

6.19 Program development project: Extend the program developed in Problem 2.17 to compute nodal values of stress and strain. Follow the procedure given in Sec. 6.4 to project element values to nodes.

Test your program using (a) the patch test of Problem 2.17 and (b) the curved beam problem shown in Fig. 2.11.

6.20 Program development project: Add a module to the program developed in Problem 2.17 to plot contours of stress and strain components for plane stress, plane strain and axisymmetric solids. Use the capability developed in Problem 6.19 to obtain nodal values and the contour routine developed in Problem 2.18.

Test your program system by plotting contours of stress components for the curved beam meshes described in Problem 2.18.

6.21 Program development project: Add a 4-node quadrilateral element to the program system developed in Problem 2.17. Use shape functions and numerical integration to compute the element stiffness matrix. Also include the force vector from a constant element body force (you may need to add **b** to your input module).

Test your program on the curved beam problems described in Problem 2.18. Compare the accuracy to that obtained using triangular elements.

References

1. M.J. Turner, R.W. Clough, H.C. Martin, and L.J. Topp. Stiffness and deflection analysis of complex structures. *J. Aero. Sci.*, 23:805–823, 1956.
2. R.W. Clough. The finite element method in plane stress analysis. In *Proc. 2nd ASCE Conf. on Electronic Computation*, Pittsburgh, Pa., Sept. 1960.
3. R.H. Gallagher, J. Padlog, and P.P. Bijlaard. Stress analysis of heated complex shapes. *ARS J.*, 29:700–707, 1962.
4. R.J. Melosh. Structural analysis of solids. *J. Struct. Eng., ASCE*, 4:205–223, Aug. 1963.
5. J.H. Argyris. Matrix analysis of three-dimensional elastic media – small and large displacements. *J. AIAA*, 3:45–51, Jan. 1965.
6. J.H. Argyris. Three-dimensional anisotropic and inhomogeneous media – matrix analysis for small and large displacements. *Ingenieur Archiv*, 34:33–55, 1965.
7. Y.R. Rashid and W. Rockenhauser. Pressure vessel analysis by finite element techniques. In *Proc. Conf. Prestressed Concrete Pressure Vessels*, Institute of Civil Engineering, 1968.
8. J.H. Argyris. Continua and discontinua. In *Proc. 1st Conf. Matrix Methods in Structural Mechanics*, volume AFFDL-TR-66-80, pages 11–189, Wright Patterson Air Force Base, Ohio, Oct. 1966.

9. B.M. Irons. Engineering applications of numerical integration in stiffness methods. *J. AIAA*, 4:2035–2037, 1966.
10. J.G. Ergatoudis, B.M. Irons, and O.C. Zienkiewicz. Three dimensional analysis of arch dams and their foundations. In *Proc. Symp. Arch Dams*, Inst. Civ. Eng., London, 1968.
11. J.H. Argyris and J.C. Redshaw. Three dimensional analysis of two arch dams by a finite element method. In *Proc. Symp. Arch Dams*, Inst. Civ. Eng., 1968.
12. S. Fjeld. Three dimensional theory of elastics. In I. Holand and K. Bell, editors, *Finite Element Methods in Stress Analysis*, Trondheim, 1969. Tech. Univ. of Norway, Tapir Press.
13. A.E.H. Love. *A Treatise on the Mathematical Theory of Elasticity*. Cambridge University Press, Cambridge, 4th edition, 1927.
14. N.I. Muskhelishvili. *Some Basic Problems of the Mathematical Theory of Elasticity*. Noordhoff, Groningen, 3rd edition, 1953. English translation by J.R.M. Radok.
15. S.P. Timoshenko and J.N. Goodier. *Theory of Elasticity*. McGraw-Hill, New York, 3rd edition, 1969.
16. S.G. Lekhnitskii. *Theory of Elasticity of an Anisotropic Elastic Body*. Holden Day, San Francisco, 1963. (Translation from Russian by P. Fern.)
17. I.S. Sokolnikoff. *The Mathematical Theory of Elasticity*. McGraw-Hill, New York, 2rd edition, 1956.
18. P.G. Ciarlet. *Mathematical Elasticity. Volume 1: Three-dimensional Elasticity*. North-Holland, Amsterdam, 1988.
19. O.C. Zienkiewicz and F.C. Scott. On the principle of repeatability and its application in analysis of turbine and pump impellers. *Int. J. Numer. Meth. Eng.*, 9:445–452, 1972.
20. R.F.S. Hearmon. *An Introduction to Applied Anisotropic Elasticity*. Oxford University Press, Oxford, 1961.
21. T.C.-T Ting. *Anisotropic Elasticity: Theory and Applications*. Oxford University Press, New York, 1996.
22. G. Strang and G.J. Fix. *An Analysis of the Finite Element Method*. Prentice-Hall, Englewood Cliffs, N.J., 1973.
23. H.J. Brauchli and J.T. Oden. On the calculation of consistent stress distributions in finite element applications. *Int. J. Numer. Meth. Eng.*, 3:317–325, 1971.
24. E. Hinton and J. Campbell. Local and global smoothing of discontinuous finite element function using a least squares method. *Int. J. Numer. Meth. Eng.*, 8:461–480, 1974.
25. O.C. Zienkiewicz and Y.K. Cheung. Buttress dams on complex rock foundations. *Water Power*, 16:193, 1964.
26. O.C. Zienkiewicz and Y.K. Cheung. Finite element procedures in the solution of plate and shell problems. In O.C. Zienkiewicz and G.S. Holister, editors, *Stress Analysis*, Chapter 8. John Wiley & Sons, Chichester, 1965.
27. O.C. Zienkiewicz, A.H.C. Chan, M. Pastor, B.A. Schrefler, and T. Shiomi. *Computational Geomechanics: With Special Reference to Earthquake Engineering*. John Wiley & Sons, Chichester, 1999.
28. K. Terzhagi. *Theoretical Soil Mechanics*. John Wiley & Sons, New York, 1943.
29. O.C. Zienkiewicz, C. Humpheson, and R.W. Lewis. A unified approach to soil mechanics problems, including plasticity and visco-plasticity. In *Int. Symp. on Numerical Methods in Soil and Rock Mechanics*, Karlsruhe, 1975. See also Chapter 4 of *Finite Elements in Geomechanics* (ed. G. Gudehus), pages 151–78. Wiley, 1977.
30. O.C. Zienkiewicz and R.L. Taylor. *The Finite Element Method for Solid and Structural Mechanics*. Butterworth-Heinemann, Oxford, 6th edition, 2005.

7

Field problems – heat conduction, electric and magnetic potential and fluid flow

7.1 Introduction

The general procedures discussed in the previous chapters can be applied to a variety of physical problems. Indeed, some such possibilities have been indicated in Chapter 3 and here more detailed attention will be given to a particular, but wide class, of such situations.

Primarily we shall deal with situations governed by the general 'quasi-harmonic' equation, the particular cases of which are the well-known Laplace and Poisson equations.[1–6] The range of physical problems falling into this category is large. To list but a few frequently encountered in engineering practice we have:

- Heat conduction
- Seepage through porous media
- Irrotational flow of ideal fluids
- Distribution of electrical (or magnetic) potential
- Torsion of prismatic shafts
- Lubrication of pad bearings, etc.

The formulation developed in this chapter is equally applicable to all, and hence only limited reference will be made to the actual physical quantities. In all the above classes of problems, the behaviour can be represented in terms of a scalar variable for which we will generally use the symbol ϕ. In the applications to specific problems, however, we shall generally introduce the physical variable describing the behaviour. For instance, in discussing heat conduction applications we use the symbol T to denote the temperature.

In Chapter 3 we indicated both the 'weak form' and a variational principle applicable to the Poisson and Laplace equations (see Secs 3.2 and 3.8.1). In the following sections we shall apply these approaches to a general, quasi-harmonic equation and indicate the ranges of applicability of a *single*, *unified*, *approach* by which one computer program can solve a large variety of physical problems. It will be observed that the C_0 'shape functions' presented in Chapters 4 and 5 can be directly applied and that both isotropic and anisotropic behaviour can be treated with equal ease.

7.2 General quasi-harmonic equation

7.2.1 Governing equations

In many physical situations we are concerned with the *diffusion* or *flow* of some quantity such as heat, mass, concentration, etc. In such problems, the rate of transfer per unit area (flux), $\mathbf{q}$, can be written in terms of its cartesian components as

$$\mathbf{q} = \left[q_x, \quad q_y, \quad q_z\right]^{\mathrm{T}} \tag{7.1}$$

If the rate at which the relevant quantity is generated (or removed) per unit volume is Q, then for steady-state flow the balance or continuity requirement gives

$$\frac{\partial q_x}{\partial x} + \frac{\partial q_y}{\partial y} + \frac{\partial q_z}{\partial z} + Q = 0 \tag{7.2}$$

Introducing the gradient operator

$$\boldsymbol{\nabla} = \left[\frac{\partial}{\partial x}, \quad \frac{\partial}{\partial y}, \quad \frac{\partial}{\partial z}\right]^{\mathrm{T}} \tag{7.3}$$

we can write (7.2) as

$$\boldsymbol{\nabla}^{\mathrm{T}}\mathbf{q} + Q = 0 \tag{7.4}$$

Generally the rates of flow will be related to the *gradient* of some potential quantity ϕ. This may be temperature in the case of heat flow, etc. A very general linear relationship will be of the form

$$\mathbf{q} = \begin{Bmatrix} q_x \\ q_y \\ q_z \end{Bmatrix} = -\begin{bmatrix} k_{xx}, & k_{xy}, & k_{xz} \\ k_{yx}, & k_{yy}, & k_{yz} \\ k_{zx}, & k_{zy}, & k_{zz} \end{bmatrix} \begin{Bmatrix} \dfrac{\partial \phi}{\partial x} \\ \dfrac{\partial \phi}{\partial y} \\ \dfrac{\partial \phi}{\partial z} \end{Bmatrix} = -\mathbf{k}\,\boldsymbol{\nabla}\phi \tag{7.5}$$

where $\mathbf{k}$ is a symmetric form due to energy arguments (i.e., $k_{xy} = k_{yx}$, etc.) and is variously referred to as Fourier's, Fick's, or Darcy's law depending on the physical problem.

The final governing differential equation for the 'potential' ϕ is obtained by substitution of Eq. (7.5) into (7.4), leading to

$$-\boldsymbol{\nabla}^{\mathrm{T}}\left(\mathbf{k}\,\boldsymbol{\nabla}\phi\right) + Q = 0 \tag{7.6}$$

which has to be solved in a domain Ω. On the boundaries of such a domain we shall usually encounter one of the following conditions:

1. On Γ_ϕ,

$$\phi = \bar{\phi} \tag{7.7a}$$

   i.e., the potential is specified (Dirichlet condition).
2. On Γ_q the normal component of flow (or flux), q_n, is given as (Neumann condition)

$$q_n = \bar{q} - H(\phi - \phi_0)$$

   where H is a transfer or radiation coefficient, ϕ_0 is a known equilibrium value and $\bar{q}$ is a specified value. Here q_n is defined as

$$q_n = \mathbf{n}^{\mathrm{T}}\mathbf{q} \quad \text{with} \quad \mathbf{n} = \left[n_x\,, n_y\,, n_z\right]^{\mathrm{T}}$$

   where $\mathbf{n}$ is a vector of direction cosines of the normal to the boundary surface. Accordingly, we may write the second boundary condition

$$\bar{q} + \mathbf{n}^{\mathrm{T}}(\mathbf{k}\,\nabla\phi) + H(\phi - \phi_0) = 0 \tag{7.7b}$$

   which holds on Γ_q.

7.2.2 Anisotropic and isotropic forms for k

If we consider the general statement of Eq. (7.5) as being determined for an arbitrary set of coordinate axes x, y, z we shall find that it is always possible to determine locally another set of axes x', y', z' with respect to which the matrix $\mathbf{k}'$ becomes diagonal, as shown in

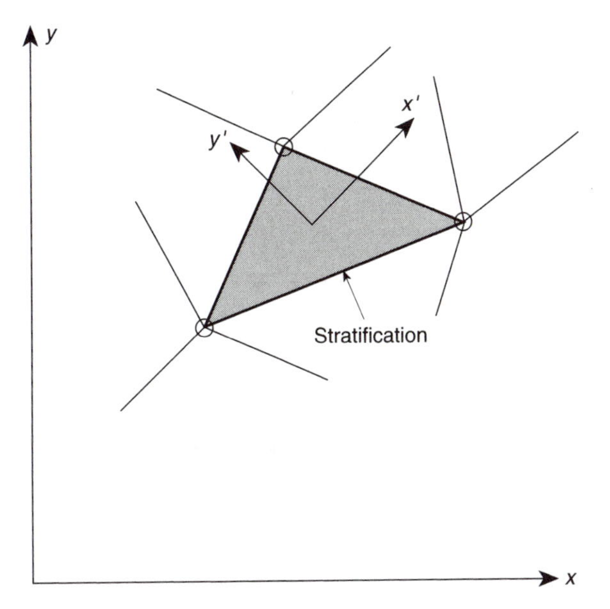

Fig. 7.1 Anisotropic material. Local coordinates coincide with the principal directions of stratification.

Fig. 7.1. With respect to such axes we have

$$\mathbf{k}' = \begin{bmatrix} k_{x'x'} & 0 & 0 \\ 0 & k_{y'y'} & 0 \\ 0 & 0 & k_{z'z'} \end{bmatrix} \tag{7.8}$$

Thus, the general form of the **k** has only three components which are associated with three orthogonal axes. Such materials are called *anisotropic* or *orthotropic*.

The governing differential equation (7.6) for these axes can be written

$$\begin{gathered} -\left[\frac{\partial}{\partial x'}\left(k_{x'x'}\frac{\partial \phi}{\partial x'}\right) + \frac{\partial}{\partial y'}\left(k_{y'y'}\frac{\partial \phi}{\partial y'}\right) + \frac{\partial}{\partial z'}\left(k_{z'z'}\frac{\partial \phi}{\partial z'}\right)\right] + Q = 0 \\ -\left(\boldsymbol{\nabla}'\right)^{\mathrm{T}}\left(\mathbf{k}'\,\boldsymbol{\nabla}'\phi\right) + Q = 0 \end{gathered} \tag{7.9}$$

where

$$\boldsymbol{\nabla}' = \left[\frac{\partial}{\partial x'}, \quad \frac{\partial}{\partial y'}, \quad \frac{\partial}{\partial z'}\right]^{\mathrm{T}}$$

defines the gradient operator for the 'prime' coordinate system.

Alternatively, knowing $\mathbf{k}'$ and the orientation of the axes x', y', z' a transformation of coordinates is given by

$$\mathbf{x}' = \mathbf{T}\mathbf{x}$$

in which **T** are *direction cosines* defined as

$$\mathbf{T} = \begin{bmatrix} \cos(x', x), & \cos(x', y), & \cos(x', z) \\ \cos(y', x), & \cos(y', y), & \cos(y', z) \\ \cos(z', x), & \cos(z', y), & \cos(z', z) \end{bmatrix}$$

where $\cos(x', x)$ is the cosine of the angle between the x' direction and the x direction. The inverse of **T** is equal to its transpose; hence

$$\mathbf{x} = \mathbf{T}^{\mathrm{T}}\mathbf{x}'$$

In addition we may write the gradient with respect to the prime axes as

$$\boldsymbol{\nabla}'(\cdot) = \mathbf{T}\boldsymbol{\nabla}(\cdot)$$

or alternatively

$$\boldsymbol{\nabla}(\cdot) = \mathbf{T}^{\mathrm{T}}\,\boldsymbol{\nabla}'(\cdot)$$

Using the above we obtain the expression

$$\left(\boldsymbol{\nabla}'\right)^{\mathrm{T}}\left(\mathbf{k}'\boldsymbol{\nabla}'\phi\right) = (\boldsymbol{\nabla})^{\mathrm{T}}\,\mathbf{T}^{\mathrm{T}}\left(\mathbf{k}'\mathbf{T}\boldsymbol{\nabla}\phi\right) \tag{7.10a}$$

or

$$\mathbf{k} = \mathbf{T}^{\mathrm{T}}\mathbf{k}'\mathbf{T} \quad \text{or} \quad \mathbf{k}' = \mathbf{T}\mathbf{k}\mathbf{T}^{\mathrm{T}} \tag{7.10b}$$

Lastly for an isotropic material we can write

$$\mathbf{k} = k\mathbf{I} \tag{7.11}$$

where **I** is an identity matrix. In two dimensions this leads to the simple form of Eq. (3.8) as discussed in Chapter 3.

7.2.3 Weak form and variational principle for the general quasi-harmonic equation

Following the principles of Chapter 3, Sec. 3.2, we can obtain the weak form of Eqs (7.6) and (7.7b) by writing (using $v = \delta\phi$)

$$\int_\Omega \delta\phi \left[-\boldsymbol{\nabla}^{\mathrm{T}} (\mathbf{k}\, \boldsymbol{\nabla}\phi) + Q\right] \mathrm{d}\Omega + \int_{\Gamma_q} \delta\phi \left[\bar{q} + \mathbf{n}^{\mathrm{T}} (\mathbf{k}\, \boldsymbol{\nabla}\phi) + H\,(\phi - \phi_0)\right] \mathrm{d}\Gamma = 0 \tag{7.12}$$

for arbitrary functions $\delta\phi$. Integration by parts (see Appendix G) will result in the following weak statement

$$\int_\Omega \left[(\boldsymbol{\nabla}\delta\phi)^{\mathrm{T}} (\mathbf{k}\, \boldsymbol{\nabla}\phi) + \delta\phi\, Q\right] \mathrm{d}\Omega + \int_{\Gamma_q} \delta\phi \left[\bar{q} + H\,(\phi - \phi_0)\right] \mathrm{d}\Gamma + \int_{\Gamma_\phi} \delta\phi\, q_n\, \mathrm{d}\Gamma = 0 \tag{7.13}$$

Generally, the last term is omitted by requiring $\delta\phi = 0$ and imposing the *forced* (Dirichlet) boundary condition (7.7a) on Γ_q.

It is also possible to express an integral form for the quasi-harmonic equation as a variational principle. The functional

$$\Pi = \int_\Omega \left[\frac{1}{2}\, (\boldsymbol{\nabla}\phi)^{\mathrm{T}} (\mathbf{k}\, \boldsymbol{\nabla}\phi) + \phi\, Q\right] \mathrm{d}\Omega + \int_{\Gamma_q} \left[\phi\, \bar{q} + H\,(\frac{1}{2}\,\phi^2 - \phi\phi_0)\right] \mathrm{d}\Gamma \tag{7.14}$$

gives on minimization [subject to the constraint of Eq. (7.7a)] the original problem in Eqs (7.6) and (7.7b). The algebraic manipulations required to verify the above principle follow precisely the lines of Sec. 3.8.

Clearly material properties defined by the **k** matrix can vary from element to element in a discontinuous manner. This is implied in both the weak and variational statements of the problem.

7.3 Finite element solution process

7.3.1 Finite element discretization

The finite element solution process follows the standard solution methodology and for the quasi-harmonic equation approximates the trial function using any of the C_0 shape function expressions given in Chapters 4 and 5. Accordingly, we use

$$\phi \approx \hat{\phi} = \sum_a N_a \tilde{\phi}_a = \mathbf{N}\,\tilde{\boldsymbol{\phi}} \tag{7.15}$$

in either the weak formulation of Eq. (7.13) or the variational statement of Eq. (7.14). If, in the weak statement, we take

$$\delta\phi \approx \delta\hat{\phi} = \sum_a W_a\, \delta\tilde{\phi}_a = \mathbf{W}\, \delta\tilde{\boldsymbol{\phi}} \quad \text{with} \quad \mathbf{W} = \mathbf{N} \tag{7.16}$$

according to the Galerkin principle, an identical form will arise with that obtained from the minimization of the variational principle.

The gradient of ϕ is now given by the approximation

$$
\begin{aligned}
\nabla\hat{\phi} &= \sum_a (\nabla N_a)\,\tilde{\phi}_a \\
&= \sum_a \left[\frac{\partial N_a}{\partial x}, \quad \frac{\partial N_a}{\partial y}, \quad \frac{\partial N_a}{\partial z}\right]^{\mathrm{T}} \tilde{\phi}_a = \sum_a \mathbf{b}_a \tilde{\phi}_a
\end{aligned}
\tag{7.17}
$$

where now $\mathbf{b}_a$ denotes the *gradient matrix* of shape functions.

Substituting Eqs (7.15) to (7.17) into (7.13), we have a typical statement for an arbitrary $\delta\tilde{\phi}_a$ giving, for each a (assuming summation convention for b),

$$
\left\{\left[\int_\Omega \mathbf{b}_a^{\mathrm{T}}\mathbf{k}\mathbf{b}_b\,\mathrm{d}\Omega + \int_{\Gamma_q} N_a H N_b\,\mathrm{d}\Gamma\right]\tilde{\phi}_b + \int_\Omega N_a Q\,\mathrm{d}\Omega + \int_{\Gamma_q} N_a\,(\bar{q} - H\phi_0)\,\mathrm{d}\Gamma\right\} = \mathbf{0}
\tag{7.18}
$$

Evaluating the integrals for all elements leads to the set of standard discrete equations of the form

$$
\mathbf{H}\tilde{\boldsymbol{\phi}} + \mathbf{f} = \mathbf{0}
\tag{7.19}
$$

with

$$
H_{ab} = \int_\Omega \mathbf{b}_a^{\mathrm{T}}\mathbf{k}\mathbf{b}_b\,\mathrm{d}\Omega + \int_{\Gamma_q} N_a H N_b\,\mathrm{d}\Gamma \quad \text{and} \quad f_a = \int_\Omega N_a Q\,\mathrm{d}\Omega + \int_{\Gamma_q} N_a\,(\bar{q} - H\phi_0)\,\mathrm{d}\Gamma
\tag{7.20}
$$

to which prescribed values of $\bar{\phi}$ have to be imposed on boundaries Γ_ϕ. We note that an additional 'stiffness' is contributed on boundaries for which a radiation constant H is specified.

Indeed, standard operations are followed to evaluate the above arrays using quadrature. In the general three-dimensional case using Lagrange or serendipity-type 'brick' elements, use of Gauss quadrature results in

$$
\begin{aligned}
H_{ab} &= \sum_{l=1}^{L_3} \mathbf{b}_a(\xi_l, \eta_l, \zeta_l)^{\mathrm{T}}\mathbf{k}\mathbf{b}_b(\xi_l, \eta_l, \zeta_l)\, J(\xi_l, \eta_l, \zeta_l)\, W_l \\
&\quad + \sum_{l=1}^{L_2} N_a(\xi_l, \eta_l)\, H N_b(\xi_l, \eta_l)\, j(\xi_l, \eta_l)\, W_l
\end{aligned}
$$

with a similar expression for f_a.

Indeed in a computer program the same standard operations are followed to evaluate the fluxes using

$$
\mathbf{q} \equiv -\mathbf{k}\,\nabla\phi \approx -\mathbf{k}\sum_b \mathbf{b}_b\,\tilde{\phi}_b
\tag{7.21}
$$

The fluxes may be computed within the elements; however, it is often desirable to obtain their values at nodes. This is best accomplished by the procedure summarized in Sec. 6.4 and discussed in more detail later in Chapter 13.

7.3.2 Two-dimensional plane and axisymmetric problem

The two-dimensional *plane case* is obtained by taking the gradient in the form

$$\nabla = \left[\frac{\partial}{\partial x}, \quad \frac{\partial}{\partial y}\right]^{\mathrm{T}} \tag{7.22}$$

and taking the flux as

$$\mathbf{q} = \begin{Bmatrix} q_x \\ q_y \end{Bmatrix} = -\begin{bmatrix} k_{xx} & k_{xy} \\ k_{yx} & k_{yy} \end{bmatrix} \begin{Bmatrix} \dfrac{\partial \phi}{\partial x} \\ \dfrac{\partial \phi}{\partial y} \end{Bmatrix} \tag{7.23}$$

On discretization by Eqs (7.15) to (7.17) a slightly simplified form of the matrices will now be found with $\mathbf{b}_a$ in Eq. (7.17) replaced by

$$\mathbf{b}_a = \left[\frac{\partial N_a}{\partial x}, \quad \frac{\partial N_a}{\partial y}\right]^{\mathrm{T}} \tag{7.24}$$

and the volume element by

$$\mathrm{d}\Omega = t\,\mathrm{d}x\,\mathrm{d}y$$

where t is the slab thickness. Alternatively the formulation may be specialized to cylindrical coordinates and used for the solution of *axisymmetric* situations by introducing the gradient

$$\nabla = \left[\frac{\partial}{\partial r}, \quad \frac{\partial}{\partial z}\right]^{\mathrm{T}} \tag{7.25}$$

where r, z replace x, y to describe both the gradient and $\mathbf{b}_a$. With the flux now given by

$$\mathbf{q} = \begin{Bmatrix} q_r \\ q_z \end{Bmatrix} = -\begin{bmatrix} k_{rr} & k_{rz} \\ k_{zr} & k_{zz} \end{bmatrix} \begin{Bmatrix} \dfrac{\partial \phi}{\partial r} \\ \dfrac{\partial \phi}{\partial z} \end{Bmatrix} \tag{7.26}$$

the discretization of Eq. (7.18) is now performed with the volume element expressed by

$$\mathrm{d}\Omega = 2\pi r\,\mathrm{d}r\,\mathrm{d}z$$

and integration carried out using quadrature as described above.

Example 7.1: Plane triangular element with 3 nodes. We particularize here to the simplest triangular element (Fig. 7.2).

With shape functions written in the alternative forms

$$N_a = L_a = \frac{a_a + b_a x + c_a y}{2\Delta}$$

in which Δ is defined in (4.26) and a_a, b_a, c_a in (4.28), we can compute the derivatives as

$$\frac{\partial N_a}{\partial x} = \frac{\partial L_a}{\partial x} = \frac{b_a}{2\Delta}; \quad \frac{\partial N_a}{\partial y} = \frac{\partial L_a}{\partial y} = \frac{c_a}{2\Delta}$$

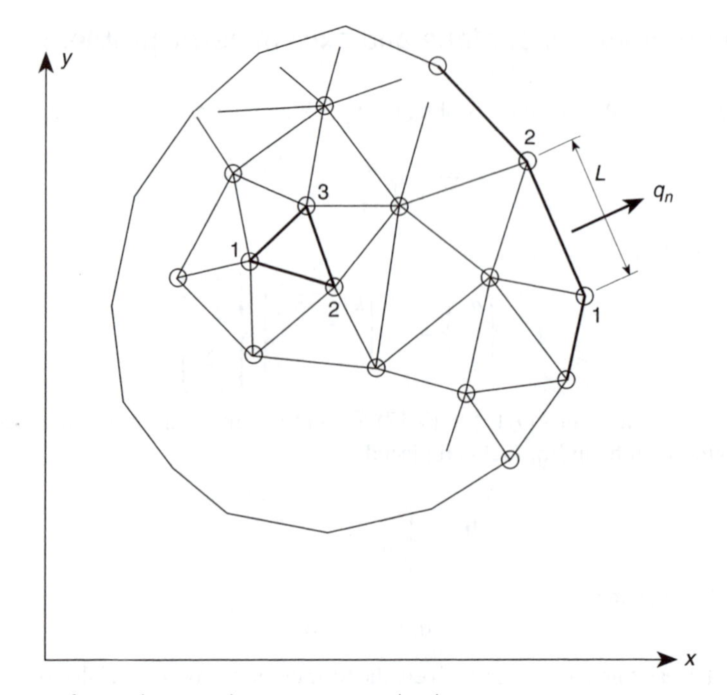

Fig. 7.2 Division of a two-dimensional region into triangular elements.

giving the gradient matrix

$$\mathbf{b}_a = \frac{1}{2\Delta}\begin{bmatrix} b_a & c_a \end{bmatrix}^{\mathrm{T}}$$

Since the gradient matrix is constant the element 'stiffness' matrix (ignoring the H boundary term) is given by

$$\mathbf{H}^e = \frac{k_{xx}t}{4\Delta}\begin{bmatrix} b_1b_1 & b_1b_2 & b_1b_3 \\ b_2b_1 & b_2b_2 & b_2b_3 \\ b_3b_1 & b_3b_2 & b_3b_3 \end{bmatrix} + \frac{k_{yy}t}{4\Delta}\begin{bmatrix} c_1c_1 & c_1c_2 & c_1c_3 \\ c_2c_1 & c_2c_2 & c_2c_3 \\ c_3c_1 & c_3c_2 & c_3c_3 \end{bmatrix}$$
$$+ \frac{k_{xy}t}{4\Delta}\begin{bmatrix} b_1c_1 & b_1c_2 & b_1c_3 \\ b_2c_1 & b_2c_2 & b_2c_3 \\ b_3c_1 & b_3c_2 & b_3c_3 \end{bmatrix} + \frac{k_{yx}t}{4\Delta}\begin{bmatrix} c_1b_1 & c_1b_2 & c_1b_3 \\ c_2b_1 & c_2b_2 & c_2b_3 \\ c_3b_1 & c_3b_2 & c_3b_3 \end{bmatrix}$$

The load matrices follow a similar simple pattern and thus, for instance, due to constant Q and using a 1-point quadrature from Table 5.3 we have $L_a = 1/3$ so that

$$f_a^e = L_a Q t \Delta = \frac{1}{3} Q t \Delta$$

This is a very simple (almost 'obvious') result.

Example 7.2: 'Stiffness' matrix for axisymmetric triangular element with 3 nodes. The computation of the arrays for an axisymmetric problem may be performed using area coordinates as described in Sec. 4.7.1 and quadrature in Sec. 5.11. Since the integrals for the 'stiffness' matrix only involve a linear function in r (from the volume element) a 1-point integration from Table 5.3 still is exact and results in

$$\mathbf{H}^e = \left\{ \frac{k_{rr}}{4\Delta} \begin{bmatrix} b_1b_1 & b_1b_2 & b_1b_3 \\ b_2b_1 & b_2b_2 & b_2b_3 \\ b_3b_1 & b_3b_2 & b_3b_3 \end{bmatrix} + \frac{k_{zz}}{4\Delta} \begin{bmatrix} c_1c_1 & c_1c_2 & c_1c_3 \\ c_2c_1 & c_2c_2 & c_2c_3 \\ c_3c_1 & c_3c_2 & c_3c_3 \end{bmatrix} \right.$$

$$\left. + \frac{k_{rz}}{4\Delta} \begin{bmatrix} b_1c_1 & b_1c_2 & b_1c_3 \\ b_2c_1 & b_2c_2 & b_2c_3 \\ b_3c_1 & b_3c_2 & b_3c_3 \end{bmatrix} + \frac{k_{zr}}{4\Delta} \begin{bmatrix} c_1b_1 & c_1b_2 & c_1b_3 \\ c_2b_1 & c_2b_2 & c_2b_3 \\ c_3b_1 & c_3b_2 & c_3b_3 \end{bmatrix} \right\} 2\pi\bar{r}$$

where $\bar{r} = (r_1 + r_2 + r_3)/3$.

Example 7.3: Load matrix for axisymmetric triangular element with 3 nodes. The nodal forces from a constant source term Q are computed from

$$f_a^e = \int_\Delta L_a Q 2\pi r_b L_b \, \mathrm{d}r \, \mathrm{d}z \quad \text{sum on } b$$

and thus now has quadratic terms. From Table 5.3 use of a 3-point formula is adequate to obtain an exact result. For node 1 this gives

$$f_1^e = \tfrac{1}{2} 2\pi Q \left[\tfrac{1}{2}(r_1 + r_2) + \tfrac{1}{2}(r_1 + r_3) \right] \Delta \, \tfrac{1}{3} = \tfrac{1}{6}(2r_1 + r_2 + r_3)\, \pi Q \Delta$$

with results for f_2^e and f_3^e obtained by cyclic permutation. The use of a 1-point formula gives results which are of the same accuracy as that of the basic linear functions in the approximation of ϕ, namely, $O(h^2)$ where h is the diameter of an element. Using this we obtain the force array

$$f_a^e \approx \tfrac{1}{3}\, 2\pi \bar{r} Q \Delta = \tfrac{2}{3}\, \pi \bar{r} Q \Delta$$

7.4 Partial discretization – transient problems

The above developments have assumed that the solution to the problem is independent of time. Many problems, however, require the solution to depend explicitly on time, both in the loading and in the differential equation.

An example of a problem which is *time dependent* is a heat conduction problem in which the loading varies with time. The solution for the temperature now requires use of the differential equation given by

$$c \frac{\partial T}{\partial t} - \boldsymbol{\nabla}^{\mathrm{T}} (\mathbf{k} \, \boldsymbol{\nabla} T) + Q = 0 \tag{7.27}$$

where T is temperature (which now replaces ϕ), $\mathbf{k}$ the thermal conductivity, c the specific heat per unit volume and Q a heat source term. In addition to boundary conditions of the form given in (7.7a) and (7.7b) it is now necessary to provide the distribution of temperature at the initial time

$$T(x, y, z, 0) = T_0(x, y, z) \tag{7.28}$$

Extending the method used to develop (7.13), a weak form† of the time dependent problem is given by

$$\int_\Omega \left[\delta T\, c \frac{\partial T}{\partial t} + (\nabla \delta T)^{\mathrm{T}} (\mathbf{k}\, \nabla T) + \delta T Q \right] \mathrm{d}\Omega + \int_{\Gamma_q} \delta T\, (\bar{q} + H(T - T_0))\, \mathrm{d}\Gamma = 0 \tag{7.29}$$

where we require $\delta T = 0$ and $T = \bar{T}$ on Γ_{T}.

7.4.1 Finite element discretizations

A finite element solution of (7.29) is constructed using an approximation of the type given in Sec. 3.5 where now we assume the separable form

$$T(x, y, z, y) \approx \hat{T}(x, y, z, t) = N_a(x, y, z)\, \tilde{T}_a(t) \tag{7.30}$$

With this form the spatial derivatives are associated with the shape functions N_a and the time derivative with the parameters $\tilde{T}_a$. Substituting (7.30) into (7.29) yields the *semi-discrete* set of ordinary differential equations

$$\mathbf{C} \frac{\mathrm{d}\tilde{\mathbf{T}}}{\mathrm{d}t} + \mathbf{H}\tilde{\mathbf{T}} + \mathbf{f} = \mathbf{0} \tag{7.31}$$

or for node a

$$C_{ab} \frac{\mathrm{d}\tilde{T}_b}{\mathrm{d}t} + H_{ab}\tilde{T}_b + f_a = 0$$

where H_{ab} and f_a are given by (7.20) and

$$C_{ab} = \int_\Omega N_a c N_b \,\mathrm{d}\Omega$$

In Chapters 16 and 17 we shall discuss in more detail methods of solution for large sets of equations of the form (7.31). Here, however, we consider a simple procedure in which the time dependence is given by a finite difference approximation. We will approximate the nodal temperatures at a time t_n by

$$\tilde{\mathbf{T}}(t_n) \approx \tilde{\mathbf{T}}_n$$

and the time derivative by

$$\left. \frac{\mathrm{d}\tilde{\mathbf{T}}}{\mathrm{d}t} \right|_{t=t_n} \approx \frac{1}{\Delta t} (\tilde{\mathbf{T}}_n - \tilde{\mathbf{T}}_{n-1})$$

where $\Delta t = t_n - t_{n-1}$. An approximate solution to the semi-discrete equations at each time t_n is obtained by solving the set of equation

$$\left[\frac{1}{\Delta t}\mathbf{C} + \mathbf{H} \right] \tilde{\mathbf{T}}_n = \frac{1}{\Delta t}\mathbf{C}\, \tilde{\mathbf{T}}_{n-1} - \mathbf{f} \tag{7.32}$$

If the initial condition is approximated as

$$T(\mathbf{x}, 0) \approx \mathbf{N}(\mathbf{x})\, \tilde{\mathbf{T}}(0) \quad \text{with} \quad \tilde{\mathbf{T}}(0) = \tilde{\mathbf{T}}_0$$

a solution for $\tilde{\mathbf{T}}_1$ is immediately available from (7.32) by solving a set of *algebraic equations*. For each subsequent time step the solution process is identical to the time independent

† Note that no variational principle of the type (7.14) exists.

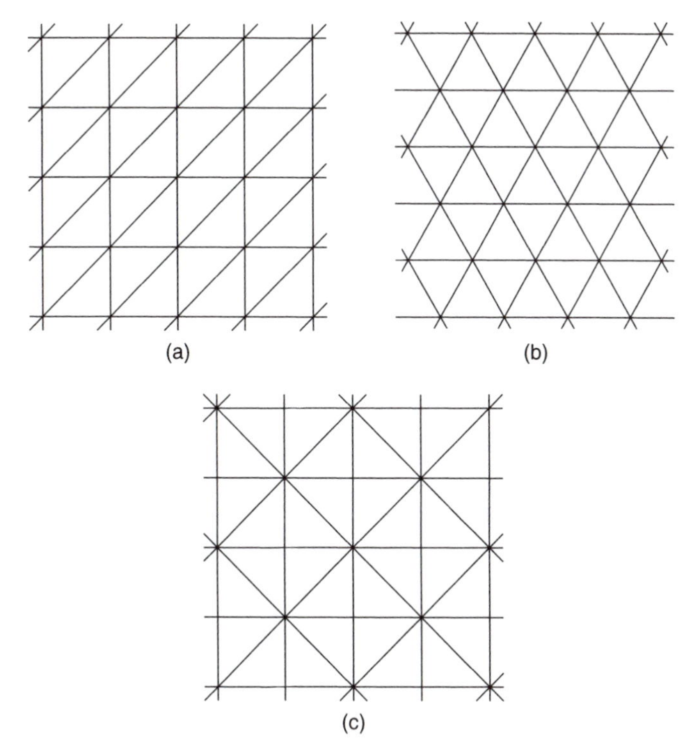

Fig. 7.3 'Regular' and 'irregular' subdivision patterns.

problem except for the modified force vector and a need to use a coefficient matrix which has a term inversely proportional on the size of the time increment.

7.5 Numerical examples – an assessment of accuracy

In Sec. 3.3, Example 3.6, we showed that by assembling explicitly worked-out 'stiffnesses' of triangular elements for the 'regular' mesh pattern shown in Fig. 7.3(a) the discretized equations are *identical* with those that are derived by well-known finite difference methods. The same result holds for the mesh pattern shown in Fig. 7.3(b).[7] For cases where all boundary conditions are given as prescribed values

$$\phi = \bar{\phi} \quad \text{on } \Gamma_\phi$$

the solutions obtained by the two methods obviously will be identical, and so also will be the orders of approximation.

However, if the mesh shown in Fig. 7.3(c) which is also based on a square arrangement of nodes but with 'irregular' element pattern is used a difference between the two approaches for the 'load' vector $\mathbf{f}^e$ will be evident. The assembled equations will have the same 'stiffness' matrix as in Fig. 7.3(a) but will show 'loads' which differ by small amounts from node to node, but the sum of which is still the same as that due to the finite difference expressions. The solutions therefore differ only locally and will represent the same averages.

Further advantages of the finite element process are:

1. It can deal simply with non-homogeneous and anisotropic situations (particularly when the direction of anisotropy is variable).
2. The elements can be graded in shape and size to follow arbitrary boundaries and to allow for regions of rapid variation of the function sought, thus controlling the errors in a most efficient way (viz. Chapters 13 and 14).
3. Specified gradient or 'radiation' boundary conditions are introduced naturally and with a better accuracy than in standard finite difference procedures.
4. Higher order elements can be readily used to improve accuracy without complicating boundary conditions – a difficulty always arising with finite difference approximations of a higher order.
5. Finally, but of considerable importance in the computer age, standard programs may be used for assembly and solution.

7.5.1 Torsion of prismatic bars

The torsion of prismatic elastic bars may be solved using a quasi-harmonic equation formulation. Here either a *warping function* or a *stress function* approach may be used. In Fig. 7.4(a) we show a rectangular bar loaded by an end torque M_t. The analysis is performed on the cross-section as shown in Fig. 7.4(b).

The use of a warping function is governed by the formulation in which displacements are given as

$$u = -yz\theta; \quad v = xz\theta \quad \text{and} \quad w = \psi(x, y)\,\theta \tag{7.33}$$

where x, y are coordinates in the cross-section and z is a coordinate of the bar axis; θ is the rate of twist and ψ the warping function. The non-zero strain components resulting from these displacements are given by

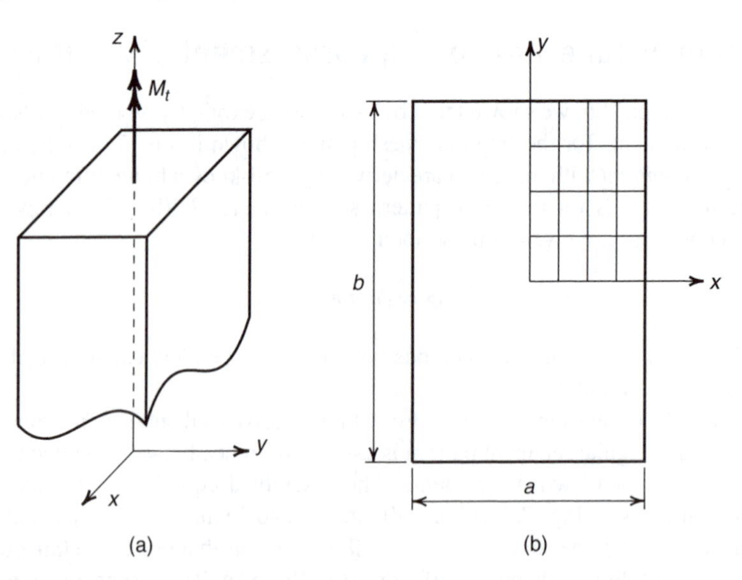

Fig. 7.4 Torsion of rectangular prismatic bar.

$$\gamma_{xz} = \theta \left(\frac{\partial \psi}{\partial x} - y \right) \quad \text{and} \quad \gamma_{yz} = \theta \left(\frac{\partial \psi}{\partial y} + x \right) \tag{7.34}$$

giving, for an isotropic elastic material, the stresses

$$\tau_{xz} = G\,\gamma_{xz} \quad \text{and} \quad \tau_{yz} = G\,\gamma_{yz} \tag{7.35}$$

Inserting the stresses into the equilibrium equation gives the governing differential equation

$$\frac{\partial}{\partial x}\left(G \frac{\partial \psi}{\partial x} \right) + \frac{\partial}{\partial y}\left(G \frac{\partial \psi}{\partial y} \right) = 0 \tag{7.36}$$

and for stress-free boundary conditions

$$\tau_{nz} = n_x\,\tau_{xz} + n_y\,\tau_{yz} = 0 \tag{7.37}$$

in which n_x and n_y are the direction cosines for the outward normal to the boundary of the rectangular section. At least one value of the warping function must be specified to have a unique solution.

The total torque acting on a cross-section is given by

$$\begin{aligned} M_t &= \int_A \left[-\tau_{xz}\, y + \tau_{yz}\, x \right] \mathrm{d}A \\ &= \int_A G \left[x^2 + y^2 - y \frac{\partial \psi}{\partial x} + x \frac{\partial \psi}{\partial y} \right] \mathrm{d}A\,\theta = \overline{GJ}_{\psi}\,\theta \end{aligned} \tag{7.38}$$

where $\overline{GJ}_{\psi}$ is the effective torsional stiffness.

A stress function formulation is deduced using the representation for stresses

$$\tau_{xz} = -\frac{\partial \phi}{\partial y} \quad \text{and} \quad \tau_{yz} = \frac{\partial \phi}{\partial x} \tag{7.39}$$

Combining (7.36) and (7.37) with (7.39) and eliminating the warping function ψ gives the differential equation

$$\frac{\partial}{\partial x}\left(\frac{1}{G} \frac{\partial \phi}{\partial x} \right) + \frac{\partial}{\partial y}\left(\frac{1}{G} \frac{\partial \phi}{\partial y} \right) + 2\theta = 0 \tag{7.40}$$

with

$$\phi(s) = \text{Constant} \quad \text{on } \Gamma_q \tag{7.41}$$

representing a stress-free boundary condition.

The total torque acting on a cross-section is now given by

$$M_t = \int_A G \left[x \frac{\partial \phi}{\partial x} + y \frac{\partial \phi}{\partial y} \right] \mathrm{d}A\,\theta = \overline{GJ}_{\phi}\,\theta \tag{7.42}$$

where $\overline{GJ}_{\phi}$ is the effective torsional stiffness.

The two solutions provide a bound on the torsional stiffness with the warping function solution giving an upper bound, $\overline{GJ}_{\psi}$, and the stress function a lower bound, $\overline{GJ}_{\phi}$.

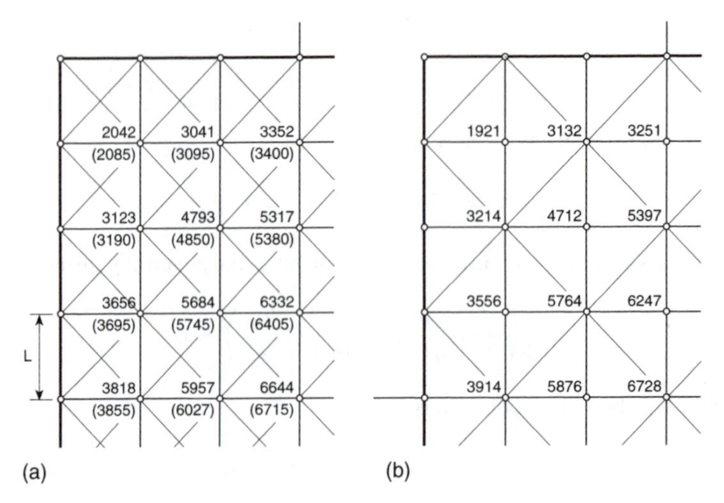

Fig. 7.5 Torsion of a rectangular shaft. Numbers in parentheses show a more accurate solution due to Southwell using a 12×16 mesh (values of $\phi/G\theta L^2$).

Example 7.4: Torsion of rectangular shaft. In Fig. 7.5 a test comparing the results obtained on an 'irregular' mesh of 3-node triangular elements with a relaxation solution of the lowest order finite difference approximation is shown. Both give results of similar accuracy, as indeed would be anticipated. In general superior accuracy is available with the finite element discretization. Furthermore, it is possible to get bounds on the torsional stiffness, as indicated above. To illustrate this latter aspect we consider a square bar which is solved using 4-node rectangular elements and a range of $n \times n$ meshes in which n is the number of spaces between nodes on each side. The results for the computed torsional stiffness values are plotted in Fig. 7.6.

The improvement in the rate of convergence for higher order elements may also be illustrated by comparing the total error using 4-node and 9-node elements of lagrangian type. A very accurate solution is computed from the series solution given in reference 8 and used to compute the error in the finite element solution (see Fig. 7.7).

Example 7.5: Torsion of hollow bimetallic shaft. The pure torsion of a non-homogeneous rectangular shaft with a circular hole is illustrated in Fig. 7.8. In the finite element solution presented, the hollow section is represented by a material for which G has a value of the order of 10^{-3} compared with the other materials.† The results compare well with the contours derived from an accurate finite difference solution.[9]

7.5.2 Transient heat conduction

Example 7.6: Transient heat conduction of a rectangular bar. In this example we consider the transient heat conduction in a long square prism with sides $L \times L$ and subjected to a rate of heat generation

†This was done to avoid difficulties due to the 'multiple connection' of the region and to permit the use of a standard program.

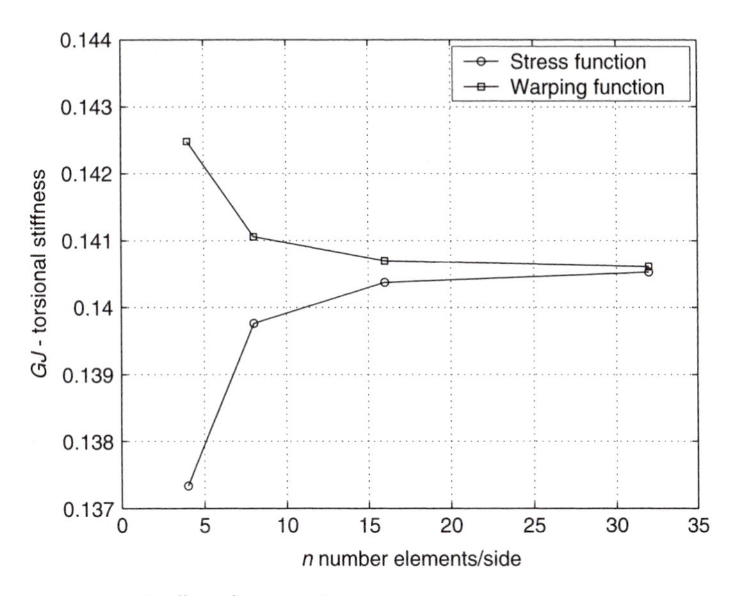

Fig. 7.6 Bound on torsional stiffness for square bar.

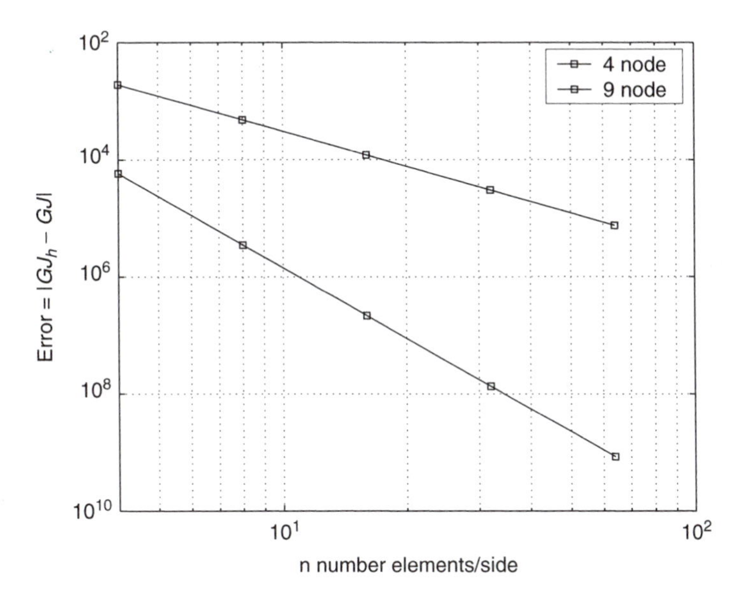

Fig. 7.7 Rate of convergence for square bar. 4- and 9-node lagrangian elements.

$$Q = Q_0 \, \mathrm{e}^{-\alpha t}$$

The problem is identical to the one considered in Sec. 3.5 where shape functions are assumed in a cosine form given by Eq. (3.57). Here, however, we use a standard finite element solution with 4-node square elements. The transient solution is performed using

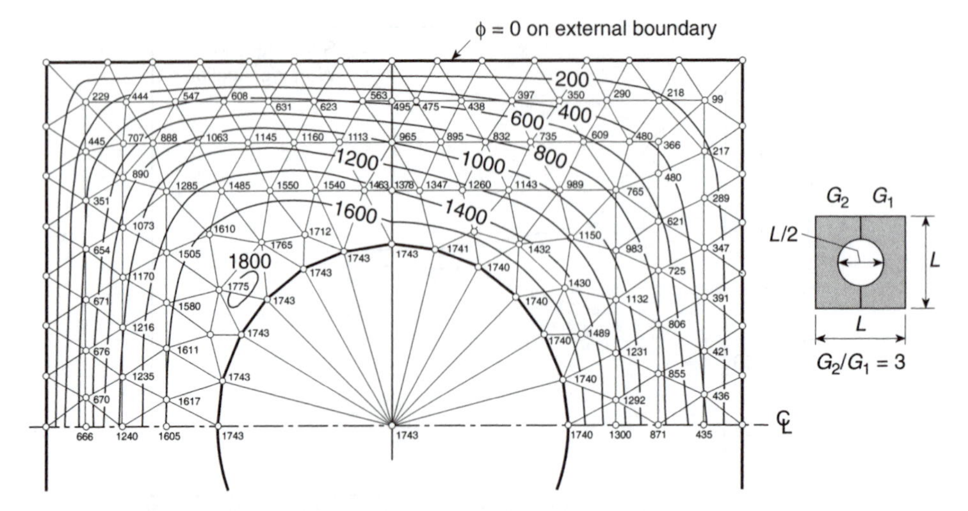

Fig. 7.8 Torsion of a hollow bimetallic shaft. $\phi/G\theta L^2 \times 10^4$.

the procedure given in Sec. 7.4.1. For the analysis we assume the following parameters:

$$L = c = Q_0 = \alpha = 1 \quad \text{and} \quad k = \frac{0.75}{\pi^2}$$

Using symmetry conditions, a mesh of 20 × 20 4-node elements is used to approximate one quadrant of the domain. A constant increment in time, $\Delta t = 0.01$, is used to perform the solution. Results for the temperature at the centre of the prism are given in Fig. 7.9 and compared to the series solutions computed in Sec. 3.5, Fig. 3.9.

Transient heat conduction of a rotor blade

In Fig. 7.10 we show some results for the transient temperature distribution in a turbine rotor blade. The blade is subjected to a hot gas at 1145C° applied to the outer boundary in which a variable radiation constant $H = \alpha$ is employed. Cooling is introduced in the internal ducts. The analysis is performed using cubic elements of serendipity type which permit the representation of the boundaries using very few elements.

7.5.3 Anisotropic seepage

The next problem is concerned with the flow through highly non-homogeneous, anisotropic, and contorted strata. The basic governing equation is

$$\frac{\partial}{\partial x'}\left(k_{x'x'}\frac{\partial H}{\partial x'}\right) + \frac{\partial}{\partial y'}\left(k_{y'y'}\frac{\partial H}{\partial y'}\right) = 0 \tag{7.43}$$

in which H is the hydraulic head and $k_{x'x'}$ and $k_{y'y'}$ represent the permeability coefficients in the direction of the (inclined) principal axes. However, a special feature has to be incorporated to allow for changes of x' and y' principal directions from element to element.

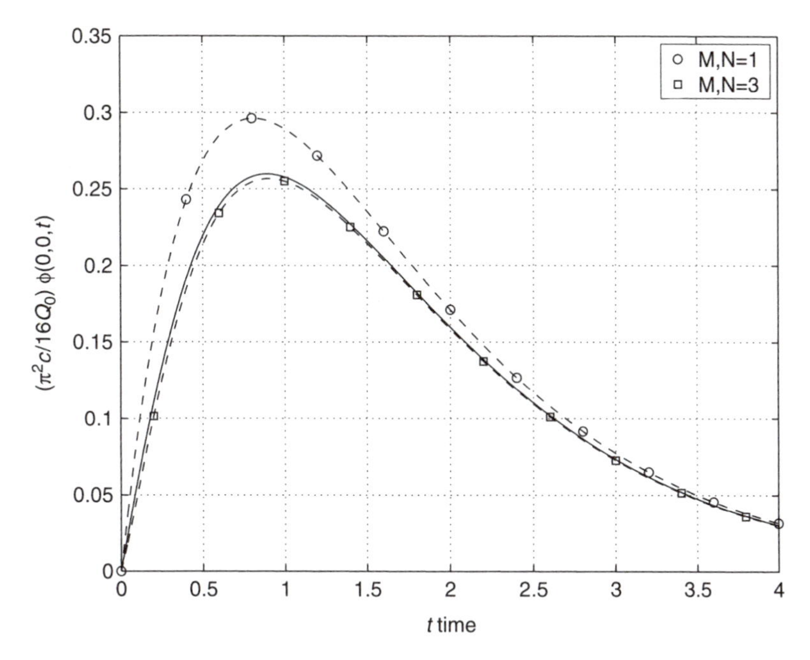

Fig. 7.9 Transient heat development in a square prism – plot of temperature at centre.

No difficulties are encountered in computation, and the problem together with its solution is given in Fig. 7.11.[3]

7.5.4 Electrostatic and magnetostatic problems

In this area of activity frequent need arises to determine appropriate field strengths and the governing equations are usually of the standard quasi-harmonic type discussed here. Thus the formulations are directly transferable. One of the first applications made as early as 1967[4] was to fully three-dimensional electrostatic field distributions governed by simple Laplace equations (Fig. 7.12).

In Fig. 7.13 a similar use of triangular elements was made in the context of magnetic two-dimensional fields by Winslow[6] in 1966. These early works stimulated considerable activity in this area and much additional work has been published.[11–14]

The magnetic problem is of particular interest as its formulation usually involves the introduction of a *vector potential* with three components which leads to a formulation different from those discussed in this chapter. It is, therefore, worthwhile introducing a variant which allows the standard programs of this section to be utilized for this problem.[15–17]

In electromagnetic theory for steady-state fields the problem is governed by Maxwell's equations which are

$$\begin{aligned} \nabla \times \mathbf{H} &= -\mathbf{J} \\ \mathbf{B} &= \mu \mathbf{H} \\ \nabla^{\mathrm{T}} \mathbf{B} &= 0 \end{aligned} \tag{7.44}$$

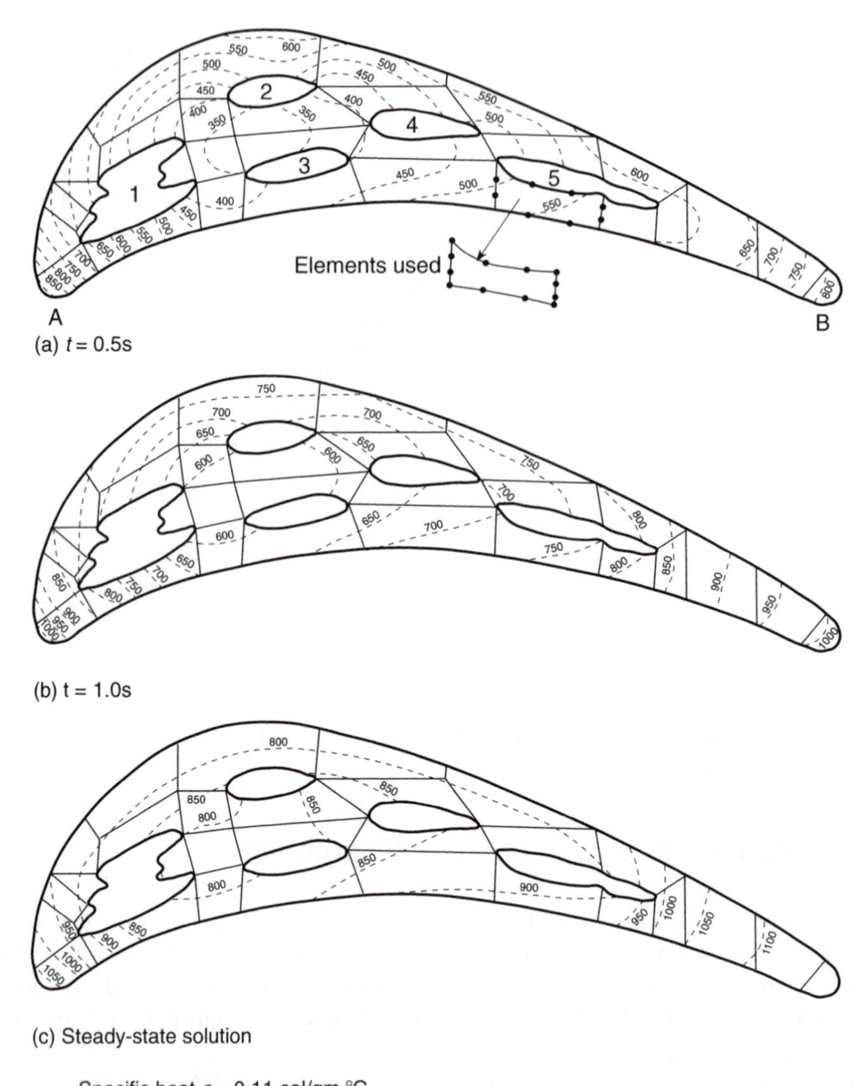

Specific heat $c = 0.11$ cal/gm °C

Density = 7.99 gm/cm3

Conductivity $k = 0.05$ cal/s cm °C

Gas temperature around blade = 1145 °C

Heat transfer coefficient α varies from 0.390 to 0.056 on the outside surfaces of the blade (A–B)

Hole number	Cooling hole temperature	α around perimeter of each hole
1	545 °C	0.0980
2	587 °C	0.0871

Fig. 7.10 Temperature distribution in a cooled rotor blade, initially at zero temperature.

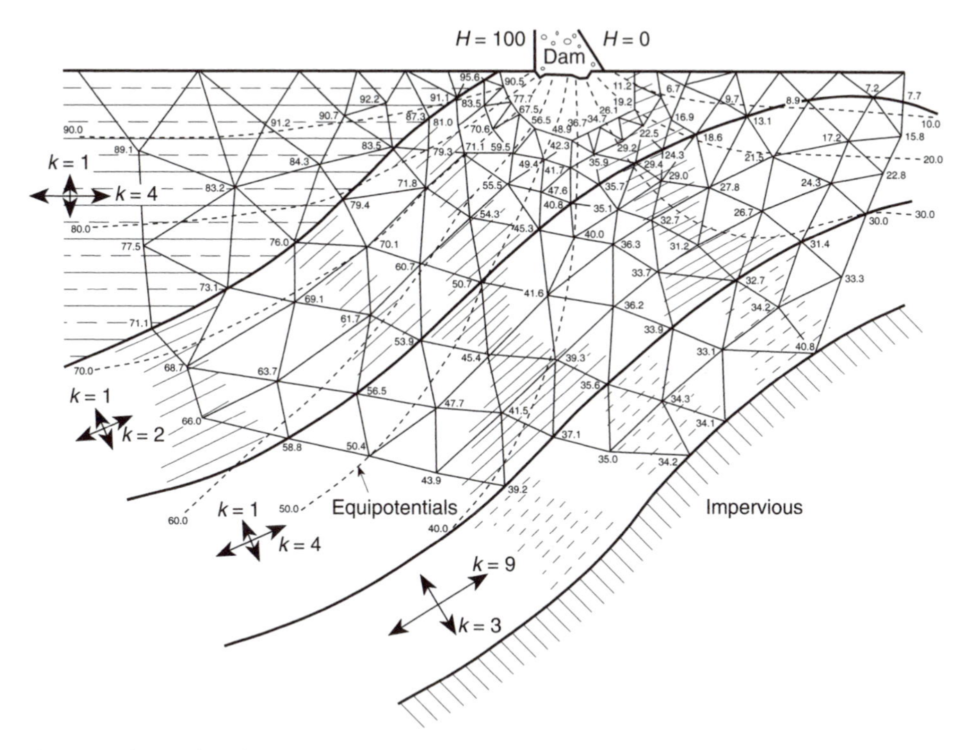

Fig. 7.11 Flow under a dam through a highly non-homogeneous and contorted foundation.

with the boundary condition specified at an infinite distance from the disturbance, requiring **H** and **B** to tend to zero there. In the above **J** is a prescribed electric current density confined to conductors, **H** and **B** are vector quantities with three components denoting the magnetic field strength and flux density respectively, μ is the magnetic permeability which varies (in an absolute set of units) from unity *in vacuo* to several thousand in magnetizing materials and $\times$ denotes the vector (cross) product, defined in Appendix F.

The formulation presented here depends on the fact that it is a relatively simple matter to determine the field $\mathbf{H}_s$ which exactly solves Eq. (7.44) when $\mu \equiv 1$ everywhere. This is given at any point defined by a vector coordinate $\mathbf{r}$ by an integral:

$$\mathbf{H}_s = \frac{1}{4\pi}\int_\Omega \frac{\mathbf{J}\times(\mathbf{r}-\mathbf{r}')}{|\mathbf{r}-\mathbf{r}'|^3}\,\mathrm{d}\Omega; \qquad |\mathbf{r}-\mathbf{r}'| = \sqrt{(\mathbf{r}-\mathbf{r}')^{\mathrm{T}}(\mathbf{r}-\mathbf{r}')} \tag{7.45}$$

In the above, $\mathbf{r}'$ refers to the coordinates of $\mathrm{d}\Omega$ and obviously the integration domain only involves the electric conductors where $\mathbf{J} \neq 0$.

With $\mathbf{H}_s$ known we can write

$$\mathbf{H} = \mathbf{H}_s + \mathbf{H}_m$$

and, on substitution into Eq. (7.44), we have a system

$$\begin{aligned} \boldsymbol{\nabla} \times \mathbf{H}_m &= \mathbf{0} \\ \mathbf{B} &= \mu(\mathbf{H}_s + \mathbf{H}_m) \\ \boldsymbol{\nabla}^{\mathrm{T}}\mathbf{B} &= 0 \end{aligned} \tag{7.46}$$

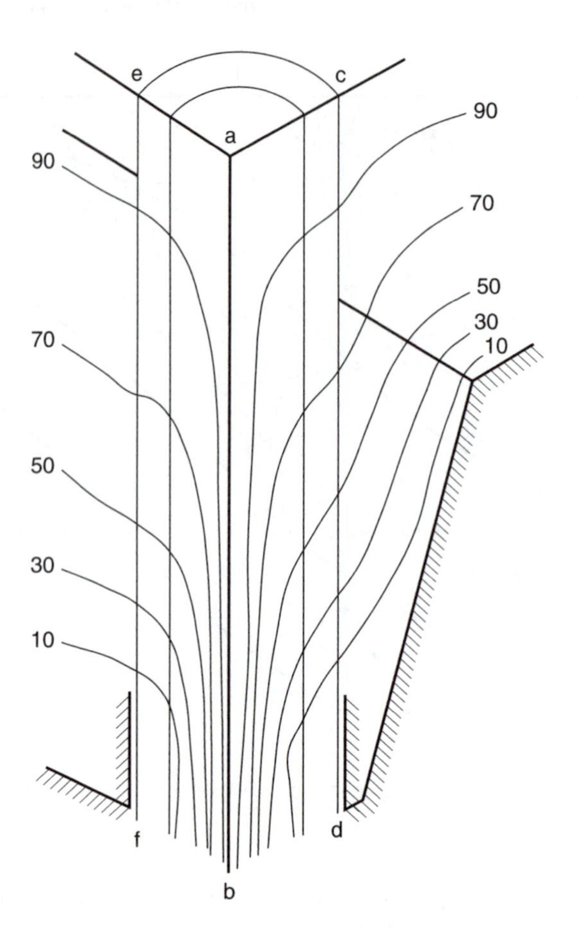

Fig. 7.12 A three-dimensional distribution of electrostatic potential around a porcelain insulator in an earthed trough.[10]

If we now introduce a *scalar* potential ϕ, defining $\mathbf{H}_m$ as

$$\mathbf{H}_m \equiv \boldsymbol{\nabla}\phi \tag{7.47}$$

we find the first of Eqs (7.46) to be automatically satisfied and, on eliminating $\mathbf{B}$ in the other two, the governing equation becomes

$$\boldsymbol{\nabla}^{\mathrm{T}}(\mu\boldsymbol{\nabla}\phi) + \boldsymbol{\nabla}^{\mathrm{T}}(\mu\mathbf{H}_s) = 0 \tag{7.48}$$

with $\phi \rightarrow 0$ at infinity. This is precisely of the standard form discussed in this chapter [Eq. (7.6)] with the second term, which is now specified, replacing Q.

An apparent difficulty exists, however, if μ has a discontinuity, as indeed we would expect it to do on the interfaces of two materials. Here the term Q is now undefined and, in the standard discretization of Eq. (7.18) or (7.19), the term (for node a)

$$\int_\Omega N_a\, Q\, \mathrm{d}\Omega \equiv -\int_\Omega N_a\, \boldsymbol{\nabla}^{\mathrm{T}}(\mu\mathbf{H}_s)\, \mathrm{d}\Omega \tag{7.49}$$

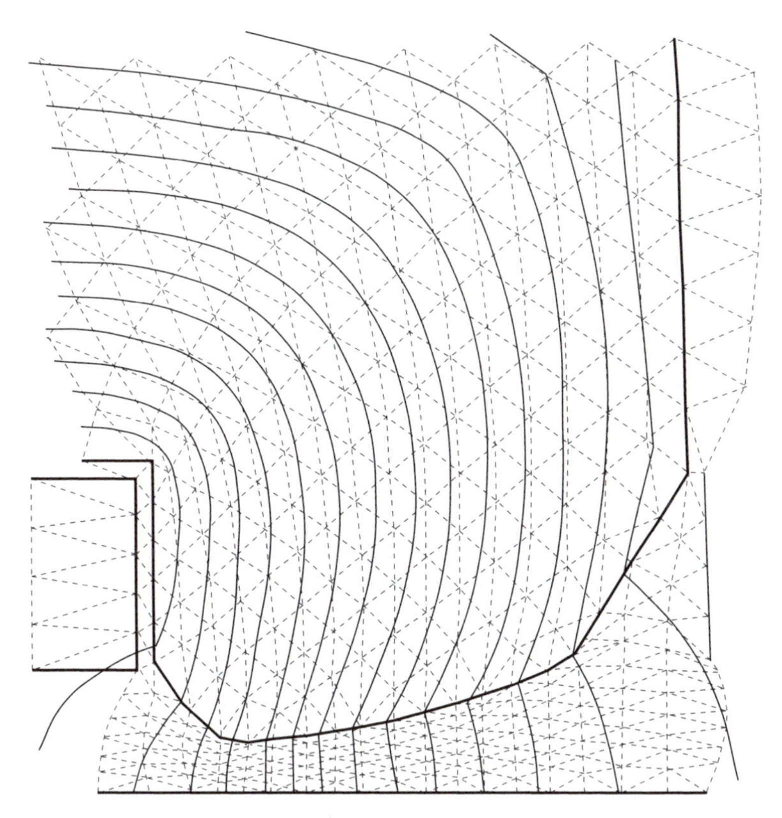

Fig. 7.13 Field near a magnet (after Winslow[6]).

apparently has no meaning. Integration by parts comes once again to the rescue and we note that

$$\int_{\Omega} N_a \boldsymbol{\nabla}^{\mathrm{T}} (\mu \mathbf{H}_s)\, \mathrm{d}\Omega \equiv -\int_{\Omega} (\boldsymbol{\nabla} N_a)^{\mathrm{T}} \mu \mathbf{H}_s\, \mathrm{d}\Omega + \int_{\Gamma} N_a \mathbf{n}^{\mathrm{T}} (\mu \mathbf{H}_s)\, \mathrm{d}\Gamma \tag{7.50}$$

In subregions of constant μ, $\boldsymbol{\nabla}^{\mathrm{T}}\mathbf{H}_s \equiv 0$, the only contribution to the forcing terms comes as a line integral of the second term at discontinuity interfaces.

Introduction of the scalar potential makes both two- and three-dimensional magnetostatic problems solvable by a standard program used for all the problems in this chapter. Figure 7.14 shows a typical three-dimensional solution for a transformer. Here isoparametric quadratic brick elements of the type which were described in Chapter 5 were used.[15]

In typical magnetostatic problems a high non-linearity exists with

$$\mu = \mu(|\mathbf{H}|) \qquad \text{where} \quad |\mathbf{H}| = \sqrt{H_x^2 + H_y^2 + H_z^2} \tag{7.51}$$

The treatment of such non-linearities is outside the scope of this volume; however, the solution of such problems generally uses an iterative approach in which a sequence of linearized problems is solved.[18]

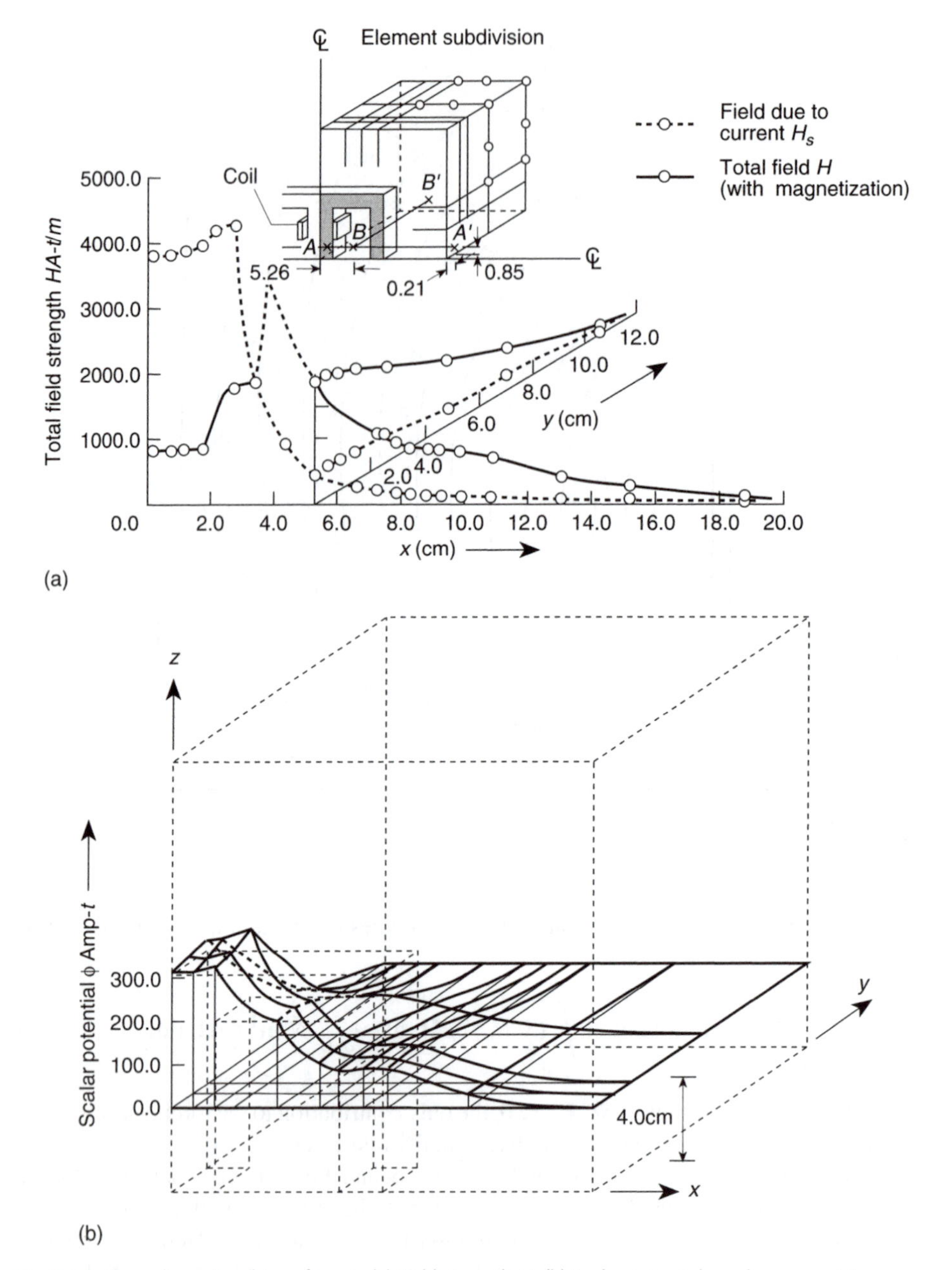

Fig. 7.14 Three-dimensional transformer. (a) Field strength H. (b) Scalar potential on plane $z = 4.0$ cm.

Considerable economy in this and other problems of infinite extent can be achieved by the use of the *infinite* elements discussed in Chapter 5.

Many examples of practical applications of computing magnetic and electric field solutions have been given by Binns *et al.* and some are included in his recent book.[19] Plate 3 given in the front of this book presents one such example.

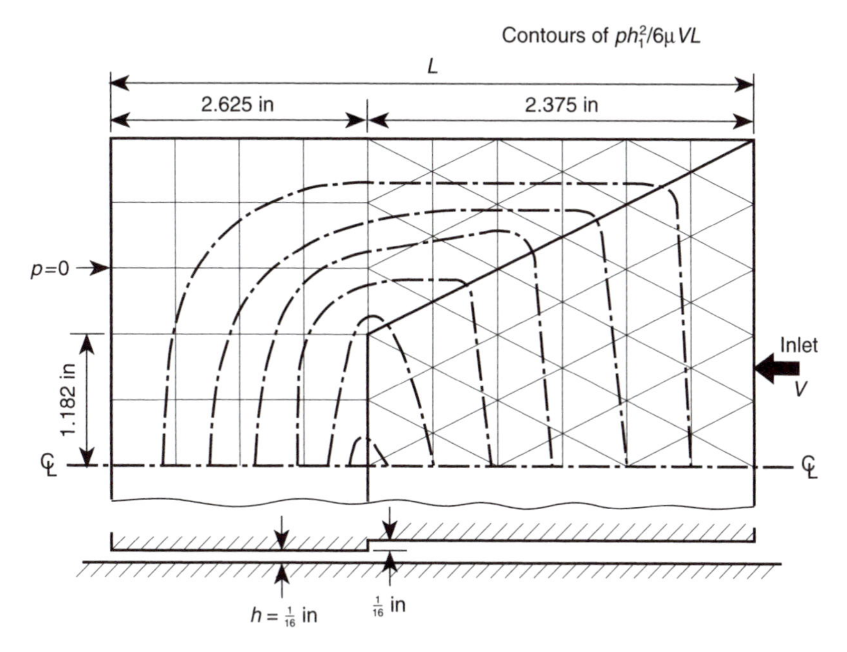

Fig. 7.15 A stepped pad bearing. Pressure distribution.

7.5.5 Lubrication problems

Once again a standard Poisson type of equation is encountered in the two-dimensional domain of a bearing pad. In the simplest case of constant lubricant density and viscosity the equation to be solved is the Reynolds equation

$$\frac{\partial}{\partial x}\left(h^3\frac{\partial p}{\partial x}\right)+\frac{\partial}{\partial y}\left(h^3\frac{\partial p}{\partial y}\right)=6\mu V\frac{\partial h}{\partial x} \tag{7.52}$$

where h is the film thickness, p the pressure developed, μ the viscosity and V the velocity of the pad in the x direction.

Figure 7.15 shows the pressure distribution in a typical finite width stepped pad.[20] The boundary condition is simply that of zero pressure and it is of interest to note that the step causes an equivalent of a 'line load' on integration by parts of the right-hand side of Eq. (7.52), just as in the case of magnetic discontinuity mentioned above.

More general cases of lubrication problems, including vertical pad movements (squeeze films) and compressibility, can obviously be dealt with, and much work has been done here.[21–28]

7.5.6 Irrotational and free surface flows

The basic Laplace equation which governs the flow of viscous fluid in seepage problems is also applicable in the problem of irrotational fluid flow outside the boundary layer created by viscous effects. The seepage example given above is adequate to illustrate the general applicability in this context. Further examples for this class of problems are cited by Martin[29] and others.[30–36]

If no viscous effects exist, then it can be shown that for a fluid starting at rest the motion must be irrotational, i.e.,

$$\omega_z \equiv \frac{\partial u}{\partial y} - \frac{\partial v}{\partial x} = 0 \tag{7.53}$$

where u and v are appropriate velocity components.

This implies the existence of a velocity potential, giving

$$u = -\frac{\partial \phi}{\partial x} \qquad v = -\frac{\partial \phi}{\partial y} \tag{7.54a}$$

or

$$\mathbf{u} = -\boldsymbol{\nabla}\phi \tag{7.54b}$$

If, further, the flow is incompressible the continuity equation [which is similar to Eq. (7.4)] has to be satisfied, i.e.,

$$\boldsymbol{\nabla}^{\mathrm{T}}\mathbf{u} = 0 \tag{7.55}$$

and therefore

$$\boldsymbol{\nabla}^{\mathrm{T}}(\boldsymbol{\nabla}\phi) = \nabla^2\phi = 0 \tag{7.56}$$

Alternatively, for two-dimensional flow a stream function may be introduced defining the velocities as

$$u = -\frac{\partial \psi}{\partial y} \qquad v = \frac{\partial \psi}{\partial x} \tag{7.57}$$

and this identically satisfies the continuity equation. The irrotationality condition must now ensure that

$$\boldsymbol{\nabla}^{\mathrm{T}}(\boldsymbol{\nabla}\psi) = \nabla^2\psi = 0 \tag{7.58}$$

and thus problems of ideal fluid flow can be posed in either form. As the standard formulation is again applicable, there is little more that needs to be added, and for examples the reader can well consult the literature cited. We also discuss this problem in more detail in reference 37.

The similarity with problems of seepage flow, which has already been discussed, is obvious.[38, 39]

A particular class of fluid flow deserves mention. This is the case when a free surface limits the extent of the flow and this surface is not known *a priori*.

The class of problem is typified by two examples – that of a freely overflowing jet [Fig. 7.16(a)] and that of flow through an earth dam [Fig. 7.16(b)]. In both, the free surface represents a streamline and in both the position of the free surface is unknown *a priori* but has to be determined so that an *additional condition* on this surface is satisfied. For instance, in the second problem, if formulated in terms of the potential for the hydraulic head H, Eq. (7.43) governs the problem.

The free surface, being a streamline, imposes the condition

$$\frac{\partial H}{\partial n} = 0 \tag{7.59}$$

be satisfied there. In addition, however, the pressure must be zero on the surface as this is exposed to atmosphere. As

$$H = \frac{p}{\gamma} + y \tag{7.60}$$

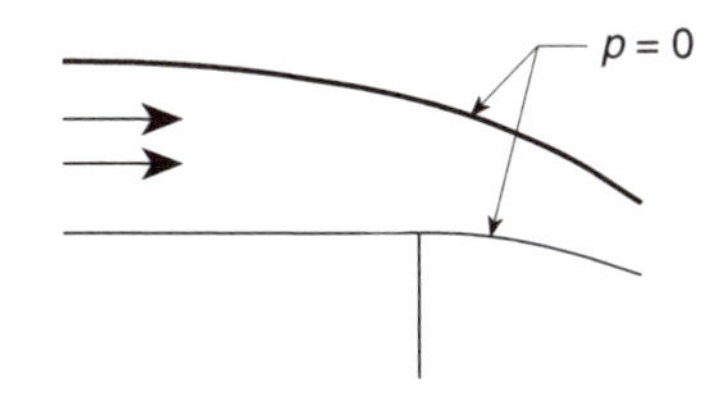

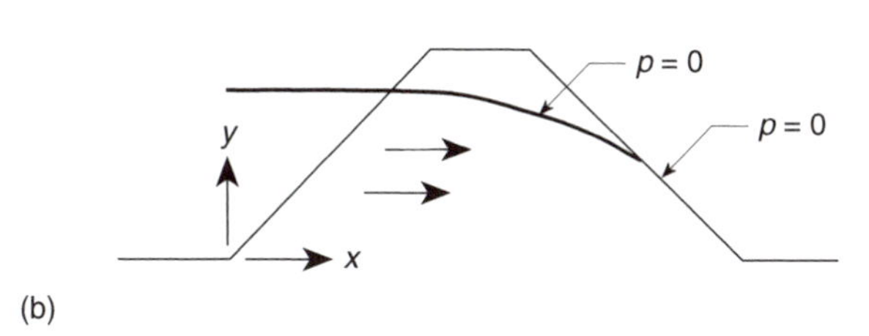

Fig. 7.16 Typical free surface problems with a streamline also satisfying an additional condition of pressure = 0. (a) Jet overflow. (b) Seepage through an earth dam.

where γ is the fluid specific weight, p is the fluid pressure, and y the elevation above some (horizontal) datum, we must have on the surface

$$H = y \tag{7.61}$$

The solution may be approached iteratively. Starting with a prescribed free surface streamline the standard problem is solved. A check is carried out to see if Eq. (7.61) is satisfied and, if not, an adjustment of the surface is carried out to make the new y equal to the H just found. A few iterations of this kind show that convergence is reasonably rapid. Taylor and Brown[40] show such a process. Alternative methods including special variational principles for dealing with this problem have been devised over the years and interested readers can consult references 41–49.

7.6 Concluding remarks

We have shown how a general formulation for the solution of a steady-state quasi-harmonic problem can be written, and how a single program of such a form can be applied to a wide variety of physical situations. Indeed, the selection of problems dealt with here is by no means exhaustive and many other examples of application are of practical interest. Readers will doubtless find appropriate analogies for their own problems.

7.7 Problems

7.1 The anisotropic properties for $\mathbf{k}$ are $k_{x'} = 0.4$, $k_{y'} = 2.1$ and $k_{z'} = 1.0$. The axes are oriented as shown in Fig. 7.17. For $\theta = 30^\circ$ compute the terms in the matrix $\mathbf{k}$ (e.g., k_{xx}, k_{xy}, etc.) with respect to the axes x, y, z.

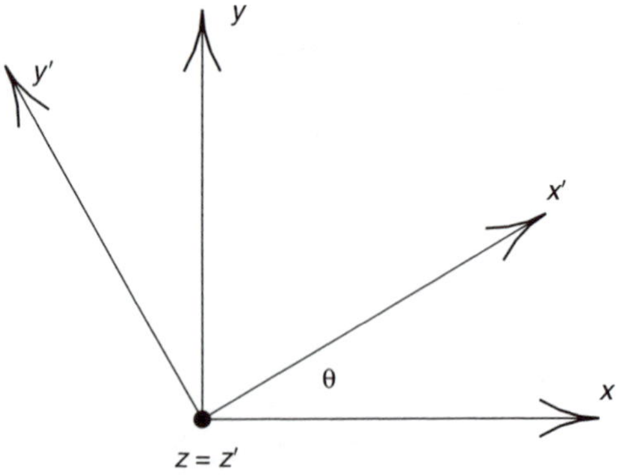

Fig. 7.17 Orientation of axes for Problem 7.1.

7.2 A two-dimensional heat equation has its surface located in the x–y plane. The problem is allowed to convect heat from the *surface* of the surrounding region according to

$$Q(x, y) = -\beta[\phi(x, y) - \phi_0]$$

where β is a convection parameter and ϕ_0 the temperature of the surrounding medium.

Construct a weak form for the problem by modifying Eq. (7.13).

For a finite element approximation to ϕ and $\delta\phi$ deduce the form of the matrices which result from the modified weak form.

7.3 For the quasi-harmonic equation consider a square 8-node serendipity element with unit side lengths in the x and y directions. Using *FEAPpv* (or any other available program) determine the rank of the element matrix $\mathbf{H}$ for the case where $\mathbf{k} = \mathbf{I}$ (i.e., isotropic with $k = 1$) and $H = 0$ using 1×1 Gaussian quadrature. Repeat the calculation using 2×2, 3×3 and 4×4 quadrature.

(a) What is the lowest order quadrature that gives a matrix $\mathbf{H}$ with full rank?

(b) What is the lowest order quadrature that evaluates the matrix $\mathbf{H}$ exactly?

[Hint: The rank of $\mathbf{H}$ may be determined from the eigenproblem given by:

$$\mathbf{H}\,\mathbf{v}_i = \lambda_i \mathbf{v}_i \quad \text{with} \quad \mathbf{v}_i^T \mathbf{v}_j = \delta_{ij}$$

where δ_{ij} is the Kronnecker delta. The rank of $\mathbf{H}$ is the number of non-zero eigenvalues λ_i (a zero is any value below the round-off limit).]

7.4 Solve Problem 7.3 for a 9-node lagrangian element.

Using the eigenvector for the zero eigenvalue of the fully integrated element array $\mathbf{H}$ determine and sketch the shape of eigenvectors from any additional (spurious) zero eigenvalues. (Note: The fully integrated element has one zero eigenvalue, λ_0.)

(Hint: For the case where two zero eigenvectors $\mathbf{v}_1$ and $\mathbf{v}_2$ exist they may be expressed in terms of $\mathbf{v}_0$ and another orthogonal unit vector $\mathbf{w}_0$ as:

$$\mathbf{v}_0 = \alpha_1 \mathbf{v}_1 + \alpha_2 \mathbf{v}_2 \quad \text{where } \alpha_i = \mathbf{v}_0^T \mathbf{v}_i$$
$$\mathbf{w}_0 = \alpha_2 \mathbf{v}_1 - \alpha_1 \mathbf{v}_2$$

number of non-zero eigenvalues λ_i (a zero being a value below the computer round-off). The vectors $\mathbf{v}_0$, $\mathbf{w}_0$ and the vectors $\mathbf{v}_1$, $\mathbf{w}_2$ are both eigenvectors of the same subspace.

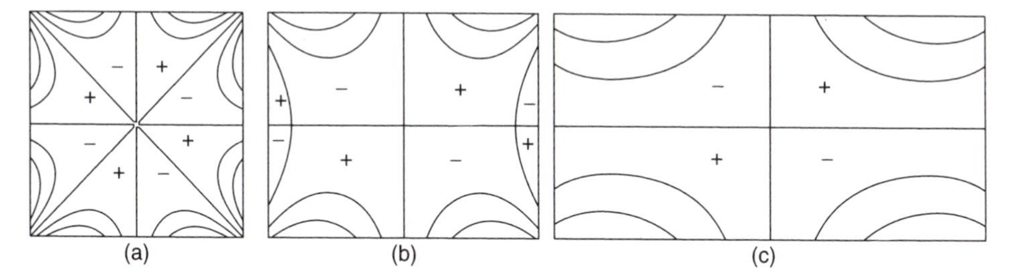

Fig. 7.18 Warping function for torsion of rectangular bar. Problem 7.5.

7.5 Consider the torsion of a rectangular bar by the warping function formulation discussed in Sec. 7.5.1. Let a and b be the side lengths in the x and y directions, respectively. For a homogeneous section with shear modulus G the warping function has the behaviour shown in Fig. 7.18 for a/b ratios of 1, 1.25 and 2. Note that the behaviour transitions from eight to four regions of $\pm$ variation. Estimate the a/b ratio where this transition just occurs.

To make your estimate use *FEAPpv* (or any other available program) with a fine mesh of quadratic lagrangian elements. Set the boundary conditions to make the warping function zero along the x and y axes. The transition will occur at the smallest a/b for which all the values on the perimeter of one quadrant of the cross-section have the same sign or are 'numerically' zero.

7.6 A cross-section of a long prismatic section is shown in Fig. 7.19 and subjected to constant uniform temperatures 370 C^o on the left boundary and 66 C^o on the right boundary. The top and bottom edges are assumed to be insulated so that $q_n = 0$.

The cross-section is a composite of fir (A), concrete (B), glass wool (C) and yellow pine (D). The thermal conductivity for each of the parts is: $k_A = 0.11$, $k_B = 0.78$, $k_C = 0.04$ and $k_D = 0.147$ in consistent units for the geometry of the section shown.

(a) Estimate the heat flow through the cross-section assuming $q_y = 0$ and q_x constant

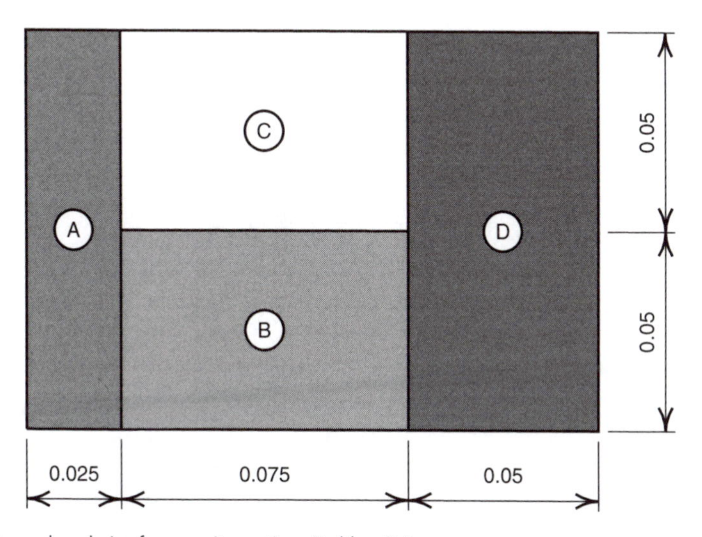

Fig. 7.19 Thermal analysis of composite section. Problem 7.6.

in each part. Let the temperatures at each junction be $T(0) = 370$, $T(0.025) = T_1$, $T(0.10) = T_2$ and $T(0.15) = 66$. (Hint: Assume T is a function of x only.)

(b) Use *FEAPpv* (or any other available program) to compute a finite element solution using 4-node and 9-node quadrilateral elements. First perform a solution on a coarse mesh and use this to design a mesh using a finer discretization. Let your final 4-node element mesh have nodal locations which coincide with those used for the corresponding 9-node element mesh.

Plot a distribution of heat flow q_n across each of the internal boundaries.

7.7 The cross-section of two tubular sections is shown in Fig. 7.20. The parts are to be assembled by heating the outer part until it just passes over the inner part as shown in the figure. Let $r_i = 10$ cm, $t = 5$ cm and $h = 10$ cm and take elastic properties as $E = 200$ Gpa, $\nu = 0.3$ and $\alpha = 12 \times 10^{-6}$ per C°. The parts are stress free at room temperature 20C°. The parts just fit when the outer bar is heated to 220C° (while the inner part is maintained at room temperature).

(a) What is the correct inner radius of the outer part at room temperature?

(b) Solve the problem using *FEAPpv* (or any other available program). Use a mesh of 4-node quadrilateral elements to compute the final solution for the assembled part at room temperature assuming complete contact at the mating surface and no slip during cooling. Plot the radial displacement at the interaction surface.

(c) Compute an estimate of the traction components at the interaction surface. Do you think there will be slip? Why?

7.8 Company X&Y plans to produce a rectangular block which needs to be processed by a thermal quench in a medium which is 100°C above room temperature. The block shown in Fig. 7.21(a) has $a = 10$ and $b = 20$ (i.e., the block is $10 \times 10 \times 20$). It has been determined that the thermal properties of the block may be specified by an isotropic Fourier model in which $k = 1$ and $c = 1$. The surface convection constant H is 0.05.

The quench must be maintained until the minimum temperature in the block reaches 99°C above room temperature. Use *FEAPpv* (or any other available program) to perform a transient analysis to estimate the required quench time.

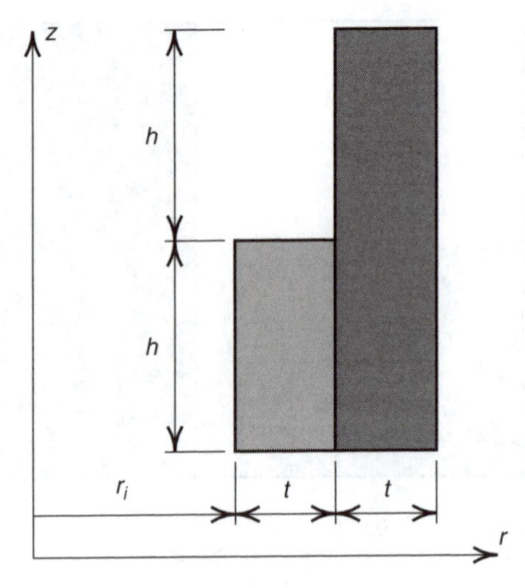

Fig. 7.20 Thermal assembly of tubular sections. Problem 7.7.

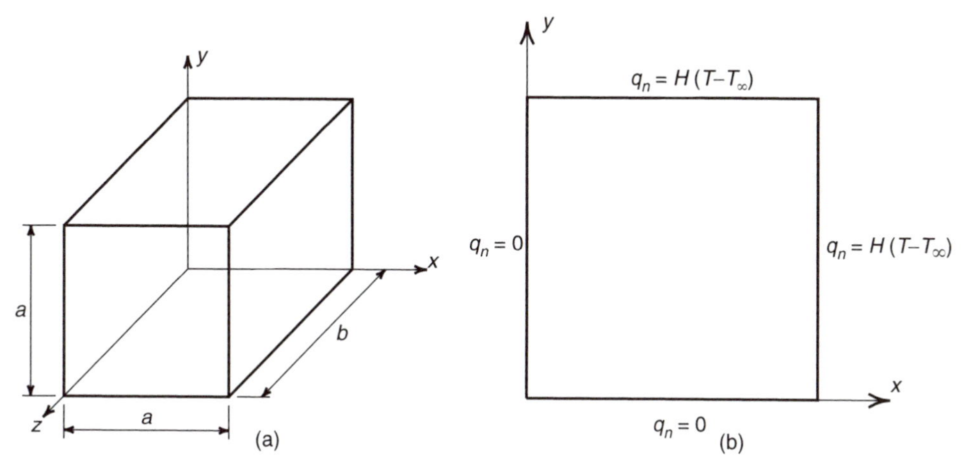

Fig. 7.21 Thermal quench in 2 and 3 dimensions. Problem 7.8.

(a) First perform a two-dimensional plane analysis on a 10×10 cross-section using a uniform mesh of 4-node quadrilateral elements. Use symmetry to reduce the size of the domain analysed. The surface convection will be modelled by 2-node line elements along the outer perimeter. The analysis region is shown in Fig. 7.21(b) with the boundary conditions to be imposed. Locate the node where the minimum temperature occurs and plot the behaviour vs time (a good option is to use MATLAB to perform plots).

Estimate the duration of time needed for the minimum temperature to reach the desired value. (Hint: One approach to selecting time increments is to select a very small value, e.g. $\Delta t = 10^{-8}$ and perform 10 steps of the solution. Multiply the time increment by 10 and perform 9 more steps. Repeat the multiplication until the desired time is reached.)

(b) Using the time duration estimated in (a) perform a three-dimensional analysis using a uniform mesh of 8-node hexahedral elements. Use symmetry to reduce the size of the region analysed. (Note: The convection condition applies to all outer surfaces.)

Estimate the duration of quench time needed for the minimum temperature to reach the desired value.

(c) What analyses would you perform if the block was $10 \times 10 \times 5$?

(d) Comment on use of a two-dimensional solution to estimate the required quench times for other shaped parts.

7.9 The distribution of shear stresses on the cross-section of a cantilever beam shown in Fig. 7.22(a) may be determined by solving the quasi-harmonic equation[50]

$$\frac{\partial^2 \phi}{\partial x^2} + \frac{\partial^2 \phi}{\partial y^2} = 0$$

with boundary condition

$$\phi = \frac{P}{2I}\left[\int y^2 \, \mathrm{d}x - \frac{\nu}{3(1+\nu)}\, y^3\right]$$

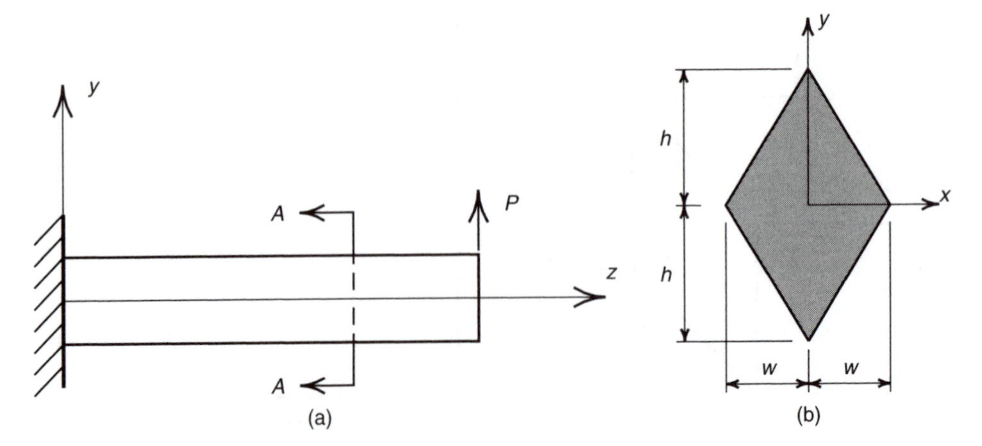

Fig. 7.22 End loaded cantilever beam. Problem 7.9.

where P is the end load, I is the moment of inertia of the cross-section, ν is the Poisson ratio of an isotropic elastic material and ϕ is a stress function. The shear stresses are determined from

$$\tau_{xz} = -\frac{\partial \phi}{\partial y} \quad \text{and} \quad \tau_{yz} = \frac{\partial \phi}{\partial x} + \frac{P}{2I}\left[\frac{\nu}{1+\nu}x^2 - y^2\right]$$

See reference 50 for details on the formulation.

(a) Show that the stress function satisfies the equilibrium equation when the bending stress is computed from

$$\sigma_z = -\frac{P\,(L-z)\,y}{I}$$

and $\sigma_x = \sigma_y = \tau_{xy} = 0$. L is the length of the beam.

(b) Develop a weak form for the problem in terms of the stress function ϕ.

(c) For a finite element formulation develop the relation to compute the boundary condition for the case when either 3-node triangular or 4-node quadrilateral elements are used.

(d) Write a program to determine the boundary values for the cross-section shown in Fig. 7.22(b). Let $w = 2$ and $h = 3$. Use the quasi-harmonic thermal element in *FEAPpv* (or any other available program) to solve for the stress function ϕ. Plot the distribution for ϕ on the cross section.

(e) Modify the expressions in *FEAPpv* (or any other program for which source code is available) to compute the stress distribution on the cross-section. Solve and plot their distribution. Compare your results to those computed from the classical strength of materials approach.

(Hint: Normalize your solution by the factor $P/2I$ to simplify expressions.)

7.10 A long sheet pile is placed in soil as shown in Fig. 7.23. The anisotropic properties of the soil are oriented so that $x = x'$, and $y = y'$. The governing differential equation is given in Sec. 7.5.3. The soil has the properties $k_x = 2$ and $k_y = 3$. Use *FEAPpv* (or any other available program) to determine the distribution of head and the flow in the region shown. Solve the problem using a mesh of 4-node, 8-node, and 9-node

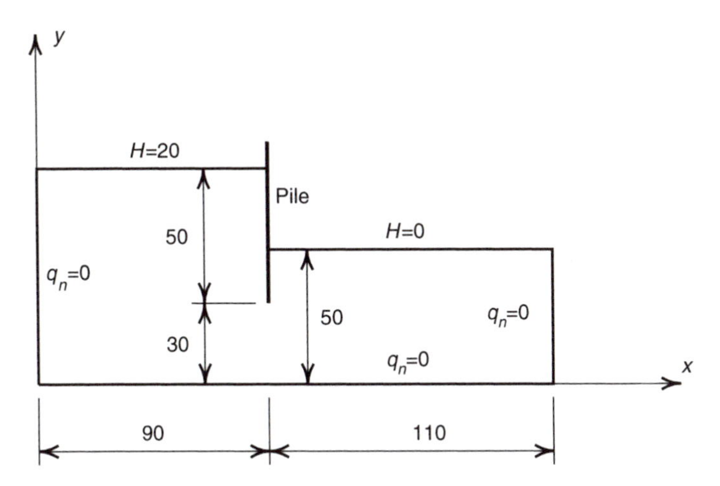

Fig. 7.23 Seepage under a sheet pile. Problem 7.10.

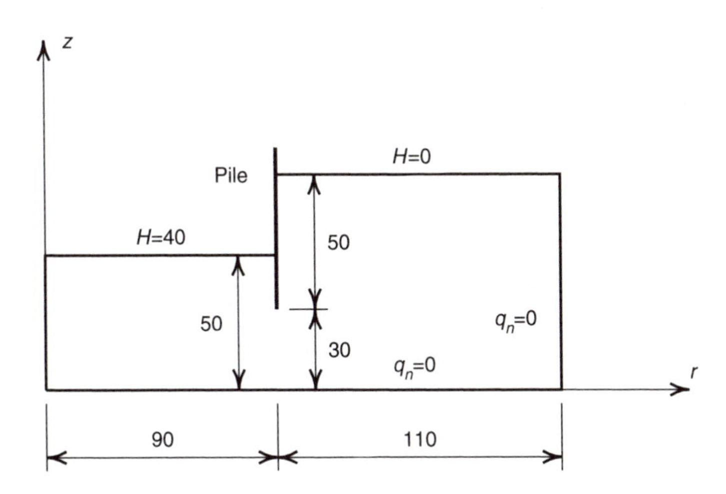

Fig. 7.24 Seepage under an axisymmetric sheet pile. Problem 7.11.

quadrilateral elements. Model the problem so that there are about four times as many 4-node elements as used for the 8- and 9-node models (and thus approximately an equal number of nodes for each model). Compare total flow obtained from each analysis.

7.11 An axisymmetric sheet pile is placed in soil as shown in Fig. 7.24. The anisotropic properties of the soil are oriented so that $r = r'$ and $z = z'$. The governing equation for plane flow is given in Sec. 7.5.3. Deduce the Euler differential equation for the axisymmetric problem from the weak form given in Secs 7.2.3 and 7.3.2 suitably modified for the seepage problem.

Assuming isotropic properties with $k = 3$, use *FEAPpv* (or any other available program) to determine the distribution of head and the flow in the region shown. Solve the problem using a mesh of 4-node, 8-node, and 9-node quadrilateral elements. Model the problem so that there are about four times as many 4-node elements as used for the 8- and 9-node models (and thus approximately an equal number of nodes for each model). Compare total flow obtained from each analysis.

7.12 A membrane occupies a region in the x–y plane and is stretched by a uniform tension T. When subjected to a transient load $q(x, y, t)$ acting normal to the surface the governing differential equation is given by

$$-T\left[\frac{\partial^2 u}{\partial x^2} + \frac{\partial^2 u}{\partial y^2}\right] + m\,\frac{\partial^2 u}{\partial t^2} = q(x, y, t)$$

(a) Construct a weak form for the differential equation for the case when boundary conditions are given by $u(s, t) = 0$ for s on Γ.

(b) Show that the solution by a finite element method may be constructed using C^0 functions.

(c) Approximate the u and δu by C^0 shape functions $N_a(x, y)$ and determine the semi-discrete form of the equations.

(d) For the case of steady harmonic motion, u may be replaced by

$$u(x, y, t) = w(x, y)\,\exp i\omega t$$

where $i = \sqrt{-1}$ and ω is the frequency of excitation.

Using this approximation, deduce the governing equation for w. Construct a weak form for this equation. Using C^0 approximations for w determine the form of the discretized problem.

7.13 Program development project: Modify the program developed for solution of linear elasticity problems to solve problems described by the quasi-harmonic equation for heat conduction. Include capability to solve both plane and axisymmetric geometry.

Specify the material properties by anisotropic values k'_x, k'_y and β (where β is the angle x' makes with the x axis).

Use your program to solve the problem described in Problem 7.6. Plot contours for temperature and heat flows q_x and q_y.

7.14 Program development project: Extend the program developed in Problem 7.13 to solve transient problems.

Include an input module to specify the initial temperatures.

Also add a capability to consider time-dependent source terms for Q.

Test your program by solving the problem described in Example 7.6 of Sec. 7.5.2.

7.15 Program development project: Extend the program developed in Problem 2.17 to compute nodal values of fluxes from the quasi-harmonic equation. Follow the procedure given in Sec. 6.4 to project element values to nodes.

Test your program using (a) a patch test of your design and (b) the problem described in Example 6.6.

References

1. O.C. Zienkiewicz and Y.K. Cheung. Finite elements in the solution of field problems. *The Engineer*, pages 507–510, Sept. 1965.
2. W. Visser. A finite element method for the determination of non-stationary temperature distribution and thermal deformation. In *Proc. 1st Conf. Matrix Methods in Structural Mechanics*, volume AFFDL-TR-66-80, Wright Patterson Air Force Base, Ohio, Oct. 1966.
3. O.C. Zienkiewicz, P. Mayer, and Y.K. Cheung. Solution of anisotropic seepage problems by finite elements. *J. Eng. Mech., ASCE*, 92(EM1):111–120, 1966.
4. O.C. Zienkiewicz, P.L. Arlett, and A.K. Bahrani. Solution of three-dimensional field problems by the finite element method. *The Engineer*, Oct. 1967.
5. L.R. Herrmann. Elastic torsion analysis of irregular shapes. *J. Eng. Mech., ASCE*, 91(EM6):11–19, 1965.
6. A.M. Winslow. Numerical solution of the quasi-linear Poisson equation in a non-uniform triangle 'mesh'. *J. Comp. Phys.*, 1:149–172, 1966.
7. D.N. de G. Allen. *Relaxation Methods*. McGraw-Hill, London, 1955.
8. S.P. Timoshenko and J.N. Goodier. *Theory of Elasticity*. McGraw-Hill, New York, 3rd edition, 1969.
9. J.F. Ely and O.C. Zienkiewicz. Torsion of compound bars – a relaxation solution. *Int. J. Mech. Sci.*, 1:356–365, 1960.
10. O.C. Zienkiewicz and Y.K. Cheung. *The Finite Element Method in Structural Mechanics*. McGraw-Hill, London, 1967.
11. P. Silvester and M.V.K. Chari. Non-linear magnetic field analysis of D.C. machines. *Trans. IEEE*, No. 7:5–89, 1970.
12. P. Silvester and M.S. Hsieh. Finite element solution of two dimensional exterior field problems. *Proc. IEEE*, 118, 1971.
13. B.H. McDonald and A. Wexler. Finite element solution of unbounded field problems. *Proc. IEEE*, MTT-20(12), 1972.
14. E. Munro. Computer design of electron lenses by the finite element method. In *Image Processing and Computer Aided Design in Electron Optics*, page 284. Academic Press, New York, 1973.
15. O.C. Zienkiewicz, J.F. Lyness, and D.R.J. Owen. Three-dimensional magnetic field determination using a scalar potential. A finite element solution. *IEEE Trans. Mag.*, MAG-13(5):1649–1656, 1977.
16. J. Simkin and C.W. Trowbridge. On the use of the total scalar potential in the numerical solution of field problems in electromagnets. *Int. J. Numer. Meth. Eng.*, 14:423–440, 1979.
17. J. Simkin and C.W. Trowbridge. Three-dimensional non-linear electromagnetic field computations using scalar potentials. *Proc. Inst. Elec. Eng.*, 127(B(6)):423–440, 1980.
18. O.C. Zienkiewicz and R.L. Taylor. *The Finite Element Method for Solid and Structural Mechanics*. Butterworth-Heinemann, Oxford, 6th edition, 2005.
19. K.J. Binns, P.J. Lawrenson, and C.W. Trowbridge. *The Analytical and Numerical Solution of Electric and Magnetic Fields*. John Wiley & Sons, Chichester, 1992.
20. D.V. Tanesa and I.C. Rao. Student project report on lubrication. Royal Naval College, 1966.
21. M.M. Reddi. Finite element solution of the incompressible lubrication problem. *Trans. Am. Soc. Mech. Eng.*, 91(Ser. F):524, 1969.
22. M.M. Reddi and T.Y. Chu. Finite element solution of the steady state compressible lubrication problem. *Trans. Am. Soc. Mech. Eng.*, 92(Ser. F):495, 1970.
23. J.H. Argyris and D.W. Scharpf. The incompressible lubrication problem. *J. Roy. Aero. Soc.*, 73:1044–1046, 1969.
24. J.F. Booker and K.H. Huebner. Application of finite element methods to lubrication: an engineering approach. *J. Lubr. Techn., Trans, ASME*, 14((Ser. F)):313, 1972.

25. K.H. Huebner. Application of finite element methods to thermohydrodynamic lubrication. *Int. J. Numer. Meth. Eng.*, 8:139–168, 1974.
26. S.M. Rohde and K.P. Oh. Higher order finite element methods for the solution of compressible porous bearing problems. *Int. J. Numer. Meth. Eng.*, 9:903–912, 1975.
27. A.K. Tieu. Oil film temperature distributions in an infinitely wide glider bearing: an application of the finite element method. *J. Mech. Eng. Sci.*, 15:311, 1973.
28. K.H. Huebner. Finite element analysis of fluid film lubrication – a survey. In R.H. Gallagher, J.T. Oden, C. Taylor, and O.C. Zienkiewicz, editors, *Finite Elements in Fluids*, volume II, pages 225–254. John Wiley & Sons, New York, 1975.
29. H.C. Martin. Finite element analysis of fluid flows. In *Proc. 2nd Conf. Matrix Methods in Structural Mechanics*, volume AFFDL-TR-68-150, Wright Patterson Air Force Base, Ohio, Oct. 1968.
30. G. de Vries and D.H. Norrie. Application of the finite element technique to potential flow problems. Technical Reports 7 and 8, Dept. Mech. Eng., Univ. of Calgary, Alberta, Canada, 1969.
31. J.H. Argyris, G. Mareczek, and D.W. Scharpf. Two and three dimensional flow using finite elements. *J. Roy. Aero. Soc.*, 73:961–964, 1969.
32. L.J. Doctors. An application of finite element technique to boundary value problems of potential flow. *Int. J. Numer. Meth. Eng.*, 2:243–252, 1970.
33. G. de Vries and D.H. Norrie. The application of the finite element technique to potential flow problems. *J. Appl. Mech., ASME*, 38:978–802, 1971.
34. S.T.K. Chan, B.E. Larock, and L.R. Herrmann. Free surface ideal fluid flows by finite elements. *J. Hydraulics Division, ASCE*, 99(HY6), 1973.
35. B.E. Larock. Jets from two dimensional symmetric nozzles of arbitrary shape. *J. Fluid Mech.*, 37:479–483, 1969.
36. A. Curnier and R.L. Taylor. A thermomechanical formulation and solution of lubricated contacts between deformable solids. *J. Lub. Tech., ASME*, 104:109–117, 1982.
37. O.C. Zienkiewicz, R.L. Taylor, and P. Nithiarasu. *The Finite Element Method for Fluid Dynamics*. Butterworth-Heinemann, Oxford, 6th edition, 2005.
38. C.S. Desai. Finite element methods for flow in porous media. In J.T. Oden, O.C. Zienkiewicz, R.H. Gallagher, and C. Taylor, editors, *Finite Elements in Fluids*, volume 1, pages 157–182. John Wiley & Sons, New York, 1976.
39. I. Javandel and P.A. Witherspoon. Applications of the finite element method to transient flow in porous media. *Trans. Soc. Petrol. Eng.*, 243:241–251, 1968.
40. R.L. Taylor and C.B. Brown. Darcy flow solutions with a free surface. *J. Hydraulics Division, ASCE*, 93(HY2):25–33, 1967.
41. J.C. Luke. A variational principle for a fluid with a free surface. *J. Fluid Mech.*, 27:395–397, 1957.
42. K. Washizu. *Variational Methods in Elasticity and Plasticity*. Pergamon Press, New York, 3rd edition, 1982.
43. J.C. Bruch. A survey of free-boundary value problems in the theory of fluid flow through porous media. *Adv. Water Res.*, 3:65–80, 1980.
44. C. Baiocchi, V. Comincioli, and V. Maione. Unconfined flow through porous media. *Meccanice, Ital. Ass. Theor. Appl. Mech.*, 10:51–60, 1975.
45. J.M. Sloss and J.C. Bruch. Free surface seepage problem. *J. Eng. Mech., ASCE*, 108(EM5):1099–1111, 1978.
46. N. Kikuchi. Seepage flow problems by variational inequalities. *Int. J. Numer. Anal. Meth. Geomech.*, 1:283–290, 1977.
47. C.S. Desai. Finite element residual schemes for unconfined flow. *Int. J. Numer. Meth. Eng.*, 10:1415–1418, 1976.

48. C.S. Desai and G.C. Li. A residual flow procedure and application for free surface, and porous media. *Adv. Water Res.*, 6:27–40, 1983.
49. K.J. Bathe and M. Koshgoftar. Finite elements from surface seepage analysis without mesh iteration. *Int. J. Numer. Anal. Meth. Geomech.*, 3:13–22, 1979.
50. S.P. Timoshenko and J.N. Goodier. *Theory of Elasticity*. McGraw-Hill, New York, 2nd edition, 1951.

8

Automatic mesh generation

8.1 Introduction

In the previous chapters we have introduced various forms of elements and the procedures of using these elements in the computation of approximate solutions of a wide range of engineering problems. It is now obvious that the first step in the finite element computation is to discretize the problem domain into a union of elements. These elements could be of any one type or a combination of different types of those described in Chapters 4 and 5. The union of these elements is the so-called finite element mesh. The process of creating a finite element mesh is often termed as *mesh generation*.

Mesh generation has always been a time-consuming and error-prone process. This is especially true in the practical science and engineering computations, where meshes have to be generated for three-dimensional geometries of various levels of complexity. The attempt to create a fully automatic mesh generator, which is a particular mesh generation algorithm that is capable of generating valid finite element meshes over arbitrary domains and needs only the information of the specified geometric boundary of the domain and the required distribution of the element size, started from the work of Zienkiewicz and Phillips[1] in the early 1970s. Since then many methodologies have been proposed and different algorithms have been devised in the development of automatic mesh generators.

The early proposed mesh generation methods, such as the isoparametric mapping method by Zienkiewicz and Phillips,[1] the transfinite mapping method by Gordon and Hall[2] discussed in Sec. 5.13, and the method of generating a mesh by solving various types of partial differential equations as described by Thompson *et al.*,[3,4] are often regarded as semi-automatic mesh generation methods. This is because in the mesh generation process the model domain has to be subdivided manually into simple subregions, i.e., multi-blocks, which are then mapped onto regular grids to produce a mesh. This manual process is tedious and occasionally difficult, particularly in the case of three-dimensional complex geometries. Such mesh generation procedures are complicated further by the requirement of varying mesh size distributions since the element sizes are controlled by the subdivision of the simple subregions. Thus, more subregions are needed to generate a mesh which can accommodate changes in the desired element sizes from region to region. However, one of the main features of these mapping techniques is that, once the domain is decomposed into mappable subregions, the generation of the elements is much easier than any other methods. In addition, the elements generated by mapping methods usually have good shape and regular orientation. Mapping methods are often used to generate quadrilateral elements

in two dimensions and hexahedral elements in three dimensions. Triangles, tetr
and all other types of elements can be obtained by dividing quadrilaterals and hex
accordingly. These generated meshes are sometimes called *structured meshes*. O
years, continuous efforts have been made to automate the mapping methods,[5–10] although automatic decomposition of a complex domain into subregions seems to be a non-trivial task. Today, no fully automatic mesh generator using a mapping method has been achieved.

In contrast with mapping methods, in recent years concrete achievements have been made in the development of various algorithms for the automatic generation of the so-called *unstructured meshes*. Most of the unstructured mesh generation methods are designed for generating triangular elements in two dimensions and tetrahedral elements in three dimensions (known as simplex forms). These simplex forms lead to the simplest discretization of two- and three-dimensional domains of any shape, especially when meshes with varying element sizes in different regions of the domain are requested. A large number of automatic unstructured mesh generation algorithms have been proposed in the literature, but the most widely used algorithms are based on one or some kind of combination of the three fundamentally distinctive methods, which are the Delaunay triangulation method,[11–20] the advancing front method[21–24, 26–28, 43] and tree methods [29–31] (the finite quadtree method in two dimensions and the finite octree method in three dimensions). By observing the fact that a quadrilateral can be formed by two triangles which share a common edge, the above-mentioned methods can be extended to automatically generate unstructured quadrilateral meshes in two dimensions. However, automatically generating a hexahedral mesh[10, 32–35] encounters the almost identical difficulties as that in the mapping methods, much research is still needed in this direction.

The automatic mesh generation process has been an active research subject since the early 1970s. The research literature on the subject is vast and different methodologies and techniques have been proposed. In this chapter, we are mainly concerned with the automatic mesh generation methods based on the advancing front method and the Delaunay triangulation method. These are the basis of many existing mesh generation programs and the basis of current research. We shall discuss the algorithmic procedures of the advancing front method in two dimensions and the Delaunay triangulation method in three dimensions. We shall also discuss curve and surface mesh generations. The reader is referred to Thompson *et al.*[5, 36] for discussions on the development of semi-automatic multi-block mesh generation methods.

Before proceeding further on mesh generation schemes it is necessary to specify the kind of mesh we desire. Here we should give the following information:

1. The type of element and the number of nodes required on each;
2. The size of the desired element, here the minimum size of each element generally is specified;
3. Specification of regions of different material types or characteristics to be attached to a given element; and
4. In some cases, the so-called stretch ratio if we wish to present elements which are elongated in some preferential direction. This is often needed for problems in fluid mechanics in regions where boundary layers and shocks are encountered.

Any and all of the above information has to be available at all points of the space in which elements are to be generated. It is often convenient to present this information as numbers attached to a *background mesh*, consisting say of elements of a linear kind, from which these values can be interpolated to any point in space. The procedure is important particularly if the use of adaptive refinement is considered – and here all mesh generation schemes must ensure that the input data contains this information. In adaptive refinement, in fact in general analysis, the background mesh will simply be the last mesh used for analysis of the problem and the refinement will proceed from there as this is the starting point for any new mesh to be developed.

8.2 Two-dimensional mesh generation – advancing front method

Conceptually, the advancing front method is one of the simplest generation processes. The element generation algorithm, starting from an initial 'front' formed from the specified boundary of the domain, generates elements, one by one, as the front advances into the region to be discretized until the whole domain is completely covered by elements.

The representative element generation algorithms of the advancing front method include the procedure introduced by Lo,[21] which constructs a triangulation over a set of *a priori* generated points inside of the domain, and the methodology developed by Peraire *et al.*,[22] which generates points and triangular elements at the same time.

One of the main distinctions of the mesh generation algorithm of Peraire *et al.* is that the geometrical characteristics of the mesh, such as the location of the newly generated point, the shape of the element and the size of the element, can be controlled during the mesh generation process, due to the fact that individual points and elements are generated simultaneously. With the assistance of a background mesh, which is utilized to define the geometrical characteristics of the mesh, non-uniform distribution of element sizes, often required in highly graded meshes, can be achieved throughout the domain according to particular specifications. Any directional orientation of the elements can also be realized by introducing stretches in certain specified directions. These features are particularly desirable for the nearly optimal mesh design in adaptive analysis (viz. Chapter 14) and adaptive computations of fluid dynamics as discussed in reference 37.

The mesh generation procedure includes three main steps:

- Node generation along boundary edges to form a discretized boundary of the domain.
- Element (and node) generation within the discretized boundary.
- Element shape enhancement to improve the quality of the mesh.

Before we proceed to the discussion of mesh generation procedures, the geometrical representation of the two-dimensional domain is introduced.

8.2.1 Geometrical characteristics of the mesh

The geometrical characteristics of the mesh such as element size, element shape and element orientation are represented by means of mesh parameters which are spatial functions. The mesh parameters include two orthogonal directions defined by unit vectors $\boldsymbol{\alpha}_i$ $(i = 1, 2)$ and the associated element sizes h_i $(i = 1, 2)$ as illustrated in Fig. 8.1. The orthogonal directions $\boldsymbol{\alpha}_i$ $(i = 1, 2)$ describe the directions of element stretching. A mesh with stretched elements in certain directions is only necessary when a non-isotropic mesh is desired, otherwise the

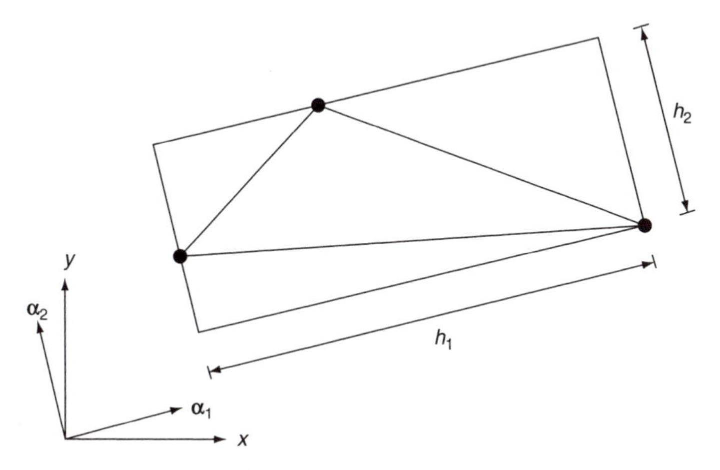

Fig. 8.1 Mesh parameters in two dimensions.

stretching directions are set to be constant unit vectors in the coordinate directions and the related element sizes are set to be equal, i.e., $h_1 = h_2$. In this case, the generated elements will not be stretched in any direction and an isotropic mesh type will be generated.

Background mesh

The background mesh may be represented by simple triangular elements and is employed to accurately control the distribution of the geometrical characteristics on the new mesh. A piecewise linear distribution of the mesh parameters (mainly element sizes and stretching as discussed above) is represented by data assigned to nodes of the background mesh. Values of the mesh parameters at any point inside the domain or on the boundary of the domain can be obtained by linear interpolation. There is no requirement that the background mesh precisely represent the geometry, but it should completely cover the domain to be meshed. The number of the elements and the position of the nodes in the background mesh are chosen so that the mesh parameters can be approximated in a satisfactory manner. One or two background elements will be sufficient if a uniform (isotropic) distribution of the element sizes $h_i (i = 1, 2)$ is required. Examples of background meshes for a given domain are illustrated in Fig. 8.2.

8.2.2 Geometrical representation of the domain

A general two-dimensional domain, which is covered by a background mesh (viz. Fig. 8.2), is defined by its boundary which consists of a closed loop of curved boundary segments (viz. Fig. 8.3).

Boundary curve representation

The curvilinear boundary segments are in general represented by composite parametric spline curves. A curved boundary segment in two dimensions, can be expressed by a vector valued function, using a parameter t, as

$$\mathbf{x}(t) = \{x(t) \quad y(t)\}, \qquad 0 \leq t \leq 1 \tag{8.1}$$

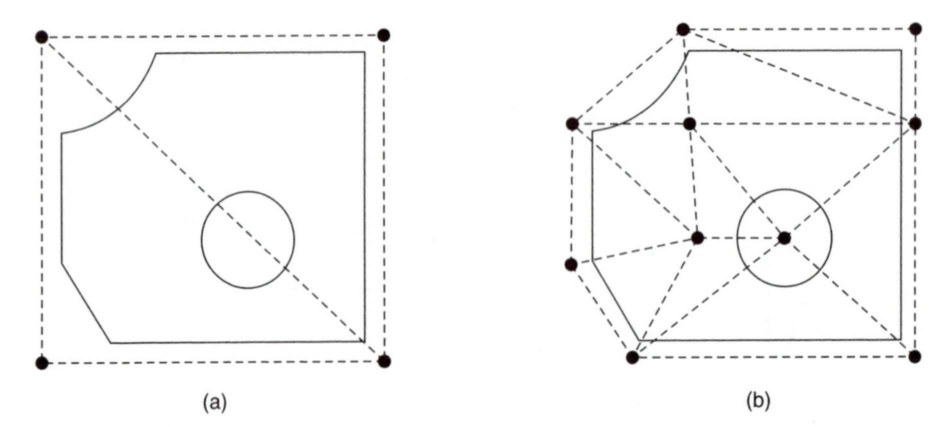

Fig. 8.2 Background meshes for a typical domain. (a) Two triangles are used in the background mesh to represent a uniform distribution of the mesh parameters. (b) Eleven triangles are used in the background mesh to generate a graded mesh.

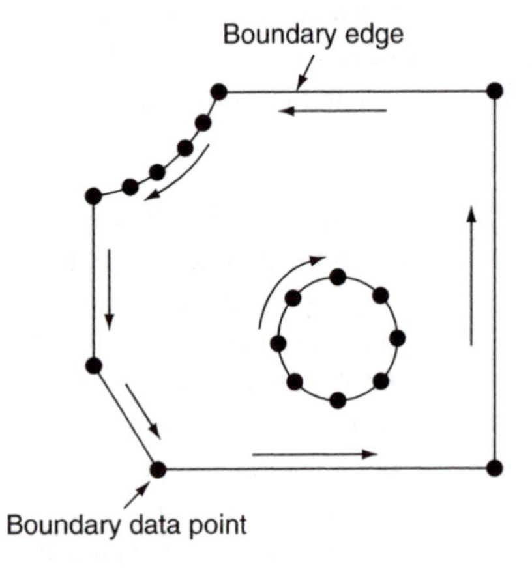

Fig. 8.3 Boundary segments, boundary data points and orientation of a typical domain.

In general, a composite spline curve is required to be at least C^1 continuous to preserve the smoothness of the boundary curve and to satisfy the continuity conditions required by mesh generation algorithms. A Hermite cubic spline is used in the following; however, there are many other types of parametric spline curves that can be used to represent the curved boundary edges.[38, 39] The parametric description of a Hermite cubic spline is given by the form

$$\mathbf{x}(t) = \{H_0(t)\ H_1(t)\ G_0(t)\ G_1(t)\} \begin{Bmatrix} \mathbf{x}(0) \\ \mathbf{x}(1) \\ \mathbf{x}_{,t}(0) \\ \mathbf{x}_{,t}(1) \end{Bmatrix}, \quad 0 \le t \le 1 \tag{8.2}$$

in which $\mathbf{x}(0), \mathbf{x}(1)$ are the coordinates of the end points and $\mathbf{x}_{,t}(0), \mathbf{x}_{,t}(1)$ are their respective tangent vectors defined as

$$\mathbf{x}_{,t}(t) = \frac{\mathrm{d}\mathbf{x}(t)}{\mathrm{d}t} \tag{8.3}$$

Cubic Hermite polynomials are expressed as

$$\begin{aligned} H_0(t) &= 1 - 3t^2 + 2t^3; \qquad & H_1(t) &= 3t^2 - 2t^3 \\ G_0(t) &= t - 2t^2 + t^3; \qquad & G_1(t) &= -t^2 + t^3 \end{aligned} \tag{8.4}$$

and depicted in Fig. 8.4. It is easy to verify that

$$H_a(b) = G'_a(b) = \delta_{ab} \quad \text{and} \quad H'_a(b) = G_a(b) = 0; \quad a, b = 0, 1 \tag{8.5}$$

Substituting (8.4) into (8.2) we obtain

$$\mathbf{x}(t) = \left\{1 \quad t \quad t^2 \quad t^3\right\} \mathbf{M} \begin{Bmatrix} \mathbf{x}(0) \\ \mathbf{x}(1) \\ \mathbf{x}_{,t}(0) \\ \mathbf{x}_{,t}(1) \end{Bmatrix}, \quad 0 \leq t \leq 1 \tag{8.6}$$

where

$$\mathbf{M} = \begin{bmatrix} 1 & 0 & 0 & 0 \\ 0 & 0 & 1 & 0 \\ -3 & 3 & -2 & -1 \\ 2 & -2 & 1 & 1 \end{bmatrix} \tag{8.7}$$

All boundary edges are transformed to their spline representations. For each boundary edge, a set of ordered data points $\mathbf{x}_i$ $(i = 0, 1, \ldots, n)$ are located. An interpolation by piecewise cubic Hermite polynomials through pairs of these points forms a composite parametric spline curve. The number and distribution of the points should be chosen such that the resulting piecewise cubic spline accurately represents the geometry of the boundary. For a curved edge segment $[u_{i-1}, u_i]$ of length $\Delta_i = u_i - u_{i-1}$ $(i = 1, 2, \ldots, n)$ of the interpolated cubic composite curve, the cubic spline has the form of

$$\begin{aligned} \mathbf{x}(u) &= H_0(t)\mathbf{x}(u_{i-1}) + H_1(t)\mathbf{x}(u_i) + \Delta_i G_0(t)\mathbf{x}_{,t}(u_{i-1}) + \Delta_i G_1(t)\mathbf{x}_{,t}(u_i) \\ &= \left\{1 \quad t \quad t^2 \quad t^3\right\} \mathbf{M} \begin{Bmatrix} \mathbf{x}(u_{i-1}) \\ \mathbf{x}(u_i) \\ \Delta_i \mathbf{x}_{,t}(u_{i-1}) \\ \Delta_i \mathbf{x}_{,t}(u_i) \end{Bmatrix} \qquad 0 \leq t \leq 1 \end{aligned} \tag{8.8}$$

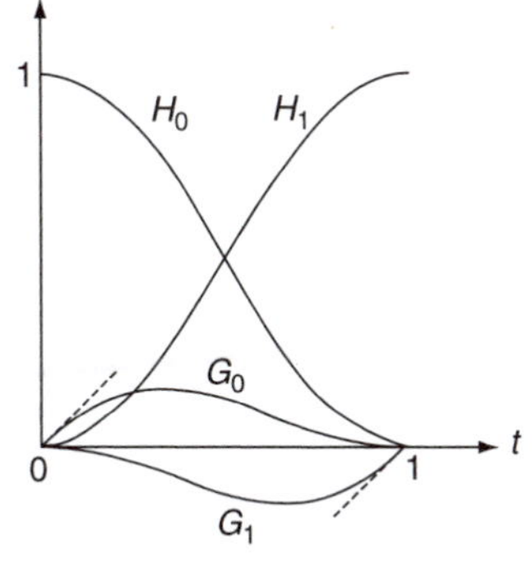

Fig. 8.4 Cubic Hermite interpolation functions.

where $t = (u - u_{i-1})/\Delta_i$ is the local parameter of the segment $[u_{i-1}, u_i]$ and $u \in [u_0, u_n]$. The unknown tangents of the composite curve can be computed by the standard cubic spline interpolation procedures, provided that the final parametric curve is C^2 continuous, with proper end conditions.[38, 40] The implementation of the spline interpolation can be simplified when the parametric coordinates of u_i are taken to be $u_i = i(i = 1, 2, \ldots, n)$. For segment $[u_{i-1}, u_i] = [i - 1, i]$, the global parametric coordinates u is then related to the local coordinate t according to

$$u = u_{i-1} + t = i - 1 + t \tag{8.9}$$

The global mapping of the region $u \in [u_0, u_n]$ in parametric space and the cubic composite curve provided by $\mathbf{x}(u)$ is depicted in Fig. 8.5.

The collection of all boundary edges, following a specific sequence that is convenient for mesh generation, forms the complete boundary of the domain. For the advancing front method, the sequence of exterior boundary edges is usually in a counterclockwise order, but, for interior boundary edges, is set in a clockwise order, i.e., the domain to be discretized always has an interior area situated to the left of the boundary edges. Figure 8.3 shows the direction of the boundary edges together with the boundary data points, of a typical domain.

8.2.3 Triangular mesh generation

Among all the steps in the mesh generation process, we are particularly concerned with the procedure for element generation, which include node and side generation on the boundary curve and triangular element generation inside of the two-dimensional domain.[22]

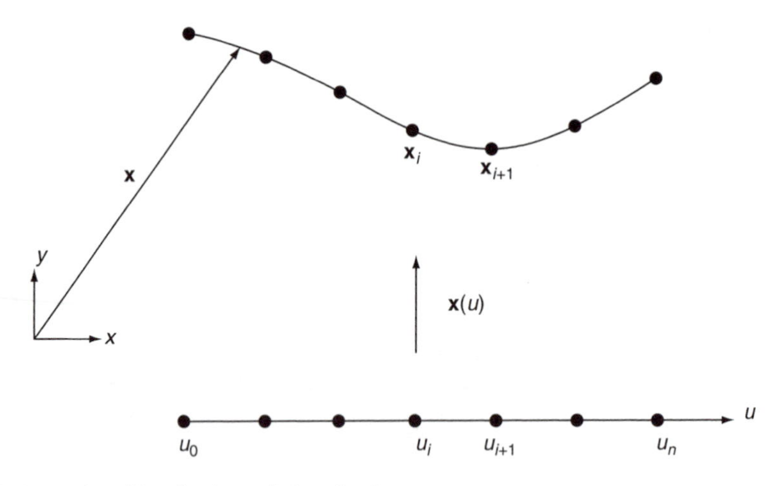

Fig. 8.5 Composite cubic spline interpolation of a planar curve.

Geometrical transformation of the mesh

In order to simplify the mesh generation process, a symmetric transformation matrix $\mathbf{T}$, defined by the mesh parameters, is introduced and has the form

$$\mathbf{T}(\mathbf{x}) = \sum_{i=1}^{2} \frac{1}{h_i} \boldsymbol{\alpha}_i \boldsymbol{\alpha}_i^{\mathrm{T}}; \quad \boldsymbol{\alpha}_i = \begin{Bmatrix} \alpha_{1i} \\ \alpha_{2i} \end{Bmatrix} \tag{8.10}$$

It is easy to verify that the local transformation

$$\mathbf{x}' = \mathbf{T}\mathbf{x} \tag{8.11}$$

is in fact imposing two scaling operations with factor $1/h_i$ in each of the corresponding directions $\boldsymbol{\alpha}_i$. Figure 8.6 illustrates the effect of the transformation $\mathbf{T}$ on a triangle formed by nodes abc in the coordinate system (x, y) to form the triangle $a'b'c'$ in the coordinate system (x', y'). This demonstrates that, at a particular point, the transformation $\mathbf{T}$ maps triangle with element size $h_i (i = 1, 2)$ formed in the neighbourhood of the point into a *normalized space* (x', y'), in which the triangular elements are approximately equilateral.

Example 8.1: Transformation of a triangle. As an example, the details of the coordinates transformation shown in Fig. 8.6 are given as follows.
From the coordinates of nodes a, b and c, we can easily find that at node b

$$\boldsymbol{\alpha}_1 = \begin{pmatrix} \frac{\sqrt{2}}{2} \\ \frac{\sqrt{2}}{2} \end{pmatrix} \quad \text{and} \quad \boldsymbol{\alpha}_2 = \begin{pmatrix} -\frac{\sqrt{2}}{2} \\ \frac{\sqrt{2}}{2} \end{pmatrix}$$

with the associated element sizes $h_1 = 4\sqrt{2}$ and $h_2 = \sqrt{2}$.

The transformation matrix $\mathbf{T}$ at node b is computed, using Eq. (8.10), as

$$\mathbf{T} = \frac{\sqrt{2}}{16} \begin{pmatrix} 1 & 1 \\ 1 & 1 \end{pmatrix} + \frac{\sqrt{2}}{4} \begin{pmatrix} 1 & -1 \\ -1 & 1 \end{pmatrix}$$

Applying $\mathbf{T}$ to nodes a, b, c results in

$$\mathbf{x}_{a'} = \mathbf{T}\mathbf{x}_a = \frac{\sqrt{2}}{8} \begin{Bmatrix} -1 \\ 7 \end{Bmatrix}, \quad \mathbf{x}_{b'} = \mathbf{T}\mathbf{x}_b = \frac{\sqrt{2}}{4} \begin{Bmatrix} 1 \\ 1 \end{Bmatrix} \quad \text{and} \quad \mathbf{x}_{c'} = \mathbf{T}\mathbf{x}_c = \frac{\sqrt{2}}{4} \begin{Bmatrix} 3 \\ 3 \end{Bmatrix}$$

These are the nodal positions for the triangle $a'b'c'$ in the coordinate system (x', y').

Boundary node generation

The boundary of the domain will be discretized into a polygon which will form the element edges, herein called ‘sides’. The sides are defined by nodes generated on the composite spline curves that represent the boundary edges. The nodes will be generated along the curve edge and expressed by their parametric positions. The coordinates of the nodes in the two-dimensional domain are determined using Eq. (8.8). The algorithmic procedure of the boundary node generation is described in the following.

1. For a curved edge with typical length L, a set of sampling points $\mathbf{x}_l = \mathbf{x}(u_l)$ $(l = 0, 1, 2, \ldots, m)$ is first placed along the curve with parameters u_l uniformly

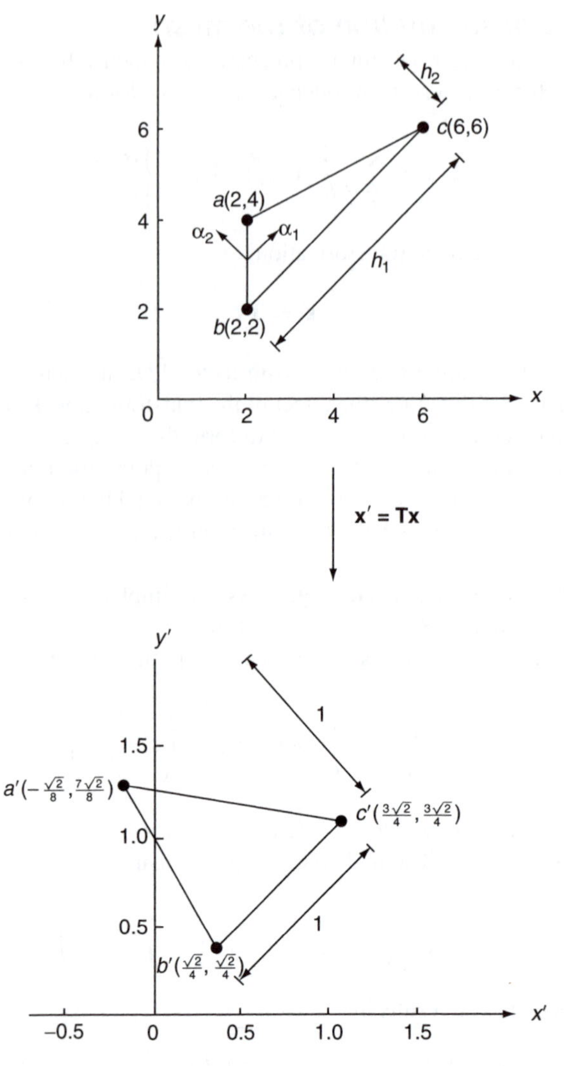

Fig. 8.6 An irregular triangle *abc* in coordinate system (x, y) mapped to a regular triangle $a'b'c'$ in coordinate system (x', y').

distributed as shown in Fig. 8.7. The unit tangent vector of the curve is determined at each of the sampling points as

$$\mathbf{t}_l = (t_{1_l}, t_{2_l}) \tag{8.12}$$

where

$$t_{1_l} = \frac{x_{,u_l}}{\sqrt{x^2_{,u_l} + y^2_{,u_l}}} \quad \text{and} \quad t_{2_l} = \frac{y_{,u_l}}{\sqrt{x^2_{,u_l} + y^2_{,u_l}}} \tag{8.13}$$

and

$$\mathbf{x}_{,u_l} = (x_{,u_l}, y_{,u_l}); \quad x_{,u_l} = \left.\frac{\mathrm{d}x}{\mathrm{d}u}\right|_{u_l} \tag{8.14}$$

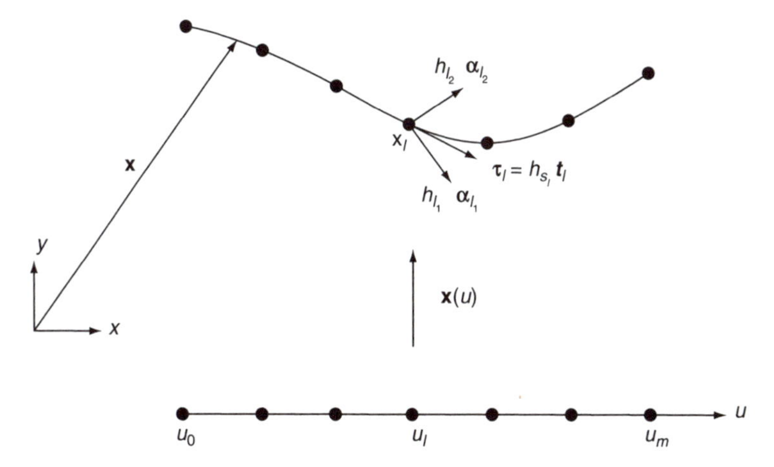

Fig. 8.7 Sampling points and mesh parameters on a composite curve segment.

2. The mesh parameters α_{l_i} and h_{l_i} $(i = 1, 2)$ are computed at each sampling point by interpolation from their values assigned on the background mesh. The transformation matrix $\mathbf{T}_l$ is formed accordingly at each of the sampling point.
3. In order to find the position of the new nodes on the curve, an element size distribution function needs to be determined along the curve edge.
   Let h_{s_l} denote the element size at the arc length s_l corresponding to the sampling point $\mathbf{x}_l$, a vector of length h_{s_l} in the tangent direction is defined as

$$\boldsymbol{\tau}_l = h_{s_l}\mathbf{t}_l \tag{8.15}$$

   Applying transformation $\mathbf{T}_l$ and assuming that $\mathbf{T}_l$ maps $\boldsymbol{\tau}_l$ to a vector $\mathbf{T}\boldsymbol{\tau}_l$ in the normalized space with unit length, i.e.,

$$\mathbf{T}_l\boldsymbol{\tau}_l = \mathbf{T}_l(h_{s_l}\mathbf{t}_l) = h_{s_l}\mathbf{T}_l\mathbf{t}_l \tag{8.16}$$

   with

$$\sqrt{(\mathbf{T}_l\boldsymbol{\tau}_l)^{\mathrm{T}}(\mathbf{T}_l\boldsymbol{\tau}_l)} = h_{s_l}\sqrt{(\mathbf{T}_l\mathbf{t}_l)^{\mathrm{T}}(\mathbf{T}_l\mathbf{t}_l)} = 1 \tag{8.17}$$

   Thus, the curvilinear element size $h_{s_l}(l = 0, 1, 2, \ldots, m)$ at the sampling points along the curve in the direction of the tangent is

$$h_{s_l} = \frac{1}{\sqrt{(\mathbf{T}_l\mathbf{t}_l)^{\mathrm{T}}(\mathbf{T}_l\mathbf{t}_l)}} = \frac{1}{\sqrt{\mathbf{t}_l^{\mathrm{T}}\mathbf{C}_l\mathbf{t}_l}} \tag{8.18}$$

   where

$$\mathbf{C} = \mathbf{T}^{\mathrm{T}}\mathbf{T} = \sum_{i=1}^{2}\frac{1}{h_i^2}\boldsymbol{\alpha}_i\boldsymbol{\alpha}_i^{\mathrm{T}} \tag{8.19}$$

   is the two-dimensional matrix of *Euclidean metric tensor* in the normalized space.
4. Assume that $\mathbf{T}_l$ is a constant matrix in the neighbourhood of $\mathbf{x}_l$, it is observed, from Eq. (8.13), that h_{s_l} is a function of the parameter u. A continuous element size distribution

function may be achieved by a piecewise linear interpolation of the nodal values h_{s_l} using (lagrangian) finite element shape functions along the arc length of the curve

$$h(u) = \sum_{l=0}^{m} h_{s_l} N_l(u) \tag{8.20}$$

The element density function, i.e., the number of elements per length scale along the curve, is defined as $1/h(u)$.

5. The total number of the sides N to be generated along the curved edge needs to be consistent with the specified element size, which is now represented by the element density function. Therefore, N is taken to be the nearest integer to

$$A = \int_0^L \frac{1}{h(u)}\,\mathrm{d}s = \int_{u_0}^{u_m} \frac{1}{h(u)} \sqrt{(x_u)^2 + (y_u)^2}\ \mathrm{d}u \tag{8.21}$$

where A is the ideal number of the sides that should have been created on the boundary curve edge. However, A is in general not an integer. To measure how close N is to A, a *consistency index* is defined as

$$\theta = \frac{A}{N} = \frac{1}{N} \int_{u_0}^{u_m} \frac{1}{h(u)} \sqrt{(x_u)^2 + (y_u)^2}\ \mathrm{d}u \tag{8.22}$$

Because the position of the nodes at the end of the edge $\mathbf{x}_0 = \mathbf{x}(u_0)$ and $\mathbf{x}_m = \mathbf{x}(u_m)$ are already known, there will be $(N-1)$ new nodes to be generated.

6. Assume that every node on the boundary edge is generated with the same consistency index θ, the position of a particular new node $n_k (k = 1, 2, \ldots, (N-1))$, represented by its parametric position u_k on the boundary curve can be computed as

$$\theta = \theta_k = \frac{1}{k} \int_{u_0}^{u_k} \frac{1}{h(u)} \sqrt{(x_u)^2 + (y_u)^2}\ \mathrm{d}u \tag{8.23}$$

and similarly, the position of new node n_{k+1} is given by

$$\theta = \theta_{k+1} = \frac{1}{k+1} \int_{u_0}^{u_{k+1}} \frac{1}{h(u)} \sqrt{(x_u)^2 + (y_u)^2}\ \mathrm{d}u \tag{8.24}$$

From $\theta_k = \theta_{k+1} = \theta$, we obtain the parametric position of node n_{k+1} computed consecutively as

$$\theta = \int_{u_k}^{u_{k+1}} \frac{1}{h(u)} \sqrt{(x_u)^2 + (y_u)^2}\ \mathrm{d}u \tag{8.25}$$

where $k = 0, 1, 2, \ldots, (N-2)$. In general, Eq. (8.25) can be solved iteratively for u_{k+1}. For example, writing Eq. (8.25) in the form of

$$F(u_{k+1}) = \int_{u_k}^{u_{k+1}} \frac{1}{h(u)} \sqrt{(x_u)^2 + (y_u)^2}\ \mathrm{d}u - \theta = 0 \tag{8.26}$$

A Newton iterative process results in

$$u_{k+1}^{j+1} = u_{k+1}^{j} - \frac{h(u_{k+1}^{j})}{\sqrt{(x_{u_{k+1}^{j}})^2 + (y_{u_{k+1}^{j}})^2}} \left[F(u_{k+1}^{j}) \right] \tag{8.27}$$

for $j = 0, 1, 2, \ldots$ with initial value $u_{k+1}^{0} = u_k$.

7. Finally, the position of the new points is mapped onto the boundary curve using Eq. (8.8).

The generation of the boundary nodes will be performed edge by edge following the above procedure. The boundary of the domain is finally discretized and transformed to a union of straight line sides formed by connecting the consecutive boundary nodes.

Example 8.2: To verify the above procedure, the node generation technique is applied to a curve edge representing a quarter of a circle shown in Fig. 8.8(a). To simplify the presentation and also to demonstrate that the node generation procedure is independent of the curve representation of the boundary edge, we chose not to use Hermite spline to represent the curve, but use its parametric expression in the form of

$$x = 4\cos(u), \qquad y = 4\sin(u); \qquad 0 \leq u \leq \frac{\pi}{2} \tag{8.28}$$

The mesh parameters $\boldsymbol{\alpha}_i$ and h_i $(i = 1, 2)$ are chosen to be independent of their spatial positions, the background mesh is therefore not required. We shall follow each step of the boundary node generation procedure to create nodes in accordance with the specified element size.

1. Three sampling points $\mathbf{x}(u_l)$ $(l = 0, 1, 2)$ are located with u_l equally distributed at

$$u_0 = 0; \qquad u_1 = \frac{\pi}{4}; \qquad u_2 = \frac{\pi}{2}$$

The unit tangent vector to the curve is, by noting Eq. (8.13),

$$\mathbf{t} = (-\sin(u),\ \cos(u))^{\mathrm{T}}$$

and

$$\mathbf{t}(u_0) = (0,\ 1)^{\mathrm{T}}; \qquad \mathbf{t}(u_1) = \left(-\frac{\sqrt{2}}{2},\ \frac{\sqrt{2}}{2}\right)^{\mathrm{T}}; \qquad \mathbf{t}(u_2) = (1,\ 0)^{\mathrm{T}}$$

2. The stretching directions are chosen as depicted in Fig. 8.8(a), i.e.,

$$\boldsymbol{\alpha}_1 = (1,\ 0)^{\mathrm{T}}; \qquad \boldsymbol{\alpha}_2 = (0,\ 1)^{\mathrm{T}}$$

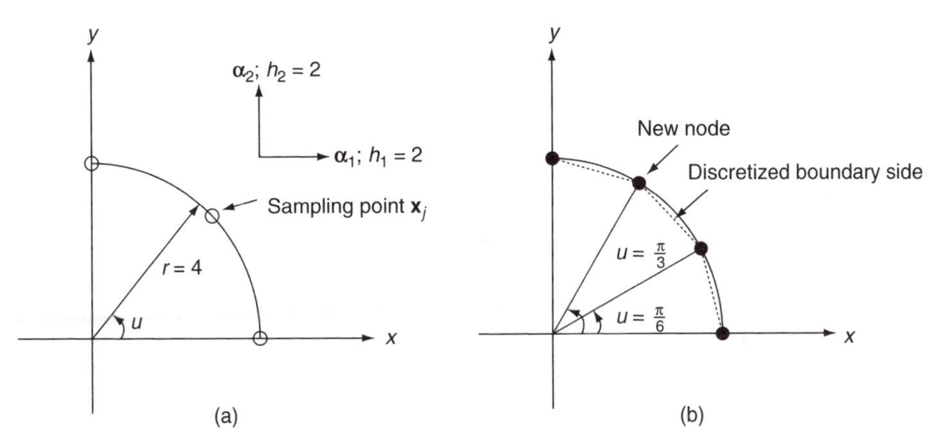

Fig. 8.8 Node generation on a curve. (a) Description of the curve and mesh parameters. (b) Nodes generated on the curve and sides formed by nodes.

The corresponding element sizes are set to be constant and have the values of $h_1 = h_2 = 2$. That is, we are looking for a uniform discretization on the edge.

As a result of our choice of mesh parameters, the transformation matrix $\mathbf{T}$ is a constant matrix and at all sampling points

$$\mathbf{T} = \frac{1}{2}\begin{pmatrix} 1 & 0 \\ 0 & 0 \end{pmatrix} + \frac{1}{2}\begin{pmatrix} 0 & 0 \\ 0 & 1 \end{pmatrix} = \frac{1}{2}\begin{pmatrix} 1 & 0 \\ 0 & 1 \end{pmatrix}$$

3. Applying $\mathbf{T}$ to tangent vector $\mathbf{t}(u_0)$, we have

$$\mathbf{Tt}(u_0) = \frac{1}{2}\begin{pmatrix} 1 & 0 \\ 0 & 1 \end{pmatrix}\begin{pmatrix} 0 \\ 1 \end{pmatrix} = \begin{pmatrix} 0 \\ \frac{1}{2} \end{pmatrix}$$

The mesh size at $\mathbf{x}(u_0)$ along the curve is, using Eq. (8.18), $h_{s_0} = 2$.

Similarly, as expected, we have

$$\mathbf{Tt}(u_1) = \begin{pmatrix} -\frac{\sqrt{2}}{4} \\ \frac{\sqrt{2}}{4} \end{pmatrix}; \quad h_{s_1} = 2; \quad \text{and} \quad \mathbf{Tt}(u_2) = \begin{pmatrix} -\frac{1}{2} \\ 0 \end{pmatrix}; \quad h_{s_2} = 2$$

4. Each element size is obtained by linear interpolation

$$h(u) = \sum_{l=0}^{2} h_{s_l} N_l(u) = 2$$

because h_{s_l} $(l = 0, 1, 2)$ are constant. The element density function is

$$\frac{1}{h(u)} = \frac{1}{2}$$

5. The integration of the element density function, using Eq. (8.21), gives the ideal number of sides, as

$$A = \int_0^{\frac{\pi}{2}} \frac{1}{2}\sqrt{16\sin^2(u) + 16\cos^2(u)}\ du = \pi$$

The nearest integer N to π is 3, i.e., there should be 3 sides being generated along the curve. In addition to the nodes at the end of the curve, 2 new nodes are required. The consistency index has the value of

$$\theta = \frac{A}{N} = \frac{\pi}{3}$$

6. The parametric position of the first new node is computed using Eq. (8.25)

$$\theta_1 = \frac{\pi}{3} = \int_0^{u_1} \frac{1}{2}\sqrt{16\sin^2(u) + 16\cos^2(u)}\, du = \int_0^{u_1} 2\, du = 2u_1$$

which gives $u_1 = \pi/6$. Here Eq. (8.25) can be solved exactly, no iterative scheme needs to be invoked. Using the above result, the parametric position u_2 of the second node is calculated as

$$\theta_2 = \frac{\pi}{3} = \int_{\frac{\pi}{6}}^{u_2} 2\, du = 2\left(u_2 - \frac{\pi}{6}\right)$$

We have $u_2 = \pi/3$.

7. Finally, the coordinates of the new nodes are obtained from the parametric equations of the circle. Substituting u_1 and u_2 into Eq. (8.28), we have for node n_1,

$$x_1 = 4\cos\left(\frac{\pi}{6}\right) = 2\sqrt{3}; \qquad y_1 = 4\sin\left(\frac{\pi}{6}\right) = 2$$

and for node n_2

$$x_2 = 4\cos\left(\frac{\pi}{3}\right) = 2; \qquad y_2 = 4\sin\left(\frac{\pi}{3}\right) = 2\sqrt{3}$$

The curved edge is discretized by three sides after linking each of the nodes as shown in Fig. 8.8(b).

Generation front

A generation front is established prior to starting the triangular element generation. The initial generation front is a collection of all of the sides which form the discretized boundary edges of the domain. Thus, it consists of a set of closed loops of boundary sides. If the domain is composed of multiple connected regions, such as regions with different material properties, an initial generation front will be formed for each of the regions.

Each side in the generation front is defined by its two end points. The sequence of the sides is also arranged such that the regions to be meshed are always situated to the left of the generation front. The initial generation front for a simple rectangular domain is shown in Fig. 8.9(a). At any stage of the element generation process, the generation front always forms the boundary of the region to be discretized as depicted in Fig. 8.9(b). In the process of element generation, a side from the generation front is chosen as a base to form a new element with either a newly generated node or an existing node from the generation front. Once a new element is formed, the generation front is updated. Any side that has been used to create a new element is removed from the generation front and the newly created side is added, as illustrated in Fig. 8.9(c) and Fig. 8.9(d). The updating procedure ensures that the generation front always forms the boundary of the region to be meshed. The sides and the nodes in the generation front are referred to as *active sides* and *active nodes* respectively. Processing of the generation front continues until the entire region is filled with elements and nodes.

Element generation

The process of generating a triangular element is illustrated in Fig. 8.10 and includes the following steps:

1. An active side ab connecting nodes a and b is selected from the generation front as a base to form the new element. To produce a mesh with smooth transition of the element size, the smallest side is considered first. The sides in the generation front are sorted and updated according to their length during the element generation process to increase the efficiency of the mesh generation algorithm.
2. At the middle point m of side ab compute the local mesh parameters α_{m_i} and h_{m_i} $(i = 1, 2)$ for the new element by interpolating from the background mesh.
3. The element creation process can be significantly simplified when point m and all the nodes in the generation front with respect to coordinates (x, y) are mapped to the normalized coordinate system (x', y') by $\mathbf{x}' = \mathbf{T}\mathbf{x}$. The element generation process will be conducted in the coordinate system (x', y') to construct a triangle that is as regular as possible.

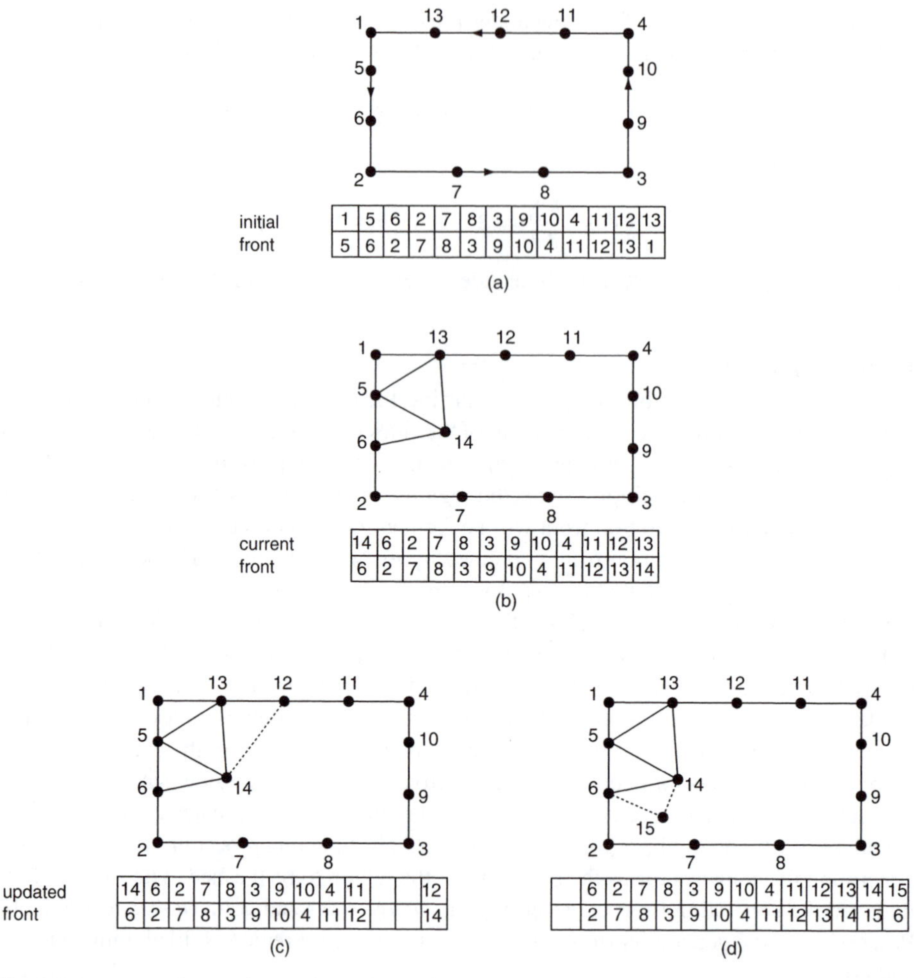

Fig. 8.9 Generation front and its updating during the element generation process: (a) The initial generation front. (b) Generation front at a certain stage. (c) and (d) Updated generation front after the creation of a new element.

4. Determine the ideal position of new node c' to form the new triangle. Node c' is constructed in the direction normal to side $a'b'$ and located at a distance h'_l from node a' and node b' as shown in Fig. 8.10. Here the normal of side $a'b'$ is pointing to the region to be meshed and h'_l is chosen as

$$h'_l = \begin{cases} 0.55l_{a'b'} & 1 \leq 0.55l_{a'b'} \\ 1 & 0.55l_{a'b'} < 1 < 2l_{a'b'} \\ 2l_{a'b'} & 2l_{a'b'} \leq 1 \end{cases} \tag{8.29}$$

to ensure that an element with excessive distortion is not created. The constants appearing in the expression are empirical but have been shown to work well in practice.

5. Additional points are located to create a list of the potential node to generate the new element. These include

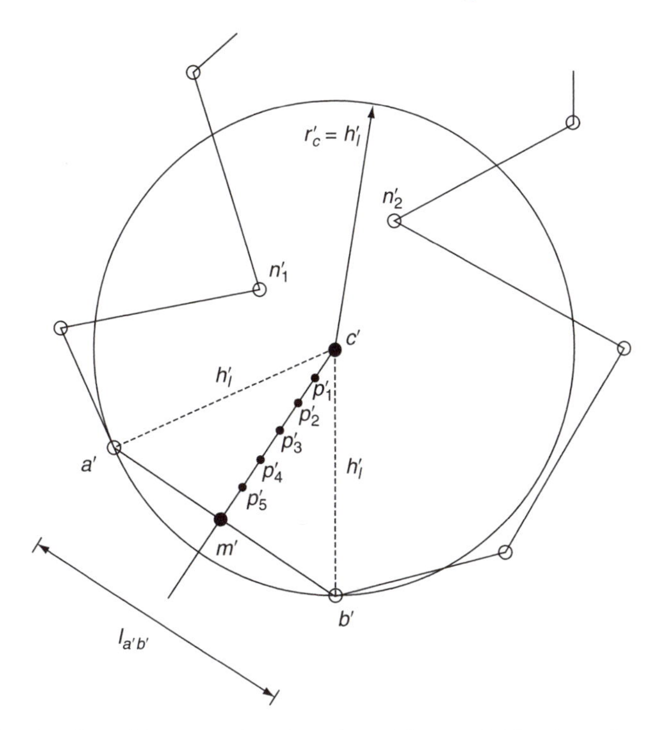

Fig. 8.10 The ideal position of the new node and locations of the potential forming nodes.

- all the nodes from the generation front that fall into the circle centred at c' with radius

$$r_{c'} = N\, h'_l \tag{8.30}$$

  These nodes are ordered by their distance from c' and denoted by $n'_1, n'_2, \ldots, n'_M$ with n'_1 being the closest to c'. The value of constant N is given here as $N = 1$. The inclusion of such nodes creates the opportunity for the new element being formed from existing nodes on the generation front.

- A collection of L points are generated along the straight line between points c' and m'. These points, denoted by $p'_1, p'_2, \ldots, p'_L$ are also ordered according to how close they are to node c'. Their addition ensures that a new element can always be generated. Here $L = 5$ is commonly used.

It is noted that the empirical values given to N and L can be automatically modified during the mesh generation process to include a sufficient number of nodes.

6. Nodes n'_j $(j = 1, 2, \ldots, M)$ from the potential node list and node c' are considered sequentially to form the new triangle with side $a'b'$. Such ordering allows the existing nodes to be considered first when they are not too far away from c'. The point that forms the new triangle with side $a'b'$ is taken to be the first node that satisfies the criterion (8.29) such that the newly formed sides of the triangle do not intersect any of the existing sides in the generation front.

   If neither nodes n'_j $(j = 1, 2, \ldots, M)$ nor node c' can form the new triangle, nodes p'_j $(j = 1, 2, \ldots, L)$ are tested. The first p'_j that verifies the criterion is taken to be the new node.

7. New element is formed. If the new node c' is adopted to generate the new triangle, its coordinates are transformed back to the original coordinate system by

$$\mathbf{x}_c = \mathbf{T}^{-1}\mathbf{x}_c' \tag{8.31}$$

8. The generation front list is updated after each new element is added and any new node is added to the node list.

   The element generation process continues until the number of the sides in the generation front reduces to zero. The domain is then discretized completely by triangular elements.

8.2.4 Mesh quality enhancement for triangles

Mesh quality enhancement is indispensable to all mesh generation algorithms, because the shape of the triangles generated directly is not always optimal, particularly for a strongly graded mesh with element size varying rapidly. To improve the shape of the elements, at the final stage of the mesh generation, various mesh quality enhancement techniques, such as mesh smoothing and mesh modification, are employed.

Mesh smoothing

In the process of mesh smoothing, the topological structure of the mesh is fixed, i.e., the nodal connections of the elements will not be altered, but the interior nodes are repositioned to produce triangles with somewhat improved shapes. The computationally most efficient smoothing algorithm is the well-known Laplacian smoothing[41] which repositions the internal node at the centroid of the polygon formed by its neighbouring nodes. The new position of an internal node i is computed as

$$x_i = \frac{1}{N}\sum_{j=1}^{N} x_j \quad \text{and} \quad y_i = \frac{1}{N}\sum_{j=1}^{N} y_j \tag{8.32}$$

where N is the number of the nodes linked to node i.

The mesh smoothing process consists of several (usually three to five) iterations. The technique has proved to be effective and generally adjusts the mesh into one with better shaped elements as shown in Fig. 8.11. However, the algorithm may fail if some of the neighbouring nodes of interior node i are boundary nodes and the polygon formed by these nodes is concave. The following example demonstrates this possibility.

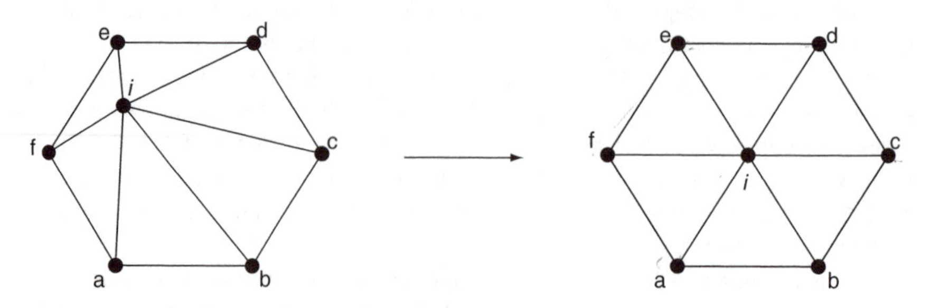

Fig. 8.11 Laplacian smoothing. Node *i* is repositioned.

Example 8.3: A simple triangulated concave domain with boundary nodes a, b, c, d, e, f and their coordinates is shown in Fig. 8.12. The new position of the only interior node i, calculated by Eq. (8.32), is at

$$x_{i'} = \frac{1}{6}(-1+4+4+0+0-1) = 1$$
$$y_{i'} = \frac{1}{6}(-1-1+0+0+4+4) = 1$$

which clearly is outside the domain.

To prevent such failure, various constraints can be added to Laplacian smoothing. One such constraint is to reposition node i only if the maximum interior angle of all the elements linked by node i is decreasing. This constraint is in fact only necessary for those interior nodes which have neighbouring nodes being boundary nodes.

Mesh modification

The topological structures of the mesh, such as the node–element relation, the side–element relation and the node–node relation, are established once the process of element generation is completed. These relations, to some extent, reflect the regularity of the mesh. For instance, the optimal value of node–element adjacency number *NE*, which shows how many elements are connected to a node in a triangular mesh, is defined as

$$NE_{op} = \begin{cases} 6 & \text{for interior nodes} \\ max(\lfloor \frac{3\theta}{\pi} + \frac{1}{2} \rfloor, 1) & \text{for boundary nodes} \end{cases} \tag{8.33}$$

where θ is the internal angle formed by boundary edges joined at the boundary node and $\lfloor c \rfloor$ is the integer part of the value c. When NE_{op} is attained for all the nodes, most of the elements in the resulting mesh are approximately equilateral. However, if an interior node has a node–element adjacency number far bigger or smaller than NE_{op}, the surrounding elements of the node may be very distorted.

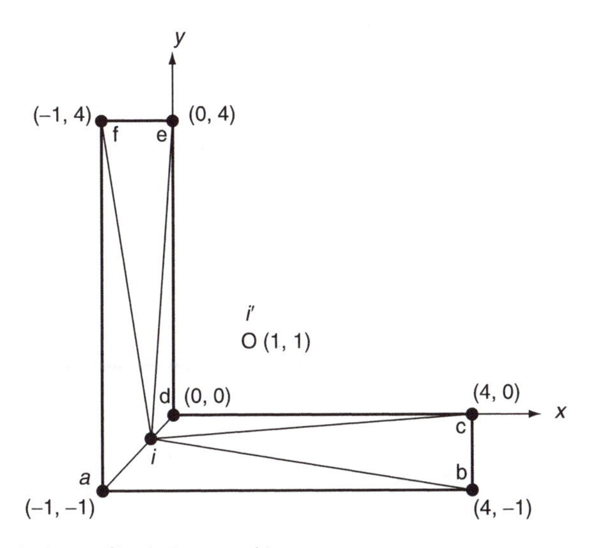

Fig. 8.12 A pathological case of Laplacian smoothing.

Distorted elements are, in general, inevitable for a mesh with varying element sizes as less regular transition elements are created during the element generation process. Distorted elements are also produced because the element size h is considered only locally when each element is formed. The distortion of the element caused by non-optimal topological structure of the mesh cannot be corrected by smoothing alone. In the following, techniques that alter the topology of the mesh in order to reduce the element distortion to a minimum are described.

Node elimination. A node elimination process consists of a loop over all the interior nodes. A node i will be eliminated if:

- i is linked with three elements, i.e., $NE_i = 3$. i is removed together with the three elements connected to it and replaced by a single element e_1' as illustrated in Fig. 8.13.
- i is shared by four elements, i.e., $NE_i = 4$. i is deleted from the mesh and its four related elements are reduced to two.

The possibility of such operations are depicted in Fig. 8.14.

It is noted that the element size distribution is almost unchanged in the process of node elimination.

Diagonal swapping. The process of diagonal swapping examines all the element sides common to two elements. Element sides that are part of a material interface should not be altered. Considering a side shared by two triangles e_1 and e_2 shown in Fig. 8.15, the edge ac will be replaced by edge bd together with elements e_1 and e_2 and be substituted by elements e_1' and e_2' when one of the following frequently used criteria is satisfied:

- The maximum internal angle of the new elements e_1' and e_2' should be smaller than that of elements e_1 and e_2.

  or
- The node–element adjacency number improves to be closer to the optimal value after swapping.

Both criteria work well in practice. Figure 8.16 shows two diagonal swapping steps that satisfy the criterion of reducing the maximum angle in a region with seven elements. The quality of the elements is obviously improved.

Swapping of diagonals is not allowed if it results in a negative area for one of the newly created elements.

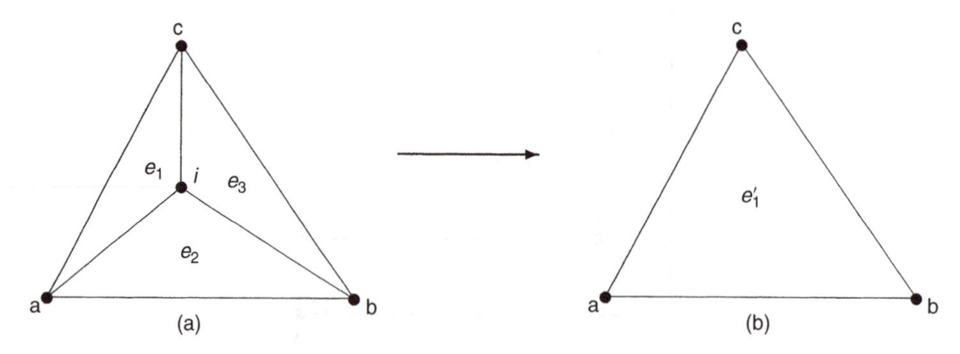

Fig. 8.13 Elimination of node i with $NE_i = 3$.

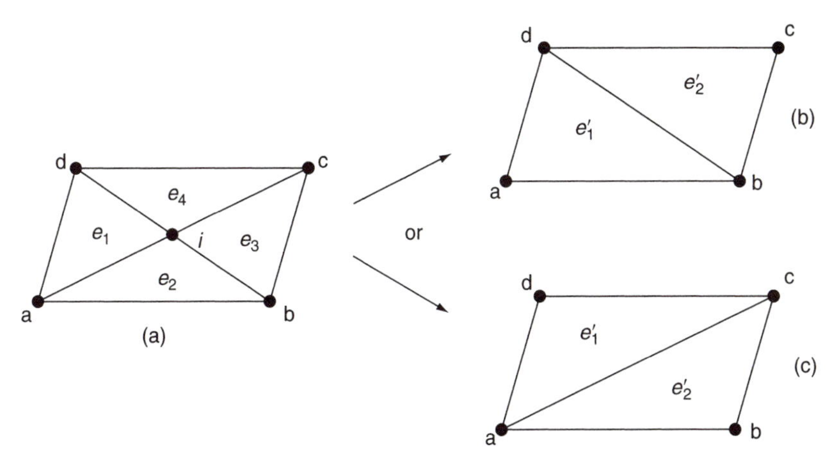

Fig. 8.14 Elimination of node e with $NE_e = 4$.

The mesh quality, such as the smoothness of the mesh and the regularity of the elements, can be significantly improved after combined applications of mesh quality enhancement techniques.

Example 8.4: Two triangular meshes generated by a mesh generator using the advancing front method are shown in Fig. 8.17 and Fig. 8.18. The mesh plotted in Fig. 8.17 is used for finite element analysis of a dam. The mesh parameters are arbitrarily given. The mesh of Fig. 8.18 was generated in the adaptive analysis of fluid dynamics. The mesh reflects the distribution of the specified mesh parameters which are computed from an *a posteriori* error estimator.[42]

8.2.5 Higher order elements

Higher order elements can be created easily by adding additional intermediate nodes to each element edge. For an interior edge, the position of an intermediate node is determined directly by interpolation using the positions of the nodes at each end of the edge. For a boundary edge, however, the parametric position of the intermediate node between nodes n_k and n_{k+1} may be computed either by Eq. (8.25) or by interpolating the parametric position of n_k and n_{k+1} directly, so that the position of the intermediate node can be mapped onto the

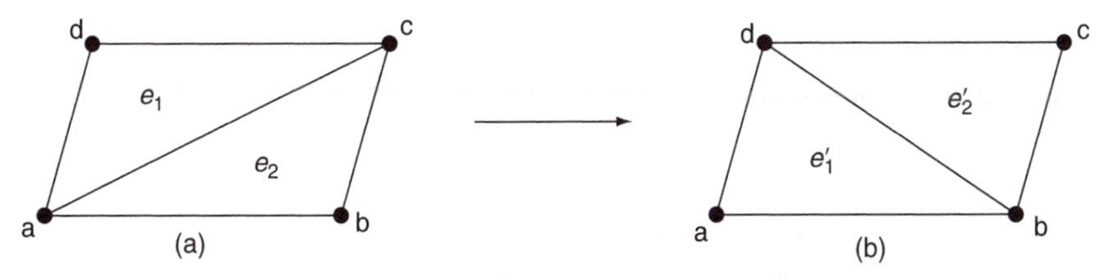

Fig. 8.15 Diagonal swapping. Diagonal *ac* replaced by diagonal *bd*. Elements e_1 and e_2 changed to elements e_1' and e_2'.

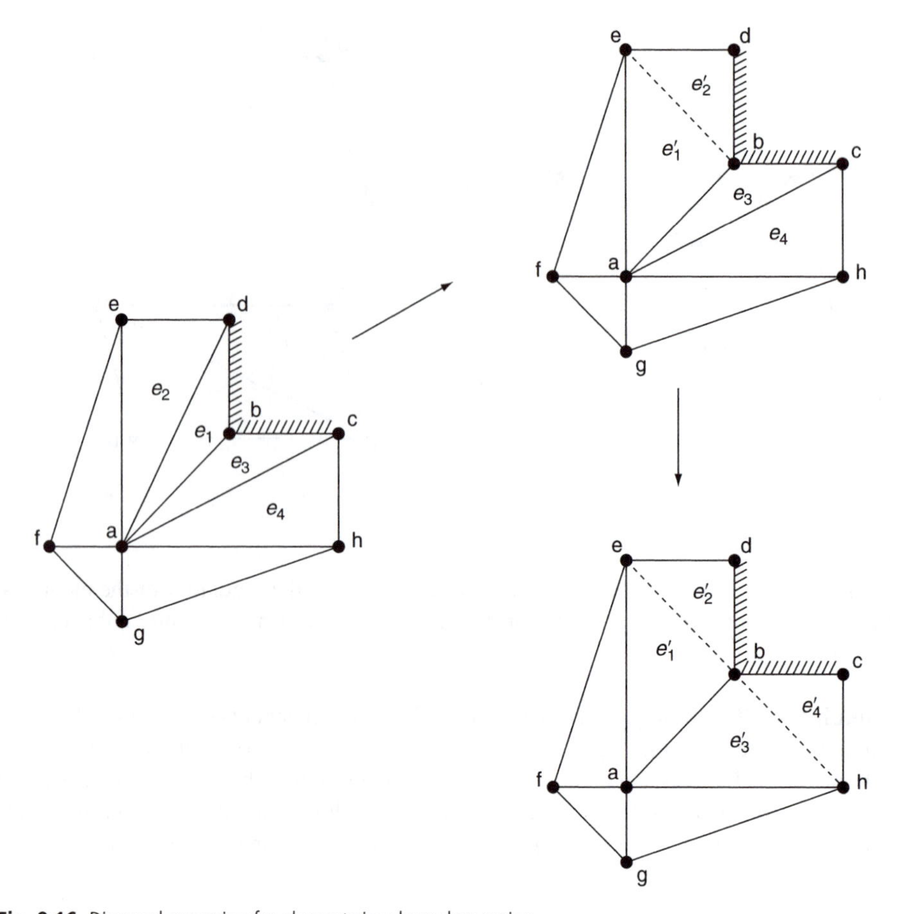

Fig. 8.16 Diagonal swapping for elements in a boundary region.

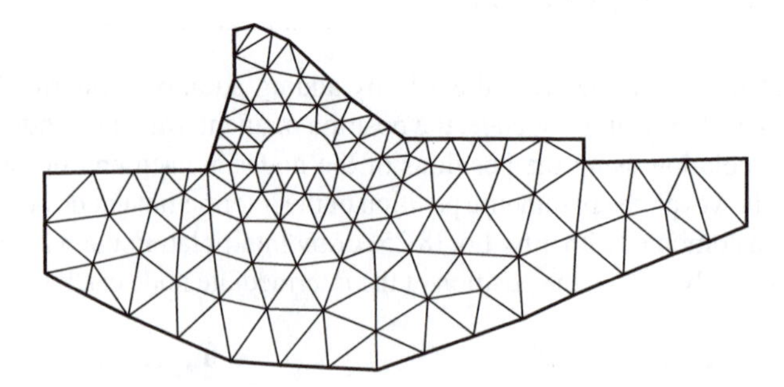

Fig. 8.17 Triangular mesh for a dam.

curvilinear boundary. These nodes are generated at the boundary node generation stage and placed on the boundary curve after the completion of the element generation process. The position of any interior nodes can be interpolated by the position of the element perimeter nodes. Figure 8.19 shows the locations of vertex nodes, edge nodes and interior nodes for some quadratic elements.

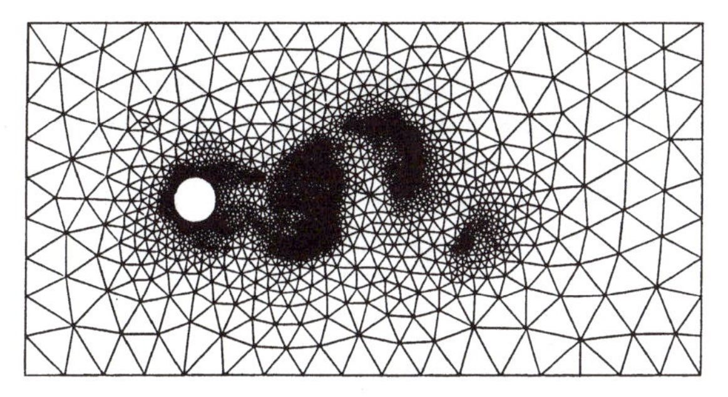

Fig. 8.18 Triangular mesh with mesh size distribution given by an error estimator.

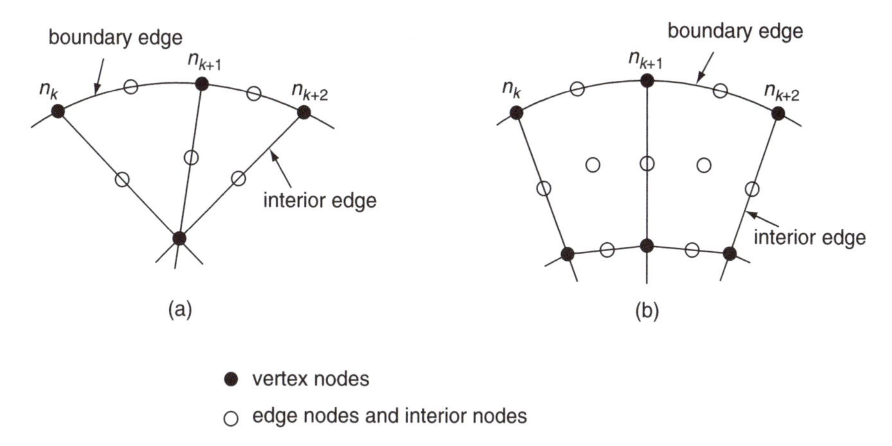

Fig. 8.19 Quadratic elements.

8.2.6 Remarks

Before we proceed to the discussion of surface mesh generation, several remarks are given on issues related to the topics discussed in this section.

Remark 8.1. The algorithm for the advancing front method has been shown to be robust in two-dimensional mesh generation of triangles and, although not discussed here, can easily be extended to generate quadrilaterals.[87] However, in the process of generating triangles, several empirical constants have been adopted, e.g., those used in Eqs (8.29) and (8.30). The optimal values of these constants are still unknown. In addition, what kind of correlation between the value of these constants and the structure of the resulting mesh is also an open question. Because of the lack of the mathematical rigour, a robust element generation process for the advancing front method in three dimensions is much more complicated than that described here for two dimensions.[23–26, 28]

Remark 8.2. In order to implement the search algorithms required in the element generation process efficiently, use of special data structures such as those proposed in references 44–46 for the advancing front method are advantageous.

Remark 8.3. Other methodologies for generating quadrilateral mesh can be found in references 30, 47–52 and 88. The various algorithms that convert an existing triangular mesh to a quadrilateral mesh[48–50, 52] are all robust, although the element size and element orientation of the resulting mesh are influenced by the pre-existing triangular mesh. Among these algorithms, the one proposed by Owen *et al.*,[52] uses the advancing front method in the process of converting triangles to quadrilaterals.

8.3 Surface mesh generation

Surface mesh generation is a prerequisite for many three-dimensional mesh generation algorithms, such as those based on the advancing front method and the Delaunay triangulation method, but it is not required for the finite octree method. However, generating a surface mesh prior to three-dimensional mesh generation has its advantages:

(a) The quality of three-dimensional meshes is strongly dependent on the quality of the surface mesh. A distorted triangle in the surface mesh will almost certainly result in a tetrahedron with poor quality.
(b) For many applications, an accurate description of the surface of a three-dimensional domain is essential. This can be readily realized by increasing the accuracy of the parametric representation of the surface and by assigning proper surface mesh element size distribution during the generation of the surface mesh.
(c) In modern engineering design, nearly all three-dimensional geometries are created by computer aided design (CAD) systems. The boundary representation (B-Rep) of the three-dimensional geometry exported by CAD systems often contains defects, e.g., gaps between connecting surfaces and discontinuities of boundary edges of the surface. These defects can be corrected before and during the surface mesh generation process. Consequently, accurate surface mesh generation prevents three-dimensional mesh generation algorithms from failing due to presence of defects.

Although essential to many three-dimensional mesh generation algorithms, surface mesh generation also has its own application in finite element analysis, this is especially true in the solution of shell problems. In the process of generating a surface mesh for a three-dimensional geometry, each face of the geometry is discretized individually, the complete surface mesh of the geometry will be formed by a final assembly of the faces. Since we are mainly concerned with the algorithmic procedure of the surface mesh generation, we shall discuss the mesh generation algorithm for an individual face. The basic idea of the algorithm described below, proposed by Peraire *et al.*[53] and Peiró,[54] is to perform mesh generation, according to the prescribed element size distribution, in the two-dimensional parametric plane and map the two-dimensional mesh onto the three-dimensional surface. From a computational standpoint, generating triangles (or quadrilaterals) on a plane is much simpler than that on a three-dimensional surface. The two-dimensional element generation algorithm described in the previous section can be readily applied. Nevertheless, in order to obtain a surface mesh that respects the prescribed geometrical characteristics, such as element size and element shape, the mesh parameters given to the three-dimensional surface mesh need to be transformed to the parametric plane. The mesh generation procedure includes four major steps:

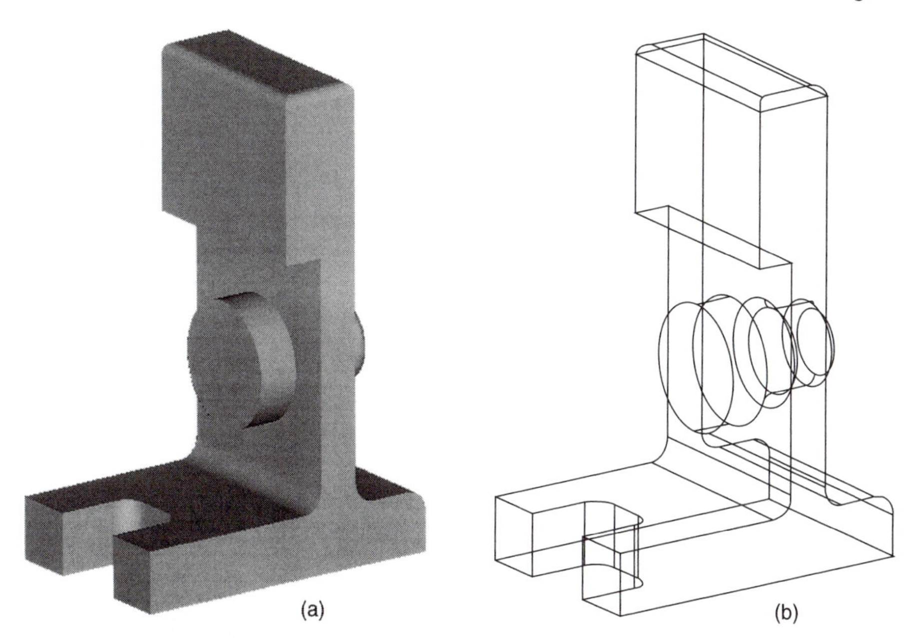

Fig. 8.20 (a) A machine part. (b) Boundary faces and boundary edges of the machine part.

- Perform node generation along the curved boundary edge to form the discretized boundary of the surface.
- Transform boundary nodes, therefore the discretized surface boundary, to the parametric plane.
- Perform element generation in the parametric space within the discretized boundary.
- Map the mesh in parametric space onto the surface using its parametric representation.

In the following we will mainly be concerned with the processes of boundary node generation and element generation in the parametric space, starting with the parametric representation of three-dimensional curves and surfaces.

8.3.1 Geometrical representation

In the boundary representation of CAD systems for three-dimensional solids, the surfaces and curves are usually given in parametric forms represented by a variety of composite spline surfaces and curves [e.g., in the form of Bézier, B-spline or NURBS (Non-Uniform Rational B-Splines)]. The faces of the solid are sections of the surfaces on which they are defined. The edges that connect the faces are portions of the spline curves and are the boundary of the faces. Figure 8.20 illustrates a machine part, its boundary faces and their connecting edges. Figure 8.21 shows two of the faces of the same part and the composite spline surfaces that include the faces as sections.

Although the surface mesh generation algorithm described in this section is independent of the parametric forms of the spline function that represents the curves and surfaces, we choose for simplicity to have boundary edges and surfaces expressed by the interpolatory cubic Hermite parametric spline curves and surfaces, respectively. It is worth noting that

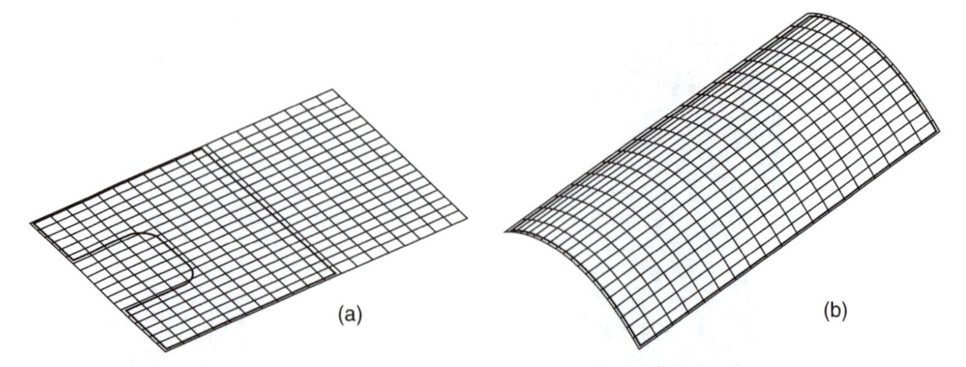

Fig. 8.21 Surfaces of the machine part. (a) A flat surface. (b) A curved surface. Darker lines are the boundaries of the machine part surfaces.

transforming various curve and surface representations created by CAD systems to a single convenient form can simplify the development of a mesh generation program.

Curve representation

The same curve representation described in Sec. 8.2.2 can be used for curves in three dimensions except that the vector valued function is now in the form of

$$\mathbf{x}(t) = \{x(t) \quad y(t) \quad z(t)\}, \qquad 0 \leq t \leq 1 \tag{8.34}$$

with t again as the parameter.

An interpolation by piecewise cubic Hermite polynomials through a set of ordered data points $\mathbf{x}_i$ $(i = 0, 1, \ldots, n)$ located on the curve generated by CAD systems for each boundary edge constructs a composite parametric spline curve in the same format as described by Eq. (8.8), except that now $\mathbf{x}(u)$ represents a position in three-dimensional space. Similarly, the number of the interpolation points and the distribution of the points should be chosen such that the surface edge can be accurately represented. The global mapping between the region $u \in [u_0, u_n]$ in parametric space and a three-dimensional cubic composite curve provided by $\mathbf{x}(u)$ is depicted in Fig. 8.22.

Surface representation

Surfaces are represented by composite Hermite surfaces. A parametric bicubic Hermite surface patch can be obtained from the tensor product of two cubic Hermite parametric segments. It is represented by the four cubic curves that form its boundary and the twist vectors at its four corner points. In parametric form, the surface patch is expressed as

$$\begin{aligned}\mathbf{x}(s,t) &= \{H_0(s) \quad H_1(s) \quad G_0(s) \quad G_1(s)\}\, \mathbf{B}_c \begin{Bmatrix} H_0(t) \\ H_1(t) \\ G_0(t) \\ G_1(t) \end{Bmatrix} \\ &= \{1 \quad s \quad s^2 \quad s^3\}\, \mathbf{M}\mathbf{B}_c\mathbf{M}^{\mathrm{T}} \begin{Bmatrix} 1 \\ t \\ t^2 \\ t^3 \end{Bmatrix}, \qquad 0 \leq s, t \leq 1\end{aligned} \tag{8.35}$$

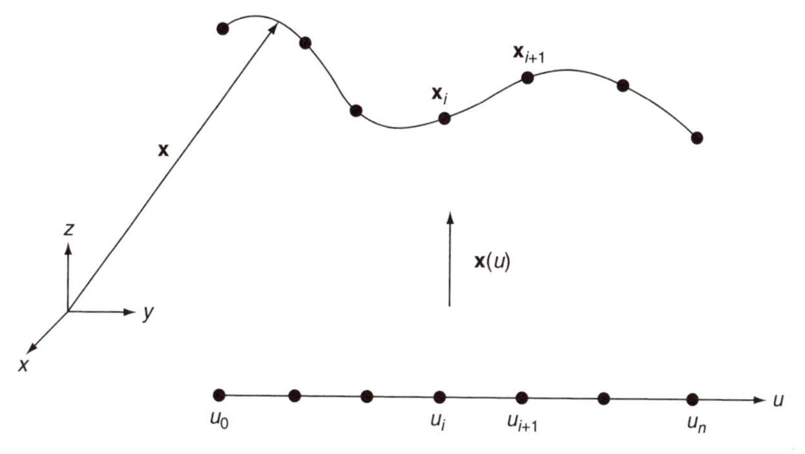

Fig. 8.22 Three-dimensional composite cubic spline interpolation.

where **M** is the matrix defined in Eq. (8.7) and

$$\mathbf{B}_c = \begin{bmatrix} \mathbf{x}(0,0) & \mathbf{x}(0,1) & \mathbf{x}_{,t}(0,0) & \mathbf{x}_{,t}(0,1) \\ \mathbf{x}(1,0) & \mathbf{x}(1,1) & \mathbf{x}_{,t}(1,0) & \mathbf{x}_{,t}(1,1) \\ \mathbf{x}_{,s}(0,0) & \mathbf{x}_{,s}(0,1) & \mathbf{x}_{,st}(0,0) & \mathbf{x}_{,st}(0,1) \\ \mathbf{x}_{,s}(1,0) & \mathbf{x}_{,s}(1,1) & \mathbf{x}_{,st}(1,0) & \mathbf{x}_{,st}(1,1) \end{bmatrix} \tag{8.36}$$

is often called the boundary condition matrix, in which $\mathbf{x}(0,0)$, $\mathbf{x}(0,1)$, $\mathbf{x}(1,0)$, $\mathbf{x}(1,1)$ are the four corners of the patch. The tangents $\mathbf{x}_{,s}$ and $\mathbf{x}_{,t}$, and the twists $\mathbf{x}_{,st}$ at the corners are given by

$$\mathbf{x}_{,s} = \frac{\partial \mathbf{x}}{\partial s}, \quad \mathbf{x}_{,t} = \frac{\partial \mathbf{x}}{\partial t}, \quad \mathbf{x}_{,st} = \frac{\partial^2 \mathbf{x}}{\partial s \partial t} \tag{8.37}$$

A piecewise composite surface is obtained by interpolation through a topologically rectangular set of data points $\mathbf{x}_{ij}$ $(i = 0, \ldots, m;\ j = 0, \ldots, n)$ and their corresponding parameter values (v_i, w_j) $(i = 0, \ldots, m;\ j = 0, \ldots, n)$ generated from the CAD representation of the surfaces. It consists of a network of $m \times n$ quadrilateral surface patches. The surface patch $[v_{i-1}, v_i] \times [w_{j-1}, w_j]$ $(i = 1, 2, \ldots, m;\ j = 1, \ldots, n)$ in the network is described by

$$\mathbf{x}(v, w) = \left\{1 \quad s \quad s^2 \quad s^3\right\} \mathbf{M}\mathbf{B}_c\mathbf{M}^{\mathrm{T}} \begin{Bmatrix} 1 \\ t \\ t^2 \\ t^3 \end{Bmatrix} \qquad 0 \le s, t \le 1 \tag{8.38}$$

Here $s = (v - v_{i-1})/\Delta_i$ and $t = (w - w_{j-1})/\Delta_j$ are local coordinates of the corresponding intervals $[v_{i-1}, v_i]$ and $[w_{j-1}, w_j]$, $\Delta_i = v_i - v_{i-1}$ and $\Delta_j = w_j - w_{j-1}$ are the respective lengths of the intervals, and the boundary condition matrix is of the form

$$\mathbf{B}_c = \begin{bmatrix} \mathbf{x}(v_{i-1}, w_{j-1}) & \mathbf{x}(v_{i-1}, w_j) & \Delta_j \mathbf{x}_{,w}(v_{i-1}, w_{j-1}) & \Delta_j \mathbf{x}_{,w}(v_{i-1}, w_j) \\ \mathbf{x}(v_i, w_{j-1}) & \mathbf{x}(v_i, w_j) & \Delta_j \mathbf{x}_{,w}(v_i, w_{j-1}) & \Delta_j \mathbf{x}_{,w}(v_i, w_j) \\ \Delta_i \mathbf{x}_{,v}(v_{i-1}, w_{i-1}) & \Delta_i \mathbf{x}_{,v}(v_{i-1}, w_j) & \Delta_i \Delta_j \mathbf{x}_{,vw}(v_{i-1}, w_{j-1}) & \Delta_i \Delta_j \mathbf{x}_{,vw}(v_{i-1}, w_j) \\ \Delta_i \mathbf{x}_{,v}(v_i, w_{j-1}) & \Delta_i \mathbf{x}_{,v}(v_i, w_j) & \Delta_i \Delta_j \mathbf{x}_{,vw}(v_i, w_{j-1}) & \Delta_i \Delta_j \mathbf{x}_{,vw}(v_i, w_j) \end{bmatrix} \tag{8.39}$$

The unknown tangents $\mathbf{x}_{,v}$ and $\mathbf{x}_{,w}$ and twist $\mathbf{x}_{,vw}$ of the composite surface can be determined by following the standard procedure for piecewise parametric surface interpolation requiring the resulting composite surface to be at least C^1 continuous[39, 40] (a C^2 continuous composite surface is usually adopted in CAD surface mesh generation).

Similar to that used in the implementation of curve interpolation described in Sec. 8.2.2, the parametric coordinates of v_i and w_j are usually taken as $v_i = i$ and $w_j = j$ respectively. For surface patch $[v_{i-1}, v_i] \times [w_{j-1}, w_j] = [i-1, i] \times [j-1, j]$, the global parametric coordinates v and w are related to the local coordinates s and t through

$$\begin{aligned} v &= v_{i-1} + s = i - 1 + s, \qquad i = 1, 2, \ldots, n; \\ w &= w_{j-1} + t = j - 1 + t, \qquad j = 1, 2, \ldots, m \end{aligned} \tag{8.40}$$

The global one-to-one mapping between the parametric plane and the composite surface provided by $\mathbf{x}(v, w)$ is illustrated in Fig. 8.23. Indeed, a region with curved boundary, as depicted in Fig. 8.24, within the parametric plane $[v_0, v_m] \times [w_0, w_n]$ constitutes a portion of the composite spline surface in a three-dimensional space.

8.3.2 Geometrical characteristics of the surface mesh

A surface mesh generally is three dimensional, thus, the spatial distribution of the shape and size of the surface elements needs to be specified in three dimensions. Because the mesh generation is performed in the two-dimensional parametric plane, proper spatial distribution of the element shape and size in the parametric space also needs to be defined.

Mesh control function in three dimensions

As was adopted in two-dimensional mesh generation, the geometrical characteristics of the surface mesh such as the distribution of the element shapes and sizes on the surface are controlled by mesh parameters. For surface mesh, mesh parameters include a set of three mutually orthogonal directions α_i $(i = 1, 2, 3)$, and their associated element sizes h_i $(i = 1, 2, 3)$ as shown in Fig. 8.25.

Mesh parameters are defined at the nodes of a three-dimensional background mesh, which usually consists of a small number of tetrahedral elements. The background mesh can be constructed to cover the entire surface of the three-dimensional geometry or to cover each face of the geometry individually. In either case, the background mesh will be created automatically by dividing one or several hexahedra (tetrahedra or prisms) into tetrahedra. Figure 8.26 shows a tetrahedral background mesh generated from a single hexahedron for a single surface. The spatial distribution of the mesh parameters is furnished by the background mesh. At a particular point on the surface, the mesh parameters are computed by a linear interpolation of the values assigned at the nodes of the background mesh. The geometrical characteristics at a point of the surface mesh is therefore attained. For instance, when all three element sizes are found to be equal at the point, the tetrahedral elements in the surrounding area of the point will be approximately equilateral. Since the faces of the tetrahedral elements form the surface elements they are therefore approximately equilateral triangles.

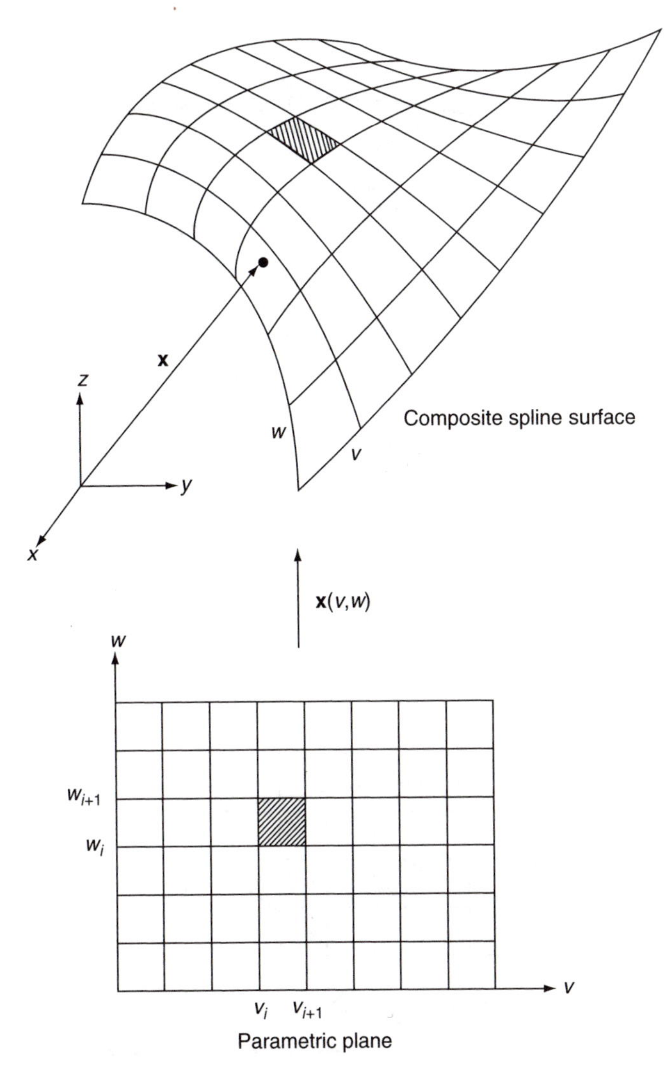

Fig. 8.23 Composite cubic surface interpolation.

A three-dimensional geometrical transformation matrix **T**, similar to that defined in two-dimensional mesh generation, is in the form of

$$\mathbf{T}(\mathbf{x}) = \sum_{i=1}^{3} \frac{1}{h_i} \boldsymbol{\alpha}_i \boldsymbol{\alpha}_i^{\mathrm{T}} \tag{8.41}$$

and constitutes a scaling factor $1/h_i$ in each of the $\boldsymbol{\alpha}_i$ directions ($i = 1, 2, 3$). When it is applied to tetrahedra with element sizes h_i in directions $\boldsymbol{\alpha}_i$ at a given point in the coordinate system (x, y, z), the tetrahedra will be mapped to equilateral tetrahedra in the normalized space (x', y', z'). Consequently, the surface element will be mapped to equilateral triangles in the normalized space. Indeed, transformation **T** also provides a mapping relationship

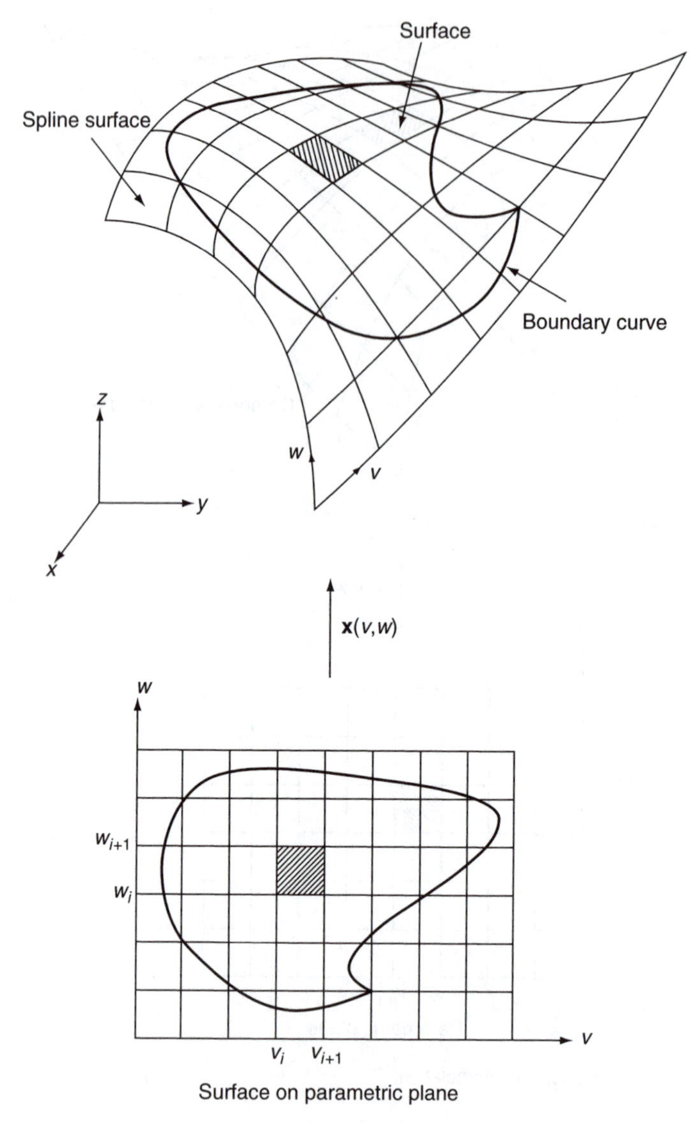

Fig. 8.24 A region in parametric plane and its image on spline surface.

between parametric coordinates and the three-dimensional normalized coordinates

$$\mathbf{x}'(v, w) = \mathbf{T}\mathbf{x}(v, w) \tag{8.42}$$

As indicated, surface mesh generation will be performed in the parametric space (v, w) and mapped onto the surface in three dimensions. In order to accomplish the planar mesh generation on the parametric plane, appropriate planar mesh parameters have to be assigned. These mesh parameters must be given in such a way that, after being mapped onto the surface, the final surface mesh respects the specified geometrical characteristics. Such

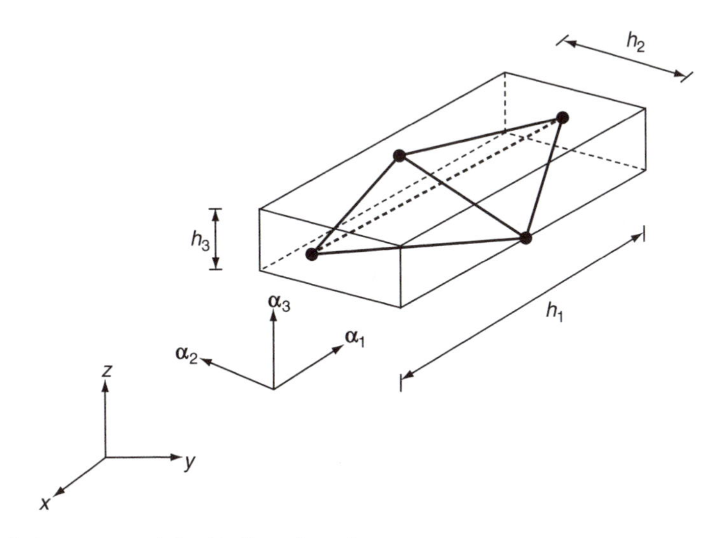

Fig. 8.25 Mesh parameters defined in three dimensions.

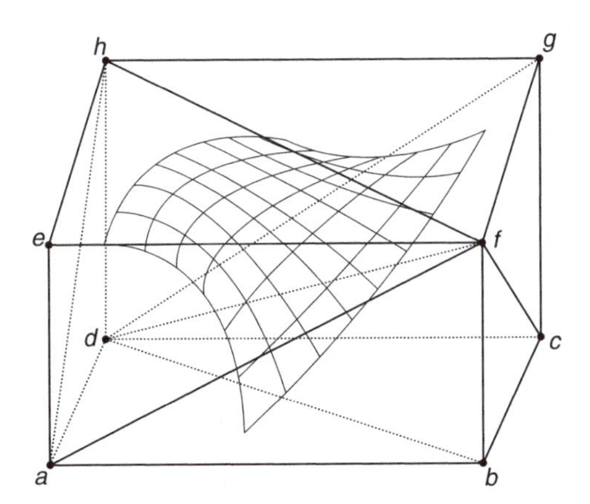

Fig. 8.26 A six tetrahedra background mesh derived from a hexahedron. The six tetrahedra are (e, f, h, a), (a, b, f, d), (f, h, d, a), (b, f, d, c), (d, g, h, f), (d, c, g, f).

a requirement can be achieved by deriving the planar mesh parameters from the three-dimensional surface mesh parameters.

Mesh parameters in parametric plane

We start by examining a curve in the parametric plane expressed by parameter ξ and illustrated in Fig. 8.27, i.e.,

$$v = v(\xi), \qquad w = w(\xi) \tag{8.43}$$

or in vector form

$$\mathbf{u}(\xi) = (v(\xi), w(\xi)) \tag{8.44}$$

At a particular point $\mathbf{u}(\xi_p)$, the square of arc length element ζ along the curve is expressed as

$$(\mathrm{d}\zeta)^2 = (\mathrm{d}v)^2 + (\mathrm{d}w)^2 = \left(\frac{\mathbf{du}}{\mathrm{d}\xi}\mathrm{d}\xi\right)^{\mathrm{T}}\left(\frac{\mathbf{du}}{\mathrm{d}\xi}\mathrm{d}\xi\right) = \mathbf{t}_\xi^{\mathrm{T}}\mathbf{t}_\xi(l_\xi d\xi)^2 = (l_\xi \mathrm{d}\xi)^2 \tag{8.45}$$

where l_ξ is the length of the tangent vector

$$l_\xi = \sqrt{\left(\frac{\mathbf{du}}{\mathrm{d}\xi}\right)^{\mathrm{T}}\left(\frac{\mathbf{du}}{\mathrm{d}\xi}\right)} \tag{8.46}$$

and $\mathbf{t}_\xi$ is the unit tangent vector

$$\mathbf{t}_\xi = \frac{1}{l_\xi}\frac{\mathbf{du}}{\mathrm{d}\xi} \tag{8.47}$$

This shows that the arc length element in the direction of a unit tangent along the curve in the parametric plane can be expressed by

$$\mathrm{d}\zeta = l_\xi\, \mathrm{d}\xi \tag{8.48}$$

We now consider the image of the planar curve at point $\mathbf{x}(v(\xi_p), w(\xi_p))$ on the surface represented by

$$\mathbf{x}(v, w) = (x(v, w),\ y(v, w),\ z(v, w)) \tag{8.49}$$

The square of the arc length element in the direction of the unit tangent $\mathbf{t}$ along the curve on the surface is given by

$$(\mathrm{d}s)^2 = (\mathrm{d}x)^2 + (\mathrm{d}y)^2 + (\mathrm{d}z)^2 = (\mathbf{dx})^{\mathrm{T}}(\mathbf{dx}) \tag{8.50}$$

where

$$\mathbf{dx} = \begin{Bmatrix} \mathrm{d}x \\ \mathrm{d}y \\ \mathrm{d}z \end{Bmatrix} = \begin{bmatrix} \dfrac{\partial x}{\partial v} & \dfrac{\partial x}{\partial w} \\ \dfrac{\partial y}{\partial v} & \dfrac{\partial y}{\partial w} \\ \dfrac{\partial z}{\partial v} & \dfrac{\partial z}{\partial w} \end{bmatrix} \begin{Bmatrix} \mathrm{d}v \\ \mathrm{d}w \end{Bmatrix} = \left[\mathbf{x}_{,v},\quad \mathbf{x}_{,w}\right]\mathbf{du} \tag{8.51}$$

Assume that the transformation matrix $\mathbf{T}_{\mathrm{d}s}$, correlated to $\mathrm{d}s$ at $\mathbf{x}(v(\xi_p), w(\xi_p))$, maps $\mathbf{dx}$ to the normalized space with a unit length, i.e.,

$$(\mathrm{d}s')^2 = (\mathrm{d}x')^2 + (\mathrm{d}y')^2 + (\mathrm{d}z')^2 = (\mathbf{dx})^{\mathrm{T}}\mathbf{C}_{ds}(\mathbf{dx}) = 1 \tag{8.52}$$

where

$$\mathbf{C}_{\mathrm{d}s} = \mathbf{T}_{\mathrm{d}s}^{\mathrm{T}}\mathbf{T}_{\mathrm{d}s} = \sum_{i=1}^{3}\frac{1}{h_{\mathrm{d}s_i}^2}\boldsymbol{\alpha}_i\boldsymbol{\alpha}_i^{\mathrm{T}} \tag{8.53}$$

corresponding to the *Euclidean metric tensor* in the normalized space, and $h_{\mathrm{d}s_i}$ can be viewed as element size associated with $\mathrm{d}s$ in the direction of $\boldsymbol{\alpha}_i$.

Substitute $\mathbf{dx}$ into Eq. (8.52), we have

$$\left([\mathbf{x}_{,v},\ \mathbf{x}_{,w}]\,\mathbf{du}\right)^{\mathrm{T}}\mathbf{C}_{\mathrm{d}s}\left([\mathbf{x}_{,v},\ \mathbf{x}_{,w}]\,\mathbf{du}\right) = 1 \tag{8.54}$$

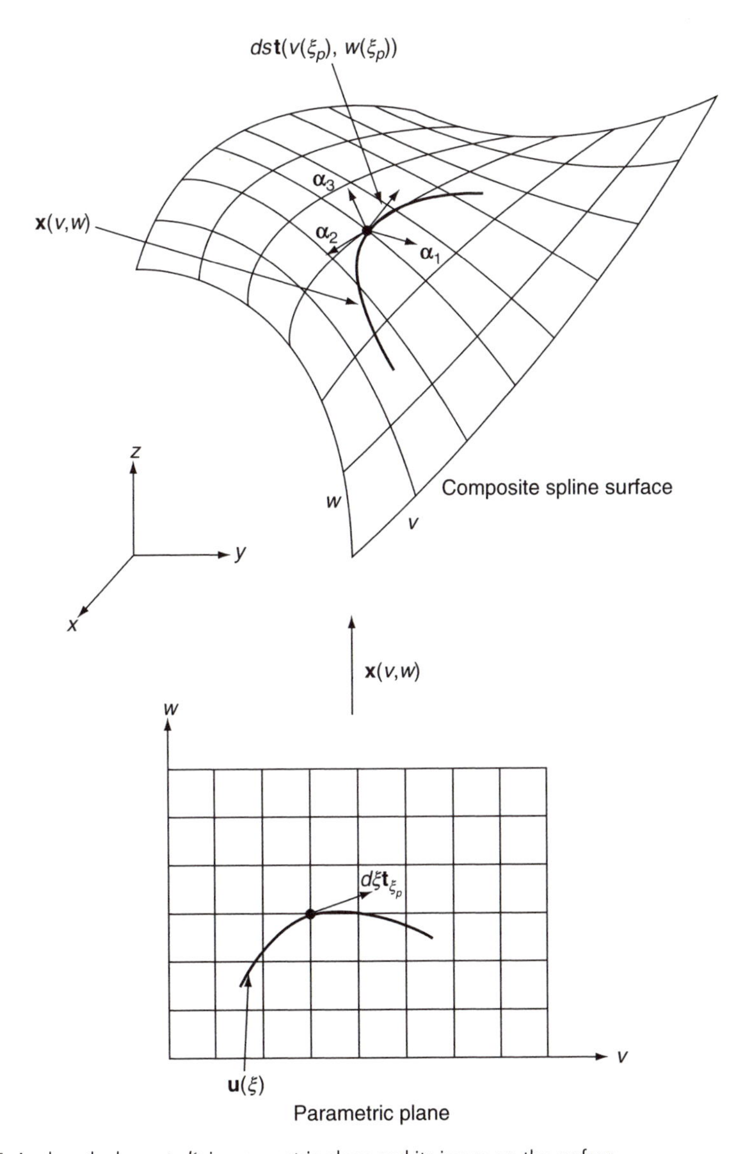

Fig. 8.27 Arc length element $d\xi$ in parametric plane and its image on the surface.

From Eqs (8.47) and (8.48), we know that

$$\mathrm{d}\mathbf{u} = l_\xi \, \mathrm{d}\xi \, \mathbf{t}_\xi = \mathrm{d}\zeta \, \mathbf{t}_\xi \tag{8.55}$$

which when substituted into Eq. (8.54) gives

$$\left([\mathbf{x}_{,v}, \ \mathbf{x}_{,w}] \, \mathbf{t}_\xi\right)^{\mathrm{T}} \mathbf{C}_{\mathrm{d}s} \left([\mathbf{x}_{,v}, \ \mathbf{x}_{,w}] \, \mathbf{t}_\xi\right) (\mathrm{d}\zeta)^2 = 1 \tag{8.56}$$

This shows that for an arc length element ds along a curve on the surface, the arc length along its curve in the parametric plane is

$$\mathrm{d}\zeta = \frac{1}{\sqrt{\left([\mathbf{x}_{,v},\ \mathbf{x}_{,w}]\,\mathbf{t}_\xi\right)^{\mathrm{T}} \mathbf{C}_{\mathrm{d}s}\left([\mathbf{x}_{,v},\ \mathbf{x}_{,w}]\,\mathbf{t}_\xi\right)}} \tag{8.57}$$

Replacing ds by the curvilinear element size h_s and $\mathbf{T}_{\mathrm{d}s}$ by $\mathbf{T}$ on the surface. Using h_ξ, the element size along the planar curve on the parametric plane, in place of dζ, we have

$$\left([\mathbf{x}_{,v},\ \mathbf{x}_{,w}]\,\mathbf{t}_\xi\right)^{\mathrm{T}} \mathbf{C}\left([\mathbf{x}_{,v},\ \mathbf{x}_{,w}]\,\mathbf{t}_\xi\right)(h_\xi)^2 = 1 \tag{8.58}$$

or

$$h_\xi = \frac{1}{\sqrt{\left([\mathbf{x}_{,v},\ \mathbf{x}_{,w}]\,\mathbf{t}_\xi\right)^{\mathrm{T}} \mathbf{C}\left([\mathbf{x}_{,v},\ \mathbf{x}_{,w}]\,\mathbf{t}_\xi\right)}} = \frac{1}{\sqrt{\mathbf{t}_\xi^{\mathrm{T}} \mathbf{G}\,\mathbf{t}_\xi}} \tag{8.59}$$

The matrix of the metric tensor is now expressed as

$$\mathbf{C} = \mathbf{T}^{\mathrm{T}}\mathbf{T} = \sum_{i=1}^{3} \frac{1}{h_i^2} \boldsymbol{\alpha}_i \boldsymbol{\alpha}_i^{\mathrm{T}} \tag{8.60}$$

and

$$\mathbf{G} = [\mathbf{x}_{,v},\ \mathbf{x}_{,w}]^{\mathrm{T}} \mathbf{C} [\mathbf{x}_{,v},\ \mathbf{x}_{,w}] \tag{8.61}$$

is the matrix of the metric tensor in the parametric space.

If we assume that $\mathbf{T}$ is a constant matrix in the neighbourhood of point $\mathbf{x}(v(\xi_p),\ w(\xi_p))$, substitute Eq. (8.60) into Eq. (8.59) and note the mapping relationship of Eq. (8.42), we obtain h_ξ in a somewhat different form

$$h_\xi = \frac{1}{\sqrt{\mathbf{t}_\xi^{\mathrm{T}} \mathbf{g} \mathbf{t}_\xi}} \tag{8.62}$$

where

$$\mathbf{g} = \begin{bmatrix} g_{vv} & g_{vw} \\ g_{wv} & g_{ww} \end{bmatrix} \tag{8.63}$$

is the *first fundamental matrix* of the surface in the normalized space,[38] and

$$\begin{aligned} g_{vv} &= \left(\frac{\partial x'}{\partial v}\right)^2 + \left(\frac{\partial y'}{\partial v}\right)^2 + \left(\frac{\partial z'}{\partial v}\right)^2 \\ g_{vw} &= 2\left(\frac{\partial x'}{\partial v}\frac{\partial x'}{\partial w} + \frac{\partial y'}{\partial v}\frac{\partial y'}{\partial w} + \frac{\partial z'}{\partial v}\frac{\partial z'}{\partial w}\right) = g_{wv} \\ g_{ww} &= \left(\frac{\partial x'}{\partial w}\right)^2 + \left(\frac{\partial y'}{\partial w}\right)^2 + \left(\frac{\partial z'}{\partial w}\right)^2 \end{aligned} \tag{8.64}$$

Consequently, we have established the relationship between $\mathbf{G}$ and $\mathbf{g}$. When transformation $\mathbf{T}$ is a constant matrix,

$$\mathbf{G} = \mathbf{g}$$

This shows that the matrix of the metric tensor in parametric space is the same as the first fundamental matrix of the surface in the normalized space.

The two-dimensional mesh parameters $\boldsymbol{\alpha}_i(v(\xi_p), w(\xi_p))$ and $h_i(v(\xi_p), w(\xi_p))$ for the planar mesh on the parametric plane are computed from the directions in which h_ξ attains an extremum. To this end, Eq. (8.59) is rewritten in the form of

$$\mathbf{t}_\xi^{\mathrm{T}} \mathbf{G} \mathbf{t}_\xi = \frac{1}{h_\xi^2} \tag{8.65}$$

Let

$$\mathbf{G} = \begin{bmatrix} G_{11} & G_{12} \\ G_{21} & G_{22} \end{bmatrix} \tag{8.66}$$

where

$$G_{11} = \mathbf{x}_{,v}^{\mathrm{T}} \mathbf{C} \mathbf{x}_{,v}, \quad G_{12} = \mathbf{x}_{,v}^{\mathrm{T}} \mathbf{C} \mathbf{x}_{,w}, \quad G_{22} = \mathbf{x}_{,w}^{\mathrm{T}} \mathbf{C} \mathbf{x}_{,w} \tag{8.67}$$

and $G_{21} = G_{12}$.

From Eq. (8.65), we know that finding the direction in which $1/h_\xi^2$ reaches an extremum involves the solution of an eigenproblem for the symmetric matrix $\mathbf{G}$. To this end we let λ_1, λ_2 with $\lambda_1 \geq \lambda_2$ denote the eigenvalues and $\mathbf{a}_1, \mathbf{a}_2$ the eigenvectors of $\mathbf{G}$. The mesh parameters in the parametric plane at point $(v(\xi_p), w(\xi_p))$ are given as

$$h_1(v, w) = \frac{1}{\sqrt{\lambda_1}}, \qquad h_2(v, w) = \frac{1}{\sqrt{\lambda_2}} \tag{8.68}$$

and

$$\boldsymbol{\alpha}_1(v, w) = \mathbf{a}_1, \qquad \boldsymbol{\alpha}_2(v, w) = \mathbf{a}_2 \tag{8.69}$$

where the eigenvalues are computed as

$$\lambda_1 = \frac{G_{11} + G_{22}}{2} + \sqrt{\frac{(G_{11} - G_{22})^2}{4} + G_{12}^2}, \quad \lambda_2 = \frac{G_{11} + G_{22}}{2} - \sqrt{\frac{(G_{11} - G_{22})^2}{4} + G_{12}^2} \tag{8.70}$$

the eigenvectors are

$$\mathbf{a}_1 = (\cos\theta, \sin\theta), \qquad \mathbf{a}_2 = (-\sin\theta, \cos\theta) \tag{8.71}$$

with

$$\theta = \frac{1}{2} \tan^{-1} \left(\frac{2G_{12}}{G_{11} - G_{22}} \right) \tag{8.72}$$

8.3.3 Discretization of three-dimensional curves

In order to perform mesh generation on the parametric plane, the boundary curves of the surface will be discretized in three-dimensional space and then projected to the parametric plane by inverse mapping.

Node generation on the curves

The algorithmic procedure for the node generation on three-dimensional curves is identical to that described in Sec. 8.2.2 for the two-dimensional curve. The curves are of course now represented by Eq. (8.34). The procedure listed below is the same as that discussed in Sec. 8.2.2 but now expressed in its three-dimensional form.

1. A set of sampling points $\mathbf{x}_l = \mathbf{x}(u_l)$ $(l = 0, 1, 2, \ldots, m)$ is first placed along the curve with parameters u_l uniformly distributed as shown in Fig. 8.28. The unit tangent vector for the three-dimensional curve is computed at the sampling point as

$$\mathbf{t}_l = (t_{1_l}, t_{2_l}, t_{3_l}) \tag{8.73}$$

where

$$t_{1_l} = \frac{x_{,u_l}}{l_{u_l}}; \quad t_{2_l} = \frac{y_{,u_l}}{l_{u_l}}; \quad t_{3_l} = \frac{z_{,u_l}}{l_{u_l}} \tag{8.74}$$

with $l_{u_l} = \sqrt{x^2_{,u_l} + y^2_{,u_l} + z^2_{,u_l}}$ and $\mathbf{x}_{,u_l} = (x_{,u_l}, y_{,u_l}, z_{,u_l})$ the tangent vector.

2. Using the interpolated values from the background mesh, the mesh parameters $\boldsymbol{\alpha}_{l_i}$, h_{l_i} $(i = 1, 2, 3)$ and the associated transformation matrix $\mathbf{T}_l$ are computed at each sampling point.
3. The element size h_{s_l} at the sampling point $\mathbf{x}_l$ $(l = 0, 1, 2, \ldots, m)$ in the tangent direction is calculated by

$$h_{s_l} = \frac{1}{\sqrt{(\mathbf{T}_l\mathbf{t}_l)^{\mathrm{T}}(\mathbf{T}_l\mathbf{t}_l)}} = \frac{1}{\sqrt{\mathbf{t}_l^{\mathrm{T}}\mathbf{C}_l\mathbf{t}_l}} \tag{8.75}$$

where the metric tensor $\mathbf{C}$ is defined in Eq. (8.60).

4. A continuous element size distribution function is obtained by linear interpolation of the nodal values h_{s_l} at sampling points using finite element shape functions

$$h(u) = \sum_{l=0}^{m} h_{s_l} N_l(u) \tag{8.76}$$

The element density function is set to be $1/h(u)$.

5. The total number of sides N to be generated along the curve is taken to be the nearest integer to

$$A = \int_0^L \frac{1}{h(u)}\,\mathrm{d}s = \int_{u_0}^{u_m} \frac{1}{h(u)} \sqrt{(x_u)^2 + (y_u)^2 + (z_u)^2}\,\mathrm{d}u \tag{8.77}$$

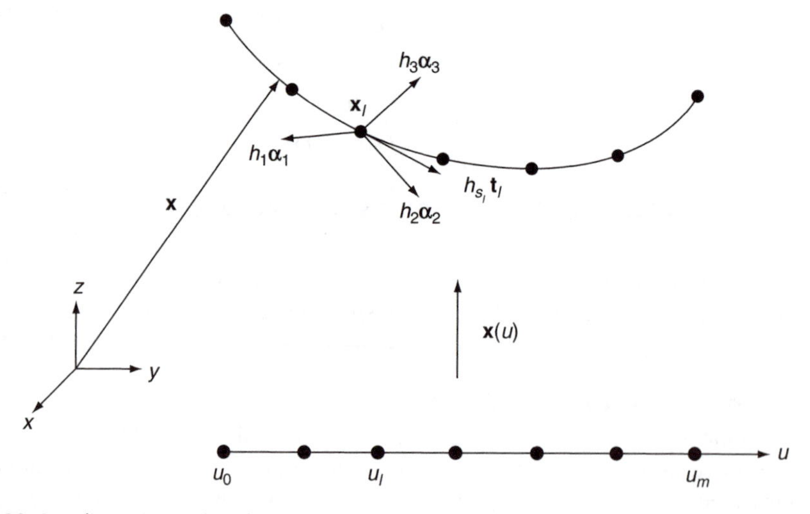

Fig. 8.28 Sampling points and mesh parameters on a three-dimensional composite curve segment.

where L is the length of the curve. The consistency index is calculated as

$$\theta = \frac{A}{N} \tag{8.78}$$

In addition to the already known end points $\mathbf{x}(u_0)$ and $\mathbf{x}(u_m)$ of the curve, there will be $(N - 1)$ new nodes created along the curve.

6. The parametric positions u_{k+1} $(k = 0, 1, 2, \ldots, (N-2))$ of the new nodes are computed consecutively from

$$\theta = \int_{u_k}^{u_{k+1}} \frac{1}{h(u)} \sqrt{(x_u)^2 + (y_u)^2 + (z_u)^2}\, du \tag{8.79}$$

by an iterative method, such as a Newton method.

7. Finally, the position of all the new points are mapped onto the boundary curve using Eq. (8.34).

Place boundary nodes to parametric plane

In order to perform surface mesh generation in the parametric plane the discretized boundary curves must be placed to the parametric plane to form the boundary of the region to be meshed. However, the inverse mapping

$$\mathbf{u}(\mathbf{x}) = (v(\mathbf{x}),\ w(\mathbf{x})) \tag{8.80}$$

is in general not expressed explicitly. When a composite cubic spline surface is used to represent the surface of the geometry, the above inverse mapping is clearly non-linear.

In addition, in the boundary representation of CAD systems and in our discussion of the geometrical representation of curves and surfaces, the composite parametric spline curve that represents the boundary is in fact an approximation to the edge of the surface, which is often formed by the intersection of two or more surfaces. Similarly, the composite spline surface is also an approximation to the surface of the geometry. As a result of such approximations, the boundary curve and its nodes are not exactly located on the surface.

The parametric position of a boundary node on the surface can be found by assuming that its parametric coordinates are the same as its closest point on the surface, which is to find a point $\mathbf{x}(v, w)$ on the surface that is the closest point to boundary node $\mathbf{x}(u_k)$. The problem of finding the closest point can be formulated as: find the parametric coordinates (v, w) of a surface point such that

$$D = \|\mathbf{x}(v, w) - \mathbf{x}(u_k)\| = \text{minimum} \tag{8.81}$$

where $\|\cdot\|$ denotes the Euclidean norm.

This problem is usually non-linear and may be solved by various iterative methods.[55, 56] The initial approximation of the parametric coordinates of a boundary node is taken as the computed position of a previous boundary point.

After all the boundary nodes are placed on the parametric plane, they are linked by straight lines to form the discretized boundary of the region to be meshed.

8.3.4 Element generation in parametric plane

The element generation procedure in the parametric plane is the same as that described in Sec. 8.2.3 except that when creating a new element, the mesh parameters $\alpha_i(v, w)$

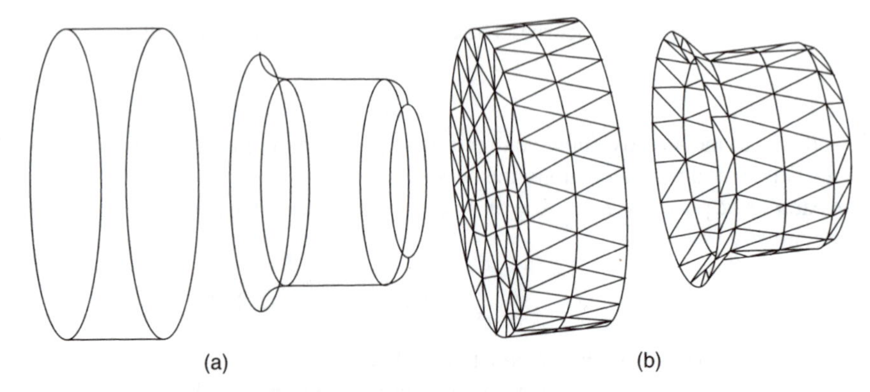

Fig. 8.29 (a) Faces and their edges of the machine part of Fig. 8.20. (b) Coarse meshes.

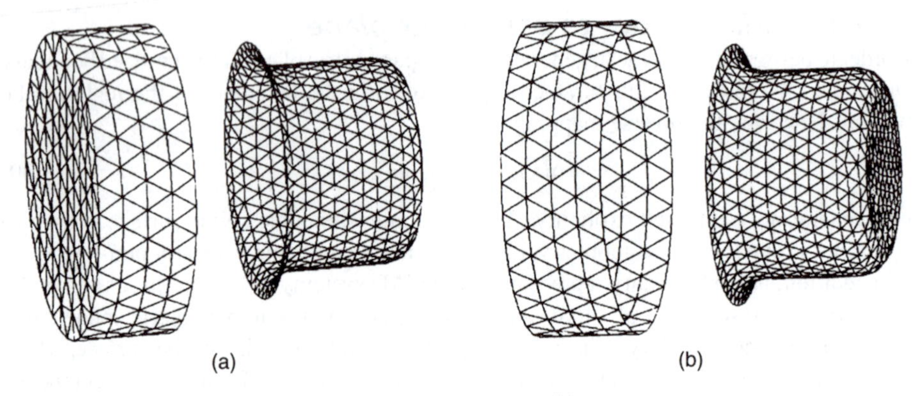

Fig. 8.30 Finer meshes of different element sizes. (a) View one. (b) View two.

[Eq. (8.69)] and $h_i(v, w)$ [Eq. (8.68)] computed from the specified mesh parameters for the surface mesh have to be utilized. After the completion of the element generation quality enhancement techniques need to be applied which include a constraint of preserving the curvature of the surface to improve the quality of the mesh.[54,57] The final mesh is then mapped onto the surface by $\mathbf{x}(v,w)$ to obtain the required surface mesh.

The complete surface of a three-dimensional solid can be achieved after assembling the surface mesh of all the faces. Such surface mesh may be used as the discretized boundary for three-dimensional mesh generation, which we shall discuss in the next section.

Example 8.5: Some of the faces of the machine part shown in Fig. 8.20 are discretized by a surface mesh generator using the algorithms discussed in this section. Figure 8.29(a) shows the boundary representation of two of the surfaces. Coarse meshes of the surfaces are illustrated in Fig. 8.29(b). Finer meshes of the surfaces with different element size distribution are shown in Fig. 8.30(a) and Fig. 8.30(b) with a view from a different angle. The surface mesh of the part with side faces removed is shown in Fig. 8.31.

Example 8.6: The boundary edges of the surface representation of a gearbox part are shown in Fig. 8.32(a). The geometry of the part is realistic and hence somewhat complex.

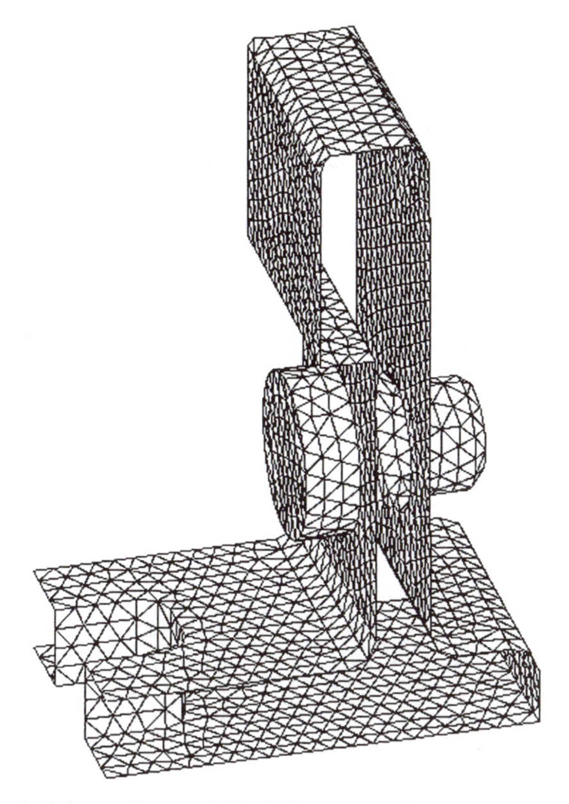

Fig. 8.31 Surface mesh of the machine part with side faces removed.

A complete surface mesh of the part generated automatically using the algorithms described in this section is illustrated in Fig. 8.32(b).

8.3.5 Higher order surface elements

One of the advantages of generating a surface mesh in the parametric plane is that it can produce higher order elements without additional difficulties. When a mesh of higher order elements is generated in the parametric plane, it will be mapped onto the surface to form a boundary fit surface mesh with all the nodes on the surface. To preserve the boundary curves, it is important to generate the intermediate boundary edge nodes by following the node generation procedure for curves. Figure 8.33 demonstrates a surface mesh of quadratic elements for a mechanical part.

8.3.6 Remarks

Several remarks on issues related to the surface mesh generation are given below.

Remark 8.5. Once the discretized boundary of the region in the parametric plane is available, any two-dimensional element generation algorithm could be used to generate a valid

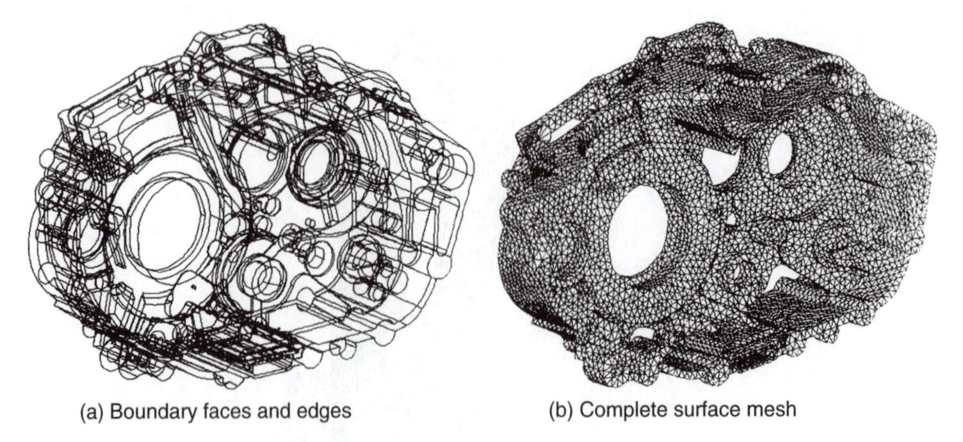

Fig. 8.32 Mesh generation for a gearbox part.

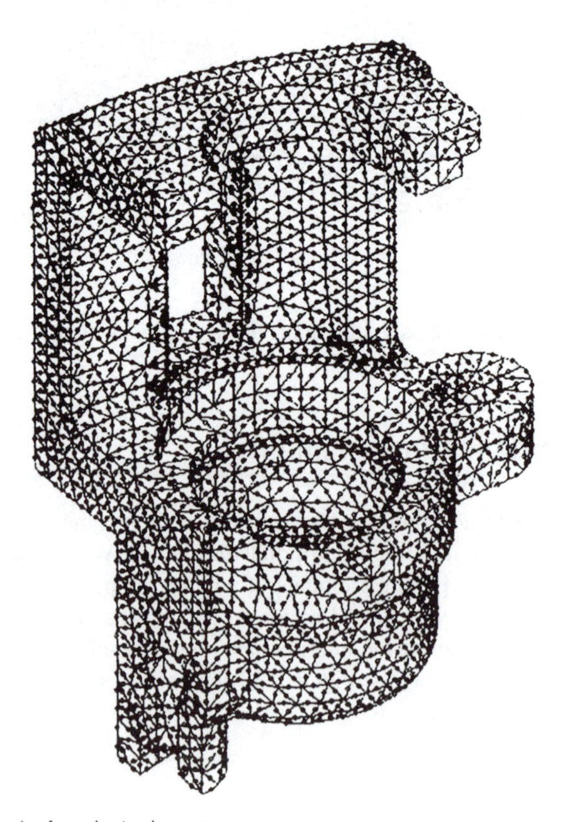

Fig. 8.33 Surface mesh of quadratic elements.

mesh on the parametric plane. For convenience, we only mentioned the two-dimensional advancing front method described in Sec. 8.2.

Remark 8.6. Equations (8.59) and (8.62) have revealed that the mesh parameters in the parametric plane is a function of the parameterization of the surface. In order to satisfy the specified geometrical characteristics of the surface, the two-dimensional mesh generation algorithm is required to be capable of generating the mesh strictly following the computed mesh parameters for the parametric plane. Otherwise, serious distortion in the surface mesh may occur, when the planar mesh is mapped onto the surface. We refer to references 58–63 for additional discussion on issues of surface mesh generation.

Remark 8.7. Besides the parametric representation, the surface of the three-dimensional geometry can also be represented in a discrete form. Surface mesh generation algorithms for surfaces represented in such form are topics of active research. Several surface mesh generation algorithms based on discrete representation of surfaces can be found in references 64–68.

Remark 8.8. The surface mesh generation algorithms discussed in this section assume that the geometrical representation and topological representation of the surfaces are correct, i.e., there are no defects in the surface representation of the three-dimensional geometry. In practical computations, this is often not the case. The boundary representation of the geometry provided by the CAD systems sometimes contains errors or undesirable features that will either cause the mesh generation algorithm to fail or the quality of the surface mesh become unacceptable for three-dimensional mesh generation algorithms. Although methodologies that automatically remove defects and detrimental features from the boundary representations of the surface are not in the scope of surface mesh generation, they critically affect the success of automatic surface mesh generation and therefore deserve further research.

Remark 8.9. Finally, surface mesh generation is often used as the boundary discretization of a three-dimensional geometry. The quality of the surface mesh not only affects the quality of the three-dimensional mesh, it also affects the robustness of any three-dimensional mesh generation algorithm. This is particularly true for the three-dimensional advancing front mesh generation method. Although the robustness of the three-dimensional Delaunay triangulation method is less dependent on the quality of the surface mesh, the quality of the final three-dimensional mesh is certainly affected. To have a successful three-dimensional mesh generation algorithm, the quality of the surface mesh must be insured before the interior mesh generation process starts.

8.4 Three-dimensional mesh generation – Delaunay triangulation

Many practical finite element computations are carried out on complex three-dimensional domains. The level of difficulty to automatically generate valid meshes for arbitrary three-dimensional domains is much greater than in two dimensions. In principle, a Delaunay triangulation, advancing front and finite octree method are all applicable to three-dimensional mesh generation. However, the Delaunay triangulation method has attracted most of the attention in theoretical research and software development, due to its conceptual simplicity, mathematical rigour and algorithmic robustness. In this section, we shall be concerned with the Delaunay triangulation method and its application to three-dimensional mesh generation. We shall also introduce mesh quality enhancement methods which are crucial to ensure the final mesh can be used in the finite element computations.

8.4.1 Voronoi diagram and Delaunay triangulation

Delaunay triangulation[69] is the dual of the Voronoi diagram.[70] The properties of the Voronoi diagram and Delaunay triangulation provide the theoretical foundation for all the mesh generation methods based on the Delaunay method. In order to facilitate the description of the Delaunay triangulation method for mesh generation, a brief review of the basic properties of the Voronoi diagram and Delaunay triangulation is presented in a two-dimensional setting for visualization convenience, but these properties are equally valid in three dimensions.

Let $P = \{p_i, i = 1, 2, \ldots, N\}$ be a set of distinct points in the two-dimensional Euclidean plane R^2. They are referred to as the *forming points* in the mesh generation literature. The *Voronoi region* $V(p_i)$ is defined as the set of points $x \in R^2$ that are at least as close to p_i as to any other forming point, i.e.,

$$V(p_i) = \{x \in R^2 : \|x - p_i\| \leq \|x - p_j\|, \forall j \neq i\} \qquad (8.82)$$

Figure 8.34(a) depicts ten Voronoi regions, with two interior regions bounded by eight others, the total being defined by an equal number of forming points. It follows that the Voronoi region $V(p_i)$ represents a convex polygonal region, possibly unbounded; and any point x inside $V(p_i)$ is nearer to p_i than any other forming point in P. The points that belong to more than one region form the edges of the Voronoi regions and the edges of the Voronoi region $V(p_i)$ are portions of the perpendicular bisectors separating the segment joining forming points p_i and p_j when $V(p_i)$ and $V(p_j)$ are contiguous. The union of the Voronoi regions is called the Voronoi diagram of the forming point set P.

The dual graph of the Voronoi diagram is produced by connecting the forming points of the neighbouring Voronoi regions sharing a common edge with straight lines. It forms the Delaunay triangulation $D(P)$ of the Voronoi forming points P. Figure 8.34(b) illustrates the Delaunay triangulation and its corresponding Voronoi diagram.

In addition to those already mentioned, several properties of the Delaunay triangulations and Voronoi diagrams that are most relevant to the mesh generation algorithms of Delaunay triangulation are listed below:[71, 72]

i. Delaunay triangulation is formed by triangles if no four points of the forming points P are co-circular. These triangles are called *Delaunay triangles*.
ii. Each Delaunay triangle corresponds to a Voronoi vertex, which is the centre of the circumcircle of the triangle, as depicted in Fig. 8.35.
iii. The interior of the circumcircle contains no forming points of P.
iv. The boundary of the Delaunay triangulation is the convex hull of the forming points.

In the Delaunay triangulation-based mesh generation algorithm, property (i) is used to avoid the degeneracy; property (ii) is often used to construct data structures; property

(iii) forms the well-known Delaunay criterion, the *empty circle criterion*, or the *in-circle criterion* when verifying whether it is violated by the new point introduced to the Delaunay triangulation; property (iv) is the theoretical origin of using a convex hull, which contains all the mesh points, in mesh generation.

Example 8.7: Each Voronoi diagram corresponds to a set of forming points which forms Delaunay triangulation. Adding a new forming point will inevitably result in a modification of the Voronoi diagram and the Delaunay triangulation. The process of constructing a new Voronoi diagram and Delaunay triangulation after the insertion of a new node is frequently used in automatic mesh generation and is illustrated here in the same two-dimensional setting shown in Fig. 8.34(b).

Let the new forming point n be inserted in the Delaunay triangulation shown in Fig. 8.36(a). It falls into the circumcircles of Delaunay triangles afg, abf and bef, therefore

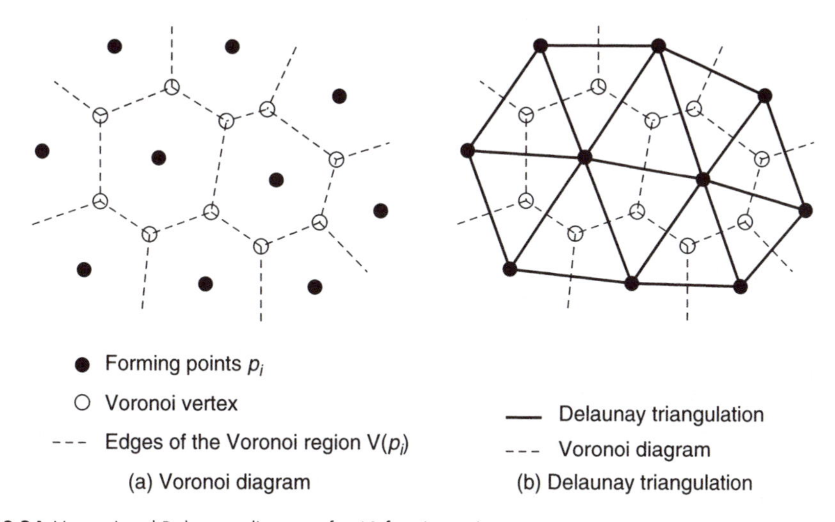

Fig. 8.34 Voronoi and Delaunay diagrams for 10 forming points.

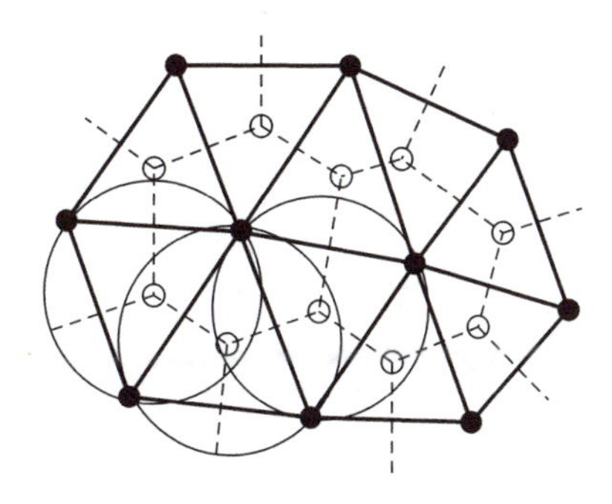

O Voronoi vertex and centre of the circumcircle

Fig. 8.35 Circumcircles of the Delaunay triangles. Only three are shown.

violating property (iii). This causes the removal of the three Voronoi vertices which are the centres of the circumcircles and their corresponding Delaunay triangles, as illustrated in Fig. 8.36(b). The new Delaunay triangulation is constructed by linking the new forming point n and its contiguous forming points that form a face of the neighbouring triangle followed by the construction of the new Voronoi diagram as shown in Fig. 8.36(c).

As we have indicated previously, the process used in the last example is applicable to three dimensions.

8.4.2 Three-dimensional mesh generation by Delaunay triangulation

Although by definition Delaunay triangulation decomposes the convex hull of the forming points into triangles in two dimensions and tetrahedra in three dimensions, it does not address the issues of how Delaunay triangulation can be formed effectively; how to generate those points that will be inserted in the triangulation; and how to preserve the boundary of a region when the forming points are from the boundary of a concave region. These issues

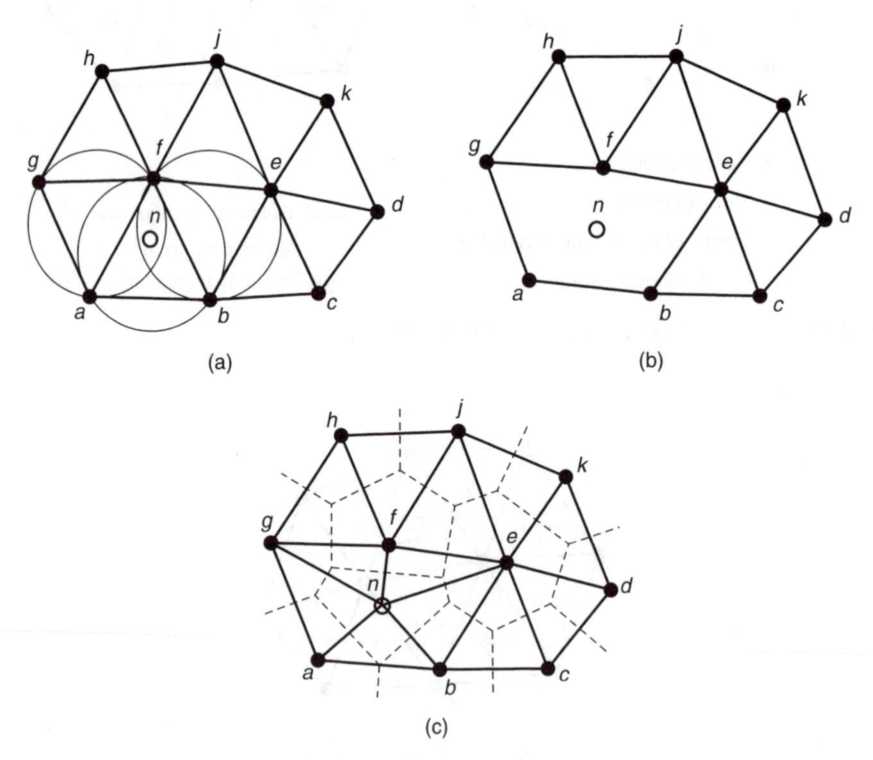

Fig. 8.36 (a) Insertion of new forming point n into Delaunay triangulation. (b) Removal of Delaunay triangles, deleted Voronoi vertices are not shown. (c) New Delaunay triangulation and Voronoi diagram.

are the three most important components of the automatic mesh generation algorithms of Delaunay type. A large body of literature exists on research of these three subjects. The most representative ones are the work of Bowyer[11] and Watson[12] on the efficient Delaunay triangulation algorithms, which were introduced to mesh generation by Cavendish *et al.*,[13] Weatherill,[14] Schroeder and Shephard,[15] Baker[16] and George *et al.*;[17] the early work of Rebay,[18] Weatherill and Hassan,[19] Marcum and Weatherill[20] on automatic point generation algorithms; and the work of Weatherill,[73] George *et al.*,[17] Weatherill and Hassan[19] on preserving the integrity of the domain boundary.

In the following, we shall introduce the three-dimensional mesh generation procedure of Weatherill and Hassan, which is one of the first Delaunay mesh generation procedures that contains all three necessary components for a robust three-dimensional Delaunay mesh generation algorithm. It includes a Delaunay triangulation algorithm; a node generation algorithm based on specified mesh size distribution; and a surface mesh recovery procedure that ensures the integrity of the boundary surface.

The global procedure of the three-dimensional mesh generation algorithm is as follows:

1. Input the triangular surface mesh and derive the topological data of the surface mesh, such as edges of surface elements and node–element connections. (Figure 8.37 shows the surface mesh of a simple three-dimensional geometry.)
2. Build a convex hull that contains all the mesh points. (An eight node convex hull is shown in Fig. 8.38.)
3. Perform Delaunay triangulation using nodes of the surface mesh to form tetrahedra. (Figure 8.39(a) illustrates the Delaunay triangulation of the surface nodes. A cross-section of the triangulation is shown in Fig. 8.39(b).)
4. Create interior points, following the specified element size distribution function, and perform Delaunay triangulation to form tetrahedra. (The results are shown in Fig. 8.40(a) and Fig. 8.40(b).)
5. Recover any missing edges and triangular faces of the surface mesh to ensure the input surface triangulation being contained in the volume triangulation.

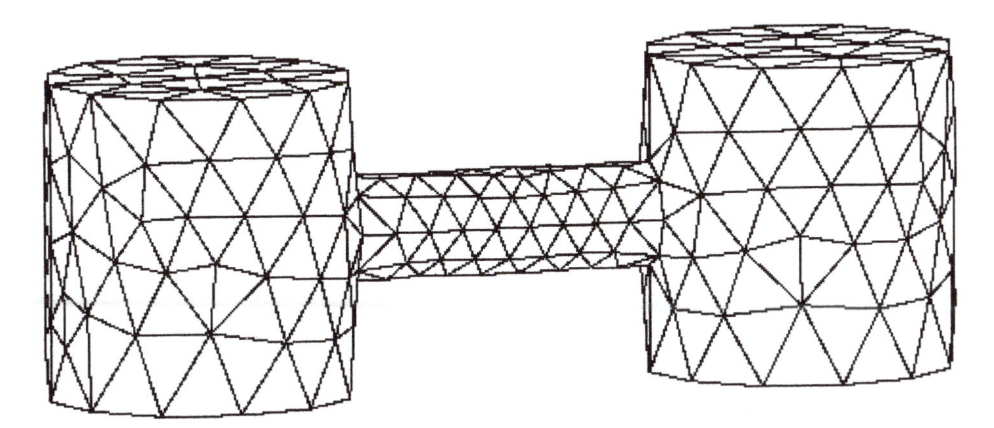

Fig. 8.37 Surface mesh of a simple three-dimensional geometry.

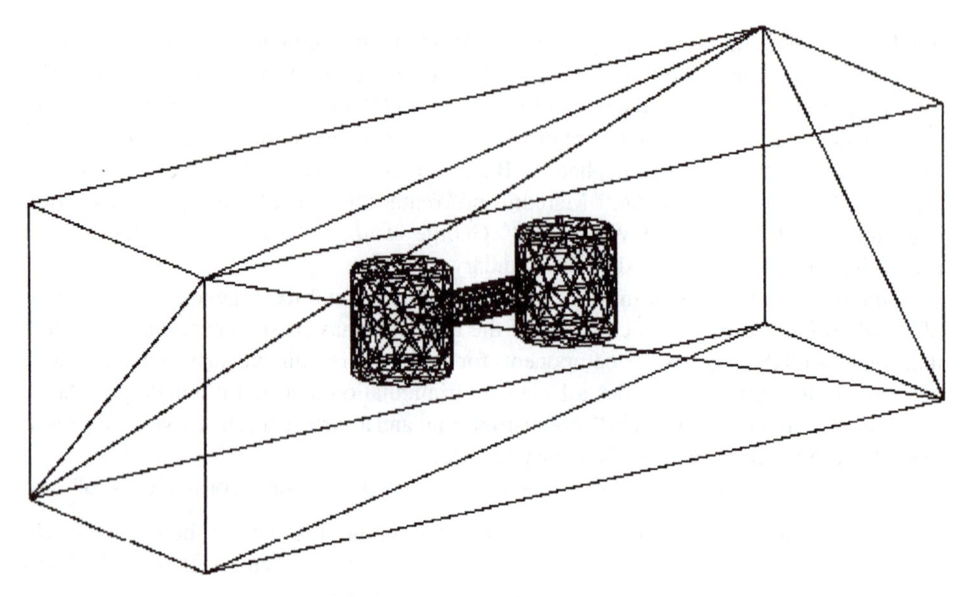

Fig. 8.38 Convex hull and surface mesh.

6. Identify and remove all the tetrahedra outside the domain of interest to give the final three-dimensional mesh. (Figure 8.41(a) shows the recovered surface mesh, which is identical to the input mesh, and Fig. 8.41(b) demonstrates the interior of the tetrahedral mesh at a cross-section of the geometry.)

The mesh generation procedure has been shown to be computationally efficient and, probably more importantly, very robust. Indeed, the robustness of the algorithm is independent on the complexity of the three-dimensional geometry. We shall, in the following, describe in more detail the three main components of the procedure, i.e., the Delaunay triangulation algorithm, the node creation algorithm and the surface mesh recovery methods.

Delaunay triangulation algorithm

The Delaunay triangulation algorithm discussed below is based on the algorithm proposed by Bowyer,[11] but it can be readily replaced by the similar algorithm of Watson[12] which differs only in its data structure.

The process of generating Delaunay triangulation is sequential. Each point is introduced into an existing structure of the Voronoi diagram and the Delaunay triangulation, which will be reformulated based on the in-circle criterion to form a new Delaunay triangulation. The process is similar to that described in Example 8.7, but of course now is in three dimensions. The main steps of the procedure are as follows:

1. Define a set of points which form a convex hull that encloses all the points to be used in the tetrahedral mesh.
2. Introduce a new point into the convex hull.
3. Determine all vertices of the Voronoi diagram to be deleted. A vertex will be deleted if the circumsphere, centred at the vertex, of four forming points contains the new point. This follows from the in-circle criterion.

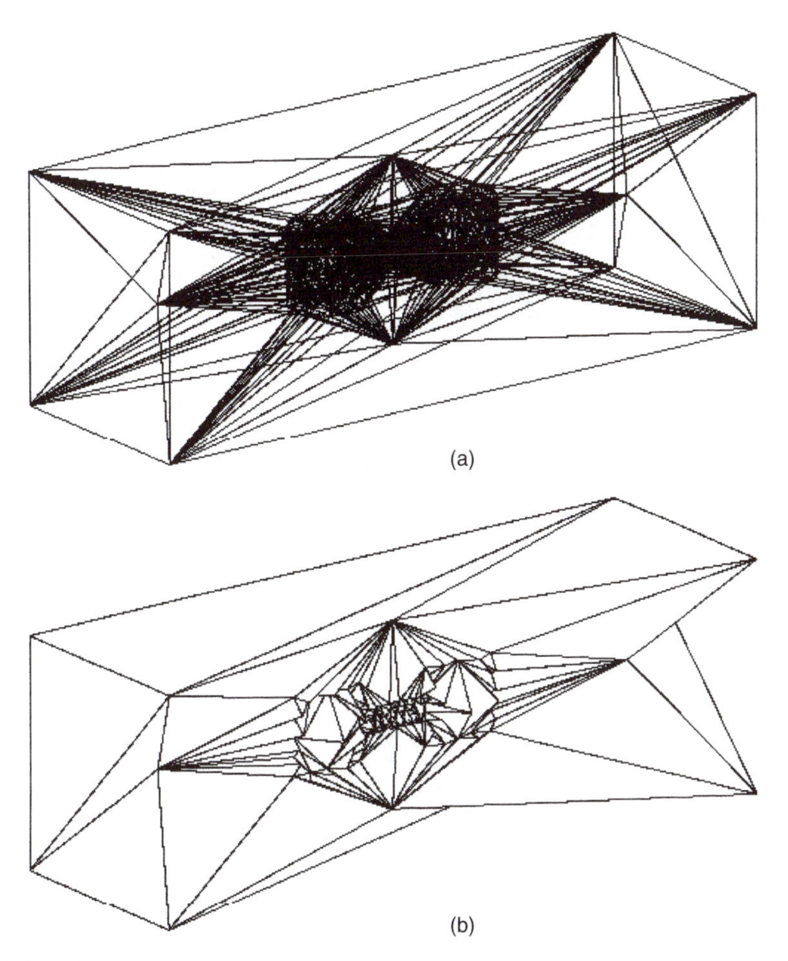

Fig. 8.39 (a) Delaunay triangulation of the surface nodes. (b) A cross-section of the surface nodes Delaunay triangulation. Six additional points are inserted before surface nodes for efficiency.

4. Find the forming points of all the deleted Voronoi vertices. These form the contiguous points to the new point.
5. Determine the neighbouring Voronoi vertices to the deleted vertices which have not themselves been deleted. This data provides the necessary information to enable valid combinations of the contiguous points to be constructed.
6. Determine the forming points of the new Voronoi vertices. The forming points of new vertices include the new point together with three points which are contiguous to the new point and form a face of a neighbouring tetrahedra. This forms the new Delaunay triangulation.
7. Determine the neighbouring Voronoi vertices to the new vertices.

   In Step 6, the forming points of all new vertices have been computed. For each new vertex, perform a search through the forming points of the neighbouring vertices, as found in Step 5, to identify common triples of forming points. When a common combination

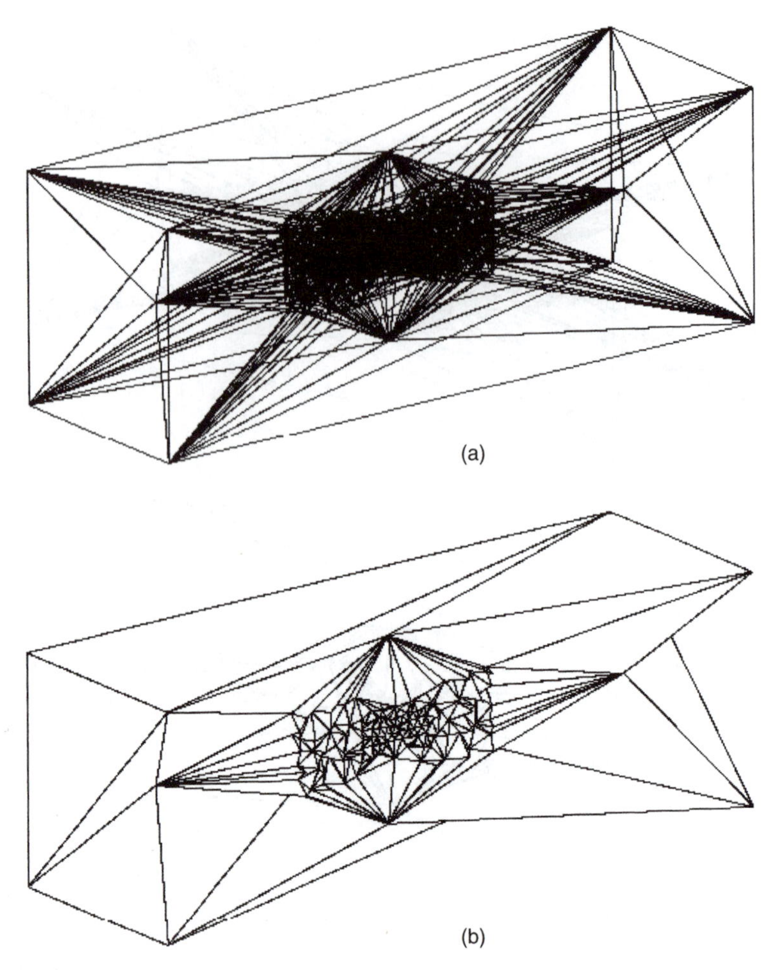

Fig. 8.40 (a) Delaunay triangulation of the interior nodes. (b) A cross-section of the interior nodes Delaunay triangulation.

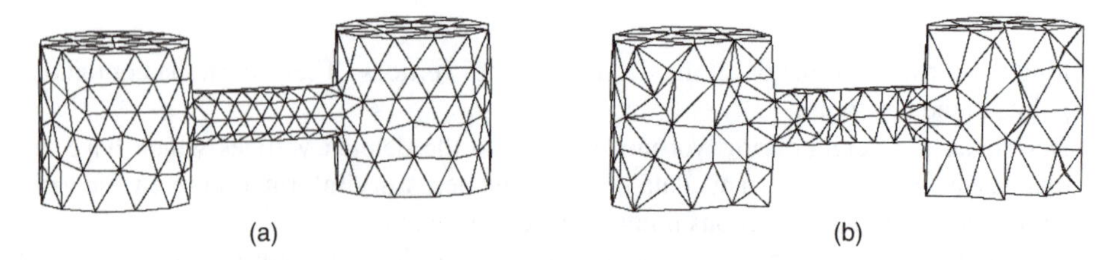

Fig. 8.41 (a) Recovered surface. (b) Tetrahedral mesh shown at a cross-section.

occurs, the neighbour of the new Voronoi vertex has been found. This forms the new Voronoi diagram.

8. Reorder the Voronoi diagram data structure and replace (overwrite) the entries of the deleted vertices.
9. Repeat Steps 2–8 until all the points have been inserted.

In the mesh generation process, the new points to be inserted into the Delaunay triangulation and Voronoi diagram are the surface mesh nodes and the mesh points generated automatically following the adopted node generation algorithm.

Automatic node generation

The detailed step-by-step node generation algorithm described below creates points based on the element size distribution of the surface mesh:

1. The node spacing function for each surface mesh node a of position $\mathbf{x}_a$ is taken as the average of the surface element edge length,

$$h_a = \frac{1}{M} \sum_{b=1}^{M} \|\mathbf{x}_b - \mathbf{x}_a\| \tag{8.83}$$

   where $\mathbf{x}_b (b = 1, 2, \ldots, M)$ are positions of the surface nodes connected to node a.
2. Perform Delaunay triangulation using surface mesh nodes.
3. Initialize the number of new interior points to be created, set $N = 0$.
4. For each tetrahedron within the domain:
   (a) Locate a prospective point c at the centroid of the tetrahedron.
   (b) Derive the node spacing function h_c, for point c, by interpolating the node spacing function h_m $(m = 1, 2, 3, 4)$ from the nodes of the tetrahedron.
   (c) Compute the distances d_m $(m = 1, 2, 3, 4)$ from the prospective point c to each of the four nodes of the tetrahedron.

   If $\{d_m < \alpha h_c\}$ for any $m = 1, 2, 3, 4$ then
   reject the point and return to the beginning of Step 4 for the next tetrahedron.
   Else
   compute the distance d_j from the prospective point c to other already created nodes p_j $(j = 1, 2, \ldots, N)$.

   If $\{d_j < \beta h_c\}$ then
   reject the point and return to the beginning of Step 4 for the next tetrahedron.
   Else
   accept the point c and add it to the interior node list p_j $(j = 1, 2, \ldots, N)$ and update N.

   (d) Assign point distribution function h_c to new node c.
   (e) Go to the next tetrahedron.
5. If $N = 0$, i.e., no new point is created, exit the node generation process.
6. Perform the Delaunay triangulation of the derived points p_j $(j = 1, 2, \ldots, N)$. Then, go to Step 3.

The parameter α controls the element density by changing the allowable shape of the formed tetrahedra, while β has an influence on the regularity of the triangulation by not allowing points within a specified distance of each other. Both parameters can be adjusted to control the mesh density. In practical computations, α can be chosen in the range of 0.85–1.1, and β in the range of 0.6–1.0 for an isotropic mesh. An effort to find the optimal value of these parameters has been made in reference 74.

It is noted that, with minor modification, the node generation algorithm can also create nodes-based element size distribution defined by a three-dimensional background mesh.

Surface mesh recovery

The property (iv) of the Voronoi diagram and Delaunay triangulation presented in Sec. 8.4.1 implies that the surface mesh and the surface boundary of a general three-dimensional geometry, which is seldom convex, will not be respected during the mesh generation process. Very often, some of the surface triangles and their edges are not present in the resulting Delaunay triangulation due to penetrations by other tetrahedra. The loss of completeness of the original surface mesh causes the loss of integrity of the surface boundary of the geometry. In order to derive a valid three-dimensional mesh for the given geometry, the integrity of the surface boundary of the geometry must be respected, which can be realized by recovering the original surface mesh.

In the surface mesh recovery procedure, the surface triangles and surface edges that are missing from the Delaunay triangulation are first identified and then restored by the following procedure:

Edge swapping. Edge swapping is illustrated in Fig. 8.42. If faces abd and bcd appear in the Delaunay triangulation, but faces abc and acd exist in the surface mesh, replacing edge bd by edge ac recovers two surface triangles. This process is attempted for each surface edge because it is the most efficient method to recover the missing edges and triangular faces.

Boundary edge recovery. Consider the case when surface edge joining points a and b are missing from the Delaunay triangulation, a line ab is formed and its intersection with the faces, edges and points of the Delaunay triangulation are identified, as shown in Fig. 8.43, with all the possible types of intersections depicted in Fig. 8.44. Local transformations with the newly added nodes at the intersection as shown in Fig. 8.45 are performed to all the involved tetrahedra to recover the edge, segment by segment. This process is executed for every missing edge.

When the combined intersection involves a node-to-face type and a face-to-node type, edge ab can be recovered by directly linking nodes a and b, with the two involved tetrahedra transformed to three tetrahedra as shown in Fig. 8.46.

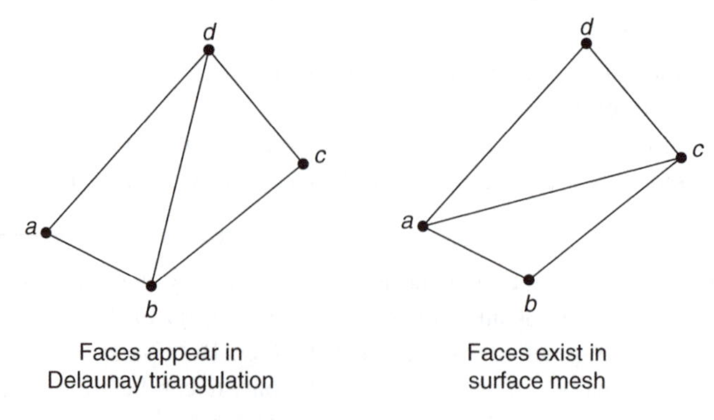

Fig. 8.42 Edge swapping to recover the surface edge and faces.

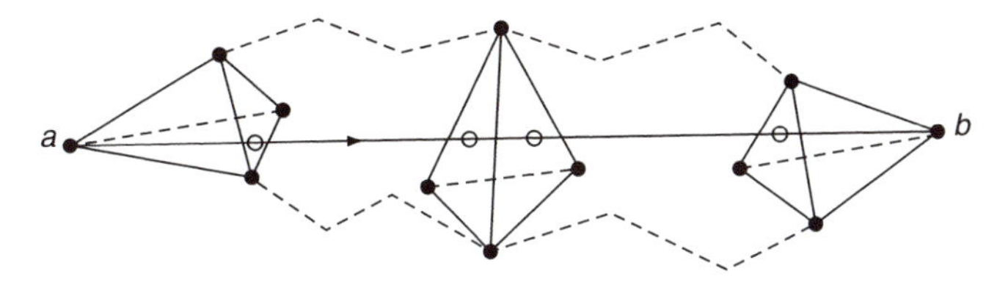

Fig. 8.43 Edge *ab* of a surface triangle missing in the Delaunay triangulation.

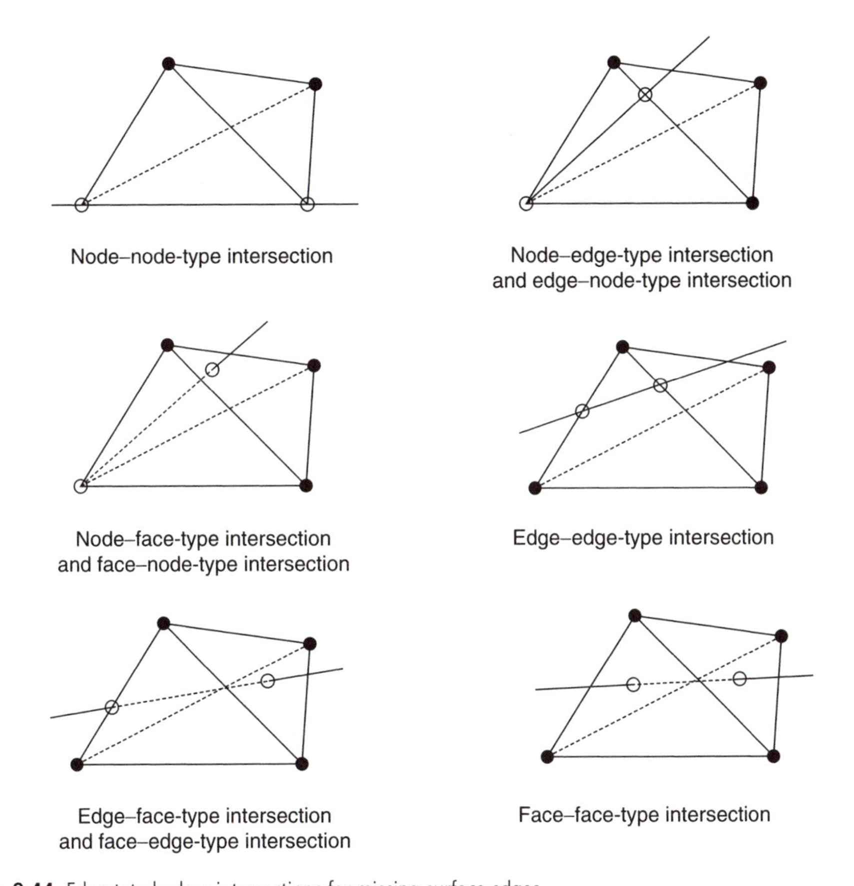

Fig. 8.44 Edge-tetrahedron intersections for missing surface edges.

Boundary face recovery. The recovery of the surface triangles is conducted after the completion of the edge recovery. A surface triangle may still be missing from the tetrahedral mesh even though all its edges are present, because the interior of the triangle face is penetrated by other tetrahedra. There are a total of four possibilities that a face can be intersected by a tetrahedra as illustrated in Fig. 8.47. Every missing face can be recovered after all the intersecting tetrahedra are determined and transformed, with newly added points, according to their intersection type. The intersecting tetrahedron shown in Fig. 8.47 is transformed to a combined shape of a tetrahedron, a pyramid or a prism which can be further divided into tetrahedra as illustrated in Fig. 8.48 and Fig. 8.49.

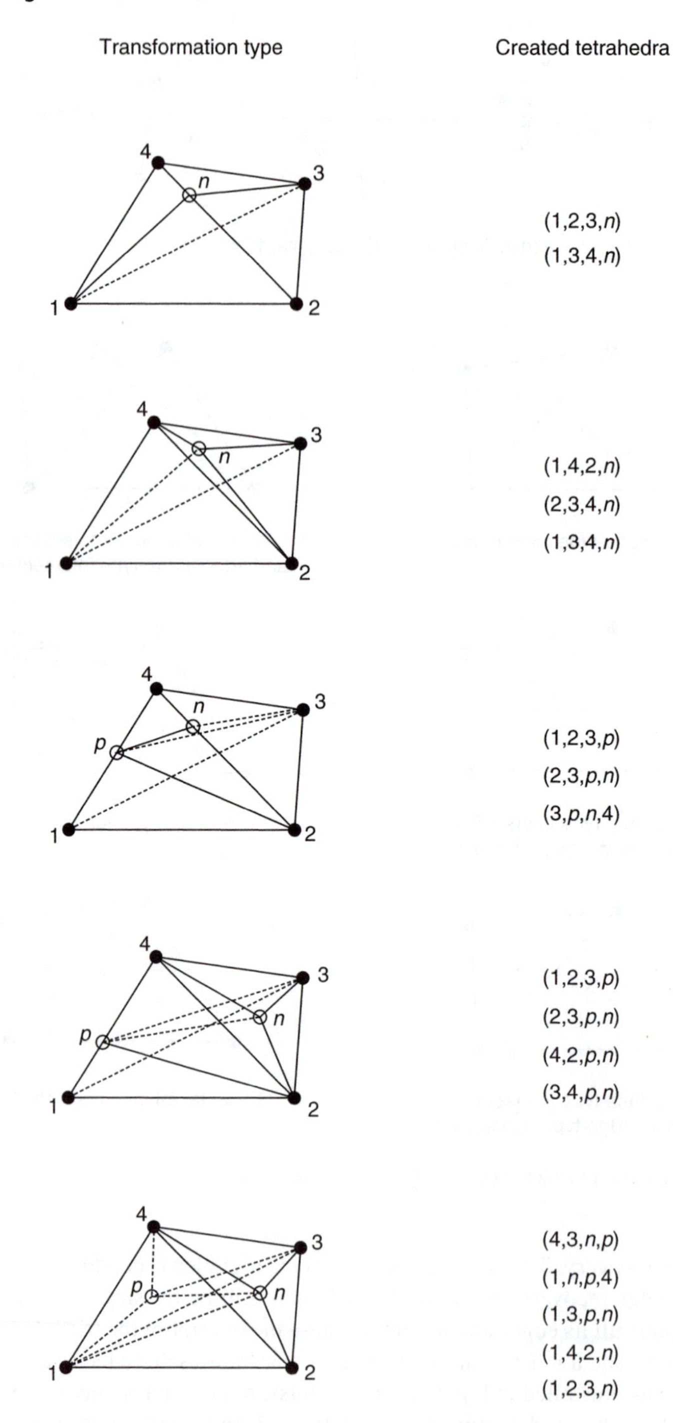

Fig. 8.45 Tetrahedral transformation to recover a segment of the missing edges.

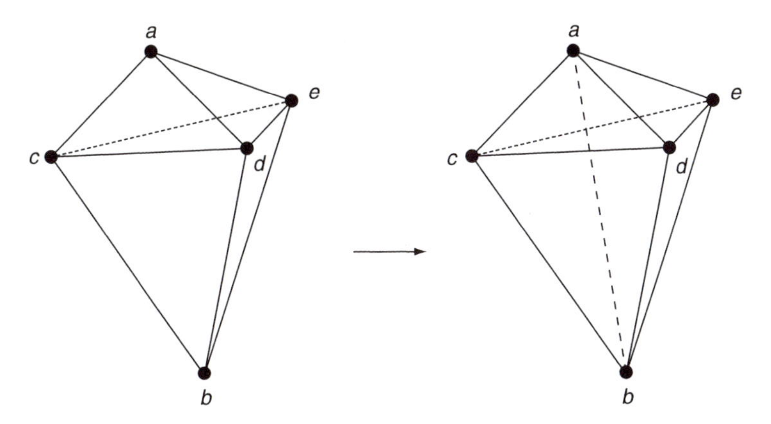

Fig. 8.46 Recovery of boundary edge *ab* by the deletion of face *cde*.

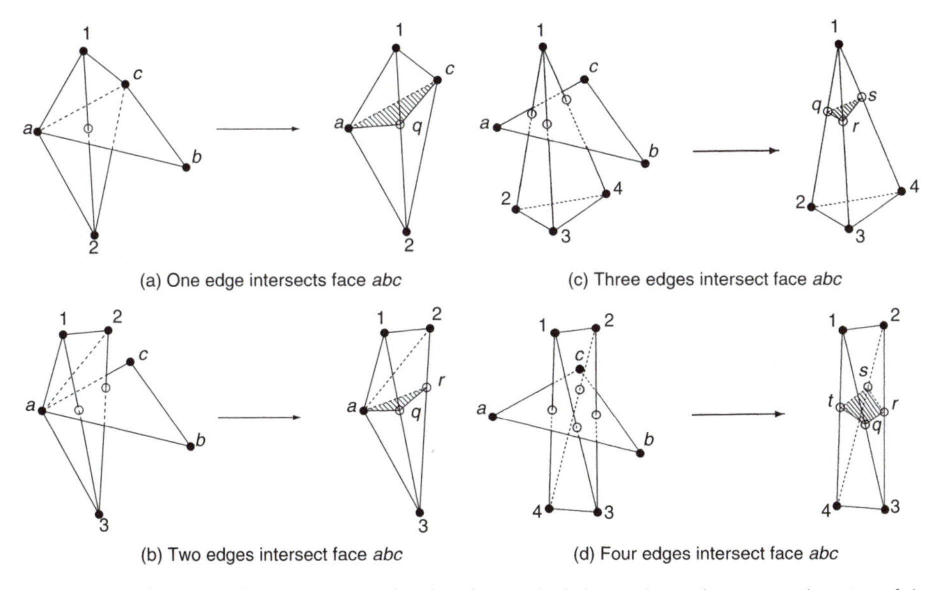

Fig. 8.47 Transformations for the recovery of surface faces. Shaded area shows the recovered portion of the boundary face.

When a face is intersected by only one edge and the edge is common to three tetrahedra, the face can be recovered directly by deleting the edge, with three tetrahedra transformed to two as shown in Fig. 8.50.

Removal of added points. The points that are added in the process of recovering the boundary edges and faces will be removed one by one together with the connected tetrahedra. The empty polyhedron left after the deletion of each added point and its connected tetrahedra will be triangulated directly, often with additional interior points.

It is noted, with the reference to the global mesh generation procedure, that once Step 5 of the global procedure for surface recovery starts, the tetrahedral mesh, in general, does

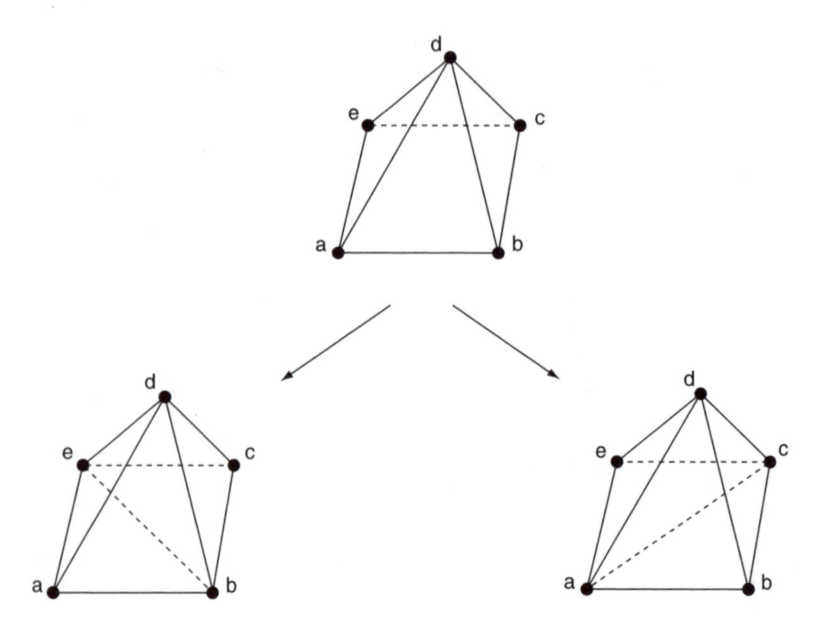

Fig. 8.48 Different patterns of dividing a pyramid to tetrahedra.

not continue to be a Delaunay triangulation, which may seriously damage the quality of the mesh near the boundary. Therefore, mesh quality enhancement becomes indispensable after the completion of the mesh generation procedure.

8.4.3 Mesh quality enhancement

In three-dimensional mesh generation, mesh quality enhancement is almost as important as element generation itself. This is because some poorly shaped tetrahedral elements are created either in the Delaunay triangulation process, due to the position of the inserting points, or in the surface mesh recovery process. Without applying certain mesh quality enhancement procedures to improve the element quality, these poorly shaped elements may render the three-dimensional mesh unusable in finite element computation. Unlike in two-dimensional mesh generation, the process of improving the quality of a three-dimensional mesh is much more complicated and tedious. The quality of a tetrahedral element may be evaluated by different measures. A wide range of measures for the quality of tetrahedral elements are presented in references 75–78, and any one of these measures can be employed as criterion in the mesh quality enhancement procedure. Here we shall not get into the details of a particular quality measure, but are mainly concerned with the methodologies that can be used to improve the quality of the tetrahedral elements, under any specified quality measure. Several effective element quality enhancement methods are described below.

Element transformation

Modifying the topological structure of the mesh is probably the most effective way to improve the quality of the mesh in three dimensions and is realized by performing element transformations of the following form:

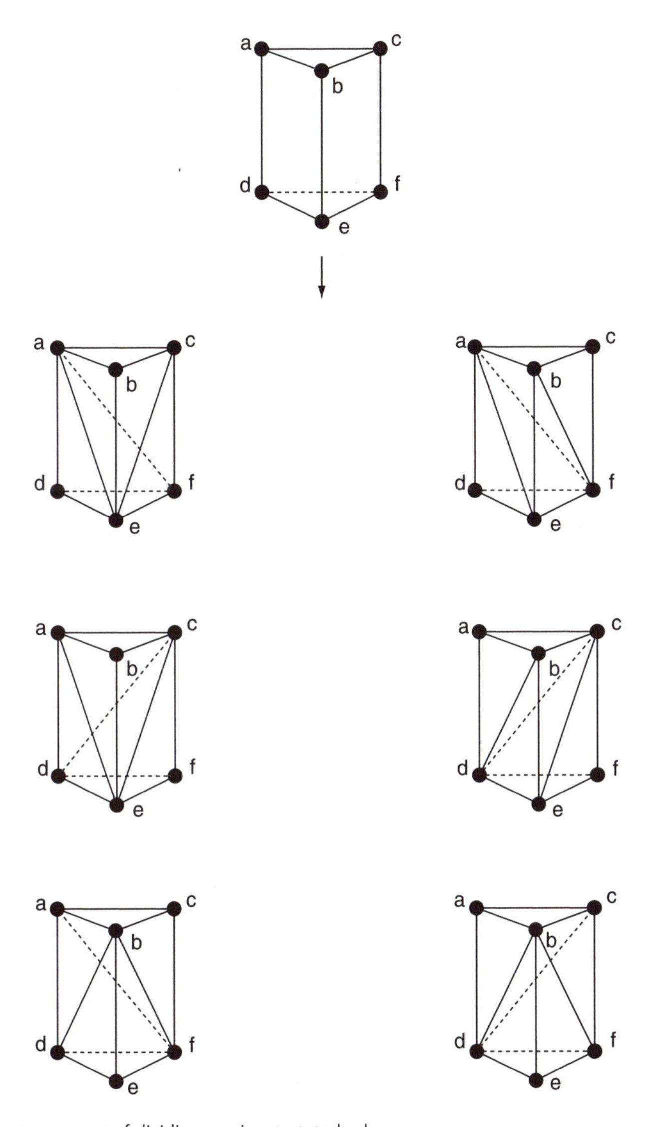

Fig. 8.49 Different patterns of dividing a prism to tetrahedra.

Two elements transformation. Two elements common to a face can be transformed to three elements as shown in Fig. 8.51, if one of the elements does not satisfy the quality criterion. To ensure that the new elements are valid, the new edge ab must intersect the removed face cde.

Three elements transformation. As an inverse of the two element transformation, three elements common to an edge are transformed to two elements as illustrated in Fig. 8.51, if one of the elements does not meet the quality criterion.

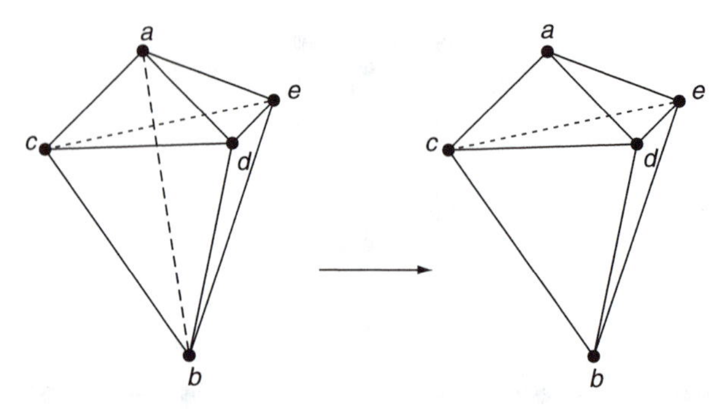

Fig. 8.50 Recovery of boundary face *cde* by the deletion of edge *ab*.

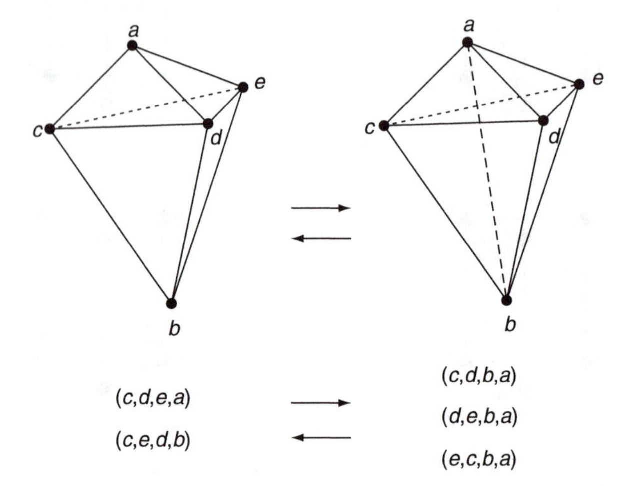

Fig. 8.51 Two elements transformed to three elements and three elements transformed to two elements.

Four elements transformation. Four elements common to an edge can be transformed to two topologically different patterns of four elements common to an edge as depicted in Fig. 8.52, when one of the elements fails to satisfy the quality criterion.

Five or more elements transformation. A split-collapse procedure is used for element transformation when an edge is common to five or more elements. For an element that fails the quality criterion, e.g., element *abcd* as shown in Fig. 8.53(a), find its edge *ab* and all the elements common to the edge (five elements are shown in Fig. 8.53(a)). A node *n* is added to the middle of the edge *ab* and splitting the elements as illustrated in Fig. 8.53(b). The new node *n* is then collapsed to one of its connecting nodes, except nodes *a* and *b*, to form new elements as demonstrated in Fig. 8.53(c) (six new elements are formed). This procedure can be used for an edge common to any number of elements and is attempted for each edge of the element.

All the transformations are carried out under the condition that the worst quality measure improves after the transformation.

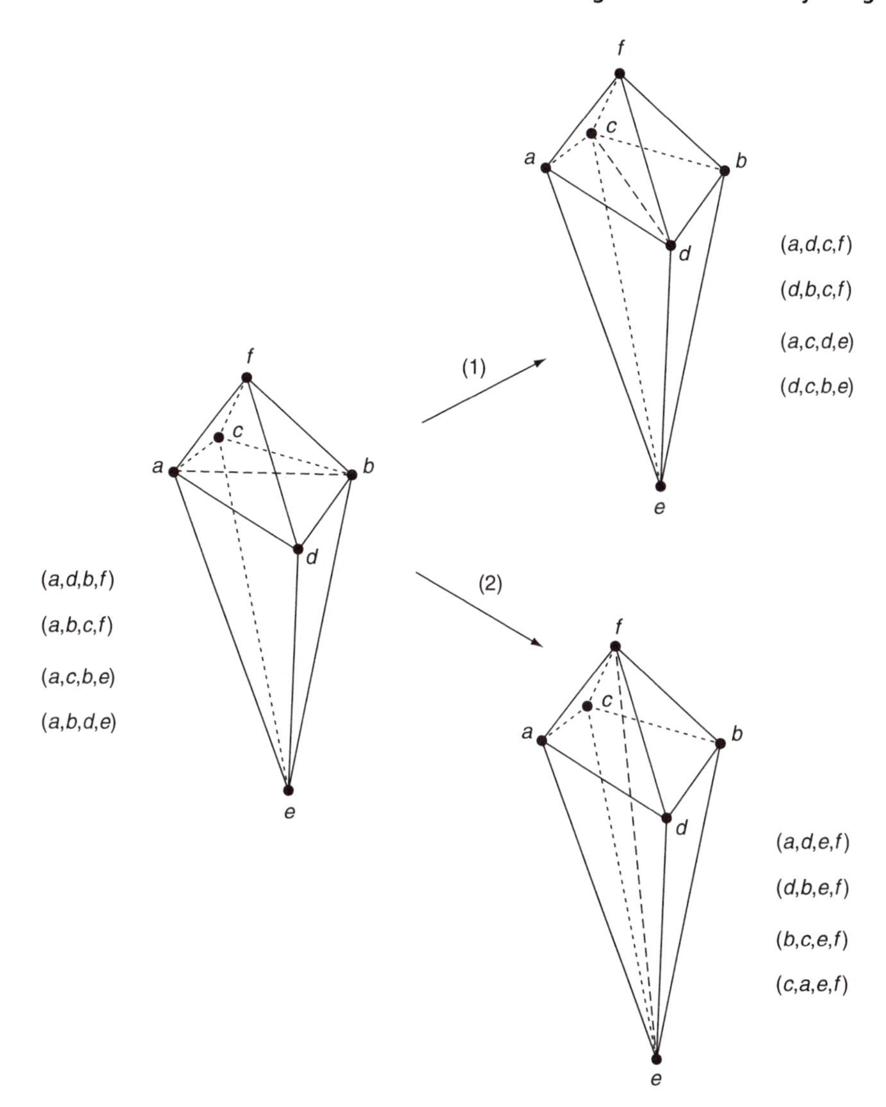

Fig. 8.52 Two patterns of four elements transformed to four elements. (1) Edge *ab* changed to edge *cd*. (2) Edge *ab* changed to edge *ef*.

Node addition and node elimination

Unlike element transformations which will only change the topological structure of the mesh, node addition and node elimination will locally change the node density of the mesh.

Node addition. A node is added to an edge of the element if the edge is deemed too long. This requires all the elements common to the edge being split, the process is shown in Fig. 8.53(a) and (b).

Node elimination. A node of the element that fails the quality criterion and all the elements connected to it are shown in Fig. 8.54(a). The node is collapsed to one of its connecting

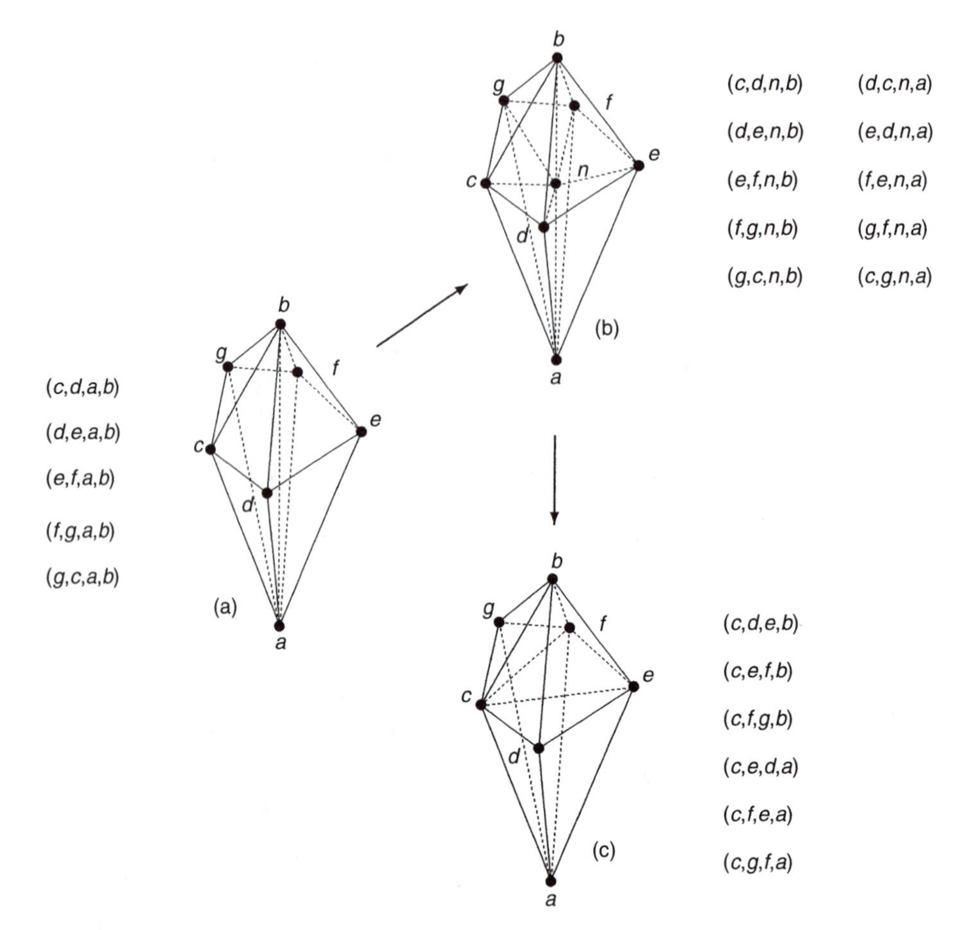

Fig. 8.53 Five or more elements transformation. (a) Edge *ab* and all the elements sharing it. (b) Node *n* is added to edge *ab*, all the common elements are split. (c) New node *n* is collapsed to node *c*.

nodes, as illustrated in Fig. 8.54(b), so that the quality of the resulting elements improves. The procedure is attempted for each node of the element.

The quality of the elements can be further improved if the positions of the interior nodes are repositioned, which leads to the so-called mesh smoothing algorithm.

Mesh smoothing

The standard Laplacian smoothing cannot be applied directly to a tetrahedral mesh. It in fact reduces the quality of the mesh. The procedure is modified to move a node incrementally and iteratively towards each of its connecting nodes and is placed at the position that will increase the quality of the worst element. The procedure stops when quality of the connected elements does not improve. Several combined applications of the quality enhancement method usually results in a mesh with much improved quality.

It is noted that the condition attached to all the quality enhancement methods, which requires the worst element quality to improve according to an element quality criterion,

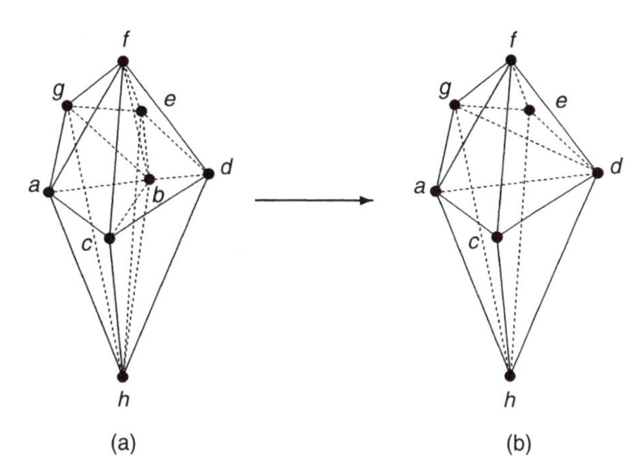

Fig. 8.54 Node collapsing. (a) Node *b* and its connecting elements. (b) Collapse node *b* to node *d*.

corresponds to an optimization problem. Its application guarantees improvement of the element quality, but could also be computationally expensive. For additional information on mesh enhancement methods using a specific quality measure as the objective function in the optimization process we refer to references 79–81.

8.4.4 Higher order elements

With higher order surface mesh available, higher order tetrahedral elements can be readily obtained by finding the positions of the intermediate nodes using linear interpolation.

8.4.5 Numerical examples

Tetrahedral meshes are generated using the mesh generation procedure of Delaunay triangulation described in this section. Figure 8.55(a) shows the mesh for a flask body casting.[89] A cross-section of the tetrahedral mesh is illustrated in Fig. 8.55(b) to demonstrate the regularity of the mesh. Figure 8.56 presents a tetrahedral mesh for a complete V8 engine block.

8.4.6 Remarks

Remark 8.10. The automatic node generation procedure described in this section was introduced by Weatherill and Hassan, many other node generation methods also exist. Indeed, the procedure by which new nodes are generated is the main difference between various Delaunay mesh generation algorithms reported in the literature. From our discussion, it is clear that once the points are available, a mesh can always be generated following a Delaunay triangulation algorithm. Therefore, it is important to adopt a suitable node generation method that can meet the specific requirements for a particular application, so that

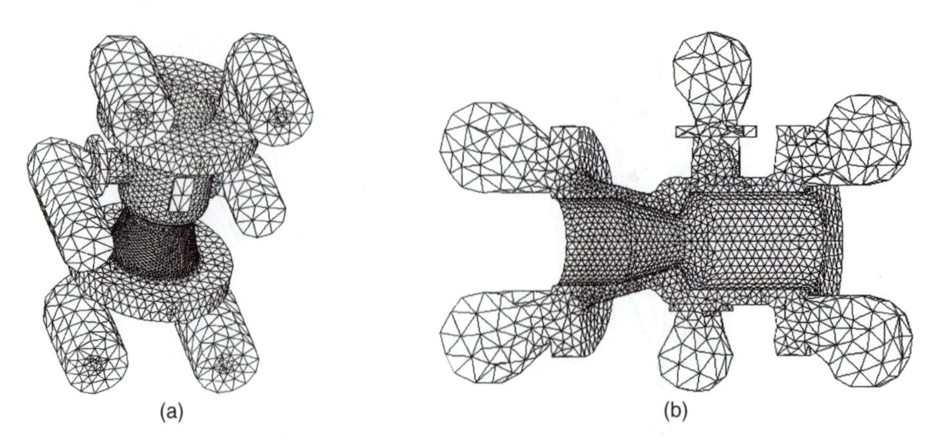

Fig. 8.55 (a) Tetrahedral mesh of casting part of flask body. (b) A cross-section of the tetrahedral mesh.

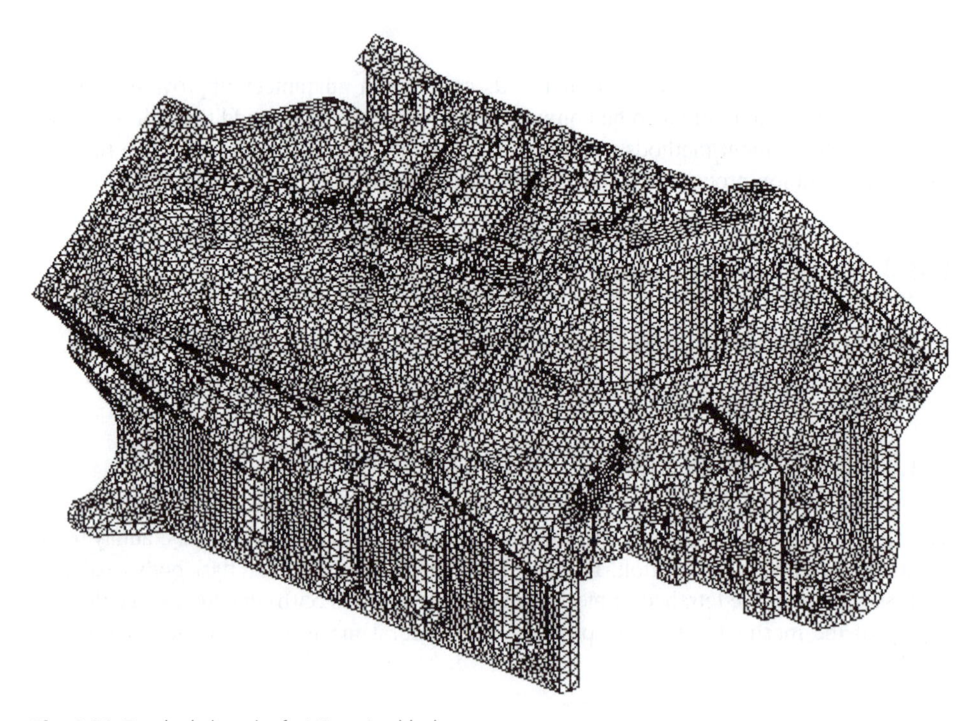

Fig. 8.56 Tetrahedral mesh of a V8 engine block.

an optimal mesh can be obtained for finite element computation. We refer to references 20, 82–85 for additional information.

Remark 8.11. As we have mentioned at the beginning of the chapter, it is still a demanding and challenging task to generate structured or unstructured hexahedral meshes automatically. In Chapter 11, we will observe that some hexahedral elements have advantages in the finite element computation for incompressible materials, in addition they generally have better accuracy compared to tetrahedral elements of the same order. Developing an automatic mesh generation algorithm for hexahedral elements certainly deserves further research.

8.5 Concluding remarks

We have shown in this chapter how to generate mesh on curves, in arbitrary two-dimensional domains, on curved surfaces and for realistic three-dimensional geometries. We presented detailed discussion as well as algorithmic procedure for curve and surface mesh generation. We also described the advancing front method in two-dimensional mesh generation and the Delaunay triangulation method in three-dimensional mesh generation. The algorithms and methodologies presented in this chapter are not only robust and have been implemented for science and engineering applications, but also provide a basis for further research in the development of various aspects of automatic mesh generation methods. Additional applications of the automatic mesh generation methods discussed in this chapter will be presented again in Chapter 14 for adaptive finite element analysis and also appear in reference 37 for fluid dynamics applications.

8.6 Problems

8.1 Write expressions for two-dimensional boundary curves in which the Hermite parametric segments are replaced by a cubic Bézier spline.

8.2 Write a MATLAB program[86] to implement the boundary node generation procedure described in Sec. 8.2.3.

8.3 Develop an algorithm to update the generation front for a two-dimensional advancing front method.

8.4 For triangular meshing in two dimensions formulate a diagonal swapping criterion such that the node–element adjacency number NE_{op} is closer to the optimal value after each swap.

8.5 When using the advancing front method to generate a quadrilateral mesh, prove that a necessary condition is that *initial front must contain an even number of sides.*

8.6 Devise a quadrilateral mesh generation algorithm for the advancing front method that forms each quadrilateral element from two neighbouring triangular elements with a common edge. Assume the triangular element mesh already exists.

8.7 Devise a quadrilateral mesh generation algorithm for the advancing front method that forms quadrilateral elements by subdividing a triangle as shown in Fig. 8.57. Assume the triangular element mesh already exists.

8.8 Define the optimal value of node–element adjacency number NE_{op} for a quadrilateral mesh.

8.9 Write expressions for three-dimensional boundary curves in which the Hermite parametric segments are replaced by a cubic Bézier spline.

8.10 Write expressions for three-dimensional boundary surfaces in which the bi-Hermite parametric surface segments are replaced by bicubic Bézier splines.

8.11 Devise an algorithm to generate a quadrilateral surface mesh using the advancing front method.

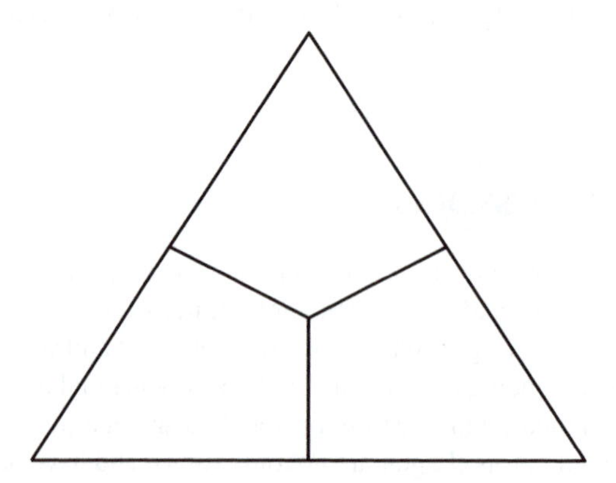

Fig. 8.57 Triangle subdivided into three quadrilaterals. Problem 8.7.

8.12 When a new point is inserted into a Delaunay triangulation in three-dimensional mesh generation, the new point is sometimes found to lie on a co-sphere with four other forming points and, thus, violates the Delaunay triangulation property i of Sec. 8.4.1. Devise an algorithm to avoid such violation.

8.13 Devise an automatic numbering algorithm to generate nodes in an advancing front method for a Delaunay triangulation mesh generation procedure.

8.14 Show that the *three elements transformation* and *four elements transformation* used to improve the quality of the tetrahedral mesh in Sec. 8.4.3 are special cases of the *split–collapse procedure* used for the *five or more elements transformation*.

8.15 In element transformations in a tetrahedral mesh show that, after applying the *split–collapse procedure* when an edge is common to six or more elements, the mesh quality can be further improved by performing edge swapping to the newly formed edges.

References

1. O.C. Zienkiewicz and D.V. Phillips. An automatic mesh generation scheme for plane and curved surfaces by isoparametric coordinates. *Int. J. Numer. Meth. Eng.*, 3:519–528, 1971.
2. W.J. Gordon and C.A. Hall. Construction of curvilinear co-ordinate systems and application to mesh generation. *Int. J. Numer. Meth. Eng.*, 3:461–477, 1973.
3. J.F. Thompson, F.C. Thames, and C.W. Martin. Automatic numerical generation of body-fitted curvilinear coordinates for a field containing any number of arbitrary two-dimensional bodies. *J. Comp. Phys.*, 15:299–319, 1974.
4. J.F. Thompson and Z.U.A. Warsi. Boundary-fitted coordinate systems for numerical solution of partial differential equations. *J. Comp. Phys.*, 47:1–108, 1982.
5. J.F. Thompson, Z.U.A. Warsi, and C.W. Martin. *Numerical Grid Generation: Foundations and Applications*. North-Holland, Dordrecht, 1987.
6. T.K.H. Tam and C.G. Armstrong. 2d finite element mesh generation by medial axis subdivision. *Adv. Eng. Soft.*, 13:313–324, 1991.
7. M.A. Price, C.G. Armstrong, and M.A. Sabin. Hexahedral mesh generation by medial axis subdivision, I: Solids with convex edges. *Int. J. Numer. Meth. Eng.*, 38:3335–3359, 1995.

8. M.A. Price, C.G. Armstrong, and M.A. Sabin. Hexahedral mesh generation by medial axis subdivision, II: Solids with flat and concave edges. *Int. J. Numer. Meth. Eng.*, 40:111–136, 1997.
9. N. Chiba, I. Nishigaki, Y. Yamashita, C. Takizawa, and K. Fujishiro. A flexible automatic hexahedral mesh generation by boundary-fit method. *Comp. Meth. Appl. Mech. Eng.*, 161:145–154, 1998.
10. A. Sheffer and M. Bercovier. Hexahedral meshing of non-linear volumes using Voronoi faces and edges. *Int. J. Numer. Meth. Eng.*, 49:329–351, 2000.
11. A. Bowyer. Computing Dirichlet tessellations. *Comp. J.*, 24(2):162–166, 1981.
12. D.F. Watson. Computing the n-dimensional Delaunay tessellation with application to Voronoi polytopes. *Comp. J.*, 24:167–172, 1981.
13. J.C. Cavendisha, D.A. Field, and W.H. Frey. An approach to automatic three dimensional finite element mesh generation. *Int. J. Numer. Meth. Eng.*, 21:329–347, 1985.
14. N.P. Weatherill. A method for generating irregular computation grids in multiply connected planar domains. *Int. J. Numer. Meth. Eng.*, 8:181–197, 1988.
15. W.J. Schroeder and M.S. Shephard. Geometry-based fully automatic mesh generation and the Delaunay triangulation. *Int. J. Numer. Meth. Eng.*, 26:2503–2524, 1988.
16. T.J. Baker. Automatic mesh generation for complex three-dimensional regions using a constrained Delaunay triangulation. *Eng. Comp.*, 5:161–175, 1989.
17. P.L. George, F. Hecht, and E. Saltel. Automatic mesh generator with specified boundary. *Comp. Meth. Appl. Mech. Eng.*, 92:269–288, 1991.
18. S. Rebay. Efficient unstructured mesh generation by means of Delaunay triangulation and Bowyer–Watson algorithm. *J. Comp. Phys.*, 106:125–138, 1993.
19. N.P. Weatherill and O. Hassan. Efficient 3-dimensional Delaunay triangulation with automatic point generation and imposed boundary constraints. *Int. J. Numer. Meth. Eng.*, 37:2005–2039, 1994.
20. D.L. Marcum and N.P. Weatherill. Unstructured grid generation using iterative point insertion and local reconnection. *AIAA J.*, 33(9):1619–1625, 1995.
21. S.H. Lo. A new mesh generation scheme for arbitrary planar domains. *Int. J. Numer. Meth. Eng.*, 21:1403–1426, 1985.
22. J. Peraire, M. Vahdati, K. Morgan, and O.C. Zienkiewicz. Adaptive remeshing for compressible flow computations. *J. Comp. Phys.*, 72:449–466, 1987.
23. J. Peraire, J. Peiró, L. Formaggio, K. Morgan, and O.C. Zienkiewicz. Finite element Euler computations in three dimensions. *Int. J. Numer. Meth. Eng.*, 26:2135–2159, 1988.
24. R. Löhner and P. Parikh. Three-dimensional grid generation by the advancing-front method. *Int. J. Numer. Meth. Eng.*, 8:1135–1149, 1988.
25. S.H. Lo. Volume discretization into tetrahedra – I. Verification and orientation of boundary surfaces. *Comp. Struct.*, 39(5):493–500, 1991.
26. H. Jin and R.I. Tanner. Generation of unstructured tetrahedral meshes by advancing front technique. *Int. J. Numer. Meth. Eng.*, 36:1805–1823, 1993.
27. P.L. George and E. Seveno. The advancing-front mesh generation method revisited. *Int. J. Numer. Meth. Eng.*, 37:3605–3619, 1994.
28. P. Möller and P. Hansbo. On advancing front mesh generation in three dimensions. *Int. J. Numer. Meth. Eng.*, 38:3551–3569, 1995.
29. M.A. Yerry and M.S. Shephard. Automatic three-dimensional mesh generation by the modified octree technique. *Int. J. Numer. Meth. Eng.*, 20:1965–1990, 1984.
30. P.L. Baehmann, S.L. Wittchen, M.S. Shephard, K.R. Grice, and M.A. Yerry. Robust geometrically-based, automatic two-dimensional mesh generation. *Int. J. Numer. Meth. Eng.*, 24:1043–1078, 1987.
31. M.S. Shephard and M.K. Georges. Automatic three-dimensional mesh generation by the finite octree technique. *Int. J. Numer. Meth. Eng.*, 32:709–749, 1991.

32. T.D. Blacker and R.J. Meyers. Seams and wedges in plastering: a 3D hexahedral mesh generation algorithm. *Eng. Comp.*, 2(9):83–93, 1993.
33. T.J. Tautges, T.D. Blacker, and S.A. Mitchell. The Whisker–Weaving algorithm: a connectivity-based method for constructing all-hexahedral finite element meshes. *Int. J. Numer. Meth. Eng.*, 39:3327–3349, 1996.
34. N.A. Calvo and S.R. Idelsohn. All-hexahedral element meshing: generation of the dual mesh by recurrent subdivision. *Comp. Meth. Appl. Mech. Eng.*, 182:371–378, 2000.
35. G. Dhondt. A new automatic hexahedral mesher based on cutting. *Int. J. Numer. Meth. Eng.*, 50:2109–2126, 2001.
36. J.F. Thompson, B.K. Soni, and N.P. Weatherill, editors. *Handbook of Grid Generation*. CRC Press, Jan. 1999.
37. O.C. Zienkiewicz, R.L. Taylor, and P. Nithiarasu. *The Finite Element Method for Fluid Dynamics*. Butterworth-Heinemann, Oxford, 6th edition, 2005.
38. I.D. Faux and M.J. Pratt. *Computational Geometry for Design and Manufacture*. Ellis Horwood, 1985.
39. G. Farin. *Curves and Surfaces for Computer Aided Geometric Design*. Academic Press, 1990.
40. F. Yamaguchi. *Curves and Surfaces in Computer Aided Geometric Design*. Springer-Verlag, Berlin, 1988.
41. J.C. Cavendish. Automatic triangulation of arbitrary planar domains for the finite element method. *Int. J. Numer. Meth. Eng.*, 8:679–696, 1974.
42. J. Wu, J.Z. Zhu, J. Szmelter, and O.C. Zienkiewicz. Error estimation and adaptivity in Navier–Stokes incompressible flows. *Comp. Mech.*, 6:259–270, 1990.
43. S.H. Lo. Volume discretization into tetrahedra – II. 3D triangulation by advancing front approach. *Comp. Struct.*, 39(5):501–511, 1991.
44. R. Löhner. Some useful data structures for the generation of unstructured grids. *Comm. Appl. Numer. Meth.*, 4:123–135, 1988.
45. J. Bonet and J. Peraire. An alternating digit tree (ADT) algorithm for 3d geometric search and intersection problems. *Int. J. Numer. Meth. Eng.*, 31:1–17, 1990.
46. W. Kwok. An efficient data structure for the advancing front triangular mesh generation technique. *Comm. Num. Meth. Eng.*, 11:465–473, 1995.
47. J.A. Talbert and A.R. Parkinson. Development of an automatic, two-dimensional finite element mesh generator using quadrilateral elements and Bezier curve boundary definitions. *Int. J. Numer. Meth. Eng.*, 29:1551–1567, 1990.
48. B.P. Johnston, J.M. Sullivan, and A. Kwasnik. Automatic conversion of triangular finite element meshes to quadrilateral elements. *Int. J. Numer. Meth. Eng.*, 31:67–84, 1991.
49. E. Rank, M. Schweingruber, and M. Sommer. Adaptive mesh generation. *Comm. Numer. Meth. Eng.*, 9:121–129, 1993.
50. C.K. Lee and S.H. Lo. A new scheme for the generation of a graded quadrilateral mesh. *Comp. Struct.*, 52:847–857, 1994.
51. B. Joe. Quadrilateral mesh generation in polygonal regions. *Comp. Aided Design*, 27:209–222, 1995.
52. S.J. Owen, M.L. Staten, S.A. Canann, and S. Saigal. Q-morph: an indirect approach to advancing front quad meshing. *Int. J. Numer. Meth. Eng.*, 44:1317–1340, 1999.
53. J. Peraire, J. Peiró, and K. Morgan. Adaptive remeshing for 3-dimensional compressible flow computations. *J. Comp. Phys.*, 103:269–285, 1992.
54. J. Peiró. Surface grid generation. In *Handbook of Grid Generation*, Chapter 19, pages 19.1 –19.20. CRC Press, 1999.
55. W.H. Press *et al.*, editor. *Numerical Recipes in Fortran: The Art of Scientific Computing*. Cambridge University Press, Cambridge, 2nd edition, 1992.
56. O.C. Zienkiewicz and R.L. Taylor. *The Finite Element Method for Solid and Structural Mechanics*. Butterworth-Heinemann, Oxford, 6th edition, 2005.

57. P. Hansbo. Generalized Laplacian smoothing of unstructured grids. *Comm. Num. Meth. Eng.*, 11:455–464, 1995.
58. Y. Zheng, R.W. Lewis, and D.T. Gethin. Three-dimensional unstructured mesh generation: Part 2. Surface meshes. *Comp. Meth. Appl. Mech. Eng.*, 134:269–284, 1996.
59. M. Suzuki. Surface grid generation with linkage to geometric generation. *Int. J. Numer. Meth. Eng.*, 34:163–176, 1993.
60. C.K. Lee. Automatic metric advancing front triangulation over curved surfaces. *Eng. Comp.*, 17(1):48–74, 2000.
61. H. Borouchaki, P. Laug, and P.L. George. Parametric surface meshing using combined advancing-front generalized Delaunay approach. *Int. J. Numer. Meth. Eng.*, 49:233–259, 2000.
62. S.J. Sherwin and J. Peiró. Mesh generation in curvilinear domain using high-order elements. *Int. J. Numer. Meth. Eng.*, 53:207–223, 2002.
63. Y.K. Lee and C.K. Lee. Automatic generation of anisotropic quadrilateral meshes on three-dimensional surfaces using metric specifications. *Int. J. Numer. Meth. Eng.*, 53:2673–2700, 2002.
64. R. Löhner. Regridding surface triangulations. *J. Comp. Phys.*, 126:1–10, 1996.
65. A. Rassineux, P. Villon, J.-M. Savignat, and O. Stab. Surface remeshing by local Hermite diffuse interpolation. *Int. J. Numer. Meth. Eng.*, 49:31–49, 2000.
66. Y. Ito and K. Nakahashi. Surface triangulation for polygonal models based on CAD data. *Int. J. Numer. Meth. Fluids*, 39:75–96, 2002.
67. C.K. Lee. Automatic metric 3D surface mesh generation using subdivision surface geometrical model. Part 1: Construction of underlying geometrical model. *Int. J. Numer. Meth. Eng.*, 56:1593–1614, 2003.
68. C.K. Lee. Automatics metric 3D surface mesh generation using subdivision surface geometrical model. Part 2: Mesh generation algorithm and examples. *Int. J. Numer. Meth. Eng.*, 56:1615–1646, 2003.
69. B. Delaunay. Sur la sphère vide. *Izv. Akad. Nauk SSSR, Otdelenie Matematicheskii i Estestvennyka Nauk*, 7:793–800, 1934.
70. G. Voronoi. Nouvelles applications des paramètres continus à la théorie des formes quadratiques. *J. Reine Angew. Math*, 133:97–178, 1907.
71. F.P. Preparata and M.I. Shamos. *Computational Geometry*. Springer-Verlag, New York, 1988.
72. J. O'Rourke. *Computational Geometry in C*. Cambridge University Press, 2nd edition, 2001.
73. N.P. Weatherill. The integrity of geometrical boundaries in the two-dimensional Delaunay triangulation. *Comm. Appl. Numer. Meth.*, 6:101–109, 1990.
74. Y. Zheng, R.W. Lewis, and D.T. Gethin. Three-dimensional unstructured mesh generation: Part 1. Fundamental aspects of triangulation and point creation. *Comp. Meth. Appl. Mech. Eng.*, 134:249–268, 1996.
75. V.N. Parthasarathy, C.M. Graichen, and A.F. Hathaway. A comparison of tetrahedron quality measures. *Fin. Elem. Anal. Design*, 15:255–261, 1993.
76. N.P. Weatherill, P.R. Eiseman, J. Hause, and J.F. Thompson. *Numerical Grid Generation in Computational Fluid Dynamics and Related Fields*. Pineridge Press, Swansea, 1994.
77. A. Liu and B. Joe. Relationship between tetrahedron shape measures. *BIT*, 34:268–287, 1994.
78. R.W. Lewis, Y. Zheng, and D.T. Gethin. Three-dimensional unstructured mesh generation: Part 3. Volume meshes. *Comp. Meth. Appl. Mech. Eng.*, 134:285–310, 1996.
79. L.A. Freitag and P.M. Knupp. Tetrahedral mesh improvement via optimization of element condition number. *Int. J. Numer. Meth. Eng.*, 53:1377–1391, 2002.
80. J.M. Escobar, E. Rodríguez, R. Montenegro, G. Montero, and J.M. González-Yuste. Simultaneous untangling and smoothing of tetrahedral meshes. *Comp. Meth. Appl. Mech. Eng.*, 192:2772–2787, 2003.
81. C.L. Bottasso. Anisotropic mesh adaption by metric-driven optimization. *Int. J. Numer. Meth. Eng.*, 60:597–639, 2004.

82. D.L. Marcum. Adaptive unstructured grid generation for viscous flow applications. *AIAA J.*, 34:2440, 1996.
83. D.J. Mavriplis. An advancing front Delaunay triangulation algorithm designed for robustness. *J. Comp. Phys.*, 117:90–101, 1995.
84. H. Borouchaki, P.L. George, F. Hecht, P. Laug, and E. Saltel. Delaunay mesh generation governed by metric specifications. Part I. Algorithms. *Fin. Elem. Anal. Design*, 25:61–83, 1997.
85. H. Borouchaki, P.L. George, and B. Mohammadi. Delaunay mesh generation governed by metric specifications. Part II. Applications. *Fin. Elem. Anal. Design*, 25:85–109, 1997.
86. MATLAB. www.mathworks.com, 2003.
87. J.Z. Zhu, O.C. Zienkiewicz, E. Hinton and J. Wu. A new approach to the development of automatic quadrilateral mesh generation. *Int. J. Numer. Meth. Eng.* 32:849–866, 1991.
88. T.D. Blacker and M.B. Stephenson. Paving: A new approach to automatic quadrilateral mesh generation. *Int. J. Numer. Meth. Eng.* 32:811–847, 1991.
89. J.Z. Zhu, O.C. Zienkiewicz. 'A posteriori' error estimation and three dimensional automatic mesh generation. *Fin. Elem. Anal. Design.* 25:167–184, 1997.

9

The patch test, reduced integration, and non-conforming elements

9.1 Introduction

We have briefly referred in Chapter 2 to the patch test as a means of assessing convergence of displacement-type elements for elasticity problems in which the shape functions violate continuity requirements. In this chapter we shall deal in more detail with this test which is *applicable to all finite element forms* and will show that

(a) it is a *necessary* condition for assessing the convergence of any finite element approximation and further that, if properly extended and interpreted, it can provide
(b) a *sufficient* requirement for convergence,
(c) an assessment of the (asymptotic) convergence rate of the element tested,
(d) a check on the robustness of the algorithm, and
(e) a means of developing new finite element forms which can violate compatibility (continuity) requirements.

While for elements which *a priori* satisfy all the continuity requirements, have correct polynomial expansions, and are exactly integrated such a test is superfluous in principle, but it is nevertheless useful as it gives

(f) a check that correct programming was achieved.

For all the reasons cited above the patch test has been, since its inception, and continues to be the most important check for practical finite element codes.

The original test was introduced by Irons *et al.*[1–3] in a physical way and could be interpreted as a check which ascertained whether a patch of elements (Fig. 9.1) subject to a constant strain reproduced exactly the constitutive behaviour of the material and resulted in correct stresses when it became infinitesimally small. If it did, it could then be argued that the finite element model represented the real material behaviour and, in the limit, as the size of the elements decreased would therefore reproduce exactly the behaviour of the real structure.

Clearly, although this test would only have to be passed when the size of the element patch became infinitesimal, for most elements in which polynomials are used the patch size did not in fact enter the consideration and the requirement that the patch test be passed for any element size became standard.

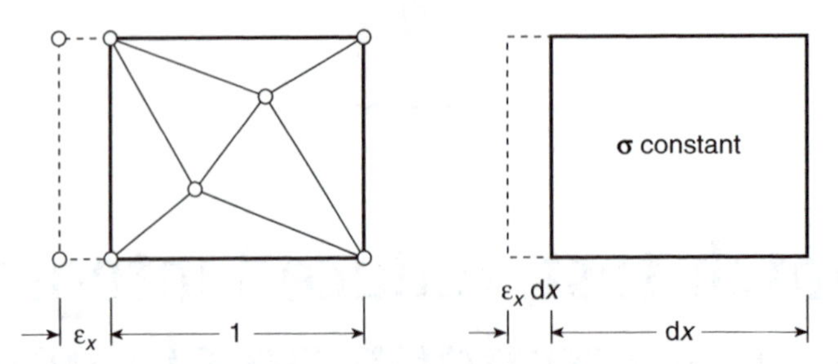

Fig. 9.1 A patch of element and a volume of continuum subject to constant strain ε_x. A physical interpretation of the constant strain or linear displacement field patch test.

Quite obviously a rigid body displacement of the patch would cause no strain, and if the proper constitutive laws were reproduced no stress changes would result. The patch test thus guarantees that no rigid body motion straining will occur.

When curvilinear coordinates are used the patch test is still required to be passed in the limit but generally will not do so for a finite size of the patch. (An exception here is the isoparametric coordinate system in problems discussed in Chapter 5 since it is guaranteed to contain linear polynomials in the global coordinates.) Thus for many problems such as shells, where local curvilinear coordinates are used, this test has to be restricted to infinitesimal patch sizes and, on physical grounds alone, appears to be a *necessary and sufficient condition* for convergence.

Numerous publications on the theory and practice of the test have followed the original publications cited[4–6] and mathematical respectability was added to those by Strang.[7,8] Although some authors have cast doubts on its validity[9,10] these have been fully refuted[11–13] and if the test is used as described here it fulfils the requirements (a)–(f) stated above.

In the present chapter we consider the patch test applied to irreducible forms (see Chapter 3) but an extension to mixed forms is more important. This has been studied in references 13, 14 and 15 and made use of in many subsequent publications. The matter of mixed form patch tests will be fully discussed in the next chapter; however, the consistency and stability tests developed in the present chapter are *always* required.

One additional use of the patch test was suggested by Babuška *et al.*[16] with a shorter description given by Boroomand and Zienkiewicz.[17] This test can establish the efficiency of gradient (stress) recovery processes which are so important in error estimation as will be discussed in Chapter 13.

9.2 Convergence requirements

We shall consider in the following the patch test as applied to a finite element solution of a set of differential equations

$$\mathcal{A}(\mathbf{u}) \equiv \mathcal{L}\mathbf{u} + \mathbf{b} = \mathbf{0} \tag{9.1}$$

in the domain Ω together with the conditions

$$\mathcal{B}(\mathbf{u}) = \mathbf{0} \tag{9.2}$$

on the boundary of the domain, Γ.

The finite element approximation is given in the form

$$\mathbf{u} \approx \hat{\mathbf{u}} = \mathbf{N}\tilde{\mathbf{u}} \tag{9.3}$$

where $\mathbf{N}$ are shape functions defined in each element, Ω_e, and $\tilde{\mathbf{u}}$ are unknown parameters.

By applying standard procedures of finite element approximation the problem reduces in a linear case to a set of algebraic equations

$$\mathbf{K}\tilde{\mathbf{u}} = \mathbf{f} \tag{9.4}$$

which when solved give an approximation to the differential equation and its boundary conditions.

What is meant by 'convergence' in the approximation sense is that the approximate solution, $\hat{\mathbf{u}}$, should tend to the exact solution $\mathbf{u}$ when the size of the elements h approaches zero (with some specified subdivision pattern). Stated mathematically we must find that the error at any point becomes (when h is sufficiently small)

$$|\mathbf{u} - \hat{\mathbf{u}}| = \mathrm{O}(h^q) \leq Ch^q \tag{9.5}$$

where $q > 0$ and C is a positive constant, depending on the position. This must also be true for all the derivatives of $\mathbf{u}$ defined in the approximation.

By the order of convergence in the variable $\mathbf{u}$ we mean the value of the index q in the above definition. To ensure convergence it is necessary that the approximation fulfil both consistency and stability conditions.[18]

The *consistency requirement* ensures that as the size of the elements h tends to zero, the approximation equation (9.4) will represent the exact differential equation (9.1) and the boundary conditions (9.2) (at least in the weak sense).

The *stability condition* is simply translated as a requirement that the solution of the discrete equation system (9.4) be unique and avoid spurious mechanisms which may pollute the solution for all sizes of elements. For linear problems in which we solve the system of algebraic equations (9.4) as

$$\tilde{\mathbf{u}} = \mathbf{K}^{-1}\mathbf{f} \tag{9.6}$$

this means simply that the matrix $\mathbf{K}$ must be non-singular for all possible element assemblies (subject to imposing minimum stable boundary conditions).

The patch test traditionally has been used as a procedure for verifying the consistency requirement; the stability was checked independently by ensuring non-singularity of matrices.[19] Further, it generally tested only the consistency in satisfaction of the differential equation (9.1) but not of its natural boundary conditions. In what follows we shall show how all the necessary requirements of convergence can be tested by a properly conceived patch test.

A 'weak' singularity of a single element may on occasion be permissible and some elements exhibiting it have been, and still are, successfully used in practice. One such case is given by the eight-node isoparametric element with a 2×2 Gauss quadrature, to which we shall refer later here. This element is on occasion observed to show peculiar behaviour (though its use has advantages as discussed in Chapter 10). An element that occasionally fails is termed *non-robust* and the patch test provides a means of assessing the *degree of robustness*.

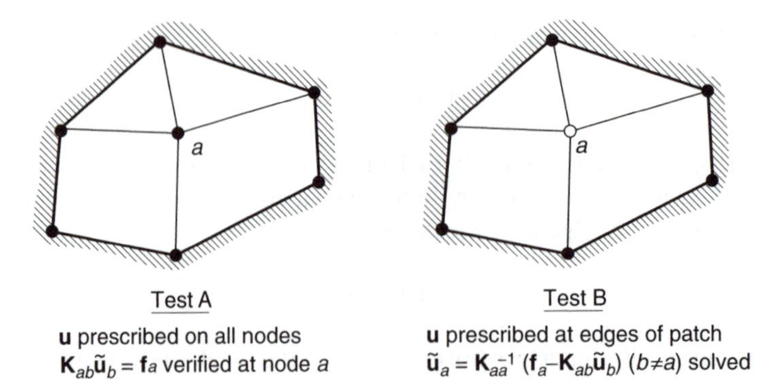

Fig. 9.2 Patch test of forms A and B.

9.3 The simple patch test (tests A and B) – a necessary condition for convergence

We shall first consider the consistency condition which requires that in the limit (as h tends to zero) the finite element approximation of Eq. (9.4) should model exactly the differential equation (9.1) and the boundary conditions (9.2). If we consider a 'small' region of the domain (of size $2h$) we can expand the unknown function $\mathbf{u}$ and the essential derivatives entering the weak approximation in a Taylor series. From this we conclude that for convergence of the function and its first derivative in typical problems of a second-order equation and two dimensions, we require that around a point a assumed to be at the coordinate origin,

$$\mathbf{u} = \mathbf{u}_a + \left(\frac{\partial \mathbf{u}}{\partial x}\right)_a x + \left(\frac{\partial \mathbf{u}}{\partial y}\right)_a y + \cdots + \mathrm{O}(h^p)$$

$$\frac{\partial \mathbf{u}}{\partial x} = \left(\frac{\partial \mathbf{u}}{\partial x}\right)_a + \cdots + \mathrm{O}(h^{p-1}) \qquad (9.7)$$

$$\frac{\partial \mathbf{u}}{\partial y} = \left(\frac{\partial \mathbf{u}}{\partial y}\right)_a + \cdots + \mathrm{O}(h^{p-1})$$

with $p \geq 2$. The finite element approximation should therefore reproduce exactly the problem posed for *any linear forms* of $\mathbf{u}$ as h tends to zero. Similar conditions can obviously be written for higher order problems. This requirement is tested by the current interpretation of the patch test illustrated in Fig. 9.2. We refer to this as the *base solution*.

For problems involving C_0 approximation we compute first an arbitrary solution of the differential equation using a linear polynomial as the base solution and set the corresponding parameters $\tilde{\mathbf{u}}$ [see Eq. (9.3)] at all 'nodes' of a *patch* which assembles completely the nodal variable $\tilde{\mathbf{u}}_a$ (i.e., provides all the equation terms corresponding to it).

In *test A* we simply insert the exact value of the parameters $\tilde{\mathbf{u}}$ into the node a equations and verify that

$$\mathbf{K}_{ab}\tilde{\mathbf{u}}_b - \mathbf{f}_a \equiv \mathbf{0} \qquad (9.8)$$

where $\mathbf{f}_a$ is a force which results from any 'body force' required to satisfy the differential equation (9.1) for the base solution. Generally in problems given in cartesian coordinates the required body force is zero; however, in curvilinear coordinates (e.g., axisymmetric elasticity problems) it can be non-zero.

In *test B* only the values of $\tilde{\mathbf{u}}$ corresponding to the boundaries of the 'patch' are inserted and $\tilde{\mathbf{u}}_a$ is found as

$$\tilde{\mathbf{u}}_a = \mathbf{K}_{aa}^{-1}(\mathbf{f}_a - \mathbf{K}_{ab}\tilde{\mathbf{u}}_b) \qquad b \neq a \tag{9.9}$$

and compared against the exact value.

Both patch tests verify only the satisfaction of the basic differential equation and not of the boundary approximations, as these have been explicitly excluded here.

We mentioned earlier that the test is, in principle, required only for an infinitesimally small patch of elements; however, for differential equations with constant coefficients the size of the patch is immaterial and the test can be carried out on a patch of arbitrary dimensions.

Indeed, if the coefficients are not constant the same size independence exists providing that a constant set of such coefficients is used in the formulation of the test. This applies, for instance, in axisymmetric problems where coefficients of the type $1/r$ (radius) enter the equations and when the patch test is here applied, it is simply necessary to enter the computation with such quantities assumed constant. Alternatively, a body force can be computed which allows the base solution to satisfy the differential equation exactly.

If mapped curvilinear elements are used it is not obvious that the patch test posed in global coordinates needs to be satisfied. Here, in general, convergence in the mapping coordinates may exist but a finite patch test may not be satisfied. However, once again if we specify the nature of the subdivision without changing the mapping function, in the limit the jacobian becomes locally constant and the previous remarks apply. To illustrate this point consider, for instance, a set of elements in which local coordinates are simply the polar coordinates as shown in Fig. 9.3. With shape functions using polynomial expansions in the r, θ terms the patch test of the kind we have described above will not be satisfied with elements of finite size – nevertheless in the limit as the element size tends to zero it will become true. Thus it is evident that patch test satisfaction is a *necessary condition* which has always to be achieved *providing the size of the patch is infinitesimal.*

This proviso which we shall call *weak patch test satisfaction* is not always simple to verify, particularly if the element coding does not easily permit the insertion of constant

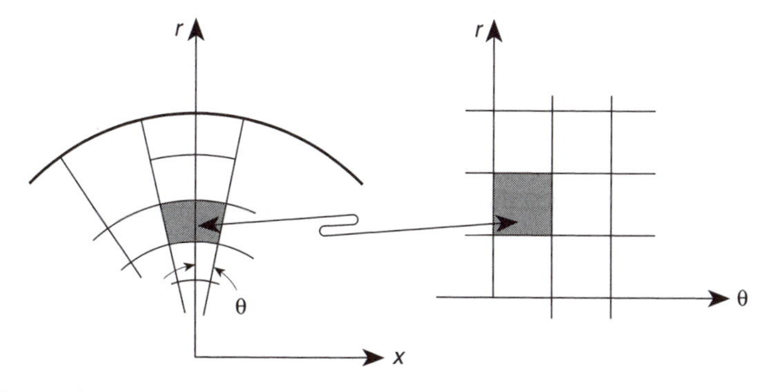

Fig. 9.3 Polar coordinate mapping.

coefficients or a constant jacobian. It is indeed fortunate that the standard isoparametric element form reproduces exactly the linear polynomial global coordinates (see Chapter 5) and for this reason does not require special treatment unless some other *crime* (such as selective or reduced integration) is introduced.

9.4 Generalized patch test (test C) and the single-element test

The patch test described in the preceding section was shown to be a necessary condition for convergence of the formulation but did not establish sufficient conditions for it. In particular, it omitted the testing of the boundary 'load' approximation for the case when the 'natural' (e.g., 'traction of elasticity') conditions are specified. Further it did not verify the stability of the approximation. A test including a check on both of the above conditions is easily constructed. We show this in Fig. 9.4 for a two-dimensional plane problem as *test C*. In this the patch of elements is assembled as before but subject to prescribed natural boundary conditions (or tractions around its perimeter) corresponding to the base function. The assembled matrix of the whole patch is written as

$$\mathbf{K}\tilde{\mathbf{u}} = \mathbf{f}$$

Fixing only the minimum number of parameters $\tilde{\mathbf{u}}$ necessary to obtain a physically valid solution (e.g., eliminating the rigid body motion in an elasticity example or a single value of temperature in a heat conduction problem), a solution is sought for the remaining $\tilde{\mathbf{u}}$ values and compared with the exact base solution assumed.

Now any singularity of the $\mathbf{K}$ matrix will be immediately observed and, as the vector $\mathbf{f}$ includes all necessary source and boundary traction terms, the formulation will be completely tested (providing of course a sufficient number of test states is used). The test described is now not only *necessary* but *sufficient* for convergence.

With boundary traction included it is of course possible to reduce the size of the patch to a single element and an alternative form of test C is illustrated in Fig. 9.4(b), which

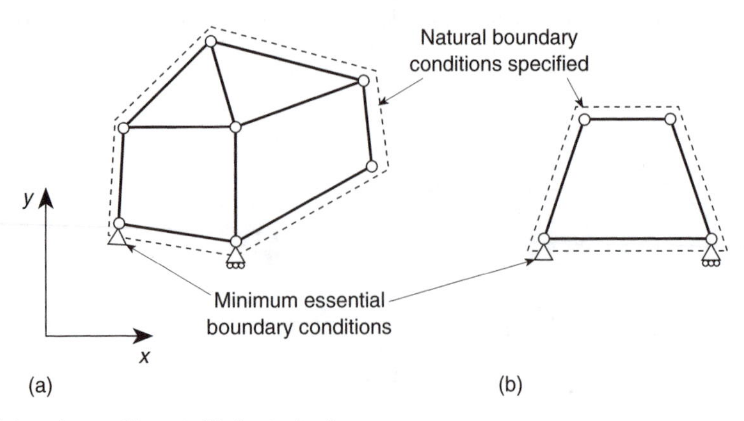

Fig. 9.4 (a) Patch test of form C. (b) The single-element test.

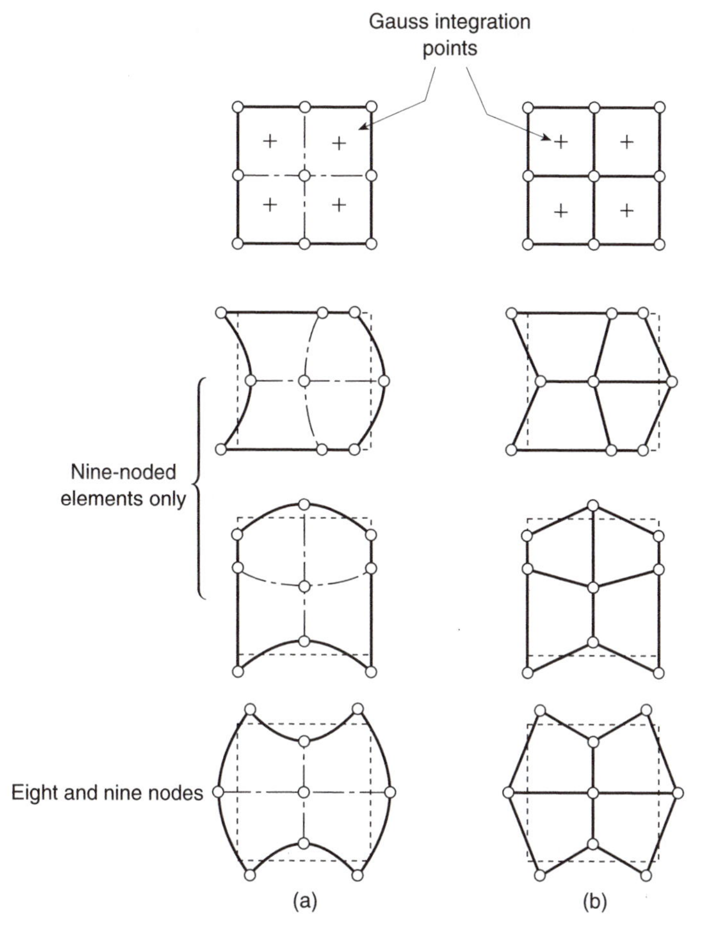

Fig. 9.5 (a) Zero energy (singular) modes for 8- and 9-noded quadratic elements and (b) for a patch of bilinear elements with single integration points.

is termed the *single-element test*.[11] This test is indeed one requirement of a good finite element formulation as, on occasion, a larger patch may not reveal the inherent instabilities of a single element. This happens in the well-documented case of the plane strain–stress 8-noded isoparametric element with (reduced) four-point Gauss quadrature, i.e., where the singular deformation mode of a single element (see Fig. 9.5) disappears when several elements are assembled.† *It should be noted, however, that satisfaction of a single element test is not a sufficient condition for convergence. For sufficiency we require at least one internal element boundary to test that consistency of a patch solution is maintained between elements.*

† This figure also shows a similar singularity for a patch of four 4-noded (bilinear interpolation) elements with single-point quadrature, and we note the similar shape of zero energy modes (see Chapter 5, Sec. 5.12.3).

9.5 The generality of a numerical patch test

In the previous section we have defined in some detail the procedures for conducting a patch test. We have also asserted the fact that such tests if passed guarantee that convergence will occur. However, all the tests are numerical and it is impractical to test all possible combinations.

In particular let us consider the base solutions used. These will invariably be a set of polynomials given in two dimensions as

$$\mathbf{u} = \sum_i \alpha_i P_i(x, y) \tag{9.10}$$

where P_i are a suitable set of low order polynomials (e.g., $1, x, y$ for Galerkin forms possessing only first order derivatives) and α_i are parameters. It is fairly obvious that if patch tests are conducted on each of these polynomials individually any base function of the form given in Eq. (9.10) can be reproduced and the generality preserved for the particular combination of elements tested. This must always be done and is almost a standard procedure in engineering tests, necessitating only a limited number of combinations.

However, as various possible patterns of elements can occur and it is possible to increase the size without limit the reader may well ask whether the test is complete from the geometrical point of view. We believe it is necessary in a numerical test to consider the possibility of several pathological arrangements of elements but that if the test is purely limited to a single element and a complete patch around a node we can be confident about the performance on more general geometric patterns.

Indeed even mathematical assessments of convergence are subject to limits often imposed *a posteriori*. Such limits may arise if for instance a singular mapping is used.

The procedures referred to in this section should satisfy most readers as to the validity and generality of the test.

On some limited occasions it is possible to perform the test purely algebraically and then its validity cannot be doubted. Some such algebraic tests will be referred to later in connection with incompatible elements.

In this chapter we have only considered linear differential equations and linear material behaviour; however, the patch test can well be used and extended to cover non-linear problems.

9.6 Higher order patch tests

While the patch tests discussed in the last three sections ensure (when satisfied) that convergence will occur, they did not test the order of this convergence, beyond assuring us that in the case of Eq. (9.7) the errors were, at least, of order $O(h^2)$ in $\mathbf{u}$. It is an easy matter to determine the actual highest asymptotic rate of convergence of a given element by simply imposing, instead of a linear solution, exact higher order polynomial solutions.[6, 8] The highest value of such polynomials for which complete satisfaction of the patch test is achieved automatically evaluates the corresponding convergence rate. It goes without saying that for such exact solutions generally non-zero source (e.g., body force) terms in the original equation (9.1) will need to be involved.

In addition, test C in conjunction with a higher order patch test may be used to illustrate any tendency for 'locking' to occur (see Chapter 10). Accordingly, element robustness with regard to various parameters (e.g., Poisson's ratios near one-half for elasticity problems in plane strain) may be established.

In such higher order patch tests it will of course first be assumed that the patch is subject to the base expansion solution as described. Thus, for higher order terms it will be necessary to start and investigate solutions of the type

$$\alpha_3 x^2 + \alpha_4 xy + \alpha_5 y^2 + \cdots$$

each of which should be applied individually or as linearly independent combinations and for each the solution should be appropriately tested.

In particular, we shall expect higher order elements to exactly satisfy certain order solutions. However, in Chapter 13 we shall use this idea to find the error between the exact solution and the recovery using precisely the same type of formulation.

9.7 Application of the patch test to plane elasticity elements with 'standard' and 'reduced' quadrature

In the next few sections we consider several applications of the patch test in the evaluation of finite element models. In each case we consider only one of the necessary tests which need to be implemented. For a complete evaluation of a formulation it is necessary to consider all possible independent base polynomial solutions as well as a variety of patch configurations which test the effects of element distortion or alternative meshing interconnections which will be commonly used in analysis. As we shall emphasize, it is important that both consistency and stability be evaluated in a properly conducted test.

In Chapter 5 (Sec. 5.12) we have discussed the minimum required order of numerical integration for various finite element problems which results in no loss of convergence rate. However, it was also shown that for some elements such a minimum integration order results in singular matrices. If we define the *standard* integration as one which evaluates the stiffness of an element exactly† (at least in the undistorted form) then any lower order of integration is called *reduced*.

Such *reduced* integration has some merits in certain problems for reasons which we shall discuss in Sec. 11.5, but it can cause singularities which should be discovered by a patch test (which supplements and verifies the arguments of Sec. 5.12.3).

Application of the patch test to some typical problems will now be shown.

Example 9.1: Patch test for base solution. We consider first a plane stress problem on the patch shown in Fig. 9.6(a). The material is linear, isotropic elastic with properties $E = 1000$ and $\nu = 0.3$. The finite element procedure used is based on the displacement form using 4-noded isoparametric shape functions and numerical integration as described in Chapter 5. Since the stiffness computation includes only first derivatives of displacements, the formulation converges provided that the patch test is satisfied for all linear polynomial

† An alternate definition for standard integration is the lowest order of integration for which the rank of the stiffness matrix does not increase.

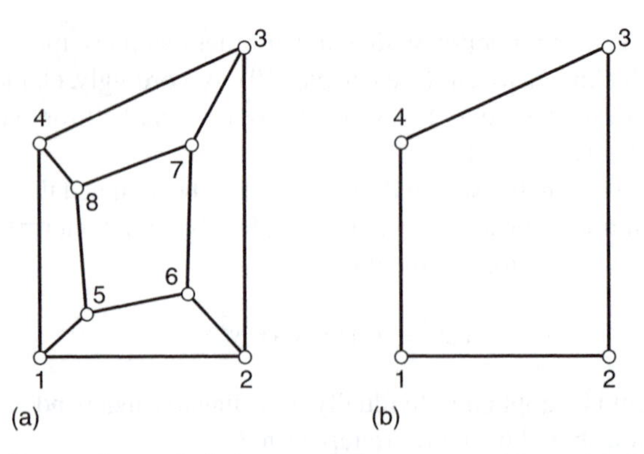

Fig. 9.6 Patch for evaluation of numerically integrated plane stress problems. (a) Five-element patch. (b) One-element patch.

solutions of displacements in the base solution. Here we consider only one of the six independent linear polynomial solutions necessary to verify satisfaction of the patch test. The solution considered is

$$\begin{aligned} u &= 0.0020\,x \\ v &= -0.0006\,y \end{aligned} \tag{9.11a}$$

which produces zero body forces and zero stresses except for

$$\sigma_x = 2 \tag{9.11b}$$

The solution given in Table 9.1 is obtained for the nodal displacements and satisfies Eq. (9.11a) exactly.

The patch test is performed first using 2×2 gaussian 'standard' quadrature to compute each element stiffness and resulting reaction forces at nodes. For patch test A all nodes are restrained and nodal displacement values are specified according to Table 9.1. Stresses are computed at specified Gauss points (1×1, 2×2, and 3×3 Gauss points were sampled) and all are exact to within round-off error (double precision was used which produced round-off errors less than 10^{-15} in the quantities computed). Reactions were also computed at all nodes and again produced the force values shown in Table 9.1 to within round-off limits. This approximation satisfies all conditions required for a finite element procedure (i.e., conforming shape functions and standard order quadrature). Accordingly, the patch test

Table 9.1 Patch solution for Fig. 9.6

	Coordinates		Computed displacements		Forces	
Node a	x_a	y_a	u_a	v_a	F_{x_a}	F_{y_a}
1	0.0	0.0	0.0	0.0	−2	0
2	2.0	0.0	0.0040	0.0	3	0
3	2.0	3.0	0.0040	−0.00180	2	0
4	0.0	2.0	0.0	−0.00120	−3	0
5	0.4	0.4	0.0008	−0.00024	0	0
6	1.4	0.6	0.0028	−0.00036	0	0
7	1.5	2.0	0.0030	−0.00120	0	0
8	0.3	1.6	0.0006	−0.00096	0	0

merely verifies that the programming steps used contain no errors. Patch test A does not require explicit use of the stiffness matrix to compute results; consequently the above patch test was repeated using patch test B where only nodes 1 to 4 are restrained with their displacements specified according to Table 9.1. This tests the accuracy of the stiffness matrix and, as expected, exact results are once again recovered to within round-off errors. Finally, patch test C was performed with node 1 fully restrained and node 4 restrained only in the x direction. Nodal forces were applied to nodes 2 and 3 in accordance with the values generated through the boundary tractions by σ_x (i.e., nodal forces shown in Table 9.1). This test also produced exact solutions for all other nodal quantities in Table 9.1 and recovered σ_x of 2 at all Gauss points in each element.

The above test was repeated for patch tests A, B, and C but using a 1×1 'reduced' Gauss quadrature to compute the element stiffness and nodal force quantities. Patch test C indicated that the global stiffness matrix contained two global 'zero energy modes' (i.e., the global stiffness matrix was rank deficient by 2), thus producing incorrect nodal displacements whose results depend solely on the round-off errors in the calculations. These in turn produced incorrect stresses except at the 1×1 Gauss point used in each element to compute the stiffness and forces. Thus, based upon stability considerations, the use of 1×1 quadrature on 4-noded elements produces a failure in the patch test. The element does satisfy consistency requirements, however, and provided a proper stabilization scheme is employed (e.g., stiffness or viscous methods are used in practice) this element may be used for practical calculations.[20, 21]

It should be noted that a one-element patch test may be performed using the mesh shown in Fig. 9.6(b). The results are given by nodes 1 to 4 in Table 9.1. For the one-element patch, patch tests A and B coincide and neither evaluates the accuracy or stability of the stiffness matrix. On the other hand, patch test C leads to the conclusions reached using the five-element patch: namely, 2×2 gaussian quadrature passes a patch test whereas 1×1 quadrature fails the stability part of the test (as indeed we would expect by the arguments of Chapter 5, Sec. 5.12).

A simple test on cancellation of a diagonal during the triangular decomposition step is sufficient to warn of rank deficiencies in the stiffness matrix.

Example 9.2: Patch test for quadratic elements: quadrature effects. In Fig. 9.7 we show a two-element patch of quadratic isoparametric quadrilaterals. Both 8-noded serendipity and 9-noded lagrangian types are considered and a basic patch test type C is performed for load case 1. For the 8-noded element both 2×2 ('reduced') and 3×3 ('standard') gaussian quadrature satisfy the patch test, whereas for the 9-noded element only 3×3 quadrature is satisfactory, with 2×2 reduced quadrature leading to failure in rank of the stiffness matrix. However, if we perform a one-element test for the 8-noded and 2×2 quadrature element, we discover the spurious zero-energy mode shown in Fig. 9.5 and thus the one-element test has failed. We consider such elements suspect and to be used only with the greatest of care. To illustrate what can happen in practice we consider the simple problem shown in Fig. 9.8(a). In this example the 'structure' modelled by a single element is considered rigid and interest is centred on the 'foundation' response. Accordingly only one element is used to model the structure. Use of 2×2 quadrature throughout leads to answers shown in Fig. 9.8(b) while results for 3×3 quadrature are shown in Fig. 9.8(c). It should be noted that no zero-energy mode exists since more than one element is used. There is here, however, a spurious response due to the large modulus variation between

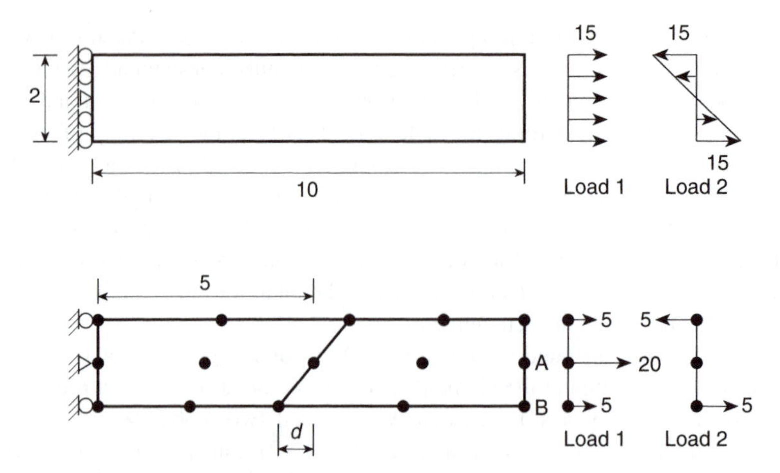

Fig. 9.7 Patch test for 8- and 9-noded isoparametric quadrilaterals.

structure and foundation. This suggests that problems in which non-linear response may lead to a large variation in material parameters could also induce such performance, and thus use of the 8-noded 2×2 integrated element should always be closely monitored to detect such anomalous behaviour.

Indeed, support or loading conditions may themselves induce very suspect responses for elements in which near singularity occurs. Figure 9.9 shows some amusing peculiarities which can occur for reduced integration elements and which disappear entirely if full integration is used.[22] In all cases the *assembly* of elements is non-singular even though individual elements are rank deficient.

Example 9.3: Higher order patch test – assessment of order. In order to demonstrate a higher order patch test we consider the two-element plane stress problem shown in Fig. 9.7 and subjected to bending loading shown as Load 2. As above, two different types of element are considered: (a) an 8-noded serendipity quadrilateral element and (b) a 9-noded lagrangian quadrilateral element. In our test we wish to demonstrate a feature for nine-noded element mapping discussed in Chapter 5 (see Sec. 5.7) and first shown by Wachspress.[23] In particular we restrict the mapping into the xy plane to be that produced by the 4-noded isoparametric bilinear element, but permit the dependent variable to assume the full range of variations consistent with the 8- or 9-noded shape functions. In Chapter 5 we showed that the 9-noded element can approximate a complete quadratic displacement function in x, y whereas the eight-noded element cannot. Thus we expect that the nine-noded element when restricted to the isoparametric mappings of the 4-noded element will pass a higher order patch test for all arbitrary quadratic displacement fields. The pure bending solution in elasticity is composed of polynomial terms up to quadratic order. Furthermore, no body force loadings are necessary to satisfy the equilibrium equations. For the mesh considered the nodal loadings are equal and opposite on the top and bottom nodes as shown in Fig. 9.7. The results for the two elements are shown in Table 9.2 for the indicated quadratures with $E = 100$ and $\nu = 0.3$.

From this test we observe that the 9-noded element does pass the higher order test performed. Indeed, provided the mapping is restricted to the 4-noded shape it will always

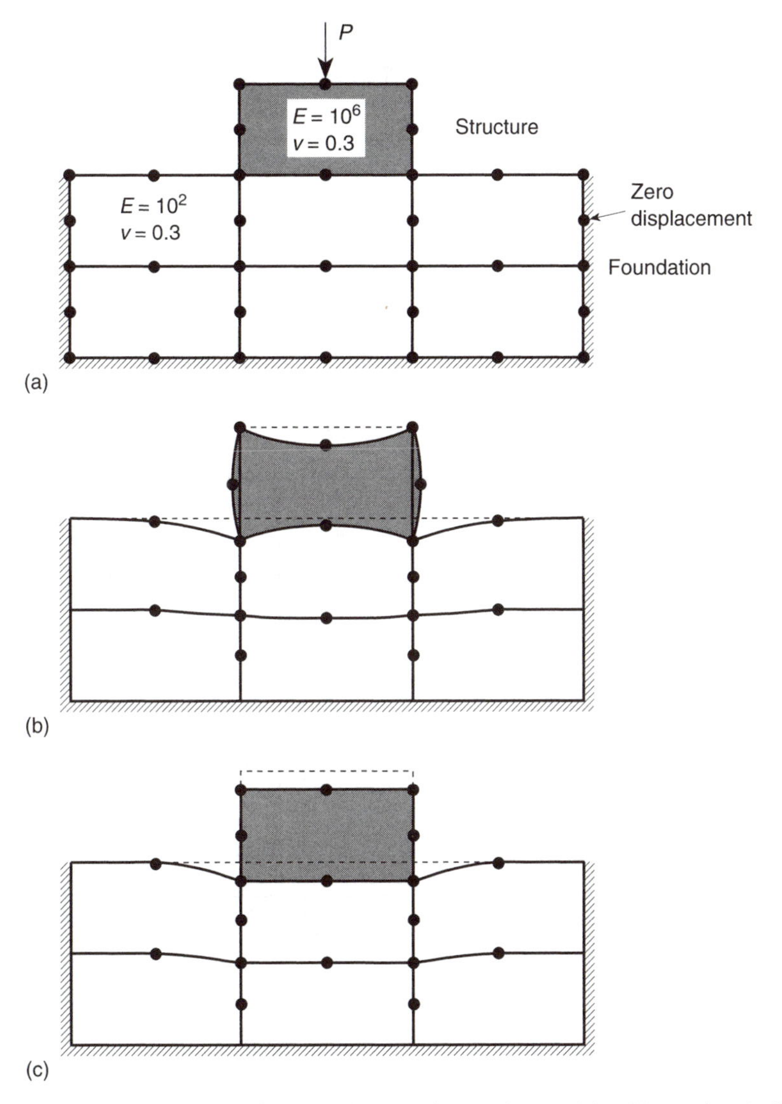

Fig. 9.8 A propagating spurious mode from a single unsatisfactory element. (a) Problem and mesh. (b) 2×2 integration. (c) 3×3 integration.

pass a patch test for displacements with terms no higher than quadratic. On the other hand, the 8-noded element passes the higher order patch test performed only for rectangular element (or constant jacobian) mappings. Moreover, the accuracy of the 8-noded element deteriorates very rapidly with increased distortions defined by the parameter d in Fig. 9.7.

The use of 2×2 reduced quadrature improves results for the higher order patch test performed. Indeed, two of the points sampled give exact results and the third is only slightly in error. As noted previously, however, a single-element test for the 2×2 integrated 8-noded element will fail the stability part of the patch test and it should thus be used with great care.

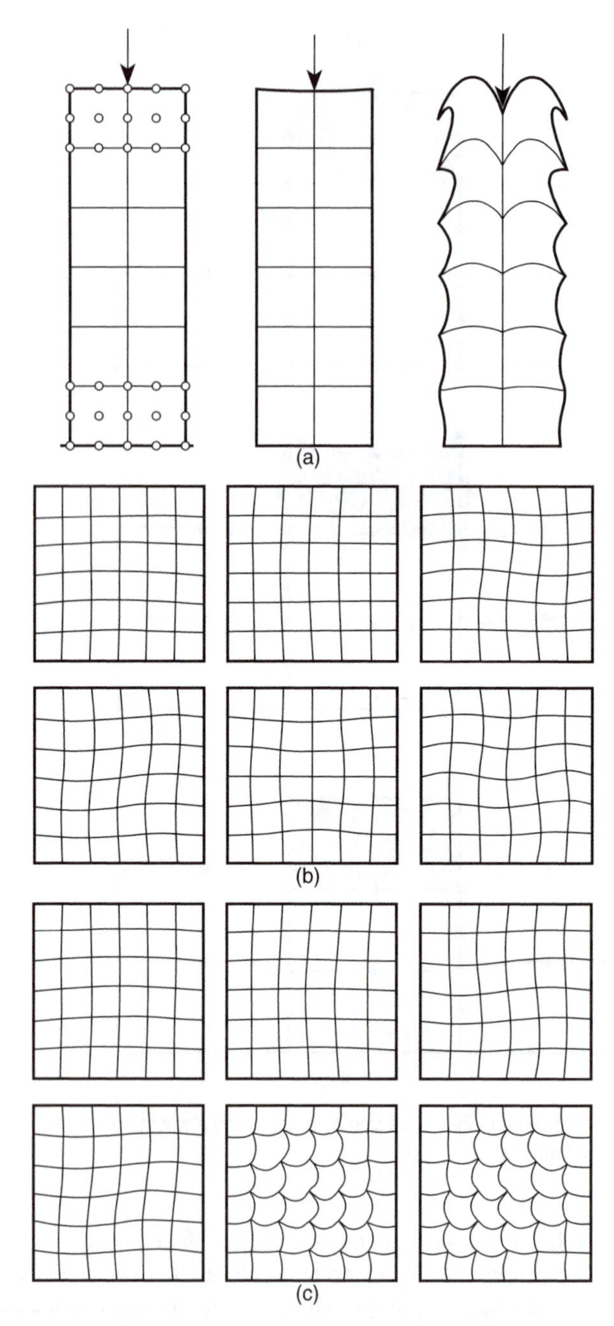

Fig. 9.9 Peculiar response of near singular assemblies of elements.[22] (a) A column of 9-noded elements with point load response of full 3×3 and 2×2 integration. The whole assembly is non-singular but singular element modes are apparent. (b) A fully constrained assembly of 9-noded elements with no singularity – first six eigenmodes with full (3×3) integration. (c) Same as (b) but with 2×2 integration. Note the appearance of 'wild' modes called 'Escher' modes named so in reference 22 after this graphic artist.

Table 9.2 Bending load case ($E = 100$, $\nu = 0.3$)

Element	Quadrature	d	v_A	u_B	v_B
8-node	3 × 3		0.750	0.150	0.75225
8-node	2 × 2	0	0.750	0.150	0.75225
9-node	3 × 3		0.750	0.150	0.75225
8-node	3 × 3		0.7448	0.1490	0.74572
8-node	2 × 2	1	0.750	0.150	0.75100
9-node	3 × 3		0.750	0.150	0.75225
8-node	3 × 3		0.6684	0.1333	0.66364
8-node	2 × 2	2	0.750	0.150	0.75225
9-node	3 × 3		0.750	0.150	0.75225
Exact	–	–	0.750	0.150	0.75225

9.8 Application of the patch test to an incompatible element

In order to demonstrate the use of the patch test for a finite element formulation which violates the usually stated requirements for shape function continuity, we consider the plane strain incompatible modes first introduced by Wilson *et al.*[24] and discussed by Taylor *et al.*[25] The specific incompatible formulation considered uses the element displacement approximations:

$$\hat{\mathbf{u}} = \mathbf{N}_a \tilde{\mathbf{u}}_a + N_1^n \boldsymbol{\alpha}_1 + N_2^n \boldsymbol{\alpha}_2 \tag{9.12}$$

where $\mathbf{N}_a$ $(a = 1, \ldots, 4)$ are the usual conforming bilinear shape functions and the last two terms are *incompatible modes of deformation* defined by the hierarchical functions

$$N_1^n = 1 - \xi^2 \quad \text{and} \quad N_2^n = 1 - \eta^2 \tag{9.13}$$

defined independently for each element.

The shape functions used are illustrated in Fig. 9.10. The first, a set of standard bilinear type, gives a displacement pattern which, as shown in Fig. 9.10(b), introduces spurious shear strains in pure bending. The second, in which the parameters $\boldsymbol{\alpha}_1$ and $\boldsymbol{\alpha}_2$ are strictly associated with a specific element, therefore introduces incompatibility but assures correct bending behaviour in an individual rectangular element. The excellent performance of this element in the bending situation is illustrated in Fig. 9.11.

In reference 25 the finite element approximation is computed by summing the potential energies of each element and computing the nodal loads due to boundary tractions from the conforming part of the displacement field only. Thus for the purposes of conducting patch tests we compute the strains using all parts of the displacement field leading to a generalization of (9.4) which may be written as

$$\begin{bmatrix} \mathbf{K}_{11} & \mathbf{K}_{12} \\ \mathbf{K}_{21} & \mathbf{K}_{22} \end{bmatrix} \begin{Bmatrix} \tilde{\mathbf{u}} \\ \boldsymbol{\alpha} \end{Bmatrix} = \begin{Bmatrix} \mathbf{f}_1 \\ \mathbf{f}_2 \end{Bmatrix} \tag{9.14}$$

Here $\mathbf{K}_{11}$ and $\mathbf{f}_1$ are the stiffness and loads of the 4-noded (conforming) bilinear element, $\mathbf{K}_{12}$ and $\mathbf{K}_{21}$ $(= \mathbf{K}_{12}^{\mathrm{T}})$ are coupling stiffnesses between the conforming and non-conforming displacements, and $\mathbf{K}_{22}$ and $\mathbf{f}_2$ are the stiffness and loads of the non-conforming displacements. We note that, according to the algorithm of reference 24, $\mathbf{f}_2$ must vanish from the patch test solutions.

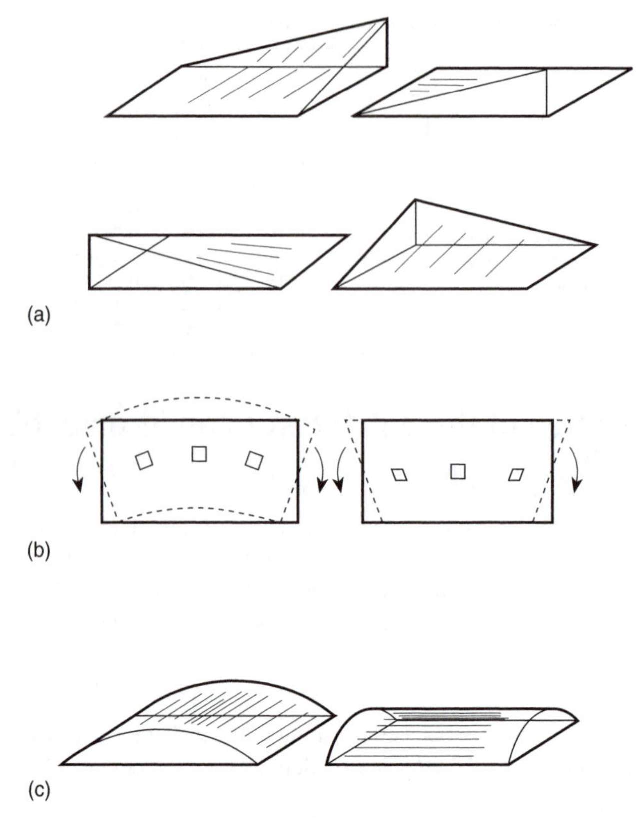

Fig. 9.10 (a) Linear quadrilateral with auxiliary incompatible shape functions. (b) Pure bending and linear displacements causing shear. (c) Auxiliary 'bending' shape functions with internal variables.

For a patch test in plane strain or plane stress, only linear polynomials need be considered for which all non-conforming displacements must vanish. Thus for a successful patch test we must have

$$\mathbf{K}_{11}\tilde{\mathbf{u}} = \mathbf{f}_1 \tag{9.15a}$$

and

$$\mathbf{K}_{21}\tilde{\mathbf{u}} = \mathbf{f}_2 \tag{9.15b}$$

If we carry out a patch test for the mesh shown in Fig. 9.12(a) we find that all three forms (i.e., patch tests A, B, and C) satisfy these conditions and thus pass the patch test. If we consider the patch shown in Fig. 9.12(b), however, the patch test is not satisfied. The lack of satisfaction shows up in different ways for each form of the patch test. Patch test A produces non-zero $\mathbf{f}_2$ values when $\boldsymbol{\alpha}$ is set to zero and $\tilde{\mathbf{u}}$ according to the displacements considered. In form B the values of the nodal displacements $\tilde{\mathbf{u}}_5$ are in error and $\boldsymbol{\alpha}$ are non-zero, also leading to erroneous stresses in each element. In form C all unspecified displacements are in error as well as the stresses.

It is interesting to note that when a patch is constructed according to Fig. 9.12(c) in which all elements are parallelograms all three forms of the patch test are once again satisfied.

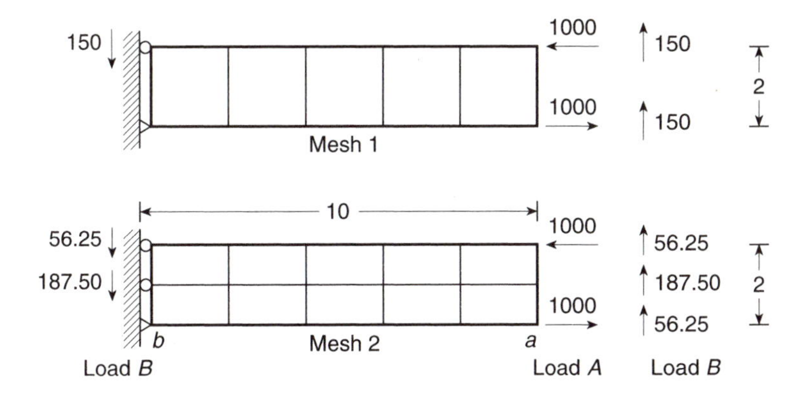

	Displacement at *a*		Displacement at *b*	
	Load *A*	Load *B*	Load *A*	Load *B*
Beam theory	10.00	103.0	300.0	4050
(a) Mesh 1	6.81	70.1	218.2	2945
(a) Mesh 2	7.06	72.3	218.8	2954
(b) Mesh 1	10.00	101.5	300.0	4050
(b) Mesh 2	10.00	101.3	300.0	4050

Fig. 9.11 Performance of the non-conforming quadrilateral in beam bending treated as plane stress. (a) Conforming linear quadrilateral. (b) Non-conforming quadrilateral.

Accordingly we can note that if any mesh is systematically refined by subdivision of each element into four elements whose sides are all along ξ, η lines in the original element with values of -1, 0, or 1 (i.e., by bisections) the mesh converges to constant jacobian approximations of the type shown in Fig. 9.12(c). Thus, in this special case the incompatible mode element satisfies a weak patch test and will converge. In general, however, it may be necessary to use a very fine discretization to achieve sufficient accuracy, and hence the element probably has no practical (or efficient) engineering use.

A simple artifice to ensure that an element passes the patch test is to replace the derivatives of the incompatible modes by

$$\begin{Bmatrix} \dfrac{\partial N_a^n}{\partial x} \\ \dfrac{\partial N_a^n}{\partial y} \end{Bmatrix} = \frac{J_0}{J(\xi,\eta)} \mathbf{J}_0^{-1} \begin{Bmatrix} \dfrac{\partial N_a^n}{\partial \xi} \\ \dfrac{\partial N_a^n}{\partial \eta} \end{Bmatrix} \tag{9.16}$$

where $J(\xi, \eta)$ is the determinant of the jacobian matrix $\mathbf{J}(\xi, \eta)$ and $\mathbf{J}_0$ and J_0 are the values of the inverse jacobian matrix and jacobian evaluated at the element centre ($\xi = \eta = 0$). This ensures satisfaction of the patch test for all element shapes, and with this alteration of the algorithm the incompatible element proves convergent and quite accurate.[25]

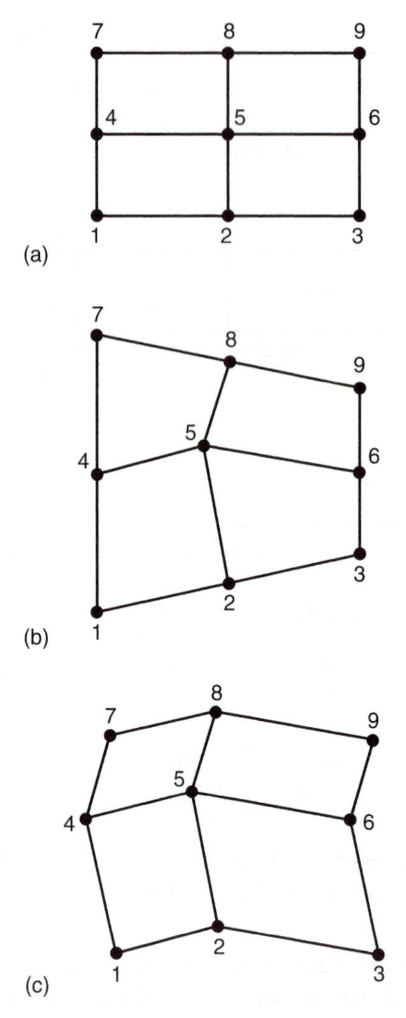

Fig. 9.12 Patch test for an incompatible element form. (a) Regular discretization. (b) Irregular discretization about node 5. (c) Constant jacobian discretization about node 5.

An alternative approach which also passes the patch test constructs the derivatives used in the strains as

$$
\begin{aligned}
\frac{\partial N_a^n}{\partial x} &\Leftarrow \frac{\partial N_a^n}{\partial x} - \frac{1}{\Omega_e}\int_{\Omega_e} \frac{\partial N_a^n}{\partial x}\, \mathrm{d}\Omega \\
\frac{\partial N_a^n}{\partial y} &\Leftarrow \frac{\partial N_a^n}{\partial y} - \frac{1}{\Omega_e}\int_{\Omega_e} \frac{\partial N_a^n}{\partial y}\, \mathrm{d}\Omega
\end{aligned}
\qquad (9.17)
$$

where Ω_e is the volume of the element.[26] Indeed, this form may also be used to deduce terms in the strain matrix of enhanced strain forms (e.g., see Sec. 10.5.3) and the justification of the modification follows from the mixed approach used there. When the shape functions (9.13) are inserted into (9.17) and the jacobian is constant (as it will be for any parallelogram shape element) we immediately find that the integral term is zero. However, when the element has a non-constant jacobian which for the two-dimensional element has the form

$$J(\xi, \eta) = J_0 + J_\xi \xi + J_\eta \eta,$$

where J_0, J_ξ, J_η are constants depending on nodal coordinates of the element, the integral term is non-zero. Thus, the effectiveness of the modification is clearly evident in producing elements which pass the constant stress patch test for all element shapes.

9.9 Higher order patch test – assessment of robustness

A higher order patch test may also be used to assess element 'robustness'. An element is termed robust if its performance is not sensitive to physical parameters of the differential equation. For example, the performance of many elements for solution of plane strain linear elasticity problems is sensitive to Poisson's ratio values near 0.5 (called 'near incompressibility'). Indeed, for Poisson ratios near 0.5 the energy stored by a unit volumetric strain is many orders larger than the energy stored by a unit deviatoric strain. Accordingly finite elements which exhibit a strong coupling between volumetric and deviatoric strains often produce poor results in the nearly incompressible range, a problem discussed further in Chapter 11.

This may be observed using a four-noded element to solve a problem with a quadratic displacement field (i.e., a higher order patch test). If we again consider a pure bending example and an eight-element mesh shown in Fig. 9.13 we can clearly observe the deterioration of results as Poisson's ratio approaches a value of one-half. Also shown in Fig. 9.13 are results for the incompatible modes described in Sec. 9.8. It is evident that the response is considerably improved by adding these modes, especially if 2×2 quadrature is used.

If we consider the regular mesh and 4-noded elements and further keep the domain constant and successively refine the problem using meshes of 8, 32, 128, and 512 elements, we observe that the answers do converge as guaranteed by the patch test. However, as shown in Fig. 9.14, the rate of convergence in energy for Poisson ratio values of 0.25 and 0.4999 is quite different. For 0.25 the rate of convergence is nearly a straight line for all meshes, whereas for 0.4999 the rate starts out quite low and approaches an asymptotic value of 2 as h tends towards zero. For ν near 0.25 the element is called robust, whereas for ν near 0.5 it is not. If we use selective reduced integration (which for the plane strain case passes strong patch tests) and repeat the experiment, both values of ν produce a similar response and thus the element becomes robust for all values of Poisson's ratio less than 0.5.

The use of higher order patch tests can thus be very important to separate robust elements from non-robust elements. For methods which seek to automatically refine a mesh adaptively in regions with high errors, as discussed in Chapter 14, it is extremely important to use robust elements.

9.10 Concluding remarks

In the preceding sections we have described the patch test and its use in practice by considering several example problems. The patch test described has two essential parts: (a) *a consistency evaluation* and (b) *a stability check.* In the consistency test a set of linearly

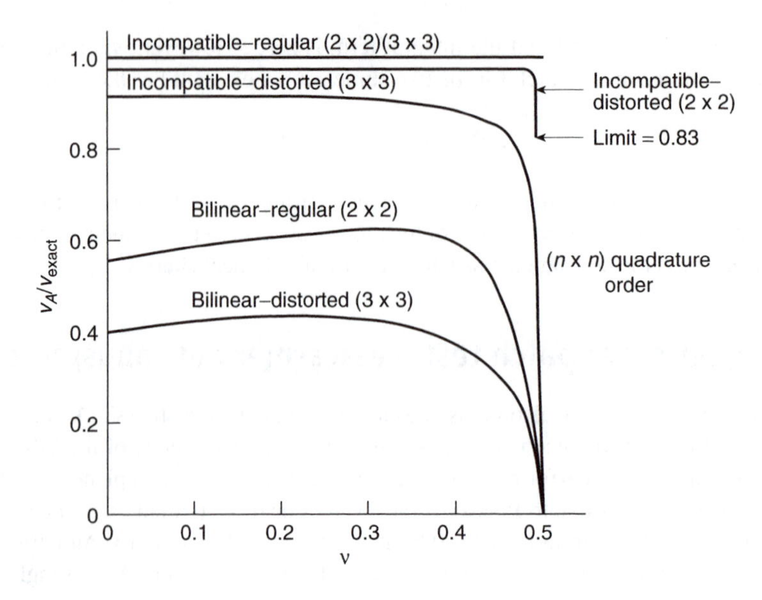

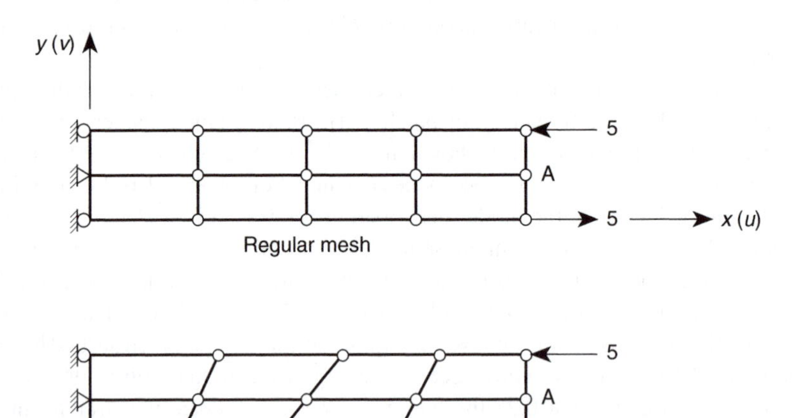

Fig. 9.13 Plane strain 4-noded quadrilaterals with and without incompatible modes (higher order patch test for performance evaluation).

independent essential polynomials (i.e., all independent terms up to the order needed to describe the finite element model) is used as a solution to the differential equations and boundary conditions, and in the limit as the size of a patch tends to zero the finite element model must exactly satisfy each solution. We presented three forms to perform this portion of the test which we call forms A, B, and C.

The use of form C, where all boundary conditions are the natural ones (e.g., tractions for elasticity) except for the minimum number of essential conditions needed to ensure a unique solution to the problem (e.g., rigid body modes for elasticity), is recommended to test consistency and stability simultaneously. Both one-element and more-than-one-

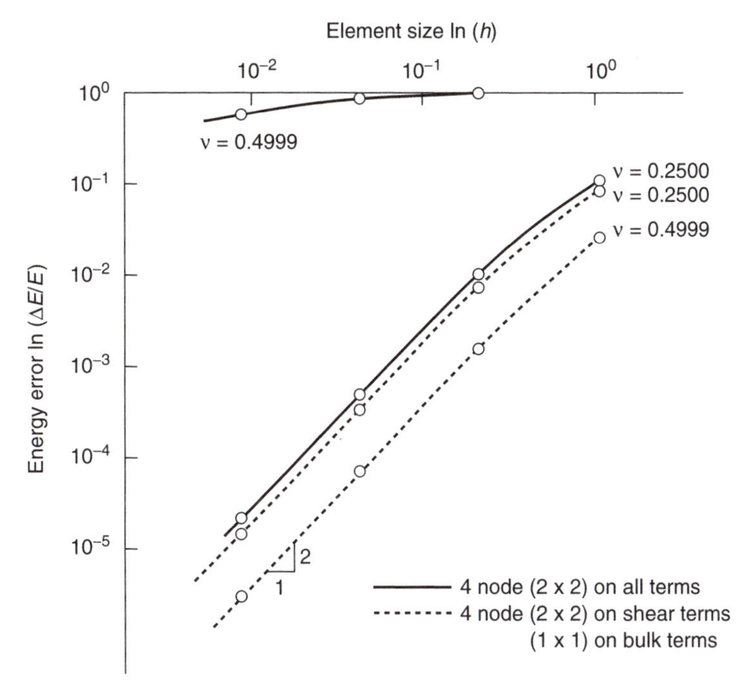

Fig. 9.14 Higher order patch test on element robustness (see Fig. 9.13) – convergence test under subdivision of elements.

element tests are necessary to ensure that the patch test is satisfied. With these conditions and assuming that the solution procedure used can detect any possible rank deficiencies the stability of solution is also tested. If no such condition is included in the program a stability test must be conducted independently. This can be performed by computing the number of zero eigenvalues in the coefficient matrix for methods that use a solution of linear equations to compute the finite element parameters, $\tilde{\mathbf{u}}$. Alternatively, the loading used for the patch solution may be perturbed at one point by a small value (say square root of the round-off limit, e.g., by 10^{-8} for round-off of order 10^{-15}) and the solution tested to ensure that it does not change by a large amount.

Once an element has been shown to pass all of the essential patch tests for both consistency and stability, convergence is assured as the size of elements tends to zero. However, in some situations (e.g., the nearly incompressible elastic problem) convergence may be very slow until a very large number of elements are used. Accordingly, we recommend that higher order patch tests be used to establish element robustness. Higher order patch tests involve the use of polynomial solutions of the differential equation and boundary conditions with the order of terms larger than the basic polynomials used in a patch test. Indeed, the order of polynomials used should be increased until the patch test is satisfied only in a weak sense (i.e., as h tends to zero). The advantage of using a higher order patch test, as opposed to other boundary value problems, is that the exact solution may be easily computed everywhere in the model.

In some of the examples we have tested the use of incompatible function and inexact numerical integration procedures (reduced and selective integration). Some of these violations of the rules previously stipulated have proved justified not only by yielding improved performance but by providing methods for which convergence is guaranteed. We shall discuss in Chapter 11 some of the reasons for such improved performance.

9.11 Problems

9.1 A Type C patch test for a plane strain problem is to be performed using the single element shown in Fig. 9.15(a).

Assume an element has the dimensions $a = 15$, $b = 12$, $c = 10$ with elastic properties $E = 200$, $\nu = 0.25$. Nodes 1 and 2 are placed on the x axis and $u_1 = v_1 = v_2 = 0$ are applied as boundary restraints.

(a) Compute all nodal forces necessary to compute the test for a stress state $\sigma_x = 8$ with all other stresses zero.

(b) Compute the displacements $u(x, y)$ and $v(x, y)$ for the solution.

(c) Use *FEAPpv* (or any other available program) to perform the test. Is it passed?

9.2 Solve Problem 9.1 for an axisymmetric geometry† with node 1 satisfying $v_1 = 0$ and all other nodes free to displace. For $\sigma_r = 8$ consider the cases:

(a) Node 1 placed at $r = 0$.

(b) Node 1 placed at $r = 15$.

9.3 Solve Problem 9.1 for a plane stress problem with an orthotropic material given by

$$\begin{Bmatrix} \sigma_{x'} \\ \sigma_{y'} \\ \tau_{x'y'} \end{Bmatrix} = \begin{bmatrix} 200 & 50 & 0 \\ 50 & 100 & 0 \\ 0 & 0 & 75 \end{bmatrix} \begin{Bmatrix} \varepsilon_{x'} \\ \varepsilon_{y'} \\ \gamma_{x'y'} \end{Bmatrix}$$

Let nodes 1 and 2 lie on the x axis with node 1 placed at the origin.

(a) Compute the nodal forces acting on all nodes when the orthotropic axes are aligned as shown in Fig. 9.15(b) with $\theta = 30°$ and a single stress $\sigma'_x = 5$ is applied.

(b) Compute the displacement field for the case $u_1 = v_1 = u_4 = 0$.

(c) Use *FEAPpv* (or any other available program) to perform the patch test. Is it passed?

9.4 The example described in Problem 9.1 is used to perform a patch test on an element with incompatible modes. The element matrix is given by

$$\begin{bmatrix} \mathbf{K} & \mathbf{C} \\ \mathbf{C}^{\mathrm{T}} & \mathbf{V} \end{bmatrix} \begin{Bmatrix} \tilde{\mathbf{u}} \\ \boldsymbol{\alpha} \end{Bmatrix} = \begin{Bmatrix} \mathbf{f}_u \\ \mathbf{f}_\alpha \end{Bmatrix}$$

where

† Note: *FEAPpv* computes all axisymmetric arrays on a 1-radian sector in the θ direction, thus avoiding the 2π factor in a complete ring sector.

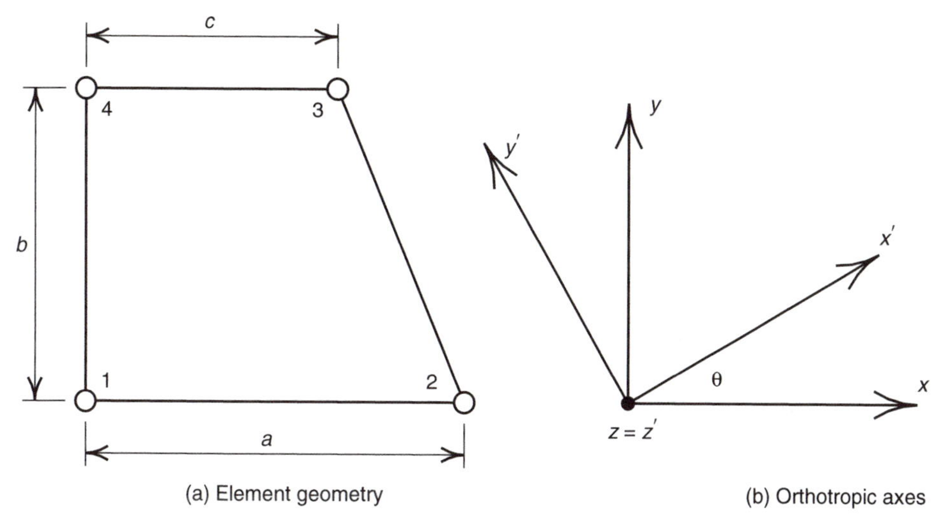

Fig. 9.15 One element patch test for 4-node quadrilateral. Problems 9.1 to 9.4.

$$\mathbf{K} = \int_{\Omega_e} \mathbf{B}_u^{\mathrm{T}} \mathbf{D} \mathbf{B}_u \, \mathrm{d}\Omega$$

$$\mathbf{C} = \int_{\Omega_e} \mathbf{B}_u^{\mathrm{T}} \mathbf{D} \mathbf{B}_\alpha \, \mathrm{d}\Omega$$

$$\mathbf{V} = \int_{\Omega_e} \mathbf{B}_\alpha^{\mathrm{T}} \mathbf{D} \mathbf{B}_\alpha \, \mathrm{d}\Omega$$

The element passes the patch test for all constant stress states when $a = c$; however, it fails when $a \neq c$.

Suggest a correction which will ensure the patch test is satisfied.

9.5 Perform a patch test for the 8-node element shown in Fig. 9.16(a) for the assumed displacements

$$u = 0.1\,x \quad \text{and} \quad v = 0$$

Let the origin be at the lower left corner of the element. The dimensions are $a = b = 3$ and $c = 3.3$. The material is linear isotropic elastic with $E = 200$ and $\nu = 0.3$ and plane strain conditions are assumed.

(a) Use 2×2 gaussian quadrature to compute the element arrays and conduct a type A, B, and C patch test.
(b) Repeat the calculation using 3×3 quadrature.
(c) Consider a higher order displacement

$$u = 0.2xy \quad \text{and} \quad v = 0$$

and repeat (a) and (b).
(d) Discuss any differences noted.

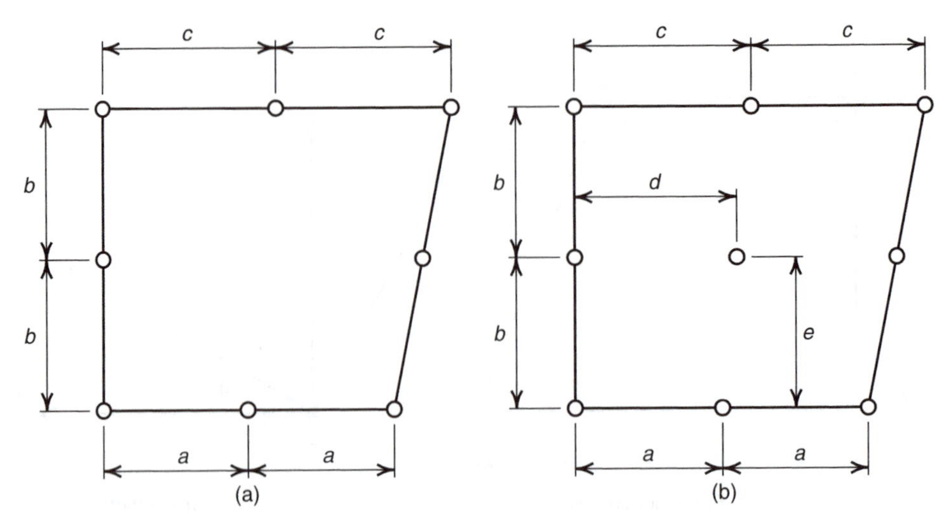

Fig. 9.16 One-element patch test for 8- and 9-node quadrilateral. Problems 9.5 to 9.8.

9.6 Perform a patch test for the 9-node element shown in Fig. 9.16(b) for the assumed displacements

$$u = 0.1x \quad \text{and} \quad v = 0$$

Let the origin be at the lower left corner of the element. The dimensions are $a = b = e = 3$, $c = 3.3$ and $d = 3.15$. Material is linear isotropic elastic with $E = 200$ and $\nu = 0.3$ and plane strain conditions are assumed.

(a) Use 2×2 gaussian quadrature to compute the element arrays and conduct a type A, B, and C patch test.
(b) Repeat the calculation using 3×3 quadrature.
(c) Set $d = 3.4$ and $e = 2.9$ and repeat (a) and (b).
(d) Consider a higher order displacement

$$u = 0.2xy \quad \text{and} \quad v = 0$$

and repeat (a) to (c).
(e) Discuss any differences noted.

9.7 Solve Problem 9.5 for an axisymmetric geometry (replace x, y by r, z).
(a) Let the inner radius be located at $r = 0$.
(b) Let the inner radius be located at $r = 3$.

9.8 Solve Problem 9.6 for an axisymmetric geometry (replace x, y by r, z).
(a) Let the inner radius be located at $r = 0$.
(b) Let the inner radius be located at $r = 3$.

9.9 For the 4-element mesh configurations shown in Fig. 9.17 devise a set of patch tests for a plane strain problem in which individual constant stress components are evaluated. Choose appropriate dimensions and isotropic elastic properties with $\nu \neq 0$.

Use *FEAPpv* (or any available program) to perform Type A, B, and C tests for the arrays evaluated by (a) 1×1 quadrature and (b) 2×2 quadrature. Discuss your findings.

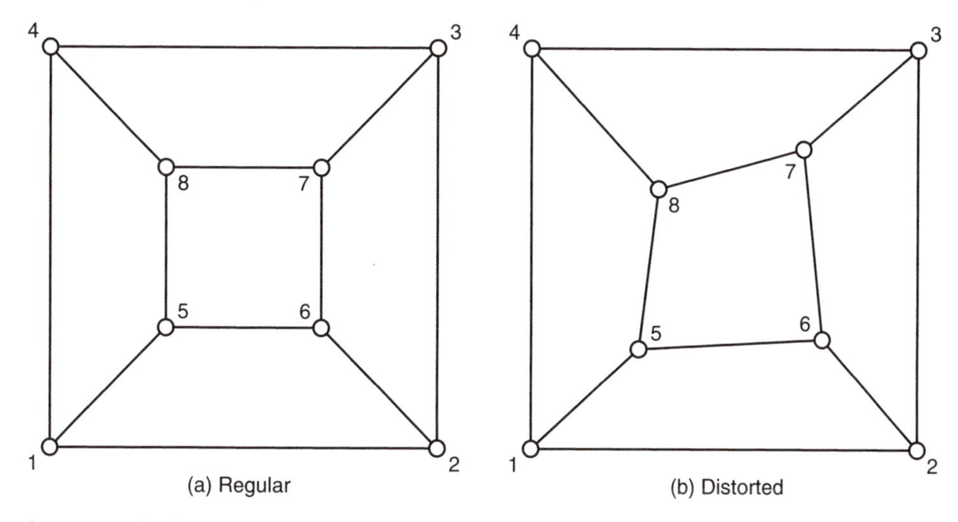

Fig. 9.17 Multi-element patch test. Problems 9.9 to 9.13.

9.10 Each quadrilateral subregion in Fig. 9.17 is to be represented by an 8-node isoparametric serendipity element. Each side of the region has a length of 10 units. A higher order patch test of a plane strain problem with isotropic material with $E = 200$ and $\nu = 0$ has the displacements

$$u = -0.1xy \quad \text{and} \quad v = 0.05x^2$$

(a) Compute the state of stress for the given displacement field.
(b) Select appropriate positions for nodes 5 to 8 for configurations (a) and (b). Specify appropriate nodal boundary conditions to prevent rigid body motion. (Hint: Place the origin of coordinates at the mid-point between nodes 1 and 4.)
(c) Compute appropriate nodal forces and perform a Type C patch test for each configuration using 3×3 gaussian quadrature to compute arrays.

Briefly discuss your findings.

9.11 Solve Problem 9.10 using 9-node isoparametric lagrangian elements.

9.12 Replace x by r and y by z and solve Problem 9.10 using 8-node isoparametric serendipity elements on an axisymmetric geometry.

9.13 Replace x by r and y by z and solve Problem 9.12 using 9-node isoparametric lagrangian elements.

9.14 Construct the generalization of the mesh configuration shown in Fig. 9.17 to a three-dimensional problem. For $E = 200$, $\nu = 0.25$ and equal side lengths of 10 units use 8-node isoparametric hexagonal elements to perform a Type C patch test for the single stress $\sigma_z = 5$. Use both regular and distorted positions for the internal nodes.
(Hint: Check that there are no negative jacobian determinants at the nodes of each element.)

9.15 Select dimensions and use *FEAPpv* (or any available program) to verify the results shown in Fig. 9.8.

9.16 Select dimensions and use *FEAPpv* (or any available program) to verify the results shown in Fig. 9.9(a).
9.17 Select dimensions and use *FEAPpv* (or any available program) to verify the results shown in Fig. 9.9(b) and (c).

References

1. B.M. Irons. Numerical integration applied to finite element methods. In *Proc. Conf. on Use of Digital Computers in Structural Engineering*, University of Newcastle, 1966.
2. G.P. Bazeley, Y.K. Cheung, B.M. Irons, and O.C. Zienkiewicz. Triangular elements in bending – conforming and non-conforming solutions. In *Proc. 1st Conf. Matrix Methods in Structural Mechanics*, volume AFFDL-TR-66-80, pages 547–576, Wright Patterson Air Force Base, Ohio, Oct. 1966.
3. B.M. Irons and A. Razzaque. Experience with the patch test for convergence of finite elements. In A.K. Aziz, editor, *The Mathematics of Finite Elements with Application to Partial Differential Equations*, pages 557–587. Academic Press, New York, 1972.
4. B. Fraeijs de Veubeke. Variational principles and the patch test. *Int. J. Numer. Meth. Eng.*, 8:783–801, 1974.
5. G. Sander. Bournes supérieures et inérieures dans l'analyse matricielle des plates en flexion-torsion. *Bull. Soc. Royale des Sci. de Liège*, 33:456–494, 1974.
6. E.R. de Arantes Oliveira. The patch test and the general convergence criteria of the finite element method. *Int. J. Solids Struct.*, 13:159–178, 1977.
7. G. Strang. Variational crimes and the finite element method. In A.K. Aziz, editor, *Proc. Foundations of the Finite Element Method*, pages 689–710. Academic Press, New York, 1972.
8. G. Strang and G.J. Fix. *An Analysis of the Finite Element Method.* Prentice-Hall, Englewood Cliffs, N.J., 1973.
9. F. Stummel. The limitations of the patch test. *Int. J. Numer. Meth. Eng.*, 15:177–188, 1980.
10. J. Robinson *et al.* Correspondence on patch test. *Fin. Elem. News*, 1:30–34, 1982.
11. R.L. Taylor, O.C. Zienkiewicz, J.C. Simo, and A.H.C. Chan. The patch test – a condition for assessing FEM convergence. *Int. J. Numer. Meth. Eng.*, 22:39–62, 1986.
12. R.E. Griffiths and A.R. Mitchell. Non-conforming elements. In *Mathematical Basis of Finite Element Methods*, pages 41–69, Oxford, 1984. Clarendon Press. Inst. Math. and Appl. Conference series.
13. O.C. Zienkiewicz and R.L. Taylor. The finite element patch test revisited: a computer test for convergence, validation and error estimates. *Comp. Meth. Appl. Mech. Eng.*, 149:523–544, 1997.
14. O.C. Zienkiewicz, S. Qu, R.L. Taylor, and S. Nakazawa. The patch test for mixed formulations. *Int. J. Numer. Meth. Eng.*, 23:1873–1883, 1986.
15. W.X. Zhong. FEM patch test and its convergence. Technical Report 97–3001, Research Institute Engineering Mechanics, Dalian University of Technology, 1997 (in Chinese).
16. I. Babuška, T. Strouboulis, and C.S. Upadhyay. A model study of the quality of a posteriori error estimators for linear elliptic problems. Error estimation in the interior of patchwise uniform grids of triangles. *Comp. Meth. Appl. Mech. Eng.*, 114:307–378, 1994.
17. B. Boroomand and O.C. Zienkiewicz. An improved REP recovery and the effectivity robustness test. *Int. J. Numer. Meth. Eng.*, 40:3247–3277, 1997.
18. A. Ralston. *A First Course in Numerical Analysis*. McGraw-Hill, New York, 1965.
19. B.M. Irons and S. Ahmad. *Techniques of Finite Elements*. Horwood, Chichester, 1980.
20. D. Kosloff and G.A. Frasier. Treatment of hour glass patterns in low order finite element codes. *Int. J. Numer. Anal. Meth. Geomech.*, 2:57–72, 1978.

21. T. Belytschko and W.E. Bachrach. The efficient implementation of quadrilaterals with high coarse mesh accuracy. *Comp. Meth. Appl. Mech. Eng.*, 54:276–301, 1986.
22. N. Biĉaniĉ and E. Hinton. Spurious modes in two dimensional isoparametric elements. *Int. J. Numer. Meth. Eng.*, 14:1545–1557, 1979.
23. E.L. Wachspress. High order curved finite elements. *Int. J. Numer. Meth. Eng.*, 17:735–745, 1981.
24. E.L. Wilson, R.L. Taylor, W.P. Doherty, and J. Ghaboussi. Incompatible displacement models. In S.T. Fenves *et al.*, editor, *Numerical and Computer Methods in Structural Mechanics*, pages 43–57. Academic Press, New York, 1973.
25. R.L. Taylor, P.J. Beresford, and E.L. Wilson. A non-conforming element for stress analysis. *Int. J. Numer. Meth. Eng.*, 10:1211–1219, 1976.
26. A. Ibrahimbegovic and E.L. Wilson. A modified method of incompatible modes. *Comm. Numer. Meth. Eng.*, 7:187–194, 1991.

10

Mixed formulation and constraints – complete field methods

10.1 Introduction

The set of differential equations from which we start the discretization process will determine whether we refer to the formulation as *mixed* or *irreducible*. Thus if we consider an equation system with several dependent variables $\mathbf{u}$ written as [see Eqs (3.1) and (3.2)]

$$\mathcal{A}(\mathbf{u}) = \mathbf{0} \qquad \text{in domain } \Omega \tag{10.1a}$$

and

$$\mathcal{B}(\mathbf{u}) = \mathbf{0} \qquad \text{on boundary } \Gamma \tag{10.1b}$$

in which none of the components of $\mathbf{u}$ can be eliminated still leaving a well-defined problem, then the formulation will be termed *irreducible*. If this is not the case the formulation will be called *mixed*. These definitions were given in Chapter 3 (p. 56).

This definition is not the only one possible[1] but appears to the authors to be widely applicable[2, 3] if in the elimination process referred to we are allowed to introduce penalty functions. Further, for any given physical situation we shall find that more than one irreducible form is usually possible.

As an example we shall consider the simple problem of heat conduction (or the quasi-harmonic equation) to which we have referred in Chapters 3 and 7. In this we start with a physical constitutive relation defining the flux [see Eq. (7.5)] in terms of the potential (temperature) gradients, i.e.,

$$\mathbf{q} = -\mathbf{k}\,\boldsymbol{\nabla}\phi \qquad \mathbf{q} = \begin{Bmatrix} q_x \\ q_y \end{Bmatrix} \tag{10.2}$$

The continuity equation can be written as [see Eq. (7.7)]

$$\boldsymbol{\nabla}^{\mathrm{T}}\mathbf{q} \equiv \frac{\partial q_x}{\partial x} + \frac{\partial q_y}{\partial y} = -Q \tag{10.3}$$

If the above equations are satisfied in Ω and the boundary conditions

$$\phi = \bar{\phi} \quad \text{on } \Gamma_\phi \qquad \text{or} \qquad q_n = \bar{q}_n \quad \text{on } \Gamma_q \tag{10.4}$$

are obeyed then the problem is solved.

Clearly elimination of the vector $\mathbf{q}$ is possible and simple substitution of Eq. (10.2) into Eq. (10.3) leads to

$$-\boldsymbol{\nabla}^{\mathrm{T}}(\mathbf{k}\,\boldsymbol{\nabla}\phi) + Q = 0 \qquad \text{in } \Omega \tag{10.5}$$

with appropriate boundary conditions expressed in terms of ϕ or its gradient.

In Chapter 7 we showed discretized solutions starting from this point and clearly, as no further elimination of variables is possible, the formulation is *irreducible*.

On the other hand, if we start the discretization from Eqs (10.2)–(10.4) the formulation would be *mixed*.

An alternative irreducible form is also possible in terms of the variables $\mathbf{q}$. Here we have to introduce a penalty form and write in place of Eq. (10.3)

$$\boldsymbol{\nabla}^{\mathrm{T}}\mathbf{q} + Q = \frac{\phi}{\alpha} \tag{10.6}$$

where α is a penalty number which tends to infinity. Clearly in the limit both equations are the same and in general if α is very large but finite the solutions should be approximately the same.

Now substitution into Eq. (10.2) gives the single governing equation

$$\boldsymbol{\nabla}(\boldsymbol{\nabla}^{\mathrm{T}}\mathbf{q}) + \frac{1}{\alpha}\mathbf{k}^{-1}\mathbf{q} + \boldsymbol{\nabla}Q = \mathbf{0} \tag{10.7}$$

which again could be used for the start of a discretization process as a possible irreducible form.[4]

The reader should observe that, by the definition given, the formulations so far used in this book were *irreducible*. In subsequent sections we will show how elasticity problems can be dealt with in *mixed* form and indeed will show how such formulations are essential in certain problems typified by the incompressible elasticity example to which we have referred in Chapter 6. In Chapter 3 (Sec. 3.9) we have shown how discretization of a mixed problem can be accomplished.

Before proceeding to a discussion of such discretization (which will reveal the advantages and disadvantages of mixed methods) it is important to observe that if the operator specifying the mixed form is *symmetric* or *self-adjoint* (see Sec. 3.9) the formulation can proceed from the basis of a *variational principle* which can be directly obtained for linear problems. We invite the reader to prove by using the methods of Chapter 3 that stationarity of the *variational principle* given below is equivalent to the differential equations (10.2) and (10.3) together with the boundary conditions (10.4):

$$\Pi = \tfrac{1}{2}\int_{\Omega}\mathbf{q}^{\mathrm{T}}\mathbf{k}^{-1}\mathbf{q}\,\mathrm{d}\Omega + \int_{\Omega}\mathbf{q}^{\mathrm{T}}\boldsymbol{\nabla}\phi\,\mathrm{d}\Omega - \int_{\Omega}\phi Q\,\mathrm{d}\Omega - \int_{\Gamma_q}\phi\bar{q}_n\,\mathrm{d}\Gamma \tag{10.8}$$

for

$$\phi = \bar{\phi} \qquad \text{on } \Gamma_\phi$$

The establishment of such variational principles is a worthy academic pursuit and had led to many famous forms given in the classical work of Washizu.[5] However, we also know (see Sec. 3.7) that if symmetry of weighted residual matrices is obtained in a linear problem then a variational principle exists and can be determined. As such symmetry can be established by inspection we shall, in what follows, proceed with such weighting directly and thus avoid some unwarranted complexity.

10.2 Discretization of mixed forms – some general remarks

We shall demonstrate the discretization process on the basis of the mixed form of the heat conduction equations (10.2) and (10.3). Here we start by assuming that each of the unknowns is approximated in the usual manner by appropriate shape functions and corresponding unknown parameters. Thus,

$$\mathbf{q} \approx \hat{\mathbf{q}} = \mathbf{N}_q \tilde{\mathbf{q}} \qquad \text{and} \qquad \phi \approx \hat{\phi} = \mathbf{N}_\phi \tilde{\boldsymbol{\phi}} \tag{10.9}$$

where $\tilde{\mathbf{q}}$ and $\tilde{\boldsymbol{\phi}}$ are the nodal (or element) parameters that have to be determined. Similarly the weighting functions are given by

$$\mathbf{v}_q \approx \hat{\mathbf{v}}_q = \mathbf{W}_q \delta\tilde{\mathbf{q}} \qquad \text{and} \qquad v_\phi \approx \hat{v}_\phi = \mathbf{W}_\phi \delta\tilde{\boldsymbol{\phi}} \tag{10.10}$$

where $\delta\tilde{\mathbf{q}}$ and $\delta\tilde{\boldsymbol{\phi}}$ are arbitrary parameters.

Assuming that the boundary conditions for $\phi = \bar{\phi}$ are satisfied by the choice of the expansion, the weighted statement of the problem is, for Eq. (10.2) after elimination of the arbitrary parameters,

$$\int_\Omega \mathbf{W}_q^{\mathrm{T}}(\mathbf{k}^{-1}\hat{\mathbf{q}} + \boldsymbol{\nabla}\hat{\phi})\,\mathrm{d}\Omega = \mathbf{0} \tag{10.11}$$

and, for Eq. (10.3) and the 'natural' boundary conditions,

$$-\int_\Omega \mathbf{W}_\phi^{\mathrm{T}}(\boldsymbol{\nabla}^{\mathrm{T}}\hat{\mathbf{q}} + Q)\,\mathrm{d}\Omega + \int_{\Gamma_q} \mathbf{W}_\phi^{\mathrm{T}}(\hat{q}_n - \bar{q}_n)\,\mathrm{d}\Gamma = \mathbf{0} \tag{10.12}$$

The reason we have premultiplied Eq. (10.2) by $\mathbf{k}^{-1}$ is now evident as the choice

$$\mathbf{W}_q = \mathbf{N}_q \qquad \mathbf{W}_\phi = \mathbf{N}_\phi \tag{10.13}$$

will yield symmetric equations [using Green's theorem to perform integration by parts on the gradient term in Eq. (10.12)] of the form

$$\begin{bmatrix} \mathbf{A} & \mathbf{C} \\ \mathbf{C}^{\mathrm{T}} & \mathbf{0} \end{bmatrix} \begin{Bmatrix} \tilde{\mathbf{q}} \\ \tilde{\boldsymbol{\phi}} \end{Bmatrix} = \begin{Bmatrix} \mathbf{f}_1 \\ \mathbf{f}_2 \end{Bmatrix} \tag{10.14}$$

with

$$\begin{aligned} \mathbf{A} &= \int_\Omega \mathbf{N}_q^{\mathrm{T}}\mathbf{k}^{-1}\mathbf{N}_q\,\mathrm{d}\Omega \qquad & \mathbf{C} &= \int_\Omega \mathbf{N}_q^{\mathrm{T}}\,\boldsymbol{\nabla}\mathbf{N}_\phi\,\mathrm{d}\Omega \\ \mathbf{f}_1 &= \mathbf{0} & \mathbf{f}_2 &= \int_\Omega \mathbf{N}_\phi^{\mathrm{T}} Q\,\mathrm{d}\Omega + \int_{\Gamma_q} \mathbf{N}_\phi^{\mathrm{T}}\bar{q}_n\,\mathrm{d}\Gamma \end{aligned} \tag{10.15}$$

This problem, which we shall consider as typifying a large number of mixed approximations, illustrates the main features of the mixed formulation, including its advantages and disadvantages. We note that:

1. The continuity requirements on the shape functions chosen are different. It is easily seen that those given for $\mathbf{N}_\phi$ can be C_0 continuous while those for $\mathbf{N}_q$ can be discontinuous in or between elements (C_{-1} continuity) as no derivatives of this are present. Alternatively,

this discontinuity can be transferred to $\mathbf{N}_\phi$ (using Green's theorem on the integral in $\mathbf{C}$) while maintaining C_0 continuity for $\mathbf{N}_q$.

This relaxation of continuity is of particular importance in plate and shell bending problems (see reference 6) and indeed many important early uses of mixed forms have been made in that context.[7–10]

2. If interest is focused on the variable $\mathbf{q}$ rather than ϕ, use of an improved approximation for this may result in higher accuracy than possible with the irreducible form previously discussed. *However, we must note that if the approximation function for* $\mathbf{q}$ *is capable of reproducing precisely the same type of variation as that determinable from the irreducible form then no additional accuracy will result and, indeed, the two approximations will yield identical answers.*

   Thus, for instance, if we consider the mixed approximation to the field problems discussed using a linear triangle to determine $\mathbf{N}_\phi$ and piecewise constant $\mathbf{N}_q$, as shown in Fig. 10.1, we will obtain precisely the same results as those obtained by the irreducible formulation with the same $\mathbf{N}_\phi$ applied directly to Eq. (10.5), *providing* $\mathbf{k}$ *is constant within each element*. This is evident as the second of Eqs (10.14) is precisely the weighted continuity statement used in deriving the irreducible formulation in which the first of the equations is identically satisfied.

   Indeed, should we choose to use a linear but discontinuous approximation form of $\mathbf{N}_q$ in the interior of such a triangle, we would still obtain precisely the same answers, with the additional coefficients becoming zero. This discovery was made by Fraeijs de Veubeke[11] and is called the *principle of limitation*, showing that under some circumstances no additional accuracy is to be expected from a mixed formulation. In a more general case where $\mathbf{k}$ is, for instance, discontinuous and variable within an element, the results of the mixed approximation will be different and on occasion superior.[2] Note that a C_0 continuous approximation for $\mathbf{q}$ does not fall into this category as it is not capable of reproducing the discontinuous ones.

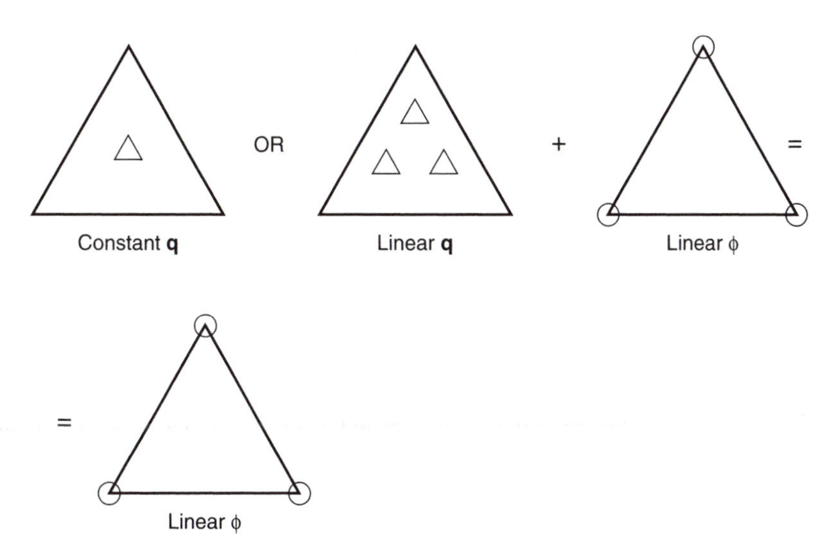

Fig. 10.1 A mixed approximation to the heat conduction problem yielding identical results as the corresponding irreducible form (the constant $\mathbf{k}$ is assumed in each element).

3. The equations resulting from mixed formulations frequently have zero diagonal terms as indeed in the case of Eq. (10.14).

   We noted in Chapter 3 that this is a characteristic of problems constrained by a Lagrange multiplier variable. Indeed, this is the origin of the problem, which adds some difficulty to a standard gaussian elimination process used in equation solving. As the form of Eq. (10.14) is typical of many two-field problems we shall refer to the first variable (here $\tilde{\mathbf{q}}$) as the *primary variable* and the second (here $\tilde{\phi}$) as the *constraint variable*.
4. The added number of variables means that generally larger size algebraic problems have to be dealt with.

The characteristics so far discussed did not mention one vital point which we elaborate in the next section.

10.3 Stability of mixed approximation. The patch test

10.3.1 Solvability requirement

Despite the relaxation of shape function continuity requirements in the mixed approximation, for certain choices of the individual shape functions the mixed approximation will not yield meaningful results. This limitation is indeed much more severe than in an *irreducible* formulation where a very simple 'constant gradient' (or constant strain) condition sufficed to ensure a convergent form once continuity requirements were satisfied.

The mathematical reasons for this difficulty are discussed by Babuška[12, 13] and Brezzi,[14] who formulated a mathematical criterion associated with their names. However, some sources of the difficulties (and hence ways of avoiding them) follow from quite simple reasoning.

If we consider the equation system (10.14) to be typical of many mixed systems in which $\tilde{\mathbf{q}}$ is the *primary variable* and $\tilde{\phi}$ is the *constraint variable* (equivalent to a lagrangian multiplier), we note that the solution can proceed by eliminating $\tilde{\mathbf{q}}$ from the first equation and by substituting into the second to obtain

$$(\mathbf{C}^{\mathrm{T}}\mathbf{A}^{-1}\mathbf{C})\tilde{\phi} = -\mathbf{f}_2 + \mathbf{C}^{\mathrm{T}}\mathbf{A}^{-1}\mathbf{f}_1 \tag{10.16}$$

which requires the matrix $\mathbf{A}$ to be non-singular (or $\mathbf{A}\tilde{\mathbf{q}} \neq \mathbf{0}$ for all $\tilde{\mathbf{q}} \neq \mathbf{0}$). To calculate $\tilde{\phi}$ it is necessary to ensure that the bracketed matrix, i.e.,

$$\mathbf{H} = \mathbf{C}^{\mathrm{T}}\mathbf{A}^{-1}\mathbf{C} \tag{10.17}$$

is non-singular.

Singularity of the $\mathbf{H}$ matrix will always occur if the number of unknowns in the vector $\tilde{\mathbf{q}}$, which we call n_q, is less than the number of unknowns n_ϕ in the vector $\tilde{\phi}$. Thus for avoidance of singularity

$$n_q \geq n_\phi \tag{10.18}$$

is *necessary* though not *sufficient* as we shall find later.

The reason for this is evident as the rank of the matrix (10.17), which needs to be n_ϕ, cannot be greater than n_q, i.e., the rank of $\mathbf{A}^{-1}$.

In some problems the matrix $\mathbf{A}$ may well be singular. It can normally be made non-singular by addition of a multiple of the second equation, thus changing the first equation to

$$\bar{\mathbf{A}} = \mathbf{A} + \gamma \mathbf{C}\mathbf{C}^{\mathrm{T}}$$
$$\bar{\mathbf{f}}_1 = \mathbf{f}_1 + \gamma \mathbf{C}\mathbf{f}_2$$

where γ is an arbitrary number. We note that the solution to (10.14) is not changed by this modification.

Although both the matrices $\mathbf{A}$ and $\mathbf{C}\mathbf{C}^{\mathrm{T}}$ are singular their combination $\bar{\mathbf{A}}$ should not be, providing we ensure that for all vectors $\tilde{\mathbf{q}} \neq \mathbf{0}$ either

$$\mathbf{A}\tilde{\mathbf{q}} \neq \mathbf{0} \quad \text{or} \quad \mathbf{C}^{\mathrm{T}}\tilde{\mathbf{q}} \neq \mathbf{0}$$

In mathematical terminology this means that $\mathbf{A}$ is non-singular in the null space of $\mathbf{C}\mathbf{C}^{\mathrm{T}}$.

The requirement of Eq. (10.18) is a necessary but not sufficient condition for non-singularity of the matrix $\mathbf{H}$. An additional requirement evident from Eq. (10.16) is

$$\mathbf{C}\tilde{\phi} \neq \mathbf{0} \quad \text{for all} \quad \tilde{\phi} \neq \mathbf{0}$$

If this is not the case the solution would not be unique.

The above requirements are inherent in the Babuška–Brezzi condition previously mentioned, but can always be verified algebraically.

10.3.2 Locking

The condition (10.18) ensures that non-zero answers for the variables $\tilde{\mathbf{q}}$ are possible. If it is violated *locking* or non-convergent results will occur in the formulation, giving near-zero answers for $\tilde{\mathbf{q}}$ [see Chapter 3, Eq. (3.137) ff.].

To show this, we shall replace Eq. (10.14) by its penalized form:

$$\begin{bmatrix} \mathbf{A} & \mathbf{C} \\ \mathbf{C}^{\mathrm{T}} & -\dfrac{1}{\alpha}\mathbf{I} \end{bmatrix} \begin{Bmatrix} \tilde{\mathbf{q}} \\ \tilde{\phi} \end{Bmatrix} = \begin{Bmatrix} \mathbf{f}_1 \\ \mathbf{f}_2 \end{Bmatrix} \qquad \begin{array}{l} \text{with } \alpha \to \infty \\ \text{and } \mathbf{I} = \text{identity matrix} \end{array} \tag{10.19}$$

Elimination of $\tilde{\phi}$ leads to

$$(\mathbf{A} + \alpha \mathbf{C}\mathbf{C}^{\mathrm{T}})\tilde{\mathbf{q}} = \mathbf{f}_1 + \alpha \mathbf{C}\mathbf{f}_2 \tag{10.20}$$

As $\alpha \to \infty$ the above becomes simply

$$(\mathbf{C}\mathbf{C}^{\mathrm{T}})\tilde{\mathbf{q}} = \mathbf{C}\mathbf{f}_2 \tag{10.21}$$

Non-zero answers for $\tilde{\mathbf{q}}$ should exist even when $\mathbf{f}_2$ is zero and hence the matrix $\mathbf{C}\mathbf{C}^{\mathrm{T}}$ *must be singular*. This singularity will always exist if $n_q > n_\phi$, but can exist also when $n_q = n_\phi$ if the rank of $\mathbf{C}$ is less than n_q.

The stability conditions derived on the particular example of Eq. (10.14) are generally valid for any problem exhibiting the standard Lagrange multiplier form. In particular the necessary count condition will in many cases suffice to determine element acceptability; however, final conclusions for successful elements which pass all count conditions and the full test to ensure consistency must be evaluated by rank tests on the full matrix.

In the example just quoted $\tilde{\mathbf{q}}$ denotes flux and $\tilde{\phi}$ temperature and perhaps the concept of locking was not clearly demonstrated. It is much more definite where the first primary variable is a displacement and the second constraining one is a stress or a pressure. There locking is more evident physically and simply means an occurrence of zero displacements throughout as the solution approaches a numerical instability limit. This unfortunately will happen on occasion.

10.3.3 The patch test

The patch test for mixed elements can be carried out in exactly the way we have described in the previous chapter for irreducible elements. As *consistency* is easily assured by taking a polynomial approximation for each of the variables, only *stability* needs generally to be investigated. *Most answers* to this can be obtained by simply ensuring that *count condition* (10.18) is satisfied for any isolated patch on the boundaries of which we constrain the *maximum* number of primary variables and the *minimum* number of constraint variables.[15]

Example 10.1: A single-element test. In Fig. 10.2 we illustrate a single-element test for two possible formulations with C_0 continuous N_ϕ (quadratic) and discontinuous N_q, assumed to be either constant or linear within an element of triangular form. As no values of $\tilde{\mathbf{q}}$ can here be specified on the boundaries, on the patch (which is here simply that of a single element) we shall fix a single value of $\tilde{\phi}$ only, as is necessary to ensure uniqueness. A count shows that only one of the formulations, i.e., that with linear flux variation, satisfies condition (10.18) and therefore may be acceptable (but will always determine elements which fail!).

Example 10.2: A single-element test with C_0, $\tilde{\mathbf{q}}$ and ϕ. In Fig. 10.3 we illustrate a similar patch test on the same element but with identical C_0 continuous shape functions specified for both $\tilde{\mathbf{q}}$ and $\tilde{\phi}$ variables. This example shows satisfaction of the basic condition of Eq. (10.18)

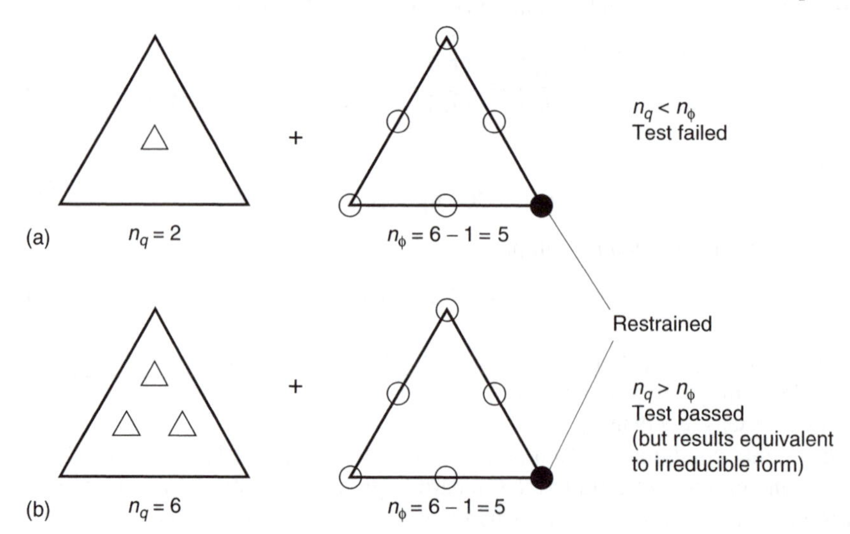

Fig. 10.2 Single-element patch test for mixed approximations to the heat conduction problem with discontinuous flux $\mathbf{q}$ assumed. (a) Quadratic C_0, ϕ; constant $\mathbf{q}$. (b) Quadratic C_0, ϕ; linear $\mathbf{q}$.

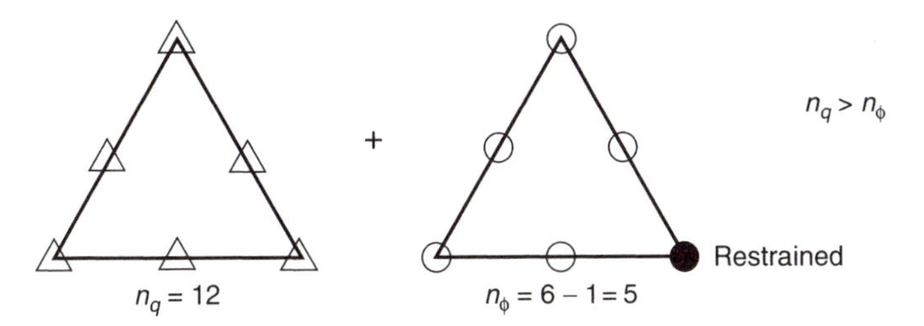

Fig. 10.3 As Fig. 10.2 but with quadratic C_0 continuous **q**.

and therefore is apparently a permissible formulation. The permissible formulation must always be subjected to a numerical rank test.

Clearly condition (10.18) will need to be satisfied and many useful conclusions can be drawn from such counts. These eliminate elements which will not function and on many occasions will give guidance to elements which will.

Even if the patch test is satisfied occasional difficulties can arise, and these are indicated mathematically by the Babuška–Brezzi condition already referred to.[16] These difficulties can be due to *excessive continuity* imposed on the problem by requiring, for instance, the flux condition to be of C_0 continuity class. In Fig. 10.4 we illustrate some cases in which the imposition of such continuity is *physically incorrect* and therefore can be expected to produce erroneous (and usually highly oscillating) results. In all such problems we recommend that *the continuity be relaxed on all surfaces where a physical discontinuity can occur*.

We shall discuss this problem further in Sec. 10.4.3.

10.4 Two-field mixed formulation in elasticity

10.4.1 General

In all the previous formulations of elasticity problems in this book we have used an irreducible formulation, using the displacement **u** as the primary variable. In earlier chapters,

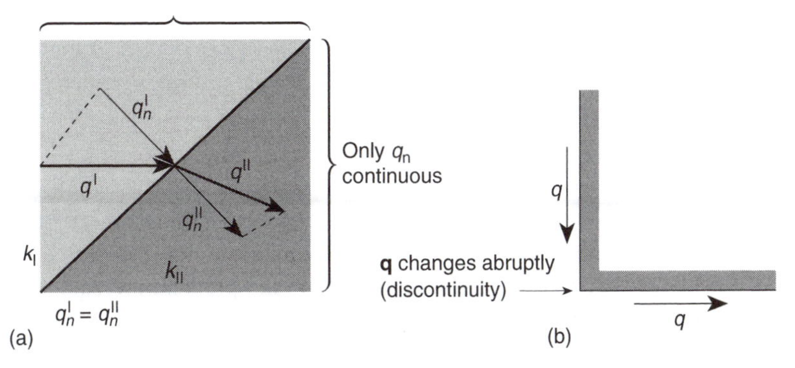

Fig. 10.4 Some situations for which C_0 continuity of flux **q** is inappropriate. (a) Discontinuous change of material properties. (b) Singularity.

the virtual work principle was used to establish the equilibrium conditions and was written as

$$\int_{\Omega} \delta\boldsymbol{\varepsilon}^{\mathrm{T}}\boldsymbol{\sigma}\,\mathrm{d}\Omega - \int_{\Omega} \delta\mathbf{u}^{\mathrm{T}}\mathbf{b}\,\mathrm{d}\Omega - \int_{\Gamma_t} \delta\mathbf{u}^{\mathrm{T}}\bar{\mathbf{t}}\,\mathrm{d}\Gamma = 0 \tag{10.22}$$

where $\bar{\mathbf{t}}$ are the tractions prescribed on Γ_t and with

$$\boldsymbol{\sigma} = \mathbf{D}\boldsymbol{\varepsilon} \tag{10.23}$$

as the constitutive relation (omitting here initial strains and stresses for simplicity).

We recall that statements such as Eq. (10.22) are equivalent to weighted residual forms (see Chapter 3) and in what follows we shall use these frequently. In the above the strains are related to displacement by the matrix operator $\boldsymbol{\mathcal{S}}$ introduced in Chapter 2, giving

$$\begin{aligned} \boldsymbol{\varepsilon} &= \boldsymbol{\mathcal{S}}\mathbf{u} \\ \delta\boldsymbol{\varepsilon} &= \boldsymbol{\mathcal{S}}\,\delta\mathbf{u} \end{aligned} \tag{10.24}$$

with the displacement expansions constrained to satisfy the prescribed displacements on Γ_u. This is, of course, equivalent to Galerkin-type weighting.

With the displacement $\mathbf{u}$ approximated as

$$\mathbf{u} \approx \hat{\mathbf{u}} = \mathbf{N}_u\tilde{\mathbf{u}} \tag{10.25}$$

the required stiffness equations were obtained in terms of the unknown displacement vector $\tilde{\mathbf{u}}$ and the solution obtained.

It is possible to use mixed forms in which either $\boldsymbol{\sigma}$ or $\boldsymbol{\varepsilon}$, or, indeed, both these variables, are approximated independently. We shall discuss such formulations below.

10.4.2 The u–σ mixed form

In this we shall assume that Eq. (10.22) is valid but that we approximate $\boldsymbol{\sigma}$ independently as

$$\boldsymbol{\sigma} \approx \hat{\boldsymbol{\sigma}} = \mathbf{N}_\sigma\tilde{\boldsymbol{\sigma}} \tag{10.26}$$

and approximately satisfy the constitutive relation

$$\boldsymbol{\sigma} = \mathbf{D}\boldsymbol{\mathcal{S}}\mathbf{u} \tag{10.27}$$

which replaces (10.23) and (10.24). The approximate integral form is written as

$$\int_{\Omega} \delta\boldsymbol{\sigma}^{\mathrm{T}}(\boldsymbol{\mathcal{S}}\mathbf{u} - \mathbf{D}^{-1}\boldsymbol{\sigma})\,\mathrm{d}\Omega = 0 \tag{10.28}$$

where the expression in the brackets is simply Eq. (10.27) premultiplied by $\mathbf{D}^{-1}$ to establish symmetry and $\delta\boldsymbol{\sigma}$ is introduced as a weighting variable.

Indeed, Eqs (10.22) and (10.28) which now define the problem are equivalent to the stationarity of the functional

$$\Pi_{\mathrm{HR}} = \int_{\Omega} \boldsymbol{\sigma}^{\mathrm{T}}\boldsymbol{\mathcal{S}}\mathbf{u}\,\mathrm{d}\Omega - \tfrac{1}{2}\int_{\Omega} \boldsymbol{\sigma}^{\mathrm{T}}\mathbf{D}^{-1}\boldsymbol{\sigma}\,\mathrm{d}\Omega - \int_{\Omega} \mathbf{u}^{\mathrm{T}}\mathbf{b}\,\mathrm{d}\Omega - \int_{\Gamma_t} \mathbf{u}^{\mathrm{T}}\bar{\mathbf{t}}\,\mathrm{d}\Gamma \tag{10.29}$$

where the boundary displacement

$$\mathbf{u} = \bar{\mathbf{u}}$$

is enforced on Γ_u, as the reader can readily verify. This is the well-known Hellinger–Reissner[17, 18] variational principle, but, as we have remarked earlier, it is unnecessary in deriving approximate equations. Using

$$\begin{aligned} \mathbf{N}_u \delta\tilde{\mathbf{u}} &\quad \text{in place of} \quad \delta\mathbf{u} \\ \mathbf{B}\delta\tilde{\mathbf{u}} \equiv \mathcal{S}\mathbf{N}_u \delta\tilde{\mathbf{u}} &\quad \text{in place of} \quad \delta\boldsymbol{\varepsilon} \\ \mathbf{N}_\sigma \delta\tilde{\boldsymbol{\sigma}} &\quad \text{in place of} \quad \delta\boldsymbol{\sigma} \end{aligned}$$

we write the approximate equations (10.28) and (10.22) in the standard form [see Eq. (10.14)]

$$\begin{bmatrix} \mathbf{A} & \mathbf{C} \\ \mathbf{C}^{\mathrm{T}} & \mathbf{0} \end{bmatrix} \begin{Bmatrix} \tilde{\boldsymbol{\sigma}} \\ \tilde{\mathbf{u}} \end{Bmatrix} = \begin{Bmatrix} \mathbf{f}_1 \\ \mathbf{f}_2 \end{Bmatrix} \tag{10.30}$$

with

$$\begin{aligned} \mathbf{A} &= -\int_\Omega \mathbf{N}_\sigma^{\mathrm{T}} \mathbf{D}^{-1} \mathbf{N}_\sigma \, \mathrm{d}\Omega \qquad & \mathbf{C} &= \int_\Omega \mathbf{N}_\sigma^{\mathrm{T}} \mathbf{B} \, \mathrm{d}\Omega \\ \mathbf{f}_1 &= \mathbf{0} & \mathbf{f}_2 &= \int_\Omega \mathbf{N}_u^{\mathrm{T}} \mathbf{b} \, \mathrm{d}\Omega + \int_{\Gamma_t} \mathbf{N}_u^{\mathrm{T}} \bar{\mathbf{t}} \, \mathrm{d}\Gamma \end{aligned} \tag{10.31}$$

In the form given above the $\mathbf{N}_u$ shape functions have still to be of C_0 continuity, though $\mathbf{N}_\sigma$ can be discontinuous. However, integration by parts of the expression for $\mathbf{C}$ allows a reduction of such continuity and indeed this form has been used by Herrmann[7, 19, 20] for problems of plates and shells.

10.4.3 Stability of two-field approximation in elasticity (u–σ)

Before attempting to formulate practical mixed approach approximations in detail, identical stability problems to those discussed in Sec. 10.3 have to be considered.

For the **u**–$\boldsymbol{\sigma}$ forms it is clear that $\boldsymbol{\sigma}$ is the *primary variable* and **u** the *constraint variable* (see Sec. 10.2), and for the total problem as well as for element patches we must have as a necessary, though not sufficient, condition

$$n_\sigma \geq n_u \tag{10.32}$$

where n_σ and n_u stand for numbers of degrees of freedom in appropriate variables.

In Fig. 10.5 we consider a two-dimensional plane problem and show a series of elements in which $\mathbf{N}_\sigma$ is discontinuous while $\mathbf{N}_u$ has C_0 continuity. We note again, by invoking the Veubeke 'principle of limitation', that all the elements that pass the single-element test here will in fact yield identical results to those obtained by using the equivalent irreducible form, providing the **D** matrix and the determinant of the jacobian matrix are constant within each element. They are therefore of little interest. However, we note in passing that the Q 4/8, which fails in a single-element test, passes that patch test for assemblies of two or more elements, and performs well in many circumstances. We shall see later that this is equivalent to using four-point Gauss, *reduced* integration (see Sec. 11.5), and as we have mentioned in Chapter 9 such elements will not always be robust.

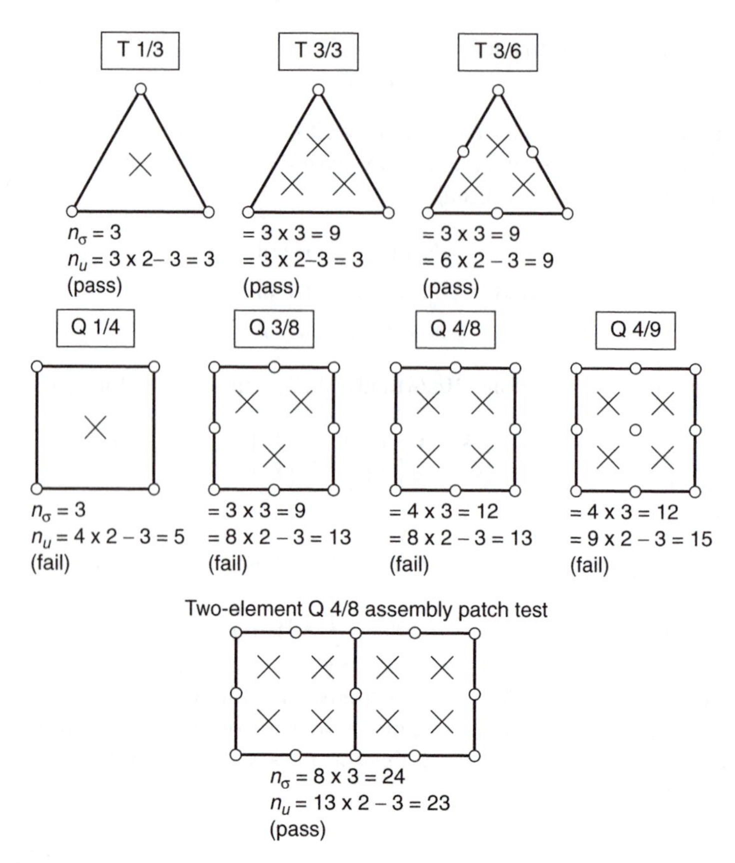

Fig. 10.5 Elasticity by the mixed $\boldsymbol{\sigma}$–$\mathbf{u}$ formulation. Discontinuous stress approximation. Single-element patch test. No restraint on $\tilde{\boldsymbol{\sigma}}$ variables but three $\tilde{\mathbf{u}}$ degrees of freedom restrained on patch. Test condition $n_\sigma \geq n_u$ [$\times$ denotes $\tilde{\boldsymbol{\sigma}}$ (3 DOF) and $\circ$ the $\tilde{\mathbf{u}}$ (2 DOF) variables].

It is of interest to note that if a higher order of interpolation is used for $\boldsymbol{\sigma}$ than for $\mathbf{u}$ the patch test is still satisfied, but in general the results will not be improved because of the principle of limitation.

We do not show the similar patch test for the C_0 continuous $\mathbf{N}_\sigma$ assumption but state simply that, similarly to the example of Fig. 10.3, identical interpolation of $\mathbf{N}_\sigma$ and $\mathbf{N}_u$ is acceptable from the point of view of stability. However, as in Fig. 10.4, restriction of *excessive continuity* for stresses has to be avoided at singularities and at abrupt material property change interfaces, where only the normal and tangential tractions are continuous.

The disconnection of stress variables at corner nodes can only be accomplished for all the stress variables. For this reason an alternative set of elements with continuous stress nodes at element interfaces can be introduced (see Fig. 10.6).[21]

In such elements excessive continuity can easily be avoided by disconnecting only the direct stress components parallel to an interface at which material changes occur. It should be noted that even in the case when all stress components are connected at a mid-side node such elements do not ensure stress continuity along the whole interface. Indeed, the amount of such discontinuity can be useful as an error measure. However, we observe that for the linear element [Fig. 10.6(a)] the interelement stresses are continuous *in the mean*.

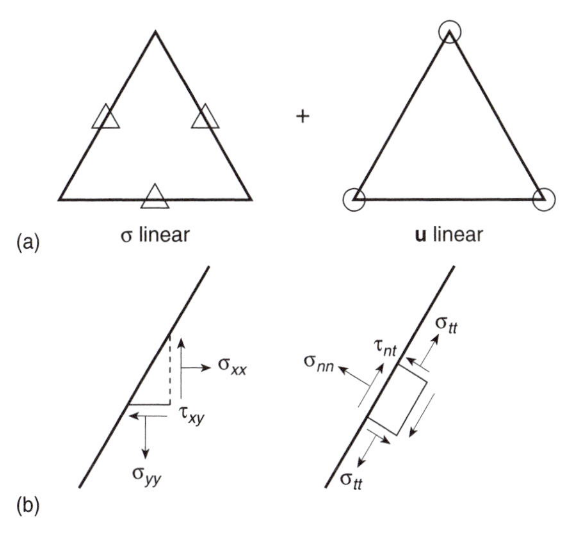

Fig. 10.6 Elasticity by the mixed σ–**u** formulation. Partially continuous σ (continuity at nodes only). (a) σ linear, **u** linear. (b) Possible transformation of interface stresses with σ_{tt} disconnected.

It is, of course, possible to derive elements that exhibit complete continuity of the appropriate components along interfaces and indeed this was achieved by Raviart and Thomas[22] in the case of the heat conduction problem discussed previously. Extension to the full stress problem is difficult[23] and as yet such elements have not been successfully noted.

Example 10.3: Pian–Sumihara rectangle. Today very few two-field elements based on interpolation of the full stress and displacement fields are used. One, however, deserves to be mentioned. We begin by first considering a rectangular element where interpolations may be given directly in terms of cartesian coordinates. A 4-node plane rectangular element with side lengths $2a$ in the x direction and $2b$ in the y direction, shown in Fig. 10.7, has

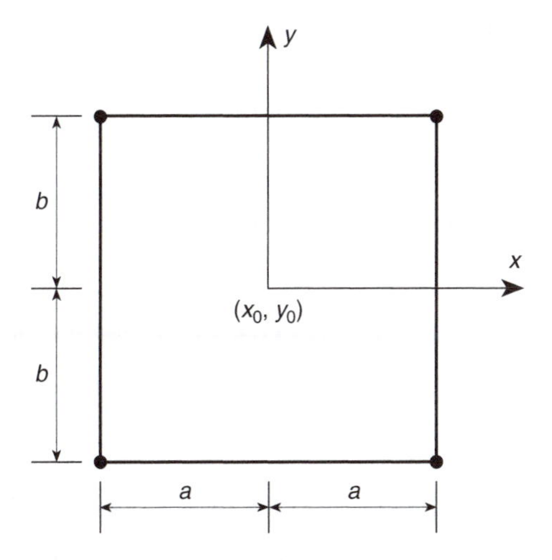

Fig. 10.7 Geometry of rectangular σ–**u** element.

displacement interpolation given by

$$\mathbf{u} = \sum_{a=1}^{4} N_a(x, y)\tilde{\mathbf{u}}_a$$

The shape functions are given by

$$N_1(x, y) = \tfrac{1}{4}\left(1 - \frac{x - x_0}{a}\right)\left(1 - \frac{y - y_0}{b}\right); \; N_2(x, y) = \tfrac{1}{4}\left(1 + \frac{x - x_0}{a}\right)\left(1 - \frac{y - y_0}{b}\right)$$

$$N_3(x, y) = \tfrac{1}{4}\left(1 + \frac{x - x_0}{a}\right)\left(1 + \frac{y - y_0}{b}\right); \; N_4(x, y) = \tfrac{1}{4}\left(1 - \frac{x - x_0}{a}\right)\left(1 + \frac{y - y_0}{b}\right)$$

in which x_0 and y_0 are the cartesian coordinates at the element centre. The strains generated from this interpolation will be such that

$$\varepsilon_x = \eta_1 + \eta_2 y; \quad \varepsilon_y = \eta_3 + \eta_4 x; \quad \gamma_{xy} = \eta_5 + \eta_6 x + \eta_7 y$$

where η_j are expressed in terms of $\tilde{\mathbf{u}}$. For isotropic linear elasticity problems these strains will lead to stresses which have a complete linear polynomial variation in each element (except for the special case when $\nu = 0$).

Here the stress interpolation is restricted to each element individually and, thus, can be discontinuous between adjacent elements. The limitation principle restricts the possible choices which lead to different results from the standard displacement solution. Namely, the approximation must be less than a complete linear polynomial. To satisfy the stability condition given by Eq. (10.18) we need at least five stress parameters in each element. A viable choice for a five-term approximation is one which has the same variation in each element as the normal strains given above but only a constant shear stress. Accordingly,

$$\begin{Bmatrix} \sigma_x \\ \sigma_y \\ \tau_{xy} \end{Bmatrix} = \begin{bmatrix} 1 & 0 & 0 & y - y_0 & 0 \\ 0 & 1 & 0 & 0 & x - x_0 \\ 0 & 0 & 1 & 0 & 0 \end{bmatrix} \begin{Bmatrix} \alpha_1 \\ \alpha_2 \\ \alpha_3 \\ \alpha_4 \\ \alpha_5 \end{Bmatrix}$$

Indeed, this approximation satisfies Eq. (10.18) and leads to excellent results for a rectangular element.

Example 10.4: Pian–Sumihara quadrilateral. We now rewrite the formulation given in Example 10.3 to permit a general quadrilateral shape to be used. The element coordinate and displacement field are given by a standard bilinear isoparametric expansion

$$\mathbf{x} = \sum_{a=1}^{4} N_a(\xi, \eta)\tilde{\mathbf{x}}_a \quad \text{and} \quad \hat{\mathbf{u}} = \sum_{a=1}^{4} N_a(\xi, \eta)\,\tilde{\mathbf{u}}_a$$

where now

$$N_a(\xi, \eta) = \tfrac{1}{4}(1 + \xi_a \xi)(1 + \eta_a \eta)$$

in which ξ_a and η_a are the values of the parent coordinates at node a.

The problem remains to deduce an approximation for stresses for the general quadrilateral element. Here this is accomplished by first assuming stresses Σ on the parent element (for convenience in performing the coordinate transformation the tensor form is used, see Appendix B) in an analogous manner as the rectangle above:

$$\boldsymbol{\Sigma}(\xi,\eta) = \begin{bmatrix} \Sigma_{\xi\xi} & \Sigma_{\xi\eta} \\ \Sigma_{\eta\xi} & \Sigma_{\eta\eta} \end{bmatrix} = \begin{bmatrix} \alpha_1 + \alpha_4\eta & \alpha_3 \\ \alpha_3 & \alpha_2 + \alpha_5\xi \end{bmatrix}$$

In the above the parent normal stresses again produce constant and bending terms while shear stress is only constant. These stresses are then transformed to cartesian space using

$$\boldsymbol{\sigma} = \mathbf{T}^{\mathrm{T}}\boldsymbol{\Sigma}(\xi,\eta)\mathbf{T}$$

It remains now only to select an appropriate form for $\mathbf{T}$. The transformation must

1. produce stresses in cartesian space which satisfy the patch test (i.e., can produce constant stresses and be stable);
2. be independent of the orientation of the initially chosen element coordinate system and numbering of element nodes (invariance requirement).

Pian and Sumihara[24] use a constant array (to preserve constant stresses) deduced from the jacobian matrix at the centre of the element. Accordingly, with

$$\mathbf{J}_0 = \begin{bmatrix} J_{0,11} & J_{0,12} \\ J_{0,21} & J_{0,22} \end{bmatrix} = \begin{bmatrix} \dfrac{\partial x}{\partial \xi} & \dfrac{\partial y}{\partial \xi} \\ \dfrac{\partial x}{\partial \eta} & \dfrac{\partial y}{\partial \eta} \end{bmatrix}_{\xi,\eta=0}$$

the elements of the jacobian matrix at the centre are given by [see Eq. (5.11)]

$$J_{0,11} = \tfrac{1}{4}x_a\,\xi_a \qquad J_{0,12} = \tfrac{1}{4}x_a\,\eta_a$$
$$J_{0,21} = \tfrac{1}{4}y_a\,\xi_a \qquad J_{0,22} = \tfrac{1}{4}y_a\,\eta_a$$

Using $\mathbf{T} = \mathbf{J}_0$ gives the stresses (in matrix form)

$$\begin{Bmatrix} \sigma_x \\ \sigma_y \\ \tau_{xy} \end{Bmatrix} = \begin{Bmatrix} \bar{\alpha}_1 \\ \bar{\alpha}_2 \\ \bar{\alpha}_3 \end{Bmatrix} + \begin{bmatrix} J_{0,11}^2\eta & J_{0,12}^2\xi \\ J_{0,21}^2\eta & J_{0,22}^2\xi \\ J_{0,12}J_{0,21}\eta & J_{0,12}J_{0,22}\xi \end{bmatrix} \begin{Bmatrix} \alpha_4 \\ \alpha_5 \end{Bmatrix}$$

where the parameters $\bar{\alpha}_i$, $i = 1, 2, 3$, replace the transformed quantities for the constant part of the stresses. This approximation satisfies the constant stress condition (Condition 1) and can also be shown to satisfy the invariance condition (Condition 2). The development is now complete and the arrays indicated in Eq. (10.31) may be computed. We note that the integrals are computed exactly for all quadrilateral elements (with constant $\mathbf{D}$) using 2×2 gaussian quadrature.

An alternative to the above definition for $\mathbf{T}$ is to use the transpose of the jacobian inverse at the centre of the element (i.e., $\mathbf{T} = \mathbf{J}_0^{-\mathrm{T}}$). This has also been suggested recently by several authors as an invariant transformation. However, as shown in Fig. 10.8, the sensitivity to element distortion is much greater for this form than the original one given by Pian and Sumihara for the above two-field approximation. The other two options (e.g., $\mathbf{T} = \mathbf{J}_0^{\mathrm{T}}$ and $\mathbf{T} = \mathbf{J}_0^{-1}$) do not satisfy the frame invariance requirement, thus giving elements which depend on the orientation of the element with respect to the global coordinates.

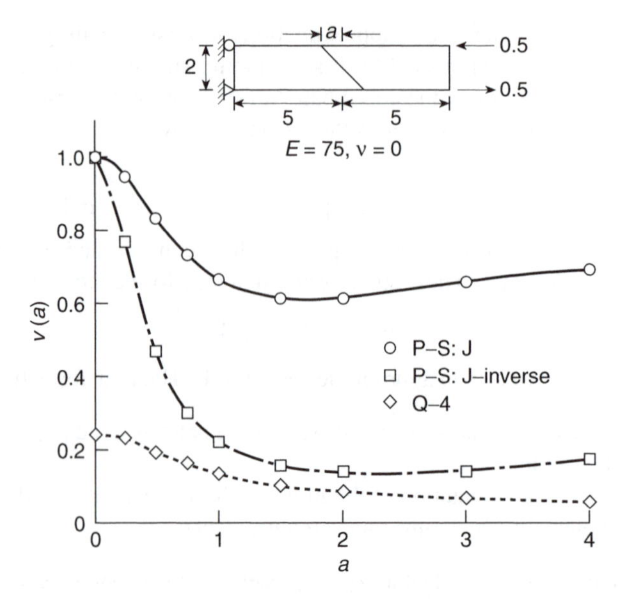

Fig. 10.8 Pian–Sumihara quadrilateral (P–S) compared with displacement quadrilateral (Q-4). Effect of element distortion (Exact = 1.0).

10.5 Three-field mixed formulations in elasticity

10.5.1 The $\mathbf{u}$–$\boldsymbol{\sigma}$–$\boldsymbol{\varepsilon}$ mixed form

It is, of course, possible to use an independent approximation to all the essential variables entering the elasticity problem. We can then write the three equations (10.22), (10.23), and (10.24) in their weak form as

$$\begin{aligned} \int_\Omega \delta\boldsymbol{\varepsilon}^{\mathrm{T}}(\mathbf{D}\boldsymbol{\varepsilon}-\boldsymbol{\sigma})\,\mathrm{d}\Omega &= 0 \\ \int_\Omega \delta\boldsymbol{\sigma}^{\mathrm{T}}(\boldsymbol{\mathcal{S}}\mathbf{u}-\boldsymbol{\varepsilon})\,\mathrm{d}\Omega &= 0 \\ \int_\Omega \delta(\boldsymbol{\mathcal{S}}\mathbf{u})^{\mathrm{T}}\boldsymbol{\sigma}\,\mathrm{d}\Omega - \int_\Omega \delta\mathbf{u}^{\mathrm{T}}\mathbf{b}\,\mathrm{d}\Omega - \int_{\Gamma_t}\delta\mathbf{u}^{\mathrm{T}}\bar{\mathbf{t}}\,\mathrm{d}\Gamma &= 0 \end{aligned} \qquad (10.33)$$

where $\mathbf{u} \equiv \bar{\mathbf{u}}$ on Γ_u is enforced.† The variational principle equivalent to Eq. (10.33) is known by the name of Hu–Washizu[5] (see Problem 10.1).

Introducing the approximations

$$\mathbf{u} \approx \hat{\mathbf{u}} = \mathbf{N}_u\tilde{\mathbf{u}} \qquad \boldsymbol{\sigma} \approx \hat{\boldsymbol{\sigma}} = \mathbf{N}_\sigma\tilde{\boldsymbol{\sigma}} \qquad \text{and} \qquad \boldsymbol{\varepsilon} \approx \hat{\boldsymbol{\varepsilon}} = \mathbf{N}_\varepsilon\tilde{\boldsymbol{\varepsilon}} \qquad (10.34)$$

with corresponding 'variations' (i.e., the Galerkin form $\mathbf{W}_u = \mathbf{N}_u$, etc.) into Eq. (10.33), and writing the approximating equations in a similar fashion as we have in the previous

† It is possible to include the displacement boundary conditions in Eq. (10.33) as a natural rather than imposed constraint; however, most finite element applications of the principle are in the form shown.

section yields an equation system of the following form:

$$\begin{bmatrix} \mathbf{A} & \mathbf{C} & \mathbf{0} \\ \mathbf{C}^{\mathrm{T}} & \mathbf{0} & \mathbf{E} \\ \mathbf{0} & \mathbf{E}^{\mathrm{T}} & \mathbf{0} \end{bmatrix} \begin{Bmatrix} \tilde{\boldsymbol{\varepsilon}} \\ \tilde{\boldsymbol{\sigma}} \\ \tilde{\mathbf{u}} \end{Bmatrix} = \begin{Bmatrix} \mathbf{f}_1 \\ \mathbf{f}_2 \\ \mathbf{f}_3 \end{Bmatrix} \tag{10.35}$$

where

$$\begin{aligned} \mathbf{A} &= \int_\Omega \mathbf{N}_\varepsilon^{\mathrm{T}} \mathbf{D} \mathbf{N}_\varepsilon \, \mathrm{d}\Omega; \quad \mathbf{E} = \int_\Omega \mathbf{N}_\sigma^{\mathrm{T}} \mathbf{B} \, \mathrm{d}\Omega; \quad \mathbf{C} = -\int_\Omega \mathbf{N}_\varepsilon^{\mathrm{T}} \mathbf{N}_\sigma \, \mathrm{d}\Omega \\ \mathbf{f}_1 &= \mathbf{f}_2 = 0; \quad \mathbf{f}_3 = \int_\Omega \mathbf{N}_u^{\mathrm{T}} \mathbf{b} \, \mathrm{d}\Omega + \int_{\Gamma_t} \mathbf{N}_u^{\mathrm{T}} \bar{\mathbf{t}} \, \mathrm{d}\Gamma \end{aligned} \tag{10.36}$$

The reader will observe that in this section we have developed all the approximations directly without using a variational principle. In Problem 10.2 we suggest that the reader show the equivalence of a development from the variational principle.

10.5.2 Stability condition of three-field approximation ($\mathbf{u}$–σ–ε)

The stability condition derived in Sec. 10.3 [Eq. (10.18)] for two-field problems, which we later used in Eq. (10.32) for the simple mixed elasticity form, needs to be modified when three-field approximations of the form given in Eq. (10.35) are considered.

Many other problems fall into a similar category (for instance, plate bending) and hence the conditions of stability are generally useful. The requirement now is that

$$\begin{aligned} n_\varepsilon + n_u &\geq n_\sigma \\ n_\sigma &\geq n_u \end{aligned} \tag{10.37}$$

This was first stated in reference 25 and follows directly from the two-field criterion as shown below.

The system of Eq. (10.35) can be 'regularized' by adding $\gamma\mathbf{E}$ times the third equation to the second, with γ being an arbitrary constant. We now have

$$\begin{bmatrix} \mathbf{A} & \mathbf{C} & \mathbf{0} \\ \mathbf{C}^{\mathrm{T}} & \gamma\mathbf{E}\mathbf{E}^{\mathrm{T}} & \mathbf{E} \\ \mathbf{0} & \mathbf{E}^{\mathrm{T}} & \mathbf{0} \end{bmatrix} \begin{Bmatrix} \tilde{\boldsymbol{\varepsilon}} \\ \tilde{\boldsymbol{\sigma}} \\ \tilde{\mathbf{u}} \end{Bmatrix} = \begin{Bmatrix} \mathbf{f}_1 \\ \mathbf{f}_2 + \gamma\mathbf{E}\mathbf{f}_3 \\ \mathbf{f}_3 \end{Bmatrix}$$

On elimination of ε using the first of the above we have

$$\begin{bmatrix} (\gamma\mathbf{E}\mathbf{E}^{\mathrm{T}} - \mathbf{C}^{\mathrm{T}}\mathbf{A}^{-1}\mathbf{C}), & \mathbf{E} \\ \mathbf{E}^{\mathrm{T}}, & \mathbf{0} \end{bmatrix} \begin{Bmatrix} \tilde{\boldsymbol{\sigma}} \\ \tilde{\mathbf{u}} \end{Bmatrix} = \begin{Bmatrix} \mathbf{f}_2 + \gamma\mathbf{E}\mathbf{f}_3 - \mathbf{C}^{\mathrm{T}}\mathbf{A}^{-1}\mathbf{f}_1 \\ \mathbf{f}_3 \end{Bmatrix}$$

From the two-field requirement [Eq. (10.18)] it follows that we require

$$n_\sigma \geq n_u \tag{10.38}$$

for the equation system to have a solution.

To establish the second condition we rearrange Eq. (10.35) as

$$\begin{bmatrix} \mathbf{A} & \mathbf{0} & \mathbf{C} \\ \mathbf{0} & \mathbf{0} & \mathbf{E}^{\mathrm{T}} \\ \mathbf{C}^{\mathrm{T}} & \mathbf{E} & \mathbf{0} \end{bmatrix} \begin{Bmatrix} \tilde{\boldsymbol{\varepsilon}} \\ \tilde{\mathbf{u}} \\ \tilde{\boldsymbol{\sigma}} \end{Bmatrix} = \begin{Bmatrix} \mathbf{f}_1 \\ \mathbf{f}_3 \\ \mathbf{f}_2 \end{Bmatrix}$$

This again can be regularized by adding multiples $\gamma\mathbf{C}$ and $\gamma\mathbf{E}^{\mathrm{T}}$ of the third of the above equations to the first and second respectively obtaining

$$\begin{bmatrix} \mathbf{A}+\gamma\mathbf{C}\mathbf{C}^{\mathrm{T}}, & \gamma\mathbf{C}\mathbf{E} & \mathbf{C} \\ \gamma\mathbf{E}^{\mathrm{T}}\mathbf{C}^{\mathrm{T}}, & \gamma\mathbf{E}^{\mathrm{T}}\mathbf{E} & \mathbf{E}^{\mathrm{T}} \\ \mathbf{C}^{\mathrm{T}}, & \mathbf{E} & \mathbf{0} \end{bmatrix} \begin{Bmatrix} \tilde{\boldsymbol{\varepsilon}} \\ \tilde{\mathbf{u}} \\ \tilde{\boldsymbol{\sigma}} \end{Bmatrix} = \begin{Bmatrix} \mathbf{f}_1+\gamma\mathbf{C}\mathbf{f}_2 \\ \mathbf{f}_3+\gamma\mathbf{E}^{\mathrm{T}}\mathbf{f}_2 \\ \mathbf{f}_2 \end{Bmatrix}$$

By partitioning as above it is evident that we require

$$n_\varepsilon + n_u \geq n_\sigma \tag{10.39}$$

We shall not discuss in detail any of the possible approximations to the $\boldsymbol{\varepsilon}$–$\boldsymbol{\sigma}$–$\mathbf{u}$ formulation or their corresponding patch tests as the arguments are similar to those of two-field problems.

In some practical applications of the three-field form the approximation of the second and third equations in (10.33) is used directly to eliminate all but the displacement terms. This leads to a special form of the displacement method which has been called a $\bar{\mathbf{B}}$ ($\mathbf{B}$-bar) form.[26, 27] In the $\bar{\mathbf{B}}$ form the shape function derivatives are replaced by approximations resulting from the mixed form. We shall illustrate this concept with an example of a *nearly incompressible* material in Sec. 11.4.

10.5.3 The $\mathbf{u}$–$\boldsymbol{\sigma}$–$\boldsymbol{\varepsilon}_{\mathbf{en}}$ form. Enhanced strain formulation

In the previous two sections the general form and stability conditions of the three-field formulation for elasticity problems are given in Eqs (10.32) and (10.37). Here we consider a special case of this form from which several useful elements may be deduced.

In the special form considered the strain approximation is split into two parts: one the usual displacement-gradient term and, second, an added or *enhanced strain* part. Accordingly, we write

$$\boldsymbol{\varepsilon} = \boldsymbol{\mathcal{S}}\mathbf{u} + \boldsymbol{\varepsilon}_{\mathrm{en}} \qquad \delta\boldsymbol{\varepsilon} = \delta(\boldsymbol{\mathcal{S}}\mathbf{u}) + \delta\boldsymbol{\varepsilon}_{\mathrm{en}} \tag{10.40}$$

Substitution into Eq. (10.33) yields the weak forms as

$$\begin{aligned} \int_\Omega \delta(\boldsymbol{\mathcal{S}}\mathbf{u})^{\mathrm{T}}\mathbf{D}(\boldsymbol{\mathcal{S}}\mathbf{u}+\boldsymbol{\varepsilon}_{\mathrm{en}})\,\mathrm{d}\Omega - \int_\Omega \delta\mathbf{u}^{\mathrm{T}}\mathbf{b}\,\mathrm{d}\Omega - \int_{\Gamma_t} \delta\mathbf{u}^{\mathrm{T}}\bar{\mathbf{t}}\,\mathrm{d}\Gamma &= 0 \\ \int_\Omega \delta\boldsymbol{\varepsilon}_{\mathrm{en}}^{\mathrm{T}}\left(\mathbf{D}(\boldsymbol{\mathcal{S}}\mathbf{u}+\boldsymbol{\varepsilon}_{\mathrm{en}}) - \boldsymbol{\sigma}\right)\mathrm{d}\Omega &= 0 \\ \int_\Omega \delta\boldsymbol{\sigma}^{\mathrm{T}}\boldsymbol{\varepsilon}_{\mathrm{en}}\,\mathrm{d}\Omega &= 0 \end{aligned} \tag{10.41}$$

where, as before, $\mathbf{u} = \bar{\mathbf{u}}$ is enforced on Γ_u.

We can directly discretize Eq. (10.41) by taking the following approximations

$$\mathbf{u} \approx \hat{\mathbf{u}} = \mathbf{N}_u\tilde{\mathbf{u}} \qquad \boldsymbol{\sigma} \approx \hat{\boldsymbol{\sigma}} = \mathbf{N}_\sigma\tilde{\boldsymbol{\sigma}} \qquad \boldsymbol{\varepsilon}_{\mathrm{en}} \approx \hat{\boldsymbol{\varepsilon}}_{\mathrm{en}} = \mathbf{N}_{\mathrm{en}}\tilde{\boldsymbol{\varepsilon}}_{\mathrm{en}} \tag{10.42}$$

with corresponding expressions for variations. Substituting the approximations into Eq. (10.41) yields the discrete equation system

$$\begin{bmatrix} \mathbf{A} & \mathbf{C} & \mathbf{G} \\ \mathbf{C}^{\mathrm{T}} & \mathbf{0} & \mathbf{0} \\ \mathbf{G}^{\mathrm{T}} & \mathbf{0} & \mathbf{K} \end{bmatrix} \begin{Bmatrix} \tilde{\boldsymbol{\varepsilon}}_{\mathrm{en}} \\ \tilde{\boldsymbol{\sigma}} \\ \tilde{\mathbf{u}} \end{Bmatrix} = \begin{Bmatrix} \mathbf{f}_1 \\ \mathbf{f}_2 \\ \mathbf{f}_3 \end{Bmatrix} \tag{10.43}$$

where

$$\mathbf{A} = \int_\Omega \mathbf{N}_{\text{en}}^{\text{T}} \mathbf{D}\, \mathbf{N}_{\text{en}}\, \text{d}\Omega; \quad \mathbf{C} = -\int_\Omega \mathbf{N}_{\text{en}}^{\text{T}} \mathbf{N}_\sigma\, \text{d}\Omega; \quad \mathbf{G} = \int_\Omega \mathbf{N}_{\text{en}}^{\text{T}} \mathbf{D}\, \mathbf{B}\, \text{d}\Omega$$
$$\mathbf{K} = \int_\Omega \mathbf{B}^{\text{T}} \mathbf{D}\, \mathbf{B}\, \text{d}\Omega; \quad \mathbf{f}_1 = \mathbf{f}_2 = \mathbf{0}; \quad \mathbf{f}_3 = \int_\Omega \mathbf{N}_u^{\text{T}} \mathbf{b}\, \text{d}\Omega + \int_{\Gamma_t} \mathbf{N}_u^{\text{T}} \bar{\mathbf{t}}\, \text{d}\Gamma \tag{10.44}$$

In this form there is only one zero diagonal term and the stability condition reduces to the single condition

$$n_u + n_{\text{en}} \geq n_\sigma \tag{10.45}$$

Further, the use of the strains deduced from the displacement interpolation leads to a matrix which is identical to that from the irreducible form and we have thus included this in Eq. (10.44) as $\mathbf{K}$.

Example 10.5: Simo–Rifai quadrilateral. An enhanced strain formulation for application to problems in plain elasticity was introduced by Simo and Rifai.[28] The element has 4 nodes and employs isoparametric interpolation for the displacement field. The derivatives of the shape functions yield a form

$$\begin{Bmatrix} \dfrac{\partial N_a}{\partial x} \\ \dfrac{\partial N_a}{\partial y} \end{Bmatrix} = \frac{1}{J(\xi,\eta)} \begin{Bmatrix} a_{x,a}(y_b) + b_{x,a}(y_b)\xi + c_{x,a}(y_b)\eta \\ a_{y,a}(x_b) + b_{y,a}(x_b)\xi + c_{y,a}(x_b)\eta \end{Bmatrix}$$

where a_a, b_a and c_a depend on the nodal coordinates, and the jacobian determinant for the 4-node quadrilateral is given by†

$$\det \mathbf{J} = J(\xi,\eta) = J_0 + J_\xi \xi + J_\eta \eta$$

The enhanced strains are first assumed in the parent coordinate frame and transformed to the cartesian frame using a transformation similar to that used in developing the Pian–Sumihara quadrilateral in Example 10.4. Due to the presence of the jacobian determinant in the strains computed from the displacements (as well as the requirement to later pass the patch test for constant stress states) the enhanced strains are computed from

$$\varepsilon_{\text{en}} = \frac{1}{J(\xi,\eta)} \mathbf{T}^{\text{T}} \mathbf{E}(\xi,\eta) \mathbf{T}$$

where

$$\mathbf{E} = \begin{bmatrix} E_{\xi\xi} & E_{\xi\eta} \\ E_{\eta\xi} & E_{\eta\eta} \end{bmatrix}$$

In matrix form this may be written as

$$\begin{Bmatrix} \varepsilon_x \\ \varepsilon_y \\ \gamma_{xy} \end{Bmatrix}_{en} = \frac{1}{J(\xi,\eta)} \begin{bmatrix} T_{11}^2 & T_{21}^2 & T_{11}T_{21} \\ T_{12}^2 & T_{22}^2 & T_{12}T_{22} \\ 2T_{11}T_{12} & 2T_{21}T_{22}^2 & T_{11}T_{22} + T_{12}T_{21} \end{bmatrix} \begin{Bmatrix} E_{\xi\xi} \\ E_{\eta\eta} \\ 2E_{\xi\eta} \end{Bmatrix}$$

† In general, the determinant of the jacobian for the two-dimensional Lagrange family of elements will not contain the term with the product of the highest order polynomial, e.g., $\xi\eta$ for the 4-node element, $\xi^2\eta^2$ for the 9-node element, etc.

The parent strains (strains with components in the parent element frame) are assumed as

$$\begin{Bmatrix} E_{\xi\xi} \\ E_{\eta\eta} \\ 2E_{\xi\eta} \end{Bmatrix} = \begin{bmatrix} \xi & 0 & 0 & 0 \\ 0 & \eta & 0 & 0 \\ 0 & 0 & \xi & \eta \end{bmatrix} \begin{Bmatrix} \eta_1 \\ \eta_2 \\ \eta_3 \\ \eta_4 \end{Bmatrix}$$

The above is motivated by the fact that the derivatives of the shape functions with respect to parent coordinates yields

$$\frac{\partial N_a}{\partial \xi} = a_\xi + b_\xi \eta \qquad \frac{\partial N_a}{\partial \eta} = a_\eta + b_\eta \xi$$

and these may be combined to form strains in the usual manner, but in the parent frame. Thus, by design, the above enhanced strains are specified to generate complete polynomials in the parent coordinates for each strain component. References 29 and 30 discuss the relationship between the design of assumed stress elements using the two-field form and the selection of enhanced strain modes so as to produce the same result.

Remarks

1. The above enhanced strains are defined so that the **C** array is identically zero for constant assumed stresses in each element.
2. Parent normal strains have linearly independent terms added. However, the assumed parent shear strains are linearly dependent. Due to this linear dependence the final shearing strain will usually be nearly constant in each element. Accordingly, to be more explicit, normal strains are *enhanced* while shearing strain is *de-enhanced*.

Since the **C** array vanishes, the equation set to be solved becomes

$$\begin{bmatrix} \mathbf{A} & \mathbf{G} \\ \mathbf{G}^{\mathrm{T}} & \mathbf{K} \end{bmatrix} \begin{Bmatrix} \tilde{\boldsymbol{\varepsilon}}_{\mathrm{en}} \\ \tilde{\mathbf{u}} \end{Bmatrix} = \begin{Bmatrix} \mathbf{f}_1 \\ \mathbf{f}_3 \end{Bmatrix}$$

and in this form no additional count conditions are apparently needed. The solution may be accomplished partly at the element level by eliminating the equation associated with the enhanced strain parameters. Accordingly,

$$\mathbf{K}^* \tilde{\mathbf{u}} = \mathbf{f}_3^*$$

where

$$\mathbf{K}^* = \mathbf{K} - \mathbf{G}^{\mathrm{T}} \mathbf{A}^{-1} \mathbf{G} \quad \text{and} \quad \mathbf{f}_3^* = \mathbf{f}_3 - \mathbf{G}^{\mathrm{T}} \mathbf{A}^{-1} \mathbf{f}_1$$

The sensitivity of the enhanced strain element to geometric distortion is evaluated using the problem shown in Fig. 10.9. The transformation from the parent to the global frame is assessed using $\mathbf{T} = \mathbf{J}_0$ and $\mathbf{T} = \mathbf{J}_0^{-\mathrm{T}}$. These are the only options which maintain frame invariance for the element. As observed in Fig. 10.9 the results are now better using the inverse transpose. Since the stress and strain are conjugates in an energy sense, this result could be anticipated from the equivalence relationship

$$E = \frac{1}{2} \int_\Omega \boldsymbol{\sigma}^{\mathrm{T}} \boldsymbol{\varepsilon} \, \mathrm{d}\Omega \equiv \frac{1}{2} \int_\Box \boldsymbol{\Sigma}^{\mathrm{T}} \mathbf{E} \, \mathrm{d}\Box$$

where E is energy and $\Box$ denotes the domain of the element in the parent coordinate system (i.e., the bi-unit square for a quadrilateral element).

The performance of the enhanced element is compared to the Pian–Sumihara element for a shear loading on the mesh shown in Fig. 10.10. In Fig. 10.11 the convergence results for various order meshes are shown for linear elastic, plane strain conditions with: (a) $E = 70$ and $\nu = 1/3$ and (b) for $E = 70$ and $\nu = 0.499995$. The results shown in Fig. 10.11 clearly show the strong dependence of the displacement formulation on Poisson's ratio – namely the tendency for the element to *lock* for values which approach the incompressibility limit of $\nu = 1/2$. On the other hand, the performance of both the enhanced strain and the Pian–Sumihara element are nearly insensitive to the value of Poisson's ratio selected, with somewhat better performance of the enhanced element on coarse meshing.

10.6 Complementary forms with direct constraint

10.6.1 General forms

In the introduction to this chapter we defined the irreducible and mixed forms and indicated that on occasion it is possible to obtain more than one 'irreducible' form. To illustrate this in the problem of heat transfer given by Eqs (10.2) and (10.3) we introduced a penalty function α in Eq. (10.6) and derived a corresponding single governing equation (10.7) given in terms of $\mathbf{q}$. This penalty function here has no obvious physical meaning and served simply as a device to obtain a *close enough* approximation to the satisfaction of the continuity of flow equations.

On occasion it is possible to solve the problem as an irreducible one assuming *a priori* that the choice of the variable satisfies one of the equations. We call such forms *directly constrained* and obviously the choice of the shape function becomes difficult.

We shall consider two examples.

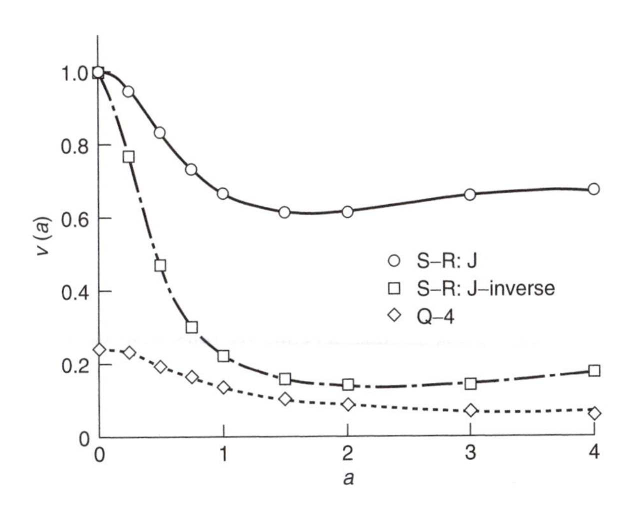

Fig. 10.9 Simo–Rifai enhanced strain quadrilateral (S–R) compared with displacement quadrilateral (Q-4). Effect of element distortion (Exact = 1.0).

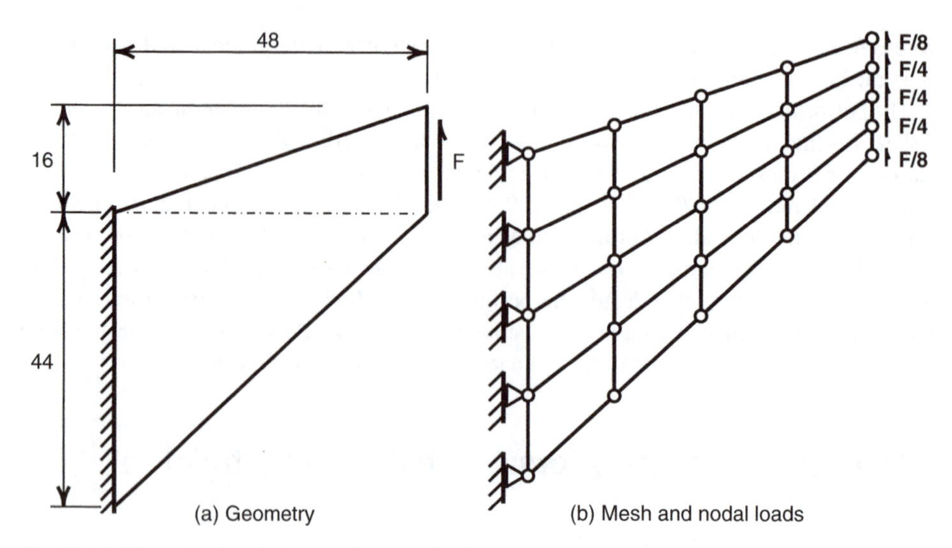

Fig. 10.10 Mesh with 4 × 4 elements for shear load.

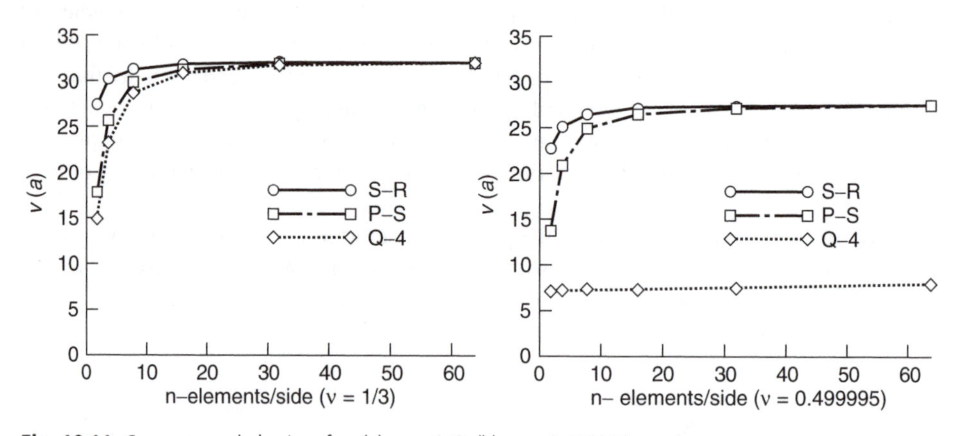

Fig. 10.11 Convergence behaviour for: (a) $\nu = 1/3$; (b) $\nu = 0.499995$.

The complementary heat transfer problem

In this we assume *a priori* that the choice of $\mathbf{q}$ is such that it satisfies Eq. (10.3) and the natural boundary conditions

$$\boldsymbol{\nabla}^{\mathrm{T}}\mathbf{q} = -Q \text{ in } \Omega \qquad \text{and} \qquad q_n = \mathbf{q}^{\mathrm{T}}\mathbf{n} = \bar{q}_n \text{ on } \Gamma_q \tag{10.46}$$

where $\mathbf{n}$ is the unit normal to the boundary. Thus we only have to satisfy the constitutive relation (10.2), i.e.,

$$\mathbf{k}^{-1}\mathbf{q} + \boldsymbol{\nabla}\phi = \mathbf{0} \text{ in } \Omega \qquad \text{with} \qquad \phi = \bar{\phi} \text{ on } \Gamma_\phi \tag{10.47}$$

A weak statement of the above is

$$\int_\Omega \delta\mathbf{q}^{\mathrm{T}}(\mathbf{k}^{-1}\mathbf{q} + \boldsymbol{\nabla}\phi)\,\mathrm{d}\Omega - \int_{\Gamma_\phi} \delta q_n(\phi - \bar{\phi})\,\mathrm{d}\Gamma = 0 \tag{10.48}$$

in which $\delta q_n = \delta \mathbf{q}^{\mathrm{T}} \mathbf{n}$ represents the variation of normal flux on the boundary.

Use of Green's theorem transforms the above into

$$\int_{\Omega} \delta \mathbf{q}^{\mathrm{T}} \mathbf{k}^{-1} \mathbf{q} \, \mathrm{d}\Omega - \int_{\Omega} \boldsymbol{\nabla}^{\mathrm{T}} \delta \mathbf{q} \phi \, \mathrm{d}\Omega + \int_{\Gamma_\phi} \delta q_n \bar{\phi} \, \mathrm{d}\Gamma + \int_{\Gamma_q} \delta q_n \phi \, \mathrm{d}\Gamma = 0 \tag{10.49}$$

If we further assume that $\boldsymbol{\nabla}^{\mathrm{T}} \delta \mathbf{q} \equiv 0$ in Ω and $\delta q_n = 0$ on Γ_q, i.e., that the weighting functions are simply the variations of $\mathbf{q}$, the equation reduces to

$$\int_{\Omega} \delta \mathbf{q}^{\mathrm{T}} \mathbf{k}^{-1} \mathbf{q} \, \mathrm{d}\Omega + \int_{\Gamma_\phi} \delta q_n \bar{\phi} \, \mathrm{d}\Gamma = 0 \tag{10.50}$$

This is in fact the variation of a complementary flux principle

$$\Pi = \int_{\Omega} \tfrac{1}{2} \mathbf{q}^{\mathrm{T}} \mathbf{k}^{-1} \mathbf{q} \, \mathrm{d}\Omega + \int_{\Gamma_\phi} q_n \bar{\phi} \, \mathrm{d}\Gamma \tag{10.51}$$

Numerical solutions can obviously be started from either of the above equations but the difficulty is the choice of the trial function satisfying the constraints. We shall return to this problem in Sec. 10.6.2.

The complementary elastic energy principle

In the elasticity problem specified in Sec. 10.4 we can proceed similarly, assuming stress fields which satisfy the equilibrium conditions both on the boundary Γ_t and in the domain Ω.

Thus in an analogous manner to that of the previous example we impose on the permissible stress field the constraints which we assume to be satisfied by the approximation identically, i.e.,

$$\boldsymbol{\mathcal{S}}^{\mathrm{T}} \boldsymbol{\sigma} + \mathbf{b} = \mathbf{0} \text{ in } \Omega \qquad \text{and} \qquad \mathbf{t} = \bar{\mathbf{t}} \text{ on } \Gamma_t \tag{10.52}$$

Thus only the constitutive relations and displacement boundary conditions remain to be satisfied, i.e.,

$$\mathbf{D}^{-1} \boldsymbol{\sigma} - \boldsymbol{\mathcal{S}} \mathbf{u} = \mathbf{0} \text{ in } \Omega \qquad \text{and} \qquad \mathbf{u} = \bar{\mathbf{u}} \text{ on } \Gamma_u \tag{10.53}$$

The weak statement of the above can be written as

$$\int_{\Omega} \delta \boldsymbol{\sigma}^{\mathrm{T}} (\mathbf{D}^{-1} \boldsymbol{\sigma} - \boldsymbol{\mathcal{S}} \mathbf{u}) \, \mathrm{d}\Omega + \int_{\Gamma_u} \delta \mathbf{t}^{\mathrm{T}} (\mathbf{u} - \bar{\mathbf{u}}) \, \mathrm{d}\Gamma = 0 \tag{10.54}$$

which on integration by Green's theorem gives

$$\int_{\Omega} \delta \boldsymbol{\sigma}^{\mathrm{T}} \mathbf{D}^{-1} \boldsymbol{\sigma} \, \mathrm{d}\Omega + \int_{\Omega} (\boldsymbol{\mathcal{S}}^{\mathrm{T}} \delta \boldsymbol{\sigma})^{\mathrm{T}} \mathbf{u} \, \mathrm{d}\Omega - \int_{\Gamma_u} \delta \mathbf{t}^{\mathrm{T}} \bar{\mathbf{u}} \, \mathrm{d}\Gamma - \int_{\Gamma_t} \delta \mathbf{t}^{\mathrm{T}} \mathbf{u} \, \mathrm{d}\Gamma = 0 \tag{10.55}$$

Again assuming that the test functions are complete variations satisfying the homogeneous equilibrium equation, i.e.,

$$\boldsymbol{\mathcal{S}}^{\mathrm{T}} \delta \boldsymbol{\sigma} = \mathbf{0} \text{ in } \Omega \qquad \text{and} \qquad \delta \mathbf{t} = \mathbf{0} \text{ on } \Gamma_t \tag{10.56}$$

we have as the weak statement

$$\int_{\Omega} \delta \boldsymbol{\sigma}^{\mathrm{T}} \mathbf{D}^{-1} \boldsymbol{\sigma} \, \mathrm{d}\Omega - \int_{\Gamma_u} \delta \mathbf{t}^{\mathrm{T}} \bar{\mathbf{u}} \, \mathrm{d}\Gamma = 0 \tag{10.57}$$

The corresponding complementary energy variational principle is

$$\Pi = \tfrac{1}{2} \int_{\Omega} \boldsymbol{\sigma}^{\mathrm{T}} \mathbf{D}^{-1} \boldsymbol{\sigma} \, \mathrm{d}\Omega - \int_{\Gamma_u} \mathbf{t}^{\mathrm{T}} \bar{\mathbf{u}} \, \mathrm{d}\Gamma \tag{10.58}$$

Once again in practical use the difficulties connected with the choice of the approximating function arise but on occasion a direct choice is possible.[31]

10.6.2 Solution using auxiliary functions

Both the complementary forms can be solved using auxiliary functions to ensure the satisfaction of the constraints.

Example 10.6: Heat transfer solution by potential function. In the *heat transfer problem* it is easy to verify that the homogeneous equation

$$\boldsymbol{\nabla}^{\mathrm{T}}\mathbf{q} \equiv \frac{\partial q_x}{\partial x} + \frac{\partial q_y}{\partial y} = 0 \tag{10.59}$$

is automatically satisfied by defining a function ψ such that

$$q_x = \frac{\partial \psi}{\partial y} \qquad q_y = -\frac{\partial \psi}{\partial x} \tag{10.60}$$

Thus we define

$$\mathbf{q} = \boldsymbol{\mathcal{L}}\psi + \mathbf{q}_0 \qquad \text{and} \qquad \delta\mathbf{q} = \boldsymbol{\mathcal{L}}\delta\psi \tag{10.61}$$

where $\mathbf{q}_0$ is any flux chosen so that

$$\boldsymbol{\nabla}^{\mathrm{T}}\mathbf{q}_0 = -Q \tag{10.62}$$

and

$$\boldsymbol{\mathcal{L}} = \left[\frac{\partial}{\partial y}, \ -\frac{\partial}{\partial x}\right]^{\mathrm{T}} \tag{10.63}$$

the formulations of Eqs (10.50) and (10.51) can be used without any constraints and, for instance, the stationarity

$$\Pi = \int_{\Omega} \tfrac{1}{2}(\boldsymbol{\mathcal{L}}\psi + \mathbf{q}_0)^{\mathrm{T}}\mathbf{k}^{-1}(\boldsymbol{\mathcal{L}}\psi + \mathbf{q}_0)\,\mathrm{d}\Omega - \int_{\Gamma_\phi} \left(\frac{\partial \psi}{\partial s}\right)\bar{\phi}\,\mathrm{d}\Gamma \tag{10.64}$$

will suffice to so formulate the problem (here s is the tangential direction to the boundary).

The above form will require shape functions for ψ satisfying C_0 continuity.

Example 10.7: Elasticity solution by Airy stress function. In the elasticity problem a two-dimensional form can be obtained by the use of the so-called Airy stress function ψ.[32]

Now the equilibrium equations

$$\boldsymbol{\mathcal{S}}^{\mathrm{T}}\boldsymbol{\sigma} + \mathbf{b} \equiv \begin{Bmatrix} \dfrac{\partial \sigma_x}{\partial x} + \dfrac{\partial \tau_{xy}}{\partial y} + b_x \\ \dfrac{\partial \tau_{xy}}{\partial x} + \dfrac{\partial \sigma_y}{\partial y} + b_y \end{Bmatrix} = \mathbf{0} \tag{10.65}$$

are identically solved by choosing

$$\boldsymbol{\sigma} = \boldsymbol{\mathcal{L}}\psi + \boldsymbol{\sigma}_0 \tag{10.66}$$

where

$$\boldsymbol{\mathcal{L}} = \left[\frac{\partial^2}{\partial y^2}, \ \frac{\partial^2}{\partial x^2}, \ -\frac{\partial^2}{\partial x\,\partial y}\right]^{\mathrm{T}} \tag{10.67}$$

and $\boldsymbol{\sigma}_0$ is an arbitrary stress chosen so that

$$\boldsymbol{\mathcal{S}}^{\mathrm{T}}\boldsymbol{\sigma}_0 + \mathbf{b} = \mathbf{0} \tag{10.68}$$

Again the substitution of (10.66) into the weak statement (10.57) or the complementary variational problem (10.58) will yield a direct formulation to which no additional constraints need be applied. However, use of the above forms does lead to further complexity in multiply connected regions where further conditions are needed. The reader will note that in Chapter 7 we encountered this in a similar problem in torsion and suggested a very simple procedure of avoidance (see Sec. 7.5).

The use of this stress function formulation in the two-dimensional context was first made by de Veubeke and Zienkiewicz[33] and Elias,[34] but the reader should note that now with second order operators present, C_1 continuity of shape functions is needed in a similar manner to the problems which we have to consider in plate bending (see reference 6).

Incidentally, analogies with plate bending go further here and indeed it can be shown that some of these can be usefully employed for other problems.[35]

10.7 Concluding remarks – mixed formulation or a test of element 'robustness'

The mixed form of finite element formulation outlined in this chapter opens a new range of possibilities, many with potentially higher accuracy and robustness than those offered by irreducible forms. However, an additional advantage arises even in situations where, by the *principle of limitation*, the irreducible and mixed forms yield identical results. Here the study of the behaviour of the mixed form can frequently reveal weaknesses or lack of 'robustness' in the irreducible form which otherwise would be difficult to determine.

The mixed approximation, if properly understood, expands the potential of the finite element method and presents almost limitless possibilities of detailed improvement. Some of these will be discussed further in the next two chapters, and others in references 6 and 36.

10.8 Problems

10.1 Show that the stationarity of the variational principle given by

$$\Pi_{\mathrm{HW}} = \int_\Omega \tfrac{1}{2}\boldsymbol{\varepsilon}^{\mathrm{T}}\mathbf{D}\boldsymbol{\varepsilon}\,\mathrm{d}\Omega - \int_\Omega \boldsymbol{\sigma}^{\mathrm{T}}(\boldsymbol{\varepsilon} - \boldsymbol{\mathcal{S}}\mathbf{u})\,\mathrm{d}\Omega - \int_\Omega \mathbf{u}^{\mathrm{T}}\mathbf{b}\,\mathrm{d}\Omega - \int_{\Gamma_t} \mathbf{u}^{\mathrm{T}}\bar{\mathbf{t}}\,\mathrm{d}\Gamma$$

where $\mathbf{u} \equiv \bar{\mathbf{u}}$ on Γ_u is equivalent to Eq. (10.33).

10.2 Using the variational principle of Problem 10.1 with the approximations (10.34) show that the stationarity condition gives (10.35) and (10.36).

10.3 Show that the variational principle given by stationarity of

$$\Pi_{\mathrm{en}} = \int_\Omega \tfrac{1}{2}\,(\boldsymbol{\mathcal{S}}\mathbf{u} + \boldsymbol{\varepsilon}_{\mathrm{en}})^{\mathrm{T}}\,\mathbf{D}\,(\boldsymbol{\mathcal{S}}\mathbf{u} + \boldsymbol{\varepsilon}_{\mathrm{en}})\,\mathrm{d}\Omega + \int_\Omega \boldsymbol{\sigma}^{\mathrm{T}}\boldsymbol{\varepsilon}_{\mathrm{en}}\,\mathrm{d}\Omega - \int_\Omega \mathbf{u}^{\mathrm{T}}\mathbf{b}\,\mathrm{d}\Omega - \int_{\Gamma_t} \mathbf{u}^{\mathrm{T}}\bar{\mathbf{t}}\,\mathrm{d}\Gamma$$

with $\mathbf{u} = \bar{\mathbf{u}}$ enforced on Γ_u is equivalent to Eq. (10.41).

10.4 For the rectangular element shown in Fig. 10.7 develop the expressions for η_i for the Pian–Sumihara element described in Sec. 10.4.3. For an isotropic elastic material and a plane stress problem compute the expressions for the stresses which result from the strains (these are those of the displacement model described in Chap. 6). How do these differ from those assumed for the mixed element?

10.5 For the enhanced strain formulation described in Sec. 10.5.3 use the constant stress patch test for a plane strain problem to show that $\boldsymbol{\varepsilon}_{en} = \mathbf{0}$.

Show that a necessary condition to satisfy this requirement is

$$\int_{\Omega_e} \mathbf{N}_{en} \, d\Omega = 0$$

10.6 Generalize the Simo–Rifai quadrilateral given as Example 10.5 in Sec. 10.5.3 for a three-dimensional solid modelled by 8-node hexahedral elements.

10.7 Generalize the Simo–Rifai quadrilateral given as Example 10.5 in Sec. 10.5.3 for an axisymmetric geometry.

10.8 A plane stress problem has the geometry shown in Fig. 10.11 and is loaded by a uniformly distributed shear traction (i.e., t_y = const.). Use *FEAPpv* to solve the problem using a series of 3-node triangular meshes. The first mesh should be as shown with each quadrilateral divided into two triangles. Consider two values for the elastic properties: (a) $E = 70$, $\nu = 1/3$ and (b) $E = 70$, $\nu = 0.499995$. Let the thickness of the slab be one unit.

Next, perform the solution using 4-node quadrilaterals based on (a) the displacement solution described in Chap. 6; (b) the Simo-Rifai enhanced element described in Sec. 10.5.3. Plot the displacement convergence for the top and bottom points at the loaded end. Plot contours for displacement and principal stresses.

Repeat the calculations assuming plane strain conditions.

Briefly discuss your findings.

References

1. S.N. Atluri, R.H. Gallagher, and O.C. Zienkiewicz, editors. *Hybrid and Mixed Finite Element Methods*. John Wiley & Sons, New York, 1983.
2. O.C. Zienkiewicz, R.L. Taylor, and J.A.W. Baynham. Mixed and irreducible formulations in finite element analysis. In S.N. Atluri, R.H. Gallagher, and O.C. Zienkiewicz, editors, *Hybrid and Mixed Finite Element Methods*, pages 405–431. John Wiley & Sons, 1983.
3. I. Babuška and J.E. Osborn. Generalized finite element methods and their relations to mixed problems. *SIAM J. Num. Anal.*, 30:510–536, 1983.
4. R.L. Taylor and O.C. Zienkiewicz. Complementary energy with penalty function in finite element analysis. In R. Glowinski, E.Y. Rodin, and O.C. Zienkiewicz, editors, *Energy Methods in Finite Element Analysis*, Chapter 8. John Wiley & Sons, Chichester, 1979.
5. K. Washizu. *Variational Methods in Elasticity and Plasticity*. Pergamon Press, New York, 3rd edition, 1982.
6. O.C. Zienkiewicz and R.L. Taylor. *The Finite Element Method for Solid and Structural Mechanics*. Butterworth-Heinemann, Oxford, 6th edition, 2005.
7. L.R. Herrmann. Finite element bending analysis of plates. In *Proc. 1st Conf. Matrix Methods in Structural Mechanics*, AFFDL-TR-66-80, pages 577–602, Wright-Patterson Air Force Base, Ohio, 1965.

8. K. Hellan. Analysis of elastic plates in flexure by a simplified finite element method. Technical Report Civ. Eng. Series 46, Acta Polytechnica Scandinavia, Trondheim, 1967.
9. R.S. Dunham and K.S. Pister. A finite element application of the Hellinger–Reissner variational theorem. In *Proc. 1st Conf. Matrix Methods in Structural Mechanics*, volume AFFDL-TR-66-80, Wright Patterson Air Force Base, Ohio, Oct. 1966.
10. R.L. Taylor and O.C. Zienkiewicz. Mixed finite element solution of fluid flow problems. In R.H. Gallagher, G.F. Carey, J.T. Oden, and O.C. Zienkiewicz, editors, *Finite Elements in Fluids*, volume 1, Chapter 4, pages 1–20. John Wiley & Sons, 1982.
11. B. Fraeijs de Veubeke. Displacement and equilibrium models in finite element method. In O.C. Zienkiewicz and G.S. Holister, editors, *Stress Analysis*, Chapter 9, pages 145–197. John Wiley & Sons, Chichester, 1965.
12. I. Babuška. Error bounds for finite element methods. *Numer. Math.*, 16:322–333, 1971.
13. I. Babuška. The finite element method with lagrangian multipliers. *Numer. Math.*, 20:179–192, 1973.
14. F. Brezzi. On the existence, uniqueness and approximation of saddle-point problems arising from Lagrange multipliers. *Rev. Française d'Automatique Inform. Rech. Opér., Ser. Rouge Anal. Numér.*, 8(R-2):129–151, 1974.
15. O.C. Zienkiewicz, S. Qu, R.L. Taylor, and S. Nakazawa. The patch test for mixed formulations. *Int. J. Numer. Meth. Eng.*, 23:1873–1883, 1986.
16. J.T. Oden and N. Kikuchi. Finite element methods for constrained problems in elasticity. *Int. J. Numer. Meth. Eng.*, 18:701–725, 1982.
17. E. Hellinger. Die allgemeine Aussetze der Mechanik der Kontinua. In F. Klein and C. Muller, editors, *Encyclopedia der Mathematishen Wissnschaften*, volume 4. Tebner, Leipzig, 1914.
18. E. Reissner. On a variational theorem in elasticity. *J. Math. Phys.*, 29(2):90–95, 1950.
19. L.R. Herrmann. Finite element bending analysis of plates. *J. Eng. Mech., ASCE*, 94(EM5): 13–25, 1968.
20. L.R. Herrmann and D.M. Campbell. Finite element analysis for thin shells. *J. AIAA*, 6:1842–1847, 1968.
21. O.C. Zienkiewicz and D. Lefebvre. Mixed methods for FEM and the patch test. Some recent developments. In F. Murat and O. Pirenneau, editors, *Analyse Mathematique of Application*. Gauthier Villars, Paris, 1988.
22. P.A. Raviart and J.M. Thomas. A mixed finite element method for second order elliptic problems. In *Lect. Notes in Math.*, no. 606, pages 292–315. Springer-Verlag, Berlin, 1977.
23. D.N. Arnold, F. Brezzi, and J. Douglas. PEERS, a new mixed finite element for plane elasticity. *Japan J. Appl. Math.*, 1:347–367, 1984.
24. T.H.H. Pian and K. Sumihara. Rational approach for assumed stress finite elements. *Int. J. Numer. Meth. Eng.*, 20:1685–1695, 1985.
25. O.C. Zienkiewicz and D. Lefebvre. Three field mixed approximation and the plate bending problem. *Comm. Appl. Num. Meth.*, 3:301–309, 1987.
26. T.J.R. Hughes. Generalization of selective integration procedures to anisotropic and non-linear media. *Int. J. Numer. Meth. Eng.*, 15:1413–1418, 1980.
27. J.C. Simo, R.L. Taylor, and K.S. Pister. Variational and projection methods for the volume constraint in finite deformation plasticity. *Comp. Meth. Appl. Mech. Eng.*, 51:177–208, 1985.
28. J.C. Simo and M.S. Rifai. A class of mixed assumed strain methods and the method of incompatible modes. *Int. J. Numer. Meth. Eng.*, 29:1595–1638, 1990.
29. U. Andelfinger and E. Ramm. EAS-elements for two-dimensional, three-dimensional, plate and shell structures and their equivalence to HR-elements. *Int. J. Numer. Meth. Eng.*, 36:1311–1337, 1993.
30. M. Bischoff, E. Ramm, and D. Braess. A class of equivalent enhanced assumed strain and hybrid stress finite elements. *Comput. Mech.*, 22:443–449, 1999.

31. C. Loubignac, G. Cantin, and C. Touzot. Continuous stress fields in finite element analysis. *J. AIAA*, 15:1645–1647, 1978.
32. S.P. Timoshenko and J.N. Goodier. *Theory of Elasticity*. McGraw-Hill, New York, 3rd edition, 1969.
33. B. Fraeijs de Veubeke and O.C. Zienkiewicz. Strain energy bounds in finite element analysis by slab analogy. *J. Strain Anal.*, 2:265–271, 1967.
34. Z.M. Elias. Duality in finite element methods. *Proc. Am. Soc. Civ. Eng.*, 94(EM4):931–946, 1968.
35. R.V. Southwell. On the analogues relating flexure and displacement of flat plates. *Quart. J. Mech. Appl. Math.*, 3:257–270, 1950.
36. O.C. Zienkiewicz, R.L. Taylor, and P. Nithiarasu. *The Finite Element Method for Fluid Dynamics*. Butterworth-Heinemann, Oxford, 6th edition, 2005.

11

Incompressible problems, mixed methods and other procedures of solution

11.1 Introduction

We have noted earlier that the standard displacement formulation of elastic problems fails when Poisson's ratio ν becomes 0.5 or when the material becomes incompressible. Indeed, problems arise even when the material is nearly incompressible with $\nu > 0.4$ and the simple linear approximation with triangular elements gives highly oscillatory results in such cases.

The application of a mixed formulation for such problems can avoid the difficulties and is of great practical interest as *nearly* incompressible behaviour is encountered in a variety of real engineering problems ranging from soil mechanics to aerospace engineering. Identical problems also arise when the flow of incompressible fluids is encountered.

In this chapter we shall discuss fully the mixed approaches to incompressible problems, generally using a two-field manner where displacement (or fluid velocity) $\mathbf{u}$ and the pressure p are the variables. Such formulation will allow us to deal with full incompressibility as well as near incompressibility as it occurs. However, what we will find is that the interpolations used will be very much limited by the stability conditions of the mixed patch test. For this reason much interest has been focused on the development of so-called *stabilized* procedures in which the violation of the mixed patch test (or Babuška–Brezzi conditions) is artificially compensated. A part of this chapter will be devoted to such stabilized methods.

11.2 Deviatoric stress and strain, pressure and volume change

The main problem in the application of a 'standard' displacement formulation to incompressible or nearly incompressible problems lies in the determination of the mean stress or pressure which is related to the volumetric part of the strain (for isotropic materials). For this reason it is convenient to separate this from the total stress field and treat it as an independent variable. Using the 'vector' notation of stress, the mean stress or pressure is given by

$$p = \tfrac{1}{3}\left(\sigma_x + \sigma_y + \sigma_z\right) = \tfrac{1}{3}\mathbf{m}^{\mathrm{T}}\boldsymbol{\sigma} \tag{11.1}$$

where $\mathbf{m}$ for the general three-dimensional state of stress is given by

$$\mathbf{m} = \begin{bmatrix}1, & 1, & 1, & 0, & 0, & 0\end{bmatrix}^{\mathrm{T}}$$

For isotropic behaviour the 'pressure' is related to the volumetric strain, ε_v, by the bulk modulus of the material, K. Thus,

$$\varepsilon_v = \varepsilon_x + \varepsilon_y + \varepsilon_z = \mathbf{m}^{\mathrm{T}}\boldsymbol{\varepsilon} = \frac{p}{K} \tag{11.2}$$

For an incompressible material $K = \infty$ ($\nu \equiv 0.5$) and the volumetric strain is simply zero.

The deviatoric strain $\boldsymbol{\varepsilon}^d$ is defined by

$$\boldsymbol{\varepsilon}^d = \boldsymbol{\varepsilon} - \tfrac{1}{3}\mathbf{m}\varepsilon_v \equiv \left(\mathbf{I} - \tfrac{1}{3}\mathbf{m}\mathbf{m}^{\mathrm{T}}\right)\boldsymbol{\varepsilon} = \mathbf{I}_d\boldsymbol{\varepsilon} \tag{11.3}$$

where $\mathbf{I}_d$ is a deviatoric projection matrix which also proves useful in problems with more general constitutive relations.[1] In isotropic elasticity the deviatoric strain is related to the deviatoric stress by the shear modulus G as

$$\boldsymbol{\sigma}^d = \mathbf{I}_d\boldsymbol{\sigma} = 2G\mathbf{I}_0\boldsymbol{\varepsilon}^d = 2G\left(\mathbf{I}_0 - \tfrac{1}{3}\mathbf{m}\mathbf{m}^{\mathrm{T}}\right)\boldsymbol{\varepsilon} \tag{11.4}$$

where the diagonal matrix

$$\mathbf{I}_0 = \frac{1}{2}\begin{bmatrix} 2 & & & & & \\ & 2 & & & & \\ & & 2 & & & \\ & & & 1 & & \\ & & & & 1 & \\ & & & & & 1 \end{bmatrix}$$

is introduced because of the vector notation. A deviatoric form for the elastic moduli of an isotropic material is written as

$$\mathbf{D}_d = 2G\left(\mathbf{I}_0 - \tfrac{1}{3}\mathbf{m}\mathbf{m}^{\mathrm{T}}\right) \tag{11.5}$$

for convenience in writing subsequent equations.

The above relationships are but an alternate way of determining the stress–strain relations shown in Chapters 2 and 6, with the material parameters related through

$$\begin{aligned} G &= \frac{E}{2(1+\nu)} \\ K &= \frac{E}{3(1-2\nu)} \end{aligned} \tag{11.6}$$

and indeed Eqs (11.4) and (11.2) can be used to define the standard $\mathbf{D}$ matrix in an alternative manner.

11.3 Two-field incompressible elasticity (u–*p* form)

In the mixed form considered next we shall use as variables the displacement $\mathbf{u}$ and the pressure p.

Now the equilibrium equation (10.22) is rewritten using (11.4), treating p as an independent variable, as

$$\int_\Omega \delta\varepsilon^{\mathrm{T}}\mathbf{D}_d\varepsilon\,\mathrm{d}\Omega + \int_\Omega \delta\varepsilon^{\mathrm{T}}\mathbf{m}\,p\,\mathrm{d}\Omega - \int_\Omega \delta\mathbf{u}^{\mathrm{T}}\mathbf{b}\,\mathrm{d}\Omega - \int_{\Gamma_t} \delta\mathbf{u}^{\mathrm{T}}\bar{\mathbf{t}}\,\mathrm{d}\Gamma = 0 \tag{11.7}$$

and in addition we shall impose a weak form of Eq. (11.2), i.e.,

$$\int_\Omega \delta p\left[\mathbf{m}^{\mathrm{T}}\varepsilon - \frac{p}{K}\right]\mathrm{d}\Omega = 0 \tag{11.8}$$

with $\varepsilon = \mathcal{S}\mathbf{u}$. Independent approximation of $\mathbf{u}$ and p as

$$\mathbf{u} \approx \hat{\mathbf{u}} = \mathbf{N}_u\tilde{\mathbf{u}} \quad \text{and} \quad p \approx \hat{p} = \mathbf{N}_p\tilde{\mathbf{p}} \tag{11.9}$$

immediately gives the mixed approximation in the form

$$\begin{bmatrix} \mathbf{A} & \mathbf{C} \\ \mathbf{C}^{\mathrm{T}} & -\mathbf{V} \end{bmatrix} \begin{Bmatrix} \tilde{\mathbf{u}} \\ \tilde{\mathbf{p}} \end{Bmatrix} = \begin{Bmatrix} \mathbf{f}_1 \\ \mathbf{f}_2 \end{Bmatrix} \tag{11.10}$$

where

$$\begin{aligned} \mathbf{A} &= \int_\Omega \mathbf{B}^{\mathrm{T}}\mathbf{D}_d\,\mathbf{B}\,\mathrm{d}\Omega; \quad \mathbf{C} = \int_\Omega \mathbf{B}^{\mathrm{T}}\mathbf{m}\mathbf{N}_p\,\mathrm{d}\Omega \\ \mathbf{V} &= \int_\Omega \mathbf{N}_p^{\mathrm{T}}\frac{1}{K}\mathbf{N}_p\,\mathrm{d}\Omega; \quad \mathbf{f}_1 = \int_\Omega \mathbf{N}_u^{\mathrm{T}}\mathbf{b}\,\mathrm{d}\Omega + \int_{\Gamma_t} \mathbf{N}_u^{\mathrm{T}}\bar{\mathbf{t}}\,\mathrm{d}\Gamma \qquad \mathbf{f}_2 = \mathbf{0} \end{aligned} \tag{11.11}$$

We note that for incompressible situations the equations are of the 'standard' form, see Eq. (10.14) with $\mathbf{V} = \mathbf{0}$ (as $K = \infty$), but the formulation is useful in practice when K has a high value (or $\nu \to 0.5$).

A formulation similar to that above and using the corresponding variational theorem was first proposed by Herrmann[2] and later generalized by Key[3] for anisotropic elasticity. The arguments concerning stability (or singularity) of the matrices which we presented in Sec. 10.3 are again of great importance in this problem.

Clearly the mixed patch condition about the number of degrees of freedom now yields [see Eq. (10.18)]

$$n_u \geq n_p \tag{11.12}$$

and *is necessary* for prevention of locking (or instability) with the pressure acting now as the constraint variable of the lagrangian multiplier enforcing zero volumetric strain.

In the form of a patch test this condition is most critical and we show in Figs 11.1 and 11.2 a series of such patch tests on elements with C_0 continuous interpolation of $\mathbf{u}$ and either discontinuous or continuous interpolation of p. For each we have included all combinations of constant, linear and quadratic functions.

In the test we prescribe *all* the displacements on the boundaries of the patch and one pressure variable as it is well known that in fully incompressible situations pressure will be indeterminate by a constant for the problem with all boundary displacements prescribed.†

†Alternatively, it is possible to omit all boundary conditions on pressure if one displacement with a component normal to the boundary is allowed to exist.

The single-element test is very stringent and eliminates most continuous pressure approximations whose performance is known to be acceptable in many situations. For this reason we attach more importance to the assembly test and it would appear that the following elements could be permissible according to the criteria of Eq. (11.12) (indeed all pass the B-B condition fully):

Triangles: T6/1; T10/3; T6/C3

Quadrilaterals: Q9/3; Q8/C4; Q9/C4

We note, however, that in practical applications quite adequate answers have been reported with Q4/1, Q8/3 and Q9/4 quadrilaterals, although severe oscillations of p may occur. If full robustness is sought the choice of the elements is limited.[4]

It is unfortunate that in the present 'acceptable' list, the linear triangle and quadrilateral are missing. This appreciably restricts the use of these simplest elements. A possible and indeed effective procedure here is not to apply the pressure constraint at the level of a single element but on an assembly. This was done by Herrmann in his original presentation[2] where four elements were chosen for such a constraint as shown in Fig. 11.3(a). This composite 'element' passes the single-element (and multiple-element) patch tests but apparently so do several others fitting into this category. In Fig. 11.3(b) we show how a single triangle can be internally subdivided into three parts by the introduction of a central node. This coupled with constant pressure on the assembly allows the necessary count condition to be satisfied and a standard element procedure applies to the original triangle treating the central node as an internal variable. Indeed, the same effect could be achieved by the introduction of any other internal element function which gives zero value on the main triangle perimeter. Such a *bubble function* can simply be written in terms of the area coordinates (see Chapter 4) as

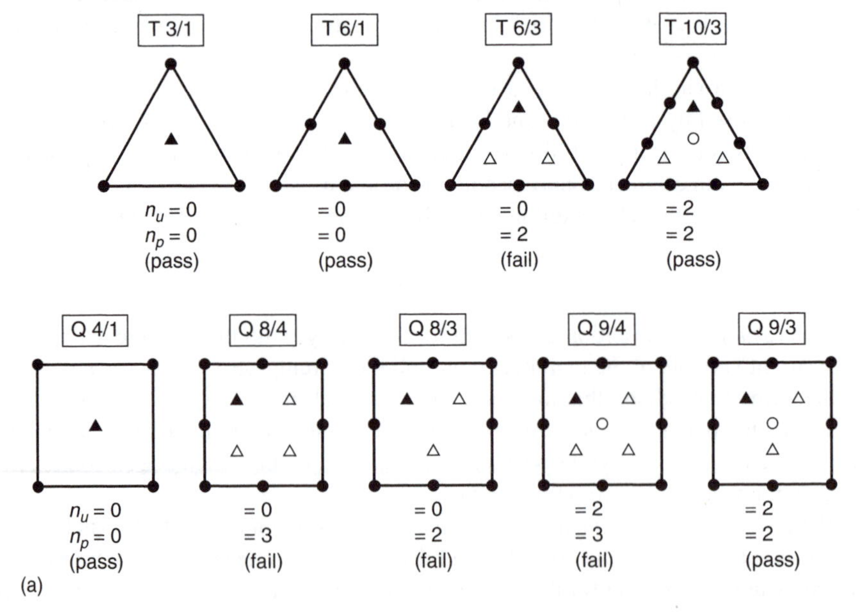

Fig. 11.1 Incompressible elasticity **u**–p formulation. Discontinuous pressure approximation. (a) Single-element patch tests. **u** variable (● restrained, ○ free) 2 DOF, p variable (▲ restrained, △ free) 1 DOF.

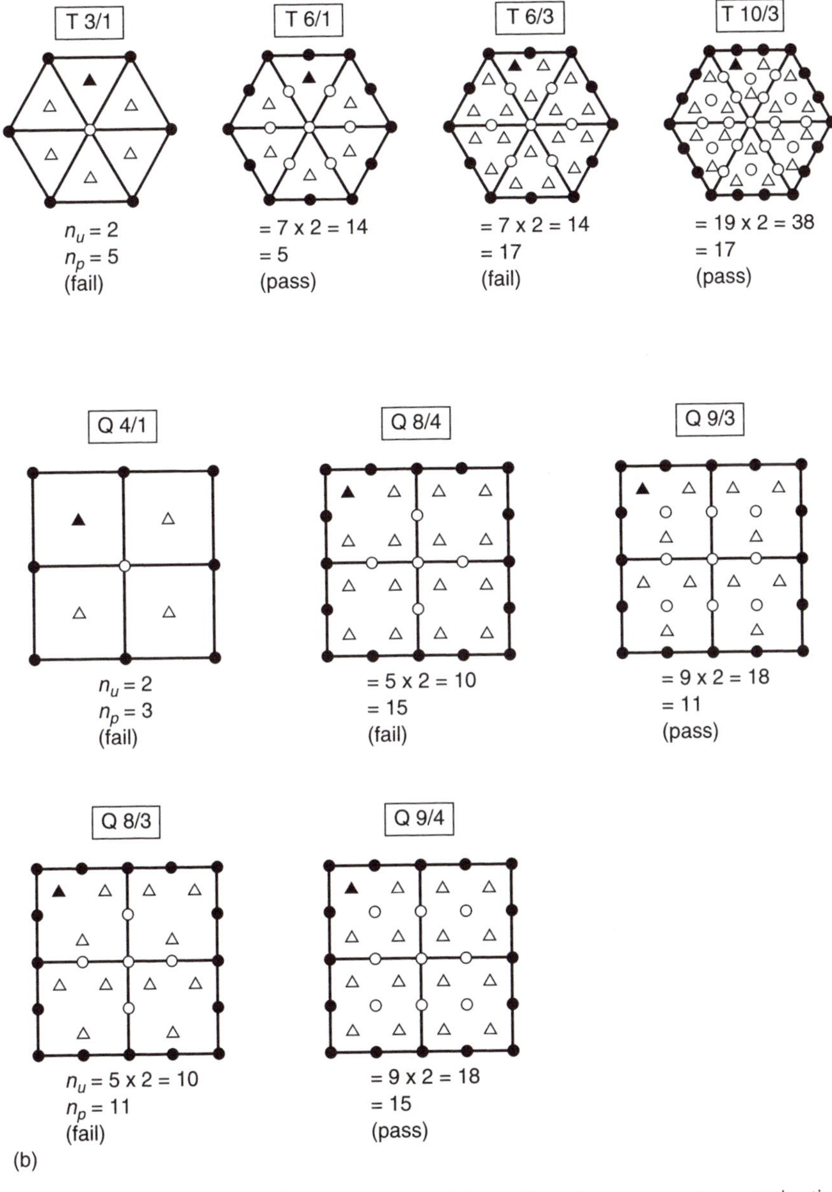

Fig. 11.1 (Cont.) Incompressible elasticity **u**–p formulation. Discontinuous pressure approximation. (b) Multiple-element patch tests.

$L_1L_2L_3$. However, as we have stated before, the degree of freedom count is a necessary but not sufficient condition for stability and a direct rank test is always required. In particular it can be verified by algebra that the conditions stated in Sec.10.3 are not fulfilled for this triple subdivision of a linear triangle (or the case with the bubble function) and thus

$$\mathbf{C}\tilde{\mathbf{p}} = \mathbf{0} \text{ for some non-zero values of } \tilde{\mathbf{p}}$$

indicating instability.

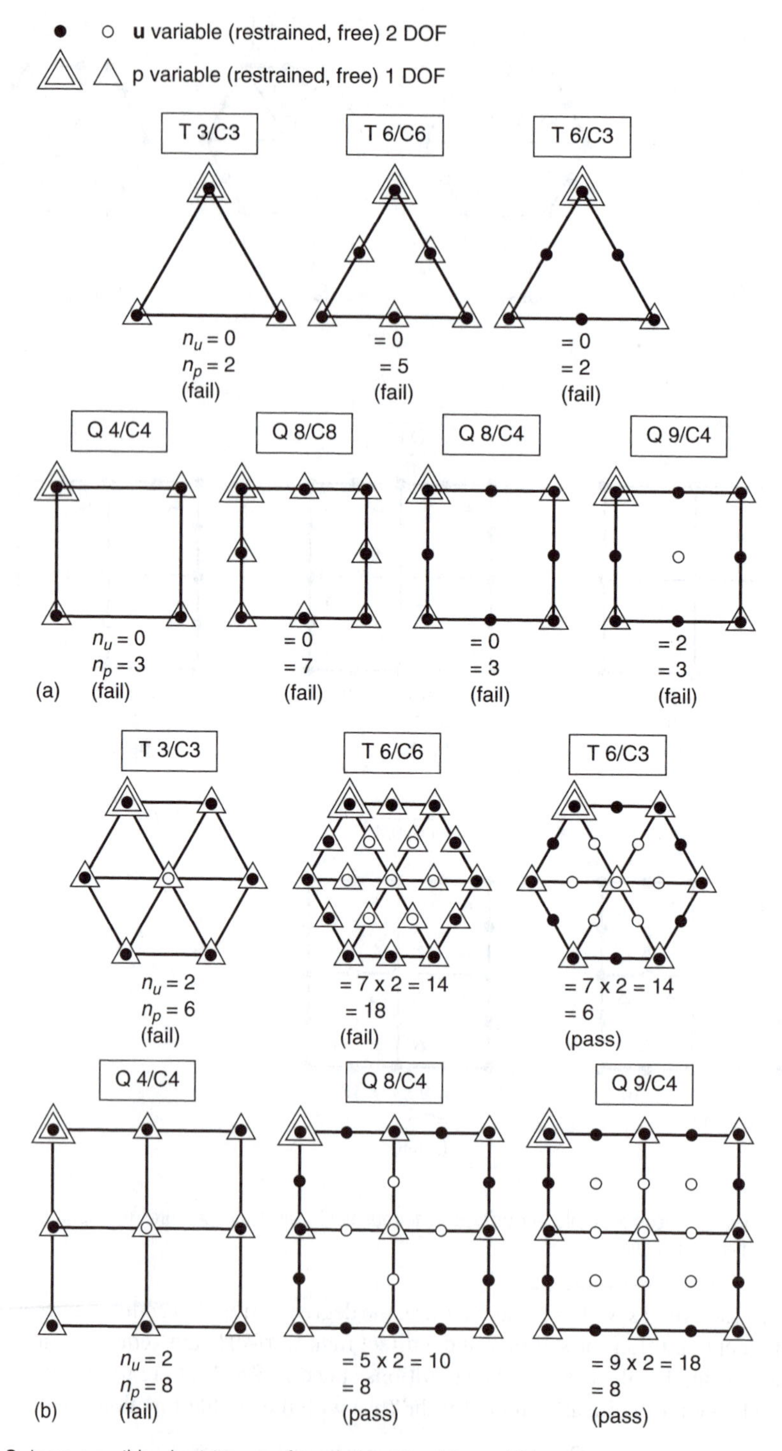

Fig. 11.2 Incompressible elasticity **u**–p formulation. Continuous (C_0) pressure approximation. (a) Single-element patch tests. (b) Multiple-element patch tests.

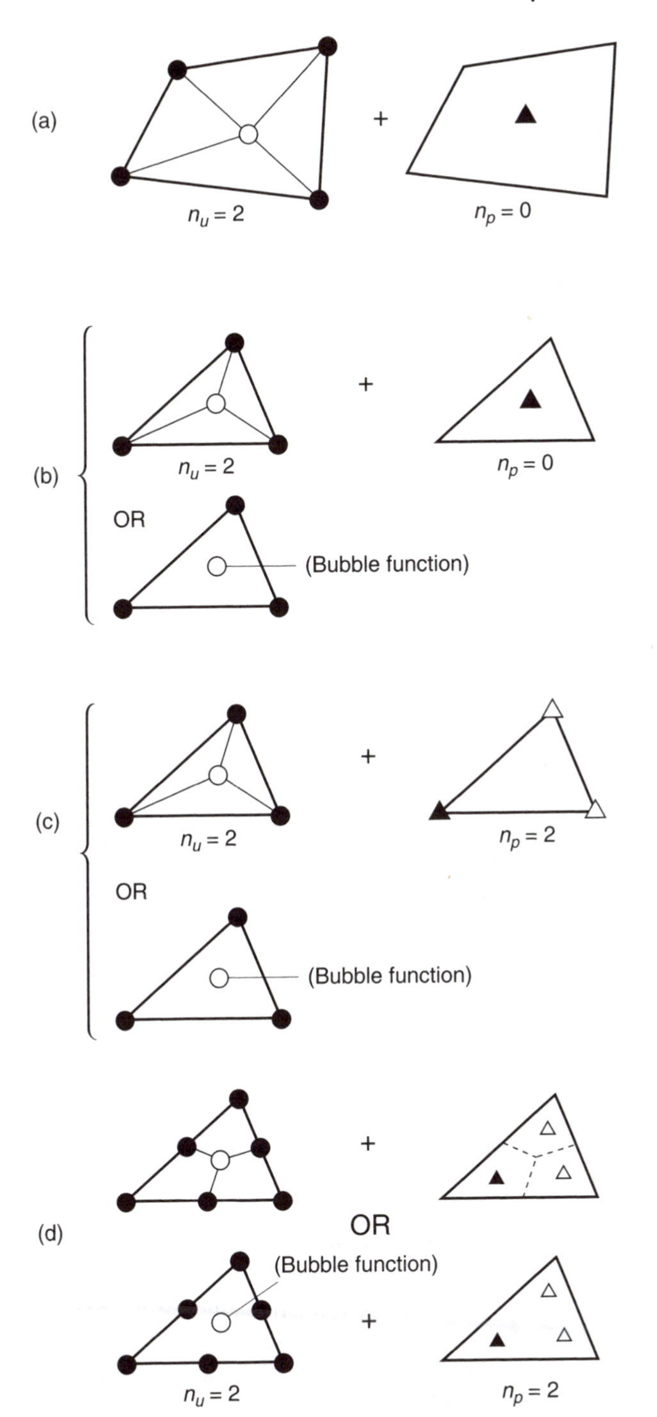

Fig. 11.3 Some simple combinations of linear triangles and quadrilaterals that pass the necessary patch test counts. Combinations (a), (c), and (d) are successful but (b) is still singular and not usable.

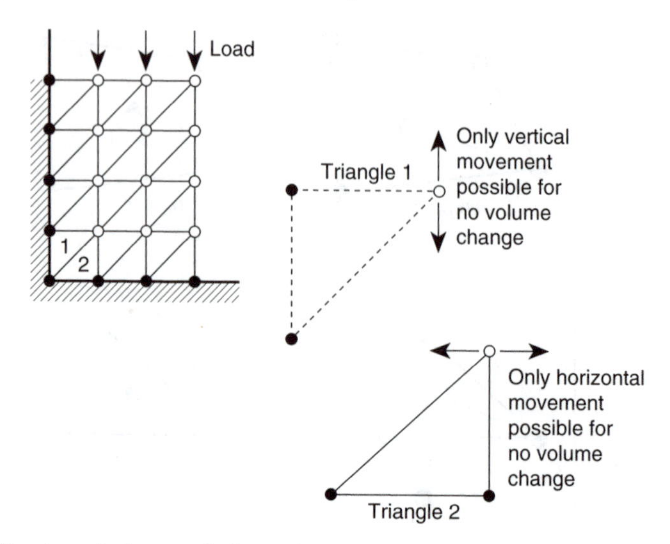

Fig. 11.4 Locking (zero displacements) of a simple assembly of linear triangles for which incompressibility is fully required ($n_p = n_u = 24$).

In Fig. 11.3(c) we show, however, that the same concept can be used with good effect for C_0 continuous p.[5] Similar internal subdivision into quadrilaterals or the introduction of bubble functions in quadratic triangles can be used, as shown in Fig. 11.3(d), with success.

The performance of all the elements mentioned above has been extensively discussed[6–11] but detailed comparative assessment of merit is difficult. As we have observed, it is essential to have $n_u \geq n_p$ but if near equality is only obtained in a large problem no meaningful answers will result for $\mathbf{u}$ as we observe, for example, in Fig. 11.4 in which linear triangles for $\mathbf{u}$ are used with the element constant p. Here the only permissible answer is of course $\mathbf{u} = \mathbf{0}$ as the triangles have to preserve constant volumes.

The ratio n_u/n_p which occurs as the field of elements is enlarged gives some indication of the relative performance, and we show this in Fig. 11.5. This approximates to the behaviour of a very large element assembly, but of course for any practical problem such a ratio will depend on the boundary conditions imposed.

We see that for the discontinuous pressure approximation this ratio for 'good' elements is 2–3 while for C_0 continuous pressure it is 6–8. All the elements shown in Fig. 11.5 perform very well, though two (Q4/1 and Q9/4) can on occasion lock when most boundary conditions are on $\mathbf{u}$.

Example 11.1: Simple triangle with bubble – MINI element. In Fig. 11.3(c) we indicate that the simple triangle with C_0 linear interpolation and an added bubble for the displacements $\mathbf{u}$ together with continuous C_0 linear interpolation for the pressure p satisfied the count test part of the mixed patch test and, verifying the consistency condition, can be used with success.[5] Here we consider this element further to develop some understanding about its performance at the incompressible limit.

The displacement field with the bubble is written in hierarchical form as

$$\mathbf{u} \approx \hat{\mathbf{u}} = \sum_a N_a \tilde{\mathbf{u}}_a + N_{bub} \tilde{\mathbf{u}}_{bub} \tag{11.13}$$

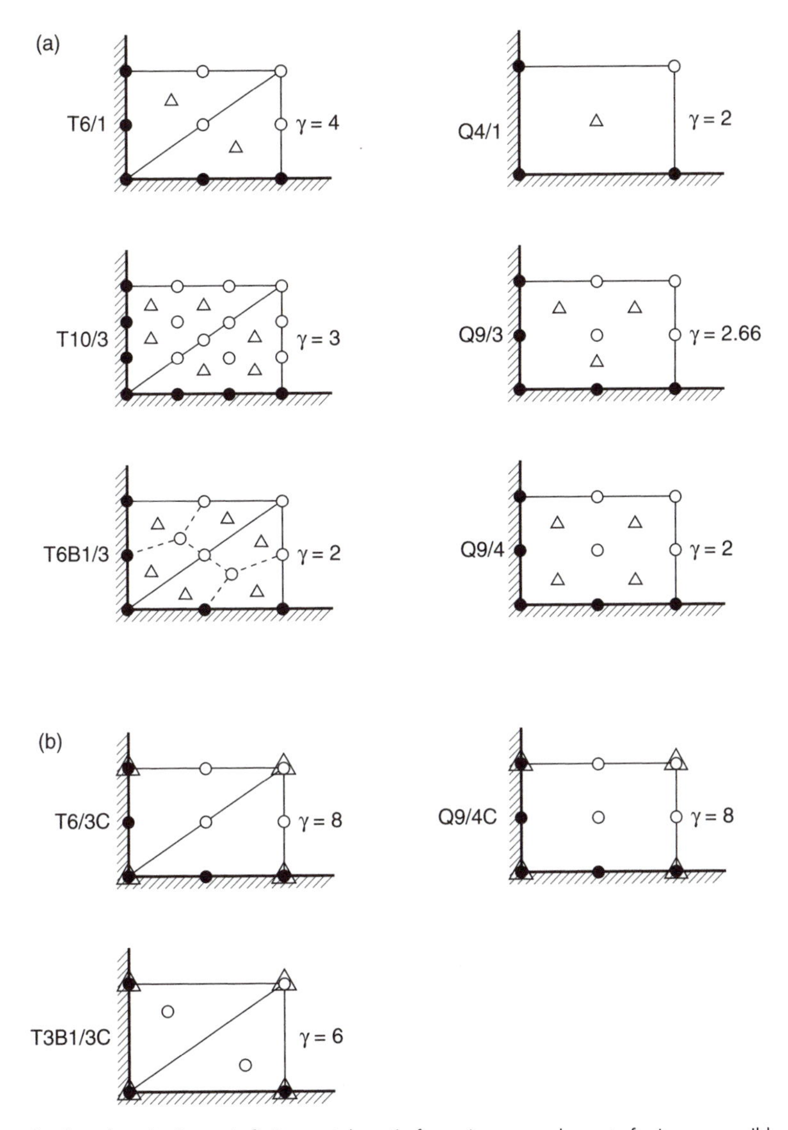

Fig. 11.5 The *freedom index* or *infinite patch ratio* for various **u**–*p* elements for incompressible elasticity ($\gamma = n_u/n_p$). (a) Discontinuous pressure. (b) Continuous pressure. B–bubble, C–continuous.

where here

$$N_{bub} = L_1 L_2 L_3 \tag{11.14}$$

$\tilde{\mathbf{u}}_a$ are nodal parameters of displacement and $\tilde{\mathbf{u}}_{bub}$ are parameters of the hierarchical bubble function. The pressures are similarly given by

$$p \approx \hat{p} = \sum_a N_a \tilde{p}_a \tag{11.15}$$

where $\tilde{p}_a$ are nodal parameters of the pressure. In the above the shape functions are given by (e.g., see Eqs (4.26) and (4.29))

$$N_a = L_a = \frac{1}{2\Delta}(a_a + b_a x + c_a y) \tag{11.16}$$

where

$$a_a = x_b y_c - x_c y_b; \qquad b_a = y_b - y_c; \qquad c_a = x_c - y_b$$

b, c are cyclic permutations of a and

$$2\Delta = \det \begin{bmatrix} 1 & x_1 & y_1 \\ 1 & x_2 & y_2 \\ 1 & x_3 & y_3 \end{bmatrix} = a_1 + a_2 + a_3$$

The derivatives of the shape functions are thus given by

$$\frac{\partial N_a}{\partial x} = \frac{b_a}{2\Delta} \quad \text{and} \quad \frac{\partial N_a}{\partial y} = \frac{c_a}{2\Delta}$$

Similarly the derivatives of the bubble are given by

$$\begin{aligned} \frac{\partial N_{bub}}{\partial x} &= \frac{1}{2\Delta}(b_1 L_2 L_3 + b_2 L_3 L_1 + b_3 L_1 L_2) \\ \frac{\partial N_{bub}}{\partial y} &= \frac{1}{2\Delta}(c_1 L_2 L_3 + c_2 L_3 L_1 + c_3 L_1 L_2) \end{aligned}$$

The strains may be expressed in terms of the above and the nodal parameters as†

$$\boldsymbol{\varepsilon} = \sum_a \frac{1}{2\Delta} \begin{bmatrix} b_a & 0 \\ 0 & c_a \\ c_a & b_a \end{bmatrix} \tilde{\mathbf{u}}_a + \sum_a \frac{L_b L_c}{2\Delta} \begin{bmatrix} b_a & 0 \\ 0 & c_a \\ c_a & b_a \end{bmatrix} \tilde{\mathbf{u}}_{bub} \tag{11.17}$$

where again b, c are cyclic permutations of a.

Substituting the above strains into Eq. (11.11) and evaluating the integrals give

$$\mathbf{A} = \begin{bmatrix} \mathbf{A}_{11} & \mathbf{A}_{12} & \mathbf{A}_{13} & \mathbf{0} \\ \mathbf{A}_{21} & \mathbf{A}_{22} & \mathbf{A}_{23} & \mathbf{0} \\ \mathbf{A}_{31} & \mathbf{A}_{32} & \mathbf{A}_{33} & \mathbf{0} \\ \mathbf{0} & \mathbf{0} & \mathbf{0} & \mathbf{A}_{bb} \end{bmatrix} \tag{11.18}$$

where

$$\begin{aligned} \mathbf{A}_{ab} &= \frac{G}{6\Delta} \begin{bmatrix} (4b_a b_b + 3c_a c_b) & (3c_a b_b - 2b_a c_b) \\ (3b_a c_b - 2c_a b_b) & (3b_a b_b + 4c_a c_b) \end{bmatrix} \\ \mathbf{A}_{bubbub} &= \frac{G}{2160\Delta} \begin{bmatrix} (4\mathbf{b}^{\mathrm{T}}\mathbf{b} + 3\mathbf{c}^{\mathrm{T}}\mathbf{c}) & \mathbf{b}^{\mathrm{T}}\mathbf{c} \\ \mathbf{b}^{\mathrm{T}}\mathbf{c} & (3\mathbf{b}^{\mathrm{T}}\mathbf{b} + 4\mathbf{c}^{\mathrm{T}}\mathbf{c}) \end{bmatrix} \end{aligned}$$

and

$$\mathbf{b} = \begin{bmatrix} b_1, & b_2, & b_3 \end{bmatrix}^{\mathrm{T}} \quad \text{and} \quad \mathbf{c} = \begin{bmatrix} c_1, & c_2, & c_3 \end{bmatrix}^{\mathrm{T}}$$

† At this point it is also possible to consider the term added to the derivatives to be *enhanced modes* and delete the bubble mode from displacement terms.

Note in the above that all terms except $\mathbf{A}_{bubbub}$ are standard displacement stiffnesses for the deviatoric part. Similarly,

$$\mathbf{C} = \begin{bmatrix} \mathbf{C}_{11} & \mathbf{C}_{12} & \mathbf{C}_{13} \\ \mathbf{C}_{21} & \mathbf{C}_{22} & \mathbf{C}_{23} \\ \mathbf{C}_{31} & \mathbf{C}_{32} & \mathbf{C}_{33} \\ \mathbf{C}_{bub1} & \mathbf{C}_{bub2} & \mathbf{C}_{bub3} \end{bmatrix} \tag{11.19}$$

where

$$\mathbf{C}_{ab} = \frac{1}{6}\begin{bmatrix} b_b \\ c_b \end{bmatrix} \quad \text{and} \quad \mathbf{C}_{bubb} = -\frac{1}{120}\begin{bmatrix} b_b \\ c_b \end{bmatrix}$$

In all the above arrays a and b have values from 1 to 3 and bub denotes the bubble mode.

We note that the bubble mode is decoupled from the other entries in the **A** array – it is precisely for this reason that the discontinuous constant pressure case shown in Fig. 11.3(b) cannot be improved by the addition of the internal parameters associated with $\tilde{\mathbf{u}}_{bub}$. Also, the parameters $\tilde{\mathbf{u}}_{bub}$ are defined separately for each element. Consequently, we may perform a partial solution at the element level[12] to obtain the set of equations in the form Eq. (11.10) where now

$$\mathbf{A} = \begin{bmatrix} \mathbf{A}_{11} & \mathbf{A}_{12} & \mathbf{A}_{13} \\ \mathbf{A}_{21} & \mathbf{A}_{22} & \mathbf{A}_{23} \\ \mathbf{A}_{31} & \mathbf{A}_{32} & \mathbf{A}_{33} \end{bmatrix}; \quad \mathbf{C} = \begin{bmatrix} \mathbf{C}_{11} & \mathbf{C}_{12} & \mathbf{C}_{13} \\ \mathbf{C}_{21} & \mathbf{C}_{22} & \mathbf{C}_{23} \\ \mathbf{C}_{31} & \mathbf{C}_{32} & \mathbf{C}_{33} \end{bmatrix}; \quad \mathbf{V} = \begin{bmatrix} V_{11} & V_{12} & V_{13} \\ V_{21} & V_{22} & V_{23} \\ V_{31} & V_{32} & V_{33} \end{bmatrix}$$

with

$$V_{ab} = \begin{bmatrix} \dfrac{b_a}{2\Delta} & \dfrac{c_a}{2\Delta} \end{bmatrix} \begin{bmatrix} \tau_{11} & \tau_{12} \\ \tau_{21} & \tau_{22} \end{bmatrix} \begin{Bmatrix} \dfrac{b_b}{2\,\Delta} \\ \dfrac{c_b}{2\,\Delta} \end{Bmatrix} \Delta \tag{11.20a}$$

and

$$\boldsymbol{\tau} = \frac{3\Delta^2}{10Gc}\begin{bmatrix} (3\mathbf{b}^{\mathrm{T}}\mathbf{b} + 4\mathbf{c}^{\mathrm{T}}\mathbf{c}) & -\mathbf{b}^{\mathrm{T}}\mathbf{c} \\ -\mathbf{b}^{\mathrm{T}}\mathbf{c} & (4\mathbf{b}^{\mathrm{T}}\mathbf{b} + 3\mathbf{c}^{\mathrm{T}}\mathbf{c}) \end{bmatrix} \tag{11.20b}$$

in which

$$c = 12\left(\mathbf{b}^{\mathrm{T}}\mathbf{b}\right)^2 + 25\left(\mathbf{b}^{\mathrm{T}}\mathbf{b}\right)\left(\mathbf{c}^{\mathrm{T}}\mathbf{c}\right) + 12\left(\mathbf{c}^{\mathrm{T}}\mathbf{c}\right)^2 - \left(\mathbf{b}^{\mathrm{T}}\mathbf{c}\right)^2$$

The reader may recognize the **V** array given above as that for the two-dimensional, steady heat equation with conductivity $\mathbf{k} = \boldsymbol{\tau}$ and discretized by linear triangular elements. The direct reduction of the bubble matrix $\mathbf{A}_{bubbub}$ as given above leads to a full matrix $\boldsymbol{\tau}$. Some numerical experiments including the above formulation are presented in Sec. 11.7.

11.4 Three-field nearly incompressible elasticity (u–p–ε_v form)

A direct approximation of the three-field form leads to an important method in finite element solution procedures for nearly incompressible materials which has sometimes been called the **B**-bar method. The methodology can be illustrated for the nearly incompressible isotropic problem. For this problem the method often reduces to the same two-field form previously discussed. However, for more general anisotropic or inelastic materials and in finite deformation problems the method has distinct advantages as are discussed in reference 1. The usual irreducible form (displacement method) has been shown to 'lock' for the nearly incompressible problem. As shown in Sec. 11.3, the use of a two-field mixed method can avoid this locking phenomenon when properly implemented (e.g., using the Q9/3 two-field form). Below we present an alternative which leads to an efficient and accurate implementation in many situations. For the development shown we shall assume that the material is isotropic linear elastic but it may be extended easily to include anisotropic materials.

Assuming an independent approximation to ε_v and p we can formulate the problem by use of Eq. (11.7) and the weak statement of relation (11.2) written as

$$\int_\Omega \delta p\left[\mathbf{m}^{\mathrm{T}}\boldsymbol{\mathcal{S}}\mathbf{u} - \varepsilon_v\right] \,\mathrm{d}\Omega = 0 \tag{11.21}$$

and

$$\int_\Omega \delta\varepsilon_v [K\varepsilon_v - p] \,\mathrm{d}\Omega = 0 \tag{11.22}$$

If we approximate the **u** and p fields by Eq. (11.9) and

$$\varepsilon_v \approx \hat{\varepsilon}_v = \mathbf{N}_v \tilde{\boldsymbol{\varepsilon}}_v \tag{11.23}$$

we obtain a mixed approximation in the form of Sec. (10.5.3) but now only for p and ε_v

$$\begin{bmatrix} \mathbf{A} & \mathbf{C} & \mathbf{0} \\ \mathbf{C}^{\mathrm{T}} & \mathbf{0} & -\mathbf{E} \\ \mathbf{0} & -\mathbf{E}^{\mathrm{T}} & \mathbf{H} \end{bmatrix} \begin{Bmatrix} \tilde{\mathbf{u}} \\ \tilde{\mathbf{p}} \\ \tilde{\boldsymbol{\varepsilon}}_v \end{Bmatrix} = \begin{Bmatrix} \mathbf{f}_1 \\ \mathbf{f}_2 \\ \mathbf{f}_3 \end{Bmatrix} \tag{11.24}$$

where $\mathbf{A}$, $\mathbf{C}$, $\mathbf{f}_1$, $\mathbf{f}_2$ are given by Eq. (11.11) and

$$\mathbf{H} = \int_\Omega \mathbf{N}_v^{\mathrm{T}} K \mathbf{N}_v \,\mathrm{d}\Omega; \qquad \mathbf{E} = \int_\Omega \mathbf{N}_v^{\mathrm{T}} \mathbf{N}_p \,\mathrm{d}\Omega; \qquad \mathbf{f}_3 = \mathbf{0} \tag{11.25}$$

For completeness we give the variational theorem whose first variation gives Eqs (11.7), (11.21) and (11.22). First we define the strain deduced from the standard displacement approximation as

$$\varepsilon_u = \boldsymbol{\mathcal{S}}\mathbf{u} \approx \mathbf{B}\tilde{\mathbf{u}} \tag{11.26}$$

The variational theorem is then given as

$$\begin{aligned} \Pi = \frac{1}{2}\int_\Omega \left(\varepsilon_u^{\mathrm{T}} \mathbf{D}_d \varepsilon_u + \varepsilon_v K \varepsilon_v\right) \mathrm{d}\Omega + \int_\Omega p\left(\mathbf{m}^{\mathrm{T}}\varepsilon_u - \varepsilon_v\right) \mathrm{d}\Omega \\ - \int_\Omega \mathbf{u}^{\mathrm{T}}\mathbf{b} \,\mathrm{d}\Omega - \int_{\Gamma_t} \mathbf{u}^{\mathrm{T}}\bar{\mathbf{t}} \,\mathrm{d}\Gamma \end{aligned} \tag{11.27}$$

Example 11.2: An enhanced strain triangle. In Example 11.1 we presented a two-field formulation using continuous **u** and p approximations together with an added hierarchical bubble mode to the displacements. For more general applications this form is not the most convenient. For example, if transient problems are considered the accelerations will also involve the bubble mode and affect the inertial terms. We will also find in the Sec. 11.7 that use of the above bubble is not fully effective in eliminating pressure oscillations in solutions. An alternative form is discussed in which we use a three-field approximation involving **u**, p and ε_v discussed above, together with an enhanced strain formulation as discussed in Sec. 10.5.3.

The enhanced strains are added to those computed from displacements as

$$\breve{\boldsymbol{\varepsilon}} = \boldsymbol{\varepsilon}_u + \boldsymbol{\varepsilon}_e \tag{11.28}$$

in which $\boldsymbol{\varepsilon}_e$ represents a set of enhanced strain terms. The internal strain energy is represented by

$$W(\breve{\boldsymbol{\varepsilon}}, \varepsilon_v) = \tfrac{1}{2}\left(\breve{\boldsymbol{\varepsilon}}^{\mathrm{T}}\mathbf{D}_d\breve{\boldsymbol{\varepsilon}} + \varepsilon_v K \varepsilon_v\right) \tag{11.29}$$

Using the above notation a Hu–Washizu-type variational theorem for the deviatoric–spherical split may be written as

$$\Pi_{HW} = \int_\Omega \left[W(\breve{\boldsymbol{\varepsilon}}, \varepsilon_v) + p\left(\mathbf{m}^{\mathrm{T}}\breve{\boldsymbol{\varepsilon}} - \varepsilon_v\right) + \boldsymbol{\sigma}^{\mathrm{T}}\left(\boldsymbol{\varepsilon}_u - \breve{\boldsymbol{\varepsilon}}\right)\right] \mathrm{d}\Omega + \Pi_{ext} \tag{11.30}$$

where Π_{ext} represents the terms associated with body and traction forces. After substitution for the mixed enhanced strain the last term simplifies to

$$\int_\Omega \boldsymbol{\sigma}^{\mathrm{T}}\left(\boldsymbol{\varepsilon}_u - \breve{\boldsymbol{\varepsilon}}\right) \mathrm{d}\Omega = -\int_\Omega \boldsymbol{\sigma}^{\mathrm{T}}\boldsymbol{\varepsilon}_e \,\mathrm{d}\Omega \tag{11.31}$$

Taking variations with respect to **u**, p, ε_v, $\boldsymbol{\varepsilon}_e$ and $\boldsymbol{\sigma}$ the principle yields the weak form

$$\begin{aligned}\delta\Pi_{HW} = &\int_\Omega \delta\mathbf{u}^{\mathrm{T}}\mathbf{B}^{\mathrm{T}}\left[\mathbf{D}_d\breve{\boldsymbol{\varepsilon}} + \mathbf{m}p\right] \mathrm{d}\Omega + \delta\Pi_{ext}\\ &+ \int_\Omega \delta\varepsilon_v\left[K\varepsilon_v - p\right] \mathrm{d}\Omega + \int_\Omega \delta p\left[\mathbf{m}^{\mathrm{T}}\breve{\boldsymbol{\varepsilon}} - \varepsilon_v\right] \mathrm{d}\Omega\\ &+ \int_\Omega \delta\boldsymbol{\varepsilon}_e^{\mathrm{T}}\left[\mathbf{D}_d\breve{\boldsymbol{\varepsilon}} + \mathbf{m}p - \boldsymbol{\sigma}\right] \mathrm{d}\Omega + \int_\Omega \delta\boldsymbol{\sigma}^{\mathrm{T}}\boldsymbol{\varepsilon}_e \,\mathrm{d}\Omega = 0\end{aligned} \tag{11.32}$$

Equal order interpolation with shape functions **N** are used to approximate **u**, p and ε_v as

$$\begin{aligned}\mathbf{u} &\approx \hat{\mathbf{u}} = \mathbf{N}\tilde{\mathbf{u}}\\ p &\approx \hat{p} = \mathbf{N}\tilde{\mathbf{p}}\\ \varepsilon_v &\approx \hat{\varepsilon}_v = \mathbf{N}\tilde{\boldsymbol{\varepsilon}}_v\end{aligned} \tag{11.33}$$

However, only approximations for **u** and p are C_0 continuous between elements. The approximation for ε_v may be discontinuous between elements. The stress $\boldsymbol{\sigma}$ in each element is assumed constant. Thus, only the approximation for $\boldsymbol{\varepsilon}_e$ remains to be constructed in such a way that the third equation in (10.41) is satisfied. For the present we shall assume that this approximation may be represented by

$$\boldsymbol{\varepsilon}_e \approx \hat{\boldsymbol{\varepsilon}}_e = \mathbf{B}_e\tilde{\boldsymbol{\alpha}}_e \tag{11.34}$$

so that the terms involving $\boldsymbol{\sigma}$ and its variation in Eq. (11.32) are zero and thus do not appear in the final discrete equations.

With the above approximations, Eq. (11.32) may be evaluated as

$$\begin{bmatrix}\mathbf{A}_{uu} & \mathbf{A}_{ue} & \mathbf{C}_u & \mathbf{0}\\ \mathbf{A}_{eu} & \mathbf{A}_{ee} & \mathbf{C}_e & \mathbf{0}\\ \mathbf{C}_u^{\mathrm{T}} & \mathbf{C}_e^{\mathrm{T}} & \mathbf{0} & -\mathbf{E}\\ \mathbf{0} & \mathbf{0} & -\mathbf{E}^{\mathrm{T}} & \mathbf{H}\end{bmatrix}\begin{Bmatrix}\tilde{\mathbf{u}}\\ \tilde{\boldsymbol{\alpha}}_e\\ \tilde{\mathbf{p}}\\ \tilde{\boldsymbol{\varepsilon}}_v\end{Bmatrix} = \begin{Bmatrix}\mathbf{f}_1\\ \mathbf{0}\\ \mathbf{f}_2\\ \mathbf{f}_3\end{Bmatrix} \tag{11.35}$$

where $\mathbf{A}_{uu} = \mathbf{A}$, $\mathbf{C}_u = \mathbf{C}$, $\mathbf{f}_i$, $\mathbf{E}$ and $\mathbf{H}$ are as defined in Eqs (11.11), (11.25) and

$$
\begin{aligned}
\mathbf{A}_{ue} &= \int_\Omega \mathbf{B}\mathbf{D}_d\mathbf{B}_e \,\mathrm{d}\Omega = \mathbf{A}_{eu}^{\mathrm{T}} \\
\mathbf{A}_{ee} &= \int_\Omega \mathbf{B}_e\mathbf{D}_d\mathbf{B}_e \,\mathrm{d}\Omega \\
\mathbf{C}_e &= \int_\Omega \mathbf{B}_e\mathbf{m}\mathbf{N} \,\mathrm{d}\Omega
\end{aligned}
\tag{11.36}
$$

Since the approximations for ε_v and ε_e are discontinuous between elements we can again perform a partial solution for $\tilde{\varepsilon}_v$ and $\tilde{\alpha}_e$ using the second and fourth row of (11.35). After eliminating these variables from the first and third equation we again, as in Example 11.1, obtain a form identical to Eq. (11.10).

As an example we consider again the 3-noded triangular element with linear approximations for $\mathbf{N}$ in terms of area coordinates L_i. We will construct enhanced strain terms from the derivatives of an assumed function.

Here we consider three enhanced functions given by

$$
N_e^i = \beta L_i + L_j L_k \tag{11.37}
$$

in which $i,\ j,\ k$ is a cyclic permutation and β is a parameter to be determined. Note that this form only involves quadratic terms and thus gives linear strains which are fully consistent with the linear interpolations for p and ε_v. The derivatives of the enhanced function are given by

$$
\begin{aligned}
\frac{\partial N_e^i}{\partial x} &= \frac{1}{2\Delta}\left[\beta b_i + L_j b_k + L_k b_j\right] \\
\frac{\partial N_e^i}{\partial y} &= \frac{1}{2\Delta}\left[\beta c_i + L_j c_k + L_k c_j\right]
\end{aligned}
\tag{11.38}
$$

where

$$
b_i = y_j - y_k \qquad \text{and} \qquad c_i = x_k - x_j
$$

and Δ is the area of a triangular element. For constant p the requirement imposed by Eq. 11.21 gives $\beta = 1/3$. The derivatives are inserted in the usual strain–displacement matrix

$$
\mathbf{B}_e^i = \begin{bmatrix} \dfrac{\partial N_e^i}{\partial x} & 0 \\ 0 & \dfrac{\partial N_e^i}{\partial y} \\ \dfrac{\partial N_e^i}{\partial y} & \dfrac{\partial N_e^i}{\partial x} \end{bmatrix} \tag{11.39}
$$

While the use of added enhanced modes leads to increased cost (over use of a simple bubble mode, as in Example 11.1) in eliminating the $\tilde{\varepsilon}_v$ and α_e parameters in Eq. (11.35) the results obtained are improved considerably, as indicated in the numerical results presented in Sec. 11.7. Furthermore, this form leads to improved consistency between the pressure and strain.

11.4.1 The B-bar method for nearly incompressible problems

The second of (11.24) has the solution

$$\tilde{\varepsilon}_v = \mathbf{E}^{-1}\mathbf{C}^{\mathrm{T}}\tilde{\mathbf{u}} = \mathbf{W}\tilde{\mathbf{u}} \tag{11.40}$$

In the above we assume that $\mathbf{E}$ may be inverted, which implies that $\mathbf{N}_v$ and $\mathbf{N}_p$ have the same number of terms. Furthermore, the approximations for the volumetric strain and pressure are constructed for each element individually and are not continuous across element boundaries. Thus, the solution of Eq. (11.40) may be performed for each individual element. In practice $\mathbf{N}_v$ is normally assumed identical to $\mathbf{N}_p$ so that $\mathbf{E}$ is symmetric positive definite. The solution of the third equation of (11.24) yields the pressure parameters in terms of the volumetric strain parameters and is given by

$$\tilde{\mathbf{p}} = \mathbf{E}^{-\mathrm{T}}\mathbf{H}\tilde{\varepsilon}_v \tag{11.41}$$

Substitution of (11.40) and (11.41) into the first of (11.24) gives a solution that is in terms of displacements only. Accordingly,

$$\bar{\mathbf{A}}\tilde{\mathbf{u}} = \mathbf{f}_1 \tag{11.42}$$

where for isotropy

$$\begin{aligned}\bar{\mathbf{A}} &= \int_\Omega \mathbf{B}^{\mathrm{T}}\mathbf{D}_d\mathbf{B}\,\mathrm{d}\Omega + \mathbf{W}^{\mathrm{T}}\mathbf{H}\mathbf{W}\\ &= \mathbf{A} + \mathbf{W}^{\mathrm{T}}\mathbf{H}\mathbf{W}\end{aligned} \tag{11.43}$$

The solution of (11.42) yields the nodal parameters for the displacements. Use of (11.40) and (11.41) then gives the approximations for the volumetric strain and pressure.

The result given by (11.43) may be further modified to obtain a form that is similar to the standard displacement method. Accordingly, we write

$$\bar{\mathbf{A}} = \int_\Omega \bar{\mathbf{B}}^{\mathrm{T}}\mathbf{D}\bar{\mathbf{B}}\,\mathrm{d}\Omega \tag{11.44}$$

where the strain–displacement matrix is now

$$\bar{\mathbf{B}} = \mathbf{I}_d\mathbf{B} + \tfrac{1}{3}\mathbf{m}\mathbf{N}_v\mathbf{W} \tag{11.45}$$

For isotropy the modulus matrix is

$$\mathbf{D} = \mathbf{D}_d + K\mathbf{m}\mathbf{m}^{\mathrm{T}} \tag{11.46}$$

We note that the above form is identical to a standard displacement model except that $\mathbf{B}$ is replaced by $\bar{\mathbf{B}}$. The method has been discussed more extensively in references 13, 14 and 15.

The equivalence of (11.43) and (11.44) can be verified by simple matrix multiplication. Extension to treat general small strain formulations can be simply performed by replacing the isotropic $\mathbf{D}$ matrix by an appropriate form for the general material model. The formulation shown above has been implemented into an element included as part of the program available on the web site. The elegance of the method is more fully utilized when considering non-linear problems, such as plasticity and finite deformation elasticity (see reference 1).

We note that elimination starting with the third equation of (11.24) could be accomplished leading to a $\mathbf{u}$–p two-field form using K as a penalty number. This is convenient for the case where p is continuous but ε_v remains discontinuous – as already discussed in Example 11.2. Such an elimination, however, points out that precisely the same stability criteria operate here as in the two-field approximation discussed earlier.

11.5 Reduced and selective integration and its equivalence to penalized mixed problems

In Chapter 5 we mentioned the lowest order numerical integration rules that still preserve the required convergence order for various elements, but at the same time pointed out the possibility of a singularity in the resulting element matrices. In Chapter 9 we again referred to such low order integration rules, introducing the name 'reduced integration' for those that did not evaluate the stiffness exactly for simple elements and pointed out some dangers of its indiscriminate use due to resulting instability. Nevertheless, such reduced integration and selective integration (where the low order integration is only applied to certain parts of the matrix) has proved its worth in practice, often yielding much more accurate results than the use of more precise integration rules. This was particularly noticeable in nearly incompressible elasticity (or Stokes fluid flow which is similar)[16–18] and in problems of plate and shell flexure dealt with as a case of a degenerate solid[19, 20] (see reference 1 for more information on plate and shell problems).

The success of these procedures derived initially by heuristic arguments proved quite spectacular – though some consider it somewhat verging on immorality to obtain improved results while doing less work! Obviously fuller justification of such processes is required.[21] The main reason for success is associated with the fact that it provides the necessary singularity of the constraint part of the matrix [viz. Eqs (10.19)–(10.21)] which avoids locking. Such singularity can be deduced from a count of integration points,[22, 23] but it is simpler to show that there is a complete equivalence between reduced (or selective) integration procedures and the mixed formulation already discussed in Sec. 11.3. This equivalence was first shown by Malkus and Hughes[24] and later in a general context by Zienkiewicz and Nakazawa.[25]

We shall demonstrate this equivalence on the basis of the nearly incompressible elasticity problem for which the mixed weak Galerkin integral statement is given by Eqs (11.7) and (11.8). It should be noted, however, that equivalence holds only for the discontinuous pressure approximation.

The corresponding irreducible form can be written by satisfying the second of Eq. (11.8) exactly, implying

$$p = K\mathbf{m}^{\mathrm{T}}\varepsilon \tag{11.47}$$

and substituting above into (11.7) as

$$\int_\Omega \delta\boldsymbol{\varepsilon}^{\mathrm{T}} 2G\left(\mathbf{I}_0 - \frac{1}{3}\mathbf{m}^{\mathrm{T}}\mathbf{m}\right)\boldsymbol{\varepsilon}\,\mathrm{d}\Omega + \int_\Omega \delta\boldsymbol{\varepsilon}^{\mathrm{T}}\mathbf{m}K\mathbf{m}^{\mathrm{T}}\boldsymbol{\varepsilon}\,\mathrm{d}\Omega \\ - \int_\Omega \delta\mathbf{u}^{\mathrm{T}}\mathbf{b}\,\mathrm{d}\Omega - \int_{\Gamma_t} \delta\mathbf{u}^{\mathrm{T}}\bar{\mathbf{t}}\,\mathrm{d}\Gamma = 0 \tag{11.48}$$

On substituting

$$\mathbf{u} \approx \hat{\mathbf{u}} = \mathbf{N}_u\tilde{\mathbf{u}} \quad \text{and} \quad \boldsymbol{\varepsilon} \approx \hat{\boldsymbol{\varepsilon}} = \boldsymbol{\mathcal{S}}\mathbf{N}_u\tilde{\mathbf{u}} = \mathbf{B}\tilde{\mathbf{u}} \tag{11.49}$$

we have

$$\left(\mathbf{A} + \bar{\mathbf{A}}\right)\tilde{\mathbf{u}} = \mathbf{f}_1 \tag{11.50}$$

where $\mathbf{A}$ and $\mathbf{f}_1$ are exactly as given in Eq. (11.11) and

$$\bar{\mathbf{A}} = \int_\Omega \mathbf{B}^{\mathrm{T}}\mathbf{m}K\mathbf{m}^{\mathrm{T}}\mathbf{B}\,\mathrm{d}\Omega \tag{11.51}$$

The solution of Eq. (11.50) for $\tilde{\mathbf{u}}$ allows the pressures to be determined at all points by Eq. (11.47). In particular, if we have used an integration scheme for evaluating (11.51) which samples at points (ξ_k) we can write

$$p(\xi_k) = K\mathbf{m}^{\mathrm{T}}\boldsymbol{\varepsilon}(\xi_k) = K\mathbf{m}^{\mathrm{T}}\mathbf{B}(\xi_k)\tilde{\mathbf{u}} = \sum_a N_{pa}(\xi_k)\tilde{p}_a \tag{11.52}$$

Now if we turn our attention to the penalized mixed form of Eqs (11.7)–(11.11) we note that the second of Eq. (11.10) is explicitly

$$\int_\Omega \mathbf{N}_p^{\mathrm{T}}\left(\mathbf{m}^{\mathrm{T}}\mathbf{B}\tilde{\mathbf{u}} - \frac{1}{K}\mathbf{N}_p\tilde{\mathbf{p}}\right)\mathrm{d}\Omega = \mathbf{0} \tag{11.53}$$

If a numerical integration is applied to the above sampling at the pressure nodes located at coordinate (ξ_l), previously defined in Eq. (11.52), we can write for each scalar component of $\mathbf{N}_p$

$$\sum_l N_{pa}(\xi_l)\left(\mathbf{m}^{\mathrm{T}}\mathbf{B}(\xi_l)\tilde{\mathbf{u}} - \frac{1}{K}\mathbf{N}_p(\xi_l)\tilde{\mathbf{p}}\right)w_l = 0 \tag{11.54}$$

in which the summation is over all integration points (ξ_l) and W_l are the appropriate weights and jacobian determinants. Now as

$$N_{pa}(\xi_l) = \delta_{al}$$

if ξ_l is located at the pressure node a and zero at other pressure nodes, Eq. (11.54) reduces simply to the requirement that at all pressure nodes

$$\mathbf{m}^{\mathrm{T}}\mathbf{B}(\xi_l)\tilde{\mathbf{u}} = \frac{1}{K}\mathbf{N}_p(\xi_l)\tilde{\mathbf{p}} \tag{11.55}$$

This is precisely the same condition as that given by Eq. (11.52) and the equivalence of the procedures is proved, *providing the integrating scheme used for evaluating* $\bar{\mathbf{A}}$ *gives an identical integral of the mixed form of Eq. (11.53).*

This is true in many cases and for these the reduced integration-mixed equivalence is exact. In all other cases this equivalence exists for a mixed problem in which an inexact rule of integration has been used in evaluating equations such as (11.53).

For curved isoparametric elements the equivalence is in fact inexact, and slightly different results will be obtained using reduced integration and mixed forms. This is illustrated in examples given in reference 26.

We can conclude without detailed proof that this type of equivalence is quite general and that with any problem of a similar type the application of numerical quadrature at n_p points in evaluating the matrix $\bar{\mathbf{A}}$ within each element is equivalent to a mixed problem in which the variable p is interpolated element by element using as p-nodal values the same integrating points.

The equivalence is only complete for the selective integration process, i.e., application of reduced numerical quadrature only to the matrix $\bar{\mathbf{A}}$, and ensures that this matrix is singular, i.e., no locking occurs if we have satisfied the previously stated conditions ($n_u > n_p$).

The full use of reduced integration on the remainder of the matrix determining $\tilde{\mathbf{u}}$, i.e., $\mathbf{A}$, is only permissible if that remains non-singular – the case which we have discussed previously for the Q8/4 element.

It can therefore be concluded that all the elements with discontinuous interpolation of p which we have verified as applicable to the mixed problem (viz. Fig. 11.1, for instance) can be implemented for nearly incompressible situations by a penalized irreducible form using corresponding selective integration.†

In Fig. 11.6 we show an example which clearly indicates the improvement of displacements achieved by such reduced integration as the compressibility modulus K increases (or the Poisson ratio tends to 0.5). We note also in this example the dramatically improved performance of such points for stress sampling.

For problems in which the p (constraint) variable is continuously interpolated (C_0) the arguments given above fail as quantities such as $\mathbf{m}^{\mathrm{T}}\varepsilon$ are not interelement continuous in the irreducible form.

A very interesting corollary of the equivalence just proved for (nearly) incompressible behaviour is observed if we note the rapid increase of order of integrating formulae with the number of quadrature points (viz. Chapter 5). *For high order elements the number of quadrature points equivalent to the p constraint permissible for stability rapidly reaches that required for exact integration and hence their performance in nearly incompressible situations is excellent, even if exact integration is used.* This was observed on many occasions[27–29] and Sloan and Randolf[30] have shown good performance with the quintic triangle. Unfortunately such high order elements pose other difficulties and are seldom used in practice.

A final remark concerns the use of 'reduced' integration in particular and of penalized, mixed, methods in general. As we have pointed out in Sec. 10.3.1 it is possible in such forms to obtain sensible results for the *primary variable* (**u** in the present example) even though the general stability conditions are violated, providing some of the *constraint equations* are linearly dependent. Now of course the *constraint variable* (p in the present example) is not determinate in the limit.

This situation occurs with some elements that are occasionally used for the solution of incompressible problems but which do not pass our mixed patch test, such as Q8/4 and Q9/4 of Fig. 11.1. If we take the latter number to correspond to the integrating points these will yield acceptable **u** fields, though not p.

†The Q9/3 element would involve three-point quadrature which is somewhat unnatural for quadrilaterals. It is therefore better to simply use the mixed form here – and, indeed, in any problem which has non-linear behaviour between p and **u** (see reference 1).

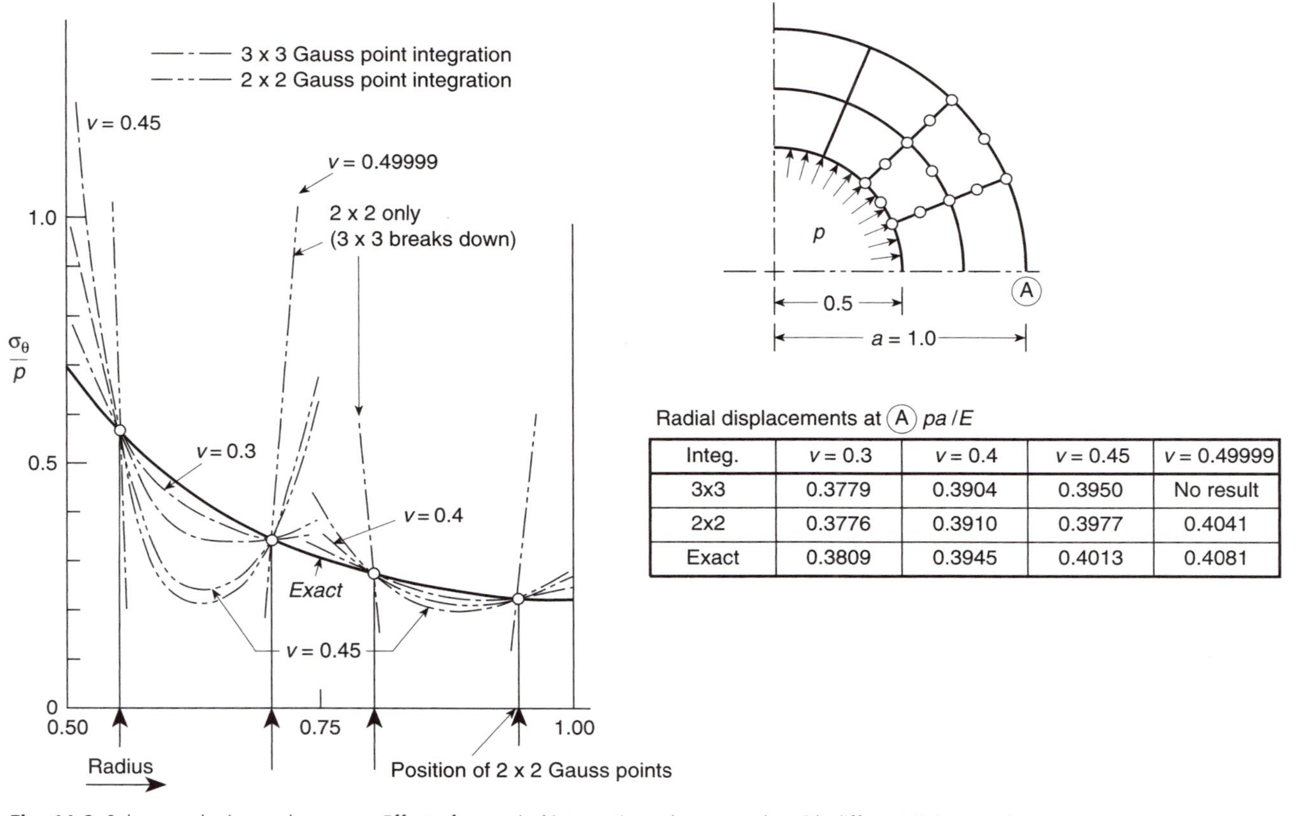

Radial displacements at (A) pa/E

Integ.	$v = 0.3$	$v = 0.4$	$v = 0.45$	$v = 0.49999$
3x3	0.3779	0.3904	0.3950	No result
2x2	0.3776	0.3910	0.3977	0.4041
Exact	0.3809	0.3945	0.4013	0.4081

Fig. 11.6 Sphere under internal pressure. Effect of numerical integration rules on results with different Poisson ratios.

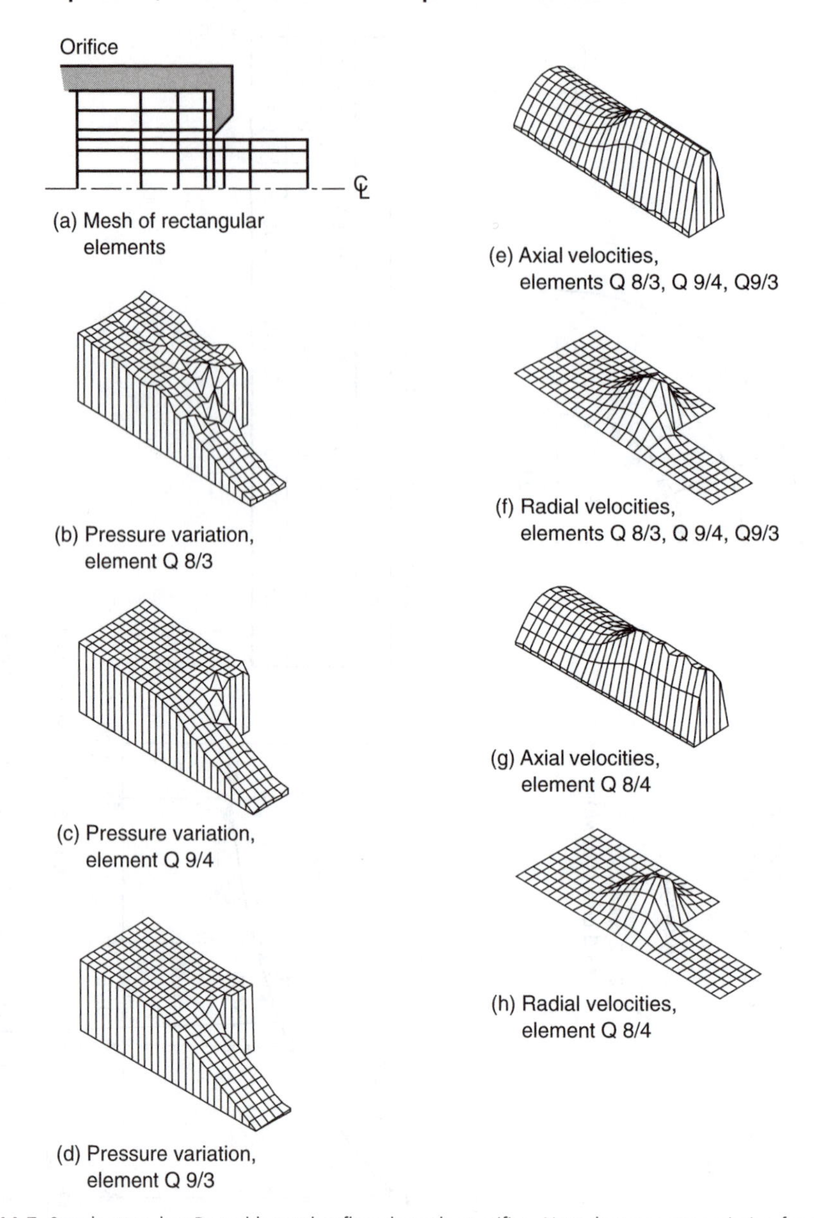

Fig. 11.7 Steady-state, low Reynolds number flow through an orifice. Note that pressure variation for element Q8/4 is so large it cannot be plotted. Solution with **u**/*p* elements Q8/3, Q8/4, Q9/3, Q9/4.

Figure 11.7 illustrates the point on an application involving slow viscous flow through an orifice – a problem that obeys identical equations to those of incompressible elasticity. Here elements Q8/4, Q8/3, Q9/4 and Q9/3 are compared although only the last completely satisfies the stability requirements of the mixed patch test. All elements are found to give a reasonable velocity (**u**) field but pressures are acceptable only for the last one, with element Q8/4 failing to give results which can be plotted.[4]

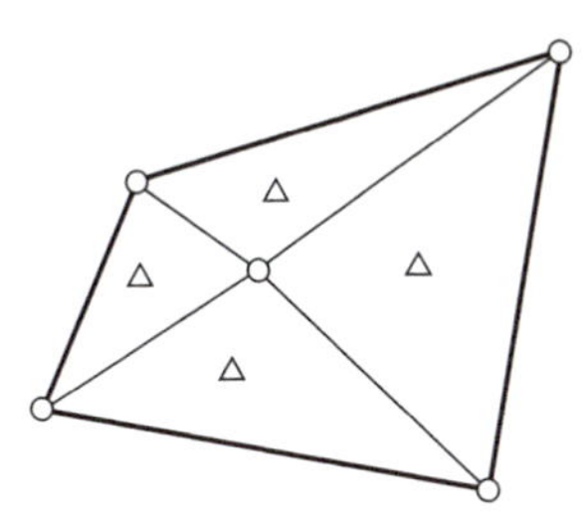

Fig. 11.8 A quadrilateral with intersecting diagonals forming an assembly of four T3/1 elements. This allows displacements to be determined for nearly incompressible behaviour but does not yield pressure results.

It is of passing interest to note that a similar situation develops if four triangles of the T3/1 type are assembled to form a quadrilateral in the manner of Fig. 11.8. Although the original element locks, as we have previously demonstrated, a linear dependence of the constraint equation allows the assembly to be used quite effectively in many incompressible situations, as shown in reference 31.

Example 11.3: A weak patch test – selective integration. In order to illustrate the performance of an element which only satisfies a weak patch test we consider an axisymmetric linear elastic problem modelled by 4-noded isoparametric elements. The material is assumed isotropic and the finite element stiffness and reaction force matrices are computed using a selective integration method where terms associated with the bulk modulus are evaluated by a single-point Gauss quadrature, whereas all other terms are computed using a 2×2 (standard) gaussian quadrature. It may be readily verified that the stiffness matrix is of proper rank and thus stability of solutions is not an issue. On the other hand, consistency must still be evaluated.

In order to assess the performance of a selective reduced quadrature formulation we consider the patch of elements shown in Fig. 11.9. The patch is not as generally shaped as desirable and is only used to illustrate performance of an element that satisfies a weak patch test. The polynomial solution considered is

$$\begin{aligned} u &= 2r \\ v &= 0 \end{aligned} \tag{11.56}$$

and material constants $E = 1$ and $\nu = 0$ are used in the analysis. The resulting stress field is given by

$$\sigma_r = \sigma_\theta = 2 \tag{11.57}$$

with other components identically zero. The exact solution for the nodal quantities of the mesh shown in Fig. 11.9 are summarized in Table 11.1. Patch tests have been performed for this problem using the selective reduced integration scheme described above and values of h of 0.8, 0.4, 0.2, 0.1, and 0.05. The result for the radial displacement at nodes 2 and 5 (reported to six digits) is given in Table 11.2. All other quantities (displacements, strains, and stresses) have a similar performance with convergence rates of at least $O(h)$ or more. Based on this assessment we conclude the element passes a weak patch test. A similar result will be found for elements which are not rectangular and thus the element produces convergent results.

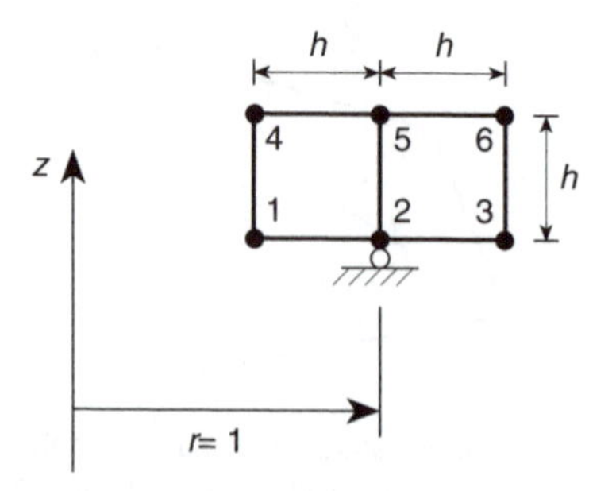

Fig. 11.9 Patch for selective, reduced quadrature on axisymmetric 4-noded elements.

Table 11.1 Exact solution for patch

Node a	Radius r_a	Displacement $\tilde{u}_a$	Displacement $\tilde{v}_a$	Force F_{ra}	Force F_{za}
1, 4	$1-h$	$2(1-h)$	0	$-(1-h)h$	0
2, 5	1	2	0	0	0
3, 6	$1+h$	$2(1+h)$	0	$(1+h)h$	0

Table 11.2 Radial displacement at nodes 2 and 5

h	$\hat{u}$
0.8	2.01114
0.4	2.00049
0.2	2.00003
0.1	2.00000
0.05	2.00000

11.6 A simple iterative solution process for mixed problems: Uzawa method

11.6.1 General

In the general remarks on the algebraic solution of mixed problems characterized by equations of the type [viz. Eq. (10.14)]

$$\begin{bmatrix} \mathbf{A} & \mathbf{C} \\ \mathbf{C}^{\mathrm{T}} & \mathbf{0} \end{bmatrix} \begin{Bmatrix} \mathbf{x} \\ \mathbf{y} \end{Bmatrix} = \begin{Bmatrix} \mathbf{f}_1 \\ \mathbf{f}_2 \end{Bmatrix} \tag{11.58}$$

we have remarked on the difficulties posed by the zero diagonal and the increased number of unknowns $(n_x + n_y)$ as compared with the irreducible form (n_x or n_y).

A general iterative form of solution is possible, however, which substantially reduces the cost.[32] In this we solve successively

$$\mathbf{y}^{(k+1)} = \mathbf{y}^{(k)} + \rho \mathbf{r}^{(k)} \tag{11.59}$$

where $\mathbf{r}^{(k)}$ is the residual of the second equation computed as

$$\mathbf{r}^{(k)} = \mathbf{C}^{\mathrm{T}} \mathbf{x}^{(k)} - \mathbf{f}_2 \tag{11.60}$$

and follow with solution of the first equation, i.e.,

$$\mathbf{x}^{(k+1)} = \mathbf{A}^{-1}(\mathbf{f}_1 - \mathbf{C}\mathbf{y}^{(k+1)}) \tag{11.61}$$

In the above ρ is a 'convergence accelerator matrix' and is chosen to be efficient and simple to use.

The algorithm is similar to that described initially by Uzawa[33] and has been widely applied in an optimization context.[28, 34–38]

Its relative simplicity can best be grasped when a particular example is considered.

11.6.2 Iterative solution for incompressible elasticity

In this case we start from Eq. (11.10) now written with $\mathbf{V} = \mathbf{0}$, i.e., complete incompressibility is assumed. The various matrices are defined in (11.11), resulting in the form

$$\begin{bmatrix} \mathbf{A} & \mathbf{C} \\ \mathbf{C}^{\mathrm{T}} & \mathbf{0} \end{bmatrix} \begin{Bmatrix} \tilde{\mathbf{u}} \\ \tilde{\mathbf{p}} \end{Bmatrix} = \begin{Bmatrix} \mathbf{f}_1 \\ \mathbf{0} \end{Bmatrix} \tag{11.62}$$

Now, however, for three-dimensional problems the matrix $\mathbf{A}$ is singular (as volumetric changes are not restrained) and it is necessary to *augment* it to make it non-singular. We can do this in the manner described in Sec. 10.3.1, or equivalently by the addition of a fictitious compressibility matrix, thus replacing $\mathbf{A}$ by

$$\bar{\mathbf{A}} = \mathbf{A} + \int_{\Omega} \mathbf{B}^{\mathrm{T}}(\lambda G \mathbf{m}\mathbf{m}^{\mathrm{T}})\mathbf{B}\, \mathrm{d}\Omega \tag{11.63}$$

If the second matrix uses an integration consistent with the number of discontinuous pressure parameters assumed, then this is precisely equivalent to writing

$$\bar{\mathbf{A}} = \mathbf{A} + \lambda G \mathbf{C}\mathbf{C}^{\mathrm{T}} \tag{11.64}$$

and is simpler to evaluate. Clearly this addition does not change the equation system.

The iteration of the algorithm (11.59)–(11.61) is now conveniently taken with the 'convergence accelerator' being simply defined as

$$\rho = \lambda G \mathbf{I} \tag{11.65}$$

We now have the iterative system given as

$$\tilde{\mathbf{p}}^{(k+1)} = \tilde{\mathbf{p}}^{(k)} + \lambda G \mathbf{r}^{(k)} \tag{11.66}$$

where

$$\mathbf{r}^{(k)} = \mathbf{C}^{\mathrm{T}}\tilde{\mathbf{u}}^{(k)} \tag{11.67}$$

the residual of the incompressible constraint, and

$$\tilde{\mathbf{u}}^{(k+1)} = \bar{\mathbf{A}}^{-1}(\mathbf{f}_1 - \mathbf{C}\tilde{\mathbf{p}}^{(k+1)}) \tag{11.68}$$

In this $\bar{\mathbf{A}}$ can be interpreted as the stiffness matrix of a compressible material with bulk modulus $K = \lambda G$ and the process may be interpreted as the successive addition of volumetric 'initial' strains designed to reduce the volumetric strain to zero. Indeed this simple approach led to the first realization of this algorithm.[39–41] Alternatively the process can be visualized as an amendment of the original equation (11.62) by subtracting the term $\mathbf{p}/(\lambda G)$ from each side of the second to give (this is often called an *augmented lagrangian form*)[32, 37, 38]

$$\begin{bmatrix} \mathbf{A} & \mathbf{C} \\ \mathbf{C}^{\mathrm{T}} & -\dfrac{1}{\lambda G}\mathbf{I} \end{bmatrix} \begin{Bmatrix} \tilde{\mathbf{u}} \\ \tilde{\mathbf{p}} \end{Bmatrix} = \begin{Bmatrix} \mathbf{f}_1 \\ -\dfrac{1}{\lambda G}\tilde{\mathbf{p}} \end{Bmatrix} \tag{11.69}$$

and adopting the iteration

$$\begin{bmatrix} \bar{\mathbf{A}} & \mathbf{C} \\ \mathbf{C}^{\mathrm{T}} & -\dfrac{1}{\lambda G}\mathbf{I} \end{bmatrix} \begin{Bmatrix} \tilde{\mathbf{u}} \\ \tilde{\mathbf{p}} \end{Bmatrix}^{(k+1)} = \begin{Bmatrix} \mathbf{f}_1 \\ -\dfrac{1}{\lambda G}\tilde{\mathbf{p}}^{(k)} \end{Bmatrix} \tag{11.70}$$

With this, on elimination, a sequence similar to Eqs (11.66)–(11.68) will be obtained provided $\bar{\mathbf{A}}$ is defined by Eq. (11.64).

Starting the iteration from

$$\tilde{\mathbf{u}}^{(0)} = \mathbf{0} \quad \text{and} \quad \tilde{\mathbf{p}}^{(0)} = \mathbf{0}$$

in Fig. 11.10 we show the convergence of the maximum div $\mathbf{u}$ computed at any of the integrating points used. We note that this convergence becomes quite rapid for large values of $\lambda = (10^3–10^4)$.

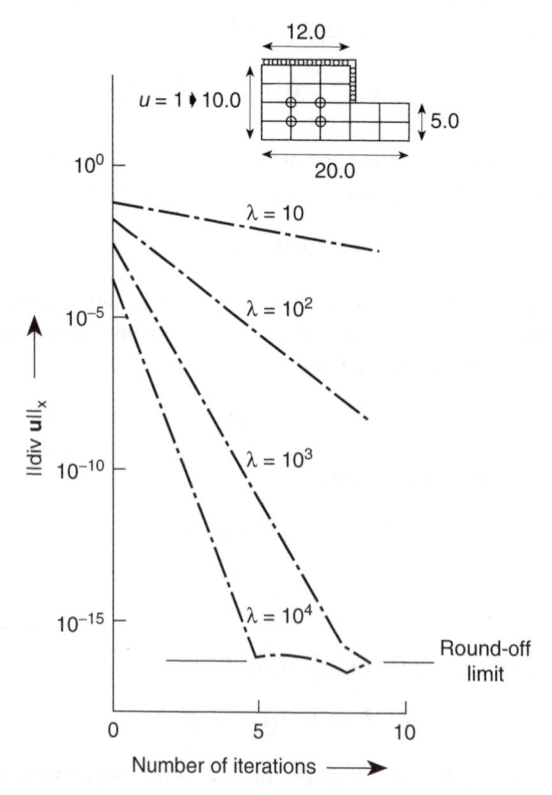

Fig. 11.10 Convergence of iterations in an extrusion problem for different values of the parameter λ.

For smaller λ values the process can be accelerated by using different ρ[32] but for practical purposes the simple algorithm shown above suffices for many problems, including applications in large strain.[42] Clearly much better satisfaction of the incompressibility constraint can now be obtained by the simple use of a 'large enough' bulk modulus or penalty parameter. With $\lambda = 10^4$, for instance, in five iterations the initial div $\mathbf{u}$ is reduced from the value $\sim 10^{-4}$ to 10^{-16}, which is at the round-off limit of the particular computer used.

Finally, we remind the reader that the above iterative process solves the equations of a mixed problem. Accordingly, it is fully effective only when the element used satisfies the stability and consistency conditions of the mixed patch test.

11.7 Stabilized methods for some mixed elements failing the incompressibility patch test

It has been observed earlier in this chapter that many of the two-field $\mathbf{u}$–p elements do not pass the stability conditions imposed by the mixed patch test at the incompressible limit (or the Babuška–Brezzi conditions). Here in particular we have such methods in which the displacement and pressure are interpolated in an identical manner (for instance, linear triangles, linear quadrilaterals, quadratic triangles, etc.) and many attempts for *stabilization* of such elements have been introduced. Indeed one may view the bubble introduced in Example 11.1 and the enhanced strain treatment of Example 11.2 as stabilized methods. However, several alternative categories to these exist. The first category is the introduction of non-zero diagonal terms of the constraint equation by adding a least squares form to the Galerkin formulation. This was first suggested by Courant[43] as a means of improving accuracy in solutions. It appears that Brezzi and Pitkaranta in 1984[44] were the first to add terms to the Galerkin solution in an attempt to stabilize results. Numerous further suggestions have been proposed by Hughes *et al.* between 1986 and 1989 with the final form again a least squares approach called the *Galerkin least squares method*.[45–47] An alternative proposal of achieving similar answers has been proposed by Oñate[48] which gains the addition of diagonal terms by the introduction of so-called *finite increment calculus* to the formulation. More recently, a very simple stabilization has been proposed by Dohrmann and Bochev[49] in which a stabilization involving the difference between the interpolated pressure and a direct projection of pressure is appended to the Galerkin equations in a least squares form.

There is, however, an alternative possibility introduced by time integration of the full incompressible formulation. Here many of the algorithms will yield, when steady-state conditions are recovered, a stabilized form. A number of such algorithms have been discussed by Zienkiewicz and Wu in 1991[50] and a very efficient method has appeared as a by-product of a fluid mechanics algorithm named the *characteristic-based split* (CBS) procedure[51–55] (which is discussed at length in reference 56).

In the latter algorithm there exists a free parameter. This parameter depends on the size of the time increment. In the other methods there is a weighting parameter applied to the additional terms introduced. We shall discuss each of these algorithms in the following subsections and compare the numerical results obtainable.

One may question, perhaps, that resort to stabilization procedures is not worthwhile in view of the relative simplicity of the full mixed form. But this is a matter practice will decide and is clearly in the hands of the analyst applying the necessary solutions.

11.7.1 Laplacian pressure stabilization

In the first part of this chapter we separated the stress into the deviatoric and pressure components as

$$\boldsymbol{\sigma} = \boldsymbol{\sigma}^d + \mathbf{m}p$$

Using the tensor form described in Appendix B this may be written in index form as

$$\sigma_{ij} = \sigma_{ij}^d + \delta_{ij} p$$

The deviatoric stresses are related to the deviatoric strains through the relation

$$\sigma_{ij}^d = 2G\varepsilon_{ij}^d = G\left(\frac{\partial u_i}{\partial x_j} + \frac{\partial u_j}{\partial x_i} - \frac{2}{3}\delta_{ij}\frac{\partial u_k}{\partial x_k}\right) \tag{11.71}$$

The equilibrium equations (in the absence of inertial forces) are:

$$\frac{\partial \sigma_{ij}^d}{\partial x_i} + \frac{\partial p}{\partial x_j} + b_j = 0$$

Substituting the constitutive equations for the deviatoric part yields the equilibrium form (assuming G is constant)

$$G\left[\frac{\partial^2 u_j}{\partial x_i\,\partial x_i} + \frac{1}{3}\frac{\partial^2 u_i}{\partial x_i\,\partial x_j}\right] + \frac{\partial p}{\partial x_j} + b_j = 0 \tag{11.72}$$

In vector form this is given as

$$G[\nabla^2\mathbf{u} + \tfrac{1}{3}\boldsymbol{\nabla}(\text{div } \mathbf{u})] + \boldsymbol{\nabla} p + \mathbf{b} = \mathbf{0}$$

where ∇^2 is the laplacian operator and $\boldsymbol{\nabla}$ the gradient operator. The constitutive equation (11.2) is expressed in terms of the displacement as

$$\varepsilon_v = \frac{\partial u_i}{\partial x_i} = \text{div } \mathbf{u} = \frac{1}{K} p \tag{11.73}$$

where div $(\cdot)$ is the divergence of the quantity. A single equation for pressure may be deduced from the divergence of the equilibrium equation. Accordingly, from Eq. (11.72) we obtain

$$\frac{4G}{3}\nabla^2\,(\text{div } \mathbf{u}) + \nabla^2 p + \text{div } \mathbf{b} = 0 \tag{11.74}$$

where upon noting (11.73) we obtain finally

$$\left(1 + \frac{4G}{3K}\right)\nabla^2 p + \text{div } \mathbf{b} = 0 \tag{11.75}$$

Thus, in general, the pressure must satisfy a Poisson equation, or in the absence of body forces, a Laplace equation.

We have noted the dangers of artificially raising the order of the differential equation in introducing spurious solutions; however, in the context of constructing approximate solutions to the incompressible problem the above is useful in providing additional terms to the weak form which otherwise would be zero. Brezzi and Pitkaranta[44] suggested adding Eq. (11.75) to Eq. (11.8) and (on setting the body force to zero for simplicity) obtain

$$\int_{\Omega} \delta p \left(\mathbf{m}^{\mathrm{T}} \boldsymbol{\varepsilon} - \frac{1}{K} p \right) \mathrm{d}\Omega + \sum_e \beta \int_{\Omega_e} \delta p \nabla^2 p \, \mathrm{d}\Omega = 0 \tag{11.76}$$

where β is a parameter introduced to control accuracy. The last term may be integrated by parts to yield a form which is more amenable to computation as

$$\int_{\Omega} \delta p \left(\mathbf{m}^{\mathrm{T}} \boldsymbol{\varepsilon} - \frac{1}{K} p \right) \mathrm{d}\Omega - \sum_e \beta \int_{\Omega_e} \frac{\partial \delta p}{\partial x_i} \frac{\partial p}{\partial x_i} \mathrm{d}\Omega = 0 \tag{11.77}$$

in which the resulting boundary terms are ignored. Upon discretization using equal order linear interpolation on triangles for $\mathbf{u}$ and p we obtain a form identical to that for the bubble in Example 11.1 with the exception that $\boldsymbol{\tau}$ [viz. Eq. (11.20b)] is now given by

$$\boldsymbol{\tau} = \beta \mathbf{I} \tag{11.78}$$

On dimensional considerations with the first term in Eq. (11.77) the parameter β should have a value proportional to L^4/F, where L is length and F is force. We defer discussion on the particular value until after presenting the Galerkin least squares method.

11.7.2 Galerkin least squares method

The Galerkin least squares (GLS) approach is a general scheme for solving the differential equations (3.1) by a finite element method. We may write the GLS form as

$$\int_{\Omega} \delta \mathbf{u}^{\mathrm{T}} \mathcal{A}(\mathbf{u}) \, \mathrm{d}\Omega + \sum_e \int_{\Omega_e} \delta \mathcal{A}(\mathbf{u})^{\mathrm{T}} \boldsymbol{\tau} \mathcal{A}(\mathbf{u}) \, \mathrm{d}\Omega = 0 \tag{11.79}$$

where the first term represents the normal Galerkin form and the added terms are computed for each element individually including a weight $\boldsymbol{\tau}$ to provide dimensional balance and scaling. Generally, the $\boldsymbol{\tau}$ will involve parameters which have to be selected for good performance. Discontinuous terms on boundaries between elements that arise from higher order terms in $\mathcal{A}(\mathbf{u})$ are commonly omitted.

The form given above has been used by Hughes[47] as a means of stabilizing the fluid flow equations, which for the case of the incompressible Stokes problem coincide with those for incompressible linear elasticity. For this problem only the momentum equation is used in the least squares terms. After substituting Eq. (11.73) into Eq. (11.72) the momentum equation may be written as (assuming that G and K are constant in each element and body forces are ignored)

$$G \frac{\partial^2 u_j}{\partial x_i^2} + \left(1 + \frac{G}{3K} \right) \frac{\partial p}{\partial x_j} = 0 \tag{11.80}$$

A more convenient form results by using a single parameter defined as

$$\bar{G} = \frac{G}{1 + G/3K} \tag{11.81}$$

With this form the least squares term to be appended to each element may be written as

$$\int_{\Omega_e} \left(\bar{G}\frac{\partial^2 \delta u_i}{\partial x_k^2} + \frac{\partial \delta p}{\partial x_i} \right) \tau_{ij} \left(\bar{G}\frac{\partial^2 u_j}{\partial x_m^2} + \frac{\partial p}{\partial x_j} \right) \mathrm{d}\Omega \tag{11.82}$$

This leads to terms to be added to the standard Galerkin equations and is expressed as

$$\begin{bmatrix} \mathbf{A}^s & \mathbf{C}^s \\ \mathbf{C}^{s,T} & \mathbf{V}^s \end{bmatrix} \begin{Bmatrix} \tilde{\mathbf{u}} \\ \tilde{\mathbf{p}} \end{Bmatrix}$$

where

$$\mathbf{A}^s_{ab} = \int_{\Omega_e} \bar{G}^2 \nabla^2 N_a \boldsymbol{\tau} \nabla^2 N_b \, \mathrm{d}\Omega$$

$$\mathbf{C}^s_{ab} = \int_{\Omega_e} \bar{G} \nabla^2 N_a \boldsymbol{\tau} \boldsymbol{\nabla} N_b \, \mathrm{d}\Omega$$

$$V^s_{ab} = \int_{\Omega_e} (\boldsymbol{\nabla} N_a)^{\mathrm{T}} \boldsymbol{\tau} \boldsymbol{\nabla} N_b \, \mathrm{d}\Omega$$

and the operators on the shape functions are given in two dimensions by

$$\nabla^2 N_a = \frac{\partial^2 N_a}{\partial x_1^2} + \frac{\partial^2 N_a}{\partial x_2^2} \quad \text{and} \quad \boldsymbol{\nabla} N_a = \begin{bmatrix} \dfrac{\partial N_a}{\partial x_1} & \dfrac{\partial N_a}{\partial x_2} \end{bmatrix}^{\mathrm{T}}$$

Note again that all *infinite* terms between elements are ignored (i.e., those arising from second derivatives when C_0 functions are used).

For linear triangular elements the second derivatives of the shape functions are identically zero within the element and only the **V** term remains and is now nearly identical to the Brezzi–Pitkaranta form if β coincides with the definition of $\boldsymbol{\tau}$. In the work of Hughes *et al.*, $\boldsymbol{\tau}$ is given by

$$\boldsymbol{\tau} = -\frac{\alpha h^2}{2G}\mathbf{I} \tag{11.83}$$

where α is a parameter which is recommended to be of $O(1)$ for linear triangles and quadrilaterals and h is the size of the element.

11.7.3 Direct pressure stabilization

In the previous two sections we have discussed the procedures which needed certain disregard for consistency to be introduced. In particular, in both methodologies certain integrals were allowed over the individual elements with high order derivatives and interelement values being omitted especially if these reached infinity, as happens for instance in the GLS method when second derivatives on the interface between elements or boundary terms in the Brezzi–Pitkaranta method are ignored.

In this section we introduce another process proposed in a recent paper by Dohrmann and Bochev[49] which seems to be totally correct and is arrived at without ignoring any terms in the overall integrals. In this procedure we try to ensure that the difference between the C_0 interpolated pressures gives answers consistent with those in which a lower order, discontinuous approximation is used – i.e., one which is consistent with the general approximation for stresses. Thus, for instance, in triangular elements in which linear displacements are used the stresses are only allowed to be constant within any element and the assumption of any component being also linear is not consistent. For this reason the method looks at the difference between the interpolated pressure which is of the same order as the displacement and its projection onto one order lower expansion consistent with that of the stresses.

The work of Dohrmann and Bochev considers a two-field mixed approximation given by

$$\begin{aligned}
&\int_\Omega \delta\boldsymbol{\varepsilon}^{\mathrm{T}}\mathbf{D}_d\boldsymbol{\varepsilon}\,\mathrm{d}\Omega + \int_\Omega \delta\boldsymbol{\varepsilon}^{\mathrm{T}}\mathbf{m}p\,\mathrm{d}\Omega - \int_\Omega \delta\mathbf{u}^{\mathrm{T}}\mathbf{b}\,\mathrm{d}\Omega - \int_{\Gamma_t} \delta\mathbf{u}^{\mathrm{T}}\bar{\mathbf{t}}\,\mathrm{d}\Gamma = 0 \\
&\int_\Omega \delta p\left[\mathbf{m}\boldsymbol{\varepsilon} - \frac{1}{K}p\right]\mathrm{d}\Omega - \sum_e \int_{\Omega_e} (\delta p - \delta\breve{p})\,\frac{\alpha}{G}\,(p - \breve{p})\,\mathrm{d}\Omega = 0
\end{aligned} \tag{11.84}$$

in which the displacements $\mathbf{u}$ and pressure p are approximated by k order continuous polynomial shape functions, $\breve{p}$ is a discontinuous projection of p onto a polynomial space of order $k-1$ and α is a parameter to be selected for stability. When $K \to \infty$ the above form represents a stable approximation for the incompressible problem for all order of elements provided α is set at a nominal value. In the examples we use $\alpha = 2$.†

We note that the form (11.84) requires no integration by parts in which terms are ignored. Thus, the method has considerable theoretical advantages over the previously discussed stabilization methods.

The pressure stabilization is computed for *each element* individually using

$$\int_{\Omega_e} \delta\breve{p}(p - \breve{p})\,\mathrm{d}\Omega = 0 \tag{11.85}$$

and, thus, has low additional cost. Due to this form, which also holds when the variation is interchanged on pressures, the stabilization term may also be written as

$$\int_{\Omega_e} (\delta p - \delta\breve{p})\,\frac{\alpha}{G}\,(p - \breve{p})\,\mathrm{d}\Omega = \int_{\Omega_e} \frac{\alpha}{G}\,(\delta p p - \delta\breve{p}\breve{p})\,\mathrm{d}\Omega \tag{11.86}$$

which is now in the form of a difference of two 'mass'-type arrays. If we approximate the pressure by

$$p \approx \hat{p} = \sum_a N_a\,\tilde{p}_a = \mathbf{N}\tilde{\mathbf{p}}$$

in which N_a contain the set of polynomials of order k and

$$\breve{p} = \sum_b h_b(\mathbf{x})\,\beta_b = \mathbf{h}(\mathbf{x})\,\boldsymbol{\beta}$$

where $h_b(\mathbf{x})$ are the polynomials of order $k-1$ the solution to (11.85) is determined from

$$\begin{aligned}
\int_{\Omega_e} \mathbf{h}^{\mathrm{T}}\mathbf{h}\,\mathrm{d}\Omega\boldsymbol{\beta} &= \int_{\Omega_e} \mathbf{h}^{\mathrm{T}}\mathbf{N}\,\mathrm{d}\Omega\tilde{\mathbf{p}} \\
\mathbf{H}\,\boldsymbol{\beta} &= \mathbf{G}\tilde{\mathbf{p}}
\end{aligned}$$

Thus, the pressure projection is given by

$$\breve{p} = \mathbf{h}(\mathbf{x})\mathbf{H}^{-1}\mathbf{G}\tilde{\mathbf{p}} \tag{11.87}$$

† Reference 49 uses $\alpha = 1$. While this leads to convergence we find this value is somewhat small for our examples.

Using the usual finite element approximation for the displacement

$$\mathbf{u} = \sum_a N_a \, \tilde{\mathbf{u}}_a = \mathbf{N}\tilde{\mathbf{u}}$$

the stabilized weak form may be written in matrix form as

$$\begin{bmatrix} \mathbf{K}_d & \mathbf{C} \\ \mathbf{C}^T & -\mathbf{V} \end{bmatrix} \begin{Bmatrix} \tilde{\mathbf{u}} \\ \tilde{\mathbf{p}} \end{Bmatrix} = \begin{Bmatrix} \mathbf{f} \\ \mathbf{0} \end{Bmatrix} \tag{11.88}$$

where the arrays are given by

$$\mathbf{K}_d = \int_\Omega \mathbf{B}^{\mathrm{T}} \mathbf{D}_d \mathbf{B} \, \mathrm{d}\Omega; \quad \mathbf{C} = \int_\Omega \mathbf{B}^{\mathrm{T}} \mathbf{m} \mathbf{N} \, \mathrm{d}\Omega$$

$$\mathbf{V} = \int_\Omega \mathbf{N}^{\mathrm{T}} \left[\frac{1}{K} + \frac{\alpha}{G} \right] \mathbf{N} \, \mathrm{d}\Omega - \frac{\alpha}{G} \mathbf{G}^{\mathrm{T}} \mathbf{H}^{-1} \mathbf{G}$$

and $\mathbf{f}$ is the usual force due to boundary traction and body loads. It is clear from the definition of $\mathbf{V}$ that, when G/α is much smaller than K, the effect of the direct pressure stabilization is a *penalty* form on the difference of the interpolated and projected pressure. The patch test only requires pressures of order $k-1$ (i.e., the order of the projected pressures) to satisfy the consistency condition. This partly explains why the above approach is successful. Further validation is provided by the numerical experiments given below.

Example 11.4: Direct stabilization for 3-node triangular element. As an example consider the problem of the two-dimensional plane strain problem in which the solution is performed using linear triangles ($k = 1$) with shape functions given by

$$N_a = L_a; \quad a = 1, 2, 3$$

Here the projection for $\check{p}$ is given by a constant ($k - 1 = 0$) value

$$\check{p} = \tfrac{1}{3} (\tilde{p}_1 + \tilde{p}_2 + \tilde{p}_3)$$

Note that numerical integration of the stabilizing term may not be performed using one-point quadrature at the element baricentre as then no contribution to the stabilizing term would be found.

Performing the integrations for the stabilizing term (11.86) gives the result

$$\mathbf{V}_{stab} = \frac{\Delta\alpha}{12G} \begin{bmatrix} 2 & 1 & 1 \\ 1 & 2 & 1 \\ 1 & 1 & 2 \end{bmatrix} - \frac{\Delta\alpha}{9G} \begin{bmatrix} 1 & 1 & 1 \\ 1 & 1 & 1 \\ 1 & 1 & 1 \end{bmatrix} = \frac{\Delta\alpha}{18G} \begin{bmatrix} 2 & -1 & -1 \\ -1 & 2 & -1 \\ -1 & -1 & 2 \end{bmatrix} \tag{11.89}$$

where Δ is the area of the triangular element. We recognize this result to have the same form as the deviatoric projection array which is positive semi-definite. The singular nature of this array permits the constant values of p to be unaffected by the stabilizing terms, thus, maintaining optimal accuracy for the method.

If we assemble the stabilization array given in (11.89) for the four-element patch shown in Fig. 11.11(a) we obtain an equation for node 0

$$\frac{\Delta\alpha}{18G} \left[4\, p_0 - \sum_1^4 p_i \right]$$

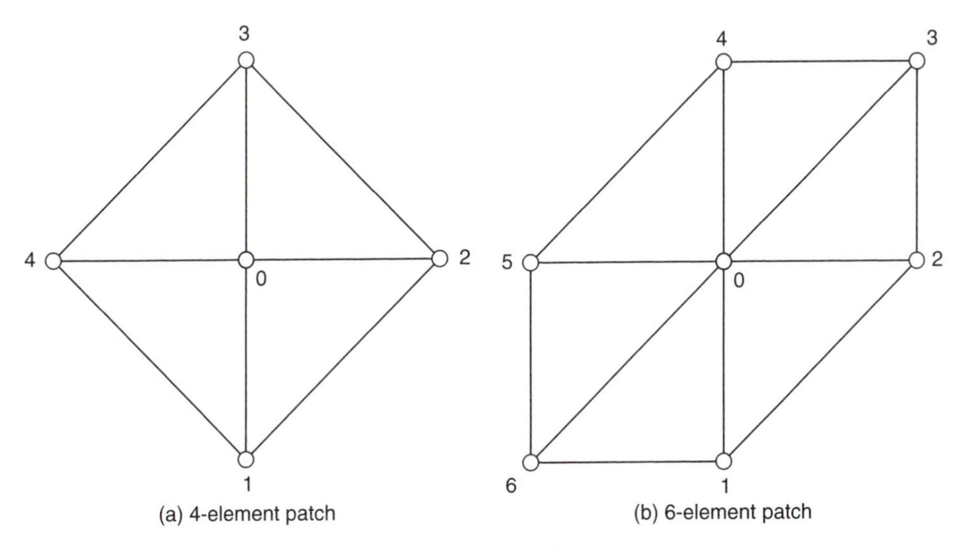

Fig. 11.11 Mesh patterns for pressure stabilization matrix evaluation.

which we recognize as a laplacian-type form. Similarly, for the six-element patch shown in Fig. 11.11(b) we obtain

$$\frac{\Delta\alpha}{18G}\left[4\,p_0 - \sum_1^6 p_i\right]$$

which has a similar form but is not the same as the laplacian operator for this mesh pattern.

The simplicity of the direct pressure stabilization is one of its main advantages. However, it also permits applications on elements of other order and shape without significant complication. For example, if degenerate element forms for quadratic elements are used as discussed in Chapter 5 the direct approach provides a means of stabilizing computations for incompressible forms without the need to add any second derivative terms (as, for example, needed for the GLS form).

We will show later on numerical examples of how well the direct stabilization approach works.

11.7.4 Incompressibility by time stepping

The fully incompressible case (i.e., $K = \infty$) has been studied by Zienkiewicz and Wu[50] using various time stepping procedures. Their applications concern the solution of fluid problems in which the rate effects for the Stokes equation appear as first derivatives of time. We can consider such a method here as a procedure to obtain the static solutions of elasticity problems in the limit as the rate terms become zero. Thus, this approach is considered here as a method for either the Stokes equation or the case of static incompressible elasticity.

The governing equations for slightly compressible Stokes flow may be written as

$$\rho_0 \frac{\partial u_i}{\partial t} - \frac{\partial \sigma_{ij}^d}{\partial x_j} - \frac{\partial p}{\partial x_i} = 0$$
$$\frac{1}{\rho_0 c^2} \frac{\partial p}{\partial t} - \frac{\partial u_i}{\partial x_i} = 0 \tag{11.90}$$

where ρ_0 is density (taken as unity in subsequent developments), $c = (K/\rho_0)^{1/2}$ is the speed of compressible waves, p is the pressure (here taken as positive in tension), and u_i is a velocity (or for elasticity interpretations a displacement) in the i-coordinate direction. Note that the above form assumes some compressibility in order to introduce the pressure rate term. At the steady limit this term is not involved, consequently, the solution will correspond to the incompressible case. Deviatoric stresses σ_{ij}^d are related to deviatoric strains (or strain rates for fluids) as described by Eq. (11.71).

Zienkiewicz and Wu consider many schemes for integrating the above equations in time. Here we introduce only one of the forms, which is widely used in the solution of the fluid equations which include transport effects (see reference 56). For the full fluid equations the algorithm is part of the *characteristic-based split* (CBS) method.[51, 52, 54–57]

The equations are discretized in time using the approximations $u(t_n) \approx u^n$ and time derivatives

$$\frac{\partial u_i}{\partial t} \approx \frac{u^{n+1} - u^n}{\Delta t} \tag{11.91}$$

where $\Delta t = t_{n+1} - t_n$. The time discretized equations are given by

$$\frac{u_i^{n+1} - u_i^n}{\Delta t} = \frac{\partial \sigma_{ij}^{d,n}}{\partial x_j} + \frac{\partial p^n}{\partial x_i} + \theta_2 \frac{\partial \Delta p}{\partial x_i} \tag{11.92a}$$

and

$$\frac{1}{c^2} \frac{p^{n+1} - p^n}{\Delta t} = \frac{\partial u_i^n}{\partial x_i} + \theta_1 \frac{\partial \Delta u_i}{\partial x_i} \tag{11.92b}$$

where $\Delta p = p^{n+1} - p^n$; $\Delta u_i = u_i^{n+1} - u_i^n$; θ_1 can vary between $1/2$ and 1; and θ_2 can vary between 0 and 1. In all that follows we shall use $\theta_1 = 1$.

The form to be considered uses a split of the equations by defining an intermediate approximate velocity u_i^* at time t_{n+1} when integrating the equilibrium equation (11.92a). Accordingly, we consider

$$\frac{u_i^* - u_i^n}{\Delta t} = \frac{\partial \sigma_{ij}^{d,n}}{\partial x_j} \tag{11.93a}$$

and

$$\frac{u_i^{n+1} - u_i^*}{\Delta t} = \frac{\partial p^n}{\partial x_i} + \theta_2 \frac{\partial \Delta p}{\partial x_i} \tag{11.93b}$$

Differentiating the second of these with respect to x_i to get the divergence of u_i^{n+1} and combining with the discrete pressure equation (11.92b) results in

$$\frac{1}{c^2} \frac{\Delta p}{\Delta t} - \theta_2 \Delta t \frac{\partial^2 \Delta p}{\partial x_i \, \partial x_i} = \Delta t \frac{\partial^2 p^n}{\partial x_i \, \partial x_i} + \frac{\partial u_i^*}{\partial x_i} \tag{11.93c}$$

Thus, the original problem has been replaced by a set of three equations which need to be solved successively.

Equations (11.93a), (11.93b) and (11.93c) may be written in a weak form using as weighting functions $\delta\mathbf{u}^*$, $\delta\mathbf{u}$ and δp, respectively (viz. Chapter 3). They are then discretized in space using the approximations

$$
\begin{aligned}
\mathbf{u}^n &\approx \hat{\mathbf{u}}^n = \mathbf{N}_u\tilde{\mathbf{u}}^n &\quad \text{and} \quad & \delta\mathbf{u} \approx \delta\hat{\mathbf{u}} = \mathbf{N}_u\delta\tilde{\mathbf{u}} \\
\mathbf{u}^* &\approx \hat{\mathbf{u}}^* = \mathbf{N}_u\tilde{\mathbf{u}}^* &\quad \text{and} \quad & \delta\mathbf{u}^* \approx \delta\hat{\mathbf{u}}^* = \mathbf{N}_u\delta\tilde{\mathbf{u}}^* \\
p^n &\approx \hat{p}^n = \mathbf{N}_p\tilde{\mathbf{p}}^n &\quad \text{and} \quad & \delta p \approx \delta\hat{p} = \mathbf{N}_p\delta\tilde{\mathbf{p}}
\end{aligned}
$$

with similar expressions for $\mathbf{u}^{n+1}$ and p^{n+1}. The final discrete form is given by the three equation sets

$$
\begin{aligned}
\frac{1}{\Delta t}\mathbf{M}_u\left(\tilde{\mathbf{u}}^* - \tilde{\mathbf{u}}^n\right) &= -\mathbf{A}\tilde{\mathbf{u}}^n + \mathbf{f}_1 \\
\left[\frac{1}{\Delta t}\mathbf{M}_p + \theta_2\Delta t\mathbf{H}\right]\Delta\tilde{\mathbf{p}} &= -\mathbf{C}\tilde{\mathbf{u}}^* - \Delta t\mathbf{H}\tilde{\mathbf{p}}^n + \mathbf{f}_2 \\
\frac{1}{\Delta t}\mathbf{M}_u\left(\tilde{\mathbf{u}}^{n+1} - \tilde{\mathbf{u}}^*\right) &= -\mathbf{C}^{\mathrm{T}}\left(\tilde{\mathbf{p}}^n + \theta_2\Delta\tilde{\mathbf{p}}\right)
\end{aligned}
\tag{11.94}
$$

In the above we have integrated by parts all the terms which involve derivatives on deviator stress (σ_{ij}^d), pressure (p) and displacements (velocities). In addition we consider only the case where $u_i^{n+1} = u_i^* = \bar{u}_i$ on the boundary Γ_u (thus requiring $\delta u_i = \delta u_i^* = 0$ on Γ_u). Accordingly, the matrices are defined as

$$
\begin{aligned}
\mathbf{M}_u &= \int_\Omega \mathbf{N}_u^{\mathrm{T}}\mathbf{N}_u\,\mathrm{d}\Omega & \mathbf{M}_p &= \int_\Omega \frac{1}{c^2}\mathbf{N}_p^{\mathrm{T}}\mathbf{N}_p\,\mathrm{d}\Omega \\
\mathbf{A} &= \int_\Omega \mathbf{B}^{\mathrm{T}}\mathbf{D}_d\mathbf{B}\,\mathrm{d}\Omega & \mathbf{C} &= \int_\Omega \frac{\partial\mathbf{N}_p}{\partial x_i}\mathbf{N}_u\,\mathrm{d}\Omega \\
\mathbf{H} &= \int_\Omega \frac{\partial\mathbf{N}_p^{\mathrm{T}}}{\partial x_i}\frac{\partial\mathbf{N}_p}{\partial x_i}\,\mathrm{d}\Omega & \mathbf{f}_1 &= \int_{\Gamma_t} \mathbf{N}_u^{\mathrm{T}}(\bar{\mathbf{t}} - k\mathbf{n}p^n)\,\mathrm{d}\Gamma \\
\mathbf{f}_2 &= \int_{\Gamma_u} \mathbf{N}_p^{\mathrm{T}}\mathbf{n}^{\mathrm{T}}\bar{\mathbf{u}}\,\mathrm{d}\Gamma
\end{aligned}
\tag{11.95}
$$

in which $\mathbf{D}_d$ are the deviatoric moduli defined previously. The parameter k denotes an option on alternative methods to split the boundary traction term and is taken as either zero or unity. We note that a choice of zero simplifies the computation of boundary contributions; however, some would argue that unity is more consistent with the integration by parts.

The boundary pressure acting on Γ_t is computed from the specified surface tractions ($\bar{t}_i$) and the 'best' estimate for the deviator stress at step $(n+1)$ which is given by $\sigma_{ij}^{d,*}$. Accordingly,

$$
\bar{p}^{n+1} \approx n_i\bar{t}_i - n_i\sigma_{ij}^{d,*}n_j
$$

is imposed at each node on the boundary Γ_t.

In general we require that $\Delta t < \Delta t_{\mathrm{crit}}$ where the critical time step is $h^2/2G$ (in which h is the element size). Such a quantity is obviously calculated independently for each element and the lowest value occurring in any element governs the overall stability. It is possible and useful to use here the value of Δt calculated for each element separately when calculating incompressible stabilizing terms in the pressure calculation and the overall time step elsewhere (we shall label the time increments multiplying $\mathbf{H}$ in Eq. $(11.94)_3$ as Δt_{int}).

A ratio of $\gamma = \Delta t_{\text{int}}/\Delta t$ greater than unity improves considerably the stabilizing properties. As Eq. $(11.94)_3$ has greater stability than the first two equations in (11.94), and for $\theta_2 \geq 1/2$ is unconditionally stable, we recommend that the time step used in this equation be $\gamma\,\Delta t_{\text{cr}}$ for each node. Generally a value of 2 is good as we shall show in the examples (for additional details see reference 54).

Equation (11.94) defines a value of $\tilde{\mathbf{u}}^*$ entirely in terms of known quantities at the n-step. If the mass matrix $\mathbf{M}_u$ is made diagonal by lumping (see Chapter 16 and Appendix I) the solution is thus trivial. Such an equation is called *explicit*. The equation for $\Delta\tilde{\mathbf{p}}$, on the other hand, depends on both $\mathbf{M}_p$ and $\mathbf{H}$ and it is not possible to make the latter diagonal easily.† It is possible to make $\mathbf{M}_p$ diagonal using a similar method as that employed for $\mathbf{M}_u$. Thus, if θ_2 is zero this equation will also be explicit, otherwise it is necessary to solve a set of algebraic equations and the method for this equation is called *implicit*. Once the value of $\Delta\tilde{\mathbf{p}}$ is known the solution for $\tilde{\mathbf{u}}^{n+1}$ is again explicit. In practice the above process is quite simple to implement; however, it is necessary to satisfy *stability* requirements by limiting the size of the time increment. This is discussed further in Chapter 17 and in reference 51. Here we only wish to show the limit result as the changes in time go to zero (i.e., for a constant in time load value) and when full incompressibility is imposed.

At the steady limit the solutions become

$$\tilde{\mathbf{u}}^n = \tilde{\mathbf{u}}^{n+1} = \tilde{\mathbf{u}} \quad \text{and} \quad \tilde{\mathbf{p}}^n = \tilde{\mathbf{p}}^{n+1} = \tilde{\mathbf{p}} \tag{11.96}$$

Eliminating $\mathbf{u}^*$ the discrete equations reduce to the mixed problem

$$\begin{bmatrix} \mathbf{A} & \mathbf{C} \\ \mathbf{C}^{\mathrm{T}} & \Delta t\left(\mathbf{C}^{\mathrm{T}}\mathbf{M}_u^{-1}\mathbf{C} - \theta_1\mathbf{H}\right) \end{bmatrix} \begin{Bmatrix} \tilde{\mathbf{u}} \\ \tilde{\mathbf{p}} \end{Bmatrix} + \begin{Bmatrix} \mathbf{f} \\ \mathbf{0} \end{Bmatrix} = \mathbf{0} \tag{11.97}$$

At the steady limit we again recover a term on the diagonal which stabilizes the solution. This term is again of a Laplace equation type – indeed, it is now the difference between two discrete forms for the Laplace equation. The term $\mathbf{C}^{\mathrm{T}}\mathbf{M}_u^{-1}\mathbf{C}$ makes the bandwidth of the resulting equations larger – thus this form is different from all the previously discussed methods.

11.7.5 Numerical comparisons

To provide some insight into the behaviour of the above methods we consider two example problems. The first is a problem often used to assess the performance of codes to solve steady-state Stokes flow problems – which is identical to the case for incompressible linear elasticity. The second example is a problem in nearly incompressible linear elasticity.

Example 11.5: Driven cavity. A two-dimensional plane (strain) case is considered for a square domain with unit side lengths. The material properties are assumed to be fully incompressible ($\nu = 0.5$) with unit viscosity (elastic shear modulus, G, of unity). All boundaries of the domain are restrained in the x and y directions with the top boundary having a unit tangential velocity (displacement) at all nodes except the corner ones. Since

† It is possible to diagonalize the matrix by solving an eigenproblem as shown in Chapter 16 – for large problems this requires more effort than is practical.

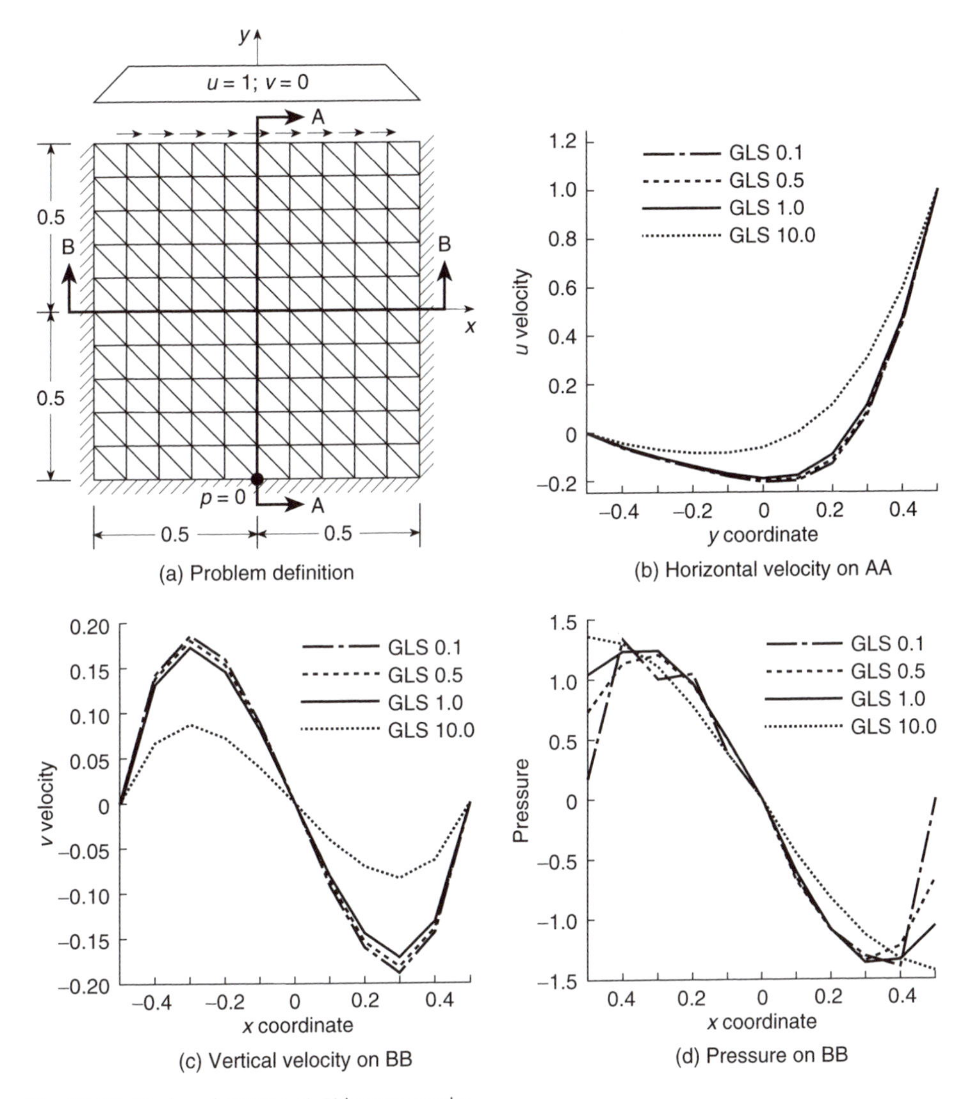

Fig. 11.12 Mesh and GLS/Brezzi–Pitkaranta results.

the problem is incompressible it is necessary to prescribe the pressure at one point in the mesh – this is selected as the centre node along the bottom edge. The 10×10 element mesh of triangular elements (200 elements total) used for the comparison is shown in Fig. 11.12(a). The elements used for the analysis use linear velocity (displacement) and pressure on 3-noded triangles. Results are presented for the horizontal velocity along the vertical centre-line AA and for vertical velocity and pressure along the horizontal centre-line BB. Three forms of stabilization are considered:

1. Galerkin least squares (GLS)/Brezzi–Pitkaranta (BP) where the effect of α on τ is assessed. The results for the horizontal velocity are given in Fig. 11.12(b) and for the vertical velocity and pressure in Figs 11.12(c) and (d), respectively. From the analysis it

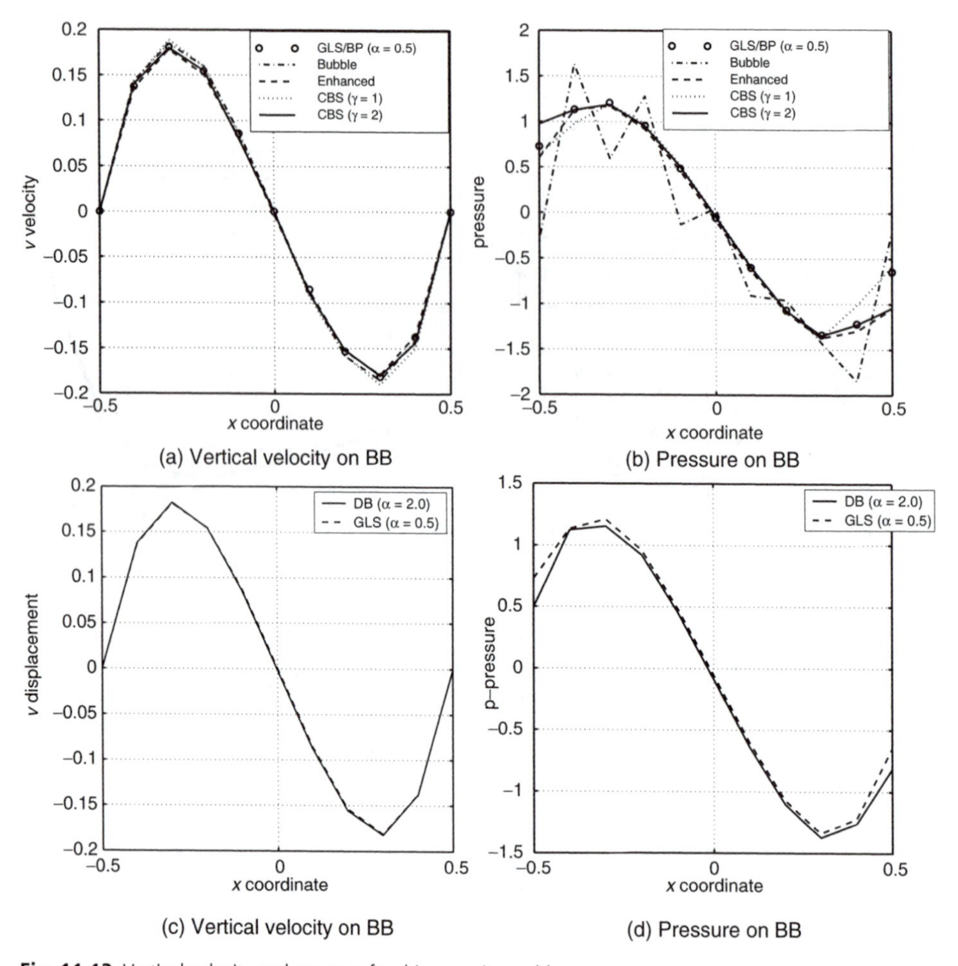

Fig. 11.13 Vertical velocity and pressure for driven cavity problem.

is assessed that the stabilization parameter α should be about 0.5 to 1 (as also indicated by Hughes *et al.*[47]). Use of lower values leads to excessive oscillation in pressure and use of higher values to strong dissipation of pressure results.

2. Cubic bubble (MINI) element stabilization. Results for vertical velocity are nearly indistinguishable from the GLS results as indicated in Fig. 11.13; however, those for pressure show oscillation. Such oscillation has also been observed by others along with some suggested boundary modifications.[58] No free parameters exist for this element (except possible modification of the bubble mode used), thus, no artificial 'tuning' is possible. Use of more refined meshes leads to a strong decrease in the oscillation.
3. Direct pressure stabilization (DB). Results for vertical velocity are again well captured as shown in Fig. 11.13; pressure results are also smooth and give good peak answers. We have not explored the range of α which may be used for the stabilization.
4. The CBS algorithm. Finally in Fig. 11.13 we present results using the CBS solution which may be compared with GLS, $\alpha = 0.5$. Once again the reader will observe that with $\gamma = 2$, the results of CBS reproduce very closely those of GLS, $\alpha = 0.5$. However,

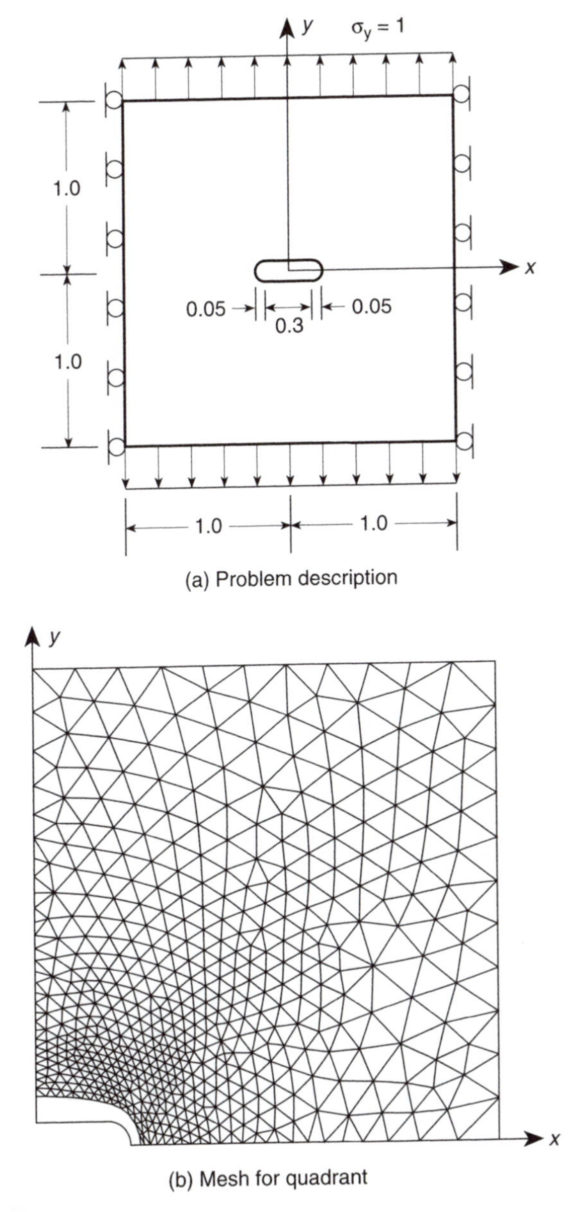

Fig. 11.14 Region and mesh used for slotted tension strip.

in results for $\gamma = 1$ no oscillations are observed and they are quite reasonable. This ratio for γ is where the algorithm gives excellent results in incompressible flow modelling as will be demonstrated further in results presented in reference 56.

Example 11.6: Tension strip with slot. As our next example we consider a plane strain linear problem on a square domain with a central slot. The domain is two units square

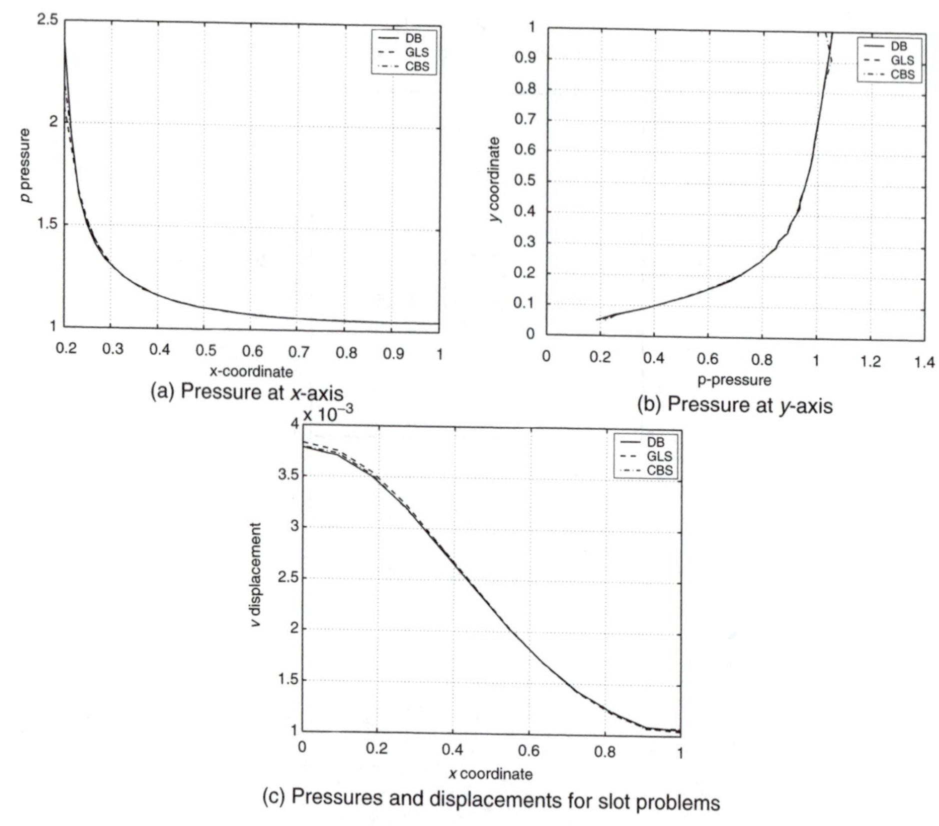

Fig. 11.15 Pressures and displacements for slot problems.

and the central slot has a total width of 0.4 units and a height of 0.1 units. The ends of the slot are semicircular. Lateral boundaries have specified normal displacement and zero tangential traction. The top and bottom boundaries are uniformly stretched by a uniform axial loading and lateral boundaries are maintained at zero horizontal displacement. We consider the linear elastic problem with elastic properties $E = 24$ and $\nu = 0.5$; thus, giving an incompressible situation. An unstructured mesh of triangles is constructed as shown in Fig. 11.14(b). The problem is solved using direct pressure (DB), Galerkin least squares (GLS) and characteristic-based split (CBS) stabilization methods. Results for pressure along the horizontal and vertical centre-lines (i.e., the x and y axes) are presented in Figs 11.15(a) and 11.15(b) and in Tables 11.3 and 11.4. The distribution of the vertical displacement is shown in Fig. 11.15(c). We note that the results for this problem cause very strong gradients in stress near the ends of the slot. The mesh used for the analysis is not highly refined in this region and hence results from different analyses can be expected to differ in this region. The results obtained elsewhere using all three formulations are indistinguishable on the plot. In general the results achieved with all forms are satisfactory and indicate that stabilized methods may be considered for use in problems where constraints, such as incompressibility, are encountered.

Table 11.3 Pressure for slot problem along $x = 0$

	Pressure at $x = 0$		
Coord. y	DB	GLS	CBS
0.0500	0.1841	0.2185	0.1964
0.0693	0.2516	0.2593	0.2619
0.0886	0.3563	0.3541	0.3537
0.1079	0.4339	0.4337	0.4373
0.1273	0.5091	0.5050	0.5083
0.1466	0.5744	0.5693	0.5709
0.1661	0.6311	0.6260	0.6261
0.1860	0.6857	0.6770	0.6756
0.2065	0.7243	0.7230	0.7226
0.2275	0.7592	0.7589	0.7587
0.2491	0.7943	0.7921	0.7936
0.2714	0.8174	0.8187	0.8185
0.2944	0.8503	0.8452	0.8470
0.3182	0.8589	0.8636	0.8624
0.3429	0.8918	0.8861	0.8886
0.3686	0.8982	0.9016	0.9029
0.3954	0.9163	0.9159	0.9157
0.4235	0.9354	0.9301	0.9303
0.4532	0.9360	0.9398	0.9386
0.4846	0.9503	0.9506	0.9498
0.5181	0.9605	0.9612	0.9610
0.5543	0.9720	0.9712	0.9716
0.5938	0.9813	0.9801	0.9803
0.6380	0.9871	0.9886	0.9889
0.6887	0.9976	0.9988	0.9974
0.7505	1.0098	1.0096	1.0097
0.8262	1.0222	1.0240	1.0222
0.9090	1.0383	1.0390	1.0502
1.0000	1.0534	1.0516	1.0275

11.8 Concluding remarks

In this chapter we have considered in some detail the application of mixed methods to incompressible problems and also we have indicated some alternative procedures. The extension to non-isotropic problems and non-linear problems is presented in reference 1, but will follow similar lines. Here we note how important the problem is in the context of fluid mechanics and it is there that much of the attention to it has been given.[56]

In concluding this chapter we would like to point out three matters:

1. The mixed formulation discovers immediately the non-robustness of certain irreducible (displacement) elements and, indeed, helps us to isolate those which perform well from those that do not. Thus, it has merit which as a test is applicable to many irreducible forms at all times.
2. In elasticity, certain mixed forms work quite well at the near incompressible limit without resort to splits into deviatoric and mean parts. These include the two-field quadrilateral element of Pian–Sumihara and the enhanced strain quadrilateral element of Simo–Rifai which were presented in the previous chapter. There we noted how well such elements work for Poisson's ratio approaching one-half as compared to the standard irreducible element of a similar type.

Table 11.4 Pressure for slot problem along $y=0$

	Pressure at $y=0$		
Coord. x	DB	GLS	CBS
0.2000	2.4997	2.0775	2.1983
0.2152	1.9449	1.8805	1.9405
0.2313	1.6527	1.6682	1.6539
0.2485	1.5073	1.5314	1.5219
0.2668	1.4098	1.4291	1.4253
0.2864	1.3392	1.3542	1.3510
0.3062	1.2952	1.2957	1.2927
0.3265	1.2532	1.2543	1.2523
0.3474	1.2206	1.2209	1.2202
0.3691	1.1964	1.1910	1.1902
0.3915	1.1685	1.1702	1.1692
0.4149	1.1514	1.1515	1.1508
0.4392	1.1322	1.1348	1.1349
0.4647	1.1195	1.1209	1.1205
0.4916	1.1068	1.1082	1.1074
0.5200	1.0988	1.0977	1.0975
0.5502	1.0895	1.0879	1.0874
0.5828	1.0793	1.0796	1.0795
0.6183	1.0702	1.0722	1.0723
0.6578	1.0645	1.0657	1.0655
0.7027	1.0587	1.0596	1.0591
0.7566	1.0525	1.0539	1.0536
0.8268	1.0479	1.0489	1.0486
0.9084	1.0444	1.0453	1.0450
1.0000	1.0438	1.0443	1.0439

3. Use of stabilizing forms such as the direct pressure or time stepping form allows use of mixed $\mathbf{u}$–p elements with equal order interpolation – a form which otherwise fails the mixed patch test (or Babuška–Brezzi condition).

11.9 Problems

11.1 Show that the variational theorem

$$\Pi_{HR} = \int_\Omega \tfrac{1}{2}\boldsymbol{\varepsilon}^{\mathrm{T}}\mathbf{D}_d\boldsymbol{\varepsilon}\,\mathrm{d}\Omega - \int_\Omega \mathbf{u}^{\mathrm{T}}\mathbf{b}\,\mathrm{d}\Omega - \int_\Gamma \mathbf{u}^{\mathrm{T}}\bar{\mathbf{t}}\,\mathrm{d}\Gamma + \int_\Omega \left[p\mathbf{m}^{\mathrm{T}}\boldsymbol{\varepsilon} - \frac{1}{2K}p^2 \right] \mathrm{d}\Omega$$

generates the problem given in Eq. (11.10) as its first variation.

11.2 Show that the variational theorem given in Eq. (11.27) generates the problem given by (11.24).

11.3 If the approximation for p contains all the terms that are in the approximation to $\mathbf{m}^{\mathrm{T}}\boldsymbol{\varepsilon}$ the limitation principle yields the result that the formulation will be identical to the standard displacement approximation given in Chapter 6.

If the number of internal degrees of freedom for the displacement $\mathbf{u}$ in an element is equal to the number of parameters in the pressure p the mixed patch count condition is passed and the consistency condition is also passed an element will not lock.

Consider the case where $\mathbf{u}$ is C^0 continuous and p is discontinuous.

(a) The first three members of the rectangular lagrangian family of two-dimensional plane strain elements is shown in Fig. 4.7. Consider the general member of this class in which the displacement $\mathbf{u}$ is of order n (i.e., has $n + 1$ nodes in each direction) and show on the Pascal triangle (viz. Fig. 4.8) the polynomial terms contained in the divergence term for volumetric strain $\mathbf{m}^T \boldsymbol{\varepsilon}$.

(b) If the pressure p is approximated by an order m lagrangian interpolation determine the lowest order for m which will contain all the polynomial terms found in (a).

(c) Determine the lowest order of n for which the limitation principle is satisfied and the element will not lock at the nearly incompressible limit.

11.4 Repeat Problem 11.3 for the triangular family of elements in plane strain (viz. Fig. 4.13). Let the displacement $\mathbf{u}$ be approximated by an order n polynomial (i.e., has $n+1$ nodes on each edge).

(a) Show on the Pascal triangle (viz. Fig. 4.8) the polynomial terms contained in the divergence term $\mathbf{m}^T \boldsymbol{\varepsilon}$.

(b) What order approximation for p will contain all the polynomial terms in (a)?

(c) Determine the lowest order of n for which the limitation principle is satisfied. Note that this order of approximation will yield a displacement formulation which will not 'lock' near the incompressible limit.

(d) Is the result valid for an axisymmetric geometry? Explain your answer.

11.5 For a plane strain problem consider a linear triangular element in which the displacement approximation is given by

$$\mathbf{u} \approx \sum_{a=1}^{3} L_a \tilde{\mathbf{u}}$$

together with a constant approximation for the pressure p.

Using Eq. (11.24) compute $\bar{\mathbf{B}}$. Can this formulation be used to model nearly incompressible problems? Justify your answer.

11.6 For an axisymmetric problem consider a linear triangular element in which the displacement approximation is given by

$$\mathbf{u} \approx \sum_{a=1}^{3} L_a \tilde{\mathbf{u}}$$

together with a constant approximation for the pressure p.

Using Eq. (11.24) compute $\bar{\mathbf{B}}$. Can this formulation be used to model nearly incompressible problems? Justify your answer.

11.7 Consider a plane strain problem which is to be solved using a linear quadrilateral element with the displacement approximation

$$\mathbf{u} \approx \sum_{a=1}^{4} N_a \tilde{\mathbf{u}}$$

where $N_a = 1/4(1 + \xi_a \xi)(1 + \eta_a \eta)$ together with a constant approximation for the pressure p.

Let the element have a rectangular form with sides a and b in the x and y directions, respectively. Using Eq. (11.24) compute $\bar{\mathbf{B}}$. Can this formulation be used to model nearly incompressible problems? Justify your answer.

11.8 Consider an axisymmetric problem which is to be solved using a linear quadrilateral element with the displacement approximation

$$\mathbf{u} \approx \sum_{a=1}^{4} N_a \tilde{\mathbf{u}}$$

where $N_a = 1/4(1 + \xi_a \xi)(1 + \eta_a \eta)$ together with a constant approximation for the pressure p.

Let the element have a rectangular form with sides a and b in the r and z directions, respectively. Using Eq. (11.24) compute $\bar{\mathbf{B}}$. Can this formulation be used to model nearly incompressible problems? Justify your answer.

11.9 Consider a rectangular plane element with sides a and b in the x and y directions, respectively.

(a) Compute the matrix $\mathbf{V}$ for GLS stabilization. Ignore second derivatives of $\mathbf{u}$.
(b) Consider four elements of equal size with a central node c and compute the assembled equation for the pressure p at this node.

11.10 Consider a rectangular axisymmetric element with sides a and b in the r and z directions, respectively. Let the inner radius of the element be located at $r_i > a/2$.

(a) Compute the matrix $\mathbf{V}$ for GLS stabilization. Ignore second derivatives of $\mathbf{u}$.
(b) Consider four elements of equal size with a central node c located at r_i and compute the assembled equation for this node.

11.11 Consider a rectangular plane element with sides a and b in the x and y directions, respectively.

(a) Compute the matrix $\mathbf{V}$ for direct pressure stabilization.
(b) Consider four elements of equal size with a central node c and compute the assembled equation for this node.

11.12 Consider a rectangular axisymmetric element with sides a and b in the r and z directions, respectively. Let the inner radius of the element be located at $r_i > a/2$.

(a) Compute the matrix $\mathbf{V}$ for direct pressure stabilization.
(b) Consider four elements of equal size with a central node c located at r_i and compute the assembled equation for this node.

11.13 The steel–rubber composite bearing shown in Fig. 11.16(a) is used to support a heavy machine. The bearing is to have high vertical stiffness but be flexible in shear (similar bearings are also used to support structures in seismic regions). Consider a typical layer where $t_r = 1$ cm and $t_s = 0.1$ cm with a width $w = 5$ cm. Let the properties be $E_s = 200$ GPa, $\nu_s = 0.3$ and $E_r = 5$ GPa, $\nu_r = 0.495$. For a state of plane strain use *FEAPpv* (or any appropriate available program) to compute the stiffness for a vertical and a horizontal applied loading.

Use 4-node and 9-node mixed $\mathbf{u}$–p–ε_v elements to compute the vertical and horizontal stiffness of a single layer [viz. Fig. 11.16(b)]. Compare your solution to answers from a standard displacement formulation in $\mathbf{u}$.

11.14 Program development project: Extend your program system started in Problem 2.17 to permit solution of a stabilized method as described in Sec. 11.7. You may select a stabilization scheme from either the GLS method of Sec. 11.7.2 or the direct pressure stabilization method of Sec. 11.7.3.

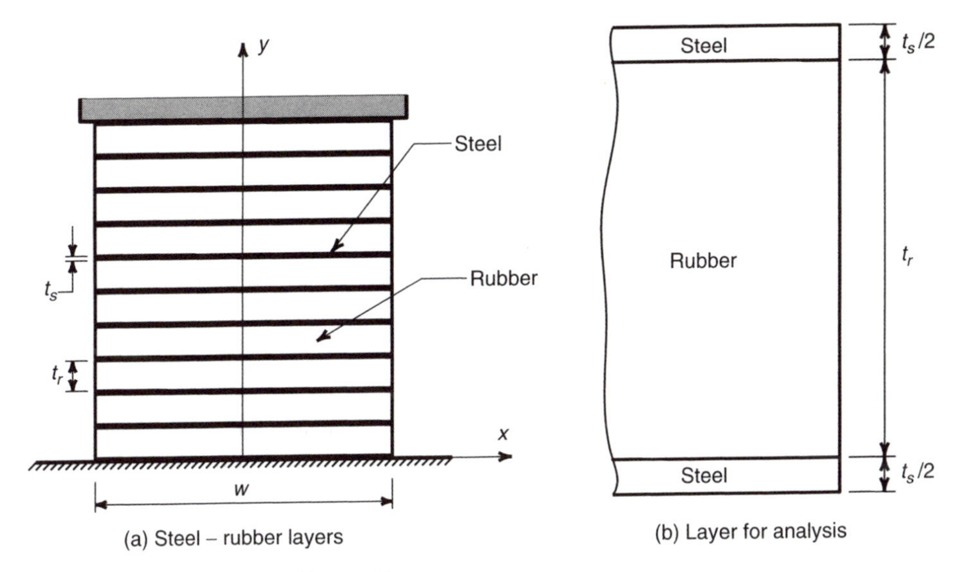

Fig. 11.16 Support bearing. Problem 12.13.

Use your program to solve the driven cavity problem described in Example 11.5. Set the boundary velocity as shown in Fig. 11.12(a) and the nodal pressure to zero at the centre of the bottom. Plot values shown in Figs 11.12 and 11.13. Also plot contours for velocity components and pressure. Briefly discuss your findings.

References

1. O.C. Zienkiewicz and R.L. Taylor. *The Finite Element Method for Solid and Structural Mechanics*. Butterworth-Heinemann, Oxford, 6th edition, 2005.
2. L.R. Herrmann. Finite element bending analysis of plates. In *Proc. 1st Conf. Matrix Methods in Structural Mechanics*, AFFDL-TR-66-80, pages 577–602, Wright-Patterson Air Force Base, Ohio, 1965.
3. S.W. Key. Variational principle for incompressible and nearly incompressible anisotropic elasticity. *Int. J. Solids Struct.*, 5:951–964, 1969.
4. O.C. Zienkiewicz, R.L. Taylor, and J.A.W. Baynham. Mixed and irreducible formulations in finite element analysis. In S.N. Atluri, R.H. Gallagher, and O.C. Zienkiewicz, editors, *Hybrid and Mixed Finite Element Methods*, pages 405–431. John Wiley & Sons, 1983.
5. D.N. Arnold, F. Brezzi, and M. Fortin. A stable finite element for the Stokes equations. *Calcolo*, 21:337–344, 1984.
6. M. Fortin and N. Fortin. Newer and newer elements for incompressible flow. In R.H. Gallagher, G.F. Carey, J.T. Oden, and O.C. Zienkiewicz, editors, *Finite Elements in Fluids*, volume 6, Chapter 7, pages 171–188. John Wiley & Sons, 1985.
7. J.T. Oden. R.I.P. methods for Stokesian flow. In R.H. Gallagher, D.N. Norrie, J.T. Oden, and O.C. Zienkiewicz, editors, *Finite Elements in Fluids*, volume 4, Chapter 15, pages 305–318. John Wiley & Sons, 1982.

8. M. Crouzcix and P.A. Raviart. Conforming and non-conforming finite element methods for solving stationary Stokes equations. *RAIRO*, 7–R3:33–76, 1973.
9. D.S. Malkus. Eigenproblems associated with the discrete LBB condition for incompressible finite elements. *Int. J. Eng. Sci.*, 19:1299–1370, 1981.
10. M. Fortin. Old and new finite elements for incompressible flow. *Int. J. Numer. Meth. Fluids*, 1:347–364, 1981.
11. C. Taylor and P. Hood. A numerical solution of the Navier–Stokes equations using the finite element technique. *Comp. Fluids*, 1:73–100, 1973.
12. R.E. Bank and B.D. Welfert. A comparison between the mini-element and the Petrov–Galerkin formulations for the generalized Stokes problem. *Comp. Meth. Appl. Mech. Eng.*, 83:61–68, 1990.
13. T.J.R. Hughes. Generalization of selective integration procedures to anisotropic and non-linear media. *Int. J. Numer. Meth. Eng.*, 15:1413–1418, 1980.
14. J.C. Simo, R.L. Taylor, and K.S. Pister. Variational and projection methods for the volume constraint in finite deformation plasticity. *Comp. Meth. Appl. Mech. Eng.*, 51:177–208, 1985.
15. J.C. Simo and T.J.R. Hughes. *Computational Inelasticity*, volume 7 of *Interdisciplinary Applied Mathematics*. Springer-Verlag, Berlin, 1998.
16. D.J. Naylor. Stresses in nearly incompressible materials for finite elements with application to the calculation of excess pore pressures. *Int. J. Numer. Meth. Eng.*, 8:443–460, 1974.
17. O.C. Zienkiewicz and P.N. Godbole. Viscous incompressible flow with special reference to non-Newtonian (plastic) flows. In R.H. Gallagher *et al.*, editor, *Finite Elements in Fluids*, volume 1, Chapter 2, pages 25–55. John Wiley & Sons, 1975.
18. T.J.R. Hughes, R.L. Taylor, and J.F. Levy. High Reynolds number, steady, incompressible flows by a finite element method. In R.H. Gallagher *et al.*, editor, *Fin. Elem. Fluids*, volume 3. John Wiley & Sons, 1978.
19. O.C. Zienkiewicz, J. Too, and R.L. Taylor. Reduced integration technique in general analysis of plates and shells. *Int. J. Numer. Meth. Eng.*, 3:275–290, 1971.
20. S.F. Pawsey and R.W. Clough. Improved numerical integration of thick slab finite elements. *Int. J. Numer. Meth. Eng.*, 3:575–586, 1971.
21. O.C. Zienkiewicz and E. Hinton. Reduced integration, function smoothing and non-conformity in finite element analysis. *J. Franklin Inst.*, 302:443–461, 1976.
22. O.C. Zienkiewicz and P. Bettess. Infinite elements in the study of fluid-structure interaction problems. In *2nd Int. Symp. on Computing Methods in Applied Science and Engineering*, Versailles, France, Dec. 1975.
23. O.C. Zienkiewicz. *The Finite Element Method*. McGraw-Hill, London, 3rd edition, 1977.
24. D.S. Malkus and T.J.R. Hughes. Mixed finite element methods in reduced and selective integration techniques: a unification of concepts. *Comp. Meth. Appl. Mech. Eng.*, 15:63–81, 1978.
25. O.C. Zienkiewicz and S. Nakazawa. On variational formulations and its modification for numerical solution. *Comp. Struct.*, 19:303–313, 1984.
26. M.S. Engleman, R.L. Sani, P.M. Gresho, and H. Bercovier. Consistent vs. reduced integration penalty methods for incompressible media using several old and new elements. *Int. J. Numer. Meth. Fluids*, 2:25–42, 1982.
27. D.N. Arnold. Discretization by finite elements of a model parameter dependent problem. *Num. Meth.*, 37:405–421, 1981.
28. M.J.D. Powell. A method for nonlinear constraints in minimization problems. In R. Fletcher, editor, *Optimization*. Academic Press, London, 1969.
29. M. Vogelius. An analysis of the *p*-version of the finite element method for nearly incompressible materials; uniformly optimal error estimates. *Num. Math.*, 41:39–53, 1983.
30. S.W. Sloan and M.F. Randolf. Numerical prediction of collapse loads using finite element methods. *Int. J. Numer. Anal. Meth. Geomech.*, 6:47–76, 1982.

31. J.C. Nagtegaal, D.M. Parks, and J.R. Rice. On numerical accurate finite element solutions in the fully plastic range. *Comp. Meth. Appl. Mech. Eng.*, 4:153–177, 1974.
32. O.C. Zienkiewicz, J.P. Vilotte, S. Toyoshima, and S. Nakazawa. Iterative method for constrained and mixed approximation. An inexpensive improvement of FEM performance. *Comp. Meth. Appl. Mech. Eng.*, 51:3–29, 1985.
33. K.J. Arrow, L. Hurwicz, and H. Uzawa. *Studies in Non-Linear Programming*. Stanford University Press, Stanford, CA, 1958.
34. M.R. Hestenes. Multiplier and gradient methods. *J. Opt. Theory Appl.*, 4:303–320, 1969.
35. C.A. Felippa. Iterative procedure for improving penalty function solutions of algebraic systems. *Int. J. Numer. Meth. Eng.*, 12:165–185, 1978.
36. M. Fortin and F. Thomasset. Mixed finite element methods for incompressible flow problems. *J. Comp. Phys.*, 31:113–145, 1973.
37. M. Fortin and R. Glowinski. *Augmented Lagrangian Methods: Applications to Numerical Solution of Boundary-Value Problems*. North-Holland, Amsterdam, 1983.
38. D.G. Luenberger. *Linear and Nonlinear Programming*. Addison-Wesley, Reading, Mass., 1984.
39. J.H. Argyris. Three-dimensional anisotropic and inhomogeneous media – matrix analysis for small and large displacements. *Ingenieur Archiv*, 34:33–55, 1965.
40. O.C. Zienkiewicz and S. Valliappan. Analysis of real structures for creep plasticity and other complex constitutive laws. In M. Te'eni, editor, *Structure of Solid Mechanics and Engineering Design*, volume, Part 1, pages 27–48, 1971.
41. O.C. Zienkiewicz. *The Finite Element Method in Engineering Science*. McGraw-Hill, London, 2nd edition, 1971.
42. J.C. Simo and R.L. Taylor. Quasi-incompressible finite elasticity in principal stretches: continuum basis and numerical algorithms. *Comp. Meth. Appl. Mech. Eng.*, 85:273–310, 1991.
43. R. Courant. Variational methods for the solution of problems of equilibrium and vibration. *Bull. Amer. Math Soc.*, 49:1–61, 1943.
44. F. Brezzi and J. Pitkäranta. On the stabilization of finite element approximations of the Stokes problem. In W. Hackbusch, editor, *Efficient Solution of Elliptic Problems, Notes on Numerical Fluid Mechanics*, volume 10. Vieweg, Wiesbaden, 1984.
45. T.J.R. Hughes, L.P. Franca, and M. Balestra. A new finite element formulation for computational fluid dynamics: V. Circumventing the Babuška–Brezzi condition: a stable Petrov–Galerkin formulation of the Stokes problem accommodating equal-order interpolations. *Comp. Meth. Appl. Mech. Eng.*, 59:85–99, 1986.
46. T.J.R. Hughes and L.P. Franca. A new finite element formulation for computational fluid dynamics: VII. The Stokes problem with various well-posed boundary conditions: symmetric formulation that converge for all velocity/pressure spaces. *Comp. Meth. Appl. Mech. Eng.*, 65:85–96, 1987.
47. T.J.R. Hughes, L.P. Franca, and G.M. Hulbert. A new finite element formulation for computational fluid dynamics: VIII. The Galerkin/least-squares method for advective-diffusive equations. *Comp. Meth. Appl. Mech. Eng.*, 73:173–189, 1989.
48. E. Oñate. Derivation of stabilized equations for numerical solution of advective–diffusive transport and fluid flow problems. *Comp. Meth. Appl. Mech. Eng.*, 151:233–265, 1998.
49. C.R. Dohrmann and P.B. Bochev. A stabilized finite element method for the Stokes problem based on polynomial pressure projections. *Int. J. Numer. Meth. Fluids*, to appear, 2005.
50. O.C. Zienkiewicz and J. Wu. Incompressibility without tears! How to avoid restrictions of mixed formulations. *Int. J. Numer. Meth. Eng.*, 32:1184–1203, 1991.
51. O.C. Zienkiewicz and R. Codina. A general algorithm for compressible and incompressible flow – Part I: The split, characteristic-based scheme. *Int. J. Numer. Meth. Fluids*, 20:869–885, 1995.
52. O.C. Zienkiewicz, P. Nithiarasu, R. Codina, M. Vasquez, and P. Ortiz. The characteristic-based-split (CBS) procedure: an efficient and accurate algorithm for fluid problems. *Int. J. Numer. Meth. Fluids*, 31:359–392, 1999.

53. O.C. Zienkiewicz, K. Morgan, B.V.K. Satya Sai, R. Codina, and M. Vasquez. A general algorithm for compressible and incompressible flow – Part II: Tests on the explicit form. *Int. J. Numer. Meth. Fluids*, 20:887–913, 1995.
54. P. Nithiarasu and O.C. Zienkiewicz. On stabilization of the CBS algorithm. Internal and external time steps. *Int. J. Numer. Meth. Eng.*, 48:875–880, 2000.
55. P. Nithiarasu. An efficient artificial compressibility (AC) scheme based on the characteristic based split (CBS) method for incompressible flows. *Int. J. Numer. Meth. Eng.*, 56:1815–1845, 2003.
56. O.C. Zienkiewicz, R.L. Taylor, and P. Nithiarasu. *The Finite Element Method for Fluid Dynamics*. Butterworth-Heinemann, Oxford, 6th edition, 2005.
57. O.C. Zienkiewicz. Origins, milestones and directions of the finite element method. A personal view. In P.G. Ciarlet and J.L. Lyons, editors, *Handbook of Numerical Analysis*, volume IV, pages 3–65. North Holland, 1996.
58. R. Pierre. Simple C^0 approximations for the computation of incompressible flows. *Comp. Meth. Appl. Mech. Eng.*, 68:205–227, 1988.

12

Multidomain mixed approximations – domain decomposition and 'frame' methods

12.1 Introduction

In the previous chapters we have assumed in the approximations that all the variables were defined in the same manner throughout the domain of the analysis. This process can, however, be conveniently abandoned on occasion with the same or different formulations adopted in different subdomains of the problem. In this case some variables are only approximated on surfaces joining such subdomains.

There are two motivations for separating the whole domain into several subdomain regions. In the first of these the concept of parallel computation is paramount. Such parallel computation has become very important in many fields of engineering and allows us to use completely different methodologies for solving the problem in each individual part and even if this is not used allows us very much to increase the computer power by having separate operations going on simultaneously. In general the process we have just mentioned is referred to as *domain decomposition* and we shall devote the first part of this chapter to domain decomposition methodologies.

As this volume is not concerned in detail with the process of calculation and therefore does not discuss the subject of parallel computing in extended form, we refer the reader to references on the subject.[1–4]

Indeed the whole problem of domain decomposition associated with parallel computation is today an active field in which many conferences are held at frequent intervals and which seem to stir the imagination of many mathematicians and engineers. We shall discuss the problems of this kind in part one of this chapter.

It is of interest to note, however, that the methodologies for connecting separate subdomains can have other outcomes and objectives. In particular here the so-called *frame methods* have proved successful and, for instance, the introduction of *hybrid elements* by Pian *et al.* pioneered this type of approximation in the 1960s.[5, 6] More recently, other forms of frame approximation have been introduced and of particular interest is one in which so-called boundary approximations are used within an element with standard displacements specified on a frame. This allows for the introduction of many complex elements, capable of dealing with interesting problems on their own, which can be linked with more standard finite element computations.[7–14]

Domain decomposition methods

12.2 Linking of two or more subdomains by Lagrange multipliers

In this section we deal with the problem of connecting two or more subdomains in which standard finite element approximations of one form or another have been used. In particular we shall give as examples the process in which 'irreducible' formulations are used but, of course, other approximations could be introduced. The linking of such subdomains can be easily accomplished by the introduction of Lagrange multiplier methods to which we already referred to in Chapter 3 and elsewhere. The Lagrange multipliers for this case are defined on the boundary interface of the connecting subdomains.

In the present case we consider two subdomains, Ω^1 and Ω^2, which are to be joined together along an interface Γ_I. The generalization to multiple domains follows the same pattern. Independently approximated *Lagrange multipliers* (fluxes or tractions) are used on the interface to join the subdomains, as in Fig. 12.1.

In the first problem considered we treat the quasi-harmonic equation expressed in terms of the scalar potential function ϕ. This is followed by treatment for the elasticity problem.

12.2.1 Linking subdomains for quasi-harmonic equations

In Chapter 7 we considered the general problem for steady-state field problems. This problem resulted in a weak form in terms of a potential function ϕ. The approximation in a domain Ω^1 may be expressed as [viz. see Eq. (7.13)] (we ignore ϕ_0^1 for simplicity)

$$\int_{\Omega^1} \left[(\nabla\delta\phi^1)^{\mathrm{T}} \left(\mathbf{k}^1 \nabla\phi^1\right) + \delta\phi^1 Q^1 \right] \mathrm{d}\Omega + \int_{\Gamma_q} \delta\phi^1 \left(\bar{q}^1 + H^1\phi^1\right) \mathrm{d}\Gamma + \int_{\Gamma_I} \delta\phi^1 \lambda \,\mathrm{d}\Gamma = 0 \tag{12.1}$$

where the normal flux has been replaced by a Lagrange multiplier function λ defined on the interface Γ_I. Similarly, for the domain Ω^2 we have

$$\int_{\Omega^2} \left[(\nabla\delta\phi^2)^{\mathrm{T}} \left(\mathbf{k}^2 \nabla\phi^2\right) + \delta\phi^2 Q^2 \right] \mathrm{d}\Omega + \int_{\Gamma_q} \delta\phi^2 \left(\bar{q}^2 + H^2\phi^2\right) \mathrm{d}\Gamma - \int_{\Gamma_I} \delta\phi^2 \lambda \,\mathrm{d}\Gamma = 0 \tag{12.2}$$

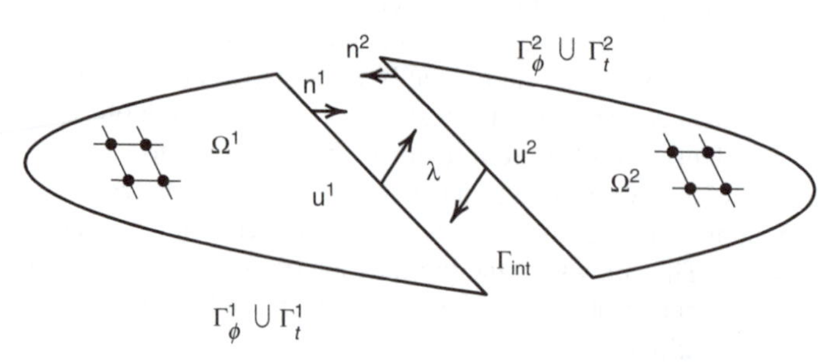

Fig. 12.1 Linking two (or more) domains by traction variables defined only on the interfaces. Variables in each domain are displacement **u** (irreducible form).

in which we have used $q_n^2 = -\lambda$ to satisfy flux continuity on the interface. The two subdomain equations are completed by a weak statement of continuity of the potential between the two subdomains. Thus we have

$$\int_{\Gamma_I} \delta\lambda \left(\phi^1 - \phi^2\right) \mathrm{d}\Gamma = 0 \tag{12.3}$$

Discretization of the potential in each domain and the Lagrange multiplier on the interface yields the final set of equations. Thus expressing the independent approximations as

$$\phi^1 = \mathbf{N}_1\tilde{\boldsymbol{\phi}}^1; \quad \phi^2 = \mathbf{N}_2\tilde{\boldsymbol{\phi}}^2 \quad \text{and} \quad \lambda = \mathbf{N}_\lambda\tilde{\boldsymbol{\lambda}} \tag{12.4}$$

we have

$$\begin{bmatrix} \mathbf{H}^1 & \mathbf{0} & \mathbf{Q}^1 \\ \mathbf{0} & \mathbf{H}^2 & \mathbf{Q}^2 \\ \mathbf{Q}^{1\mathrm{T}} & \mathbf{Q}^{2\mathrm{T}} & \mathbf{0} \end{bmatrix} \begin{Bmatrix} \tilde{\boldsymbol{\phi}}^1 \\ \tilde{\boldsymbol{\phi}}^2 \\ \tilde{\boldsymbol{\lambda}} \end{Bmatrix} + \begin{Bmatrix} \mathbf{f}^1 \\ \mathbf{f}^2 \\ \mathbf{0} \end{Bmatrix} = \mathbf{0} \tag{12.5}$$

where

$$\begin{aligned}
\mathbf{H}^1 &= \int_{\Omega_1} (\boldsymbol{\nabla}\mathbf{N}_1)^{\mathrm{T}}(\mathbf{k}^1\boldsymbol{\nabla}\mathbf{N}_1)\,\mathrm{d}\Omega, & \mathbf{H}^2 &= \int_{\Omega^2} (\boldsymbol{\nabla}\mathbf{N}_2)^{\mathrm{T}}(\mathbf{k}^2\boldsymbol{\nabla}\mathbf{N}_2)\,\mathrm{d}\Omega, \\
&\quad + \int_{\Gamma_q^1} \mathbf{N}_1^{\mathrm{T}} H^1 \mathbf{N}_1\,\mathrm{d}\Gamma, & &\quad + \int_{\Gamma_q^2} \mathbf{N}_2^{\mathrm{T}} H^2 \mathbf{N}_2\,\mathrm{d}\Gamma, \\
\mathbf{Q}^1 &= \int_{\Gamma_I} \mathbf{N}_1^{\mathrm{T}}\mathbf{N}_\lambda\,\mathrm{d}\Gamma, & \mathbf{Q}^2 &= -\int_{\Gamma_I} \mathbf{N}_2^{\mathrm{T}}\mathbf{N}_\lambda\,\mathrm{d}\Gamma, \\
\mathbf{f}^1 &= \int_{\Omega^1} \mathbf{N}_1^{\mathrm{T}} Q^1\,\mathrm{d}\Omega + \int_{\Gamma_q^1} \mathbf{N}_1^{\mathrm{T}} \bar{q}^1\,\mathrm{d}\Gamma, & \mathbf{f}^2 &= \int_{\Omega^2} \mathbf{N}_2^{\mathrm{T}} Q^2\,\mathrm{d}\Omega + \int_{\Gamma_q^2} \mathbf{N}_2^{\mathrm{T}} \bar{q}^2\,\mathrm{d}\Gamma
\end{aligned} \tag{12.6}$$

The formulation outlined above for two domains can obviously be extended to many subdomains and in many cases of practical analysis is useful in ensuring a better matrix conditioning and allowing the solution to be obtained with reduced computational effort.[15]

The variables $\tilde{\boldsymbol{\phi}}^1$ and $\tilde{\boldsymbol{\phi}}^2$ appear as internal and boundary variables within each subdomain (or superelement) and can be eliminated locally providing the matrices $\mathbf{H}^1$ and $\mathbf{H}^2$ are non-singular. Such non-singularity presupposes, however, that each of the subdomains has enough prescribed values to prevent the singular modes. If this is not the case partial elimination is always possible, retaining the singular modes until the complete solution is achieved.

We note that in the derivation of the matrices in Eq. (12.6) the shape function $\mathbf{N}_\lambda$ and hence λ itself are only specified along the interface surface. The choice of appropriate functions for the $\mathbf{N}_\lambda$ must, of course, satisfy the mixed patch requirement with counts performed for the interface degree of freedoms. Here the count condition can be more difficult to satisfy when multiple subdomains are connected at a point or along a line due to presence of multiple λ functions at these locations. One procedure to satisfy the condition is to use *mortar* or *dual mortar* methods. This matter is taken up later in this section. However, prior to this we consider the use of the above to include the Dirichlet boundary condition $\phi - \bar{\phi} = 0$ as part of the weak solution to the problem.

Treatment for forced boundary conditions

We note that the above form also may be used to satisfy the forced boundary condition $\phi = \bar{\phi}$ on Γ_ϕ. For this we let $\Gamma_I = \Gamma_\phi$ and from Eq. (12.1) (dropping the superscript '1') obtain

$$\int_\Omega \left[(\boldsymbol{\nabla} \delta\phi)^{\mathrm{T}} (\mathbf{k}\, \boldsymbol{\nabla}\phi) + \delta\phi\, Q \right] \mathrm{d}\Omega + \int_{\Gamma_q} \delta\phi\, (\bar{q} + H\phi)\, \mathrm{d}\Gamma + \int_{\Gamma_\phi} \delta\phi\, \lambda\, \mathrm{d}\Gamma = 0 \quad (12.7)$$

Similarly, from Eq. (12.3) with $\phi^2 = \bar{\phi}$ we have

$$\int_{\Gamma_\phi} \delta\lambda(\phi - \bar{\phi})\, \mathrm{d}\Gamma = 0 \quad (12.8)$$

The discrete form of the equations becomes

$$\begin{bmatrix} \mathbf{H} & \mathbf{Q} \\ \mathbf{Q}^{\mathrm{T}} & \mathbf{0} \end{bmatrix} \begin{Bmatrix} \tilde{\phi} \\ \tilde{\lambda} \end{Bmatrix} + \begin{Bmatrix} \mathbf{f} \\ \mathbf{f}_\lambda \end{Bmatrix} = 0 \quad (12.9)$$

where

$$\begin{aligned} \mathbf{H} &= \int_\Omega (\boldsymbol{\nabla}\mathbf{N})^{\mathrm{T}} (\mathbf{k}\boldsymbol{\nabla}\mathbf{N})\, \mathrm{d}\Omega + \int_{\Gamma_q} \mathbf{N}^{\mathrm{T}} H \mathbf{N}\, \mathrm{d}\Gamma \\ \mathbf{Q} &= \int_{\Gamma_\phi} \mathbf{N}^{\mathrm{T}} \mathbf{N}_\lambda\, \mathrm{d}\Gamma \\ \mathbf{f} &= \int_\Omega \mathbf{N}^{\mathrm{T}} Q\, \mathrm{d}\Omega + \int_{\Gamma_q} \mathbf{N}^{\mathrm{T}} \bar{q}\; \mathrm{d}\Gamma, \;\; \mathbf{f}_\lambda = -\int_{\Gamma_\phi} \mathbf{N}_\lambda^{\mathrm{T}} \phi\, \mathrm{d}\Gamma \end{aligned} \quad (12.10)$$

in which

$$\phi = \mathbf{N}\tilde{\phi} \quad \text{and} \quad \lambda = \mathbf{N}_\lambda \tilde{\lambda} \quad (12.11)$$

Mortar and dual mortar methods

The *mortar method* is a procedure which is used to join multiple subdomains.[16] Consider as an example a two-dimensional problem in which we use 4-noded (bilinear) quadrilaterals in Ω^1 and 9-noded (biquadratic) quadrilaterals in Ω^2. To connect subdomains the Lagrange multiplier may be approximated as shown in Fig. 12.2 for a subdomain with five segments along the interface. The use of the constant part at an end is required if multiple subdomains exist at the end point, otherwise the interpolation may be continued with normal linear interpolation as shown for the left end in Fig. 12.2. Along the interface we may connect subdomains with a different number of segments as shown in Fig. 12.3(a). Thus, if we

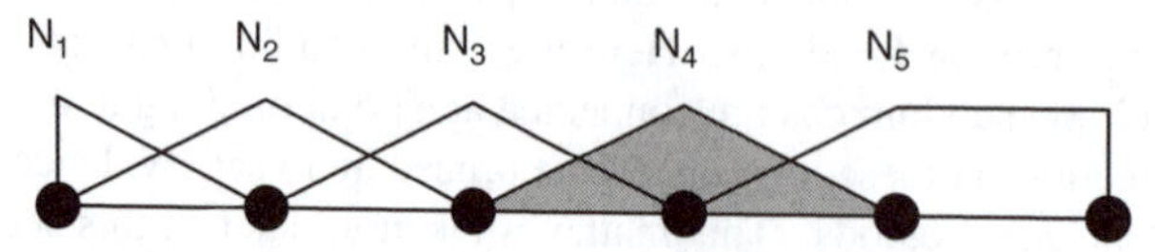

Fig. 12.2 Mortar function for Lagrange multiplier. Form for linear edges on Ω^1 elements.

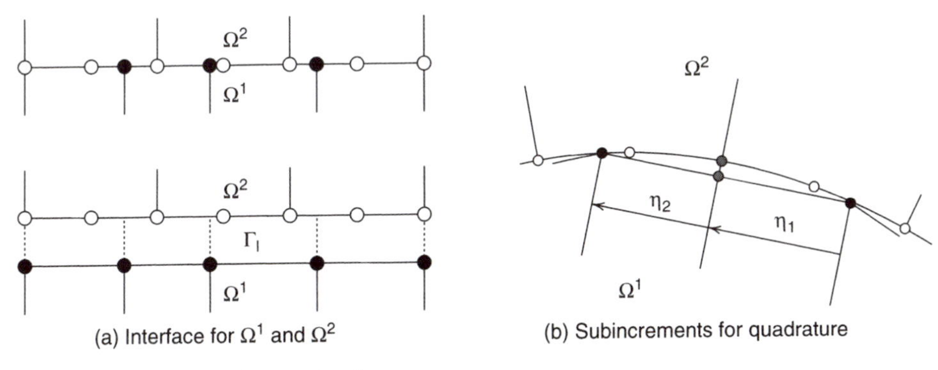

(a) Interface for Ω^1 and Ω^2

(b) Subincrements for quadrature

Fig. 12.3 Two-dimensional mortar interface.

assume the Lagrange multiplier interpolation uses $\mathbf{N}_\lambda = \mathbf{N}_1$ (except at end points)† the interface term resulting from (12.3) yields

$$\mathbf{Q}^1 = \int_{\Gamma_I} \mathbf{N}_1^T \mathbf{N}_1 \, \mathrm{d}\Gamma \quad \text{and} \quad \mathbf{Q}^2 = -\int_{\Gamma_I} \mathbf{N}_2^T \mathbf{N}_1 \, \mathrm{d}\Gamma \tag{12.12}$$

The integral for $\mathbf{Q}^1$ may be evaluated for each element edge using quadrature described in Chapter 4; however, evaluation by quadrature of the integral for $\mathbf{Q}^2$ requires further subdivision into *subincrements* along the element edges as indicated in Fig. 12.3(b).

The *dual mortar method* is an alternate form of the mortar method which has advantages for Lagrange multiplier and penalty forms. The dual shape functions are defined to satisfy

$$\int_{\Gamma_e} \widehat{N}_a N_b \, \mathrm{d}\Omega = \delta_{ab} \int_{\Gamma_e} N_b \, \mathrm{d}\Omega \tag{12.13}$$

where $\widehat{N}_a$ denotes a dual shape function and δ_{ab} is a Kronecker delta function. Figure 12.4 shows the dual functions computed for the standard linear functions shown in Fig. 12.2. The dual functions are *discontinuous* between elements, which is permitted since no derivatives appear for the Lagrange multipliers $\boldsymbol{\lambda}$.

The dual functions may be computed for each element edge separately. For linear edges the result is shown in Fig. 12.5. The process may be repeated for higher order functions without difficulty; however, for higher order edges nodes appear between the ends and, thus, with arbitrary spacing the computation must be computed for each case separately.

The advantage of the dual functions is evident from the definition of $\mathbf{Q}^1$ (assuming we start with $\mathbf{N}_\lambda = \mathbf{N}_1$). Here we observe that

$$\mathbf{Q}^1 = \int_{\Gamma_I} \mathbf{N}_1^T \widehat{\mathbf{N}}_\lambda \, \mathrm{d}\Gamma = \widehat{\mathbf{Q}}^1 \tag{12.14}$$

where $\widehat{\mathbf{Q}}^1$ is *diagonal* by the properties of Eq. (12.13).

†An alternative to avoiding modification at end points is to use a stabilization method such as one equivalent to the direct pressure stabilization presented in Sec. 11.7.3.

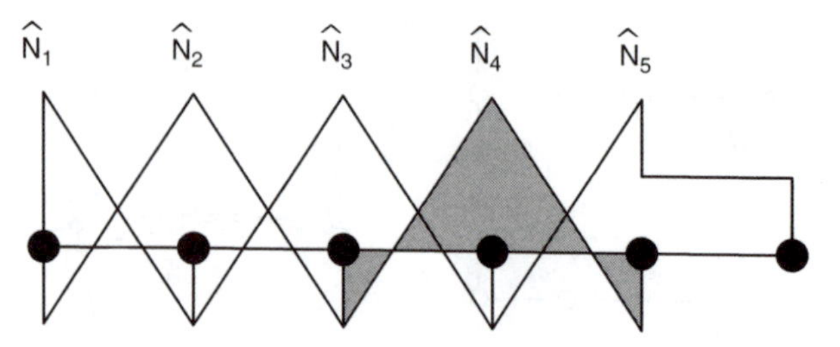

Fig. 12.4 Dual mortar function for Lagrange multiplier. Form for linear edges on Ω^1 elements.

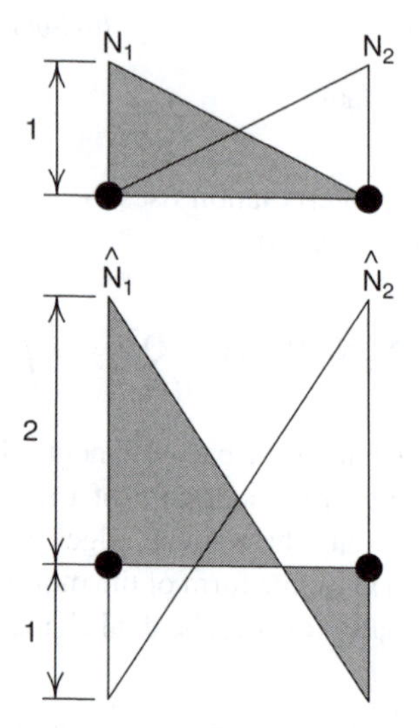

Fig. 12.5 Mortar and dual mortar shape functions for two-dimensional linear edge.

12.2.2 Linking subdomains for elasticity equations

In this problem we formulate the approximation in domain Ω^1 in terms of displacements $\mathbf{u}^1$ resulting from an irreducible (displacement) form of the elasticity equations. The traction $\mathbf{t}^1$ on the interface is denoted by $\boldsymbol{\lambda}$. With the weak form using the standard virtual work expression [see Eq. (6.52)] ignoring $\boldsymbol{\sigma}_0$ and ε_0 we have†

$$\int_{\Omega^1} \delta(\boldsymbol{\mathcal{S}}\mathbf{u}^1)^{\mathrm{T}}\mathbf{D}^1\boldsymbol{\mathcal{S}}\mathbf{u}^1 \,\mathrm{d}\Omega - \int_{\Gamma_I} \delta\mathbf{u}^{1\mathrm{T}}\boldsymbol{\lambda}\,\mathrm{d}\Gamma - \int_{\Omega^1} \delta\mathbf{u}^{1\mathrm{T}}\mathbf{b}\,\mathrm{d}\Omega - \int_{\Gamma_t^1} \delta\mathbf{u}^{1\mathrm{T}}\bar{\mathbf{t}}\,\mathrm{d}\Gamma = 0 \quad (12.15)$$

in which as usual we assume that the satisfaction of the prescribed displacement on Γ_u^1 is implied by the approximation for $\mathbf{u}^1$. Similarly in domain Ω^2 we can write, now putting the interface traction as $\mathbf{t}^2 = -\boldsymbol{\lambda}$ to ensure equilibrium between the two domains,

$$\int_{\Omega^2} \delta(\boldsymbol{\mathcal{S}}\mathbf{u}^2)^{\mathrm{T}}\mathbf{D}^2\boldsymbol{\mathcal{S}}\mathbf{u}^2 \,\mathrm{d}\Omega + \int_{\Gamma_I} \delta\mathbf{u}^{2\mathrm{T}}\boldsymbol{\lambda}\,\mathrm{d}\Gamma - \int_{\Omega^2} \delta\mathbf{u}^{2\mathrm{T}}\mathbf{b}\,\mathrm{d}\Omega - \int_{\Gamma_t^2} \delta\mathbf{u}^{2\mathrm{T}}\bar{\mathbf{t}}\,\mathrm{d}\Gamma = 0 \quad (12.16)$$

The two subdomain equations are completed by a weak statement of displacement continuity on the interface between the two domains, i.e.,

† Here we use (6.4) to replace ε and $\delta\varepsilon$ by $\boldsymbol{\mathcal{S}}\mathbf{u}$ and $\boldsymbol{\mathcal{S}}\delta\mathbf{u}$, respectively.

$$\int_{\Gamma_I} \delta\boldsymbol{\lambda}^{\mathrm{T}}(\mathbf{u}^2 - \mathbf{u}^1)\,\mathrm{d}\Gamma = 0 \tag{12.17}$$

Discretization of displacements in each domain and of the Lagrange multipliers (tractions) $\boldsymbol{\lambda}$ on the interface yields the final system of equations. Thus putting the independent approximations as

$$\mathbf{u}^1 = \mathbf{N}_1\tilde{\mathbf{u}}^1; \quad \mathbf{u}^2 = \mathbf{N}_2\tilde{\mathbf{u}}^2; \quad \boldsymbol{\lambda} = \mathbf{N}_\lambda\tilde{\boldsymbol{\lambda}} \tag{12.18}$$

we have

$$\begin{bmatrix} \mathbf{K}^1 & \mathbf{0} & \mathbf{Q}^1 \\ \mathbf{0} & \mathbf{K}^2 & \mathbf{Q}^2 \\ \mathbf{Q}^{1\mathrm{T}} & \mathbf{Q}^{2\mathrm{T}} & \mathbf{0} \end{bmatrix} \begin{Bmatrix} \tilde{\mathbf{u}}^1 \\ \tilde{\mathbf{u}}^2 \\ \tilde{\boldsymbol{\lambda}} \end{Bmatrix} = \begin{Bmatrix} \mathbf{f}^1 \\ \mathbf{f}^2 \\ \mathbf{0} \end{Bmatrix} \tag{12.19}$$

where

$$\begin{aligned} \mathbf{K}^1 &= \int_{\Omega^1} \mathbf{B}^{1\mathrm{T}}\mathbf{D}^1\mathbf{B}^1\,\mathrm{d}\Omega, \quad \mathbf{K}^2 = \int_{\Omega^2} \mathbf{B}^{2\mathrm{T}}\mathbf{D}^2\mathbf{B}^2\,\mathrm{d}\Omega \\ \mathbf{Q}^1 &= -\int_{\Gamma_I} \mathbf{N}_1^{\mathrm{T}}\mathbf{N}_\lambda\,\mathrm{d}\Gamma, \quad \mathbf{Q}^2 = \int_{\Gamma_I} \mathbf{N}_2^{\mathrm{T}}\mathbf{N}_\lambda\,\mathrm{d}\Gamma \\ \mathbf{f}^1 &= \int_{\Omega^1} \mathbf{N}_1^{\mathrm{T}}\mathbf{b}^1\,\mathrm{d}\Omega + \int_{\Gamma^1} \mathbf{N}_1^{\mathrm{T}}\bar{\mathbf{t}}^1\,\mathrm{d}\Gamma, \quad \mathbf{f}^2 = \int_{\Omega^2} \mathbf{N}_2^{\mathrm{T}}\mathbf{b}^2\,\mathrm{d}\Omega + \int_{\Gamma^2} \mathbf{N}_2^{\mathrm{T}}\bar{\mathbf{t}}^2\,\mathrm{d}\Gamma \end{aligned} \tag{12.20}$$

The process described here is very similar to that introduced by Kron [17] at a very early date and, more recently, used by Farhat *et al.* in the FETI (finite element tearing and interconnecting) method[18] which uses the process on many individual element partitions as a means of iteratively solving large problems.

The formulation is, of course, subject to limitations imposed by the stability and consistency conditions of the mixed patch test for selection of appropriate number of $\boldsymbol{\lambda}$ variables.

The formulation just used can, of course, be applied to a single field displacement formulation in which we are required to specify the displacement on the boundaries in a weak sense – rather than imposing these directly on displacement shape functions.

This problem can be approached directly or can be derived simply using (12.15) and (12.17) in which we put $\mathbf{u}^2 = \bar{\mathbf{u}}$, the specified displacement on $\Gamma_I \equiv \Gamma_u$.

Now the equation system is simply

$$\begin{bmatrix} \mathbf{K}^1 & \mathbf{Q}^1 \\ \mathbf{Q}^{1\mathrm{T}} & \mathbf{0} \end{bmatrix} \begin{Bmatrix} \tilde{\mathbf{u}}^1 \\ \tilde{\boldsymbol{\lambda}} \end{Bmatrix} = \begin{Bmatrix} \mathbf{f}_1 \\ \mathbf{f}_\lambda \end{Bmatrix} \tag{12.21}$$

where

$$\mathbf{f}_\lambda = -\int_{\Gamma_I} \mathbf{N}_\lambda^{\mathrm{T}}\bar{\mathbf{u}}\,\mathrm{d}\Gamma \tag{12.22}$$

This formulation is often convenient for imposing a prescribed displacement on an element field when the boundary values cannot fit the shape function form.

We have approached the above formulation directly via weak or weighted residual forms. A variational principle could be given here simply as the minimization of total potential energy (see Chapter 2) subject to a Lagrange multiplier $\boldsymbol{\lambda}$ imposing subdomain continuity. The stationarity of

$$\Pi = \sum_{i=1}^{2}\left[\frac{1}{2}\int_{\Omega^i} (\mathcal{S}\mathbf{u}^i)^{\mathrm{T}}\mathbf{D}^i\mathcal{S}\mathbf{u}^i\,\mathrm{d}\Omega - \int_{\Omega^i} \mathbf{u}^{i\mathrm{T}}\mathbf{b}\,\mathrm{d}\Omega - \int_{\Gamma_t^i} \mathbf{u}^{i\mathrm{T}}\bar{\mathbf{t}}\,\mathrm{d}\Gamma\right] + \int_{\Gamma_I} \boldsymbol{\lambda}^{\mathrm{T}}(\mathbf{u}^2 - \mathbf{u}^1)\,\mathrm{d}\Gamma \tag{12.23}$$

would result in the equation set (12.15) to (12.17). The formulation is, of course, subject to limitations imposed by the stability and consistency conditions of the mixed patch test for selection of the appropriate number of $\boldsymbol{\lambda}$ variables.

Example 12.1: A mortar method for two-dimentional elasticity. Mortar and dual mortar methods may also be used in the solution of elasticity problems. The formulation follows that given for the quasi-harmonic equation with appropriate change in variables. To indicate the type of result which occurs using mortar or dual mortar forms we consider the problem of a strip loaded by a uniform pressure along a short segment. The problem is solved as a single region using a fine mesh over the whole domain and also by a two subdomain form in which fine elements are in the top layer only (see Fig. 12.6). Contours for the vertical displacement are presented in Fig. 12.7 for the two cases. It is evident that the mortar treatment produces excellent continuity in displacement. A comparison for the vertical stress, σ_y, is not shown here again the results exhibit very small discontinuity.

12.3 Linking of two or more subdomains by perturbed lagrangian and penalty methods

In the previous section we have shown how linking can be achieved using Lagrange multipliers. A disadvantage of the Lagrange multiplier approach is the addition of extra unknowns (the Lagrange multipliers $\boldsymbol{\lambda}$) and the creation of equations which have zero on the diagonal. As we have shown previously (viz. Chapter 3) it is possible to avoid both of these situations using a *perturbed lagrangian* or *penalty* form.

The perturbed lagrangian form of the equations may be achieved by modifying Eq. (12.17) to

$$\int_{\Gamma_I} \delta\boldsymbol{\lambda}^{\mathrm{T}}(\mathbf{u}^2 - \mathbf{u}^1)\,\mathrm{d}\Gamma - \frac{1}{\alpha}\int_{\Gamma_I} \delta\boldsymbol{\lambda}^{\mathrm{T}}\boldsymbol{\lambda}\,\mathrm{d}\Gamma = 0 \tag{12.24}$$

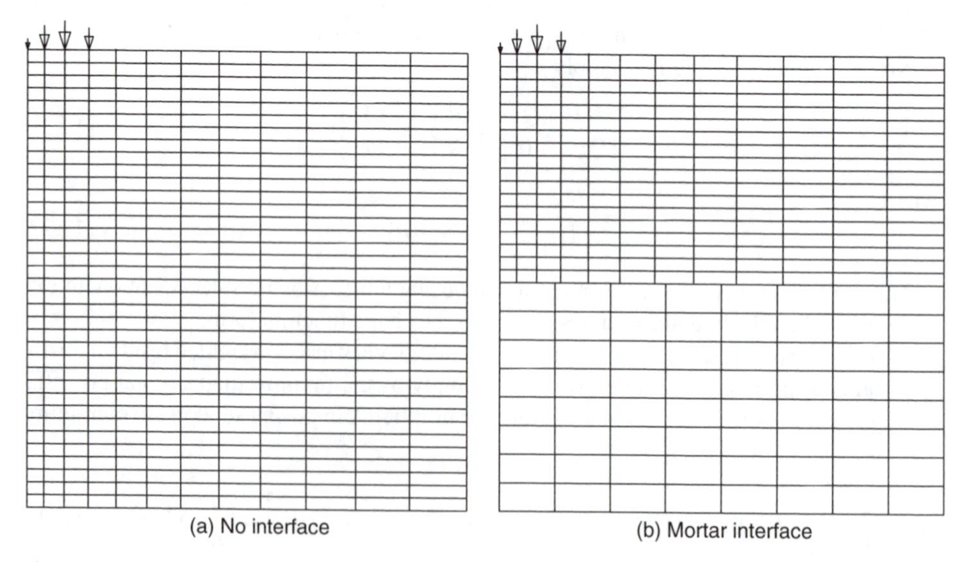

Fig. 12.6 Mesh and nodal loading for vertically loaded strip.

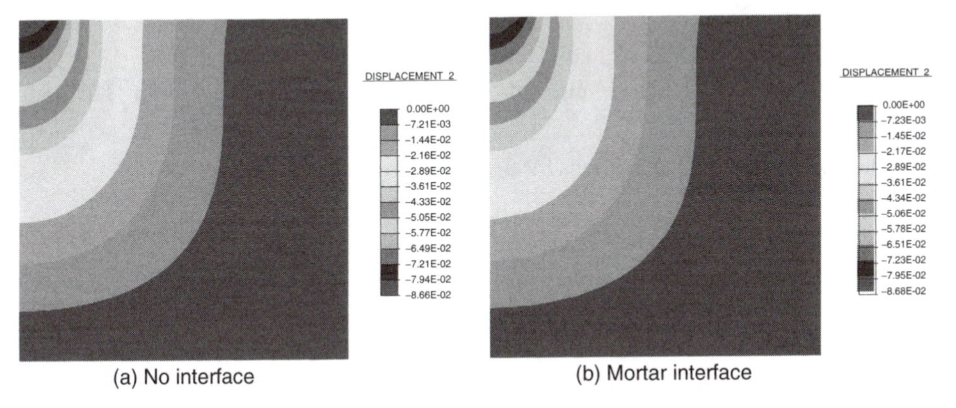

(a) No interface (b) Mortar interface

Fig. 12.7 Vertical displacement for strip loaded over short segment of top.

in which α is a large (penalty) parameter. Inserting the approximation (12.18) into (12.15), (12.16) and (12.24) results in the form

$$\begin{bmatrix} \mathbf{K}^1 & \mathbf{0} & \mathbf{Q}^1 \\ \mathbf{0} & \mathbf{K}^2 & \mathbf{Q}^2 \\ \mathbf{Q}^{1\mathrm{T}} & \mathbf{Q}^{2\mathrm{T}} & -\frac{1}{\alpha}\mathbf{V} \end{bmatrix} \begin{Bmatrix} \tilde{\mathbf{u}}^1 \\ \tilde{\mathbf{u}}^2 \\ \tilde{\boldsymbol{\lambda}} \end{Bmatrix} = \begin{Bmatrix} \mathbf{f}^1 \\ \mathbf{f}^2 \\ \mathbf{0} \end{Bmatrix} \tag{12.25}$$

where in addition to the arrays defined in Eq. (12.20)

$$\mathbf{V} = \int_{\Gamma_I} \mathbf{N}_\lambda^{\mathrm{T}} \mathbf{N}_\lambda \, \mathrm{d}\Gamma \tag{12.26}$$

Clearly, as the parameter α tends to infinity the result becomes identical to the Lagrange multiplier form. Such approximation thus behaves as a *penalty*-type form. Formally, we can eliminate the Lagrange multiplier parameters from (12.25) to obtain

$$\begin{bmatrix} (\mathbf{K}^1 + \alpha \mathbf{Q}^1 \mathbf{V}^{-1} \mathbf{Q}^{1\mathrm{T}}) & \alpha \mathbf{Q}^1 \mathbf{V}^{-1} \mathbf{Q}^{2\mathrm{T}} \\ \alpha \mathbf{Q}^2 \mathbf{V}^{-1} \mathbf{Q}^{1\mathrm{T}} & (\mathbf{K}^2 + \alpha \mathbf{Q}^2 \mathbf{V}^{-1} \mathbf{Q}^{2\mathrm{T}}) \end{bmatrix} \begin{Bmatrix} \tilde{\mathbf{u}}^1 \\ \tilde{\mathbf{u}}^2 \end{Bmatrix} = \begin{Bmatrix} \mathbf{f}^1 \\ \mathbf{f}^2 \end{Bmatrix} \tag{12.27}$$

which we recognize as a *penalty*-type form

$$[\mathbf{K}_1 + \alpha \mathbf{K}_2]\, \tilde{\mathbf{u}} = \mathbf{f}$$

[viz. Eq. (3.137)].

An alternative to the above solves (12.24) *for each point on the boundary* Γ_I yielding

$$\boldsymbol{\lambda} = \alpha \, \left(\mathbf{u}^2 - \mathbf{u}^1\right) \tag{12.28}$$

Substituting this into (12.15) and (12.16) then gives

$$\int_{\Omega^1} \delta(\mathcal{S}\mathbf{u}^1)^{\mathrm{T}} \mathbf{D}^1 \mathcal{S}\mathbf{u}^1 \, \mathrm{d}\Omega - \alpha \int_{\Gamma_I} \delta\mathbf{u}^{1\mathrm{T}} (\mathbf{u}^2 - \mathbf{u}^1) \, \mathrm{d}\Gamma - \int_{\Omega^1} \delta\mathbf{u}^{1\mathrm{T}} \mathbf{b} \, \mathrm{d}\Omega - \int_{\Gamma_t^1} \delta\mathbf{u}^{1\mathrm{T}} \bar{\mathbf{t}} \, \mathrm{d}\Gamma = 0 \tag{12.29}$$

and

$$\int_{\Omega^2} \delta(\mathcal{S}\mathbf{u}^2)^{\mathrm{T}}\mathbf{D}^2\mathcal{S}\mathbf{u}^2 \,\mathrm{d}\Omega + \alpha \int_{\Gamma_I} \delta\mathbf{u}^{2\mathrm{T}}(\mathbf{u}^2 - \mathbf{u}^1)\,\mathrm{d}\Gamma - \int_{\Omega^2} \delta\mathbf{u}^{2\mathrm{T}}\mathbf{b}\,\mathrm{d}\Omega - \int_{\Gamma_t^2} \delta\mathbf{u}^{2\mathrm{T}}\bar{\mathbf{t}}\,\mathrm{d}\Gamma = 0 \tag{12.30}$$

Introducing now the approximations for $\mathbf{u}^1$ and $\mathbf{u}^2$ produces the penalty form

$$\begin{bmatrix} (\mathbf{K}^1 + \alpha\mathbf{K}_2^{11}) & -\alpha\mathbf{K}_2^{12} \\ -\alpha\mathbf{K}_2^{21} & (\mathbf{K}^2 + \alpha\mathbf{K}_2^{22}) \end{bmatrix} \begin{Bmatrix} \tilde{\mathbf{u}}^1 \\ \tilde{\mathbf{u}}^2 \end{Bmatrix} = \begin{Bmatrix} \mathbf{f}^1 \\ \mathbf{f}^2 \end{Bmatrix} \tag{12.31}$$

where

$$\mathbf{K}_2^{ij} = \int_{\Gamma_I} \mathbf{N}_i^{\mathrm{T}}\mathbf{N}_j \,\mathrm{d}\Gamma; \; i, j = 1, 2 \tag{12.32}$$

which we again recognize as a penalty form as given in Eq. (3.137).

The differences between the penalty form (12.27) and that of (12.31) are significant:

(a) The form given by (12.27) will not exhibit *locking* provided the choice for the $\mathbf{N}_\lambda$ satisfies the conditions for the mixed patch test.
(b) The form given by (12.31) and (12.32) usually requires use of *reduced quadrature* on $\mathbf{K}_2^{ij}$ in order to avoid locking for reasons we discussed in Chapter 11.
(c) Using standard or dual mortar methods the form (12.27) satisfies consistency conditions (e.g., constant stress) across the interface.[19] Generally, the form (12.31) does not transmit a constant stress condition correctly at the interface unless perfect matching of meshes occurs on Γ_I.

The above remarks clearly favour the form (12.27); however, this form requires the inversion of the matrix $\mathbf{V}$ (or a solution process equivalent to such inversion) and this can present difficulties. For the dual mortar method discussed in Sec. 12.2.1, the Lagrange multiplier can be eliminated by a perturbed lagrangian approach using the discretized form of (12.26) approximated as

$$\mathbf{V} \approx \int_{\Gamma_i} \hat{\mathbf{N}}_\lambda^{\mathrm{T}}\mathbf{N}_\lambda \,\mathrm{d}\Gamma \tag{12.33}$$

In this case the matrix to be inverted is also diagonal and the Lagrange multiplier may be locally eliminated to give a penalty form.

12.3.1 Nitsche method and discontinuous Galerkin approximation

An alternative to Lagrange multiplier and penalty methods for including the Dirichlet boundary condition was introduced by Nitsche.[20] Here we consider the procedure to include the condition $\phi = \bar{\phi}$ in the weak form of the quasi-harmonic equation. We first add together Eqs (12.7) and (12.8) to obtain

$$\int_\Omega \left[(\boldsymbol{\nabla}\delta\phi)^{\mathrm{T}}(\mathbf{k}\,\boldsymbol{\nabla}\phi) + \delta\phi Q\right] \mathrm{d}\Omega + \int_{\Gamma_q} \delta\phi\,(\bar{q} + H\phi)\,\mathrm{d}\Gamma + \int_{\Gamma_\phi} \delta\phi\,\lambda\,\mathrm{d}\Gamma + \int_{\Gamma_\phi} \delta\lambda\,(\phi - \bar{\phi})\,\mathrm{d}\Gamma = 0 \tag{12.34}$$

The normal flux q_n on the boundary part Γ_ϕ is replaced by

$$\lambda = q_n(\phi) = -\mathbf{n}^{\mathrm{T}}(\mathbf{k}\boldsymbol{\nabla}\phi) \qquad \delta\lambda = q_n(\delta\phi) = -\mathbf{n}^{\mathrm{T}}(\mathbf{k}\boldsymbol{\nabla}\delta\phi)$$

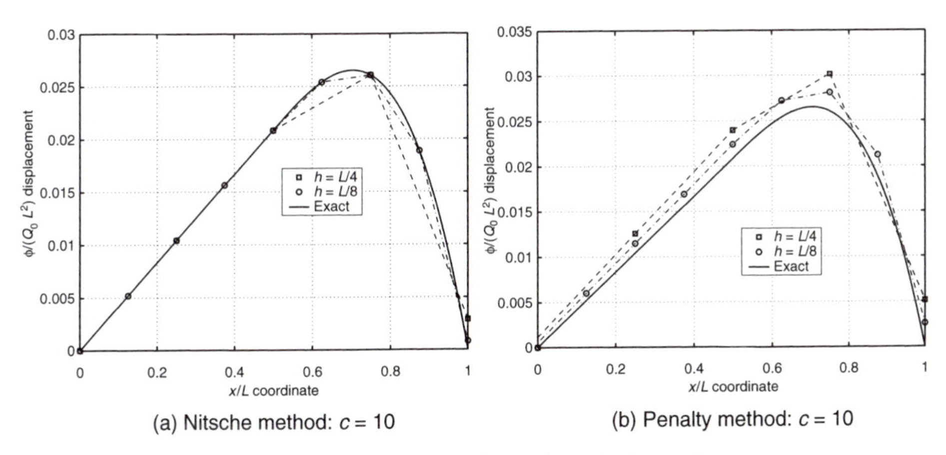

(a) Nitsche method: $c = 10$ (b) Penalty method: $c = 10$

Fig. 12.8 Solution of one-dimensional heat equation of Example 3.5 in Chapter 3.

thus eliminating the appearance of the Lagrange multiplier and giving a weak form expressed entirely in terms of ϕ. One now can note that these two terms on Γ_ϕ can be zero for ϕ and $\delta\phi$ having constant values. Thus, to make the method *stable* Nitsche adds a penalty-like term

$$\int_{\Gamma_\phi} \delta\phi\, \alpha\, \left(\phi - \bar{\phi}\right) \mathrm{d}\Gamma$$

However, it is not required that α be large to ensure a good satisfaction of the boundary condition. The value recommended by Nitsche for linear elements is

$$\alpha = c\,\frac{|\mathbf{k}|}{h} \qquad c = O(10)$$

where h is an element size and $|\mathbf{k}|$ a norm of the diffusion matrix.

The above steps give the weak form

$$\begin{aligned}
&\int_\Omega \left[(\boldsymbol{\nabla}\delta\phi)^{\mathrm{T}}(\mathbf{k}\boldsymbol{\nabla}\phi) + \delta\phi Q\right] \mathrm{d}\Omega + \int_{\Gamma_q} \delta\phi\,(\bar{q} + H\phi)\,\mathrm{d}\Gamma + \int_{\Gamma_\phi} \delta\phi q_n(\phi)\,\mathrm{d}\Gamma \\
&\quad + \underbrace{\int_{\Gamma_\phi} q_n(\delta\phi)\,[\phi - \bar{\phi}]\,\mathrm{d}\Gamma}_{\text{Symmetry}} + \underbrace{\int_{\Gamma_\phi} \delta\phi\alpha\,[\phi - \bar{\phi}]\,\mathrm{d}\Gamma}_{\text{Stability}} = 0
\end{aligned} \tag{12.35}$$

Substituting the approximation for ϕ given by Eq. (12.11) into (12.35) gives

$$\mathbf{H}\,\tilde{\phi} = \mathbf{f} \tag{12.36}$$

where

$$\begin{aligned}\mathbf{H} &= \int_\Omega (\boldsymbol{\nabla}\mathbf{N})^\mathrm{T}\mathbf{k}(\boldsymbol{\nabla}\mathbf{N})\,\mathrm{d}\Omega + \int_{\Gamma_q} \mathbf{N}^\mathrm{T} H \mathbf{N}\,\mathrm{d}\Gamma \\ &\quad - \int_{\Gamma_\phi} \mathbf{N}^\mathrm{T}\left(\mathbf{n}^\mathrm{T}[\mathbf{k}(\boldsymbol{\nabla}\mathbf{N})]\right)\,\mathrm{d}\Gamma - \int_{\Gamma_\phi}\left(\mathbf{n}^\mathrm{T}[\mathbf{k}(\boldsymbol{\nabla}\mathbf{N})]\right)^\mathrm{T}\mathbf{N}\,\mathrm{d}\Gamma + \int_{\Gamma_\phi}\mathbf{N}^\mathrm{T}\alpha\,\mathbf{N}\,\mathrm{d}\Gamma \\ \mathbf{f} &= -\int_\Omega \mathbf{N}^\mathrm{T} Q\,\mathrm{d}\Omega - \int_{\Gamma_q}\mathbf{N}^\mathrm{T}\bar{q}\,\mathrm{d}\Gamma \\ &\quad - \int_{\Gamma_\phi}\left(\mathbf{n}^\mathrm{T}[\mathbf{k}(\boldsymbol{\nabla}\mathbf{N})]\right)^\mathrm{T}\bar{\phi}\,\mathrm{d}\Gamma + \int_{\Gamma_\phi}\mathbf{N}^\mathrm{T}\alpha\bar{\phi}\,\mathrm{d}\Gamma\end{aligned}$$

The Nitsche method results in a form in terms of the original primary variables of the problem. We can easily extend this to consider the connection of multiple subdomains.

Example 12.2: Dirichlet boundary condition. To indicate the performance of the Nitsche method in satisfaction of the Dirichlet boundary condition, we consider the one-dimensional problem given in Chapter 3 as Example 3.5. There the differential equation was given as

$$A(\phi) = -\frac{\mathrm{d}^2\phi}{\mathrm{d}x^2} + Q(x) = 0 \quad 0 \le x \le L$$

with boundary conditions $\phi(0) = \phi(L) = 0$. We shall consider two domains: Ω^1 for $0 \le x \le L/2$ and Ω^2 for $L/2 < x \le L$. The loading on Ω^2 is a linear continuous function and that on Ω^1 is zero. Using the Lagrange multiplier solution for this problem results in exact satisfaction of the boundary conditions $\phi(0) = \phi(L) = 0$, and, consequently, the same solution as given for the standard finite element solution in Chapter 3, Fig. 3.5. Using the Nitsche method with $c = 10$ and $h = L/4$ and $L/8$ ($k = 1$) gives the solution shown in Fig. 12.8(a). For comparison we drop the terms on the boundary with $q(\phi)$, and $q(\delta\phi)$ (i.e., use the penalty form alone) but keep the same value for c. This solution is shown in Fig. 12.8(b). Of course, increasing the size of c with either approach will improve the satisfaction of the boundary condition – but with an increased sensitivity in equation solution. The overall improvement of the Nitsche method is clearly evident and is accomplished without an increase in equation condition number. The results using quadratic results are even better as shown in Fig. 12.9.

Multiple subdomain problems

We again return to the problem of connecting two subdomains defined in Ω^1 and Ω^2 in which the common interface is Γ_I. The weak form of the problem may be written now as

$$\begin{aligned}G &= \sum_{i=1}^{2}\int_{\Omega^i}\left[\left(\boldsymbol{\nabla}\delta\phi^i\right)^\mathrm{T}\left(\mathbf{k}^i\boldsymbol{\nabla}\phi^i\right) + \delta\phi^i Q^i\right]\mathrm{d}\Omega + \int_{\Gamma_q^i}\delta\phi^i\left(\bar{q}^i + H^i\phi^i\right)\mathrm{d}\Gamma \\ &\quad + \int_{\Gamma_I}[\delta\phi^1 - \delta\phi^2]q_n(\phi^1,\phi^2)\,\mathrm{d}\Gamma + \int_{\Gamma_I} q_n(\delta\phi^1,\delta\phi^2)\,[\phi^1-\phi^2]\,\mathrm{d}\Gamma \\ &\quad + \int_{\Gamma_I}[\delta\phi^1 - \delta\phi^2]\,\alpha\,[\phi^1-\phi^2]\,\mathrm{d}\Gamma = 0\end{aligned} \tag{12.37}$$

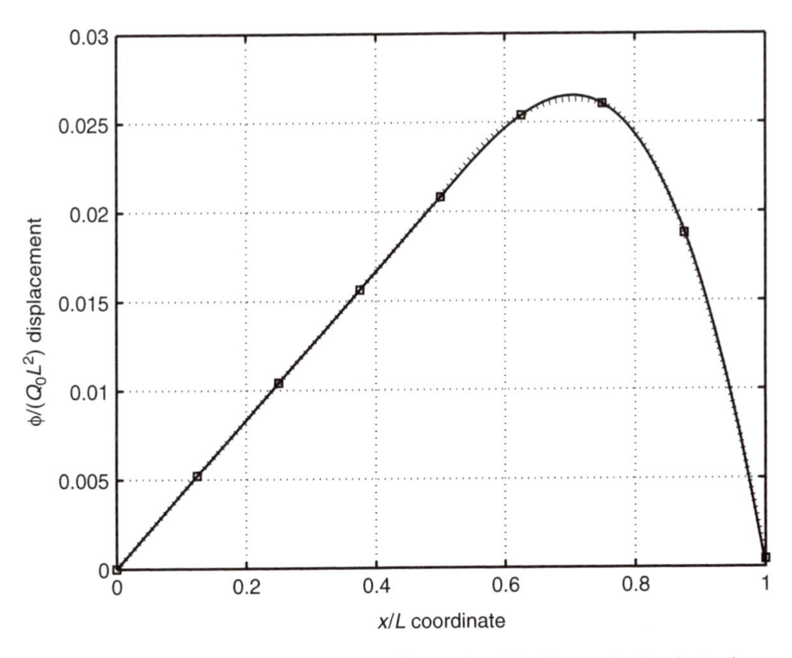

Fig. 12.9 Solution of one-dimensional heat equation of Example 3.5 in Chapter 3. Quadratic elements.

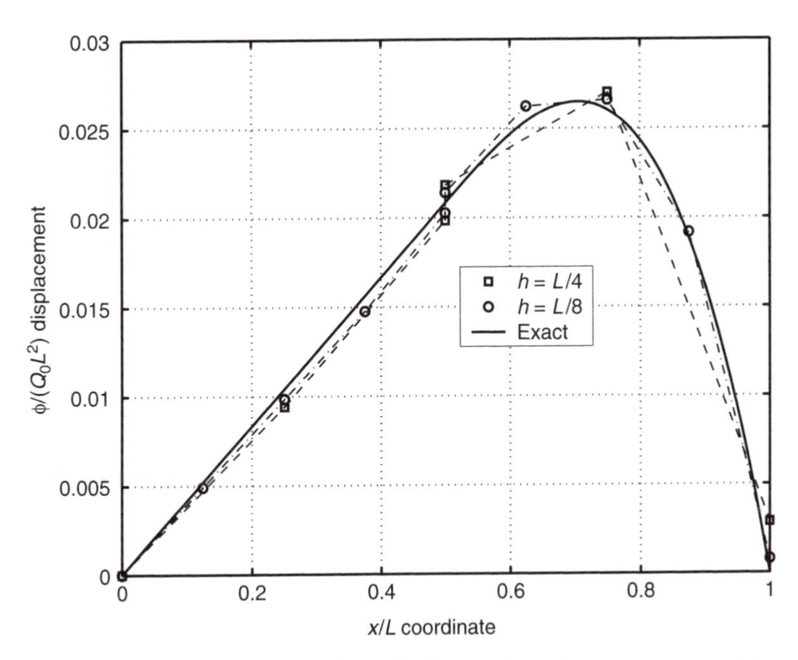

Fig. 12.10 Two subdomain solution using Nitsche method for one-dimensional heat equation of Example 3.5 in Chapter 3. Linear elements.

which results from adding Eqs (12.1) and (12.2) and setting the Lagrange multiplier to

$$\lambda = q_n(\phi^1, \phi^2) \quad \text{and} \quad \delta\lambda = q_n(\delta\phi^1, \delta\phi^2) \tag{12.38}$$

which now becomes a function of the flux from both sides of the interface.

This form is an extension of the concept of Nitsche and, of course, can be effectively used to consider multiple subdomains in an obvious manner. When extended to the case where each element becomes a subdomain the problem assumes a form known as the *discontinuous Galerkin method.*[21–25]

The discontinuous Galerkin method was first introduced by Reed and Hill[26] for analysis of neutron transport problems. It was analysed by Lesaint and Raviart for its mathematical properties.[27] As shown in the paper by Zienkiewicz *et al.* the method is most effective in problems which have significant convection effects – and is less accurate than standard (continuous) finite elements for problems which possess only diffusion effects.[28] Here we are interested in the method primarily for connecting subdomains which contain either a large number of standard elements or have high order expansions with significant number of parameters not associated with the boundary.

Example 12.3: Two domain problem. To indicate the performance in the presence of multiple domains we again consider the one-dimensional of Example 12.2. We shall consider two domains: Ω^1 for $0 \leq x \leq L/2$ and Ω^2 for $L/2 < x \leq L$. The loading on Ω^2 is a linear continuous function and that on Ω^1 is zero. The Nitsche method is used with four and eight elements (two and four in each subdomain, respectively) and a value of $c = 10$. The solution is shown in Fig. 12.10 and again indicates quite rapid convergence with increased number of elements. We also show results for the same problem with quadratic elements, Fig. 12.11, in which no discernible jump exists for the eight-element case.

Frame methods

12.4 Interface displacement 'frame'

12.4.1 General remarks

In the preceding examples we have used traction as the Lagrange multiplier interface variable linking two or more subdomains of elasticity problems. Due to lack of rigid body constraints the elimination of local subdomain displacements has generally been impossible. For this and other reasons it is convenient to accomplish the linking of subdomains via a displacement field *defined only on the interface* [Fig. 12.12(a)] and to eliminate all the interior variables so that this linking can be accomplished via a standard stiffness matrix procedure using only the interface variables.

A *displacement frame* can be made to surround the subdomain completely and if all internal variables are eliminated will yield a stiffness matrix of a new 'element' which can be used directly in coupling with any other element with similar displacement assumptions on the interface, irrespective of the procedure used for deriving such an element [Fig. 12.12(b)].

In all the examples of this section we shall approximate the frame displacements as

$$\mathbf{v} = \mathbf{N}_v \tilde{\mathbf{v}} \quad \text{on} \quad \Gamma_I \tag{12.39}$$

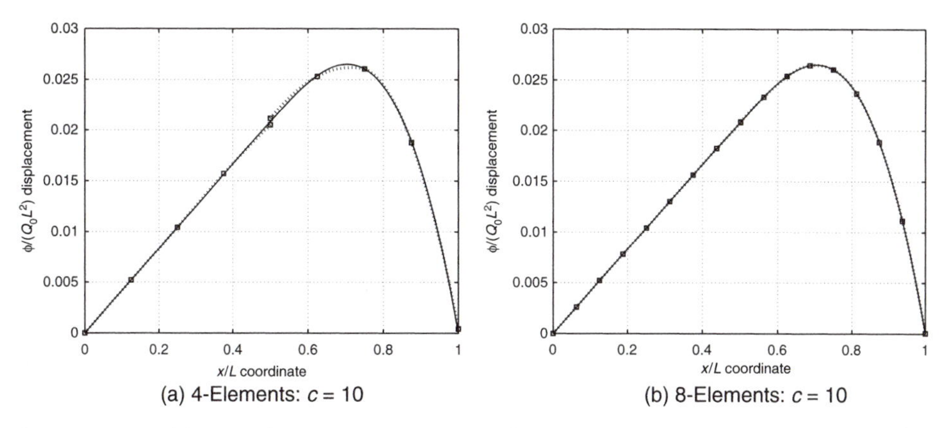

Fig. 12.11 Two subdomain solution using Nitsche method for one-dimensional heat equation of Example 3.5 in Chapter 3. Quadratic elements.

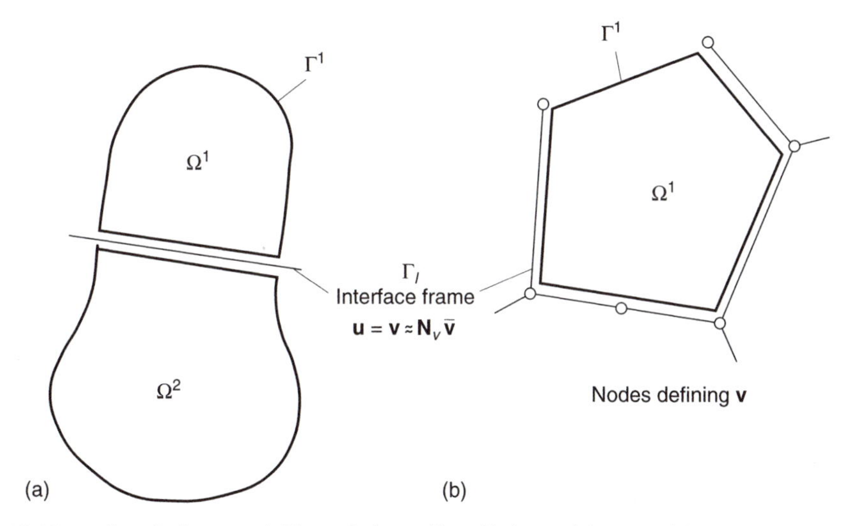

Fig. 12.12 Interface displacement field specified on a 'frame' linking subdomains. (a) Two-domain link. (b) A 'superelement' (hybrid) which can be linked to many other similar elements.

and consider the 'nodal forces' contributed by a single subdomain Ω^1 to the 'nodes' on this frame. Using virtual work (or weak) statements we have with discretization

$$\int_{\Gamma_i^1} \mathbf{N}_v^{\mathrm{T}} \mathbf{t} \, \mathrm{d}\Gamma = \mathbf{q}^1 \tag{12.40}$$

where $\mathbf{t}$ are the tractions the interior exerts on the imaginary frame and $\mathbf{q}^1$ are the nodal forces developed. The balance of the nodal forces contributed by each subdomain now provides the weak condition for traction continuity.

As finally the tractions $\mathbf{t}$ can be expressed in terms of the frame parameters $\tilde{\mathbf{v}}$ only, we shall arrive at

$$\mathbf{q}^1 = \mathbf{K}^1 \tilde{\mathbf{v}} + \mathbf{f}_0^1 \tag{12.41}$$

where $\mathbf{K}^1$ is the stiffness matrix of the subdomain Ω^1 and $\mathbf{f}_0^1$ its internally contributed 'forces'.

From this point onwards the standard assembly procedures are valid and the subdomain can be treated as a standard element which can be assembled with others by ensuring that

$$\sum_j \mathbf{q}^j = \mathbf{0} \tag{12.42}$$

where the sum includes all subdomains (or elements). We thus have only to consider a single subdomain Ω_e in what follows.

12.4.2 Linking displacement frame on equilibrating form subdomains

In this form we shall assume *a priori* the stress field expansion is given by

$$\boldsymbol{\sigma}_T = \boldsymbol{\sigma} + \boldsymbol{\sigma}_0 \tag{12.43}$$

and that the equilibrium equations are identically satisfied as

$$\boldsymbol{\mathcal{S}}^{\mathrm{T}}\boldsymbol{\sigma} \equiv \mathbf{0};\ \boldsymbol{\mathcal{S}}^{\mathrm{T}}\boldsymbol{\sigma}_0 \equiv \mathbf{b} \text{ in } \Omega \qquad \text{and} \qquad \mathbf{G}\boldsymbol{\sigma} = \mathbf{0};\ \mathbf{G}\boldsymbol{\sigma}_0 = \bar{\mathbf{t}} \text{ on } \Gamma_{t^e}$$

The equation

$$\int_{\Omega^e} \delta(\boldsymbol{\mathcal{S}}\mathbf{u})^{\mathrm{T}}\boldsymbol{\sigma}\, \mathrm{d}\Omega - \int_{\Omega^e} \delta\mathbf{u}^{\mathrm{T}}\mathbf{b}\, \mathrm{d}\Omega - \int_{\Gamma_{t^e}} \delta\mathbf{u}^{\mathrm{T}}\bar{\mathbf{t}}\, \mathrm{d}\Gamma = 0 \tag{12.44}$$

is identically satisfied and we write a weak form of the constitutive equation and interface condition as (see Chapter 10, Sec. 10.6)

$$\int_{\Omega^e} \delta\boldsymbol{\sigma}^{\mathrm{T}}(\mathbf{D}^{-1}\boldsymbol{\sigma}_T - \boldsymbol{\mathcal{S}}\mathbf{u})\, \mathrm{d}\Omega + \int_{\Gamma_{Ie}} \delta\mathbf{t}^{\mathrm{T}}(\mathbf{u} - \mathbf{v})\, \mathrm{d}\Gamma$$
$$\equiv \int_{\Omega^e} \delta\boldsymbol{\sigma}^{\mathrm{T}}\mathbf{D}^{-1}(\boldsymbol{\sigma} + \boldsymbol{\sigma}_0)\, \mathrm{d}\Omega - \int_{\Gamma_{Ie}} (\mathbf{G}\delta\boldsymbol{\sigma})^{\mathrm{T}}\mathbf{v}\, \mathrm{d}\Gamma = 0$$

On discretization, noting that the field $\mathbf{u}$ does not enter the problem

$$\boldsymbol{\sigma} = \mathbf{N}_\sigma\tilde{\boldsymbol{\sigma}} \quad \text{and} \quad \mathbf{v} = \mathbf{N}_v\tilde{\mathbf{v}}$$

we have, on including Eq. (12.40),

$$\begin{bmatrix} \mathbf{A}^e & \mathbf{Q}^e \\ \mathbf{Q}^{e\mathrm{T}} & \mathbf{0} \end{bmatrix} \begin{Bmatrix} \tilde{\boldsymbol{\sigma}} \\ \tilde{\mathbf{v}} \end{Bmatrix} = \begin{Bmatrix} \mathbf{f}_1^e \\ \mathbf{q}^e - \mathbf{f}_2^e \end{Bmatrix} \tag{12.45}$$

where

$$\mathbf{A}^e = \int_{\Omega^e} \mathbf{N}_\sigma\mathbf{D}^{-1}\mathbf{N}_\sigma\, \mathrm{d}\Omega \qquad \mathbf{Q}^e = -\int_{\Gamma_{Ie}} (\mathbf{G}\mathbf{N}_\sigma)^{\mathrm{T}}\mathbf{N}_v\, \mathrm{d}\Gamma$$

$$\mathbf{f}_1^e = -\int_{\Omega^e} \mathbf{N}_\sigma\mathbf{D}^{-1}\boldsymbol{\sigma}_0\, \mathrm{d}\Omega \quad \text{and} \quad \mathbf{f}_2^e = \int_{\Gamma_{Ie}} \mathbf{N}_v\mathbf{G}\boldsymbol{\sigma}_0\, \mathrm{d}\Gamma$$

Here elimination of $\tilde{\boldsymbol{\sigma}}$ is simple and we can write directly

$$\mathbf{K}^e\tilde{\mathbf{v}} = \mathbf{q}^e - \mathbf{f}_2^e + \mathbf{Q}^{eT}(\mathbf{A}^e)^{-1}\mathbf{f}_1^e \quad \text{and} \quad \mathbf{K}^e = \mathbf{Q}^{eT}(\mathbf{A}^e)^{-1}\mathbf{Q}^e \tag{12.46}$$

In Sec10.6 we have discussed the possible equilibration fields and have indicated the difficulties in choosing such fields for a finite element, subdivided, field. In the present case the situation is quite simple as the parameters describing the equilibrating stresses inside the element can be chosen arbitrarily in a polynomial expression.

Example 12.4: Equilibrium field. If we use a simple polynomial expression in two dimensions:

$$\begin{aligned} \sigma_x &= \alpha_0 + \alpha_1 x + \alpha_2 y \\ \sigma_y &= \beta_0 + \beta_1 x + \beta_2 y \\ \tau_{xy} &= \gamma_0 + \gamma_1 x + \gamma_2 y \end{aligned}$$

we note that to satisfy the equilibrium we require

$$\boldsymbol{\mathcal{S}}^{\mathrm{T}}\boldsymbol{\sigma} = \begin{bmatrix} \dfrac{\partial}{\partial x} & 0 & \dfrac{\partial}{\partial y} \\ 0 & \dfrac{\partial}{\partial y} & \dfrac{\partial}{\partial x} \end{bmatrix} \boldsymbol{\sigma} = \begin{Bmatrix} \alpha_1 + \gamma_2 \\ \beta_2 + \gamma_1 \end{Bmatrix} = \mathbf{0}$$

and this simply means

$$\begin{aligned} \gamma_2 &= -\alpha_1 \\ \gamma_1 &= -\beta_2 \end{aligned}$$

Thus a linear expansion in terms of $9 - 2 = 7$ independent parameters is easily achieved. Similar expansions can of course be used with higher order terms.

It is interesting to observe that:

1. $n_\sigma \geq n_v - 3$ is needed to preserve stability.
2. By the principle of limitation, the accuracy of this approximation cannot be better than that achieved by a simple displacement formulation with compatible expansion of $\mathbf{v}$ throughout the element, providing similar polynomial expressions arise in stress component variations.

In practice two advantages of such elements, known as *hybrid-stress elements*, are obtained. In the first place it is not necessary to construct compatible displacement fields throughout the element (a point useful in their application to, say, a plate bending problem). In the second for distorted (isoparametric) elements it is easy to use stress fields varying with the global coordinates.

The first use of such elements was made by Pian[5] and many successful variants are in use today.[6, 29–41]

12.5 Linking of boundary (or Trefftz)-type solution by the 'frame' of specified displacements

Boundary methods in which the chosen fields for both displacement and stress fields satisfy *a priori* the homogeneous equations of equilibrium and constitutive equations (and indeed on occasion some prescribed boundary traction or displacement conditions) have been considered by Trefftz.[42] Here such methods are called *Trefftz-type* solutions.

Thus in Eqs (12.45) and (12.44) the subdomain (element e) Ω_e integral terms disappear and, as the internal $\delta\mathbf{t}$ and $\delta\mathbf{u}$ variations are linked, we combine all into a single statement (in the absence of body force terms) as

$$-\int_{\Gamma_{Ie}} \delta\mathbf{t}^{\mathrm{T}}(\mathbf{u}-\mathbf{v})\,\mathrm{d}\Gamma - \int_{\Gamma_{te}} \delta\mathbf{u}^{\mathrm{T}}(\mathbf{t}-\bar{\mathbf{t}})\,\mathrm{d}\Gamma = 0 \tag{12.47}$$

This coupled with the boundary statement (12.40) provides the means of devising stiffness matrix statements of such subdomains.

For instance, if we express the approximate fields as

$$\mathbf{u} = \mathbf{N}\tilde{\mathbf{u}} \tag{12.48}$$

implying

$$\boldsymbol{\sigma} = \mathbf{D}(\mathcal{S}\mathbf{N})\tilde{\mathbf{u}} \qquad \text{and} \qquad \mathbf{t} = \mathbf{G}\boldsymbol{\sigma} = \mathbf{G}\mathbf{D}(\mathcal{S}\mathbf{N})\tilde{\mathbf{u}}$$

we can write

$$\begin{bmatrix} -\mathbf{H}^e & \mathbf{Q}^e \\ \mathbf{Q}^{e\mathrm{T}} & \mathbf{0} \end{bmatrix} \begin{Bmatrix} \tilde{\mathbf{u}} \\ \tilde{\mathbf{v}} \end{Bmatrix} = \begin{Bmatrix} \mathbf{f}_1^e \\ \mathbf{q} \end{Bmatrix} \tag{12.49}$$

where

$$\begin{aligned} \mathbf{H}^e &= \int_{\Gamma_{Ie}} [\mathbf{G}\mathbf{D}(\mathcal{S}\mathbf{N})]^{\mathrm{T}}\mathbf{N}\,\mathrm{d}\Gamma + \int_{\Gamma_{te}} \mathbf{N}^{\mathrm{T}}\mathbf{G}\mathbf{D}(\mathcal{S}\mathbf{N})\,\mathrm{d}\Gamma \\ \mathbf{Q}^e &= \int_{\Gamma_{Ie}} [\mathbf{G}\mathbf{D}(\mathcal{S}\mathbf{N})]^{\mathrm{T}}\mathbf{N}_v\mathrm{d}\Gamma \\ \mathbf{f}_1^e &= -\int_{\Gamma_{te}} \mathbf{N}^{\mathrm{T}}\bar{\mathbf{t}}\,\mathrm{d}\Gamma \end{aligned} \tag{12.50}$$

In Eqs (12.49) and (12.50) we have omitted the domain integral of the particular solution $\boldsymbol{\sigma}_0$ corresponding to the body forces $\mathbf{b}$ but have allowed a portion of the boundary Γ_{t^e} to be subject to prescribed tractions. Full expressions including the particular solution can easily be derived.

Equation (12.49) is immediately available for solution of a single boundary problem in which $\mathbf{v}$ and $\bar{\mathbf{t}}$ are described on portions of the boundary. More importantly, however, it results in a very simple stiffness matrix for a full element enclosed by the frame. We now have

$$\mathbf{K}^e\tilde{\mathbf{v}} = \mathbf{q} + \mathbf{f}^e \tag{12.51}$$

in which

$$\begin{aligned} \mathbf{K}^e &= \mathbf{Q}^{e\mathrm{T}}(\mathbf{H}^e)^{-1}\mathbf{Q}^e \\ \mathbf{f}^e &= \mathbf{Q}^{e\mathrm{T}}(\mathbf{H}^e)^{-1}\mathbf{f}_1^e \end{aligned} \tag{12.52}$$

This form is very similar to that of Eq. (12.46) except that now only integrals on the boundaries of the subdomain element need to be evaluated.

Much has been written about so-called 'boundary elements' and their merits and disadvantages.[9–11, 13, 43–51] Very frequently singular Green's functions are used to satisfy the governing field equations in the domain.[46–50] The singular function distributions used do not lend themselves readily to the derivation of symmetric coupling forms of the type given in Eq. (12.49). Zienkiewicz *et al.*[51–54] show that it is possible to obtain symmetry at a cost of two successive integrations. Further it should be noted that the singular distributions always involve difficult integration over a point of singularity and special procedures need to be used for numerical implementation. For this reason the use of generally non-singular Trefftz functions is preferable and it is possible to derive complete sets of functions satisfying the governing equations without introducing singularities,[51–54] and simple integration then suffices.

While boundary solutions are confined to linear homogeneous domains these give very accurate solutions for a limited range of parameters, and their combination with 'standard' finite elements has been occasionally described. Several coupling procedures have been developed in the past,[51–54] but the form given here coincides with the work of Zielinski and Zienkiewicz,[55] Jirousek[7, 56–58] and Piltner.[14] Jirousek *et al.* have developed very general two-dimensional elasticity and plate bending elements which can be enclosed by a many-sided polygonal domain (element) that can be directly coupled to standard elements providing that same-displacement interpolation along the edges is involved, as shown in Fig. 12.13. Here both *interior* elements with a frame enclosing an element volume and *exterior* elements satisfying tractions at free surface and infinity are illustrated.

Rather than combining in a finite element mesh the standard and the Trefftz-type elements ('T-elements'[13]) it is often preferable to use the T-elements alone. This results in the whole domain being discretized by elements of the same nature and offering each about the same degree of accuracy. The subprogram of such elements can include an arsenal of homogeneous 'shape functions' $\mathbf{N}^e$ [see Eq. (12.48)] which are exact solutions to different types of singularities as well as those which automatically satisfy traction boundary conditions on internal boundaries, e.g., circles or ellipses inscribed within large elements as shown in Fig. 12.14. Moreover, by completing the set of homogeneous shape functions by suitable 'load terms' representing the non-homogeneous differential equation solution, $\mathbf{u}_0$,

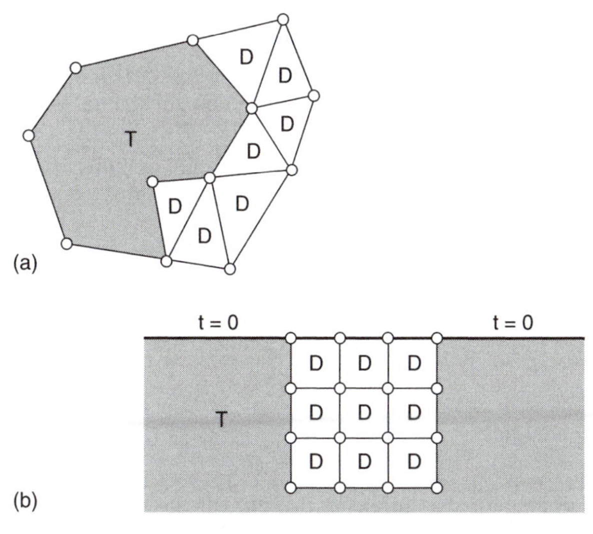

Fig. 12.13 Boundary–Trefftz-type elements (T) with complex-shaped 'frames' allowing combination with standard, displacement elements (D). (a) An *interior* element. (b) An *exterior* element.

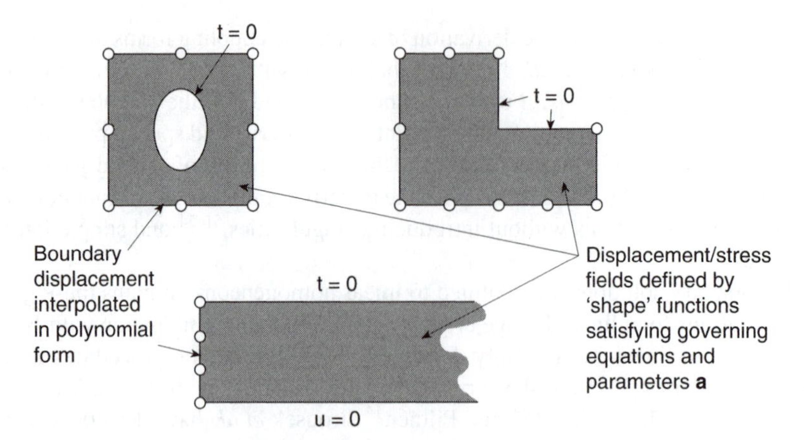

Fig. 12.14 Boundary–Trefftz-type elements. Some useful general forms.[58]

one may account accurately for various discontinuous or concentrated loads without laborious adjustment of the finite element mesh.

Clearly such elements can perform very well when compared with standard ones, as the nature of the analytical solution has been essentially included. Figure 12.15 shows excellent results which can be obtained using such complex elements. The number of degrees of freedom is here much smaller than with a standard displacement solution but, of course, the bandwidth is much larger.[58]

Two points come out clearly in the general formulation of Eqs (12.47)–(12.50).

First, the displacement field, $\mathbf{u}$ given by parameters $\tilde{\mathbf{u}}$, can only be determined by excluding any rigid body modes. These can only give strains $\boldsymbol{\mathcal{S}}\mathbf{N}$ identically equal to zero and hence make no contribution to the $\mathbf{H}$ matrix.

Second, stability conditions require that (in two dimensions)

$$n_u \geq n_v - 3$$

and thus the minimum n_u can be readily found (viz. Chapter 10). Once again there is little point in increasing the number of internal parameters substantially above the minimum number as additional accuracy may not be gained.

We have said earlier that the 'translation' of the formulation discussed to problems governed by the quasi-harmonic equations is almost evident. Now identical relations will hold if we replace

$$\begin{aligned} \mathbf{u} &\rightarrow \phi \\ \sigma &\rightarrow \mathbf{q} \\ \mathbf{t} &\rightarrow q_n \\ \boldsymbol{\mathcal{S}} &\rightarrow \nabla \end{aligned} \tag{12.53}$$

For the Poisson equation

$$\nabla^2 \phi = Q \tag{12.54}$$

a complete series of analytical solutions in two dimensions can be written as

$$\begin{aligned} \text{Re}\,(z^n) &= 1, x, x^2 - y^2, x^3 - 3xy^3, \ldots \\ \text{Im}\,(z^n) &= y, 2xy, \ldots \end{aligned} \qquad \text{for } z = x + \mathrm{i}y \tag{12.55}$$

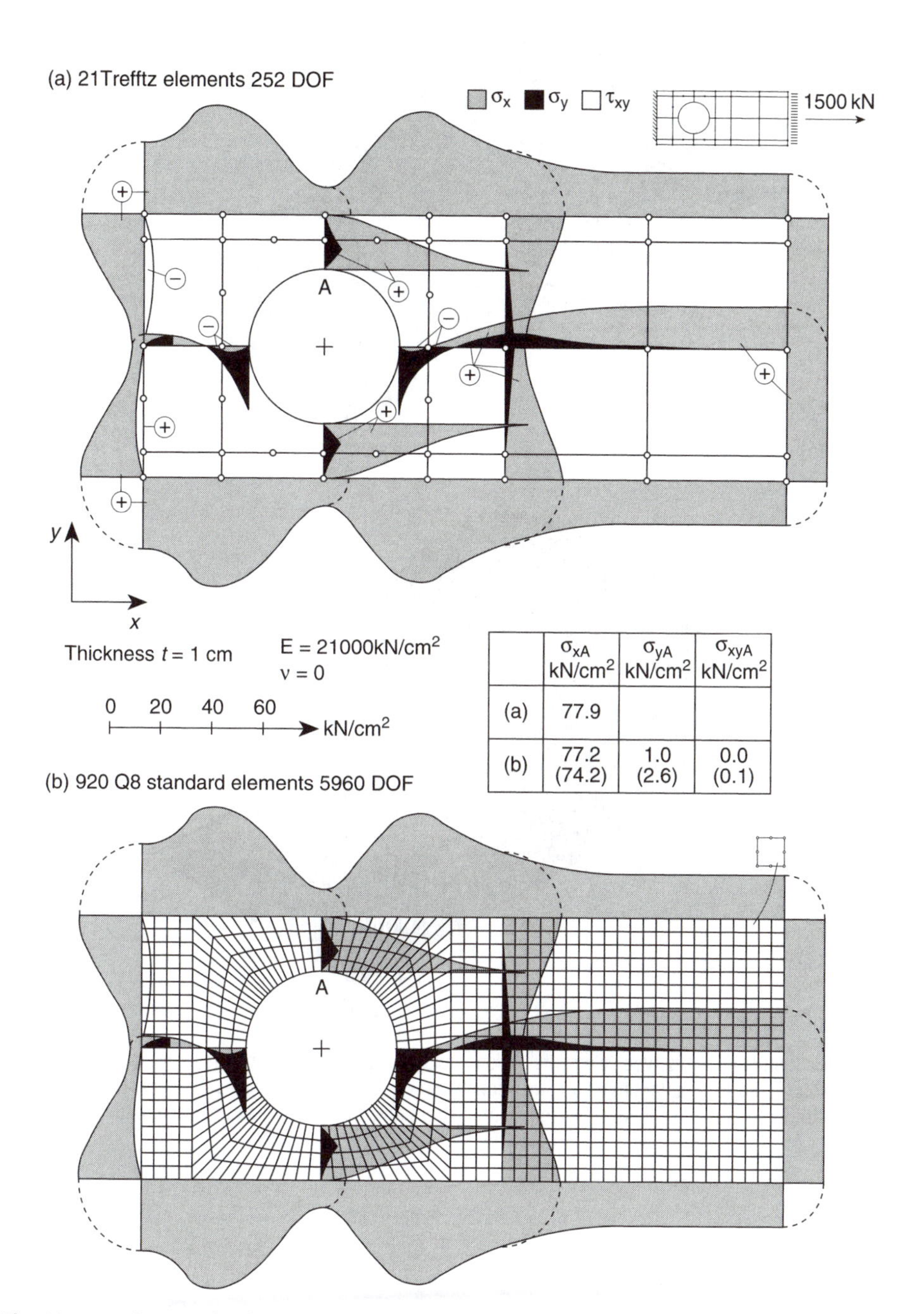

	σ_{xA} kN/cm^2	σ_{yA} kN/cm^2	σ_{xyA} kN/cm^2
(a)	77.9		
(b)	77.2 (74.2)	1.0 (2.6)	0.0 (0.1)

Fig. 12.15 Application of Trefftz-type elements to a problem of a plane stress tension bar with a circular hole. (a) Trefftz element solution. (b) Standard displacement element solution. (Numbers in parentheses indicate standard solution with 230 elements, 1600 DOF.)

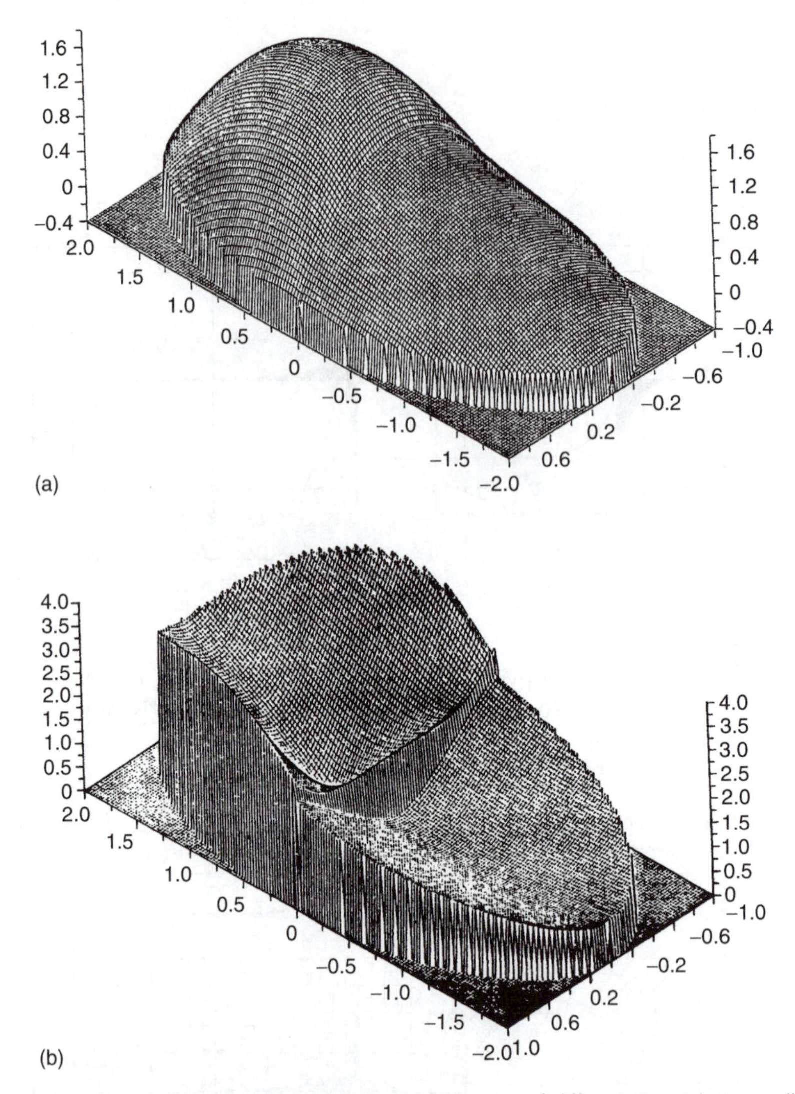

Fig. 12.16 Boundary–Trefftz-type 'elements' linking two domains of different materials in an elliptic bar subject to torsion (Poisson equations).[55] (a) Stress function given by internal variables showing almost complete continuity. (b) x component of shear stress (gradient of stress function showing abrupt discontinuity of material junction).

With the above we get

$$\mathbf{N}^e = \left[1, \quad x, \quad y, \quad x^2 - y^2, \quad 2xy, \quad x^3 - 3xy^2, \quad 3x^2y, \quad \ldots\right] \tag{12.56}$$

A simple solution involving two subdomains with constant but different values of Q and a linking on the boundary is shown in Fig. 12.16, indicating the accuracy of the linking procedures.

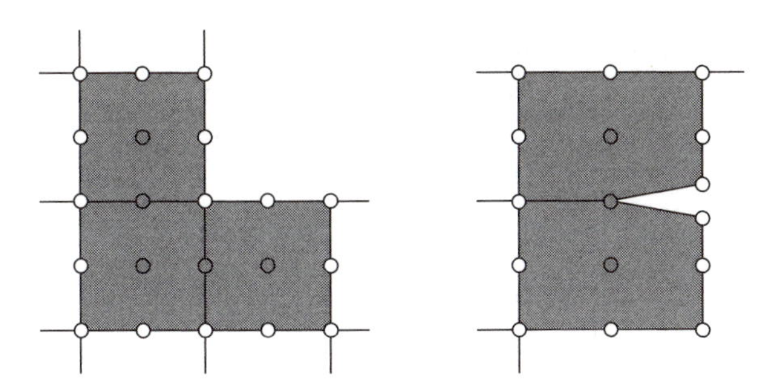

Fig. 12.17 'Superelements' built from assembly of standard displacement elements with global functions eliminating singularities confined to the assembly.

12.6 Subdomains with 'standard' elements and global functions

The procedure just described can be conveniently used with approximations made internally with standard (displacement) elements and global functions helping to deal with singularities or other internal problems. Now simply an additional term will arise inside nodes placed internally in the subdomain but the effect of global functions can be contained inside the subdomain. The formulation is somewhat simpler as complicated Trefftz-type functions need not be used.

We leave details to the reader and in Fig. 12.17 show some possible, useful, subdomain assemblies. We shall return to this again in Chapter 15.

12.7 Concluding remarks

The possibilities of elements or 'superelements' constructed by the mixed-incomplete field methods of this chapter are numerous. Many have found practical use in existing computer codes as 'hybrid elements'; others are only now being made widely available. The use of a frame of specified displacements is only one of the possible methods for linking Trefftz-type solutions. As an alternative, a frame of specified boundary tractions **t** has also been successfully investigated.[10, 45] In addition, the so-called 'frameless formulation'[9, 11] has been found to be another efficient solution (for a review see reference 13) in the Trefftz-type element approach. All of the above-mentioned alternative approaches may be implemented into standard finite element computer codes. Much further research will elucidate the advantages of some of the forms discovered and we expect the use of such developments to continue to increase in the future.

12.8 Problems

12.1 Compute explicit relations for linear one-dimensional dual shape functions using Eq. (12.13). Verify the results shown in Fig. 12.5.

12.2 Compute explicit relations for quadratic one-dimensional dual shape functions using Eq. (12.13). Assume the element side is straight and the interior node is at the centre of the edge.

12.3 Compute an explicit relation at node a for 4-node dual shape functions. Use Eq. (12.13) and assume the surface mesh for elements is as shown in Fig. 12.18. Sketch the shape of the global dual function at node a (e.g., as shown for a two-dimensional edge in Fig. 12.4).

12.4 The mesh segment shown in Fig. 12.19 occurs in a problem in which the two sides are to be joined using a standard mortar method. If node a is located at $0.4h$ from node b perform the integrals necessary to construct the contributions to the $\mathbf{Q}_i$ arrays appearing in Eq. (12.12).

12.5 The mesh segment shown in Fig. 12.19 occurs in a problem in which the two sides are to be joined using a dual mortar method. If node a is located at $0.4h$ from node b perform the integrals necessary to construct the contributions to the $\mathbf{Q}_i$ arrays appearing in Eq. (12.12). (Note: It is necessary to replace one N by $\hat{N}$ for the dual approach.)

12.6 Write a MATLAB program to solve the one-dimensional problem of Example 3.5 in Chapter 3. Modify the program to enforce the boundary conditions using the Nitsche method described in Sec. 12.3.1. Verify your program by solving the example illustrated in Fig. 12.8(a).

12.7 Perform the derivations given in Sec. 12.5 which include the effects of a non-zero body force $\mathbf{b}$ to define σ_0.

12.8 For the quasi-harmonic equation given by $\nabla^2\phi = Q$ construct the linking of Trefftz-type solutions by a 'frame' of specified values for ϕ. (Hint: Follow the suggestions given in Eq. (12.53).)

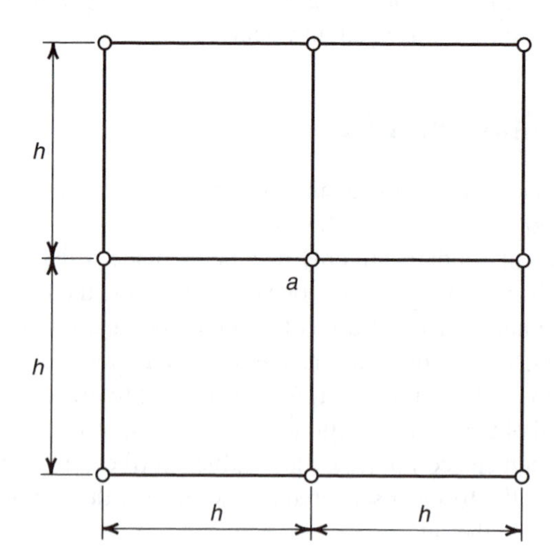

Fig. 12.18 Surface description for Problem 12.3.

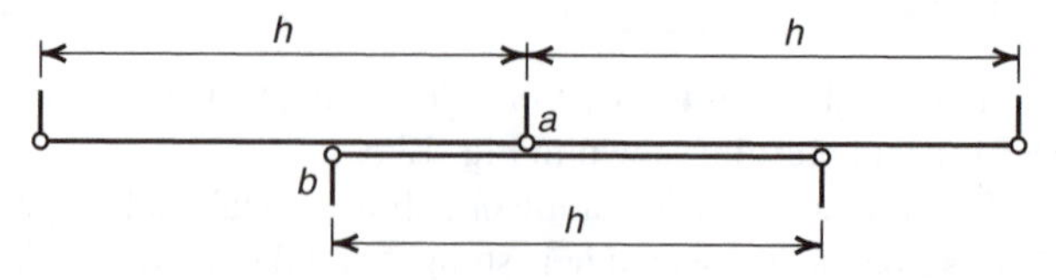

Fig. 12.19 Tied segment for Problems 12.4 and 12.5.

References

1. Ch. Farhat and F.-X. Roux. Optimal convergence properties of the FETI domain decomposition method. *Comp. Meth. Appl. Mech. Eng.*, 115:365–385, 1994.
2. C. Farhat and J. Mandel. The two-level FETI method. I. An optimal iterative solver for biharmonic systems. *Comp. Meth. Appl. Mech. Eng.*, 155:129–151, 1998.
3. C. Farhat, Chen Po-shu, J. Mandel, and F.X. Roux. The two-level FETI method. II. Extension to shell problems, parallel implementation and performance results. *Comp. Meth. Appl. Mech. Eng.*, 155:153–179, 1998.
4. C. Farhat, K. Pierson, and M. Lesoinne. The second generation FETI methods and their application to the parallel solution of large-scale linear and geometrically non-linear structural analysis problems. *Comp. Meth. Appl. Mech. Eng.*, 184:333–374, 2000.
5. T.H.H. Pian. Derivation of element stiffness matrices by assumed stress distribution. *J. AIAA*, 2:1332–1336, 1964.
6. T.H.H. Pian and P. Tong. Basis of finite element methods for solid continua. *Int. J. Numer. Meth. Eng.*, 1:3–28, 1969.
7. J. Jirousek and L. Guex. The hybrid-Trefftz finite element model and its application to plate bending. *Int. J. Numer. Meth. Eng.*, 23:651–693, 1986.
8. J. Jirousek. Improvement of computational efficiency of the 9 dof triangular hybrid-Trefftz plate bending element. *Int. J. Numer. Meth. Eng.*, 23:2167–2168, 1986. (Letter to Editor.)
9. J. Jirousek and A.P. Zieliński. Study of two complementary hybrid-Trefftz p-element formulations. In *Numerical Methods in Engineering 92*, pages 583–590. Elsevier, 1992.
10. J. Jirousek and A.P. Zieliński. Dual hybrid-Trefftz element formulation based on independent boundary traction frame. *Int. J. Numer. Meth. Eng.*, 36:2955–2980, 1993.
11. J. Jirousek and A. Wróblewski. Least-squares T-elements: equivalent FE and BE forms of a substructure-oriented boundary solution approach. *Comm. Numer. Meth. Eng.*, 10:21–32, 1994.
12. J. Jirousek and A. Wróblewski. T-elements: a finite element approach with advantages of boundary solution methods. *Adv. Eng. Soft.*, 24:71–88, 1995.
13. J. Jirousek and A. Wróblewski. T-elements: state of the art and future trends. *Arch. Comput. Meth. Eng.*, 3(4), 1996.
14. R. Piltner. Special elements with holes and internal cracks. *Int. J. Numer. Meth. Eng.*, 21:1471–1485, 1985.
15. N.-E. Wiberg. Matrix structural analysis with mixed variables. *Int. J. Numer. Meth. Eng.*, 8:167–194, 1974.
16. B.I. Wohlmuth. *Discretization Methods and Iterative Solvers Based on Domain Decomposition*. Springer-Verlag, Heidelberg, 2001.
17. G. Kron. *Tensor Analysis of Networks*. John Wiley & Sons, New York, 1939.
18. Ch. Farhat and F.-X. Roux. A method of finite element tearing and interconnecting and its parallel solution algorithm. *Int. J. Numer. Meth. Eng.*, 32:1205–1227, 1991.
19. M.A. Puso and T.A. Laursen. Mesh tying on curved interfaces in 3d. *Eng. Comput.*, 20:305–319, 2003.
20. J.A. Nitsche. Über ein Variationsprinzip zur Lösung Dirichlet-Problemen bei Verwendung von Teilräumen, die keinen Randbedingungen uneworfen sind. *Abh. Math. Sem. Univ. Hamburg*, 36:9–15, 1971.
21. C.G. Makridakis and I. Babuška. On the stability of the discontinuous Galerkin method for the heat equation. *SIAM J. Num. Anal.*, 34:389–401, 1997.
22. J.T. Oden, I. Babuška, and C.E. Baumann. A discontinuous *hp* finite element method for diffusion problems. *J. Comp. Phys.*, 146(2):491–519, 1998.
23. T.J.R. Hughes, G. Engel, L. Mazzei, and M.G. Larson. A comparison of discontinuous and continuous Galerkin methods based on error estimates, conservation, robustness and efficiency.

In *Discontinuous Galerkin Methods: Theory, Computation and Applications*, pages 135–146. Springer-Verlag, Berlin, 2000.

24. B. Cockburn, G.E. Karniadakis, and Chi-Wang Shu. *Discontinuous Galerkin Methods: Theory, Computation and Applications*. Springer-Verlag, Berlin, 2000.
25. D.N. Arnold, F. Brezzi, B. Cockburn, and D. Marini. Unified analysis of discontinuous Galerkin methods for elliptic problems. *SIAM J. Numer. Anal.*, 39:1749–1779, 2002.
26. W.H. Reed and T.R. Hill. Triangular mesh methods for the neutron transport equation. Technical Report LA-UR-73–479, Los Alamos Scientific Laboratory, 1973.
27. P. Lesaint and P.-A. Raviart. On a finite element method for solving the neutron transport equation. In C. de Boor, editor, *Mathematical Aspects of Finite Elements in Partial Differential Equations*. Academic Press, New York, 1974.
28. O.C. Zienkiewicz, R.L. Taylor, S.J. Sherwin, and J. Peiró. On discontinuous Galerkin methods. *Int. J. Numer. Meth. Eng.*, 58:1119–1148, 2003.
29. S.N. Atluri, R.H. Gallagher, and O.C. Zienkiewicz, editors. *Hybrid and Mixed Finite Element Methods*. John Wiley & Sons, New York, 1983.
30. T.H.H. Pian. Element stiffness matrices for boundary compatibility and for prescribed boundary stresses. In *Proc. 1st Conf. Matrix Methods in Structural Mechanics*, volume AFFDL-TR-66-80, pages 457–478, Wright Patterson Air Force Base, Ohio, Oct. 1966.
31. R.D. Cook and J. At-Abdulla. Some plane quadrilateral 'hybrid' finite elements. *J. AIAA*, 7:2184–2185, 1969.
32. S.N. Atluri. A new assumed stress hybrid finite element model for solid continua. *J. AIAA*, 9:1647–1649, 1971.
33. R.D. Henshell. On hybrid finite elements. In J.R. Whiteman, editor, *The Mathematics of Finite Elements and Applications*, pages 299–312. Academic Press, London, 1973.
34. R. Dungar and R.T. Severn. Triangular finite elements of variable thickness. *J. Strain Anal.*, 4:10–21, 1969.
35. R.J. Allwood and G.M.M. Cornes. A polygonal finite element for plate bending problems using the assumed stress approach. *Int. J. Numer. Meth. Eng.*, 1:135–160, 1969.
36. T.H.H. Pian. Hybrid models. In S.J. Fenves *et al.*, editors, *Numerical and Computer Methods in Applied Mechanics*. Academic Press, New York, 1971.
37. Y. Yoshida. A hybrid stress element for thin shell analysis. In V. Pulmano and A. Kabailia, editors, *Finite Element Methods in Engineering*, pages 271–286. University of New South Wales, Australia, 1974.
38. R.D. Cook and S.G. Ladkany. Observations regarding assumed-stress hybrid plate elements. *Int. J. Numer. Meth. Eng.*, 8:513–520, 1974.
39. J.A. Wolf. Generalized hybrid stress finite element models. *J. AIAA*, 11:385–388, 1973.
40. P.L. Gould and S.K. Sen. Refined mixed method finite elements for shells of revolution. In *Proc. 3rd Conf. Matrix Methods in Structural Mechanics*, volume AFFDL-TR-71-160, Wright-Patterson Air Force Base, Ohio, 1972.
41. P. Tong. New displacement hybrid models for solid continua. *Int. J. Numer. Meth. Eng.*, 2:73–83, 1970.
42. E. Trefftz. Ein Gegenstruck zum Ritz'schem Verfohren. In *Proc. Int. Cong. Appl. Mech.*, Zurich, 1926.
43. P.K. Banerjee and R. Butterfield. *The Boundary Element Methods in Engineering Science*. McGraw-Hill, London, 1981.
44. J.A. Ligget and P.L-F. Liu. *The Boundary Integral Equation Method for Porous Media Flow*. Allen and Unwin, London, 1983.
45. C.A. Brebbia and S. Walker. *Boundary Element Technique in Engineering*. Newnes-Butterworth, London, 1980.
46. I. Herrera. Boundary methods: a criteria for completeness. In *Proc. Nat. Acad. Sci.*, volume 77, pages 4395–4398, USA, Aug. 1980.

47. I. Herrera. Boundary methods for fluids. In R.H. Gallagher, H.D. Norrie, J.T. Oden, and O.C. Zienkiewicz, editors, *Finite Elements in Fluids*, volume 4, Chapter 19. John Wiley & Sons, New York, 1982.
48. I. Herrera. Trefftz method. In C.A. Brebbia, editor, *Progress in Boundary Element Methods*, volume 3. John Wiley & Sons, New York, 1983.
49. I. Herrera and H. Gourgeon. Boundary methods, *c*-complete system for Stokes problems. *Comp. Meth. Appl. Mech. Eng.*, 30:225–244, 1982.
50. I. Herrera and F.J. Sabina. Connectivity as an alternative to boundary integral equations construction of bases. In *Proc. Nat. Acad. Sci.*, volume 75, pages 2059–2063, USA, May 1978.
51. O.C. Zienkiewicz, D.W. Kelley, and P. Bettess. The coupling of the finite element and boundary solution procedures. *Int. J. Numer. Meth. Eng.*, 11:355–375, 1977.
52. O.C. Zienkiewicz, D.W. Kelly, and P. Bettess. Marriage a la mode – the best of both worlds (finite elements and boundary integrals). In R. Glowinski, E.Y. Rodin, and O.C. Zienkiewicz, editors, *Energy Methods in Finite Element Analysis*, Chapter 5, pages 81–107. John Wiley & Sons, London and New York, 1979.
53. O.C. Zienkiewicz and K. Morgan. *Finite Elements and Approximation*. John Wiley & Sons, London, 1983.
54. O.C. Zienkiewicz. The generalized finite element method – state of the art and future directions. *J. Appl. Mech., ASME*, 1983. 50th anniversary issue.
55. A.P. Zielinski and O.C. Zienkiewicz. Generalized finite element analysis with *t* complete boundary solution functions. *Int. J. Numer. Meth. Eng.*, 21:509–528, 1985.
56. J. Jirousek. A powerful finite element for plate bending. *Comp. Meth. Appl. Mech. Eng.*, 12:77–96, 1977.
57. J. Jirousek. Basis for development of large finite elements locally satisfying all field equations. *Comp. Meth. Appl. Mech. Eng.*, 14:65–92, 1978.
58. J. Jirousek and P. Teodorescu. Large finite elements for the solution of problems in the theory of elasticity. *Comp. Struct.*, 15:575–587, 1982.

13

Errors, recovery processes and error estimates

13.1 Definition of errors

We have stressed from the beginning of this book the approximate nature of the finite element method and on many occasions we have compared it with known exact solutions. Also, in reference to the 'accuracy' of the procedures, we suggested and discussed the manner by which this accuracy could be improved. Indeed one of the objectives of this chapter is concerned with the question of accuracy and a possible improvement on it by *a posteriori* treatments of the finite element data. We refer to such processes as *recovery*. We shall also consider the discretization error of the finite element approximation and *a posteriori* estimates of such error. In particular, we describe two distinct types of *a posteriori* error estimators, *recovery-based error estimators* and *residual-based error estimators*. The importance of highly accurate recovery methods in the computation of the recovery-based error estimators is discussed. We also demonstrate how various recovery methods can be used in the construction of residual-based error estimators.

Before proceeding further it is necessary to define what we mean by error. This we consider to be the difference between the exact solution and the approximate one. This can apply to the basic function, such as displacement which we have called $\mathbf{u}$, and is given as

$$\mathbf{e} = \mathbf{u} - \hat{\mathbf{u}} \tag{13.1}$$

where, as before, $\hat{\mathbf{u}}$ denotes a finite element solution and $\mathbf{u}$ the exact solution. In a similar way, however, we could focus on the error in the strains (i.e., gradients in the solution), such as ε or stresses $\boldsymbol{\sigma}$ and describe the error in these quantities as

$$\begin{aligned} \mathbf{e}_\varepsilon &= \varepsilon - \hat{\varepsilon} \\ \mathbf{e}_\sigma &= \boldsymbol{\sigma} - \hat{\boldsymbol{\sigma}} \end{aligned} \tag{13.2}$$

The specification of local error in the manner given in Eqs (13.1) and (13.2) is generally not convenient and occasionally misleading. For instance, under a point load both errors in displacements and stresses will be locally infinite but the overall solution may well be acceptable. Similar situations will exist near re-entrant corners where, as is well known, stress singularities exist in elastic analysis and gradient singularities develop in field problems. For this reason various 'norms' representing some *integral scalar quantity* are often introduced to measure the error.

13.1.1 Norms of errors

If, for instance, we are concerned with a general linear equation of the form of Eq. (3.6) (cf. Chapter 3), i.e.,

$$\boldsymbol{\mathcal{L}}\mathbf{u} + \mathbf{b} = \mathbf{0} \tag{13.3}$$

we can define an *energy norm* written for the error as

$$\|\mathbf{e}\| = \left| \int_\Omega \mathbf{e}^{\mathrm{T}} \boldsymbol{\mathcal{L}} \mathbf{e} \, \mathrm{d}\Omega \right|^{\frac{1}{2}} \equiv \left| \int_\Omega (\mathbf{u} - \hat{\mathbf{u}})^{\mathrm{T}} \boldsymbol{\mathcal{L}} (\mathbf{u} - \hat{\mathbf{u}}) \, \mathrm{d}\Omega \right|^{\frac{1}{2}} \tag{13.4}$$

where $|\cdot|$ denotes the absolute value of the argument.

This scalar measure corresponds in fact to the square root of the quadratic functional such as we have discussed in Sec. 3.9 of Chapter 3 and where we sought its minimum in the case of a self-adjoint operator $\boldsymbol{\mathcal{L}}$.

For elasticity problems the energy norm is defined in the same manner and yields,

$$\|\mathbf{e}\| = \left[\int_\Omega (\boldsymbol{\mathcal{S}}\mathbf{e})^{\mathrm{T}} \mathbf{D} \boldsymbol{\mathcal{S}} \mathbf{e} \, \mathrm{d}\Omega \right]^{\frac{1}{2}} \tag{13.5}$$

(with symbols as used in Chapters 2 and 6).

Here $\mathbf{e}$ is given by Eq. (13.1), the operator $\boldsymbol{\mathcal{S}}$ defines the strains as

$$\varepsilon = \boldsymbol{\mathcal{S}}\mathbf{u} \qquad \text{and} \qquad \hat{\varepsilon} = \boldsymbol{\mathcal{S}}\hat{\mathbf{u}} \tag{13.6a}$$

and $\mathbf{D}$ is the elasticity matrix (see Chapters 2 or 6), giving the stress as

$$\sigma = \mathbf{D}\varepsilon \qquad \text{and} \qquad \hat{\sigma} = \mathbf{D}\hat{\varepsilon} \tag{13.6b}$$

in which for simplicity we ignore initial stresses and strains.

Using the above relations the energy norm of Eq. (13.5) can be written alternatively as

$$\begin{aligned} \|\mathbf{e}\| &= \left[\int_\Omega (\varepsilon - \hat{\varepsilon})^{\mathrm{T}} \mathbf{D} (\varepsilon - \hat{\varepsilon}) \, \mathrm{d}\Omega \right]^{\frac{1}{2}} \\ &= \left[\int_\Omega (\varepsilon - \hat{\varepsilon})^{\mathrm{T}} (\sigma - \hat{\sigma}) \, \mathrm{d}\Omega \right]^{\frac{1}{2}} \\ &= \left[\int_\Omega (\sigma - \hat{\sigma})^{\mathrm{T}} \mathbf{D}^{-1} (\sigma - \hat{\sigma}) \, \mathrm{d}\Omega \right]^{\frac{1}{2}} \end{aligned} \tag{13.7}$$

and its relation to strain energy is evident.

Other scalar norms can easily be devised. For instance, the L_2 norm of displacement error can be written as

$$\|\mathbf{e}\|_{L_2} = \left[\int_\Omega (\mathbf{u} - \hat{\mathbf{u}})^{\mathrm{T}} (\mathbf{u} - \hat{\mathbf{u}}) \, \mathrm{d}\Omega \right]^{\frac{1}{2}} \tag{13.8a}$$

and that for stresses error as

$$\|\mathbf{e}_\sigma\|_{L_2} = \left[\int_\Omega (\sigma - \hat{\sigma})^{\mathrm{T}} (\sigma - \hat{\sigma}) \, \mathrm{d}\Omega \right]^{\frac{1}{2}} \tag{13.8b}$$

Such norms allow us to focus on a particular quantity of interest and indeed it is possible to evaluate 'root mean square' (RMS) values of its error. For instance, the RMS error in displacement, $\Delta\mathbf{u}$, becomes for the domain Ω

$$|\Delta\mathbf{u}| = \left(\frac{\|\mathbf{e}\|_{L_2}^2}{\Omega}\right)^{\frac{1}{2}} \tag{13.9}$$

Similarly, the RMS error in stress, $\Delta\boldsymbol{\sigma}$, becomes for the domain Ω

$$|\Delta\boldsymbol{\sigma}| = \left(\frac{\|\mathbf{e}_\sigma\|_{L_2}^2}{\Omega}\right)^{\frac{1}{2}} \tag{13.10}$$

Any of the above norms can be evaluated over the whole domain, any subdomain, or even an individual element.

We note that

$$\|\mathbf{e}\| = \left(\sum_{K=1}^{m} \|\mathbf{e}\|_K^2\right)^{\frac{1}{2}} \tag{13.11}$$

where K refers to individual elements Ω_K such that their sum (union) is Ω.

We note further that the energy norm given in terms of the stresses, the L_2 norm of stress and the RMS stress error have a very similar structure and that these are similarly approximated.

Effect of a singularity

At this stage it is of interest to invoke the discussion of Chapter 2 (Sec. 2.6) concerning the rates of convergence. We noted there that with trial functions in the displacement formulation of degree p, the errors in the stresses were of the order $O(h^p)$. This order of error should therefore apply to the energy norm error $\|\mathbf{e}\|$. While the arguments are correct for well-behaved problems with no singularity, it is of interest to see how the above rule is violated when singularities exist.

To describe the behaviour of stress analysis problems we define the variation of the *relative energy norm error* (percentage) as

$$\eta = \frac{\|\mathbf{e}\|}{\|\mathbf{u}\|} \times 100\% \tag{13.12}$$

where

$$\|\mathbf{u}\| = \left[\int_\Omega \boldsymbol{\varepsilon}^{\mathrm{T}}\mathbf{D}\boldsymbol{\varepsilon}\,\mathrm{d}\Omega\right]^{\frac{1}{2}} \tag{13.13}$$

is the energy norm of the solution. In Figs 13.1 and 13.2 we consider two similar stress analysis problems. In the first a strong singularity is present, however, in the second the singularity is removed by introducing a rounded corner. In both figures we show the relative energy norm error for an h refinement constructed by uniform subdivision of the initial mesh and for a p refinement in which polynomial order is increased throughout the original mesh.

We note two interesting facts. First, the h convergence rate for various polynomial orders of the shape functions is nearly the same in the example with singularity (Fig. 13.1) and is well below the theoretically predicted optimal order $O(h^p)$, [or $O(\mathrm{NDF})^{-p/2}$ as the NDF (number of degrees of freedom) is approximately inversely proportional to h^2 for a two-dimensional problem].

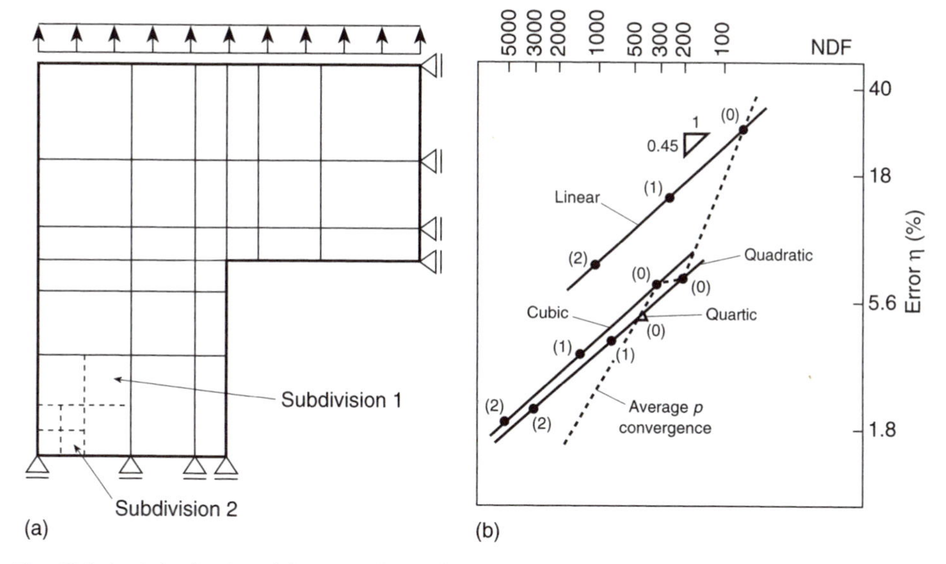

Fig. 13.1 Analysis of L-shaped domain with singularity.

Second, in the case shown in Fig. 13.2, where the singularity is avoided by rounding the re-entrant corner, the h convergence rate improves for elements of higher order, although again the theoretical (asymptotic) rate is not quite achieved.

The reason for this behaviour is clearly the singularity, and in general it can be shown that the rate of convergence for problems with singularity is

$$O(\text{NDF})^{-[\min(\lambda, p)]/2} \tag{13.14}$$

where λ is a number associated with the intensity of the singularity. For elasticity problems λ ranges from 0.5 for a nearly closed crack to 0.711 for a 90° corner. The rate of convergence illustrated in Fig. 13.2 approaches the theoretically optimal order for all values of p used in the elements.

13.2 Superconvergence and optimal sampling points

In this section we shall consider the location of points at which the stresses, or displacements, give their most accurate values in typical problems of a self-adjoint kind. We shall note that on many occasions the displacements, or the function itself, are most accurately sampled at the nodes defining an element and that the gradients or stresses are best sampled at some interior points. Indeed, in one dimension at least, we find that such points often exhibit the quality known as *superconvergence* (i.e., the values sampled at these points show an error which decreases more rapidly than elsewhere). Obviously, the user of finite element analysis should be encouraged to employ such points but at the same time note that the errors overall may be much larger. To clarify these ideas we start with a typical problem of second order in one dimension.

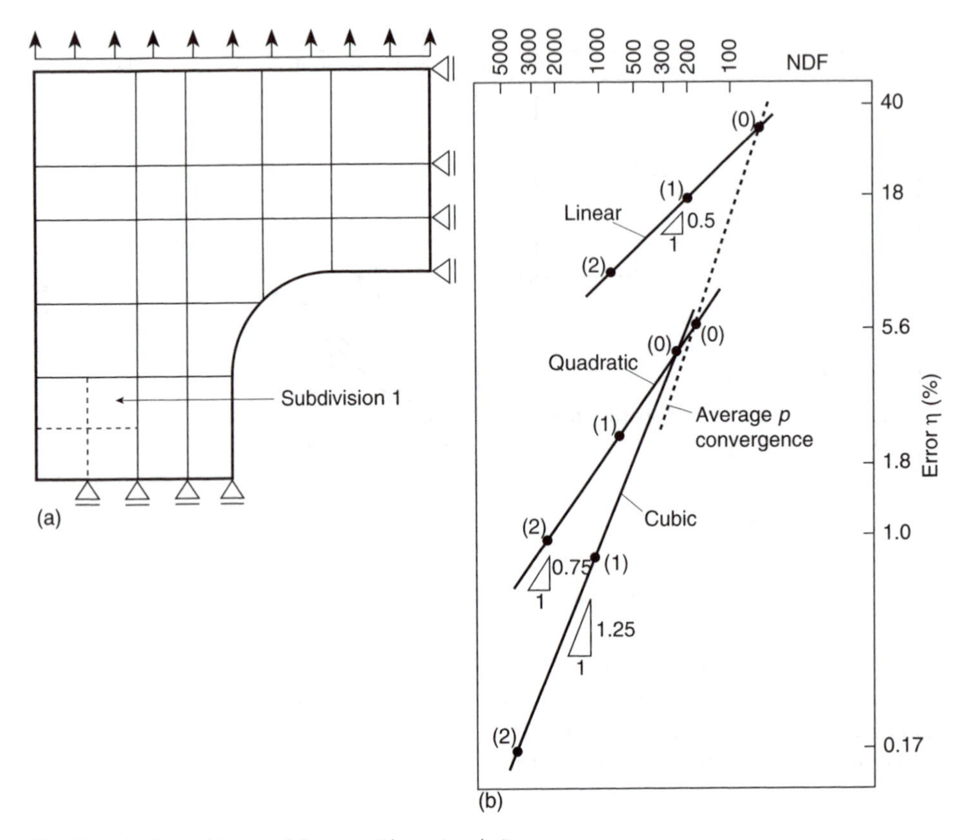

Fig. 13.2 Analysis of L-shaped domain without singularity.

13.2.1 A one-dimensional example

Here we consider a problem of a second order equation such as we have discussed in Chapter 3 and which may be typical of either one-dimensional heat conduction or the displacements of an elastic bar with varying cross-section. This equation can readily be written as

$$\frac{\mathrm{d}}{\mathrm{d}x}\left(k\frac{\mathrm{d}u}{\mathrm{d}x}\right) + \beta u + Q = 0 \tag{13.15}$$

with the boundary conditions either defining the values of the function u or of its gradients at the ends of the domain.

Let us consider a typical problem as illustrated in Fig. 13.3. Here we show an exact solution for u and $\mathrm{d}u/\mathrm{d}x$ for a span of several elements and indicate the type of solution which will result from a finite element calculation using linear elements. We have already noted that on occasions we shall obtain exact solutions for u at nodes (see Figs 3.5 and 3.6). This will happen when the weighting function contains the exact solution of the homogeneous differential equation (Appendix H) – a situation which happens for Eq. (13.15) when $\beta = 0$, k is constant in each element and polynomial shape functions are used. In all cases, even when β is non-zero and linear shape functions are used, the nodal values generally will be much more accurate than those elsewhere, Fig. 13.3(a). For the gradients shown

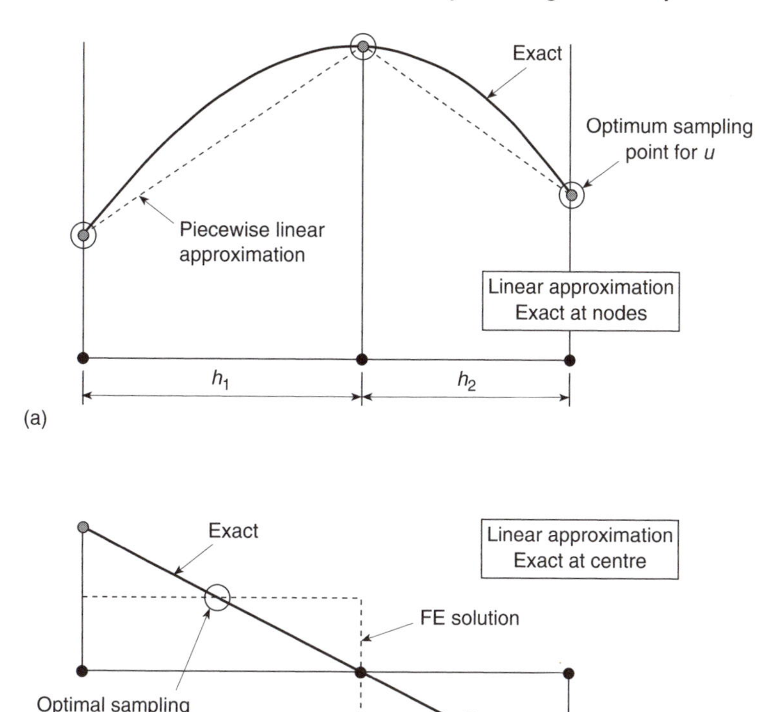

Fig. 13.3 Optimal sampling points for the function (a) and its gradient (b) in one dimension (linear elements).

in Fig. 13.3(b) we observe large discrepancies of the finite element solution from the exact solution but we note that somewhere within each element the results are exact.

It would be useful to locate such points and indeed we have already remarked in the context of two-dimensional analysis that values obtained within the elements tend to be more accurate for gradients (strains and stresses) than those values calculated at nodes. Clearly, for the problem illustrated in Fig. 13.3(b) we should sample somewhere near the centre of each element.

Pursuing this problem further in a heuristic manner we note that if higher order elements (e.g., quadratic elements) are used the solution still remains exact or nearly exact at the end nodes of an element but may depart from exactness at the interior nodes, as shown in Fig. 13.4(a). The stresses, or gradients, in this case will be optimal at points which correspond to the two Gauss quadrature points for each element as indicated in Fig. 13.4(b). This fact was observed experimentally by Barlow.[1]

We shall now state in an axiomatic manner that:

(a) the displacements are best sampled at the nodes of the element, whatever the order of element used, and
(b) the best accuracy for gradients or stresses is obtained at the Gauss points corresponding to the order of polynomial used in the solution.

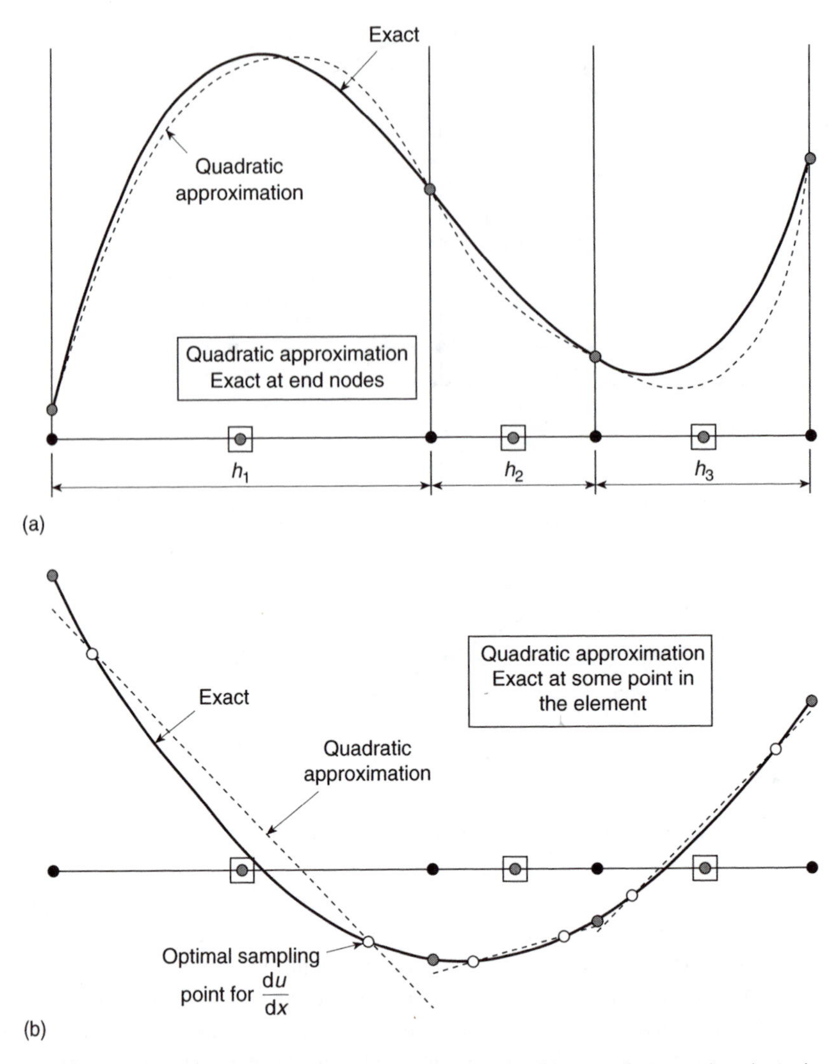

Fig. 13.4 Optimal sampling points for the function (a) and its gradient (b) in one dimension (quadratic elements).

At such points the order of the convergence of the function or its gradients is at least one order higher than that which would be anticipated from the appropriate polynomial and thus such points are known as *superconvergent*. The reason for such superconvergence will be shown in the next section where we introduce the reader to a theorem developed by Herrmann.[2]

13.2.2 The Herrmann theorem and optimal sampling points

The concept of least squares fitting has additional justification in self-adjoint problems in which an energy functional is minimized. In such cases, typical of a displacement formulation of elasticity, it can be readily shown that the minimization is equivalent to a least squares fit of the approximate stresses to the exact ones. Thus quite generally we can start from a theory given by the differential equation

$$\mathcal{L}\mathbf{u} = \mathcal{S}^{\mathrm{T}}(\mathbf{A}\mathcal{S}\mathbf{u}) = \mathbf{p} \tag{13.16}$$

In the above, $\mathcal{L}$ is a self-adjoint operator defined by $\mathcal{S}$ and $\mathbf{A}$ (symmetric) and $\mathbf{p}$ are prescribed matrices of position. The *minimization of an energy functional* Π defined as

$$\Pi = \tfrac{1}{2}\int_\Omega (\mathcal{S}\mathbf{u})^{\mathrm{T}}\mathbf{A}\mathcal{S}\mathbf{u}\,\mathrm{d}\Omega - \int_\Omega \mathbf{u}^{\mathrm{T}}\mathbf{p}\,\mathrm{d}\Omega \tag{13.17}$$

gives at an absolute minimum the exact solution $\mathbf{u} = \bar{\mathbf{u}}$, is equivalent to minimization of another functional Π^* defined as

$$\Pi^* = \tfrac{1}{2}\int_\Omega [\mathcal{S}(\mathbf{u}-\bar{\mathbf{u}})]^{\mathrm{T}}\mathbf{A}\mathcal{S}(\mathbf{u}-\bar{\mathbf{u}})\,\mathrm{d}\Omega \tag{13.18}$$

The above quadratic functional [Eq. (13.17)] arises in all linear self-adjoint problems.

For elasticity problems this theorem is given by Herrmann[2] and shows that the approximate solution for $\mathcal{S}\mathbf{u}$ approaches the exact one $\mathcal{S}\bar{\mathbf{u}}$ as a *weighted least squares approximation*.

The proof of the Herrmann theorem is as follows. The variation of Π defined in Eq. (13.17) gives, at $\mathbf{u} = \bar{\mathbf{u}}$ (the exact solution),

$$\delta\Pi = \tfrac{1}{2}\int_\Omega (\mathcal{S}\delta\mathbf{u})^{\mathrm{T}}\mathbf{A}\mathcal{S}\bar{\mathbf{u}}\,\mathrm{d}\Omega + \tfrac{1}{2}\int_\Omega (\mathcal{S}\bar{\mathbf{u}})^{\mathrm{T}}\mathbf{A}\mathcal{S}\delta\mathbf{u}\,\mathrm{d}\Omega - \int_\Omega \delta\mathbf{u}^{\mathrm{T}}\mathbf{p}\,\mathrm{d}\Omega = 0$$

or as $\mathbf{A}$ is symmetric

$$\delta\Pi = \int_\Omega (\mathcal{S}\delta\mathbf{u})^{\mathrm{T}}\mathbf{A}\mathcal{S}\bar{\mathbf{u}}\,\mathrm{d}\Omega - \int_\Omega \delta\mathbf{u}^{\mathrm{T}}\mathbf{p}\,\mathrm{d}\Omega = 0$$

in which $\delta\mathbf{u}$ is any arbitrary variation. Thus we can select

$$\delta\mathbf{u} = \mathbf{u}$$

and

$$\int_\Omega (\mathcal{S}\mathbf{u})^{\mathrm{T}}\mathbf{A}\mathcal{S}\bar{\mathbf{u}}\,\mathrm{d}\Omega - \int_\Omega \mathbf{u}^{\mathrm{T}}\mathbf{p}\,\mathrm{d}\Omega = 0$$

Subtracting the above from Eq. (13.17) and noting the symmetry of the $\mathbf{A}$ matrix, we can write

$$\Pi = \tfrac{1}{2}\int_\Omega [\mathcal{S}(\mathbf{u}-\bar{\mathbf{u}})]^{\mathrm{T}}\mathbf{A}\mathcal{S}(\mathbf{u}-\bar{\mathbf{u}})\,\mathrm{d}\Omega - \tfrac{1}{2}\int_\Omega (\mathcal{S}\bar{\mathbf{u}})^{\mathrm{T}}\mathbf{A}\mathcal{S}\bar{\mathbf{u}}\,\mathrm{d}\Omega \tag{13.19}$$

where the last term is not subject to variation. Thus

$$\Pi^* = \Pi + \text{constant} \tag{13.20}$$

and its stationarity is equivalent to the stationarity of Π.

It follows directly from the Herrmann theorem that, for one dimension and by a well-known property of the Gauss–Legendre quadrature points, if the approximate gradients

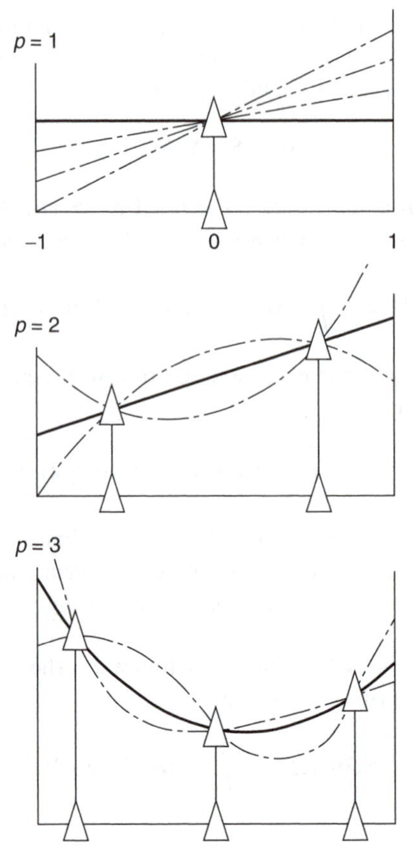

Fig. 13.5 The integration property of Gauss points: $p = 1$, $p = 2$, and $p = 3$ which guarantees superconvergence.

are defined by a polynomial of degree $p - 1$, where p is the degree of the polynomial used for the unknown function u, then stresses taken at these quadrature points must be superconvergent. The single point at the centre of an element integrates precisely all linear functions passing through that point and, hence, if the stresses are exact to the linear form they will be exact at that point of integration. For any higher order polynomial of order p, the Gauss–Legendre points numbering p will also provide points of superconvergent sampling. We see this from Fig. 13.5 directly. Here we indicate one, two, and three point Gauss–Legendre quadrature showing why exact results are recovered there for gradients and stresses.

For points based on rectangles and products of polynomial functions it is clear that the exact integration points will exist at the product points as shown in Fig. 13.6 for various rectangular elements assuming that the weighting matrix $\mathbf{A}$ is diagonal. In the same figure we show some triangles and what appear to be 'good' but are not necessarily superconvergent sampling points. Though we find that superconvergent points do not exist in triangles, the points shown in Fig. 13.6 are optimal. In Fig. 13.6 we contrast these points with the minimum number of quadrature points necessary for obtaining an accurate (though not always stable) stiffness representation and find these to be almost coincident at all times.

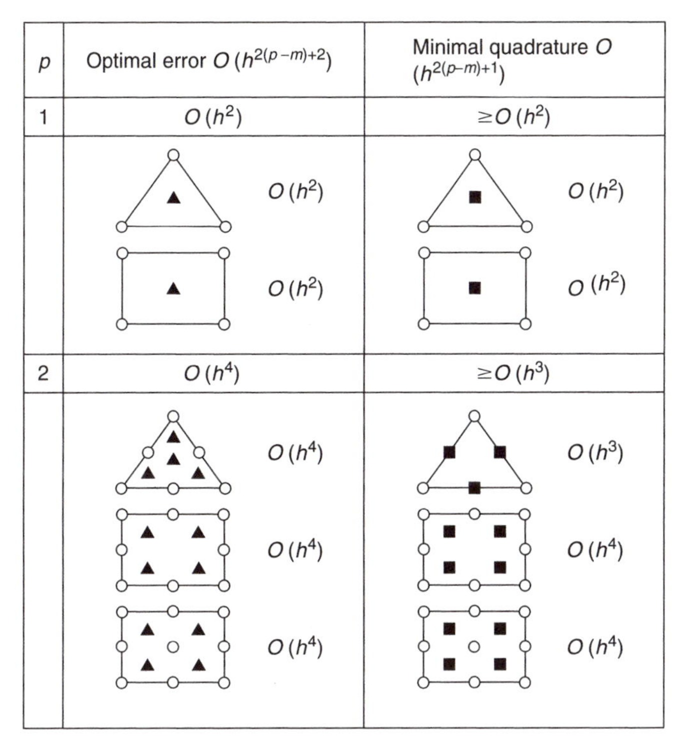

Fig. 13.6 Optimal superconvergent sampling and minimum integration points for some C_0 elements.

In Fig. 13.7 representing an analysis of a cantilever beam by four rectangular quadratic serendipity elements we see how well the stresses sampled at superconvergent points behave compared to the overall stress pattern computed in each element.

The extension of the idea of superconvergent points from one-dimensional elements to two-dimensional rectangles is fairly obvious. However, the full order of superconvergence is lost when isoparametric distortion of elements occurs. We have shown, however, that results at the pth order Gauss–Legendre points still remain excellent and we suggest that superconvergent properties of the integration points continue to be used for sampling.

In all of the above discussion we have assumed that the weighting matrix $\mathbf{A}$ is diagonal. If a diagonal structure does not exist the existence of superconvergent points is questionable. However, excellent results are still available through the sampling points defined as above.

Finally, we refer readers to references 3–8 for surveys on the superconvergence phenomenon and its detailed analyses.

13.3 Recovery of gradients and stresses

In the previous section we have shown that sampling of the gradients and stresses at certain points within an element is optimal and higher order accuracy can be achieved. However, we would also like to have similarly accurate quantities elsewhere within each element for general analysis purposes, and in particular we need such highly accurate displacements,

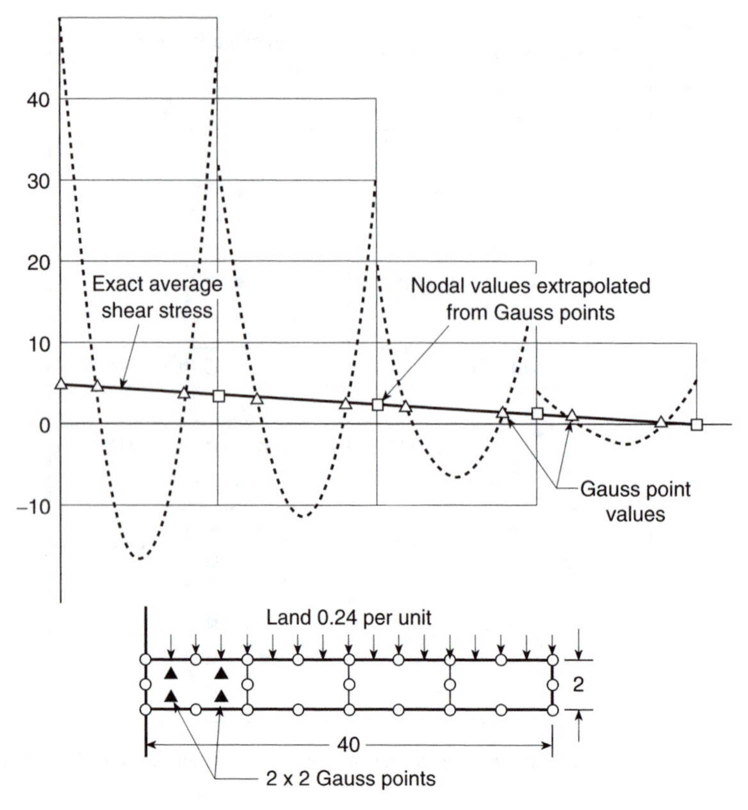

Fig. 13.7 Cantilever beam with four quadratic (Q8) elements. Stress sampling at cubic order (2 × 2) Gauss points with extrapolation to nodes.

gradients and stresses when energy norm or other norms representing the particular quantity of interest have to be evaluated in error estimates. We have already shown how with some elements very large errors exist beyond the superconvergent point and attempts have been made from the earliest days to obtain a complete picture of stresses which is more accurate overall. Here attempts are generally made to recover the nodal values of stresses and gradients from those sampled internally and then to assume that throughout the element the recovered stresses $\boldsymbol{\sigma}^*$ are obtained by interpolation in the same manner as the displacements

$$\boldsymbol{\sigma}^* = \mathbf{N}_u \tilde{\boldsymbol{\sigma}}^* \tag{13.21}$$

We have already suggested a process used almost from the beginning of finite element calculations for triangular elements, where elements are sampled at the centroid (assuming linear shape functions have been used) and then the stresses are averaged at nodes. We have referred to such recovery in Chapter 6. However, this is not the best for triangles and for higher order elements such averaging is inadequate. Here other procedures were necessary, for instance Hinton and Campbell[9] suggested a method in which stresses at all nodes were calculated by extrapolating the Gauss point values. A method of a similar kind was suggested by Brauchli and Oden[10] who used the stresses in the manner given by Eq. (13.21) and assumed that these stresses should represent in a least squares sense the actual finite element stresses. This is therefore an L_2 projection. Although this has a

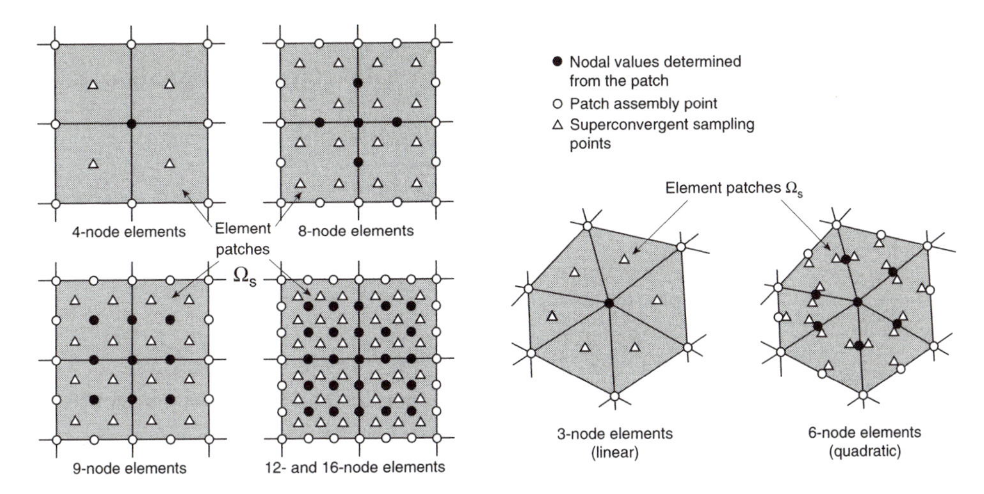

Fig. 13.8 Interior superconvergent patches for quadrilateral elements (linear, quadratic, and cubic) and triangles (linear and quadratic).

similarity with the ideas contained in the Herrmann theorem it reverses the order of least squares application and has not proved to be always stable and accurate, especially for even order elements. In the following presentation we will show that highly improved results can be obtained by direct polynomial 'smoothing' of the optimal values. Here the first method of importance is called *superconvergent patch recovery.*[11–13]

13.4 Superconvergent patch recovery – SPR

13.4.1 Recovery for gradients and stresses

We have noted above that the stresses sampled at certain points in an element possess a superconvergent property (i.e., converge at a rate comparable to that of displacement) and have errors of order $O(h^{p+1})$. A fairly obvious procedure for utilizing such sampled values seems to the authors to be that of involving a smoothing of such values by a polynomial of order p within a *patch of elements* for which the number of sampling points can be taken as greater than the number of parameters in the polynomial. In Fig. 13.8 we show several such patches each assembled around an interior vertex (corner) node. The first four represent rectangular elements where the superconvergent points are well defined. The last two give patches of triangles where the 'optimal' sampling points used are not quite superconvergent.

If we accept the superconvergence of $\hat{\sigma}$ at certain points k in each element then it is a simple matter (which also turns out computationally much less expensive than the L_2 projection) *to compute σ^* which is superconvergent at all points within the element.* The procedure is illustrated for two dimensions in Fig. 13.8, where we shall consider interior patches (assembling all elements at interior nodes) as shown.

At each superconvergent point the values of $\hat{\sigma}$ are accurate to order $p+1$ (not p as is true elsewhere). However, we can easily obtain an approximation $\bar{\sigma}^*$ given by a polynomial of degree p, with identical order to those occurring in the shape function for displacement, which has superconvergent accuracy everywhere when this polynomial is made to fit the superconvergent points in a least squares manner.

Thus we proceed for each component $\hat{\sigma}_i$ of $\hat{\boldsymbol{\sigma}}$ as follows: writing the recovered solution as

$$\bar{\sigma}_i^* = \mathbf{p}(x, y)\mathbf{a}_i \tag{13.22a}$$

in which

$$\begin{aligned} \mathbf{p}(x, y) &= \begin{bmatrix} 1, & \bar{x}, & \bar{y}, & \cdots, & \bar{y}^p \end{bmatrix} \\ \mathbf{a}_i &= \begin{bmatrix} a_1, & a_2, & \cdots, & a_m \end{bmatrix}^{\mathrm{T}} \end{aligned} \tag{13.22b}$$

with $\bar{x} = x - x_c$, $\bar{y} = y - y_c$ where x_c, y_c are the coordinates of the interior vertex node describing the patch.

For each element patch we minimize a least squares functional with n sampling points,

$$\Pi = \frac{1}{2} \sum_{k=1}^{n} \left[\hat{\sigma}_i(x_k, y_k) - \mathbf{p}_k \mathbf{a}_i\right]^2 \tag{13.23}$$

where

$$\mathbf{p}_k = \mathbf{p}(x_k, y_k)$$

[(x_k, y_k) correspond to the coordinates of the sampling superconvergent point k)] obtaining immediately the coefficient $\mathbf{a}_i$ as

$$\mathbf{a}_i = \mathbf{A}^{-1}\mathbf{b}_i \tag{13.24}$$

where

$$\mathbf{A} = \sum_{k=1}^{n} \mathbf{p}_k^{\mathrm{T}} \mathbf{p}_k \quad \text{and} \quad \mathbf{b}_i = \sum_{k=1}^{n} \mathbf{p}_k^{\mathrm{T}} \hat{\sigma}_i(x_k, y_k) \tag{13.25}$$

The availability of $\bar{\boldsymbol{\sigma}}^*$ allows superconvergent values of $\tilde{\boldsymbol{\sigma}}^*$ to be determined at all nodes. For example, each component of the recovered solution at node a in the element patch is obtained by

$$(\tilde{\sigma}_i^*)_a = \bar{\sigma}_i^*(x_a, y_a) = \mathbf{p}(x_a, y_a)\mathbf{a}_i \tag{13.26}$$

It should be noted that on external boundaries or indeed on interfaces where stresses are discontinuous the nodal values should be calculated from interior patches and evaluated in the manner shown in Fig. 13.9. As some nodes belong to more than one patch, average values of $\tilde{\boldsymbol{\sigma}}^*$ are best obtained. The superconvergence of $\boldsymbol{\sigma}^*$ throughout each element is established by Eq. (13.21).

In Fig. 13.10 we show in a one-dimensional example how the superconvergent patch recovery reproduces *exactly* the stress (gradient) solutions of order $p + 1$ for linear or quadratic elements. Following the arguments of Chapter 9 on the patch test it is evident that superconvergent recovery is now achieved at all points. Indeed, the same figure shows why averaging (or L_2 projection) is inferior (particularly on boundaries).

Figure 13.11 shows experimentally determined convergence rates for a one-dimensional problem (stress distribution in a bar of length $L = 1$; $0 \leq x \leq 1$ and prescribed body forces). A uniform subdivision is used here to form the elements, and the convergence rates for the stress error at $x = 0.5$ are shown using the direct stress approximation $\hat{\sigma}$, the L_2 recovery σ_L and σ^* obtained by the SPR procedure using elements from order $p = 1$

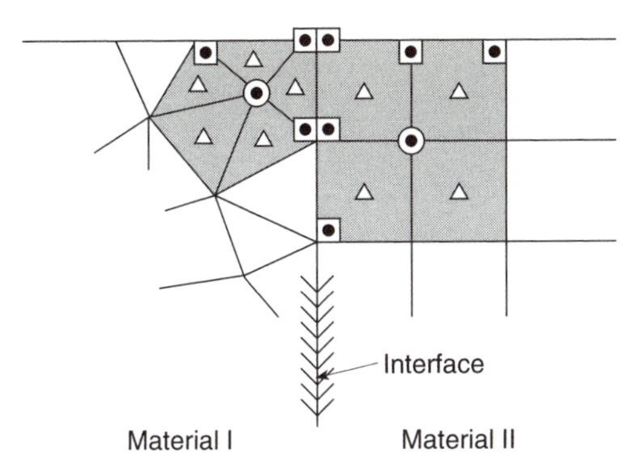

Fig. 13.9 Recovery of boundary or interface gradients.

to $p = 6$. It is immediately evident that σ^* is superconvergent with a rate of convergence being at least one order higher than that of $\hat{\sigma}$. However, as anticipated, the L_2 recovery gives much poorer answers, showing superconvergence only for odd values of p and almost no improvement for even values of p, while σ^* shows a two-order increase of convergence rate for even order elements (tests on higher order polynomials are reported in reference 14). This *ultraconvergence* has been verified mathematically.[15–17] Although it is not observed when elements of varying size are used, the important tests shown in Figs 13.12 and 13.13 indicate how well the recovery process works for problems in two dimensions.

In the first of these, Fig. 13.12, a field problem is solved in two dimensions using a very irregular mesh for which the existence of superconvergent points is only inferred heuristically. The very small error in σ_x^* is compared with the error of $\hat{\sigma}_x$ and the improvement is obvious. Here $\sigma_x = \partial u/\partial x$ where u is the field variable.

In the second, i.e., Fig. 13.13, a problem of stress analysis, for which an exact solution is known, is solved using three different recovery methods. Once again the recovered solution σ^* (SPR) shows much improved values compared with σ_L. It is clear that the SPR process *should be included in all codes if simply to present improved stress values*, to which we have already alluded in Chapters 6 and 7.

The SPR procedure which we have just outlined has proved to be a very powerful tool leading to superconvergent results on regular meshes and much improved results (nearly superconvergent) on irregular meshes. It has been shown numerically that it produces superconvergent recovery even for triangular elements which do not have superconvergent points within the element. Recent mathematical proofs confirm these capabilities of SPR.[16–21] It is also found, for linear elements on irregular meshes, that SPR produces superconvergence of order $O(h^{1+\alpha})$ with α greater than zero.[22] The SPR procedure, introduced by Zienkiewicz and Zhu in 1992,[11–13] is recommended as the best recovery procedure which is simple to use. However, the procedure has been modified by various investigators.[23–27] Some of the modifications have been shown to produce improved results in certain instances but with additional computational costs. One such modification appends satisfaction of discrete equilibrium equations and/or boundary conditions to the functional where the least squares fit is performed. While the satisfaction of known boundary tractions can on occasion

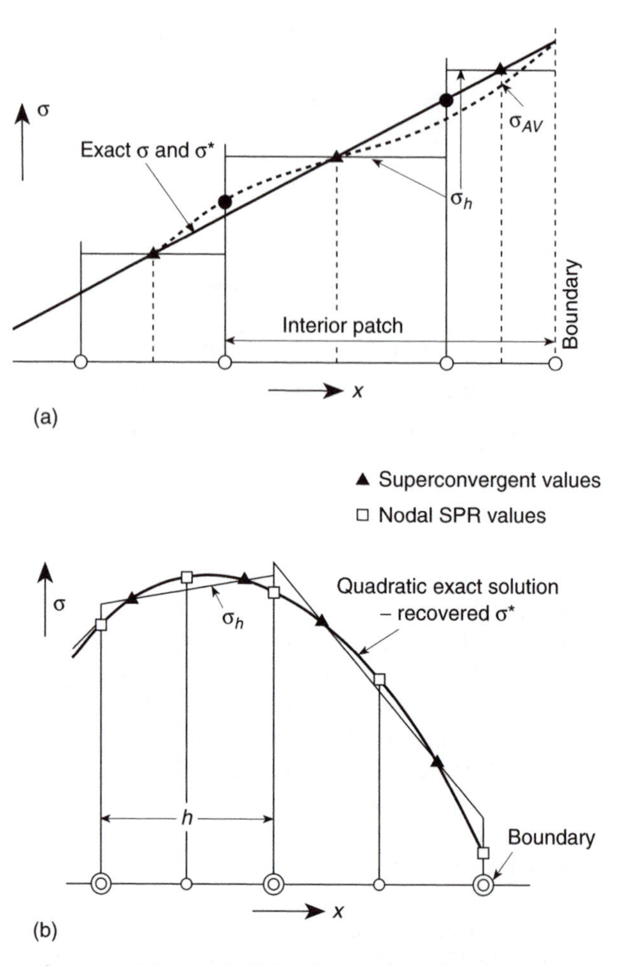

Fig. 13.10 Recovery of exact σ of degree p by linear elements ($p = 1$) and quadratic elements ($p = 2$).

be useful most of the additional constraints introduced have affected the superconvergent properties adversely and in general the modified versions of SPR such as those by Wiberg *et al.*[23, 24] and by Blacker and Belytschko[25] have not proved to be effective.

Example 13.1: SPR stress projection for rectangular element patch. As an example we consider the SPR projection for a stress component σ_i on the patch of rectangular elements shown in Fig. 13.14. The elements are 4-node rectangles in which shape functions are given by bilinear interpolations. Thus, the optimal sampling points are given by the points at the centre of each element.

The recovered solution is given by a linear polynomial expressed as

$$\sigma_i^* = \begin{bmatrix} 1, & (x - x_1), & (y - y_1) \end{bmatrix} \begin{Bmatrix} \bar{a}_1 \\ \bar{a}_2 \\ \bar{a}_3 \end{Bmatrix}$$

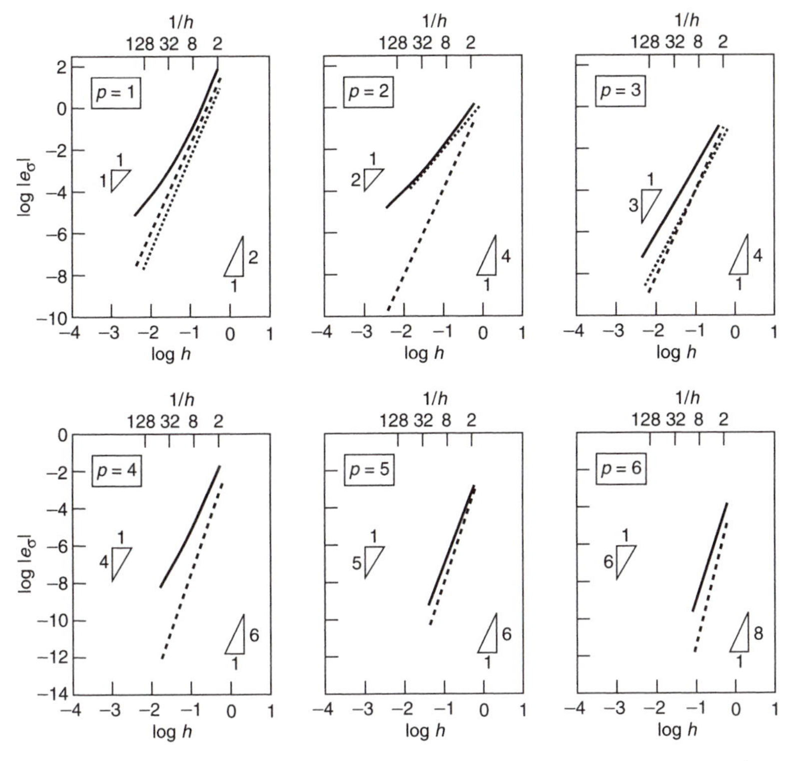

Fig. 13.11 Problem of a stressed bar. Rates of convergence (error) of stress, where $x = 0.5$ ($0 \leq x \leq 1$). ($\hat{\sigma}$ ——; σ_L · · · ·; σ^* - - - - -).

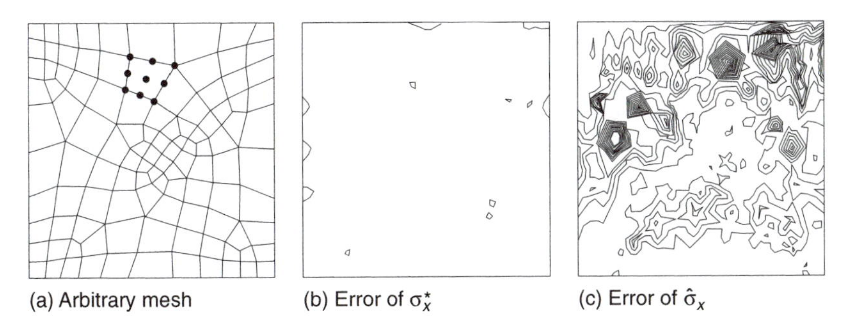

(a) Arbitrary mesh (b) Error of σ_x^* (c) Error of $\hat{\sigma}_x$

Fig. 13.12 Poisson equation in two dimensions solved using arbitrary-shaped quadratic quadrilaterals.

For this patch of elements, (13.23) is given by

$$\Pi = \frac{1}{2} \sum_{k=1}^{4} \left[\hat{\sigma}_i(x_k, y_k) - \mathbf{p}_k \mathbf{a} \right]^2$$

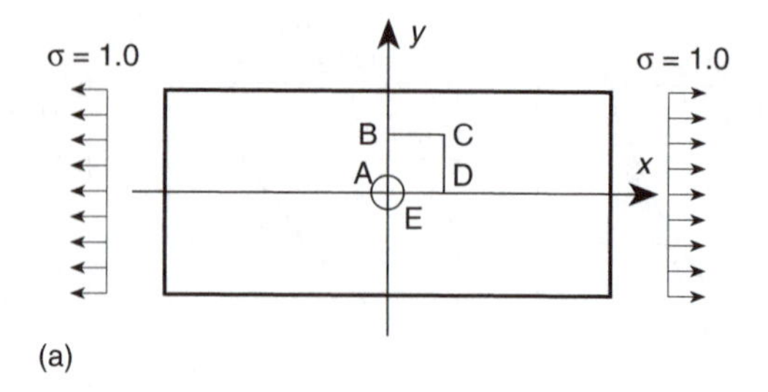

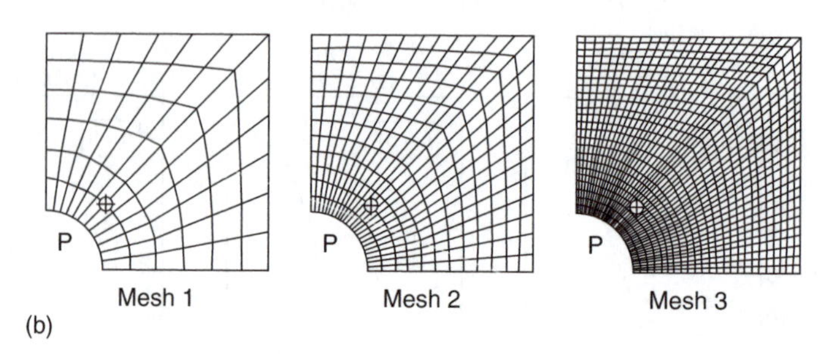

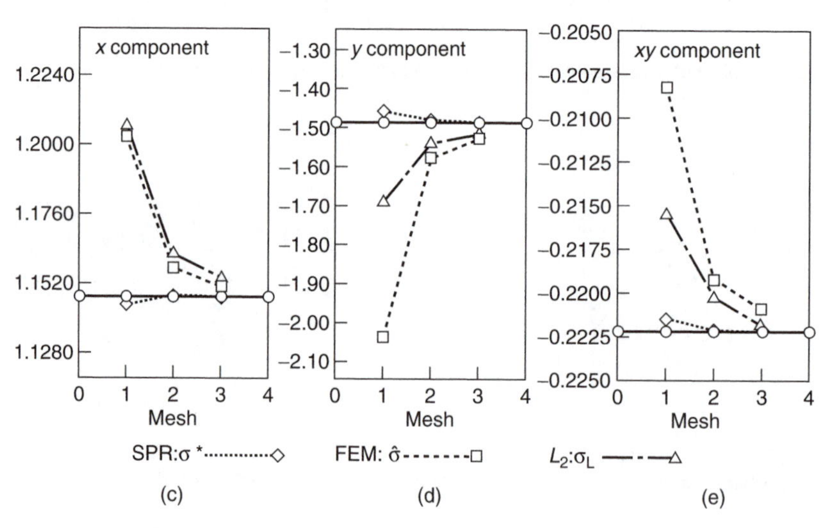

Fig. 13.13 Plane stress analysis of stresses around a circular hole in a uniaxial field.

where

$$
\mathbf{p}_k = \begin{cases} [1 \quad -a/2 \quad -b/2] & \text{for } k = 1 \\ [1 \quad a/2 \quad -b/2] & \text{for } k = 2 \\ [1 \quad a/2 \quad b/2] & \text{for } k = 3 \\ [1 \quad -a/2 \quad b/2] & \text{for } k = 4 \end{cases}
$$

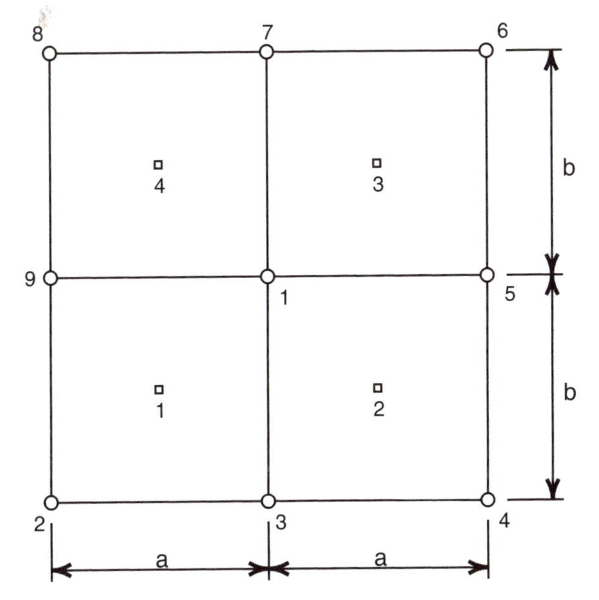

Fig. 13.14 Patch of rectangular elements for SPR projection. Optimal points to sample stresses indicated by □.

Evaluating the minimum for Π and performing the sum gives the equations

$$\mathbf{A}\mathbf{a} = \mathbf{b}$$

where

$$\mathbf{A} = \begin{bmatrix} 4 & 0 & 0 \\ 0 & a^2 & 0 \\ 0 & 0 & b^2 \end{bmatrix}$$

and

$$\mathbf{b} = \left(\begin{bmatrix} 1 \\ -a/2 \\ -b/2 \end{bmatrix} \hat{\sigma}_{i1} + \begin{bmatrix} 1 \\ a/2 \\ -b/2 \end{bmatrix} \hat{\sigma}_{i2} + \begin{bmatrix} 1 \\ a/2 \\ b/2 \end{bmatrix} \hat{\sigma}_{i3} + \begin{bmatrix} 1 \\ -a/2 \\ b/2 \end{bmatrix} \hat{\sigma}_{i4} \right)$$

The solution for the parameters is given by

$$\begin{aligned} a_1 &= \frac{1}{4}[\ \hat{\sigma}_{i1} + \hat{\sigma}_{i2} + \hat{\sigma}_{i3} + \hat{\sigma}_{i4}] \\ a_2 &= \frac{1}{2a}[-\hat{\sigma}_{i1} + \hat{\sigma}_{i2} + \hat{\sigma}_{i3} - \hat{\sigma}_{i4}] \\ a_3 &= \frac{1}{2b}[-\hat{\sigma}_{i1} - \hat{\sigma}_{i2} + \hat{\sigma}_{i3} + \hat{\sigma}_{i4}] \end{aligned}$$

Inserting the parameters into the equation for the recovered stress gives

$$\sigma_i^* = \left[\frac{1}{4}, \quad \frac{(x - x_1)}{2a}, \quad \frac{(y - y_1)}{2b} \right] \begin{Bmatrix} \hat{\sigma}_{i1} + \hat{\sigma}_{i2} + \hat{\sigma}_{i3} + \hat{\sigma}_{i4} \\ -\hat{\sigma}_{i1} + \hat{\sigma}_{i2} + \hat{\sigma}_{i3} - \hat{\sigma}_{i4} \\ -\hat{\sigma}_{i1} - \hat{\sigma}_{i2} + \hat{\sigma}_{i3} + \hat{\sigma}_{i4} \end{Bmatrix}$$

We note that the above yields SPR values at an internal node of a regular mesh which are the same as that obtained by averaging. Unfortunately, this is not the case when the mesh is irregular or boundary nodes are considered, as the reader can easily establish, where SPR will retain high accuracy but averaging will not.

13.4.2 SPR for displacements and stresses

The superconvergent patch recovery can be extended to produce superconvergent displacements. The procedure for the displacements is quite simple if we assume the superconvergent points to be at nodes of the patch. However, as we have already observed it is always necessary to have more data than the number of coefficients in the particular polynomial to be able to execute a least squares minimization. Here of course we occasionally need a patch which extends further than before, particularly since the displacements will be given by a polynomial one order higher than that used for the shape functions. In Fig. 13.8, however, we show for most assemblies that an identical patch to that used for stresses will suffice. Larger element patches have also been suggested in reference 28 but it does not seem anything is gained.

The recovered solution $\mathbf{u}^*$ has on occasion been used in dynamic problems (e.g., Wiberg[28, 29]), since in this class of problems the displacements themselves are often important. We also find such recovery useful in problems of fluid dynamics.

When both recovered displacements and stresses are desired, it is advantageous to compute the recovered stresses directly using the derivatives of the recovered displacements. The advantage of computing recovered stresses directly from displacements means that we have now obtained fully superconvergent results for all element types. Indeed, a recent study by Zhang and Naga,[30] for field problems, has found that SPR using nodal field variable sampling produces better recovered gradients in certain instances. For example, although both SPR using gradient sampling and SPR using field variable sampling achieve ultraconvergence in the recovered gradient at vertex nodes of quadratic triangles, ultraconvergence of the recovered gradient at the mid-edge nodes can only be obtained by SPR using field variable sampling. A similar procedure to that studied in reference 30 has been used by Wiberg and Hager[31] in eigenfrequency computations. Thus, field variable recovery should probably always be used for triangular and tetrahedral elements, as well as for other element types when both superconvergent displacements and stresses or strains are required.

The SPR recovery technique described in this section takes advantage of the superconvergence property of the finite element solutions and/or the availability of optimal sampling points. A recovery method which does not need such information has been devised and will be discussed in the next section.

13.5 Recovery by equilibration of patches – REP

Although SPR has proved to work well generally and much research has been devoted to its mathematical analyses, the reason behind its capability of producing an accurate recovered solution even when superconvergent points do not in fact exist remains an open question. We have therefore sought to determine viable recovery alternatives. One of these, known by the acronym REP (recovery by equilibrium of patches), will be described next. This procedure was first presented in reference 32 and later improved in reference 33.

To some extent the motivation is similar to that of Ladevèze *et al.*[34, 35] who sought to establish (for somewhat different reasons) a fully equilibrating stress field which can replace that of the finite element approximation. However, we believe that the process presented here and in reference 33 is simpler although equilibrium is satisfied in an approximate manner.

The starting point for REP is the governing equilibrium equation

$$\boldsymbol{\mathcal{S}}^{\mathrm{T}}\boldsymbol{\sigma} + \mathbf{b} = \mathbf{0} \tag{13.27}$$

In a finite element approximation this becomes

$$\int_{\Omega_p} \mathbf{B}^{\mathrm{T}}\hat{\boldsymbol{\sigma}}\,\mathrm{d}\Omega - \int_{\Omega_p} \mathbf{N}^{\mathrm{T}}\mathbf{b}\,\mathrm{d}\Omega - \int_{\Gamma_p} \mathbf{N}^{\mathrm{T}}\mathbf{t}\,\mathrm{d}\Gamma = \mathbf{0} \tag{13.28}$$

where $\hat{\boldsymbol{\sigma}}$ are the stresses from the finite element solution. In the above Ω_p is the domain of a patch and the last term comes from the tractions on the boundary of the patch domain Γ_p. These can, of course, represent the whole problem, a patch of a few elements or a single element.

As is well known the stresses $\hat{\boldsymbol{\sigma}}$ which result from the finite element analysis will in general be discontinuous and we shall seek to replace them in *every element patch* by a recovered system which is smooth and continuous.

To achieve the recovery we proceed in an analogous way to that used in the SPR procedure, *first* approximating the stress in each patch by a polynomial of appropriate order $\bar{\boldsymbol{\sigma}}^*$, *second* using this approximation to obtain nodal values of $\tilde{\boldsymbol{\sigma}}^*$ and *finally* interpolating these values by standard shape functions.

The stress $\boldsymbol{\sigma}$ is taken as a vector of appropriate components, which for convenience we write as:

$$\boldsymbol{\sigma} = \begin{Bmatrix} \sigma_1 \\ \sigma_2 \\ \vdots \\ \sigma_n \end{Bmatrix} \tag{13.29}$$

The above notation is general with, for instance, $\sigma_1 = \sigma_x$, $\sigma_2 = \sigma_y$ and $\sigma_3 = \tau_{xy}$ describing a two-dimensional plane elastic analysis.

We shall write each component of the above as a polynomial expansion of the form:

$$\bar{\sigma}_i^* = \begin{bmatrix} 1, & \bar{x}, & \bar{y}, & \cdots \end{bmatrix} \mathbf{a}_i = \mathbf{p}(x, y)\mathbf{a}_i \tag{13.30}$$

where $\mathbf{p}$ is a vector of polynomials, $\mathbf{a}_i$ is a set of unknown coefficients for the ith component of stress and $\bar{x}$, $\bar{y}$ are as described for (13.22b).

For equilibrium we shall always attempt to ensure that the smoothed stress $\bar{\boldsymbol{\sigma}}^*$ satisfies in a least squares sense the same patch equilibrium conditions as the finite element solution. Accordingly,

$$\int_{\Omega_p} \mathbf{B}^{\mathrm{T}}\hat{\boldsymbol{\sigma}}\,\mathrm{d}\Omega \approx \int_{\Omega_p} \mathbf{B}^{\mathrm{T}}\bar{\boldsymbol{\sigma}}^*\,\mathrm{d}\Omega \tag{13.31}$$

where

$$\bar{\boldsymbol{\sigma}}^* = \mathbf{Pa} = \begin{bmatrix} \mathbf{p} & \mathbf{0} & \mathbf{0} \\ \mathbf{0} & \mathbf{p} & \mathbf{0} \\ \mathbf{0} & \mathbf{0} & \mathbf{p} \end{bmatrix} \begin{Bmatrix} \mathbf{a}_1 \\ \mathbf{a}_2 \\ \mathbf{a}_3 \end{Bmatrix} \tag{13.32}$$

written here again for the case of three stress components. Obvious modifications are made for more or fewer components.

It has been found in practice that the constraints provided by Eq. (13.31) are not sufficient to always produce non-singular least squares minimization. Accordingly, the equilibrium constraints are split into an alternative form in which each component of stress is subjected to equilibrium requirements. This may be achieved by expressing the stress as

$$\begin{aligned}\bar{\boldsymbol{\sigma}}^* &= \sum_i \mathbf{1}_i \bar{\sigma}_i^* = \sum_i \bar{\boldsymbol{\sigma}}_i^* \\ \hat{\boldsymbol{\sigma}} &= \sum_i \mathbf{1}_i \hat{\sigma}_i = \sum_i \hat{\boldsymbol{\sigma}}_i\end{aligned} \tag{13.33}$$

in which

$$\mathbf{1}_1 = \begin{bmatrix}1, & 0, & 0\end{bmatrix}^{\mathrm{T}}; \qquad \mathbf{1}_2 = \begin{bmatrix}0, & 1, & 0\end{bmatrix}^{\mathrm{T}} \quad \text{etc.} \tag{13.34}$$

The equations are now obtained by imposing the set of constraints

$$\int_{\Omega_p} \mathbf{B}^{\mathrm{T}} \hat{\boldsymbol{\sigma}}_i \, \mathrm{d}\Omega \approx \int_{\Omega_p} \mathbf{B}^{\mathrm{T}} \bar{\boldsymbol{\sigma}}_i^* \, \mathrm{d}\Omega = \int_{\Omega_p} \mathbf{B}^{\mathrm{T}} \mathbf{1}_i \mathbf{p} \, \mathrm{d}\Omega \, \mathbf{a}_i \tag{13.35}$$

The imposition of the approximate equation (13.35) allows each set of coefficients $\mathbf{a}_i$ to be solved independently reducing considerably the solution cost and here repeating a procedure used with success in SPR.

A least squares minimization of Eq. (13.35) is expressed as

$$\Pi = \frac{1}{2} \left(\mathbf{H}_i \mathbf{a}_i - \mathbf{f}_i^p\right)^{\mathrm{T}} \left(\mathbf{H}_i \mathbf{a}_i - \mathbf{f}_i^p\right) \tag{13.36}$$

where

$$\mathbf{H}_i = \int_{\Omega_p} \mathbf{B}^{\mathrm{T}} \mathbf{1}_i \mathbf{p} \, \mathrm{d}\Omega \quad \text{and} \quad \mathbf{f}_i^p = \int_{\Omega_p} \mathbf{B}^{\mathrm{T}} \hat{\boldsymbol{\sigma}}_i \, \mathrm{d}\Omega \tag{13.37}$$

The minimization condition results in

$$\mathbf{a}_i = \left[\mathbf{H}_i^{\mathrm{T}} \mathbf{H}_i\right]^{-1} \mathbf{H}_i^{\mathrm{T}} \mathbf{f}_i^p \tag{13.38}$$

Nodal values $\tilde{\boldsymbol{\sigma}}^*$ are obtained from Eq. (13.30) and the final recovered solution is given by Eq. (13.21).

The REP procedure follows precisely the details of SPR near boundaries and gives overall an approximation which does not require knowledge of any superconvergent points. The accuracy of both processes is comparable.

13.6 Error estimates by recovery

One of the most important applications of the recovery methods is its use in the computation of *a posteriori* error estimators. With the recovered solutions available, we can now evaluate errors simply by replacing the exact values of quantities such as $\mathbf{u}$, $\boldsymbol{\sigma}$, etc., which are in general unknown, in Eqs (13.1) and (13.2), by the recovered values which are more accurate than the direct finite element solution. We write the error estimators in various norms such as

$$\begin{aligned} \|\mathbf{e}\| &\approx \|\hat{\mathbf{e}}\| &= \|\mathbf{u}^* - \hat{\mathbf{u}}\| \\ \|\mathbf{e}\|_{L_2} &\approx \|\hat{\mathbf{e}}\|_{L_2} &= \|\mathbf{u}^* - \hat{\mathbf{u}}\|_{L_2} \\ \|\mathbf{e}_\sigma\|_{L_2} &\approx \|\hat{\mathbf{e}}_\sigma\|_{L_2} &= \|\boldsymbol{\sigma}^* - \hat{\boldsymbol{\sigma}}\|_{L_2} \end{aligned} \tag{13.39}$$

For example, an error estimator of the energy norm for elasticity problems has the form

$$\|\hat{\mathbf{e}}\| = \left[\int_\Omega \left(\boldsymbol{\sigma}^* - \hat{\boldsymbol{\sigma}}\right)^{\mathrm{T}} \mathbf{D}^{-1} \left(\boldsymbol{\sigma}^* - \hat{\boldsymbol{\sigma}}\right) \mathrm{d}\Omega \right]^{\frac{1}{2}} \tag{13.40}$$

Similarly, estimates of the RMS error in displacement and stress can be obtained through Eqs (13.9)–(13.10). Error estimators formulated by replacing the exact solution with the recovered solution are sometimes called *recovery-based error estimators*. This type of error estimator was first introduced by Zienkiewicz and Zhu.[36]

The accuracy or the quality of the error estimators is measured by the *effectivity index* θ, which is defined as

$$\theta = \frac{\|\hat{\mathbf{e}}\|}{\|\mathbf{e}\|} \tag{13.41}$$

A theorem presented by Zienkiewicz and Zhu[12] shows that for all estimators based on recovery we can establish the following bounds for the effectivity index:

$$1 - \frac{\|\mathbf{e}^*\|}{\|\mathbf{e}\|} \le \theta \le 1 + \frac{\|\mathbf{e}^*\|}{\|\mathbf{e}\|} \tag{13.42}$$

where $\mathbf{e}$ is the actual error and $\mathbf{e}^*$ is the error of the recovered solution, e.g.,

$$\|\mathbf{e}^*\| = \|\mathbf{u} - \mathbf{u}^*\| \tag{13.43}$$

The proof of the above theorem is straightforward if we write Eq. (13.40) as

$$\|\hat{\mathbf{e}}\| = \|\mathbf{u}^* - \hat{\mathbf{u}}\| = \|\left(\mathbf{u} - \hat{\mathbf{u}}\right) - \left(\mathbf{u} - \mathbf{u}^*\right)\| = \|\mathbf{e} - \mathbf{e}^*\| \tag{13.44}$$

Using now the triangle inequality we have

$$\|\mathbf{e}\| - \|\mathbf{e}^*\| \le \|\hat{\mathbf{e}}\| \le \|\mathbf{e}\| + \|\mathbf{e}^*\| \tag{13.45}$$

from which the inequality (13.42) follows after division by $\|\mathbf{e}\|$. Obviously, the theorem is also true for error estimators of other norms. Two important conclusions follow:

1. *any recovery process* which results in reduced error will give a reasonable error estimator and, more importantly,
2. if the recovered solution converges at a higher rate than the finite element solution we shall always have asymptotically exact estimation.

To prove the second point we consider a typical finite element solution with shape functions of order p where we know that the error (in the energy norm) is

$$\|\mathbf{e}\| = O(h^p) \tag{13.46}$$

If the recovered solution gives an error of a higher order, e.g.,

$$\|\mathbf{e}^*\| = O(h^{p+\alpha}) \qquad \alpha > 0 \tag{13.47}$$

then the bounds of the effectivity index are:

$$1 - O(h^\alpha) \leq \theta \leq 1 + O(h^\alpha) \tag{13.48}$$

and the error estimator is asymptotically exact, that is

$$\theta \to 1 \quad \text{as} \quad h \to 0 \tag{13.49}$$

This means that the error estimator converges to the true error. This is a very important property of error estimators based on recovery and is not generally shared by residual-based estimators which we discuss in the next section.

13.7 Residual-based methods

Other methods to obtain error estimators have been proposed by many investigators working in the field.[37–49] Most of these make use of the residuals of the finite element approximation, either explicitly or implicitly. Error estimators based on these methods are often called *residual error estimators*. Those using residuals explicitly are termed explicit residual error estimators; the others are called implicit residual error estimators.

In this section we are concerned with both explicit and implicit residual error estimators. To simplify the presentation, we use the quasi-harmonic equation in a two-dimensional domain as the model problem. The governing equation of the problem is given by

$$-\boldsymbol{\nabla}^{\mathrm{T}}(k\,\boldsymbol{\nabla}\phi) + Q = 0 \quad \text{in } \Omega \tag{13.50}$$

with boundary conditions

$$\phi = \bar{\phi} \quad \text{on } \Gamma_\phi$$
$$\mathbf{q}^{\mathrm{T}}\mathbf{n} = q_n = \bar{q} \quad \text{on } \Gamma_q$$

In the above

$$\mathbf{q} = -k\,\boldsymbol{\nabla}\phi = \begin{bmatrix} q_x, & q_y \end{bmatrix}^{\mathrm{T}} \tag{13.51}$$

is a flux, $\mathbf{n}$ is the unit outward normal to the boundary Γ and q_n is the flux normal to the boundary (see Chapters 3 and 7).

The error of the finite element solution $\hat{\phi}$ is written as

$$e = \phi - \hat{\phi} \tag{13.52}$$

The global energy norm error for domain Ω [viz. Eq. (13.11)] is

$$\|e\| = \left(\sum_{K=1}^{m} \|e\|_K^2 \right)^{\frac{1}{2}} \tag{13.53}$$

where for each element K

$$\begin{aligned} \|e\|_K^2 &= \int_{\Omega_K} (\boldsymbol{\nabla} e)^{\mathrm{T}} k\, \boldsymbol{\nabla} e \, \mathrm{d}\Omega \\ &= \int_{\Omega_K} \frac{1}{k} [(q_x - \hat{q}_x)^2 + (q_y - \hat{q}_y)^2] \, \mathrm{d}\Omega \end{aligned} \tag{13.54}$$

In what follows, we shall first discuss the explicit residual error estimator.

13.7.1 Explicit residual error estimator

The energy norm for an explicit residual error estimator has been derived by various authors[48, 49] and has a general form

$$\|\hat{e}\| = \left(\sum_{K=1}^{m} \|\hat{e}\|_{r_K}^2\right)^{\frac{1}{2}} \tag{13.55}$$

with element contributions

$$\|\hat{e}\|_{r_K}^2 = C_1 \int_{\Omega_K} r_K^2 \, \mathrm{d}\Omega + C_2 \int_{\Gamma_K} J^2 \, \mathrm{d}\Gamma \tag{13.56}$$

where

$$r_K = -\nabla^{\mathrm{T}} \left(k \nabla \hat{\phi}\right) + Q \tag{13.57}$$

is the element interior residual and J is the discontinuity in the normal flux q_n at each edge of element K, which we call a *jump discontinuity*. For example, at an edge shared by element K and its neighbouring element I, we have

$$J = \hat{q}_{n_K} + \hat{q}_{n_I} \tag{13.58}$$

where

$$\hat{q}_{n_K} = \hat{\mathbf{q}}^{\mathrm{T}} \mathbf{n}_K \quad \text{and} \quad \hat{q}_{n_I} = \hat{\mathbf{q}}^{\mathrm{T}} \mathbf{n}_I$$

are the finite element normal fluxes.

The constants C_1 and C_2 that appear in (13.56) are mesh dependent parameters and generally are unknown. This renders the explicit residual error estimators in the form of Eq. (13.56) less useful in practical computations.

For the particular case of constant k an explicit form for C_1 and C_2 has been obtained for a 4-node quadrilateral element.[38, 39] This element explicit residual error estimator has the form

$$\|\hat{e}\|_{r_K}^2 = \frac{h^2}{24k} \int_{\Omega_K} r^2 \, \mathrm{d}\Omega + \frac{h}{24k} \int_{\Gamma_K} J^2 \, \mathrm{d}\Gamma \tag{13.59}$$

The derivation of Eq. (13.59) was achieved following some heuristic assumptions on the error distribution and manipulations of the element residuals. It was found that the major contribution to the error estimator is from the term involving the jump discontinuities and that the term for the element interior residual is of higher order. Therefore, in practice the following form

$$\|\hat{e}\|_{r_K}^2 = \frac{h}{24k} \int_{\Gamma_K} J^2 \, \mathrm{d}\Gamma \tag{13.60}$$

is often used. Indeed, this form of the explicit residual error estimator has been most widely used.

In the following, we shall show, as an example, that the explicit residual error estimator of Eq. (13.60) can also be derived from a particular recovery-based error estimator.

Example 13.2: Deriving explicit residual error estimator. For simplicity we consider a square element Ω_K and its neighbouring elements as shown in Fig. 13.15. The element contribution of the recovery-based error estimator is in the form

$$\|\hat{e}\|_K^2 = \int_{\Omega_K} \frac{1}{k} [(q_x^* - \hat{q}_x)^2 + (q_y^* - \hat{q}_y)^2] \, \mathrm{d}\Omega \tag{13.61}$$

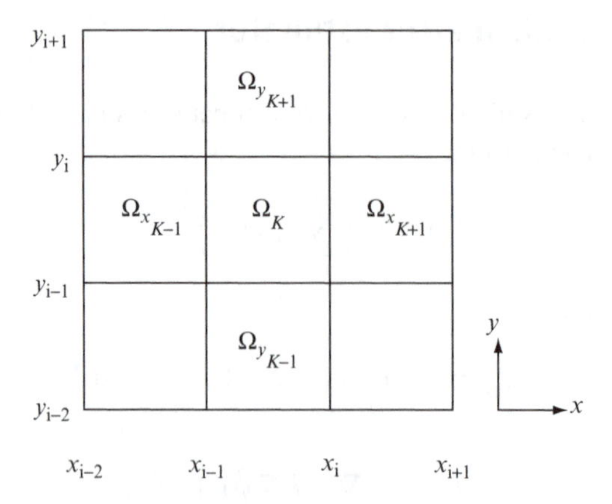

Fig. 13.15 An element patch. Element Ω_K and its neighbors.

The main steps involved in the derivation of the residual error estimator of Eq. (13.60) are as follows:

1. Construct a recovered solution for element Ω_K from elements Ω_K, $\Omega_{x_{K+1}}$ and $\Omega_{y_{K+1}}$ and forming recovery-based error estimator $\|\hat{e}\|^2_{K_1}$.
2. Construct a recovered solution for element Ω_K from elements Ω_K, $\Omega_{x_{K-1}}$ and $\Omega_{y_{K-1}}$ and forming recovery-based error estimator $\|\hat{e}\|^2_{K_2}$.
3. Average $\|\hat{e}\|^2_{K_1}$ and $\|\hat{e}\|^2_{K_2}$ to obtain the final recovery-based error estimator which results in the explicit residual error estimator of Eq. (13.60).

In the first step, consider Ω_K and its two neighbouring elements $\Omega_{x_{K+1}}$ and $\Omega_{y_{K+1}}$, the recovered solutions are expressed as

$$\begin{aligned} q^*_{x_1} &= \hat{q}_{x_K} + \alpha_1 Z_x(x) \\ q^*_{y_1} &= \hat{q}_{y_K} + \beta_1 Z_y(y) \end{aligned} \tag{13.62}$$

where Z_x and Z_y are linear functions in x and y respectively, i.e.,

$$\begin{aligned} Z_x(x) &= 1 - 2\left(\frac{x_i - x}{h}\right) \\ Z_y(y) &= 1 - 2\left(\frac{y_i - y}{h}\right) \end{aligned} \tag{13.63}$$

with h the edge length of the square element

$$h = x_i - x_{i-1} = y_i - y_{i-1}$$

and α_1, β_1 are unknown parameters to be determined by a recovery process. A recovery process for $q^*_{x_1}$ is shown in Fig. 13.16. A similar result holds for $q^*_{y_1}$.

The recovery method of averaging is used by requiring the recovered solution to be the average of the finite element solution at the boundary of the element, i.e., at the edge shared by Ω_K and $\Omega_{x_{K+1}}$

$$q^*_{x_1}(x_i) = \frac{1}{2}[\hat{q}_{x_K}(x_i) + \hat{q}_{x_{K+1}}(x_i)] \tag{13.64a}$$

and at the edge shared by element Ω_K and $\Omega_{y_{K+1}}$

$$q^*_{y_1}(y_i) = \frac{1}{2}[\hat{q}_{y_K}(y_i) + \hat{q}_{y_{K+1}}(y_i)] \tag{13.64b}$$

Substituting Eq. (13.62) into Eqs (13.64a)–(13.64b), α_1 and β_1 have the solution

$$\begin{aligned}
\alpha_1 &= \frac{1}{2Z_x(x_i)}(\hat{q}_{x_{K+1}}(x_i) - \hat{q}_{x_K}(x_i)) = -\frac{1}{2}J(x_i) \\
\beta_1 &= \frac{1}{2Z_y(y_i)}(\hat{q}_{y_{K+1}}(y_i) - \hat{q}_{y_K}(y_i)) = -\frac{1}{2}J(y_i)
\end{aligned} \tag{13.65}$$

where $J(x_i)$ is the jump discontinuity along edge x_i (viz. Fig. 13.16) and $J(y_i)$ is the jump discontinuity along edge y_i. In the above, we have used the fact that at x_i

$$\hat{q}_{n_K} = \hat{q}_{x_K} \qquad \text{and} \qquad \hat{q}_{n_{K+1}} = -\hat{q}_{x_{K+1}}$$

and at y_i

$$\hat{q}_{n_K} = \hat{q}_{y_K} \qquad \text{and} \qquad \hat{q}_{n_{K+1}} = -\hat{q}_{y_{K+1}}$$

The determined recovered solutions are now in the form of

$$\begin{aligned}
q^*_{x_1} &= \hat{q}_{x_K} - \frac{1}{2}J(x_i)Z_x \\
q^*_{y_1} &= \hat{q}_{y_K} - \frac{1}{2}J(y_i)Z_y
\end{aligned} \tag{13.66}$$

Error estimator $\|\hat{e}\|^2_{K_1}$ for element Ω_K is attained by substituting the above $q^*_{x_1}$ and $q^*_{y_1}$ into Eq. (13.61)

$$\|\hat{e}\|^2_{K_1} = \frac{1}{4k}\int_{\Omega_K}(J(x_i)^2 Z_x^2 + J(y_i)^2 Z_y^2)\,\mathrm{d}\Omega \tag{13.67}$$

Notice that $J(x_i)$ and Z_y are the only function of y and $J(y_i)$ and Z_x are the only function of x, we have the first recovery-based error estimator for element Ω_K

$$\begin{aligned}
\|\hat{e}\|^2_{K_1} &= \frac{1}{4k}\left(\int_{x_{i-1}}^{x_i} Z_x^2\,\mathrm{d}x \int_{y_{i-1}}^{y_i} J(x_i)^2\,\mathrm{d}y + \int_{y_{i-1}}^{y_i} Z_y^2\,\mathrm{d}y \int_{x_{i-1}}^{x_i} J(y_i)^2\,\mathrm{d}x\right) \\
&= \frac{h}{12k}\left(\int_{\Gamma_{y_i}} J(x_i)^2\,\mathrm{d}\Gamma + \int_{\Gamma_{x_i}} J(y_i)^2\,\mathrm{d}\Gamma\right)
\end{aligned} \tag{13.68}$$

where Γ_{x_i} denotes the limits from x_{i-1} to x_i and Γ_{y_i} from y_{i-1} to y_i. In the above we have used the following results

$$\int_{x_{i-1}}^{x_i} Z_x^2\,\mathrm{d}x = \int_{y_{i-1}}^{y_i} Z_y^2\,\mathrm{d}y = \frac{h}{3}$$

Similarly, in the second step consider elements Ω_K, $\Omega_{x_{K-1}}$ and $\Omega_{y_{K-1}}$ with the recovered solutions written as

$$\begin{aligned}
q^*_{x_2} &= \hat{q}_{x_K} + \alpha_2 Z_x \\
q^*_{y_2} &= \hat{q}_{y_K} + \beta_2 Z_y
\end{aligned} \tag{13.69}$$

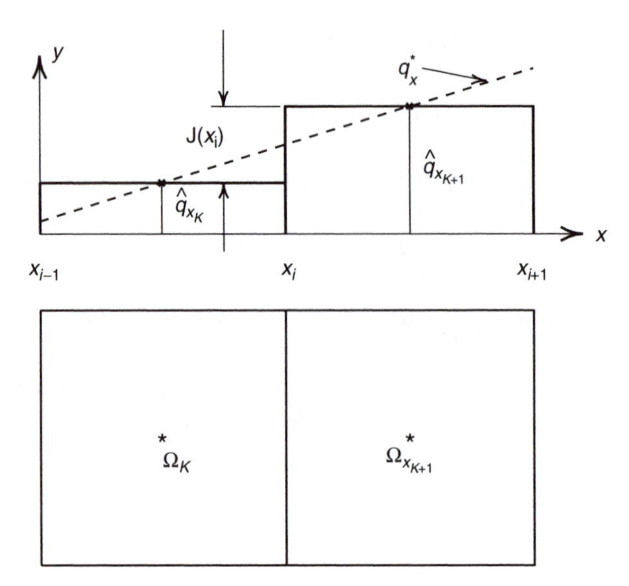

Fig. 13.16 Recovered solution and jump for Ω_{x_K} and $\Omega_{x_{K+1}}$.

To determine the unknown parameters α_2 and β_2 we again use the recovery method of averaging and require that at the edge shared by Ω_K and $\Omega_{x_{K-1}}$

$$q^*_{x_2}(x_{i-1}) = \frac{1}{2}(\hat{q}_{x_{K-1}}(x_{i-1}) + \hat{q}_{x_K}(x_{i-1})) \tag{13.70}$$

and at the edge shared by element Ω_K and $\Omega_{y_{K-1}}$

$$q^*_{y_2}(y_{i-1}) = \frac{1}{2}(\hat{q}_{y_{K-1}}(y_{i-1}) + \hat{q}_{y_K}(y_{i-1})) \tag{13.71}$$

Following the exact procedure used in step one, α_2 and β_2 are solved as

$$\begin{aligned} \alpha_2 &= \frac{1}{2Z_x(x_{i-1})}(\hat{q}_{x_{K-1}}(x_{i-1}) - \hat{q}_{x_K}(x_{i-1})) = -\frac{1}{2}J(x_{i-1}) \\ \beta_2 &= \frac{1}{2Z_y(y_{i-1})}(\hat{q}_{y_{K-1}}(y_{i-1}) - \hat{q}_{y_K}(y_{i-1})) = -\frac{1}{2}J(y_{i-1}) \end{aligned} \tag{13.72}$$

Here $J(x_{i-1})$ and $J(y_{i-1})$ are jump discontinuities along edges x_{i-1} and y_{i-1} respectively. The recovered solutions are now written as

$$\begin{aligned} q^*_{x_2} &= \hat{q}_{x_K} - \frac{1}{2}J(x_{i-1})Z_x \\ q^*_{y_2} &= \hat{q}_{y_K} - \frac{1}{2}J(y_{i-1})Z_y \end{aligned} \tag{13.73}$$

Substituting Eq. (13.73) into Eq. (13.61) the second recovery-based error estimator can be obtained

$$\|\hat{e}\|^2_{K_2} = \frac{h}{12k}\left(\int_{\Gamma_{x_{i-1}}} J(y_{i-1})^2\,\mathrm{d}\Gamma + \int_{\Gamma_{y_{i-1}}} J(x_{i-1})^2\,\mathrm{d}\Gamma\right) \tag{13.74}$$

Finally, to include the influence of all the neighbouring elements, it is only natural to let the recovery error estimator for element Ω_K be taken as the average of $\|\hat{e}\|_{K_1}^2$ and $\|\hat{e}\|_{K_2}^2$, i.e.,

$$\|\hat{e}\|_K^2 = \frac{1}{2}(\|\hat{e}\|_{K_1}^2 + \|\hat{e}\|_{K_2}^2) \tag{13.75}$$

Substituting the expressions of $\|\hat{e}\|_{K_1}^2$ and $\|\hat{e}\|_{K_2}^2$, we have

$$\begin{aligned}\|\hat{e}\|_K^2 &= \frac{h}{24k}\left(\int_{\Gamma_{y_i}} J(x_i)^2 \,\mathrm{d}\Gamma + \int_{\Gamma_{x_i}} J(y_i)^2 \,\mathrm{d}\Gamma + \int_{\Gamma_{x_{i-1}}} J(y_{i-1})^2 \,\mathrm{d}\Gamma + \int_{\Gamma_{y_{i-1}}} J(x_{i-1})^2 \,\mathrm{d}\Gamma\right) \\ &= \frac{h}{24k}\int_{\Gamma_K} J^2 \,\mathrm{d}\Gamma\end{aligned} \tag{13.76}$$

This is exactly the explicit residual error estimator of Eq. (13.60).

We have demonstrated, by the above example, that the explicit residual error estimator for the bilinear element can be derived from a recovery-based error estimator using averaging as the recovery method. For more general discussions on the relationship between recovery-based error estimators and explicit residual error estimators we refer to references 50 and 51; for discussion on the equivalence of recovery-based error estimators with certain explicit residual estimators we refer to references 48 and 52; for using a recovery method in the computation of the explicit residual error estimator the reader is referred to reference 53. We shall now turn our attention to how to use a recovery method in the computation of implicit residual error estimators.

13.7.2 Implicit residual error estimators

The computation of implicit residual error estimators requires solving an auxiliary boundary value problem with residuals as input data for the approximation error. Among all the existing implicit residual error estimators, the *equilibrated residual estimator* has been shown to be the most robust.[54–56]

In what follows we restrict our discussion to the equilibrated residual error estimator for the model problem of Eq. (13.50). The construction of an equilibrated residual error estimator for other problems, such as elasticity problems, proceeds in an analogous manner.[57]

We again consider an interior element K. Substituting the finite element solution $\hat{\phi}$ into Eq. (13.50) results in, for element K,

$$-\boldsymbol{\nabla}^{\mathrm{T}}(k\,\boldsymbol{\nabla}\hat{\phi}) + Q = r_K \quad \text{in } \Omega_K \tag{13.77}$$

and

$$-(k\,\boldsymbol{\nabla}\hat{\phi})^{\mathrm{T}}\mathbf{n} = \hat{q}_n \quad \text{on } \Gamma_K$$

Subtracting the above equations from Eq. (13.50) gives an element boundary value problem for error e as

$$-\boldsymbol{\nabla}^{\mathrm{T}}(k\,\boldsymbol{\nabla}e) + r_K = 0 \quad \text{in } \Omega_K \tag{13.78}$$

with boundary condition

$$-(k\,\boldsymbol{\nabla}e)^{\mathrm{T}}\,\mathbf{n} = q_n - \hat{q}_n \quad \text{on } \Gamma_K$$

We notice immediately that Eq. (13.78) is not solvable because the exact normal flux q_n on the element boundary is in general unknown. A natural strategy to overcome this difficulty is to replace the exact solution by a recovered solution q_n^* which can be computed from the finite element flux in element K and its surrounding elements (as we did in the computation of a recovery-based error estimator).

We can now write the Neumann boundary value problem for the element error as

$$-\boldsymbol{\nabla}^{\mathrm{T}}(k\,\boldsymbol{\nabla} e) + r_K = 0 \quad \text{in } \Omega_K \tag{13.79}$$

with boundary condition

$$-(k\,\boldsymbol{\nabla} e)^{\mathrm{T}}\,\mathbf{n} = q_n^* - \hat{q}_n \quad \text{on } \Gamma_K$$

An approximate solution to the above equation for $\hat{e}$ appearing in the energy norm, $\|\hat{e}\|_K$, defines an *implicit element residual error estimator*.

Various recovery techniques can be used to compute the normal flux q_n^*.[34, 42, 43] However, the Neumann problem of Eq. (13.79) will have a solution if q_n^* is computed such that the residuals satisfy the *equilibrium condition*

$$\int_{\Omega_K} N_c r_K \,\mathrm{d}\Omega + \int_{\Gamma_K} N_c \left(q_n^* - \hat{q}_n\right)\,\mathrm{d}\Gamma = 0 \tag{13.80}$$

where N_c is the shape function for node c of element K. Although N_c can be a shape function of any order, a linear shape function seems to be the most practical in the following computation.

The residuals which satisfy Eq. (13.80) are said to be equilibrated, thus the recovered solution q_n^* satisfying Eq. (13.80) is called the equilibrated flux. An error estimator which uses the solution of the element error problem of Eq. (13.80) with the equilibrated flux q_n^* is termed an *equilibrated residual error estimator*. This type of residual error estimator was first introduced by Bank and Weiser[42] and later more rigorously pursued by Ainsworth and Oden.[46]

It is apparent that the most important step in the computation of the equilibrated residual error estimator is to achieve the recovered normal flux q_n^* which satisfies Eq. (13.80). Once q_n^* is determined, the error problem Eq. (13.79) can be readily solved for an element following a standard finite element procedure. Therefore we shall focus our attention on the recovery process.

The technique of recovering normal flux by equilibrated residuals was first proposed by Ladevéze *et al.*,[34] Kelly[41] and followed by Ohtsubo and Mitamura.[58] A different version of this technique was later used by Ainsworth and Oden[49] where a detailed description of the application to various mesh patterns can be found. Here we shall consider a typical element patch of triangles as shown in Fig. 13.17.

To determine q_n^*, we first substitute the residual r_K of Eq. (13.77) into Eq. (13.80) and upon integrating by parts obtain

$$\int_{\Omega_K} N_c Q \,\mathrm{d}\Omega + \int_{\Omega_K} (\boldsymbol{\nabla} N_c)^{\mathrm{T}} \left(k\,\boldsymbol{\nabla}\hat{\phi}\right)\,\mathrm{d}\Omega + \int_{\Gamma_K} N_c q_n^* \,\mathrm{d}\Gamma = 0 \tag{13.81}$$

Let the recovered interelement boundary normal flux have the form

$$q_n^* = \tfrac{1}{2}\left(\hat{\mathbf{q}}_K + \hat{\mathbf{q}}_I\right)^{\mathrm{T}} \mathbf{n}_s + Z_s \tag{13.82}$$

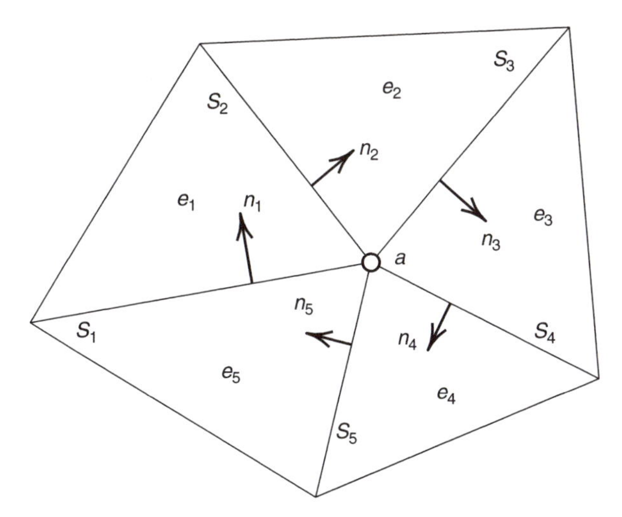

Fig. 13.17 Typical patch with interior vertex node a showing a local numbering of elements e_i and edges S_i.

where the first term on the right-hand side is the average of the normal flux of the finite element solution from element K and its neighbour element I as shown in Fig. 13.18; $\mathbf{n}_s$ is the outward normal on the edge s of element K; and Z_s is a linear function defined on the edge s, shared by elements K and I, with end nodes a and b and

$$Z_s = \hat{N}_a^s a_a^s + \hat{N}_b^s a_b^s \tag{13.83}$$

where $\hat{N}_a^s$, $\hat{N}_b^s$ are the dual shape functions introduced in Sec. 12.2.1 and in the present case are given by

$$\hat{N}_a^s = \frac{2}{|h_s|}\,[2\,N_a^s - N_b^s] \quad \text{and} \quad \hat{N}_b^s = \frac{2}{|h_s|}\,[2\,N_b^s - N_a^s] \tag{13.84}$$

where N_a^s and N_b^s are the linear shape functions defined for edge S and $|h_s|$ is the length of the edge. The unknown parameters a_a^s and a_b^s are to be determined from the residual equilibrium equation (13.81).

It is easy to verify that

$$\int_s N_a^s \hat{N}_b^s \,\mathrm{d}\Gamma = \delta_{ab} \tag{13.85}$$

where δ_{ab} is the Kronecker delta given by:

$$\delta_{ab} = \begin{cases} 1, & a = b \\ 0, & a \neq b \end{cases} \tag{13.86}$$

Let a denote a typical interior vertex node. Choose $N_c = N_a$ in Eq. (13.81) and consider the element patch associated with the linear shape function N_a as shown in Fig. 13.17. Assign element 1 as element K in the patch (i.e., $e_1 = e_K$).

It is obvious that N_a is zero at the exterior boundary of the element patch. A local numbering for the elements and edges connected to node a in the patch is given. The edge normals shown here are the result of a global edge orientation.

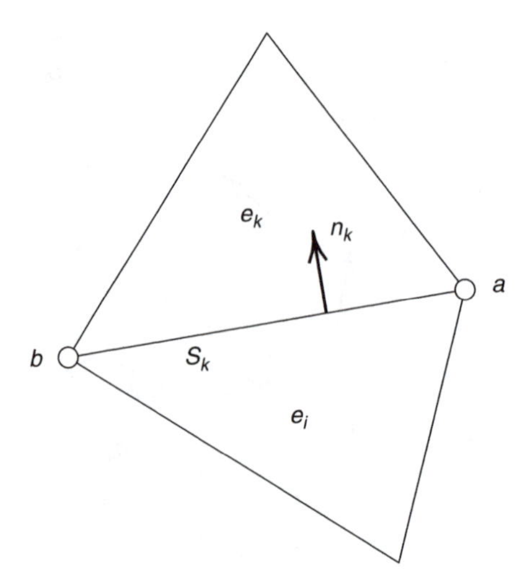

Fig. 13.18 Element interface for equilibrated flux recovery.

For element e_1 in the patch, substituting Eq. (13.82) into Eq. (13.81) for each edge and observing that N_a is non-zero only on the edges s_1 and s_2 and in the n_1 and n_2 directions we have

$$\begin{aligned}\int_{\Omega_{e_1}} N_a Q \, \mathrm{d}\Omega + \int_{\Omega_{e_1}} (\boldsymbol{\nabla} N_a)^{\mathrm{T}} (k \, \boldsymbol{\nabla} \hat{\phi}) \, \mathrm{d}\Omega - \int_{s_1} \tfrac{1}{2} N_a (\hat{\mathbf{q}}_{e_1} + \hat{\mathbf{q}}_{e_5})^{\mathrm{T}} \mathbf{n}_{s_1} \, \mathrm{d}\Gamma \\ + \int_{s_2} \tfrac{1}{2} N_a (\hat{\mathbf{q}}_{e_1} + \hat{\mathbf{q}}_{e_2})^{\mathrm{T}} \mathbf{n}_{s_2} \, \mathrm{d}\Gamma - \int_{s_1} N_a Z_{s_1} \, \mathrm{d}\Gamma + \int_{s_2} N_a Z_{s_2} \, \mathrm{d}\Gamma = 0\end{aligned} \tag{13.87}$$

where a boundary integral takes a negative sign if the edge normal shown in Figs 13.17 and 13.18 points inward to the element.

Let f_{e_1} denote the first four, computable, terms of the above equation and notice that [using Eq. (13.85)]

$$\int_{s_1} N_a Z_{s_1} \, \mathrm{d}\Gamma = \int_{s_1} N_a \left(\hat{N}_a^s a_a^{s_1} + \hat{N}_b^s a_b^{s_1} \right) \mathrm{d}\Gamma = a_a^{s_1} \tag{13.88a}$$

and

$$\int_{s_2} N_a Z_{s_2} \, \mathrm{d}\Gamma = \int_{s_2} N_a \left(\hat{N}_a^s a_a^{s_2} + \hat{N}_b^s a_b^{s_2} \right) \mathrm{d}\Gamma = a_a^{s_2} \tag{13.88b}$$

Equation (13.87) now becomes

$$-a_a^{s_1} + a_a^{s_2} = -f_{e_1} \tag{13.89}$$

Similar equations result for element e_2 to e_5 of the patch in Fig. 13.17 giving the equation set

$$\mathbf{A}\mathbf{a} = \mathbf{f} \tag{13.90}$$

where

$$\mathbf{A} = \begin{bmatrix} -1 & 1 & 0 & 0 & 0 \\ 0 & -1 & 1 & 0 & 0 \\ 0 & 0 & -1 & 1 & 0 \\ 0 & 0 & 0 & -1 & 1 \\ 1 & 0 & 0 & 0 & -1 \end{bmatrix}$$

$$\mathbf{a} = \left[a_a^{s_1}, \quad a_a^{s_2}, \quad a_a^{s_3}, \quad a_a^{s_4}, \quad a_a^{s_5}\right]^{\mathrm{T}}$$

and

$$\mathbf{f} = \left[-f_{e_1}, \quad -f_{e_2}, \quad -f_{e_3}, \quad -f_{e_4}, \quad -f_{e_5}\right]^{\mathrm{T}}$$

It is easy to verify that these equations are linearly dependent but have solutions determined up to an arbitrary constant. A procedure to obtain an *optimal* particular solution is described as follows.[35, 42, 49] First, a particular solution $\mathbf{a}_0$ of Eq. (13.90) is found by choosing, for example, $a_a^{s_5} = 0$. Second, the corresponding homogeneous equation

$$\mathbf{A}\mathbf{b} = \mathbf{0} \tag{13.91}$$

with $\mathbf{b} = [b_1, b_2, b_3, b_4, b_5]^{\mathrm{T}}$ is solved for a non-zero particular solution with the choice of, corresponding to $a_a^{s_5}$, $b_5 = 1$. It is easy to verify that b_i is either 1 or -1 due to the structure of $\mathbf{A}$. In the element patch considered here $\mathbf{b} = [1, 1, 1, 1, 1]^{\mathrm{T}}$.

The final particular solution of Eq. (13.90) takes the form of

$$\mathbf{a} = \mathbf{a}_0 + \gamma\mathbf{b} \tag{13.92}$$

where the constant γ is determined by the minimization of

$$\Pi = \mathbf{a}^{\mathrm{T}}\mathbf{a} \tag{13.93}$$

The minimization condition gives

$$\gamma = -\frac{\mathbf{b}^{\mathrm{T}}\mathbf{a}_0}{\mathbf{b}^{\mathrm{T}}\mathbf{b}} \tag{13.94}$$

The solution gives the nodal value $a_a^{s_i}$ at node a for each connected edge of the element patch.

Boundary nodes and their related element patches can be considered in the same fashion except that we can take $q_n^* = \bar{q}_n$, the known flux, for the element edge being part of Γ_q. For edges coincident with Γ_ϕ, we let the first term on the right-hand side of Eq. (13.82) be zero. By considering each vertex node of the mesh and its associated element patch, we will be able to determine a_a^s and a_b^s in Eq. (13.83) for every edge, thus the recovered normal flux q_n^* in the form defined by Eq. (13.82) on the element boundary is achieved. The procedure described above for recovering the normal flux is a *recovery by element residuals*.

We note that the non-uniqueness of the solution of Eq. (13.90) represents the non-uniqueness of the equilibrium status of the element residuals. The choice of the arbitrary constant in solving Eq. (13.90) will certainly affect the accuracy of the recovered solution q_n^*, and therefore the accuracy of the error estimator.

With q_n^* determined, the local error problem Eq. (13.79) is usually solved by a higher order (e.g., $p + 1$ or even $p + 2$) approximation. The solution of the problem is then employed in the equilibrated residual error estimator $\|\hat{e}\|_{r_K}$. The global error estimator $\|\hat{e}\|$ is obtained through Eq. (13.55). The global error estimator has been shown to be an upper bound of the exact error,[46] although it is not a trivial task to prove its convergence.

We have shown here that a proper recovery method is the key to the computation of equilibrated residual error estimators. Indeed, carefully chosen recovery methods are very important in the computation of all the implicit residual error estimators. Numerical performance of residual-based error estimators was tested by Babuška *et al.*[54–56] and Carstensen *et al.*[59] and compared with that of recovery-based error estimators.

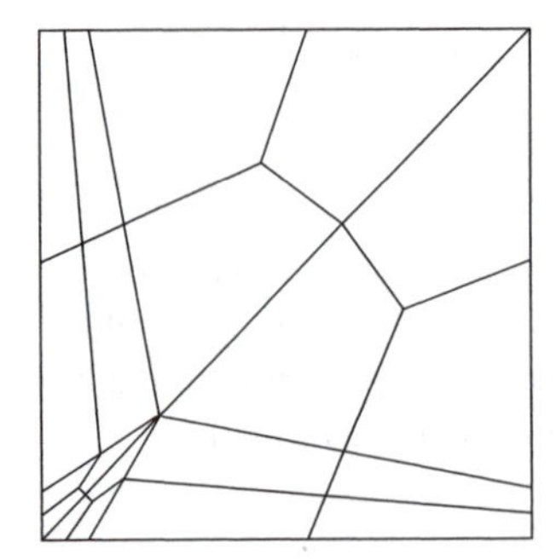

Fig. 13.19 Repeating patch of irregular and quadrilateral elements.

13.8 Asymptotic behaviour and robustness of error estimators – the Babuška patch test

It is well known that elements in which polynomials of order p are used to represent the unknown $\mathbf{u}$ will reproduce exactly any problem for which the exact solution is also defined by such a polynomial. Indeed the verification of this behaviour is an essential part of the 'patch test' which has to be satisfied by all elements to ensure convergence, as we have discussed in Chapter 9.

Thus if we are attempting to determine the error in a general smooth solution we will find that this error is dominated by terms of order $p + 1$. The response of any patch to an exact solution of order $p+1$ will therefore determine the asymptotic behaviour when both the size of the patch and of all the elements tends to zero. If the patch is assumed to be one of a repeatable kind, its behaviour when subjected to an exact solution of order $p + 1$ will give the exact asymptotic error of the finite element solution. Thus, any estimator can be compared with this exact value and the asymptotic effectivity index can be established. Figure 13.19 shows such a repeatable patch of quadrilateral elements which evaluate the performance of the error estimators for quite irregular meshes.

We have indeed shown how true superconvergent behaviour reproduces exactly such higher order solutions and thus leads to an effectivity index of unity in the asymptotic limit. In the papers presented by Babuška *et al.*[54–56] the procedure of dealing with such repeatable patches for various patterns of two-dimensional elements is developed. Thus, if we are interested in solving the differential equation

$$\mathcal{L}u + f = 0 \tag{13.95}$$

where $\mathcal{L}$ is a linear differential operator of order $2p$, we consider *exact solutions* (harmonic solutions) to the homogeneous equation ($f = 0$) of the form

$$u_{\text{ex}} = \sum_m a_m x^m y^n = \mathbf{P}(x, y)\mathbf{a}; \qquad n = p + 1 - m \tag{13.96}$$

The boundary conditions are taken as

$$u_{\text{ex}}|_{x+L_x} = u_{\text{ex}}|_x \qquad \text{and} \qquad u_{\text{ex}}|_{y+L_y} = u_{\text{ex}}|_y \tag{13.97}$$

where L_x and L_y are periodic distances in the x and y directions, respectively (viz. repeatability, Sec. 6.2.4, page 192). In general, the individual terms of Eq. (13.96) do not satisfy the differential equation and it is necessary to consider linear combinations in terms of the parameters in $\mathcal{L}$ as

$$\mathbf{a}' = \mathbf{T}\mathbf{a} \tag{13.98}$$

This solution serves as the basis for conducting a patch test in which the boundary conditions are assigned to be periodic and to prevent constant changes to u.† The correct constant value may be computed from

$$\int_{\text{patch}} (\mathbf{N}\tilde{\mathbf{u}} + C)\, \mathrm{d}\Omega = \int_{\text{patch}} u_{\text{ex}}\, \mathrm{d}\Omega \tag{13.99}$$

To compute upper and lower bounds (θ_U and θ_L) on the possible effectivity indices of the error estimators, all possible combinations of the harmonic solution must be considered. This may be achieved by constructing an error norm of the solutions, for example the L_2 norm of the flux (or stress)

$$\|\mathbf{e}_q\|^2_{L_2} = \int_{\text{patch}} (\mathbf{q}_{\text{ex}} - \hat{\mathbf{q}})^{\mathrm{T}} (\mathbf{q}_{\text{ex}} - \hat{\mathbf{q}})\, \mathrm{d}\Omega = \left(\mathbf{a}'\right)^{\mathrm{T}} \mathbf{T}^{\mathrm{T}}\mathbf{E}_{\text{ex}}\mathbf{T}\mathbf{a}' \tag{13.100}$$

and

$$\|\hat{\mathbf{e}}_q\|^2_{L_2} = \int_{\text{patch}} \left(\mathbf{q}^* - \hat{\mathbf{q}}\right)^{\mathrm{T}} \left(\mathbf{q}^* - \hat{\mathbf{q}}\right)\, \mathrm{d}\Omega = \left(\mathbf{a}'\right)^{\mathrm{T}} \mathbf{T}^{\mathrm{T}}\mathbf{E}^*\mathbf{T}\mathbf{a}' \tag{13.101}$$

and solving the eigenproblem

$$\mathbf{T}^{\mathrm{T}}\mathbf{E}^*\mathbf{T}\mathbf{a}' = \theta^2\mathbf{T}^{\mathrm{T}}\mathbf{E}_{\text{ex}}\mathbf{T}\mathbf{a}' \tag{13.102}$$

to determine the minimum (lower bound) and maximum (upper bound) effectivity indices. Further details of the process summarized here are given in Boroomand and Zienkiewicz[32, 33] and by Zienkiewicz *et al.*[60]

These bounds on the effectivity index are very useful for comparing various error estimators and their behaviour for different mesh and element patterns. However, a single parameter called the *robustness index* has also been devised[54] and is useful as a guide to the robustness of any particular estimator

$$R = \max\left(|1 - \theta_L| + |1 - \theta_U|, \quad |1 - \frac{1}{\theta_L}| + |1 - \frac{1}{\theta_U}| \right) \tag{13.103}$$

A large value of this index obviously indicates a poor performance. Conversely the best behaviour is that in which

$$\theta_L = \theta_U = 1 \tag{13.104}$$

and this gives

$$R = 0 \tag{13.105}$$

In the series of tests reported in references 54–56 various estimators have been compared. Table 13.1 shows the highest robustness index value of an equilibrating residual-based error estimator, ERpB, and the SPR recovery error estimator for a set of particular patches of triangular elements.[54]

† For elasticity-type problems the periodic boundary conditions prevent rigid rotations.

Table 13.1 Robustness index R for equilibrated residual ERpB and SPR (ZZ-discrete) estimators for a variety of anisotropic situations and element patterns, $p = 2$

Estimator form	Robustness index – R
ERpB	10.21
SPR (ZZ-discrete)	0.02

This performance comparison is quite remarkable and it seems that in all the tests quoted by Babuška *et al.*[54–56] and summarized in Babuška and Strouboulis[61] the recovery estimator using SPR performs best. Indeed we shall observe that in many cases of regular subdivision, when full superconvergence occurs the ideal, asymptotically exact solution characterized by $R = 0$ will be obtained.

In Table 13.2 we show some results obtained for regular meshes of triangles and rectangles with linear and quadratic elements. In the rectangular elements used for problems of heat conduction type, superconvergent points are exact and the ideal result is obtained for both linear and quadratic elements. It is surprising that this also occurs in elasticity where the proof of superconvergent points is lacking [since for $\nu > 0$ $\mathbf{A}$ in (13.17) is not diagonal]. Further, the REP procedure also seems to yield superconvergence except for elasticity with quadratic elements.

For regular meshes of quadratic triangles generally superconvergence is not expected and it does not occur for either heat conduction or elasticity problems. However, the robustness index has very small values ($R < 0.10$ for SPR and $R < 0.12$ for REP) and these estimators are therefore very accurate.

In Fig. 13.20 and Table 13.3 very irregular meshes of triangular and quadrilateral elements are analysed in repeatable patterns. It is of course not possible to present here all tests conducted by the effectivity patch test. The results shown are, however, typical – others are given in reference 32. It is interesting to observe that the performance measured by the robustness index on quadrilateral elements is always superior to that measured on triangles.

13.9 Bounds on quantities of interest

Although we have shown that excellent estimators of errors exist today, many are striving to know that these estimators are not only close but that they are *bounded*. The strain energy was one of the first quantities in which bounds could be established. Here the classical work of Fraeijs de Veubeke in the mid-1960s is of vital importance.[62, 63]

It was quickly realized by Fraeijs de Veubeke that the standard (displacement) procedures from which structural analysis usually started would provide a lower bound of the strain energy contained in the structure and thus always underestimated the value of strain energy. He therefore sought procedures which could solve the same structural problem by concentrating on so-called complementary energy which would allow to be established solutions in which the strain energy would always be overestimated. This process proved very difficult as equilibrating solutions have to be established at all stages. A possible way, useful for many two-dimensional problems, was suggested in reference 64 in which stress functions and the slab analogy were used. Nevertheless the methodology never succeeded as a practical way of providing the bounds of strain energy in an actual analysis.

Table 13.2 Effectivity bounds and robustness of SPR and REP recovery estimator for regular meshes of triangles and rectangles with linear and quadratic shape function (applied to heat conduction and elasticity problems). Aspect ratio = length(L)/height(H) of elements in patch tested

	Linear triangles and rectangles (heat conduction/elasticity)					
	SPR			REP		
Aspect ratio L/H	θ_L	θ_U	R	θ_L	θ_U	R
1/1	1.0000	1.0000	0.0000	1.0000	1.0000	0.0000
1/2	1.0000	1.0000	0.0000	1.0000	1.0000	0.0000
1/4	1.0000	1.0000	0.0000	1.0000	1.0000	0.0000
1/8	1.0000	1.0000	0.0000	1.0000	1.0000	0.0000
1/16	1.0000	1.0000	0.0000	1.0000	1.0000	0.0000
1/32	1.0000	1.0000	0.0000	1.0000	1.0000	0.0000
1/64	1.0000	1.0000	0.0000	1.0000	1.0000	0.0000
	Quadratic rectangles (heat conduction)					
	θ_L	θ_U	R	θ_L	θ_U	R
1/1	1.0000	1.0000	0.0000	1.0000	1.0000	0.0000
1/2	1.0000	1.0000	0.0000	1.0000	1.0000	0.0000
1/4	1.0000	1.0000	0.0000	1.0000	1.0000	0.0000
1/8	1.0000	1.0000	0.0000	1.0000	1.0000	0.0000
1/16	1.0000	1.0000	0.0000	1.0000	1.0000	0.0000
1/32	1.0000	1.0000	0.0000	1.0000	1.0000	0.0000
1/64	1.0000	1.0000	0.0000	1.0000	1.0000	0.0000
	Quadratic rectangles (elasticity)					
	θ_L	θ_U	R	θ_L	θ_U	R
1/1	1.0000	1.0000	0.0000	0.9991	1.0102	0.0111
1/2	1.0000	1.0000	0.0000	0.9991	1.0181	0.0189
1/4	1.0000	1.0000	0.0000	0.9991	1.0136	0.0145
1/8	1.0000	1.0000	0.0000	0.9991	1.0030	0.0039
1/16	1.0000	1.0000	0.0000	0.9968	1.0001	0.0033
1/32	1.0000	1.0000	0.0000	0.9950	1.0000	0.0050
1/64	1.0000	1.0000	0.0000	0.9945	1.0000	0.0055
	Quadratic triangles (elasticity)					
	θ_L	θ_U	R	θ_L	θ_U	R
1/1	0.9966	1.0929	0.0963	0.9562	1.0503	0.0940
1/2	0.9966	1.0931	0.0965	0.9559	1.0481	0.0923
1/4	0.9967	1.0937	0.0970	0.9535	1.0455	0.0924
1/8	0.9967	1.0943	0.0976	0.9522	1.0603	0.1081
1/16	0.9966	1.0946	0.0980	0.9518	1.0666	0.1148
1/32	0.9966	1.0947	0.0981	0.9517	1.0684	0.1167
1/64	0.9965	1.0947	0.0982	0.9516	1.0688	0.1172

Much later when the residual method was being applied to determine error in structural analysis it was realized that once again opportunity existed for establishing bounds. The residuals are nothing else but a measure by which the numerical solution fails to satisfy the differential equations of the problem. By using a local solution, generally based on a few elements or even a single element, the error can be estimated locally and the total error obtained by combining these estimates for all elements. The completely independent solution for the displacement, stresses, etc. established by the residuals provides the measure of the error. This solution can be carried out in a number of ways and here a departure from

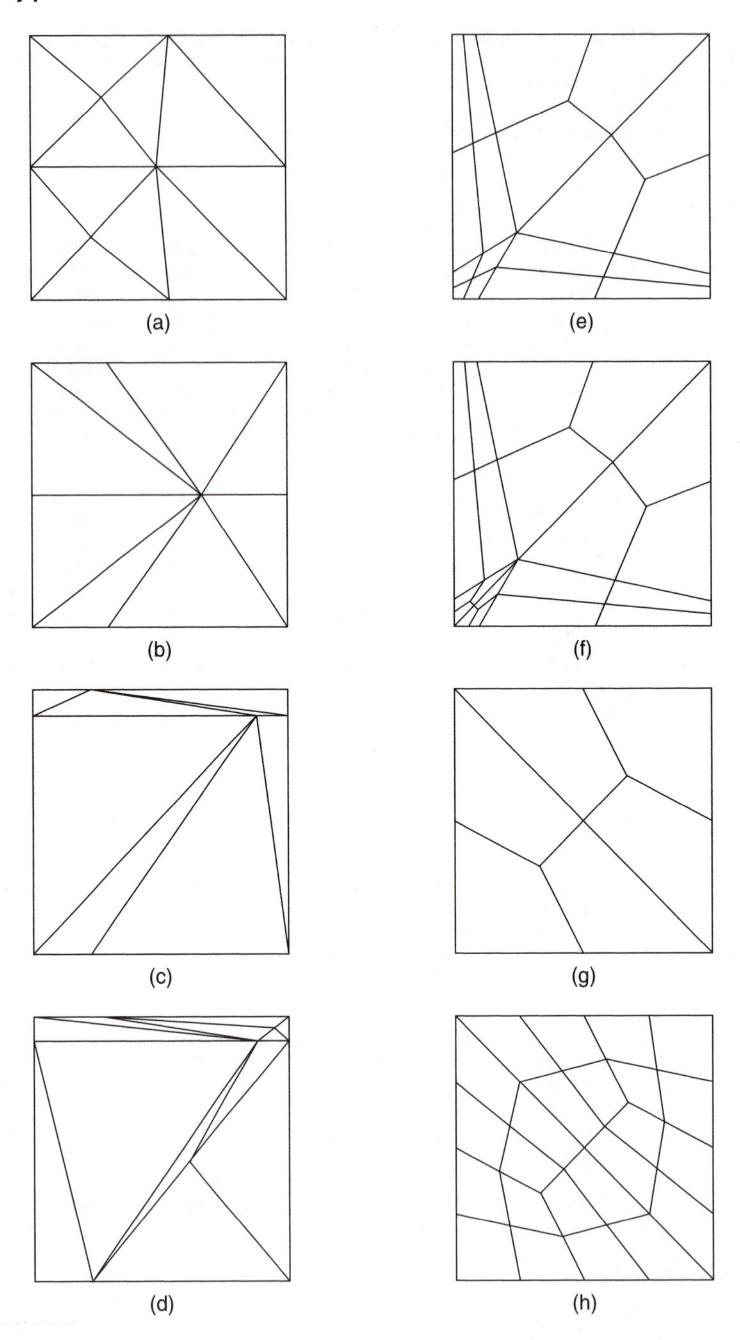

Fig. 13.20 Repeating patch types.

the original, say displacement, method can be made. As many local problems are solved, it is assumed the total error is a combination of local (patch) solutions. It seems that one of the first to extend the concept of establishing upper bounds is Kelly in 1984[41] and subsequent work.[65] He endeavoured to obtain solutions for the residual placed as a load with

Table 13.3 Effectivity bounds and robustness of SPR and REP recovery estimator for irregular meshes of (a, b, c, d) and quadrilaterals (e, f, g, h)

	Linear element (heat conduction)					
	SPR			REP		
Mesh pattern	θ_L	θ_U	R	θ_L	θ_U	R
a	0.9626	1.0054	0.0442	0.9709	1.0145	0.0443
b	0.9715	1.0156	0.0447	0.9838	1.0167	0.0329
c	0.9228	1.4417	0.5189	0.8938	1.8235	0.9297
d	0.8341	1.2027	0.3685	0.9463	1.9272	0.9810
e	0.9943	1.0175	0.0232	0.9800	1.0589	0.0789
f	0.9969	1.0152	0.0183	0.9849	1.0582	0.0733
g	0.9987	1.0175	0.0188	0.9987	1.0175	0.0188
h	0.9991	1.0068	0.0077	0.9979	1.0062	0.0083
	Linear elements (elasticity)					
	SPR			REP		
	θ_L	θ_U	R	θ_L	θ_U	R
a	0.9404	1.0109	0.0741	0.9468	1.0148	0.0707
b	0.8869	1.0250	0.1520	0.9392	1.0275	0.0915
c	0.8550	1.6966	0.8415	0.8037	2.0522	1.2486
d	0.7945	1.2734	0.4788	0.7576	1.9416	1.1840
e	0.9946	1.0247	0.0301	0.9579	1.0508	0.0928
f	1.0038	1.0281	0.0318	0.9612	1.0467	0.0855
g	0.9959	1.0300	0.0341	0.9960	1.0298	0.0338
h	0.9972	1.0139	0.0168	0.9965	1.0122	0.0157
	Quadratic elements (heat conduction)					
	θ_L	θ_U	R	θ_L	θ_U	R
a	0.9443	1.0295	0.0877	0.9339	1.0098	0.0805
b	0.8146	1.0037	0.2313	0.9256	1.0028	0.0832
c	0.7640	1.0486	0.3000	0.9559	1.2229	0.2670
d	0.8140	1.0141	0.2423	0.9091	1.2808	0.3717
e	0.9762	1.0053	0.0296	0.9901	1.0177	0.0276
f	0.9691	1.0045	0.0363	0.9901	1.0322	0.0421
g	0.9692	1.0004	0.0322	0.9833	1.0024	0.0195
h	0.9906	1.0113	0.0207	1.0045	1.0261	0.0307
	Quadratic elements (elasticity)					
	θ_L	θ_U	R	θ_L	θ_U	R
a	0.9144	1.0353	0.1277	0.9197	1.0244	0.1111
b	0.7302	1.0355	0.4038	0.8643	1.0346	0.1905
c	0.7556	1.1024	0.4163	0.8387	1.2422	0.4035
d	0.7624	1.0323	0.3430	0.8244	1.2632	0.4388
e	0.9702	1.0102	0.0408	0.9682	1.0058	0.0386
f	0.9651	1.0085	0.0446	0.9749	1.0286	0.0537
g	0.9457	1.0115	0.0688	0.9807	1.0125	0.0321
h	0.9852	1.0141	0.0290	0.9996	1.0522	0.0526

equilibrating methodologies. Two similar alternative approaches, though stemming from completely different origins, were proposed by Ladevèze[34, 66] and almost simultaneously by Bank and Weiser.[42] These ideas were later adopted by Ainsworth and Oden forming much of the basis to their book.[49]

The methodologies so far produced are very effective when the basic quantity of interest is a simple one, such as strain energy or the energy norm. However, when the quantity of interest is more localized and if for instance it is the displacement at some part of the structure, rather than an overall measure of stresses as it is in the case of energy, different procedures arise and pure examination of energy errors does not suffice or it is not very selective in showing how to obtain the answers. For this reason much effort has been given in recent years in discussing the possibilities of bounds and the manner by which such localized goals of analysis can be solved. Much of the recent work in this field concentrates on such methodologies.

The first to give attention to the possible extension of norms to other quantities of interest appeared in a series of papers presented by Babuška and Miller in 1984.[67–69] These papers laid the foundations for much of the work continued some ten years later and which today occupies much interest. Here it appears that the first full extension of the methodology is due to Peraire *et al.* first published in 1997 with many papers following.[70–75]

By extending the ideas introduced in the Babuška and Miller papers, Peraire *et al.* show that whatever the quantity of interest is, it is always possible to establish an adjoint problem which can be solved on the same mesh of the original problem but now with different loads for dealing with the accuracy. Such adjoint problems may be called differently and names such as extraction problems and dual problems are also used. Although many people are now entering the field and the methodology has been followed by Oden and Prudhomme in a series of papers, it appears that only Peraire so far has extended the approach to non-self-adjoint problems such as fluid dynamics.[73–75]

The equilibrated methods have always provided upper bounds for such quantities as strain energy. A similar bounding occurs if we look at the energy in the adjoint approaches. However, some interest now goes back to placing satisfactorily the lower bounds and thus bracketing the solution. Of course, lower bounds of zero values have been used and this is rather defeatist because they do not provide tight estimates. It is our belief that more precise bounds are required and here the work of Díez and Huerta seems to lead the way.[47, 76, 77]

13.10 Which errors should concern us?

In this chapter we have shown how various recovery procedures can accurately estimate the overall error of the finite element approximation and thus provide a very accurate error estimating method. We have also shown that estimators based on SPR recovery are superior to those based on residual computation. The error estimation discussed here concerns, however, only the original solution and if the user takes advantage of the recovered values a much better solution is already available. In the next chapter we shall be concerned with adaptivity processes which are aimed at reduction of the original finite element error. Here again we shall show the excellent values of the effectivity index which can be obtained with SPR-type methods on examples for which an 'exact' solution is available from very fine mesh computations. What perhaps we should also be concerned with are the errors remaining in the recovered solutions, if indeed these are to be made use of. This problem is still unsolved and at the moment all the adaptive methods simply aim at the reduction of various norms of error in the finite element solution directly provided.

13.11 Problems

13.1 Let the assumed stress for Example 13.1 in Sec. 13.4.1 be given as

$$\sigma_i^* = \begin{bmatrix} 1, & (x-x_1), & (y-y_1), & (x-x_1)(y-y_1) \end{bmatrix} \begin{Bmatrix} \bar{a}_1 \\ \bar{a}_2 \\ \bar{a}_3 \\ \bar{a}_4 \end{Bmatrix}$$

Compute the recovered stress and compare the result with that of Example 13.1. Is this result superconvergent?

13.2 Show all the relations necessary to extend the SPR algorithm to three-dimensional elastic problems.

What is the expression for σ_i^* that should be used for 8-node hexahedral elements?

13.3 Program development project: Implement the SPR procedure in the solver system developed in Problem 2.17 and subsequent chapters. Assume the problem is modelled by the quasi-harmonic equation using 4-node quadrilateral elements. (Hint: Extend the result from Problem 6.19.)

13.4 Program development project: Implement the SPR procedure in the solver developed in Problem 2.17 and subsequent additions. Assume a linear elastic problem that is modelled using 4-node quadrilateral elements. (Hint: Extend the result from Problem 6.19.)

13.5 The element size h appearing in the explicit residual error estimator given by Eq. (13.59) is often taken as a constant for a particular element of certain shape. Consider results from Example 13.2 and explain why the accuracy of the explicit residual error estimator will deteriorate when the aspect ratio of the element increases, i.e., when the mesh becomes more anisotropic.

13.6 Extend the technique of *recovering normal flux by equilibrated residuals* described in Sec. 13.7.2 to two-dimensional elastic problems. Consider both plane and axisymmetric geometry.

13.7 Extend the technique of *recovering normal flux by equilibrated residuals* described in Sec. 13.7.2 to three-dimensional elastic problems.

13.8 Program development project: Implement a *recovery-based error estimator* or a *residual-based error estimator* in the solver system developed in Problem 2.17 and subsequent exercises.

13.9 Program development project: Extend the program developed in Problem 2.17 to compute the SPR solution for displacements. Use the recovered displacements to compute strains and from these stresses.

Follow the procedure given in Sec. 13.4.2 to project 3-node triangular and 4-node quadrilateral element values to nodes.

Test your program using (a) the patch test of Problem 2.17 and (b) the curved beam problem shown in Fig. 2.11.

Report results for both displacements and stresses.

References

1. J. Barlow. Optimal stress locations in finite element models. *Int. J. Numer. Meth. Eng.*, 10: 243–251, 1976.
2. L.R. Herrmann. Interpretation of finite element procedures in stress error minimization. *Proc. Am. Soc. Civ. Eng.*, 98(EM5):1331–1336, 1972.
3. M. Krizek and P. Neitaanmaki. On superconvergence techniques. *Acta. Appl. Math.*, 9:75–198, 1987.
4. Q.D. Zhu and Q. Lin. *Superconvergence Theory of the Finite Element Methods*. Hunan Science and Technology Press, Hunan, China, 1989.
5. L.B. Wahlbin. *Superconvergence in Galerkin Finite Element Methods, Lecture Notes in Mathematics Vol. 1605*. Springer, Berlin, 1995.
6. C.M. Chen and Y. Huang. *High Accuracy Theory of Finite Element Methods*. Hunan Science and Technology Press, Hunan, China, 1995.
7. Q. Lin and N. Yan. *Construction and Analyses of Highly Effective Finite Elements*. Hebei University Press, Hebei, China, 1996.
8. M. Krizek, P. Neitaanmaki, and R. Stenberg, editors. *Finite Element Methods: Superconvergence, Post-processing, and A Posteriori Estimates*. Lecture Notes in Mathematics, No. 363. Marcel Dekker, New York, 1997.
9. E. Hinton and J. Campbell. Local and global smoothing of discontinuous finite element function using a least squares method. *Int. J. Numer. Meth. Eng.*, 8:461–480, 1974.
10. H.J. Brauchli and J.T. Oden. On the calculation of consistent stress distributions in finite element applications. *Int. J. Numer. Meth. Eng.*, 3:317–325, 1971.
11. O.C. Zienkiewicz and J.Z. Zhu. The superconvergent patch recovery and *a posteriori* error estimates. Part 1: The recovery technique. *Int. J. Numer. Meth. Eng.*, 33:1331–1364, 1992.
12. O.C. Zienkiewicz and J.Z. Zhu. The superconvergent patch recovery and *a posteriori* error estimates. Part 2: Error estimates and adaptivity. *Int. J. Numer. Meth. Eng.*, 33:1365–1382, 1992.
13. O.C. Zienkiewicz and J.Z. Zhu. The superconvergent patch recovery (SPR) and adaptive finite element refinement. *Comp. Meth. Appl. Mech. Eng.*, 101:207–224, 1992.
14. O.C. Zienkiewicz, J.Z. Zhu, and J. Wu. Superconvergent recovery techniques – some further tests. *Commun. Num. Meth. Eng.*, 9:251–258, 1993.
15. Z. Zhang. Ultraconvergence of the patch recovery technique. *Math. Comput.*, 65:1431–1437, 1996.
16. Q.D. Zhu and Q. Zhao. SPR technique and finite element correction. *Numer. Math.*, 96:185–196, 2003.
17. Z. Zhang. Ultraconvergence of the patch recovery technique II. *Math. Comput.*, 69:141–158, 2000.
18. Z. Zhang and H.D. Victory Jr. Mathematical analysis of Zienkiewicz–Zhu's derivative patch recovery techniques. *Num. Meth. Partial Diff. Eq.*, 12:507–524, 1996.
19. Z. Zhang and J.Z. Zhu. Analysis of the superconvergent patch recovery technique and *a posteriori* error estimator in the finite element method (II). *Comp. Meth. Appl. Mech. Eng.*, 163:159–170, 1998.
20. B. Li and Z. Zhang. Analysis of a class of superconvergence patch recovery techniques for linear and bilinear finite elements. *Num. Meth. Partial Diff. Eq.*, 15:151–167, 1999.
21. Z. Zhang and R. Lin. Ultraconvergence of ZZ patch recovery at mesh symmetry points. *Numer. Math.*, 95:781–801, 2003.
22. J. Xu and Z. Zhang. Analysis of recovery type *a posteriori* error estimators for mildly structured grids. *Math. Comput.*, 73:1139–1152, 2004.

23. N.-E. Wiberg and F. Abdulwahab. Patch recovery based on superconvergent derivatives and equilibrium. *Int. J. Numer. Meth. Eng.*, 36:2703–2724, 1993.
24. N.-E. Wiberg, F. Abdulwahab, and S. Ziukas. Enhanced superconvergent patch recovery incorporating equilibrium and boundary conditions. *Int. J. Numer. Meth. Eng.*, 37:3417–3440, 1994.
25. T.D. Blacker and T. Belytschko. Superconvergent patch recovery with equilibrium and conjoint interpolation enhancements. *Int. J. Numer. Meth. Eng.*, 37:517–536, 1994.
26. T. Lee, H.C. Park, and S.W. Lee. A superconvergent stress recovery technique with equilibrium constraint. *Int. J. Numer. Meth. Eng.*, 40:1139–1160, 1997.
27. B. Heimsund, X. Tai, and J. Wang. Superconvergence for gradient of finite element approximations by l^2 projections. *SIAM Numer. Anal.*, 40:1263–1280, 2002.
28. X.D. Li and N.-E. Wiberg. *A posteriori* error estimate by element patch postprocessing, adaptive analysis in energy and l_2 norm. *Comp. Struct.*, 53:907–919, 1994.
29. N.-E. Wiberg and X.D. Li. Superconvergent patch recovery of finite element solutions and *a posteriori* l_2 norm error estimate. *Commun. Numer. Meth. Eng.*, 10:313–320, 1994.
30. Z. Zhang and A. Naga. A new finite element gradient recovery method. Part I: Superconvergence property. *SIAM J. Sci. Comp.*, to appear.
31. N.-E. Wiberg and P. Hager. Error estimation and adaptivity for h-version eigenfrequency analysis. In P. Ladevèze and J.T. Oden, editors, *Advances in Adaptive Computational Methods in Mechanics*, pages 461–476. Elsevier, 1998.
32. B. Boroomand and O.C. Zienkiewicz. Recovery by equilibrium patches (REP). *Int. J. Numer. Meth. Eng.*, 40:137–154, 1997.
33. B. Boroomand and O.C. Zienkiewicz. An improved REP recovery and the effectivity robustness test. *Int. J. Numer. Meth. Eng.*, 40:3247–3277, 1997.
34. P. Ladevèze and D. Leguillon. Error estimate procedure in the finite element method and applications. *SIAM J. Num. Anal.*, 20(3):485–509, 1983.
35. P. Ladevèze, J.P. Pelle, and P. Rougeot. Error estimation and mesh optimization for classical finite elements. *Eng. Comp.*, 8:69–80, 1991.
36. O.C. Zienkiewicz and J.Z. Zhu. A simple error estimator and adaptive procedure for practical engineering analysis. *Int. J. Numer. Meth. Eng.*, 24:337–357, 1987.
37. I. Babuška and C. Rheinboldt. *A-posteriori* error estimates for the finite element method. *Int. J. Numer. Meth. Eng.*, 12:1597–1615, 1978.
38. I. Babuška and W.C. Rheinboldt. Reliable error estimation and mesh adaptation for the finite element method. In J.T. Oden, editor, *Computational Methods in Nonlinear Mechanics*, pages 67–1084. North Holland, Amsterdam, 1980.
39. D.W. Kelly, J.P. De S.R. Gago, O.C. Zienkiewicz, and I. Babuška. *A posteriori* error analysis and adaptive processes in the finite element method: Part I – Error analysis. *Int. J. Numer. Meth. Eng.*, 19:1593–1619, 1983.
40. O.C. Zienkiewicz, J.P. De S.R. Gago, and D.W. Kelly. The hierarchical concept in finite element analysis. *Comp. Struct.*, 16:53–65, 1983.
41. D.W. Kelly. The self-equilibration of residuals and complementary a posteriori error estimates in the finite element method. *Int. J. Numer. Meth. Eng.*, 20:1491–1506, 1984.
42. R.E. Bank and A. Weiser. Some a posteriori error estimators for elliptic partial differential equations. *Math. Comp.*, 44:283–301, 1985.
43. J.T. Oden, L. Demkowicz, W. Rachowicz, and T.A. Westermann. Toward a universal h-p adaptive finite element strategy. Part 2: A posteriori error estimation. *Comp. Meth. Appl. Mech. Eng.*, 77:113–180, 1989.
44. R. Verfurth. *A posteriori* error estimators for the Stokes equations. *Numer. Math.*, 55:309–325, 1989.
45. C. Johnson and P. Hansbo. Adaptive finite element methods in computational mechanics. *Comp. Meth. Appl. Mech. Eng.*, 101:143–181, 1992.

46. M. Ainsworth and J.T. Oden. A unified approach to a posteriori error estimation using element residual methods. *Numerische Mathematik*, 65:23–50, 1993.
47. P. Díez, J.J. Egozcue, and A. Huerta. *A posteriori* error estimation for standard finite element analysis. *Comp. Meth. Appl. Mech. Eng.*, 163:141–157, 1998.
48. R. Verfurth. *A Review of A Posteriori Error Estimation and Adaptive Mesh Refinement Techniques*. Wiley-Teubner, New York, 1996.
49. M. Ainsworth and J.T. Oden. *A Posteriori Error Estimation in Finite Element Analysis*. John Wiley & Sons, New York, 2000.
50. J.Z. Zhu and O.C. Zienkiewicz. Superconvergence recovery technique and *a posteriori* error estimators. *Int. J. Numer. Meth. Eng.*, 30:1321–1339, 1990.
51. J.Z. Zhu. *A posteriori* error estimation – the relationship between different procedures. *Comp. Meth. Appl. Mech. Eng.*, 150:411–422, 1997.
52. R. Rodriguez. Some remarks on the Zienkiewicz–Zhu estimator. *Numer. Meth. Partial Diff. Eq.*, 10:625–635, 1994.
53. M. Picasso. An anisotropic error indicator based on Zienkiewicz–Zhu error estimator: application to elliptic and parabolic problems. *SIAM J. Sci. Comp.*, 4:1328–1355, 2003.
54. I. Babuška, T. Strouboulis, and C.S. Upadhyay. A model study of the quality of a posteriori error estimators for linear elliptic problems. Error estimation in the interior of patchwise uniform grids of triangles. *Comp. Meth. Appl. Mech. Eng.*, 114:307–378, 1994.
55. I. Babuška, T. Strouboulis, C.S. Upadhyay, S.K. Gangaraj, and K. Copps. Validation of *a posteriori* error estimators by numerical approach. *Int. J. Numer. Meth. Eng.*, 37:1073–1123, 1994.
56. I. Babuška, T. Strouboulis, and C.S. Upadhyay. A model study of the quality of a posteriori error estimators for finite element solutions of linear elliptic problems, with particular reference to the behavior near the boundary. *Int. J. Numer. Meth. Eng.*, 40:2521–2577, 1997.
57. S. Ohnimus, E. Stein, and E. Walhorn. Local error estimates of FEM for displacements and stresses in linear elasticity by solving local Neumann problems. *Int. J. Numer. Meth. Eng.*, 52:727–746, 2001.
58. H. Ohtsubo and M. Kitamura. Numerical investigation of element-wise *a posteriori* error estimation in two and three-dimensional elastic problem. *Int. J. Numer. Meth. Eng.*, 34:969–977, 1992.
59. C. Carstensen and J. Alberty. Averaging techniques for reliable *a posteriori* FE-error control in elastoplasticity with hardening. *Comp. Meth. Appl. Mech. Eng.*, 192:1435–1450, 2003.
60. O.C. Zienkiewicz, B. Boroomand, and J.Z. Zhu. Recovery procedures in error estimation and adaptivity: adaptivity in linear problems. *Comp. Meth. Appl. Mech. Eng.*, 176:111–125, 1999.
61. I. Babuška and T. Strouboulis. *The Finite Element Method and its Reliability*. Clarendon Press, Oxford, 2001.
62. B. Fraeijs de Veubeke. Displacement and equilibrium models in finite element method. In O.C. Zienkiewicz and G.S. Holister, editors, *Stress Analysis*, Chapter 9, pages 145–197. John Wiley & Sons, Chichester, 1965.
63. B. Fraeijs de Veubeke. Bending and stretching of plates. Special models for upper and lower bounds. In *Proc. 1st Conf. Matrix Methods in Structural Mechanics*, volume AFFDL-TR-66-80, pages 863–886, Wright Patterson Air Force Base, Ohio, Oct. 1966.
64. B. Fraeijs de Veubeke and O.C. Zienkiewicz. Strain energy bounds in finite element analysis by slab analogy. *J. Strain Anal.*, 2:265–271, 1967.
65. M. Ainsworth, D. Kelly, S. Sloan, and S. Wang. Post processing with computable error bounds for finite element approximation of a non-linear heat conduction problem. *IMA J. Numer. Anal.*, 17:547–561, 1997.
66. P. Ladevèze, G. Coffignal, and J.P. Pelle. Accuracy of elastoplastic and dynamic analysis. In I. Babuška, O.C. Zienkiewicz, J. Gago, and E.R. de A. Oliviera, editors, *Accuracy Estimates and*

Adaptive Refinements in Finite Element Computations, Chapter 11. John Wiley & Sons, New York, 1986.

67. I. Babuška and A. Miller. The post-processing approach in the finite element method: Part 1. Calculation of displacements, stresses and other higher derivatives of displacements. *Int. J. Numer. Meth. Eng.*, 20:1085–1109, 1984.
68. I. Babuška and A. Miller. The post-processing approach in the finite element method: Part 2. The calculation of stress intensity factors. *Int. J. Numer. Meth. Eng.*, 20:1111–1129, 1984.
69. I. Babuška and A. Miller. The post-processing approach in the finite element method: Part 3. A posteriori error estimates and adaptive mesh selection. *Int. J. Numer. Meth. Eng.*, 20:2311–2324, 1984.
70. M. Paraschivoiu, J. Peraire, and A. Patera. *A posteriori* finite element bounds for linear-functional outputs of elliptic partial differential equations. *Comp. Meth. Appl. Mech. Eng.*, 150(1–4): 289–312, 1997.
71. J. Peraire and A.T. Patera. Asymptotic a posteriori finite element bounds for the outputs of noncoercive problems: the Helmholtz and Burgers equations. *Comp. Meth. Appl. Mech. Eng.*, 171(1–2):77–86, 1999.
72. Y. Maday, A.T. Patera, and J. Peraire. A general formulation for a posteriori bounds for output functionals of partial differential equations; application to the eigenvalue problem. *C. R. Acad. Sci. Paris Sér. I Math.*, 328(9):823–828, 1999.
73. L. Machiels, J. Peraire, and A.T. Patera. Output bound approximations for partial differential equations; application to the incompressible Navier–Stokes equations. In *Industrial and Environmental Applications of Direct and Large-eddy Simulation (Istanbul, 1998)*, volume 529 of *Lecture Notes in Phys.*, pages 93–108. Springer, Berlin, 1999.
74. L. Machiels, J. Peraire, and A.T. Patera. A posteriori finite-element output bounds for the incompressible Navier–Stokes equations: application to a natural convection problem. *J. Comp. Phys.*, 172(2):401–425, 2001.
75. A.T. Patera and J. Peraire. A general Lagrangian formulation for the computation of a posteriori finite element bounds. In *Error Estimation and Adaptive Discretization Methods in Computational Fluid Dynamics*, volume 25 of *Lect. Notes Comput. Sci. Eng.*, pages 159–206. Springer, Berlin, 2003.
76. A. Huerta and P. Díez. Error estimation including pollution assessment for nonlinear finite element analysis. *Comp. Meth. Appl. Mech. Eng.*, 181:21–41, 2000.
77. P. Díez, N. Parés, and A. Huerta. Reconverting lower bounds of the error by postprocessing implicit residual *a posteriori* error estimates. *Int. J. Numer. Meth. Eng.*, 56:1465–1488, 2003.

14

Adaptive finite element refinement

14.1 Introduction

In the previous chapter we have discussed at some length various methods of recovery by which the finite element solution results could be made more accurate and this led us to devise various procedures for error estimation. In this chapter we are concerned with methods which can be used to reduce the errors once a finite element solution has been obtained. As the process depends on previous results at all stages it is called adaptive. Such adaptive methods were first introduced to finite element calculations by Babuška and Rheinboldt in the late 1970s.[1, 2]

Before proceeding further it is necessary to clarify the objectives of refinement and specify 'permissible error magnitudes' and here the engineer or user must have very clear aims. For instance the naive requirement that all displacements or all stresses should be given within a specified tolerance is not acceptable. The reasons for this are obvious as at singularities, for example, stresses will always be infinite and therefore no finite tolerance could be specified. The same difficulty is true for displacements if point or knife edge loads are considered.

The most common criterion in general engineering use is that of prescribing a total limit of the error computed in the energy norm. Often this error is required not to exceed a specified percentage of the total energy norm of the solution and in the many examples presented later we shall use this simple criterion. However, using a recovery type of error estimator it is possible to adaptively refine the mesh so that the accuracy of a certain quantity of interest, such as the RMS error in displacement and/or RMS error in stress (see Chapter 13, Eqs (13.9) and (13.10)), satisfies some user-specified criterion. We should recognize that mesh refinement based on reducing the RMS error in displacement is in effect reducing the average displacement error in a user-specified region (e.g., in each element); similarly mesh refinement based on reducing the RMS error in stress is the same as reducing the average stress error in a user-specified region. Here we could, for instance, specify directly the permissible error in stresses or displacements at any location. Some investigators (e.g., Zienkiewicz and Zhu[3]) have used RMS error in stress in the adaptive mesh refinement to obtain more accurate stress solutions. Others (e.g., Oñate and Bugeda[4]) have used the requirement of constant energy norm density in the adaptive analysis, which is in fact equivalent to specifying a uniform distribution of RMS error in stress in each element. We note that the recovery type of error estimators are particularly useful and convenient in designing adaptive analysis procedures for the quantities of interest. For other methodologies of designing adaptive analysis procedures based on error estimation of the quantities of interest, we refer to references 5–7.

As we have already remarked in the previous chapter we will at all times consider the error in the actual finite element solution rather than the error in the recovered solution. It may indeed be possible in special problems for the error in the recovered solution to be zero, even if the error in the finite element solution itself is quite substantial. (Consider here for instance a problem with a linear stress distribution being solved by linear elements which result in constant element stresses. Obviously the element error will be quite large. But if recovered stresses are used, exact results can be obtained and no errors will exist.) The problem of which errors to consider still needs to be answered. At the present time we shall consider the question of recovery as that of providing a very substantial margin of safety in the definition of errors.

Various procedures exist for the refinement of finite element solutions. Broadly these fall into two categories:

1. The h-refinement in which the same class of elements continue to be used but are changed in size, in some locations made larger and in others made smaller, to provide maximum economy in reaching the desired solution.
2. The p-refinement in which we continue to use the same element size and simply increase, generally hierarchically, the order of the polynomial used in their definition.

It is occasionally useful to divide the above categories into subclasses, as the h-refinement can be applied and thought of in different ways. In Fig. 14.1 we illustrate three typical methods of h-refinement:

1. The first of these h-refinement methods is *element subdivision* (enrichment) [Fig. 14.1(b)]. Here refinement can be conveniently implemented and existing elements, if they show too much error, are simply divided into smaller ones keeping the original element boundaries intact. Such a process is cumbersome as many *hanging points* are created where an element with mid-side nodes is joined to a linear element with no such nodes. On such occasions it is necessary to provide local constraints at the hanging points and the calculations become more involved. In addition, the implementation of de-refinement requires rather complex data management which may reduce the efficiency of the method. Nevertheless, the method of element subdivision is quite widely used.
2. The second method is that of a complete *mesh regeneration* or *remeshing* [Fig. 14.1(c)]. Here, on the basis of a given solution, a new element size is predicted in all the domains and a totally new mesh is generated. Thus a refinement and de-refinement are simultaneously allowed. This of course can be expensive, especially in three dimensions where mesh generation is difficult for certain types of elements, and it also presents a problem of transferring data from one mesh to another. However, the results are generally much superior and this method will be used in most of the examples shown in this chapter. For many practical engineering problems, particularly of those for which the element shape will be severely distorted during the analysis, adaptive mesh regeneration is a natural choice.
3. The final method, sometimes known as *r-refinement* [Fig. 14.1(d)], keeps the total number of nodes constant and adjusts their position to obtain an optimal approximation.[5–7] While this procedure is theoretically of interest it is difficult to use in practice and there is little to recommend it. Further it is not a true refinement procedure as a prespecified accuracy cannot generally be reached.

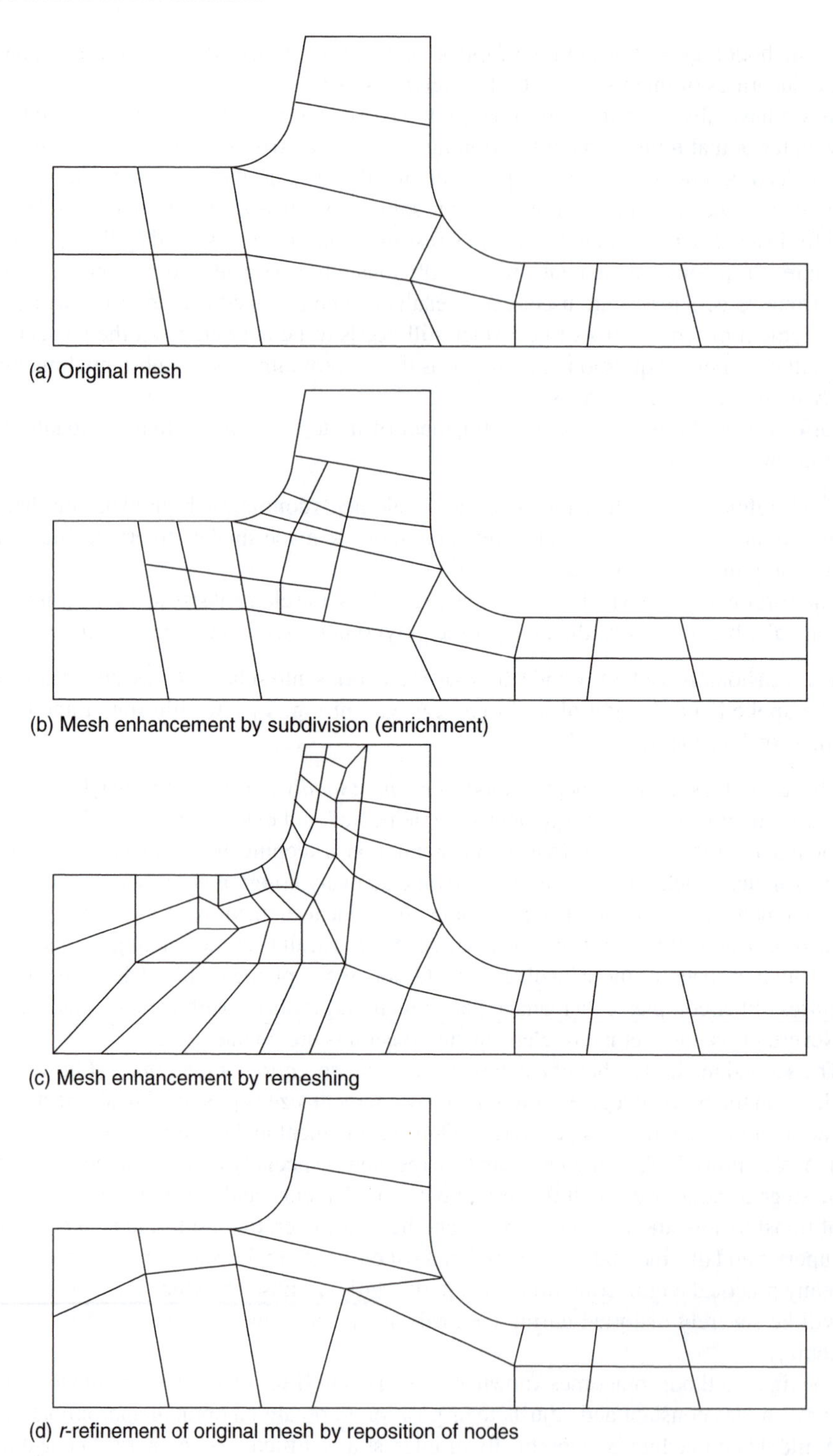

Fig. 14.1 Various procedures by *h*-refinement.

We shall see that with energy norms specified as the criterion, it is a fairly simple matter to predict the element size required for a given degree of approximation. Thus very few re-solutions are generally necessary to reach the objective.

With p-refinement the situation is different. Here two subclasses exist:

1. One in which the polynomial order is increased uniformly throughout the whole domain;
2. One in which the polynomial order is increased locally using hierarchical refinement.

In neither of these has a direct procedure been developed which allows the prediction of the best refinement to be used to obtain a given error. Here the procedures generally require more resolutions and tend to be more costly. However, the convergence for a given number of variables is more rapid with p-refinement and it has much to recommend it.

On occasion it is possible to combine efficiently the h- and p-refinements and call it the hp-refinement. In this procedure both the size of elements h and their degree of polynomial p are altered. Much work has been reported in the literature by Babuška, Oden and others and the interested reader is referred to the references.[8–18]

In Secs 14.2 and 14.3 we shall discuss both the h- and the p-refinement methods. In Sec. 14.3 we also include some details of the very simple and yet efficient hp-refinement process introduced by Zienkiewicz, Zhu and Gong.[19]

14.2 Adaptive *h*-refinement

14.2.1 Predicting the required element size in *h* adaptivity

In the introduction to this chapter we have mentioned several alternative processes of h-adaptivity and we suggested that the process in which the complete mesh is regenerated is in general the most efficient. Such a procedure allows elements to be de-refined (or enlarged) as well as refined (made smaller) and invariably starts at each stage of the analysis from a specification of the mesh size h defined at each nodal point of the previous mesh. Standard interpolation is used to find the size of elements required at any point in the domain. As the refinement process proceeds for each subsequent stage of analysis the computed mesh sizes h are based on a prescribed accuracy at the nodes of the previous mesh.

The error estimators discussed in the previous chapter allow the global energy (or similar) norm of the error to be determined and the errors occurring locally (at the element level) are usually also well represented. If these errors are within the limits prescribed by the analyst then clearly the work is completed. More frequently these limits are exceeded and refinement is necessary. The question which this section addresses is how best to effect this refinement. Here obviously many strategies are possible and much depends on the *objectives* or *goals* to be achieved.

In the simplest case we shall seek, for instance, to make the relative energy norm error η [viz. Eq. (13.12)] less than some specified value $\bar{\eta}$ (say 5% for many engineering applications). Thus

$$\eta \leq \bar{\eta} \tag{14.1}$$

is to be achieved.

In an 'optimal mesh' it is desirable that the distribution of element energy norm error (i.e., $\|\mathbf{e}\|_K$) should be equal for all elements. Thus if the total permissible error is determined (assuming that it is given by the result of the approximate analysis) as

$$\text{Permissible error} \equiv \bar{\eta}\|\mathbf{u}\| = \bar{\eta}\left(\|\hat{\mathbf{u}}\|^2 + \|\mathbf{e}\|^2\right)^{1/2} \tag{14.2}$$

here we have used[20]

$$\|\mathbf{e}\|^2 = \|\mathbf{u}\|^2 - \|\hat{\mathbf{u}}\|^2 \tag{14.3}$$

We could pose a requirement that the error in any element k should be

$$\|\mathbf{e}\|_K < \bar{\eta}\left(\frac{\|\hat{\mathbf{u}}\|^2 + \|\mathbf{e}\|^2}{m}\right)^{1/2} \equiv \bar{e}_m \tag{14.4}$$

where m is the number of elements involved.

Elements in which the above is not satisfied are obvious candidates for refinement. Thus if we define a refinement ratio by

$$\xi_K = \frac{\|\mathbf{e}\|_K}{\bar{e}_m} \tag{14.5}$$

we shall refine whenever†

$$\xi_K > 1 \tag{14.6}$$

The refinement ratio ξ_K can be approximated, of course, by replacing the true error in Eqs (14.4) and (14.5) with the error estimators.

The refinement could be carried out progressively by refining only a certain number of elements in which ξ is higher than a specified limit. This type of element subdivision process is also known as *mesh enrichment* as depicted in Fig. 14.1(b). This process of refinement though ultimately leading to a satisfactory solution being obtained with a relatively small number of total degrees of freedom, is in general not economical as the total number of trial solutions is usually excessive.

It is more efficient to try to design a completely new mesh which satisfies the requirement that

$$\xi_K \leq 1 \tag{14.7}$$

in all elements.

One possibility here is to invoke the asymptotic convergence rate criteria to predict the element size distribution. For instance, if we assume

$$\|\mathbf{e}\|_K \propto h_K^p \tag{14.8}$$

where h_K is the current element size and p the polynomial order of approximation, then to satisfy the requirement of Eq. (14.4) the new generated element size should be no larger than

$$h_{\text{new}} = \xi_K^{-1/p} h_K \tag{14.9}$$

Mesh generation programs in which the local element size can be specified are available now as we have already discussed in Chapter 8 and these can be used to design a new mesh for which the reanalysis is carried out.[21, 22] In the figures we show how starting from a relatively coarse solution a single mesh prediction often allows a solution (almost) satisfying the specified accuracy requirement to be achieved.

The reason for the success of the mesh regeneration based on the simple assumption of asymptotic convergence rate implied in Eq. (14.8) is the fact that with refinement the mesh

† We can indeed 'de-refine' or use a larger element spacing where $\xi_K < 1$ if computational economy is desired.

tends to be 'optimal' and the localized singularity influence no longer affects the overall convergence.

Of course the effects of any singularity will still remain present in the elements adjacent to it. An improved mesh results if in such elements we use the appropriate convergence and replace p by λ in Eqs (14.8) and (14.9), to obtain

$$h_{\text{new}} = \xi_K^{-1/\lambda} h_K \tag{14.10}$$

in which λ is the singularity strength, see Chapter 13, Eq. (13.14). A conservative number to use here is $\lambda = 0.5$ as most singularity parameters lie in the range 0.5–1.0. With this procedure, added to the refinement strategy, we frequently achieve accuracies better than the prescribed limit in one remeshing.

14.2.2 Numerical examples

In the examples which follow we will show in general a process of refinement in which the total number of degrees of freedom increases with each stage, even though the mesh is redesigned. This need not necessarily be the case as a fine but badly structured mesh can show much greater error than a near-optimal one. To illustrate this point we show in Fig. 14.2 a refinement designed to reach 5% accuracy in one step starting from uniform mesh subdivisions. We note that now, in at least one refinement, a decrease of total error occurs with a reduction of total degrees of freedom (starting from a uniform 8×8 subdivision with 544 equations and $\eta = 9.8\%$ to $\eta = 3.1\%$ with 460 equations).

We shall now present further typical examples of h-refinement with mesh adaptivity. In all of these, full mesh regeneration is used at every step.

Example 14.1: Short cantilever beam. This problem refers to a short cantilever beam in which two very high singularities exist at the corners attached to a rigid wall. The beam is loaded by a uniform load along the top boundary as illustrated in Figs 14.3 and 14.4. In the refinement process we use both the mesh criteria of Eqs (14.9) and (14.10).[23] In Figs 14.3 and 14.4 we show three stages of an adaptive solution and in Fig. 14.5 we indicate how rapidly these converge, although all uniform refinements converge at a very slow rate (due to the singularities).

The same problem is also solved by both mesh enrichment and mesh regeneration using linear quadrilateral elements to achieve 5% accuracy. The prescribed accuracy is obtained with optimal rate of convergence being reached by both adaptive refinement processes (Fig. 14.6). However, the mesh enrichment method requires seven refinements, as shown in Fig. 14.7, while mesh regeneration requires only three (see Fig. 14.8). Here the refinement criterion, Eq. (14.6), is used for the mesh enrichment process.

Example 14.2: Stressed cylinder. As we mentioned earlier, the value of the energy norm error is not necessarily the best criterion for practical refinement. Limits on the local stress error can be used effectively. Such errors are quite simply obtained by the recovery processes described in the previous chapter (SPR in Sec. 13.4 and REP in Sec. 13.5). In Fig. 14.9 we show a simple exercise recently conducted by Oñate and Bugeda[4] in which a refinement of a stressed cylinder is made using various criteria as described in the caption

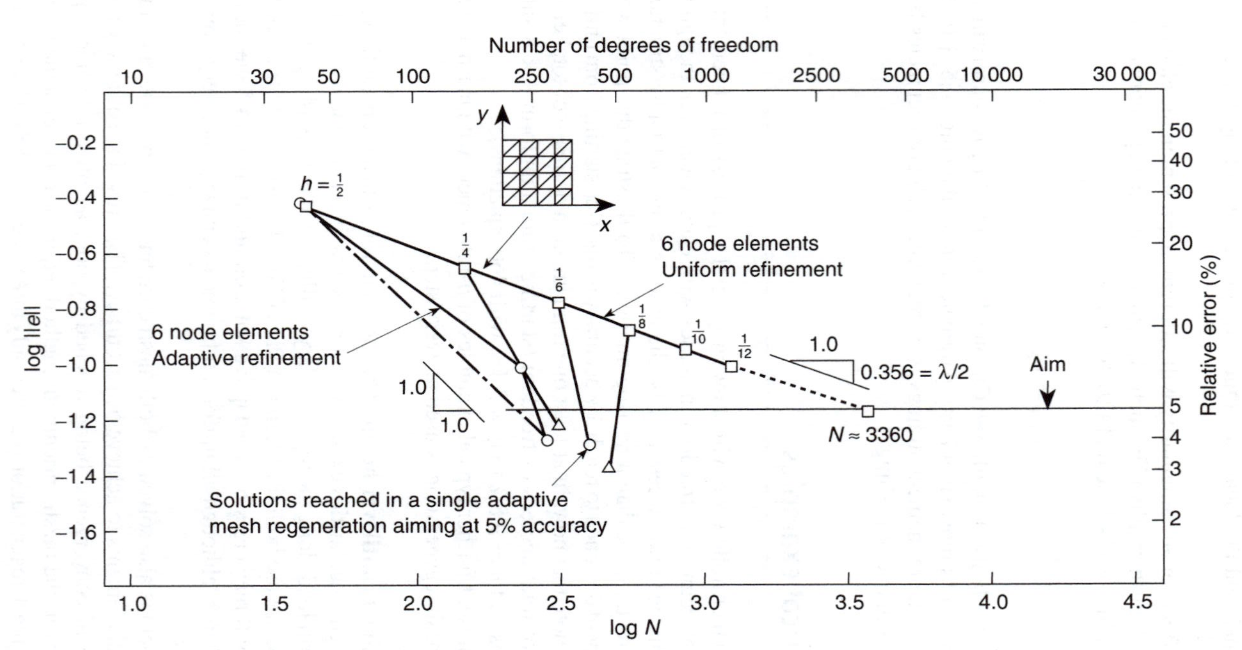

Fig. 14.2 The influence of initial mesh to convergence rates in h version. Adaptive refinement using quadratic triangular elements. Problem of Fig. 14.3. Note that if initial mesh is finer than $h = 1/8$ adaptive refinement reducesthe number of equations.

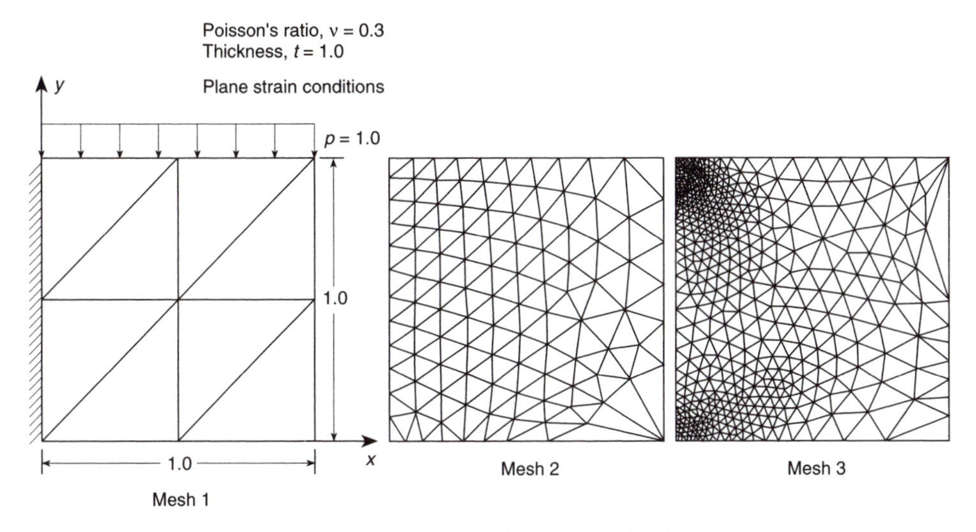

Fig. 14.3 Short cantilever beam and adaptive meshes of linear triangular elements.

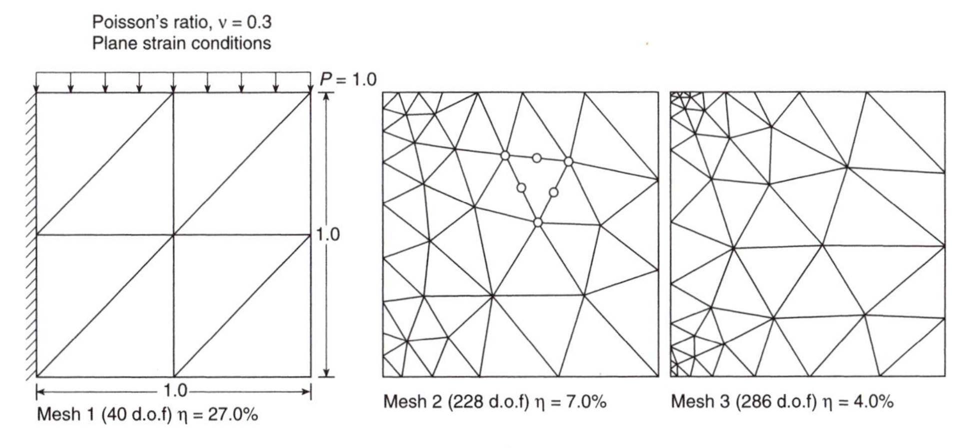

Fig. 14.4 Adaptive mesh of quadratic triangular elements for short cantilever beam.

of Fig. 14.9. It will be observed that the stress tolerance method generally needs a much finer mesh.

Example 14.3: A Poisson equation in a square domain. This example is fairly straightforward and starts from a simple square domain in which suitable loading terms exist in a Poisson equation to give the solution shown in Fig. 14.10.[12] In Fig. 14.11 we show the first subdivision of this domain into regular linear and quadratic elements and the subsequent refinements. The elements are of both triangular and quadrilateral shape and for the linear ones a target error of 10% in total energy has been set, while for quadratic elements the target error is 1% of total energy. In practically all cases three refinements suffice to reach a very accurate solution satisfying the requirements despite the fact that the original mesh cannot capture in any way the high intensity region illustrated in the previous figure. It is of interest to note that the effectivity indices in all cases are very close to one – this is true even for the original refinement. Figure 14.12 shows the convergence for various elements

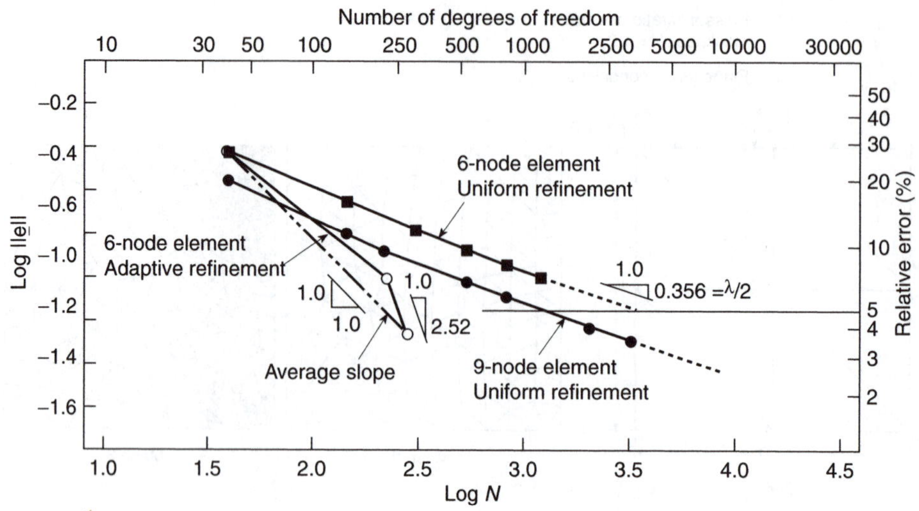

Fig. 14.5 Experimental rates of convergence for short cantilever beam.

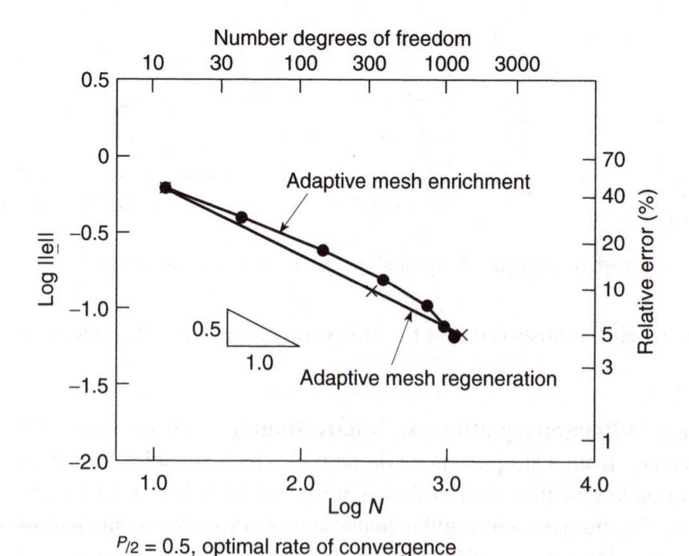

Fig. 14.6 Short cantilever beam. Mesh enrichment versus mesh regeneration using linear quadrilateral elements.

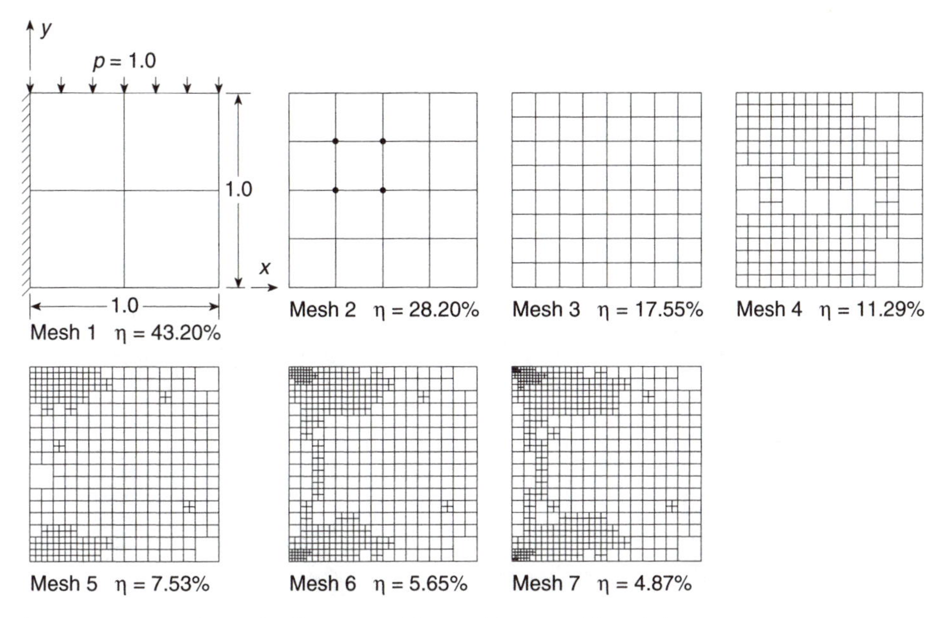

Fig. 14.7 Short cantilever solved by mesh enrichment. Linear quadrilateral elements.

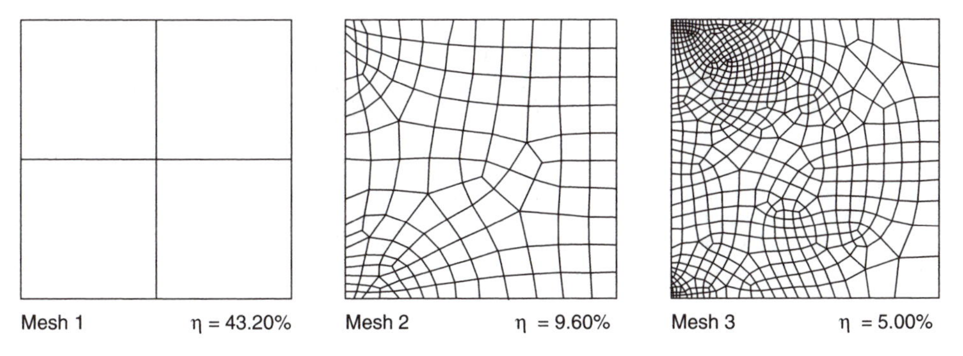

Fig. 14.8 Short cantilever solved by mesh regeneration. Linear quadrilateral elements.

with the error plotted against the total number of degrees of freedom. The reader should note that the asymptotic rate of convergence is exceeded when the refinement gets closer to its final objective.

Example 14.4: An L-shaped domain. It is of interest to note the results in Fig. 14.13 which come from an analysis of a re-entrant corner using isoparametric quadratic quadrilaterals. Here two meshes are shown together with the convergence data of the solution.

Example 14.5: A machine part. For this machine part problem plane strain conditions are assumed. A prescribed accuracy of 5% relative error is achieved in one adaptive refinement (see Fig. 14.14) with linear quadrilateral elements. The convergence of the shear stress τ_{xy} is shown in Fig. 14.15.

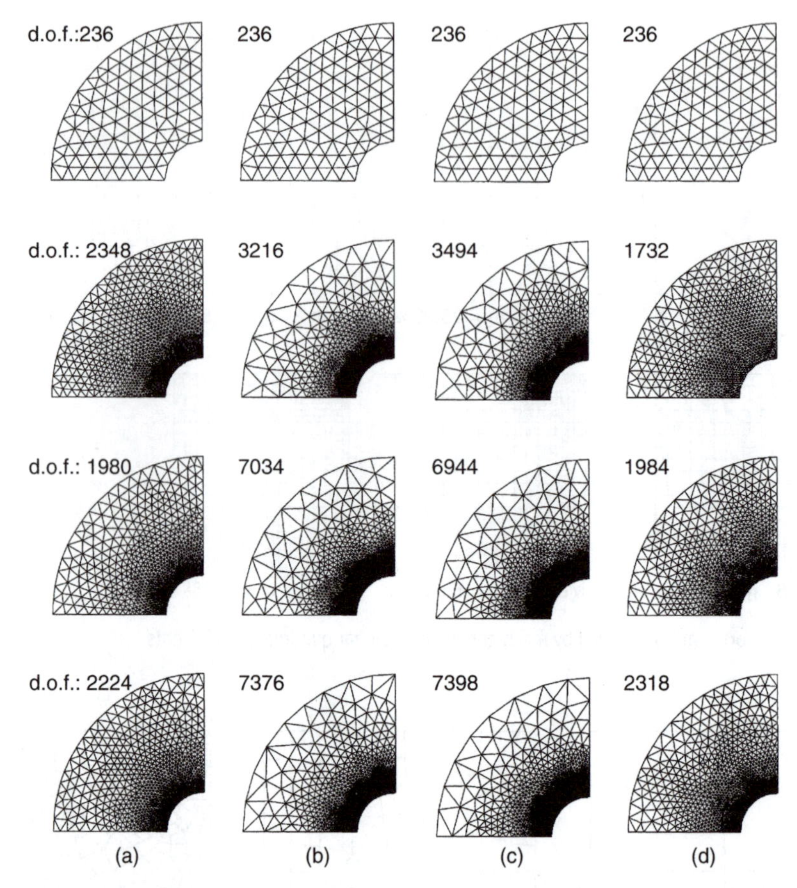

Fig. 14.9 Sequence of adaptive mesh refinement strategies based on (a) equal distribution of the global energy error between all the elements, (b) equal distribution of the density of energy error, (c) equal distribution of the maximum error in stresses at each point, and (d) equal distribution of the maximum percentage of the error in stresses at each point. All final meshes have less than 5% energy norm error.

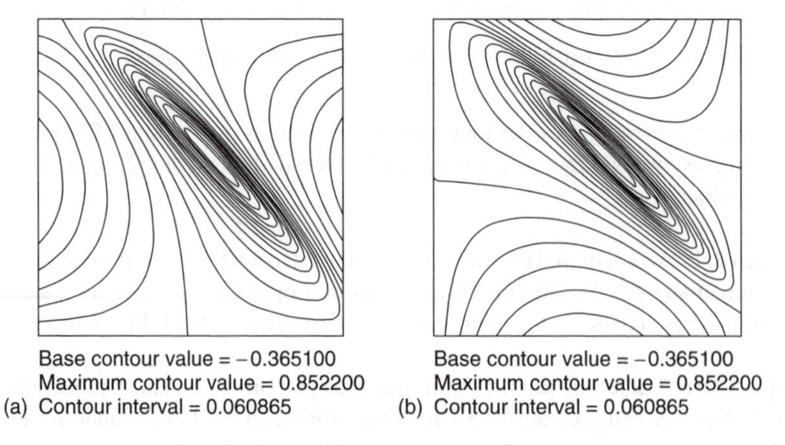

Fig. 14.10 Poisson equation 'exact' solutions. (a) $\partial u/\partial x$ contours. (b) $\partial u/\partial y$ contours.

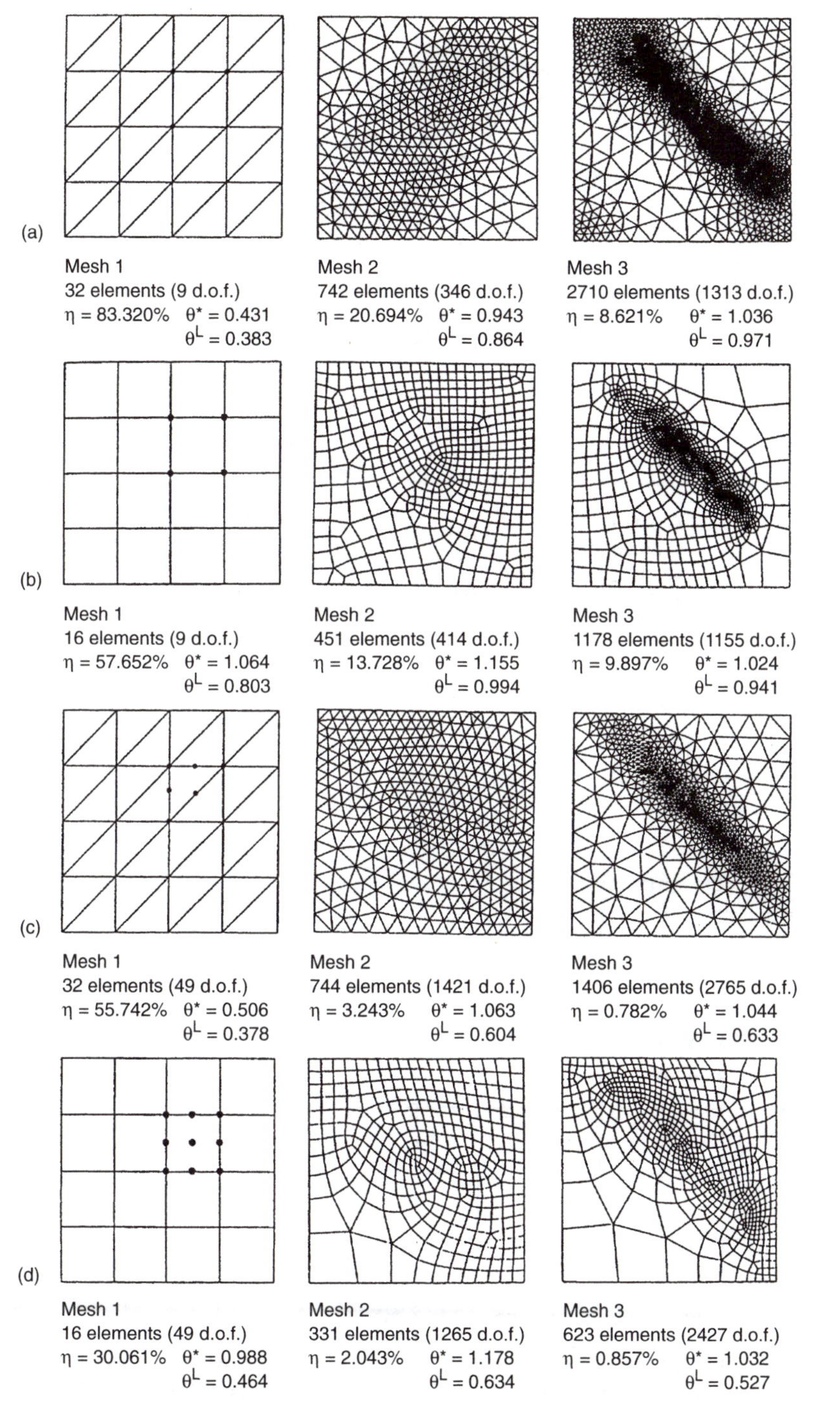

Fig. 14.11 Poisson problem of Fig. 14.10. Adaptive solutions for: (a) linear triangles; (b) linear quadrilaterals; (c) quadratic triangles; (d) quadratic quadrilaterals. θ^* based on SPR, θ^L based on L_2 projection. Target error 10% for linear elements and 1% for quadratic elements.

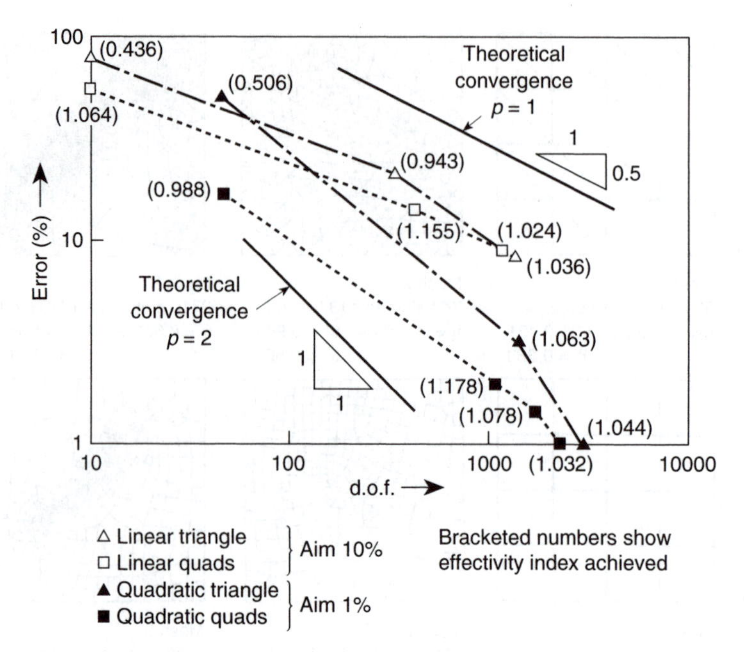

Fig. 14.12 Adaptive refinement for Poisson problem of Fig. 14.10.

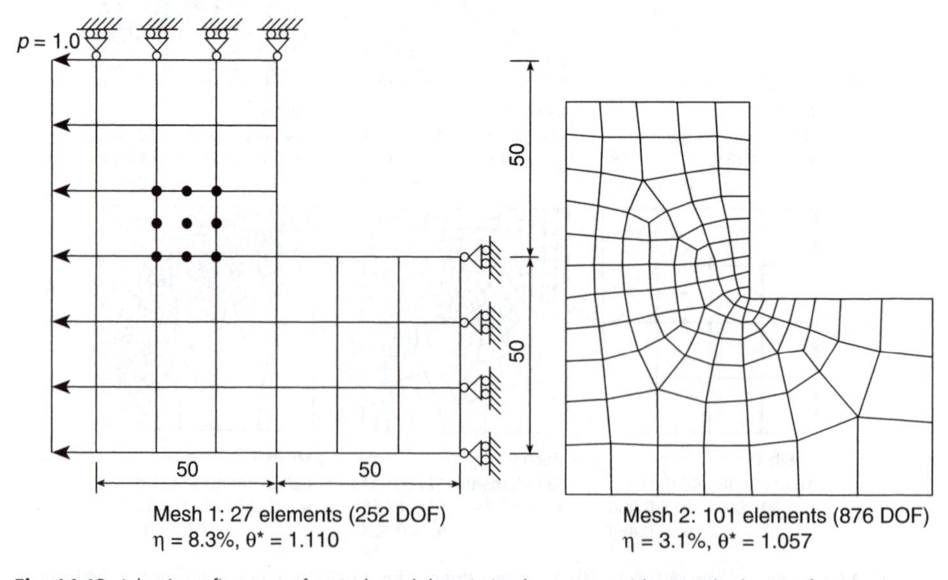

Fig. 14.13 Adaptive refinement of an L-shaped domain in plane stress with prescribed error of 1%.

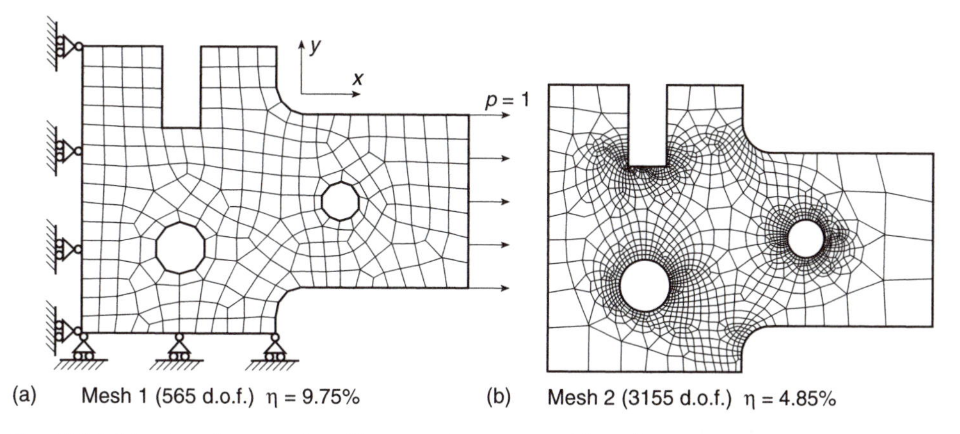

Fig. 14.14 Adaptive refinement of machine part using linear quadrilateral elements. Target error 5%.

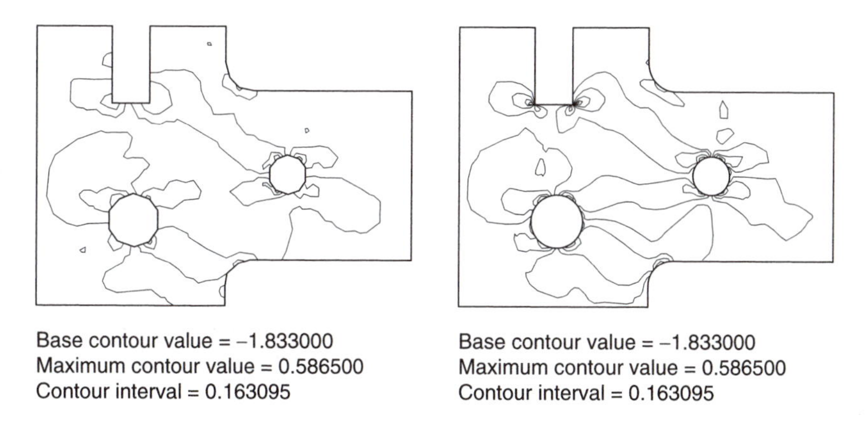

Fig. 14.15 Adaptive refinement of machine part. Contours of shear stress for original and final mesh.

Example 14.6: A perforated gravity dam. The final example of this section shows a more practical engineering problem of a perforated dam. This dam was analysed in the late 1960s during its construction. The problem was revisited to choose a suitable mesh of quadratic triangles. Figure 14.16(a) shows the mesh chosen. Despite the high order of elements the error is quite high, being around 17%. One stage of adaptive refinement reaches the specified value of 5% error in energy norm. As we have seen in previous examples such convergence is not always possible but it is achieved here. We believe this typical example shows the advantages of adaptivity and the ease with which a final good mesh can be arrived at automatically.

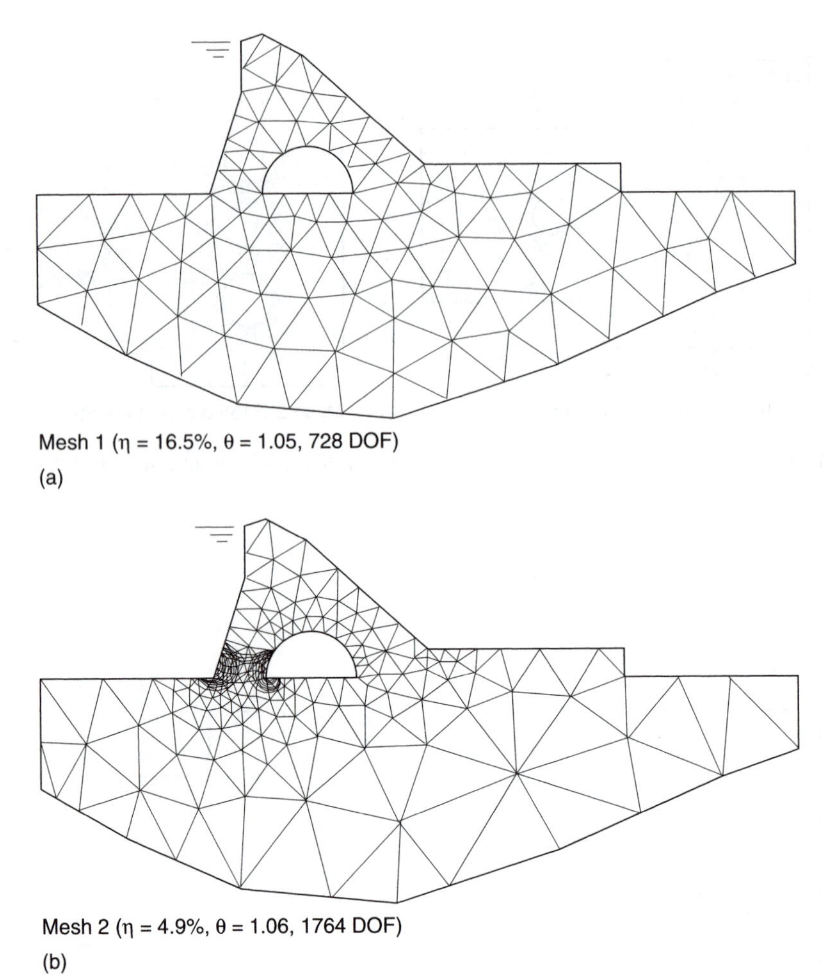

Fig. 14.16 Quadratic triangle. Automatic mesh generation to achieve 5% accuracy. Plane strain analysis of a dam with perforation, water loading only. (a) Original mesh. (b) Refined mesh.

14.3 *p*-refinement and *hp*-refinement

The use of non-uniform p-refinement is of course possible if done hierarchically and many attempts have been made to do this efficiently. Some of this was done as early as 1983.[24, 25] However, the general process is difficult and necessitates many assumptions about the decrease of error. Certainly, the desired accuracy can seldom be obtained in a single step and most of the work on this requires a sequence of steps. We illustrate such a refinement process in Fig. 14.17 for the perforated dam problem presented in the previous section.

The same applies to hp-processes in which much work has been done during the last two decades.[8–18] We shall quote here only one particular attempt at hp-refinement which seems to be particularly efficient and where the number of resolutions is quite small. The

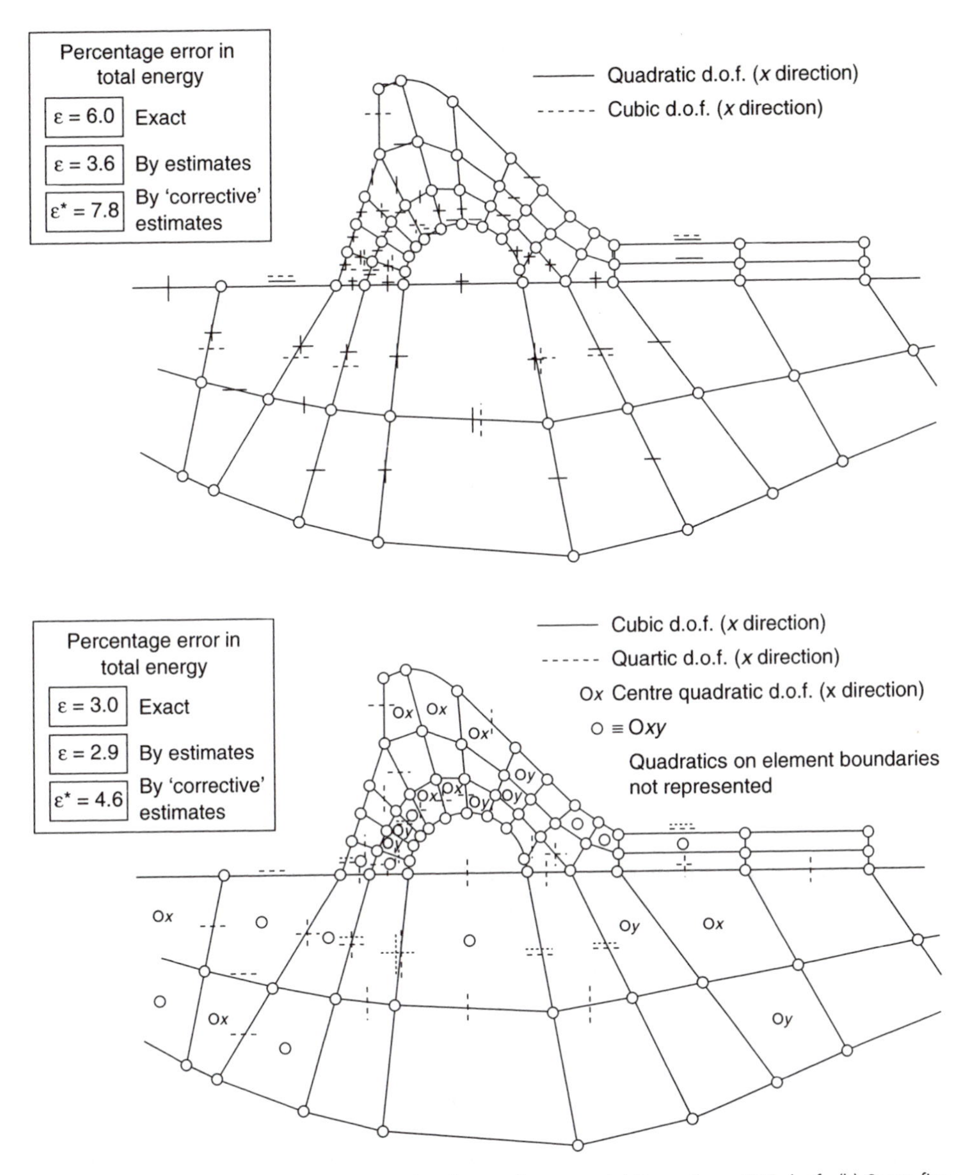

Fig. 14.17 Adaptive solution of perforated dam by p-refinement. (a) Stage three, 206 d.o.f. (b) Stage five, 365 d.o.f.

methodology was introduced by Zienkiewicz *et al.* in 1989[19] and we shall quote here some of the procedures suggested.

The first procedure is that of pursuing an h-refinement with lower order elements (e.g., linear or quadratic elements) to obtain, say, a 5% accuracy, at which stage the energy norm error is nearly uniformly distributed throughout all elements. From there a p-refinement is applied in a uniform manner (i.e., the same p is used in all elements). This has very substantial computational advantages as programming is easy and can be readily accomplished, especially if hierarchical functions are used. The uniform p-refinement also allows the global energy norm error to be approximately extrapolated by three consecutive solutions.[26]

The convergence of the p-refinement finite element solution can be written as[27]

$$\|\mathbf{e}\| \leq CN^{-\beta} \tag{14.11}$$

where C and β are positive constants depending on the solution of the problem and N is the number of degrees of freedom. We assume that for each refinement the error is, observing Eq. (14.3),

$$\|\mathbf{u}\|^2 - \|\hat{\mathbf{u}}_q\|^2 = CN_q^{-2\beta} \tag{14.12}$$

with $q = p-2, p-1, p$ for the three solutions. Eliminating the two constants C and β from the above three equations, $\|\mathbf{u}\|^2$ can be solved by

$$\frac{\|\mathbf{u}\|^2 - \|\hat{\mathbf{u}}_p\|^2}{\|\mathbf{u}\|^2 - \|\hat{\mathbf{u}}_{p-1}\|^2} = \left(\frac{\|\mathbf{u}\|^2 - \|\hat{\mathbf{u}}_{p-1}\|^2}{\|\mathbf{u}\|^2 - \|\hat{\mathbf{u}}_{p-2}\|^2}\right)^{\frac{\log(N_{p-1}/N_p)}{\log(N_{p-2}/N_{p-1})}} \tag{14.13}$$

The global energy norm error for the final solution and indeed the error at any stage of the p-refinement can be determined using

$$\|\mathbf{e}\|^2 = \|\mathbf{u}\|^2 - \|\hat{\mathbf{u}}_q\|^2 \tag{14.14}$$

$q = 1, 2, \ldots, p$.

Example 14.7: *h–p*-refinement of L-shaped domain and short cantilever beam. Generally the high accuracy is gained rapidly by refinement, at least from examples performed to date. In Figs 14.18 and 14.19 we show two examples for which we have previously used an h-refinement. The first illustrates the L-shaped domain with one singularity and the second the short cantilever beam with two strong singularities. Both problems are solved first using h-refinement until target 5% accuracy is reached using quadratic triangles. At this stage the p is increased to third and fourth order so that three solutions are available. At the end of the third solution the error is less than 1%.

In the same paper[19] an alternative procedure is suggested. This uses a very coarse mesh at the outset followed by p-refinement. In this case the error at the element level is estimated at the last stage of the p-refinement as the difference between the last two refinements (e.g., the third and fourth order when the maximum p is 4). The global error estimator is calculated by the extrapolation procedure used in the previous example. The element error estimator is for order $p-1$ rather than the highest order p. It is, however, very accurate. The element error estimator is subsequently used to compute the optimal mesh size as described in Sec. 14.2.1. Nearly optimal rate of convergence is expected to be achieved because the optimal mesh is designed for $p-1$ order elements. Details of this process will be found again in the reference and will not be discussed further.

At no stage of the hp-refinements have we used here any of the estimators quoted in the previous chapter. However, their use would make the optimal mesh design at order p

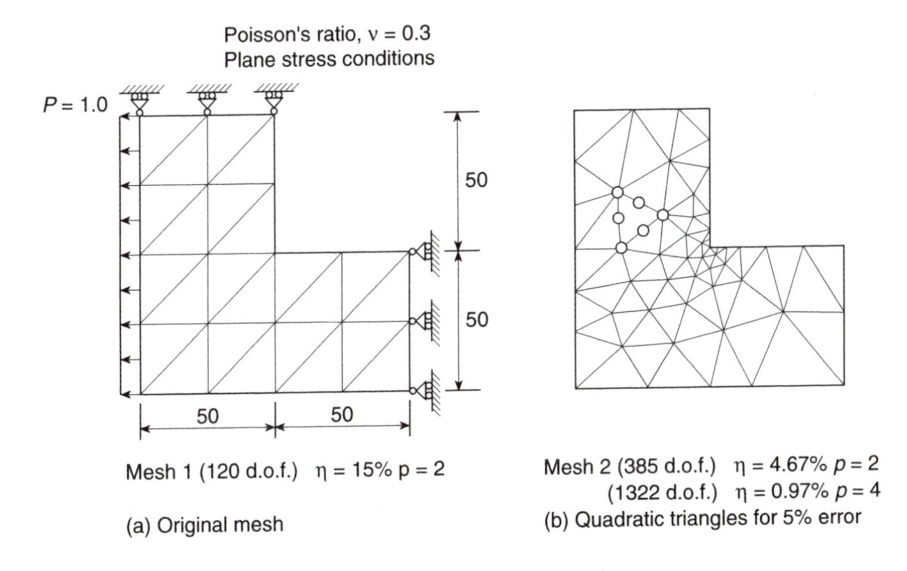

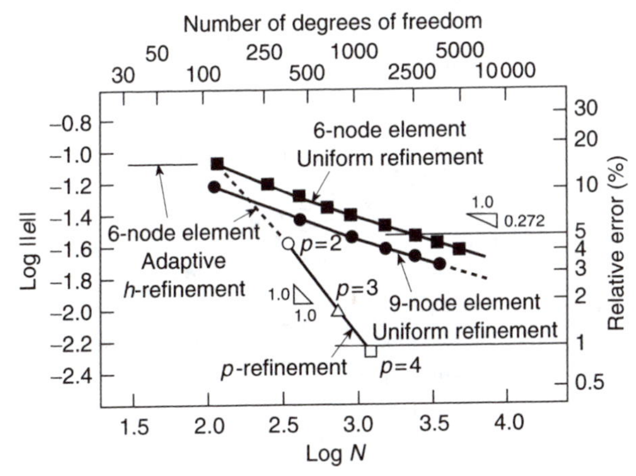

Fig. 14.18 Solution of L-shaped domain by h–p-refinement (as defined in Example 14.4 of previous section) using procedure one of reference 19.

possible, because the element error can be accurately estimated at order p. It will result in an optimal hp-refinement.

The two examples we have quoted above are reanalysed using the alternative process described above and presented in Figs 14.20 and 14.21. In both cases the final accuracy shows an error of less than 1% but it is noteworthy that the total number of degrees of freedom used with the second method is considerably less than that in the first and still achieves a nearly optimal rate of convergence.

We can conclude this section on hp-refinement with a final example where a highly singular crack domain is studied. Once again the second procedure is used showing in Fig. 14.22 a remarkable rate of convergence.

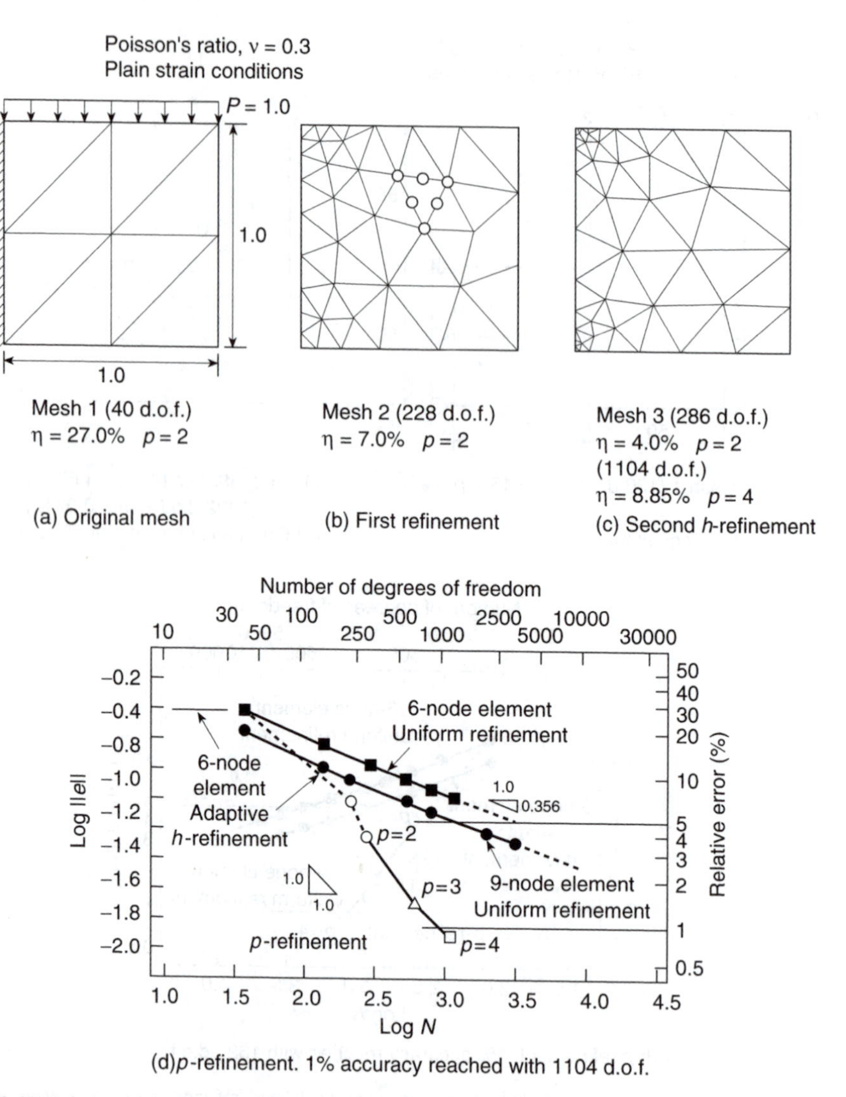

Fig. 14.19 Solution of short cantilever by h–p-adaptive refinement using procedure one of reference 19.

14.4 Concluding remarks

The methods of estimating errors and adaptive refinement which are described in this and the previous chapter constitute a very important tool for practical application of finite element methods. The range of applications is large and we have only touched here upon the relatively simple range of linear elasticity and similar self-adjoint problems. A recent survey shows many more areas of application[28] and the reader is referred to this publication for interesting details. At this stage we would like to reiterate that many different norms or measures of error can be used and that for some problems the energy norm is not in fact 'natural'. A good example of this is given by problems of high-speed gas flow, where very

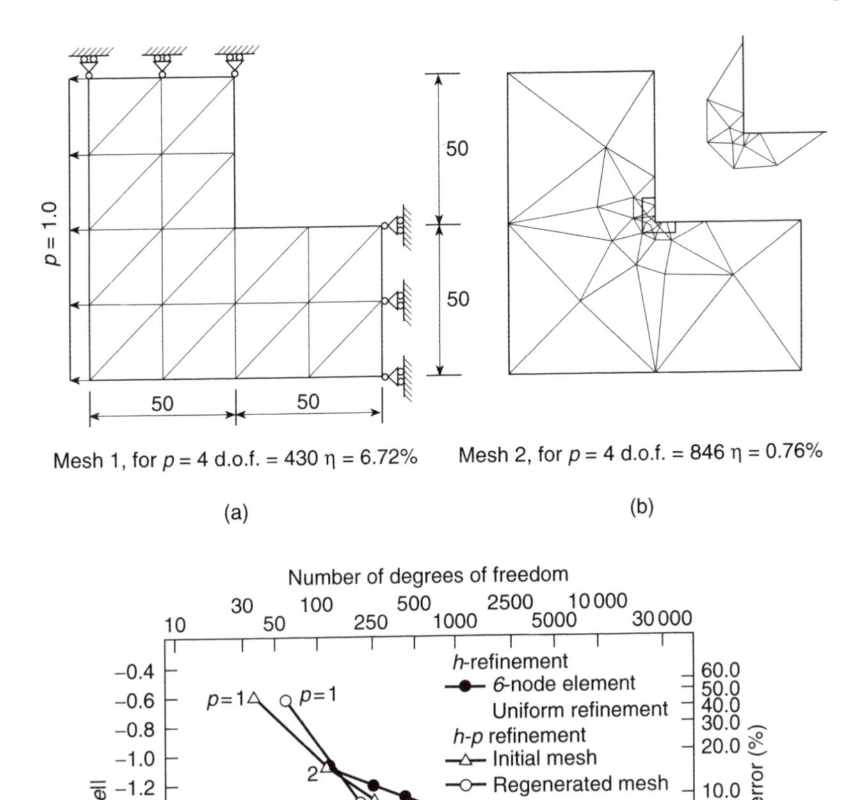

Fig. 14.20 Solution of L-shaped domain by h–p-adaptive refinement using alternative procedure of reference 19.

steep gradients (shocks) can develop. The formulation of such problems is complex, but this is not necessary for the present argument.

For problems in fluid mechanics discussed in reference 29 and similarly for problems of strain localization in plastic softening discussed in reference 30 no global norms can be used effectively. In such situations it is convenient to base the refinement on the value of the maximum curvatures developed by the solution of u. On occasion an elongation of the elements will be used to refine the mesh appropriately. Figure 14.23 shows a typical problem of shock capturing solved adaptively.

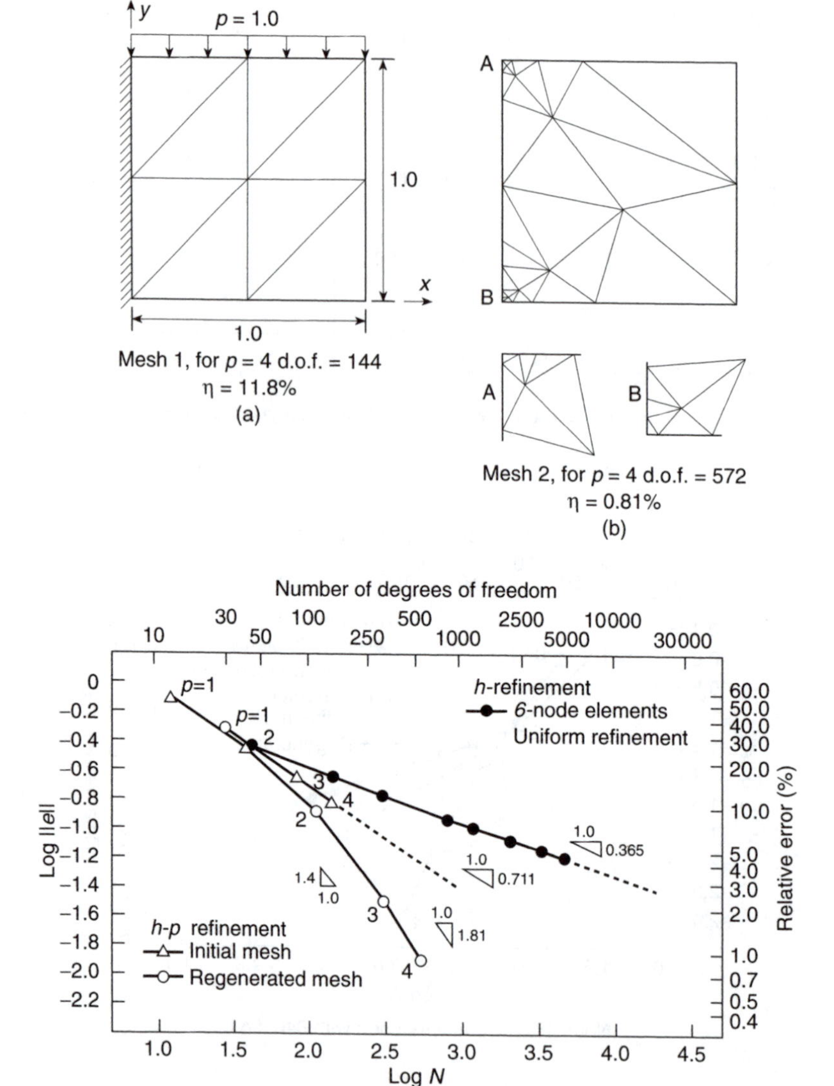

Fig. 14.21 Solution of short cantilever by h–p-adaptive refinement using alternative procedure of reference 19.

14.5 Problems

14.1 Program development project: Implement a *mesh enrichment* algorithm (as described in Sec. 14.2.1) in the solution system started in Problem 2.17 and extended in subsequent exercises. Assume the problem is given by the quasi-harmonic equation and modelled using 3-node triangular elements. (Hint: Adapt the mesh generation program developed in Problems 5.17 and 5.18 to generate the mesh using the new coordinates resulting from enrichment.)

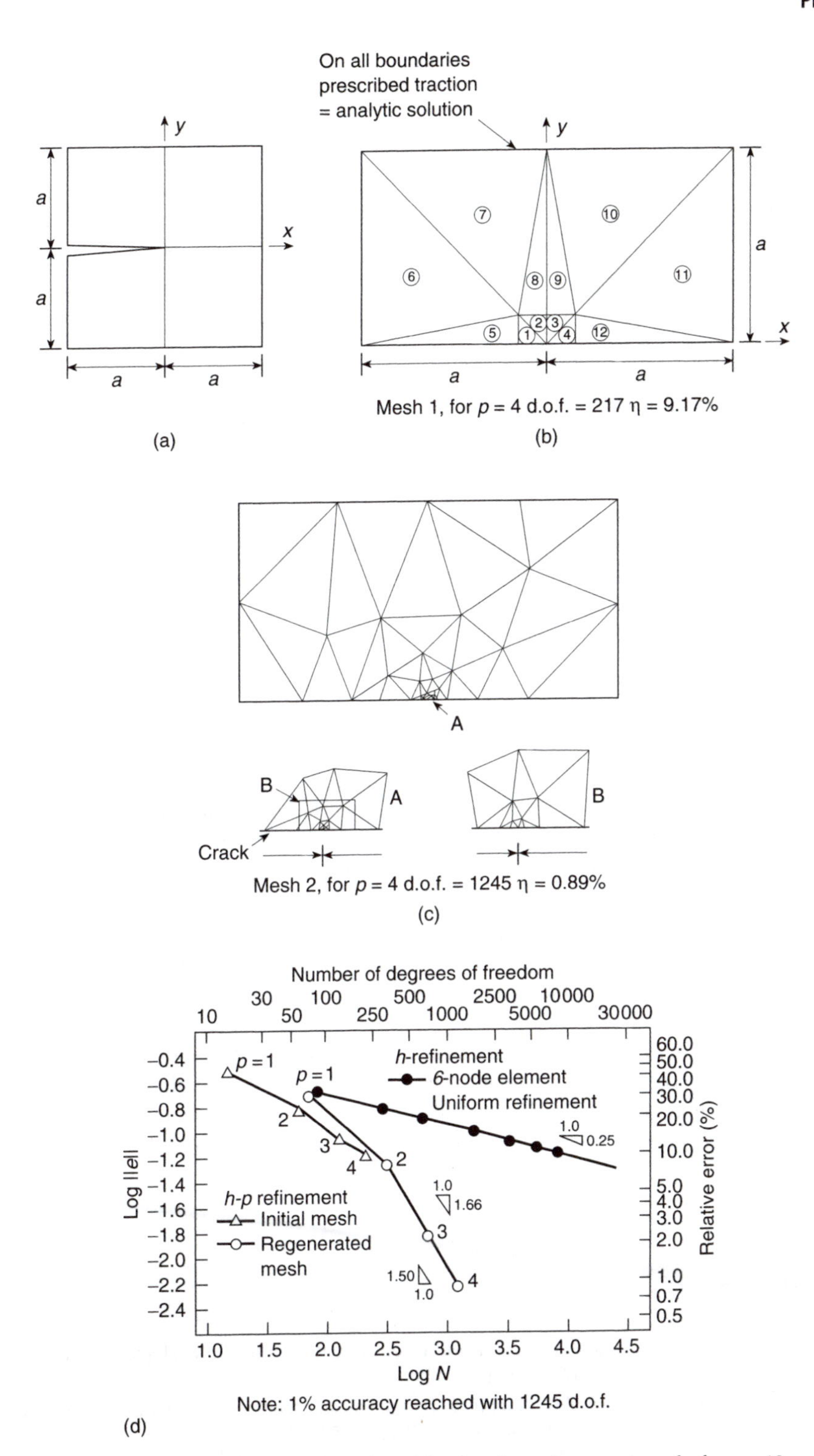

Fig. 14.22 Adaptive *h–p*-refinement for a singular crack using alternative procedure of reference 19.

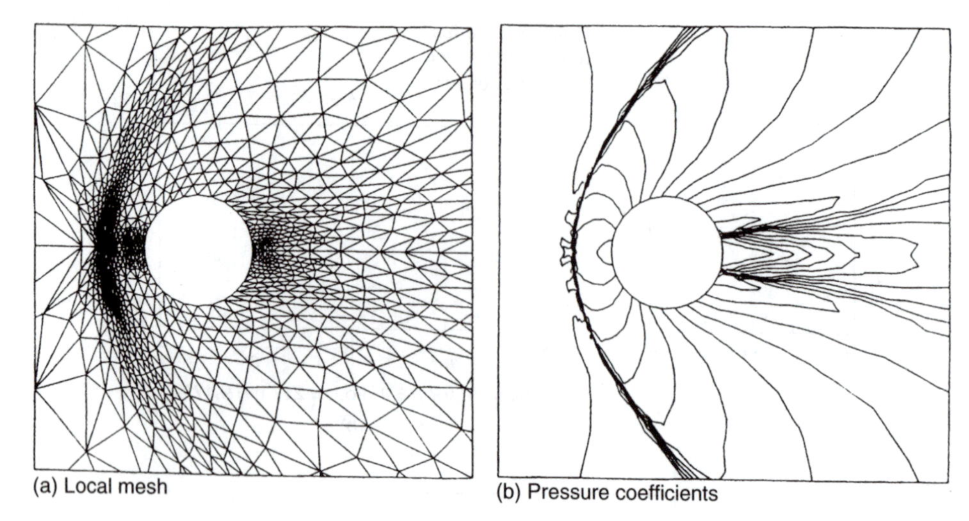

Fig. 14.23 Directional mesh refinement. Gas flow past a circular cylinder – Mach number 3. Third refinement mesh 709 nodes (1348 elements).

14.2 Program development project: Implement an *adaptive mesh regeneration* algorithm (as described in Sec. 14.2) in the solution system started in Problem 2.17 and extended in subsequent exercises. Assume the problem is given by the quasi-harmonic equation and modelled using 3-node triangular elements.

14.3 Solve Example 2.3 using linear (3-node) elements. Follow the mesh refinement procedure described in Sec. 14.2 for adaptive h refinement and show that the optimal rate of convergence of the finite element method can be attained when a prescribed accuracy is achieved.

14.4 Solve Problem 2.3 using quadratic (6-node) elements. Follow the mesh refinement procedure described in Sec. 14.2 for adaptive h refinement and show that the optimal rate of convergence of the finite element method can be attained when prescribed accuracy is achieved.

14.5 Program development project: Devise and implement in the solution system started in Problem 2.17 an hp refinement strategy (see Sec. 14.3) to attain a prescribed accuracy. Assume the problem is given by the quasi-harmonic equation and hierarchical triangular elements are used to define the finite element p-models.

References

1. I. Babuška and C. Rheinboldt. *A-posteriori* error estimates for the finite element method. *Int. J. Numer. Meth. Eng.*, 12:1597–1615, 1978.
2. I. Babuška and C. Rheinboldt. Adaptive approaches and reliability estimates in finite element analysis. *Comp. Meth. Appl. Mech. Eng.*, 17/18:519–540, 1979.
3. O.C. Zienkiewicz and J.Z. Zhu. A simple error estimator and adaptive procedure for practical engineering analysis. *Int. J. Numer. Meth. Eng.*, 24:337–357, 1987.

4. E. Oñate and G. Bugeda. A study of mesh optimality criteria in adaptive finite element analysis. *Eng. Comput.*, 10:307–321, 1993.
5. E.R. de Arantes e Oliveira. Theoretical foundations of the finite element method. *Int. J. Solids Struct.*, 4:929–952, 1968.
6. E.R. de Arantes e Oliveira. Optimization of finite element solutions. In *Proc. 3rd Conf. Matrix Methods in Structural Mechanics*, volume AFFDL-TR-71-160, pages 423–446, Wright-Patterson Air Force Base, Ohio, 1972.
7. R.L. Taylor and R. Iding. Applications of extended variational principles to finite element analysis. In C.A. Brebbia and H. Tottenham, editors, *Proc. of the International Conference on Variational Methods in Engineering*, volume II, pages 2/54–2/67. Southampton University Press, 1973.
8. W. Gui and I. Babuška. The h, p and h-p version of the finite element method in 1 dimension. Part 1: The error analysis of the p-version. Part 2: The error analysis of the h- and h-p version. Part 3: The adaptive h-p version. *Numerische Math.*, 48:557–683, 1986.
9. B. Guo and I. Babuška. The h-p version of the finite element method. Part 1: The basic approximation results. Part 2: General results and applications. *Comp. Mech.*, 1:21–41, 203–226, 1986.
10. I. Babuška and B. Guo. The h-p version of the finite element method for domains with curved boundaries. *SIAM J. Numer. Anal.*, 25:837–861, 1988.
11. I. Babuška and B.Q. Guo. Approximation properties of the hp version of the finite element method. *Comp. Meth. Appl. Mech. Eng.*, 133:319–349, 1996.
12. L. Demkowicz, J.T. Oden, W. Rachowicz, and O. Hardy. Toward a universal h–p adaptive finite element strategy. Part 1: Constrained approximation and data structure. *Comp. Meth. Appl. Mech. Eng.*, 77:79–112, 1989.
13. W. Rachowicz, J.T. Oden, and L. Demkowicz. Toward a universal h–p adaptive finite element strategy. Part 3: Design of h–p meshes. *Comp. Meth. Appl. Mech. Eng.*, 77:181–211, 1989.
14. K.S. Bey and J.T. Oden. hp-version discontinuous Galerkin methods for hyperbolic conservation laws. *Comp. Meth. Appl. Mech. Eng.*, 133:259–286, 1996.
15. C.E. Baumann and J.T. Oden. A discontinuous hp finite element method for convection-diffusion problems. *Comp. Meth. Appl. Mech. Eng.*, 175:311–341, 1999.
16. P. Monk. On the p and hp extension of Nedelecs curl-conforming elements. *J. Comput. Appl. Math.*, 53:117–137, 1994.
17. L.K. Chilton and M. Suri. On the selection of a locking-free hp element for elasticity problems. *Int. J. Numer. Meth. Eng.*, 40:2045–2062, 1997.
18. L. Vardapetyan and L. Demkowicz. hp-Adaptive finite elements in electromagnetics. *Comp. Meth. Appl. Mech. Eng.*, 169:331–344, 1999.
19. O.C. Zienkiewicz, J.Z. Zhu, and N.G. Gong. Effective and practical h-p-version adaptive analysis procedures for the finite element method. *Int. J. Numer. Meth. Eng.*, 28:879–891, 1989.
20. P.G. Ciarlet. *The Finite Element Method for Elliptic Problems*. North-Holland, Amsterdam, 1978.
21. J. Peraire, M. Vahdati, K. Morgan, and O.C. Zienkiewicz. Adaptive remeshing for compressible flow computations. *J. Comput. Phys.*, 72:449–466, 1987.
22. J.Z. Zhu, O.C. Zienkiewicz, E. Hinton, and J. Wu. A new approach to the development of automatic quadrilateral mesh generation. *Int. J. Numer. Meth. Eng.*, 32:849–866, 1991.
23. J.Z. Zhu and O.C. Zienkiewicz. Adaptive techniques in the finite element method. *Comm. Appl. Num. Math.*, 4:197–204, 1988.
24. D.W. Kelly, J.P. De S.R. Gago, O.C. Zienkiewicz, and I. Babuška. *A posteriori* error analysis and adaptive processes in the finite element method: Part I – Error analysis. *Int. J. Numer. Meth. Eng.*, 19:1593–1619, 1983.
25. J.P. De S.R. Gago, D.W. Kelly, O.C. Zienkiewicz, and I. Babuška. *A posteriori* error analysis and adaptive processes in the finite element method: Part II – Adaptive mesh refinement. *Int. J. Numer. Meth. Eng.*, 19:1621–1656, 1983.

26. B.A. Szabo. Mesh design for the p version of the finite element. *Comp. Meth. Appl. Mech. Eng.*, 55:181–197, 1986.
27. I. Babuška, B.A. Szabo, and I.N. Katz. The p version of the finite element method. *SIAM J. Numer. Anal.*, 18:512–545, 1981.
28. P. Ladevèze and J.T. Oden, editors. *Advances in Adaptive Computational Methods in Mechanics. Studies in Applied Mechanics 47*. Elsevier, 1998.
29. O.C. Zienkiewicz, R.L. Taylor, and P. Nithiarasu. *The Finite Element Method for Fluid Dynamics*. Butterworth-Heinemann, Oxford, 6th edition, 2005.
30. O.C. Zienkiewicz and R.L. Taylor. *The Finite Element Method for Solid and Structural Mechanics*. Butterworth-Heinemann, Oxford, 6th edition, 2005.

15

Point-based and partition of unity approximations. Extended finite element methods

15.1 Introduction

In all of the preceding chapters, the finite element method was characterized by the subdivision of the total domain of the problem into a set of subdomains called elements. The union of such elements gave the total domain. The subdivision of the domain into such components is of course laborious and difficult necessitating mesh generation as discussed in Chapter 8. Further if adaptivity processes are used, generally large areas of the problem have to be remeshed. For this reason, much attention has been given to devising approximation methods which are based on points without necessity of forming elements.

When we discussed the matter of generalized finite element processes in Chapter 3, we noted that point collocation or in general finite differences did in fact satisfy the requirement of the pointwise definition. However, the early finite differences were always based on a regular arrangement of nodes which severely limited their applications. To overcome this difficulty, since the late 1960s the proponents of the finite difference method have worked on establishing the possibility of finite difference calculus being based on an arbitrary disposition of collocation points. Here the work of Girault,[1] Pavlin and Perrone,[2] and Snell *et al.*[3] should be mentioned. However, a full realization of the possibilities was finally offered by Liszka and Orkisz,[4,5] and Krok and Orkisz[6] who introduced the use of least squares methods to determine the appropriate shape functions.

At this stage Orkisz and coworkers realized not only that collocation methods could be used but also the full finite element, weak formulation could be adopted by performing integration. Questions of course arose as to what areas such integration should be applied. Liszka and Orkisz[4] suggested determining a 'tributary area' to each node providing these nodes were triangulated as shown in Fig. 15.1(a). On the other hand in a somewhat different context Nay and Utku[7] also used the least squares approximation including triangular vertices and points of other triangles placed outside a triangular element thus simply returning to the finite element concept. We show this kind of approximation in Fig. 15.1(b). Whichever form of tributary area was used the direct least squares approximation centred at each node will lead to discontinuities of the function between the chosen integration areas and thus will violate the rules which we have imposed on the finite element method. However, it turns out that such rules could be violated and here the patch test will show that convergence is still preserved.

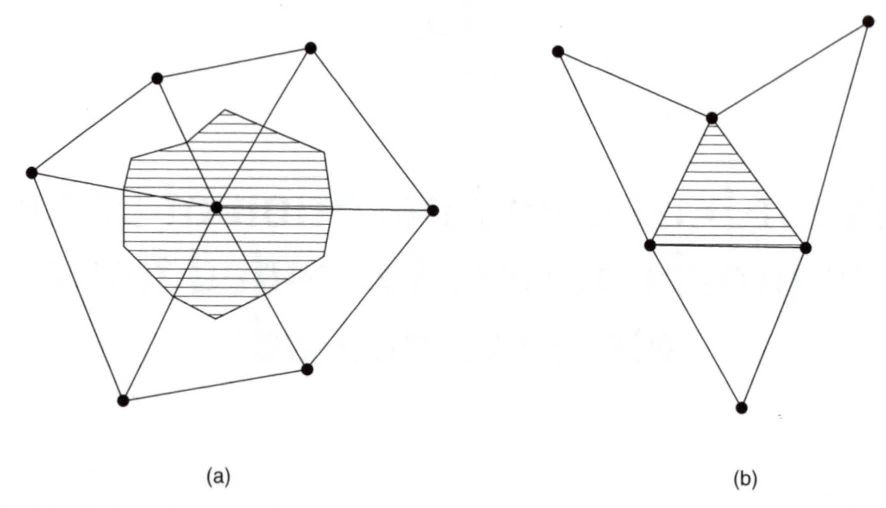

Fig. 15.1 Patches of triangular elements and tributary areas.

However, the possibility of determining a completely compatible form of approximation existed. This compatible form in which continuity of the function and of its slope if required and even higher derivatives could be accomplished by the use of so-called moving least squares methods. Such methods were originated in another context (Shepard,[8] Lancaster and Salkauskas[9, 10]). The use of such interpolation in the meshless approximation was first suggested by Nayroles *et al.*[11–13] This formulation was named by the authors as the *diffuse finite element method.*

Belytschko and coworkers[14, 15] realized the advantages offered by such an approach especially when dealing with the development of cracks and other problems for which standard elements presented difficulties. His so-called 'element-free Galerkin' method led to many seminal publications.

An alternative use of moving least squares procedures, called *hp*-cloud methods, was suggested by Duarte and Oden.[16, 17] They introduced at the same time a concept of hierarchical forms by noting that all shape functions derived by least squares possess the partition of unity property (viz. Chapter 4). Thus higher order interpolations could be added at each node rather than each element, and the procedures of element-free Galerkin or of the diffuse element method could be extended.

The use of all the above methods still, necessitates integration. Now, however, this integration need not be carried out over complex areas. A background grid for integration purposes is introduced though internal boundaries are no longer required. Thus such numerical integration on regular grids is used by Belytschko[18, 19] and other approaches are being explored. However, another interesting possibility was suggested by Babuška and Melenk.[20, 21]

Babuška and Melenk use a partition of unity but now the first set of basic shape functions is derived on a standard finite element, say the linear triangle. Most of the approximations then arise through addition of hierarchical variables centred at nodes. We feel that this kind of approach which necessitates very few elements for integration purposes combines well the methodologies of 'element-free' and 'standard element' approximation procedures. We shall demonstrate a few examples later for the application of such methods which seem to present a very useful extension of the hierarchical approach.

Incidentally the procedures based on local elements also have the additional advantage that global functions can be introduced in addition to the basic ones to represent special phenomena, for instance the presence of a singularity or waves. Both of these are important and the idea presented by this can be exploited. This is especially useful in solution of certain wave phenomena[22] and Belytschko and coworkers have coined the term 'XFEM' (extended finite element method) and exploited the approach to insert cracks in domains.[23–34]

This chapter will conclude with reference to other similar procedures which we do not have time to discuss.

15.2 Function approximation

We consider here a local set of n points in two (or three) dimensions defined by the coordinates x_k, y_k, z_k; $k = 1, 2, \ldots, n$ or simply $\mathbf{x_k} = [x_k, y_k, z_k]$ at which a set of data values of the unknown function $\tilde{u}_k$ are given. It is desired to fit a specified function form to the data points. In order to make a fit it is necessary to:

1. Specify the form of the functions, $\mathbf{p}(\mathbf{x})$, to be used for the approximation. Here as in the standard finite element method, it is essential to include low order polynomials necessary to model all the derivatives contained in the differential equation or in the weak form approximation being used. Certainly a complete linear and sometimes quadratic polynomial will always be necessary.
2. Define the procedure for establishing the fit.

Here we will consider some *least squares fit* methods as the basis for performing the fit. The functions will mostly be assumed to be polynomials; however, in addition other functions can be considered if these are known to model well the solution expected (e.g., see reference 22 on use of 'wave' functions).

15.2.1 Least squares fit

We shall first consider a least squares fit scheme which minimizes the square of the distance between n data values $\tilde{u}_k$ defined at the points $\mathbf{x}_k$ and an approximating function evaluated at the same points $\hat{u}(\mathbf{x_k})$. We assume the approximation function is given by a linearly independent set of m polynomials $p_j(\mathbf{x})$

$$\hat{u}(\mathbf{x}) = \sum_{j=1}^{m} p_j(\mathbf{x})\alpha_j \equiv \mathbf{p}(\mathbf{x})\boldsymbol{\alpha} \tag{15.1}$$

in which $\boldsymbol{\alpha}$ is a set of parameters to be determined. A least squares scheme is introduced to perform the fit to a set of n data points and this is written as (see Chapter 13 for similar operations). Minimize

$$J = \tfrac{1}{2}\sum_{k=1}^{n} (\hat{u}(\mathbf{x_k}) - \tilde{u}_k)^2 = \min \tag{15.2}$$

where the minimization is to be performed with respect to the values of $\boldsymbol{\alpha}$. Substituting the values of $\hat{u}$ at the points $\mathbf{x}_k$ we obtain

$$\frac{\partial J}{\partial \alpha_j} = \sum_{k=1}^{n} \frac{\partial \hat{u}_k}{\partial \alpha_j} \cdot (\hat{u}(\mathbf{x}_k) - \tilde{u}_k) = 0; \qquad j = 1, 2, \ldots, m \tag{15.3}$$

where

$$\hat{u}_k = \sum_{j=1}^{m} p_j(\mathbf{x}_k)\alpha_j = \mathbf{p}_k\,\boldsymbol{\alpha}$$

in which $\mathbf{p}_k \equiv \mathbf{p}(\mathbf{x}_k)$. This set of equations may be written in a compact matrix form as

$$\frac{\partial J}{\partial \boldsymbol{\alpha}} = \sum_{k=1}^{n} \mathbf{p}_k^{\mathrm{T}}\,(\mathbf{p}_k\boldsymbol{\alpha} - \tilde{u}_k) = \mathbf{0}\ . \tag{15.4}$$

We can define the result of the sums as

$$\mathbf{H} = \sum_{k=1}^{n} \mathbf{p}_k^{\mathrm{T}}\mathbf{p}_k = \mathbf{P}^{\mathrm{T}}\mathbf{P} \quad \text{and} \quad \mathbf{g} = \sum_{k=1}^{n} \mathbf{p}_k^{\mathrm{T}}\tilde{u}_k = \mathbf{P}^{\mathrm{T}}\tilde{\mathbf{u}} \tag{15.5}$$

in which

$$\mathbf{P} = \begin{bmatrix} \mathbf{p}_1 \\ \mathbf{p}_2 \\ \vdots \\ \mathbf{p}_n \end{bmatrix} \qquad \text{and} \qquad \tilde{\mathbf{u}} = \begin{Bmatrix} \tilde{u}_1 \\ \tilde{u}_2 \\ \vdots \\ \tilde{u}_n \end{Bmatrix}$$

The above process yields the set of linear algebraic equations

$$\mathbf{H}\boldsymbol{\alpha} = \mathbf{g}$$

which, provided $\mathbf{H}$ is non-singular, has the solution

$$\boldsymbol{\alpha} = \mathbf{H}^{-1}\mathbf{g} \tag{15.6}$$

We can now write the approximation for the function as

$$\hat{u} = \mathbf{p}(\mathbf{x})\,\mathbf{H}^{-1}\mathbf{P}^{\mathrm{T}}\tilde{\mathbf{u}} = \mathbf{N}(\mathbf{x})\tilde{\mathbf{u}}$$

where $\mathbf{N}(\mathbf{x})$ are the appropriate shape or basis functions. In general

$$\mathbf{N}_i(\mathbf{x}_j) \neq \delta_{ij}$$

as it always has been for standard finite element shape functions. However, the partition of unity [viz. Eq. (4.4)] is always preserved provided $\mathbf{p}(\mathbf{x})$ contains a constant.

Example 15.1: Fit of a linear polynomial. To make the process clear we first consider a dataset, $\tilde{u}_k$, defined at four points, $\mathbf{x_k}$, to which we desire to fit an approximation given by a linear polynomial

$$\hat{u}(\mathbf{x}) = \alpha_1 + x\alpha_2 + y\alpha_3 = \mathbf{p}(\mathbf{x})\boldsymbol{\alpha}$$

If we consider the set of data defined by

$$x_k = \begin{bmatrix}-4.0 & -1.0 & 0.0 & 6.0\end{bmatrix}$$
$$y_k = \begin{bmatrix}\ 5.0 & -5.0 & 0.0 & 3.0\end{bmatrix}$$
$$\tilde{u}_k = \begin{bmatrix}-1.5 & 5.1 & 3.5 & 4.3\end{bmatrix}$$

we can write the arrays as

$$\mathbf{P} = \begin{bmatrix} 1 & -4 & 5 \\ 1 & -1 & -5 \\ 1 & 0 & 0 \\ 1 & 6 & 3 \end{bmatrix} \quad \text{and} \quad \tilde{\mathbf{u}} = \begin{Bmatrix} -1.5 \\ 5.1 \\ 3.5 \\ 4.3 \end{Bmatrix}$$

Using Eq. (15.5) we obtain the values

$$\mathbf{H} = \mathbf{P}^{\mathrm{T}}\mathbf{P} = \begin{bmatrix} 4 & 1 & 3 \\ 1 & 53 & 3 \\ 3 & 3 & 59 \end{bmatrix} \quad \text{and} \quad \mathbf{g} = \mathbf{P}^{\mathrm{T}}\tilde{\mathbf{u}} = \begin{Bmatrix} 11.4 \\ 26.7 \\ -20.1 \end{Bmatrix}$$

which from Eq. (15.6) has the solution $\boldsymbol{\alpha} = [3.1241\,,\, 0.4745\,,\, -0.5237]^{\mathrm{T}}$. The least squares fit for these data points together with the difference between the data points and the values of the fit at $\mathbf{x}_k$ is given in Table 15.1.

15.2.2 Weighted least squares fit

Let us now assume that the point at the origin, $\mathbf{x}_0 = \mathbf{0}$ ($k = 3$ of Example 15.1), is the point about which we are making the expansion and, therefore, the one where we would like to have the best accuracy. Based on the linear approximation above we observe that the direct least squares fit yields at the point in question the *largest* discrepancy. In order to improve the fit we can modify our least squares fit for weighting the data in a way that emphasizes the effect of distance from a chosen point. We can write such a *weighted least squares fit* as the minimization of

$$J = \tfrac{1}{2} \sum_{k=1}^{n} w(\mathbf{x}_k - \mathbf{x}_0)\,(\hat{u}(\mathbf{x}_k) - \tilde{u}_k)^2 = \min \tag{15.7}$$

Table 15.1 Data and least squares fit for Example 15.1

k	1	2	3	4
x_k	−4	−1	0	6
y_k	5	−5	0	3
$\tilde{u}_k$	−1.500	5.100	3.500	4.300
$\hat{u}_k$	−1.392	5.268	3.124	4.400
Difference	−0.108	−0.168	0.376	−0.100

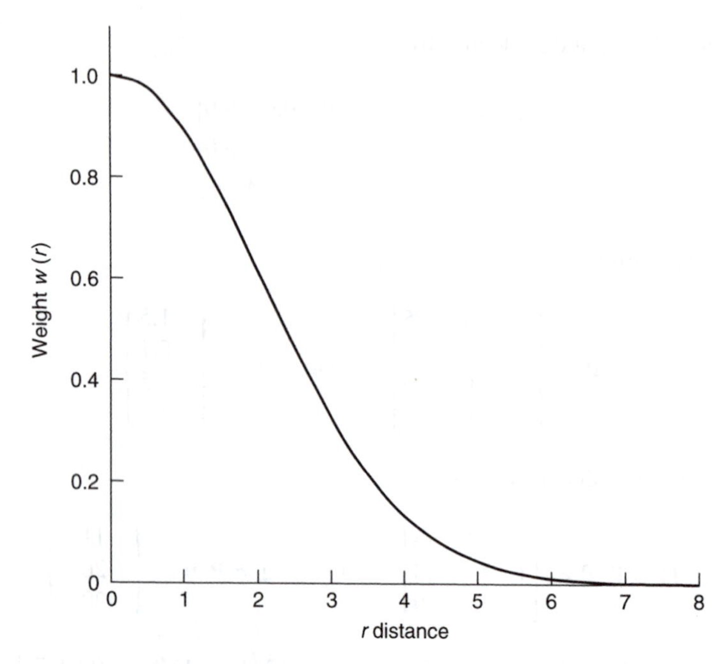

Fig. 15.2 Weighting function for Eq. (15.8): $c = 0.125$.

where w is the weighting function. Many choices may be made for the shape of the function w. If we assume that the weight function depends on a radial distance, r, from the chosen point we have

$$w = w(r); \qquad r^2 = (\mathbf{x} - \mathbf{x}_0) \cdot (\mathbf{x} - \mathbf{x}_0)$$

One functional form for $w(r)$ is the exponential Gauss function:

$$w(r) = \exp(-cr^2); \qquad c > 0 \text{ and } r \geq 0 \tag{15.8}$$

For $c = 0.125$ this function has the shape shown in Fig. 15.2 and when used with the previously given four data points yields the linear fit shown in Table 15.2.

15.2.3 Interpolation domains and shape functions

In what follows we shall invariably use the least squares procedure to interpolate the unknown function in the vicinity of a particular node i. The first problem is that when approximating to the function it is necessary to include a number of nodes equal at least

Table 15.2 Difference between weighted least squares fit and data

x_k	−4	−1	0	6
y_k	5	−5	0	3
$\tilde{u}_k$	−1.500	5.100	3.500	4.300
$\hat{u}_k$	−0.880	5.247	3.487	5.246
Error	−0.620	−0.147	0.013	−0.946

to the number of parameters of α sought to represent a given polynomial. This number, for instance, in two dimensions is three for linear polynomials and six for quadratic ones. As always the number of nodal points has to be greater than or equal to the bare minimum which is the number of parameters required. We should note in passing that it is always possible to develop a singularity in the equation used for solving α, i.e., Eq. (15.6) if the data points lie for instance on a straight line in two or three dimensions. However, in general we shall try to avoid such difficulties by reasonable spacing of nodes. The domain of influence can well be defined by making sure that the weighting function is limited in extent so that any point lying beyond a certain distance r_m is weighted by zero and therefore is not taken into account. Commonly used weighting functions are, for instance, in direction r, given by

$$w(r) = \begin{cases} \dfrac{\exp(-cr^2) - \exp(-cr_m^2)}{1 - \exp(-cr_m^2)}; & c > 0 \text{ and } 0 \le r \le r_m \\ 0; & r > r_m \end{cases} \tag{15.9}$$

which represents a truncated Gauss function. Another alternative is to use a Hermitian interpolation function as employed for the beam example in Sec. 2.9:

$$w(r) = \begin{cases} 1 - 3\left(\dfrac{r}{r_m}\right)^2 + 2\left(\dfrac{r}{r_m}\right)^3; & 0 \le r \le r_m \\ 0; & r > r_m \end{cases} \tag{15.10}$$

or alternatively the function

$$w(r) = \begin{cases} \left[1 - \left(\dfrac{r}{r_m}\right)^2\right]^n; & 0 \le r \le r_m \text{ and } n \ge 2 \\ 0; & r > r_m \end{cases} \tag{15.11}$$

is simple and has been effectively used. For circular domains, or spherical ones in three dimensions, a simple limitation of r_m suffices as shown in Fig. 15.3(a). However, occasionally use of rectangular or hexahedral subdomains is useful as also shown in that figure and now of course the weighting function takes on a different form:

$$w(x, y) = \begin{cases} X_i(x)Y_j(y); & 0 \le x \le x_m; \quad 0 \le y \le y_m; \quad \text{and } i, j \ge 2 \\ 0; & x > x_m, \, y > y_m \end{cases} \tag{15.12}$$

with

$$X_i(x) = \left[1 - \left(\frac{x}{x_m}\right)^2\right]^i; \qquad Y_j(y) = \left[1 - \left(\frac{y}{y_m}\right)^2\right]^j$$

The above two possibilities are shown in Fig. 15.3. Extensions to three dimensions using these methods is straightforward.

Clearly the domains defined by the weighting functions will overlap and it is necessary if any of the integral procedures are used such as the Galerkin method to avoid such an overlap by defining the areas of integration. We have suggested a couple of possible ideas in Fig. 15.1 but other limitations are clearly possible. In Fig. 15.4, we show an approximation to a series of points sampled in one dimension. The weighting function here always

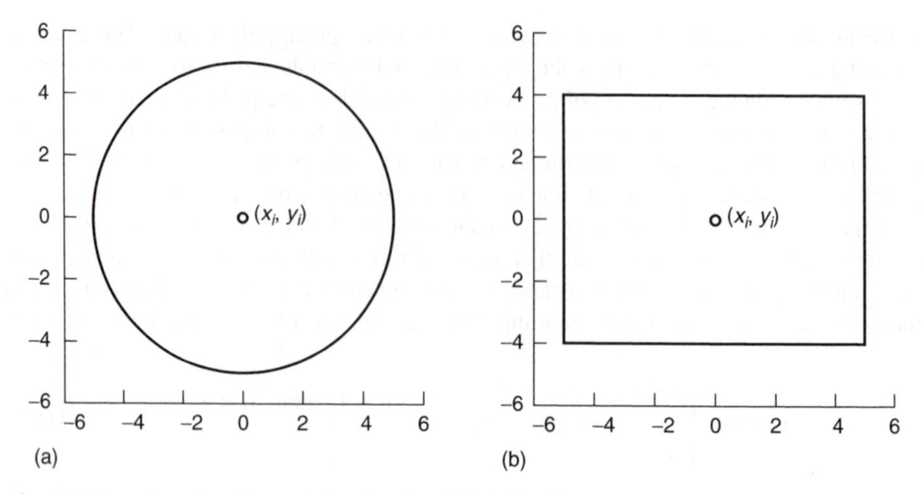

Fig. 15.3 Two-dimensional interpolation domains. (a) Circular. (b) Rectangular.

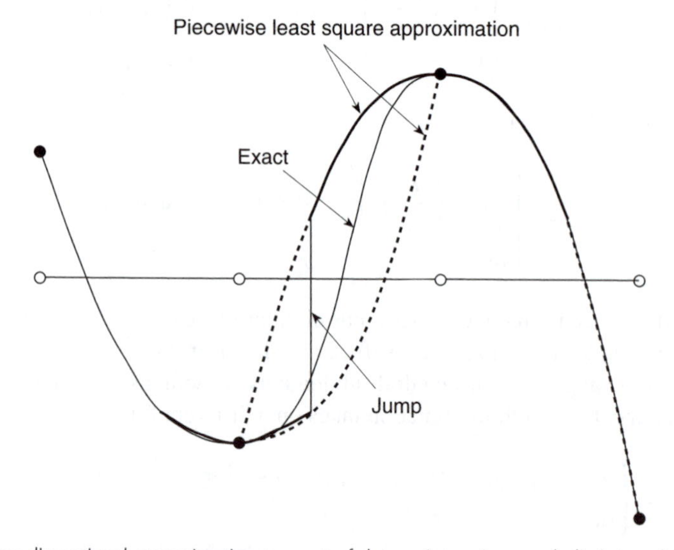

Fig. 15.4 A one-dimensional approximation to a set of data points using parabolic interpolation and direct least squares fit to adjacent points.

embraces three or four nodes. Limiting, however, the domains of their validity to a distance which is close to each of the points provides a unique definition of interpolation. The reader will observe that this interpolation is discontinuous. We have already pointed out such a discontinuity in Chapter 3, but if strictly finite difference approximations are used this does not matter. It can, however, have serious consequences if integral procedures are used and for this reason it is convenient to introduce a modification to the definition of weighting and method of calculation of the shape function which is given in the next section.

15.3 Moving least squares approximations – restoration of continuity of approximation

The method of moving least squares was introduced in the late 1960s by Shepard[8] as a means of generating a smooth surface interpolating between various specified point values. The procedure was later extended for the same reasons by Lancaster and Salkauskas[9, 10] to deal with very general surface generation problems but again it was not at that time considered of importance in finite elements. Clearly in the present context the method of moving least squares could be used to replace the local least squares we have so far considered and make the approximation fully continuous.

In the moving least squares methods, the weighted least squares approximation is applied in exactly the same manner as we have discussed in the preceding section but is established for every point at which the interpolation is to be evaluated. The result of course completely smooths the weighting functions used and it also presents smooth derivatives noting of course that such derivatives will depend on the locally specified polynomial.

To describe the method, we again consider the problem of fitting an approximation to a set of data items $\tilde{u}_i$, $i = 1, \ldots, n$ defined at the n points $\mathbf{x}_i$. We again assume the approximating function is described by the relation

$$u(\mathbf{x}) \approx \hat{u}(\mathbf{x}) = \sum_{j=1}^{m} p_j(\mathbf{x})\alpha_j = \mathbf{p}(\mathbf{x})\boldsymbol{\alpha} \tag{15.13}$$

where p_j are a set of linearly independent (polynomial) functions and α_j are unknown quantities to be determined by the fit algorithm. A generalization to the weighted least squares fit given by Eq. (15.7) may be defined for each point $\mathbf{x}$ in the domain by solving the problem

$$J(\mathbf{x}) = \tfrac{1}{2}\sum_{k=1}^{n} w_x(\mathbf{x}_k - \mathbf{x})[\tilde{u}_k - \mathbf{p}(\mathbf{x}_k)\boldsymbol{\alpha}]^2 = \min \tag{15.14}$$

In this form the weighting function is defined for *every* point in the domain and thus can be considered as translating or *moving* as shown in Fig. 15.5. This produces a *continuous* interpolation throughout the whole domain.

Figure 15.6 illustrates the problem previously presented in Fig. 15.4 now showing continuous interpolation. We should note that it is now no longer necessary to specify 'domains of influence' as the shape functions are defined in the whole domain.

The main difficulty with this form is the generation of a *moving* weight function which can change size continuously to match any given distribution of points $\mathbf{x}_k$ with a limited number of points entering each calculation. One expedient method to accomplish this is to assume the function is *symmetric* so that

$$w_x(\mathbf{x}_k - \mathbf{x}) = w_x(\mathbf{x} - \mathbf{x}_k)$$

and use a weighting function associated with each data point $\mathbf{x}_k$ as

$$w_x(\mathbf{x}_k - \mathbf{x}) = w_k(\mathbf{x} - \mathbf{x}_k)$$

The function to be minimized now becomes

$$J(\mathbf{x}) = \tfrac{1}{2}\sum_{k=1}^{n} w_k(\mathbf{x} - \mathbf{x}_k)[\tilde{u}_k - \mathbf{p}(\mathbf{x}_k)\boldsymbol{\alpha}]^2 = \min \tag{15.15}$$

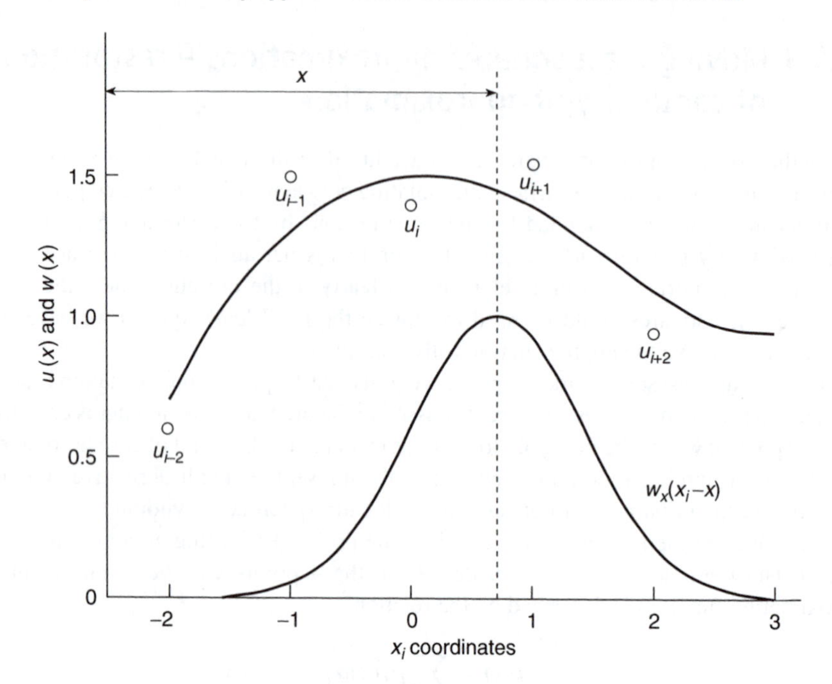

Fig. 15.5 Moving weighting function approximation in MLS.

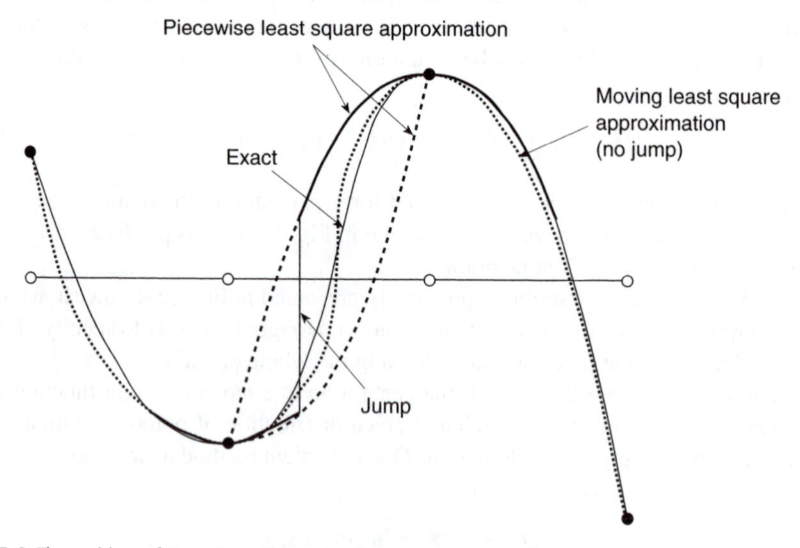

Fig. 15.6 The problem of Fig. 15.4 with moving least squares interpolation.

In this form the weighting function is *fixed* at a data point $\mathbf{x}_k$ and evaluated at the point $\mathbf{x}$ as shown in Fig. 15.7. Each weighting function may be defined such that

$$w_x(r) = \begin{cases} f_k(r), & \text{if } |r| \leq r_k \\ 0, & \text{otherwise} \end{cases} \tag{15.16}$$

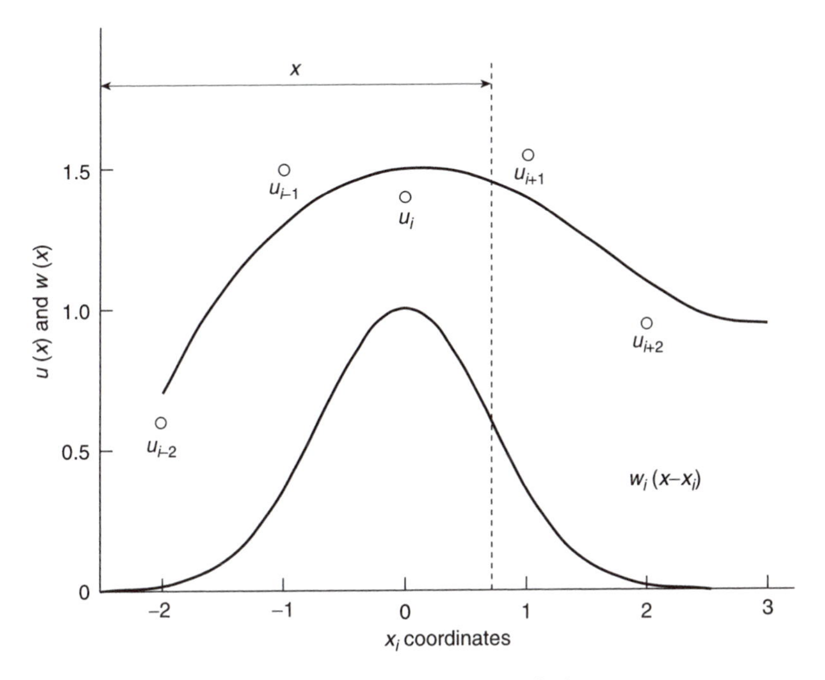

Fig. 15.7 A 'fixed' weighting function approximation to the MLS method.

and the terms in the sum are zero whenever $r^2 = (\mathbf{x} - \mathbf{x}_k)^{\mathrm{T}}(\mathbf{x} - \mathbf{x}_k)$ and $|r| > r_k$. The parameter r_k defines the radius of a ball around each point, $\mathbf{x}_k$; inside the ball the weighting function is non-zero while outside the radius it is zero. Each point may have a different weighting function and/or radius of the ball around its defining point. The weighting function should be defined such that it is zero on the boundary of the ball. This class of function may be denoted as $C_q^0(r_k)$, where the superscript denotes the boundary value and the subscript the highest derivative for which C_0 continuity is achieved. Other options for defining the weighting function are available as discussed in the previous section. The solution to the least squares problem now leads to

$$\alpha(\mathbf{x}) = \mathbf{H}^{-1}(\mathbf{x}) \sum_{j=1}^{n} \mathbf{g}_j(\mathbf{x})\tilde{u}_j = \mathbf{H}^{-1}(\mathbf{x})\mathbf{g}(\mathbf{x})\tilde{\mathbf{u}} \tag{15.17}$$

where

$$\mathbf{H}(\mathbf{x}) = \sum_{k=1}^{n} w_k(\mathbf{x} - \mathbf{x}_k)\mathbf{p}(\mathbf{x}_k)^{\mathrm{T}}\mathbf{p}(\mathbf{x}_k) \tag{15.18a}$$

and

$$\mathbf{g}_j(\mathbf{x}) = w_j(\mathbf{x} - \mathbf{x}_j)\mathbf{p}(\mathbf{x}_j)^{\mathrm{T}} \tag{15.18b}$$

In matrix form the arrays $\mathbf{H}(\mathbf{x})$ and $\mathbf{g}(\mathbf{x})$ may be written as

$$\begin{aligned} \mathbf{H}(\mathbf{x}) &= \mathbf{P}^{\mathrm{T}}\mathbf{w}(\Delta\mathbf{x})\mathbf{P} \\ \mathbf{g}(\mathbf{x}) &= \mathbf{P}^{\mathrm{T}}\mathbf{w}(\Delta\mathbf{x}) \end{aligned} \tag{15.19}$$

in which

$$\mathbf{w}(\Delta\mathbf{x}) = \begin{bmatrix} w_1(\mathbf{x}-\mathbf{x}_1) & 0 & \cdots & \cdots \\ 0 & w_2(\mathbf{x}-\mathbf{x}_2) & 0 & \cdots \\ \vdots & \vdots & \ddots & \vdots \\ \cdots & \cdots & 0 & w_n(\mathbf{x}-\mathbf{x}_n) \end{bmatrix} \tag{15.20}$$

The moving least squares algorithm produces solutions for $\boldsymbol{\alpha}$ which depend continuously on the point selected for each fit. The approximation for the function $u(\mathbf{x})$ now may be written as

$$\hat{u}(\mathbf{x}) = \sum_{j=1}^{n} N_j(\mathbf{x})\tilde{u}_j \tag{15.21}$$

where

$$N_j(\mathbf{x}) = \mathbf{p}(\mathbf{x})\mathbf{H}^{-1}(\mathbf{x})\mathbf{g}_j(\mathbf{x}) \tag{15.22}$$

define interpolation functions for each data item $\tilde{u}_j$. We note that in general these 'shape functions' do not possess the Kronecker delta property which we noted previously for finite element methods – that is

$$N_j(\mathbf{x}_i) \neq \delta_{ji} \tag{15.23}$$

It must be emphasized that all least squares approximations generally have values at the defining points $\mathbf{x}_j$ in which

$$\tilde{u}_j \neq \hat{u}(\mathbf{x}_j) \tag{15.24}$$

i.e., the local values of the approximating function do not fit the nodal unknown values. Indeed $\hat{u}$ will be the approximation used in seeking solutions to differential equations and boundary conditions and $\tilde{u}_j$ are simply the unknown parameters defining this approximation.

The main drawback of the least squares approach is that the approximation rapidly deteriorates if the number of points used, n, largely exceeds that of the m polynomial terms in $\mathbf{p}$. This is reasonable since a least squares fit usually does not match the data points exactly.

A moving least squares interpolation as defined by Eq. (15.21) can approximate *globally* all the functions used to define $\mathbf{p}(\mathbf{x})$. To show this we consider the set of approximations

$$\mathbf{U} = \sum_{j=1}^{n} N_j(x)\tilde{\mathbf{U}}_j \tag{15.25}$$

where

$$\mathbf{U} = \begin{bmatrix}\hat{u}_1(x) & \hat{u}_2(x) & \ldots & \hat{u}_n(x)\end{bmatrix}^{\mathrm{T}} \tag{15.26a}$$

and

$$\tilde{\mathbf{U}}_j = \begin{bmatrix}\tilde{u}_{j1} & \tilde{u}_{j2} & \ldots & \tilde{u}_{jn}\end{bmatrix}^{\mathrm{T}} \tag{15.26b}$$

Next, assign to each $\tilde{u}_{jk}$ the value of the polynomial $p_k(\mathbf{x}_j)$ (i.e., the kth entry in $\mathbf{p}$) so that

$$\tilde{\mathbf{U}}_j = \mathbf{p}(x_j) \tag{15.27}$$

Using the definition of the interpolation functions given by Eqs (15.21) and (15.22) we have

$$\mathbf{U} = \sum_{j=1}^{n} N_j(\mathbf{x})\mathbf{p}(\mathbf{x}_j) = \sum_{j=1}^{n} \mathbf{p}(\mathbf{x})\mathbf{H}^{-1}(\mathbf{x})\mathbf{g}_j(\mathbf{x})\mathbf{p}(\mathbf{x}_j) \tag{15.28}$$

which after substitution of the definition of $\mathbf{g}_j(x)$ yields

$$\begin{aligned}\mathbf{U} &= \sum_{j=1}^{n} \mathbf{p}(\mathbf{x})\mathbf{H}^{-1}(\mathbf{x})w_j(\mathbf{x}-\mathbf{x}_j)\mathbf{p}(\mathbf{x}_j)^{\mathrm{T}}\mathbf{p}(\mathbf{x}_j)\\ &= \mathbf{p}(\mathbf{x})\mathbf{H}^{-1}\sum_{j=1}^{n} w_j(\mathbf{x}-\mathbf{x}_j)\mathbf{p}(\mathbf{x}_j)^{\mathrm{T}}\mathbf{p}(\mathbf{x}_j)\\ &= \mathbf{p}(\mathbf{x})\mathbf{H}^{-1}\mathbf{H}(\mathbf{x}) = \mathbf{p}(\mathbf{x})\end{aligned}$$

Equation (15.29) shows that a moving least squares form can exactly interpolate *any function* included as part of the definition of $\mathbf{p}(\mathbf{x})$. If polynomials are used to define the functions, the interpolation always includes exact representations for each included polynomial. Inclusion of the zero order polynomial (i.e., 1) implies that

$$\sum_{j=1}^{n} N_j(\mathbf{x}) = 1 \tag{15.29}$$

This is called a *partition of unity* (provided it is true for all points, $\mathbf{x}$, in the domain).[35] It is easy to recognize that this is the same requirement as applies to standard finite element shape functions.

Derivatives of moving least squares interpolation functions may be constructed from the representation

$$N_j(\mathbf{x}) = \mathbf{p}(\mathbf{x})\mathbf{v}_j(\mathbf{x}) \tag{15.30}$$

where

$$\mathbf{H}(\mathbf{x})\mathbf{v}_j(\mathbf{x}) = \mathbf{g}_j(\mathbf{x}) \tag{15.31}$$

For example, the first derivatives with respect to x is given by

$$\frac{\partial N_j}{\partial x} = \frac{\partial \mathbf{p}}{\partial x}\mathbf{v}_j + \mathbf{p}\frac{\partial \mathbf{v}_j}{\partial x} \tag{15.32a}$$

and

$$\mathbf{H}\frac{\partial \mathbf{v}_j}{\partial x} + \frac{\partial \mathbf{H}}{\partial x}\mathbf{v}_j = \frac{\partial \mathbf{g}_j}{\partial x} \tag{15.32b}$$

where

$$\frac{\partial \mathbf{H}}{\partial x} = \sum_{k=1}^{n} \frac{\partial w_k(\mathbf{x}-\mathbf{x}_k)}{\partial x}\mathbf{p}(\mathbf{x}_k)^{\mathrm{T}}\mathbf{p}(\mathbf{x}_k) \tag{15.33a}$$

and

$$\frac{\partial \mathbf{g}_j}{\partial x} = \frac{\partial w_j(\mathbf{x}-\mathbf{x}_j)}{\partial x}\mathbf{p}(\mathbf{x}_j) \tag{15.33b}$$

Higher derivatives may be computed by repeating the above process to define the higher derivatives of $\mathbf{v}_j$. An important finding from higher derivatives is the order at which the interpolation becomes discontinuous between the interpolation subdomains. This will be controlled by the continuity of the weight function only. For weight functions which are C_q^0 continuous in each subdomain the interpolation will be continuous for all derivatives up to order q. For the truncated Gauss function given by Eq. (15.9) only the approximated function will be continuous in the domain, no matter how high the order used for the $\mathbf{p}$ basis functions. On the other hand, use of the Hermitian interpolation given by Eq. (15.10) produces C_1 continuous interpolation and use of Eq. (15.11) produces C_n continuous interpolation. This generality can be utilized to construct approximations for high order differential equations.

Nayroles *et al.* suggest that approximations ignoring the derivatives of $\boldsymbol{\alpha}$ may be used to define the derivatives of the interpolation functions.[11–13] While this approximation simplifies the construction of derivatives as it is no longer necessary to compute the derivatives for $\mathbf{H}$ and $\mathbf{g}_j$, there is little additional effort required to compute the derivatives of the weighting function. Furthermore, for a constant in $\mathbf{p}$ no derivatives are available. Consequently, there is little to recommend the use of this approximation.

15.4 Hierarchical enhancement of moving least squares expansions

The moving least squares approximation of the function $u(\mathbf{x})$ was given in the previous section as

$$\hat{u}(\mathbf{x}) = \sum_{j=1}^{n} N_j(\mathbf{x})\tilde{u}_j \tag{15.34}$$

where $N_j(\mathbf{x})$ defined the interpolation or shape functions based on linearly independent functions prescribed by $\mathbf{p}(\mathbf{x})$ as given by Eq. (15.22). Here we shall restrict attention to one-dimensional forms and employ polynomial functions to describe $\mathbf{p}(x)$ up to degree k. Accordingly, we have

$$\mathbf{p}(x) = \begin{bmatrix} 1 & x & x^2 & \dots & x^k \end{bmatrix} \tag{15.35}$$

For this case we will denote the resulting interpolation functions using the notation $N_j^k(x)$, where j is associated with the location of the point where the parameter $\tilde{u}_j$ is given and k denotes the order of the polynomial approximating functions. Duarte and Oden suggest using Legendre polynomials instead of the form given above;[16] however, conceptually the two are equivalent and we use the above form for simplicity. A hierarchical construction based on $N_j^k(x)$ can be established which increases the order of the complete polynomial to degree p. The hierarchical interpolation is written as

$$\begin{aligned}\hat{u}(x) &= \sum_{j=1}^{n} \left(N_j^k(x)\tilde{u}_j + N_j^k(x) \begin{bmatrix} x^{k+1} & x^{k+2} & \dots & x^p \end{bmatrix} \begin{Bmatrix} \tilde{b}_{j1} \\ \tilde{b}_{j2} \\ \vdots \\ \tilde{b}_{jq} \end{Bmatrix} \right) \\ &= \sum_{j=1}^{n} N_j^k(x) \left(\tilde{u}_j + \mathbf{q}(x)\tilde{\mathbf{b}}_j \right) = \sum_{j=1}^{n} N_j^k(x) \begin{bmatrix} 1 & \mathbf{q}(x) \end{bmatrix} \begin{Bmatrix} \tilde{u}_j \\ \tilde{\mathbf{b}}_j \end{Bmatrix}\end{aligned}$$

where $q = p - k$ and $\tilde{b}_{jm}$, $m = 1, \dots, q$, are additional parameters for the approximation. Derivatives of the interpolation function may be constructed using the method described by Eqs (15.30)–(15.33b).

The advantage of the above method lies in the reduced cost of computing the interpolation function $N_j^k(x)$ compared to that required to compute the p-order interpolations $N_j^p(x)$.

Shepard interpolation

For example, use of the functions $N_j^0(x)$, which are called Shepard interpolations,[8] leads to a scalar matrix $\mathbf{H}$ which is trivial to invert to define the N_j^0. Specifically, the Shepard interpolations are

$$N_j^0(x) = H^{-1}(x) g_j(x) \tag{15.36}$$

where

$$H(x) = \sum_{k=1}^{n} w_k(x - x_k) \tag{15.37a}$$

and

$$g_j(x) = w_j(x - x_j) \tag{15.37b}$$

The fact that the hierarchical interpolations include polynomials up to order p is easy to demonstrate. Based on previous results from standard moving least squares the interpolation with $\tilde{\mathbf{b}}_j = \mathbf{0}$ contains all the polynomials up to degree k. Higher degree polynomials may be constructed from

$$\hat{u}(x) = \sum_{j=1}^{n} \left(N_j^k(x) \tilde{u}_j + N_j^k(x) \begin{bmatrix} x^{k+1} & x^{k+2} & \dots & x^p \end{bmatrix} \begin{Bmatrix} \tilde{b}_{j1} \\ \tilde{b}_{j2} \\ \vdots \\ \tilde{b}_{jq} \end{Bmatrix} \right) \tag{15.38}$$

by setting all $\tilde{u}_j$ to zero and for each interpolation term setting one of the $\tilde{b}_{jk}$ to unity with the remaining values set to zero. For example, setting $\tilde{b}_{j1}$ to unity results in the expansion

$$\hat{u}(x) = \sum_{j=1}^{n} N_j^k(x) x^{k+1} = x^{k+1} \tag{15.39}$$

This result requires only the partition of unity property

$$\sum_{j=1}^{n} N_j^k(x) = 1 \tag{15.40}$$

The remaining polynomials are obtained by setting the other values of $\tilde{b}_{jk}$ to unity one at a time. We note further that the same order approximation is obtained using $k = 0, 1$ or p.[16]

The above hierarchical form has parameters which do not relate to approximate values of the interpolation function. For the case where $k = 0$ (i.e., Shepard interpolation), Babuška and Melenk[36] suggest an alternate expression be used in which $\mathbf{q}$ in Eq. (15.38) is taken as $\begin{bmatrix} 1 & x & x^2 & \dots & x^p \end{bmatrix}$ and the interpolation written as

$$\hat{u}(x) = \sum_{j=1}^{n} N_j^0(x) \left(\sum_{k=0}^{p} l_k^p(x) \tilde{u}_{jk} \right) \tag{15.41}$$

In this form the $l_k^p(x)$ are Lagrange interpolation polynomials (e.g., see Sec. 4.5) and $\tilde{u}_{jk}$ are parameters with dimensions of u for the jth term at point x_k of the Lagrange interpolation. The above result follows since Lagrange interpolation polynomials have the property

$$l_k(x_i) = \delta_{ki} = \begin{cases} 1, & \text{if } k = i \\ 0, & \text{otherwise} \end{cases} \tag{15.42}$$

We should also note that options other than polynomials may be used for the $\mathbf{q}(x)$. Thus, for any function $q_i(x)$ we can set the associated $\tilde{b}_{ji}$ to unity (with all others and $\tilde{u}_j$ set to zero) and obtain

$$\hat{u}(x) = \sum_{j=1}^{n} N_j^k(x) q_i(x) = q_i(x) \tag{15.43}$$

Again the only requirement is that

$$\sum_{j=1}^{n} N_j^k(x) = 1$$

Thus, for any basic functions satisfying the partition of unity a hierarchical enrichment may be added using any type of functions. For example, if one knows that the structure of the solution involves exponential functions in x it is possible to include them as members of the $\mathbf{q}(x)$ functions and thus capture the essential part of the solution with just a few terms. This is especially important for problems which involve solutions with different length scales. A large length scale can be included in the basic functions, $N_j^k(x)$, while other smaller length scales may be included in the functions $\mathbf{q}(x)$. This will be illustrated further in Volume 3 in the chapter dealing with waves.

The above discussion has been limited to functions in one space variable; however, extensions to two and three dimensions can be easily constructed. In the process of this extension we shall encounter some difficulties which we address in more detail in the section on partition-of-unity finite element methods. Before doing this we explore in the next section the direct use of least squares methods to solve differential equations using collocation methods.

15.5 Point collocation – finite point methods

Finite difference methods based on Taylor formula expansions on regular grids can, as explained in Chapter 3, Sec. 3.14, always be considered as *point collocation methods* applied to the differential equation. They have been used to solve partial differential equations for many decades.[37–39] Classical finite difference methods commonly restrict applications to regular grids. This limits their use in obtaining accurate solutions to general engineering problems which have curved (irregular) boundaries and/or multiple material interfaces. To overcome the boundary approximation and interface problem curvilinear mapping may be used to define the finite difference operators.[40]

The extension of the finite difference methods from regular grids to general arbitrary and irregular grids or sets of points has received considerable attention (Girault,[1] Pavlin and Perrone,[2] Snell *et al.*[3]). An excellent summary of the current state of the art may be found in a recent paper by Orkisz[40] who himself has contributed very much to the subject since the late 1970s (Liszka and Orkisz[4]).

More recently such finite difference approximations on irregular grids have been proposed by Batina[41] in the context of aerodynamics and by Oñate *et al.*[42–44] who introduced the name 'finite point method'. Here both elasticity and fluid mechanics problems have been addressed.

In point collocation methods the set of differential equations, which here is taken in the form described in Sec. 3.1, is used directly without the need to construct a weak form or perform domain integrals. Accordingly, we consider

$$\mathcal{A}(\mathbf{u}) = \mathbf{0} \tag{15.44a}$$

as a set of governing differential equations in a domain Ω subject to boundary conditions

$$\mathcal{B}(\mathbf{u}) = \mathbf{0} \tag{15.44b}$$

applied on the boundaries Γ. An approximation to the dependent variable $\mathbf{u}$ may be constructed using either a weighted or moving least squares approximation since at each collocation point the methods become identical. In this we must first describe the (collocation) *points* and the *weighting function*. The approximation is then constructed from Eq. (15.21) by assuming a sufficient order polynomial for $\mathbf{p}$ in Eq. (15.13) such that *all* derivatives appearing in Eqs (15.44a) and (15.44b) may be computed. Generally, it is advantageous to use the same order of interpolation to approximate both the differential and boundary conditions.[40] The resulting discrete form for the differential equations at each collocation point becomes

$$\mathcal{A}(\mathbf{N}(\mathbf{x}_i)\tilde{\mathbf{u}}_i) = \mathbf{0}; \qquad i = 1, 2, \ldots, n_e \tag{15.45a}$$

and the discrete form for each boundary condition is

$$\mathcal{B}(\mathbf{N}(\mathbf{x}_i)\tilde{\mathbf{u}}_i) = \mathbf{0}; \qquad i = 1, 2, \ldots, n_b \tag{15.45b}$$

The total number of equations must equal the number of collocation points selected. Accordingly,

$$n_e + n_b = n \tag{15.46}$$

It would appear that little difference will exist between continuous approximations involving moving least squares and discontinuous ones as in both locally the same polynomial will be used. This may well account for the convergence of standard least squares approximations which we have observed in Chapter 3 for discontinuous least squares forms but in view of our previous remarks about differentiation, a slight difference will in fact exist if moving least squares are used and in the work of Onãte *et al.*[42–44] which we mentioned before such moving least squares are adopted.

In addition to the choice for $\mathbf{p}(\mathbf{x})$, a key step in the approximation is the choice of the weighting function for the least squares method and the domain over which the weighting function is applied. In the work of Orkisz[45] and Liszka[46] two methods are used:

1. A 'cross' criterion in which the domain at a point is divided into quadrants in a cartesian coordinate system originating at the 'point' where the equation is to be evaluated. The domain is selected such that each quadrant contains a fixed number of points, n_q. The product of n_q and the number of quadrants, q, must equal or exceed the number of polynomial terms in $\mathbf{p}$ less one (the central node point). An example is shown in Fig. 15.8(a) for a two-dimensional problem ($q = 4$ quadrants) and $n_q = 2$.
2. A 'Voronoi neighbour' criterion in which the closest nodes are selected as shown for a two-dimensional example in Fig. 15.8(b).

There are advantages and disadvantages to both approaches – namely, the cross criterion leads to dependence on the orientation of the global coordinate axes while the Voronoi method gives results which are sometimes too few in number to get appropriate order

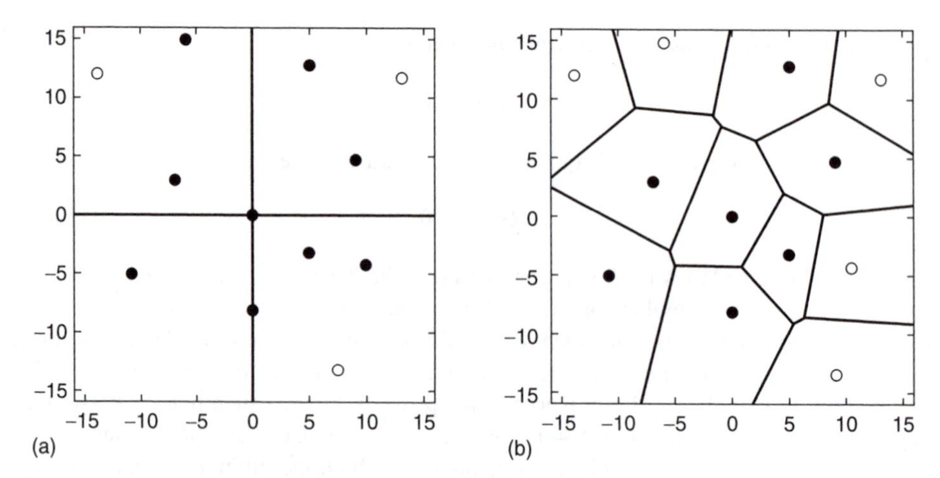

Fig. 15.8 Methods for selecting points. (a) Cross. (b) Voronoi.

approximations. The Voronoi method is, however, effective for use in Galerkin solution methods or finite volume (subdomain collocation) methods in which only first derivatives are needed.

The interested reader can consult reference 40 for examples of solutions obtained by this approach. Additional results for finite point solutions may be found in work by Onãte *et al.*[42] and Batina.[41]

One advantage of considering moving least squares approximations instead of simple fixed point weighted least squares is that approximations at points other than those used to write the differential equations and boundary conditions are also continuously available. Thus, it is possible to perform a full post-processing to obtain the contours of the solution and its derivatives.

In the next part of this section we consider the application of the moving least squares method to solve a second-order ordinary differential equation using point collocation.

Example 15.2: Collocation (point) solution of ordinary differential equations. We consider the solution of ordinary differential equations using a point collocation method. The differential equation in our examples is taken as

$$-a\frac{\mathrm{d}^2u}{\mathrm{d}x^2}+b\frac{\mathrm{d}u}{\mathrm{d}x}+cu-f(x)=0 \tag{15.47}$$

on the domain $0 < x < L$ with constant coefficients a, b, c, subject to the boundary conditions $u(0) = g_1$ and $u(L) = g_2$. The domain is divided into an equally spaced set of points located at x_i, $i = 1, \ldots, n$. The moving least squares approximation described in Sec. 15.3 is used to write difference equations at each of the interior points (i.e., $i = 2, \ldots, n-1$). The boundary conditions are also written in terms of discrete approximations using the moving least squares approximation. Accordingly, for the approximate solution using p-order polynomials to define the $\mathbf{p}(x)$ in the interpolations

$$\hat{u}(x)=\sum_{j=1}^{n}N_i^p(x)\tilde{u}_i \tag{15.48}$$

we have the set of n equations in n unknowns:

$$\sum_{i=1}^{n} N_i^p(x_1)\tilde{u}_i = g_1 \tag{15.49a}$$

$$\sum_{i=1}^{n} \left(-a\frac{\mathrm{d}^2 N_i^p}{\mathrm{d}x^2} + b\frac{\mathrm{d}N_i^p}{\mathrm{d}x} + cN_i^p \right)_{x=x_j} \tilde{u}_i - f(x_j) = 0; \qquad j = 2, \ldots, n-1 \tag{15.49b}$$

and

$$\sum_{i=1}^{n} N_i^p(x_n)u_i = g_2 \tag{15.49c}$$

The above equations may be written compactly as:

$$\mathbf{K}\tilde{\mathbf{u}} + \mathbf{f} = \mathbf{0} \tag{15.50}$$

where $\mathbf{K}$ is a square coefficient matrix, $\mathbf{f}$ is a load vector consisting of the entries from g_i and $f(x_j)$, and $\tilde{\mathbf{u}}$ is the vector of unknown parameters defining the approximate solution $\hat{u}(x)$. A unique solution to this set of equations requires $\mathbf{K}$ to be non-singular (i.e., rank$(\mathbf{K}) = n$). The rank of $\mathbf{K}$ depends both on the weighting function used to construct the least squares approximation as well as the number of functions used to define the polynomials $\mathbf{p}$. In order to keep the least squares matrices as well conditioned as possible, a different approximation is used at each node with

$$\mathbf{p}^{(j)}(x) = [1 \quad x - x_j \quad (x - x_j)^2 \quad \ldots \quad (x - x_j)^p] \tag{15.51}$$

defining the interpolations associated with $N_j^p(x)$. The matrix $\mathbf{K}$ will be of correct rank provided the weighting function can generate linearly independent equations.

The accurate approximation of second derivatives in the differential equation requires the use of quadratic or higher order polynomials in $\mathbf{p}(x)$.[40] In addition, the span of the weighting function must be sufficient to keep the least squares matrix $\mathbf{H}$ non-singular *at every collocation point*. Thus, the minimum span needed to define quadratic interpolations of $\mathbf{p}(x)$ (i.e., $p = k = 2$) must include at least three mesh points with non-zero contributions. At the problem boundaries only half of the weighting function span will be used (e.g., the right half at the left boundary). Consequently, for weighting functions which go smoothly to zero at their boundary, a span larger than four mesh spaces is required. The span should not be made too large, however, since the sparse structure of $\mathbf{K}$ will then be lost and overdiffuse solutions may result.

Use of hierarchical interpolations reduces the required span of the weighting function. For example, use of interpolations with $k = 0$ requires only a span at each point for which the domain is just covered (since any span will include its defining point, x_k, the $\mathbf{H}$ matrix will always be non-singular). For a uniformly spaced set of points this is any span greater than one mesh spacing.

For the example we use the weighting function described by Eq. (15.11) with a weight span 4.4 ($r_m = 2.2h$) times the largest adjacent mesh space for the quadratic interpolations with $k = p = 2$ and a weight 2.01 times the mesh space for the hierarchical quadratic interpolations with $k = 0$, $p = 2$.

We consider the example of a string on an elastic foundation with the differential equation

$$-a\frac{\mathrm{d}^2 u}{\mathrm{d}x^2} + cu + f = 0; \qquad 0 < x < 1 \tag{15.52}$$

with the boundary conditions $u(0) = u(1) = 0$. This is a special form of Eq. (15.47). The parameters for solution are selected as

$$a = 0.01 \qquad c = 1 \qquad f = -1$$

The exact solution is given by

$$u(x) = 1 - \cosh(mx) - (1 - \cosh(m))\frac{\sinh(mx)}{\sinh(m)}, \qquad m = \left(\frac{c}{a}\right)^{1/2}$$

The problem is solved using 27 points and $k = p = 2$ producing the results shown in Fig. 15.9.

The process was repeated using the hierarchical interpolations with $k = 0$ and $p = 2$ using nine points (which results in 27 parameters, the same as for the first case). The results are shown in Fig. 15.10.

The hierarchical interpolation permits the solution to be obtained using as few as two points. A solution with two points and interpolations with $k = 0$ and $p = 3$ and 5 is shown in Figs 15.11 and 15.12, respectively. Note, however, that with the hierarchical form additional collocation points have to be introduced to achieve a sufficient number of equations. We show such collocation points in Fig. 15.10.

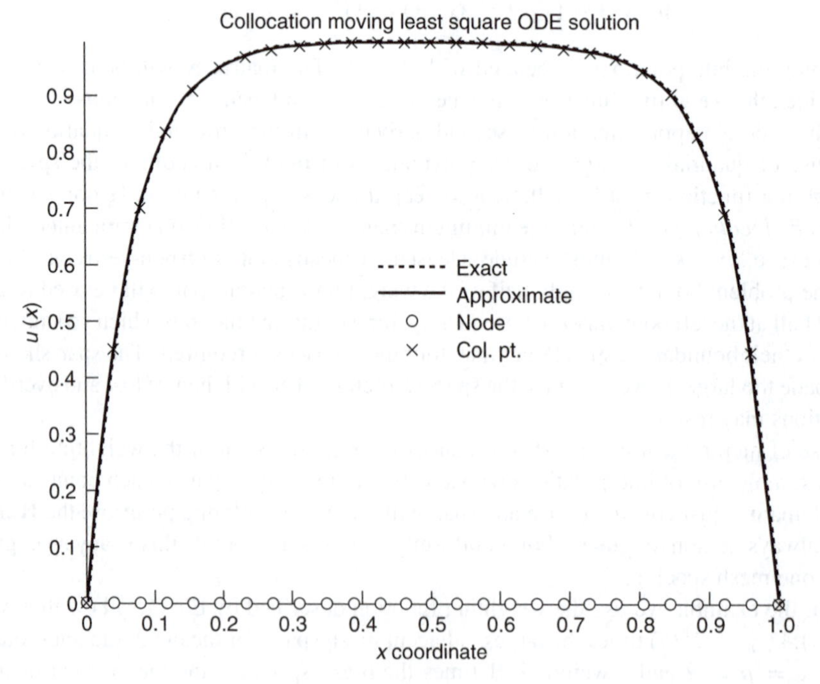

Fig. 15.9 String on elastic foundation solution using MLS form based on nodes: 27 points, $k = 2, p = 2$.

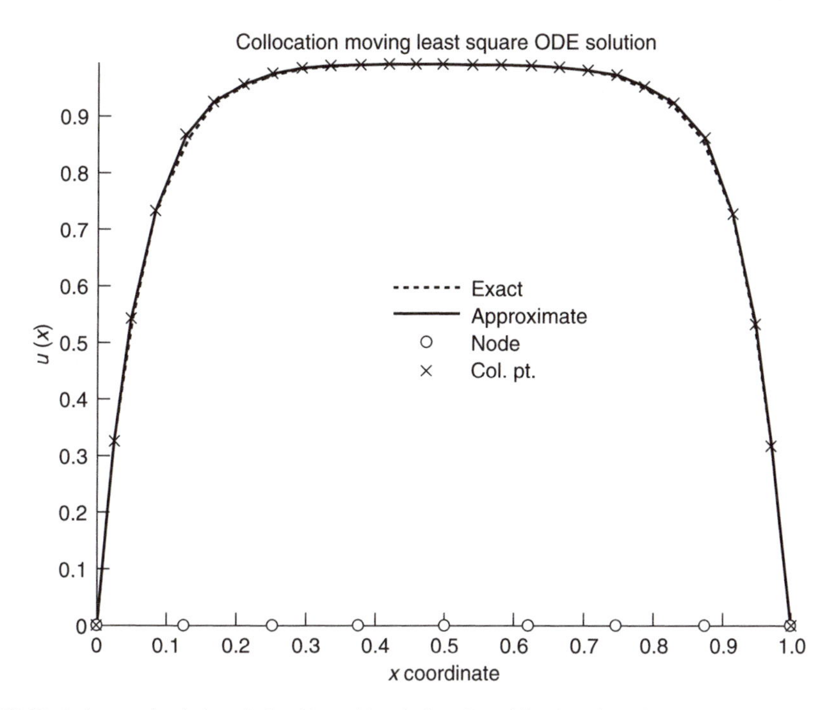

Fig. 15.10 String on elastic foundation hierarchic solution: 9 nodal points, $k = 0, p = 2$.

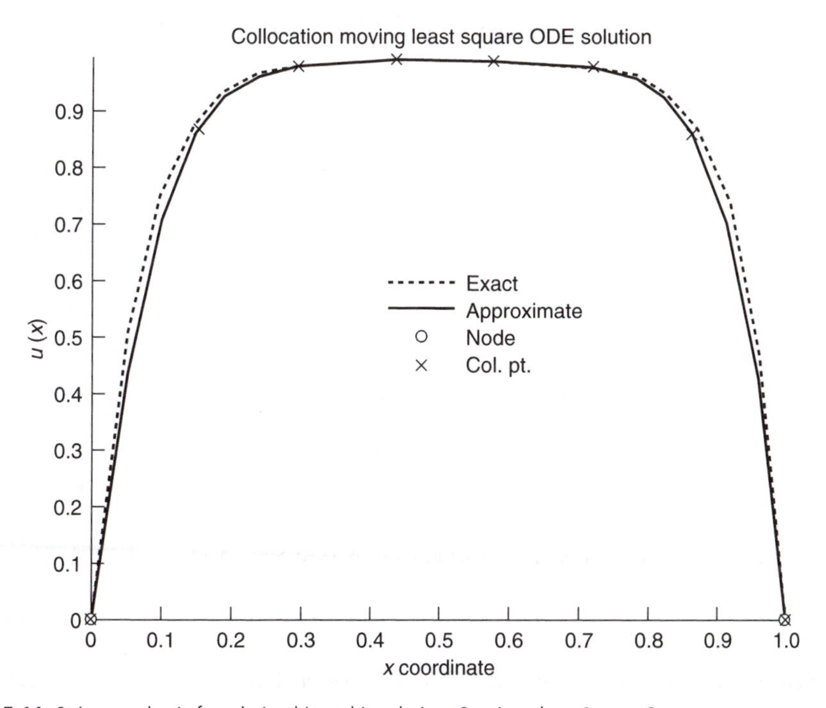

Fig. 15.11 String on elastic foundation hierarchic solution: 2 points, $k = 0, p = 3$.

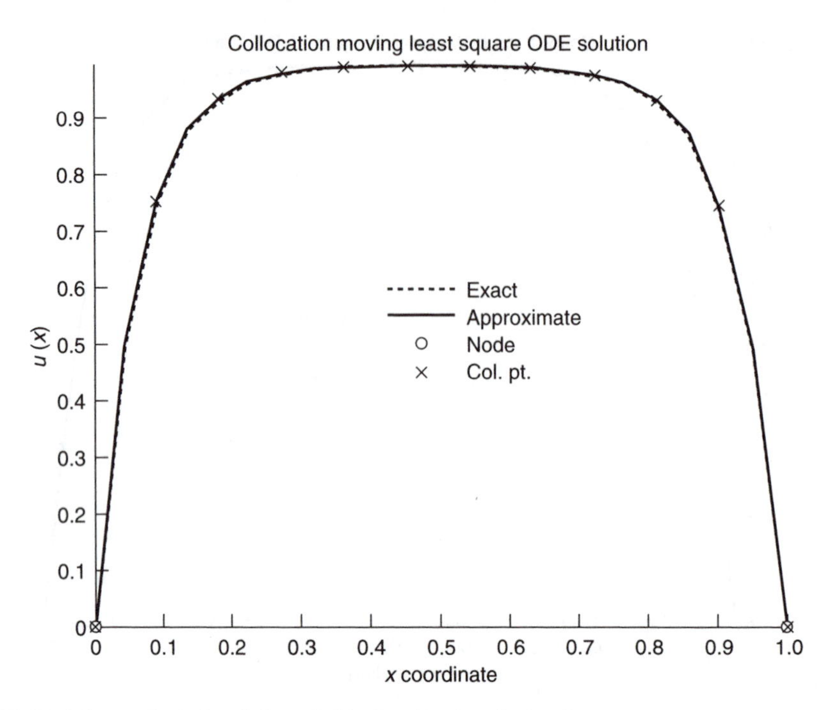

Fig. 15.12 String on elastic foundation solution: 2 points, $k = 0, p = 5$.

15.6 Galerkin weighting and finite volume methods

15.6.1 Introduction

Point collocation methods are straightforward and quite easy to implement, the main task being only the selection of the subdomain on which to perform the fit of the function from which the derivatives are computed. Disadvantages arise, however, in the need to use high order interpolations such that accurate derivatives of the order of the differential equation may be computed. Further the treatment of boundaries and material interfaces present difficulties.

An alternative, as we have discussed in Chapter 3, is the use of 'weak' or 'variational' forms which are equivalent to the differential equation. Approximations then require functions which have lower order than in the differential equation. In addition, boundary conditions often appear as 'natural' conditions in the weak form – especially for flux (derivative or Neumann)-type boundary conditions. This advantage now is balanced by a need to perform integration over the whole domain.

Here, we consider problems of the form given by (see Sec. 3.2)†

$$\int_{\Omega} \mathcal{C}(\mathbf{v})^{\mathrm{T}} \mathcal{D}(\mathbf{u})\, \mathrm{d}\Omega + \int_{\Gamma} \mathcal{E}(\mathbf{v})^{\mathrm{T}} \mathcal{F}(\mathbf{u})\, \mathrm{d}\Gamma = 0 \tag{15.53}$$

in which the operators $\mathcal{C}$, $\mathcal{D}$, $\mathcal{E}$ and $\mathcal{F}$ contain lower derivatives than those occurring in operators $\mathcal{A}$ and $\mathcal{B}$ given in Eqs (15.44a) and (15.44b), respectively. For example,

† We assume that the boundary terms are described such that $\bar{\mathbf{v}} = \mathbf{v}$.

the solution of second order differential equations (such as those occurring in the quasi-harmonic or linear elasticity equation) have differential operators for $\mathcal{C}$ to $\mathcal{F}$ with derivatives no higher than first order.

The approximate solution to forms given by Eq. (15.53) may be achieved using moving least squares and alternative methods for performing the domain integrals.

15.6.2 Subdomain collocation – finite volume method

A simple extension of the point collocation method is to use subdomains (elements) defined by the Voronoi neighbour criterion. The integrals for each subdomain are approximated as a *constant* evaluated at the originating point as

$$\sum_{i}^{n_d} \mathcal{C}(\mathbf{v}_i)^{\mathrm{T}} \mathcal{D}(\mathbf{u}_i) \Omega_i + \sum_{i}^{n_b} \mathcal{E}(\mathbf{v}_i)^{\mathrm{T}} \mathcal{F}(\mathbf{u}_i) \Gamma_i = 0 \tag{15.54}$$

where $n_d + n_b = n$, the total number of unknown parameters appearing in the approximations of $\mathbf{u}$ and $\mathbf{v}$.

The validity of the above approximation form can be established using patch tests (see Chapter 9). This approach is often called *subdomain collocation* or the *finite volume* method. This approach has been used extensively in constructing approximations for fluid flow problems.[47–53] It has also been employed with some success in the solution of problems in structural mechanics.[54]

15.6.3 Galerkin methods – diffuse elements

Moving least squares approximations have been used with weak forms to construct Galerkin-type approximations. The origin of this approach can be traced to the work of Liszka[46] and Orkisz.[40] Additional work, originally called the diffuse element approximation, was presented in the early 1990s by Nayroles *et al.*[11–13] Beginning in the mid-1990s the method has been extensively developed and improved by Belytschko and coauthors under the name *element-free Galerkin.*[14, 15, 55, 56] A similar procedure, called 'hp-clouds', was also presented by Oden and Duarte.[16, 17, 57] Each of the methods is also said to be 'meshless'; however, in order to implement a true Galerkin process it is necessary to carry out integrations over the domain. What distinguishes each of the above processes is the manner in which these integrations are carried out. In the element-free Galerkin method a background 'grid' is often used to define the integrals whereas in the hp-cloud method circular subdomains are employed. Differing weights are also used as a means of generating the moving least squares approximation. The interested reader is referred to the appropriate literature for more details. Another source to consult for implementation of the EFG method is reference 19. Here we present only a simple implementation for solution of an ordinary differential equation.

Example 15.3: Galerkin solution of ordinary differential equations. The moving least squares approximation described in Sec. 15.3 is now used as a Galerkin method to solve a second order ordinary differential equation. For an arbitrary function $W(x)$ satisfying

$W(0) = W(L) = 0$, a weak form for the differential equation may be deduced using the procedures presented in Chapter 3. Accordingly, we obtain

$$\int_0^L \left[\frac{dW}{dx} a \frac{du}{dx} + W \left(b \frac{du}{dx} + cu - f(x) \right) \right] dx = 0 \qquad (15.55)$$

subject to the boundary conditions $u(0) = g_1$ and $u(L) = g_2$. Using a *hierarchical moving least squares form* a p-order polynomial approximation to the dependent variable may be written as

$$\hat{u}(x) = \sum_{j=1}^{n} N_j^0(x) \bar{\mathbf{q}}_{jp}(x) \tilde{\mathbf{u}}_j^p \qquad (15.56)$$

where

$$\bar{\mathbf{q}}_{jp} = [1 \quad x - x_j \quad (x - x_j)^2 \quad \dots \quad (x - x_j)^p] \qquad (15.57)$$

Note that in the above form we have used the representation

$$\bar{\mathbf{q}}_{jp}(x) \tilde{\mathbf{u}}_j^p = \tilde{u}_j + \mathbf{q}(x) \tilde{\mathbf{b}}_j$$

The approximation to the weight function is similarly taken as

$$\hat{W}(x) = \sum_{j=1}^{n} N_j^0(x) \bar{\mathbf{q}}_{jp}(x) \tilde{\mathbf{W}}_j^p \qquad (15.58)$$

in which $\tilde{\mathbf{W}}_j^p$ are arbitrary parameters satisfying $W(0) = W(L) = 0$. The approximation yields the discrete problem

$$\sum_{i=1}^{n} (\tilde{\mathbf{W}}_i^p)^{\mathrm{T}} \sum_{j=1}^{n} \left\{ \int_0^L \left[\frac{d(\bar{\mathbf{q}}_{ip}^{\mathrm{T}} N_i^0)}{dx} a \frac{d(\bar{\mathbf{q}}_{jp} N_j^0)}{dx} + \bar{\mathbf{q}}_{ip}^{\mathrm{T}} N_i^0 \left(b \frac{d(\bar{\mathbf{q}}_{jp} N_j^0)}{dx} + c \bar{\mathbf{q}}_{jp} N_j^0 \right) \right] dx \right\} \tilde{\mathbf{u}}_j$$
$$= \sum_{i=1}^{n} (\tilde{\mathbf{W}}_i^p)^{\mathrm{T}} \int_0^L \bar{\mathbf{q}}_{ip}^{\mathrm{T}} N_i^0 f(x) \, dx \qquad (15.59)$$

Since $\tilde{\mathbf{W}}_i^p$ is arbitrary, the solution to the approximate weak form yields the set of equations

$$\sum_{j=1}^{n} \left\{ \int_0^L \left[\frac{d(\bar{\mathbf{q}}_{ip}^{\mathrm{T}} N_i^0)}{dx} a \frac{d(\bar{\mathbf{q}}_{jp} N_j^0)}{dx} + \bar{\mathbf{q}}_{ip}^{\mathrm{T}} N_i^0 \left(b \frac{d(\bar{\mathbf{q}}_{jp} N_j^0)}{dx} + c \bar{\mathbf{q}}_{jp} N_j^0 \right) \right] dx \right\} \tilde{\mathbf{u}}_j$$
$$= \int_0^L \bar{\mathbf{q}}_{ip}^{\mathrm{T}} N_i^0 f(x) \, dx; \qquad i = 1, 2, \dots, n \qquad (15.60)$$

The set of equations only needs to be modified to satisfy the essential boundary equations. This is accomplished by replacing the equations corresponding to $W_1 = W_n = 0$ by $\tilde{u}_1 = g_1$ and $\tilde{u}_n = g_2$.

The Galerkin form requires only first derivatives of the approximating functions as opposed to the second derivatives required for the point collocation method. This reduction, however, is accompanied by a need to perform integrals over the domain. For weighting functions given by Eq. (15.11) all functions entering the approximation are polynomial and rational polynomial expressions, thus, a closed form evaluation is impractical. Accordingly, we evaluate integrals using Gauss and Gauss–Lobatto quadrature over each *interval* generated by the basis points in the moving least squares representation (i.e., x_j for $j = 1, 2, \ldots, n$). As an example of the type of solutions possible we consider the string on elastic foundation problem given in the previous section. For the parameters $a = 0.001$, $c = 1$ with loading $f = -1$ and zero boundary conditions a Galerkin solution using 3- and 4-point Gauss quadrature and 4- and 5-point Gauss–Lobatto quadrature is shown in Figs 15.13–15.16. A mesh consisting of nine equally spaced points is used to define the intervals for the solution and quadrature. The weight function is generated for $k = 0$, $p = 2$ with a span of 2.1 mesh points.

Based upon this elementary example it is evident that the answers for a 9-point mesh depend on accurate evaluation of integrals to produce high-quality answers.

15.7 Use of hierarchic and special functions based on standard finite elements satisfying the partition of unity requirement

15.7.1 Introduction

In Sec. 15.4, we discussed the possibility of introducing hierarchical variables to shape functions based on moving least squares interpolations. However, a simpler approach to

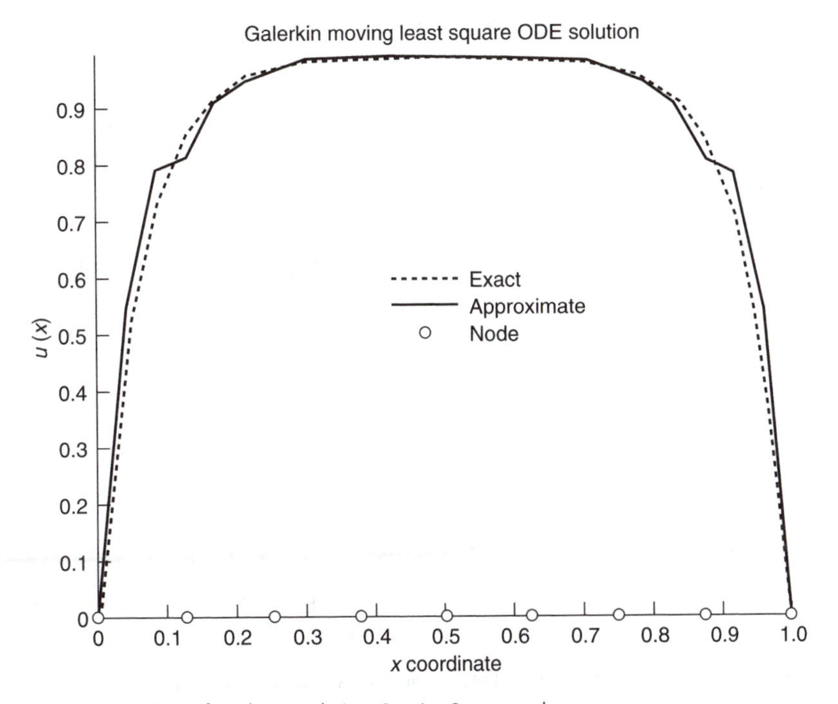

Fig. 15.13 String on elastic foundation solution: 3-point Gauss quadrature.

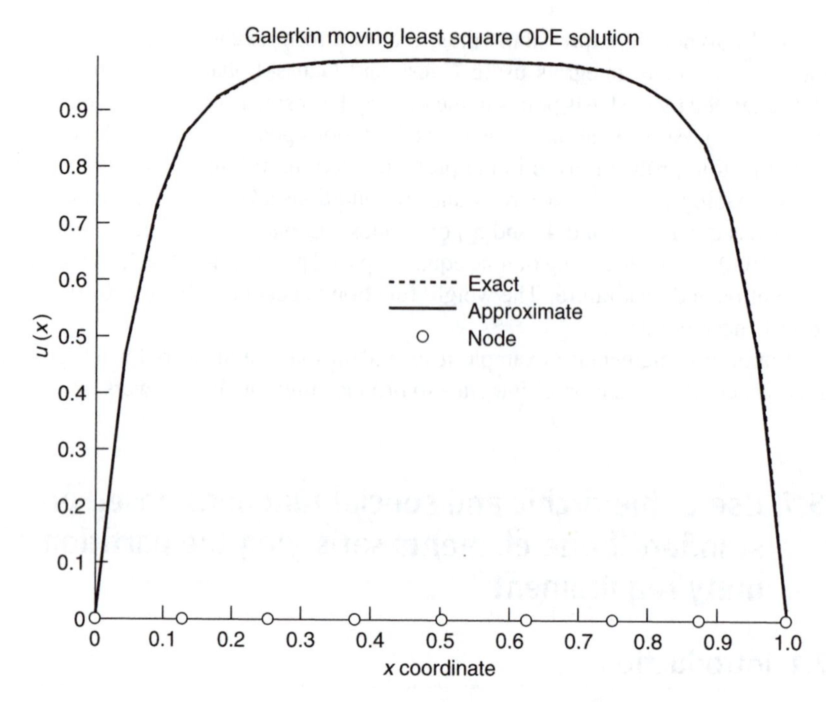

Fig. 15.14 String on elastic foundation solution: 4-point Gauss quadrature.

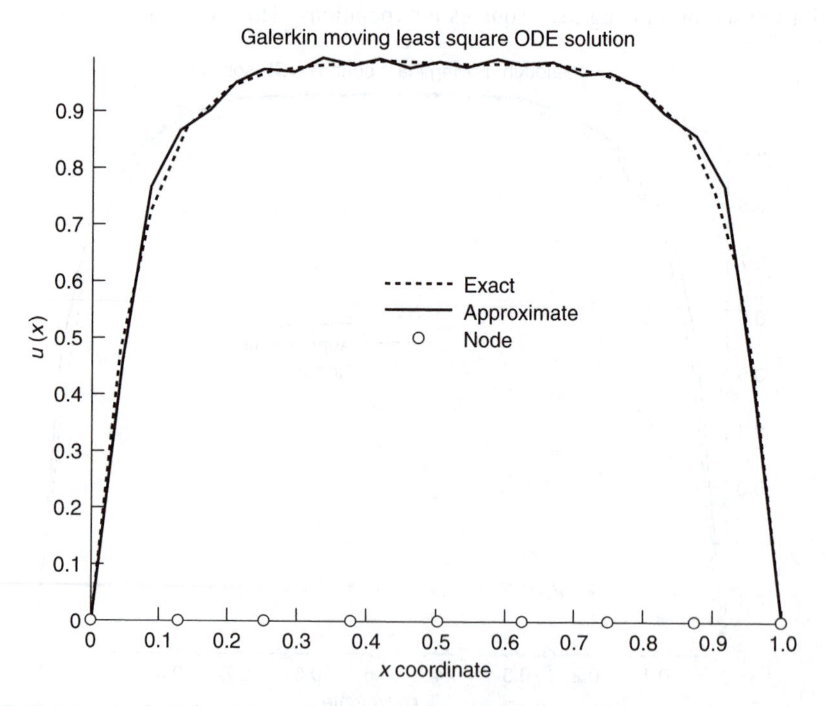

Fig. 15.15 String on elastic foundation solution: 4-point Gauss–Lobatto quadrature.

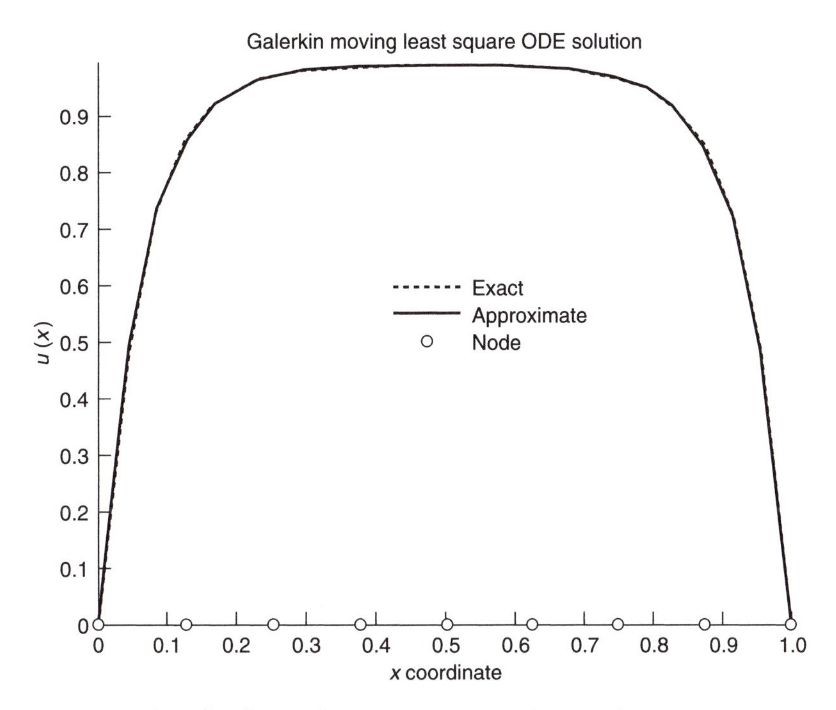

Fig. 15.16 String on elastic foundation solution: 5-point Gauss–Lobatto quadrature.

hierarchical forms and indeed to extensions by other functions can be based on simple finite element shape functions.

One important application of the partition of unity method starts from a set of finite element basis functions, $N_i(\mathbf{x})$. An approximation to $u(x)$ is now given by

$$u(\mathbf{x}) \approx \hat{u}(\mathbf{x}) = \sum_i N_i(\mathbf{x}) \left[\tilde{u}_i + \sum_\alpha q_\alpha^{(i)}(\mathbf{x}) b_{\alpha i} \right] \tag{15.61}$$

where $N_i(\mathbf{x})$ is the conventional (possibly isoparametric) finite element shape function at node i, $q_\alpha^{(i)}$ are *global functions* associated with node i, and $\tilde{u}_i$, and $b_{\alpha i}$ are parameters associated with the added global hierarchical functions. We must note that as before $\tilde{u}_i$ will not represent a local value of the function unless the function q^i become zero at the node i.

Here we assume that conventional shape functions which satisfy the partition of unity condition

$$\sum_i N_i = 1$$

are used. Thus, the above form is a *hierarchic finite element method based on the partition of unity*.[21, 58, 59] We note in particular that the function $q_\alpha^{(i)}$ may be different for each node and thus the form may be effectively used in an adaptive finite element procedure as described in Chapter 14.

Equation (15.61) provides options for a wide choice of functions for $q_\alpha^{(i)}$:

1. Polynomial functions. In this case the method becomes an alternative hierarchical scheme to that presented in Part 2 of Chapter 4.

2. Harmonic 'wave' functions. This is a multiscale method and will be discussed in detail in Volume 3.
3. Singular functions. These can be used to introduce re-entrant corner or singular load effects in elliptic problems (e.g., heat conduction or elasticity forms).

Derivatives of Eq. (15.61) are computed directly as

$$\frac{\partial \hat{u}}{\partial x_k} = \sum_i \left[\frac{\partial N_i}{\partial x_k} \tilde{u}_i + \sum_\alpha \left(\frac{\partial N_i}{\partial x_k} q_\alpha^{(i)} + N_i \frac{q_\alpha^{(i)}}{\partial x_k} \right) b_{\alpha i} \right] \tag{15.62}$$

The reader will note that the narrow band structure of the standard finite element method will always be maintained as it is determined by the connectivity of N_i. Note also that the standard element on which the shape functions N_i were generated can be used for all subsequent integrations. Such a formulation is very easy to fit into any finite element program.

15.7.2 Polynomial hierarchical method

To give more details of the above hierarchical finite element method we first consider the one-dimensional approximation in a 2-noded element where

$$\hat{u} = N_1 \left[\tilde{u}_1 + \mathbf{q}^{(1)} \mathbf{b}_1 \right] + N_2 \left[\tilde{u}_2 + \mathbf{q}^{(2)} \mathbf{b}_2 \right] \tag{15.63}$$

in which

$$N_1 = \frac{x_2 - x}{x_2 - x_1}; \qquad N_2 = \frac{x - x_1}{x_2 - x_1}$$

and

$$\begin{aligned} \mathbf{q}^{(1)} = \mathbf{q}^{(2)} &= \left[x^k, \quad x^{k+1}, \quad \cdots \right] \\ \mathbf{b}_i &= \left[b_{i1}, \quad b_{i2}, \quad \cdots \right]^{\mathrm{T}} \end{aligned} \tag{15.64}$$

We recall that $N_1 + N_2 = 1$ and $N_1 x_1 + N_2 x_2 = x$.

Investigation of the term x^k in the approximation

$$\hat{u} = N_1(x) \left[\tilde{u}_1 + x^k b_{k1} \right] + N_2(x) \left[\tilde{u}_2 + x^k b_{k2} \right] \tag{15.65}$$

we observe that a linear dependence with the usual finite element approximation occurs when $\tilde{u}_i = x_i \tilde{b}_0$ and $k = 1$ with $b_{11} = b_{12} = \tilde{b}_1$. In this case Eq. (15.65) becomes

$$\begin{aligned} \hat{u} &= [N_1 x_1 + N_2 x_2] \tilde{b}_0 + [N_1 + N_2] x \tilde{b}_1 \\ &= x \tilde{b}_0 + x \tilde{b}_1 \end{aligned}$$

In one dimension linear dependence can be avoided by setting k to 2 in Eqs (15.63) and (15.64). However, in two- and three-dimensional problems the linear dependence cannot be completely avoided, and we address this next.[58, 60]

An approximation over two-dimensional triangles may be expressed as

$$u(x, y) \approx \hat{u}(x, y) = \sum_{i=1}^{3} L_i \left[\tilde{u}_i + \mathbf{q}^{(i)} \mathbf{b}_i \right] \tag{15.66}$$

where L_i are the area coordinates defined in Chapter 4. We consider the case where complete quadratic functions are added as

$$\mathbf{q}^{(i)} = \left[x^2, \quad xy, \quad y^2\right] \tag{15.67}$$

to give a complete second-order polynomial approximation for u. Although this gives a complete second-order polynomial approximation there are two ways in which the cubic term x^2y can be obtained.

1. The first sets

$$\tilde{u}_i = b_{i1} = b_{i3} = 0 \qquad \text{and} \qquad b_{i2} = x_i\tilde{\alpha}$$

giving

$$\hat{u} = \sum_{i=1}^{3} L_i \cdot [xy] \cdot x_i\tilde{\alpha} = x^2y\tilde{\alpha}$$

2. The second alternative to compute the same term sets

$$\tilde{u}_i = b_{i2} = b_{i3} = 0 \qquad \text{and} \qquad b_{i1} = y_i\tilde{\alpha}$$

giving

$$\hat{u} = \sum_{i=1}^{3} L_i \cdot [x^2] \cdot y_i\tilde{\alpha} = x^2y\tilde{\alpha}$$

A similar construction may be made for the polynomial term xy^2.

An alternative is to construct the interpolation to depend on each node as

$$\mathbf{q}^{(i)} = \left[(x - x_i)^2 \quad (x - x_i)(y - y_i) \quad (y - y_i)^2\right] \tag{15.68}$$

This form, while conceptually the same as the original formulation, appears to be better conditioned and also avoids some of the problems of linear dependency.[60] In Sec. 15.7.4 we will discuss in more detail a methodology to deal with the problem of linear dependency; however, before doing so we illustrate the use of the hierarchical finite element method by an application to two-dimensional problems in linear elasticity.

15.7.3 Application to linear elasticity

In the previous section the form for polynomial interpolation in two dimensions was given. Here we consider the use of the interpolation to model the behaviour of problems in linear elasticity. For simplicity only the displacement model for plane strain as discussed in Chapters 2 and 6 is considered; however, the use of the hierarchic interpolations can easily be extended to other forms and to mixed models.

For a displacement model the finite element arrays may be computed using the formulation given in Chapter 6. For two-dimensional plane strain problems, the strain–displacement relations may be written in matrix form as

$$\boldsymbol{\varepsilon} = \begin{bmatrix} \dfrac{\partial u}{\partial x} \\ \dfrac{\partial v}{\partial y} \\ \dfrac{\partial u}{\partial y} + \dfrac{\partial v}{\partial x} \end{bmatrix} \tag{15.69}$$

Inserting the interpolations for u and v given by Eq. (15.61) and using Eq. (15.62) to compute derivatives, the strain–displacement relations become

$$\varepsilon = \sum_{i=1}^{N} \begin{bmatrix} \dfrac{\partial N_i^k}{\partial x} & 0 \\ 0 & \dfrac{\partial N_i^k}{\partial y} \\ \dfrac{\partial N_i^k}{\partial y} & \dfrac{\partial N_i^k}{\partial x} \end{bmatrix} \begin{bmatrix} \tilde{u}_i \\ \tilde{v}_i \end{bmatrix} + \sum_{i=1}^{N} \begin{bmatrix} \left(\dfrac{\partial N_i^k}{\partial x}\mathbf{q}_i^k + N_i^k \dfrac{\partial \mathbf{q}_i^k}{\partial x}\right) & 0 \\ 0 & \left(\dfrac{\partial N_i^k}{\partial y}\mathbf{q}_i^k + N_i^k \dfrac{\partial \mathbf{q}_i^k}{\partial y}\right) \\ \left(\dfrac{\partial N_i^k}{\partial y}\mathbf{q}_i^k + N_i^k \dfrac{\partial \mathbf{q}_i^k}{\partial y}\right) & \left(\dfrac{\partial N_i^k}{\partial x}\mathbf{q}_i^k + N_i^k \dfrac{\partial \mathbf{q}_i^k}{\partial x}\right) \end{bmatrix} \begin{bmatrix} \tilde{\mathbf{b}}_i^u \\ \tilde{\mathbf{b}}_i^v \end{bmatrix}$$

where N is the number of nodes for an element. The first term is identical to the usual finite element strain–displacement matrices [see Eq. (6.57)] and the second term has identical structure to the usual arrays. Thus, the development of all element arrays follows standard procedures.

Example 15.4: A quadratic triangular element. For a triangular element with linear interpolation the shape functions and quadratic polynomial hierarchic terms are given by $N_i = L_i$ and Eq. (15.68), respectively. Using isoparametric concepts the coordinates are given by

$$\mathbf{x} = \sum_{i=1}^{3} N_i^1 \tilde{\mathbf{x}}_i = \sum_{i=1}^{3} L_i \tilde{\mathbf{x}}_i \tag{15.70}$$

and are used to construct all polynomials appearing in hierarchical form (15.68).

A set of patch tests is first performed to assess the stability and consistency of the above hierarchic form. The set consists of one-, two-, four-, and eight-element patches as shown in Fig. 15.17. First, we perform a stability assessment by determining the number of zero eigenvalues for each patch. The results for hierarchical interpolation are shown in Table 15.3.

The eigenproblem assessment reveals that the hierarchic interpolation has excess zero eigenvalues (i.e., spurious zero energy modes) only for meshes consisting of one or two elements. Furthermore, only two element meshes in which one side is a straight line through both elements have excess zero values. Once the mesh has no straight intersections the number of zero modes becomes correct (e.g., contain only the three rigid body modes).

Consistency tests verify that all meshes contain terms of up to quadratic polynomial order – thus also validating the correctness of the coding.

As a simple test problem using the hierarchical finite element method we consider a finite width strip containing a circular hole with diameter half the width of the strip. The strip is subjected to axial extension in the vertical direction and, due to symmetry of the loading and geometry, only one quadrant is discretized as shown in Figs 15.18 and 15.19.

The meshes in Fig. 15.18 employ the hierarchical interpolation considered above; whereas those in Fig. 15.19 use standard 6-node isoparametric quadratic triangles with two degrees of freedom per node (i.e., u and v). The material is taken as linear elastic with $E = 1000$

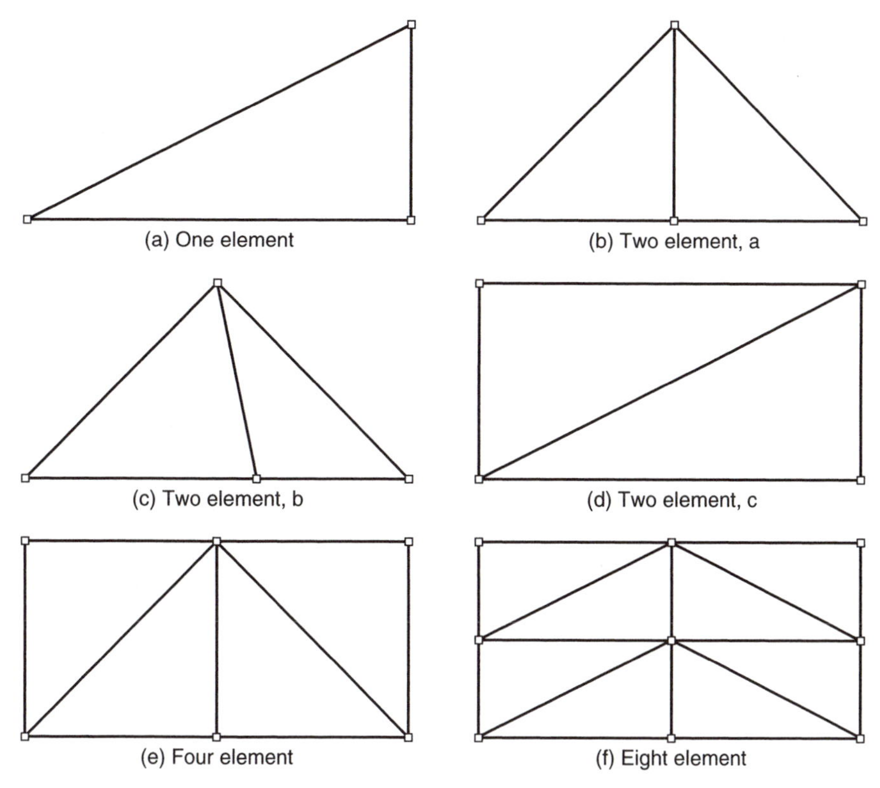

Fig. 15.17 Patches for eigenproblem assessment.

Table 15.3 Triangle element patch tests: number of eigenvalues, minimum non-zero value, and maximum value ($k = 2$) – quadratic hierarchical terms

Mesh	No. zero	Min. value	Max. value
1	7	4.7340E + 01	2.0560E + 06
2a	5	4.0689E + 01	2.1543E + 05
2b	5	4.1971E + 02	2.2648E + 05
2c	3	1.5728E + 02	2.3883E + 06
4	3	1.0446E + 02	2.9027E + 05
8	3	9.5560E + 01	3.4813E + 05

and $\nu = 0.25$. The half-width of the strip is 10 units and the half-height is 18 units. The hole has radius 5.

The problem size and computed energy (which indicates solution accuracy) are shown in Table 15.4 for the hierarchical method, in Table 15.5 for the 6-node isoparametric formulation and in Table 15.6 for 3-node linear triangular elements.

The 6-node isoparametric method gives overall the best accuracy; however, the hierarchical element is considerably better than the 3-node triangular element and offers great advantages when used in adaptive analysis.[60]

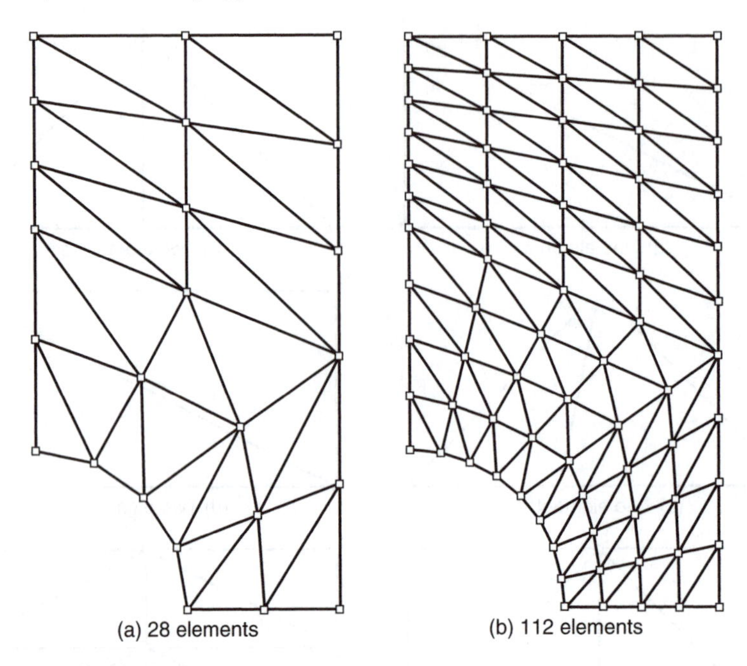

Fig. 15.18 Hierarchic elements: tension strip.

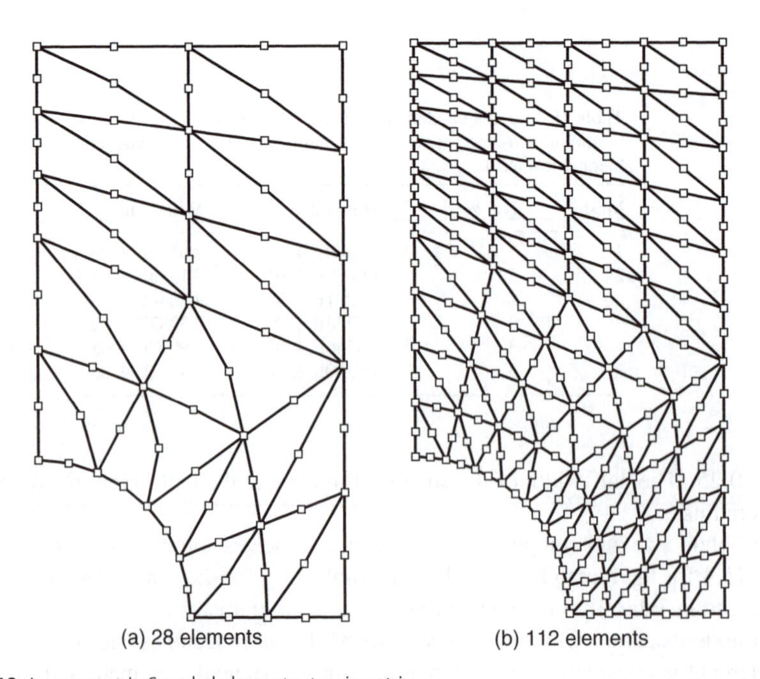

Fig. 15.19 Isoparametric 6-noded elements: tension strip.

Table 15.4 Hierarchical element. Boundary segments straight

Nodes	Elements	Equations	Energy
30	28	156	131.7088
85	112	537	127.8260
279	448	1971	126.7641
1003	1792	7527	126.5908

Table 15.5 Isoparametric element. Boundary segments have curved sides

Nodes	Elements	Equations	Energy
30	28	129	127.3350
279	112	483	126.6483
1003	448	1863	126.5661
3795	1792	7311	126.5593

Table 15.6 Linear triangular element

Nodes	Elements	Equations	Energy
30	28	36	137.652
85	112	129	131.065
279	448	483	128.008
1003	1792	1863	126.958
3795	7168	7311	126.662

15.7.4 Solution of forms with linearly dependent equations

A typical problem for a steady-state analysis in which the algebraic equations are generated from the hierarchical finite element form described above, such as given by Eqs (15.66) and (15.67), produces algebraic equations in the standard form, i.e.,

$$\mathbf{K}\tilde{\mathbf{a}} + \mathbf{f} = \mathbf{0} \tag{15.71}$$

where the parameters $\tilde{\mathbf{a}}$ include both nodal $\tilde{u}_i$ and hierarchical parameters $\mathbf{b}_i$. We assume that occasionally the 'stiffness matrix' $\mathbf{K}$ and 'force' vector $\mathbf{f}$ include equations which are linearly dependent with other equations in the system and, thus, $\mathbf{K}$ can be singular.

If the system is solved by a direct elimination scheme (e.g., as described in Chapter 2 or in books on linear algebra such as references 61 or 62) it is possible to set a tolerance for the pivot below which an equation is assumed to be linearly dependent and can be omitted from the calculations (e.g., see references 63 and 64).

An alternative to the above is to perturb Eq. (15.71) to

$$[\mathbf{K} + \varepsilon\mathbf{D}_K]\,\Delta\tilde{\mathbf{a}}^k = \mathbf{f} - \mathbf{K}\tilde{\mathbf{a}}^k \tag{15.72}$$

where $\mathbf{D}_K$ are diagonal entries of $\mathbf{K}$, ε is a specified value and

$$\tilde{\mathbf{a}}^{k+1} = \tilde{\mathbf{a}}^k + \Delta\tilde{\mathbf{a}}^k \tag{15.73}$$

is used to define an iterative strategy. An initial guess of zero may be used to start the solution process. Certainly a choice of a small value for ε (e.g., 10^{-6}) leads to rapid convergence.[60]

15.8 Concluding remarks

In this chapter we have considered a number of methods which eliminate or reduce our dependence on meshing the total domain. There are a number of other approaches having the same aim which have been pursued with success. These include the *smooth particle hydrodynamics* (SPH) method (Lucy,[65] Gingold and Monaghan,[66] Benz[67]) and the *reproducing kernel* (RPK) method (Liu *et al.*[68,69]) applied to problems in solid and fluid mechanics. Bonet and coworkers[70] improve the method of SPH and show its possibilities. Another approach has recently been introduced by Yagawa.[71,72] These are not described here and the reader is referred to the literature for details.

15.9 Problems

15.1 Data for a distributed loading to be applied to an analysis is tabulated in Table 15.7. The data is to be fit using the least squares analysis described in Sec. 15.2.1 in which

$$\mathbf{p(x)} = \begin{bmatrix} 1 & x & x^2 & x^3 & x^4 \end{bmatrix}$$

Determine and plot the solution obtained with this data.

15.2 Data for a distributed loading to be applied to an analysis is tabulated in Table 15.7. The data is to be fit using the moving least squares analysis described in Sec.15.3. Use the weight function given by Eq. (15.11) with $r_m = 2$ and a Shepard approximation,

$$p(x) = 1$$

Write a MATLAB program to compute the fit at intervals on x of 0.1 units. Plot the solution obtained.

15.3 Data for a distributed loading to be applied to an analysis is tabulated in Table 15.7. The data is to be fit using the moving least squares analysis described in Sec.15.3. Use the weight function given by Eq. (15.11) with $r_m = 2$ and a linear approximation,

$$\mathbf{p(x)} = \begin{bmatrix} 1 & x - x_j \end{bmatrix} \quad \text{for } x_j - r_m < x < x_j + r_m$$

Write a MATLAB program to compute the fit at intervals on x of 0.1 units. Plot the solution obtained.

15.4 Solve the differential equation

$$-\frac{\mathrm{d}^2 u}{\mathrm{d}x^2} + u + f = 0; \quad 0 < x < 1$$

Table 15.7 Data for force in Problems 15.1 to 15.3

k	1	2	3	4	5	6	7	8
x_k	0.0	1.0	2.0	3.5	5.0	6.5	8.3	10.0
$\tilde{f}_k$	6.0	3.0	1.6	−1.4	−1.3	0.3	0.9	0.0

where the boundary conditions are $u(0) = u(1) = 1$ and f is a concentrated unit load at $x = 0.4$ units.

Write a MATLAB program to solve the problem. Use a moving least square method with Shepard interpolation and the point collocation method described in Sec. 15.5. Let the points be located at 0.2 intervals (6 points). Repeat the solution using points spaced at 0.1 intervals (11 points). Plot the two solutions and comment on the behaviour obtained.

15.5 Solve the differential equation

$$-\frac{d^2u}{dx^2} + u + f = 0; \quad 0 < x < 1$$

where the boundary conditions are $u(0) = u(1) = 1$ and f is a concentrated unit load at $x = 0.4$ units.

Write a MATLAB program to solve the problem. Use a moving least square method with Shepard interpolation and the Galerkin method described in Sec. 15.6. Let the points be located at 0.2 intervals (6 points). Repeat the solution using points spaced at 0.1 intervals (11 points). Plot the two solutions and comment on the behaviour obtained.

15.6 Repeat Problem 15.4 using a hierarchical interpolation with

$$q^{(j)}(x) = x - x_j$$

where x_j is the coordinate of the point.

15.7 Repeat Problem 15.5 using a hierarchical interpolation with

$$q^{(j)}(x) = x - x_j$$

where x_j is the coordinate of the point.

15.8 In the moving least squares approximation, the shape function $N_j(\mathbf{x})$ of Eq. (15.22) is derived by minimization of the function $J(\mathbf{x})$ in Eq. (15.14). Derive a shape function for the moving least squares approximation by minimization of the function

$$J(\mathbf{x}) = \tfrac{1}{2} \int_{\Omega} w(\mathbf{x} - \mathbf{y})[u(\mathbf{y}) - \mathbf{p}(\mathbf{y})\alpha]^2 d\mathbf{y} = \min$$

15.9 In the least square fit and the construction of shape functions for the moving least squares approximation the matrix $\mathbf{H}$ in Eqs (15.5) and (15.18a) may be singular for a given set of points if the monomials used in the polynomial basis are not chosen properly. Devise an algorithm that can choose the terms used in the polynomial basis automatically so that matrix $\mathbf{H}$ is always non-singular.

References

1. V. Girault. Theory of a finite difference method on irregular networks. *SIAM J. Num. Anal.*, 11:260–282, 1974.
2. V. Pavlin and N. Perrone. Finite difference energy techniques for arbitrary meshes. *Comp. Struct.*, 5:45–58, 1975.

3. C. Snell, D.G. Vesey, and P. Mullord. The application of a general finite difference method to some boundary value problems. *Comp. Struct.*, 13:547–552, 1981.
4. T. Liszka and J. Orkisz. Finite difference methods of arbitrary irregular meshes in non-linear problems of applied mechanics. In *Proceedings 4th Int. Conference on Structural Mechanics in Reactor Technology*, San Francisco, California, 1977.
5. T. Liszka and J. Orkisz. The finite difference method at arbitrary irregular grids and its applications in applied mechanics. *Comp. Struct.*, 11:83–95, 1980.
6. J. Krok and J. Orkisz. A unified approach to the FE generalized variational FD method in nonlinear mechanics. Concept and numerical approach. In *Discretization Methods in Structural Mechanics*, pages 353–362. Springer-Verlag, Berlin-Heidelberg, 1990. IUTAM/IACM Symposium, Vienna, 1989.
7. R.A. Nay and S. Utku. An alternative for the finite element method. *Vari. Meth. Eng.*, 1, 1972.
8. D. Shepard. A two-dimensional function for irregularly spaced data. In *ACM National Conference*, pages 517–524, 1968.
9. P. Lancaster and K. Salkauskas. Surfaces generated by moving least squares methods. *Math. Comput.*, 37:141–158, 1981.
10. P. Lancaster and K. Salkauskas. *Curve and Surface Fitting*. Academic Press, New York, 1990.
11. B. Nayroles, G. Touzot, and P. Villon. La méthode des éléments diffuse. *C.R. Acad. Sci. Paris*, 313:133–138, 1991.
12. B. Nayroles, G. Touzot, and P. Villon. L'approximation diffuse. *C.R. Acad. Sci. Paris*, 313: 293–296, 1991.
13. B. Nayroles, G. Touzot, and P. Villon. Generalizing the FEM: diffuse approximation and diffuse elements. *Comput. Mech.*, 10:307–318, 1992.
14. T. Belytschko, Y. Lu, and L. Gu. Element free Galerkin methods. *Int. J. Numer. Meth. Eng.*, 37:397–414, 1994.
15. T. Belytschko, Y. Lu, and L. Gu. Crack propagation by element-free Galerkin methods. *Eng. Fracture Mech.*, 51:295–315, 1995.
16. C.A. Duarte and J.T. Oden. $h-p$ clouds – a meshless method to solve boundary-value problems. Technical Report TICAM Report 95–05, The University of Texas, May 1995.
17. C.A. Duarte and J.T. Oden. An $h-p$ adaptive method using clouds. *Comp. Meth. Appl. Mech. Eng.*, 139(1–4):237–262, 1996.
18. T. Belytschko, J. Fish, and A. Bayless. The spectral overlay on finite elements for problems with high gradients. *Comp. Meth. Appl. Mech. Eng.*, 81:71–89, 1990.
19. J. Dolbow and T. Belytschko. An introduction to programming the meshless element free Galerkin method. *Arch. Comput. Meth. Eng.*, 5(3):207–241, 1998.
20. I. Babuška and J.M. Melenk. The partition of unity finite element method. Technical Report Technical Note BN-1185, Institute for Physical Science and Technology, University of Maryland, April 1995.
21. J.M. Melenk and I. Babuška. The partition of unity finite element method: basic theory and applications. *Comp. Meth. Appl. Mech. Eng.*, 139:289–314, 1996.
22. O.C. Zienkiewicz, R.L. Taylor, and P. Nithiarasu. *The Finite Element Method for Fluid Dynamics*. Butterworth-Heinemann, Oxford, 6th edition, 2005.
23. C. Daux, N. Moes, J. Dolbow, N. Sukumar, and T. Belytschko. Arbitrary branched and intersecting cracks with the extended finite element method. *Int. J. Numer. Meth. Eng.*, 48:1741–1760, 2000.
24. N. Sukumar, N. Moes, B. Moran, and T. Belytschko. Extended finite element method for three-dimensional crack modelling. *Int. J. Numer. Meth. Eng.*, 48:1549–1570, 2000.
25. N. Sukumar, D.L. Chopp, N. Moes, and T. Belytschko. Modeling holes and inclusions by level sets in the extended finite element method. *Comp. Meth. Appl. Mech. Eng.*, 190:6183–6846, 2001.

26. J. Dolbow, N. Moes, and T. Belytschko. An extended finite element method for modeling crack growth with frictional contact. *Comp. Meth. Appl. Mech. Eng.*, 190:6825–6846, 2000.
27. N. Moes, A. Gravouil, and T. Belytschko. Non-planar 3D crack growth by the extended finite element and level sets. Part I: Mechanical model. *Int. J. Numer. Meth. Eng.*, 53:2549–2568, 2002.
28. A. Gravouil, N. Moes, and T. Belytschko. Non-planar 3D crack growth by the extended finite element and level sets. Part II: Level set update. *Int. J. Numer. Meth. Eng.*, 53:2569–2586, 2002.
29. N. Moes and T. Belytschko. Extended finite element method for cohesive crack growth. *Eng. Frac. Mech.*, 69:813–833, 2002.
30. T. Belytschko, C. Parimi, N. Moes, N. Sukumar, and S. Usui. Structured extended finite element methods for solids defined by implicit surfaces. *Int. J. Numer. Meth. Eng.*, 56:609–635, 2002.
31. J. Chessa, P. Smolinski, and T. Belytschko. The extended finite element method (XFEM) for solidification problems. *Int. J. Numer. Meth. Eng.*, 53:1959–1977, 2002.
32. G. Zi and T. Belytschko. New crack-tip elements for XFEM and applications to cohesive cracks. *Int. J. Numer. Meth. Eng.*, 57:2221–2240, 2003.
33. F.L. Stazi, E. Budyn, J. Chessa, and T. Belytschko. An extended finite element method with higher-order elements for curved cracks. *Comput. Mech.*, 31:38–48, 2003.
34. J. Chessa, H. Wang, and T. Belytschko. On the construction of blending elements for local partition of unity enriched finite elements. *Int. J. Numer. Meth. Eng.*, 57:1015–1038, 2003.
35. W. Rudin. *Principles of Mathematical Analysis*. McGraw-Hill, 3rd edition, 1976.
36. I. Babuška and J.M. Melenk. The partition of unity method. *Int. J. Numer. Meth. Eng.*, 40:727–758, 1997.
37. L. Collatz. *The Numerical Treatment of Differential Equations*. Springer, Berlin, 1966.
38. G.E. Forsythe and W.R. Wasow. *Finite Difference Methods for Partial Differential Equations*. John Wiley & Sons, New York, 1960.
39. R.D. Richtmyer and K.W. Morton. *Difference Methods for Initial Value Problems*. Wiley (Interscience), New York, 1967.
40. J. Orkisz. Finite difference method. In M. Kleiber, editor, *Handbook of Computational Solid Mechanics*. Springer-Verlag, Berlin, 1998.
41. J. Batina. A gridless Euler/Navier–Stokes solution algorithm for complex aircraft applications. In *AIAA 93–0333*, Reno, NV, Jan. 1993.
42. E. Oñate, S. Idelsohn, and O.C. Zienkiewicz. Finite point methods in computational mechanics. Technical Report CIMNE Report 67, Int. Center for Num. Meth. Engr., Barcelona, July 1995.
43. E. Oñate, S.R. Idelsohn, O.C. Zienkiewicz, R.L. Taylor, and C. Sacco. A stabilized finite point method for analysis of fluid mechanics problems. *Comp. Meth. Appl. Mech. Eng.*, 139:315–346, 1996.
44. E. Oñate, S.R. Idelsohn, O.C. Zienkiewicz, and R.L. Taylor. A finite point method in computational mechanics. Applications to convective transport and fluid flow. *Int. J. Numer. Meth. Eng.*, 39:3839–3866, 1996.
45. J. Orkisz. Computer approach to the finite difference method (in Polish). *Mech. i Komp.*, 2:7–69, 1979.
46. T. Liszka. An interpolation method for an irregular net of nodes. *Int. J. Numer. Meth. Eng.*, 20:1599–1612, 1984.
47. R.B. Pelz and A. Jameson. Transonic flow calculations using triangular finite elements. *J. AIAA*, 23:569–576, 1985.
48. J.T. Batina. Vortex-dominated conical-flow computations using unstructured adaptively-refined meshes. *J. AIAA*, 28(11):1925–1932, 1990.
49. J.T. Batina. Unsteady Euler airfoil solutions using unstructured dynamic meshes. *J. AIAA*, 28(8):1381–1388, 1990.
50. D.J. Mavriplis and A. Jameson. Multigrid solution of the Navier–Stokes equations on triangular meshes. *J. AIAA*, 28:1415–1425, 1990.

51. R.D. Rausch, J.T. Batina, and H.T.Y. Yang. Spatial adaptation of unstructured meshes for unsteady aerodynamic flow computations. *J. AIAA*, 30(5):1243–1251, 1992.
52. R.D. Rausch, J.T. Batina, and H.T.Y. Yang. Three-dimensional time-marching aeroelastic analyses using an unstructured-grid Euler method. *J. AIAA*, 31(9):1626–1633, 1993.
53. K. Xu, L. Martinelli, and A. Jameson. Gas-kinetic finite volume methods. In S.M. Deshpande, S.S. Desai, and R. Narasimha, editors, *Proc. 14th Int. Conf. Num. Meth. Fluid Dynamics*, pages 106–111, 1995.
54. E. Oñate, F. Zarate, and F. Flores. A simple triangular element for thick and thin plate and shell analysis. *Int. J. Numer. Meth. Eng.*, 37:2569–2582, 1994.
55. Y. Lu, T. Belytschko, and L. Gu. A new implementation of the element-free Galerkin method. *Comp. Meth. Appl. Mech. Eng.*, 113:397–414, 1994.
56. M. Tabbara, T. Blacker, and T. Belytschko. Finite element derivative recovery by moving least square interpolates. *Comp. Meth. Appl. Mech. Eng.*, 117:211–223, 1994.
57. C.A. Duarte. A review of some meshless methods to solve partial differential equations. Technical Report TICAM Report 95–06, The University of Texas, May 1995.
58. R.L. Taylor, O.C. Zienkiewicz, and E. Oñate. A hierarchical finite element method based on the partition of unity. *Comp. Meth. Appl. Mech. Eng.*, 152:73–84, 1998.
59. J.T. Oden, C.A.M. Duarte, and O.C. Zienkiewicz. A new cloud-based *hp* finite element method. *Comp. Meth. Appl. Mech. Eng.*, 153(1–2):117–126, 1998.
60. C.A. Duarte, I. Babuška, and J.T. Oden. Generalized finite element methods for three dimensional structural mechanics problems. *Comp. Struct.*, 77:215–232, 2000.
61. G. Strang. *Linear Algebra and its Application*. Academic Press, New York, 1976.
62. J. Demmel. *Applied Numerical Linear Algebra*. Society for Industrial and Applied Mathematics, Philadelphia, PA, 1997.
63. I.S. Duff and J.K. Reid. Exploiting zeros on the diagonal in the direct solution of indefinite sparse linear systems. *ACM Trans. Math. Soft.*, 22:227–257, 1996.
64. I.S. Duff and J.A. Scott. Ma62 – a frontal code for sparse positive-definite symmetric systems from finite element applications. In M. Papadrakakis and B.H.V. Topping, editors, *Innovative Computational Methods for Structural Mechanics*, pages 1–25, June 1999.
65. L.B. Lucy. A numerical approach to the testing of fusion process. *The Astron. J.*, 88, 1977.
66. R.A. Gingold and J.J. Monaghan. Smoothed particle hydrodynamics: theory and application to non-spherical stars. *Monthly Notices of Royal Astron. Sci.*, 181, 1977.
67. W. Benz. Smoothed particle hydrodynamics: a review. Preprint 2884, 1989.
68. W.K. Liu, S. Jun, S. Li, J. Adee, and T. Belytschko. Reproducing kernel particle methods for structural dynamics. *Comp. Meth. Appl. Mech. Eng.*, 38:1655–1679, 1995.
69. W.K. Liu, S. Jun, and Y.F. Zhang. Reproducing kernel particle methods. *Int. J. Numer. Meth. Eng.*, 20:1081–1106, 1995.
70. J. Bonet and S. Kulasegaram. Correction and stabilization of smooth particle hydrodynamics methods with applications in metal forming simulations. *Int. J. Numer. Meth. Eng.*, 47:1189–1214, 2000.
71. G. Yagawa and T. Yamada. Free mesh method. A kind of meshless finite element method. *Comput. Mech.*, 18:383–386, 1996.
72. G. Yagawa, T. Yamada, and T. Furukawa. Parallel computing with free mesh method: virtually meshless FEM. In H.A. Mang and F.G. Rammerstorfer, editors, *IUTAM Sym., Solid Mechanics and its Applications*, pages 165–172. Kluwer Acd. Pub., 1997.

16

The time dimension – semi-discretization of field and dynamic problems and analytical solution procedures

16.1 Introduction

In most of the problems considered so far in this text conditions that do not vary with time were generally assumed. There is little difficulty in extending the finite element idealization to situations that are time dependent as indicated briefly in Chapters 3 and 7.

The range of practical problems in which the time dimension has to be considered is great. Transient heat conduction, wave transmission in fluids and dynamic behaviour of structures are typical examples. While it is usual to consider these various problems separately – sometimes classifying them according to the mathematical structure of the governing equations as 'parabolic' or 'hyperbolic'[1] – we shall group them into one category to show that the formulation is identical.

In the first part of this chapter we shall formulate, by a simple extension of the methods used so far, matrix differential equations governing such problems for a variety of physical situations. Here a finite element discretization in the space dimension only will be used and a semi-discretization process followed (see Chapter 3). In the remainder of this chapter various analytical procedures of the solution for the resulting ordinary linear differential equation system will be dealt with. These form the basic arsenal of steady-state and transient analysis.

Chapter 17 will be devoted to the discretization of the time domain itself.

16.2 Direct formulation of time-dependent problems with spatial finite element subdivision

16.2.1 The 'quasi-harmonic' equation with time differential

In many physical problems the quasi-harmonic equation takes the form in which time derivatives of the unknown function ϕ occur. In the three-dimensional case typically we might have [viz. Eq. (7.6)]

$$-\frac{\partial}{\partial x}\left(k\frac{\partial \phi}{\partial x}\right)-\frac{\partial}{\partial y}\left(k\frac{\partial \phi}{\partial y}\right)-\frac{\partial}{\partial z}\left(k\frac{\partial \phi}{\partial z}\right)+\left(\bar{Q}+\mu\frac{\partial \phi}{\partial t}+\rho\frac{\partial^2 \phi}{\partial t^2}\right)=0 \quad (16.1)$$

In the above, quite generally, all the parameters may be prescribed functions of time, or in non-linear cases of ϕ, as well as of space $\mathbf{x}$, i.e.,

$$k = k(\mathbf{x}, \phi, t) \qquad \bar{Q} = \bar{Q}(\mathbf{x}, \phi, t) \quad \text{etc.} \quad (16.2)$$

If a situation at a particular instant of time is considered, the time derivatives of ϕ and all the parameters can be treated as *prescribed functions of space coordinates*. Thus, at that instant the problem is precisely identified with those treated in Chapter 7 if the whole of the quantity in the last parentheses of Eq. (16.1) is identified as the source term Q.

The finite element discretization of this in terms of *space* elements has already been fully discussed and we found that with the prescription

$$\phi = \sum N_a \tilde{\phi}_a = \mathbf{N}\tilde{\mathbf{u}} \quad \text{with} \quad \mathbf{N} = \mathbf{N}(x, y, z) \quad \text{and} \quad \tilde{\mathbf{u}} = \tilde{\mathbf{u}}(t) \quad (16.3)$$

for each element, the standard form of assembled equations†

$$\mathbf{K}\tilde{\mathbf{u}} + \mathbf{f} = \mathbf{0} \quad (16.4)$$

was obtained. Element contributions to the above matrices are defined in Chapter 7 and need not be repeated here except for that representing the 'load' term due to Q. This is given by

$$\mathbf{f} = \int_\Omega \mathbf{N}^{\mathrm{T}} Q \, \mathrm{d}\Omega \quad (16.5)$$

Replacing Q by the last bracketed term of Eq. (16.1) we have

$$\mathbf{f} = \int_\Omega \mathbf{N}^{\mathrm{T}}\left(\bar{Q}+\mu\frac{\partial \phi}{\partial t}+\rho\frac{\partial^2 \phi}{\partial t^2}\right)\mathrm{d}\Omega \quad (16.6)$$

However, from Eq. (16.3) it is noted that ϕ is approximated in terms of the nodal parameters $\tilde{\mathbf{u}}$. On substitution of this approximation we have

$$\mathbf{f} = \int_\Omega \mathbf{N}^{\mathrm{T}}\bar{Q}\,\mathrm{d}\Omega + \left(\int_\Omega \mathbf{N}^{\mathrm{T}}\mu\mathbf{N}\,\mathrm{d}\Omega\right)\frac{\mathrm{d}\tilde{\mathbf{u}}}{\mathrm{d}t} + \left(\int_\Omega \mathbf{N}^{\mathrm{T}}\rho\mathbf{N}\,\mathrm{d}\Omega\right)\frac{\mathrm{d}^2\tilde{\mathbf{u}}}{\mathrm{d}t^2} \quad (16.7)$$

and on expanding Eq. (16.4) in its final assembled form we get the following *matrix differential equation*:

$$\mathbf{M}\ddot{\tilde{\mathbf{u}}} + \mathbf{C}\dot{\tilde{\mathbf{u}}} + \mathbf{K}\tilde{\mathbf{u}} + \mathbf{f} = \mathbf{0}$$
$$\dot{\tilde{\mathbf{u}}} \equiv \frac{\mathrm{d}\tilde{\mathbf{u}}}{\mathrm{d}t} \qquad \ddot{\tilde{\mathbf{u}}} \equiv \frac{\mathrm{d}^2\tilde{\mathbf{u}}}{\mathrm{d}t^2} \quad (16.8)$$

in which all the matrices are assembled from element submatrices in the standard manner with submatrices $\mathbf{K}^e$ and $\mathbf{f}^e$ still given by relations (7.20) in Chapter 7 and

$$C_{ab}^e = \int_\Omega N_a \mu \, N_b \, \mathrm{d}\Omega \quad \text{and} \quad M_{ab}^e = \int_\Omega N_a \rho \, N_b \, \mathrm{d}\Omega \quad (16.9)$$

† We have replaced the matrix $\mathbf{H}$ of Chapter 7 by $\mathbf{K}$ and $\tilde{\phi}$ by $\tilde{\mathbf{u}}$ to facilitate later comparison with other transient equations.

Once again these matrices are symmetric as seen from the above relations.

Boundary conditions imposed at any time instant are treated in the standard manner.

The variety of physical problems governed by Eq. (16.1) is so large that a comprehensive discussion of them is beyond the scope of this book. A few typical examples will, however, be quoted.

Equation (16.1) with $\rho = 0$

This is the standard *transient heat conduction equation*[1, 2] which has been discussed in the finite element context in Sec. 7.4 and by several authors.[3–6] This same equation is applicable in other physical situations – one of these being the *soil consolidation equations*[7] associated with *transient seepage forms*.[8]

Equation (16.1) with $\mu = 0$

Now the relationship becomes the famous *Helmholz wave equation* governing a wide range of physical phenomena. Electromagnetic waves,[9] fluid surface waves[10] and compression waves[11] are but a few cases to which the finite element process has been applied.

Equation (16.1) with $\mu \neq \rho \neq 0$

This damped wave equation is of yet more general applicability and has particular significance in fluid mechanics (wave) problems.

The reader will recognize that what we have done here is simply an application of the process of partial discretization described in Sec. 3.5. It is convenient, however, to perform the operations in the manner suggested above as all the matrices and discretization expressions obtained from steady-state analysis are immediately available.

16.2.2 Dynamic behaviour of elastic structures with linear damping

While in the previous section we have been concerned with, apparently, a purely mathematical problem, identical reasoning can be applied directly to the wide class of dynamic behaviour of elastic structures following precisely the general lines of Chapters 2 and 6.

When displacements of an elastic body vary with time two sets of additional forces are called into play. The first is the inertia force, which for an acceleration characterized by $\ddot{\mathbf{u}}$ can be replaced by its static equivalent $-\rho\ddot{\mathbf{u}}$ using the well-known d'Alembert principle. This is a force with components in directions identical to those of the displacement $\mathbf{u}$ and (generally) given per unit of volume. In this context ρ is simply the mass per unit volume.

The second force is that due to (frictional) resistances opposing the motion. These may be due to microstructure movements, air resistance, etc., and are often related in a non-linear way to the velocity $\dot{\mathbf{u}}$. For simplicity of treatment, however, only a linear viscous-type resistance will be considered, resulting again in unit volume forces in an equivalent static problem of magnitude $-\mu\dot{\mathbf{u}}$. In the above μ is a set of viscosity parameters which can presumably be given numerical values.[12]

The equivalent static problem, at any instant of time, is now discretized precisely in the manner of Chapters 2 and 6, but replacing the distributed body force $\mathbf{b}$ by its equivalent

$$\bar{\mathbf{b}} - \rho\ddot{\mathbf{u}} - \mu\dot{\mathbf{u}}$$

The element (nodal) forces given by Eq. (6.62) now become (excluding initial stress and strain contributions)

$$\mathbf{f}^e = -\int_{\Omega^e} \mathbf{N}^{\mathrm{T}}\mathbf{b}\,\mathrm{d}\Omega = -\int_{\Omega^e} \mathbf{N}^{\mathrm{T}}\bar{\mathbf{b}}\,\mathrm{d}\Omega + \int_{\Omega^e} \mathbf{N}^{\mathrm{T}}\rho\ddot{\mathbf{u}}\,\mathrm{d}\Omega + \int_{\Omega^e} \mathbf{N}^{\mathrm{T}}\boldsymbol{\mu}\dot{\mathbf{u}}\,\mathrm{d}\Omega \qquad (16.10)$$

in which the first force is that due to an external distributed body load and need not be considered further.

Substituting Eq. (16.10) into the general equilibrium equations we obtain finally, on assembly, the following matrix differential equation:

$$\mathbf{M}\ddot{\tilde{\mathbf{u}}} + \mathbf{C}\dot{\tilde{\mathbf{u}}} + \mathbf{K}\tilde{\mathbf{u}} + \mathbf{f} = \mathbf{0} \qquad (16.11)$$

in which **K** and **f** are assembled stiffness and force matrices obtained by the usual addition of element stiffness coefficients and of element forces due to external specified loads, initial stresses, etc., in the manner fully described before. The new matrices **C** and **M** are assembled by the usual rule from element submatrices given by

$$\mathbf{C}^e_{ab} = \int_{\Omega^e} \mathbf{N}_a^{\mathrm{T}}\boldsymbol{\mu}\mathbf{N}_b\,\mathrm{d}\Omega \qquad (16.12a)$$

and

$$\mathbf{M}^e_{ab} = \int_{\Omega^e} \mathbf{N}_a^{\mathrm{T}}\rho\,\mathbf{N}_b\,\mathrm{d}\Omega \qquad (16.12b)$$

The matrix $\mathbf{M}^e$ is known as the *element mass matrix* and the assembled matrix **M** as the system mass matrix. Similarly, the matrix $\mathbf{C}^e$ is known as the *element damping matrix* and the assembled matrix **C** as the system damping matrix.

It is of interest to note that in early attempts to deal with dynamic problems of this nature the mass of the elements was usually arbitrarily ‘lumped’ at nodes, always resulting in a diagonal mass matrix even if no actual concentrated masses existed. The fact that such a procedure was, in fact, unnecessary and apparently inconsistent was simultaneously recognized by Archer[13] and independently by Leckie and Lindberg[14] in 1963. The general presentation of the results given in Eq. (16.12b) is due to Zienkiewicz and Cheung.[15] The name *consistent mass matrix* has been coined for the mass matrix defined here, a term which may be considered to be unnecessary since it is the logical and natural consequence of the discretization process. By analogy the matrices $\mathbf{C}^e$ and **C** may be called *consistent damping matrices*.†

For many computational processes the lumped mass matrix is, however, more convenient and economical. Many practitioners are today using such matrices exclusively – sometimes showing good accuracy. While with simple elements a physically obvious methodology of lumping is easy to devise, this is not the case with higher order elements and we shall return to the process of ‘lumping’ later.

Determination of the damping matrix **C** is in practice difficult as knowledge of the viscous matrix $\boldsymbol{\mu}$ is lacking. It is often assumed, therefore, that the damping matrix is a linear combination of stiffness and mass matrices, i.e.,

$$\mathbf{C} = \alpha\mathbf{M} + \beta\mathbf{K} \qquad (16.13)$$

Here the parameters α and β are determined experimentally.[12,16] Such damping is known as ‘Rayleigh damping’ and has certain mathematical advantages which we shall discuss later. On occasion **C** may be completely specified and such approximation devices are not necessary.

† For simplicity we shall only consider *distributed* inertia – concentrated damping forces being a limiting case.

16.2.3 'Mass' or 'damping' matrices for some typical elements

It is impractical to present in an explicit form all the mass matrices for the various elements discussed in previous chapters. Some selected examples only will be discussed here.

Example 16.1: Plane stress and plane strain. Using triangular elements discussed in Chapter 2 the matrix $\mathbf{N}^e$ is defined as

$$\mathbf{N}^e = \begin{bmatrix}\mathbf{N}_1^e & \mathbf{N}_2^e & \mathbf{N}_3^e\end{bmatrix} \text{ where } \mathbf{N}_a^e = N_a\mathbf{I}, \quad a = 1, 2, 3 \text{ and } \mathbf{I} = \begin{bmatrix}1 & 0\\ 0 & 1\end{bmatrix}$$

Equation (2.8) gives the shape functions as

$$N_a = \frac{a_a + b_a x + c_a y}{2\Delta}, \quad a = 1, 2, 3$$

where Δ is the area of the triangular element.

If the thickness of the element is h and this is assumed to be constant within the element, we have, for the mass matrix, Eq. (16.12b),

$$\mathbf{M}^e = \rho h \iint \mathbf{N}^{\mathrm{T}}\mathbf{N}\,\mathrm{d}x\,\mathrm{d}y \quad \text{or} \quad \mathbf{M}_{ab}^e = \rho\, h\, \mathbf{I} \iint N_a\, N_b\,\mathrm{d}x\,\mathrm{d}y$$

If the relationships of Eq. (2.8) are substituted, it is easy to verify that

$$\iint N_a N_b\,\mathrm{d}x\,\mathrm{d}y = \begin{cases}\frac{1}{6}\Delta & \text{when } a = b\\ \frac{1}{12}\Delta & \text{when } a \neq b\end{cases} \tag{16.14}$$

Thus taking the total mass of the element as

$$m = \rho h \Delta$$

the (consistent) mass matrix becomes

$$\mathbf{M}^e = \frac{m}{12}\left[\begin{array}{cc|cc|cc} 2 & 0 & 1 & 0 & 1 & 0\\ 0 & 2 & 0 & 1 & 0 & 1\\ \hline 1 & 0 & 2 & 0 & 1 & 0\\ 0 & 1 & 0 & 2 & 0 & 1\\ \hline 1 & 0 & 1 & 0 & 2 & 0\\ 0 & 1 & 0 & 1 & 0 & 2 \end{array}\right] \tag{16.15}$$

If the mass is physically lumped at the nodes in three equal parts the 'lumped' mass matrix contributed by the element is

$$\mathbf{M}^e = \frac{m}{3}\left[\begin{array}{cc|cc|cc} 1 & 0 & 0 & 0 & 0 & 0\\ 0 & 1 & 0 & 0 & 0 & 0\\ \hline 0 & 0 & 1 & 0 & 0 & 0\\ 0 & 0 & 0 & 1 & 0 & 0\\ \hline 0 & 0 & 0 & 0 & 1 & 0\\ 0 & 0 & 0 & 0 & 0 & 1 \end{array}\right] \tag{16.16}$$

Certainly both matrices differ considerably and yet in applications the results of the analysis can be almost identical.

Example 16.2: Mass for isoparametric elements. The mass matrix for an isoparametric element may be computed by numerical integration as described in Chapter 5. For example, for two-dimensional elements the mass is given by

$$\mathbf{M}^e_{ab} \approx \sum_l N_a(\xi_l, \eta_l)\rho N_a(\xi_l, \eta_l)\, J_l\, w_l \tag{16.17}$$

for plane problems and

$$\mathbf{M}^e_{ab} \approx \sum_l N_a(\xi_l, \eta_l)\rho N_a(\xi_l, \eta_l)\, r_l\, J_l\, w_l \tag{16.18}$$

for axisymmetric problems.

Now since it is the shape functions which are integrated, the order of quadrature needs to be selected according to the requirements given in Sec. 5.12. Generally, the order used for standard integration will suffice to accurately compute the mass. Reduced quadrature should not be used since spurious results can be obtained due to loss in rank of the mass matrix.

16.2.4 Mass 'lumping' or diagonalization

We have referred to the computational convenience of lumping of mass matrices and presenting these in diagonal form. On some occasions such lumping is physically obvious (see the linear triangle for instance), in others this is not the case and a 'rational' procedure is required. For matrices of the type given in Eq. (16.12b) several alternative approximations have been developed, as discussed in Appendix I. In all of these the essential requirement of mass preservation is satisfied, i.e.,

$$\sum_a \tilde{M}_{aa} = \int_\Omega \rho \,\mathrm{d}\Omega \tag{16.19}$$

where $\tilde{M}_{aa}$ is the diagonal for a component of the lumped mass matrix $\tilde{\mathbf{M}}$.

Three main procedures exist (see Fig. 16.1):

1. The row sum method in which
$$\tilde{M}_{aa} = \sum_b M_{ab}$$
2. Diagonal scaling in which
$$\tilde{M}_{aa} = c\, M_{aa}$$
with c adjusted so that Eq. (16.19) is satisfied,[17, 18] and
3. Evaluation of M using a quadrature involving only the nodal points and thus automatically yielding a diagonal matrix for standard finite element shape functions[19, 20] in which $N_a = 0$ for $\mathbf{x} = \mathbf{x}_b,\ b \neq a$.

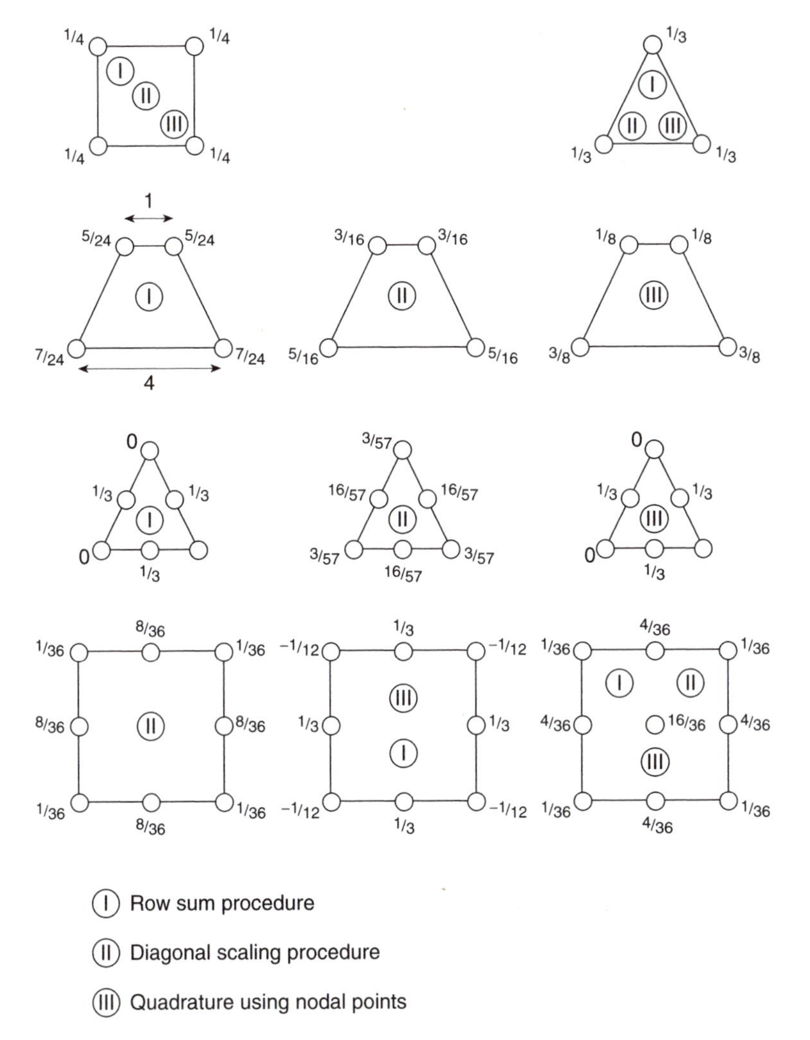

Fig. 16.1 Mass lumping for some two-dimensional elements.

It should be remarked that Eq. (16.19) does not hold for hierarchical shape functions where no lumping procedure appears satisfactory.

The quadrature (numerical integration) process is mathematically most appealing but frequently leads to negative or zero lumped masses. Such a loss of positive definiteness is undesirable in some solution processes and cancels out the advantages of lumping. In Fig. 16.1 we show the effect of various lumping procedures on triangular and quadrilateral elements of linear and quadratic type. It is clear from these that the optimal choice to lump the mass is by no means unique.

In general we would recommend the use of lumped matrices only as a convenient numerical device generally paid for by some loss of accuracy. An exception to this is for 'explicit' time integration of dynamics problems where the considerable efficiency of their use more than compensates for any loss in accuracy (see Chapter 17). However, we note that it has occasionally been shown that lumping can *improve* accuracy of some problem by error cancellation. It can be shown that in the transient approximation the lumping process introduces additional dissipation of the 'stiffness matrix' form and this can help in cancelling out numerical oscillation.

To demonstrate the nature of lumped and consistent mass matrices it is convenient to consider a typical one-dimensional problem specified by the equation

$$\frac{\partial \phi}{\partial t} - \frac{\partial}{\partial x}\left[\mu \frac{\partial}{\partial x}\left(\frac{\partial \phi}{\partial t}\right)\right] - \frac{\partial}{\partial x}\left(k\frac{\partial \phi}{\partial x}\right) = 0$$

Semi-discretization here gives a typical nodal equation a as

$$(M_{ab} + H_{ab})\,\dot{\tilde{u}}_b + K_{ab}\tilde{u}_b = 0$$

where

$$M_{ab} = \int_\Omega N_a N_b \,\mathrm{d}x,\; H_{ab} = \int_\Omega \frac{\mathrm{d}N_a}{\mathrm{d}x}\mu\frac{\mathrm{d}N_b}{\mathrm{d}x}\,\mathrm{d}x,\; K_{ab} = \int_\Omega \frac{\mathrm{d}N_a}{\mathrm{d}x}k\frac{\mathrm{d}N_b}{\mathrm{d}x}\,\mathrm{d}x$$

and it is observed that **H** and **K** have identical structure. With linear elements of constant size h the approximating equation at a typical node a (and surrounding nodes $a-1$ and $a+1$) can be written as follows

$$\begin{aligned}
M_{ab}\dot{\tilde{u}}_b &\equiv \frac{h}{6}\left(\dot{\tilde{u}}_{a-1} + 4\dot{\tilde{u}}_a + \dot{\tilde{u}}_{a+1}\right)\\
H_{ab}\dot{\tilde{u}}_b &\equiv \frac{\mu}{h}\left(-\dot{\tilde{u}}_{a-1} + 2\dot{\tilde{u}}_a - \dot{\tilde{u}}_{a+1}\right)\\
K_{ab}\tilde{u}_b &\equiv \frac{k}{h}\left(-\tilde{u}_{a-1} + 2\tilde{u}_a - \tilde{u}_{a+1}\right)
\end{aligned}$$

If a lumped approximation is used for **M**, that is $\tilde{\mathbf{M}}$, we have, simply by adding coefficients using the row sum method,

$$\tilde{M}_{ab}\dot{\tilde{u}}_b = h\dot{\tilde{u}}_a$$

The difference between the two expressions is

$$\tilde{M}_{ab}\dot{\tilde{u}}_b - M_{ab}\dot{\tilde{u}}_b \equiv \frac{h}{6}\left(-\dot{\tilde{u}}_{a-1} + 2\dot{\tilde{u}}_a - \dot{\tilde{u}}_{a+1}\right)$$

and is clearly identical to that which would be obtained by replacing μ by $h^2/6$. As μ in the above example can be considered as a viscous dissipation we note that the effect of using a lumped matrix is that of adding an extra amount of such viscosity and can often result in smoother (though possibly less accurate) solutions.

Eigenvalues and analytical solution procedures

16.3 General classification

We have seen that as a result of semi-discretization many time-dependent problems can be reduced to a system of ordinary differential equations of the characteristic form given by

$$\mathbf{M}\ddot{\tilde{\mathbf{u}}} + \mathbf{C}\dot{\tilde{\mathbf{u}}} + \mathbf{K}\tilde{\mathbf{u}} + \mathbf{f} = \mathbf{0} \tag{16.20}$$

In this, in general, all the matrices are symmetric. Cases involving non-symmetric matrices are also found in some fluid problems.[21] This second-order system often becomes first order if $\mathbf{M}$ is zero as, for instance, in transient heat conduction problems. We shall now discuss some methods of solution of such ordinary differential equation systems. In general, the above equations can be non-linear (if, for instance, stiffness matrices are dependent on non-linear material properties or if large deformations are involved) but here we shall concentrate on linear cases only.

Systems of ordinary linear differential equations can always in principle be solved analytically without the introduction of additional approximations. The remainder of this chapter will be concerned with such analytical processes. While such solutions are possible they may be so complex that further recourse has to be taken to the process of approximation; we shall deal with this matter in the next chapter. The analytical approach provides, however, an insight into the behaviour of the system which the authors always find helpful.

Some of the matter in this chapter will be an extension of standard well-known procedures used for the solution of differential equations with constant coefficients that are encountered in most studies of dynamics or mathematics. In the following we shall deal successively with:

1. Determination of free response ($\mathbf{f} = \mathbf{0}$)
2. Determination of steady-state periodic response ($\mathbf{f}(t)$ periodic)
3. Determination of transient response ($\mathbf{f}(t)$ arbitrary).

In the first two, initial conditions of the system are not required and a general solution is simply sought. The transient response initial conditions are required and we will devote considerable attention to this type in Sec. 16.8.

16.4 Free response – eigenvalues for second-order problems and dynamic vibration

16.4.1 Free dynamic vibration – real eigenvalues

If no damping or forcing terms exist in the dynamic problem of Eq. (16.20) it reduces to

$$\mathbf{M}\ddot{\tilde{\mathbf{u}}} + \mathbf{K}\tilde{\mathbf{u}} = \mathbf{0} \tag{16.21}$$

A general solution of such an equation may be written as

$$\tilde{\mathbf{u}} = \bar{\mathbf{u}}\exp(\mathrm{i}\omega t)$$

the real part of which simply represents a harmonic response as $\exp(\mathrm{i}\omega t) \equiv \cos\omega t + \mathrm{i}\sin\omega t$. Then on substitution we find that ω can be determined from

$$(-\omega^2\mathbf{M} + \mathbf{K})\bar{\mathbf{u}} = \mathbf{0} \tag{16.22}$$

This is a *general linear eigenvalue* or *characteristic value problem* and for non-zero solutions the determinant of the above coefficient matrix must be zero:

$$|-\omega^2\mathbf{M} + \mathbf{K}| = 0 \tag{16.23}$$

Such a determinant will in general give n positive values of ω^2 (or $\omega_j^2,\ j = 1, 2, \ldots, n$) when the size of the matrices $\mathbf{K}$ and $\mathbf{M}$ is $n \times n$, providing the matrices $\mathbf{K}$ and $\mathbf{M}$ are symmetric positive definite.†

While the solution of Eq. (16.23) cannot determine the actual values of $\bar{\mathbf{u}}$ we can find n vectors $\bar{\mathbf{u}}_j$ that give the proportions for the various terms. Such vectors are known as the *normal modes of the system* or *eigenvectors* and are made unique by normalizing so that

$$\bar{\mathbf{u}}_j^{\mathrm{T}} \mathbf{M} \bar{\mathbf{u}}_j = 1; \quad j = 1, 2, \ldots, n \tag{16.24}$$

At this stage it is useful to note the property of *modal orthogonality*, i.e., that

$$\bar{\mathbf{u}}_i^{\mathrm{T}} \mathbf{M} \bar{\mathbf{u}}_j = 0; \quad (i \neq j) \quad \text{and} \quad \bar{\mathbf{u}}_i^{\mathrm{T}} \mathbf{K} \bar{\mathbf{u}}_j = 0; \quad (i \neq j) \tag{16.25}$$

The proof of the above statement is simple. As Eq. (16.22) is valid for any mode we can write

$$\omega_i^2 \mathbf{M} \bar{\mathbf{u}}_i = \mathbf{K} \bar{\mathbf{u}}_i \quad \text{and} \quad \omega_j^2 \mathbf{M} \bar{\mathbf{u}}_j = \mathbf{K} \bar{\mathbf{u}}_j$$

Premultiplying the first by $\bar{\mathbf{u}}_j^{\mathrm{T}}$ and the second by $\bar{\mathbf{u}}_i^{\mathrm{T}}$ and noting the symmetry of $\mathbf{M}$ and $\mathbf{K}$ so that

$$\bar{\mathbf{u}}_j^{\mathrm{T}} \mathbf{M} \bar{\mathbf{u}}_i = \bar{\mathbf{u}}_i^{\mathrm{T}} \mathbf{M} \bar{\mathbf{u}}_j \quad \text{and} \quad \bar{\mathbf{u}}_j^{\mathrm{T}} \mathbf{K} \bar{\mathbf{u}}_i = \bar{\mathbf{u}}_i^{\mathrm{T}} \mathbf{K} \bar{\mathbf{u}}_j$$

the difference becomes

$$(\omega_i^2 - \omega_j^2) \bar{\mathbf{u}}_i^{\mathrm{T}} \mathbf{M} \bar{\mathbf{u}}_j = 0$$

and if $\omega_i \neq \omega_j$‡ the orthogonality condition for the matrix $\mathbf{M}$ has been proved. From this the orthogonality of the vectors with $\mathbf{K}$ follows immediately. The final condition

$$\bar{\mathbf{u}}_i^{\mathrm{T}} \mathbf{K} \bar{\mathbf{u}}_i = \omega^2$$

follows from Eq. (16.24) and a premultiplication of Eq. (16.22) for equation i by $\bar{\mathbf{u}}_i$.

16.4.2 Determination of eigenvalues

To find the actual eigenvalues it is seldom practical to write the polynomial expanding the determinant given in Eq. (16.23) and alternative techniques have to be developed. Many extremely efficient procedures are available and the reader can find some interesting matter in references 22–28.

In some processes the starting point is the *standard eigenvalue problem* given by

$$\mathbf{H}\mathbf{x} = \lambda \mathbf{x} \tag{16.26}$$

in which $\mathbf{H}$ is a symmetric matrix and hence has real eigenvalues. Equation (16.22) can be written as

$$\mathbf{M}^{-1} \mathbf{K} \bar{\mathbf{u}} = \omega^2 \bar{\mathbf{u}} \tag{16.27}$$

† A symmetric matrix is positive definite if all the diagonals of the triangular factors are positive, this is a usual case with structural problems – all roots of Eq. (16.23) are real positive numbers (for a proof see reference 1). These are known as the natural frequencies of the system. If only the $\mathbf{M}$ matrix is symmetric positive definite while $\mathbf{K}$ is symmetric positive semidefinite the roots are real and positive or zero.

‡ For any case where repeated frequencies occur we merely enforce the orthogonality by construction.

on inverting $\mathbf{M}$ with $\lambda = \omega^2$, but symmetry is in general lost.

If, however, we write in triangular form (i.e., the Cholesky factors)

$$\mathbf{M} = \mathbf{L}\mathbf{L}^{\mathrm{T}} \quad \text{and} \quad \mathbf{M}^{-1} = \mathbf{L}^{-\mathrm{T}}\mathbf{L}^{-1}$$

in which $\mathbf{L}$ is a lower triangular matrix (i.e., has all zero coefficients above the diagonal), Eq. (16.22) may now be written as

$$\mathbf{K}\bar{\mathbf{u}} = \omega^2 \mathbf{L}\mathbf{L}^{\mathrm{T}}\bar{\mathbf{u}}$$

Calling

$$\mathbf{L}^{\mathrm{T}}\bar{\mathbf{u}} = \mathbf{x} \tag{16.28}$$

and multiplying by $\mathbf{L}^{-1}$ we have finally

$$\mathbf{H}\mathbf{x} = \omega^2 \mathbf{x} \tag{16.29}$$

in which

$$\mathbf{H} = \mathbf{L}^{-1}\mathbf{K}\mathbf{L}^{-\mathrm{T}} \tag{16.30}$$

which is of the standard form of Eq. (16.26), as $\mathbf{H}$ is now symmetric.

Having determined ω^2 (all, or only a few, of the selected smallest values corresponding to fundamental periods) the modes of $\mathbf{x}$ are found, and hence by use of Eq. (16.28) the modes of $\bar{\mathbf{u}}$.

If the matrix $\mathbf{M}$ is diagonal – as it will be if the masses have been 'lumped' – the procedure of deriving the standard eigenvalue problem is simplified and here appears the first advantage of the diagonalization, which we have discussed in Sec. 16.2.4.

16.4.3 Free vibration with the singular *K* matrix

In static problems we have always introduced a suitable number of *support* conditions to allow the stiffness matrix $\mathbf{K}$ to be inverted, or what is equivalent to solve the static equations uniquely. If such 'support' conditions are in fact not specified, as may well be the case with a rocket travelling in space, the arbitrary fixing of a minimum number of support conditions allows a static solution to be obtained without affecting the stresses. In dynamic situations such a fixing is not permissible and frequently one is faced with the problem of a free oscillation for which $\mathbf{K}$ is singular and therefore does not possess unique triangular factors or an inverse.

To preserve the applicability of methods which require an inverse (e.g., methods based on inverse power iteration[27]) a simple artifice is possible. Equation (16.22) is modified to

$$[(\mathbf{K} + \alpha\mathbf{M}) - (\omega^2 + \alpha)\mathbf{M}]\bar{\mathbf{u}} = \mathbf{0} \tag{16.31}$$

in which α is an arbitrary constant of the same order as the typical ω^2 sought. The new matrix $(\mathbf{K} + \alpha\mathbf{M})$ is no longer singular and can be factored (or inverted) for use in the standard eigensolution procedure to find $(\omega^2 + \alpha)$ and hence ω^2.

This simple but effective avoidance of an otherwise serious difficulty was first suggested by Cox[29] and Jennings.[30] Alternative methods of dealing with the above problem are given in references 31 and 32.

16.4.4 Reduction of the eigenvalue system

Independent of which technique is used to determine the eigenpairs of the system (16.22), the effort for $n \times n$ matrices is at least one order greater than that involved in an equivalent static situation. Further, while the number of eigenvalues of the real system is infinite, in practice, we are generally interested only in a relatively small number of the lower frequencies and it is possible to simplify the computation by reducing the size of the problem.

To achieve a reduced problem we assume that the unknown $\bar{\mathbf{u}}$ can be expressed in terms of m ($\ll n$) vectors $\mathbf{t}_1, \mathbf{t}_2, \ldots, \mathbf{t}_m$ and corresponding participating factors x_i. We now write

$$\bar{\mathbf{u}} = \mathbf{t}_1 x_1 + \mathbf{t}_2 x_2 + \cdots + \mathbf{t}_m x_m = \mathbf{T}\mathbf{x} \tag{16.32}$$

Inserting Eq. (16.32) into Eq. (16.22) and premultiplying by $\mathbf{T}^{\mathrm{T}}$ we have a reduced problem with only m eigenpairs:

$$(\omega^*)^2 \mathbf{M}^* \mathbf{x} = \mathbf{K}^* \mathbf{x} \tag{16.33}$$

where

$$\mathbf{M}^* = \mathbf{T}^{\mathrm{T}} \mathbf{M} \mathbf{T} \qquad \mathbf{K}^* = \mathbf{T}^{\mathrm{T}} \mathbf{K} \mathbf{T}$$

and ω^* are now eigenvalues of the *reduced* system, which for the appropriate choice of the $\mathbf{t}_i$ vectors can be good approximations to the eigenvalues of the original system.

If by good fortune the trial vectors were to be chosen as eigenvectors of the original matrix the system would become diagonal and all eigenvalues (i.e., in this case $\omega^* = \omega$) could be determined by a trivial calculation. This indeed is what some iterative eigenproblem strategies attempt (e.g., subspace or Lanczos methods[27, 33]). It is also of course possible by physical insight to find vectors $\mathbf{t}$ that correspond closely to the principal modes of the movement (e.g., see reference 34).

16.4.5 Some examples

There are a variety of problems for which practical solutions exist, so only a few simple examples will be shown.

Example 16.3: Vibration of a simply supported beam. Figure 16.2 shows the first three vibration modes of a simply supported beam with length 40 and rectangular cross-section of width 1 and depth 2 units. The elastic properties are $E = 30\,000$, $\nu = 0$, and $\rho = 0.1$ units. The beam is modelled using 9-noded quadrilateral elements of lagrangian type with the central node at the left end restrained in the x and y direction and the central node at the right end restrained only in the y direction. The problem is also solved using a mesh with 1000 2-noded beam elements which include effects of transverse shearing deformation. In Table 16.1 we present the values for the first three frequencies obtained from the finite element analysis and compare to the value obtained from an exact solution for the Euler–Bernoulli beam without shear deformation. It is evident that transverse shearing strains affect the frequencies computed for this problem and, thus, illustrate the importance of using a correct theory for calculations.

Table 16.1 Frequencies for a simply supported beam

Solution form	ω_1	ω_2	ω_3
9-noded element	3.7785	59.2236	290.0804
2-noded element	3.7787	59.2338	290.1774
Beam theory	3.8050	60.8807	308.2080

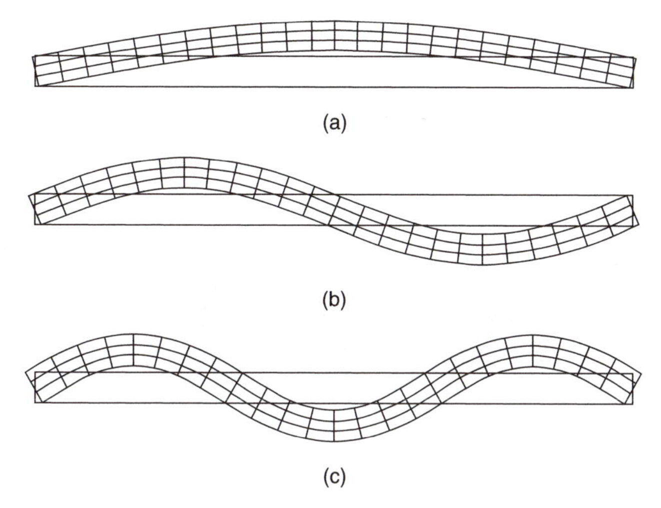

Fig. 16.2 Simply supported beam. (a) $\omega_1 = 3.7785$. (b) $\omega_2 = 59.2236$. (c) $\omega_3 = 290.0804$.

Example 16.4: Vibration of an earth dam. Figure 16.3 shows the vibration of a two-dimensional earth dam resting on a rigid foundation. The earth dam is modelled by linear triangular elements and includes the effects of different material layers.

Example 16.5: Electromagnetic fields. The basic dynamic equation (16.8) can be derived for a variety of non-structural problems. The eigenvalue problem once again occurs with 'stiffness' and 'mass' matrices now having alternate physical meanings.

A particular form of the more general equations discussed earlier is the well-known Helmholz wave equation which, in two-dimensional form, is

$$\frac{\partial^2 \phi}{\partial x^2} + \frac{\partial^2 \phi}{\partial y^2} + \frac{1}{\bar{c}^2} \frac{\partial^2 \phi}{\partial t^2} = 0 \tag{16.34}$$

If the boundary conditions do not force a response, an eigenvalue problem results which has significance in several fields of physical science.

The first application is to *electromagnetic fields*. Figure 16.4 shows a modal shape of a field for a *waveguide problem*. Simple linear triangular elements are used here. More complex three-dimensional oscillations are also discussed in reference 9.

Example 16.6: Waves in shallow water. A similar equation also describes to a reasonable approximation the behaviour of shallow water waves in a body of water:

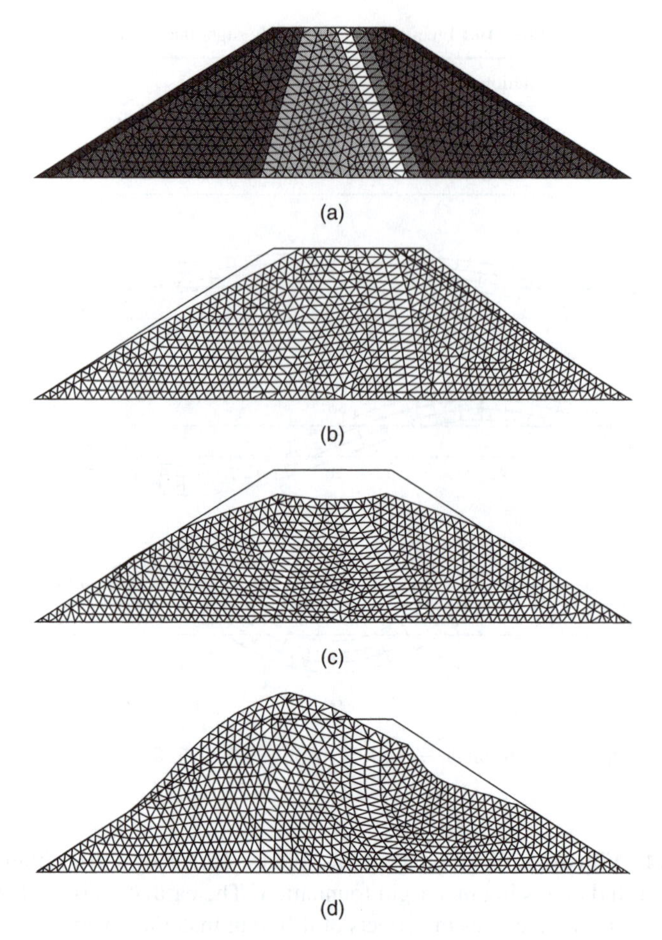

Fig. 16.3 (a) Mesh showing layers considered. (b) Earth dam, Mode 1. (c) Earth dam, Mode 2. (d) Earth dam, Mode 3.

$$\frac{\partial}{\partial x}\left(h\,\frac{\partial \psi}{\partial x}\right)+\frac{\partial}{\partial y}\left(h\,\frac{\partial \psi}{\partial y}\right)+\frac{1}{g}\,\frac{\partial^2 \psi}{\partial t^2}=0 \tag{16.35}$$

in which h is the average water depth, ψ the surface elevation above average and g the gravity acceleration.[21]

Thus natural frequencies of bodies of water contained in harbours of varying depths may easily be found.[10] Figure 16.5 shows the modal shape for a particular harbour.

16.5 Free response – eigenvalues for first-order problems and heat conduction, etc.

If in Eq. (16.20) $\mathbf{M}=\mathbf{0}$, we have a form typical of the transient heat conduction equation [see Eq. (16.1)]. For free response we seek a solution of the homogeneous equation

$$\mathbf{C}\dot{\tilde{\mathbf{u}}}+\mathbf{K}\tilde{\mathbf{u}}=\mathbf{0} \tag{16.36}$$

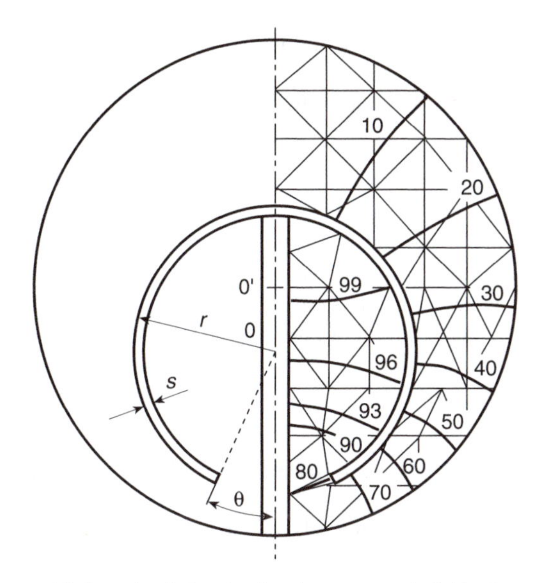

Fig. 16.4 A 'lunar' waveguide;[9] mode of vibration for electromagnetic field. Outer diameter $= d$, $OO' = 0.13d$, $r = 0.29d$, $S = 0.055d$, $\theta = 22^\circ$.

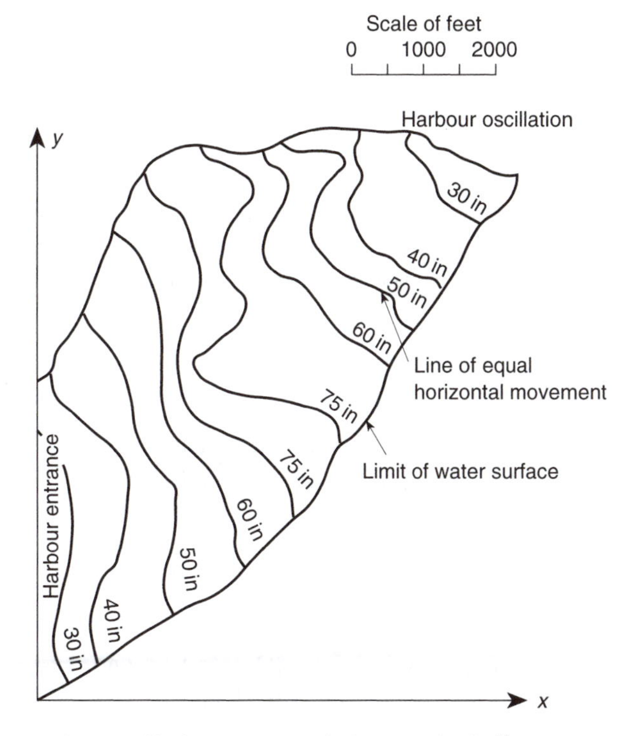

Fig. 16.5 Oscillations of a natural harbour: contours of velocity amplitudes.[10]

Once again an exponential form can be used:

$$\tilde{\mathbf{u}} = \bar{\mathbf{u}} \exp(-\lambda t)$$

Substituting we have

$$(-\lambda \mathbf{C} + \mathbf{K})\bar{\mathbf{u}} = \mathbf{0} \tag{16.37}$$

which again gives an eigenvalue problem identical to that of Eq. (16.22). As **C** and **K** are usually positive definite, λ will be positive and real. The solution therefore represents simply an exponential decay term and is not really steady state. Combination of such terms, however, can be useful in the solution of initial value transient problems but is of little value *per se*.

16.6 Free response – damped dynamic eigenvalues

We shall now consider the full equation (16.20) for free response conditions. Writing

$$\mathbf{M}\ddot{\tilde{\mathbf{u}}} + \mathbf{C}\dot{\tilde{\mathbf{u}}} + \mathbf{K}\tilde{\mathbf{u}} = \mathbf{0} \tag{16.38}$$

and substituting

$$\tilde{\mathbf{u}} = \bar{\mathbf{u}} \exp(\alpha t) \tag{16.39}$$

we have the characteristic equation

$$(\alpha^2 \mathbf{M} + \alpha \mathbf{C} + \mathbf{K})\bar{\mathbf{u}} = \mathbf{0} \tag{16.40}$$

where α and $\bar{\mathbf{u}}$ will in general be found to be complex. The real part of the solution represents a decaying vibration.

The eigenvalue problem involved in Eq. (16.39) is more difficult than that arising in the previous sections. In solutions to date the problem is usually solved by splitting Eq. (16.38) into two first-order equations. This is accomplished by defining

$$\dot{\tilde{\mathbf{u}}} = \mathbf{v}$$

and writing the split form as

$$\begin{bmatrix} \mathbf{M} & \mathbf{0} \\ \mathbf{0} & -\mathbf{M} \end{bmatrix} \begin{Bmatrix} \dot{\mathbf{v}} \\ \dot{\tilde{\mathbf{u}}} \end{Bmatrix} + \begin{bmatrix} \mathbf{C} & \mathbf{K} \\ \mathbf{M} & \mathbf{0} \end{bmatrix} \begin{Bmatrix} \mathbf{v} \\ \tilde{\mathbf{u}} \end{Bmatrix} = \begin{Bmatrix} \mathbf{0} \\ \mathbf{0} \end{Bmatrix} \tag{16.41}$$

Now substituting

$$\tilde{\mathbf{u}} = \bar{\mathbf{u}} \exp(\alpha t) \qquad \mathbf{v} = \bar{\mathbf{v}} \exp(\alpha t)$$

gives the general linear eigenproblem

$$\left(\alpha \begin{bmatrix} \mathbf{M} & \mathbf{0} \\ \mathbf{0} & -\mathbf{M} \end{bmatrix} + \begin{bmatrix} \mathbf{C} & \mathbf{K} \\ \mathbf{M} & \mathbf{0} \end{bmatrix} \right) \begin{Bmatrix} \bar{\mathbf{v}} \\ \bar{\mathbf{u}} \end{Bmatrix} = \begin{Bmatrix} \mathbf{0} \\ \mathbf{0} \end{Bmatrix} \tag{16.42}$$

This form has been studied by Chen *et al.*[35–37] Similar to the first-order problem, no steady-state solution exists and once more the concept of eigenvalues of the above kind is generally of importance only in modal analysis, as we shall see later.

16.7 Forced periodic response

If the forcing term in Eq. (16.20) is periodic or, more generally, if we can express it as

$$\mathbf{f} = \bar{\mathbf{f}} \exp(\alpha t) \tag{16.43}$$

where α is complex, i.e.

$$\alpha = \alpha_1 + \mathrm{i}\,\alpha_2 \tag{16.44}$$

then a general solution can once more be written as

$$\tilde{\mathbf{u}} = \bar{\mathbf{u}} \exp(\alpha t) \tag{16.45}$$

Substituting the above in Eq. (16.20) gives

$$\left(\alpha^2 \mathbf{M} + \alpha \mathbf{C} + \mathbf{K}\right) \bar{\mathbf{u}} \equiv \bar{\mathbf{K}} \bar{\mathbf{u}} = -\bar{\mathbf{f}} \tag{16.46}$$

which is no longer an eigenvalue problem but can be solved formally by inverting the matrix $\bar{\mathbf{K}}$ as

$$\bar{\mathbf{u}} = -\bar{\mathbf{K}}^{-1} \bar{\mathbf{f}} \tag{16.47}$$

The solution is thus precisely of the same form as that used for static problems but now, however, has to be determined in terms of complex quantities.

With periodic input the solution after an initial transient is not sensitive to the initial conditions and the above solution represents the finally established response. It is valid for problems of dynamic structural and fluid-structure responses as well as for problems typical of heat conduction in which we simply put $\mathbf{M} = \mathbf{0}$.

16.8 Transient response by analytical procedures

16.8.1 General

In the previous sections we have been concerned with steady-state general solutions which took no account of the initial conditions of the system or of the non-periodic form of the forcing terms. The response taking these features into account is essential if we consider, for instance, the earthquake behaviour of structures or the transient behaviour of the heat conduction problem. The solution of such general cases requires either a full-time discretization, which we shall discuss in detail in the next chapter, or the use of special analytical procedures.

16.8.2 Frequency response procedures

In Sec. 16.7 we have shown how the response of the system to any forcing terms of the general periodic type or in particular to a periodic forcing function

$$\mathbf{f} = \bar{\mathbf{f}} \exp(\mathrm{i}\,\omega t) \tag{16.48}$$

can be obtained by solving a simple equation system. As a completely arbitrary forcing function can be represented approximately by a Fourier series or in the limit, exactly, as a Fourier integral, the response to such an input can be obtained by a synthesis of a curve representing the response of any quantity of interest, e.g., the displacement at a particular point, etc., to all frequencies ranging from zero to infinity. In fact only a limited number of such forcing frequencies has to be considered and a result can be synthesized efficiently by fast Fourier transform techniques.[38] We shall not discuss the mathematical details for such procedures which can be found in standard texts on structural dynamics.[12, 16]

The technique of frequency response is readily adapted to problems where the damping matrix **C** is of an arbitrary specified form. This is not the case with the more widely used modal decomposition procedures which are to be described in the next section.

By way of illustration we show in Fig. 16.6 the frequency response of an artificial harbour [see Eq. (16.35)] to an input of waves with different frequencies and damping due to the radiation of reflected waves which imposes a very particular form on the damping matrix. Details of this problem are given elsewhere.[21, 39, 40] Similar techniques are frequently used in the analysis for the foundation response of structures where radiation of energy occurs.[41]

16.8.3 Modal decomposition analysis

This procedure is probably the most important and widely used in practice. Further, it provides an insight into the behaviour of the whole system, which is of value where strictly numerical processes are used. We shall therefore describe it in detail in the context of the general problem of Eq. (16.20), i.e.,

$$\mathbf{M}\ddot{\tilde{\mathbf{u}}} + \mathbf{C}\dot{\tilde{\mathbf{u}}} + \mathbf{K}\tilde{\mathbf{u}} + \mathbf{f} = \mathbf{0} \tag{16.49}$$

where $\mathbf{f}$ is an arbitrary function of time.

We have seen that the general solution for the free response is of the form

$$\tilde{\mathbf{u}} = \sum_{i=1}^{n} \bar{\mathbf{u}}_i \exp(\alpha_i t) \tag{16.50}$$

where α_i are the (complex) eigenvalues and $\bar{\mathbf{u}}_i$ are the (complex) eigenvectors (Sec. 16.6). For forced response we shall assume that the problem is linear such that the solution can be written as a linear combination of the modes

$$\tilde{\mathbf{u}} = \sum_{i=1}^{n} \bar{\mathbf{u}}_i y_i(t) = \begin{bmatrix} \bar{\mathbf{u}}_1, & \bar{\mathbf{u}}_2, & \ldots \end{bmatrix} \mathbf{y}(t) \tag{16.51}$$

where the scalar modal participation factor y_i is now a function of time. This shows in a clear manner the proportions of each mode occurring. Such a decomposition of an arbitrary vector presents no restriction as all the modes are linearly independent vectors (with those for repeated frequencies being constructed to be linearly independent as mentioned in Sec. 16.4).

If expression (16.51) is substituted into Eq. (16.49) and the result is premultiplied by the complex conjugate transposed, $\bar{\mathbf{u}}_i^{\mathrm{T}}$ $(i = 1, \ldots, n)$, then the result is simply a set of scalar, independent, equations

$$m_i \ddot{y}_i + c_i \dot{y}_i + k_i y_i + f_i = 0 \tag{16.52}$$

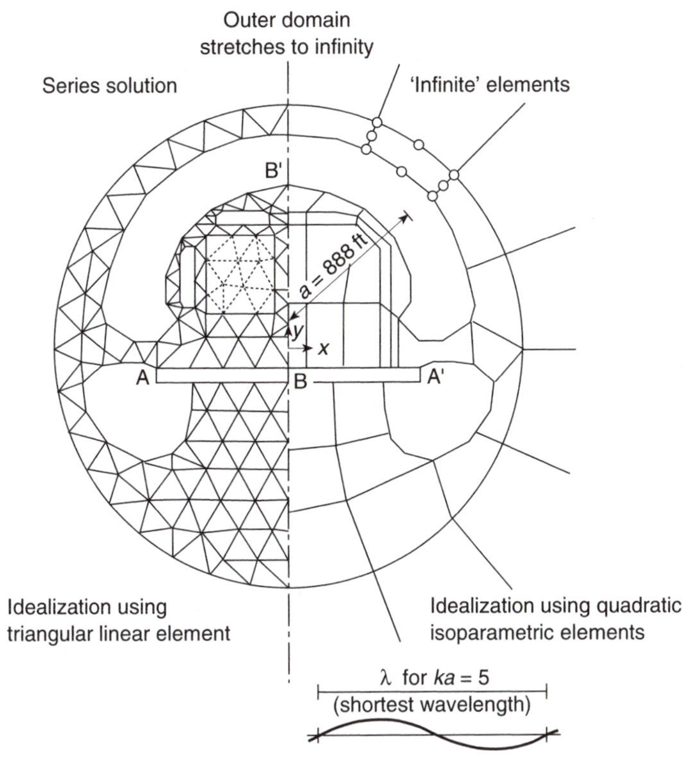

(a) Geometric details and FEM idealization.
Wave forcing frequency $\omega = k\sqrt{gh} = ka$, h = depth of water

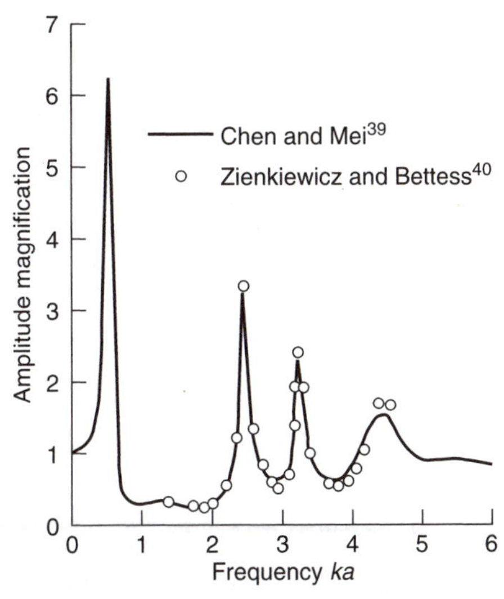

(b) Amplitude magnification response of mean depth in harbour for various frequencies

Fig. 16.6 Frequency response of an artificial harbour to an input of periodic wave.

where

$$m_i = \bar{\mathbf{u}}_i^{\mathrm{T}}\mathbf{M}\bar{\mathbf{u}}_i, \; c_i = \bar{\mathbf{u}}_i^{\mathrm{T}}\mathbf{C}\bar{\mathbf{u}}_i, \; k_i = \bar{\mathbf{u}}_i^{\mathrm{T}}\mathbf{K}\bar{\mathbf{u}}_i \;\text{ and }\; f_i = \bar{\mathbf{u}}_i^{\mathrm{T}}\mathbf{f}$$

as for true eigenvectors $\bar{\mathbf{u}}_i$

$$\bar{\mathbf{u}}_i^{\mathrm{T}}\mathbf{M}\bar{\mathbf{u}}_j = \bar{\mathbf{u}}_i^{\mathrm{T}}\mathbf{C}\bar{\mathbf{u}}_j = \bar{\mathbf{u}}_i^{\mathrm{T}}\mathbf{K}\bar{\mathbf{u}}_j \doteq 0$$

when $i \neq j$ (this result was proved in Sec. 16.4 for real eigenpairs but is valid generally for complex pairs, as could be verified by the reader).

Each scalar equation of (16.52) can be solved by elementary procedures independently and the total vector of response obtained by superposition following Eq. (16.51). In the general case, as we have shown in Sec. 16.6, the eigenpairs are complex and their determination is not simple.[31] The more usual procedure is to use real eigenpairs corresponding to the solution of Eq. (16.21):

$$\mathbf{K}\bar{\mathbf{u}} = \omega^2\mathbf{M}\bar{\mathbf{u}} \tag{16.53}$$

Decoupled equations with real variables $\mathbf{y}$ exist only if

$$\bar{\mathbf{u}}_i^{\mathrm{T}}\mathbf{C}\bar{\mathbf{u}}_j = 0; \qquad i \neq j$$

which generally does not occur as the eigenvectors now guarantee only orthogonality with $\mathbf{M}$ and $\mathbf{K}$ and not of the damping matrix. However, if the damping matrix $\mathbf{C}$ is of the form of Eq. (16.13), i.e., a linear combination of $\mathbf{M}$ and $\mathbf{K}$, such orthogonality will obviously occur. Unless the damping is of a definite form which requires special treatment, an assumption of orthogonality is made and Eq. (16.52) is assumed valid in terms of such eigenvectors.

From Eq. (16.53) we have

$$\mathbf{K}\bar{\mathbf{u}}_i = \omega_i^2\mathbf{M}\bar{\mathbf{u}}_i \tag{16.54a}$$

and on premultiplying by $\bar{\mathbf{u}}_i^{\mathrm{T}}$ we obtain

$$k_i = \omega_i^2 m_i \tag{16.54b}$$

Writing the modal damping in the form

$$c_i = 2\omega\xi_i \tag{16.54c}$$

(where ξ_i represents the ratio of damping to its critical value) and assuming that the modes have been normalized so that $m_i = 1$ [see Eq. (16.24)], Eq. (16.52) can be rewritten in standard second order form:

$$\ddot{y}_i + 2\omega_i\xi_i\dot{y}_i + \omega_i^2 y_i + f_i = 0 \tag{16.54d}$$

A general solution is then obtained as

$$\begin{aligned} y_i = {} & \exp(-\xi_i\omega_i t)\left[\frac{\dot{y}_{i0} + \xi_i\omega_i y_{i0}}{\bar{\omega}}\sin\bar{\omega}_i t + y_{i0}\cos\bar{\omega}_i t\right] \\ & + \frac{1}{\bar{\omega}_i}\int_0^t \exp(-\xi_i\omega_i[t-\tau])\sin\bar{\omega}_i(t-\tau)f_i(\tau)\,\mathrm{d}\tau \end{aligned} \tag{16.55}$$

in which $\bar{\omega}_i = \omega_i\sqrt{1-\xi_i^2}$ and y_{i0}, $\dot{y}_{i0}$ are initial conditions computed from

$$y_{i0} = \bar{\mathbf{u}}_i^{\mathrm{T}}\mathbf{M}\tilde{\mathbf{u}}(0) \quad\text{and}\quad \dot{y}_{i0} = \bar{\mathbf{u}}_i^{\mathrm{T}}\mathbf{M}\dot{\tilde{\mathbf{u}}}(0) \tag{16.56}$$

The solution of Eq. (16.55) can be carried out by assuming the forcing function is given by linear interpolation between discrete time points t_k and then evaluating the resulting integrals exactly. Alternatively, a numerical solution can be carried out and the response obtained. In practice, often a single calculation is carried out for each mode to determine the maximum responses and a suitable addition of these results is used. Such processes are described in standard texts and are used as procedures to calculate the bounds on behaviour of structures subjected to seismic loading.[12, 16, 28]

16.8.4 Damping and participation of modes

The type of calculation implied in modal decomposition apparently necessitates the determination of all modes and eigenvalues, a task of considerable magnitude. In fact only a limited number of modes usually need to be taken into consideration as often the response to higher frequency is critically damped and insignificant.

To show that this is true consider the form of the damping matrices. In Sec. 16.2 [Eq. (16.13)] we have indicated that the damping matrix is often assumed as

$$\mathbf{C} = \alpha \mathbf{M} + \beta \mathbf{K} \tag{16.57}$$

Indeed a form of this type is necessary for the use of modal decomposition, although other generalizations are possible.[42, 43] From the definition of ξ_i, the ratio to critical damping ratio in Eq. (16.54c), we see that this can now be written as

$$\xi_i = \frac{1}{2\omega_i} \bar{\mathbf{u}}_i^{\mathrm{T}} (\alpha \mathbf{M} + \beta \mathbf{K}) \bar{\mathbf{u}}_i = \frac{1}{2\omega_i} \left(\alpha + \beta \omega_i^2 \right) \tag{16.58}$$

Thus if the coefficient β is of greater importance, as is the case with most structural damping, ξ_i grows with ω_i and at high frequency an overdamped condition will arise.[12] This is indeed fortunate as, in general, an infinite number of high frequencies exist which are not modelled by any finite element discretization.

We shall see in the next chapter that in the step-by-step recurrence computation the high frequencies often control the problem, and this effect needs to be 'filtered out' for realistic results.

16.9 Symmetry and repeatability

In concluding this chapter it is worth remarking that in dynamic calculation we have once again encountered all the general principles of assembly, etc., that are applicable to static problems. However, some aspects of symmetry and repeatability which were used previously (see Sec. 6.2.4) need amending. It is obviously possible for symmetric structures to vibrate in an unsymmetrical manner, for instance, and similarly a repeatable structure contains modes which are themselves non-repeatable. However, even here considerable simplification can still be made; details of this are discussed by Williams,[44] Thomas[45] and Evensen.[46]

16.10 Problems

16.1 Specialize the problem given in Sec. 16.2 for the case where $\rho = 0$. Construct $\mathbf{C}^e$ and $\mathbf{K}^e$ for a typical 3-node triangular element and a 6-node hierarchical triangle in which coordinates are given by

$$\mathbf{x} = \sum_{a=1}^{3} L_a \tilde{\mathbf{x}}_a$$

16.2 An axial bar under transient loading is governed by

$$\frac{\partial}{\partial x}\left(EA\frac{\partial u}{\partial x}\right) + q = \rho A \frac{\partial^2 u}{\partial t^2}$$

and boundary conditions

$$u(x,t) = \bar{g}(t), \; x \text{ on } \Gamma_1 \quad \text{or} \quad EA\frac{\partial u}{\partial x} = \bar{P}(t), \; x \text{ on } \Gamma_2$$

where $\bar{g}(t)$ and $\bar{P}(t)$ are specified displacement and force, respectively.

(a) Construct a weak form for the problem. Is there a variational theorem for the problem?

(b) Consider an isoparametric element interpolation

$$x^e(\xi) = N_1(\xi)\,\tilde{x}_1^e + N_2(\xi)\,\tilde{x}_2^e$$
$$u^e(\xi,t) = N_1(\xi)\,\tilde{u}_1^e(t) + N_2(\xi)\,\tilde{u}_2^e(t)$$

with $N_1 = (1-\xi)/2$ and $N_2 = (1+\xi)/2$ and show the *stiffness* and *mass* element arrays are given by

$$\mathbf{K}^e = \frac{EA}{h}\begin{bmatrix} 1 & -1 \\ -1 & 1 \end{bmatrix} \quad \text{and} \quad \mathbf{M}^e = \tfrac{1}{6}\,\rho A h \begin{bmatrix} 2 & 1 \\ 1 & 2 \end{bmatrix},$$

respectively (where $h = \tilde{x}_2^e - \tilde{x}_1^e$).

(c) Let

$$u(x,t) = \sum_i \bar{u}_i(x) \exp i\omega_i t$$

and determine the discrete eigenproblem resulting from the weak form developed in (a). Is there a variational theorem for the problem?

(d) Consider a two-element problem shown in Fig. 16.7 and solve the eigenproblem developed in (c). Let $u(0,t) = \bar{P}(L,t) = 0$ and use material properties $E = A = \rho = L = 1$ and $q = 0$.

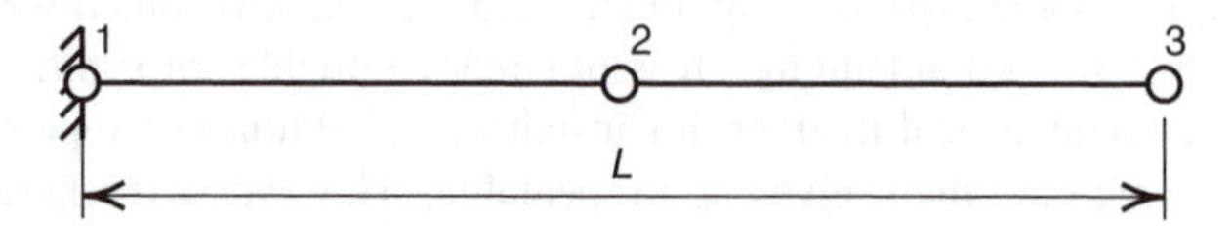

Fig. 16.7 Two-element bar for Problem 16.2.

(e) Write a MATLAB program to solve the discrete eigenproblem. Check your program using the two-element solution; then solve the problem using 4, 8, and 16 elements. Plot the first two eigenvalues vs the number of elements.

(f) Obtain an exact value for the first two eigenvalues and plot the error for each vs the number of elements on a log–log plot. What is the rate of convergence?

(g) Replace the element mass matrix by a *lumped* form given by

$$\mathbf{M}^e = \tfrac{1}{2}\,\rho h \begin{bmatrix} 1 & 0 \\ 0 & 1 \end{bmatrix}$$

and repeat parts (d) and (e)

16.3 Compute the lumped mass matrix by row sum (Method I) for a cubic order serendipity element which is a square with side length a. Note it is only necessary to compute one vertex and one mid-side value.

16.4 Compute the lumped mass matrix by diagonal scaling (Method II) for a cubic order serendipity element which is a square with side length a. Note it is only necessary to compute one vertex and one mid-side value.

16.5 Compute the lumped mass matrix by row sum (Method I) for a cubic order lagrangian element which is a square with side length a. Note it is only necessary to compute one vertex, one mid-side value and the interior node value.

16.6 For a three-dimensional cube with side lengths a in each direction compute the lumped mass matrix by row sum (Method I) for a quadratic order lagrangian element. Note it is only necessary to compute one vertex, one mid-side value and the interior node value.

16.7 Show that the degeneration of a 9-node lagrangian quadrilateral in ξ, η coordinates into a 6-node triangle as described in Example 5.2 of Sec. 5.8.1 yields the same mass matrix as that derived for the triangle using L_1, L_2, L_3 area coordinates. Note it is only necessary to compute one vertex and one mid-side value for the triangle.

16.8 The bar shown in Fig. 16.8 is divided into four elements and has the right end attached to a damper. A weak form for the problem may be written as

$$\mathbf{M}\ddot{\tilde{\mathbf{u}}} + \mathbf{C}\dot{\tilde{\mathbf{u}}} + \mathbf{K}\tilde{\mathbf{u}} + \mathbf{f} = \mathbf{0}$$

where $\mathbf{M}$ and $\mathbf{K}$ are obtained using the element arrays given in Problem 16.1. For $\tilde{\mathbf{u}} = \begin{bmatrix} \tilde{u}_2 & \tilde{u}_3 & \tilde{u}_4 & \tilde{u}_5 \end{bmatrix}^{\mathrm{T}}$:

(a) Construct $\mathbf{M}$ and $\mathbf{K}$ for the problem.

(b) Construct $\mathbf{C}$ for the problem.

(c) Use MATLAB to compute the eigensolution for the problem. Plot real and imaginary parts for the problem. On a separate plot show as vectors the real and imaginary parts of the complex frequencies. Are they proportional?

16.9 Use *FEAPpv* to verify the results given for Example 16.3 in Table 16.1 and Fig. 16.2. Use a consistent mass matrix for the computation. Repeat the analysis using a lumped mass. If the mesh is refined several times, do you expect the results to converge? Why?

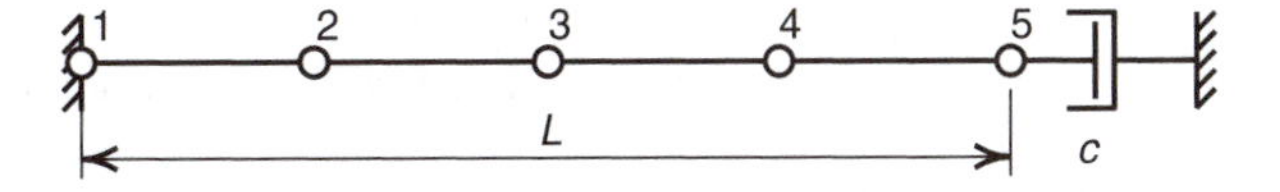

Fig. 16.8 Four-element bar with end damper for Problem 16.8.

16.10 Use *FEAPpv* to compute the first three eigenpairs for the rectangular beam problem described in Example 2.3 of Sec. 2.9. Use the same properties for E and ν and let $\rho = 0.001$.

Using the first mode and $P = 1$ for $0 < t < 2$ obtain the solution for the first 3 seconds using $\Delta t = 0.1$. Plot the results for the vertical displacement at the tip where loading is applied.

Repeat the solution using three modes.

16.11 The curved beam problem described in Example 2.4 of Sec. 2.9 is to be solved for the case where the boundary condition at $y = 0$ is specified by the shear stress of the exact solution. For the case of linear variation of displacements on the boundary edge write a MATLAB program to compute the consistent nodal loads for a unit force P and four equally spaced segments.

For the mesh shown in Fig. 2.11(b) use *FEAPpv* to compute the first eigenpair (the one where ω_i is smallest) for the problem assuming the same properties for E and ν and take $\rho = 1$.

Using the first mode and $u(x, 0) = \sin^2 t$ for $0 < t < \pi$ obtain the solution for the first 5 seconds using $\Delta t = 0.1$. Plot the results for the vertical displacement at the tip where loading is applied.

Repeat the solution using three modes.

16.12 Program development project: Extend the program system started in Problem 2.17 to compute a lumped and a consistent mass matrix for 3-node triangular and 4-node quadrilateral elements. Use the generalized eigenproblem $[V, D] = EIG(K, M)$ from MATLAB to compute the eigenvectors ($V = \bar{u}_i$) and eigenvalues ($D = \omega_i$).

Use your program to determine the eigenvalues and eigenvectors for the curved beam analysed in Problem 16.11. Results may be checked using *FEAPpv*.

16.13 Program development project: Extend the program system developed for Problem 16.12 to perform mode superposition as described in Sec. 16.8.3. You may omit the modal damping factors ξ_i for simplicity.

For the rectangular beam considered as Example 2.3 (using triangular elements as shown in Fig. 2.8), assume the end shear is applied suddenly at time zero and held constant for 2 seconds at which time it is suddenly removed.

Perform a modal solution in which only the lowest eigenvalue mode is used. For a time increment of $\Delta t = 0.001$ determine and plot the first 5 seconds of response for the vertical displacement at the tip centre-line. Repeat the solution using the lowest three eigenvalue modes. Compare your solutions with that in which all modes are included (which is the exact solution for the *semi-discrete* equations). Comment on differences obtained.

References

1. S. Crandall. *Eng. Analysis*. McGraw-Hill, New York, 1956.
2. H.S. Carslaw and J.C. Jaeger. *Conduction of Heat in Solids*. Clarendon Press, Oxford, 2nd edition, 1959.
3. W. Visser. A finite element method for the determination of non-stationary temperature distribution and thermal deformation. In *Proc. 1st Conf. Matrix Methods in Structural Mechanics*, volume AFFDL-TR-66-80, Wright Patterson Air Force Base, Ohio, Oct. 1966.

4. O.C. Zienkiewicz and Y.K. Cheung. *The Finite Element Method in Structural Mechanics*. McGraw-Hill, London, 1967.
5. E.L. Wilson and R.E. Nickell. Application of finite element method to heat conduction analysis. *Nuclear Eng. Design*, 4:1–11, 1966.
6. O.C. Zienkiewicz and C.J. Parekh. Transient field problems – two and three dimensional analysis by isoparametric finite elements. *Int. J. Numer. Meth. Eng.*, 2:61–71, 1970.
7. K. Terzhagi and R.B. Peck. *Soil Mechanics in Engineering*. John Wiley & Sons, New York, 1948.
8. D.K. Todd. *Ground Water Hydrology*. John Wiley & Sons, New York, 1959.
9. P.L. Arlett, A.K. Bahrani, and O.C. Zienkiewicz. Application of finite elements to the solution of Helmholz's equation. *Proc. IEE*, 115:1762–1764, 1968.
10. C. Taylor, B.S. Patil, and O.C. Zienkiewicz. Harbour oscillation: a numerical treatment for undamped natural modes. *Proc. Inst. Civ. Eng.*, 43:141–156, 1969.
11. O.C. Zienkiewicz and R.E. Newton. Coupled vibration of a structure submerged in a compressible fluid. In *Proc. Int. Symp. on Finite Element Techniques*, pages 359–371, Stuttgart, 1969.
12. A.K. Chopra. *Dynamics of Structures*. Prentice-Hall, Upper Saddle River, N.J., 1995.
13. J.S. Archer. Consistent mass matrix for distributed systems. *Proc. Am. Soc. Civ. Eng.*, 89(ST4):161, 1963.
14. F.A. Leckie and G.M. Lindberg. The effect of lumped parameters on beam frequencies. *Aero. Q.*, 14:234, 1963.
15. O.C. Zienkiewicz and Y.K. Cheung. The finite element method for analysis of elastic isotropic and orthotropic slabs. *Proc. Inst. Civ. Eng.*, 28:471–488, 1964.
16. R.W. Clough and J. Penzien. *Dynamics of Structures*. McGraw-Hill, New York, 2nd edition, 1993.
17. S.W. Key and Z.E. Beisinger. The transient dynamic analysis of thin shells in the finite element method. In *Proc. 1st Conf. Matrix Methods in Structural Mechanics*, volume AFFDL-TR-66-80, pages 667–710, Wright Patterson Air Force Base, Ohio, Oct. 1966.
18. E. Hinton, T. Rock, and O.C. Zienkiewicz. A note on mass lumping and related processes in the finite element method. *Earth. Eng. Struct. Dyn.*, 4:245–249, 1976.
19. P. Tong, T.H.H. Pian, and L.L. Bociovelli. Mode shapes and frequencies by the finite element method using consistent and lumped matrices. *Comp. Struct.*, 1:623–638, 1971.
20. I. Fried and D.S. Malkus. Finite element mass matrix lumping by numerical integration with the convergence rate loss. *Int. J. Solids Struct.*, 11:461–465, 1975.
21. O.C. Zienkiewicz, R.L. Taylor, and P. Nithiarasu. *The Finite Element Method for Fluid Dynamics*. Butterworth-Heinemann, Oxford, 6th edition, 2005.
22. J.H. Wilkinson. *The Algebraic Eigenvalue Problem*. Clarendon Press, Oxford, 1965.
23. I. Fried. Gradient methods for finite element eigen problems. *J. AIAA*, 7:739–741, 1969.
24. J.H. Wilkinson and C. Reinsch. *Linear Algebra. Handbook for Automatic Computation*, volume II. Springer-Verlag, Berlin, 1971.
25. K.K. Gupta. Solution of eigenvalue problems by Sturm sequence method. *Int. J. Numer. Meth. Eng.*, 4:379–404, 1972.
26. A. Jennings. Mass condensation and similarity iterations for vibration problems. *Int. J. Numer. Meth. Eng.*, 6:543–552, 1973.
27. B.N. Parlett. *The Symmetric Eigenvalue Problem*. Prentice Hall, Englewood Cliffs, N.J., 1980.
28. K.-J. Bathe. *Finite Element Procedures*. Prentice Hall, Englewood Cliffs, N.J., 1996.
29. H.L. Cox. Vibration of missiles. *Aircraft Eng.*, 33:2–7 and 48–55, 1961.
30. A. Jennings. Natural vibration of a free structure. *Aircraft Eng.*, 34:8, 1962.
31. W.C. Hurty and M.F. Rubinstein. *Dynamics of Structures*. Prentice Hall, Englewood Cliffs, N.J., 1974.
32. A. Craig and M.C.C. Bampton. On the iterative solution of semi definite eigenvalue problems. *Aero. J.*, 75:287–290, 1971.

33. J. Demmel. *Applied Numerical Linear Algebra*. Society for Industrial and Applied Mathematics, Philadelphia, PA, 1997.
34. O.C. Zienkiewicz and R.L. Taylor. *The Finite Element Method*, volume 2. McGraw-Hill, London, 4th edition, 1991.
35. H.-C. Chen and R.L. Taylor. Using Lanczos vectors and Ritz vectors for computing dynamic responses. *Eng. Comp.*, 6:151–157, 1989.
36. A. Ibrahimbegovic, H.-C. Chen, E.L. Wilson, and R.L. Taylor. Ritz method for dynamic analysis of large discrete linear systems with non-proportional damping. *Earth. Eng. Struct. Dyn.*, 19:877–889, 1990.
37. H.-C. Chen and R.L. Taylor. Properties and solutions of the eigensystem of non-proportionally damped linear dynamic systems. Technical Report UCB/SEMM-86/10, University of California, Berkeley, Nov. 1986.
38. E.O. Brigham. *The Fast Fourier Transform*. Prentice Hall, Englewood Cliffs, N.J., 1974.
39. H.S. Chen and C.C. Mei. Hybrid-element method for water waves. In *Proc. Modelling Techniques Conf. (Modelling 1975)*, volume 1, pages 63–81, San Francisco, 1975.
40. O.C. Zienkiewicz and P. Bettess. Infinite elements in the study of fluid-structure interaction problems. In *2nd Int. Symp. on Computing Methods in Applied Science and Engineering*, Versailles, France, Dec. 1975.
41. J. Penzien. Frequency domain analysis including radiation damping and water load coupling. In O.C. Zienkiewicz, R.W. Lewis, and K.G. Stagg, editors, *Numerical Methods in Offshore Engineering*. John Wiley & Sons, 1978.
42. E.L. Wilson and J. Penzien. Evaluation of orthogonal damping matrices. *Int. J. Numer. Meth. Eng.*, 4:5–10, 1972.
43. H.T. Thomson, T. Collins, and P. Caravani. A numerical study of damping. *Earth. Eng. Struct. Dyn.*, 3:97–103, 1974.
44. F.W. Williams. Natural frequencies of repetitive structures. *Q. J. Mech. Appl. Math.*, 24:285–310, 1971.
45. D.L. Thomas. Standing waves in rotationally periodic structures. *J. Sound Vibr.*, 37:288–290, 1974.
46. D.A. Evensen. Vibration analysis of multi-symmetric structures. *J. AIAA*, 14:446–453, 1976.

17

The time dimension – discrete approximation in time

17.1 Introduction

In the last chapter we have shown how semi-discretization of dynamic or transient field problems leads, in linear cases, to sets of ordinary differential equations of the form†

$$\mathbf{M}\ddot{\mathbf{u}} + \mathbf{C}\dot{\mathbf{u}} + \mathbf{K}\mathbf{u} + \mathbf{f} = \mathbf{0} \qquad \text{where} \qquad \frac{\mathrm{d}\mathbf{u}}{\mathrm{d}t} \equiv \dot{\mathbf{u}}, \text{ etc.} \tag{17.1}$$

subject to initial conditions

$$\mathbf{u}(0) = \mathbf{u}_0 \qquad \text{and} \qquad \dot{\mathbf{u}}(0) = \dot{\mathbf{u}}_0$$

or for transient field problems (e.g., heat conduction) to

$$\mathbf{C}\dot{\mathbf{u}} + \mathbf{K}\mathbf{u} + \mathbf{f} = \mathbf{0} \tag{17.2}$$

subject to the initial condition

$$\mathbf{u}(0) = \mathbf{u}_0$$

In many practical situations non-linearities exist, typically altering the above equations by making

$$\mathbf{M} = \mathbf{M}(\mathbf{u}) \qquad \mathbf{C} = \mathbf{C}(\mathbf{u}) \qquad \mathbf{K}\mathbf{u} = \mathbf{P}(\mathbf{u}) \tag{17.3}$$

The analytical solutions previously discussed, while providing much insight into the behaviour patterns (and indispensable in establishing such properties as natural system frequencies), are in general not economical for the solution of transient problems in linear cases and not applicable when non-linearity exists. In this chapter we shall therefore revert to discretization processes applicable directly to the time domain.

For such time discretization the finite element method, including in its definition the finite difference approximation, is of course widely applicable and provides the greatest possibilities, though much of the classical literature on the subject uses only the latter.[1–6] We shall demonstrate here how the finite element method provides a useful generalization unifying many existing algorithms and providing a variety of new ones.

As the time domain is infinite we shall inevitably curtail it to a finite time increment Δt and relate the initial conditions at t_n (and sometimes before) to those at time $t_{n+1} = t_n + \Delta t$, obtaining so-called *recurrence relations*. In all of this chapter, the starting point will be that of the semi-discrete equations (17.1) or (17.2), though, of course, the full space–time

† To simplify notation, we omit the 'tilde' on approximations in time of the independent variable, thus, $\tilde{\mathbf{u}}(t_n) \approx \mathbf{u}_n$.

domain discretization could be considered simultaneously. This, however, usually offers no advantage, for, with the regularity of the time domain, irregular space–time elements are not required. Indeed, if product-type shape functions are chosen, the process will be identical to that obtained by using first semi-discretization in space followed by time discretization. An exception here is provided in convection dominated problems where simultaneous discretization may be desirable (as is discussed in reference 7).

The first concepts of space–time elements were introduced in 1969–70[8–11] and the development of processes involving semi-discretization is presented in references 12–21. Full space–time elements are described for convection-type equations in references 22, 23 and 24 and for elastodynamics in references 25, 26 and 27.

The presentation of this chapter will be divided into three parts. In the first we shall derive a set of *single-step* recurrence relations for the linear first and second order problems of Eqs (17.2) and (17.1). Such schemes have a very general applicability and are preferable to *multistep schemes* described in the second part as the time step can be easily and adaptively varied. In the third part we briefly describe a *discontinuous Galerkin scheme* and show its application in some simple problems.

When discussing stability problems we shall often revert to the concept of modally uncoupled equations introduced in the previous chapter. Here we recall that the equation systems (17.1) and (17.2) can be written as a set of scalar equations:

$$m_i \ddot{y}_i + c_i \dot{y}_i + k_i y_i + f_i = 0 \tag{17.4}$$

or

$$c_i \dot{y}_i + k_i y_i + f_i = 0 \tag{17.5}$$

in the respective eigenvalue participation factors y_i. We shall find that the stability requirements here are dependent on the eigenvalues associated with such equations, ω_i. It turns out, however, fortunately, that it is never necessary to obtain the system eigenvalues or eigenvectors due to a powerful theorem first stated for finite element problems by Irons and Treharne.[28]

The theorem states simply that the system eigenvalues can be bounded by the eigenvalues of individual elements ω^e. Thus

$$\min_j(\omega_j)^2 \geq \min_e(\omega^e)^2 \quad \text{and} \quad \max_j(\omega_j)^2 \leq \max_e(\omega^e)^2 \tag{17.6}$$

The stability limits can thus (as will be shown later) be related to Eq. (17.4) or (17.5) written for a single element.

Single-step algorithms

17.2 Simple time-step algorithms for the first-order equation

17.2.1 Weighted residual finite element approach

We shall now consider Eq. (17.2) which may represent a semi-discrete approximation to a particular physical problem or simply be itself a discrete system. The objective is to obtain an approximation for $\mathbf{u}_{n+1}$ given the value of $\mathbf{u}_n$ and the forcing vector $\mathbf{f}$ acting in the interval of time Δt. It is clear that in the first interval $\mathbf{u}_n$ is the initial condition $\mathbf{u}_0$, thus we have an *initial value problem*. In subsequent time intervals $\mathbf{u}_n$ will always be a known quantity determined from the previous step.

In each interval, in the manner used in all finite element approximations, we assume that $\mathbf{u}$ varies as a polynomial and take here the lowest (linear) expansion as shown in Fig. 17.1 writing

$$\mathbf{u} \approx \hat{\mathbf{u}}(t) = \mathbf{u}_n + \frac{\tau}{\Delta t}(\mathbf{u}_{n+1} - \mathbf{u}_n) \tag{17.7}$$

with $\tau = t - t_n$.

This can be translated to the standard finite element expansion giving

$$\hat{\mathbf{u}}(t) = \sum N_i \mathbf{u}_i = \left(1 - \frac{\tau}{\Delta t}\right)\mathbf{u}_n + \left(\frac{\tau}{\Delta t}\right)\mathbf{u}_{n+1} \tag{17.8}$$

in which the unknown parameter is $\mathbf{u}_{n+1}$.

The equation by which this unknown parameter is provided will be a weighted residual approximation to Eq. (17.2). Accordingly, we write the variational problem

$$\int_0^{\Delta t} \mathbf{w}(\tau)^{\mathrm{T}} \left[\mathbf{C}\dot{\mathbf{u}} + \mathbf{K}\mathbf{u} + \mathbf{f}\right] \mathrm{d}\tau = 0 \tag{17.9}$$

in which $\mathbf{w}(\tau)$ is an arbitrary weighting function. We write the approximate form

$$\mathbf{w}(\tau) = W(\tau)\delta\mathbf{u}_{n+1} \tag{17.10}$$

in which $\delta\mathbf{u}_{n+1}$ is an arbitrary parameter. With this approximation the weighted residual equation to be solved is given by

$$\int_0^{\Delta t} W(\tau)[\mathbf{C}\dot{\hat{\mathbf{u}}} + \mathbf{K}\hat{\mathbf{u}} + \mathbf{f}]\,\mathrm{d}\tau = \mathbf{0} \tag{17.11}$$

Introducing θ as a weighting parameter given by

$$\theta = \frac{1}{\Delta t}\frac{\int_0^{\Delta t} W(\tau)\tau\,\mathrm{d}\tau}{\int_0^{\Delta t} W(\tau)\,\mathrm{d}\tau} \tag{17.12}$$

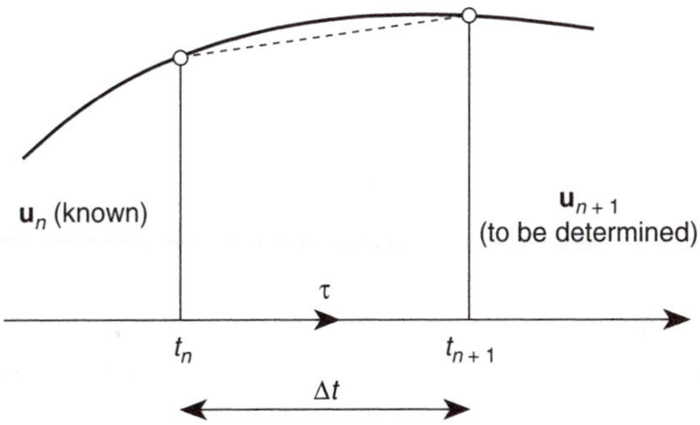

Fig. 17.1 Approximation to **u** in the time domain.

we can immediately write

$$\frac{1}{\Delta t}\mathbf{C}(\mathbf{u}_{n+1} - \mathbf{u}_n) + \mathbf{K}[\mathbf{u}_n + \theta(\mathbf{u}_{n+1} - \mathbf{u}_n)] + \bar{\mathbf{f}} = \mathbf{0} \tag{17.13}$$

where $\bar{\mathbf{f}}$ represents an average value of $\mathbf{f}$ given by

$$\bar{\mathbf{f}} = \frac{\int_0^{\Delta t} W\mathbf{f}\,\mathrm{d}\tau}{\int_0^{\Delta t} W\,\mathrm{d}\tau} \tag{17.14}$$

or

$$\bar{\mathbf{f}} = \mathbf{f}_n + \theta(\mathbf{f}_{n+1} - \mathbf{f}_n) \tag{17.15}$$

if a linear variation of $\mathbf{f}$ is assumed within the time increment.

Equation (17.13) is in fact almost identical to a finite difference approximation to the governing equation (17.2) at time $t_n + \theta\Delta t$, and in this example little advantage is gained by introducing the finite element approximation. However, the averaging of the forcing term is important, as shown in Fig. 17.2, where a constant W (that is $\theta = 1/2$) is used and a finite difference approximation presents difficulties.

Figure 17.3 shows how different weight functions can yield alternate values of the parameter θ. The solution of Eq. (17.13) yields

$$\mathbf{u}_{n+1} = (\mathbf{C} + \theta\Delta t\mathbf{K})^{-1}[(\mathbf{C} - (1-\theta)\Delta t\mathbf{K})\mathbf{u}_n - \Delta t\bar{\mathbf{f}}] \tag{17.16}$$

and it is evident that in general at each step of the computation a full equation system needs to be solved though of course a single inversion (or factorization using a Gauss-type solution process) is sufficient for linear problems in which the time increment Δt is held constant. Methods requiring such an inversion are called *implicit*. However, when $\theta = 0$ and the matrix $\mathbf{C}$ is approximated by its lumped equivalent $\mathbf{C}_L$ the solution is called *explicit* and is exceedingly cheap for each time interval. We shall show later that explicit algorithms are *conditionally stable* (requiring the Δt to be less than some critical value Δt_{crit}) whereas implicit methods may be made *unconditionally stable* for some choices of the parameters.

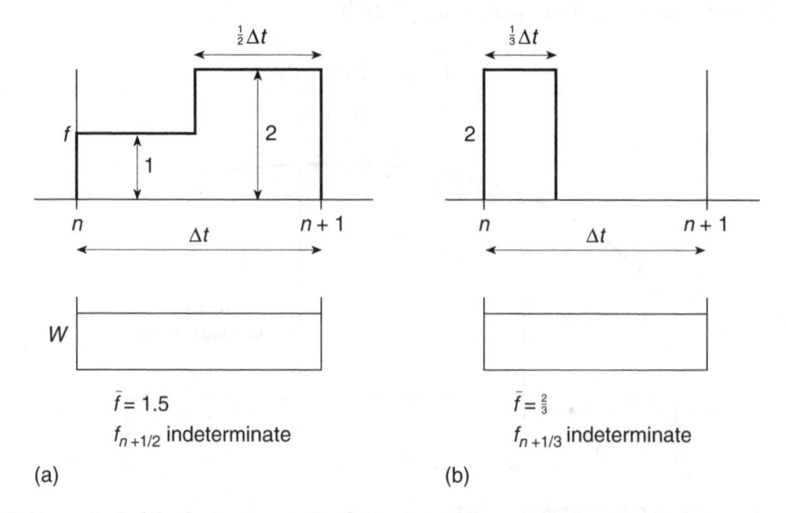

Fig. 17.2 'Averaging' of the forcing term in the finite-element-time approach.

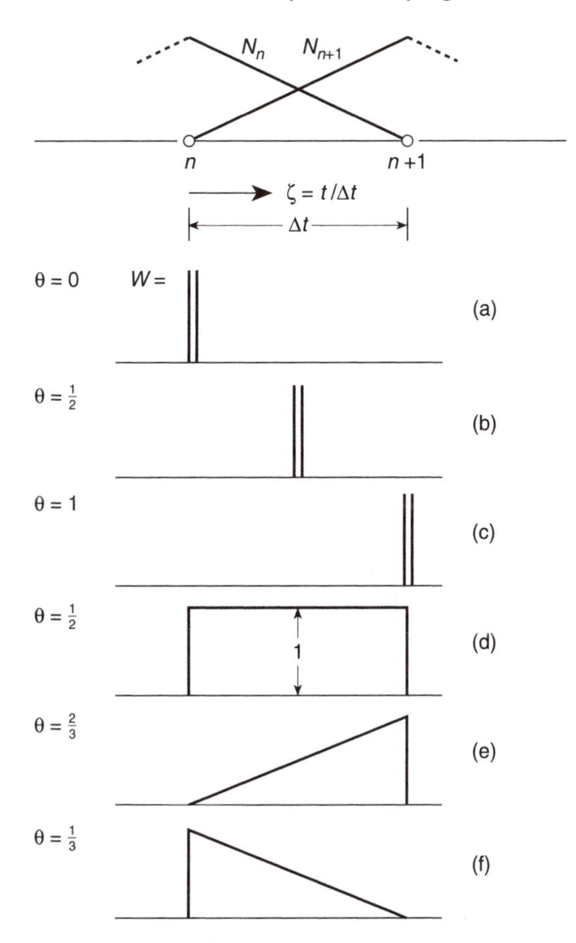

Fig. 17.3 Shape functions and weight functions for two-point recurrence formulae.

17.2.2 Taylor series collocation

A frequently used alternative to the algorithm presented above is obtained by approximating separately $\mathbf{u}_{n+1}$ and $\dot{\mathbf{u}}_{n+1}$ by truncated Taylor series. We can write, assuming that $\mathbf{u}_n$ and $\dot{\mathbf{u}}_n$ are known:

$$\mathbf{u}_{n+1} = \mathbf{u}_n + \Delta t \dot{\mathbf{u}}_n + \beta \Delta t (\dot{\mathbf{u}}_{n+1} - \dot{\mathbf{u}}_n) \tag{17.17a}$$

and use collocation to satisfy the governing equation at t_{n+1} [or alternatively using the weight function shown in Fig. 17.3(*c*)] which gives

$$\mathbf{C}\dot{\mathbf{u}}_{n+1} + \mathbf{K}\mathbf{u}_{n+1} + \mathbf{f}_{n+1} = \mathbf{0} \tag{17.17b}$$

In the above β is a parameter, $0 \leq \beta \leq 1$, such that the last term of Eq. (17.17a) represents a suitable difference approximation to the truncated expansion.

Substitution of Eq. (17.17a) into Eq. (17.17b) yields a recurrence relation for $\dot{\mathbf{u}}_{n+1}$:

$$\dot{\mathbf{u}}_{n+1} = -(\mathbf{C} + \beta \Delta t \mathbf{K})^{-1}[\mathbf{K}(\mathbf{u}_n + (1-\beta)\Delta t \dot{\mathbf{u}}_n) + \mathbf{f}_{n+1}] \tag{17.18}$$

where $\mathbf{u}_{n+1}$ is now computed by substitution of Eq. (17.18) into Eq. (17.17a).

We remark that:

(a) the scheme is not *self-starting*† and requires the satisfaction of Eq. (17.2) at $t = 0$ [whereas the finite element in the time scheme given by (17.13) is self-starting];
(b) the computation requires, with identification of the parameters $\beta = \theta$, an identical equation-solving problem to that in the finite element scheme of Eq. (17.16) and, finally, as we shall see later, stability considerations are identical.

The procedure is introduced here as it has some advantages in non-linear computations.

17.2.3 Other single-step procedures

As an alternative to the weighted residual process other possibilities of deriving finite element approximations exist, as discussed in Chapter 3. For instance, variational principles in time could be established and used for the purpose. This was indeed done in the early approaches to finite element approximation using Hamilton's or Gurtin's variational principle.[29–32] However, as expected, the final algorithms turn out to be identical. A variant on the above procedures is the use of a least square approximation for minimization of the equation residual.[13, 14] This is obtained by insertion of the approximation (17.7) into (17.2). The reader can verify that the recurrence relation becomes

$$\begin{aligned}&\left[\frac{1}{\Delta t}\mathbf{C}^{\mathrm{T}}\mathbf{C}+\frac{1}{2}\left(\mathbf{K}^{\mathrm{T}}\mathbf{C}+\mathbf{C}^{\mathrm{T}}\mathbf{K}\right)+\frac{1}{3}\Delta t\mathbf{K}^{\mathrm{T}}\mathbf{K}\right]\mathbf{u}_{n+1}\\&\quad-\left[\frac{1}{\Delta t}\mathbf{C}^{\mathrm{T}}\mathbf{C}+\frac{1}{2}\left(\mathbf{K}^{\mathrm{T}}\mathbf{C}-\mathbf{C}^{\mathrm{T}}\mathbf{K}\right)-\frac{1}{6}\Delta t\mathbf{K}^{\mathrm{T}}\mathbf{K}\right]\mathbf{u}_{n}\\&\quad+\frac{1}{\Delta t^2}\mathbf{C}^{\mathrm{T}}\int_0^{\Delta t}\mathbf{f}\,\mathrm{d}\tau+\frac{1}{\Delta t}\mathbf{K}^{\mathrm{T}}\int_0^{\Delta t}\mathbf{f}\tau\,\mathrm{d}\tau=\mathbf{0}\end{aligned} \tag{17.19}$$

requiring a more complex equation solution and always remaining 'implicit'. For this reason the algorithm is largely of purely theoretical interest, though as expected its accuracy is nearly exact for results shown in Fig. 17.4, in which a single degree of freedom equation (17.2) is used with

$$\mathbf{K}\rightarrow K=1 \qquad \mathbf{C}\rightarrow C=1 \qquad \mathbf{f}\rightarrow f=0$$

with initial condition $u_0 = 1$. Here, the various algorithms previously discussed are compared. Now we see from this example that the $\theta = 1/2$ algorithm performs almost as well as the least squares one. It is popular for this reason and is known as the Crank–Nicolson scheme after its originators.[33]

17.2.4 Consistency and approximation error

For the convergence of any finite element approximation, it is necessary and sufficient that it be *consistent* and *stable*. We have discussed these two conditions in Chapter 9

† By 'self-starting' we mean an algorithm is directly applicable without solving any subsidiary equations. Other definitions are also in use.

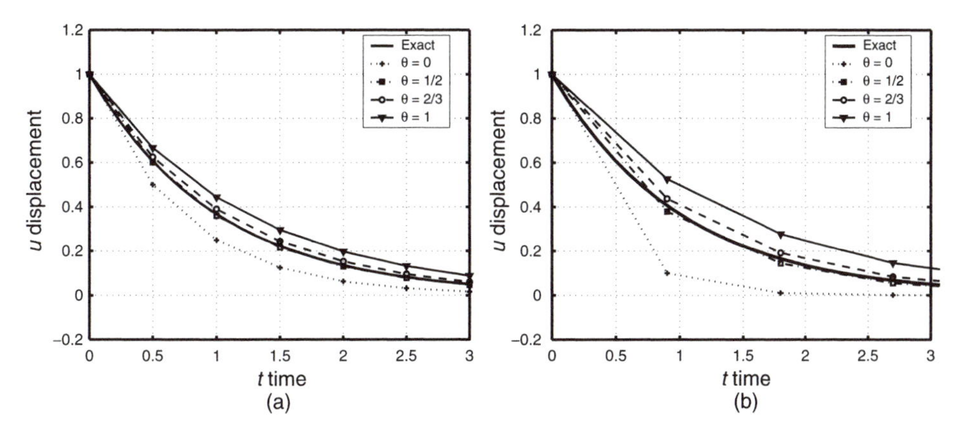

Fig. 17.4 Comparison of various time-stepping schemes on a first-order initial value problem.

and introduced appropriate requirements for boundary value problems. In the temporal approximation similar conditions apply though the stability problem is more delicate.

Clearly the function $\mathbf{u}$ itself and its derivatives occurring in the equation have to be approximated with a truncation error of $O(\Delta t^{\alpha})$, where $\alpha \geq 1$ is needed for consistency to be satisfied. For the first-order equation (17.2) it is thus necessary to use an approximating polynomial of order $p \geq 1$ which is capable of approximating $\dot{\mathbf{u}}$ to at least $O(\Delta t)$.

The *truncation error in the local approximation* of $\mathbf{u}$ with such an approximation is $O(\Delta t^2)$ and all the algorithms we have presented here using the $p = 1$ approximation of Eq. (17.7) will have at least that *local accuracy*,[34] as at a given time, $t = n\Delta t$, the total error can be magnified n times and the final accuracy at a given time for schemes discussed here is of order $O(\Delta t)$ in general.

We shall see later that the arguments used here lead to $p \geq 2$ for the second-order equation (17.1) and that an increase of accuracy can generally be achieved by use of higher order approximating polynomials.

It would of course be possible to apply such a polynomial increase to the approximating function (17.7) by adding higher order degrees of freedom. For instance, we could write in place of the original approximation a quadratic expansion:

$$\mathbf{u} \approx \hat{\mathbf{u}}(t) = \mathbf{u}_n + \frac{\tau}{\Delta t}(\mathbf{u}_{n+1} - \mathbf{u}_n) + \frac{\tau}{\Delta t}\left(1 - \frac{\tau}{\Delta t}\right)\breve{\mathbf{u}}_{n+1} \tag{17.20}$$

where $\breve{\mathbf{u}}$ is a hierarchic internal variable. Obviously now both $\mathbf{u}_{n+1}$ and $\breve{\mathbf{u}}_{n+1}$ are unknowns and will have to be solved for simultaneously. This is accomplished by using the weighting function

$$\mathbf{w} = W(\tau)\delta\mathbf{u}_{n+1} + \breve{W}(\tau)\delta\breve{\mathbf{u}}_{n+1} \tag{17.21}$$

where $W(\tau)$ and $\breve{W}(\tau)$ are two independent weighting functions. This will obviously result in an increased size of the problem.

It is of interest to consider the first of these obtained by using the weighting W alone in the manner of Eq. (17.11). It is easy to verify that we now have to add to Eq. (17.13) a term involving $\breve{\mathbf{u}}_{n+1}$ which is

$$\left[\frac{1}{\Delta t}(1 - 2\theta)\mathbf{C} + (\theta - \tilde{\theta})\mathbf{K}\right]\breve{\mathbf{u}}_{n+1} \tag{17.22}$$

where

$$\tilde{\theta} = \frac{1}{\Delta t^2} \frac{\int_0^{\Delta t} W \tau^2 \mathrm{d}\tau}{\int_0^{\Delta t} W \,\mathrm{d}\tau}$$

It is clear that the choice of $\theta = \tilde{\theta} = 1/2$ eliminates the quadratic term and regains the previous scheme, thus showing that the values so obtained have a local truncation error of $O(\Delta t^3)$. This explains why the Crank–Nicolson scheme possesses higher accuracy.

In general the addition of higher order internal variables makes recurrence schemes too expensive and we shall later show how an increase of accuracy can be more economically achieved.

In a later section of this chapter we shall refer to some currently popular schemes in which often sets of **u**s have to be solved for simultaneously. In such schemes a discontinuity is assumed at the initial condition and additional parameters ($\tilde{\mathbf{u}}$) can be introduced to keep the same linear conditions we assumed previously. In this case an additional equation appears as a weighted satisfaction of continuity in time.

The procedure is therefore known as the *discontinuous Galerkin process* and was introduced initially by Reed and Hill[35] to solve neutron transport problems. An analysis of the method was given by Lesaint and Raviart.[36] It has subsequently been applied to solve problems in fluid mechanics and heat transfer[23, 37, 38] and to problems in structural dynamics.[25–27] As we have already stated, the introduction of additional variables is expensive, so somewhat limited use of the concept has so far been made. However, one interesting application is in error estimation and adaptive time stepping.[39]

17.2.5 Stability

If we consider any of the recurrence algorithms so far derived, we note that for the homogeneous form (i.e., with $\mathbf{f} = \mathbf{0}$) all can be written in the form

$$\mathbf{u}_{n+1} = \mathbf{A}\mathbf{u}_n \tag{17.23}$$

where **A** is known as the *amplification matrix*.

The form of this matrix for the first algorithm derived is, for instance, evident from Eq. (17.16) as

$$\mathbf{A} = (\mathbf{C} + \theta\Delta t\mathbf{K})^{-1}(\mathbf{C} - (1-\theta)\Delta t\mathbf{K}) \tag{17.24}$$

Any errors present in the solution will of course be subject to amplification by precisely the same factor.

A general solution of any recurrence scheme can be written as[40]

$$\mathbf{u}_{n+1} = \mu\mathbf{u}_n \tag{17.25}$$

and by insertion into Eq. (17.23) we observe that μ is given by eigenvalues of the matrix as

$$(\mathbf{A} - \mu\mathbf{I})\,\mathbf{u}_n = \mathbf{0} \tag{17.26}$$

Clearly if any eigenvalue μ is such that

$$|\mu| > 1 \tag{17.27}$$

all initially small errors will increase without limit and the solution will be unstable. In the case of complex eigenvalues the above is modified to the requirement that the modulus of μ satisfies Eq. (17.27).

As the determination of system eigenvalues is a large undertaking it is useful to consider only a scalar equation of the form (17.5) (representing, say, one-element performance). The bounding theorems of Irons and Treharne[28] will show why we do so and the results will provide general stability bounds if maximums are used. Thus for the case of the algorithm discussed in Eq. (17.26) we have for the scalar form of (17.24)

$$A = \frac{c - (1-\theta)\Delta t k}{c + \theta \Delta t k} = \frac{1-(1-\theta)\omega\Delta t}{1+\theta\omega\Delta t} = \mu \tag{17.28}$$

where $\omega = k/c$ and μ is evaluated from Eq. (17.26) simply as $\mu = A$ to allow non-trivial u_n. (This is equivalent to making the determinant of $\mathbf{A} - \mu\mathbf{I}$ zero in the more general case.)

In Fig. 17.5 we show how μ (or A) varies with $\omega\Delta t$ for various θ values. We observe immediately that:

(a) for $\theta \geq 1/2$

$$|\mu| \leq 1 \tag{17.29}$$

and such algorithms are *unconditionally stable*;

(b) for $\theta < 1/2$ we require

$$\omega\Delta t \leq \frac{2}{1-2\theta} \tag{17.30}$$

for stability. Such algorithms are therefore only *conditionally stable*. Here of course the explicit form with $\theta = 0$ is typical.

The critical value of Δt below which the scheme is stable with $\theta < 1/2$ needs the determination of the maximum value of μ from a typical element. For instance, in the case of the thermal conduction problem in which we have the coefficients c_{aa} and k_{aa} defined by expressions

$$c_{aa} = \int_\Omega \tilde{c} N_a^2 \, d\Omega \quad \text{and} \quad k_{aa} = \int_\Omega \nabla N_a \tilde{k} \nabla N_a \, d\Omega \tag{17.31}$$

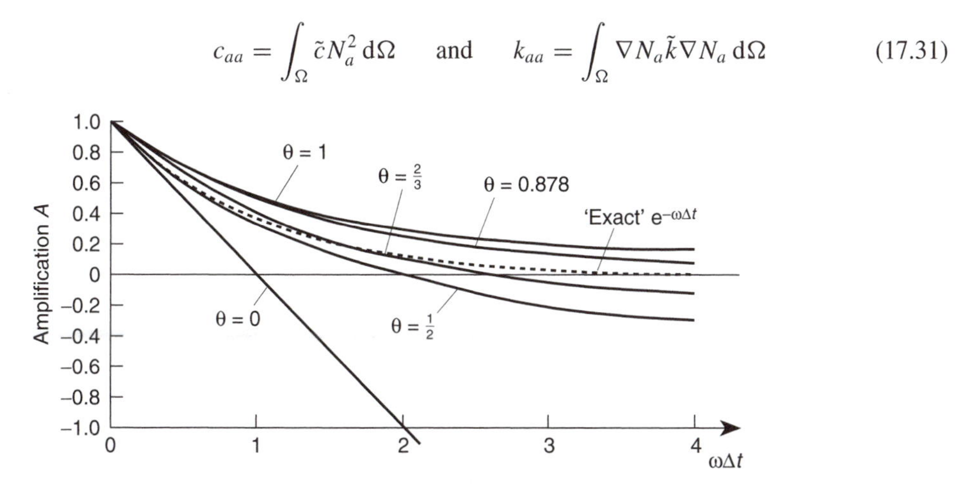

Fig. 17.5 The amplification A for various versions of the θ algorithm.

we can presuppose uniaxial behaviour with a single degree of freedom and write for a linear element

$$N = \frac{h-x}{h} \qquad c = \int_0^h \tilde{c} N^2 \, \mathrm{d}x = \frac{1}{3}\tilde{c}h \qquad k = \int_0^h \tilde{k}\left(\frac{\mathrm{d}N}{\mathrm{d}x}\right)^2 \mathrm{d}x = \frac{\tilde{k}}{h}$$

Now

$$\omega = \frac{k}{c} = \frac{3\tilde{k}}{\tilde{c}h^2}$$

This gives

$$\Delta t \leq \frac{2}{1-2\theta}\,\frac{\tilde{c}h^2}{3\tilde{k}} = \Delta t_{\mathrm{crit}} \tag{17.32}$$

which of course means that the smallest element size, $h_{\min}$, dictates overall stability. We note from the above that:

(a) in first-order problems the critical time step is proportional to h^2 and thus decreases rapidly with element size making explicit computations difficult;
(b) if mass lumping is assumed where $\tilde{c}_{lump} > \tilde{c}_{cons}$ the critical time step is larger than that obtained using a consistent mass.

In Fig. 17.6 we show the performance of the scheme described in Sec. 17.2.1 for various values of θ and Δt in the example we have already illustrated in Fig. 17.4, but now using larger values of Δt. We note now that the conditionally stable scheme with $\theta = 0$ and a stability limit of $\Delta t = 2$ shows oscillations as this limit is approached ($\Delta t = 1.5$) and diverges when exceeded.

Stability computations which were presented for the algorithm of Sec. 17.2.1 can of course be repeated for the other algorithms which we have discussed.

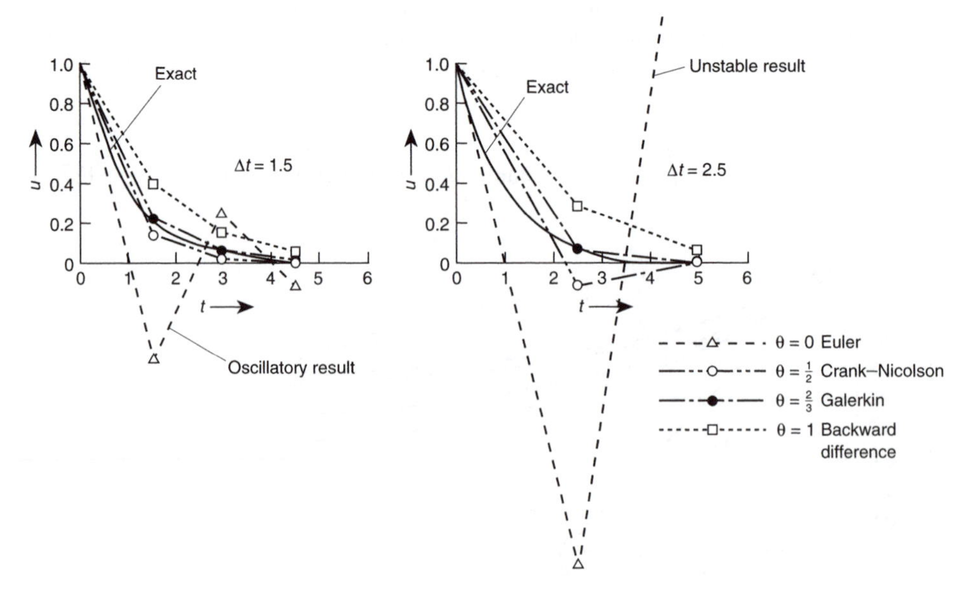

Fig. 17.6 Performance of some θ algorithms in the problem of Fig. 17.4 and larger time steps. Note oscillation and instability.

If identical procedures are used, for instance on the algorithm of Sec. 17.2.2, we shall find that the stability conditions, based on the determinant of the amplification matrix $(\mathbf{A} - \mu\mathbf{I})$, are identical with the previous one providing we set $\theta = \beta$. Algorithms that give such identical determinants will be called *similar* in the following presentations.

In general, it is possible for different amplification matrices $\mathbf{A}$ to have identical determinants of $(\mathbf{A} - \mu\mathbf{I})$ and hence identical stability conditions, but differ otherwise. If in addition the amplification matrices are the same, the schemes are known as *identical*. In the two cases described here such an identity can be shown to exist despite different derivations.

17.2.6 Some further remarks. Initial conditions and examples

The question of choosing an optimal value of θ is not always obvious from theoretical accuracy considerations. In particular with $\theta = 1/2$ oscillations are sometimes present,[14] as we observe in Fig. 17.6 ($\Delta t = 2.5$), and for this reason some prefer to use $\theta = 2/3$, which is considerably 'smoother' (and which incidentally corresponds to a standard Galerkin approximation in time[41]). In Table 17.1 we show the results for a one-dimensional finite element problem where a bar at uniform initial temperature is subject to zero temperatures applied suddenly at the ends. Here 10 linear elements are used in the space dimension with $L = 1$. The oscillation errors occurring with $\theta = 1/2$ are much reduced for $\theta = 2/3$. The time step used here is much longer than that corresponding to the lowest eigenvalue period, but the main cause of the oscillation is in the abrupt discontinuity of the temperature change.

For similar reasons Liniger[42] derives θ which minimizes the error in the whole time domain and gives $\theta = 0.878$ for the simple one-dimensional case. We observe in Fig. 17.5 how well the amplification factor fits the exact solution with these values. Again this value will smooth out many oscillations. However, most oscillations are introduced by simply using a physically unrealistic initial condition.

In part at least, the oscillations which for instance occur with $\theta = 1/2$ and $\Delta t = 2.5$ (see Fig. 17.6) in the previous example are due to a sudden jump in the forcing term introduced at the start of the computation. This jump is evident if we consider this simple problem posed in the context of the whole time domain. We can take the problem as implying

$$f(t) = -1 \quad \text{for} \quad t < 0$$

Table 17.1 Percentage error for finite elements in time: $\theta = 2/3$ (Galerkin) and $\theta = 1/2$ (Crank–Nicolson) scheme; $\Delta t = 0.01$

	$x = 0.1$		$x = 0.2$		$x = 0.3$		$x = 0.4$		$x = 0.5$	
t/θ	2/3	1/2	2/3	1/2	2/3	1/2	2/3	1/2	2/3	1/2
0.01	10.8	28.2	1.6	3.2	0.5	0.7	0.6	0.1	0.5	0.2
0.02	0.5	3.5	2.1	9.5	0.1	0.0	0.5	0.7	0.7	0.4
0.03	1.3	9.9	0.5	0.7	0.8	3.1	0.5	0.2	0.5	0.6
0.05	0.5	4.5	0.4	0.2	0.5	2.3	0.4	0.8	0.5	1.0
0.10	0.1	1.4	0.1	2.0	0.1	1.4	0.1	1.9	0.1	1.6
0.15	0.3	2.2	0.3	2.1	0.3	2.2	0.3	2.1	0.3	2.2
0.20	0.6	2.6	0.6	2.6	0.6	2.6	0.6	2.6	0.6	2.6
0.30	1.4	3.5	1.4	3.5	1.4	3.5	1.4	3.5	1.4	3.5

giving the solution $u = 1$ with a sudden change at $t = 0$, resulting in

$$f(t) = 0 \quad \text{for} \quad t \geq 0$$

As shown in Fig. 17.7 this represents a discontinuity of the loading function at $t = 0$.

Although load discontinuities are permitted by the algorithm they lead to a sudden discontinuity of $\dot{u}$ and hence induce undesirable oscillations. If in place of this discontinuity we assume that f varies linearly in the first time step Δt ($-\Delta t/2 \leq t \leq \Delta t/2$) then smooth results are obtained with a much improved physical representation of the true solution, even for such a long time step as $\Delta t = 2.5$, as shown in Fig. 17.7.

Similar use of smoothing is illustrated in a multidegree of freedom system (the representation of heat conduction in a wall) which is solved using two-dimensional finite elements[43] (Fig. 17.8).

Here the problem corresponds to an instantaneous application of prescribed temperature ($T = 1$) at the wall sides with zero initial conditions. Now again troublesome oscillations are almost eliminated for $\theta = 1/2$ and improved results are obtained for other values of θ (2/3, 0.878) by assuming the step change to be replaced by a continuous one. Such smoothing is always advisable and a continuous representation of the forcing term is important.

17.3 General single-step algorithms for first- and second-order equations

17.3.1 Introduction

We shall introduce in this section two general single-step algorithms applicable to Eq. (17.1):

$$\mathbf{M}\ddot{\mathbf{u}} + \mathbf{C}\dot{\mathbf{u}} + \mathbf{K}\mathbf{u} + \mathbf{f} = \mathbf{0}$$

These algorithms will of course be applicable to the first-order problem of Eq. (17.2) simply by putting $\mathbf{M} = \mathbf{0}$.

An arbitrary degree polynomial p for approximating the unknown function $\mathbf{u}$ will be used and we must note immediately that for the second-order equations $p \geq 2$ is required for consistency as second-order derivatives have to be approximated.

The first algorithm SSpj (single step with approximation of degree p for equations of order $j = 1, 2$) will be derived by use of the weighted residual process and we shall find that the algorithm of Sec. 17.2.1 is but a special case. The second algorithm GNpj (generalized Newmark[44] with degree p and order j) will follow the procedures using a truncated Taylor series approximation in a manner similar to that described in Sec. 17.2.2.

In what follows we shall *assume* that at the start of the interval, i.e., at $t = t_n$, we know the values of the unknown function $\mathbf{u}$ and its derivatives, that is $\mathbf{u}_n$, $\dot{\mathbf{u}}_n$, $\ddot{\mathbf{u}}_n$ up to $\overset{p-1}{\mathbf{u}}_n$ and our objective will be to determine $\mathbf{u}_{n+1}$, $\dot{\mathbf{u}}_{n+1}$, $\ddot{\mathbf{u}}_{n+1}$ up to $\overset{p-1}{\mathbf{u}}_{n+1}$, where p is the order of the expansion used in the interval.

This is indeed a rather strong presumption as for first-order problems we have already stated that only a single initial condition, $\mathbf{u}(0)$, is given and for second-order problems two conditions, $\mathbf{u}(0)$ and $\dot{\mathbf{u}}(0)$, are available (i.e., the initial displacement and velocity of the

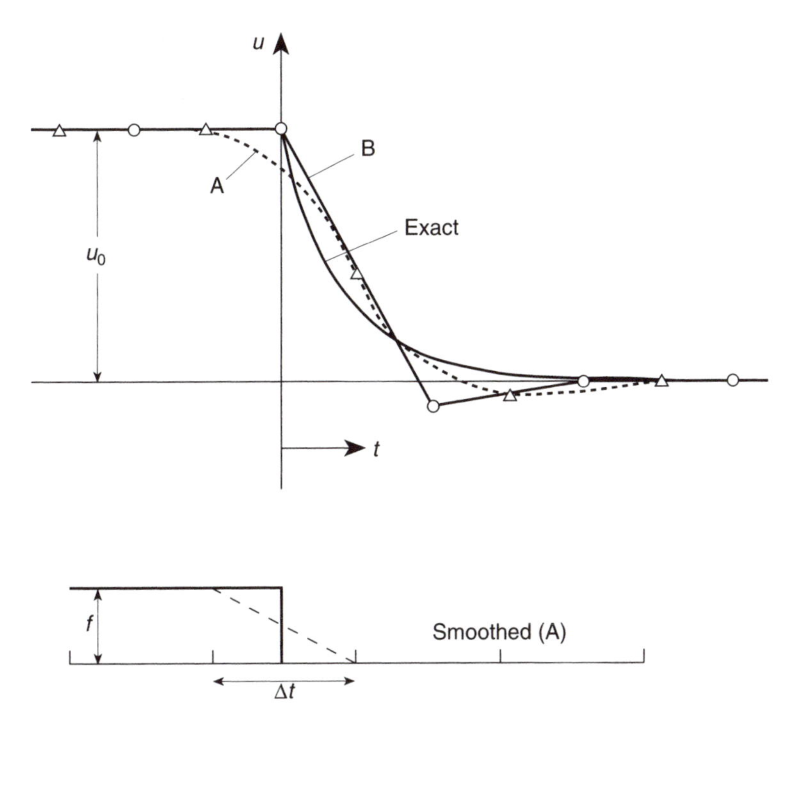

Fig. 17.7 Importance of 'smoothing' the force term in elimination of oscillations in the solution. $\Delta t = 2.5$.

system). We can, however, argue that if the system starts from rest we could take $\mathbf{u}(0)$ to $\overset{p-1}{\mathbf{u}}(0)$ as equal to zero and, providing *that suitably continuous forcing of the system occurs*, the solution will remain smooth in the higher derivatives. Alternatively, we can differentiate the differential equation to obtain the necessary starting values.

17.3.2 The weighted residual finite element form SSpj

In the SSpj algorithm the expansion of the unknown vector $\mathbf{u}$ is taken as a polynomial of degree p.[19, 20] With the *known* values of $\mathbf{u}_n$, $\dot{\mathbf{u}}_n$, $\ddot{\mathbf{u}}_n$ up to $\overset{p-1}{\mathbf{u}}_n$ at the beginning of the time step Δt, we write, as in Sec. 17.2.1,

$$\tau = t - t_n \qquad \Delta t = t_{n+1} - t_n \tag{17.33}$$

and using a polynomial expansion of degree p,

$$\mathbf{u} \approx \hat{\mathbf{u}} = \mathbf{u}_n + \tau\dot{\mathbf{u}}_n + \frac{1}{2!}\tau^2\ddot{\mathbf{u}}_n + \cdots + \frac{1}{(p-1)!}\tau^{p-1}\overset{p-1}{\mathbf{u}}_n + \frac{1}{p!}\tau^p\alpha_n^p \tag{17.34}$$

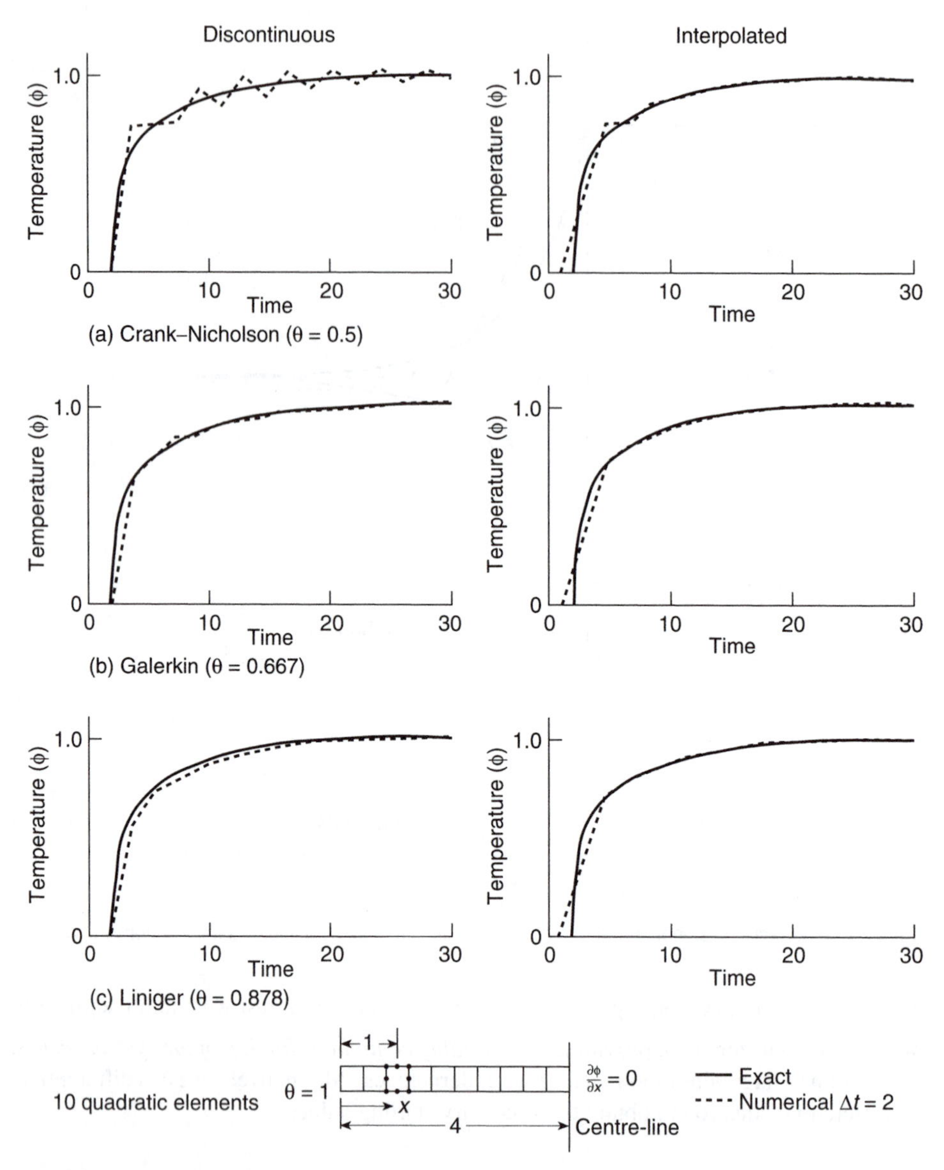

Fig. 17.8 Transient heating of a bar; comparison of discontinuous and interpolated (smoothed) initial conditions for single-step schemes.

where the only unknown is the vector $\boldsymbol{\alpha}_n^p$,

$$\boldsymbol{\alpha}_n^p \approx \overset{p}{\mathbf{u}} \equiv \frac{\mathrm{d}^p\mathbf{u}}{\mathrm{d}t^p} \tag{17.35}$$

which represents some average value of the pth derivative occurring in the interval Δt. The approximation to $\mathbf{u}$ for the case of $p = 2$ is shown in Fig. 17.9.

We recall that in order to obtain a consistent approximation to all the derivatives that occur in the differential equations (17.1) and (17.2), $p \geq 2$ is necessary for the full

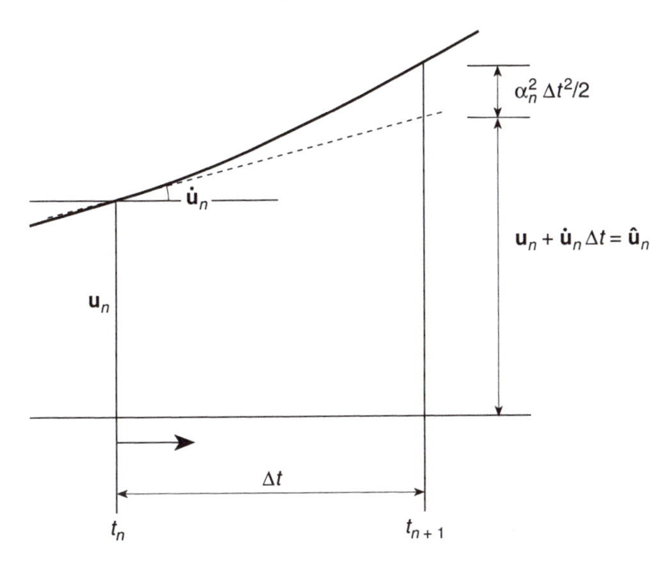

Fig. 17.9 A second-order time approximation.

dynamic equation and $p \geq 1$ is necessary for the first-order equation. Indeed the lowest approximation, that is $p = 1$, is the basis of the algorithm derived in the previous section.

The recurrence algorithm will now be obtained by inserting $\mathbf{u}$, $\dot{\mathbf{u}}$ and $\ddot{\mathbf{u}}$ [obtained by differentiating Eq. (17.34)] into Eq. (17.1) and satisfying the weighted residual equation with a single weighting function $W(\tau)$. This gives

$$\begin{aligned}\int_0^{\Delta t} W(\tau)\Bigg[&\mathbf{M}\left(\ddot{\mathbf{u}}_n + \tau\dddot{\mathbf{u}}_n + \cdots + \frac{1}{(p-2)!}\tau^{p-2}\boldsymbol{\alpha}_n^p\right)\\ &+\mathbf{C}\left(\dot{\mathbf{u}}_n + \tau\ddot{\mathbf{u}}_n + \cdots + \frac{1}{(p-1)!}\tau^{p-1}\boldsymbol{\alpha}_n^p\right)\\ &+\mathbf{K}\left(\mathbf{u}_n + \tau\dot{\mathbf{u}}_n + \cdots + \frac{1}{p!}\tau^{p}\boldsymbol{\alpha}_n^p\right) + \mathbf{f}\Bigg]\,\mathrm{d}t = \mathbf{0}\end{aligned} \tag{17.36}$$

as the basic equation for determining $\boldsymbol{\alpha}_n^p$.

Without explicitly specifying the weighting function used we can, as in Sec. 17.2.1, generalize its effects by writing

$$\begin{aligned}\theta_k &= \frac{\int_0^{\Delta t} W\tau^k\,\mathrm{d}\tau}{\Delta t^k\int_0^{\Delta t} W\,\mathrm{d}\tau} \qquad k = 0, 1, \ldots, p\\ \bar{\mathbf{f}} &= \frac{\int_0^{\Delta t} W\mathbf{f}\,\mathrm{d}\tau}{\int_0^{\Delta t} W\,\mathrm{d}\tau}\end{aligned} \tag{17.37}$$

where we note θ_0 is always unity. Equation (17.36) can now be written more compactly as

$$\mathbf{A}\boldsymbol{\alpha}_n^p + \mathbf{M}\ddot{\bar{\mathbf{u}}}_{n+1} + \mathbf{C}\dot{\bar{\mathbf{u}}}_{n+1} + \mathbf{K}\bar{\mathbf{u}}_{n+1} + \bar{\mathbf{f}} = \mathbf{0} \tag{17.38}$$

where

$$
\begin{aligned}
\mathbf{A} &= \frac{\theta_{p-2}\Delta t^{p-2}}{(p-2)!}\mathbf{M} + \frac{\theta_{p-1}\Delta t^{p-1}}{(p-1)!}\mathbf{C} + \frac{\theta_p \Delta t^p}{p!}\mathbf{K} \\
\bar{\mathbf{u}}_{n+1} &= \sum_{q=0}^{p-1} \frac{\theta_q \Delta t^q}{q!} \overset{q}{\mathbf{u}}_n \\
\dot{\bar{\mathbf{u}}}_{n+1} &= \sum_{q=1}^{p-1} \frac{\theta_{q-1} \Delta t^{q-1}}{(q-1)!} \overset{q}{\mathbf{u}}_n \\
\ddot{\bar{\mathbf{u}}}_{n+1} &= \sum_{q=2}^{p-1} \frac{\theta_{q-2} \Delta t^{q-2}}{(q-2)!} \overset{q}{\mathbf{u}}_n
\end{aligned} \tag{17.39}
$$

As $\bar{\mathbf{u}}_{n+1}$, $\dot{\bar{\mathbf{u}}}_{n+1}$ and $\ddot{\bar{\mathbf{u}}}_{n+1}$ can be computed directly from the initial values we can solve Eq. (17.38) to obtain

$$
\boldsymbol{\alpha}_n^p = -\mathbf{A}^{-1}\left[\mathbf{M}\ddot{\bar{\mathbf{u}}}_{n+1} + \mathbf{C}\dot{\bar{\mathbf{u}}}_{n+1} + \mathbf{K}\bar{\mathbf{u}}_{n+1} + \bar{\mathbf{f}}\right] \tag{17.40}
$$

It is important to observe that $\bar{\mathbf{u}}_{n+1}$, $\dot{\bar{\mathbf{u}}}_{n+1}$ and $\ddot{\bar{\mathbf{u}}}_{n+1}$ here represent some *mean predicted values* of $\mathbf{u}_{n+1}$, $\dot{\mathbf{u}}_{n+1}$ and $\ddot{\mathbf{u}}_{n+1}$ in the interval and satisfy the governing Eq. (17.1) in a weighted sense if *$\boldsymbol{\alpha}_n^p$ is chosen as zero.*

The procedure is now complete as knowledge of the vector $\boldsymbol{\alpha}_n^p$ permits the evaluation of $\mathbf{u}_{n+1}$ to $\overset{p-1}{\mathbf{u}}_{n+1}$ from the expansion originally used in Eq. (17.34) by putting $\tau = \Delta t$. This gives

$$
\begin{aligned}
\mathbf{u}_{n+1} &= \mathbf{u}_n + \Delta t \dot{\mathbf{u}}_n + \cdots + \frac{\Delta t^p}{p!}\boldsymbol{\alpha}_n^p = \hat{\mathbf{u}}_{n+1} + \frac{\Delta t^p}{p!}\boldsymbol{\alpha}_n^p \\
\dot{\mathbf{u}}_{n+1} &= \dot{\mathbf{u}}_n + \Delta t \ddot{\mathbf{u}}_n + \cdots + \frac{\Delta t^{p-1}}{(p-1)!}\boldsymbol{\alpha}_n^p = \dot{\hat{\mathbf{u}}}_{n+1} + \frac{\Delta t^{p-1}}{(p-1)!}\boldsymbol{\alpha}_n^p \\
&\;\vdots \\
\overset{p-1}{\mathbf{u}}_{n+1} &= \overset{p-1}{\mathbf{u}}_n + \Delta t \boldsymbol{\alpha}_n^p
\end{aligned} \tag{17.41}
$$

In the above $\hat{\mathbf{u}}$, $\dot{\hat{\mathbf{u}}}$, etc., are again quantities that can be written down *a priori* (before solving for $\boldsymbol{\alpha}_n^p$). These represent predicted values at the end of the interval with $\boldsymbol{\alpha}_n^p = \mathbf{0}$.

To summarize, the general algorithm necessitates the choice of values for θ_1 to θ_p and requires

(a) computation of $\bar{\mathbf{u}}_{n+1}$, $\dot{\bar{\mathbf{u}}}_{n+1}$ and $\ddot{\bar{\mathbf{u}}}_{n+1}$ using the definitions of Eqs (17.39);
(b) computation of $\boldsymbol{\alpha}_n^p$ by solution of Eq. (17.40);
(c) computation of $\mathbf{u}_{n+1}$ to $\overset{p-1}{\mathbf{u}}_{n+1}$ by Eqs (17.41).

After completion of stage (c) a new time step can be started. In first-order problems the computation of $\ddot{\mathbf{u}}$ can obviously be omitted.

If matrices $\mathbf{C}$ and $\mathbf{M}$ are diagonal the solution of Eq. (17.40) is trivial providing we choose

$$
\theta_p = 0 \tag{17.42}
$$

With this choice the algorithms are *explicit* but, as we shall find later, only sometimes *conditionally stable*.

When $\theta_p > 0$, *implicit* algorithms of various kinds will be available and some of these will be found to be *unconditionally stable*. Indeed, it is such algorithms that are of great practical use.

Important special cases of the general algorithm are the SS11 and SS22 forms given below.

Example 17.1: The SS11 algorithm. If we consider the first-order equation (that is $j = 1$) it is evident that only the value of $\mathbf{u}_n$ is necessarily specified as the initial value for any computation. For this reason the choice of a linear expansion in the time interval is *natural* ($p = 1$) and the SS11 algorithm is for that reason most widely used.

Now the approximation of Eq. (17.34) is simply

$$\mathbf{u} = \mathbf{u}_n + \tau \boldsymbol{\alpha} \qquad (\boldsymbol{\alpha}_n^1 = \boldsymbol{\alpha} = \dot{\mathbf{u}}) \tag{17.43}$$

and the approximation to the average satisfaction of Eq. (17.2) is simply

$$\mathbf{C}\boldsymbol{\alpha} + \mathbf{K}(\bar{\mathbf{u}}_{n+1} + \theta \Delta t \boldsymbol{\alpha}) + \bar{\mathbf{f}} = \mathbf{0} \tag{17.44}$$

with $\bar{\mathbf{u}}_{n+1} = \mathbf{u}_n$. Solution of Eq. (17.44) determines $\boldsymbol{\alpha}$ as

$$\boldsymbol{\alpha} = -\left(\mathbf{C} + \theta \Delta t \mathbf{K}\right)^{-1} \left(\bar{\mathbf{f}} + \mathbf{K}\mathbf{u}_n\right) \tag{17.45}$$

and finally

$$\mathbf{u}_{n+1} = \mathbf{u}_n + \Delta t \boldsymbol{\alpha} \tag{17.46}$$

The reader will verify that this process is identical to that developed in Eqs (17.7)–(17.13) and hence will not be further discussed except perhaps for noting the more elegant computation form above.

Example 17.2: The SS22 algorithm. With Eq. (17.1) we considered a second-order system ($j = 2$) in which the necessary initial conditions require the specification of two quantities, $\mathbf{u}_n$ and $\dot{\mathbf{u}}_n$. The simplest and most natural choice here is to specify the minimum value of p, that is $p = 2$, as this does not require computation of additional derivatives at the start. This algorithm, SS22, is *thus basic for dynamic equations* and we present it here in full.

From Eq. (17.34) the approximation is a quadratic

$$\mathbf{u} = \mathbf{u}_n + \tau \dot{\mathbf{u}}_n + \tfrac{1}{2}\tau^2 \boldsymbol{\alpha} \qquad (\boldsymbol{\alpha}_n^2 \equiv \boldsymbol{\alpha} \approx \ddot{\mathbf{u}}) \tag{17.47}$$

The approximate form of the 'weighted' dynamic equation is now

$$\mathbf{M}\boldsymbol{\alpha} + \mathbf{C}(\dot{\bar{\mathbf{u}}}_{n+1} + \theta_1 \Delta t \boldsymbol{\alpha}) + \mathbf{K}(\bar{\mathbf{u}}_{n+1} + \tfrac{1}{2}\theta_2 \Delta t^2 \boldsymbol{\alpha}) + \bar{\mathbf{f}} = \mathbf{0} \tag{17.48}$$

with predicted 'mean' values

$$\begin{aligned} \bar{\mathbf{u}}_{n+1} &= \mathbf{u}_n + \theta_1 \Delta t \dot{\mathbf{u}}_n \\ \dot{\bar{\mathbf{u}}}_{n+1} &= \dot{\mathbf{u}}_n \end{aligned} \tag{17.49}$$

After evaluation of $\boldsymbol{\alpha}$ from Eq. (17.40), the values of $\mathbf{u}_{n+1}$ are found by Eqs (17.41) which become simply

$$\begin{aligned} \mathbf{u}_{n+1} &= \mathbf{u}_n + \Delta t \dot{\mathbf{u}}_n + \tfrac{1}{2}\Delta t^2 \boldsymbol{\alpha} \\ \dot{\mathbf{u}}_{n+1} &= \dot{\mathbf{u}}_n + \Delta t \boldsymbol{\alpha} \end{aligned} \tag{17.50}$$

This completes the algorithm which is of much practical value in the solution of dynamics problems.

In many respects the previous example resembles the Newmark algorithm[44] which we shall discuss in the next section and which is widely used in practice with the forms

$$\begin{aligned} \mathbf{u}_{n+1} &= \mathbf{u}_n + \Delta t \dot{\mathbf{u}}_n + (\tfrac{1}{2} - \beta)\Delta t^2 \ddot{\mathbf{u}}_n + \beta \Delta t^2 \ddot{\mathbf{u}}_{n+1} \\ \dot{\mathbf{u}}_{n+1} &= \dot{\mathbf{u}}_n + (1-\gamma)\Delta t \ddot{\mathbf{u}}_n + \gamma \Delta t \ddot{\mathbf{u}}_{n+1} \end{aligned}$$

Indeed, the stability properties of the SS22 algorithm turn out to be identical with the Newmark algorithm if

$$\theta_1 = \gamma; \quad \theta_2 = 2\beta; \quad \theta_2 \geq \theta_1 \geq \tfrac{1}{2} \tag{17.51}$$

for unconditional stability. In the above γ and β are conventionally used Newmark parameters.

For $\theta_2 = 0$ the algorithm is 'explicit' (assuming both $\mathbf{M}$ and $\mathbf{C}$ to be diagonal) and can be made conditionally stable if $\theta_1 \geq 1/2$.

The algorithm is clearly applicable to first-order equations described as SS21 and we shall find that the stability conditions are identical. In this case, however, it is necessary to identify an initial condition for $\dot{\mathbf{u}}_0$ and

$$\dot{\mathbf{u}}_0 = -\mathbf{C}^{-1}\left(\mathbf{K}\mathbf{u}_0 + \bar{\mathbf{f}}_0\right)$$

is one possibility.

17.3.3 Truncated Taylor series collocation algorithm GNpj

In the derivation using collocation, we consider the satisfaction of the governing equation (17.1) only at the end points of the interval Δt [which results from the weighting function shown in Fig. 17.3(c)] and write

$$\mathbf{M}\ddot{\mathbf{u}}_{n+1} + \mathbf{C}\dot{\mathbf{u}}_{n+1} + \mathbf{K}\mathbf{u}_{n+1} + \mathbf{f}_{n+1} = \mathbf{0} \tag{17.52}$$

with appropriate approximations for the values of $\mathbf{u}_{n+1}$, $\dot{\mathbf{u}}_{n+1}$ and $\ddot{\mathbf{u}}_{n+1}$. It will be shown that again as in Sec. 17.2.2 a non-self-starting process is obtained, which in most cases, however, gives an algorithm similar to the SSpj one we have derived. The classical Newmark method[44] will be recognized as a particular case together with its derivation process in a form presented generally in existing texts.[45] Because of this similarity we shall term the new algorithm generalized Newmark (GNpj).

If we consider a truncated Taylor series expansion similar to Eq. (17.17a) for the function $\mathbf{u}$ and its derivatives, we can write

$$\begin{aligned}
\mathbf{u}_{n+1} &= \mathbf{u}_n + \Delta t \dot{\mathbf{u}}_n + \cdots + \frac{\Delta t^p}{p!} \overset{p}{\mathbf{u}}_n + \beta_p \frac{\Delta t^p}{p!} \left(\overset{p}{\mathbf{u}}_{n+1} - \overset{p}{\mathbf{u}}_n \right) \\
\dot{\mathbf{u}}_{n+1} &= \dot{\mathbf{u}}_n + \Delta t \ddot{\mathbf{u}}_n + \cdots + \frac{\Delta t^{p-1}}{(p-1)!} \overset{p}{\mathbf{u}}_n + \beta_{p-1} \frac{\Delta t^{p-1}}{(p-1)!} \left(\overset{p}{\mathbf{u}}_{n+1} - \overset{p}{\mathbf{u}}_n \right) \\
&\vdots \\
\overset{p-1}{\mathbf{u}}_{n+1} &= \overset{p-1}{\mathbf{u}}_n + \Delta t \overset{p}{\mathbf{u}}_n + \beta_1 \Delta t \left(\overset{p}{\mathbf{u}}_{n+1} - \overset{p}{\mathbf{u}}_n \right)
\end{aligned} \tag{17.53}$$

In Eqs (17.53) we have effectively allowed for a polynomial of degree p (i.e., by including terms up to Δt^p) plus a Taylor series remainder term in each of the expansions for the function and its derivatives with a parameter β_j, $j = 1, 2, \ldots, p$, which can be chosen to give good approximation properties to the algorithm.

Insertion of the first three expressions of (17.53) into Eq. (17.52) gives a single equation from which $\overset{p}{\mathbf{u}}_{n+1}$ can be found. When this is determined, $\mathbf{u}_{n+1}$ to $\overset{p-1}{\mathbf{u}}_{n+1}$ can be evaluated using Eqs (17.53). Satisfying Eq. (17.52) is almost a 'collocation' which could be obtained by inserting the expressions (17.53) into a weighted residual form (17.36) with $W = \delta(t_{n+1})$ (the Dirac delta function). However, the expansion does not correspond to a unique function $\mathbf{u}$.

In detail we can write the first three expansions of Eqs (17.53) as

$$\begin{aligned}
\mathbf{u}_{n+1} &= \breve{\mathbf{u}}_{n+1} + \beta_p \frac{\Delta t^p}{p!} \overset{p}{\mathbf{u}}_{n+1} \\
\dot{\mathbf{u}}_{n+1} &= \dot{\breve{\mathbf{u}}}_{n+1} + \beta_{p-1} \frac{\Delta t^{p-1}}{(p-1)!} \overset{p}{\mathbf{u}}_{n+1} \\
\ddot{\mathbf{u}}_{n+1} &= \ddot{\breve{\mathbf{u}}}_{n+1} + \beta_{p-2} \frac{\Delta t^{p-2}}{(p-2)!} \overset{p}{\mathbf{u}}_{n+1}
\end{aligned} \tag{17.54}$$

where

$$\begin{aligned}
\breve{\mathbf{u}}_{n+1} &= \mathbf{u}_n + \Delta t \dot{\mathbf{u}}_n + \cdots + (1 - \beta_p) \frac{\Delta t^p}{p!} \overset{p}{\mathbf{u}}_n \\
\dot{\breve{\mathbf{u}}}_{n+1} &= \dot{\mathbf{u}}_n + \Delta t \ddot{\mathbf{u}}_n + \cdots + (1 - \beta_{p-1}) \frac{\Delta t^{p-1}}{(p-1)!} \overset{p}{\mathbf{u}}_n \\
\ddot{\breve{\mathbf{u}}}_{n+1} &= \ddot{\mathbf{u}}_n + \Delta t \dddot{\mathbf{u}}_n + \cdots + (1 - \beta_{p-2}) \frac{\Delta t^{p-2}}{(p-2)!} \overset{p}{\mathbf{u}}_n
\end{aligned} \tag{17.55}$$

Inserting the above into Eq. (17.52) and solving for $\overset{p}{\mathbf{u}}_{n+1}$ gives

$$\overset{p}{\mathbf{u}}_{n+1} = -\mathbf{A}^{-1} \left[\mathbf{M} \ddot{\breve{\mathbf{u}}}_{n+1} + \mathbf{C} \dot{\breve{\mathbf{u}}}_{n+1} + \mathbf{K} \breve{\mathbf{u}}_{n+1} + \mathbf{f}_{n+1} \right] \tag{17.56}$$

where

$$\mathbf{A} = \frac{\beta_{p-2} \Delta t^{p-2}}{(p-2)!} \mathbf{M} + \frac{\beta_{p-1} \Delta t^{p-1}}{(p-1)!} \mathbf{C} + \frac{\beta_p \Delta t^p}{p!} \mathbf{K}$$

We note immediately that the above expression is formally identical to that of the SSpj algorithm, Eq. (17.40), if we make the substitutions

$$\beta_p = \theta_p \qquad \beta_{p-1} = \theta_{p-1} \qquad \beta_{p-2} = \theta_{p-2} \tag{17.57}$$

However, $\breve{\mathbf{u}}_{n+1}$, $\dot{\breve{\mathbf{u}}}_{n+1}$, etc., in the generalized Newmark, GNpj, are not identical to $\bar{\mathbf{u}}_{n+1}$, $\dot{\bar{\mathbf{u}}}_{n+1}$, etc., in the SSpj algorithms. In the SSpj algorithm these represent predicted mean values in the interval Δt while in the GNpj algorithms they represent predicted values at t_{n+1}.

The computation procedure for the GN algorithms is very similar to that for the SS algorithms, starting now with known values of $\mathbf{u}_n$ to $\overset{p}{\mathbf{u}}_n$. As before we have the given initial conditions and we can often arrange to use the differential equation and its derivatives to generate higher derivatives for $\mathbf{u}$ at $t = 0$. However, the GN algorithm requires use of $\overset{p}{\mathbf{u}}_0$ in the computation of the next time step.

An important member of this family is the GN22 algorithm.

The Newmark algorithm (GN22)

We have already mentioned the classical Newmark algorithm as it is one of the most popular for dynamic analysis. It is indeed a special case of the general algorithm of the preceding section in which a quadratic ($p = 2$) expansion is used, this being the minimum required for second-order problems. We describe here the details in view of its widespread use.

The expansion of Eq. (17.53) for $p = 2$ gives

$$\begin{aligned} \mathbf{u}_{n+1} &= \mathbf{u}_n + \Delta t\dot{\mathbf{u}}_n + \tfrac{1}{2}(1-\beta_2)\Delta t^2\ddot{\mathbf{u}}_n + \tfrac{1}{2}\beta_2\Delta t^2\ddot{\mathbf{u}}_{n+1} = \breve{\mathbf{u}}_{n+1} + \tfrac{1}{2}\beta_2\Delta t^2\ddot{\mathbf{u}}_{n+1} \\ \dot{\mathbf{u}}_{n+1} &= \dot{\mathbf{u}}_n + (1-\beta_1)\Delta t\ddot{\mathbf{u}}_n + \beta_1\Delta t\ddot{\mathbf{u}}_{n+1} = \dot{\breve{\mathbf{u}}}_{n+1} + \beta_1\Delta t\ddot{\mathbf{u}}_{n+1} \end{aligned} \tag{17.58}$$

and this together with the dynamic equation (17.52),

$$\mathbf{M}\ddot{\mathbf{u}}_{n+1} + \mathbf{C}\dot{\mathbf{u}}_{n+1} + \mathbf{K}\mathbf{u}_{n+1} + \mathbf{f}_{n+1} = \mathbf{0} \tag{17.59}$$

allows the three unknowns $\mathbf{u}_{n+1}$, $\dot{\mathbf{u}}_{n+1}$ and $\ddot{\mathbf{u}}_{n+1}$ to be determined.

We now proceed as we have already indicated and solve first for $\ddot{\mathbf{u}}_{n+1}$ by substituting (17.58) into (17.59). This yields as the first step

$$\ddot{\mathbf{u}}_{n+1} = -\mathbf{A}^{-1}\{\mathbf{f}_{n+1} + \mathbf{C}\dot{\breve{\mathbf{u}}}_{n+1} + \mathbf{K}\breve{\mathbf{u}}_{n+1}\} \tag{17.60}$$

where

$$\mathbf{A} = \mathbf{M} + \beta_1\Delta t\mathbf{C} + \tfrac{1}{2}\beta_2\Delta t^2\mathbf{K} \tag{17.61}$$

After this step the values of $\mathbf{u}_{n+1}$ and $\dot{\mathbf{u}}_{n+1}$ can be found using Eqs (17.58).

As in the general case, $\beta_2 = 0$ produces an explicit algorithm whose solution is very simple if $\mathbf{M}$ and $\mathbf{C}$ are assumed diagonal.

It is of interest to remark that the accuracy can be slightly improved and yet the advantages of the explicit form preserved for SS/GN algorithms by a simple iterative process within each time increment. In this, for the GN algorithm, we predict $\mathbf{u}^i_{n+1}$, $\dot{\mathbf{u}}^i_{n+1}$ and $\ddot{\mathbf{u}}^i_{n+1}$ using expressions (17.54) with

$$\left(\overset{p}{\mathbf{u}}_{n+1}\right)^i = \left(\overset{p}{\mathbf{u}}_{n+1}\right)^{i-1}$$

and setting for $i = 1$

$$\left(\overset{p}{\mathbf{u}}_{n+1}\right)^0 = \mathbf{0}$$

This is followed by rewriting the governing equation (17.52) as

$$\mathbf{M}\left[\ddot{\mathbf{u}}_{n+1}^{i-1} + \frac{\beta_2\,\Delta t^{p-2}}{(p-2)!}\left(\overset{p}{\mathbf{u}}_{n+1}\right)^{i}\right] + \mathbf{C}\dot{\mathbf{u}}_{n+1}^{i-1} + \mathbf{K}\mathbf{u}_{n+1}^{i-1} + \mathbf{f}_{n+1} = \mathbf{0} \qquad (17.62)$$

and solving for $\left(\overset{p}{\mathbf{u}}_{n+1}\right)^{i}$.

This predictor–corrector iteration has been successfully used for various algorithms, though of course the stability conditions remain unaltered from those of a simple explicit scheme.[46]

For implicit schemes we note that in the general case, Eqs (17.58) have scalar coefficients while Eq. (17.59) has matrix coefficients. Thus, for the implicit case some users prefer a slightly more complicated procedure than indicated above in which the first unknown determined is $\mathbf{u}_{n+1}$. This may be achieved by expressing Eqs (17.58) in terms of the $\mathbf{u}_{n+1}$ to obtain

$$\begin{aligned}\ddot{\mathbf{u}}_{n+1} &= \ddot{\hat{\mathbf{u}}}_{n+1} + \frac{2}{\beta_2\Delta t^2}\mathbf{u}_{n+1}\\ \dot{\mathbf{u}}_{n+1} &= \dot{\hat{\mathbf{u}}}_{n+1} + \frac{2\beta_1}{\beta_2\Delta t}\mathbf{u}_{n+1}\end{aligned} \qquad (17.63)$$

where

$$\begin{aligned}\ddot{\hat{\mathbf{u}}}_{n+1} &= -\frac{2}{\beta_2\Delta t^2}\mathbf{u}_n - \frac{2}{\beta_2\Delta t}\dot{\mathbf{u}}_n - \frac{1-\beta_2}{\beta_2}\ddot{\mathbf{u}}_n\\ \dot{\hat{\mathbf{u}}}_{n+1} &= -\frac{2\beta_1}{\beta_2\Delta t}\mathbf{u}_n + \left(1-\frac{2\beta_1}{\beta_2}\right)\dot{\mathbf{u}}_n + \left(1-\frac{\beta_1}{\beta_2}\right)\Delta t\ddot{\mathbf{u}}_n\end{aligned} \qquad (17.64)$$

These are now substituted into Eq. (17.59) to give the result

$$\mathbf{u}_{n+1} = -\mathbf{A}^{-1}(\mathbf{f}_{n+1} + \mathbf{C}\dot{\hat{\mathbf{u}}}_{n+1} + \mathbf{M}\ddot{\hat{\mathbf{u}}}_{n+1}) \qquad (17.65)$$

where now

$$\mathbf{A} = \frac{2}{\beta_2\Delta t^2}\mathbf{M} + \frac{2\beta_1}{\beta_2\Delta t}\mathbf{C} + \mathbf{K}$$

which again on using Eqs (17.63) and (17.64) gives $\dot{\mathbf{u}}$ and $\ddot{\mathbf{u}}$. The inversion is here identical to within a scalar multiplier and, thus, precludes use of the explicit form where β_2 is zero.

17.4 Stability of general algorithms

Consistency of the general algorithms of SS and GN type is self-evident and assured by their formulation.

In a similar manner to that used in Sec. 17.2.5 we can conclude from this that the *local truncation error* is $O(\Delta t^{p+1})$ as the expansion contains all terms up to τ^p for SS and Δt^p for GN algorithms. However, the total truncation error after n steps is only $O(\Delta t^p)$ for the first-order equation system and $O(\Delta t^{p-1})$ for the second-order one. Details of accuracy discussions and reasons for this can be found in reference 6.

The question of stability is paramount and in this section we shall discuss it in detail for the SS type of algorithms. The establishment of similar conditions for the GN algorithms follows precisely the same pattern and is left as an exercise to the reader. It is, however, important to remark here that it can be shown that

(a) the SS and GN algorithms are generally similar in performance;
(b) *their stability conditions are identical when* $\theta_p \equiv \beta_p$.

The proof of the last statement requires some elaborate algebra and is given in reference 6.

The determination of stability requirements follows precisely the pattern outlined in Sec. 17.2.5. However, for practical reasons we shall

(a) avoid writing explicitly the amplification matrix **A**;
(b) immediately consider the scalar equation system implying modal decomposition and no forcing, i.e.,

$$m\ddot{u} + c\dot{u} + ku = 0 \tag{17.66}$$

Equations (17.37), (17.40) and (17.41) written in scalar terms define the recurrence algorithms. For the homogeneous case the general solution can be written down as

$$\begin{aligned} u_{n+1} &= \mu u_n \\ \dot{u}_{n+1} &= \mu \dot{u}_n \\ &\vdots \\ \overset{p-1}{u}_{n+1} &= \mu \overset{p-1}{u}_n \end{aligned} \tag{17.67}$$

and substitution of the above into the equations governing the recurrence can be written quite generally as

$$\mathbf{S}\mathbf{X}_n = \mathbf{0} \tag{17.68}$$

where

$$\mathbf{X}_n = \begin{Bmatrix} u_n \\ \Delta t \dot{u}_n \\ \vdots \\ \Delta t^{p-1} \overset{p-1}{u}_n \\ \Delta t^p \overset{p}{u}_n \end{Bmatrix} \tag{17.69}$$

The matrix **S** is given below in a compact form which can be verified by the reader:

$$\mathbf{S} = \begin{bmatrix} b_0 & b_1 & b_2 & \cdots & b_{p-1} & b_p \\ 1-\mu & 1 & \dfrac{1}{2!} & \cdots & \dfrac{1}{(p-1)!} & \dfrac{1}{p!} \\ 0 & 1-\mu & 1 & \cdots & \dfrac{1}{(p-2)!} & \dfrac{1}{(p-1)!} \\ \vdots & \vdots & \vdots & \vdots & \vdots & \vdots \\ 0 & 0 & 0 & \cdots & 1 & \dfrac{1}{2!} \\ 0 & 0 & 0 & \cdots & 1-\mu & 1 \end{bmatrix} \tag{17.70}$$

where

$$b_0 = \theta_0 \Delta t^2 k$$
$$b_1 = \theta_0 \Delta t c + \theta_1 \Delta t^2 k$$
$$\vdots$$
$$b_q = \frac{\theta_{q-2}}{(q-2)!} m + \frac{\theta_{q-1} \Delta t}{(q-1)!} c + \frac{\theta_q \Delta t^2}{q!} k, \qquad q = 2, 3, \ldots, p$$

and $\theta_0 = 1$.

For non-trivial solutions for the vector $\mathbf{X}_n$ to exist it is necessary for

$$\det \mathbf{S} = 0 \tag{17.71}$$

This provides a *characteristic polynomial* of order p for μ which yields the eigenvalues of the amplification matrix. For stability it is sufficient and necessary that the moduli of all eigenvalues [see Eq. (17.27)] satisfy

$$|\mu| \leq 1 \tag{17.72}$$

We remark that in the case of repeated roots the equality sign does not apply. The reader will have noticed that the direct derivation of the determinant of **S** is much simpler than writing down matrix **A** and finding the eigenvalues. The results are, of course, identical.

The calculation of stability limits, even with the scalar (modal) equation system, is non-trivial. For this reason in what follows we shall only do it for $p = 2$ and $p = 3$. However, two general procedures will be introduced here.

The first of these is the so-called *z transformation*. In this we use a change of variables in the polynomial putting

$$\mu = \frac{1+z}{1-z} \tag{17.73}$$

where z as well as μ are in general complex numbers. It is easy to show that the requirement of Eq. (17.72) is identical to that demanding the *real part of z to be negative* (see Fig. 17.10).

The second procedure introduced is the well-known Routh–Hurwitz condition[47–49] which states that for a polynomial

$$c_0 z^n + c_1 z^{n-1} + \cdots + c_{n-1} z + c_n = 0 \quad \text{with} \quad c_0 > 0 \tag{17.74}$$

the real part of all roots will be negative if, for $c_1 > 0$,

$$\det \begin{bmatrix} c_1 & c_3 \\ c_0 & c_2 \end{bmatrix} > 0; \quad \det \begin{bmatrix} c_1 & c_3 & c_5 \\ c_0 & c_2 & c_4 \\ 0 & c_1 & c_3 \end{bmatrix} > 0 \tag{17.75}$$

and generally

$$\det \begin{bmatrix} c_1 & c_3 & c_5 & c_7 & \cdots & \\ c_0 & c_2 & c_4 & c_6 & \cdots & \\ 0 & c_1 & c_3 & c_5 & \cdots & \\ 0 & 0 & c_2 & c_4 & \cdots & \\ \vdots & & & & \ddots & \\ 0 & 0 & 0 & \cdots & c_{n-2} & c_n \end{bmatrix} > 0 \tag{17.76}$$

With these tools in hand we can discuss in detail the stability of specific algorithms.

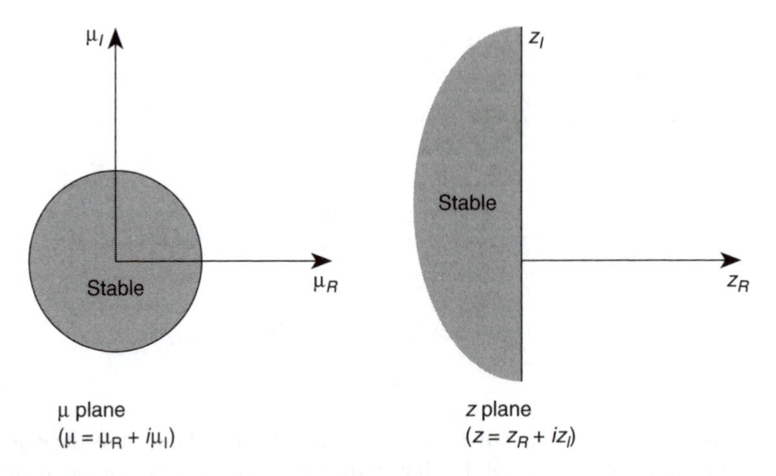

Fig. 17.10 The $\mu = (1+z)/(1-z)$ transformation.

17.4.1 Stability of SS22/SS21 algorithms

The recurrence relations for the algorithm given in Eqs (17.48) and (17.50) can be written after inserting

$$u_{n+1} = \mu u_n; \quad \dot{u}_{n+1} = \mu \dot{u}_n \quad \text{and} \quad f = 0 \tag{17.77}$$

as

$$\begin{aligned} m\alpha + c\left(\dot{u}_n + \theta_1 \Delta t \alpha\right) + k\left(u_n + \theta_1 \Delta t \dot{u}_n + \tfrac{1}{2}\theta_2 \Delta t^2 \alpha\right) &= 0 \\ -\mu u_n + u_n + \Delta t \dot{u}_n + \tfrac{1}{2}\Delta t^2 \alpha &= 0 \\ -\mu \dot{u}_n + \dot{u}_n + \Delta t \alpha &= 0 \end{aligned}$$

Changing the variable according to Eq. (17.73) results in the characteristic polynomial

$$c_0 z^2 + c_1 z + c_2 = 0 \tag{17.78}$$

with

$$\begin{aligned} c_0 &= 4m + (4\theta_1 - 2)\Delta t c + 2(\theta_2 - \theta_1)\Delta t^2 k \\ c_1 &= 2\Delta t c + (2\theta_1 - 1)\Delta t^2 k \\ c_2 &= \Delta t^2 k \end{aligned} \tag{17.79}$$

The Routh–Hurwitz requirement for stability is simply that

$$c_0 > 0 \qquad c_1 \geq 0 \qquad \det \begin{bmatrix} c_1 & 0 \\ c_0 & c_2 \end{bmatrix} > 0$$

or simply

$$c_0 > 0 \qquad c_1 \geq 0 \qquad c_2 > 0 \tag{17.80}$$

These inequalities give for *unconditional stability* the condition that

$$\theta_2 \geq \theta_1 \geq \tfrac{1}{2} \tag{17.81}$$

This condition is also generally valid when $m = 0$, i.e., for the SS21 algorithm (the first-order equation) though now $\theta_2 = \theta_1$ must be excluded.

It is possible to satisfy the inequalities (17.80) only at some values of Δt yielding conditional stability. For the explicit process $\theta_2 = 0$ with SS22/SS21 algorithms the inequalities (17.80) demand that

$$\begin{aligned} 2m + (2\theta_1 - 1)\Delta t c - \theta_1 \Delta t^2 k &\geq 0 \\ 2c + (2\theta_1 - 1)\Delta t k &\geq 0 \end{aligned} \tag{17.82}$$

The second one is satisfied whenever

$$\theta_1 \geq \tfrac{1}{2} \tag{17.83}$$

and for $\theta_1 = 1/2$ the first supplies the requirement that

$$\Delta t^2 \leq \frac{4m}{k} \tag{17.84}$$

The last condition does not permit an explicit scheme for SS21, i.e., when $m = 0$. Here, however, if we take $\theta_1 > 1/2$ we have from the first equation of Eq. (17.82)

$$\Delta t < \frac{2\theta_1 - 1}{\theta_1} \frac{c}{k} \tag{17.85}$$

It is of interest for problems of structural dynamics to consider the nature of the bounds in an elastic situation. Here we can use the same process as that described in Sec. 17.2.5 for first-order problems of heat conduction. Looking at a single element with a single degree of freedom and consistent mass yields in place of condition (17.84)

$$\Delta t \leq \frac{2}{\sqrt{3}} \frac{h}{C} = \Delta t_{\text{crit}}$$

where h is the element size and

$$C = \sqrt{\frac{E}{\rho}}$$

is the speed of elastic wave propagation. For lumped mass matrices the factor becomes $\sqrt{2}$.

Once again the ratio of the smallest element size over wave speed governs the stability but it is interesting to note that in problems of dynamics the critical time step is proportional to h while, as shown in Eq. (17.32), for first-order problems it is proportional to h^2. Clearly for decreasing mesh size explicit schemes in dynamics are more efficient than in thermal analysis and are exceedingly popular in certain classes of problems.

17.4.2 Stability of various higher order schemes and equivalence with some known alternatives

Identical stability considerations as those described in previous sections can be applied to SS32/SS31 and higher order approximations. We omit here the algebra and simply quote some results.[6]

SS32/31. Here for zero damping ($c = 0$) in SS32 we require for unconditional stability that

$$\theta_1 > \tfrac{1}{2} \qquad \theta_2 \geq \theta_1 + \tfrac{1}{6} \qquad \theta_2 \geq \tfrac{1}{4} \qquad \theta_3 \geq \tfrac{3}{2}$$
$$3\theta_1\theta_2 - 3\theta_1^2 + \theta_1 \geq \theta_3 \tag{17.86}$$

For first-order problems ($m = 0$), i.e., SS31, the first requirements are as in dynamics but the last one becomes

$$3\theta_1\theta_1^2 - 3\theta_2 + \theta_1 \geq \theta_3 - \frac{[6\theta_1(\theta_1 - 1) + 1]^2}{9(2\theta_1 - 1)} \tag{17.87}$$

With $\theta_3 = 0$, i.e., an explicit scheme when $c = 0$,

$$\Delta t^2 \leq \frac{12(2\theta_1 - 1)}{6\theta_2 - 1}\frac{m}{k} \tag{17.88}$$

and when $m = 0$,

$$\Delta t \leq \frac{\theta_2 - \theta_1}{6\theta_2 - 1}\frac{c}{k} \tag{17.89}$$

SS42/41. For this (and indeed higher orders) unconditional stability in dynamics problems $m \neq 0$ does not exist. This is a consequence of a theorem by Dahlquist.[50] The SS41 scheme can have unconditional stability but the general expressions for this are cumbersome. We quote one example that is unconditionally stable:

$$\theta_1 = \tfrac{5}{2} \qquad \theta_2 = \tfrac{35}{6} \qquad \theta_3 = \tfrac{25}{2} \qquad \theta_4 = 24$$

This set of values corresponds to a backward difference four-step algorithm of Gear.[51]

It is of general interest to remark that certain members of the SS (or GN) families of algorithms are similar in performance and identical in the stability (and hence recurrence) properties to others published in the large literature on the subject. Each algorithm claims particular advantages and properties. In Tables 17.2–17.4 we show some members of this family.[41, 50–56] Clearly many more algorithms that are applicable are present in the general formulae.

We remark here that identity of stability and recurrence always occurs with multistep algorithms, which we shall discuss briefly in the next section.

Table 17.2 SS21 equivalents

Algorithms	Theta values
Zlamal[41]	$\theta_1 = \frac{5}{6}, \theta_2 = 2$
Gear[51]	$\theta_1 = \frac{3}{2}, \theta_2 = 2$
Liniger[52]	$\theta_1 = 1.0848, \theta_2 = 1$
Liniger[52]	$\theta_1 = 1.2184, \theta_2 = 1.292$

Table 17.3 SS31 equivalents

Algorithms	Theta values
Gear[51]	$\theta_1 = 2, \theta_2 = \frac{11}{3}, \theta_3 = 6$
Liniger[52]	$\theta_1 = 1.84, \theta_2 = 3.07, \theta_3 = 4.5$
Liniger[52]	$\theta_1 = 0.8, \theta_2 = 1.03, \theta_3 = 1.29$

Table 17.4 SS32 equivalents

Algorithms	Theta values
Houbolt[53]	$\theta_1 = 2, \theta_2 = \frac{11}{3}, \theta_3 = 6$
Wilson Θ[54]	$\theta_1 = \Theta, \theta_2 = \Theta^2, \theta_3 = \Theta^3$ ($\Theta = 1.4$ commonly used)
Bossak–Newmark[55] ($m\ddot{u} + ku = 0$, $\gamma_B = \frac{1}{2} - \alpha_B$)	$\theta_1 = 1 - \alpha_B$ $\theta_2 = \frac{2}{3} - \alpha_B + 2\beta_B$ $\theta_3 = 6\beta_B$
Bossak–Newmark[55] ($m\ddot{u} + c\dot{u} + ku = 0$, $\gamma_B = \frac{1}{2} - \alpha_B$, $\beta_B = \frac{1}{6} - \frac{1}{2}\alpha_B$)	$\theta_1 = 1 - \alpha_B$ $\theta_2 = 1 - 2\alpha_B$ $\theta_3 = 1 - 3\alpha_B$
Hilber–Hughes–Taylor[56] ($m\ddot{u} + ku = 0$, $\gamma_H = \frac{1}{2} - \alpha_H$)	$\theta_1 = 1$ $\theta_2 = \frac{2}{3} + 2\beta_H - 2\alpha_H^2$ $\theta_3 = 6\beta_H(1 + \alpha_H)$

Multistep methods

17.5 Multistep recurrence algorithms

17.5.1 Introduction

In the previous sections we have been concerned with recurrence algorithms valid within a single time step and relating the values of $\mathbf{u}_{n+1}$, $\dot{\mathbf{u}}_{n+1}$, $\ddot{\mathbf{u}}_{n+1}$ to $\mathbf{u}_n$, $\dot{\mathbf{u}}_n$, $\ddot{\mathbf{u}}_n$, etc. It is possible to derive, using very similar procedures to those previously introduced, multistep algorithms in which we relate $\mathbf{u}_{n+1}$ to the values $\mathbf{u}_n$, $\mathbf{u}_{n-1}$, $\mathbf{u}_{n-2}$, etc., without explicitly introducing the derivatives. Much classical work on stability and accuracy has been introduced on such multistep algorithms and hence they deserve mention here.

We shall show in this section that a series of such algorithms may be simply derived using the weighted residual process. For constant time increments Δt, it can be shown that this set possesses identical stability and accuracy properties to the SSpj procedures.

17.5.2 The approximation procedure for a general multistep algorithm

As in Sec. 17.3.2 we shall approximate the function $\mathbf{u}$ of the second-order equation

$$\mathbf{M}\ddot{\mathbf{u}} + \mathbf{C}\dot{\mathbf{u}} + \mathbf{K}\mathbf{u} + \mathbf{f} = \mathbf{0} \tag{17.90}$$

by a polynomial expansion of the order p, now containing a single unknown $\mathbf{u}_{n+1}$. This polynomial assumes knowledge of the value of $\mathbf{u}_n, \mathbf{u}_{n-1}, \ldots, \mathbf{u}_{n-p+1}$ at appropriate times $t_n, t_{n-1}, \ldots, t_{n-p+1}$ (Fig. 17.11).

We can write this polynomial as

$$\mathbf{u}(t) = \sum_{j=1-p}^{1} N_j(t)\mathbf{u}_{n+j} \tag{17.91}$$

where Lagrange interpolation in time is given by (see Chapter 4)

$$N_j(t) = \prod_{\substack{k=1-p \\ k \neq j}}^{1} \frac{t - t_{n+k}}{t_{n+j} - t_{n+k}} \tag{17.92}$$

Substituting this approximation into Eq. (17.91) gives

$$\dot{\mathbf{u}} = \sum_{j=1-p}^{1} \dot{N}_j(t)\mathbf{u}_{n+j} \quad \text{and} \quad \ddot{\mathbf{u}} = \sum_{j=1-p}^{1} \ddot{N}_j(t)\mathbf{u}_{n+j} \tag{17.93}$$

where $\dot{N}_j$ and $\ddot{N}_j$ denote the time derivatives of the shape functions. Insertion of $\mathbf{u}$, $\dot{\mathbf{u}}$ and $\ddot{\mathbf{u}}$ into the weighted residual equation form yields

$$\int_{t_n}^{t_{n+1}} W(t) \sum_{j=1-p}^{1} \left[\left(\ddot{N}_j\mathbf{M} + \dot{N}_j\mathbf{C} + N_j\mathbf{K}\right)\mathbf{u}_{n+j} + N_j\mathbf{f}_{n+j}\right] \mathrm{d}t = 0 \tag{17.94}$$

with the forcing functions interpolated similarly from its nodal values. Using (17.92) and the definition for θ_k given by (17.37) leads to a recurrence relation which may be used to compute $\mathbf{u}_{n+1}$.

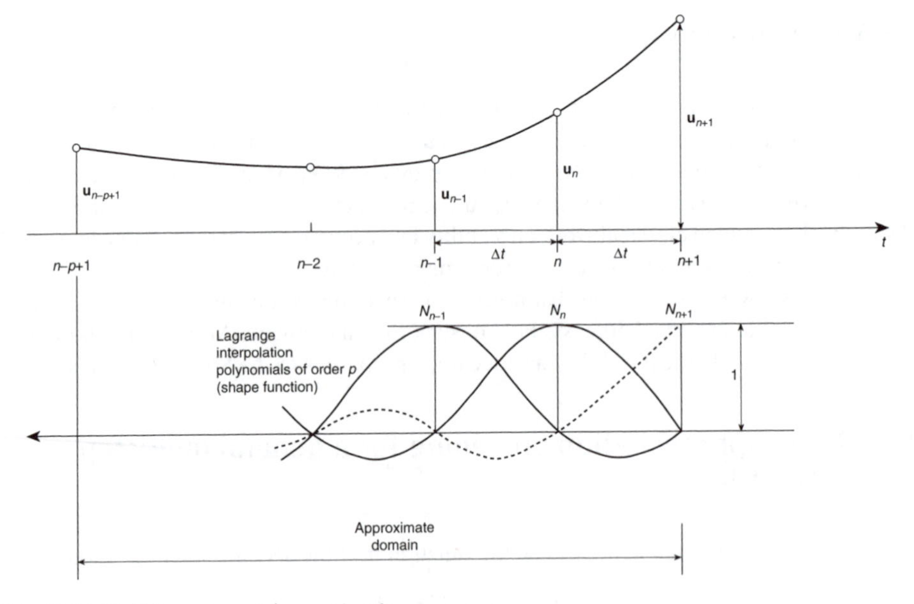

Fig. 17.11 Multistep polynomial approximation.

Example 17.3: Two-point interpolation: p = 1. Evaluating Eq. (17.92) for the two points we obtain

$$N_1 = \frac{t - t_n}{t_{n+1} - t_n} = \frac{1}{\Delta t}(t - t_n) = \frac{\tau}{\Delta t}$$
$$N_0 = \frac{t_{n+1} - t}{t_{n+1} - t_n} = \frac{1}{\Delta t}(t_{n+1} - t) = 1 - \frac{\tau}{\Delta t} \tag{17.95a}$$

where $\Delta t = t_{n+1} - t_n$ and $\tau = t - t_n$. Here the derivative is computed directly as

$$\frac{\mathrm{d}N_1}{\mathrm{d}t} = -\frac{\mathrm{d}N_0}{\mathrm{d}t} = \frac{1}{\Delta t} \tag{17.95b}$$

Second derivatives are obviously zero, hence this form may only be used for first-order equations as

$$\frac{1}{\Delta t}\mathbf{C}(\mathbf{u}_{n+1} - \mathbf{u}_n) + \mathbf{K}[(1-\theta)\mathbf{u}_n + \theta\mathbf{u}_{n+1}] + \bar{\mathbf{f}} = \mathbf{0} \tag{17.95c}$$

which is, obviously, identical to the SS11 result given previously.

Example 17.4: Three-point interpolation: p = 2. Evaluating Eq. (17.92) for the three points gives

$$N_1 = \frac{(t - t_{n-1})(t - t_n)}{(t_{n+1} - t_{n-1})(t_{n+1} - t_n)}$$
$$N_0 = \frac{(t - t_{n-1})(t - t_{n+1})}{(t_n - t_{n-1})(t_n - t_{n+1})} \tag{17.96a}$$
$$N_{-1} = \frac{(t - t_n)(t - t_{n+1})}{(t_{n-1} - t_n)(t_{n-1} - t_{n+1})}$$

The derivatives follow immediately from Eqs (17.92) and (17.93) as

$$\frac{\mathrm{d}N_1}{\mathrm{d}t} = \frac{(t - t_n) + (t - t_{n-1})}{(t_{n+1} - t_{n-1})(t_{n+1} - t_n)}$$
$$\frac{\mathrm{d}N_0}{\mathrm{d}t} = \frac{(t - t_{n+1}) + (t - t_{n-1})}{(t_n - t_{n-1})(t_n - t_{n+1})} \tag{17.96b}$$
$$\frac{\mathrm{d}N_{-1}}{\mathrm{d}t} = \frac{(t - t_{n+1}) + (t - t_n)}{(t_{n-1} - t_n)(t_{n-1} - t_{n+1})}$$

This is the lowest order which can be used for second-order equations and has second derivatives

$$\frac{\mathrm{d}^2N_1}{\mathrm{d}t^2} = \frac{2}{(t_{n+1} - t_{n-1})(t_{n+1} - t_n)}$$
$$\frac{\mathrm{d}^2N_0}{\mathrm{d}t^2} = \frac{2}{(t_n - t_{n-1})(t_n - t_{n+1})} \tag{17.96c}$$
$$\frac{\mathrm{d}^2N_{-1}}{\mathrm{d}t^2} = \frac{2}{(t_{n-1} - t_n)(t_{n-1} - t_{n+1})}$$

The recurrence relation for the two-step method with Δt constant is given by

$$\left[\frac{1}{\Delta t^2}\mathbf{M} + \frac{1}{\Delta t}(\bar{\theta}_1 + \tfrac{1}{2})\mathbf{C} + \tfrac{1}{2}(\bar{\theta}_2 + \bar{\theta}_1)\mathbf{K}\right]\mathbf{u}_{n+1} + \left[-\frac{2}{\Delta t^2}\mathbf{M} - \frac{2}{\Delta t}\bar{\theta}_1\mathbf{C} + (1 - \bar{\theta}_2)\mathbf{K}\right]\mathbf{u}_n + \left[\frac{1}{\Delta t^2}\mathbf{M} + \frac{1}{\Delta t}(\bar{\theta}_1 - \tfrac{1}{2})\mathbf{C} + \tfrac{1}{2}(\bar{\theta}_2 - \bar{\theta}_1)\mathbf{K}\right]\mathbf{u}_{n-1} + \bar{\mathbf{f}} = \mathbf{0} \qquad (17.96d)$$

where $\bar{\mathbf{f}}$ is the effect of the integrated force resultant and $\bar{\theta}_k$ is computed using (17.92), but now has different values for stability than given for θ_k in the SS22 form.

The above form is identical to the form originally derived by Newmark[44] (however, the conventional parameters are usually β and γ) and also corresponds to the SS22 and GN22 forms when parameters are related by:

$$\gamma = \bar{\theta}_1 + \tfrac{1}{2} = \theta_1 = \beta_1 \quad \text{and} \quad \beta = \tfrac{1}{2}(\bar{\theta}_2 + \bar{\theta}_1) = \tfrac{1}{2}\theta_2 = \tfrac{1}{2}\beta_2$$

The explicit form of this algorithm with $2\beta = \bar{\theta}_2 = \theta_2 = \beta_2 = 0$ and $\gamma = \bar{\theta}_1 + 1/2 = \theta_1 = \beta_1 = 1/2$ is frequently used as an alternative to the single-step explicit form. It is then known as the *central difference approximation* obtained by direct differencing. The reader can easily verify that the simplest finite difference approximation of Eq. (17.1) in fact corresponds to the above with $\bar{\theta}_2 = 0$ and $\bar{\theta}_1 = 0$.

Higher order multistep forms follow the general pattern given above for the two- and three-point forms and need not be discussed more here. In general there are no added advantages using the multistep form and, quite generally, we recommend use of the one-step forms SSpj and GNpj given above.

17.6 Some remarks on general performance of numerical algorithms

In Secs 17.2.5 and 17.3.3 we have considered the *exact* solution of the approximate recurrence algorithm given in the form

$$u_{n+1} = \mu u_n, \qquad \text{etc.} \qquad (17.97)$$

for the modally decomposed, single degree of freedom systems typical of Eqs (17.4) and (17.5). The evaluation of μ was important to ensure that its modulus does not exceed unity so that stability is preserved.

However, analytical solution of the linear homogeneous differential equations is also easy to obtain in the form

$$\tilde{u} = \bar{u}e^{\lambda t} \quad \text{or} \quad u_{n+1} = u_n e^{\lambda \Delta t} \qquad (17.98)$$

and comparison of μ with such a solution is always instructive to provide information on the performance of algorithms in the particular range of eigenvalues.

In Fig. 17.5 we plotted the exact solution $e^{-\omega \Delta t}$ and compared it with the values of μ for various θ algorithms approximating the first-order equation, noting that here

$$\lambda = -\omega = -\frac{k}{c}$$

and is real.

Immediately we see there that the performance error is very different for various values of Δt and obviously deteriorates at large values. Such values in a real multivariable problem correspond of course to the 'high frequency' responses which are often less important, and for smooth solutions we favour algorithms where μ tends to values much less than unity for such problems. However, response through the whole time range is important and attempts to choose an optimal value of θ for various time ranges has been performed by Liniger.[52] Table 17.1 of Sec. 17.2.6 illustrates how an algorithm with $\theta = 2/3$ and a higher truncation error than that of $\theta = 1/2$ can perform better in a multidimensional system because of such properties.

Similar analysis can be applied to the second-order equation. Here, to simplify matters, we consider only the homogeneous undamped equation in the form

$$m\ddot{u} + ku = 0 \tag{17.99}$$

in which the value of λ is purely imaginary and corresponds to a simple oscillator. By examining μ we can find not only the amplitude ratio (which for high accuracy should be unity) but also the phase error.

In Fig. 17.12(a) we show both the variation of the modulus of μ (which is called the *spectral radius*) and in Fig. 17.12(b) that of the relative period for the SS22/GN22 schemes, which of course are also applicable to the two-step equivalent. The results are plotted against

$$\frac{\Delta t}{T} \quad \text{where} \quad T = \frac{2\pi}{\omega}; \quad \omega^2 = \frac{k}{m}$$

In Fig. 17.13(a) and (b) similar curves are given for the SS23 and GN23 schemes frequently used in practice and discussed previously.

Here as in the first-order problem we often wish to suppress (or damp out) the response to frequencies in which $\Delta t/T$ is large (say greater than 0.1) in multidegree of freedom systems, as such a response will invariably be inaccurate. At the same time below this limit it is desirable to have amplitude ratios as close to unity as possible. It is clear that the stability limit with $\theta_1 = \theta_2 = 1/2$ giving unit response everywhere is often undesirable (unless physical damping is sufficient to damp high frequency modes) and that some *algorithmic damping* is necessary in these cases. The various schemes shown in Figs 17.12 and 17.13 can be judged accordingly and provide the reason for a search for an optimum algorithm.

We have remarked frequently that although schemes can be identical with regard to stability their performances may differ slightly. In Fig. 17.14 we illustrate the application of SS22 and GN22 to a single degree of freedom system showing results and errors in each scheme.

17.7 Time discontinuous Galerkin approximation

A time discontinuous Galerkin formulation may be deduced from the finite element in the time approximation procedure considered in this chapter. This is achieved by assuming the weight function $\mathbf{w}$ and solution variables $\mathbf{u}$ are approximated within each time interval Δt as

$$\begin{aligned} \mathbf{u} &= \mathbf{u}_n^+ + \Delta\mathbf{u}(t) \quad t_n^- \le t < t_{n+1}^- \\ \mathbf{w} &= \mathbf{w}_n^+ + \Delta\mathbf{w}(t) \quad t_n^- \le t < t_{n+1}^- \end{aligned} \tag{17.100}$$

where the time t_n^- is the limit from times smaller than t_n and t_n^+ is the limit from times larger than t_n and, thus, admit a discontinuity in the approximation to occur at each discrete

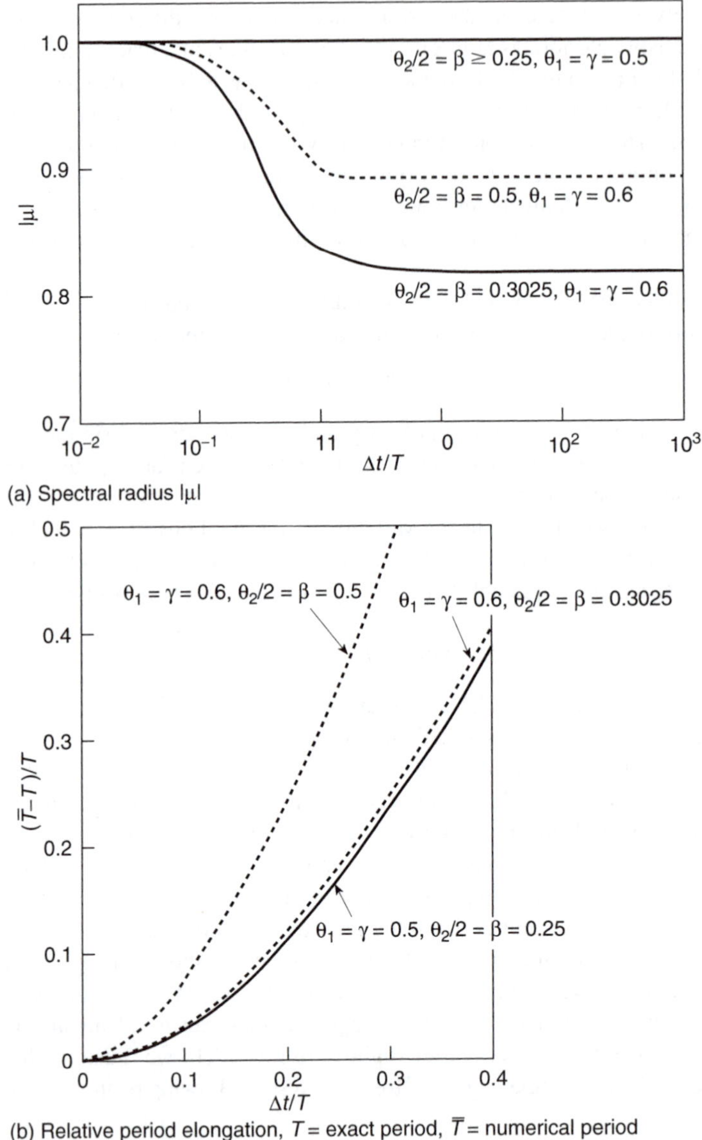

(a) Spectral radius $|\mu|$

(b) Relative period elongation, T = exact period, $\overline{T}$ = numerical period

Fig. 17.12 SS22, GN22 (Newmark) or their two-step equivalent.

time location. The functions $\Delta\mathbf{u}$ and $\Delta\mathbf{w}$ are defined to be zero at t_n and continuous up to the time t_{n+1}^- where again a discontinuity can occur during the next time interval.

The discrete form of the governing equations may be deduced starting from the time dependent partial differential equations where standard finite elements in space are combined with the time discontinuous Galerkin approximation and defining a weak form in a space–time slab. Alternatively, we may begin with the semi-discrete form as done previously in this chapter for other finite element in time methods. In this second form, for the first-order case, we write

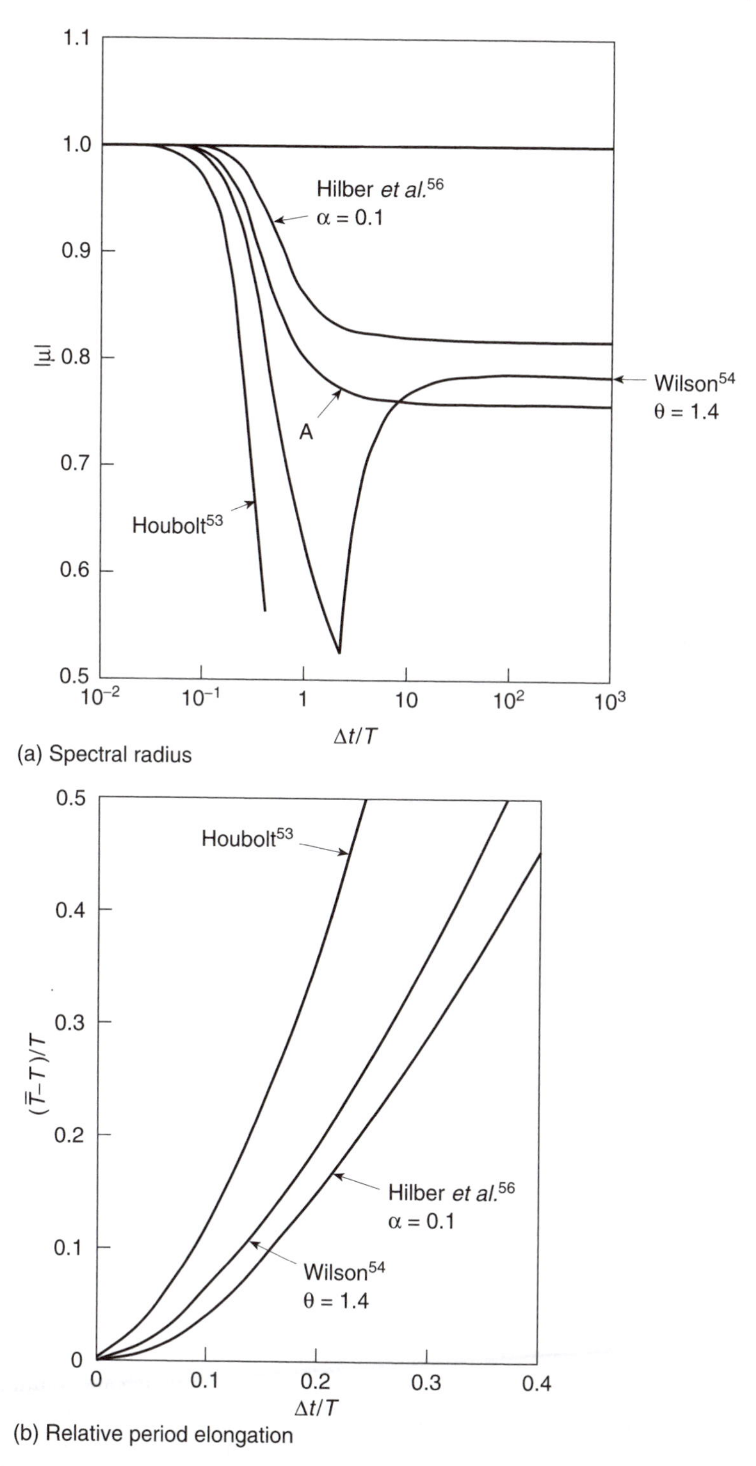

Fig. 17.13 SS23, GN23 or their two-step equivalent (see Problem 17.10 for description of α).

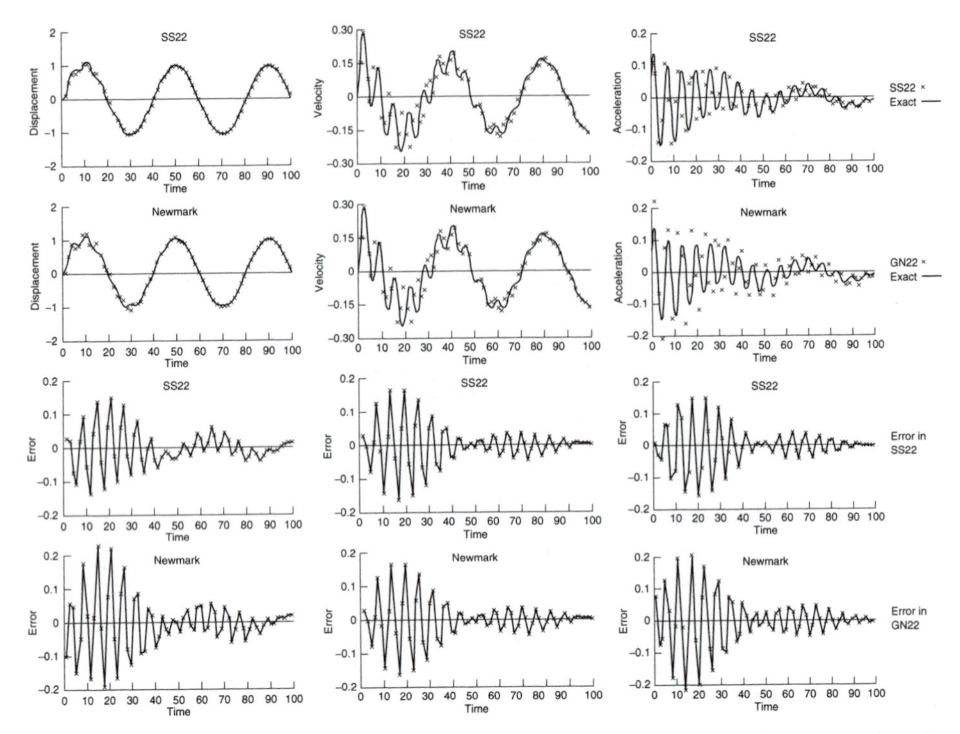

Fig. 17.14 Comparison of the SS22 and GN22 (Newmark) algorithms: a single DOF dynamic equation with periodic forcing term, $\theta_1 = \beta_1 = 1/2, \theta_2 = \beta_2 = 0$.

$$I = \int_{t_n^-}^{t_{n+1}^-} \mathbf{w}^{\mathrm{T}} \left(\mathbf{C}\dot{\mathbf{u}} + \mathbf{K}\mathbf{u} + \mathbf{f}\right) \, \mathrm{d}\tau = 0 \tag{17.101}$$

Due to the discontinuity at t_n it is necessary to split the integral into

$$I = \int_{t_n^-}^{t_n^+} \mathbf{w}^{\mathrm{T}}(\mathbf{C}\dot{\mathbf{u}} + \mathbf{K}\mathbf{u} + \mathbf{f}) \, \mathrm{d}\tau + \int_{t_n^+}^{t_{n+1}^-} \mathbf{w}^{\mathrm{T}}(\mathbf{C}\dot{\mathbf{u}} + \mathbf{K}\mathbf{u} + \mathbf{f}) \, \mathrm{d}\tau = 0 \tag{17.102}$$

which gives

$$\begin{aligned} I = (\mathbf{w}_n^+)^{\mathrm{T}}[\mathbf{C}(\mathbf{u}_n^+ - \mathbf{u}_n^-)] + (\mathbf{w}_n^+)^{\mathrm{T}} \int_{t_n^+}^{t_{n+1}^-} (\mathbf{C}\dot{\mathbf{u}} + \mathbf{K}\mathbf{u} + \mathbf{f}) \, \mathrm{d}\tau \\ + \int_{t_n^+}^{t_{n+1}^-} (\Delta \mathbf{w})^{\mathrm{T}} \left(\mathbf{C}\dot{\mathbf{u}} + \mathbf{K}\mathbf{u} + \mathbf{f}\right) \, \mathrm{d}\tau = 0 \end{aligned} \tag{17.103}$$

in which now all integrals involve approximations to functions which are continuous.

To apply the above process to a second-order equation it is necessary first to reduce the equation to a pair of first-order equations. This may be achieved by defining the momenta

$$\mathbf{p} = \mathbf{M}\dot{\mathbf{u}} \tag{17.104}$$

and then writing the pair

$$\begin{aligned}\mathbf{M}\dot{\mathbf{u}} - \mathbf{p} &= \mathbf{0} \\ \dot{\mathbf{p}} + \mathbf{C}\dot{\mathbf{u}} + \mathbf{K}\mathbf{u} + \mathbf{f} &= \mathbf{0}\end{aligned} \tag{17.105}$$

The time discrete process may now be applied by introducing two weighting functions as described in reference 39.

Example 17.5: Solution of a scalar equation. To illustrate the process we consider the simple first-order scalar equation

$$c\dot{u} + ku + f = 0 \tag{17.106}$$

We consider the specific approximations

$$\begin{aligned}u(t) &= u_n^+ + \tau \Delta u_{n+1}^- \\ w(t) &= w_n^+ + \tau \Delta w_{n+1}^-\end{aligned} \tag{17.107}$$

where $\Delta u_{n+1}^- = u_{n+1}^- - u_n^+$, etc., and

$$\tau = \frac{t - t_n}{t_{n+1} - t_n} = \frac{t - t_n}{\Delta t}$$

defines the time interval $0 < \tau < \Delta t$. This approximation gives the integral form

$$\begin{aligned}I = w_n^+ \left[c(u_n^+ - u_n^-)\right] + w_n^+ \Delta t \int_{0+}^{1-} \left[\frac{1}{\Delta t} c \Delta u_{n+1}^- + k\left(u_n^+ + \tau \Delta u_{n+1}^-\right) + f\right] \mathrm{d}\tau \\ + \Delta t \int_{0+}^{1-} \Delta w_{n+1}^- \tau \left[\frac{1}{\Delta t} c \Delta u_{n+1}^- + k\left(u_n^+ + \tau \Delta u_{n+1}^-\right) + f\right] \mathrm{d}\tau = 0\end{aligned} \tag{17.108}$$

Evaluation of the integrals gives the pair of equations

$$\begin{bmatrix}(c + k\Delta t) & \frac{1}{2}k\Delta t \\ \frac{1}{2}k\Delta t & (\frac{1}{2}c + \frac{1}{3}k\Delta t)\end{bmatrix} \begin{Bmatrix} u_n^+ \\ \Delta u_{n+1}^- \end{Bmatrix} + \begin{Bmatrix} \Delta t \bar{f} \\ \Delta t \Delta \bar{f} \end{Bmatrix} = \begin{Bmatrix} c u_n^- \\ 0 \end{Bmatrix} \tag{17.109a}$$

where

$$\begin{Bmatrix} \bar{f} \\ \Delta \bar{f} \end{Bmatrix} = \int_0^{\Delta t} \begin{Bmatrix} f \\ \tau f \end{Bmatrix} \mathrm{d}\tau \tag{17.109b}$$

Thus, with linear approximation of the variables the time discontinuous Galerkin method gives two equations to be solved for the two unknowns u_n^+ and u_{n+1}^-.

To illustrate the performance of the above scheme we compare the amplification matrix for the discontinuous Galerkin and standard Galerkin method in Fig. 17.15. In addition we use the method to solve the example described in Fig. 17.4 and present the results in Fig. 17.16. It is possible to also perform the solution with *constant* approximation. Based on the above this is achieved by setting Δu_{n+1}^- and Δw_{n+1}^- to zero yielding the single equation

$$(c + k\Delta t)u_n^+ + \Delta t \bar{f} = c u_n^- \tag{17.110}$$

and now since the approximation is constant over the entire time the u_n^+ also defines exactly the u_{n+1}^- value. This form will now be recognized as identical to the *backward difference* implicit scheme defined in Fig. 17.4 for $\theta = 1$.

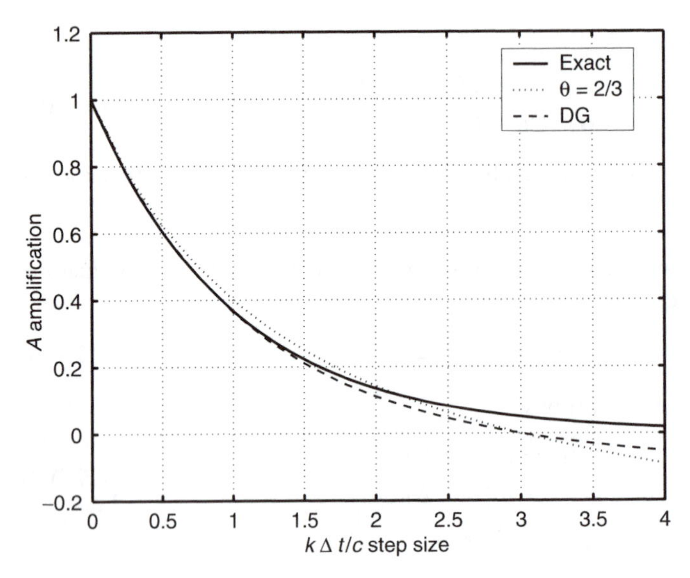

Fig. 17.15 The amplification A for standard and discontinuous Galerkin schemes.

Finally, we compare the error in the amplification matrix for different step sizes. The error defined by

$$E = \hat{A}(\Delta t) - A_{ex}(\Delta t)$$

where the $\hat{A}$ is the amplification for the approximate form and $A_{ex} = \exp(-\Delta t)$, the exact value. In Fig. 17.17(a) we present the values for the single-step algorithms and in Fig. 17.17(b) those for the discontinuous Galerkin and two-step quadratic continuous Galerkin solution. We note that the $\theta = 1/2$ (Crank–Nicolson), discontinuous Galerkin and $p = 2$ continuous Galerkin solutions are all second-order accurate (slope zero for small $\Delta t = 0$) while other values have finite slope and hence are only first-order accurate. It is also evident that the error at larger steps for the $p = 2$ continuous Galerkin is more accurate than the discontinuous Galerkin. Thus, for the same computational effort the use of the continuous form is more appropriate in this class of problems. For this reason we will not pursue use of the discontinuous Galerkin time integration procedure further here.

17.8 Concluding remarks

The derivation and examples presented in this chapter cover, we believe, the necessary toolkit for efficient solution of many transient problems governed by Eqs (17.1) and (17.2). In the next chapter we shall elaborate further on the application of the procedures discussed here and show that they can be extended to solve coupled problems which frequently arise in practice and where simultaneous solution by time stepping is often needed.

Finally, as we have indicated in Eq. (17.3), many problems have coefficient matrices or other variations which render the problem non-linear. This topic is addressed further for structural and solid mechanics problems in reference 57 and we note also that the issue of stability after many time steps is more involved than the procedures introduced here to investigate *local stability*.

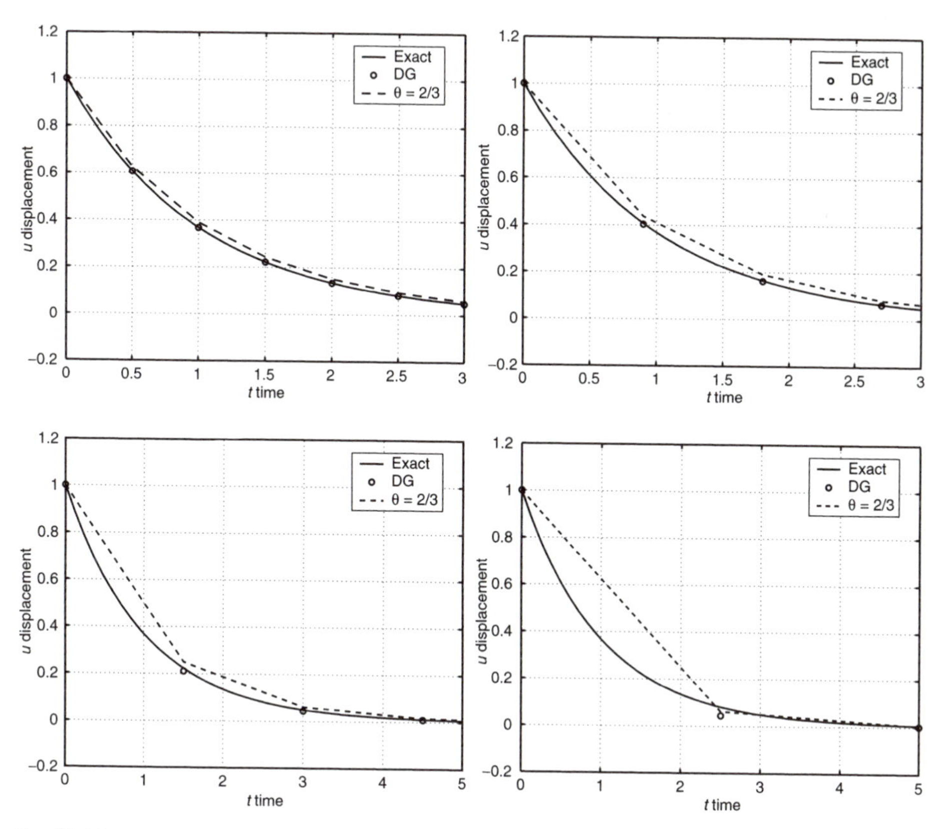

Fig. 17.16 Comparison of standard and discontinuous Galerkin schemes on a first-order initial value problem.

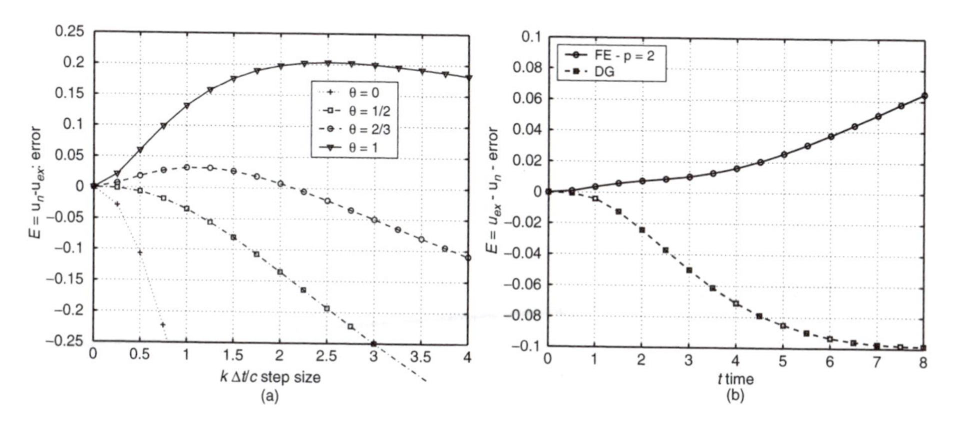

Fig. 17.17 Error in amplification matrix for single steps.

17.9 Problems

17.1 Verify the recurrence relation given in Eq. (17.19) using a least squares minimization process in which (17.7) is substituted into (17.2).

17.2 Determine the stability characteristics for a scalar form of the least squares recurrence relation given in Eq. (17.19). Plot the behaviour of the amplification matrix vs Δt.

17.3 Houbolt's method was originally developed as a multi-step method for the equation of motion written as

$$\mathbf{M}\ddot{\mathbf{u}}_{n+1} + \mathbf{C}\dot{\mathbf{u}}_{n+1} + \mathbf{K}\mathbf{u}_{n+1} + \mathbf{f}_{n+1} = \mathbf{0}$$

with updates given by

$$\dot{\mathbf{u}}_{n+1} = \frac{1}{6\,\Delta t}\,(11\mathbf{u}_{n+1} - 18\mathbf{u}_n + 9\mathbf{u}_{n-1} - 2\mathbf{u}_{n-2})$$

$$\ddot{\mathbf{u}}_{n+1} = \frac{1}{\Delta t^2}\,(2\mathbf{u}_{n+1} - 5\mathbf{u}_n + 4\mathbf{u}_{n-1} - \mathbf{u}_{n-2})$$

(a) Following the approach given in Sec. 17.2.5 determine the amplification matrix in the form

$$\mathbf{X}_{n+1} = \mathbf{A}\mathbf{X}_n \quad \text{with} \quad \mathbf{X}_{n+1} = \begin{bmatrix} u_{n+1} & u_n & u_{n-1} \end{bmatrix}^{\mathrm{T}}$$

(b) For the undamped and unloaded case (i.e., $c = f = 0$) determine and plot the spectral radius $|\mu|$ and period elongation $\Delta T/T$ vs the time increment $\Delta t/T$ as shown in Fig. 17.13. T is the period of the undamped equation.

17.4 Consider the scalar first-order equation:

$$c\dot{u} + ku + f = 0$$

Construct the discrete form for transient solution using the SS11 algorithm described in Sec. 17.3.

For the data $c = k = 1$ and $f = \sin^2 t$ obtain, by hand, the solution for the first five steps using a time step of $\Delta t = 0.1$.

Write a MATLAB program to solve the problem for $0 \le t \le 2$.

17.5 Consider the scalar second-order equation:

$$m\ddot{u} + c\dot{u} + ku + f = 0$$

Construct the discrete form for transient solution using the SS22 algorithm described in Sec. 17.3.

(a) For the data $m = k = 1$, $c = 0$ and $f = \sin^2 t$ obtain, by hand, the solution for the first five steps using a time step of $\Delta t = 0.05$.

Write a MATLAB program to solve the problem for $0 \le t \le 2$.

(b) Repeat (a) for $c = 0.05$.

17.6 Consider the scalar second-order equation:

$$m\ddot{u} + c\dot{u} + ku + f = 0$$

Construct the discrete form for transient solution using the GN22 algorithm described in Sec. 17.3.

(a) For the data $m = k = 1$, $c = 0$ and $f = \sin^2 t$ obtain, by hand, the solution for the first five steps using a time step of $\Delta t = 0.05$.
Write a MATLAB program to solve the problem for $0 \le t \le 2$.
(b) Repeat (a) for $c = 0.05$.

17.7 For the general second-order equation

$$m\ddot{u} + c\dot{u} + ku + f = 0$$

develop the discrete form for the SS32 algorithm.

17.8 For the general second-order equation

$$m\ddot{u} + c\dot{u} + ku + f = 0$$

develop the discrete form for the GN32 algorithm.

17.9 The general second-order equation may be split into the pair of first-order equations given by

$$\mathbf{M}\dot{\mathbf{v}} + \mathbf{C}\mathbf{v} + \mathbf{K}\mathbf{u} + \mathbf{f} = \mathbf{0}$$
$$\dot{\mathbf{u}} - \mathbf{v} = \mathbf{0}$$

(a) Develop the discrete form of the equations using the SS11 algorithm.
(b) For the scalar form of the equations determine the amplification matrix and the stability characteristics of the method.
(c) For the data $m = k = 1$, $c = 0$ and $f = \sin^2 t$ obtain, by hand, the solution for the first five steps using a time step of $\Delta t = 0.05$.
(d) Write a MATLAB program to solve the problem for $0 \le t \le 2$.

17.10 The Hilber–Hughes–Taylor (HHT) algorithm[56] is given by†

$$\mathbf{M}\ddot{\mathbf{u}}_{n+\alpha} + \mathbf{C}\dot{\mathbf{u}}_{n+\alpha} + \mathbf{K}\mathbf{u}_{n+\alpha} + \mathbf{f}_{n+\alpha} = \mathbf{0}$$

where $t_{n+\alpha} = (1 - \alpha_H)t_n + \alpha_H t_{n+1}$,

$$\mathbf{u}_{n+\alpha} = (1 - \alpha_H)\mathbf{u}_n + \alpha_H \mathbf{u}_{n+1}$$
$$\dot{\mathbf{u}}_{n+\alpha} = (1 - \alpha_H)\dot{\mathbf{u}}_n + \alpha_H \dot{\mathbf{u}}_{n+1}$$
$$\ddot{\mathbf{u}}_{n+\alpha} = \ddot{\mathbf{u}}_{n+1}$$

and $\mathbf{f}_{n+\alpha}$ is the force at $t_{n+\alpha}$. The algorithm is completed using the GN22 relations (17.58) with

$$\beta_1 = \frac{3}{2} - \alpha_H \quad \text{and} \quad \beta_2 = \frac{1}{2}(2 - \alpha_H)^2$$

(a) For the scalar form of the equations determine the amplification matrix and the stability characteristics of the method. (If necessary, use MATLAB to determine the roots of the stability equation.)
(b) For the data $m = k = 1$, $c = 0$ and $f = \sin^2 t$ obtain, by hand, the solution for the first five steps using a time step of $\Delta t = 0.05$.
(c) Write a MATLAB program to solve the problem for $0 \le t \le 2$.

† In the original publication $\alpha_H = 1 + \alpha$ and α had negative values. The definition used here is more consistent with other usage in this chapter and α_H is always positive.

17.11 Using the Routh–Hurwitz criterion described in Sec. 17.4 perform the stability analysis for the first-order problem described by SS11.

17.12 Using the Routh–Hurwitz criterion described in Sec. 17.4 perform the stability analysis for the second-order problem described by SS22.

17.13 Using the Routh–Hurwitz criterion described in Sec. 17.4 perform the stability analysis for the second-order problem described by GN22.

17.14 Use *FEAPpv* to solve the rectangular beam problem described in Problem 16.13. Compare the solution with that computed by modal analysis. (Note: The comparison with the full modal solution gives the error between the discrete integration and an exact integration of the semi-discrete system.)

17.15 Use *FEAPpv* to solve the curved beam problem described in Problem 16.11. Compare the solution with that computed by modal analysis.

17.16 Program development project: Extend the program system started in Problem 2.17 to perform time integration using the single-step algorithms described in Sec. 17.3. Your implementation should include:

(a) SS11 to integrate a first-order system such as encountered for thermal analysis.
(b) SS22 to integrate a second-order system for transient analysis of solids.
(c) GN22 to integrate a second-order system for transient analysis of solids.

Test your program by integrating a single degree of freedom problem for which you have a hand calculation for verification use.

Solve the rectangular beam problem described in Problem 16.13. Compare the solution with that computed by modal analysis. (Note: The comparison with the full modal solution gives the error between the discrete integration and the exact integration of the semi-discrete system.)

References

1. R.D. Richtmyer and K.W. Morton. *Difference Methods for Initial Value Problems.* Wiley (Interscience), New York, 1967.
2. T.D. Lambert. *Computational Methods in Ordinary Differential Equations.* John Wiley & Sons, Chichester, 1973.
3. P. Henrici. *Discrete Variable Methods in Ordinary Differential Equations.* John Wiley & Sons, New York, 1962.
4. F.B. Hildebrand. *Finite Difference Equations and Simulations.* Prentice-Hall, Englewood Cliffs, N.J., 1968.
5. G.W. Gear. *Numerical Initial Value Problems in Ordinary Differential Equations.* Prentice-Hall, Englewood Cliffs, N.J., 1971.
6. W.L. Wood. *Practical Time Stepping Schemes.* Clarendon Press, Oxford, 1990.
7. O.C. Zienkiewicz, R.L. Taylor, and P. Nithiarasu. *The Finite Element Method for Fluid Dynamics.* Butterworth-Heinemann, Oxford, 6th edition, 2005.
8. J.T. Oden. A general theory of finite elements. Part II. Applications. *Int. J. Numer. Meth. Eng.*, 1:247–254, 1969.
9. I. Fried. Finite element analysis of time-dependent phenomena. *J. AIAA*, 7:1170–1173, 1969.
10. J.H. Argyris and D.W. Scharpf. Finite elements in time and space. *Nucl. Eng. and Design*, 10:456–469, 1969.

11. O.C. Zienkiewicz and C.J. Parekh. Transient field problems – two and three dimensional analysis by isoparametric finite elements. *Int. J. Numer. Meth. Eng.*, 2:61–71, 1970.
12. O.C. Zienkiewicz. *The Finite Element Method in Engineering Science*. McGraw-Hill, London, 2nd edition, 1971.
13. O.C. Zienkiewicz and R.W. Lewis. An analysis of various time stepping schemes for initial value problems. *Earth. Eng. Struct. Dyn.*, 1:407–408, 1973.
14. W.L. Wood and R.W. Lewis. A comparison of time marching schemes for the transient heat conduction equation. *Int. J. Numer. Meth. Eng.*, 9:679–689, 1975.
15. O.C. Zienkiewicz. A new look at the Newmark, Houbolt and other time stepping formulas. A weighted residual approach. *Earth. Eng. Struct. Dyn.*, 5:413–418, 1977.
16. W.L. Wood. On the Zienkiewicz four-time-level scheme for numerical integration of vibration problems. *Int. J. Numer. Meth. Eng.*, 11:1519–1528, 1977.
17. O.C. Zienkiewicz, W.L. Wood, and R.L. Taylor. An alternative single-step algorithm for dynamic problems. *Earth. Eng. Struct. Dyn.*, 8:31–40, 1980.
18. W.L. Wood. A further look at Newmark, Houbolt, etc. time-stepping formulae. *Int. J. Numer. Meth. Eng.*, 20:1009–1017, 1984.
19. O.C. Zienkiewicz, W.L. Wood, N.W. Hine, and R.L. Taylor. A unified set of single-step algorithms. Part 1: General formulation and applications. *Int. J. Numer. Meth. Eng.*, 20:1529–1552, 1984.
20. W.L. Wood. A unified set of single-step algorithms. Part 2: Theory. *Int. J. Numer. Meth. Eng.*, 20:2302–2309, 1984.
21. M. Katona and O.C. Zienkiewicz. A unified set of single-step algorithms. Part 3: The beta-m method, a generalization of the Newmark scheme. *Int. J. Numer. Meth. Eng.*, 21:1345–1359, 1985.
22. E. Varoglu and N.D.L. Finn. A finite element method for the diffusion convection equations with concurrent coefficients. *Adv. Water Res.*, 1:337–341, 1973.
23. C. Johnson, U. Nävert, and J. Pitkäranta. Finite element methods for linear hyperbolic problems. *Comp. Meth. Appl. Mech. Eng.*, 45:285–312, 1984.
24. T.J.R. Hughes, L.P. Franca, and G.M. Hulbert. A new finite element formulation for computational fluid dynamics: VIII. The Galerkin/least-squares method for advective-diffusive equations. *Comp. Meth. Appl. Mech. Eng.*, 73:173–189, 1989.
25. T.J.R. Hughes and G.M. Hulbert. Space–time finite element methods in elastodynamics: formulation and error estimates. *Comp. Meth. Appl. Mech. Eng.*, 66:339–363, 1988.
26. G.M. Hulbert and T.J.R. Hughes. Space–time finite element methods for second-order hyperbolic equations. *Comp. Meth. Appl. Mech. Eng.*, 84:327–348, 1990.
27. G.M. Hulbert. Time finite element methods for structural dynamics. *Int. J. Numer. Meth. Eng.*, 33:307–331, 1992.
28. B.M. Irons and C. Treharne. A bound theorem for eigenvalues and its practical application. In *Proc. 3rd Conf. Matrix Methods in Structural Mechanics*, volume AFFDL-TR-71-160, pages 245–254, Wright-Patterson Air Force Base, Ohio, 1972.
29. K. Washizu. *Variational Methods in Elasticity and Plasticity*. Pergamon Press, New York, 3rd edition, 1982.
30. M.E. Gurtin. Variational principles for linear initial-value problems. *Q. Appl. Math.*, 22: 252–256, 1964.
31. M.E. Gurtin. Variational principles for linear elastodynamics. *Arch. Rat. Mech. Anal.*, 16:34–50, 1969.
32. E.L. Wilson and R.E. Nickell. Application of finite element method to heat conduction analysis. *Nucl. Eng. Design*, 4:1–11, 1966.
33. J. Crank and P. Nicolson. A practical method for numerical integration of solutions of partial differential equations of heat conduction type. *Proc. Camb. Phil. Soc.*, 43:50, 1947.

34. R.L. Taylor and O.C. Zienkiewicz. A note on the 'order of approximation'. *Int. J. Solids Struct.*, 21:793–798, 1985.
35. W.H. Reed and T.R. Hill. Triangular mesh methods for the neutron transport equation. Technical Report LA-UR-73–479, Los Alamos Scientific Laboratory, 1973.
36. P. Lesaint and P.-A. Raviart. On a finite element method for solving the neutron transport equation. In C. de Boor, editor, *Mathematical Aspects of Finite Elements in Partial Differential Equations*. Academic Press, New York, 1974.
37. C. Johnson. *Numerical Solutions of Partial Differential Equations by the Finite Element Method.* Cambridge University Press, Cambridge, 1987.
38. K. Eriksson and C. Johnson. Adaptive finite element methods for parabolic problems I: A linear model problem. *SIAM J. Numer. Anal.*, 28:43–77, 1991.
39. X.D. Li and N.-E. Wiberg. Structural dynamic analysis by a time-discontinuous Galerkin finite element method. *Int. J. Numer. Meth. Eng.*, 39:2131–2152, 1996.
40. M. Salvadori and M. Baron. *Numerical Methods in Engineering*. Prentice-Hall, New York, 1952.
41. M. Zlamal. Finite element methods in heat conduction problems. In J. Whiteman, editor, *The Mathematics of Finite Elements and Applications*, pages 85–104. Academic Press, London, 1977.
42. W. Liniger. Optimisation of a numerical integration method for stiff systems of ordinary differential equations. Technical Report RC2198, IBM Research, 1968.
43. J.M. Bettencourt, O.C. Zienkiewicz, and G. Cantin. Consistent use of finite elements in time and the performance of various recurrence schemes for heat diffusion equation. *Int. J. Numer. Meth. Eng.*, 17:931–938, 1981.
44. N. Newmark. A method of computation for structural dynamics. *J. Eng. Mech., ASCE*, 85:67–94, 1959.
45. T. Belytschko and T.J.R. Hughes, editors. *Computational Methods for Transient Analysis.* North-Holland, Amsterdam, 1983.
46. I. Miranda, R.M. Ferencz, and T.J.R. Hughes. An improved implicit–explicit time integration method for structural dynamics. *Earth. Eng. Struct. Dyn.*, 18:643–655, 1989.
47. E.J. Routh. *A Treatise on the Stability of a Given State or Motion.* Macmillan, London, 1977.
48. A. Hurwitz. Uber die Bedingungen, unter welchen eine Gleichung nur Würzeln mit negativen reellen teilen besitzt. *Math. Ann.*, 46:273–284, 1895.
49. F.R. Gantmacher. *The Theory of Matrices.* Chelsea, New York, 1959.
50. G.G. Dahlquist. A special stability problem for linear multistep methods. *BIT*, 3:27–43, 1963.
51. C.W. Gear. The automatic integration of stiff ordinary differential equations. In A.J.H. Morrell, editor, *Information Processing 68*. North-Holland, Dordrecht, 1969.
52. W. Liniger. Global accuracy and A-stability of one and two step integration formulae for stiff ordinary differential equations. In *Proc. Conf. on Numerical Solution of Differential Equations*, Dundee University, 1969.
53. J.C. Houbolt. A recurrence matrix solution for dynamic response of elastic aircraft. *J. Aero. Sci.*, 17:540–550, 1950.
54. K.-J. Bathe and E.L. Wilson. Stability and accuracy analysis of direct integration methods. *Earth. Eng. Struct. Dyn.*, 1:283–291, 1973.
55. W. Wood, M. Bossak, and O.C. Zienkiewicz. An alpha modification of Newmark's method. *Int. J. Numer. Meth. Eng.*, 15:1562–1566, 1980.
56. H. Hilber, T.J.R. Hughes, and R.L. Taylor. Improved numerical dissipation for the time integration algorithms in structural dynamics. *Earth. Eng. Struct. Dyn.*, 5:283–292, 1977.
57. O.C. Zienkiewicz and R.L. Taylor. *The Finite Element Method for Solid and Structural Mechanics*. Butterworth-Heinemann, Oxford, 6th edition, 2005.

18

Coupled systems

18.1 Coupled problems – definition and classification

Frequently two or more physical systems interact with each other, with the independent solution of any one system being impossible without simultaneous solution of the others. Such systems are known as coupled and of course such coupling may be weak or strong depending on the degree of interaction.

An obvious 'coupled' problem is that of dynamic fluid–structure interaction. Here neither the fluid nor the structural system can be solved independently of the other due to the unknown interface forces.

A definition of coupled systems may be generalized to include a wide range of problems and their numerical discretization as:[1]

Coupled systems and formulations are those applicable to multiple domains and dependent variables which usually (but not always) describe different physical phenomena and in which

(a) *neither domain can be solved while separated from the other;*
(b) *neither set of dependent variables can be explicitly eliminated at the differential equation level.*

The reader may well contrast this with definitions of *mixed* and *irreducible* formulations introduced in Chapter 3 and discussed fully in Chapter 10 and find some similarities. Clearly 'mixed' and 'coupled' formulations are analogous, with the main difference being that in the former elimination of some dependent variables is possible at the governing differential equation level. In the coupled system a full analytical solution or inversion of a (discretized) single system is necessary before such elimination is possible.

Indeed, a further distinction can be made. In coupled systems the solution of any single system is a well-posed problem and is possible when the variables corresponding to the other system are prescribed. This is not always the case in mixed formulations.

It is convenient to classify coupled systems into two categories:

Class I. This class contains problems in which coupling occurs on domain interfaces via the boundary conditions imposed there. Generally the domains describe different physical situations but it is possible to consider coupling between domains that are physically similar in which different discretization processes have been used.

Class II. This class contains problems in which the various domains overlap (totally or partially). Here the coupling occurs through the governing differential equations describing different physical phenomena.

Typical of the first category are the problems of fluid–structure interaction illustrated in Fig. 18.1(a) where physically different problems interact and also structure–structure interactions of Fig. 18.1(b) where the interface simply divides arbitrarily chosen regions in which different numerical discretizations are used.

The need for the use of different discretizations may arise from different causes. Here for instance:

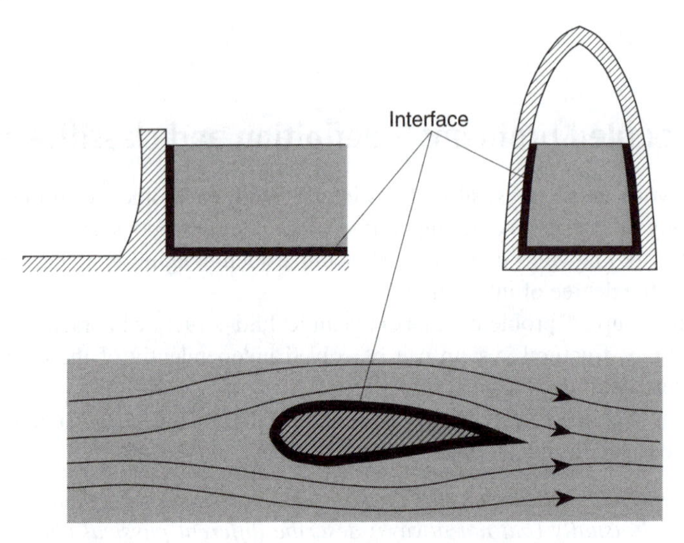

(a) Fluid–structure interaction (physically different domains)

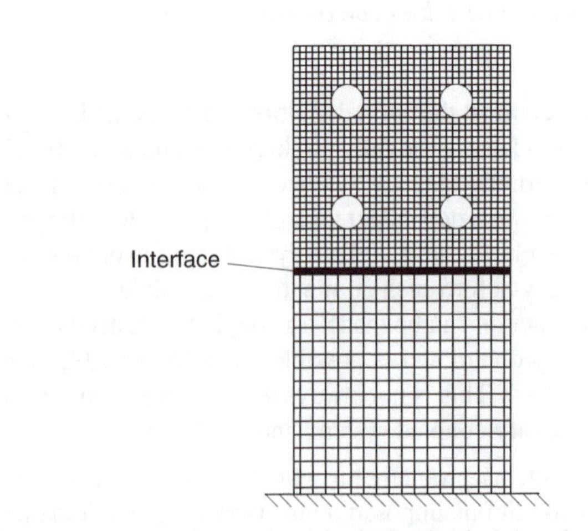

(b) Structure–structure interaction (physically identical domains)

Fig. 18.1 Class I problems with coupling via interfaces (shown as thick line).

1. Different finite element meshes may be advantageous to describe the subdomains.
2. Different procedures such as the combination of boundary method and finite elements in respective regions may be computationally desirable.
3. Domains may simply be divided by the choice of different time-stepping procedures, e.g., of an implicit and explicit kind.

In the second category, typical problems are illustrated in Fig. 18.2. One of these is that of metal extrusion where the plastic flow is strongly coupled with the temperature field while at the same time the latter is influenced by the heat generated in the plastic flow. This problem is included to illustrate a form of coupling that commonly occurs in analyses of solids. The other problem shown in Fig. 18.2 is that of soil dynamics (earthquake response of a dam) in which the seepage flow and pressures interact with the dynamic behaviour of the soil 'skeleton'.

We observe that, in the examples illustrated, motion invariably occurs. Indeed, the vast majority of coupled problems involve such transient behaviour and for this reason the present chapter will only consider this area. It will thus follow and expand the analysis techniques presented in Chapters 16 and 17.

As the problems encountered in coupled analysis of various kinds are similar, we shall focus the presentation on three examples:

1. fluid–structure interaction (confined to small amplitudes);
2. soil–fluid interaction;

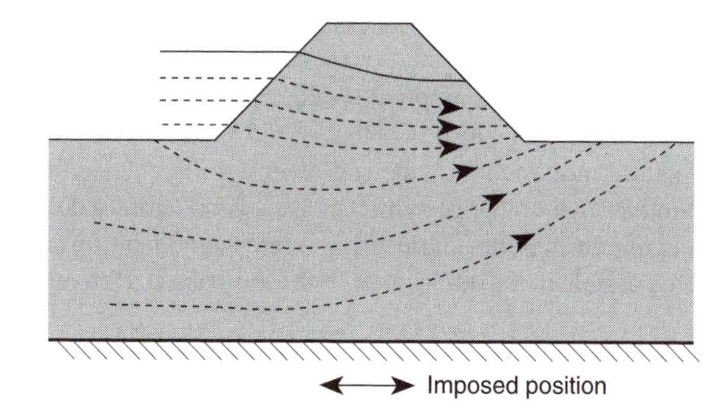

(a) Seepage through a porous medium interacts with its dynamic, structural behaviour

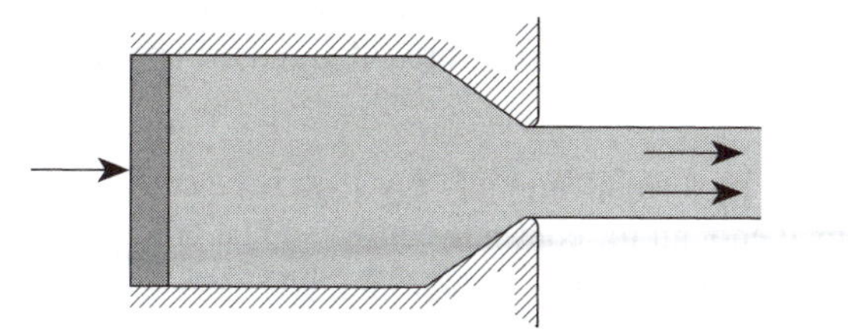

(b) Problem of metal extrusion in which the plastic flow is coupled with the thermal field

Fig. 18.2 Class II problems with coupling in overlapping domains.

3. implicit–explicit dynamic analysis of a structure where the separation involves the process of temporal discretization.

In these problems all the typical features of coupled analysis will be found and extension to others will normally follow similar lines. As a final remark, it is worthwhile mentioning that problems such as linear thermal stress analysis to which we have referred frequently in this volume are not coupled in the terms defined here. In this the stress analysis problem requires a knowledge of the temperature field but the temperature problem can be solved independently of the stress field.† Thus the problem decouples in one direction. Many examples of truly coupled problems will be found in available books.[3–5]

18.2 Fluid–structure interaction (Class I problems)

18.2.1 General remarks and fluid behaviour equations

The problem of fluid–structure interaction is a wide one and covers many forms of fluid which we do not discuss in this book. The consideration of problems in which the fluid is in substantial motion is considered in standard texts on fluid dynamics (e.g., see reference 6) and, thus, we exclude at this stage such problems as flutter where movement of an aerofoil influences the flow pattern and forces around it leading to possible instability. For the same reason we also exclude here the 'singing wire' problem in which the shedding of vortices reacts with the motion of the wire.

However, in a very considerable range of problems the fluid displacement remains small while interaction is substantial. In this category fall the first two examples of Fig. 18.1 in which the structural motions influence and react with the generation of pressures in a reservoir or a container. A number of symposia have been entirely devoted to this class of problems which is of considerable engineering interest, and here fortunately considerable simplifications are possible in the description of the fluid phase. References 7–22 give some typical studies.

In such problems it is possible to write the linearized dynamic equations of fluid behaviour about the hydrostatic state as

$$\frac{\partial(\rho \mathbf{v})}{\partial t} \approx \rho_0 \frac{\partial \mathbf{v}}{\partial t} = -\boldsymbol{\nabla} p + \mathbf{b} \tag{18.1}$$

where $\mathbf{v}$ is the fluid velocity, ρ is the fluid density (with ρ_0 the density in the hydrostatic state), p the pressure and $\mathbf{b}$ is a *constant* body force of gravity. In postulating the above we have assumed

1. that the density ρ_0 varies by a small amount only so may be considered constant;
2. that velocities are small enough for convective effects to be omitted;
3. that viscous effects by which deviatoric stresses are introduced can be neglected in the fluid.

† In a general setting the temperature field does depend upon the strain rate. However, these terms are not included in the form presented in this volume and in many instances produce insignificant changes to the solution.[2]

The reader can in fact note that with the preceding assumption Eq. (18.1) is a special form of a more general relation (described in reference 6).

The linearized continuity equation based on the same assumption is

$$\rho_0 \text{ div } \mathbf{v} \equiv \rho_0 \boldsymbol{\nabla}^{\mathrm{T}} \mathbf{v} = -\frac{\partial \rho}{\partial t} \tag{18.2}$$

and noting that

$$\frac{\partial \rho}{\partial t} \approx \frac{\rho_0}{K} \frac{\partial p}{\partial t} \tag{18.3}$$

where K is the bulk modulus of the fluid, we can write

$$\boldsymbol{\nabla}^{\mathrm{T}} \mathbf{v} = -\frac{1}{K} \frac{\partial p}{\partial t} \tag{18.4}$$

Elimination of $\mathbf{v}$ between (18.1) and (18.4) gives the well-known scalar wave equation governing the pressure p:

$$\boldsymbol{\nabla}^2 p = \frac{1}{c^2} \frac{\partial^2 p}{\partial t^2} \tag{18.5}$$

where

$$c = \sqrt{\frac{K}{\rho_0}} \tag{18.6}$$

denotes the speed of sound in the fluid.

The equations described above are the basis of *acoustic* problems.

18.2.2 Boundary conditions for the fluid. Coupling and radiation

In Fig. 18.3 we focus on the Class I problem illustrated in Fig. 18.1(a) and on the boundary conditions possible for the fluid part described by the governing equation (18.5). As we know well, either normal gradients or values of p now need to be specified.

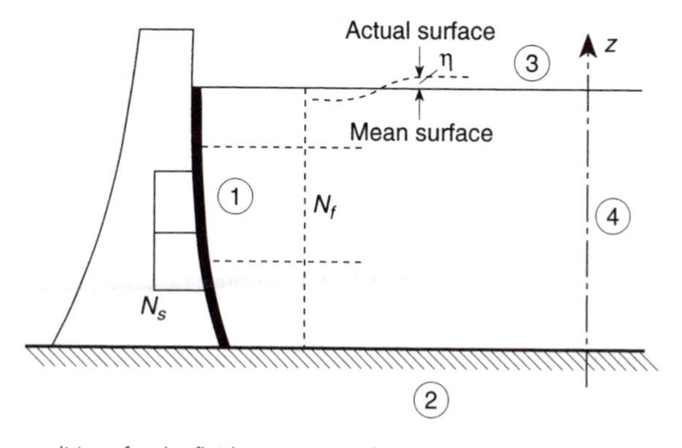

Fig. 18.3 Boundary conditions for the fluid component of the fluid–structure interaction.

Interface with solid

On the boundaries ① and ② in Fig. 18.3 the normal velocities (or their time derivatives) are prescribed. Considering the pressure gradient in the normal direction to the face n we can thus write, by Eq. (18.1),

$$\frac{\partial p}{\partial n} = -\rho_0 \dot{v}_n = -\rho_0 \mathbf{n}^{\mathrm{T}} \dot{\mathbf{v}} \tag{18.7}$$

where $\mathbf{n}$ is the direction cosine vector for an outward pointing normal to the fluid region and $\dot{v}_n$ is prescribed.

Thus, for instance, on boundary ① coupling with the motion of the structure described by displacement $\mathbf{u}$ occurs. Here we put

$$\dot{v}_n = \ddot{u}_n = \mathbf{n}^{\mathrm{T}} \ddot{\mathbf{u}} \tag{18.8}$$

while on boundary ② where only horizontal motion exists we have

$$\dot{v}_z = 0 \tag{18.9}$$

Coupling with the structure motion occurs only via boundary ①.

Free surface

On the free surface (boundary ③ in Fig. 18.3) the simplest assumption is that

$$p = 0 \tag{18.10}$$

However, this does not allow for any possibility of surface gravity waves. These can be approximated by assuming the actual surface to be at an elevation η relative to the mean surface. Now

$$p = \rho_0 g \eta \tag{18.11}$$

where g is the acceleration due to gravity. From Eq. (18.1) we have, on noting $v_z = \partial \eta / \partial t$ and assuming ρ_0 to be constant,

$$\rho_0 \frac{\partial^2 \eta}{\partial t^2} = -\frac{\partial p}{\partial z} \tag{18.12}$$

and on elimination of η, using Eq. (18.11), we have a specified normal gradient condition

$$\frac{\partial p}{\partial z} = -\frac{1}{g} \frac{\partial^2 p}{\partial t^2} = -\frac{1}{g} \ddot{p} \tag{18.13}$$

This allows for gravity waves to be approximately incorporated in the analysis and is known as the *linearized surface wave condition.*

Radiation boundary

Boundary ④ physically terminates an infinite domain and some approximation to account for the effect of such a termination is necessary. The main dynamic effect is simply that the wave solution of the governing equation (18.5) must here be composed of *outgoing waves only* as no input from the infinite domain exists.

If we consider only variations in x (the horizontal direction) we know that the general solution of Eq. (18.5) can be written as

$$p = F(x - ct) + G(x + ct) \tag{18.14}$$

where c is the wave velocity given by Eq. (18.6) and the two waves F and G travel in positive and negative directions of x, respectively.

The absence of the incoming wave G means that on boundary ④ we have only

$$p = F(x - ct) \tag{18.15}$$

Thus

$$\frac{\partial p}{\partial n} \equiv \frac{\partial p}{\partial x} = F' \tag{18.16}$$

and

$$\frac{\partial p}{\partial t} = -cF' \tag{18.17}$$

where F' denotes the derivative of F with respect to $(x - ct)$. We can therefore eliminate the unknown function F' and write

$$\frac{\partial p}{\partial n} = -\frac{1}{c}\dot{p} \tag{18.18}$$

which is a condition very similar to that of Eq. (18.13). This boundary condition was first presented in reference 7 for radiating boundaries and has an analogy with a damping element placed there. More accurate forms are possible to represent far field radiation conditions. For example, use of so-called *perfectly matched layers* (PML) is reported in references 23, 24.

18.2.3 Weak form for coupled systems

A weak form for each part of the coupled system may be written as described in Chapter 3. Accordingly, for the fluid we can write the differential equation as

$$\delta\Pi_f = \int_{\Omega_f} \delta p \left[\frac{1}{c^2}\ddot{p} - \boldsymbol{\nabla}^2 p\right] \mathrm{d}\Omega = 0 \tag{18.19}$$

which after integration by parts and substitution of the boundary conditions described above yields

$$\int_{\Omega_f} \left[\delta p \frac{1}{c^2}\ddot{p} + (\boldsymbol{\nabla}\delta p)^{\mathrm{T}}(\boldsymbol{\nabla}p)\right] \mathrm{d}\Omega + \int_{\Gamma_1} \delta p \rho_0\, \mathbf{n}^{\mathrm{T}}\ddot{\mathbf{u}}\, \mathrm{d}\Gamma + \int_{\Gamma_3} \delta p \frac{1}{g}\ddot{p}\, \mathrm{d}\Gamma + \int_{\Gamma_4} \delta p \frac{1}{c}\dot{p}\, \mathrm{d}\Gamma = 0 \tag{18.20}$$

where Ω_f is the fluid domain and Γ_i the integral over boundary part ⓘ.

Similarly for the solid the weak form after integration by parts is given by

$$\int_{\Omega} \delta\mathbf{u}^{\mathrm{T}}\left[\rho_s\ddot{\mathbf{u}} + \mu\dot{\mathbf{u}} + \boldsymbol{\mathcal{S}}^{\mathrm{T}}\mathbf{D}\boldsymbol{\mathcal{S}}\mathbf{u} - \mathbf{b}\right] \mathrm{d}\Omega - \int_{\Gamma_t} \delta\mathbf{u}^{\mathrm{T}}\bar{\mathbf{t}}\, \mathrm{d}\Gamma = 0 \tag{18.21}$$

where for pressure defined positive in compression the surface traction is defined as

$$\bar{\mathbf{t}} = -p\mathbf{n}_s = p\mathbf{n} \tag{18.22}$$

since the outward normal to the solid is $\mathbf{n}_s = -\mathbf{n}$. The traction integral in Eq. (18.21) is now expressed as

$$\int_{\Gamma_t} \delta\mathbf{u}^{\mathrm{T}}\bar{\mathbf{t}}\, \mathrm{d}\Gamma = \int_{\Gamma_t} \delta\mathbf{u}^{\mathrm{T}}\mathbf{n}p\, \mathrm{d}\Gamma \tag{18.23}$$

(1) In complex physical situations, the interaction between compressibility and internal gravity waves (interaction between acoustic modes and sloshing modes) leads to a modified scalar wave equation. Equation (18.5) should then be replaced by a more complex equation: in a stratified medium for instance, the irrotationality condition for the fluid is not totally verified (the fluid is irrotational in a plane perpendicular to the stratification axis).[16]

(2) The variational formulation defined by Eq. (18.20) is valid in the static case provided the following constraints conditions are added $\int_{\Omega_f} p \, \mathrm{d}\Omega + \rho_0 c^2 \int_{\partial\Omega_f} \mathbf{n}^{\mathrm{T}}\mathbf{u} \, \mathrm{d}\Gamma = 0$ for a compressible fluid filling a cavity, $\int_{\Gamma_1} \mathbf{n}^{\mathrm{T}}\mathbf{u} \, \mathrm{d}\Gamma + \int_{\Gamma_2} p/\rho_0 g \, \mathrm{d}\Gamma = 0$, for an incompressible liquid with a free surface contained inside a reservoir. The static behaviour is important for the modal response of coupled systems when modal truncation needs static corrections in order to accelerate the convergence of the method. This static behaviour is also of prime importance for the construction of reduced matrix models when using dynamic substructuring methods for fluid structure interaction problems.[17, 18]

18.2.4 The discrete coupled system

We shall now consider the coupled problem discretized in the standard (displacement) manner with the displacement vector approximated as

$$\mathbf{u} \approx \hat{\mathbf{u}} = \mathbf{N}_u \tilde{\mathbf{u}} \tag{18.24}$$

and the fluid similarly approximated by

$$p \approx \hat{p} = \mathbf{N}_p \tilde{\mathbf{p}} \tag{18.25}$$

where $\tilde{\mathbf{u}}$ and $\tilde{\mathbf{p}}$ are the nodal parameters of each field and $\mathbf{N}_u$ and $\mathbf{N}_p$ are appropriate shape functions.

The discrete structural problem thus becomes

$$\mathbf{M}\ddot{\tilde{\mathbf{u}}} + \mathbf{C}\dot{\tilde{\mathbf{u}}} + \mathbf{K}\tilde{\mathbf{u}} - \mathbf{Q}\tilde{\mathbf{p}} + \mathbf{f} = \mathbf{0} \tag{18.26}$$

where the coupling term arises due to the pressures (tractions) specified on the boundary as

$$\int_{\Gamma_t} \mathbf{N}_u^{\mathrm{T}} \bar{\mathbf{t}} \, \mathrm{d}\Gamma = \int_{\Gamma_t} \mathbf{N}_u^{\mathrm{T}} \mathbf{n} \mathbf{N}_p \, \mathrm{d}\Gamma \tilde{\mathbf{p}} = \mathbf{Q}\tilde{\mathbf{p}} \tag{18.27}$$

The terms of the other matrices are already well known to the reader as mass, damping, stiffness and force.

Standard Galerkin discretization applied to the weak form of the fluid equation (18.20) leads to (including the possibility of a source term, $\mathbf{q}$)

$$\mathbf{S}\ddot{\tilde{\mathbf{p}}} + \tilde{\mathbf{C}}\dot{\tilde{\mathbf{p}}} + \mathbf{H}\tilde{\mathbf{p}} + \rho_0 \mathbf{Q}^{\mathrm{T}}\ddot{\tilde{\mathbf{u}}} + \mathbf{q} = \mathbf{0} \tag{18.28}$$

where

$$\begin{aligned}
\mathbf{S} &= \int_{\Omega} \mathbf{N}_p^{\mathrm{T}} \frac{1}{c^2} \mathbf{N}_p \, \mathrm{d}\Omega + \int_{\Gamma_3} \mathbf{N}_p^{\mathrm{T}} \frac{1}{g} \mathbf{N}_p \, \mathrm{d}\Gamma \\
\tilde{\mathbf{C}} &= \int_{\Gamma_4} \mathbf{N}_p^{\mathrm{T}} \frac{1}{c} \mathbf{N}_p \, \mathrm{d}\Gamma \\
\mathbf{H} &= \int_{\Omega} (\boldsymbol{\nabla}\mathbf{N}_p)^{\mathrm{T}} \boldsymbol{\nabla}\mathbf{N}_p \, \mathrm{d}\Omega
\end{aligned} \tag{18.29}$$

and $\mathbf{Q}$ is identical to that of Eq. (18.27).

18.2.5 Free vibrations

If we consider free vibrations and omit all force and damping terms (noting that in the fluid component the damping is strictly that due to radiation energy loss) we can write the two equations (18.26) and (18.28) as a set:

$$\begin{bmatrix} \mathbf{M} & \mathbf{0} \\ \rho_0\mathbf{Q}^{\mathrm{T}} & \mathbf{S} \end{bmatrix} \begin{Bmatrix} \ddot{\tilde{\mathbf{u}}} \\ \ddot{\tilde{\mathbf{p}}} \end{Bmatrix} + \begin{bmatrix} \mathbf{K} & -\mathbf{Q} \\ \mathbf{0} & \mathbf{H} \end{bmatrix} \begin{Bmatrix} \tilde{\mathbf{u}} \\ \tilde{\mathbf{p}} \end{Bmatrix} = \mathbf{0} \tag{18.30}$$

and attempt to proceed to establish the eigenvalues corresponding to natural frequencies. However, we note immediately that the system is not symmetric (nor positive definite) and that standard eigenvalue computation methods are not directly applicable. Physically it is, however, clear that the eigenvalues are real and that free vibration modes exist.

The above problem is similar to that arising in vibration of rotating solids and special solution methods are available, though costly.[25] It is possible by various manipulations to arrive at a symmetric form and reduce the problem to a standard eigenvalue one.[14–22, 25–28]

A simple method proposed by Ohayon proceeds to achieve the symmetrization objective by putting $\tilde{\mathbf{u}} = \breve{\mathbf{u}}e^{i\omega t}$, $\tilde{\mathbf{p}} = \breve{\mathbf{p}}e^{i\omega t}$ and rewriting Eq. (18.30) as

$$\begin{aligned} \mathbf{K}\breve{\mathbf{u}} - \mathbf{Q}\breve{\mathbf{p}} - \omega^2\mathbf{M}\breve{\mathbf{u}} &= \mathbf{0} \\ \mathbf{H}\breve{\mathbf{p}} - \omega^2\mathbf{S}\breve{\mathbf{p}} - \omega^2\rho_0\mathbf{Q}^{\mathrm{T}}\breve{\mathbf{u}} &= \mathbf{0} \end{aligned} \tag{18.31}$$

and an additional variable $\breve{\mathbf{q}}$ such that

$$\breve{\mathbf{p}} = \omega^2\breve{\mathbf{q}} \tag{18.32}$$

After some manipulation and substitution we can write the new system as

$$\left\{ \begin{bmatrix} \mathbf{K} & \mathbf{0} & \mathbf{0} \\ \mathbf{0} & \frac{1}{\rho_0}\mathbf{S} & \mathbf{0} \\ \mathbf{0} & \mathbf{0} & \mathbf{0} \end{bmatrix} - \omega^2 \begin{bmatrix} \mathbf{M} & \mathbf{0} & \mathbf{Q} \\ \mathbf{0} & \mathbf{0} & \frac{1}{\rho_0}\mathbf{S} \\ \mathbf{Q}^{\mathrm{T}} & \frac{1}{\rho_0}\mathbf{S}^{\mathrm{T}} & -\frac{1}{\rho_0}\mathbf{H} \end{bmatrix} \right\} \begin{Bmatrix} \breve{\mathbf{u}} \\ \breve{\mathbf{p}} \\ \breve{\mathbf{q}} \end{Bmatrix} = \mathbf{0} \tag{18.33}$$

which is a symmetric generalized eigenproblem. Further, the variable $\breve{\mathbf{q}}$ can now be eliminated by static condensation and the final system becomes symmetric and now contains only the basic variables. The system (18.32), with static corrections, may lead to convenient reduced matrix models through appropriate dynamic substructuring methods.[19]

An alternative that has frequently been used is to introduce a new symmetrizing variable at the governing equation level, but this is clearly not necessary.[14, 15]

As an example of a simple problem in the present category we show an analysis of a three-dimensional flexible wall vibrating with a fluid encased in a 'rigid' container[29] (Fig. 18.4).

18.2.6 Forced vibrations and transient step-by-step algorithms

The reader can easily verify that the steady-state, linear response to periodic input can be readily computed in the complex frequency domain by the procedures described in Chapter 16. Here no difficulties arise due to the non-symmetric nature of equations and standard procedures can be applied. Chopra and coworkers have, for instance, done many

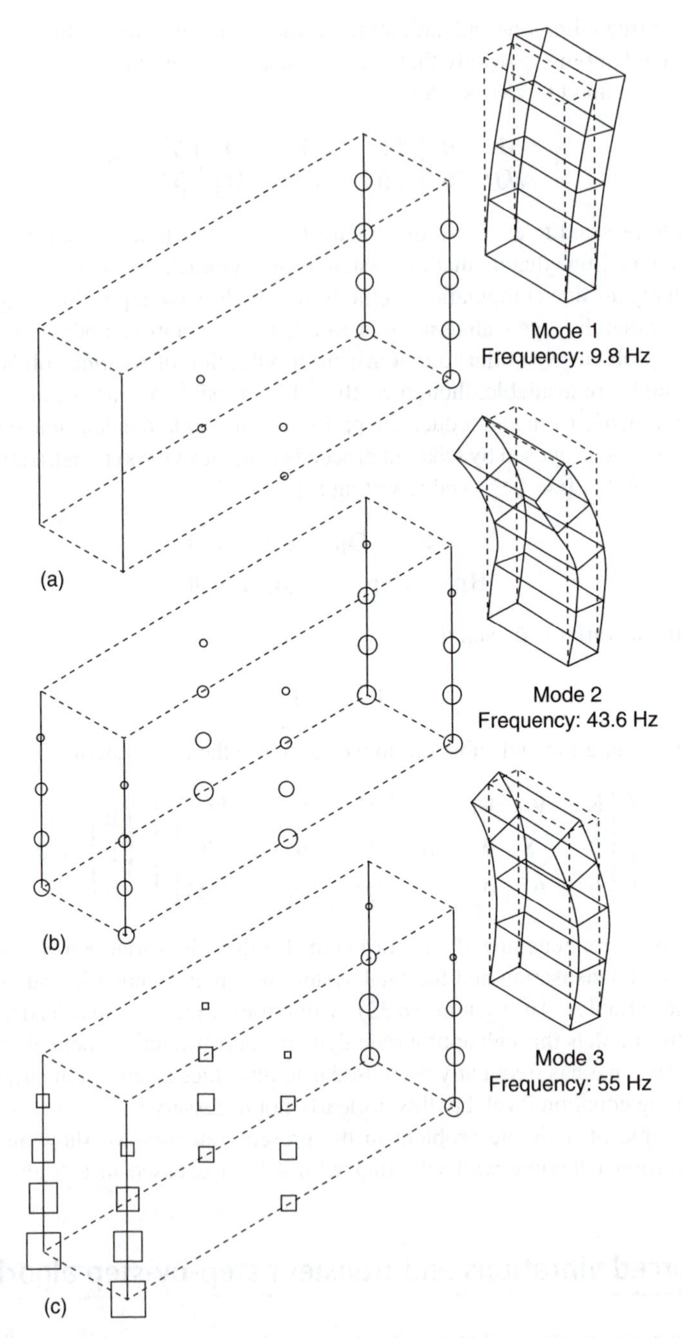

Fig. 18.4 Body of fluid with a free surface oscillating with a wall. Circles show pressure amplitude and squares indicate opposite signs. Three-dimensional approach using parabolic elements.

studies of dam/reservoir interaction using such methods.[30, 31] However, such methods are not generally economical for very large problems and fail in non-linear response studies. Here time-stepping procedures are required in the manner discussed in the previous chapter. However, simple application of methods developed there leads to an unsymmetric problem for the combined system (with $\tilde{\mathbf{u}}$ and $\tilde{\mathbf{p}}$ as variables) due to the form of the matrices appearing in (18.30) and a modified approach is desirable.[32] In this, each of the equations (18.26) and (18.28) is first *discretized in time separately* using the general approaches of Chapter 17.

Thus in the time interval Δt we can approximate $\tilde{\mathbf{u}}$ using, say, the general SS22 procedure as follows. First we write

$$\tilde{\mathbf{u}} = \mathbf{u}_n + \dot{\mathbf{u}}_n \tau + \boldsymbol{\alpha}\frac{\tau^2}{2} \tag{18.34}$$

with a similar expression for $\mathbf{p}$,

$$\tilde{\mathbf{p}} = \mathbf{p}_n + \dot{\mathbf{p}}_n \tau + \boldsymbol{\beta}\frac{\tau^2}{2} \tag{18.35}$$

where $\tau = t - t_n$.

Insertion of the above into Eqs (18.26) and (18.28) and weighting with two *separate weighting functions* results in two relations in which $\boldsymbol{\alpha}$ and $\boldsymbol{\beta}$ are the unknowns. These are

$$\begin{aligned} &\mathbf{M}\boldsymbol{\alpha} + \mathbf{C}\big(\dot{\bar{\mathbf{u}}}_{n+1} + \theta_1 \Delta t \boldsymbol{\alpha}\big) + \mathbf{K}\big(\bar{\mathbf{u}}_{n+1} + \tfrac{1}{2}\theta_2 \Delta t^2 \boldsymbol{\alpha}\big) \\ &\quad - \mathbf{Q}\big(\bar{\mathbf{p}}_{n+1} + \tfrac{1}{2}\bar{\theta}_2 \Delta t^2 \boldsymbol{\beta}\big) + \mathbf{f}_{n+1} = \mathbf{0} \end{aligned} \tag{18.36a}$$

and

$$\mathbf{S}\boldsymbol{\beta} + \mathbf{Q}^{\mathrm{T}}\boldsymbol{\alpha} + \mathbf{H}\big(\bar{\mathbf{p}}_{n+1} + \tfrac{1}{2}\bar{\theta}_2 \Delta t^2 \boldsymbol{\beta}\big) + \mathbf{q}_{n+1} = \mathbf{0} \tag{18.36b}$$

where

$$\begin{aligned} \bar{\mathbf{u}}_{n+1} &= \mathbf{u}_n + \theta_1 \Delta t \dot{\mathbf{u}}_n \\ \dot{\bar{\mathbf{u}}}_{n+1} &= \dot{\mathbf{u}}_n \\ \bar{\mathbf{p}}_{n+1} &= \mathbf{p}_n + \bar{\theta}_1 \Delta t \dot{\mathbf{p}}_n \end{aligned} \tag{18.37}$$

are the predictors for the $n+1$ time step. In the above the parameters θ_i and $\bar{\theta}_i$ are similar to those of Eq. (18.49) and can be chosen by the user. It is interesting to note that the equation system can be put in symmetric form as

$$\begin{bmatrix} (\mathbf{M} + \theta_1 \Delta t \mathbf{C} + \tfrac{1}{2}\theta_2 \Delta t^2 \mathbf{K}) & -\mathbf{Q} \\ -\mathbf{Q}^{\mathrm{T}} & -\left(\mathbf{H} + \dfrac{2}{\bar{\theta}_2 \Delta t^2}\mathbf{S}\right) \end{bmatrix} \begin{Bmatrix} \boldsymbol{\alpha} \\ \hat{\boldsymbol{\beta}} \end{Bmatrix} = \begin{Bmatrix} \mathbf{F}_1 \\ \mathbf{F}_2 \end{Bmatrix} \tag{18.38}$$

where the second equation has been multiplied by -1, the unknown $\boldsymbol{\beta}$ has been replaced by

$$\hat{\boldsymbol{\beta}} = \tfrac{1}{2}\bar{\theta}_2 \Delta t^2 \boldsymbol{\beta} \tag{18.39}$$

and the forces are given by

$$\begin{aligned} \mathbf{F}_1 &= -\mathbf{f}_{n+1} - \mathbf{C}\dot{\bar{\mathbf{u}}}_{n+1} - \mathbf{K}\bar{\mathbf{u}}_{n+1} + \mathbf{Q}\bar{\mathbf{p}}_{n+1} \\ \mathbf{F}_2 &= \mathbf{q}_{n+1} + \mathbf{H}\bar{\mathbf{p}}_{n+1} \end{aligned} \tag{18.40}$$

It is not necessary to go into detail about the computation steps as these follow the usual patterns of determining $\boldsymbol{\alpha}$ and $\boldsymbol{\beta}$ and then evaluation of the problem variables, that is $\mathbf{u}_{n+1}$,

$\mathbf{p}_{n+1}$, $\dot{\mathbf{u}}_{n+1}$ and $\dot{\mathbf{p}}_{n+1}$ at t_{n+1} before proceeding with the next time step. Non-linearity of structural behaviour can also be accommodated (e.g., see reference 33). It is, however, important to consider the stability of the linear system which will, of course, depend on the choice of θ_i and $\bar{\theta}_i$. Here we find, by using procedures described in Chapter 17, that unconditional stability is obtained when

$$\begin{aligned} \theta_2 &\geq \theta_1 \qquad & \theta_1 &\geq \tfrac{1}{2} \\ \bar{\theta}_2 &\geq \bar{\theta}_1 \qquad & \bar{\theta}_1 &\geq \tfrac{1}{2} \end{aligned} \tag{18.41}$$

It is instructive to note that precisely the same result would be obtained if GN22 approximations were used in Eqs (18.34) and (18.35).

The derivation of such stability conditions is straightforward and follows precisely the lines of Sec.17.4 of the previous chapter. However, the algebra is sometimes tedious. Nevertheless, to allow the reader to repeat such calculations for any case encountered we shall outline the calculations for the present example.

Stability of the fluid–structure time-stepping scheme[32]

For stability evaluations it is always advisable to consider the modally decomposed system with scalar variables. We thus rewrite Eqs (18.36a) and (18.36b) omitting the forcing terms and putting $\theta_i = \bar{\theta}_i$ as

$$\begin{aligned} m\alpha + c(\dot{u}_n + \theta_1 \Delta t \alpha) + k(u_n + \theta_1 \Delta t \dot{u}_n + \tfrac{1}{2}\theta_2 \Delta t^2 \alpha) \\ - q(p_n + \theta_1 \Delta t \dot{p}_n + \tfrac{1}{2}\theta_2 \Delta t^2 \beta) = 0 \end{aligned} \tag{18.42a}$$

and

$$s\beta + q\alpha + h(p_n + \theta_1 \Delta t \dot{p} + \tfrac{1}{2}\theta_2 \Delta t^2 \beta) = 0 \tag{18.42b}$$

To complete the recurrence relations we have

$$\begin{aligned} u_{n+1} &= u_n + \Delta t \dot{u}_n + \tfrac{1}{2}\Delta t^2 \alpha \\ \dot{u}_{n+1} &= \dot{u}_n + \Delta t \alpha \\ p_{n+1} &= p_n + \Delta t \dot{p}_n + \tfrac{1}{2}\Delta t^2 \beta \\ \dot{p}_{n+1} &= \dot{p}_n + \Delta t \beta \end{aligned} \tag{18.42c}$$

The exact solution of the above system will always be of the form

$$\begin{aligned} u_{n+1} &= \mu u_n \\ \dot{u}_{n+1} &= \mu \dot{u}_n \\ p_{n+1} &= \mu p_n \\ \dot{p}_{n+1} &= \mu \dot{p}_n \end{aligned} \tag{18.43}$$

and immediately we put

$$\mu = \frac{1+z}{1-z}$$

knowing that for stability we require the real part of z to be negative.

Eliminating all $n+1$ values from Eqs (18.42c) and (18.43) leads to

$$\begin{aligned} \dot{u}_n &= \frac{2z}{\Delta t} u_n \qquad & \dot{p}_n &= \frac{2z}{\Delta t} p_n \\ \alpha &= \frac{4z^2}{(1-z)\Delta t^2} u_n \qquad & \beta &= \frac{4z^2}{(1-z)\Delta t^2} p_n \end{aligned} \tag{18.44}$$

Inserting (18.44) into the system (18.42a) and (18.42b) gives

$$\begin{aligned}(a_{11}z^2 + b_{11}z + k)u_n + (a_{12}z^2 + b_{12}z - q)p_n &= 0 \\ 4qz^2u_n + (a_{22}z^2 + b_{22}z + h')p_n &= 0\end{aligned} \tag{18.45}$$

where

$$\begin{aligned}a_{11} &= 4m' - 2(1 - 2\theta_1)c' - 2k(\theta_1 - \theta_2) \\ a_{12} &= 2q(\theta_1 - \theta_2) \\ a_{22} &= 4s - 2(\theta_1 - \theta_2)h' \\ b_{11} &= 2c' - k(1 - 2\theta_1) \\ b_{12} &= (1 - 2\theta_1)q \\ b_{22} &= -(1 - 2\theta_1)h'\end{aligned} \tag{18.46}$$

in which

$$m' = \frac{m}{\Delta t^2} \qquad c' = \frac{c}{\Delta t} \qquad h' = \Delta t^2 h$$

For non-trivial solutions to exist the determinant of the coefficient matrix Eq. (18.45) has to be zero. This determinant provides the characteristic equation for z which, in the present case, is a polynomial of fourth order of the form

$$a_0 z^4 + a_1 z^3 + a_2 z^2 + a_3 z + a_4 = 0$$

Thus use of the Routh–Hurwitz conditions given in Sec. 17.4 ensures stability requirements are satisfied, i.e., that the roots of z have negative real parts. For the present case the requirements are the following

$$a_0 > 0 \qquad \text{and} \qquad a_i \geq 0, \qquad i = 1, 2, 3, 4$$

The inequality

$$a_{11}a_{22} - 8q^2(\theta_1 - \theta_2) > 0 \tag{18.47}$$

is satisfied for $m', c', k, s, h' \geq 0$ if

$$\theta_1 \geq \tfrac{1}{2} \qquad \theta_2 \geq \theta_1$$

The inequality

$$a_1 = a_{11}\left[-h'(1 - 2\theta_1)\right] + \left[2c' - k(1 - 2\theta_1)\right]a_{22} - 4qb_{12} \geq 0 \tag{18.48}$$

is also satisfied if

$$\theta_1 \geq \tfrac{1}{2} \qquad \theta_2 \geq \theta_1$$

The inequalities

$$\begin{aligned}a_2 &= a_{11}h' + b_{11}b_{22} + a_{22}k + 4q^2 \geq 0 \\ a_3 &= b_{11}h' + b_{22}k \geq 0\end{aligned} \tag{18.49}$$

are satisfied if (18.47) and (18.48) are satisfied. The inequality

$$a_4 = kh' \geq 0 \tag{18.50}$$

is automatically satisfied. Finally the two inequalities

$$\begin{aligned} a_1a_2 - a_0a_3 &\geq 0 \\ a_1a_2a_3 - a_0a_3^2 - a_4a_1^2 &\geq 0 \end{aligned} \tag{18.51}$$

are also satisfied if (18.47) and (18.48) are satisfied.

If all the equalities hold then $m's > 0$ has to be satisfied. In case $m's = 0$ and $c' = 0$ then $\theta_2 > \theta_1$ must be enforced.

18.2.7 Special case of incompressible fluids

If the fluid is incompressible as well as being inviscid, its behaviour is described by a simple laplacian equation

$$\nabla^2 p = 0 \tag{18.52}$$

obtained by putting $c = \infty$ in Eq. (18.5).

In the absence of surface wave effects and of non-zero prescribed pressures the discrete equation (18.28) becomes simply

$$\mathbf{H}\tilde{\mathbf{p}} = -\mathbf{Q}^{\mathrm{T}}\ddot{\tilde{\mathbf{u}}} \tag{18.53}$$

as wave radiation disappears. It is now simple to obtain

$$\tilde{\mathbf{p}} = -\mathbf{H}^{-1}\mathbf{Q}^{\mathrm{T}}\ddot{\tilde{\mathbf{u}}} \tag{18.54}$$

and substitution of the above into the structure equation (18.26) results in

$$\left(\mathbf{M} + \mathbf{Q}\mathbf{H}^{-1}\mathbf{Q}^{\mathrm{T}}\right)\ddot{\tilde{\mathbf{u}}} + \mathbf{C}\dot{\tilde{\mathbf{u}}} + \mathbf{K}\tilde{\mathbf{u}} + \mathbf{f} = \mathbf{0} \tag{18.55}$$

This is now a standard structural system in which the mass matrix has been augmented by an *added mass matrix* as

$$\mathbf{M}_u = \mathbf{Q}\mathbf{H}^{-1}\mathbf{Q}^{\mathrm{T}} \tag{18.56}$$

and its solution follows the standard procedures of previous chapters.

We have to remark that:

1. In general the complete inverse of **H** is not required as pressures at interface nodes only are needed.
2. In general the question of when compressibility effects can be ignored is a difficult one and will depend much on the frequencies that have to be considered in the analysis. For instance, in the analysis of the reservoir–dam interaction much debate on the subject has been recorded.[34] Here the fundamental compressible period may be of order H/c where H is a typical dimension (such as height of the dam). If this period is of the same order as that of, say, earthquake forcing motion then, of course, compressibility must be taken into account. If it is much shorter then its neglect can be justified.

18.2.8 Cavitation effects in fluids

In fluids such as water the linear behaviour under volumetric strain ceases when pressures fall below a certain threshold. This is the vapour pressure limit. When this is reached cavities or distributed bubbles form and the pressure remains almost constant. To follow such behaviour a non-linear constitutive law has to be introduced. Although this book is primarily devoted to linear problems we here indicate some of the steps which are necessary to extend analyses to account for non-linear behaviour.

A convenient variable useful in cavitation analysis was defined by Newton[35]

$$s = \mathrm{div}(\rho \mathbf{u}) \equiv \boldsymbol{\nabla}^{\mathrm{T}}(\rho \mathbf{u}) \tag{18.57}$$

where $\mathbf{u}$ is the fluid displacement. The non-linearity now is such that

$$\begin{aligned} p &= -K\,\mathrm{div}\,\mathbf{u} = c^2 s, \qquad && \text{if } s < (p_a - p_v)/c^2 \\ p &= p_a - p_v, && \text{if } s > (p_a - p_v)/c^2 \end{aligned} \tag{18.58}$$

Here p_a is the atmospheric pressure (at which $\mathbf{u} = \mathbf{0}$ is assumed), p_v is the vapour pressure and c is the sound velocity in the fluid.

Clearly monitoring strains is a difficult problem in the formulation using the velocity and pressure variables [Eq. (18.1) and (18.5)]. Here it is convenient to introduce a displacement potential ψ such that

$$\rho \mathbf{u} = -\boldsymbol{\nabla}\psi \tag{18.59}$$

From the momentum equation (18.1) we see that

$$\rho \ddot{\mathbf{u}} = -\boldsymbol{\nabla}\ddot{\psi} = -\boldsymbol{\nabla} p$$

and thus

$$\ddot{\psi} = p \tag{18.60}$$

The continuity equation (18.2) now gives

$$s = \rho\,\mathrm{div}\,\mathbf{u} = -\nabla^2\psi = \frac{1}{c^2}p = \frac{1}{c^2}\ddot{\psi} \tag{18.61}$$

in the linear case [with an appropriate change according to conditions (18.58) during cavitation].

Details of boundary conditions, discretization and coupling are fully described in reference 36 and follow the standard methodology previously given. Figure 18.5, taken from that reference, illustrates the results of a non-linear analysis showing the development of cavity zones in a reservoir.

18.3 Soil–pore fluid interaction (Class II problems)

18.3.1 The problem and the governing equations. Discretization

It is well known that the behaviour of soils (and indeed other geomaterials) is strongly influenced by the pressures of the fluid present in the pores of the material. Indeed, the

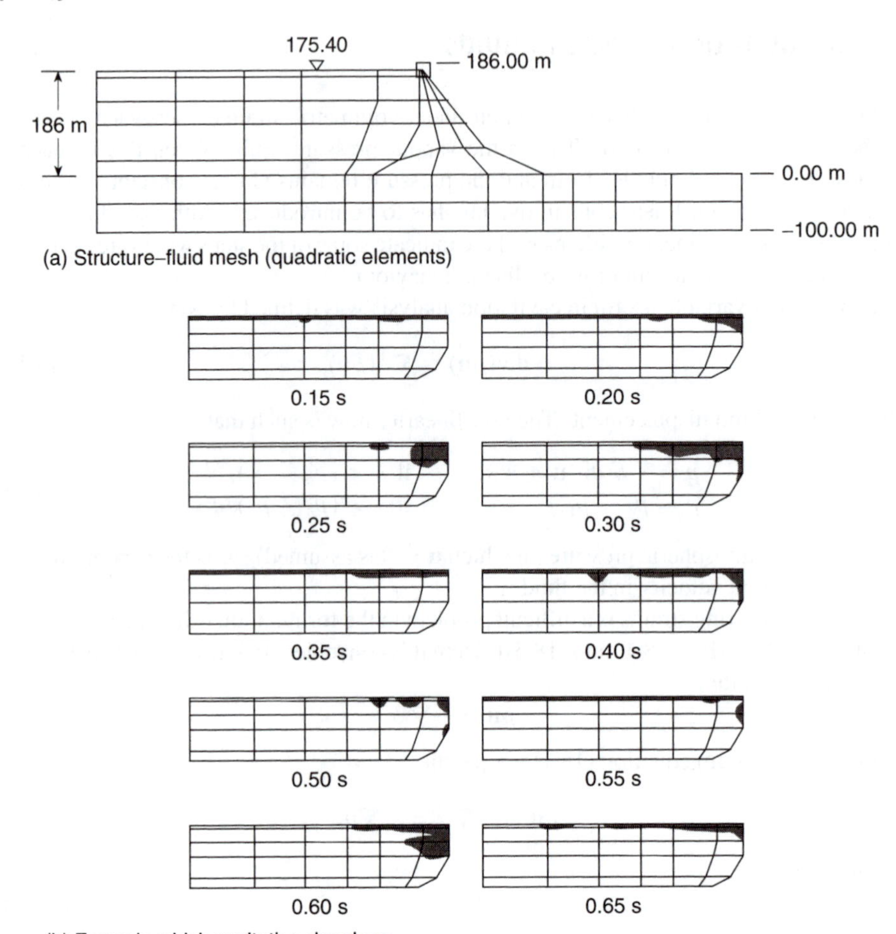

Fig. 18.5 The Bhakra dam–reservoir system.[36] Interaction during the first second of earthquake motion showing the development of cavitation.

concept of *effective stress* is here of paramount importance. Thus if $\boldsymbol{\sigma}$ describes the total stress (positive in tension) acting on the total area of the soil and the pores, and p is the pressure of the fluid (positive in compression) in the pores (generally of water), the effective stress is defined as

$$\boldsymbol{\sigma}' = \boldsymbol{\sigma} + \mathbf{m}p \tag{18.62}$$

Here $\mathbf{m}^{\mathrm{T}} = [1, 1, 1, 0, 0, 0]$ if we use the notation in Chapter 11. Now it is well known that it is only the stress $\boldsymbol{\sigma}'$ which is responsible for the deformations (or failure) of the solid skeleton of the soil (excluding here a very small volumetric grain compression which has to be included in some cases). Assuming for the development given here that the soil can be represented by a linear elastic model we have

$$\boldsymbol{\sigma}' = \mathbf{D}\boldsymbol{\varepsilon} \tag{18.63}$$

Immediately the total discrete equilibrium equations for the soil–fluid mixture can be written in exactly the same form as is done for all problems of solid mechanics:

$$\mathbf{M}\ddot{\tilde{\mathbf{u}}} + \mathbf{C}\dot{\tilde{\mathbf{u}}} + \int_{\Omega} \mathbf{B}^{\mathrm{T}}\boldsymbol{\sigma}\,\mathrm{d}\Omega + \mathbf{f} = \mathbf{0} \tag{18.64}$$

where $\tilde{\mathbf{u}}$ are the displacement discretization parameters, i.e.,

$$\mathbf{u} \approx \hat{\mathbf{u}} = \mathbf{N}\tilde{\mathbf{u}} \tag{18.65}$$

$\mathbf{B}$ is the strain–displacement matrix and $\mathbf{M}$, $\mathbf{C}$, $\mathbf{f}$ have the usual meaning of mass, damping and force matrices, respectively.

Now, however, the term involving the stress must be split as

$$\int_{\Omega} \mathbf{B}^{\mathrm{T}}\boldsymbol{\sigma}\,\mathrm{d}\Omega = \int_{\Omega} \mathbf{B}^{\mathrm{T}}\boldsymbol{\sigma}'\,\mathrm{d}\Omega - \int_{\Omega} \mathbf{B}^{\mathrm{T}}\mathbf{m}p\,\mathrm{d}\Omega \tag{18.66}$$

to allow the direct relationship between effective stresses and strains (and hence displacements) to be incorporated. For a linear elastic soil skeleton we immediately have

$$\mathbf{M}\ddot{\tilde{\mathbf{u}}} + \mathbf{C}\dot{\tilde{\mathbf{u}}} + \mathbf{K}\tilde{\mathbf{u}} - \mathbf{Q}\tilde{\mathbf{p}} + \mathbf{f} = \mathbf{0} \tag{18.67}$$

where $\mathbf{K}$ is the standard stiffness matrix written as

$$\int_{\Omega} \mathbf{B}^{\mathrm{T}}\boldsymbol{\sigma}'\,\mathrm{d}\Omega = \left(\int_{\Omega} \mathbf{B}^{\mathrm{T}}\mathbf{D}\mathbf{B}\,\mathrm{d}\Omega\right)\tilde{\mathbf{u}} = \mathbf{K}\tilde{\mathbf{u}} \tag{18.68}$$

and $\mathbf{Q}$ couples the field of pressures in the equilibrium equations assuming these are discretized as

$$p \approx \hat{p} = \mathbf{N}_p\tilde{\mathbf{p}} \tag{18.69}$$

Thus

$$\mathbf{Q} = \int_{\Omega} \mathbf{B}^{\mathrm{T}}\mathbf{m}\mathbf{N}_p\,\mathrm{d}\Omega \tag{18.70}$$

In the above discretization conventionally the same element shapes are used for the $\tilde{\mathbf{u}}$ and $\tilde{\mathbf{p}}$ variables, though not necessarily identical interpolations. With the dynamic equations coupled to the pressure field an additional equation is clearly needed from which the pressure field can be derived. This is provided by the transient seepage equation of the form

$$-\boldsymbol{\nabla}^{\mathrm{T}}(k\boldsymbol{\nabla}p) + \frac{1}{Q}\dot{p} + \dot{\varepsilon}_v = 0 \tag{18.71}$$

where Q is related to the compressibility of the fluid, k is the permeability and ε_v is the volumetric strain in the soil skeleton, which on discretization of displacements is given by

$$\varepsilon_v = \mathbf{m}^{\mathrm{T}}\boldsymbol{\varepsilon} = \mathbf{m}^{\mathrm{T}}\mathbf{B}\tilde{\mathbf{u}} \tag{18.72}$$

The equation of seepage can now be discretized in the standard Galerkin manner as

$$\mathbf{Q}^{\mathrm{T}}\dot{\tilde{\mathbf{u}}} + \mathbf{S}\dot{\tilde{\mathbf{p}}} + \mathbf{H}\tilde{\mathbf{p}} + \mathbf{q} = \mathbf{0} \tag{18.73}$$

where $\mathbf{Q}$ is precisely that of Eq. (18.70), and

$$\mathbf{S} = \int_{\Omega} \mathbf{N}_p^{\mathrm{T}}\frac{1}{Q}\mathbf{N}_p\,\mathrm{d}\Omega \qquad \mathbf{H} = \int_{\Omega} (\boldsymbol{\nabla}\mathbf{N}_p)^{\mathrm{T}}k\boldsymbol{\nabla}\mathbf{N}_p\,\mathrm{d}\Omega \tag{18.74}$$

with $\mathbf{q}$ containing the forcing and boundary terms. The derivation of coupled flow–soil equations was first introduced by Biot[37] but the present formulation is elaborated upon in references 32, 34–45 where various approximations, as well as the effect of various non-linear constitutive relations, are discussed.

We shall not comment in detail on any of the boundary conditions as these are of standard type and are well documented in previous chapters.

18.3.2 The format of the coupled equations

The solution of coupled equations often involves non-linear behaviour, as noted previously in the cavitation problem. However, it is instructive to consider the linear version of Eqs (18.67) and (18.73). This can be written as

$$\begin{bmatrix} \mathbf{M} & \mathbf{0} \\ \mathbf{0} & \mathbf{0} \end{bmatrix} \begin{Bmatrix} \ddot{\tilde{\mathbf{u}}} \\ \ddot{\tilde{\mathbf{p}}} \end{Bmatrix} + \begin{bmatrix} \mathbf{C} & \mathbf{0} \\ \mathbf{Q}^{\mathrm{T}} & \mathbf{S} \end{bmatrix} \begin{Bmatrix} \dot{\tilde{\mathbf{u}}} \\ \dot{\tilde{\mathbf{p}}} \end{Bmatrix} + \begin{bmatrix} \mathbf{K} & -\mathbf{Q} \\ \mathbf{0} & \mathbf{H} \end{bmatrix} \begin{Bmatrix} \tilde{\mathbf{u}} \\ \tilde{\mathbf{p}} \end{Bmatrix} = -\begin{Bmatrix} \mathbf{f} \\ \mathbf{q} \end{Bmatrix} \tag{18.75}$$

Once again, like in the fluid–structure interaction problem, overall asymmetry occurs despite the inherent symmetry of the $\mathbf{M}$, $\mathbf{C}$, $\mathbf{K}$, $\mathbf{S}$ and $\mathbf{H}$ matrices. As the free vibration problem is of no great interest here, we shall not discuss its symmetrization. In the transient solution algorithm we shall proceed in a similar manner to that described in Sec. 18.2.6 and again symmetry will be observed.

18.3.3 Transient step-by-step algorithm

Time-stepping procedures can be derived in a manner analogous to that presented in Sec. 18.2.6. Here we choose to use the GNpj algorithm of lowest order to approximate each variable.

Thus for $\tilde{\mathbf{u}}$ we shall use GN22, writing

$$\begin{aligned} \mathbf{u}_{n+1} &= \mathbf{u}_n + \Delta t \dot{\mathbf{u}}_n + \tfrac{1}{2}\Delta t^2 \ddot{\mathbf{u}}_n + \frac{1}{2}\beta_2 \Delta t^2 \Delta \ddot{\mathbf{u}}_{n+1} \\ &\equiv \mathbf{u}^p_{n+1} + \tfrac{1}{2}\beta_2 \Delta t^2 \Delta \ddot{\mathbf{u}}_{n+1} \\ \dot{\mathbf{u}}_{n+1} &= \dot{\mathbf{u}}_n + \Delta t \ddot{\mathbf{u}}_n + \beta_1 \Delta t \Delta \ddot{\mathbf{u}}_{n+1} \\ &\equiv \dot{\mathbf{u}}^p_{n+1} + \beta_1 \Delta t \Delta \ddot{\mathbf{u}}_{n+1} \end{aligned} \tag{18.76}$$

For the variables p that occur in first-order form we shall use GN11, as

$$\begin{aligned} \mathbf{p}_{n+1} &= \mathbf{p}_n + \Delta t \dot{\mathbf{p}}_n + \theta \Delta t \Delta \dot{\mathbf{p}}_{n+1} \\ &\equiv \mathbf{p}^p_{n+1} + \theta \Delta t \Delta \dot{\mathbf{p}}_{n+1} \end{aligned} \tag{18.77}$$

In the above $\mathbf{u}^p_{n+1}$, etc., denote values that can be 'predicted' from known parameters at time t_n and

$$\Delta \ddot{\mathbf{u}}_{n+1} = \ddot{\mathbf{u}}_{n+1} - \ddot{\mathbf{u}}_n \qquad \Delta \dot{\mathbf{p}}_{n+1} = \dot{\mathbf{p}}_{n+1} - \dot{\mathbf{p}}_n \tag{18.78}$$

are the unknowns.

To complete the recurrence algorithm it is necessary to insert the above into the coupled governing equations [(18.64) and (18.73)] written at time t_{n+1}. Thus we require the following equalities

$$\mathbf{M}\ddot{\mathbf{u}}_{n+1} + \mathbf{C}\dot{\mathbf{u}}_{n+1} + \int_{\Omega} \mathbf{B}^{\mathrm{T}}\boldsymbol{\sigma}'_{n+1} - \mathbf{Q}\mathbf{p}_{n+1} + \mathbf{f}_{n+1} = \mathbf{0}$$
$$\mathbf{Q}^{\mathrm{T}}\dot{\mathbf{u}}_{n+1} + \mathbf{S}\dot{\mathbf{p}}_{n+1} + \mathbf{H}\mathbf{p}_{n+1} + \mathbf{q}_{n+1} = \mathbf{0} \tag{18.79}$$

in which $\boldsymbol{\sigma}'_{n+1}$ is evaluated using the constitutive equation (18.63) in incremental form and knowledge of $\boldsymbol{\sigma}'_n$ as

$$\boldsymbol{\sigma}'_{n+1} = \boldsymbol{\sigma}'_n + \mathbf{D}\Delta\boldsymbol{\varepsilon}_{n+1} = \boldsymbol{\sigma}'_n + \mathbf{D}\mathbf{B}\Delta\mathbf{u}_{n+1} \tag{18.80}$$

In general the above system may be non-linear and indeed on many occasions the $\mathbf{H}$ matrix itself may be dependent on the values of $\mathbf{u}$ due to permeability variations with strain. It is of interest to look at the linear form as the non-linear system usually solves a similar form iteratively.[33]

Here insertion of Eqs (18.76), (18.77) and (18.80) into (18.79) results in the equation system

$$\begin{bmatrix} (\mathbf{M} + \beta_1\Delta t\mathbf{C} + \frac{1}{2}\beta_2\Delta t^2\mathbf{K}) & -\mathbf{Q} \\ -\mathbf{Q}^{\mathrm{T}} & -\left(\mathbf{H} + \dfrac{1}{\theta\Delta t}\mathbf{S}\right) \end{bmatrix} \begin{Bmatrix} \Delta\ddot{\mathbf{u}}_{n+1} \\ \Delta\dot{\breve{\mathbf{p}}}_{n+1} \end{Bmatrix} = \begin{Bmatrix} \mathbf{F}_1 \\ \mathbf{F}_2 \end{Bmatrix} \tag{18.81}$$

where $\mathbf{F}_1$ and $\mathbf{F}_2$ are vectors that can be evaluated from loads and solution values at t_n. Symmetry in the above is obtained by multiplying Eq. (18.36b) by -1 and defining

$$\Delta\dot{\breve{\mathbf{p}}}_{n+1} = \beta_1\Delta t\Delta\dot{\mathbf{p}}_{n+1} \tag{18.82}$$

The solution of Eq. (18.81) and the use of Eqs (18.76) and (18.77) complete the recurrence relation.

The stability of the linear scheme can be found by following identical procedures to those used in Sec. 18.2.6 and the result is that stability is unconditional when[27]

$$\beta_2 \geq \beta_1 \qquad \beta_1 \geq \tfrac{1}{2} \qquad \theta \geq \tfrac{1}{2} \tag{18.83}$$

18.3.4 Special cases and robustness requirements

Frequently the compressibility of the fluid phase, which forms the matrix $\mathbf{S}$, is such that

$$\mathbf{S} \approx \mathbf{0}$$

compared with other terms. Further, the permeability k may on occasion also be very small (as, say, in clays) and

$$\mathbf{H} \approx \mathbf{0}$$

leading to so-called 'undrained' behaviour.

Now the coefficient matrix in (18.81) becomes of the lagrangian constrained form (see Chapter 10), i.e.,

$$\begin{bmatrix} \mathbf{A} & -\mathbf{Q} \\ -\mathbf{Q}^{\mathrm{T}} & \mathbf{0} \end{bmatrix} \begin{Bmatrix} \Delta\ddot{\mathbf{u}}_{n+1} \\ \Delta\dot{\breve{\mathbf{p}}}_{n+1} \end{Bmatrix} = \begin{Bmatrix} \mathbf{F}_1 \\ \mathbf{F}_2 \end{Bmatrix} \tag{18.84}$$

and is solvable only if

$$n_u \geq n_p$$

where n_u and n_p denote the number of $\tilde{\mathbf{u}}$ and $\tilde{\mathbf{p}}$ parameters, respectively.

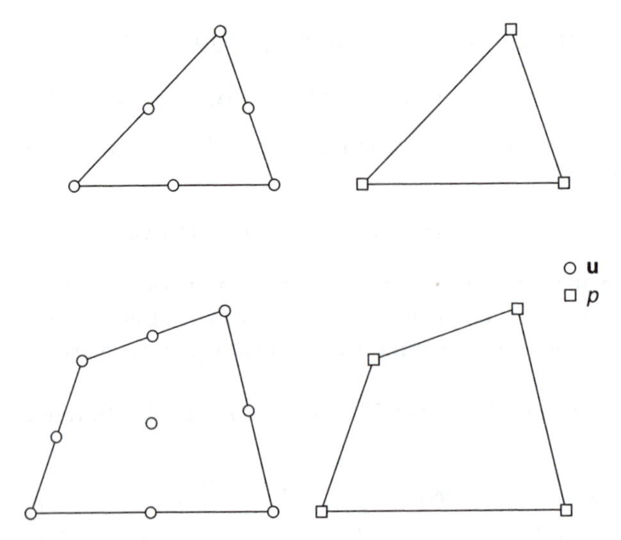

Fig. 18.6 'Robust' interpolations for the coupled soil–fluid problem.

The problem is indeed identical to that encountered in incompressible behaviour and the interpolations used for the **u** and p variables have to satisfy identical criteria. As C_0 interpolation for both variables is necessary for the general case, suitable element forms are shown in Fig. 18.6 and can be used with confidence. Alternatively, equal order interpolation may be used for **u** and p in conjunction with stabilized forms discussed in Sec.11.7.

The formulation can of course be used for steady-state solutions but it must be remarked that in such cases an uncoupling occurs as the seepage equation can be solved independently.

Finally, it is worth remarking that the formulation also solves the well-known soil consolidation problem where the phenomena are so slow that the dynamic term $\mathbf{M}\ddot{\mathbf{u}}$ tends to $\mathbf{0}$. However, no special modifications are necessary and the algorithm form is again applicable.

18.3.5 Examples – soil liquefaction

As we have already mentioned, the most interesting applications of the coupled soil–fluid behaviour is when non-linear soil properties are taken into account. In particular, it is a well-known fact that repeated straining of a granular, soil-like material in the absence of the pore fluid results in a decrease of volume (densification) due to particle rearrangement. Constitutive equations which include this effect are available;[33] however, here we only represent a typical result which they can achieve when used in a coupled soil–fluid solution. When a pore fluid is present, densification will (via the coupling terms) tend to increase the fluid pressures and hence reduce the soil strength. This, as is well known, decreases with the compressive mean effective stress.

It is not surprising therefore that under dynamic action the soil frequently loses all of its strength (i.e., liquefies) and behaves almost like a fluid, leading occasionally to catastrophic failures of structural foundations in earthquakes. The reproduction of such phenomena with computational models is not easy as a complete constitutive behaviour description for soils is imperfect. However, much effort devoted to the subject has produced good results[38–45]

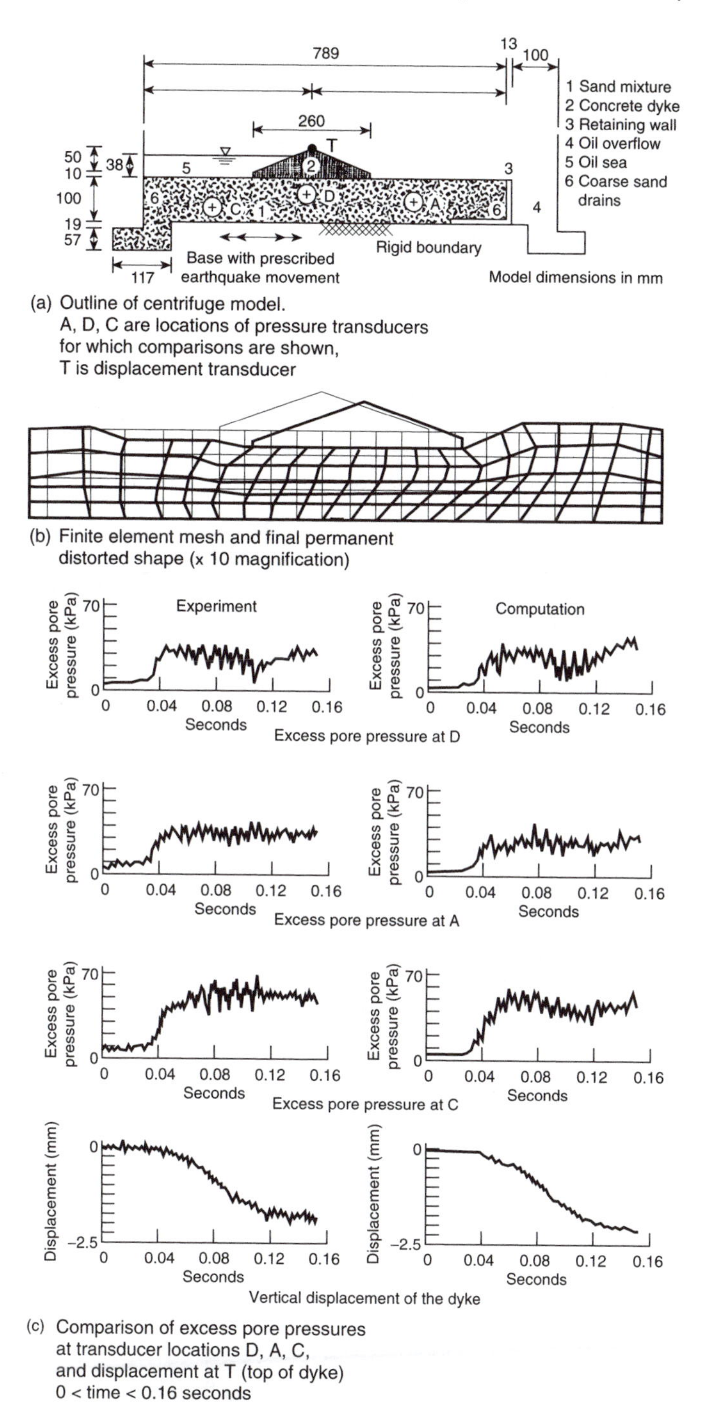

(a) Outline of centrifuge model.
A, D, C are locations of pressure transducers for which comparisons are shown,
T is displacement transducer

(b) Finite element mesh and final permanent distorted shape (x 10 magnification)

(c) Comparison of excess pore pressures at transducer locations D, A, C, and displacement at T (top of dyke)
0 < time < 0.16 seconds

Fig. 18.7 Soil–pressure water interaction. Computation and centrifuge model results compared on a problem of a dyke foundation subject to a simulated earthquake.

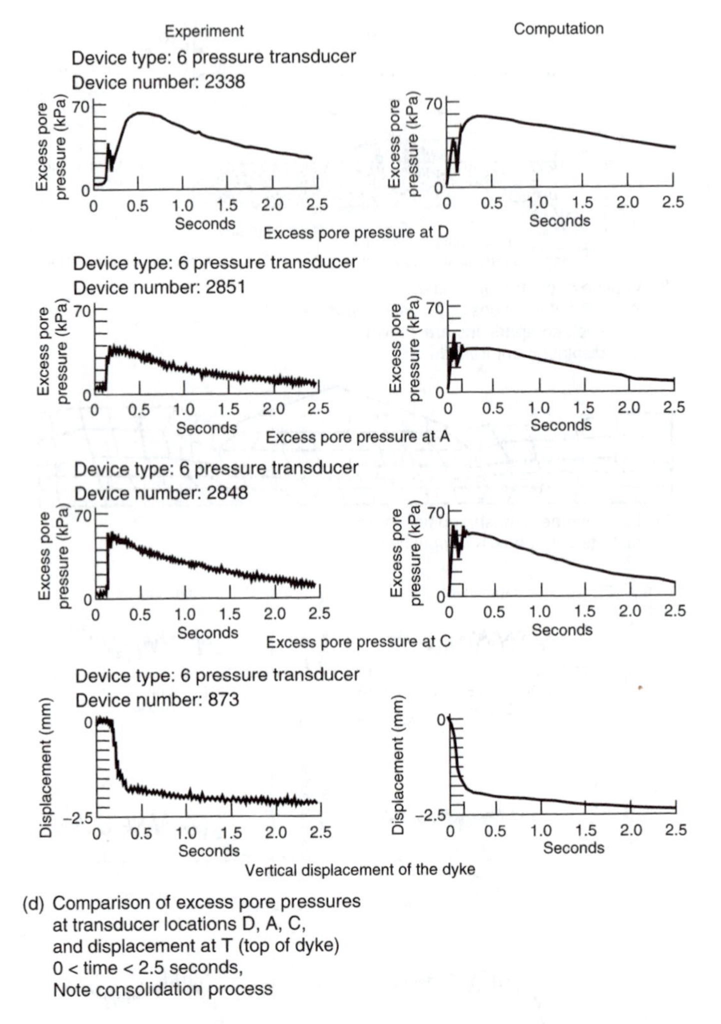

(d) Comparison of excess pore pressures at transducer locations D, A, C, and displacement at T (top of dyke) 0 < time < 2.5 seconds, Note consolidation process

Fig. 18.7 (Cont.)

and a reasonable confidence in predictions achieved by comparison with experimental studies exists. One such study is illustrated in Fig. 18.7 where a comparison with tests carried out in a centrifuge is made.[44, 45] In particular the close correlation between computed pressure and displacement with experiments should be noted.

18.3.6 Biomechanics, oil recovery and other applications

The interaction between a porous medium and interstitial fluid is not confined to soils. The same equations describe, for instance, the biomechanics problem of bone–fluid interaction *in vivo*. Applications in this field have been documented.[46, 47]

On occasion two (or more) fluids are present in the pores and here similar equations can again be written[46, 47] to describe the interaction. Problems of ground settlement in oil fields due to oil extraction, or flow of water/oil mixtures in oil recovery are good examples of application of techniques described here.

18.4 Partitioned single-phase systems – implicit–explicit partitions (Class I problems)

In Fig. 18.1(b), describing problems coupled by an interface, we have already indicated the possibility of a structure being partitioned into substructures and linked along an interface only. Here the substructures will in general be of a similar kind but may differ in the manner (or simply size) of discretization used in each or even in the transient recurrence algorithms employed. In Chapter 12 we have described special kinds of mixed formulations allowing the linking of domains in which, say, boundary-type approximations are used in one and standard finite elements in the other. We shall not return to this phase and will simply assume that the total system can be described using such procedures by a single set of equations in time. Here we shall only consider a first-order problem (but a similar approach can be extended to the second-order dynamic system):

$$\mathbf{C}\dot{\tilde{\mathbf{u}}} + \mathbf{K}\tilde{\mathbf{u}} + \mathbf{f} = \mathbf{0} \tag{18.85}$$

which can be partitioned into two (or more) components, writing

$$\begin{bmatrix} \mathbf{C}_{11} & \mathbf{C}_{12} \\ \mathbf{C}_{21} & \mathbf{C}_{22} \end{bmatrix} \begin{Bmatrix} \dot{\tilde{\mathbf{u}}}_1 \\ \dot{\tilde{\mathbf{u}}}_2 \end{Bmatrix} + \begin{bmatrix} \mathbf{K}_{11} & \mathbf{K}_{12} \\ \mathbf{K}_{21} & \mathbf{K}_{22} \end{bmatrix} \begin{Bmatrix} \tilde{\mathbf{u}}_1 \\ \tilde{\mathbf{u}}_2 \end{Bmatrix} + \begin{Bmatrix} \mathbf{f}_1 \\ \mathbf{f}_2 \end{Bmatrix} = \begin{Bmatrix} \mathbf{0} \\ \mathbf{0} \end{Bmatrix} \tag{18.86}$$

Now for various reasons it may be desirable to use in each partition a different time-step algorithm. Here we shall assume the same structure of the algorithm (SS11) and the same time step (Δt) but simply a different parameter θ in each. Proceeding thus as in the other coupled analyses we write

$$\begin{aligned} \tilde{\mathbf{u}}_1 &= \mathbf{u}_{1n} + \tau\boldsymbol{\alpha}_1 \\ \tilde{\mathbf{u}}_2 &= \mathbf{u}_{2n} + \tau\boldsymbol{\alpha}_2 \end{aligned} \tag{18.87}$$

Inserting the above into each of the partitions and using different weight functions, we obtain

$$\begin{aligned} \mathbf{C}_{11}\boldsymbol{\alpha}_1 + \mathbf{C}_{12}\boldsymbol{\alpha}_2 + \mathbf{K}_{11}(\mathbf{u}_{1n} + \theta\Delta t\boldsymbol{\alpha}_1) + \mathbf{K}_{12}(\mathbf{u}_{2n} + \theta\Delta t\boldsymbol{\alpha}_2) + \bar{\mathbf{f}}_1 = \mathbf{0} \\ \mathbf{C}_{21}\boldsymbol{\alpha}_1 + \mathbf{C}_{22}\boldsymbol{\alpha}_2 + \mathbf{K}_{21}(\mathbf{u}_{1n} + \bar{\theta}\Delta t\boldsymbol{\alpha}_1) + \mathbf{K}_{22}(\mathbf{u}_{2n} + \bar{\theta}\Delta t\boldsymbol{\alpha}_2) + \bar{\mathbf{f}}_2 = \mathbf{0} \end{aligned} \tag{18.88}$$

This system may be solved in the usual manner for $\boldsymbol{\alpha}_1$ and $\boldsymbol{\alpha}_2$ and recurrence relations obtained even if θ and $\bar{\theta}$ differ. The remaining details of the time-step calculations follow the obvious pattern but the question of coupling stability must be addressed. Details of such stability evaluation in this case are given elsewhere[48] but the result is interesting.

1. Unconditional stability of the whole system occurs if

$$\theta \geq \tfrac{1}{2} \qquad \bar{\theta} \geq \tfrac{1}{2}$$

2. Conditional stability requires that

$$\Delta t \leq \Delta t_{\text{crit}}$$

where the Δt_{crit} condition is that pertaining to each partitioned system considered *without its coupling terms.*

Indeed, similar results will be obtained for the second-order systems

$$\mathbf{M}\ddot{\tilde{\mathbf{u}}} + \mathbf{C}\dot{\tilde{\mathbf{u}}} + \mathbf{K}\tilde{\mathbf{u}} + \mathbf{f} = \mathbf{0} \tag{18.89}$$

partitioned in a similar manner with SS22 or GN22 used in each.

The reader may well ask why different schemes should be used in each partition of the domain. The answer in the case of *implicit–implicit* schemes may be simply the desire to introduce different degrees of algorithmic damping. However, much more important is the use of *implicit–explicit* partitions. As we have shown in both 'thermal' and dynamic-type problems the critical time step is inversely proportional to h^2 and h (the element size), respectively. Clearly if a single explicit scheme were to be used with very small elements (or very large material property differences) occurring in one partition, this time step may become too short for economy to be preserved in its use. In such cases it may be advantageous to use an explicit scheme (with $\theta = 0$ in first-order problems, $\theta_2 = 0$ in dynamics) for a part of the domain with larger elements while maintaining unconditional stability with the same time step in the partition in which elements are small or otherwise very 'stiff'. For this reason such implicit–explicit partitions are frequently used in practice.

Indeed, with a lumped representation of matrices $\mathbf{C}$ or $\mathbf{M}$ such schemes are in effect *staggered* as the explicit part can be advanced independently of the implicit part and immediately provides the boundary values for the implicit partition. We shall return to such staggered solutions in the next section.

The use of explicit–implicit partitions was first recorded in 1978.[49–51] In the first reference the process is given in an identical manner as presented here; in the second, a different algorithm is given based on an element split (instead of the implied nodal split above) as described next.

Implicit–explicit solution – element partition

We again consider the first-order problem given in Eq. (18.85) and split as

$$\mathbf{C}_I\dot{\tilde{\mathbf{u}}}_I + \mathbf{C}_E\dot{\tilde{\mathbf{u}}}_E + \mathbf{K}_I\tilde{\mathbf{u}}_I + \mathbf{K}_E\tilde{\mathbf{u}}_E + \mathbf{f} = \mathbf{0} \tag{18.90}$$

where the subscript I denotes an implicit partition and subscript E an explicit one. An iteration process may be used in which one or more iterations per time step are used. The recurrence relation for $\mathbf{u}$ at iteration j is written using GN11 as

$$\mathbf{u}_{n+1}^{(j)} = \mathbf{u}_{n+1}^{(j-1)} + \theta\Delta t\dot{\mathbf{u}}_{n+1}^{(j)} \tag{18.91}$$

with

$$\mathbf{u}_{n+1}^{(0)} = \mathbf{u}_n + (1-\theta)\Delta t\dot{\mathbf{u}}_n \tag{18.92}$$

Using the iteration process an approximation for the implicit–explicit split is now taken as

$$\begin{aligned}
\tilde{\mathbf{u}}_I &= \mathbf{u}_{n+1}^{(j)} \\
\tilde{\mathbf{u}}_E &= \mathbf{u}_{n+1}^{(j-1)} \\
\dot{\tilde{\mathbf{u}}}_I &= \dot{\tilde{\mathbf{u}}}_E = \dot{\mathbf{u}}_{n+1}^{(j)}
\end{aligned}$$

thus yielding the system of equations at iteration j as

$$(\mathbf{C} + \theta\Delta t\mathbf{K}_I)\dot{\mathbf{u}}_{n+1}^{(j)} + \mathbf{F}^{(j)} = \mathbf{0} \tag{18.93}$$

where $\mathbf{F}^{(j)}$ contains the loading terms which depend on known values at t_n and possibly previous iterate values $(j-1)$. The above algorithm has stability properties which depend on the choice of θ. For a linear system with $\theta \geq 0.5$ the implicit part is unconditionally stable and stability depends on the Δt_{crit} of the explicit elements.[50, 51] Performing only one iteration in each time step is normally used; however, improved accuracy in the explicit partition can occur if additional iterations are used, although the cost of each time step is obviously increased.

18.5 Staggered solution processes

18.5.1 General remarks

We have observed in the previous section that in the nodal-based implicit–explicit partitioning of time stepping it was possible to proceed in a *staggered* fashion, achieving a complete solution of the explicit scheme independently of the implicit one and then using the results to progress with the implicit partition. It is tempting to examine the possibility of such staggered procedures generally even if each uses an independent algorithm.

In such procedures the first equation would be solved with some assumed (predicted) values for the variable of the other. Once the solution for the first system was obtained its values could be substituted in the second system, again allowing its independent treatment. If such procedures can be made stable and reasonably accurate many possibilities are immediately open, for instance:

1. Completely different methodologies could be used in each part of the coupled system.
2. Independently developed codes dealing efficiently with single systems could be combined.
3. Parallel computation with its inherent advantages could be used.
4. Finally, in systems of the same physics, efficient iterative solvers could easily be developed.

The problems of such staggered solutions have been frequently discussed[36, 52–55] and on occasion unconditional stability could not be achieved without substantial modification. In the following we shall indicate some options available.

18.5.2 Staggered process of solution in single-phase systems

We shall look at this possibility first, having already mentioned it as a special form arising naturally in the implicit–explicit processes of Sec.18.4. We return here to consider the problem of Eq. (18.85) and the partitioning given in Eq. (18.86). Further, for simplicity we shall assume a diagonal form of the $\mathbf{C}$ matrix, i.e., that the problem is posed as

$$\begin{bmatrix} \mathbf{C}_{11} & \mathbf{0} \\ \mathbf{0} & \mathbf{C}_{22} \end{bmatrix} \begin{Bmatrix} \dot{\tilde{\mathbf{u}}}_1 \\ \dot{\tilde{\mathbf{u}}}_2 \end{Bmatrix} + \begin{bmatrix} \mathbf{K}_{11} & \mathbf{K}_{12} \\ \mathbf{K}_{21} & \mathbf{K}_{22} \end{bmatrix} \begin{Bmatrix} \tilde{\mathbf{u}}_1 \\ \tilde{\mathbf{u}}_2 \end{Bmatrix} + \begin{Bmatrix} \mathbf{f}_1 \\ \mathbf{f}_2 \end{Bmatrix} = \begin{Bmatrix} \mathbf{0} \\ \mathbf{0} \end{Bmatrix} \tag{18.94}$$

As we have already remarked, the use of $\theta = 0$ in the first equation and $\bar{\theta} \geq 0.5$ in the second [see Eq. (18.88)] allowed the explicit part to be solved independently of the implicit. Now, however, we shall use the same θ in both equations but in the first of the approximations, analogous to Eq. (18.88), we shall insert a predicted value for the second variable:

$$\tilde{\mathbf{u}}_2 = \mathbf{u}_2^p = \mathbf{u}_{2n} \tag{18.95}$$

This is similar to the treatment of the explicit part in the element split of the implicit–explicit scheme and gives in place of Eq. (18.88)

$$\mathbf{C}_{11}\boldsymbol{\alpha}_1 + \mathbf{K}_{11}(\mathbf{u}_{1n} + \theta\Delta t\boldsymbol{\alpha}_1) = -\mathbf{f}_1 - \mathbf{K}_{12}\mathbf{u}_{2n} \tag{18.96}$$

allowing direct solution for $\boldsymbol{\alpha}_1$.

Following this step, the second equation can be solved for $\boldsymbol{\alpha}_2$ with the previous value of $\boldsymbol{\alpha}_1$ inserted, i.e.,

$$\mathbf{C}_{22}\boldsymbol{\alpha}_2 + \mathbf{K}_{22}(\mathbf{u}_{2n} + \theta\Delta t\boldsymbol{\alpha}_2) = -\mathbf{f}_2 - \mathbf{K}_{21}(\mathbf{u}_{1n} + \theta\Delta t\boldsymbol{\alpha}_1) \tag{18.97}$$

This scheme is unconditionally stable if $\theta \geq 0.5$, i.e., the total system is stable provided each stagger is unconditionally stable. A similar condition holds for linear second-order dynamic problems.

Obviously, however, some accuracy will be lost as the approximation of Eq. (18.96) is that of the predicted value of $\mathbf{u}_2$. The approximation is consistent and hence convergence will occur as $\Delta t \to 0$.

The advantage of using the staggered process in the above is clear as the equation solving, even though not explicit, is now confined to the magnitude of each partition and computational economy occurs.

Further, it is obvious that precisely the same procedures can be used for any number of partitions and that again the same stability conditions will apply. Define the arrays

$$\mathbf{C} = \begin{bmatrix} \mathbf{C}_{11} & & & & & \\ & \mathbf{C}_{22} & & & & \\ & & \ddots & & & \\ & & & \mathbf{C}_{ii} & & \\ & & & & \ddots & \\ & & & & & \mathbf{C}_{kk} \end{bmatrix} \tag{18.98a}$$

$$\mathbf{K} = \begin{bmatrix} \mathbf{K}_{11} & \mathbf{0} & \cdots & & & \mathbf{0} \\ \mathbf{K}_{21} & \mathbf{K}_{22} & & & & \vdots \\ \vdots & & \ddots & & & \\ & & & \mathbf{K}_{ii} & & \\ \vdots & & & & \ddots & \mathbf{0} \\ \mathbf{K}_{k1} & \cdots & & & \mathbf{K}_{k,k-1} & \mathbf{K}_{kk} \end{bmatrix} + \begin{bmatrix} \mathbf{0} & \mathbf{K}_{12} & \cdots & & & \mathbf{K}_{1k} \\ \mathbf{0} & \mathbf{0} & \cdots & & & \vdots \\ \vdots & & \ddots & & & \\ & & & \mathbf{0} & & \vdots \\ & & & & \ddots & \mathbf{K}_{k-1,k} \\ \mathbf{0} & & & & \cdots & \mathbf{0} \end{bmatrix}$$

$$= \mathbf{K}_L + \mathbf{K}_U \tag{18.98b}$$

and consider the partition

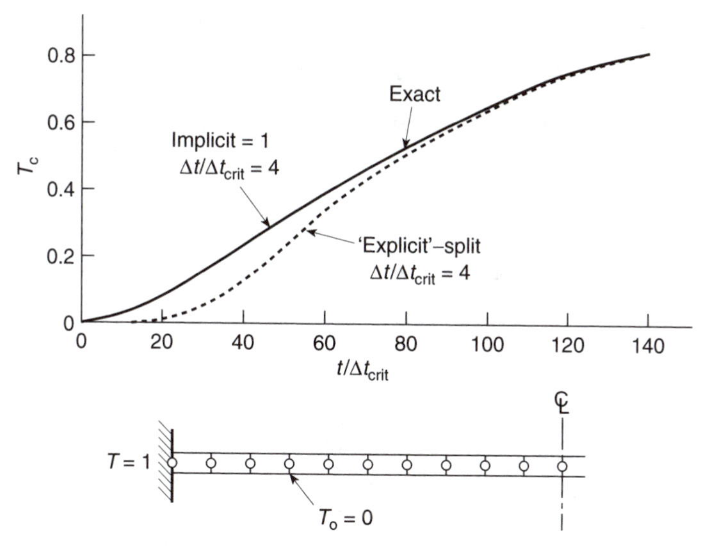

Fig. 18.8 Accuracy of an explicit-split procedure compared with a standard implicit process for heat conduction of a bar.

$$\mathbf{C}\dot{\tilde{\mathbf{u}}} + \mathbf{K}_L\tilde{\mathbf{u}} + \mathbf{K}_U\tilde{\mathbf{u}} + \mathbf{f} = \mathbf{0} \tag{18.99}$$

Introducing now the approximation

$$\tilde{\mathbf{u}}_i = \mathbf{u}_{in} + \tau\boldsymbol{\alpha}_i \tag{18.100}$$

and using Eq. (18.95) gives the discrete form

$$(\mathbf{C} + \theta\Delta t\mathbf{K}_L)\,\boldsymbol{\alpha} + \mathbf{K}_U\mathbf{u}_n + \bar{\mathbf{f}} = \mathbf{0} \tag{18.101}$$

where $\bar{\mathbf{f}}$ contains the load and effects from $\mathbf{u}_n$.

In approximating the first equation set it is necessary to use predicted values for $\mathbf{u}_2$, $\mathbf{u}_3, \cdots, \mathbf{u}_k$, writing in place of Eq. (18.96),

$$\mathbf{C}_{11}\boldsymbol{\alpha}_1 + \mathbf{K}_{11}(\mathbf{u}_{1n} + \theta\Delta t\boldsymbol{\alpha}_1) + \mathbf{K}_{12}\mathbf{u}_{2n} + \mathbf{K}_{13}\mathbf{u}_{3n} + \cdots + \mathbf{f}_1 = \mathbf{0} \tag{18.102}$$

and continue similarly to (18.97), with the predicted values now continually being replaced by better approximations as the solution progresses.

The partitioning of Eq. (18.98a) can be continued until only a single equation set is obtained. Then at each step the equation that requires solving for $\boldsymbol{\alpha}_i$ is of the form

$$(\mathbf{C}_{ii} + \theta\Delta t\mathbf{K}_{ii})\,\boldsymbol{\alpha}_i = \mathbf{F}_i \tag{18.103}$$

where $\mathbf{F}_i$ contains the effects of the load and all the previously computed $\mathbf{u}_i$. For partitions where each submatrix is a scalar Eq. (18.103) is a scalar equation and computation is thus *fully explicit and yet preserves unconditional stability* for $\theta \geq 0.5$. This type of partitioning and the derivation of an unconditionally stable explicit scheme was first proposed by Zienkiewicz *et al.*[56] An alternative and somewhat more limited scheme of a similar kind was given by Trujillo.[57]

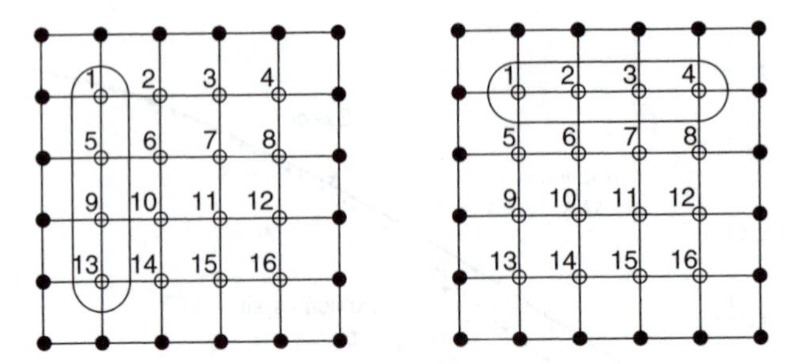

Fig. 18.9 Partitions corresponding to the well-known ADI (alternating direction implicit) finite difference scheme.

Clearly the error in the approximation in the time step decreases as the solution sweeps through the partitions and hence it is advisable to alter the sweep directions during the computation. For instance, in Fig. 18.8 we show quite reasonable accuracy for a one-dimensional heat-conduction problem in which the *explicit-split* process was used with alternating direction of sweeps. Of course the accuracy is much inferior to that exhibited by a standard implicit scheme with the same time step, though the process could be used quite effectively as an iteration to obtain steady-state solutions. Here many other options are also possible.

It is, for instance, of interest to consider the system given in Eqs (18.98a), (18.98b) and (18.99) as originating from a simple finite difference approximation to, say, a heat-conduction equation on the rectangular mesh of Fig. 18.9.

Here it is well known that the so-called alternating direction implicit (ADI) scheme[58] presents an efficient solution for both transient and steady-state problems. It is fairly obvious that the scheme simply represents the procedure just outlined with partitions representing lines of nodes such as $(1, 5, 9, 13)$, $(2, 6, 10, 14)$, etc., of Fig. 18.9 alternating with partitions $(1, 2, 3, 4)$, $(5, 6, 7, 8)$, etc.

Obviously the bigger the partition, the more accurate the scheme becomes, though of course at the expense of computational costs. The concept of the staggered partition clearly allows easy adoption of such procedures in the finite element context. Here irregular partitions arbitrarily chosen could be made but so far applications have only been recorded in regular mesh subdivisions.[58] The field of possibilities is obviously large. Use in parallel computation is obvious for such procedures.

A further possibility which has many advantages is to use hierarchical variables based on, say, linear, quadratic and higher expansions and to consider each set of these variables as a partition.[59] Such procedures are particularly efficient in iteration if coupled with suitable preconditioning[60] and form a basis of *multigrid procedures*.[61–63]

18.5.3 Staggered schemes in fluid–structure systems and stabilization processes

The application of staggered solution methods in coupled problems representing different phenomena is more obvious, though, as it turns out, more difficult.

For instance, let us consider the linear discrete fluid–structure equations with $p_o = 1$ and damping omitted, written as [see Eqs (18.26) and 18.28)]

$$\begin{bmatrix} \mathbf{M} & \mathbf{0} \\ \mathbf{Q}^T & \mathbf{S} \end{bmatrix} \begin{Bmatrix} \ddot{\mathbf{u}} \\ \ddot{\mathbf{p}} \end{Bmatrix} + \begin{bmatrix} \mathbf{K} & -\mathbf{Q} \\ \mathbf{0} & \mathbf{H} \end{bmatrix} \begin{Bmatrix} \mathbf{u} \\ \mathbf{p} \end{Bmatrix} + \begin{Bmatrix} \mathbf{f} \\ \mathbf{q} \end{Bmatrix} = \begin{Bmatrix} \mathbf{0} \\ \mathbf{0} \end{Bmatrix} \tag{18.104}$$

where we have omitted the tilde superscript for simplicity.

For illustration purposes we shall use the GN22 type of approximation for both variables and write using Eq. (18.76)

$$\begin{aligned} \mathbf{u}_{n+1} &= \mathbf{u}^p_{n+1} + \tfrac{1}{2}\beta_2 \Delta t^2 \Delta\ddot{\mathbf{u}}_{n+1} \\ \dot{\mathbf{u}}_{n+1} &= \dot{\mathbf{u}}^p_{n+1} + \beta_1 \Delta t \Delta\ddot{\mathbf{u}}_{n+1} \\ \mathbf{p}_{n+1} &= \mathbf{p}^p_{n+1} + \tfrac{1}{2}\bar{\beta}_2 \Delta t^2 \Delta\ddot{\mathbf{p}}_{n+1} \\ \dot{\mathbf{p}}_{n+1} &= \dot{\mathbf{p}}^p_{n+1} + \bar{\beta}_1 \Delta t \Delta\ddot{\mathbf{p}}_{n+1} \end{aligned} \tag{18.105}$$

which together with Eq. (18.104) written at $t = t_{n+1}$ completes the system of equations requiring simultaneous solution for $\Delta\ddot{\mathbf{u}}_{n+1}$ and $\Delta\ddot{\mathbf{p}}_{n+1}$.

Now a staggered solution of a fairly obvious kind would be to write the first set of equations (18.104) corresponding to the structural behaviour with a predicted (approximate) value of $\mathbf{p}_{n+1} = \mathbf{p}^p_{n+1}$, as this would allow an independent solution for $\Delta\ddot{\mathbf{u}}_{n+1}$ writing

$$\mathbf{M}\ddot{\mathbf{u}}_{n+1} + \mathbf{K}\mathbf{u}_{n+1} = -\mathbf{f} + \mathbf{Q}\mathbf{p}^p_{n+1} \tag{18.106}$$

This would then be followed by the solution of the fluid problem for $\Delta\ddot{\mathbf{p}}_{n+1}$ writing

$$\mathbf{S}\ddot{\mathbf{p}}_{n+1} + \mathbf{H}\mathbf{p}_{n+1} = -\mathbf{q} - \mathbf{Q}^T\ddot{\mathbf{u}}_{n+1} \tag{18.107}$$

This scheme turns out, however, to be only conditionally stable,[48] even if β_i and $\bar{\beta}_i$ are chosen so that unconditional stability of a simultaneous solution is achieved. (The stability limit is indeed the same as if a fully explicit scheme were chosen for *the fluid phase*.)

Various stabilization schemes can be used here.[27, 48] One of these is given below. In this Eq. (18.106) is augmented to

$$\mathbf{M}\ddot{\mathbf{u}}_{n+1} + \left(\mathbf{K} + \mathbf{Q}\mathbf{S}^{-1}\mathbf{Q}^T\right)\mathbf{u}_{n+1} = -\mathbf{f} + \mathbf{Q}\mathbf{p}^p_{n+1} + \mathbf{Q}\mathbf{S}^{-1}\mathbf{Q}^T\mathbf{u}^p_{n+1} \tag{18.108}$$

before solving for $\Delta\ddot{\mathbf{u}}_{n+1}$. It turns out that this scheme is now unconditionally stable provided the usual conditions

$$\beta_2 \geq \beta_1 \qquad \beta_1 \geq \tfrac{1}{2}$$

are satisfied.

Such stabilization involves the inverse of **S** but again it should be noted that this needs to be obtained only for the coupling nodes on the interface. Another stable scheme involves a similar inversion of **H** and is useful as incompressible behaviour is automatically given.

Similar stabilization processes have been applied with success to the soil–fluid system.[64, 65]

18.6 Concluding remarks

The range of problems which may be considered as coupled is very large and forms studies which are now often referred to as 'multi-physics' problems. The range of possible algorithms to solve such problems has been summarized above; however, new methods often are proposed (e.g., see reference 66).

Another class of problems which may be considered as coupled considers 'multi-scale' effects. These attempt to bridge the behaviour of materials from, for example, a micro to a macro scale. This topic is very popular today and is discussed further in reference 33.

References

1. O.C. Zienkiewicz. Coupled problems and their numerical solution. In R.W. Lewis, P. Bettess, and E. Hinton, editors, *Numerical Methods in Coupled Systems*, Chapter 1, pages 35–68. John Wiley & Sons, Chichester, 1984.
2. B.A. Boley and J.H. Weiner. *Theory of Thermal Stresses*. Dover Publications, Mineola, New York, 1997.
3. R.W. Lewis, P. Bettess, and E. Hinton, editors. *Numerical Methods in Coupled Systems*. John Wiley & Sons, Chichester, 1984.
4. R.W. Lewis, E. Hinton, P. Bettess, and B.A. Schrefler, editors. *Numerical Methods in Coupled Systems*. John Wiley & Sons, Chichester, 1987.
5. J.C. Simo and T.J.R. Hughes. *Computational Inelasticity*, volume 7 of *Interdisciplinary Applied Mathematics*. Springer-Verlag, Berlin, 1998.
6. O.C. Zienkiewicz, R.L. Taylor, and P. Nithiarasu. *The Finite Element Method for Fluid Dynamics*. Butterworth-Heinemann, Oxford, 6th edition, 2005.
7. O.C. Zienkiewicz and R.E. Newton. Coupled vibration of a structure submerged in a compressible fluid. In *Proc. Int. Symp. on Finite Element Techniques*, pages 359–371, Stuttgart, 1969.
8. P. Bettess and O.C. Zienkiewicz. Diffraction and refraction of surface waves using finite and infinite elements. *Int. J. Numer. Meth. Eng.*, 11:1271–1290, 1977.
9. O.C. Zienkiewicz, D.W. Kelly, and P. Bettess. The Sommerfield (radiation) condition on infinite domains and its modelling in numerical procedures. In *Proc. IRIA 3rd Int. Symp. on Computing Methods in Applied Science and Engineering*, Versailles, Dec. 1977.
10. O.C. Zienkiewicz, P. Bettess, and D.W. Kelly. The finite element method for determining fluid loadings on rigid structures. Two- and three-dimensional formulations. In O.C. Zienkiewicz, R.W. Lewis, and K.G. Stagg, editors, *Numerical Methods in Offshore Engineering*, pages 141–183. John Wiley & Sons, Chichester, 1978.
11. O.C. Zienkiewicz and P. Bettess. Dynamic fluid–structure interaction. Numerical modelling of the coupled problem. In O.C. Zienkiewicz, R.W. Lewis, and K.G. Stagg, editors, *Numerical Methods in Offshore Engineering*, pages 185–193. John Wiley & Sons, Chichester, 1978.
12. O.C. Zienkiewicz and P. Bettess. Fluid–structure dynamic interaction and wave forces. An introduction to numerical treatment. *Int. J. Numer. Meth. Eng.*, 13:1–16, 1978.
13. O.C. Zienkiewicz and P. Bettess. Fluid–structure dynamic interaction and some 'unified' approximation processes. In *Proc. 5th Int. Symp. on Unification of Finite Elements, Finite Differences and Calculus of Variations*. University of Connecticut, May 1980.
14. R. Ohayon. Symmetric variational formulations for harmonic vibration problems coupling primal and dual variables – applications to fluid–structure coupled systems. *La Recherche Aerospatiale*, 3:69–77, 1979.

15. R. Ohayon. True symmetric formulation of free vibrations for fluid–structure interaction in bounded media. In R.W. Lewis, P. Bettess, and E. Hinton, editors, *Numerical Methods in Coupled Systems*. John Wiley & Sons, Chichester, 1984.
16. R. Ohayon. Fluid–structure interaction. In *Proc. of the ECCM'99 Conference IACM/ECCM'99*, Munich, Germany, Sept. 1999 (on CD-ROM).
17. H. Morand and R. Ohayon. *Fluid–Structure Interaction*. John Wiley & Sons, London, 1995.
18. M.P. Paidoussis and P.P. Friedmann, editors. *4th International Symposium on Fluid–Structure Interactions, Aeroelasticity, Flow-Induced Vibration and Noise, vol. 1, 2, 3*, volume AD-vo. 52–3, Dallas, Texas, Nov. 1997. ASME/Winter Annual Meeting.
19. T. Kvamsdal *et al.*, editors. *Computational Methods for Fluid–Structure Interaction*. Tapir Publishers, Trondheim, 1997.
20. R. Ohayon and C.A. Felippa (editors). Computational Methods for Fluid–Structure Interaction and Coupled Problems. *Comp. Meth. Appl. Mech. Eng.*, 190:2977–3292 (Special issue).
21. M. Geradin, G. Roberts, and J. Huck. Eigenvalue analysis and transient response of fluid structure interaction problems. *Eng. Comput.*, 1:152–160, 1984.
22. G. Sandberg and P. Gorensson. A symmetric finite element formation of acoustic fluid–structure interaction analysis. *J. Sound Vibr.*, 123:507–515, 1988.
23. U. Basu and A.K. Chopra. Perfectly matched layers for time-harmonic elastodynamics of unbounded domains: theory and finite-element implementation. *Comp. Meth. Appl. Mech. Eng.*, 192:1337–1375, 2003.
24. U. Basu and A.K. Chopra. Perfectly matched layers for transient elastodynamics of unbounded domains. *Int. J. Numer. Meth. Eng.*, 59:1039–1074, 2004.
25. K.K. Gupta. On a numerical solution of the supersonic panel flutter eigenproblem. *Int. J. Numer. Meth. Eng.*, 10:637–645, 1976.
26. B.M. Irons. The role of part inversion in fluid–structure problems with mixed variables. *J. AIAA*, 7:568, 1970.
27. W.J.T. Daniel. Modal methods in finite element fluid–structure eigenvalue problems. *Int. J. Numer. Meth. Eng.*, 15:1161–1175, 1980.
28. C.A. Felippa. Symmetrization of coupled eigenproblems by eigenvector augmentation. *Comm. Appl. Numer. Meth.*, 4:561–563, 1988.
29. J. Holbeche. Ph.D. thesis, Department of Civil Engineering, University of Wales, Swansea, 1971.
30. A.K. Chopra and S. Gupta. Hydrodynamic and foundation interaction effects in earthquake response of a concrete gravity dam. *J. Struct. Div. Am. Soc. Civ. Eng.*, 578:1399–1412, 1981.
31. J.F. Hall and A.K. Chopra. Hydrodynamic effects in the dynamic response of concrete gravity dams. *Earth. Eng. Struct. Dyn.*, 10:333–395, 1982.
32. O.C. Zienkiewicz and R.L. Taylor. Coupled problems – a simple time-stepping procedure. *Comm. Appl. Numer. Meth.*, 1:233–239, 1985.
33. O.C. Zienkiewicz and R.L. Taylor. *The Finite Element Method for Solid and Structural Mechanics*. Butterworth-Heinemann, Oxford, 6th edition, 2005.
34. O.C. Zienkiewicz, R.W. Clough, and H.B. Seed. Earthquake analysis procedures for concrete and earth dams – state of the art. Technical Report Bulletin 32, Int. Commission on Large Dams, Paris, 1986.
35. R.E. Newton. Finite element study of shock induced cavitation. In *ASCE Spring Convention*, Portland, Oregon, 1980.
36. O.C. Zienkiewicz, D.K. Paul, and E. Hinton. Cavitation in fluid–structure response (with particular reference to dams under earthquake loading). *Earth. Eng. Struct. Dyn.*, 11:463–481, 1983.
37. M.A. Biot. Theory of propagation of elastic waves in a fluid saturated porous medium, Part I: Low frequency range; Part II: High frequency range. *J. Acoust. Soc. Am.*, 28:168–191, 1956.

38. O.C. Zienkiewicz, C.T. Chang, and E. Hinton. Non-linear seismic responses and liquefaction. *Int. J. Numer. Anal. Meth. Geomech.*, 2:381–404, 1978.
39. O.C. Zienkiewicz and T. Shiomi. Dynamic behaviour of saturated porous media, the generalized Biot formulation and its numerical solution. *Int. J. Numer. Anal. Meth. Geomech.*, 8:71–96, 1984.
40. O.C. Zienkiewicz, K.H. Leung, and M. Pastor. Simple model for transient soil loading in earthquake analysis: Part I – Basic model and its application. *Int. J. Numer. Anal. Meth. Geomech.*, 9:453–476, 1985.
41. O.C. Zienkiewicz, K.H. Leung, and M. Pastor. Simple model for transient soil loading in earthquake analysis: Part II – non-associative models for sands. *Int. J. Numer. Anal. Meth. Geomech.*, 9:477–498, 1985.
42. O.C. Zienkiewicz, A.H.C. Chan, M. Pastor, and T. Shiomi. Computational approach to soil dynamics. In A.S. Czamak, editor, *Soil Dynamics and Liquefaction*, volume *Developments in Geotechnical Engineering 42*. Elsevier, Amsterdam, 1987.
43. O.C. Zienkiewicz, A.H.C. Chan, M. Pastor, D.K. Paul, and T. Shiomi. Static and dynamic behaviour of soils: a rational approach to quantitative solutions, I. *Proc. R. Soc. London*, 429:285–309, 1990.
44. O.C. Zienkiewicz, Y.M. Xie, B.A. Schrefler, A. Ledesma, and N. Biĉaniĉ. Static and dynamic behaviour of soils: a rational approach to quantitative solutions, II. *Proc. R. Soc. London*, 429:311–321, 1990.
45. O.C. Zienkiewicz, A.H.C. Chan, M. Pastor, B.A. Schrefler, and T. Shiomi. *Computational Geomechanics: With Special Reference to Earthquake Engineering*. John Wiley & Sons, Chichester, 1999.
46. B.R. Simon, J. S-S. Wu, M.W. Carlton, L.E. Kazarian, E/P. France, J.H. Evans, and O.C. Zienkiewicz. Poroelastic dynamic structural models of rhesus spinal motion segments. *Spine*, 10(6):494–507, 1985.
47. B.R. Simon, J. S-S. Wu, and O.C. Zienkiewicz. Higher order mixed and Hermitian finite element procedures for dynamic analysis of saturated porous media. *Int. J. Numer. Meth. Eng.*, 10:483–499, 1986.
48. O.C. Zienkiewicz and A.H.C. Chan. Coupled problems and their numerical solution. In I.S. Doltsinis, editor, *Advances in Computational Non-linear Mechanics*, Chapter 3, pages 109–176. Springer-Verlag, Berlin, 1988.
49. T. Belytschko and R. Mullen. Stability of explicit–implicit time domain solution. *Int. J. Numer. Meth. Eng.*, 12:1575–1586, 1978.
50. T.J.R. Hughes and W.K. Liu. Implicit–explicit finite elements in transient analyses. Part I and Part II. *J. Appl. Mech.*, 45:371–378, 1978.
51. T. Belytschko and T.J.R. Hughes, editors. *Computational Methods for Transient Analysis*. North-Holland, Amsterdam, 1983.
52. C.A. Felippa and K.C. Park. Staggered transient analysis procedures for coupled mechanical systems: formulation. *Comp. Meth. Appl. Mech. Eng.*, 24:61–111, 1980.
53. K.C. Park. Partitioned transient analysis procedures for coupled field problems: stability analysis. *J. Appl. Mech.*, 47:370–376, 1980.
54. K.C. Park and C.A. Felippa. Partitioned transient analysis procedures for coupled field problems: accuracy analysis. *J. Appl. Mech.*, 47:919–926, 1980.
55. O.C. Zienkiewicz, E. Hinton, K.H. Leung, and R.L. Taylor. Staggered time marching schemes in dynamic soil analysis and selective explicit extrapolation algorithms. In R. Shaw *et al.*, editors, *Proc. Conf. on Innovative Numerical Analysis for the Engineering Sciences*, University of Virginia Press, 1980.
56. O.C. Zienkiewicz, C.T. Chang, and P. Bettess. Drained, undrained, consolidating dynamic behaviour assumptions in soils. *Geotechnique*, 30:385–395, 1980.

57. D.M. Trujillo. An unconditionally stable explicit scheme of structural dynamics. *Int. J. Numer. Meth. Eng.*, 11:1579–1592, 1977.
58. L.J. Hayes. Implementation of finite element alternating-direction methods on non-rectangular regions. *Int. J. Numer. Meth. Eng.*, 16:35–49, 1980.
59. A.W. Craig and O.C. Zienkiewicz. A multigrid algorithm using a hierarchical finite element basis. In D.J. Pedolon and H. Holstein, editors, *Multigrid Methods in Integral and Differential Equations*, pages 310–312. Clarendon Press, Oxford, 1985.
60. I. Babuška, A.W. Craig, J. Mandel, and J. Pitkäranta. Efficient preconditioning for the p-inversion finite element method in two dimensions. *SIAM J. Num. Anal.*, 28:624–661, 1991.
61. R. Löhner and K. Morgan. An unstructured multigrid method for elliptic problems. *Int. J. Numer. Meth. Eng.*, 24:101–115, 1987.
62. M. Adams. Heuristics for automatic construction of coarse grids in multigrid solvers for finite element matrices. Technical Report UCB//CSD-98–994, University of California, Berkeley, 1998.
63. M. Adams. Parallel multigrid algorithms for unstructured 3D large deformation elasticity and plasticity finite element problems. Technical Report UCB//CSD-99–1036, University of California, Berkeley, 1999.
64. K.C. Park. Stabilization of partitioned solution procedures for pore fluid–soil interaction analysis. *Int. J. Numer. Meth. Eng.*, 19:1669–1673, 1983.
65. O.C. Zienkiewicz, D.K. Paul, and A.H.C. Chan. Unconditionally stable staggered solution procedures for soil–pore fluid interaction problems. *Int. J. Numer. Meth. Eng.*, 26:1039–1055, 1988.
66. J.Y. Kim, N.R. Aluru, and D.A. Tortorelli. Improved multi-level Newton solvers for fully coupled multi-physics problems. *Int. J. Numer. Meth. Eng.*, 58:463–480, 2003.

19

Computer procedures for finite element analysis

19.1 Introduction

A companion program to this book is available which can carry out analyses for most of the theory presented in previous chapters. In particular the computer program discussed here may be used to solve any one-, two-, or three-dimensional linear steady-state or transient problem. The program also has capabilities to perform non-linear analysis for the type of problems discussed in reference 1.

Source listings and a user manual may be obtained at no charge from the author's internet web site (http://www.ce.berkeley.edu/~rlt) or the publisher's internet web site (http://books.elsevier.com/companions). The program is written mostly in Fortran with some routines in C (see author's web site for more information on using C for user modules). Any errors reported by readers will be corrected so that up-to-date versions are available.

The version available for download is called *FEAPpv* which is an acronym for *Finite Element Analysis Program – personal version*. It is intended mainly for use in learning finite element programming methodologies and in solving small to moderate size problems on single processor computers. A simple management scheme is employed to permit efficient use of main memory with limited need to read and write information to disk.

Finite element programs can be separated into three basic parts:

1. Data input module and pre-processor
2. Solution module
3. Results module and post-processor.

19.2 Pre-processing module: mesh creation

FEAPpv is mainly a solution module but provides simple data input and pre-processor capabilities which permit generation of meshes using the multiblock schemes of Zienkiewicz and Phillips[2] and Gordon and Hall.[3] Alternatively the data may be input from neutral files written by other pre-processing systems (e.g., GiD[4]).

Data input for the program consists of specification (or generation) of: (1) the coordinates for each node; (2) the element form and the nodal connection list for each element;

(3) boundary conditions and loads to be applied; and (4) material property data. The user manual describes the format for specifying the data to be used by *FEAPpv*.

19.2.1 Element library

As part of the input data it is necessary to describe the element formulation to be used in forming the 'stiffness' matrix and 'load' vector of each problem. This may be provided either by user written modules (see below) or using the element library provided with the program.

Currently, the element library in *FEAPpv* includes:

1. *Solid elements for two-dimensional linear elasticity.* Forms are provided for the irreducible formulation described in Chapters 2 and 6; the three-field mixed form described in Sec. 11.3 and the enhanced strain form described in Sec. 10.5.3. The elements permit consideration of elastic models which are isotropic or orthotropic as described in Chapter 6.
   (a) For the irreducible form the element shape may range from a 3-node triangle to a 9-node lagrangian quadrilateral.
   (b) For the three-field mixed form the element shape may be a 4-node, 8-node or 9-node quadrilateral form.
   (c) For the enhanced strain model the element is restricted to a 4-node quadrilateral form.
2. *Solid elements for three-dimensional linear elasticity.* Only the irreducible form for a 4-node tetrahedron or an 8-node brick may be used. The 8-node brick may be degenerated into other forms by giving the same node number to nodes used to perform the degenerate shape (see Sec.5.8). The elastic material model may be isotropic or orthotropic as described in Chapter 6.
3. *Frame (rod) elements for two- and three-dimensional elasticity.* Conventional structural elements are provided to perform analysis of elastic two- and three-dimensional frame structures. While these forms have not been discussed in this text, except as suggested problems for solution, they are useful for use in general analysis. The theory is contained in standard references for structural analysis and also in reference 1.
4. *Truss elements for two- and three-dimensional elasticity.* Similar to frame elements, the *FEAPpv* system includes conventional truss elements which may be used to analyse plane and space truss structures.
5. *Plate element for linear elasticity.* A plate bending element for use in the analysis of plates which include the primary effects of transverse shear (so-called Reissner–Mindlin theory[1]) is provided. The element form may be either a 3-node triangle or a 4-node quadrilateral. The theory is described in references 1, 5, 6.
6. *Shell element for three dimensions with linear elasticity.* A 4-node quadrilateral element form for use in modelling general shell forms is provided. The element includes membrane and bending effects only and, thus, may be used only for analysis of 'thin' shells. The theory for the element is given in reference 7. The element form should be a 4-node quadrilateral.

7. *Membrane element for linear elasticity.* A general elastic membrane form is provided which is the same as the shell element but without the bending terms. The element form should be a 4-node quadrilateral.
8. *Thermal elements for two- and three-dimensional Fourier heat conduction.* The theory described in Chapter 7 for transient heat conduction is provided in elements which solve two- and three-dimensional problems. The Fourier model may be isotropic or orthotropic.
9. *User developed elements.* Users may develop and add element modules for any problem which can be formed by the finite element approach described in this book. Details for writing modules will be found in the *Programers Manual* available at the web sites.

19.3 Solution module

The main part of *FEAPpv* is a solution module which permits users to analyse a large range of problems formulated by the finite element method. Specific solution methods are prepared by the user using a unique *command language*, which is a sequence of statements which describe each algorithm. The current version of *FEAPpv* permits both 'batch' and 'interactive' problem solution. The commands provided permit specification of problems with either symmetric or unsymmetric 'stiffness' matrices, selection of direct or iterative solution of the linear algebraic equation system, selection of different transient solution algorithms, and output of solution results in either a text or graphics format. Commands which permit solution of a symmetric generalized linear eigenproblem (see Chapter 16) using a 'subspace' method[8, 9] are also available as well as a feature to compute the eigenvalues and vectors for an element stiffness.

While the main thrust of this book is the solution of linear problems, the system *FEAPpv* is capable of solving both linear and non-linear problems. The use of special 'loop' commands permits the construction of algorithms which require iteration or time stepping. In addition features to solve problems in which load following is needed are provided in the form of 'arc-length'-type methods.[10–12] The solution of problems for which it is not possible to deduce an accurate 'stiffness' matrix may be attempted using a quasi-Newton method based on the BFGS method.[13, 14] The user manual available at the web site provides examples for several algorithms as well as a list of all available commands.

19.4 Post-processor module

As noted above the *FEAPpv* system contains capabilities to report results as text data written to an output file or in graphical form which may be displayed on the screen or written to files for processing by other systems. Files are written in *PostScript* format (in an encapsulated form which may be used by many programs – e.g., TeX or LaTeX).

The general features of graphical post-processing are limited to displaying two-dimensional objects. More complex forms require an interface to a separate pre-/post-processing system (e.g., GiD.[4]). The two-dimensional capabilities in *FEAPpv* include display of the mesh including node and element numbers, boundary conditions and loads. Contour plots for each degree of freedom of the solution system may be displayed as well as contours of element values such as stress or strain components. The user manual provides a list of all commands for constructing graphical outputs.

The available version for graphics is limited to X-window applications and compilers compatible with the current HP Fortran 95 compiler for Windows-based systems.[15]

19.5 User modules

A key ingredient of the *FEAPpv* system is the ability of a user to add their own modules to extend the capabilities of the program to other classes of problems, material models, or solution strategies. Some user developed modules are available at the authors' web site given above and include element modules for other problem forms, an interface to other linear equation solvers, etc. Experienced programmers should be able to easily adapt these routines to include additional features.

Programming additions to the system may be performed following descriptions in the *Programmer Manual* available at the web sites.

References

1. O.C. Zienkiewicz and R.L. Taylor. *The Finite Element Method for Solid and Structural Mechanics*. Butterworth-Heinemann, Oxford, 6th edition, 2005.
2. O.C. Zienkiewicz and D.V. Phillips. An automatic mesh generation scheme for plane and curved surfaces by isoparametric coordinates. *Int. J. Numer. Meth. Eng.*, 3:519–528, 1971.
3. W.J. Gordon and C.A. Hall. Transfinite element methods – blending-function interpolation over arbitrary curved element domains. *Numer. Math.*, 21:109–129, 1973.
4. GiD – The Personal Pre/Postprocessor. www.gidhome.com, 2004.
5. F. Auricchio and R.L. Taylor. A shear deformable plate element with an exact thin limit. *Comp. Meth. Appl. Mech. Eng.*, 118:393–412, 1994.
6. F. Auricchio and R.L. Taylor. A triangular thick plate finite element with an exact thin limit. *Finite Elements in Analysis and Design*, 19:57–68, 1995.
7. R.L. Taylor. Finite element analysis of linear shell problems. In J.R. Whiteman, editor, *The Mathematics of Finite Elements and Applications VI*, pages 191–203. Academic Press, London, 1988.
8. J.H. Wilkinson and C. Reinsch. *Linear Algebra. Handbook for Automatic Computation*, volume II. Springer-Verlag, Berlin, 1971.
9. K.-J. Bathe. *Finite Element Procedures*. Prentice Hall, Englewood Cliffs, N.J., 1996.
10. E. Riks. An incremental approach to the solution of snapping and buckling problems. *Int. J. Solids Struct.*, 15:529–551, 1979.
11. K. Schweizerhof. *Nitchlineare Berechnung von Tragwerken unter verformungsabhangiger belastung mit finiten Elementen*. Doctoral dissertation, U. Stuttgart, Stuttgart, Germany, 1982.
12. J.C. Simo, P. Wriggers, K.H. Schweizerhof, and R.L Taylor. Finite deformation post-buckling analysis involving inelasticity and contact constraints. *Int. J. Numer. Meth. Eng.*, 23:779–800, 1986.
13. J.E. Dennis and J. More. Quasi-Newton methods – motivation and theory. *SIAM Rev.*, 19:46–89, 1977.
14. H. Matthies and G. Strang. The solution of nonlinear finite element equations. *Int. J. Numer. Meth. Eng.*, 14:1613–1626, 1979.
15. HP Fortran home page. http://h18009.www1.hp.com/fortran, 2004.

Appendix A

Matrix algebra

The mystique surrounding matrix algebra is perhaps due to the texts on the subject requiring a student to 'swallow too much' at one time. It will be found that in order to follow the present text and carry out the necessary computation only a limited knowledge of a few basic definitions is required.

Definition of a matrix

The linear relationship between a set of variables x and b

$$\begin{aligned} a_{11}x_1 + a_{12}x_2 + a_{13}x_3 + a_{14}x_4 &= b_1 \\ a_{21}x_1 + a_{22}x_2 + a_{23}x_3 + a_{24}x_4 &= b_2 \\ a_{31}x_1 + a_{32}x_2 + a_{33}x_3 + a_{34}x_4 &= b_3 \end{aligned} \tag{A.1}$$

can be written, in a short-hand way, as

$$[A]\{x\} = \{b\} \tag{A.2}$$

or

$$\mathbf{A}\mathbf{x} = \mathbf{b} \tag{A.3}$$

where

$$\begin{aligned} \mathbf{A} \equiv [A] &= \begin{bmatrix} a_{11} & a_{12} & a_{13} & a_{14} \\ a_{21} & a_{22} & a_{23} & a_{24} \\ a_{31} & a_{32} & a_{33} & a_{34} \end{bmatrix} \\ \mathbf{x} \equiv \{x\} &= \begin{Bmatrix} x_1 \\ x_2 \\ x_3 \\ x_4 \end{Bmatrix} \\ \mathbf{b} \equiv \{b\} &= \begin{Bmatrix} b_1 \\ b_2 \\ b_3 \end{Bmatrix} \end{aligned} \tag{A.4}$$

The above notation contains within it the definition of both a matrix and the process of multiplication of two matrices. Matrices are *defined* as 'arrays of number' of the type shown in Eq. (A.4). The particular form listing a single column of numbers is often referred to as a vector or column matrix, whereas a matrix with multiple columns and rows is called a rectangular matrix. The multiplication of a matrix by a column vector is *defined* by the equivalence of the left and right sides of Eqs (A.1) and (A.2).

The use of bold characters to define both vectors and matrices will be followed throughout the text – generally lower case letters denoting vectors and capital letters matrices.

If another relationship, using the same a constants, but a different set of x and b, exists and is written as

$$\begin{aligned} a_{11}x_1' + a_{12}x_2' + a_{13}x_3' + a_{14}x_4' &= b_1' \\ a_{21}x_1' + a_{22}x_2' + a_{23}x_3' + a_{24}x_4' &= b_2' \\ a_{31}x_1' + a_{32}x_2' + a_{33}x_3' + a_{34}x_4' &= b_3' \end{aligned} \tag{A.5}$$

then we could write

$$[A]\,[X] = [B] \quad \text{or} \quad \mathbf{AX} = \mathbf{B} \tag{A.6}$$

in which

$$\mathbf{X} \equiv [X] = \begin{bmatrix} x_1, & x_1' \\ x_2, & x_2' \\ x_3, & x_3' \\ x_4, & x_4' \end{bmatrix} \qquad \mathbf{B} \equiv [B] = \begin{bmatrix} b_1, & b_1' \\ b_2, & b_2' \\ b_3, & b_3' \end{bmatrix} \tag{A.7}$$

implying both the statements (A.1) and (A.5) arranged simultaneously as

$$\begin{bmatrix} a_{11}x_1 + \cdots, & a_{11}x_1' + \cdots \\ a_{21}x_1 + \cdots, & a_{21}x_1' + \cdots \\ a_{31}x_1 + \cdots, & a_{31}x_1' + \cdots \end{bmatrix} = \mathbf{B} \equiv [B] = \begin{bmatrix} b_1, & b_1' \\ b_2, & b_2' \\ b_3, & b_3' \end{bmatrix} \tag{A.8}$$

It is seen, incidentally, that matrices can be equal only if each of the individual terms is equal.

The multiplication of full matrices is defined above, and it is obvious that it has a meaning only if the number of columns in **A** is equal to the number of rows in **X** for a relation of the type (A.6). One property that distinguishes matrix multiplication is that, in general,

$$\mathbf{AX} \neq \mathbf{XA}$$

i.e., multiplication of matrices is not commutative as in ordinary algebra.

Matrix addition or subtraction

If relations of the form from (A.1) and (A.5) are added then we have

$$\begin{aligned} a_{11}(x_1 + x_1') + a_{12}(x_2 + x_2') + a_{13}(x_3 + x_3') + a_{14}(x_4 + x_4') &= b_1 + b_1' \\ a_{21}(x_1 + x_1') + a_{22}(x_2 + x_2') + a_{23}(x_3 + x_3') + a_{24}(x_4 + x_4') &= b_2 + b_2' \\ a_{31}(x_1 + x_1') + a_{32}(x_2 + x_2') + a_{33}(x_3 + x_3') + a_{34}(x_4 + x_4') &= b_3 + b_3' \end{aligned} \tag{A.9}$$

which will also follow from

$$\mathbf{Ax} + \mathbf{Ax}' = \mathbf{b} + \mathbf{b}'$$

if we define the addition of matrices by a simple addition of the individual terms of the array. Clearly this can be done only if the size of the matrices is identical, i.e., for example,

$$\begin{bmatrix} a_{11} & a_{12} \\ a_{21} & a_{22} \\ a_{31} & a_{32} \end{bmatrix} + \begin{bmatrix} b_{11} & b_{12} \\ b_{21} & b_{22} \\ b_{31} & b_{32} \end{bmatrix} = \begin{bmatrix} a_{11}+b_{11} & a_{12}+b_{12} \\ a_{21}+b_{21} & a_{22}+b_{22} \\ a_{31}+b_{31} & a_{32}+b_{32} \end{bmatrix}$$

or

$$\mathbf{A} + \mathbf{B} = \mathbf{C} \tag{A.10}$$

implies that every term of **C** is equal to the sum of the appropriate terms of **A** and **B**.

Subtraction obviously follows similar rules.

Transpose of a matrix

This is simply a definition for reordering the terms in an array in the following manner:

$$\begin{bmatrix} a_{11} & a_{12} & a_{13} \\ a_{21} & a_{22} & a_{23} \end{bmatrix}^{\mathrm{T}} = \begin{bmatrix} a_{11} & a_{21} \\ a_{12} & a_{22} \\ a_{13} & a_{23} \end{bmatrix} \tag{A.11}$$

and will be indicated by the symbol T as shown.

Its use is not immediately obvious but will be indicated later and can be treated here as a simple prescribed operation.

Inverse of a matrix

If in the relationship (A.3) the matrix **A** is 'square', i.e., it represents the coefficients of simultaneous equations of type (A.1) equal in number to the number of unknowns **x**, then in general it is possible to solve for the unknowns in terms of the known coefficients **b**. This solution can be written as

$$\mathbf{x} = \mathbf{A}^{-1}\mathbf{b} \tag{A.12}$$

in which the matrix $\mathbf{A}^{-1}$ is known as the 'inverse' of the square matrix **A**. Clearly $\mathbf{A}^{-1}$ is also square and of the same size as **A**.

We could obtain (A.12) by multiplying both sides of (A.3) by $\mathbf{A}^{-1}$ and hence

$$\mathbf{A}^{-1}\mathbf{A} = \mathbf{I} = \mathbf{A}\mathbf{A}^{-1} \tag{A.13}$$

where **I** is an 'identity' matrix having zero on all off-diagonal positions and unity on each of the diagonal positions.

If the equations are 'singular' and have no solution then clearly an inverse does not exist.

A sum of products

In problems of mechanics we often encounter a number of quantities such as force that can be listed as a matrix 'vector':

$$\mathbf{f} = \begin{Bmatrix} f_1 \\ f_2 \\ \vdots \\ f_n \end{Bmatrix} \tag{A.14}$$

These, in turn, are often associated with the same number of displacements given by another vector, say,

$$\mathbf{u} = \begin{Bmatrix} u_1 \\ u_2 \\ \vdots \\ u_n \end{Bmatrix} \tag{A.15}$$

It is known that the work is represented as a sum of products of force and displacement

$$W = \sum_{k=1}^{n} f_k \, u_k$$

Clearly the transpose becomes useful here as we can write, by the rule of matrix multiplication,

$$W = [f_1 \quad f_2 \ldots f_n] \begin{Bmatrix} u_1 \\ u_2 \\ \vdots \\ u_n \end{Bmatrix} = \mathbf{f}^{\mathrm{T}}\mathbf{u} = \mathbf{u}^{\mathrm{T}}\mathbf{f} \tag{A.16}$$

Use of this fact is made frequently in this book.

Transpose of a product

An operation that sometimes occurs is that of taking the transpose of a matrix product. It can be left to the reader to prove from previous definitions that

$$(\mathbf{A}\,\mathbf{B})^{\mathrm{T}} = \mathbf{B}^{\mathrm{T}}\mathbf{A}^{\mathrm{T}} \tag{A.17}$$

Symmetric matrices

In structural problems symmetric matrices are often encountered. If a term of a matrix $\mathbf{A}$ is defined as a_{ij}, then for a symmetric matrix

$$a_{ij} = a_{ji} \quad \text{or} \quad \mathbf{A} = \mathbf{A}^{\mathrm{T}}$$

A symmetric matrix must be square. It can be shown that the inverse of a symmetric matrix is also symmetric

$$\mathbf{A}^{-1} = \left(\mathbf{A}^{-1}\right)^{\mathrm{T}} \equiv \mathbf{A}^{-\mathrm{T}}$$

Partitioning

It is easy to verify that a matrix product **AB** in which, for example,

$$\mathbf{A} = \left[\begin{array}{ccc|cc} a_{11} & a_{12} & a_{13} & a_{14} & a_{15} \\ a_{21} & a_{22} & a_{23} & a_{24} & a_{25} \\ \hline a_{31} & a_{32} & a_{33} & a_{34} & a_{35} \end{array}\right]$$

$$\mathbf{B} = \left[\begin{array}{cc} b_{11} & b_{12} \\ b_{21} & b_{22} \\ b_{31} & b_{32} \\ \hline b_{41} & b_{42} \\ b_{51} & b_{52} \end{array}\right]$$

could be obtained by dividing each matrix into submatrices, indicated by the lines, and applying the rules of matrix multiplication first to each of such submatrices as if it were a scalar number and then carrying out further multiplication in the usual way. Thus, if we write

$$\mathbf{A} = \begin{bmatrix} \mathbf{A}_{11} & \mathbf{A}_{12} \\ \mathbf{A}_{21} & \mathbf{A}_{22} \end{bmatrix} \quad \mathbf{B} = \begin{bmatrix} \mathbf{B}_1 \\ \mathbf{B}_2 \end{bmatrix}$$

then

$$\mathbf{AB} = \begin{bmatrix} \mathbf{A}_{11}\mathbf{B}_1 & \mathbf{A}_{12}\mathbf{B}_2 \\ \mathbf{A}_{21}\mathbf{B}_1 & \mathbf{A}_{22}\mathbf{B}_2 \end{bmatrix}$$

can be verified as representing the complete product by further multiplication.

The essential feature of partitioning is that the size of subdivisions has to be such as to make the products of the type $\mathbf{A}_{11}\mathbf{B}_1$ meaningful, i.e., the number of columns in $\mathbf{A}_{11}$ must be equal to the number of rows in $\mathbf{B}_1$, etc. If the above definition holds, then all further operations can be conducted on partitioned matrices, treating each partition as if it were a scalar.

It should be noted that any matrix can be multiplied by a scalar (number). Here, obviously, the requirements of equality of appropriate rows and columns no longer apply.

If a symmetric matrix is divided into an equal number of submatrices $\mathbf{A}_{ij}$ in rows and columns then

$$\mathbf{A}_{ij} = \mathbf{A}_{ji}^{\mathrm{T}}$$

The eigenvalue problem

An *eigenvalue* of a symmetric matrix **A** of size $n \times n$ is a scalar λ_i which allows the solution of

$$(\mathbf{A} - \lambda_i \mathbf{I})\phi_i = \mathbf{0} \text{ and } \det \mid \mathbf{A} - \lambda_i \mathbf{I} \mid = 0 \qquad \text{(A.18)}$$

where ϕ_i is called the *eigenvector*.

There are, of course, n such eigenvalues λ_i to each of which corresponds an eigenvector ϕ_i. Such vectors can be shown to be orthonormal and we write

$$\phi_i^{\mathrm{T}} \phi_j = \delta_{ij} = \begin{cases} 1 \text{ for } i = j \\ 0 \text{ for } i \neq j \end{cases}$$

The full set of eigenvalues and eigenvectors can be written as

$$\mathbf{\Lambda} = \begin{bmatrix} \lambda_1 & & 0 \\ & \ddots & \\ 0 & & \lambda_n \end{bmatrix} \qquad \mathbf{\Phi} = \begin{bmatrix} \phi_1, & \ldots & \phi_n \end{bmatrix}$$

Using these the matrix $\mathbf{A}$ may be written in its *spectral form* by noting from the orthonormality conditions on the eigenvectors that

$$\mathbf{\Phi}^{-1} = \mathbf{\Phi}^{\mathrm{T}}$$

then from

$$\mathbf{A\Phi} = \mathbf{\Phi\Lambda}$$

it follows immediately that

$$\mathbf{A} = \mathbf{\Phi\Lambda\Phi}^{\mathrm{T}} \tag{A.19}$$

The condition number κ (which is related to equation solution round-off) is defined as

$$\kappa = \frac{|\lambda_{\max}|}{|\lambda_{\min}|} \tag{A.20}$$

Appendix B

Tensor-indicial notation in the approximation of elasticity problems

Introduction

The matrix type of notation used in this volume for description of tensor quantities such as stresses and strains is compact and we believe easy to understand. However, in a computer program each quantity often will still have to be identified by appropriate indices and the conciseness of matrix notation does not always carry over to the programming steps. Further, many readers are accustomed to the use of indicial-tensor notation which is a standard tool in the study of solid mechanics. For this reason we summarize here the formulation of the finite element arrays in an indicial form.

Some advantages of such reformulation from the matrix setting become apparent when evaluation of stiffness arrays for isotropic materials is considered. Here some multiplication operations previously necessary become redundant and the element module programs can be written more economically.

When finite deformation problems in solid mechanics have to be considered the use of indicial notation is almost essential to form many of the arrays needed for the residual and tangent terms.

This appendix adds little new to the discretization ideas – it merely repeats in a different language the results already presented.

Indicial notation: summation convention

A point P in three-dimensional space may be represented in terms of its cartesian coordinates x_i, $i = 1, 2, 3$. The limits that i can take define its *range*. To define these components we must first establish an oriented orthogonal set of coordinate directions as shown in Fig. B.1. The distance from the origin of the coordinate axes to the point define a position vector $\mathbf{x}$. If along each of the coordinate axes we define the set of unit orthonormal base vectors, $\mathbf{i}_i$, $i = 1, 2, 3$ which have the property

$$\mathbf{i}_i \cdot \mathbf{i}_j = \delta_{ij} = \begin{cases} 1 \text{ for } i = j \\ 0 \text{ for } i \neq j \end{cases} \tag{B.1}$$

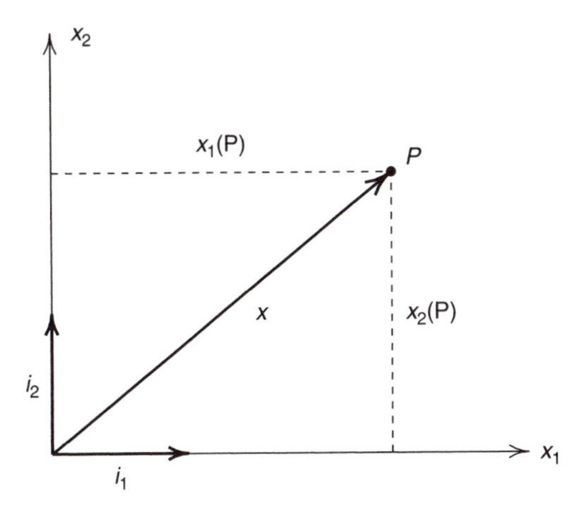

Fig. B.1 Orthogonal axes and a point: Cartesian coordinates.

where $(\,)\cdot(\,)$ denotes the vector dot product. The components of the position vector are constructed from the vector dot product

$$x_i = \mathbf{i}_i \cdot \mathbf{x}; \quad i = 1, 2, 3 \tag{B.2}$$

From this construction it is easy to observe that the vector $\mathbf{x}$ may be represented as

$$\mathbf{x} = \sum_{i=1}^{3} x_i\, \mathbf{i}_i \tag{B.3}$$

In dealing with vectors, and later tensors, the form $\mathbf{x}$ is called the *intrinsic* notation of the coordinates and $x_i\, \mathbf{i}_i$ the *indicial form*.† An intrinsic form is a physical entity which is independent of the coordinate system selected, whereas an indicial form depends on a particular coordinate system.

To simplify notation we adopt the common convention that any index which is repeated in any given term implies a summation over the range of the index. Thus, our short-hand notation for Eq. (B.3) is

$$\mathbf{x} = x_i\, \mathbf{i}_i = x_1\, \mathbf{i}_1 + x_2\, \mathbf{i}_2 + x_3\, \mathbf{i}_3 \tag{B.4}$$

For two-dimensional problems, unless otherwise stated, it will be understood that the range of the index is two.

Similarly, we can define the components of the displacement vector $\mathbf{u}$ as

$$\mathbf{u} = u_i\, \mathbf{i}_i \tag{B.5}$$

Note that the components $(u_1,\ u_2,\ u_3)$ replace the components $(u,\ v,\ w)$ used throughout most of this volume.

To avoid confusion with nodal quantities to which we previously also attached subscripts we shall simply change their position to a superscript. Thus, $\tilde{u}_2^a$ has the same meaning as $\tilde{v}_a$ used previously, etc.

† Often in an indicial form of equations the base vectors are omitted from the final equation.

Derivatives and tensorial relations

In indicial notation the derivative of any quantity with respect to a coordinate component x_i is written compactly as

$$\frac{\partial}{\partial x_i} \equiv (\,)_{,i} \tag{B.6}$$

Thus we can write the *gradient* of the displacement vector as

$$\frac{\partial u_i}{\partial x_j} \equiv u_{i,j}; \quad i, j = 1, 2, 3 \tag{B.7}$$

In a cartesian coordinate system the base vectors do not change their magnitude or direction along any coordinate direction. Accordingly their derivatives with respect to any coordinate is zero as indicated in Eq. B.8

$$\frac{\partial \mathbf{i}_i}{\partial x_j} = \mathbf{i}_{i,j} = \mathbf{0} \tag{B.8}$$

Thus, in cartesian coordinates the derivative of the intrinsic displacement $\mathbf{u}$ is given by

$$\mathbf{u}_{,j} = u_{i,j}\mathbf{i}_i + u_i\mathbf{i}_{i,j} = u_{i,j}\mathbf{i}_i \tag{B.9}$$

The collection of all the derivatives defines the *displacement gradient* which we write in intrinsic notation as

$$\boldsymbol{\nabla}\mathbf{u} = u_{i,j}\,\mathbf{i}_i \otimes \mathbf{i}_j \tag{B.10}$$

The symbol $\otimes$ denotes the *tensor product* between two base vectors and since only two vectors are involved the gradient of the displacement is called *second rank*. The notation used to define a tensor product follows that used in reference 1.

Any second rank intrinsic quantity can be split into symmetric and a skew symmetric (anti-symmetric) parts as

$$\mathbf{A} = \frac{1}{2}\left[\mathbf{A} + \mathbf{A}^{\mathrm{T}}\right] + \frac{1}{2}\left[\mathbf{A} - \mathbf{A}^{\mathrm{T}}\right] = \mathbf{A}^{(s)} + \mathbf{A}^{(a)} \tag{B.11}$$

where $\mathbf{A}$ and its transpose have cartesian components

$$\mathbf{A} = A_{ij}\,\mathbf{i}_i \otimes \mathbf{i}_j; \quad \mathbf{A}^{\mathrm{T}} = A_{ji}\,\mathbf{i}_i \otimes \mathbf{i}_j \tag{B.12}$$

The symmetric part of the displacement gradient defines the (small) strain†

$$\begin{aligned}\varepsilon = \boldsymbol{\nabla}\mathbf{u}^{(s)} &= \frac{1}{2}\left[\boldsymbol{\nabla}\mathbf{u} + (\boldsymbol{\nabla}\mathbf{u})^{\mathrm{T}}\right]\\ &= \frac{1}{2}\left[u_{i,j} + u_{j,i}\right]\,\mathbf{i}_i \otimes \mathbf{i}_j\\ &= \varepsilon_{ij}\,\mathbf{i}_i \otimes \mathbf{i}_j = \varepsilon_{ji}\,\mathbf{i}_i \otimes \mathbf{i}_j\end{aligned} \tag{B.13}$$

† Note that this definition is slightly different from that occurring in Chapters 2 to 6. Now the shearing strain is given by $\varepsilon_{ij} = 1/2\,\gamma_{ij}$ when $i \neq j$.

and the skew symmetric part gives the (small) rotation

$$\begin{aligned}\boldsymbol{\omega} &= \nabla \mathbf{u}^{(a)} = \frac{1}{2}\left[\nabla \mathbf{u} - (\nabla \mathbf{u})^{\mathrm{T}}\right] \\ &= \frac{1}{2}\left[u_{i,j} - u_{j,i}\right] \mathbf{i}_i \otimes \mathbf{i}_j \\ &= \omega_{ij}\, \mathbf{i}_i \otimes \mathbf{i}_j = -\,\omega_{ji}\, \mathbf{i}_i \otimes \mathbf{i}_j\end{aligned} \tag{B.14}$$

The strain expression is analogous to Eq. (2.13). The components ε_{ij} and ω_{ij} may be represented by a matrix as

$$\varepsilon_{ij} = \begin{bmatrix} \varepsilon_{11} & \varepsilon_{12} & \varepsilon_{13} \\ \varepsilon_{21} & \varepsilon_{22} & \varepsilon_{23} \\ \varepsilon_{31} & \varepsilon_{32} & \varepsilon_{33} \end{bmatrix} = \begin{bmatrix} \varepsilon_{11} & \varepsilon_{12} & \varepsilon_{13} \\ \varepsilon_{12} & \varepsilon_{22} & \varepsilon_{23} \\ \varepsilon_{13} & \varepsilon_{23} & \varepsilon_{33} \end{bmatrix} \tag{B.15}$$

$$\omega_{ij} = \begin{bmatrix} 0 & \omega_{12} & \omega_{13} \\ \omega_{21} & 0 & \omega_{23} \\ \omega_{31} & \omega_{32} & 0 \end{bmatrix} = \begin{bmatrix} 0 & \omega_{12} & \omega_{13} \\ -\omega_{12} & 0 & \omega_{23} \\ -\omega_{13} & -\omega_{23} & 0 \end{bmatrix} \tag{B.16}$$

Coordinate transformation

Consider now the representation of the intrinsic coordinates in a system which has different orientation than that given in Fig. B.1. We represent the components in the new system by

$$\mathbf{x} = x'_{i'}\, \mathbf{i}'_{i'} \tag{B.17}$$

Using Eq. (B.2) we can relate the components in the prime system to those in the original system as

$$x'_{i'} = \mathbf{i}'_i \cdot \mathbf{x} = \mathbf{i}'_i \cdot \mathbf{i}_j x_j = \Lambda_{i'j} x_j \tag{B.18}$$

where

$$\Lambda_{i'j} = \mathbf{i}'_{i'} \cdot \mathbf{i}_j = \cos(x'_{i'}, x_j) \tag{B.19}$$

define the direction cosines of the coordinate in a manner similar to that of Eq. (6.18).

Equation (B.18) defines how the cartesian coordinate components transform from one coordinate frame to another. Recall that summation convention implies

$$x'_{i'} = \Lambda_{i'1} x_1 + \Lambda_{i'2} x_2 + \Lambda_{i'3} x_3 \quad i' = 1, 2, 3 \tag{B.20}$$

In Eq. (B.18) i' is called a *free index* whereas j is called a *dummy index* since it may be replaced by any other unique index without changing the meaning of the term (note that the notation used does not permit an index to appear more than twice in any term). Summation convention will be employed throughout the remainder of this discussion and the reader should ensure that the concept is fully understood before proceeding. Some examples will be given occasionally to illustrate its use.

Using the notion of the direction cosines, Eq. (B.18) may be used to transform any vector with three components. Thus, transformation of the components of the displacement vector is given by

$$u'_{i'} = \Lambda_{i'j} u_j \quad i', j = 1, 2, 3 \tag{B.21}$$

Indeed we can also use the above to express the transformation for the base vectors since

$$\mathbf{i}'_{i'} = \left(\mathbf{i}'_{i'} \cdot \mathbf{i}_j\right) \mathbf{i}_j = \Lambda_{i'j}\mathbf{i}_j \tag{B.22}$$

Similarly, by interchanging the role of the base vectors we obtain

$$\mathbf{i}_j = \left(\mathbf{i}_j \cdot \mathbf{i}'_{i'}\right) \mathbf{i}'_{i'} = \Lambda_{i'j}\mathbf{i}'_{i'} \tag{B.23}$$

which indicates that the *inverse* of the direction cosine coefficient array is the same as its *transpose*.

The strain transformation follows from the intrinsic form written as

$$\boldsymbol{\varepsilon} = \varepsilon'_{i'j'}\mathbf{i}_{i'} \otimes \mathbf{i}_{j'} = \varepsilon_{kl}\mathbf{i}_k \otimes \mathbf{i}_l \tag{B.24}$$

Substitution of the base vectors from Eq. (B.23) into Eq. (B.24) gives

$$\boldsymbol{\varepsilon} = \Lambda_{i'k}\varepsilon_{kl}\Lambda_{j'l}\mathbf{i}_{i'} \otimes \mathbf{i}_{j'} \tag{B.25}$$

Comparing Eq. (B.25) with Eq. (B.24) the components of the strain transform according to the relation

$$\varepsilon'_{i'j'} = \Lambda_{i'k}\varepsilon_{kl}\Lambda_{j'l} \tag{B.26}$$

Variables that transform according to Eq. (B.21) are called *first rank cartesian tensors* whereas quantities that transform according to Eq. (B.26) are called *second rank cartesian tensors*. The use of indicial notation in the context of cartesian coordinates will lead naturally to each mechanics variable being defined in terms of a cartesian tensor of an appropriate rank.

Stress may be written in terms of its components σ_{ij} which may be written in a matrix form similar to Eq. (B.15)

$$\sigma_{ij} = \begin{bmatrix} \sigma_{11} & \sigma_{12} & \sigma_{13} \\ \sigma_{21} & \sigma_{22} & \sigma_{23} \\ \sigma_{31} & \sigma_{32} & \sigma_{33} \end{bmatrix} ; \quad i, j = 1, 2, 3 \tag{B.27}$$

In intrinsic form stress is given by

$$\boldsymbol{\sigma} = \sigma_{ij}\mathbf{i}_i \otimes \mathbf{i}_j \tag{B.28}$$

and, using similar logic as used for strain, can be shown to transform as a second rank cartesian tensor. The symmetry of the components of stress may be established by summing moments (angular momentum balance) about each of the coordinate axes to obtain

$$\sigma_{ij} = \sigma_{ji} \tag{B.29}$$

Equilibrium and energy

Introducing a body force vector

$$\mathbf{b} = b_i\mathbf{i}_i \tag{B.30}$$

we can write the static equilibrium equations (linear momentum balance) for a differential element as

$$\operatorname{div}\boldsymbol{\sigma} + \mathbf{b} \equiv \left(\sigma_{ji,j} + b_i\right)\mathbf{i}_i = \mathbf{0} \tag{B.31}$$

where the repeated index again implies summation over the range of the index, i.e.,

$$\sigma_{ji,j} \equiv \sum_{j=1}^{3} \sigma_{ji,j} = \sigma_{1i,1} + \sigma_{2i,2} + \sigma_{3i,3}$$

Note that the free index i must appear in each term for the equation to be meaningful.

As a further example of the summation convention consider an internal energy term

$$W = \sigma_{ij}\varepsilon_{ij} \tag{B.32}$$

This expression implies a double summation; hence summing first on i gives

$$W = \sigma_{1j}\varepsilon_{1j} + \sigma_{2j}\varepsilon_{2j} + \sigma_{3j}\varepsilon_{3j}$$

and then summing on j gives finally

$$\begin{aligned} W = \sigma_{11}\,\varepsilon_{11} + \sigma_{12}\,\varepsilon_{12} + \sigma_{13}\,\varepsilon_{13} \\ + \sigma_{21}\,\varepsilon_{21} + \sigma_{22}\,\varepsilon_{22} + \sigma_{23}\,\varepsilon_{23} \\ + \sigma_{31}\,\varepsilon_{31} + \sigma_{32}\,\varepsilon_{32} + \sigma_{33}\,\varepsilon_{33} \end{aligned}$$

We may use symmetry conditions on σ_{ij} and ε_{ij} to reduce the nine terms to six terms. Accordingly,

$$\begin{aligned} W &= \sigma_{11}\varepsilon_{11} + \sigma_{22}\varepsilon_{22} + \sigma_{33}\varepsilon_{33} + 2(\sigma_{12}\varepsilon_{12} + \sigma_{23}\varepsilon_{23} + \sigma_{31}\varepsilon_{31}) \\ &= \sigma_{11}\varepsilon_{11} + \sigma_{22}\varepsilon_{22} + \sigma_{33}\varepsilon_{33} + \sigma_{12}\gamma_{12} + \sigma_{23}\gamma_{23} + \sigma_{31}\gamma_{31} \end{aligned} \tag{B.33}$$

Following a similar expansion we can also show the result

$$\sigma_{ij}\omega_{ij} \equiv 0 \tag{B.34}$$

Elastic constitutive equations

For an elastic material the most general linear relationship we can write for components of the stress–strain characterization is

$$\sigma_{ij} = D_{ijkl}\left(\varepsilon_{kl} - \varepsilon^0_{kl}\right) + \sigma^0_{ij} \tag{B.35}$$

Equation (B.35) is the equivalent of Eq. (2.16) but now written in indicial notation. We note that the elastic moduli which appear in Eq. (B.35) are components of the fourth rank tensor

$$\mathbf{D} = D_{ijkl}\mathbf{i}_i \otimes \mathbf{i}_j \otimes \mathbf{i}_k \otimes \mathbf{i}_l \tag{B.36}$$

The elastic moduli possess the following symmetry conditions

$$D_{ijkl} = D_{jikl} = D_{ijlk} = D_{klij} \tag{B.37}$$

the latter arising from the existence of an internal energy density in the form[2]

$$W(\varepsilon) = \frac{1}{2}\varepsilon_{ij} D_{ijkl} \varepsilon_{kl} + \varepsilon_{ij} \left[\sigma_{ij}^0 - D_{ijkl} \varepsilon_{kl}^0\right] \tag{B.38}$$

which yields the stress from

$$\sigma_{ij} = \frac{\partial W}{\partial \varepsilon_{ij}} \tag{B.39}$$

By writing the constitutive equation with respect to $x'_{i'}$ and using properties of the base vectors we can deduce the transformation equation for moduli as

$$D'_{i'j'k'l'} = \Lambda_{i'm} \Lambda_{j'n} \Lambda_{k'p} \Lambda_{l'q} D_{mnpq} \tag{B.40}$$

A common notation for the intrinsic form of Eq. (B.35) is

$$\boldsymbol{\sigma} = \mathbf{D} : \left(\varepsilon - \varepsilon^0\right) + \boldsymbol{\sigma}^0 \tag{B.41}$$

in which : denotes the double summation (contraction) between the elastic moduli and the strains.

The elastic moduli for an isotropic elastic material may be written in indicial form as

$$D_{ijkl} = \lambda \delta_{ij} \delta_{kl} + \mu(\delta_{ik}\delta_{jl} + \delta_{il}\delta_{jk}) \tag{B.42}$$

where λ, μ are the Lamé constants. An isotropic linear elastic material is always characterized by two independent elastic constants. Instead of the Lamé constants we can use Young's modulus, E, and Poisson's ratio, ν, to characterize the material. The Lamé constants may be deduced from

$$\mu = \frac{E}{2(1+\nu)} \quad \text{and} \quad \lambda = \frac{\nu E}{(1+\nu)(1-2\nu)} \tag{B.43}$$

Finite element approximation

If we now introduce the finite element displacement approximation given by Eq. (2.1), using indicial notation we may write for a single element

$$u_i \approx \hat{u}_i = N_a \tilde{u}_i^a \quad i = 1, 2, 3; \; a = 1, 2, \ldots, n \tag{B.44}$$

where n is the total number of nodes on an element. The strain approximation in each element is given by the definition of Eq. (B.13) as

$$\hat{\varepsilon}_{ij} = \frac{1}{2}\left[N_{a,j}\tilde{u}_i^a + N_{a,i}\tilde{u}_j^a\right] \tag{B.45}$$

The internal virtual work for an element is given as

$$\delta U^I = \int_{\Omega_e} \delta\varepsilon : \boldsymbol{\sigma} \, \mathrm{d}\Omega = \int_{\Omega_e} \delta\varepsilon_{ij}\sigma_{ij} \, \mathrm{d}\Omega \tag{B.46}$$

Using Eqs (B.45) and (B.46) and noting symmetries in D_{ijkl} we may write the internal virtual work for a linear elastic material as

$$\begin{aligned}\delta U^I &= \delta\tilde{u}_i^a \int_{\Omega_e} N_{a,j} D_{ijkl} N_{b,l} \, \mathrm{d}\Omega \, \tilde{u}_k^b \\ &+ \delta\tilde{u}_i^a \int_{\Omega_e} N_{a,j} \big(\sigma_{ij}^0 - D_{ijkl}\varepsilon_{kl}^0\big) \, \mathrm{d}\Omega\end{aligned} \tag{B.47}$$

which replaces in indicial notation the matrix form presented in Chapters 2 to 6.

In describing a stiffness coefficient two subscripts have been used previously and the submatrix $\mathbf{K}_{ab}$ implied 2×2 or 3×3 entries for the ab nodal pair, depending on whether two- or three-dimension displacement components were involved. Now the scalar components

$$K_{ij}^{ab} \quad i, j = 1, 2, 3; \; a, b = 1, 2, \ldots, n \tag{B.48}$$

define completely the appropriate stiffness coefficient with ij indicating the relative submatrix position (in this case for a three-dimensional displacement).

Note that for a symmetric matrix we have previously required that

$$\mathbf{K}_{ab} = \mathbf{K}_{ba}^{\mathrm{T}} \tag{B.49}$$

In indicial notation the same symmetry is implied if

$$K_{ij}^{ab} = K_{ji}^{ba} \tag{B.50}$$

The stiffness tensor is now defined from Eq. (B.47) as

$$K_{ik}^{ab} = \int_{\Omega_e} N_{a,j} D_{ijkl} N_{b,l} \, \mathrm{d}\Omega \tag{B.51}$$

When the elastic properties are constant over the element we may separate the integration from the material constants by defining

$$W_{ij}^{ab} = \int_{\Omega_e} N_{a,i} N_{b,j} \, \mathrm{d}\Omega \tag{B.52}$$

and then perform the summations with the material moduli as

$$K_{ik}^{ab} = W_{jl}^{ab} D_{ijkl} \tag{B.53}$$

In the case of isotropy a particularly simple result is obtained

$$K_{ik}^{ab} = \lambda W_{ik}^{ab} + \mu [W_{ki}^{ab} + \delta_{ik} W_{jj}^{ab}] \tag{B.54}$$

which allows the construction of the stiffness to be carried out using fewer arithmetic operations as compared with the use of matrix form.[3]

Using indicial notation the final equilibrium equations of the system are written as

$$K_{ik}^{ab} \tilde{u}_k^b + f_i^a = 0 \quad i = 1, 2, 3 \tag{B.55}$$

and in this scalar form every coefficient is simply identified. The reader can, as a simple exercise, complete the derivation of the force terms due to the initial strain ε_{ij}^0, stress σ_{ij}^0, body force b_i and external traction $\bar{t}_i$.

Indicial notation is at times useful in clarifying individual terms, and this introduction should be helpful as a key to reading some of the current literature.

Table B.1 Mapping between matrix and tensor indices for second rank symmetric tensors

Form	Index number					
Matrix	1	2	3	4	5	6
Tensor	11 xx	22 yy	33 zz	12 & 21 xy & yx	23 & 32 yz & zy	31 & 13 zx & xz

Relation between indicial and matrix notation

The matrix form used throughout most of this volume can be deduced from the indicial form by a simple transformation between the indices. The relationship between the indices of the second rank tensors and their corresponding matrix form can be performed by an inspection of the ordering in the matrix for stress and its representation shown in Eq. (B.27). In the matrix form the stress was given in Chapter 6 as

$$\boldsymbol{\sigma} = \begin{bmatrix} \sigma_{11} & \sigma_{22} & \sigma_{33} & \sigma_{12} & \sigma_{23} & \sigma_{31} \end{bmatrix}^{\mathrm{T}} \tag{B.56}$$

This form includes use of symmetry of stress components. The mapping of the indices follows that shown in Table B.1.

Table B.1 may also be used to perform the map of the material moduli by noting that the components in the energy are associated with the index pairs from the stress and the strain. Accordingly, the moduli transform as

$$D_{1111} \to D_{11}; \quad D_{2233} \to D_{23}; \quad D_{1231} \to D_{46}; \quad \text{etc.} \tag{B.57}$$

The symmetry of the stress and strain is imbedded in Table B.1 and existence of an energy function yields symmetry of the modulus matrix, i.e., $D_{ab} = D_{ba}$.

References

1. P. Chadwick. *Continuum Mechanics*. John Wiley & Sons, New York, 1976.
2. I.S. Sokolnikoff. *The Mathematical Theory of Elasticity*. McGraw-Hill, New York, 2nd edition, 1956.
3. A.K. Gupta and B. Mohraz. A method of computing numerically integrated stiffness matrices. *Int. J. Numer. Meth. Eng.*, 5:83–89, 1972.

Appendix C

Solution of simultaneous linear algebraic equations

A finite element problem leads to a large set of simultaneous linear algebraic equations whose solution provides the nodal and element parameters in the formulation. For example, in the analysis of linear steady-state problems the direct assembly of the element coefficient matrices and load vectors leads to a set of linear algebraic equations. In this section methods to solve the simultaneous algebraic equations are summarized. We consider both *direct* methods where an *a priori* calculation of the number of numerical operations can be made, and *indirect or iterative* methods where no such estimate can be made.

Direct solution

Consider first the general problem of direct solution of a set of algebraic equations given by

$$\mathbf{K}\tilde{\mathbf{u}} = \mathbf{f} \tag{C.1}$$

where $\mathbf{K}$ is a square coefficient matrix, $\tilde{\mathbf{u}}$ is a vector of unknown parameters and $\mathbf{f}$ is a vector of known values. The reader can associate these with the quantities described previously: namely, the stiffness matrix, the nodal unknowns, and the specified forces or residuals.

In the discussion to follow it is assumed that the coefficient matrix has properties such that row and/or column interchanges are unnecessary to achieve an accurate solution. This is true in cases where $\mathbf{K}$ is symmetric positive (or negative) definite.† Pivoting may or may not be required with unsymmetric, or indefinite, conditions which can occur when the finite element formulation is based on some weighted residual methods. In these cases some checks or modifications may be necessary to ensure that the equations can be solved accurately.[1–3]

For the moment consider that the coefficient matrix can be written as the product of a lower triangular matrix with unit diagonals and an upper triangular matrix. Accordingly,

$$\mathbf{K} = \mathbf{L}\mathbf{U} \tag{C.2}$$

† For mixed methods which lead to forms of the type given in Eq. (10.14) the solution is given in terms of a positive definite part for $\tilde{\mathbf{q}}$ followed by a negative definite part for $\tilde{\phi}$.

where

$$\mathbf{L} = \begin{bmatrix} 1 & 0 & \cdots & 0 \\ L_{21} & 1 & \cdots & 0 \\ \vdots & & \ddots & \vdots \\ L_{n1} & L_{n2} & \cdots & 1 \end{bmatrix} \tag{C.3}$$

and

$$\mathbf{U} = \begin{bmatrix} U_{11} & U_{12} & \cdots & U_{1n} \\ 0 & U_{22} & \cdots & U_{2n} \\ \vdots & & \ddots & \vdots \\ 0 & 0 & \cdots & U_{nn} \end{bmatrix} \tag{C.4}$$

This form is called a *triangular decomposition* of $\mathbf{K}$. The solution to the equations can now be obtained by solving the pair of equations

$$\mathbf{L}\mathbf{y} = \mathbf{f} \tag{C.5}$$

and

$$\mathbf{U}\tilde{\mathbf{u}} = \mathbf{y} \tag{C.6}$$

where $\mathbf{y}$ is introduced to facilitate the separation, e.g., see references 1–5 for additional details.

The reader can easily observe that the solution to these equations is trivial. In terms of the individual equations the solution is given by

$$\begin{aligned} y_1 &= f_1 \\ y_i &= f_i - \sum_{j=1}^{i-1} L_{ij} y_j \qquad i = 2, 3, \ldots, n \end{aligned} \tag{C.7}$$

and

$$\begin{aligned} \tilde{u}_n &= \frac{y_n}{U_{nn}} \\ \tilde{u}_i &= \frac{1}{U_{ii}} \left(y_i - \sum_{j=i+1}^{n} U_{ij} \tilde{u}_j \right) \qquad i = n-1, n-2, \ldots, 1 \end{aligned} \tag{C.8}$$

Equation (C.7) is commonly called *forward elimination* while Eq. (C.8) is called *back substitution*.

The problem remains to construct the triangular decomposition of the coefficient matrix. This step is accomplished using variations on gaussian elimination. In practice, the operations necessary for the triangular decomposition are performed directly in the coefficient array; however, to make the steps clear the basic steps are shown in Fig. C.1 using separate arrays. The decomposition is performed in the same way as that used in the subprogram DATRI contained in the *FEAPpv* program; thus, the reader can easily grasp the details of the subprograms included once the steps in Fig. C.1 are mastered. Additional details on this step may be found in references 3–5.

In DATRI a Crout form of gaussian elimination is used to successively reduce the original coefficient array to upper triangular form. The lower portion of the array is

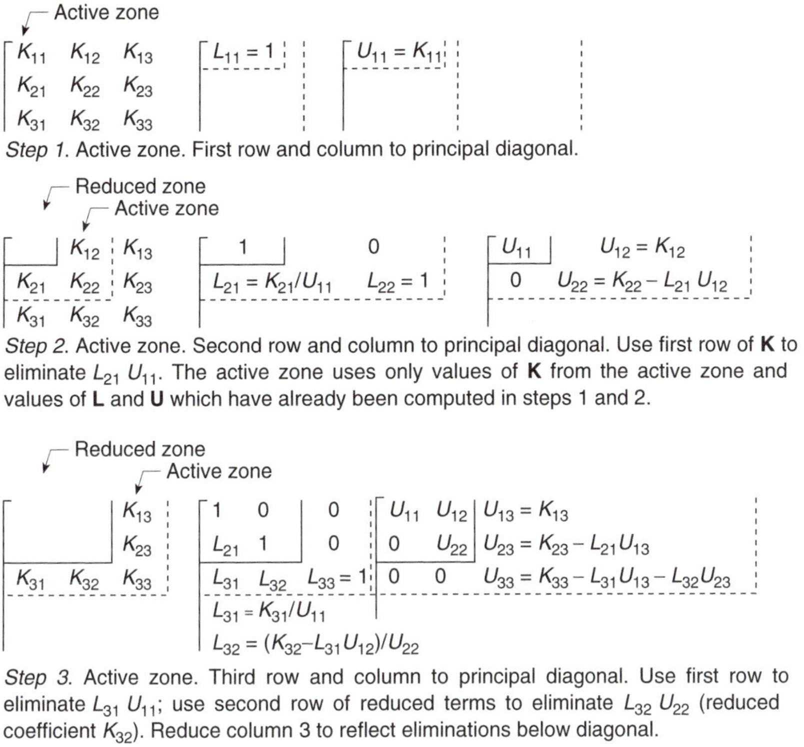

Active zone

$$\begin{bmatrix} K_{11} & K_{12} & K_{13} \\ K_{21} & K_{22} & K_{23} \\ K_{31} & K_{32} & K_{33} \end{bmatrix} \quad \begin{bmatrix} L_{11} = 1 & & \\ & & \\ & & \end{bmatrix} \quad \begin{bmatrix} U_{11} = K_{11} & & \\ & & \\ & & \end{bmatrix}$$

Step 1. Active zone. First row and column to principal diagonal.

Reduced zone
Active zone

$$\begin{bmatrix} & K_{12} & K_{13} \\ K_{21} & K_{22} & K_{23} \\ K_{31} & K_{32} & K_{33} \end{bmatrix} \quad \begin{bmatrix} 1 & 0 \\ L_{21} = K_{21}/U_{11} & L_{22} = 1 \\ & \end{bmatrix} \quad \begin{bmatrix} U_{11} & U_{12} = K_{12} \\ 0 & U_{22} = K_{22} - L_{21}\, U_{12} \\ & \end{bmatrix}$$

Step 2. Active zone. Second row and column to principal diagonal. Use first row of **K** to eliminate $L_{21}\, U_{11}$. The active zone uses only values of **K** from the active zone and values of **L** and **U** which have already been computed in steps 1 and 2.

Reduced zone
Active zone

$$\begin{bmatrix} & & K_{13} \\ & & K_{23} \\ K_{31} & K_{32} & K_{33} \end{bmatrix} \quad \begin{bmatrix} 1 & 0 & 0 \\ L_{21} & 1 & 0 \\ L_{31} & L_{32} & L_{33} = 1 \end{bmatrix} \quad \begin{bmatrix} U_{11} & U_{12} & U_{13} = K_{13} \\ 0 & U_{22} & U_{23} = K_{23} - L_{21}U_{13} \\ 0 & 0 & U_{33} = K_{33} - L_{31}U_{13} - L_{32}U_{23} \end{bmatrix}$$

$L_{31} = K_{31}/U_{11}$

$L_{32} = (K_{32} - L_{31}U_{12})/U_{22}$

Step 3. Active zone. Third row and column to principal diagonal. Use first row to eliminate $L_{31}\, U_{11}$; use second row of reduced terms to eliminate $L_{32}\, U_{22}$ (reduced coefficient K_{32}). Reduce column 3 to reflect eliminations below diagonal.

Fig. C.1 Triangular decomposition of **K**.

used to store **L** − **I** as shown in Fig. C.1. With this form, the unit diagonals for **L** are not stored.

Based on the organization of Fig. C.1 it is convenient to consider the coefficient array to be divided into three parts: part one being the region that is fully reduced; part two the region that is currently being reduced (called the active zone); and part three the region that contains the original unreduced coefficients. These regions are shown in Fig. C.2 where the jth column above the diagonal and the jth row to the left of the diagonal constitute the active zone. The algorithm for the triangular decomposition of an $n \times n$ square matrix can be deduced from Fig. C.1 and Fig. C.3 as follows:

$$U_{11} = K_{11}; \qquad L_{11} = 1 \tag{C.9}$$

For each active zone j from 2 to n,

$$L_{j1} = \frac{K_{j1}}{U_{11}}; \qquad U_{1j} = K_{1j} \tag{C.10}$$

$$\begin{aligned} L_{ji} &= \frac{1}{U_{ii}}\left(K_{ji} - \sum_{m=1}^{i-1} L_{jm}U_{mi}\right) \\ U_{ij} &= K_{ij} - \sum_{m=1}^{i-1} L_{im}U_{mj} \qquad i = 2, 3, \ldots, j-1 \end{aligned} \tag{C.11}$$

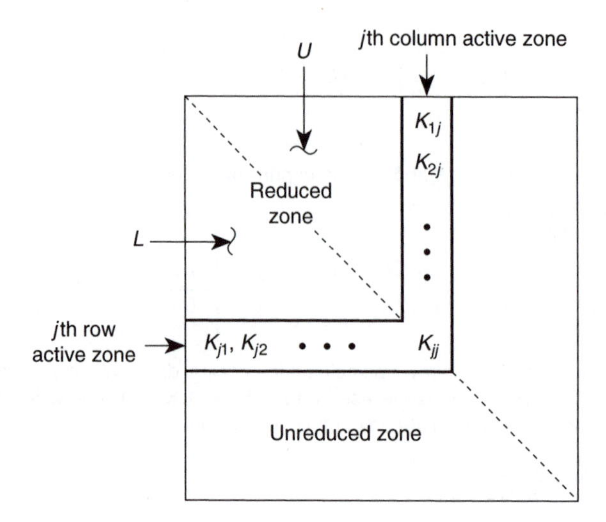

Fig. C.2 Reduced, active and unreduced parts.

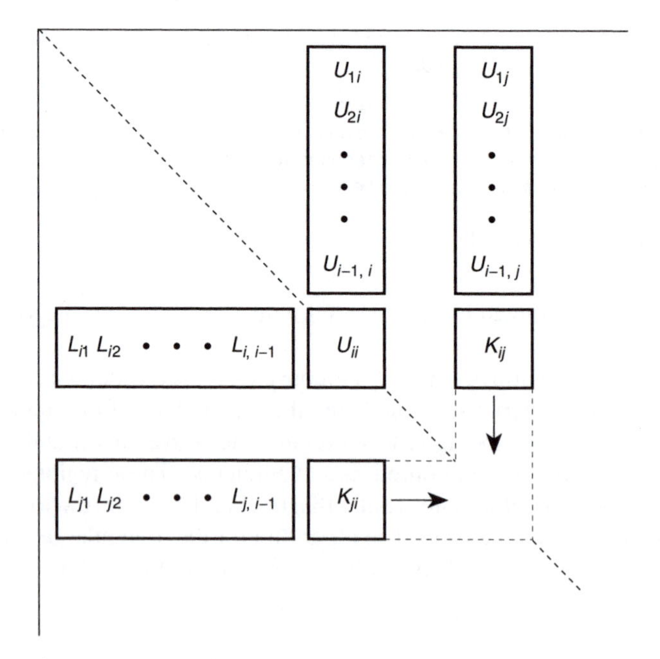

Fig. C.3 Terms used to construct U_{ij} and L_{ji}.

and finally

$$
\begin{aligned}
L_{jj} &= 1 \\
U_{jj} &= K_{jj} - \sum_{m=1}^{j-1} L_{jm} U_{mj}
\end{aligned}
\tag{C.12}
$$

Table C.1 Example: triangular decomposition of 3 × 3 matrix

K	**L**	**U**
$\begin{bmatrix} 4 & 2 & 1 \\ 2 & 4 & 2 \\ 1 & 2 & 4 \end{bmatrix}$	$\begin{bmatrix} 1 & & \\ & & \\ & & \end{bmatrix}$	$\begin{bmatrix} 4 & & \\ & & \\ & & \end{bmatrix}$
Step 1. $L_{11} = 1, U_{11} = 4$		
$\begin{bmatrix} & 2 & 1 \\ 2 & 4 & 2 \\ 1 & 2 & 4 \end{bmatrix}$	$\begin{bmatrix} 1 & & \\ 0.5 & 1 & \\ & & \end{bmatrix}$	$\begin{bmatrix} 4 & 2 & \\ & 3 & \\ & & \end{bmatrix}$
Step 2. $L_{21} = \frac{2}{4} = 0.5, U_{12} = 2, U_{22} = 1, U_{22} = 4 - 0.5 \times 2 = 3$		
$\begin{bmatrix} & & 1 \\ & & 2 \\ 1 & 2 & 4 \end{bmatrix}$	$\begin{bmatrix} 1 & & \\ 0.5 & 1 & \\ 0.25 & 0.5 & 1 \end{bmatrix}$	$\begin{bmatrix} 4 & 2 & 1 \\ & 3 & 1.5 \\ & & 3 \end{bmatrix}$
Step 3. $L_{31} = \frac{1}{4} = 0.25, U_{13} = 1, L_{32} = \dfrac{2 - 0.25 \times 2}{3} = \dfrac{1.5}{3} = 0.5$; $U_{23} = 2 - 0.5 \times 1 = 1.5, L_{33} = 1, U_{33} = 4 - 0.25 \times 1 - 0.5 \times 1.5 = 3$		
$\begin{bmatrix} 1 & & \\ 0.5 & 1 & \\ 0.25 & 0.5 & 1 \end{bmatrix} \begin{bmatrix} 4 & 2 & 1 \\ & 3 & 1.5 \\ & & 3 \end{bmatrix} = \begin{bmatrix} 4 & 2 & 1 \\ 2 & 4 & 2 \\ 1 & 2 & 4 \end{bmatrix}$		
Step 4. Check		

The ordering of the reduction process and the terms used are shown in Fig. C.3. The results from Fig. C.1 and Eqs (C.9)–(C.12) can be verified using the matrix given in the example shown in Table C.1.

Once the triangular decomposition of the coefficient matrix is computed, several solutions for different right-hand sides **f** can be computed using Eqs (C.7) and (C.8). This process is often called a *resolution* since it is not necessary to recompute the **L** and **U** arrays. For large size coefficient matrices the triangular decomposition step is very costly while a resolution is relatively cheap; consequently, a resolution capability is necessary in any finite element solution system using a direct method.

The above discussion considered the general case of equation solving (without row or column interchanges). In coefficient matrices resulting from a finite element formulation some special properties are usually present. Often the coefficient matrix is symmetric ($K_{ij} = K_{ji}$) and it is easy to verify in this case that

$$U_{ij} = L_{ji} U_{ii} \quad \text{(no sum)} \tag{C.13}$$

For this problem class it is not necessary to store the entire coefficient matrix. It is sufficient to store only the coefficients above (or below) the principal diagonal and the diagonal coefficients. Equation (C.13) may be used to construct the missing part. This reduces by almost half the required storage for the coefficient array as well as the computational effort to compute the triangular decomposition.

The required storage can be further reduced by storing only those rows and columns which lie within the region of non-zero entries of the coefficient array. Problems formulated by the finite element method and the Galerkin process normally have a symmetric profile which further simplifies the storage form. Storing the upper and lower parts in separate arrays and the diagonal entries of **U** in a third array is used in DATRI. Figure C.4 shows a typical *profile* matrix and the storage order adopted for the upper array AU, the lower array AL and the diagonal array AD. An integer array JD is used to locate the start and end of entries in each column. With this scheme it is necessary to store and compute only within the non-zero profile of the equations. This form of storage does not severely penalize the presence of a few large columns/rows and is also an easy form to program a resolution process (e.g., see subprogram DASOL in *FEAPpv* and reference 4).

The routines included in *FEAPpv* are restricted to problems for which the coefficient matrix can fit within the space allocated in the main storage array. In two-dimensional formulations, problems with several thousand degrees of freedom can be solved on today's personal computers. In three-dimensional cases, however, problems are restricted to a few thousand equations. To solve larger size problems there are several options. The first is to retain only part of the coefficient matrix in the main array with the rest saved on backing store (e.g., hard disk). This can be quite easily achieved but the size of problem is not greatly increased due to the very large solve times required and the rapid growth in the size of the profile-stored coefficient matrix in three-dimensional problems.

A second option is to use sparse solution schemes. These lead to significant program complexity over the procedure discussed above but can lead to significant savings in storage demands and compute time – especially for problems in three dimensions. Nevertheless, capacity in terms of storage and compute time is again rapidly encountered and alternatives are needed.

Iterative solution

One of the main problems in direct solutions is that terms within the coefficient matrix which are zero from a finite element formulation become non-zero during the triangular decomposition step. While sparse methods are better at limiting this fill than profile methods they still lead to a very large increase in the number of non-zero terms in the factored coefficient matrix. To be more specific consider the case of a three-dimensional linear elastic problem solved using 8-node isoparametric hexahedron elements. In a regular mesh each interior node is associated with 26 other nodes, thus, the equation of such a node has 81 non-zero coefficients – three for each of the 27 associated nodes. On the other hand, for a rectangular block of elements with n nodes on each of the sides the typical column height is approximately proportional to n^2 and the number of equations to n^3. In Table C.2 we show the size and approximate number of non-zero terms in **K** from a finite element formulation for linear elasticity (i.e., with three degrees of freedom per node). The table also indicates the size growth with column height and storage requirements for a direct solution based on a profile solution method.

From the table it can be observed that the demands for a direct solution are growing very rapidly (storage is approximately proportional to n^5) while at the same time the demands for storing the non-zero terms in the stiffness matrix grow proportional to the number of equations (i.e., proportional to n^3 for the block).

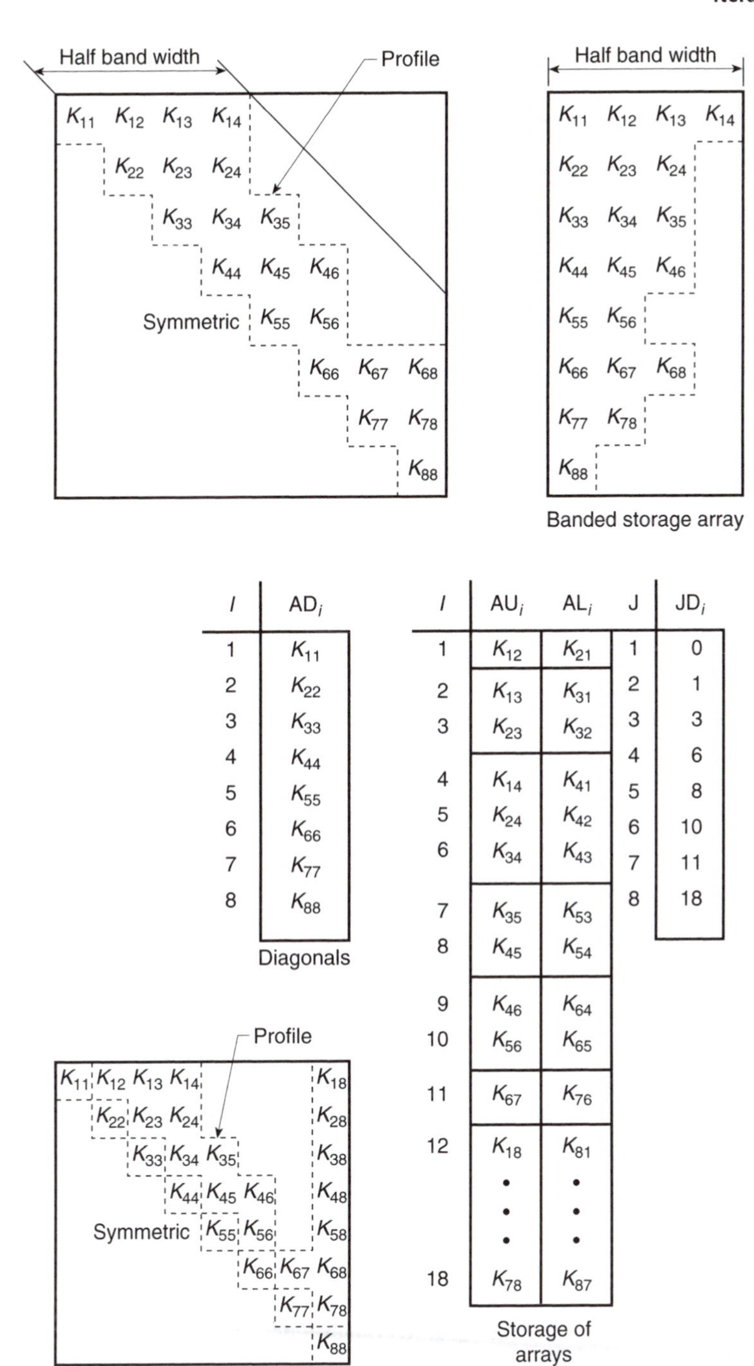

Fig. C.4 Profile storage for coefficient matrix.

Table C.2 Partial list of solutions commands

Side nodes	Number of equations	Non-zeros in K Words ($\times 10^{-6}$)	Non-zeros in K Mbytes	Col. Ht.	Profile storage data Words ($\times 10^{-6}$)	Profile storage data Mbytes
5	375	0.02	0.12	90	0.03	0.27
10	3000	0.12	0.96	330	0.99	7.92
20	24 000	0.96	7.68	1260	30.24	241.82
40	192000	7.68	61.44	4920	944.64	7557.12
80	1536000	61.44	491.52	18 440	28 323.84	226584.72

Iterative solution methods use the terms in the stiffness matrix directly and thus for large problems have the potential to be very efficient for large three-dimensional problems. On the other hand, iterative methods require the resolution of a set of equations until the residual of the linear equations, given by

$$\mathbf{r}^{(i)} = \mathbf{f} - \mathbf{K}\tilde{\mathbf{u}}^{(i)} \tag{C.14}$$

becomes less than a specified tolerance.

In order to be effective the number of iterations i to achieve a solution must be quite small – generally no larger than a few hundred. Otherwise, excessive solution costs will result. At the time of writing this book the subject of iterative solution for general finite element problems remains a topic of intense research. There are some impressive results available for the case where **K** is symmetric positive (or negative) definite; however, those for other classes (e.g., unsymmetric or indefinite forms) are generally not efficient enough for reliable use in the solution of general problems.

For the symmetric positive definite case methods based on a preconditioned conjugate gradient method have been particularly effective.[6–8] The convergence of the method depends on the condition number of the matrix **K** – the larger the condition number, the slower the convergence (see reference 3 for more discussion). The condition number for a finite element problem with a symmetric positive definite stiffness matrix **K** is defined as

$$\kappa = \frac{\lambda_n}{\lambda_1} \tag{C.15}$$

where λ_1 and λ_n are the smallest and largest eigenvalue from the solution of the eigenproblem (viz. Chapter 16)

$$\mathbf{K\Phi} = \mathbf{\Phi\Lambda} \tag{C.16}$$

in which $\mathbf{\Lambda}$ is a diagonal matrix containing the individual eigenvalues λ_i and the columns of $\mathbf{\Phi}$ are the eigenvectors ϕ_i associated with each of the eigenvalues.

Usually, the condition number for an elasticity problem modelled by the finite element method is too large to achieve rapid convergence and a *preconditioned conjugate* gradient (PCG) is used.[6] A symmetric form of preconditioned system is written as

$$\mathbf{K}_p\mathbf{z} = \mathbf{PKP}^{\mathrm{T}}\mathbf{z} = \mathbf{Pf} \tag{C.17}$$

where

$$\mathbf{P}^{\mathrm{T}}\mathbf{z} = \tilde{\mathbf{u}} \tag{C.18}$$

Now the convergence of the PCG algorithm depends on the condition number of $\mathbf{K}_p$. The problem remains to construct a preconditioner which adequately reduces the condition number. In *FEAPpv* the diagonal of $\mathbf{K}$ is used; however, more efficient schemes incorporating also multigrid methods are discussed in references 7 and 8.

References

1. A. Ralston. *A First Course in Numerical Analysis*. McGraw-Hill, New York, 1965.
2. J.H. Wilkinson and C. Reinsch. *Linear Algebra. Handbook for Automatic Computation*, volume II. Springer-Verlag, Berlin, 1971.
3. J. Demmel. *Applied Numerical Linear Algebra*. Society for Industrial and Applied Mathematics, Philadelphia, PA, 1997.
4. R.L. Taylor. Solution of linear equations by a profile solver. *Eng. Comput.*, 2:344–350, 1985.
5. G. Strang. *Linear Algebra and its Application*. Academic Press, New York, 1976.
6. R.M. Ferencz. *Element-by-element preconditioning techniques for large-scale, vectorized finite element analysis in nonlinear solid and structural mechanics*. Ph.D. thesis, Department of Mechanical Engineering, Stanford University, Stanford, California, 1989.
7. M. Adams. A parallel maximal independent set algorithm. In *Proc. 5th Copper Mountain Conference on Iterative Methods*, 1998.
8. M. Adams. Parallel multigrid solver algorithms and implementations for 3D unstructured finite element problems. In *Supercomputing '99: High Performance Networking and Computing*, volume http://www.sc99.org/proceedings, Portland, Oregon, Nov. 1999.

Appendix D

Some integration formulae for a triangle

Let a triangle be defined in the xy plane by three points (x_1, y_1), (x_2, y_2), (x_3, y_3) with the origin of the coordinates taken at the centroid (or baricentre), i.e.,

$$\frac{x_1 + x_2 + x_3}{3} = \frac{y_1 + y_2 + y_3}{3} = 0$$

Then integrating over the triangle area we obtain:

$$\int \mathrm{d}x\,\mathrm{d}y = \frac{1}{2}\begin{vmatrix} 1 & x_1 & y_1 \\ 1 & x_2 & y_2 \\ 1 & x_3 & y_3 \end{vmatrix} = \Delta = \text{area of triangle}$$

$$\int x\,\mathrm{d}x\,\mathrm{d}y = \int y\,\mathrm{d}x\,\mathrm{d}y = 0$$

$$\int x^2\,\mathrm{d}x\,\mathrm{d}y = \frac{\Delta}{12}\left(x_1^2 + x_2^2 + x_3^2\right)$$

$$\int y^2\,\mathrm{d}x\,\mathrm{d}y = \frac{\Delta}{12}\left(y_1^2 + y_2^2 + y_3^2\right)$$

$$\int x\,y\,\mathrm{d}x\,\mathrm{d}y = \frac{\Delta}{12}\left(x_1 y_1 + x_2 y_2 + x_3 y_3\right)$$

Appendix E

Some integration formulae for a tetrahedron

Let a tetrahedron be defined in the xyz coordinate system by four points (x_1, y_1, z_1), (x_2, y_2, z_2), (x_3, y_3, z_3) (x_4, y_4, z_4) with the origin of the coordinates taken at the centroid, i.e.,

$$\frac{x_1 + x_2 + x_3 + x_4}{4} = \frac{y_1 + y_2 + y_3 + y_4}{4} = \frac{z_1 + z_2 + z_3 + z_4}{4} = 0$$

Then integrating over the tetrahedron volume

$$\int \mathrm{d}x\,\mathrm{d}y\,\mathrm{d}z = \frac{1}{6}\begin{vmatrix} 1 & x_1 & y_1 & z_1 \\ 1 & x_2 & y_2 & z_2 \\ 1 & x_3 & y_3 & z_3 \\ 1 & x_4 & y_4 & z_4 \end{vmatrix} = V = \text{tetrahedron volume}$$

Provided the order of numbering the nodes is as indicated on Fig. 4.18(a) then also:

$$\int x\,\mathrm{d}x\,\mathrm{d}y\,\mathrm{d}z = \int y\,\mathrm{d}x\,\mathrm{d}y\,\mathrm{d}z = \int z\,\mathrm{d}x\,\mathrm{d}y\,\mathrm{d}z = 0$$

$$\int x^2\,\mathrm{d}x\,\mathrm{d}y\,\mathrm{d}z = \frac{V}{20}\left(x_1^2 + x_2^2 + x_3^2 + x_4^2\right)$$

$$\int y^2\,\mathrm{d}x\,\mathrm{d}y\,\mathrm{d}z = \frac{V}{20}\left(y_1^2 + y_2^2 + y_3^2 + y_4^2\right)$$

$$\int z^2\,\mathrm{d}x\,\mathrm{d}y\,\mathrm{d}z = \frac{V}{20}\left(z_1^2 + z_2^2 + z_3^2 + z_4^2\right)$$

$$\int x\,y\,\mathrm{d}x\,\mathrm{d}y\,\mathrm{d}z = \frac{V}{20}\left(x_1y_1 + x_2y_2 + x_3y_3 + x_4y_4\right)$$

$$\int y\,z\,\mathrm{d}x\,\mathrm{d}y\,\mathrm{d}z = \frac{V}{20}\left(y_1z_1 + y_2z_2 + y_3z_3 + y_4z_4\right)$$

$$\int z\,x\,\mathrm{d}x\,\mathrm{d}y\,\mathrm{d}z = \frac{V}{20}\left(z_1x_1 + z_2x_2 + z_3x_3 + z_4x_4\right)$$

Appendix F

Some vector algebra

Some knowledge and understanding of basic vector algebra is needed in dealing with complexities of elements oriented in space as occur in beams, shells, etc. Some of the operations are summarized here.

Vectors (in the geometric sense) can be described by their components along the directions of the x, y, z axes.

Thus, the vector $\mathbf{V}_{01}$ shown in Fig. F.1 can be written as

$$\mathbf{V}_{01} = x_1\mathbf{i} + y_1\mathbf{j} + z_1\mathbf{k} \tag{F.1}$$

in which $\mathbf{i}$, $\mathbf{j}$, $\mathbf{k}$ are unit vectors in the direction of the x, y, z axes.

Alternatively, the same vector could be written as

$$\mathbf{V}_{01} = \begin{Bmatrix} x_1 \\ y_1 \\ z_1 \end{Bmatrix} \tag{F.2}$$

(now a 'vector' in the matrix sense) in which the components are distinguished by positions in the column.

Addition and subtraction

Addition and subtraction is defined by addition and subtraction of components. Thus, for example,

$$\mathbf{V}_{02} - \mathbf{V}_{01} = (x_2 - x_1)\mathbf{i} + (y_2 - y_1)\mathbf{j} + (z_2 - z_1)\mathbf{k} \tag{F.3}$$

The same result is achieved by the definitions of matrix algebra; thus

$$\mathbf{V}_{02} - \mathbf{V}_{01} = \mathbf{V}_{21} = \begin{Bmatrix} x_2 - x_1 \\ y_2 - y_1 \\ z_2 - z_1 \end{Bmatrix} \tag{F.4}$$

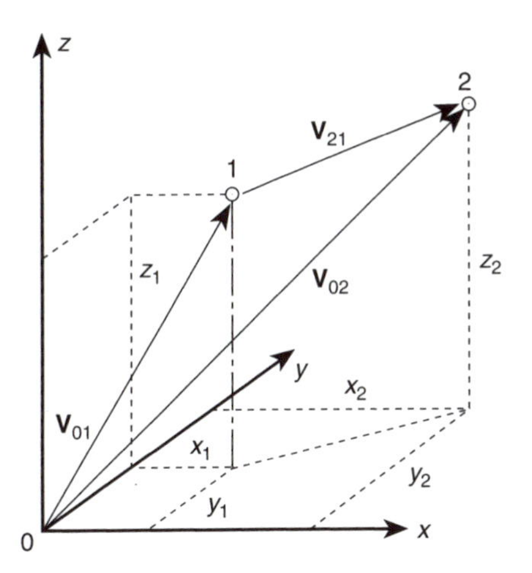

Fig. F.1 Vector addition.

'Scalar' products

A scalar product of two vectors is *defined* as

$$\mathbf{A} \cdot \mathbf{B} = \mathbf{B} \cdot \mathbf{A} = \sum_{k=1}^{3} a_k b_k \tag{F.5}$$

If

$$\begin{aligned} \mathbf{A} &= a_x\mathbf{i} + a_y\mathbf{j} + a_z\mathbf{k} \\ \mathbf{B} &= b_x\mathbf{i} + b_y\mathbf{j} + b_z\mathbf{k} \end{aligned} \tag{F.6}$$

then

$$\mathbf{A} \cdot \mathbf{B} = a_x b_x + a_y b_y + a_z b_z \tag{F.7}$$

Using the matrix notation

$$\mathbf{A} = \begin{Bmatrix} a_x \\ a_y \\ a_z \end{Bmatrix} \quad \mathbf{B} = \begin{Bmatrix} b_x \\ b_y \\ b_z \end{Bmatrix} \tag{F.8}$$

the scalar product becomes

$$\mathbf{A} \cdot \mathbf{B} = \mathbf{A}^{\mathrm{T}}\mathbf{B} = \mathbf{B}^{\mathrm{T}}\mathbf{A} \tag{F.9}$$

Length of vector

The length of the vector $\mathbf{V}_{21}$ is given, purely geometrically, as

$$l_{21} = \sqrt{(x_2 - x_1)^2 + (y_2 - y_1)^2 + (z_2 - z_1)^2} \tag{F.10}$$

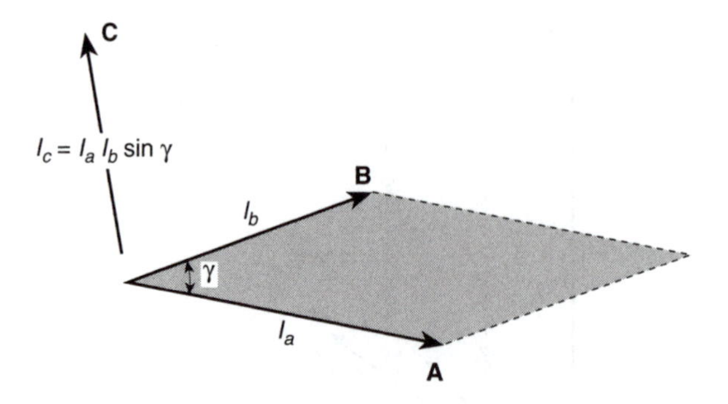

Fig. F.2 Vector multiplication (cross product).

or in terms of matrix algebra as

$$l_{21} = \sqrt{\mathbf{V}_{21} \cdot \mathbf{V}_{21}} = \sqrt{\mathbf{V}_{21}^{\mathrm{T}} \mathbf{V}_{21}} \tag{F.11}$$

Direction cosines

Direction cosines of a vector are simply, from the definition of the projected component of lengths, given as (Fig. F.1)

$$\cos \alpha_x = \frac{x_2 - x_1}{l_{21}} = \frac{\mathbf{V}_{21} \cdot \mathbf{i}}{l_{21}} \tag{F.12}$$

The scalar product may also be written as (Fig. F.2)

$$\mathbf{A} \cdot \mathbf{B} = \mathbf{B} \cdot \mathbf{A} = l_a l_b \cos\gamma \tag{F.13}$$

where γ is the angle between the two vectors **A** and **B** and l_a and l_b are their lengths, respectively.

'Vector' or cross product

Another product of vectors is defined as a vector oriented normally to the plane given by two vectors and equal in magnitude to the product of the length of the two vectors multiplied by the sine of the angle between them. Further, the direction of the normal vector follows the right-hand rule as shown in Fig. F.2 in which

$$\mathbf{A} \times \mathbf{B} = \mathbf{C} \tag{F.14}$$

is shown.

Thus, from the right-hand rule, we have

$$\mathbf{A} \times \mathbf{B} = -\mathbf{B} \times \mathbf{A} \tag{F.15}$$

It is worth noting that the magnitude (or length) of **C** is equal to the area of the parallelogram shown in Fig. F.2.

Using the definition of Eq. (F.6) and noting that

$$\begin{aligned}\mathbf{i}\times\mathbf{i} &= \mathbf{j}\times\mathbf{j} = \mathbf{k}\times\mathbf{k} = 0 \\ \mathbf{i}\times\mathbf{j} &= \mathbf{k}\,,\ \mathbf{j}\times\mathbf{k} = \mathbf{i}\,,\ \mathbf{k}\times\mathbf{i} = \mathbf{j}\end{aligned} \tag{F.16}$$

we have

$$\begin{aligned}\mathbf{A}\times\mathbf{B} &= \det\begin{vmatrix}\mathbf{i} & \mathbf{j} & \mathbf{k}\\ a_x & a_y & a_z\\ b_x & b_y & b_z\end{vmatrix}\\ &= (a_y b_z - a_z b_y)\mathbf{i} + (a_z b_x - a_x b_z)\mathbf{j} + (a_x b_y - a_y b_x)\mathbf{k}\end{aligned}$$

In matrix algebra this does not find a simple counterpart but we can use the above to define the vector **C**†

$$\mathbf{C} = \mathbf{A}\times\mathbf{B} = \begin{Bmatrix} a_y b_z - a_z b_y \\ a_z b_x - a_x b_z \\ a_x b_y - a_y b_x \end{Bmatrix} \tag{F.17}$$

The vector product will be found particularly useful when the problem of erecting a normal direction to a surface is considered.

Elements of area and volume

If ξ and η are some curvilinear coordinates, then the following vectors in two-dimensional plane

$$\mathrm{d}\boldsymbol{\xi} = \begin{Bmatrix} \dfrac{\partial x}{\partial \xi} \\ \dfrac{\partial y}{\partial \xi} \end{Bmatrix} \mathrm{d}\xi \quad \mathrm{d}\boldsymbol{\eta} = \begin{Bmatrix} \dfrac{\partial x}{\partial \eta} \\ \dfrac{\partial y}{\partial \eta} \end{Bmatrix} \mathrm{d}\eta \tag{F.18}$$

defined from the relationship between the cartesian and curvilinear coordinates, are vectors directed tangentially to the ξ and η equal constant contours, respectively. As the *length* or the vector resulting from a cross product of $\mathrm{d}\boldsymbol{\xi}\times\mathrm{d}\boldsymbol{\eta}$ is equal to the area of the elementary parallelogram we can write

$$\mathrm{d}(area) = \det\begin{bmatrix} \dfrac{\partial x}{\partial \xi} & \dfrac{\partial x}{\partial \eta} \\ \dfrac{\partial y}{\partial \xi} & \dfrac{\partial y}{\partial \eta} \end{bmatrix} \mathrm{d}\xi\,\mathrm{d}\eta \tag{F.19}$$

by Eq. (F.17).

† If we rewrite **A** as a skew symmetric matrix

$$\hat{\mathbf{A}} = \begin{bmatrix} 0 & -a_z & a_y \\ a_z & 0 & -a_x \\ -a_y & a_x & 0 \end{bmatrix}$$

then an alternative representation of the vector product in matrix form is $\mathbf{C} = \hat{\mathbf{A}}\mathbf{B}$.

Similarly, if we have three curvilinear coordinates ξ, η, ζ in the cartesian space, the 'triple' or box product defines a differential volume

$$\mathrm{d}(vol) = (\mathrm{d}\boldsymbol{\xi} \times \mathrm{d}\boldsymbol{\eta}) \cdot \mathrm{d}\boldsymbol{\zeta} = \det \begin{bmatrix} \frac{\partial x}{\partial \xi} & \frac{\partial x}{\partial \eta} & \frac{\partial x}{\partial \zeta} \\ \frac{\partial y}{\partial \xi} & \frac{\partial y}{\partial \eta} & \frac{\partial y}{\partial \zeta} \\ \frac{\partial z}{\partial \xi} & \frac{\partial z}{\partial \eta} & \frac{\partial z}{\partial \zeta} \end{bmatrix} \mathrm{d}\xi\, \mathrm{d}\eta\, \mathrm{d}\zeta \tag{F.20}$$

this follows simply from the geometry. The bracketed product, by definition, forms a vector whose length is equal to the parallelogram area with sides tangent to two of the coordinates. The second scalar multiplication by a length and the cosine of the angle between that length and the normal to the parallelogram establishes a differential volume element.

The above equations serve in changing the variables in surface and volume integrals.

Appendix G

Integration by parts in two or three dimensions (Green's theorem)

Consider the integration by parts of the following two-dimensional expression

$$\iint_{\Omega} \phi \frac{\partial \psi}{\partial x} \, \mathrm{d}x \, \mathrm{d}y \tag{G.1}$$

Integrating first with respect to x and using the well-known relation for integration by parts in one dimension

$$\int_{x_L}^{x_R} u \, \mathrm{d}v = -\int_{x_L}^{x_R} v \, \mathrm{d}u + (uv)_{x=x_R} - (uv)_{x=x_L} \tag{G.2}$$

we have, using the symbols of Fig. G.1,

$$\iint_{\Omega} \phi \frac{\partial \psi}{\partial x} \, \mathrm{d}x \, \mathrm{d}y = -\iint_{\Omega} \frac{\partial \phi}{\partial x} \psi \, \mathrm{d}x \, \mathrm{d}y + \int_{y_B}^{y_T} \left[(\phi \, \psi)_{x=x_R} - (\phi \, \psi)_{x=x_L} \right] \mathrm{d}y \tag{G.3}$$

If now we consider a direct segment of the boundary $\mathrm{d}\Gamma$ on the right-hand boundary, we note that

$$\mathrm{d}y = n_x \, \mathrm{d}\Gamma \tag{G.4}$$

where n_x is the direction cosine between the outward normal and the x direction. Similarly on the left-hand section we have

$$\mathrm{d}y = -n_x \, \mathrm{d}\Gamma \tag{G.5}$$

The final term of Eq. (G.3) can thus be expressed as the integral taken around an anticlockwise direction of the complete closed boundary:

$$\oint_{\Gamma} \phi \psi n_x \, \mathrm{d}\Gamma \tag{G.6}$$

If several closed contours are encountered this integration has to be taken around each such contour. The general expression in all cases is

$$\iint_{\Omega} \phi \frac{\partial \psi}{\partial x} \, \mathrm{d}x \, \mathrm{d}y = -\iint_{\Omega} \frac{\partial \phi}{\partial x} \psi \, \mathrm{d}x \, \mathrm{d}y + \oint_{\Gamma} \phi \psi n_x \, \mathrm{d}\Gamma \tag{G.7}$$

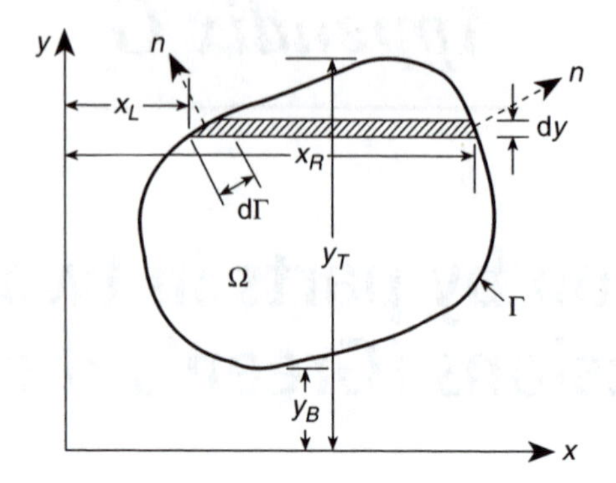

Fig. G.1 Definitions for integrations in two dimensions.

Similarly, if differentiation in the y direction arises we can write

$$\iint_{\Omega} \phi \frac{\partial \psi}{\partial y} \, \mathrm{d}x \, \mathrm{d}y = -\iint_{\Omega} \frac{\partial \phi}{\partial y} \psi \, \mathrm{d}x \, \mathrm{d}y + \oint_{\Gamma} \phi \psi n_y \, \mathrm{d}\Gamma \tag{G.8}$$

where n_y is the direction cosine between the outward normal and the y axis.

In three dimensions by identical procedure we can write

$$\iiint_{\Omega} \phi \frac{\partial \psi}{\partial y} \, \mathrm{d}x \, \mathrm{d}y \, \mathrm{d}z = -\iiint_{\Omega} \frac{\partial \phi}{\partial y} \psi \, \mathrm{d}x \, \mathrm{d}y \, \mathrm{d}z + \oint_{\Gamma} \phi \psi n_y \, \mathrm{d}\Gamma \tag{G.9}$$

where $\mathrm{d}\Gamma$ becomes the element of the surface area and the last integral is taken over the whole surface.

Appendix H

Solutions exact at nodes

The finite element solution of ordinary differential equations may be made exact at the interelement nodes by a proper choice of the *weighting* function in the weak (Galerkin) form. To be more specific, let us consider the set of ordinary differential equations given by

$$\mathbf{A}(\mathbf{u}) + \mathbf{f}(x) = \mathbf{0} \tag{H.1}$$

where $\mathbf{u}$ is the set of dependent variables which are functions of the single independent variable 'X' and $\mathbf{f}$ is a vector of specified load functions. The weak form of this set of differential equations is given by

$$\int_{x_L}^{x_R} \mathbf{v}^{\mathrm{T}} \left[\mathbf{A}(\mathbf{u}) + \mathbf{f}\right] \mathrm{d}x = 0 \tag{H.2}$$

The weak form may be integrated by parts to remove all the derivatives from $\mathbf{u}$ and place them on $\mathbf{v}$. The result of this step may be expressed as

$$\int_{x_L}^{x_R} \left[\mathbf{u}^{\mathrm{T}}\mathbf{A}^*(\mathbf{v}) + \mathbf{v}^{\mathrm{T}}\mathbf{f}\right] \mathrm{d}x + \left[\mathbf{B}^*(\mathbf{v})\right]^{\mathrm{T}} \mathbf{B}(\mathbf{u})\Big|_{x_L}^{x_R} = 0 \tag{H.3}$$

where $\mathbf{A}^*(\mathbf{v})$ is the *adjoint differential equation* and $\mathbf{B}^*(\mathbf{v})$ and $\mathbf{B}(\mathbf{u})$ are terms on the boundary resulting from integration by parts.

If we can find the general integral to the homogeneous adjoint differential equation

$$\mathbf{A}^*(\mathbf{v}) = \mathbf{0} \tag{H.4}$$

then the weak form of the problem reduces to

$$\int_{x_L}^{x_R} \mathbf{v}^{\mathrm{T}}\mathbf{f}\,\mathrm{d}x + \left[\mathbf{B}^*(\mathbf{v})\right]^{\mathrm{T}} \mathbf{B}(\mathbf{u})\Big|_{x_L}^{x_R} = 0 \tag{H.5}$$

The first term is merely an expression to generate equivalent forces from the solution to the adjoint equation and the last term is used to construct the residual equation for the problem. If the differential equation is linear these lead to a residual which depends linearly on the values of $\mathbf{u}$ at the ends x_L and x_R. If we now let these be the location of the end nodes of a typical element we immediately find an expression to generate a stiffness matrix. Since in this process we have never had to construct an *approximation* for the dependent variables $\mathbf{u}$ it is immediately evident that at the end points the discrete values of the exact solution must coincide with any admissible approximation we choose. Thus, we always obtain exact solutions at these points.

If we consider that all values of the forcing function are contained in f (i.e., no point loads at nodes), the terms in $\mathbf{B}(\mathbf{u})$ must be continuous between adjacent elements. At the boundaries the terms in $\mathbf{B}(\mathbf{u})$ include a flux term as well as displacements.

As an example problem, consider the single differential equation

$$\frac{\mathrm{d}^2 u}{\mathrm{d}x^2} + P\frac{\mathrm{d}u}{\mathrm{d}x} + f = 0 \tag{H.6}$$

with the associated weak form

$$\int_{x_L}^{x_R} v\left[\frac{\mathrm{d}^2 u}{\mathrm{d}x^2} + P\frac{\mathrm{d}u}{\mathrm{d}x} + f\right]\mathrm{d}x = 0 \tag{H.7}$$

After integration by parts the weak form becomes

$$\int_{x_L}^{x_R}\left[u\left(\frac{\mathrm{d}^2 v}{\mathrm{d}x^2} - P\frac{\mathrm{d}v}{\mathrm{d}x}\right) + vf\right]\mathrm{d}x + \left[v\left(\frac{\mathrm{d}u}{\mathrm{d}x} + Pu\right) - \frac{\mathrm{d}v}{\mathrm{d}x}u\right]_{x_L}^{x_R} = 0 \tag{H.8}$$

The adjoint differential equation is given by

$$A^*(v) = \frac{\mathrm{d}^2 v}{\mathrm{d}x^2} - P\frac{\mathrm{d}v}{\mathrm{d}x} = 0 \tag{H.9}$$

and the boundary terms by

$$\mathbf{B}^*(v) = \begin{Bmatrix} v \\ -\dfrac{\mathrm{d}v}{\mathrm{d}x} \end{Bmatrix} \tag{H.10}$$

and

$$\mathbf{B}(u) = \begin{Bmatrix} \dfrac{\mathrm{d}u}{\mathrm{d}x} + Pu \\ u \end{Bmatrix} \tag{H.11}$$

For the above example two cases may be identified:

1. P zero – where the adjoint differential equation is identical to the homogeneous equation in which case the problem is called *self-adjoint*.
2. P non-zero – where we then have the *non-self-adjoint* problem.

The finite element solution for these two cases is often quite different. In the first case an equivalent variational theorem exists, whereas for the second case no such theorem exists.†

In the first case the solution to the adjoint equation is given by

$$v = Ax + B \tag{H.12}$$

†An integrating factor often may be introduced to make the weak form generate a self-adjoint problem; however, the approximation problem will remain the same. See Sec. 3.11.2.

which may be written as conventional linear shape functions in each element as

$$N_L = \frac{x_R - x}{x_R - x_L}; \quad N_R = \frac{x - x_L}{x_R - x_L}; \tag{H.13}$$

Thus, for linear shape functions in each element used as the weighting function the interelement nodal displacements for u will always be exact (e.g., see Fig. 3.4) irrespective of the interpolation used for u.

For the second case the exact solution to the adjoint equation is

$$v = Ae^{Px} + B = Az + B \tag{H.14}$$

This yields the shape functions for the weighting function

$$N_L = \frac{z_R - z}{z_R - z_L}; \quad N_R = \frac{z - z_L}{z_R - z_L}; \tag{H.15}$$

which when used in the weak form again yield exact answers at the interelement nodes.

After constructing exact nodal solutions for u, exact solutions for the flux at the interelement nodes can also be obtained from the weak form for each element. The above process was first given by Tong for self-adjoint differential equations.[1]

References

1. P. Tong. Exact solution of certain problems by the finite element method. *J. AIAA*, 7:179–180, 1969.

Appendix I

Matrix diagonalization or lumping

Some of the algorithms discussed in this volume become more efficient if one of the global matrices can be diagonalized (also called 'lumped' by many engineers). For example, the solution of some mixed and transient problems are more efficient if a global matrix to be inverted (or equations solved) is diagonal [viz. Chapter 11, Eq. (11.94) and Chapter 16, Secs 16.2.4 and 16.4.2]. Engineers have persisted with purely physical concepts of lumping; however, there is clearly a need for devising a systematic and mathematically acceptable procedure for such lumping.

We shall define the matrix to be considered as

$$\mathbf{A} = \int_{\Omega} \mathbf{N}^{\mathrm{T}} \mathbf{c} \mathbf{N} \, \mathrm{d}\Omega \tag{I.1}$$

where $\mathbf{c}$ is a matrix with small dimension. Often $\mathbf{c}$ is a diagonal matrix (e.g., in mass or simple least square problems $\mathbf{c}$ is an identity matrix times some scalar). When $\mathbf{A}$ is computed exactly it has full rank and is not diagonal – this is called the *consistent* form of $\mathbf{A}$ since it is computed consistently with the other terms in the finite element model. The diagonalized form is defined with respect to 'nodes' or the shape functions, e.g., $\mathbf{N}_a = N_a \mathbf{I}$; hence, the matrix will have small diagonal blocks, each with the maximum dimension of $\mathbf{c}$. Only when $\mathbf{c}$ is diagonal can the matrix $\mathbf{A}$ be completely diagonalized. Four basic lines of argument may be followed in constructing a diagonal form.

The first procedure is to use different shape functions to approximate each term in the finite element discretization. For the $\mathbf{A}$ matrix we use substitute shape functions $\bar{\mathbf{N}}_a$ for the lumping process. No derivatives exist in the definition of $\mathbf{A}$; hence, for this term the shape functions may be piecewise continuous within and between elements and still lead to acceptable approximation. If the shape functions used to define $\mathbf{A}$ are piecewise constants, such that $\bar{\mathbf{N}}_a$ is a certain part of the element surrounding the node a and zero elsewhere, and such parts are not overlapping or disjoint, then clearly the matrix of Eq. (I.1) becomes nodally diagonal as

$$\int_{\Omega} \bar{\mathbf{N}}_a^{\mathrm{T}} \mathbf{c} \bar{\mathbf{N}}_b \, \mathrm{d}\Omega = \begin{cases} \int_{\Omega_a} \mathbf{c} \, \mathrm{d}\Omega & a = b \\ 0 & a \neq b \end{cases} \tag{I.2}$$

Such an approximation with different shape functions is permissible since the usual finite element criteria of integrability and completeness are satisfied. We can verify this using a patch test to show that consistency is still maintained in the approximation.

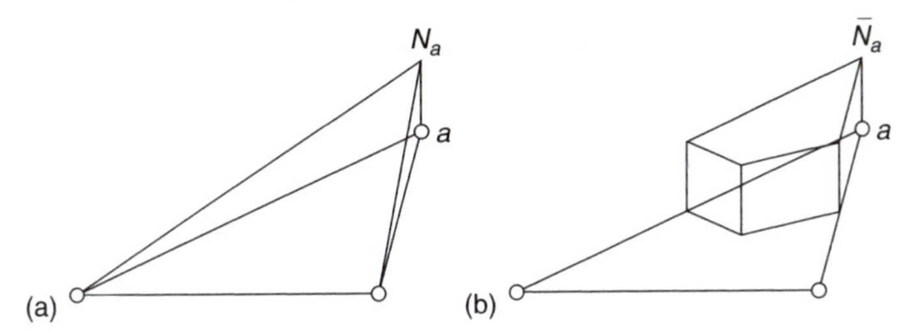

Fig. I.1 (a) Linear and (b) piecewise constant shape functions for a triangle.

The functions selected need only satisfy the condition

$$\bar{\mathbf{N}}_a = \bar{N}_a \mathbf{I} \ \text{with} \ \sum_a \bar{N}_a = 1 \tag{I.3}$$

for all points in the element and this also maintains a partition of unity property in all of Ω. In Fig. I.1 we show the functions N_a and $\bar{N}_a$ for a triangular element.

The second method to diagonalize a matrix is to note that condition (I.1) is simply a requirement that ensures conservation of the quantity $\mathbf{c}$ over the element. For structural dynamics applications this is the conservation of mass at the element level. Accordingly, it has been noted that any lumping that preserves the integral of $\mathbf{c}$ on the element will lead to convergent results, although the rate of convergence may be lower than with use of a consistent $\mathbf{A}$. Many alternatives have been proposed based upon this method. The earliest procedures performed the diagonalization using physical intuition only. Later alternative algorithms were proposed. One suggestion, often called a 'row sum' method, is to compute the diagonal matrix from

$$\mathbf{A}_{ab} = \begin{cases} \sum_c \int_{\Omega_a} \mathbf{N}_a^{\mathrm{T}} \mathbf{c} \mathbf{N}_c \, \mathrm{d}\Omega & a = b \\ \mathbf{0} & a \neq b \end{cases} \tag{I.4}$$

This simplifies to

$$\mathbf{A}_{ab} = \begin{cases} \int_{\Omega_a} \mathbf{N}_a^{\mathrm{T}} \mathbf{c} \, \mathrm{d}\Omega & a = b \\ \mathbf{0} & a \neq b \end{cases} \tag{I.5}$$

since the sum of the shape functions is unity. This algorithm makes sense only when the degrees of freedom of the problem all have the same physical interpretation. An alternative is to scale the diagonals of the consistent mass to satisfy the conservation requirement. In this case the diagonal matrix is deduced from

$$\mathbf{A}_{ab} = \begin{cases} m \int_{\Omega_a} \mathbf{N}_a^{\mathrm{T}} \mathbf{c} \mathbf{N}_b \, \mathrm{d}\Omega & a = b \\ \mathbf{0} & a \neq b \end{cases} \tag{I.6}$$

where m is selected so that

$$\sum_a \mathbf{A}_{aa} = \int_\Omega \mathbf{c} \, \mathrm{d}\Omega \tag{I.7}$$

The third procedure uses numerical integration to obtain a diagonal array without apparently introducing additional shape functions. Use of numerical integration to evaluate the **A** matrix of Eq. (I.1) yields a typical term in a summation form (following Chapter 8)

$$\mathbf{A}_{ab} = \int_\Omega \mathbf{N}_a^{\mathrm{T}} \mathbf{c} \mathbf{N}_b \, \mathrm{d}\Omega = \sum_q \left(\mathbf{N}_a^{\mathrm{T}} \mathbf{c} \mathbf{N}_b \right)_{\xi_q} J_q W_q \tag{I.8}$$

where $\boldsymbol{\xi}_q$ refers to the quadrature point at which the integrand is evaluated, J is the jacobian volume transformation at the same point and W_q gives the appropriate quadrature weight.

If the quadrature points for the numerical integration are located at nodes then, for standard shape functions (viz. Chapter 4), by Eq. (I.3) the diagonal matrix is given

$$\mathbf{A}_{ab} = \begin{cases} \mathbf{c} J_a W_a & a = b \\ \mathbf{0} & a \neq b \end{cases} \tag{I.9}$$

where J_a is the jacobian and W_a is the quadrature weight at node a.

Appropriate weighting values may be deduced by requiring the quadrature formula to exactly integrate particular polynomials in the natural coordinate system. In general the quadrature should integrate a polynomial of the highest complete order in the shape functions. Thus, for 4-noded quadrilateral elements, linear functions should be exactly integrated. Integrating additional terms may lead to improved accuracy but is not required. Indeed, only conservation of **c** is required.

For low order elements, symmetry arguments may be used to lump the matrix. It is, for instance, obvious that in a simple triangular element little improvement can be obtained by any other lumping than the simple one in which the total **c** is distributed in three equal parts. For an 8-noded two-dimensional isoparametric element no such obvious procedure is available. In Fig. I.2 we show the case of rectangular elements of 4-, 8-, and 9-noded type and lumping by Eqs (I.5), (I.6) and (I.9).

It is noted that for the 8-noded element some of the lumped quantities are negative when Eq. (I.5) or Eq. (I.9) is used. These will have some adverse effects in certain algorithms (e.g., time-stepping schemes to integrate transient problems) and preclude their use. In Fig. I.3 we show some lumped matrices for triangular elements computed by quadrature [i.e., Eq. (I.9)]. It is noted here that the cubic element has negative terms while the quadratic element has zero terms. The zero terms are particularly difficult to handle as the resulting diagonal **A** matrix no longer has full rank and thus may not be inverted.

Another aspect of lumping is the performance of the element when distorted from its parent element shape. For example, as a rectangular element is distorted and approaches a triangular shape it is desirable to have the limit triangular shape case behave appropriately. In the case of a 4-noded rectangular element the lumped matrix for all three procedures gives the same answer. However, if the element is distorted by a transformation defined by one parameter f as shown in Fig. I.4 then the three lumping procedures discussed so far give different answers. The jacobian transformation is given by

$$J = ab(1 - f) \tag{I.10}$$

and **c** is here taken as the identity matrix.

The form (I.5) gives

$$\mathbf{A}_{aa} = \begin{cases} ab\,(1 - f/3) & \text{at top nodes} \\ ab\,(1 + f/3) & \text{at bottom nodes} \end{cases} \tag{I.11}$$

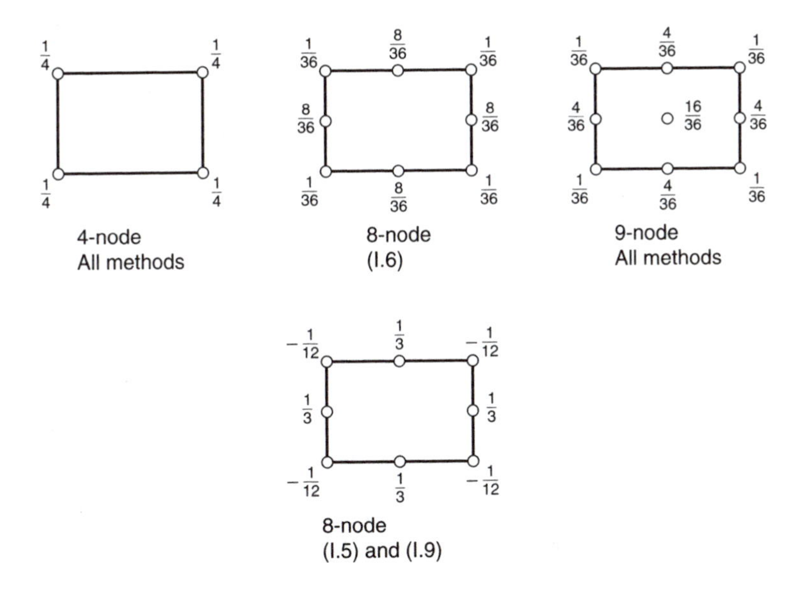

Fig. I.2 Diagonalization of rectangular elements by three methods.

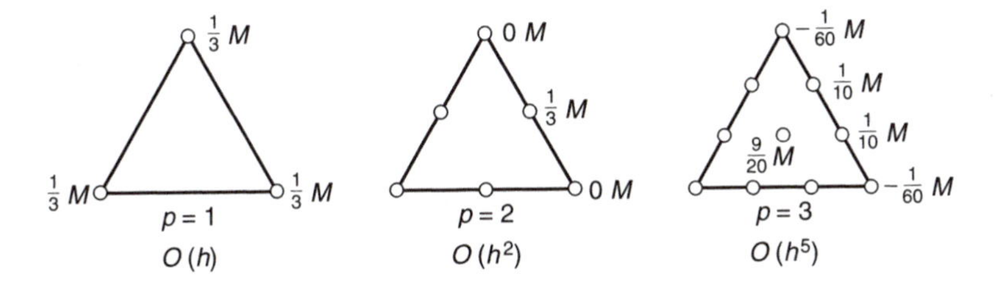

Fig. I.3 Diagonalization of rectangular elements by three methods.

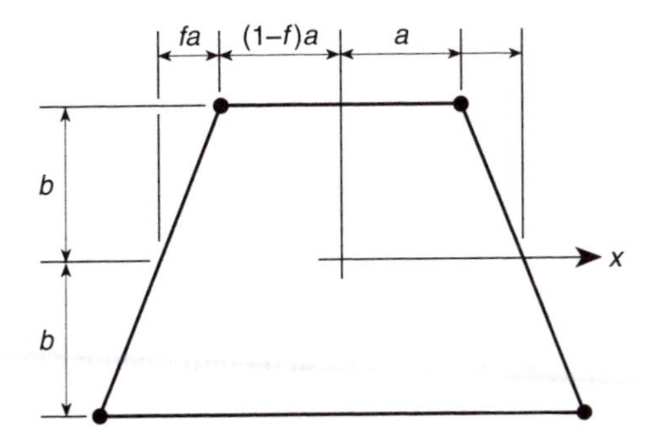

Fig. I.4 Distorted 4-noded element.

the form (I.6) gives

$$\mathbf{A}_{aa} = \begin{cases} ab(1 - f/2) & \text{at top nodes} \\ ab(1 + f/2) & \text{at bottom nodes} \end{cases} \tag{I.12}$$

and the quadrature form (I.9) yields

$$\mathbf{A}_{aa} = \begin{cases} ab(1 - f) & \text{at top nodes} \\ ab(1 + f) & \text{at bottom nodes} \end{cases} \tag{I.13}$$

The 4-noded element has the property that a triangle may be defined by coalescing 2 nodes and assigning them to the same global node in the mesh. Thus, the quadrilateral is identical to a 3-noded triangle when the parameter f is unity. The limit value for the row sum method will give equal lumped terms at the 3 nodes while method (I.6) yields a lumped value for the coalesced node which is two-thirds the value at the other nodes and the quadrature method (I.9) yields a zero lumped value at the coalesced node. Thus, methods (I.6) and (I.9) give limit cases which depend on how the nodes are numbered to form each triangular element. This lack of invariance is not desirable in computer programs; hence for the 4-noded quadrilateral, method (I.5) appears to be superior to the other two. On the other hand, we have observed above that the row sum method (I.5) leads to negative diagonal elements for the 8-noded element; hence there is no universal method for diagonalizing a matrix.

A fourth but not widely used method is available which may be explored to deduce a consistent matrix that is diagonal. This consists of making a mixed representation for the term creating the **A** matrix.

Consider a functional given by

$$\Pi_1 = \frac{1}{2} \int_\Omega \mathbf{u}^{\mathrm{T}} \mathbf{c} \mathbf{u} \, \mathrm{d}\Omega \tag{I.14}$$

The first variation of Π_1 yields

$$\delta\Pi_1 = \int_\Omega \delta\mathbf{u}^{\mathrm{T}} \mathbf{c} \mathbf{u} \, \mathrm{d}\Omega \tag{I.15}$$

Approximation using the standard form

$$\mathbf{u} \approx \hat{\mathbf{u}} = N_a \tilde{\mathbf{u}}_a = \mathbf{N}\tilde{\mathbf{u}} \tag{I.16}$$

yields

$$\delta\Pi_1 = \delta\tilde{\mathbf{u}}^{\mathrm{T}} \int_\Omega \mathbf{N}^{\mathrm{T}} \mathbf{c} \mathbf{N} \, \mathrm{d}\Omega \tilde{\mathbf{u}} \tag{I.17}$$

This yields exactly the form for **A** given by Eq. (I.1).

We can construct an alternative mixed form by introducing a momenta-type variable given by

$$\mathbf{p} = \mathbf{c}\mathbf{u} \tag{I.18}$$

A Hellinger–Reissner-type mixed form may then be expressed as

$$\Pi_2 = \int_\Omega \mathbf{u}^{\mathrm{T}} \mathbf{p} \, \mathrm{d}\Omega - \frac{1}{2} \int_\Omega \mathbf{p}^{\mathrm{T}} \mathbf{c}^{-1} \mathbf{p} \, \mathrm{d}\Omega \tag{I.19}$$

and has the first variation

$$\delta\Pi_2 = \int_\Omega \delta\mathbf{u}^{\mathrm{T}}\mathbf{p}\,\mathrm{d}\Omega + \int_\Omega \delta\mathbf{p}^{\mathrm{T}}\left(\mathbf{u} - \mathbf{c}^{-1}\mathbf{p}\right)\,\mathrm{d}\Omega \tag{I.20}$$

The term with variation on $\mathbf{u}$ will combine with other terms so is not set to zero; however, the other term will not appear elsewhere so can be solved separately.

If we now introduce an approximation for $\mathbf{p}$ as

$$\mathbf{p} \approx \hat{\mathbf{p}} = n_b\tilde{\mathbf{p}}_b = \mathbf{n}\tilde{\mathbf{p}} \tag{I.21}$$

then the variational equation becomes

$$\delta\Pi_2 = \delta\tilde{\mathbf{u}}^{\mathrm{T}}\int_\Omega \mathbf{N}^{\mathrm{T}}\mathbf{n}\,\mathrm{d}\Omega\tilde{\mathbf{p}} + \delta\tilde{\mathbf{p}}^{\mathrm{T}}\left(\int_\Omega \mathbf{n}^{\mathrm{T}}\mathbf{N}\,\mathrm{d}\Omega\tilde{\mathbf{u}} - \int_\Omega \mathbf{n}^{\mathrm{T}}\mathbf{c}^{-1}\,\mathbf{n}\,\mathrm{d}\Omega\tilde{\mathbf{p}}\right) \tag{I.22}$$

If we now define the matrices

$$\begin{aligned}\mathbf{G} &= \int_\Omega \mathbf{n}\mathbf{N}^{\mathrm{T}}\,\mathrm{d}\Omega\\ \mathbf{H} &= \int_\Omega \mathbf{n}^{\mathrm{T}}\mathbf{c}^{-1}\mathbf{n}\,\mathrm{d}\Omega\end{aligned} \tag{I.23}$$

then the weak form is

$$\delta\Pi_2 = \begin{bmatrix}\delta\tilde{\mathbf{u}}^{\mathrm{T}} & \delta\tilde{\mathbf{p}}^{\mathrm{T}}\end{bmatrix}\left(\begin{bmatrix}\mathbf{0} & \mathbf{G}^{\mathrm{T}}\\ \mathbf{G} & -\mathbf{H}\end{bmatrix}\begin{Bmatrix}\tilde{\mathbf{u}}\\ \tilde{\mathbf{p}}\end{Bmatrix} = \begin{Bmatrix}\mathbf{0}\\ \mathbf{0}\end{Bmatrix}\right) \tag{I.24}$$

Eliminating $\tilde{\mathbf{p}}$ using the second row of Eq. (I.24) gives

$$\mathbf{A} = \mathbf{G}^{\mathrm{T}}\mathbf{H}^{-1}\mathbf{G} \tag{I.25}$$

for which diagonal forms may now be sought. This form has again the same options as discussed above but, in addition, forms for the shape functions $\mathbf{n}$ can be sought which also render the matrix diagonal.

Author index

Page numbers in **bold** are for pages at the end of chapters with names of author references.

Subject index